Principles of Environmental Engineering and Science

Fourth Edition

Susan J. Masten
Michigan State University—East Lansing, MI

Mackenzie L. Davis
Emeritus, Michigan State University—East Lansing

PRINCIPLES OF ENVIRONMENTAL ENGINEERING AND SCIENCE

1 2 3 4 5 6 7 8 9 LWI 21 20 19

ISBN 978-1-260-54802-0
MHID 1-260-54802-3

Cover Image: ©*Susan Masten*

mheducation.com/highered

To our students who make it worthwhile.

Contents

Appendices

Index

Preface

Following the format of previous editions, the fourth edition of *Principles of Environmental Engineering and Science* is designed for use in an introductory sophomore-level engineering course. Basic, traditional subject matter is covered. Fundamental science and engineering principles that instructors in more advanced courses may depend upon are included. Mature undergraduate students in allied fields—such as biology, chemistry, resource development, fisheries and wildlife, microbiology, and soils science—have little difficulty with the material.

We have assumed that the students using this text have had courses in chemistry, physics, and biology, as well as sufficient mathematics to understand the concepts of differentiation and integration. Basic environmental chemistry and biology concepts are introduced at the beginning of the book.

Materials and energy balance is introduced early in the text. It is used throughout the text as a tool for understanding environmental processes and solving environmental problems. It is applied in hydrology, sustainability, water quality, water and wastewater treatment, air pollution control, as well as solid and hazardous waste management.

Each chapter concludes with a list of review items, the traditional end-of-chapter problems and discussion questions. The review items have been written in the "objective" format of the Accreditation Board for Engineering and Technology (ABET). Instructors will find this particularly helpful for directing student review for exams, for assessing continuous quality improvement for ABET and for preparing documentation for ABET curriculum review.

The fourth edition has been thoroughly revised and updated. FE formatted problems have been added to the appropriate chapters. New case studies have been added to many of the chapters as well. The following summarizes the major changes in this edition:

- Introduction
  - Data on per capita water consumption has been updated

- Biology
  - Section on harmful algae blooms has been updated
  - Addition of section on Legionellosis

- Ecosystems
  - Updated figures and charts

- Risk
  - Updated tables

- Hydrology
  - Updated figures and tables
  - Addition of discussion of effect of climate change

- Sustainability
  - Updated discussion of water resources focusing on floods and droughts with examples in the United States and in other countries
  - Updated tables and figures on energy and mineral resources
  - Expanded discussion of shale gas includes flowback and earthquakes

- Water Quality Management
  - Updated figures and charts
  - Updated section on water source protection

- Water Treatment
  - Expanded overview of treatment systems
  - Updated section on regulations
  - Updated figures and charts
  - Expansion of section on disinfection to include breakpoint chlorination and UV disinfection

- Wastewater Treatment
  - Addition of direct potable reuse discussion

- Air Pollution
  - Updated ambient air pollution standards
  - Updated discussion of acidification of lakes
  - Addition of mercury control technology for power plants
  - Update of federal motor vehicle standards
  - New discussion of CAFE standards
  - New discussion of the use of coal in power plants
  - Update of global warming potential data for selected compounds
  - Updated graphs of global surface temperature
  - New graph of CO_2 concentration

- Solid Waste
  - Updated figures and charts

Online Resources

Case Studies from the previous edition are still available for use at www.mhhe.com /davisprinciples4e. Powerpoint slides and an instructor's manual are also available. The instructor's manual includes sample course outlines, solved example exams, and detailed solutions to the end-of-chapter problems. In addition, there are suggestions for using the pedagogic aids in the next.

As always, we appreciate any comments, suggestions, corrections, and contributions for future revisions.

Susan J. Masten
Mackenzie L. Davis

Acknowledgments

As with any other text, the number of individuals who have made it possible far exceeds those whose names grace the cover. At the hazard of leaving someone out, we would like to explicitly thank the following individuals for their contribution.

The following students helped to solve problems, proofread text, prepare illustrations, raise embarrassing questions, and generally make sure that other students could understand it: Shelley Agarwal, Stephanie Albert, Deb Allen, Mark Bishop, Aimee Bolen, Kristen Brandt, Jeff Brown, Amber Buhl, Nicole Chernoby, Rebecca Cline, Linda Clowater, Shauna Cohen, John Cooley, Ted Coyer, Marcia Curran, Talia Dodak, Kimberly Doherty, Bobbie Dougherty, Lisa Egleston, Karen Ellis, Elaheh Esfahanian, Craig Fricke, Elizabeth Fry, Beverly Hinds, Edith Hooten, Brad Hoos, Kathy Hulley, Geneva Hulslander, Lisa Huntington, Angela Ilieff, Melissa Knapp, Alison Leach, Gary Lefko, Lynelle Marolf, Lisa McClanahan, Tim McNamara, Becky Mursch, Cheryl Oliver, Kyle Paulson, Marisa Patterson, Lynnette Payne, Jim Peters, Kristie Piner, Christine Pomeroy, Susan Quiring, Erica Rayner, Bob Reynolds, Laurene Rhyne, Sandra Risley, Carlos Sanlley, Lee Sawatzki, Stephanie Smith, Mary Stewart, Rick Wirsing, Ya-yun Wu. To them a hearty thank you!

The authors would also like to thank Pamela Augustine, Brandon Curtas, David Desteiger, Cheryl Edson, John Engle, Timothy Greenleaf, Erin Henderson, Robert Little, Kate Logan, Jeremy Mansell, Lorna McTaggart, Kelly Mlynarek, Brad Osinski, Alicja Pawlowska, Shannon Simpson, Lindsay Smith, Bryan Stramecki, Brad Vernier, Marcie Wawzysko, and Adam Wosneski who also helped proofread the text, check problems and make the book more readable for students. We would also like to thank Trevor Painter for his assistance with the new case studies.

We would also like to thank the following reviewers for their many helpful comments and suggestions: Max Anderson, University of Wisconsin–Platteville; Gregory Boardman, Virginia Tech; Jonathan Brant, University of Wyoming; Leonard W. Casson, University of Pittsburgh–Swanson School of Engineering; Andres Clarens, University of Virginia; Lubo Liu, California State University–Fresno; George Murgel, Boise State University; John Novak, Virginia Tech; Jonathan Sharp, Colorado School of Mines.

We give special thanks to Simon Davies for his contributions and unending support. His efforts are sincerely appreciated. And last, but certainly not least, we wish to thank our families who have put up with us during the writing of this book, especially Quentin and Jeffrey Masten-Davies, who gave up several Christmas vacations plus many other days during the year while their mom spent uncountable hours working on this book.

A special thanks to Mack's wife, Elaine, for putting up with the nonsense of book writing.

Susan J. Masten   Mackenzie L. Davis

About the Authors

Susan J. Masten is a Professor in the Department of Civil and Environmental Engineering at Michigan State University. She received her Ph.D. in environmental engineering from Harvard University in 1986. Before joining the faculty at Michigan State University in 1989, she worked for several years in environmental research at the University of Melbourne (Australia) and at the US Environmental Protection Agency's Kerr Laboratory, in Ada, Oklahoma. Professor Masten's research involves the use of chemical oxidants for the remediation of soils, water, and wastewater. Her research is presently focused on the use of ozone for reducing the concentration of disinfection by-products in drinking water, controlling fouling in membranes, and reducing the toxicity of ozonation by-products formed from the ozonation of polycyclic aromatic hydrocarbons and pesticides. She also had research projects involving the use of ozone for the reduction of odor in swine manure slurry and the elimination of chlorinated hydrocarbons and semivolatile organic chemicals from soils using in-situ ozone stripping and ozone sparging.

Dr. Masten is a member of the following professional organizations: American Water Works Association, Association of Environmental Engineering and Science Professors (AEESP), and American Chemical Society. She served on the Executive Committee of the MSU Chapter of the American Chemical Society from 1995–2005. She has served as the chair of the AEESP Publications Committee since 2013.

Professor Masten was a Lilly Teaching Fellow during the 1994–1995 academic year. She is also the recipient of the Withrow Distinguished Scholar Award, College of Engineering, MSU, March 1995, and the Teacher-Scholar Award, Michigan State University, February 1996, and the Withrow Teaching Award in 2012 and 2018. Dr. Masten was also a member of the Faculty Writing Project, Michigan State University, May 1996. In 2001, she was awarded the Association of Environmental Engineering and Science Professors/Wiley Interscience Outstanding Educator Award, and in 2013, she was awarded the Lyman A. Ripperton Environmental Educator Award by the Air and Waste Management Association. Professor Masten was awarded an Erskine Fellowship in 2018 from the University of Canterbury, Christchurch, NZ.

Dr. Masten is a registered professional engineer in the state of Michigan.

Mackenzie L. Davis, Ph.D., P.E., BCEE, is an Emeritus Professor of Environmental Engineering at Michigan State University. He received all his degrees from the University of Illinois. From 1968 to 1971 he served as a Captain in the U.S. Army Medical Service Corps. During his military service he conducted air pollution surveys at Army ammunition plants. From 1971 to 1973 he was Branch Chief of the Environmental Engineering Branch at the U.S. Army Construction Engineering Research Laboratory. His responsibilities included supervision of research on air, noise, and water pollution control and solid waste management for Army facilities. In 1973 he joined the faculty at Michigan State University. He has taught and conducted research in the areas of air pollution control and hazardous waste management.

In 1987 and 1989–1992, under an intergovernmental personnel assignment with the Office of Solid Waste of the U.S. Environmental Protection Agency, Dr. Davis performed technology assessments of treatment methods used to demonstrate the regulatory requirements for the land disposal restrictions ("land ban") promulgated under the Hazardous and Solid Waste Amendments.

Dr. Davis is a member of the following professional organizations: American Chemical Society, American Institute of Chemical Engineers, American Society for Engineering Education, American Meteorological Society, American Society of Civil Engineers, American Water

Works Association, Air & Waste Management Association, Association of Environmental Engineering and Science Professors, and the Water Environment Federation.

His honors and awards include the State-of-the-Art Award from the ASCE, Chapter Honor Member of Chi Epsilon, Sigma Xi, election as a Fellow in the Air & Waste Management Association, and election as a Diplomate in the American Academy of Environmental Engineers with certification in hazardous waste management. He has received teaching awards from the American Society of Civil Engineers Student Chapter, Michigan State University College of Engineering, North Central Section of the American Society for Engineering Education, Great Lakes Region of Chi Epsilon, and the Amoco Corporation. In 1998, he received the Lyman A. Ripperton Award for distinguished achievement as an educator from the Air & Waste Management Association. In 2007, he was recognized as the Educational Professional of the Year by the Michigan Water Environment Association. He is a registered professional engineer in Michigan.

Dr. Davis is the author of a student and professional edition of *Water and Wastewater Engineering* and co-author of *Introduction to Environmental Engineering* with Dr. David Cornwell.

In 2003, Dr. Davis retired from Michigan State University.

1

Introduction

1–1 WHAT IS ENVIRONMENTAL SCIENCE?

Natural Science

In the broadest sense, science is systematized knowledge derived from and tested by recognition and formulation of a problem, collection of data through observation, and experimentation. We differentiate between social science and natural science in that the former deals with the study of people and how they live together as families, tribes, communities, races, and nations, and the latter deals with the study of nature and the physical world. Natural science includes such diverse disciplines as biology, chemistry, geology, physics, and environmental science.

Environmental Science

Whereas the disciplines of biology, chemistry, and physics (and their subdisciplines of microbiology, organic chemistry, nuclear physics, etc.) are focused on a particular aspect of natural science, environmental science in its broadest sense encompasses all the fields of natural science. The historical focus of study for environmental scientists has been, of course, the natural environment. By this, we mean the atmosphere, the land, the water and their inhabitants as differentiated from the built environment. Modern environmental science has also found applications to the built environment or, perhaps more correctly, to the effusions from the built environment.

Quantitative Environmental Science

Science or, perhaps more correctly, the **scientific method,** deals with data, that is, with recorded observations. The data are, of course, a sample of the universe of possibilities. They may be representative or they may be skewed. Even if they are representative, they will contain some random variation that cannot be explained with current knowledge. Care and impartiality in gathering and recording data, as well as independent verification, are the cornerstones of science.

When the collection and organization of data reveal certain regularities, it may be possible to formulate a generalization or **hypothesis.** This is merely a statement that under certain circumstances certain phenomena can generally be observed. Many generalizations are statistical in that they apply accurately to large assemblages but are no more than probabilities when applied to smaller sets or individuals.

In a scientific approach, the hypothesis is tested, revised, and tested again until it is proven acceptable.

If we can use certain assumptions to tie together a set of generalizations, we formulate a theory. For example, theories that have gained acceptance over a long time are known as **laws.** Some examples are the laws of motion, which describe the behavior of moving bodies, and the gas laws, which describe the behavior of gases. The development of a **theory** is an important accomplishment because it yields a tremendous consolidation of knowledge. Furthermore, a theory gives us a powerful new tool in the acquisition of knowledge for it shows us where to look for new generalizations. "Thus, the accumulation of data becomes less of a magpie collection of facts and more of a systematized hunt for needed information. It is the existence of classification and generalization, and above all theory that makes science an organized body of knowledge" (Wright, 1964).

Logic is a part of all theories. The two types of logic are qualitative and quantitative logic. Qualitative logic is descriptive. For example we can qualitatively state that when the amount of wastewater entering a certain river is too high, the fish die. With qualitative logic we cannot identify what "too high" means—we need quantitative logic to do that.

When the data and generalizations are quantitative, we need mathematics to provide a theory that shows the quantitative relationships. For example, a quantitative statement about the river might state that "When the mass of organic matter entering a certain river equals x kilograms per day, the amount of oxygen in the stream is y."

Perhaps more importantly, quantitative logic enables us to explore "What if?" questions about relationships. For example, "If we reduce the amount of organic matter entering the stream, how much will the amount of oxygen in the stream increase?" Furthermore, theories, and in particular, mathematical theories, often enable us to bridge the gap between experimentally controlled observations and observations made in the field. For example, if we control the amount of oxygen in a fish tank in the laboratory, we can determine the minimum amount required for the fish to be healthy. We can then use this number to determine the acceptable mass of organic matter placed in the stream.

Given that environmental science is an organized body of knowledge about environmental relationships, then **quantitative environmental science** is an organized collection of mathematical theories that may be used to describe and explore environmental relationships.

In this book, we provide an introduction to some mathematical theories that may be used to describe and explore relationships in environmental science.

1–2   WHAT IS ENVIRONMENTAL ENGINEERING?

Environmental engineering is a profession that applies mathematics and science to utilize the properties of matter and sources of energy in the solution of problems of environmental sanitation. These include the provision of safe, palatable, and ample public water supplies; the proper disposal of or recycle of wastewater and solid wastes; the adequate drainage of urban and rural areas for proper sanitation; and the control of water, soil, and atmospheric pollution, and the social and environmental impact of these solutions. Furthermore, it is concerned with engineering problems in the field of public health, such as the control of arthropod-borne diseases, the elimination of industrial health hazards, the provision of adequate sanitation in urban, rural, and recreational areas, and the effect of technological advances on the environment (ASCE, 1973, 1977).

Environmental engineering is not concerned primarily with heating, ventilating, or air conditioning, nor is it concerned primarily with landscape architecture. Neither should it be confused with the architectural and structural engineering functions associated with built environments, such as homes, offices, and other workplaces.

Historically, environmental engineering has been a specialty area of civil engineering. Today it is still primarily associated with civil engineering in academic curricula. However, especially at the graduate level, students may come from a multitude of other disciplines. Examples include chemical, biosystems, electrical, and mechanical engineering as well as biochemistry, microbiology, and soil science.

Professional Development

The beginning of professional development for environmental engineers is the successful attainment of the baccalaureate degree. For continued development, a degree in engineering from a program accredited by the Accreditation Board for Engineering and Technology (ABET) provides a firm foundation for professional growth. Other steps in the progression of professional development are:

- Achievement of the title "Engineer in Training" by successful completion of the Fundamentals of Engineering (FE) examination
- Achievement of the title "Professional Engineer" by successful completion of four years of applicable engineering experience and successful completion of the Principles and Practice of Engineering (PE) exam
- Achievement of the title "Board Certified Environmental Engineer" (BCEE) by successful completion of eight years of experience and successful completion of a written certification examination or 16 years of experience and successful completion of an oral examination

The FE exam and the PE exam are developed and administered by the National Council of Examiners for Engineering and Surveying (NCEES). The BCEE exams are administered by the American Academy of Environmental Engineering (AAEE). Typically, the FE examination is taken in the last semester of undergraduate academic work.

Professions

Environmental engineers are professionals. Being a professional is more than being in or of a profession. True professionals are those who pursue their learned art in a spirit of public service (ASCE, 1973). True professionalism is defined by the following characteristics:

1. Professional decisions are made by means of general principles, theories, or propositions that are independent of the particular case under consideration.
2. Professional decisions imply knowledge in a specific area in which the person is expert. The professional is an expert only in his or her profession and not an expert at everything.
3. The professional's relations with his or her clients are objective and independent of particular sentiments about them.
4. A professional achieves status and financial reward by her or his accomplishments, not by inherent qualities such as birth order, race, religion, sex, or age or by membership in a union.
5. A professional's decisions are assumed to be on behalf of the client and to be independent of self-interest.
6. The professional relates to a voluntary association of professionals and accepts only the authority of those colleagues as a sanction on his or her own behavior (Schein, 1968).

A professional's superior knowledge is recognized. This puts the client into a very vulnerable position. The client retains significant authority and responsibility for decision making. The professional supplies ideas and information and proposes courses of action. The client's judgment and consent are required. The client's vulnerability has necessitated the development of a strong professional code of ethics. The code of ethics serves to protect not only the client but the public. Codes of ethics are enforced through the professional's peer group.

Professional Codes of Ethics. Civil engineering, from which environmental engineering is primarily, but not exclusively, derived has an established code of ethics that embodies these principles. The code is summarized in Figure 1–1. The *FE Fundamentals of Engineering Supplied-Reference Handbook,* published by the National Council of Examiners for Engineering and Surveying (NCEES) includes *Model Rules of Professional Conduct.* The NCEES amplifies the principles of the code of ethics in the *Handbook.* It is available online at www.ncees.org /Exams/Study_materials/Download_FE_supplied-Reference_Handbook.php.

1–3 HISTORICAL PERSPECTIVE

Overview

Recognizing that environmental science has its roots in the natural sciences and that the most rudimentary forms of generalization about natural processes are as old as civilizations, then environmental science is indeed very old. Certainly, the Inca cultivation of crops and the mathematics of the Maya and Sumerians qualify as early applications of natural science. Likewise the Egyptian prediction and regulation of the annual floods of the Nile demonstrate that environmental engineering works are as old as civilization. On the other hand if you asked Archimedes or Newton or Pasteur what field of environmental engineering and science they worked in, they would have given you a puzzled look indeed! For that matter, even as late as 1687 the word *science* was not in vogue; Mr. Newton's treatise alludes only to *Philosophiae Naturalis Principa Mathematics (Natural Philosophy and Mathematical Principles).*

Engineering and the sciences as we recognize them today began to blossom in the 18th century. The foundation of environmental engineering as a discipline may be considered to

**AMERICAN SOCIETY OF CIVIL ENGINEERS
CODE OF ETHICS**

Fundamental Principles

Engineers uphold and advance the integrity, honor
and dignity of the engineering profession by:

1. using their knowledge and skill for the enhancement of human welfare and the environment;
2. being honest and impartial and serving with fidel-
   ity the public, their employers and clients;
3. striving to increase the competence and prestige
   of the engineering profession; and
4. supporting the professional and technical societies of their disciplines.

Fundamental Canons

1. Engineers shall hold paramount the safety, health
   and welfare of the public and shall strive to comply with the principles of sustainable development
   in the performance of their professional duties.
2. Engineers shall perform services only in areas of
   their competence.
3. Engineers shall issue public statements only in an
   objective and truthful manner.
4. Engineers shall act in professional matters for each
   employer or client as faithful agents or trustees,
   and shall avoid conflicts of interest.
5. Engineers shall build their professional reputation
   on the merit of their services and shall not compete
   unfairly with others.
6. Engineers shall act in such a manner as to uphold
   and enhance the honor, integrity, and dignity of
   the engineering profession.
7. Engineers shall continue their professional development throughout their careers, and shall
   provide opportunities for the professional development of those engineers under their supervision.

coincide with the formation of the various societies of civil engineering in the mid-1800s (e.g., the American Society of Civil Engineers in 1852). In the first instances and well into the 20th century, environmental engineering was known as sanitary engineering because of its roots in water purification. The name changed in the late 1960s and early 1970s to reflect the broadening scope that included not only efforts to purify water but also air pollution, solid waste management, and the many other aspects of environmental protection that are included in the environmental engineer's current job description.

Although we might be inclined to date the beginnings of environmental science to the 18th century, the reality is that at any time before the 1960s there was virtually no reference to environmental science in the literature.

Although the concepts of ecology had been firmly established by the 1940s and certainly more than one individual played a role, perhaps the harbinger of environmental science as we know it today was Rachel Carson and, in particular, her book *Silent Spring* (Carson, 1962). By the mid-1970s environmental science was firmly established in academia, and by the 1980s recognized subdisciplines (environmental chemistry, environmental biology, etc.) that characterize the older disciplines of natural sciences had emerged.

Hydrology

Citations for the following section originally appeared in Chow's *Handbook of Applied Hydrology* (1964). The modern science of hydrology may be considered to have begun in the 17th century with measurements. Measurements of rainfall, evaporation, and capillarity in the Seine were taken by Perrault (1678). Mariotte (1686) computed the flow in the Seine after measuring the cross section of the channel and the velocity of the flow.

The 18th century was a period of experimentation. The predecessors for some of our current tools for measurement were invented in this period. These include Bernoulli's piezometer, the Pitot tube, Woltman's current meter, and the Borda tube. Chézy proposed his equation to describe uniform flow in open channels in 1769.

The grand era of experimental hydrology was the 19th century. The knowledge of geology was applied to hydrologic problems. Hagen (1839) and Poiseulle (1840) developed the equation to describe capillary flow, Darcy published his law of groundwater flow (1856), and Dupuit developed a formula for predicting flow from a well (1863).

During the 20th century, hydrologists moved from empiricism to theoretically based explanations of hydrologic phenomena. For example, Hazen (1930) implemented the use of statistics in hydrologic analysis, Horton (1933) developed the method for determining rainfall excess based on infiltration theory, and Theis (1935) introduced the nonequilibrium theory of hydraulics of wells. The advent of high-speed computers at the end of the 20th century led to the use of finite element analysis for predicting the migration of contaminants in soil.

Water Treatment

The provision of water and necessity of carrying away wastes were recognized in ancient civilizations: a sewer in Nippur, India, was constructed about 3750 B.C.E.; a sewer dating to the 26th century B.C.E. was identified in Tel Asmar near Baghdad, Iraq (Babbitt, 1953). Herschel (1913), in his translation of a report by Roman water commissioner Sextus Frontinus, identified nine aqueducts that carried over $3 \times 10^5 \text{ m}^3 \cdot \text{d}^{-1}$ of water to Rome in 97 A.D.

Over the centuries, the need for clean water and a means for wastewater disposal were discovered, implemented, and lost to be rediscovered again and again. The most recent rediscovery and social awakening occurred in the 19th century.

In England, the social awakening was preceded by a water filtration process installed in Paisley, Scotland, in 1804 and the entrepreneurial endeavors of the Chelsea Water Company, which installed filters for the purpose of improving the quality of the Thames River water in 1829 (Baker, 1981; Fair and Geyer, 1954). Construction of the large Parisian sewers began in 1833, and W. Lindley supervised the construction of sewers in Hamburg, Germany, in 1842 (Babbitt, 1953). The social awakening was led by physicians, attorneys, engineers, statesmen, and even the writer Charles Dickens. "Towering above all was Sir Edwin Chadwick, by training a lawyer, by calling a crusader for health. His was the chief voice in the *Report from the Poor Law Commissioners on an Inquiry into the Sanitary Conditions of the Labouring Populations of Great Britain,* 1842" (Fair and Geyer, 1954). As is the case with many leaders of the environmental movement, his recommendations were largely unheeded.

Among the first recognizable environmental scientists were John Snow (Figure 1–2) and William Budd (Figure 1–3). Their epidemiological research efforts provided a compelling demonstration of the relationship between contaminated water and disease. In 1854, Snow demonstrated the relationship between contaminated water and cholera by plotting the fatalities from cholera and their location with reference to the water supply they used (Figures 1–4 and 1–5). He found that cholera deaths in one district of London were clustered around the Broad Street Pump, which supplied contaminated water from the Thames River (Snow, 1965). In 1857, Budd began work that ultimately showed the relationship between typhoid and water contamination. His monograph, published in 1873, not only described the sequence of events in the propagation of typhoid but also provided a succinct set of rules for prevention of the spread of the disease (Budd, 1977). These rules are still valid expedients over 133 years later. The work of these two individuals is all the more remarkable in that it preceded the discovery of the germ theory of disease by Koch in 1876.

In the United States a bold but unsuccessful start on filtration was made at Richmond, Virginia, in 1832. No further installations were made in the United States until after the Civil War. Even then, they were for the most part failures. The primary means of purification from the 1830s until the 1880s was plain sedimentation.

FIGURE 1–2

Dr. John Snow. (©Pictorial Press Ltd/Alamy)

FIGURE 1–3

Dr. William Budd. (©Used with permission of the Library & Archives Service, London School of Hygiene & Tropical Medicine)

It is worthy of note that the American Water Works Association (AWWA) was established in 1881. This body of professionals joined together to share their knowledge and experience. As with other professional societies and associations formed in the late 1800s and early 1900s, the activities of the Association provide a repository for the knowledge and experience gained in purifying water. It was and is an integral part of the continuous improvement in the purification of drinking water. It serves a venue to present new ideas and challenge ineffective practices. Its journal and other publications provide a means for professionals to keep abreast of advances in the techniques for water purification.

Serious filtration research in the United States began with the establishment of the Lawrence Experiment Station by the State Board of Health in Massachusetts in 1887. On the basis of experiments conducted at the laboratory, a slow sand filter was installed in the city of Lawrence and put into operation in 1887.

At about the same time, rapid sand filtration technology began to take hold. The success here, in contrast to the failure in Britain, is attributed to the findings of Professors Austen and Wilber at Rutgers University and experiments with a full-scale plant in Cincinnati, Ohio, by George Warren Fuller. Austin and Wilber reported in 1885 that the use of alum as a coagulant when followed by plain sedimentation yielded a higher quality water than plain sedimentation alone. In 1899, Fuller reported on the results of his research. He combined the coagulation-settling process with rapid sand filtration and successfully purified Ohio River water even during its worst conditions. This work was widely disseminated.

FIGURE 1–4

Dr. Snow's map of cholera fatalities in London, August 19 to September 30, 1854. Each bar (■) represents one fatality.

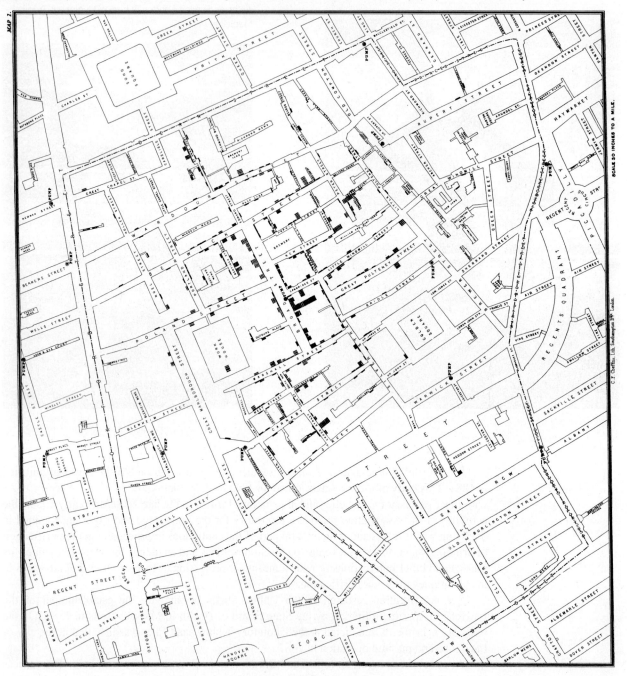

The first permanent water chlorination plant anywhere in the world was put into service in Middlekerrke, Belgium, in 1902. This was followed by installations at Lincoln, England, in 1905 and at the Boonton Reservoir for Jersey City, New Jersey, in 1908. Ozonation began about the same time as chlorination. However, until the end of the 20th century, the economics of disinfection by ozonation were not favorable.

FIGURE 1–5

Map of service areas of three water companies in London, 1854. To view the original colors, go to the UCLA website: http://www.ph.ucla.edu/epi

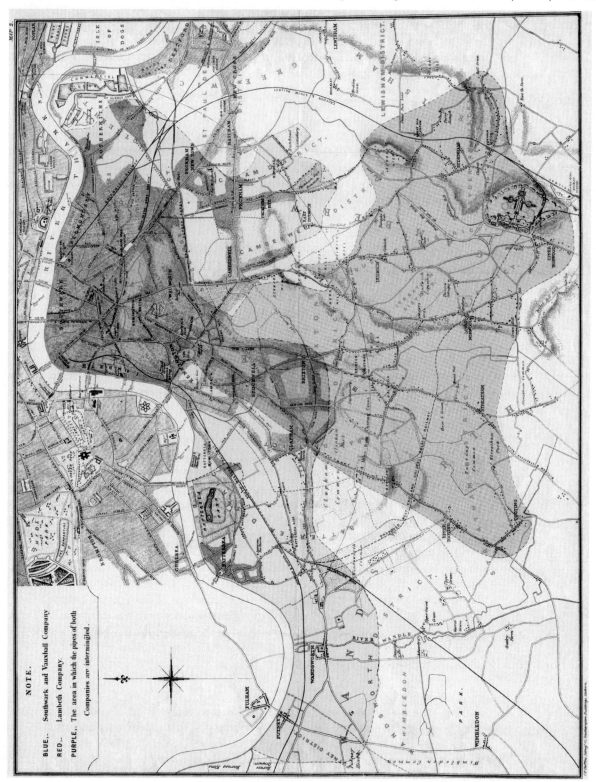

Fluoridation of water was first used for municipal water at Grand Rapids, Michigan, in 1945. The objective was to determine whether or not the level of dental cavities could be reduced if the fluoride level were raised to levels near those found in the water supplies of populations having a low prevalence of cavities. The results demonstrated that proper fluoridation results in a substantial reduction in tooth decay (AWWA, 1971).

The most recent major technological advance in water treatment is filtration with synthetic membranes. First introduced in the 1960s, membranes became economical for application in special municipal applications in the 1990s.

Wastewater Treatment

Early efforts at sewage treatment involved carrying the sewage to the nearest river or stream. Although the natural biota of the stream did indeed consume and thus treat part of the sewage, in general, the amount of sewage was too large and the result was an open sewer.

In England, the Royal Commission on Rivers Pollution was appointed in 1868. Over the course of their six reports, they provided official recognition (in decreasing order of preference) of sewage filtration, irrigation, and chemical precipitation as acceptable methods of treatment (Metcalf and Eddy, 1915).

At this point in time, events began to move rather quickly in both the United States and England. The first U.S. treatment of sewage by irrigation was attempted at the State Insane Asylum in August, Maine, in 1872.

The first experiments on aeration of sewage were carried out by W. D. Scott-Monctieff at Ashtead, England, in 1882 (Metcalf and Eddy, 1915). He used a series of nine trays over which the sewage percolated. After about 2 days operation, bacterial growths established themselves on the trays and began to effectively remove organic waste material.

With the establishment of the Lawrence Laboratory in Massachusetts in 1887, work on sewage treatment began in earnest. Among the notables who worked at the laboratory were Allen Hazen, who was in charge of the lab in its formative years, and the team of Ellen Richards and George Whipple, who were among the first to isolate the organisms that oxidized nitrogen compounds in wastewater.

In 1895, the British collected methane gas from septic tanks and used it for gas lighting in the treatment plant. After successful development by the British, the tricking filter was installed in Reading, Pennsylvania, Washington, Pennsylvania, and Columbus, Ohio, in 1908 (Emerson, 1945).

In England, Arden and Lockett conducted the first experiments that led to the development of the activated sludge process in 1914. The first municipal activated sludge plant in the United States was installed in 1916 (Emerson, 1945).

The progress of the state of the art of wastewater treatment has been recorded by the Sanitary Engineering Division (later the Environmental Engineering Division) of the American Society of Civil Engineers. It was formed in June 1922. The *Journal of the Environmental Engineering Division* is published monthly. The Federation of Sewage and Industrial Wastes Association, also known as the Water Pollution Control Federation, was established in October 1928 and publishes reports on the advancement of the state of the art. Now called the Water Environment Federation (WEF), its journal is *Water Environment Research*.

Air Pollution Control

Although there were royal proclamations and learned essays about air pollution as early as 1272, these were of note only for their historic value. The first experimental apparatus for clearing particles from the air was reported in 1824 (Hohlfeld, 1824). Hohlfeld used an electrified needle to clear fog in a jar. This effect was rediscovered in 1850 by Guitard and again in 1884 by Lodge (White, 1963).

The latter half of the 19th century and early 20th century were watershed years for the introduction of the forerunners of much of the current technology now in use: fabric filters (1852), cyclone collectors (1895), venturi scrubbers (1899), electrostatic precipitator (1907), and the plate tower for absorption of gases (1916). It is interesting to note that unlike water and wastewater treatment where disease and impure water were recognized before the advent of treatment technologies, these developments preceded the recognition of the relationship between air pollution and disease.

The Air & Waste Management Association was founded as the International Union for Prevention of Smoke in 1907. The organization grew from its initial 12 members to more than 9000 in 65 countries.

The 1952 air pollution episode in London that claimed 4000 lives (WHO, 1961), much like the cholera epidemic of 1849 that claimed more than 43,000 lives in England and Wales, finally stimulated positive legislation and technical attempts to rectify the problem.

The end of the 20th century saw advances in chemical reactor technology to control sulfur dioxide, nitrogen oxides, and mercury emissions from fossil-fired power plants. The struggle to control the air pollution from the explosive growth in use of the automobile for transportation was begun.

Environmental scientists made major discoveries about global air pollution at the end of the 20th century. In 1974, Molina and Rowland identified the chemical mechanisms that cause destruction of the ozone layer (Molina and Rowland, 1974). By 1996, the Intergovernmental Panel on Climate Change (IPCC) agreed that "(t)he balance of evidence suggests a discernable human influence on global climate" (IPCC, 1996).

Solid and Hazardous Waste

From as early as 1297, there was a legal obligation on householders in London to ensure that the pavement within the frontage of their tenements was kept clear (GLC, 1969). The authorities found it extremely difficult to enforce the regulations. In 1414, the constables and other officials had to declare their willingness to pay informers to gather evidence against the offenders who cast rubbish and dirt into the street. The situation improved for a time in 1666. The Great Fire of London had a purifying effect and for some time complaints about refuse in the streets ceased. As with previously noted advances in environmental enlightenment, not much success was achieved until the end of the 19th century (GLC, 1969).

The modern system of refuse collection and disposal instituted in 1875 has been changed little by technology. People still accompany a wheeled vehicle and load it, usually by hand, after which the material is taken to be dumped or burned. The horses formerly used have been replaced by an internal-combustion engine that has not greatly increased the speed of collection. In fact, to some extent, productivity of the crew fell because, while a horse can move on command, a motor vehicle has to have a driver, who usually does not take part in the collection process. At the end of the 20th century, one-person crews with automated loading equipment began to replace multiperson crews.

Incineration was the initial step taken in managing collected solid waste. The first U.S. incinerators were installed in 1885. By 1921, more than two hundred incinerators were operating. Waste management, with an emphasis on **sanitary landfilling,** began in the United Kingdom in the early 1930s (Jones and Owen, 1934). Sanitary landfilling included three criteria: daily soil cover, no open burning, and no water pollution problems (Hagerty and Heer, 1973).

In the 1970s, the rising environmental movement brought recognition of the need to conserve resources and to take special care of wastes that are deemed hazardous because they are ignitable, reactive, corrosive, or toxic. Incineration fell into disrepute because of the difficulty in controlling air pollution emissions. In 1976, the U.S. Congress enacted legislation to focus on resource recovery and conservation as well as the management of hazardous wastes.

1–4   HOW ENVIRONMENTAL ENGINEERS AND ENVIRONMENTAL SCIENTISTS WORK TOGETHER

There is an old saying that "Scientists discover things and engineers make them work." Like many similar old saws there is a grain of truth in the statement and part is out of date. From an educational point of view, environmental engineering is founded on environmental science. Environmental science and, in particular, quantitative environmental science provides the fundamental theories used by environmental engineers to design solutions for environmental problems. In many instances the tasks and tools of environmental scientists and environmental engineers are the same.

Perhaps the best way to explain how environmental scientists and engineers work together is to give some examples:

- Early in the 20th century, a dam was built to provide water for cooling in a power plant. The impact of the dam on the oxygen in the river and its ability to support fish life was not considered. The migration of salmon in the river was not considered. To remedy the problem, environmental engineers and scientists designed a fish ladder that not only provided a means for the fish to bypass the dam but also aerated the water to increase the dissolved oxygen. The environmental scientists provided the knowledge of the depth of water and height of the steps the fish could negotiate. The environmental engineers determined the structural requirements of the bypass to allow enough water to flow around the dam to provide the required depth.

- Storm water from city streets was carrying metal and organic contaminants from the street into a river and polluting it. Although a treatment plant could have been built, a wetland mitigation system was selected to solve the problem. The slope of the channel through the wetland was designed by the environmental engineers. The provision of limestone along the channel bed to neutralize the pH and remove metals was determined by the joint work of the environmental scientists and engineers. The selection of the plant material for the wetland was the job of the environmental scientists.

- A highway rest area septic tank system was being overloaded during holiday weekend traffic. Rather than build a bigger septic system or a conventional wastewater treatment plant, an overland flow system in the median strip of the highway was selected as the solution. The engineering system to move the wastewater from the rest area to the median strip was designed by the environmental engineer. The slope of the overland flow system and its length were jointly determined by the environmental scientist and engineer. The grass cover was selected by the environmental scientist.

In each of these instances, the environmental engineer and the environmental scientist had something to contribute. Each had to be familiar with the requirements of the other to be able to come up with an acceptable solution.

1–5   INTRODUCTION TO PRINCIPLES OF ENVIRONMENTAL ENGINEERING AND SCIENCE

Where Do We Start?

We have used the ASCE definition of an environmental engineer as a starting point for this book:

1. Provision of safe, palatable, and ample public water supplies
2. Proper disposal of or recycling of wastewater and solid wastes
3. Control of water, soil, and atmospheric pollution (including noise as an atmospheric pollutant)

To this list we have added those topics from environmental sciences that complement and round out our understanding of the environment: ecosystems, risk assessment, soil and geological resources, and agricultural effects.

A Short Outline of This Book

A short outline of this book provides an overview of the many aspects of environmental engineering and science. The first chapters present a review and introduction to the tools used in the remainder of the book. This includes a chemistry review (Chapter 2), a biology review (Chapter 3), an introduction to materials and energy balances (Chapter 4), an introduction to ecosystems (Chapter 5), and an introduction to risk assessment (Chapter 6).

The science of hydrology is introduced in Chapter 7. The principles of conservation of mass are used to describe the balance of water in nature. The physics of water behavior above and below ground will give you the quantitative tools to understand the relationships between rainfall and stream flow that you will need to understand problems of groundwater pollution.

Chapter 8 provides an overview of water, energy, mineral, and soil resources; the environmental impacts of their use; and some approaches to sustainable use of these resources.

Water quality is a dynamic process. The interrelationships between hydraulic parameters and the chemistry and biology of lakes and streams is described in Chapter 9.

The treatment of water for human consumption is founded on fundamental principles of chemistry and physics. In Chapter 10, these are used to demonstrate methods of purifying water.

Modern treatment of municipal wastewater and some industrial wastes is based on the application of fundamentals of chemistry, microbiology, and physics. These are explained in Chapter 11.

Air pollution, though occurring naturally, is most closely related to human activities. The chemical reactions that occur and the physical processes that transport air pollutants as well as their environmental effects are discussed in Chapter 12.

Chapter 13 presents an overview of the problem of solid waste generation and its environmental effect.

Hazardous waste is the topic of Chapter 14. We examine some alternatives for pollution prevention and the treatment of these wastes as an application of quantitative environmental science to environmental engineering.

It has been estimated that 1.7 million workers in the United States between the ages of 50 and 59 have enough hearing loss to be awarded compensation. The environmental insult most frequently cited in connection with highways is noise. The fundamentals of physics are used in Chapter 15 to describe and mitigate noise.

The final chapter is a brief examination of ionizing radiation with an introduction to health effects of radiation.

1–6   ENVIRONMENTAL SYSTEMS OVERVIEW

Systems

Before we begin in earnest, we thought it worth taking a look at the problems to be discussed in this text in a larger perspective. Engineers and scientists like to call this the **systems approach,** that is, looking at all the interrelated parts and their effects on one another. In environmental systems it is doubtful that mere mortals can ever hope to identify all the interrelated parts, to say nothing of trying to establish their effects on one another. The first thing the systems engineer or scientist does, then, is to simplify the system to a tractable size that behaves in a fashion similar to the real system. The simplified model does not behave in detail as the system does, but it gives a fair approximation of what is going on.

In Chapter 5 we introduce the systems of natural science called ecosystems. On the large scale shown in Figure 1–6, the ecosystem sets a framework for the topics selected for this book,

FIGURE 1–6

The Earth as an
ecosystem.

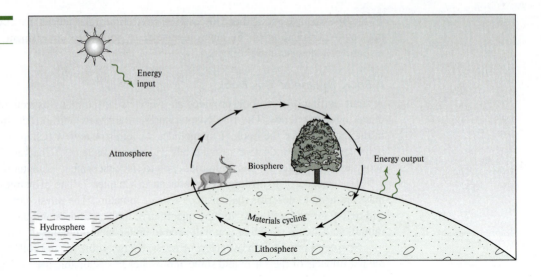

that is, the relationships and interactions of plants and animals with the water, air, and soil that make up their environment. From this large scale, we present three environmental systems that serve as the focus of this book: the water resource management system (Chapters 7, 9, 10, and 11), the air resource management system (Chapters 12 and 15), and the solid waste management system (Chapters 8, 13, and 14). Pollution problems that are confined to one of these systems are called **single-medium problems** if the medium is either air, water, or soil. Many important environmental problems are not confined to one of these simple systems but cross the boundaries from one to another. Such problems are referred to as **multimedia pollution problems.**

Water Resource Management System

Water Supply Subsystem.   The nature of the water source commonly determines the planning, design, and operation of the collection, purification, transmission, and distribution works.* The two major sources used to supply community and industrial needs are referred to as **surface-water** and **groundwater.** Streams, lakes, and rivers are the surface water sources. Groundwater sources are those pumped from wells.

Figure 1–7 depicts an extension of the water resource system to serve a small community. The source in each case determines the type of collection works and the type of treatment works. The pipe network in the city is called the **distribution system.** The pipes themselves are often referred to as **water mains.** Water in the mains generally is kept at a pressure between 200 and 860 kilopascals (kPa). Excess water produced by the treatment plant during periods of low **demand**† (usually the nighttime hours) is held in a storage reservoir. The storage reservoir may be elevated (the ubiquitous water tower), or it may be at ground level. The stored water is used to meet high demand during the day. Storage compensates for changes in demand and allows a smaller treatment plant to be built. It also provides emergency backup in case of a fire.

Population and water consumption patterns are the prime factors that govern the quantity of water required and hence the source and the whole composition of the water resource system. One of the first steps in the selection of a suitable water supply source is determining the demand that will be placed on it. The essential elements of water demand include average daily water consumption and peak rate of demand. Average daily water consumption must be

---

*Works is a noun used in the plural to mean "engineering structures." It is used in the same sense as art works.

†Demand is the use of water by consumers. This use of the word derives from the economic term meaning "the desire for a commodity." The consumers express their desire by opening the faucet or flushing the water closet (W.C.).

FIGURE 1–7

An extension of the water supply resource system.

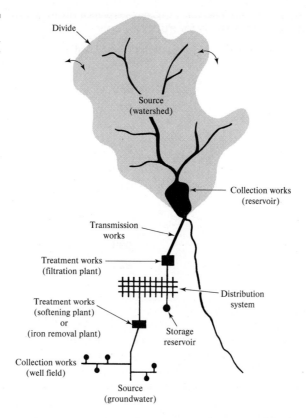

estimated for two reasons: (1) to determine the ability of the water source to meet continuing demands over critical periods when surface flows are low or groundwater tables are at minimum elevations, and (2) for purposes of estimating quantities of stored water that would satisfy demands during these critical periods. The peak demand rates must be estimated in order to determine plumbing and pipe sizing, pressure losses, and storage requirements necessary to supply sufficient water during periods of peak water demand.

Many factors influence water use for a given system. For example, the mere fact that water under pressure is available stimulates its use, often excessively, for watering lawns and gardens, for washing automobiles, for operating air-conditioning equipment, and for performing many other activities at home and in industry. The following factors have been found to influence water consumption in a major way:

1. Climate
2. Industrial activity
3. Meterage
4. System management
5. Standard of living

The following factors also influence water consumption but to a lesser degree: extent of sewers, system pressure, water price, and availability of private wells.

If the demand for water is measured on a *per capita** basis, climate is the most important factor influencing demand. This is shown dramatically in Table 1–1. The average annual

---

Per capita is a Latin term that means "by heads." Here it means "per person." This assumes that each person has one head (on the average).

TABLE 1–1	Total Fresh Water Withdrawals for Public Supply	
	State	Withdrawal, Lpcd[1]
	Wet	
	Connecticut	680
	Michigan	598
	New Jersey	465
	Ohio	571
	Pennsylvania	543
	Average	571
	Dry	
	Nevada	1,450
	New Mexico	698
	Utah	926
	Average	1,025

Source: Compiled from Kenny et al., 2009.
[1]Lpcd = liters per capita per day.

precipitation for the "wet" states is about 100 cm per year, and the average annual precipitation for the "dry" states is only about 25 cm per year. Of course, in this list, the dry states are also considerably warmer than the wet states.

The influence of industry is to increase per capita water demand. Small rural and suburban communities will use less water per person than industrialized communities.

The third most important factor in water use is whether individual consumers have water meters. Meterage imposes a sense of responsibility not found in unmetered residences and businesses. This sense of responsibility reduces per capita water consumption because customers repair leaks and make more conservative water-use decisions almost regardless of price. For residential consumers, water is so inexpensive, price is not much of a factor.

Following meterage in importance is system management. If the water distribution system is well managed, per capita water consumption is less than if it is not well managed. Well-managed systems are those in which the managers know when and where leaks in the water main occur and have them repaired promptly. Water price is extremely important for industrial and farming operations that use large volumes of water.

Climate, industrial activity, meterage, and system management are more significant factors controlling water consumption than the standard of living. The rationale for factor 5 is straightforward. Per capita water use increases with an increased standard of living. Highly developed countries use much more water than less-developed nations. Likewise, higher socioeconomic status implies greater per capita water use than lower socioeconomic status.

The total U.S. water withdrawal for all uses (agricultural, commercial, domestic, mining, and thermoelectric power) including both fresh and saline water was estimated to be approximately 5,000 liters per capita per day (Lpcd) in 2010. The amount for U.S. public supply (domestic, commercial, and industrial use) was estimated to be 590 Lpcd in 2010 (Maupin et al., 2014). The American Water Works Association estimated that the average daily household water use in the United States was 525 liters per day in 2010 (AWWA, 2016). This would amount to about 200 Lpcd. The variation in demand is normally reported as a factor of the average day. For metered dwellings the factors are as follows: maximum day equals 2.2 times

TABLE 1–2	**Examples of Variation in Per Capita Water Consumption**			
			Percent of Per Capita Consumption	
Location	**Lpcd**	**Industry**	**Commercial**	**Residential**
Lansing, Michigan	512	14	32	54
East Lansing, Michigan	310	0	10	90
Michigan State University	271	0	1	99

Source: Data from local treatment plants, 2004.

average day, while peak hour equals 5.3 times the average day (Linaweaver et al., 1967). Some mid-Michigan average daily use figures and the contribution of various sectors to demand are shown in Table 1–2.

International per capita domestic water use has been estimated by the United Nations (www.DATA360.org). For example, they report the following (all in Lpcd): Australia, 493; Bangladesh, 46; Canada, 3300; China, 86; Germany, 193; India, 135; Mexico, 366; and Nigeria, 36.

Wastewater Disposal Subsystem. Safe disposal of all human wastes is necessary to protect the health of the individual, the family, and the community, and also to prevent the occurrence of certain nuisances. To accomplish satisfactory results, human wastes must be disposed of so that:

1. They will not contaminate any drinking water supply.
2. They will not give rise to a public health hazard by being accessible to insects, rodents, or other possible disease carriers that may come into contact with food or drinking water.
3. They will not give rise to a public health hazard by being accessible to children.
4. They will not cause violation of laws or regulations governing water pollution or sewage disposal.
5. They will not pollute or contaminate the waters of any bathing beach, shellfish-breeding ground, or stream used for public or domestic water supply purposes, or for recreational purposes.
6. They will not give rise to a nuisance due to odor or unsightly appearance.

These criteria can best be met by the discharge of domestic sewage to an adequate public or community sewer system (U.S. PHS, 1970). Where no community sewer system exists, on-site disposal by an approved method is mandatory.

In its simplest form the wastewater management subsystem is composed of six parts (Figure 1–8). The source of wastewater may be either industrial wastewater or domestic sewage* or both. Industrial wastewater may be subject to some pretreatment on site if it has the potential to upset the municipal **wastewater treatment plant** (WWTP). Federal regulations refer to municipal wastewater treatment systems as **publicly owned treatment works,** or POTWs.

The quantity of sewage flowing to the WWTP varies widely throughout the day in response to water usage. A typical daily variation is shown in Figure 1–9. Most of the water used in a community will end up in the sewer. Between 5 and 10% of the water is lost in lawn watering, car washing, and other consumptive uses. In warm climates, consumptive use out of doors may be as high as 60%. Consumptive use may be thought of as the difference between the average

FIGURE 1–8

Wastewater management subsystem.

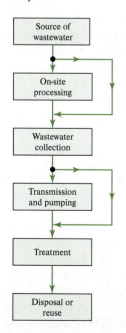

---

*Domestic sewage is sometimes called sanitary sewage, although it is far from being sanitary!

FIGURE 1–9

Typical variation in daily
wastewater flow.

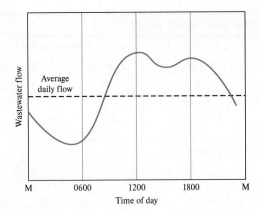

rate that water flows into the distribution system and the average rate that wastewater flows into the WWTP (excepting the effects of leaks in the pipes).

The quantity of wastewater, with one exception, depends on the same factors that determine the quantity of water required for supply. The major exception is that underground water (groundwater) conditions may strongly affect the quantity of water in the system because of leaks. Whereas the drinking water distribution system is under pressure and is relatively tight, the sewer system is gravity-operated and is relatively open. Thus, groundwater may **infiltrate,** or leak into, the system. When manholes lie in low spots, there is the additional possibility of **inflow** through leaks in the manhole cover. Other sources of inflow include direct connections from roof gutters and downspouts, as well as sump pumps used to remove water from basement footing tiles. **Infiltration** and **inflow** (I & I) are particularly important during rainstorms. The additional water from I & I may hydraulically overload the sewer, causing sewage to back up into houses, as well as to reduce the efficiency of the WWTP. New construction techniques and materials have made it possible to reduce I & I to insignificant amounts.

Sewers are classified into three categories: sanitary, storm, and combined. **Sanitary sewers** are designed to carry municipal wastewater from homes and commercial establishments. With proper pretreatment, industrial wastes may also be discharged into these sewers. **Storm sewers** are designed to handle excess rainwater and snow melt to prevent flooding of low areas. Whereas sanitary sewers convey wastewater to treatment facilities, storm sewers generally discharge into rivers and streams. **Combined sewers** are expected to accommodate both municipal wastewater and storm water. These systems were designed so that during dry periods the wastewater is carried to a treatment facility. During rain storms, the excess water is discharged directly into a river, stream, or lake without treatment. Unfortunately, the storm water is mixed with untreated sewage. The U.S. Environmental Protection Agency (EPA) has estimated that 40,000 overflows occur each year. Combined sewers are no longer being built in the United States. Many communities are in the process of replacing the combined sewers with separate systems for sanitary and storm flow.

When gravity flow is not possible or when sewer trenches become uneconomically deep, the wastewater may be pumped. When the sewage is pumped vertically to discharge into a gravity sewer at a higher elevation, the location of the sewage pump is called a **lift station.**

Sewage treatment is performed at the WWTP to stabilize the waste material, that is, to make it less **putrescible** (likely to decompose). The **effluent** from the WWTP may be discharged into an ocean, lake, or river (called the receiving body). Alternatively, it may be discharged onto (or into) the ground or be processed for reuse. The by-product sludge from the WWTP also must be disposed of in an environmentally acceptable manner.

Whether the waste is discharged onto the ground or into a receiving body, care must be exercised not to overtax the assimilative capacity of the ground or receiving body. The fact that the wastewater effluent is cleaner than the river into which it flows does not justify the discharge if it turns out to be the proverbial "straw that breaks the camel's back."

In summary, water resource management is the process of managing both the quantity and the quality of the water used for human benefit without destroying its availability and purity.

Air Resource Management System

Our air resource differs from our water resource in two important aspects. The first is in regard to quantity. Whereas engineering structures are required to provide an adequate water supply, air is delivered free of charge in whatever quantity we desire. The second aspect is in regard to quality. Unlike water, which can be treated before we use it, it is impractical to go about with a gas mask on to treat impure air or to use ear plugs to keep out the noise.

The balance of cost and benefit for obtaining a desired quality of air is termed **air resource management.** Cost-benefit analyses can be problematic for at least two reasons. First is the question of what is desired air quality. The basic objective is, of course, to protect the health and welfare of people. But how much air pollution can we stand? We know the tolerable limit is something greater than zero, but tolerance varies from person to person. Second is the question of cost versus benefit. We know that we don't want to spend the entire gross domestic product to ensure that no individual's health or welfare is impaired, but we do know that we want to spend some amount. Although the cost of control can be reasonably determined by standard engineering and economic means, the cost of pollution is still far from being quantitatively assessed.

Air resource management programs are instituted for a variety of reasons. The most defensible reasons are that (1) air quality has deteriorated and there is a need for correction, and (2) the potential for a future problem is strong.

In order to carry out an air resource management program effectively, all of the elements shown in Figure 1–10 must be employed. (Note that with the appropriate substitution of the word *water* for *air,* these elements apply to management of water resources as well.)

Solid Waste Management System

In the past, solid waste was considered a resource, and we will examine its current potential as a resource. Generally, however, solid waste is considered a problem to be solved as cheaply as possible rather than a resource to be recovered. A simplified block diagram of a solid waste management system is shown in Figure 1–11.

Typhoid and cholera epidemics of the mid-1800s spurred water resource management efforts, and, while air pollution episodes have prompted better air resource management, we have yet to feel the effect of material or energy shortages severe enough to encourage modern solid waste management. The landfill "crisis" of the 1980s appears to have abated in the early 1990s due to new or expanded landfill capacity and to many initiatives to reduce the amount of solid waste generated. By 1999, more than 9000 curbside recycling programs served roughly half of the U.S. population (U.S. EPA, 2005).

Multimedia Systems

Many environmental problems cross the air–water–soil boundary. An example is acid rain that results from the emission of sulfur oxides and nitrogen oxides into the atmosphere. These pollutants are washed out of the atmosphere, thus cleansing it, but in turn polluting water and changing the soil chemistry, which ultimately results in the death of fish and trees. Thus, our historic reliance on the natural cleansing processes of the atmosphere in designing air pollution control equipment has failed to deal with the multimedia nature of the problem. Likewise, disposal of solid waste by incineration results in air pollution, which in turn is controlled by scrubbing with water, resulting in a water pollution problem.

Three lessons have come to us from our experience with multimedia problems. First, it is dangerous to develop models that are too simplistic. Second, environmental engineers and scientists must use a multimedia approach and, in particular, work with a multidisciplinary team

A simplified block diagram of an air resource management system.

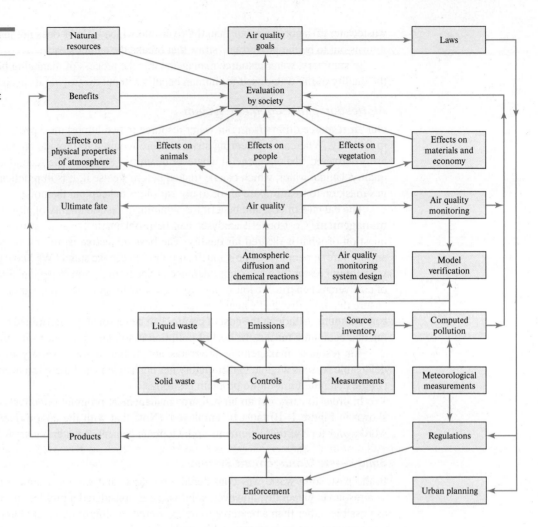

A simplified block diagram of a solid waste management system.

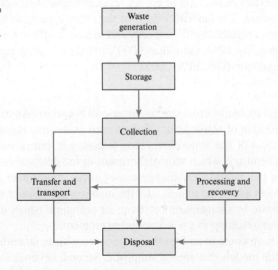

to solve environmental problems. Third, the best solution to environmental pollution is waste minimization—if waste is not produced, it does not need to be treated or disposed of.

Sustainability

"**Sustainable development** is development that meets the needs of the present without compromising the ability of future generations to meet their own needs" (WCED, 1987). Although pollution problems will remain with us for the foreseeable future, an overriding issue for the continuation of our modern living style and for the development of a similar living style for those in developing countries is the question of **sustainability.** That is, how do we maintain our ecosystem in the light of major depletion of our natural resources? If, in our systems view, we look beyond the simple idea of controlling pollution to the larger idea of sustaining our environment, we see that there are better solutions for our pollution problems. For example:

- Pollution prevention by the minimization of waste production
- Life cycle analysis of our production techniques to include built-in features for extraction and reuse of materials
- Selection of materials and methods that have a long life
- Selection of manufacturing methods and equipment that minimize energy and water consumption

1–7 ENVIRONMENTAL LEGISLATION AND REGULATION

In the United States, our publicly elected state and federal officials enact environmental laws. The laws direct the appropriate agency to develop and publish regulations to implement the requirements of the law. At the federal level, the U.S. Environmental Protection Agency (EPA) is the primary agency that develops and enforces environmental regulations. Our focus in this discussion is on federal legislation. Much of the legislation enacted at the state level is derived from federal laws.

Acts, Laws, and Regulations

The following paragraphs provide a brief introduction to the process leading to the establishment of regulations and the terms used to identify the location of information about laws and regulations. This discussion is restricted to the federal process and nomenclature.

A proposal for a new law, called a *bill,* is introduced in either the Senate or the House of Representatives (House). The bill is given a designation, for example S. 2649 in the Senate or H.R. 5959 in the House. Bills often have "companions" in that similar bills may be started in both the Senate and House at the same time. The bill is given a title, for example the "Safe Drinking Water Act," which implies an "act" of Congress. The act may be listed under one "Title" or it may be divided into several "Titles." References to the "Titles" of the act are given by roman numeral. For example, Title III of the Clean Air Act Amendments establishes a list of hazardous air pollutants. Frequently a bill directs some executive branch of the government such as the EPA to carry out an action such as setting limits for contaminants. On occasion, such a bill includes specific numbers for limits on contaminants. If the bills successfully pass the committee to which they are assigned, they are "reported out" to the full Senate (for example, Senate Report 99-56) or to the full House (House Report 99-168). The first digits preceding the dash refer to the session of Congress during which the bill is reported out. In this example, it is the 99th Congress. If bills pass the full Senate/House, they are taken by a joint committee of senators and congressmembers (conference committee) to form a single bill for action by both the Senate and House. If the bill is adopted by a majority of both houses, it goes to the president for approval or veto. When the president signs the bill, it becomes a **law** or **statute.** It is then designated, for example, as Public Law 99-339 or PL 99-339. This means it is the 339th law passed by the 99th Congress. The law or statute approved by the president's signature may alternatively be called an act that is referred to by the title assigned the bill in Congress.

The Office of the Federal Register prepares the *United States Statutes at Large* annually. This is a compilation of the laws, concurrent resolutions, reorganization plans, and proclamations issued during each congressional session. The statutes are numbered chronologically. They are not placed in order by subject matter. The shorthand reference is, for example, 104 Stat. 3000.

The *United States Code* is the compiled, written set of laws in force on the day before the beginning of the current session of Congress (*U.S. Code*, 2005). Reference is made to the *U.S. Code* by title and section number (for example, 42 USC 6901 or 42 U.S.C. §6901). The title referred to in Table 1–3 gives a sample of titles and sections of environmental interest. Note that titles of the *U.S. Code* do not match the titles of the acts of Congress.

In carrying out the directives of the Congress to develop a **regulation** or **rule,** the EPA or other executive branch of the government follows a specific set of formal procedures in a process referred to as **rule making.** The government agency (EPA, Department of Energy, Federal Aviation Agency, etc.) first publishes a **proposed rule** in the *Federal Register.* The *Federal Register* is, in essence, the government's newspaper. It is published every day that the federal government is open for business. The agency provides the logic for the rule making (called a **preamble**) as well as the proposed rule and requests comments. The preamble may be several hundred pages in length for a rule that is only a few lines long or a single page table of allowable concentrations of contaminants. Prior to the issuance of a final rule, the agency allows and considers public comment. The time period for submitting public comments varies. For rules that are not complex or controversial, it may be a few weeks. For more complicated rules, the comment period may extend for as long as a year. The reference citation to *Federal Register* publications is in the following form: 59 FR 11863. The first number is the volume number.

TABLE 1–3	U.S. Code Title and Section Numbers of Environmental Interest		
	Title	**Sections**	**Statute**
	7	136 to 136y	Federal Insecticide, Fungicide, and Rodenticide Act
	16	1531 to 1544	Endangered Species Act
	33	1251 to 1387	Clean Water Act
	33	2701 to 2761	Oil Pollution Act
	42	300f to 300j-26	Safe Drinking Water Act
	42	4321 to 4347	National Environmental Policy Act
	42	4901 to 4918	Noise Control Act
	42	6901 to 6922k	Solid Waste Disposal Act
	42	7401 to 7671q	Clean Air Act (includes noise at §7641)
	42	9601 to 9675	Comprehensive Environmental Response, Compensation, and Liability Act
	42	11001 to 11050	Emergency Planning and Community Right-to-Know Act
	42	13101 to 13109	Pollution Prevention Act
	46	3703a	Oil Pollution Act
	49	2101	Aviation Safety and Noise Abatement Act[1]
	49	2202	Airport and Airway Improvement Act[1]
	49	47501 to 47510	Airport Noise Abatement Act

[1] at *U.S. Code Annotated* (U.S.C.S.A.)

TABLE 1–4	*Code of Federal Regulation* Title Numbers of Environmental Interest	
	Title Number	**Subject**
	7	Agriculture (soil conservation)
	10	Energy (Nuclear Regulatory Commission)
	14	Aeronautics and Space (noise)
	16	Conservation
	23	Highways (noise)
	24	Housing and Urban Development (noise)
	29	Labor (noise)
	30	Mineral Resources (surface mining reclamation)
	33	Navigation and Navigable Waters (wet lands and dredging)
	40	Protection of the Environment (Environmental Protection Agency)
	42	Public Health and Welfare
	43	Public Lands: Interior
	49	Transportation (transporting hazardous waste)
	50	Wildlife and Fisheries

Volumes are numbered by year. The last number is the page number. Pages are numbered sequentially beginning with page one on the first day of business in January of each year. From the number shown, this rule making starts on page 11,863! Although one might assume this is late in the year, it may not be if a large number of rules has been published. This makes the date of publication very useful in searching for the rule.

Once a year on July 1, the rules that have been finalized in the past year are **codified.** This means they are organized and published in the *Code of Federal Regulations* (CFR, 2005). Unlike the *Federal Register,* the *Code of Federal Regulations* is a compilation of the rules/regulations of the various agencies without explanation of how the government arrived at the decision. The explanation of how the rule was developed may be found only in the *Federal Register*. The notation used for *Code of Federal Regulations* is as follows: 40 CFR 280. The first number is the title number. The second number in the citation refers to the part number. Unfortunately, this title number has no relation to either the title number in the act or the *United States Code* title number. The CFR title numbers and subjects of environmental interest are shown in Table 1–4.

1–8   ENVIRONMENTAL ETHICS

The birth of environmental ethics as a force is partly a result of concern for our own long-term survival, as well as our realization that humans are but one form of life, and that we share our earth with other forms of life (Vesilind, 1975).

Although it seems a bit unrealistic for us to set a framework for a discussion of environmental ethics in this short introduction, we have summarized a few salient points in Table 1–5.

Although these few principles seem straightforward, real-world problems offer distinct challenges. Here is an example for each of the principles listed:

• The first principle may be threatened when it comes into conflict with the need for food for a starving population and the country is overrun with locusts. Will the use of pesticides enhance and protect the environment?

TABLE 1–5	An Environmental Code of Ethics
	1. Use knowledge and skill for the enhancement and protection of the environment.
	2. Hold paramount the health, safety, and welfare of the environment.
	3. Perform services only in areas of personal expertise.
	4. Be honest and impartial in serving the public, your employers, your clients, and the environment.
	5. Issue public statements only in an objective and truthful manner.

- The EPA has stipulated that wastewater must be disinfected where people come into contact with the water. However, the disinfectant may also kill naturally occurring beneficial micro-organisms. Is this consistent with the second principle?
- Suppose your expertise is water and wastewater chemistry. Your company has accepted a job to perform air pollution analysis and asks you to perform the work in the absence of a colleague who is the company's expert. Do you decline and risk being fired?
- The public, your employers, and your client believe that dredging a lake to remove weeds and sediment will enhance the lake. However, the dredging will destroy the habitat for muskrats. How can you be impartial to *all* these constituencies?
- You believe that a new regulation proposed by EPA is too expensive to implement, but you have no data to confirm that opinion. How do you respond to a local newspaper reporter asking for your opinion? Do you violate the fifth principle even though "your opinion" is being sought?

Below are two cases that are more complex. We have not attempted to provide pat solutions but rather have left them for you and your instructor to resolve.

Case 1: To Add or Not to Add

A friend of yours has discovered that his firm is adding nitrites and nitrates to bacon to help preserve it. He also has read that these compounds are precursors to cancer-forming chemicals that are produced in the body. On the other hand, he realizes that certain disease organisms such as those that manufacture botulism toxin have been known to grow in bacon that has not been treated. He asks you whether he should (a) protest to his superiors knowing he might get fired; (b) leak the news to the press; (c) remain silent because the risk of dying from cancer is less than the absolute certainty of dying from botulism.

Note: The addition of nitrite to bacon is approved by the Food and Drug Administration. Nitrites and nitrates are, in and of themselves, not very toxic to adults. However, heating these compounds results in a reaction with the amines in proteins to form nitrosamines, which are carcinogenic compounds.

Case 2: You Can't Do Everything At Once

As an environmental scientist newly assigned to a developing country you find yourself in an isolated village with an epidemic of cholera (case 2 is adapted from Wright, 1964). You have two routes of action available:

1. You can nurse and comfort the sick.
2. You can try to clean up the water supply.

Which is the ethical choice?

We think it is important to point out that many environmentally related decisions such as those described here are much more difficult than the problems presented in the remaining chapters of this book. Frequently these problems are related more to ethics or economics than to environmental engineering and environmental science. The problems arise when several courses of action are possible with no certainty as to which is best. Decisions related to safety, health, and welfare are easily resolved. Decisions as to which course of action is in the best interest of the public are much more difficult to resolve. Furthermore, decisions as to which course of action is in the best interest of the environment are at times in conflict with those which are in the best interest of the public. Whereas decisions made in the public interest are based on professional ethics, decisions made in the best interest of the environment are based on environmental ethics.

Ethos, the Greek word from which *ethic* is derived, means the character of a person as described by his or her actions. This character was developed during the evolutionary process and was influenced by the need for adapting to the natural environment. Our ethic is our way of doing things. Our ethic is a direct result of our natural environment. During the latter stages of the evolutionary process, *Homo sapiens* began to modify the environment rather than submit to what, millennia later, became known as Darwinian natural selection. As an example, consider the cave dweller who, in the chilly dawn of prehistory, realized the value of the saber-toothed tiger's coat and appropriated it for personal use. Inevitably a pattern of appropriation developed, and our ethic became more self-modified than environmentally adapted. Thus, we are no longer adapted to our natural environment but rather to our self-made environment. In the ecological context, over millennia such maladaptation results in one of two consequences: (1) the organism (*Homo sapiens*) dies out; or (2) the organism evolves to a form and character that is once again compatible with the natural environment (Vesilind, 1975). Assuming that we choose the latter course, how can this change in character (ethic) be brought about? Each individual must change his or her character or ethic, and the social system must change to become compatible with the global ecology.

The acceptable system is one in which we learn to share our exhaustible resources to regain a balance. This requires that we reduce our needs and that the materials we use must be replenishable. We must treat all of the Earth as a sacred trust to be used so that its content is neither diminished nor permanently changed; we must release no substances that cannot be reincorporated without damage to the natural system. The recognition of the need for such adaptation (as a means of survival) has developed into what we now call the **environmental ethic** or **environmental stewardship.**

CHAPTER REVIEW

When you have completed studying the chapter, you should be able to do the following without the aid of your textbook or notes:

1. List the properties that distinguish science from other fields of inquiry.
2. Differentiate between natural science and social science.
3. List three disciplines of natural science.
4. Define environmental science.
5. Explain the advantage of a theory over a collection of ideas.
6. Describe the advantage of a quantitative theory over a qualitative one.
7. Sketch and label a water resource system, including (a) source, (b) collection works, (c) transmission works, (d) treatment works, and (e) distribution works.
8. State the proper general approach to treatment of a surface water and a groundwater (see Figure 1–6).
9. Define the word *demand* as it applies to water.
10. List the five most important factors contributing to water consumption and explain why each has an effect.
11. State the rule-of-thumb water requirement for an average city on a per-person basis and calculate the average daily water requirement for a city of a stated population.

12. Define the acronyms WWTP and POTW.
13. Explain why separate storm sewers and sanitary sewers are preferred over combined sewers.
14. Explain the purpose of a lift station.
15. Discuss the role of environmental ethics in environmental engineering and environmental science.

PROBLEMS

1-1 Estimate the total daily water withdrawal (in $m^3 \cdot d^{-1}$) including both fresh and saline water for all uses for the United States in 2000. The population was 281,421,906.

Answer: $1.52 \times 10^9 \, m^3 \cdot d^{-1}$

1-2 Estimate the per capita daily water withdrawal for public supply in the United States in 2005 (in Lpcd). Use the following population data (McGeveran, 2002) and water supply data (Hutson et al., 2001):

Year	Population	Public Supply Withdrawal, $m^3 \cdot d^{-1}$
1950	151,325,798	5.30×10^7
1960	179,323,175	7.95×10^7
1970	203,302,031	1.02×10^8
1980	226,542,203	1.29×10^8
1990	248,709,873	1.46×10^8
2000	281,421,906	1.64×10^8

(Note: This problem may be worked by hand calculation and then plotting on graph paper to extrapolate to 2005 or, it may be worked using a spreadsheet to perform the calculations, plot the graph, and extrapolate to 2005.)

1-3 A residential development of 280 houses is being planned. Assume that the American Water Works Association average daily household consumption applies, and that each house has three residents. Estimate the additional average daily water production in $L \cdot d^{-1}$ that will have to be supplied by the city.

Answer: $3.70 \times 10^5 \, L \cdot d^{-1}$

1-4 Repeat Problem 1-3 for 320 houses, but assume that low-flush valves reduce water consumption by 14%.

1-5 Using the data in Problem 1-3 and assuming that the houses are metered, determine what additional demand will be made at the peak hour.

Answer: $1.96 \times 10^6 \, L \cdot d^{-1}$

1-6 If a faucet is dripping at a rate of one drop per second and each drop contains 0.150 milliliters, calculate how much water (in liters) will be lost in one year.

1-7 Savabuck University has installed standard pressure-operated flush valves on their water closets. When flushing, these valves deliver $130.0 \, L \cdot min^{-1}$. If the delivered water costs $0.45 per cubic meter, what is the monthly cost of not repairing a broken valve that flushes continuously?

Answer: $2,527.20 or $2,530 mo^{-1}

1-8 The American Water Works Association estimates that 15% of the water that utilities process is lost each day. Assuming that the loss was from Public Supply Withdrawal in 2000 (Problem 1-2), estimate the total value of the lost water if delivered water costs $0.45 per cubic meter.

1-9 Water delivered from a public supply in western Michigan costs $0.45 per cubic meter. A 0.5 L bottle of water purchased from a dispensing machine costs $1.00. What is the cost of the bottled water on a per cubic meter basis?

Answer: $2000 m^{-3}

1–10 Using U.S. Geological Survey Circular 1268 (U.S.G.S., 2005, http://usgs.gov), estimate the daily per capita domestic withdrawal of fresh water in South Carolina in Lpcd. (Note: conversion factors, provided in Appendix D, may be helpful.)

1–11 Using the Pacific Institute for Studies in Development, Environment, and Security website (http://www .worldwater.org/table2.html), determine the lowest per capita domestic water withdrawal in the world in Lpcd and identify the country in which it occurs.

DISCUSSION QUESTIONS

1–1 A nonscientist has remarked that the laws of motion are *just* a theory. She implies that the laws are hypotheses that are not testable. Prepare a written response (explanation) to this remark that includes a definition of the following terms: *hypothesis, theory, law.*

1–2 The values for per capita water consumption in Table 1–1 are very specific to the cities noted. Determine the per capita water consumption for your college or university.

1–3 Using the Internet, answer the following questions regarding the Clean Air Act:

(a) Title II of the act addresses what kinds of pollution sources?

(b) What is the chemical name of the first hazardous air pollutant listed under Title III?

(c) Section 604 of the act lists the phase-out of production of substances that deplete the ozone layer. What is the last year that carbon tetrachloride can be produced?

1–4 Using the Internet, determine the Website address (also known as the URL) that allows you to find information on environmental regulations published in the *Federal Register* and in the Code of Federal Regulations.

1–5 The Shiny Plating Company is using about 2000.0 kg · week^{-1} of organic solvent for vapor degreasing of metal parts before they are plated. The Air Pollution Engineering and Testing Company (APET) has measured the air in the workroom and in the stack that vents the degreaser. APET has determined that 1985.0 kg · week^{-1} is being vented up the stack and that the workroom environment is within occupational standards. The 1985.0 kg · week^{-1} is well above the allowable emission rate of 11.28 kg · week^{-1}.

Elizabeth Fry, the plant superintendent, has asked J. R., the plant engineer, to review two alternative control approaches offered by APET and to recommend one of them.

The first method is to purchase a pollution control device to put on the stack. This control system will reduce the solvent emission to 1.0 kg · week^{-1}. Approximately 1950.0 kg of the solvent captured each week can be recycled back to the degreaser. Approximately 34.0 kg of the solvent must be discharged to the wastewater treatment plant (WWTP). J. R. has determined that this small amount of solvent will not adversely affect the performance of the WWTP. In addition, the capital cost of the pollution control equipment will be recovered in about 2 years as a result of savings from recovering lost solvent.

The second method is to substitute a solvent that is not on the list of regulated emissions. The price of the substitute is about 10% higher than the solvent currently in use. J. R. has estimated that the substitute solvent loss will be about 100.0 kg · week^{-1}. The substitute collects moisture and loses its effectiveness in about a month's time. The substitute solvent cannot be discharged to the WWTP because it will adversely affect the WWTP performance. Consequently, about 2000 kg must be hauled to a hazardous waste disposal site for storage each month. Because of the lack of capital funds and the high interest rate for borrowing, J. R. recommends that the substitute solvent be used. Do you agree with this recommendation? Explain your reasoning.

1–6 Ted Terrific is the manager of a leather-tanning company. In part of the tanning operation a solution of chromic acid is used. It is company policy that the spent chrome solution is put in 0.20-m^3 drums and shipped to a hazardous waste disposal facility.

On Thursday the 12th, the day shift miscalculates the amount of chrome to add to a new batch and makes it too strong. Because there is not enough room in the tank to adjust the concentration, Abe Lincoln, the shift supervisor, has the tank emptied and a new one prepared and makes a note to the manager that the bad batch needs to be reworked.

On Monday the 16th, Abe Lincoln looks for the bad batch and cannot find it. He notifies Ted Terrific that it is missing. Upon investigation, Ted finds that Rip VanWinkle, the night-shift supervisor, dumped the batch into the sanitary sewer at 3:00 A.M. on Friday the 13th. Ted makes discreet inquiries at the waste-water plant and finds that they have had no process upsets. After Ted severely disciplines Rip, he should (choose the correct answer and explain your reasoning):

(a) Inform the city and state authorities of the illegal discharge as required by law even though no apparent harm resulted.

(b) Keep the incident quiet because it will cause trouble for the company without doing the public any good. No harm was done, and the shift supervisor has been punished.

(c) Advise the president and board of directors and let them decide whether to follow alternatives (a) or (b).

1–7 The Marginal Chemical Corp. is a small outfit by Wall Street's standards, but it is one of the biggest employers and taxpayers in the little town in which it has its one and only plant. The company has an erratic earnings record, but production has been trending up at an average of 6% a year—and along with it, so has the pollution from the plant's effluents into the large stream that flows by the plant. This stream feeds a lake that has become unfit for bathing or fishing.

The number of complaints from town residents has been rising about this situation, and you, as a resident of the community and the plant's senior engineer, also have become increasingly concerned. Although the lake is a gathering place for the youth of the town, the City Fathers have applied only token pressure on the plant to clean up. Your boss, the plant manager, has other worries because the plant has been caught in a cost/price squeeze, and is barely breaking even.

After a careful study, you propose to your boss that, to have an effective pollution-abatement system, the company must make a capital investment of $1 million. This system will cost another $100,000 per year in operating expenses (e.g., for treatment chemicals, utilities, labor, laboratory support). The Boss's reaction is:

"It's out of the question. As you know, we don't have an extra million around gathering dust—we'd have to borrow it at 10% interest per year and, what with the direct operating expenses, that means it would actually cost us $200,000 a year to go through with your idea. The way things have been going, we'll be lucky if this plant *clears* $200,000 this year, and we can't raise prices. Even if we had the million bucks handy, I'd prefer to use it to expand production of our new pigment; that way it would give us a better jump on the big boys and on overseas competition. You can create a lot of new production—and new jobs—for a million bucks. And this town needs new jobs more than it needs crystal clear lakes, unless you want people to fish for a living. Besides, even if we didn't put anything in the lake, this one still wouldn't be crystal clear—there would still be all sorts of garbage in it."

During further discussion, the only concessions you can get from your boss is that you can spend $10,000 so that one highly visible (but otherwise insignificant) pollutant won't be discharged into the stream, and that if you can come up with an overall pollution-control scheme that will pay for itself via product recovery, your boss will take a hard look at it. You feel that the latter concession does not offer much hope, because not enough products with a ready market appear to be recoverable.

If you were this engineer, what do you think you *should* do? Consider the alternates below.

(a) Report the firm to your state and other governmental authorities as being a polluter (even though the possible outcome might be your dismissal, or the company deciding to close up shop).

(b) Go above your boss's head (i.e., to the president of the company). If he fails to overrule your boss, quit your job, and then take Step **a**.

(c) Go along with your boss on an interim basis, and try to improve the plant's competitive position via a rigorous cost-reduction program so that a little more money can be spent on pollution control in a year or two. In the meantime, do more studies of product-recovery systems, and keep your boss aware of your continued concern with pollution control.

(d) Relax, and let your boss tell you when to take the next antipollution step. After all, he has managerial responsibility for the plant. You have not only explained the problem to him, but have suggested a solution, so you have done your part.

Assuming the Marginal Chemical Corp. treated its engineers very well in regard to salary and working conditions, and that the firm was a responsible influence in the community in nonenvironmental matters, do you think most of its engineers would actually take Step **a, b, c,** or **d** if they became involved in a pollution problem of this type today? (Popper and Hughson, 1970)

1–8 You are the division manager of Sellwell Co.—a firm that has developed an inexpensive chemical specialty that you hope will find a huge market as a household product. You want to package this product in 1 L and 2 L sizes. A number of container materials would appear to be practical—glass, aluminum, treated paper, steel, and various types of plastic. A young engineer whom you hired recently and assigned to the packaging department has done a container-disposal study that shows that the disposal cost for 2 L containers can vary by a factor of three—depending on the weight of the container, whether it can be recycled, whether it is easy to incinerate, whether it has good landfill characteristics, etc.

Your company's marketing expert believes that the container material with the highest consumer appeal is the one that happens to present the biggest disposal problem and cost to communities. He estimates that the sales potential would be at least 10% less if the easiest-to-dispose-of, salvageable, container were used, because this container would be somewhat less distinctive and attractive.

Assuming that the actual costs of the containers were about the same, to what extent would you let the disposal problem influence your choice? Would you:

(a) Choose the container strictly on its marketing appeal, on the premise that disposal is the community's problem, not yours (and also that some communities may not be ready to use the recycling approach yet, regardless of which container-material you select).

(b) Choose the easiest-to-dispose-of container, and either accept the sales penalty, or try to overcome it by stressing the "good citizenship" angle (even though the marketing department is skeptical about whether this will work).

(c) Take the middle road, by accepting a 5% sales penalty to come up with a container that is midway on the disposability scale.

Do you think the young engineer who made the container-disposal study (but who is not a marketing expert) has any moral obligation to make strong recommendations as to which container to use?

(a) Yes. He should spare no effort in campaigning for what he believes to be socially desirable.

(b) No. He should merely point out the disposal-cost differential and not try to inject himself into decisions that also involve marketing considerations about which he may be naive. (Popper and Hughson, 1970)

1–9 Stan Smith, a young engineer with two years of experience, has been hired to assist a senior engineer in the evaluation of air and water pollution problems at a large plant—one that is considering a major expansion that would involve a new product. Local civic groups and labor unions favor this expansion, but conservation groups are opposed to it.

Smith's specific assignment is to evaluate control techniques for the effluents in accordance with state and federal standards. He concludes that the expanded plant will be able to meet these standards. However, he is not completely happy, because the aerial discharge will include an unusual byproduct whose effects are not well known, and whose control is not considered by state and federal officials in the setting of standards.

In doing further research, he comes across a study that tends to connect respiratory diseases with this type of emission in one of the few instances where such an emission took place over an extended time period. An area downwind of the responsible plant experienced a 15% increase in respiratory diseases. The study also tends to confirm that the pollutant is difficult to control by any known means.

When Smith reports these new findings to his engineering supervisor, he is told that by now the expansion project is well along, the equipment has been purchased, and it would be very expensive and embarrassing for the company to suddenly halt or change its plans.

Furthermore, the supervisor points out that the respiratory-disease study involved a different part of the country and, hence, different climatic conditions, and also that apparently only transitory diseases were increased, rather than really serious ones. This increase might have been caused by some unique combination of contaminants, rather than just the one in question, and might not have occurred at all if the other contaminants had been controlled as closely as they will be in the new facility.

If Smith still feels that there is a reasonable possibility (but not necessarily certainty) that the aerial discharge would lead to an increase in some types of ailments in the downwind area, should he:

(a) Go above his superior, to an officer of the company (at the risk of his previously good relationship with his superior)?

(b) Take it upon himself to talk to the appropriate control officials and to pass their opinions along to his superior (which entails the same risk)?

(c) Talk to the conservation groups and (in confidence) give them the type of ammunition they are looking for to halt the expansion?

(d) Accept his superior's reasoning (keeping a copy of pertinent correspondence so as to fix responsibility if trouble develops)? (Popper and Hughson, 1970)

1–10 Jerry Jones is a chemical engineer working for a large diversified company on the East Coast. For the past two years, he has been a member—the only technically trained member—of a citizens' pollution-control group working in his city.

As a chemical engineer, Jones has been able to advise the group about what can reasonably be done about abating various kinds of pollution, and he has even helped some smaller companies design and buy control equipment. (His own plant has air and water pollution under good control.) As a result of Jones' activity, he built himself considerable prestige on the pollution-control committee.

Recently, some other committee members started a drive to pressure the city administration into banning the sale of phosphate-containing detergents. They have been impressed by reports in their newspapers and magazines on the harmfulness of phosphates.

Jones feels that banning phosphates would be misdirected effort. He tries to explain that although phosphates have been attacked in regard to the eutrophication of the Great Lakes, his city's sewage flows from the sewage-treatment plant directly into the ocean. And he feels that nobody has shown any detrimental effect of phosphate on the ocean. Also, he is aware that there are conflicting theories on the effect of phosphates, even on the Great Lakes (e.g., some theories put the blame on nitrogen or carbon rather than phosphates, and suggest that some phosphate substitutes may do more harm than good).

To top it all, he points out that the major quantity of phosphate in the city's sewage comes from human wastes rather than detergent.

Somehow, all this makes no impression on the backers of the "ban phosphates" measure. During an increasingly emotional meeting, some of the committee men even accuse Jones of using stalling tactics in order to protect his employer who, they point out, has a subsidiary that makes detergent chemicals.

Jones is in a dilemma. He feels that his viewpoint makes sense, and has nothing to do with his employer's involvement with detergents (which is relatively small, anyway, and does not involve Jones' plant). Which step should he now take?

(a) Go along with the "ban phosphates" clique on the grounds that the ban won't do any harm, even if it doesn't do much good. Besides, by giving the group at least passive support, Jones can preserve his influence for future items that really matter more.

(b) Fight the phosphate foes to the end, on the grounds that their attitude is unscientific and unfair and that lending it his support would be unethical. (Possible outcomes: his ouster from the committee, or its breakup as an effective body.)

(c) Resign from the committee, giving his side of the story to the local press.

(d) None of the above. (Popper and Hughson, 1970)

FE EXAM FORMATTED PROBLEMS

1–1 A residential development of 320 houses is being planned. For planning purposes an average daily consumption of 120 gallons per capita per day (gpcd) is used. Estimate the average daily volume of water that must be supplied to this development if each house is occupied by four people.

(a) 153,600 gallons per day

(b) 38,400 gallons per day

(c) 145,300 gallons per day

(d) 581,400 gallons per day

1–2 The maximum day demand for a current population of people is 90 million gallons per day (MGD). The maximum day demand 20 years from the current estimate is 120 MGD. The following assumptions were made for the estimate: peaking factor $= 1.5$; average day demand remains constant at 150 gpcd over the 20-year period; population growth is exponential. What annual rate of population growth was used for the estimate?

(a) 15.0 percent

(b) 6.67 percent

(c) 1.44 percent

(d) 0.0144 percent

1–3 State University has a resident population of 43,000. If the average day demand is 300 Lpcd, what average flow rate per day (in m^3/d) must be supplied to the campus?

(a) $1.29 \times 10^7 \ m^3/d$

(b) $1.29 \times 10^4 \ m^3/d$

(c) $3.41 \times 10^6 \ m^3/d$

(d) $3.41 \times 10^3 \ m^3/d$

1–4 If 60 percent of the average household water use in a dry climate is used outside of the house, what is the estimated wastewater flow rate (in m^3/d) for a community of 20,100 that has an average day demand of 960 Lpc?

(a) $7.72 \times 10^6 \ m^3/d$

(b) $1.16 \times 10^7 \ m^3/d$

(c) $7.72 \times 10^3 \ m^3/d$

(d) $1.16 \times 10^4 \ m^3/d$

REFERENCES

ASCE (1973) "Statement of Purpose," *Official Record,* Environmental Engineering Division, American Society of Civil Engineers, New York.

ASCE (1977) "Statement of Purpose," *Official Register,* Environmental Engineering Division of the American Society of Civil Engineers, New York.

AWWA (2016) American Water Works Association, "Residential End Uses of Water, Version 2, Executive Report," https://www.awwa.org/portals/0/files/resources/water%20knowledge/rc%20water%20conservation/residential_end_uses_of_water.pdf

AWWA (1971) *Water Quality and Treatment,* 3rd ed., American Water Works Association, McGraw-Hill, New York, p. 11.

AWWA (2016) "Residential End Uses of Water, Version 2, Executive Report," https://www.awwa.org/portals/0/files/resources/water%20knowledge/rc%20water%20conservation/residential_end_uses_of_water.pdf

Babbitt, H. E. (1953) *Sewerage and Sewage Treatment,* John Wiley & Sons, New York, p. 3.

Baker, M. N. (1981) *The Quest for Pure Water,* vol. 1, American Water Works Association, Denver.

Budd, W. (1977) *Typhoid Fever (Its Nature, Mode of Spreading, and Prevention),* Arno Press, New York.

Carson, R. (1962) *Silent Spring,* Houghton Mifflin, Boston.

CFR (2005) U.S. Government Printing Office, Washington, DC. http://www.gpoaccess.gov/ecfr/ (in Jan 2005 this was a beta test site for searching the CFR)

Chow, V. T. (1964) "Hydrology and Its Development," in V. T. Chow, ed., *Handbook of Applied Hydrology,* McGraw-Hill, New York, pp. 1–7–1–10.

Darcy, H. (1856) *Les Fountaines Publiques de la Ville de Dijon,* Victor Dalmont, Paris, pp. 570, 590–94.

Dupuit, A. J. (1863) *Etudes Théoriques et Practiques sur le Mouvement des Eaux dans les Canaux Découverts et á Travers les Terrains Perméables,* Dunod, Paris.

Emerson, C. A. (1945) "Some Early Steps in Sewage Treatment," *Sewage Works Journal,* 17: 710–17.

Fair, G. M., and J. C. Geyer (1954) *Water Supply and Waste-Water Disposal,* John Wiley & Sons, New York, pp. 5–8.

GLC (1969) *Refuse Disposal in Greater London,* Greater London Council, London.

Hagen, G. H. L. (1839) Ueber die Bewegung des Wassers in Engen Cylindrischen Röhren, *Poggendorffs Ann. Physik Chem.,* 16: 423–42.

Hagerty, D. J., and J. E. Heer (1973) *Solid Waste Management,* Van Nostrand Reinhold, New York.

Hazen, A. (1930) *Flood Flows,* John Wiley & Sons, New York.

Herschel, C. (1913) *Frontinus and the Water Supply of the City of Rome,* Longmans, Green & Co., New York.

Hohlfeld, F. (1824) "Das Niederschlagen des Rauches durch Elekticität," *Archiv. Für die gasammte Naturlehre,* 2: 205.

Horton, R. E. (1933) "The Role of Infiltration in the Hydrologic Cycle," *Transactions of the American Geophysical Union,* 14: 446–60.

Hutson, S. S., N. L. Barber, J. F. Kenny et al. (2001) *Estimated Use of Water in the United States in 2000,* U.S. Geological Survey Circular 1268, Washington, DC.

IPCC (1996) *Climate Change, 1995,* Intergovernmental Panel on Climate Change, Cambridge University Press, Cambridge, England.

Jones, B. B., and F. Owen (1934) *Some Notes on the Scientific Aspects of Controlled Tipping,* Henry Blacklock, Manchester, England.

Kenny, J. F., N. L. Barker, S. S. Hutson et al. (2009) *Estimated Use of Water in the United States in 2005.* U.S. Geological Survey Circular 1344, Washington, DC. http://www.usgs.gov

Linaweaver, F. P., J. C. Geyer, and J. B. Wolff (1967) "Summary Report on the Residential Water Use Research Project," *Journal of the American Water Works Association,* 59: 267.

Linsley, R. K., and J. B. Fanzini (1979) *Water Resources Engineering,* McGraw-Hill, New York, p. 546.

Mariotte, E. (1686) *Traité du Movement des Eaux et Autres Corps Fluides,* E. Michallet, Paris.

Maupin, M. A., J. F. Kenny, S. S. Hutson et al. (2014) *Estimated Use of Water in the United States in 2010,* U.S. Geological Survey Circular 1405, Washington, DC.

McGeveran, W. A. (2002) *World Almanac and Book of Facts: 2002,* World Almanac Books, New York, p. 377.

Metcalf, L., and H. P. Eddy (1915) *American Sewerage Practice,* vol. III, McGraw-Hill, New York, pp. 2–3, 13.

Molina, M. J., and F. S. Rowland (1974) "Stratospheric Sink for Chlorofluoromethanes: Chlorine Atom Catalysed Destruction of Ozone," *Nature,* 248: 810–12.

Pacific Institute (2000) Pacific Institute for Studies in Development, Environment, and Security, Oakland, CA. http://www.worldwater.org/table2.html

Perrault, P. (1678) *De L'origine des Fontaines,* Pierre Le Petit, Paris.

Poiseulle, J. L. (1840) "Recherches Expérimentales sur le Movement des Liquides dans les Tubes de Très petits Diameters," *Compt. Rend.,* 11: 961, 1041.

Popper, H., and R. V. Hughson (1970) "How Would You Apply Engineering Ethics to Environmental Problems?" *Chemical Engineering,* November 2, pp. 88–93.

Schein, E. H. (1968) "Organizational Socialization and the Profession of Management," Third Douglas Murray McGregor Memorial Lecture to the Alfred P. Sloan School of Management, Massachusetts Institute of Technology.

Snow, J. (1965) *Snow on Cholera (Being a Reprint of Two Papers by John Snow, M.D. Together With a Biographical Memoir by B. W. Richardson, M.D.)* Hafner Publishing Company, New York.

Tchobanoglous, G., H. Theisen, and R. Eliassen (1977) *Solid Wastes,* McGraw-Hill, New York, p. 39.

Theis, C. V. (1935) "The Relation Between the Lowering of the Piezometric Surface and the Rate and Duration of a Well Using Ground-water Recharge," *Transactions of the American Geophysical Union,* 16: 519–24.

UCLA (2006) http://www.ph.ucla.edu/epi

U.S. EPA (2005) "Municipal Solid Waste: Reduce, Reuse, Recycle," U.S. Environmental Protection Agency, Washington, DC. http://www.epa.gov/epaoswer/non-hw/muncpl/reduce.htm

U.S.G.S. (2005) http://www.usgs.gov

U.S. House of Representatives (2005) *U.S. Code,* Washington, DC. http://uscode.house.gov/

U.S. PHS (1970) *Manual of Septic Tank Practice,* Public Health Service Publication No. 526, U.S. Department of Health Education and Welfare, Washington, DC.

Vesilind, P. A. (1975) *Environmental Pollution and Control,* Ann Arbor Science, Ann Arbor, MI, p. 214.

WCED (1987) *Our Common Future,* World Commission on Environment and Development, Oxford University Press, Oxford, England.

White, H. J. (1963) *Industrial Electrostatic Precipitation,* Addison-Wesley, Reading, MA, p. 4.

WHO (1961) *Air Pollution,* World Health Organization, Geneva Switzerland, p.180.

Wright, R. H. (1964) *The Science of Smell,* Basic Books, Inc., New York, p. 7.

2

Chemistry

Case Study

Leaded Gasoline: Corporate Greed Versus Chemistry

Students often ask why they need to study chemistry. Chemistry is the backbone for most environmental engineering and science. For example, determining the amount of chemicals needed to reduce calcium and magnesium (hardness minerals) from drinking water is based on chemical precipitation reactions. Understanding how chemicals react in the atmosphere and move through the subsurface and groundwater is based on an understanding of chemistry.

Those of you reading this book probably don't remember a time when one drove into a gasoline station and had the choice of leaded or unleaded gasoline, but for more than 60 years that was the reality in the United States, and it remains the reality in much of the developing world, Eastern Europe, and elsewhere.

Lead is a poison, a potent neurotoxin. The Romans knew it. Vitruvius wrote, "Water conducted through earthen pipes is more wholesome than that through lead; indeed that conveyed in lead must be injurious, because from it white lead [PbCO3, lead carbonate] is obtained, and this is said to be injurious to the human system" (*Lead Poisoning and Rome,* n.d.). In 1786, Benjamin Franklin wrote of the hazards of lead, which caused printers' hands to shake, and severe illness so they could no longer work (*Lead-Franklin,* n.d.).

Lead is not naturally occurring in its elemental form. It is extracted from the mineral galena, which is lead sulfide. Like all elements, it does not break down over time. An organic form of lead, tetraethyl lead (TEL), was discovered in Germany in 1854. It was known to cause hallucinations, difficulty in breathing, madness, spasms, asphyxiation, and death. For that reason, it was not commercially used until the mid-1920s (Kitman, 2000).

In 1921, an engineer, Thomas Midgley, who in his employment by General Motors (GM) Research Corporation had tested tens of thousands of chemicals for their ability to reduce the knocking or pinging in internal combustion engineers, reported to his boss, Charles Kettering, that he discovered that TEL was the solution. A year later, Pierre du Pont reported to the chairperson of DuPont that TEL is "very poisonous if absorbed through the skin, resulting in lead poisoning almost immediately." Nevertheless, GM filed and received a patent for the production of TEL, and by 1923, they began manufacturing and commercializing it (Kitman, 2000).

So why TEL? TEL was cheap: the cost was 1 cent per gallon of gasoline (Kovarik, 2005). The alternative discovered by Midgley was ethanol, but ethanol was also a fuel that could be used in the internal combustion engine and readily available; it could be produced by anyone. TEL could be patented, ethanol could not (Kitman, 2000). In addition, the industry falsely claimed that there were no alternatives to TEL, and it was necessary to ensure that the internal combustion engine ran smoothly (Kitman, 2000; Kovarik, 2005).

By 1977, the production of leaded gasoline peaked at slightly greater than 100 billion gallons of leaded gasoline/year (Great Lakes Binational Toxics Strategy, 1999). Around that time, ambient concentrations of lead ranged from 1 to 10 $\mu g/m^3$ in urban environments and 14 to 25 $\mu g/m^3$ along highways. The decline in production began not only because of environmental regulations, but because the lead poisoned the catalytic converters necessary to reduce hydrocarbon emissions. As observed in Figure 2–1, the mean lead concentration in the ambient air in the United States has decreased significantly since 1980 from 1.8 μg/m³ to 0.02 μg/m³ in 2015 (U.S. EPA, n.d.). While the ambient air levels in the United States are much lower than in previous decades, lead continues to be emitted to the atmosphere through the combustion of leaded aviation fuel. Several studies have shown that children living within 1 km of an airport where leaded aviation gasoline is used have higher blood lead levels than other children. Sixteen million Americans live close to one of 22,000 airports where leaded aviation fuel is used—and three million children go to schools near these airports (Scheer and Moss, 2012).

By 1983, the U.S. EPA estimated that the use of leaded gasoline in the United States could be responsible for well over one million cases of hypertension per year and for more than 5,000 deaths of white males aged 40 to 59 from heart attacks, strokes, and other diseases related to blood pressure (The Lead Education and Abatement Design Group, 2011). In 1995, then EPA Administrator Carol Browner reported that the dramatic reduction in ambient air lead concentrations "means that millions of children will be spared the painful consequences of lead poisoning, such as permanent nerve damage, anemia or mental retardation" (The Lead Education and Abatement Design Group, 2011). Tragically, all of this could have been avoided had the concerns of the public health and scientific community not been trumped by corporate greed and profit margins. Unfortunately, the ~7 million tons of lead that were burned in gasoline in the United States remain in the soil, in water, in air, and in the bodies of living organisms (Kitman, 2000).

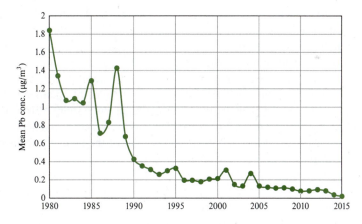

FIGURE 2–1 Mean ambient air lead concentration in the United States.
Source: https://www.epa.gov/air-trends/leadtrends.

2–1   INTRODUCTION

Environmental science is fundamentally a study of the chemistry, physics, and biology of natural systems or the environment. As such, a basic understanding of these disciplines is essential. In this chapter we will focus on a review of the fundamentals of chemistry with specific applications from environmental science.

Chemistry is the study of matter. **Matter** is any substance that has mass and occupies space. All objects, whether gas, liquid, or solids, are made up of elements. **Elements** cannot be broken down into simpler substances by any chemical reaction. **Compounds** are substances containing two or more elements that are combined chemically. A pure compound always contains the same elements in exactly the same proportions (**law of definite proportions** or **law of constant composition**). For example, 1.0000 g of sodium chloride, NaCl, always contains 0.3934 g of sodium and 0.6066 g of chlorine. The sodium and chlorine are chemically combined. The ratios and structural makeup of the compounds will significantly affect their properties. For example, the chemical compound glucose is made of carbon, hydrogen, and oxygen. The chemical formula is $C_6H_{12}O_6$, indicating that for every 6 atoms of carbon, there are 12 atoms of hydrogen and 6 atoms of oxygen. The compound CH_2O has the same ratio of carbon : hydrogen : oxygen as glucose, but very different properties. This other compound is formaldehyde, a liquid at room temperature that is toxic to humans. A **mixture** is a material, having variable composition, which can be separated by physical means. For example, table salt dissolved in water is a mixture. When you allow the water to evaporate, the table salt is left behind.

Chemical compounds made of carbon and hydrogen as the basic building blocks are considered **organic chemicals.** Organic chemicals can be formed either by living organisms or by synthetic chemical reactions. All other compounds are considered **inorganic chemicals.** A few simple compounds such as carbon monoxide (CO), carbon dioxide (CO_2), carbonates, and cyanides are considered inorganic compounds even though they contain carbon. We will discuss some of these compounds later in the chapter.

2–2 BASIC CHEMICAL CONCEPTS

Atoms, Elements, and the Periodic Table

All chemical compounds are composed of atoms of the various elements as basic structural building blocks. Elements are grouped together by their basic properties, as seen in the **periodic table.** (See Appendix C for a tabular arrangement of elements.) For example, the Group IIA (alkaline earth metals) includes calcium, magnesium, and barium, all of which are found in nature as silicate rocks. These elements also occur commonly as carbonates and sulfates. All of these metals (except beryllium) react with water if present as the zero-valent pure elemental metal. Except for beryllium, all of the hydroxides of alkaline earth metals are basic. On the other side of the periodic table are the halogens (Group VIIA): fluorine, chlorine, bromine, iodine, and astatine. All of these elements are reactive nonmetals (except for astatine, whose chemistry is still not well documented). All of the halogens form stable compounds in which the element is in the −1 oxidation state.[*]

An atom is an extremely small particle of matter that retains its identity during chemical reactions. Atoms are made up of electrons, neutrons, and protons. Neutrons and protons make up the nucleus and the vast majority of the mass of atoms. The **nucleus** may also contain one or more electrons. A **proton** is a positively charged particle having a mass more than 1800 times that of an electron. A **neutron** is a particle having a mass almost equal to that of a proton, but no charge. **Electrons** move rapidly around the nucleus, forming a cloud of negative charge.

The **atomic number** (Z) is the number of protons in the nucleus of an atom. All atoms of a particular element have the same number of protons, although the number of neutrons and electrons may vary. Therefore, an element is a substance whose atoms have the same atomic number. For example, chlorine has an atomic number of 17; it has 17 protons in the nucleus. The **mass number** of an element is the sum of the number of protons and neutrons in the nucleus of that particular element. For example, carbon-12, the most abundant form of carbon, has

---

[*]The oxidation state is the charge an atom in a chemical would have if the pairs of electrons in each bond belonged to the more electronegative atom. For chemicals that are ionicly bonded, the oxidation number equals the ionic charge. For covalently bonded compounds, the oxidation number represents a hypothetical charge assigned in accordance with a set of generally accepted rules.

six protons and six neutrons. Carbon-14, used in dating ancient objects, has the same number of protons as carbon-12 (six), but eight neutrons. These different forms of carbon are known as isotopes. **Isotopes** are chemically identical forms of an element, having different numbers of neutrons. The atomic numbers of the isotopes of the same element do not vary, whereas the mass numbers do. Some elements, such as sodium, have only one naturally occurring isotope; others, such as oxygen, carbon, and nitrogen, have several naturally occurring isotopes.

The **atomic weight** of an element is the average atomic mass for the naturally occurring element, expressed in atomic mass units (amu) or daltons. The atomic weight listed in Appendix C is determined by taking into account the fractional abundance of an isotope, that is the fraction of each naturally occurring isotope of the particular element and the atomic weight of that particular isotope.

EXAMPLE 2–1   In the following table, we have listed the isotopes of magnesium, the percentages at which they are found naturally, and the isotopic weights of each. Calculate the atomic weight of magnesium and compare the value you calculate to the value given in the list of elements in Appendix B.

Isotope	Isotopic Mass (amu)	Fractional Abundance
^{24}Mg	23.985	0.7870
^{25}Mg	24.986	0.1013
^{26}Mg	25.983	0.1117

Solution   Multiply each of the isotopic masses by its fractional abundance, then sum:

Isotope	Isotopic Mass (amu)	Fractional Abundance	Isotopic Mass × Fractional Abundance
^{24}Mg	23.985	0.787	18.876195
^{25}Mg	24.986	0.1013	2.531082
^{26}Mg	25.983	0.1117	2.902301
		Sum	24.30958

Therefore, the atomic mass of magnesium is 24.3096 amu. The slight difference between this value and that in the list of atomic weights is simply due to round-off error.

Chemical Bonds and Intermolecular Forces

All atoms that are not present in a uniatomic form are held together by chemical bonds. Molecules can be held together by intermolecular forces, which are usually weakly attractive. A **molecule** is a definite group of atoms that are chemically bonded and held together in a fixed geometric arrangement.

The two basic types of bonds are ionic and covalent. An **ionic bond** is a chemical bond formed by electrostatic attraction between positive and negative ions. With ionic bonding, one atom "gives" up or transfers at least one electron from its valence shell to the valence shell of another atom. The atom that loses electrons becomes a **cation** (positively charged ion), and the one that gains electrons becomes an **anion** (negatively charged ion). For example, the hydrogen atom in hydrofluoric acid (HF) "gives up" its electron to fluorine. In this case, the hydrogen,

FIGURE 2–2

The electronegativities of common elements (using the scale provided by Linus Pauling).

Increasing electronegativity

Increasing electronegativity

1A																	8A
H 2.1	2A											3A	4A	5A	6A	7A	
Li 1.0	Be 1.5											B 2.0	C 2.5	N 3.0	O 3.5	F 4.0	
Na 0.9	Mg 1.2	3B	4B	5B	6B	7B	⎯ 8B ⎯			1B	2B	Al 1.5	Si 1.8	P 2.1	S 2.5	Cl 3.0	
K 0.8	Ca 1.0	Sc 1.3	Ti 1.5	V 1.6	Cr 1.6	Mn 1.5	Fe 1.8	Co 1.9	Ni 1.9	Cu 1.9	Zn 1.6	Ga 1.6	Ge 1.8	As 2.0	Se 2.4	Br 2.8	
Rb 0.8	Sr 1.0	Y 1.2	Zr 1.4	Nb 1.6	Mo 1.8	Tc 1.9	Ru 2.2	Rh 2.2	Pd 2.2	Ag 1.9	Cd 1.7	In 1.7	Sn 1.8	Sb 1.9	Te 2.1	I 2.5	
Cs 0.7	Ba 0.9	La-Lu 1.0-1.2	Hf 1.3	Ta 1.5	W 1.7	Re 1.9	Os 2.2	Ir 2.2	Pt 2.2	Au 2.4	Hg 1.9	Tl 1.8	Pb 1.9	Bi 1.9	Po 2.0	At 2.2	
Fr 0.7	Ra 0.9																

essentially, becomes positively charged and the fluorine, negatively charged. Compounds that tend to have a greater ability to draw an electron toward itself are known to have **high electronegativities,**[*] or, in this case, a greater attraction for the shared electron pair than hydrogen. In general, metals are the least electronegative elements; nonmetals are the most electronegative (Figure 2–2).

Covalent bonds are formed by the sharing of a pair of electrons between atoms. For example, with hydrogen gas (H_2) the electrons in the $1s$ orbital overlap, and each electron can occupy the space around both atoms. Thus, the electrons can be described as shared by both atoms. Other chemicals that are bonded covalently include methane (CH_4), ammonia (NH_3), and ethylene (C_2H_4).

In the case of hydrogen, the bonding electrons are shared equally between the atoms. However, when the two atoms are different elements, the electrons may not be shared equally. A **polar covalent bond** is a covalent bond in which the bonding electrons are not shared equally— or the probability of finding the electrons in the vicinity of one atom is greater than the probability of finding it near the other atom. For example, with HCl, the bonding electrons are more likely to be found near the chlorine atom than the hydrogen atom. Similarly, when compounds such as hydrogen and oxygen are covalently bonded together, as in water, neither the hydrogen nor the oxygen transfers an electron, so that neither atom is essentially completely charged in nature. For example, with water, one can think of the hydrogen as being partially positive in character, with the oxygen atoms being partially negative.

So far we have discussed ionic and covalent bonding, which holds atoms together. However, there are other, usually weakly attractive, forces that can hold molecules together. These are **intermolecular forces.**

[*]Electronegativity is a measure of the ability of an atom in a molecule to attract bonding electrons to itself. The most widely used scale was developed by Linus Pauling. He derived electronegativity values from bond energies and assigned a value of 4.0 to fluorine. Lithium, at the left end of the same period, has a value of 1.0. In general, electronegativity decreases from right to left and from top to bottom in the periodic table.

Three types of intermolecular forces can exist between neutral molecules: dipole–dipole forces, London (or dispersion) forces, and hydrogen bonding. **Van der Waals** forces include both dipole–dipole interactions and dispersion forces. Van der Waals forces are weak, short-range attractive forces between neutral molecules such as Cl_2 and Br_2. Dispersion forces are also weak short-range attractive forces that result from the instantaneous dipole/induced-dipole interactions that can occur because of the varying positions of the electrons as they move around the nuclei. Because molecules with greater molecular weights tend to have more electrons, dispersion forces tend to increase with molecular weight. This increase in forces is also due to the fact that larger molecules tend to be more polarizable, resulting in a greater likelihood for the formation of induced dipoles.

Hydrogen bonding occurs in substances bonded to certain very electronegative atoms. Hydrogen bonding is very important to the environmental scientist or engineer because it gives water its unique properties.

Hydrogen bonding can be investigated by studying two chemicals: fluoromethane (CH_3F) and methanol (CH_3OH). Both substances have about the same molecular weight and dipole moment. You might expect both chemicals to have similar properties, but fluoromethane is a gas at room temperature and methanol is a liquid. The boiling points of fluoromethane and methanol are $-78°C$ and $65°C$, respectively. The reason for the very different properties is the moderately strong attractive forces that exist between a hydrogen atom covalently bonded to a very electronegative atom, X, and a lone pair of electrons on another small, electronegative atom of another molecule. For example, with methanol, the partially positive hydrogen atom covalently bonded to the partially negative oxygen atom is attracted to the partially negative oxygen atom of another molecule:

$$H_3C\text{—}\overset{\delta-}{O}\text{—}\overset{\delta+}{H}\cdots\overset{\delta-}{O}\underset{CH_3}{\overset{\overset{\delta+}{H}}{\diagup}}$$

These forces give methanol its properties. Similarly, hydrogen bonding occurs with water molecules, making water a liquid at room temperature, despite its low molecular weight.

The Mole, Molar Units, and Activity Units

A **mole** is defined as **Avogadro's number** of molecules of a substance, that is, as 6.02×10^{23} molecules of a substance. For example, a mole of benzene (a compound present in gasoline) contains 6.02×10^{23} molecules of benzene. The molecular weight can be determined by multiplying the atomic weight of each element by the number of atoms of each element present in the substance. The **molecular formula** (a chemical formula that gives the exact number of the different atoms of an element found in the molecule) for benzene is C_6H_6. The **molecular weight** of a substance is the weight of a mole of that particular substance. Because the atomic weights of carbon and hydrogen are 12.01 and 1.008 amu (atomic mass units) or daltons, respectively, the molecular weight of benzene is

$$(12.01)(6) + (1.008)(6) = 78.11 \text{ g} \cdot \text{mol}^{-1} \tag{2–1}$$

Thus a mole of benzene has a mass of 78.11 g.

An **ion** is an electrically charged atom or molecule. For example, in water, calcium ions are present as cations, Ca^{2+}, and chlorine is present as the anion chloride (Cl^-).

Molarity is the number of moles per liter of solution. For example, a 1 molar (1 M) solution of benzene has 1 mole of benzene per liter of solution. The molarity will be represented hereafter by the brackets [ ]. In environmental science, because the concentration of chemicals found in the environment is often quite small, we sometimes use millimoles per liter ($mmol \cdot L^{-1}$ or mM) or micromoles per liter ($\mu mol \cdot L^{-1}$ or μM) as a unit.

EXAMPLE 2–2   A solution of calcium chloride is prepared in a 1.00-L volumetric flask. A 60.00-g sample of $CaCl_2$ is added to a small amount of water in the flask, and then additional water is added to bring the total volume of solution to 1.00 L. What is the concentration of calcium chloride in units of molarity?

Solution   The first step in solving this problem is to determine the molecular weight of calcium chloride. The atomic weights of calcium and chlorine can be obtained from Appendix B. They are 40.078 amu and 35.4527 amu for calcium and chlorine, respectively. The molecular weight is

$$40.078 + 2(35.4527) = 110.98 \text{ g} \cdot \text{mol}^{-1}$$

The concentration in SI units is $60.00 \text{ g} \cdot \text{L}^{-1}$. Converting to molar units:

$$(60.00 \text{ g} \cdot \text{L}^{-1})/(110.98 \text{ g} \cdot \text{mol}^{-1}) = 0.5406 \text{ M}$$

Alternatively, we can write the expression as

$$\frac{60 \text{ g}}{\text{L}} \times \frac{\text{mol}}{110.98 \text{ g}} = 0.5406 \text{ M}$$

in which it is easier to see how the units cancel.

Activity is defined in terms of chemical potential[*] and is dimensionless. It will be represented hereafter by the braces, { }. For pure phases (e.g., solid, liquid, ideal one-component gas), the activity is, by definition, one. Activity can be related to molar concentrations using the activity coefficient, essentially a factor that describes the nonideal behavior of the component in the system under study.

In dilute solutions, the total ion concentration is low (generally less than 10^{-2} M). The ions in such solutions can be considered to act independently of one another. As the concentration of ions in solution increases, the interaction of their electric charges affects their equilibrium relationships. This interaction is measured in terms of ionic strength. To account for high ionic strength, the equilibrium relationships are modified by incorporating activity coefficients. These are symbolized by $\gamma_{(ion)}$ Activity is then the product of the molar concentration of the species and its activity coefficient:

$$\{i\} = [i]\gamma_{(ion)} \tag{2-2}$$

Throughout this text, we will use the notation $\{i\}$ to represent the activity of a solution, whereas $[i]$ will be used to denote molarity. For example, the activity of Ca^{2+} in a 0.1 M solution of NaCl would be related to the molar concentration of calcium by the equation

$$\{Ca^{2+}\} = \gamma_{(Ca^{2+})}[Ca^{2+}] \tag{2-3}$$

One method used to calculate activity coefficients is discussed on pages 52–54.

Chemical Reactions and Stoichiometry

Stoichiometry is that part of chemistry concerned with measuring the proportions of elements or compounds involved in a reaction. Stoichiometric calculations are an application of the principle of conservation of mass to chemical reactions.

---

[*]Potential energy is the energy an object possesses because of its position in a field of force. Chemical potential is a thermodynamic quantity that is useful as a criterion for spontaneity. Chemical potential can be thought of as the tendency for a reaction to occur. For example, a reaction that has a negative potential is analogous to a pipe full of water that is flowing downhill. In the case of the pipe, water will flow by gravity. In the case of chemical potential, the reaction is thermodynamically feasible. Chemical potential, however, tells you nothing about the rate of the reaction.

A **chemical equation** is the symbolic representation of a chemical reaction in terms of chemical formulas. For example, the most widely used automobile fuel is gasoline. One of the compounds found in gasoline is octane. If octane is burned completely, only water and carbon dioxide are formed. The equation describing this reaction is

$$2C_8H_{18} + 25O_2 \longrightarrow 16CO_2 + 18H_2O \tag{2–4}$$

The chemical compounds to the left of the reaction arrow are known as **reactants;** those to the right as **products.**

In many cases it is useful to note the states or phases of the chemicals involved in the reaction. The following labels are commonly used:

(g) = gas, (l) = liquid, (s) = solid, (aq) = aqueous (water) solution

If we use these symbols, Equation 2–4 is written as:

$$2C_8H_{18}(l) + 25O_2(g) \longrightarrow 16CO_2(g) + 18H_2O(l) \tag{2–5}$$

Balancing Chemical Reactions.   All chemical equations such as those presented here must be balanced, that is, there must be the same number of atoms of each element on both sides of the reaction arrow. The best way to emphasize this point is to show an example.

EXAMPLE 2–3   Calcium can be removed from natural waters by the addition of sodium hydroxide according to the unbalanced reaction

$$Ca(HCO_3)_2 + NaOH \rightleftharpoons Ca(OH)_2 + NaHCO_3$$

Balance this reaction.

Solution   The first step is to tally all of the elements on both sides of the reaction

Element	Reactants	Products
Calcium	1	1
Hydrogen	$(1 \times 2) + 1 = 3$	$(1 \times 2) + 1 = 3$
Carbon	$(1 \times 2) = 2$	1
Oxygen	$(3 \times 2) + 1 = 7$	$(1 \times 2) + 3 = 5$
Sodium	1	1

We are short one carbon atom on the product side, so multiply the number of moles of $NaHCO_3$ by 2:

$$Ca(HCO_3)_2 + NaOH \rightleftharpoons Ca(OH)_2 + 2NaHCO_3$$

Now we have two carbons, but we also have two sodium atoms, two hydrogen atoms and six oxygen atoms from the sodium bicarbonate (in addition to what we have from the calcium hydroxide).

Element	Reactants	Products
Calcium	1	1
Hydrogen	$(1 \times 2) + 1 = 3$	$(1 \times 2) + (1 \times 2) = 4$
Carbon	$(1 \times 2) = 2$	$\cancel{1}2$
Oxygen	$(3 \times 2) + 1 = 7$	$(1 \times 2) + (3 \times 2) = 8$
Sodium	1	$\cancel{1}2$

Now, we need to multiply the number of moles of NaOH by 2:

$$Ca(HCO_3)_2 + NaOH \rightleftharpoons Ca(OH)_2 + 2NaHCO_3$$

Which results in:

Element	Reactants	Products
Calcium	1	1
Hydrogen	$(1 \times 2) + (1 \times 2) = 4$	$(1 \times 2) + (1 \times 2) = 4$
Carbon	$(1 \times 2) = 2$	$\cancel{1}\,2$
Oxygen	$(3 \times 2) + (1 \times 2) = 8$	$(1 \times 2) + (3 \times 2) = 8$
Sodium	$\cancel{1}\,2$	$\cancel{1}\,2$

The equation is now balanced!

Chemical reactions may also involve the oxidation and reduction of species. In the section on redox reactions, we will discuss how to balance this type of reaction.

Types of Chemical Reactions. Four types of reactions are of principal importance to the environmental scientist and engineer: precipitation–dissolution, acid–base, complexation or ion-association, oxidation–reduction.

Precipitation–Dissolution Reactions. Some dissolved ions can react with other ions to form solid insoluble compounds, known as **precipitates.** The phase-change reaction by which dissolved chemicals form insoluble solids is called a **precipitation reaction.** A typical precipitation reaction is the formation of calcium carbonate when a solution of calcium chloride is mixed with a solution of sodium carbonate.

$$CaCl_2 + Na_2CO_3 \rightleftharpoons CaCO_3(s) + 2Na^+ + 2Cl^- \tag{2–6}$$

The (s) in the preceding reaction denotes that the $CaCO_3$ is a solid. When no symbol is used to designate state, it is assumed that the chemical species is dissolved. The arrows in the reaction imply that the reaction is reversible and so could proceed to the right (i.e., the ions are combining to form a solid) or to the left (i.e., the solid is dissociating into the ions). When the reaction proceeds to the left, it is referred to as a **dissolution reaction.** Dissolution reactions are important when dealing with rocks and minerals. For example, under acidic conditions, the mineral calcite ($CaCO_3$) can dissolve to release calcium ions (Ca^{2+}) and CO_3^{2-} (carbonate) into water. At equilibrium no additional calcium or carbonate can go into solution; the solution is said to be **saturated.** If the solution is not in equilibrium and additional calcium and carbonate can dissolve, the solution is said to be **undersaturated.** In some cases it is possible to obtain higher concentrations of the dissolved ions than what is predicted using the equilibrium constant; the solution is said to be **supersaturated.**

Some dissolution reactions can be said to proceed to completion, that is, they proceed essentially completely to the right or left. For example, if we add sodium chloride or calcium sulfate to water, these compounds will dissociate (break up) to release the associated ions.

$$NaCl(s) \longrightarrow Na^+ + Cl^- \tag{2–7}$$

$$CaSO_4(s) \longrightarrow Ca^{2+} + SO_4^{2-} \tag{2–8}$$

Although we might be inclined to state that the water contains NaCl and $CaSO_4$, these species are actually present in their dissociated forms (i.e., Na^+, Cl^-, Ca^{2+}, and SO_4^{2-}).

Acid–Base (or Neutralization) Reactions. The general concept of acids and bases was devised by Brønsted and Lowry. A Brønsted–Lowry acid is then defined as any substance that can donate a proton to another substance and a Brønsted–Lowry base is any substance that can accept a proton. Thus proton transfer can occur only if both an acid and a base are present.

If we use A to designate an acid and B to denote a base, then the general form of an acid–base reaction is

$$HA + H_2O \rightleftharpoons H_3O^+ + A^- \tag{2–9}$$

in which HA is a stronger acid than water and water acts as the base. The reaction results in the formation of a second (conjugate) acid (H_3O^+) and a second (conjugate) base (A^-). In this reaction

$$B^- + H_3O^+ \rightleftharpoons HB + H_2O \tag{2–10}$$

B^- is the base and water acts as the acid. Likewise, this reaction also results in the formation of a second acid (HB) and base (H_2O).

The term pH is defined as the negative log of the H^+ activity:

$$pH = -\log\{H^+\} \tag{2–11}$$

A solution having a pH of 7 is neutral; it is neither acidic nor basic. A solution having a pH < 7 is acidic, and a solution with a pH > 7 is basic. The pH of most natural waters lies in the range of 6 to 9, conditions necessary to support most life, although extremophilic bacteria can grow at pH values less than 4 and greater than 9. Biological activities involve acid–base reactions and, therefore, affect the pH of waters. As water moves through soils, acid–base reactions are likely to occur. Acid–base reactions occur in the atmosphere when acid rain is produced. Acids can enter the aquatic and terrestrial ecosystems due to their direct release from household, municipal, and industrial wastes. Unlike many precipitation reactions, acid–base reactions are very fast; with half-lives usually on the order of milliseconds.

Free protons cannot exist in water; they will combine with the H_2O molecule to form the hydronium ion (H_3O^+). If we add an acid to water it will dissociate to release a proton. For example,

$$HCl \rightleftharpoons H^+ + Cl^- \tag{2–12}$$

Because hydrochloric acid (HCl) is a strong acid, this reaction will go essentially to completion. That is, one mole of HCl will produce one mole of protons. For an acid–base reaction to occur, the proton must be transferred to a base. Water is **amphoteric,** that is, it can act as either an acid or a base. In this case water would act as a base and accept the protons released by the dissociation of hydrochloric acid.

$$H_2O + H^+ \rightleftharpoons H_3O^+ \tag{2–13}$$

If we combine the preceding reactions, we obtain the overall reaction

$$HCl + H_2O \rightleftharpoons H_3O^+ + Cl^- \tag{2–14}$$

which is the actual reaction that occurs in water, although the shorthand notation shown in Equation 2–12 is often used.

If we add a base to water, it will react with hydronium ions present in the water. For example, the addition of sodium hydroxide to water results in the consumption of H_3O^+.

$$NaOH + H_3O^+ \rightleftharpoons 2H_2O + Na^+ \tag{2–15}$$

In this case, water acts as an acid in that it donates a proton to the base.

Complexation Reactions. Complexation reactions occur in natural waters whenever the "coordination" of two (or more) atoms, molecules, or ions results in the formation of a more stable product. These reactions are important to the environmental scientist or engineer because the form of the chemical can significantly affect the toxicity, the efficiency of removal, and the biological uptake of that chemical species.

A **complex** is a compound consisting of either complex ions with other ions of opposite charge or a neutral complex species. The **complex ion** is defined as a metal ion with Lewis bases* attached to it through coordinate covalent bonding. The Lewis bases attached to the metal atom in a complex are known as **ligands.** The coordination number of a metal atom in a complex is the total number of bonds the metal atom forms with ligands. For example, with $Fe(H_2O)_6^{2+}$, Fe^{2+} is the complex ion; water is the ligand; and the coordination number of the iron in this complex is 6 because six molecules of water are bonded to the iron molecule.

Metal complexation is an important concept in environmental engineering and science because the complexation of metals greatly affects the uptake, biodegradability, and toxicity of the metal. For example, in a study in which researchers considered the complexation of several metals to nitrilotriacetic acid, used as a model organic chemical, it was found that the complexation of nitrilotriacetic acid by Cu(II), Ni(II), Co(II), and Zn(II) inhibited the biodegradation of nitrilotriacetic acid by *Chelatobacter heintzii* because the nitrilotriacetic acid decreased the bioavailability of the metal (White and Knowles, 2000). In another study, it was found that the uptake and toxicity of copper to unicellular algae decreased as the concentration of complexing agent increased and the free copper concentration decreased (Sunda and Guillard, 1976).

Oxidation–Reduction (Redox) Reactions. Without oxidation–reduction (redox) reactions, life as we know it would be impossible. Photosynthesis and respiration are essentially a sequence of redox reactions. The cycling of nutrients in the environment is also controlled by redox reactions. Redox reactions also occur when the iron in your car rusts. In general, redox reactions are quite slow, thankfully, because no one wants to see things such as their car rapidly destroyed by rust. However, this slowness presents difficulties to the environmental scientist or engineer. With redox reactions, in the natural environment, equilibrium is rarely reached.

Redox reactions involve changes in the oxidation state of an ion and the transfer of electrons. When iron metal corrodes, it releases electrons:

$$Fe^0 \rightleftharpoons Fe^{2+} + 2e^- \tag{2-16}$$

If one element releases electrons, then another must be available to accept the electrons. In iron pipe corrosion, hydrogen gas is often produced:

$$2H^+ + 2e^- \rightleftharpoons H_2(g) \tag{2-17}$$

where the symbol (g) indicates that the hydrogen is in the gas phase.

When balancing redox equations, you must also ensure that the number of electrons transferred is balanced. In Example 2–3, there was no change in the oxidation state of any of the elements. However in the next two examples, the oxidation state changes.

EXAMPLE 2–4 In the atmosphere, sulfur dioxide (released by coal-burning power plants) reacts relatively slowly with oxygen and water vapor to form sulfuric acid, a pollutant that has the potential to cause severe respiratory distress and acid rain.

The chemical equation is

$$SO_2 + O_2 + H_2O \longrightarrow H_2SO_4$$

Balance this reaction.

---

*A Lewis base has one (or more) pairs of electrons, which it can donate to the central metallic cation in the complex.

Solution   Note this reaction is not written as an equilibrium reaction. This is because the reaction proceeds essentially to completion and is not reversible. The first task is to list all of the elements present in the reaction.

Element	Reactants	Products
S (sulfur)	1	1
O (oxygen)	5	4
H (hydrogen)	2	2

Note that in this reaction the oxygen is reduced from a charge of zero (in O_2) to a charge of -2 in H_2SO_4. The sulfur is oxidized from a charge of $+4$ in SO_2 to $+6$ in H_2SO_4. Therefore, we must first balance the number of electrons transferred.

$$O_2 \rightleftharpoons 2O^{2-}$$

$$S^{4+} \rightleftharpoons S^{6+}$$

Four electrons must be transferred in the first reaction, whereas two electrons are transferred in the second. Therefore, to conserve the number of electrons transferred, we must multiply the second reaction by 2.

$$O_2 + 4e^- \rightleftharpoons 2O^{2-}$$

$$2(S^{4+} \rightleftharpoons S^{6+} + 2e^-)$$

Thus, the reaction now becomes

$$2SO_2 + O_2 + H_2O \longrightarrow 2H_2SO_4$$

The next step is to add up all of the atoms of each element on both sides of the reaction. For example, on the reactant side of the reaction there are four atoms of oxygen in $2SO_2$, two atoms of oxygen in O_2, and one atom of oxygen in water (H_2O), for a total of seven atoms. Now write 7 in the table in the cell corresponding to oxygen and reactants. You should then do the same for all other elements.

Element	Reactants	Products
S (sulfur)	2	2
O (oxygen)	7	8
H (hydrogen)	2	4

Both sides of the reaction must have the same number of atoms of all elements; however, we have seven atoms of oxygen on the reactant side and eight on the product side. We are short one oxygen atom on the reactant side of the reaction. Let's first try to balance the equation by multiplying the number of water molecules by 2.

$$2SO_2 + O_2 + 2H_2O \longrightarrow 2H_2SO_4$$

The table would then become

Element	Reactants	Products
S (sulfur)	2	2
O (oxygen)	78	8
H (hydrogen)	24	4

The equation is now balanced!

EXAMPLE 2–5   The following reaction is important in lake and river sediments that are devoid of oxygen.

$$SO_4^{2-} + H^+ + CH_2O \rightleftharpoons HS^- + CO_2 + H_2O$$

Balance this reaction.

Solution   Note that this bacterially mediated reaction is reversible. First we consider the charge on the various atoms. Oxygen almost always* has a charge of −2. If oxygen has a charge of −2, then the charge on the sulfur in SO_4^{2-} is +6, the charge on carbon in CO_2 is +4. Hydrogen almost always has a charge of +1 (except in the diatomic state, where the charge is 0). Therefore, the charge on the sulfur in HS^- is −2, and the hypothetical charge on the carbon in CH_2O is 0. If the reaction for the reduction of sulfur is

$$S(+6)O_4^{2-} \rightleftharpoons HS(-2)^-$$

we need eight electrons on the reactant side to balance the oxidation state of sulfur on the product side. The numbers in parentheses correspond to the oxidation state of sulfur.

$$S(+6)O_4^{2-} + 8e^- \rightleftharpoons HS(-2)^-$$

([+6] + [−8] = −2). Now we must balance the oxidation reaction. In this reaction, carbon goes from an oxidation state of 0 to an oxidation state of +4. Clearly we need four electrons to balance this reaction.

$$CH_2O \rightleftharpoons CO_2 + 4e^-$$

Because we need to balance the number of electrons transferred, the next step is to multiply the previous reaction by 2:

$$2CH_2O \rightleftharpoons 2CO_2 + 8e^-$$

Our original equation becomes

$$SO_4^{2-} + H^+ + 2CH_2O \rightleftharpoons HS^- + 2CO_2 + H_2O$$

with the number of electrons transferred balanced. The next steps are similar to those presented earlier. Let's now create a table showing all atoms in the reaction.

Element	Reactants	Products
S (sulfur)	1	1
O (oxygen)	6	5
C (carbon)	2	2
H (hydrogen)	5	3

Because the number of carbon and sulfur atoms are already balanced, we must avoid changing the number of molecules of any species containing carbon or sulfur. We need one more atom of oxygen and two more atoms of hydrogen on the product side of the reaction. Voila! Water! Add one more molecule of water on the product side of the reaction.

$$SO_4^{2-} + H^+ + 2CH_2O \rightleftharpoons HS^- + 2CO_2 + H_2O$$

and the table becomes

---

*Oxygen almost always has a charge of −2, except in the diatomic state, where the charge is 0, or as a peroxide, where the charge is −1.

Element	Reactants	Products
S (sulfur)	1	1
O (oxygen)	6	5̶6
C (carbon)	2	2
H (hydrogen)	5	3̶5

The preceding equation is now balanced!

Reactions Involving Gases. The transfer of gases into or from solutions is important because aquatic life could not survive if gases such as oxygen and carbon dioxide did not dissolve in water. "Pure" rainwater would not have a pH near 5.6 were it not for the fact that carbon dioxide dissolves in water. Carbon dioxide dissolves in water by the reaction

$$CO_2(g) \rightleftharpoons CO_2(aq) \qquad (2–18)$$

where the symbol (aq) indicates the aqueous phase.

Carbon dioxide can then react with water to form carbonic acid.

$$CO_2(aq) + H_2O \rightleftharpoons H_2CO_3 \qquad (2–19)$$

The dissolution of oxygen into water is important to environmental engineers and scientists and critical to life as we know it. Oxygen is only slightly soluble in water, and its solubility increases with decreasing water temperature. As the biota in water consume oxygen during respiration, that oxygen must be replenished. In a fish tank, this is accomplished by using a pump to bubble air through a diffuser into the water. In the natural environment, the oxygen from the air diffuses across the air-water interface into the bulk water solution. The dissolution reaction can be represented by the chemical reaction:

$$O_2(g) = O_2(aq) \qquad (2–20)$$

We can also use these gas transfer reactions to our advantage. For example, ammonia can be stripped from waters by raising the pH to convert NH_4^+ to NH_3 and then providing conditions in which the ammonia will be transferred to the gas phase.

$$NH_3(aq) \rightleftharpoons NH_3(g) \qquad (2–21)$$

Chemical Equilibrium

If you add calcite to a dilute hydrochloric acid solution, the calcite will dissolve, giving off carbon dioxide bubbles, until **chemical equilibrium** has been reached. At this point the rate of the reaction proceeding to the right equals that proceeding to the left.

$$CaCO_3 + 2HCl \rightleftharpoons Ca^{2+} + CO_2(g) + H_2O + 2Cl^- \qquad (2–22)$$

We define chemical equilibrium as the condition when the rate of forward reaction equals the rate of the reverse reaction.

Chemical equilibrium is always described mathematically by an equilibrium **constant.** The equilibrium constant, which is derived from chemical thermodynamic data, is numerically equivalent to the product of the concentrations of products divided by the product of the reactants, with each of the concentration terms raised to a power equivalent to the stoichiometric coefficient from the chemical reaction, where the concentrations are those at equilibrium conditions. For example, for the general reaction

$$a A + b B \rightleftharpoons c C + d D \tag{2-23}$$

the equilibrium constant, K, is described mathematically as

$$K = \frac{\{C\}^c \{D\}^d}{\{A\}^a \{B\}^b} \tag{2-24}$$

Note that for chemicals in solution, the concentrations used in Equation 2–24 must be in units of activity or molarity (if the solution is sufficiently dilute). If the chemicals are gases, then the concentrations must be in units of activity or pressure. For dilute gases, the partial pressure of the gas is often used in Equation 2–24. The **law of mass action** states that the value of the equilibrium constant expression K is a constant for a particular reaction at a given temperature, and that this value is independent of the equilibrium concentrations of the chemicals substituted into the equation.

Many environmental reactions proceed rapidly, so that equilibrium calculations can be used to predict the concentrations of chemicals after some treatment process, in a body of water, or in raindrops. This is especially true with the dissolution of chemicals in a well-mixed reactor, the dissociation of acids and bases, and the dissolution of gases in raindrops or fog. Although chemical equilibrium calculations are less useful for slower reactions like the dissolution of minerals in rocks or the dissolution of gases in large bodies of water, these calculations do provide information on the concentrations of chemical species after chemical equilibrium is reached, although the time to reach these concentrations may be long.

Solubility Calculations. All compounds are soluble in water to a certain extent. Likewise, the concentration of all compounds is limited by how much of a chemical can be dissolved in water. Some compounds, such as NaCl, are very soluble; other compounds, such as AgCl, are very insoluble, that is, only a small amount will go into solution. If you add baking soda (NaHCO$_3$), a solid compound, to distilled water, some of the compound will go into solution. After you have added a certain mass of baking soda, no more of it will go into solution (**dissolve**). At this point, equilibrium is reached. The solubility reaction for sodium bicarbonate is written as follows.

$$NaHCO_3(s) \rightleftharpoons Na^+ + HCO_3^- \tag{2-25}$$

The most general form of a precipitation–dissolution reaction is

$$A_a B_b(s) \rightleftharpoons a A^{x+} + b B^{y-} \tag{2-26}$$

As stated previously, the equilibrium expression for any reaction can be written as the product of the products (taken to the appropriate stoichiometric factors) divided by the product of the reactants. You will note that in the preceding reaction the concentration of the reactant, $A_a B_b$, does not appear in the equation. This is because the solubility products were defined using activities not molar concentrations. By definition, the activity of a pure solid is 1. As such, the activity term for the reactant, assumed to be a pure solid, drops out of the equation. Thus, for any precipitation reaction we can write a solubility product

$$K_s = \{A^{x+}\}^a \{B^{y-}\}^b \tag{2-27}$$

For the dissolution of sodium bicarbonate, shown in Equation 2–25, the solubility product can be written as

$$\{Na^+\} \{HCO_3^-\} = K_s$$

Solubility product constants (K_s) can be obtained from a variety of sources. A limited subset is presented in Table 2–1 and Appendix A–9; a more complete list is available from such sources as *The CRC Handbook of Chemistry and Physics*.

TABLE 2–1	Selected Solubility Products at 25°C			
Substance	**Equilibrium Reaction**	**pK_s**	**Application**	
Aluminum hydroxide	$Al(OH)_3\ (s) \rightleftharpoons Al^{3+} + 3OH^-$	32.9	Coagulation	
Aluminum phosphate	$AlPO_4\ (s) \rightleftharpoons Al^{3+} + PO_4^{3-}$	22.0	Phosphate removal	
Calcium carbonate (aragonite)	$CaCO_3\ (s) \rightleftharpoons Ca^{2+} + CO_3^{2-}$	8.34	Softening, corrosion control	
Ferric hydroxide	$Fe(OH)_3\ (s) \rightleftharpoons Fe^{3+} + 3OH^-$	38.57	Coagulation, iron removal	
Ferric phosphate	$FePO_4\ (s) \rightleftharpoons Fe^{3+} + PO_4^{3-}$	21.9	Phosphate removal	
Magnesium hydroxide	$Mg(OH)_2\ (s) \rightleftharpoons Mg^{2+} + 2OH^-$	11.25	Removal of calcium and magnesium	
Dolomite $(CaMg(CO_3)_2)$ (ordered)	$CaMg(CO_3)_2 \rightleftharpoons Ca^{2+} + Mg^{2+} + 2CO_3^{2-}$	17.09	Weathering of dolomitic minerals	
Kaolinite	$Al_2Si_2O_5(OH)_4 + 6H^+ \rightleftharpoons 2Al^{3+} + 2Si(OH)_4 + H_2O$	7.44	Weathering of kaolinite clays	
Gypsum	$CaSO_4 \cdot 2H_2O \rightleftharpoons Ca^{2+} + SO_4^{2-} + 2H_2O$	4.58	Weathering of gypsum minerals	

Data Source: Stumm and Morgan, 1996.

Solubility products are often reported as pK_s, that is, the negative logarithm (base 10) of K_s.

$$pK_s = -\log_{10}(K_s) \text{ or } -\log(K_s) \tag{2–28}$$

This is done because the values of K_s are often very small.

Solubility products, like all equilibrium constants, are derived from thermodynamic data and can be determined from the change in Gibbs free energy[*] for the reaction. The tendency (or driving force) for a reaction to reach equilibrium is driven by the Gibbs free energy, $\Delta G°$. The relationship between the equilibrium constant, K, and $\Delta G°$ has been defined as:

$$\Delta G° = -RT \ln K \tag{2–29}$$

where R = ideal gas constant
 K = equilibrium constant, e.g., a solubility product
 T = temperature (K)

Many solubility products have been determined empirically, although for sparingly soluble compounds, these experimental values are often highly variable and inaccurate. Solubility products are defined for a specific temperature, usually 25°C. If the temperature is other than the reference temperature for which the solubility product was determined, a temperature correction must be used. Solubility products at temperatures other than 25°C can be calculated using the basic thermodynamic expression:

$$\frac{\partial \ln K}{\partial T} = \frac{\Delta H_r^0}{RT^2} \tag{2–30}$$

where ΔH_r^0 = change in enthalpy of reaction

---

[*]Gibbs free energy is a thermodynamic quantity that gives a direct criterion for the spontaneity of a reaction.

Assuming that the change in enthalpy is constant over the temperature considered, an approximate solution for Equation 2–30 is

$$\ln \frac{K_{T_2}}{K_{T_1}} = \frac{\Delta H_r^0}{R} \left(\frac{1}{T_1} - \frac{1}{T_2} \right)$$ (2–31)

You should also remember that thermodynamic data, such as solubility products, tell you nothing about the kinetics of the reaction (how fast it proceeds). These reactions may take seconds to reach equilibrium or hundreds of thousands of years!

EXAMPLE 2–6 If you add 30 g of calcite ($CaCO_3$) to a 1.00-L volumetric flask and bring the final volume to 1.00 L, what would be the concentration of calcium (Ca^{2+}) in solution? Assume that the calcium in solution is at equilibrium with $CaCO_3(s)$ and the temperature of the solution is 25°C. The pK_s for calcite is 8.48.

Solution The pertinent reaction is

$$CaCO_3(s) \rightleftharpoons Ca^{2+} + CO_3^{2-} \qquad K_s = 10^{-pK_s} = 10^{-8.48}$$

Remember that $K_s = \{Ca^{2+}\}\{CO_3^{2-}\}$. We will assume that the solution is dilute, allowing us to approximate the activities with molar concentrations. As such, the preceding equation becomes

$$K_s = [Ca^{2+}][CO_3^{2-}]$$

For every mole of calcite that dissolves, one mole of Ca^{2+} and one mole of CO_3^{2-} are released into solution. At equilibrium, the molar concentration of Ca^{2+} and CO_3^{2-} in solution will be equal, so we may say

$$[Ca^{2+}] = [CO_3^{2-}] = s$$

Substituting s for each compound in the K_s expression,

$$10^{-8.48} = s^2$$

Solving for s (which is equal to Ca^{2+}), we can determine the concentration of calcium: $Ca^{2+} = 10^{-4.24}$ or 5.75×10^{-5} M (moles per liter) in solution. Because the concentration of calcium (and also carbonate) are so low, the assumption that the activities could be approximated by the molar concentrations was a reasonable one. Note that the amount of calcite added to the flask is irrelevant. You are asked for the equilibrium concentrations, which are independent of the "starting" conditions or the mass of calcite added, so long as the mass added exceeded the solubility.

When the background* electrolyte concentration is high (i.e., in seawater, in landfill leachates, and in activated sludge), one must take into account the effects of the ions on all equilibrium constants. Here we will investigate its effect on solubility products; however, the effects are

*The term *background* is used to denote those ions that also occur in solution.

accounted for in the same way for all types of equilibrium constants. Remember from Equation 2–27 that the solubility product is defined as

$$K_s = \{A^{x+}\}^a \{B^{y-}\}^b$$

and

$$\{i\} = \gamma_i[i] \tag{2–32}$$

where $\{i\}$ is the activity of i, γ_i is the activity coefficient of i (in the solution having a particular composition and ionic strength) and $[i]$ is the molar concentration of i. Substituting $\gamma_i[i]$ into Equation 2–27, yields

$$K_s = (\gamma_A [A^{x+}])^a (\gamma_B [B^{y-}])^b \tag{2–33}$$

Activity coefficients can be determined using a number of approximations. The Davies equation is presented here because it is valid for the widest ranges of electrolyte concentrations (where ionic strength, I, is less than 0.5 M). The Davies equation states that

$$\log \gamma = -A z^2 \left(\frac{\sqrt{I}}{1 + \sqrt{I}} - 0.2I \right) \tag{2–34}$$

where $A \approx 0.5$ for water at 25°C

z = charge of the ion

I = ionic strength of the solution = $\frac{1}{2} \Sigma C_i z_i^2$

C_i = molar concentration of each of the ith ion in solution

z_i = charge on the ith ion

EXAMPLE 2–7 Reexamine Example 2–6. Assume you added 30 g of calcite to water to make up 1.00 L of a solution containing 0.01 M NaCl. What would be the concentration of calcium (Ca^{2+}) in solution? Assume that the calcium in solution is at equilibrium with calcite (s) and the temperature of the solution is 25°C.

Solution First we must calculate the concentration of the sodium chloride solution. The two ions present in this solution would be Na^+ and Cl^-. Because NaCl would completely dissociate, the concentrations of both ions would be 0.01 M. We can now calculate the ionic strength

$$I = \frac{1}{2} [(0.01)(1)^2 + (0.01)(1)^2] = 0.01 \text{ M}$$

We must now look at the two ions we are investigating: calcium and carbonate. Because the absolute value of the charge on both calcium and carbonate is the same (i.e., two), $\gamma_{Ca} = \gamma_{CO_3}$.

$$\log \gamma_{Ca} = \log \gamma_{CO_3} = -(0.5)(2)^2 \left(\frac{\sqrt{0.01}}{1 + \sqrt{0.01}} - 0.2(0.01) \right) = -0.178$$

$$\gamma_{Ca} = \gamma_{CO_3} = 10^{-0.178} = 0.664$$

Using the activity coefficient to calculate the effect of ionic strength on the solubility product, we have

$$K_s = \{Ca^{2+}\}\{CO_3^{2-}\} = \gamma_{Ca}[Ca^{2+}]\gamma_{CO_3}[CO_3^{2-}]$$

$$\frac{K_s}{\gamma_{Ca}\gamma_{CO3}} = [Ca^{2+}][CO_3^{2-}] = \frac{10^{-8.48}}{(0.664)(0.664)} = 7.51 \times 10^{-9}$$

Solving the problem as shown in Example 2–6, we see that

$$[Ca^{2+}] = [CO_3^{2-}] = 8.7 \times 10^{-5} \text{ M}$$

EXAMPLE 2–8 Look at Example 2–6 one more time. Assume you added 30 g of calcite to 1 L of water having the composition given below. What would be the concentration of calcium (Ca^{2+}) in solution? Assume that the calcium in solution is at equilibrium with calcite (s) and the temperature of the solution is 25°C.

Ion	Concentration (mM)	Ion	Concentration (mM)
NO_3^-	2.38	Mg^{2+}	53.2
SO_4^{2-}	28.2	Na^+	468.0
Cl^-	545.0	K^+	10.2
		Cu^{2+}	10.2

$$I = \tfrac{1}{2}[(2.38 \times 10^{-3})(1)^2 + (2.82 \times 10^{-2})(2)^2 + (0.545)(1)^2 + (1.02 \times 10^{-2})(2)^2$$

$$+ (5.32 \times 10^{-2})(2)^2 + (0.468)(1)^2 + (1.02 \times 10^{-2})(1)^2] = 0.696 \text{ M}$$

$$\log \gamma_{Ca} = \log \gamma_{CO_3} = -(0.5)(2)^2 \left(\frac{\sqrt{0.696}}{1 + \sqrt{0.696}} - 0.2(0.696) \right) = -0.631$$

$$\gamma_{Ca} = \gamma_{CO_3} = 10^{-0.631} = 0.234$$

Using the activity coefficient to calculate the effect of ionic strength on the solubility product, we see that

$$K_s = \{Ca^{2+}\}\{CO_3^{2-}\} = \gamma_{Ca}[Ca^{2+}]\gamma_{CO_3}[CO_3^{2-}]$$

$$\frac{K_s}{\gamma_{Ca}\gamma_{CO3}} = [Ca^{2+}][CO_3^{2-}] = \frac{10^{-8.48}}{(0.234)(0.234)} = 6.05 \times 10^{-8}$$

Solving the problem as shown in Example 2–6, we see that

$$[Ca^{2+}] = [CO_3^{2-}] = 2.5 \times 10^{-4} \text{ M}$$

You might remember from general chemistry that "like dissolves like." Calcium and carbonate ions are highly ionic. Adding sodium chloride to distilled water makes the resulting solution more ionic. As such, calcite is more soluble in the sodium chloride solution than it was in pure water. Because the ionic strength of the aqueous solution given in Example 2–8 is greater than that of the 0.01 M NaCl solution used in Example 2–7, calcite is even more soluble in this water than it is in the sodium chloride solution.

Common Ion Effect. In natural waters, rarely does a chemical dissolve into water that is free of the ion being dissolved. For example, as groundwater flows past calcite-containing rocks, calcium will dissolve into the water by the reaction

$$CaCO_3(s) \text{ (calcite)} = Ca^{2+} + CO_3^{2-} \qquad K_s = 10^{-8.48} \tag{2–35}$$

However, rarely does this reaction occur in water that is devoid of calcium or carbonate ions at the start of the reaction. As such, the solubility of calcite decreases, and calcite will not dissolve in groundwater to the extent to which it would in pure water.

EXAMPLE 2–9   What is the solubility of dolomite in water that contains 100 mg $\cdot$ L^{-1} CO$_3^{2-}$? The solubility product of dolomite is $10^{-17.09}$. Assume that the effects of ionic strength are negligible.

Solution   The first step in solving this reaction is to determine the concentration of carbonate that exists in the water prior to the dissolution (solubilization) of dolomite. The molar concentration of carbonate must be calculated. The molecular weight of carbonate is 60.01 g $\cdot$ mol^{-1}. Therefore, the molar concentration of carbonate is

(100 mg $\cdot$ L^{-1})(10^{-3} g $\cdot$ mg^{-1})(1 mol/60.01 g) = 0.00167 M

Dolomite dissolves according to reaction

$$CaMg(CO_3)_2 \rightleftharpoons Ca^{2+} + Mg^{2+} + 2CO_3^{2-} \qquad K_s = 10^{-17.09}$$

At the start of the reaction, we have 0.00167 M CO$_3^{2-}$, 0 M Ca^{2+}, and 0 M Mg^{2+} in solution. If s moles of dolomite dissolves, then we add s moles of Ca^{2+}, s moles of Mg^{2+}, and 2s moles of CO$_3^{2-}$ to the solution. At equilibrium, the concentrations of the ions of interest are the sums of the values given here. We can write this in tabular form to summarize the calculations.

	Concentration (M)		
	Ca^{2+}	**Mg^{2+}**	**CO$_3^{2-}$**
Starting	0	0	0.00167
Change	s	s	2s
Equilibrium	s	s	0.00167 + 2s

The next step is to write the equilibrium equation.

$$[Ca^{2+}][Mg^{2+}][CO_3^{2-}]^2 = K_s = 10^{-17.09}$$

Substituting from the table, we have

$$(s)(s)(0.00167 + 2s)^2 = 10^{-17.09}$$

The preceding equation can be solved numerically using a mathematical equation-solving package such as SOLVER on EXCEL or by trial and error. By trial and error, we obtained a value of s of 1.704×10^{-6} M. As such, the concentrations of Ca^{2+}, Mg^{2+}, and CO$_3^{2-}$ are 1.704×10^{-6} M, 1.704×10^{-6} M, and 1.673×10^{-3} M, respectively.

In many situations, you must deal with conditions in which multiple common ions are present in the solution at the beginning of the reaction. In other cases, a solution can be super-saturated with the ions, and precipitation of a solid can occur. This next example illustrates how to solve a problem when the solution is supersaturated.

EXAMPLE 2–10   A solution is initially supersaturated with CO$_3^{2-}$ and Ca^{2+} such that the concentrations are both 50.0 mg $\cdot$ L^{-1}. When equilibrium is ultimately reached, what will be the final concentration of Ca^{2+}?

Solution   We are starting with a solution that is supersaturated with calcium and carbonate. Over time, as equilibrium is reached, calcium carbonate will precipitate. The reaction is

$$Ca^{2+} + CO_3^{2-} \rightleftharpoons CaCO_3(s) \qquad pK_s = 8.34$$

Remember that to solve equilibrium problems, we must use molar units! The atomic weight of Ca^{2+} is 40.08 amu and the molecular weight of CO$_3^{2-}$ is 60.01 g $\cdot$ mol^{-1}, resulting in initial molar concentrations of 1.25×10^{-3} mol $\cdot$ L^{-1} and 8.33×10^{-4} mol $\cdot$ L^{-1} for Ca^{2+} and CO$_3^{2-}$, respectively.

$$K_s = 10^{-pK_s} = 10^{-8.34} = [Ca^{2+}][CO_3^{2-}]$$

For every mole of Ca^{2+} that precipitates from solution, one mole of CO$_3^{2-}$ also precipitates. If the amount removed is given by s, then

$$10^{-8.34} = 4.57 \times 10^{-9} = [1.25 \times 10^{-3} - s][8.33 \times 10^{-4} - s]$$

$$1.037 \times 10^{-6} - (2.083 \times 10^{-3})s + s^2 = 0$$

Solving for s, using the quadratic formula, yields

$$s = \frac{-b \pm \sqrt{b^2 - 4ac}}{2a} = \frac{2.08 \times 10^{-3} \pm \sqrt{4.34 \times 10^{-6} - 4(1.037 \times 10^{-6})}}{2}$$
$$= 8.20 \times 10^{-4} \text{ M}$$

Therefore, the final Ca^{2+} concentration is

$$[Ca^{2+}] = 1.25 \times 10^{-3} \text{ M} - 8.33 \times 10^{-4} \text{ M} = 4.17 \times 10^{-4} \text{ M}$$

or

$$(4.17 \times 10^{-4} \text{ mol} \cdot \text{L}^{-1})(40 \text{ g} \cdot \text{mol}^{-1})(10^3 \text{ mg} \cdot \text{g}^{-1}) = 16.7 \text{ mg} \cdot \text{L}^{-1}$$

Note that mathematical analysis of the quadratic expression would yield two solutions, both positive, one larger than the other. The larger solution (1.25×10^{-3}) yields a value for s that is equal to the original concentrations of Ca^{2+}. This would yield a final concentration that is zero. Because this is physically impossible, we will discard the larger solution for s.

Acid–Base Equilibria. Water ionizes according to the equation

$$H_2O \rightleftharpoons H^+ + OH^- \tag{2–36}$$

The degree of ionization of water is very small and can be measured by what is called the **dissociation** (or **ionization**) **constant** of water, K_w. It is defined as

$$K_w = \{OH^-\}\{H^+\} \tag{2–37}$$

and has a value of 10^{-14} ($pK_w = 14$) at 25°C. A solution is said to be acidic if $\{H^+\}$ is greater than $\{OH^-\}$, neutral if equal, and basic if $\{H^+\}$ is less than $\{OH^-\}$. At a temperature of 25°C and in dilute solutions (where the ionic strength effects on K_w can be ignored), if $[H^+] = [OH^-] = 10^{-7}$ M, then the solution is neutral. Under the same conditions, if $[H^+]$ is greater than 10^{-7} M then the solution is acidic. Convenient expressions for the hydrogen and hydroxide ion concentrations are pH and pOH, respectively. The terms are defined as

$$pH = -\log\{H^+\} \tag{2–38}$$

$$pOH = -\log\{OH^-\} \tag{2–39}$$

where log = logarithm in base ten, as was used for pK_s values. Therefore, a neutral (and dilute) solution at 25°C has a pH of 7 (written pH 7), an acidic solution has a pH < 7, and a basic solution has a pH > 7. Also note that by taking the logarithm of Equation 2–37, we see that

$$pH + pOH = 14 \tag{2–40}$$

at a temperature of 25°C and in dilute solutions.

TABLE 2–2	**Strong and Moderately Strong Acids**			
	Acid	**Chemical Reaction**	**pK_a**	**Importance**
	Hydrochloric acid	$HCl \longrightarrow H^+ + Cl^-$	≈ -3	pH adjustment
	Nitric acid	$HNO_3 \longrightarrow H^+ + NO_3^-$	-1	Acid rain formation
	Sulfuric acid	$H_2SO_4 \longrightarrow H^+ + HSO_4^-$	≈ -3	Acid rain formation, coagulation, pH adjustment
	Bisulfate	$HSO_4^- \rightleftharpoons H^+ + SO_4^{2-}$	1.9	Formation in anoxic sediments

Acids are classified as strong acids or weak acids. As mentioned previously, strong acids have a strong tendency to donate their protons to water. For example,

$$HCl \rightleftharpoons H^+ + Cl^- \tag{2–41}$$

which is actually the simplified form of

$$HCl + H_2O \rightleftharpoons H_3O^+ + Cl^- \tag{2–42}$$

showing water as a base, which accepts a proton.

Equilibrium exists between the dissociated ions and undissociated compound. With strong acids, equilibrium is such that essentially all of the acid dissociates to form the proton and the conjugate base (Cl^- in the previous example). We can write the equilibrium expression as

$$K_a = \frac{[H_3O^+][Cl^-]}{[HCl]} \tag{2–43}$$

As with other equilibrium constants,

$$pK_a = -\log K_a \tag{2–44}$$

A list of important strong acids is in Table 2–2. Note the use of the single arrow to signify that, for all practical purposes, we may assume that the reaction proceeds completely to the right.

EXAMPLE 2–11 If 100 mg of H_2SO_4 (MW = 98) is added to water, bringing the final volume to 1.0 L, what is the final pH?

Solution Using the molecular weight of sulfuric acid, we find

$$\left(\frac{100 \text{ mg}}{1 \text{ L } H_2O} \right) \left(\frac{1 \text{ mol}}{98 \text{ g}} \right) \left(\frac{1 \text{ g}}{10^3 \text{ mg}} \right) = 1.02 \times 10^{-3} \text{ mol} \cdot L^{-1}$$

The reaction is

$$H_2SO_4 \longrightarrow 2H^+ + SO_4^{2-}$$

and sulfuric acid is a strong acid, so we can determine the pH in the following manner. If the concentration of sulfuric acid is 1.02×10^{-3} M, then during the dissociation of the acid $2(1.02 \times 10^{-3}$ M) H^+ is produced. The pH is

$$pH = -\log(2.04 \times 10^{-3}) = 2.69$$

TABLE 2–3   Selected Weak Acid Dissociation Constants at 25°C

Substance	Chemical Reaction	pK$_a$	Significance
Acetic acid	$CH_3COOH \rightleftharpoons H^+ + CH_3COO^-$	4.75	Anaerobic digestion
Carbonic acid	$H_2CO_3^* \rightleftharpoons H^+ + HCO_3^-$ $HCO_3^- \rightleftharpoons H^+ + CO_3^{2-}$	6.35 10.33	Buffering of natural waters, coagulation
Hydrogen sulfide	$H_2S \rightleftharpoons H^+ + HS^-$ $HS^- \rightleftharpoons H^+ + S^{2-}$	7.2 11.89	Aeration, odor control, anaerobic sediments
Hypochlorous acid	$HOCl \rightleftharpoons H^+ + OCl^-$	7.54	Disinfection
Phosphoric acid	$H_3PO_4 \rightleftharpoons H^+ + H_2PO_4^-$ $H_2PO_4^- \rightleftharpoons H^+ + HPO_4^{2-}$ $HPO_4^{2-} \rightleftharpoons H^+ + PO_4^{3-}$	2.12 7.20 12.32	Phosphate removal, plant nutrient, pH adjustment

*The asterisk next to H_2CO_3 signifies both true carbonic acid and dissolved carbon dioxide. Because we cannot distinguish between the two, analytically, we combine the concentrations of the two compounds and refer to the sum as H_2CO_3-star.

Weak acids are acids that do not completely dissociate in water. Equilibrium exists between the dissociated ions and undissociated compound. The reaction of a weak acid is

$$HA \rightleftharpoons H^+ + A^- \tag{2–45}$$

An equilibrium constant exists that relates the degree of dissociation to the equilibrium constant for that reaction:

$$K_a = \frac{[H^+][A^-]}{[HA]} \tag{2–46}$$

A list of weak acids important in environmental science is provided in Table 2–3. By knowing the pH of a solution (which can be easily found with a pH meter), it can be possible to get a rough idea of the degree of dissolution of the acid. For example, if the pH is equal to the pK$_a$ (i.e., $[H^+] = K_a$), then from Equation 2–46, $[HA] = [A^-]$. Because the total amount of acid species, A_T, equals $[HA] + [A^-]$ and $[HA] = [A^-]$, the acid is 50% dissociated. If the $[H^+]$ is two orders of magnitude (100 times) greater than the K_a, then

$$K_a = \frac{[H^+][A^-]}{[HA]} = \frac{100K_a[A^-]}{[HA]}$$

$$1 = \frac{100[A^-]}{[HA]}$$

$$[HA] = 100[A^-]$$

In this case pH $\ll$ pK$_a$, and the acid is found primarily in the protonated form (as HA). Conversely, if pH $\gg$ pK$_a$, then the acid is primarily in the dissociated form (A^-).

EXAMPLE 2–12   A solution of HOCl is prepared in water by adding 15 mg HOCl to a volumetric flask, and adding water to the 1.0 L mark. The final pH is measured to be 7.0. What are the concentrations of HOCl and OCl$^-$? What percent of the HOCl is dissociated? Assume the temperature is 25°C.

Solution   The dissociation reaction for HOCl is

$$HOCl \rightleftharpoons H^+ + OCl^-$$

From Table 2–3, we find the pK_a is 7.54 and

$$K_a = 10^{-7.54} = 2.88 \times 10^{-8}$$

Writing the equilibrium expression in the form of Equation 2–46 and substituting the concentration of H^+, we find

$$K_a = \frac{[H^+][OCl^-]}{[HOCl]} = \frac{[1.00 \times 10^{-7}][OCl^-]}{[HOCl]} = 2.88 \times 10^{-8}$$

Using the previous equation, we can solve for the HOCl concentration as

$$[HOCl] = 3.47[OCl^-]$$

Because, at equilibrium, the concentration of the combination of HOCl or OCl^- must equal that which was added, we can state that

$$[HOCl] + [OCl^-] = \text{molar concentration added}$$

Because we added HOCl, we need to calculate the molar concentration using the molecular weight of HOCl.

$$\text{Molar concentration} = (15 \text{ mg} \cdot L^{-1})(10^{-3} \text{ g} \cdot mg^{-1})(1 \text{ mol}/52.461 \text{ g}) = 2.86 \times 10^{-4} \text{ M}$$

Thus,

$$[HOCl] + [OCl^-] = 2.86 \times 10^{-4} \text{ M}$$

Substituting from the equation for HOCl concentration, we see

$$(3.47[OCl^-]) + [OCl^-] = 2.86 \times 10^{-4} \text{ M}$$

Therefore, $[OCl^-] = 6.39 \times 10^{-5}$ M. The concentration of HOCl can be calculated either by multiplying this concentration by 3.47 or subtracting this concentration from 2.86×10^{-4} M.

$$[HOCl] = 2.86 \times 10^{-4} - 6.39 \times 10^{-5} = 2.22 \times 10^{-4} \text{ M}$$

The percentage of OCl^- that is dissociated can be calculated as

$$\frac{[OCl^-]}{[HOCl] + [OCl^-]} = \frac{[6.4 \times 10^{-5}]}{[2.86 \times 10^{-4}]} = 22.4\%$$

Equilibrium Among Gases and Liquids. The partitioning of chemical compounds between water and air is described by Henry's law. Henry's law states that, at equilibrium, the partial pressure of a chemical in the gas phase (P_{gas}) is linearly proportional to the concentration of the chemical in the aqueous phase (C^*).

$$P_{gas} = kC^* \tag{2–47}$$

Henry's law is valid for dilute solutions (fresh water) and pressures typically found in most environmental systems. Henry's law constants are derived empirically and are reported both in dimensional and dimensionless forms.

The confusion with the dimensional form of Henry's law comes about because various concentration scales can be used for both the gas and aqueous phase concentrations and the proportionality constant can be written on either side of the equation, resulting in constants that are the inverse of one another. For example, the definition of Henry's law can be written as

$$K_H = \frac{P_{gas}}{C^*} \tag{2–48}$$

where K_H = Henry's law constant, $kPa \cdot m^3 \cdot g^{-1}$

$\quad P_{gas}$ = partial pressure of the gas at equilibrium (kPa)

$\quad C^*$ = equilibrium concentration of the gas dissolved in water $(g \cdot m^{-3})$

Similarly, one can use units of molar and atmosphere for C^* and P_{gas}, respectively. In this case, the equation can be written as

$$K'_H = \frac{C_{air}}{C^*} \tag{2-49}$$

where K'_H = Henry's law constant $(atm \cdot mol^{-1} \cdot L)$

$\quad C_{air}$ = concentration of the gas at equilibrium (atm)

$\quad C^*$ = equilibrium concentration of the gas dissolved in water $(mol \cdot L^{-1})$

In other cases, Henry's Law is written as the inverse of Equation 2–49,

$$K^{\ddagger}_H = \frac{1}{K'_H} = \frac{C^*}{C_{air}} \tag{2-50}$$

in which case $K^{\ddagger}_H$ has units of $mol \cdot L^{-1} \cdot atm$.

As stated above, Henry's Law constants can also be dimensionless. In this case, one can think of the constant as the mass of a chemical compound present in the air divided by the mass of that same chemical compound dissolved in water at equilibrium. As such,

$$H = \frac{C_{air}}{C^*} \tag{2-51}$$

where C_{air} = concentration of the chemical and air $(g \cdot m^{-3})$

$\quad C^*$ = equilibrium concentration of the chemical in water $(g \cdot m^{-3})$

If the dimensionless Henry's constant, H, is greater than one $(C_{air} > C_{water})$, then the chemical compound prefers to be in air rather than water. On the contrary, if the dimensionless Henry's constant, H, is less than one $(C_{air} < C_{water})$, then the chemical compound will tend to partition into water. The dimensional form of the Henry's constant can be obtained by multiplying the dimensionless constant by the ideal gas constant $(atm \cdot L \cdot mol^{-1} \cdot K^{-1})$ times the temperature (K), to yield a constant with units of $atm \cdot L \cdot mol^{-1}$.

The Henry's law coefficient varies both with the temperature and the concentration of other dissolved substances. Henry's law constants are given in Appendix Table A–11.

EXAMPLE 2–13 The concentration of carbon dioxide in water at 20°C is determined to be $1.00 \cdot 10^{-5}$ M. The Henry's constant for carbon dioxide dissolution in water is $3.91 \cdot 10^{-2}$ M atm^{-1} at 20°C. What is the partial pressure of CO_2 in the air?

Solution With Henry's law given in units of molar per atmosphere $(M \cdot atm^{-1})$, we can write Equation 2–50 as

$$K^{\ddagger}_H = \frac{C^*}{C_{air}}$$

Since the concentration of carbon dioxide in air is given as the partial pressure (P_{CO_2}) in units of atmosphere,

$$P_{CO_2} = C_{air} = C^*/K^{\ddagger}_H$$

$$= \frac{1.00 \times 10^{-5} \text{ M}}{3.91 \times 10^{-2} \text{ M} \cdot atm^{-1}} = 2.56 \times 10^{-4} \text{ atm}$$

EXAMPLE 2–14 You are working at a site where a leaking underground storage tank has contaminated the soil and groundwater below a gas station. The groundwater was analyzed and the concentrations of benzene and methyl tertiary butyl ether (MTBE) were found to be 45 and 500 μg/L, respectively. At the soil temperature (10°C) the dimensionless Henry's constants, H, for benzene and MTBE are 0.09 and 0.01, respectively. Calculate the soil vapor concentrations of both chemical compounds.

Solution $C_{benzene} = H \times C^*_{benzene}$

$$= (0.09)(45 \text{ μg/L}) = 0.36 \text{ μg/L}$$

$$C_{MTBE} = H \times C^*_{MTBE}$$

$$= (0.01)(500 \text{ μg/L}) = 5 \text{ μg/L}$$

Reaction Kinetics

Many reactions that occur in the environment do not reach equilibrium quickly. Some examples include the disinfection of water, gas transfer into and out of bodies of water, dissolution of rocks and minerals, and radioactive decay. The study of the speed at which these reactions proceed is called reaction kinetics. The rate of reaction, r, is used to describe the rate of formation or disappearance of a compound. Reactions that take place in a single phase (i.e., liquid, gas, or solid) are called **homogeneous reactions.** Those that occur at surfaces between phases are called **heterogeneous reactions.** For each type of reaction, the rate may be defined as follows:

For homogeneous reactions

$$r = \frac{\text{mass of chemical species}}{\text{(unit volume)(unit time)}} \tag{2–52}$$

For heterogeneous reactions

$$r = \frac{\text{mass of chemical species}}{\text{(unit surface area)(unit time)}} \tag{2–53}$$

By convention, the production of a compound is denoted by a positive sign for the reaction rate ($+r$), and a negative sign ($-r$) is used to designate the disappearance of the substance of interest. Reaction rates are a function of temperature, pressure, and the concentration of reactants. For a stoichiometric reaction of the form

$$a\text{A} + b\text{B} \longrightarrow c\text{C} \tag{2–54}$$

where a, b, and c are the proportionality coefficients for the reactants A, B, and C, the change in concentration of compound A in a batch reactor is equal to the reaction rate equation for compound A.

$$-\frac{1}{a}\frac{d[\text{A}]}{dt} = -\frac{1}{b}\frac{d[\text{B}]}{dt} = +\frac{1}{c}\frac{d[\text{C}]}{dt} \tag{2–55}$$

where [A], [B], and [C] are the concentrations of the reactants.

TABLE 2–4   **Example Reaction Orders**

Reaction Order	Rate Expression	Units on Rate Constant
Zero	$r_A = -k$	(concentration)(time)$^{-1}$
First	$r_A = -k[A]$	(time)$^{-1}$
Second	$r_A = -k[A]^2$	(concentration)$^{-1}$(time)$^{-1}$
Second	$r_A = -k[A][B]$	(concentration)$^{-1}$(time)$^{-1}$

For this general reaction, the overall reaction rate, r, and the individual reaction rates are related.

$$r = -\frac{r_a}{a} = -\frac{r_b}{b} = \frac{r_c}{c} \tag{2–56}$$

For the general equation

$$a\text{A} + b\text{B} \rightarrow c\text{C} \tag{2–57}$$

we can write the rate expression

$$\frac{d[A]}{dt} = -k[A]^\alpha[B]^\beta \tag{2–58a}$$

where α and β are empirically determined constants.

This proportionality term, k, is called the **reaction rate constant** and is generally dependent on the temperature and pressure. Because A and B are disappearing, the sign of the reaction rate equation is negative. It is positive for C because C is being formed.

The order of reaction is defined as the sum of the exponents in the reaction rate equation. The exponents may be either integers or fractions. Some sample reaction orders are shown in Table 2–4. First-order reactions are simple; the rate constant has units of inverse time. It is common, with first-order reactions, to use the half-life ($t_{1/2}$) for the reaction; that is the time required for the concentration to reach a value one-half of its initial concentration. The half-life is defined as $-\ln(0.5)k^{-1}$ or $0.693k^{-1}$.

EXAMPLE 2–15

As a result of the Aykhal Crystal detonation in 1974 (in the former Soviet Union), cesium-137 was measured (in 1993) in the soil at a concentration of 2×10^4 Bq · kg^{-1} soil.[*] If the background concentration of ^{137}Cs is 0.5 Bq · kg^{-1} soil, how many years will it take before the concentration of ^{137}Cs reaches background levels? The decay of radionuclides occurs by a first-order reaction, with a half-life of 30 years.

Solution

Because $t_{1/2} = 0.693 \cdot k^{-1}$, $k = 0.693/(30 \text{ years}) = 0.0231$ year^{-1}.
Then we can write that

$$\ln(C/C0) = -kt$$

$$\ln(0.5/2 \times 10^4) = -(0.0231 \text{ year}^{-1})t$$

$$t = 459 \text{ years}$$

---

[*]The unit becquerel (Bq) describes radioactivity. It is the number of disintegrations per second.

TABLE 2–5 **Plotting Procedure to Determine Order of Reaction by Method of Integration**

Order	Rate Equation	Integrated Equation	Linear Plot	Slope	Intercept
0	$d[A]/dt = -k$	$[A] - [A]_0 = -kt$	$[A]$ vs. t	$-k$	$[A]_0$
1	$d[A]/dt = -k[A]$	$\ln\{[A]/[A]_0\} = -kt$	$\ln[A]$ vs. t	$-k$	$\ln[A]_0$
2	$d[A]/dt = -k[A]^2$	$1/[A] - 1/[A]_0 = kt$	$1/[A]$ vs. t	k	$1/[A]_0$

Data Source: Henry and Heinke, 1989.

The reaction rate constant, k, must be determined experimentally by obtaining data on the concentrations of the reactants as a function of time and plotting on a suitable graph. The form of the graph is determined from the result of integration of the equations in Table 2–4. The integrated form and the appropriate graphical forms are shown in Table 2–5.

EXAMPLE 2–16 An environmental engineering student was very interested in the reaction of the chemical 2,4,6-chickenwire. She went into the lab and found that 2,4,6-chickenwire degrades in water. During her experiments she collected the data in the following table. Plot the data to determine if the reaction is zero-, first-, or second-order with respect to the concentration of 2,4,6-chickenwire.

Time (min)	Concentration (mg · L⁻¹)	Time (min)	Concentration (mg · L⁻¹)
0	10.0	10	5.46
1	8.56	20	4.23
2	8.14	40	1.26
4	6.96	80	0.218
8	6.77		

Solution Using the information provided above and in Table 2–5, let's first plot the data as if the reaction were zero order. In this case, we will plot $C_t - C_0$ vs. time.

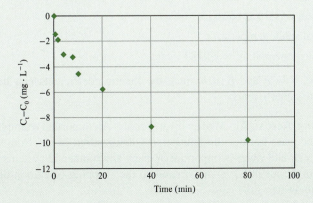

Since the resultant plot is nonlinear, the reaction of 2,4,6-chickenwire is not zero order.

Let's now plot the natural log of the ratio of the concentration of 2,4,6-chickenwire to its initial concentration versus time.

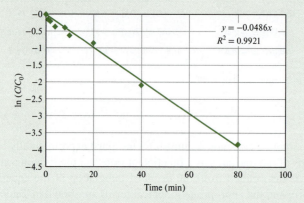

The plot is linear; a least squares regression analysis reveals a r-square value of 0.9921. The slope of the line (forced through zero) is 0.0486. This corresponds to a rate constant of 0.049 min^{-1}. Clearly, the reaction is first-order.

A special type of reaction is known as an elementary reaction. For these reactions the stoichiometric equation represents both the mass balance and the reaction process on a molecular scale, and the stoichiometric coefficients (a, b, c) are equivalent to the exponents in the reaction rate equation. For example, with the following elementary reaction:

$$a\text{A} + b\text{B} \rightarrow c\text{C} \tag{2–54}$$

we can a priori write the kinetic expression as

$$r_A = -k[\text{A}]^a[\text{B}]^b \tag{2–58b}$$

without having to develop the kinetic expression from experimental results.

The effect of temperature on elementary reactions is described by the relationship first postulated by Arrhenius (in 1889), where

$$k = Ae^{-E_a/RT} \tag{2–59}$$

where A = the Arrhenius parameter

$\quad E_a$ = the activation energy

$\quad R$ = universal gas constant

$\quad T$ = absolute temperature

$\quad e$ = exponential; $e^1 = 2.7183$

In this case, $\alpha = a$ and $\beta = b$ since the kinetic coefficients can be determined from the stoichiometric coefficients.

Gas Transfer Across Air–Water Interfaces. An important example of time-dependent reactions is the mass transfer (dissolution or volatilization) of gas from water. The transfer of a gas between air and water is important in many environmental systems from the dissolution of oxygen from the air in lakes and streams to the stripping of carbon dioxide from chemically treated groundwater.

Lewis and Whitman (1924) postulated a two-film theory to describe the mass transfer of gases. According to their theory, the boundary between the gas phase and the liquid phase (also

FIGURE 2–3

Two-film model of the interface between gas and liquid: (a) absorption mode and (b) desorption mode. C_t is the concentration at time t. C_s is the saturation concentration of the gas in the liquid.

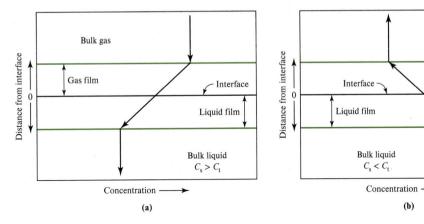

(a) (b)

called the interface) is composed of two distinct films that serve as barriers between the bulk phases (Figure 2–3). For a molecule of gas to go into solution, it must pass through the bulk of the gas, the gas film, the liquid film, and into the bulk of the liquid (Figure 2–3a). To leave the liquid, the gas molecule must follow the reverse course (Figure 2–3b).

The driving force causing the gas to move, and hence the mass transfer, is the concentration gradient: $C_s - C$. C_s is the saturation concentration of the gas in the liquid, and C is the actual aqueous concentration. The saturation concentration is a function of temperature, pressure, and gas-phase concentration (or liquid-phase concentration, depending on which phase serves as the source phase). When C_s is greater than C, the gas will go into solution. When C is greater than C_s, the gas will volatilize from solution. C_s can be determined from the relationship for Henry's constant as discussed previously.

The rate of mass transfer can be described by the following equation:

$$\frac{dC}{dt} = k_a(C_s - C) \tag{2–60}$$

where k_a = rate constant or mass transfer coefficient, $(\text{time})^{-1}$. In a batch reactor, Equation 2–60 can be written as

$$r_c = k_a(C_s - C) \tag{2–61}$$

The difference between the saturation concentration and the actual concentration, $C_s - C$, is called the deficit. Because the saturation concentration is constant with temperature and pressure, this is a first-order reaction.

Integrating the above equation yields

$$\ln\left(\frac{C_s - C_t}{C_s - C_0}\right) = -k_a t \tag{2–62}$$

where C_t = the aqueous concentration of the gas at any time, t.

EXAMPLE 2–17   A falling raindrop initially has no dissolved oxygen. The saturation concentration for oxygen in the drop of water is $9.20 \text{ mg} \cdot \text{L}^{-1}$. If, after falling for 2 s, the droplet has an oxygen concentration of $3.20 \text{ mg} \cdot \text{L}^{-1}$, how long must the droplet fall (from the start of the fall) to achieve a concentration of $8.20 \text{ mg} \cdot \text{L}^{-1}$? Assume that the rate of oxygen exchange is first-order.

Solution   We begin by calculating the deficit after 2 s, and that at a concentration of 8.20 mg · L^{-1}:

Deficit at 2 s = $9.20 - 3.20 = 6.00$ mg · L^{-1}

Deficit at t s = $9.20 - 8.20 = 1.00$ mg · L^{-1}

Now using the integrated form of the first-order rate equation from Table 2–5, noting that the rate of change is proportional to deficit and, hence, $[A] = (C_s - C)$ and that $[A]_0 = (9.20 - 0.0)$, we see that

$$\ln\left(\frac{6.00}{9.20}\right) = -k(2.00 \text{ s})$$

$$k = 0.214 \text{ s}^{-1}$$

With this value of k, we can calculate a value for t.

$$\ln\left(\frac{9.20 - 8.2}{9.20}\right) = -(0.214 \text{ s}^{-1})(t)$$

$$t = 10.4 \text{ s}$$

Reactions at the Solid-Water Interface.   As noted by Werner Stumm, often considered the father of environmental chemistry, the composition of our environment is very much controlled by the reactions that occur at the interfaces of water with naturally occurring solids. We saw the importance of reactions involving solids when we considered the dissolution of minerals. The availability of plant nutrients in soils and waters is controlled by reactions that occur at the solid-water interface. These types of reactions are also important in understanding how to inhibit corrosion, how to construct new chemical sensors and semiconductors, and how to develop better membranes for drinking-water treatment.

Reactions at the solid-water interface are a type of heterogeneous reaction, that is a reaction in which the reactants and catalysts are in different phases. Metallic oxides, a common example of a naturally occurring heterogeneous catalyst, work by adsorbing one of the reactants. Reactions that occur at the solid-water interface often depend on the reactive surface area of the solid material and on the composition of the mineral surfaces.

2–3   ORGANIC CHEMISTRY

So far much of what we have discussed in this chapter has been about inorganic chemistry. Because environmental science is very much one of the life sciences, we need to discuss organic chemistry. As mentioned previously, organic chemicals, which are those chemicals containing carbon and hydrogen, can be produced naturally or via synthetic anthropogenic reactions. This vast and diverse group of chemicals is important because the chemicals form the basis for all life; are used in many ways, as insecticides, herbicides, hormones, and antibiotics, to name a few; and can have serious deleterious effects on the environment.

Organic chemicals are divided into groups by several different approaches. For example, we can divide up organic chemicals by the types of C—C bonding present. Alkanes are those compounds that have a single bond between adjacent carbon atoms. Alkenes have a double bond between adjacent carbon atoms, whereas the highly reactive alkynes have a triple bond between adjacent carbon atoms. With alkanes, the carbon atoms involved in the single bonds share two electrons; with alkenes the double-bonded carbons share four electrons; and with alkynes six electrons are shared by the carbons involved in the triple bonds.

TABLE 2–6 **Chemical Nomenclature for Alkanes**

Molecular Formula	IUPAC Name
CH_4	methane
C_2H_6	ethane
C_3H_8	propane
C_4H_{10}	butane
C_5H_{12}	pentane
C_6H_{14}	hexane
C_7H_{16}	heptane
C_8H_{18}	octane
C_9H_{20}	nonane
$C_{10}H_{22}$	decane

Alkanes, Alkenes, and Alkynes

Alkanes (C_nH_{2n+2}) are also called paraffins or aliphatic hydrocarbons. With this group of chemicals, the carbon atoms are all linked by single covalent bonds. The nomenclature used by the IUPAC (International Union of Pure and Applied Chemists) is shown in Table 2–6. The suffix -*ane* is used to distinguish alkanes from other compounds. Alkanes can be straight-chained, branched, or cyclic. Some alkanes (and other organic compounds) may have the same molecular formula but different structures. These compounds are referred to as **structural isomers.** For example, *n*-hexane, 2-methylpentane, 3-methylpentane, and 2,3-dimethylbutane all have the molecular formula C_6H_{14}. However, the structural formulas for these chemicals all differ as shown in Figure 2–4. The chemical cyclohexane, which has six carbons, is not a structural isomer of the other hexanes listed here because its molecular formula is C_6H_{12}.

As mentioned previously, alkenes (C_nH_{2n}) contain at least one double bond. Naming alkenes is similar to naming alkanes, except that the IUPAC uses -*ene* as the suffix. Common naming calls for the use of -*ylene* as the suffix. As such, C_2H_4 has the IUPAC name *ethene* and the common name *ethylene*. Alkenes have geometric isomers because the C=C bond is rigid (i.e., it cannot rotate), unlike in alkanes. We therefore need to remember that with alkenes there are cis and trans isomers. For example, the common industrial solvent and groundwater contaminant, dichloroethene (DCE), comes in two forms, *cis*-DCE and *trans*-DCE (Figure 2–5).

Alkynes (C_nH_n) are named in a similar manner to alkanes and alkenes, except IUPAC uses the suffix -*yne*. Common naming uses acetylene as the base name. For example,

FIGURE 2–4

Structural isomers and IUPAC names for C_6H_{14}.

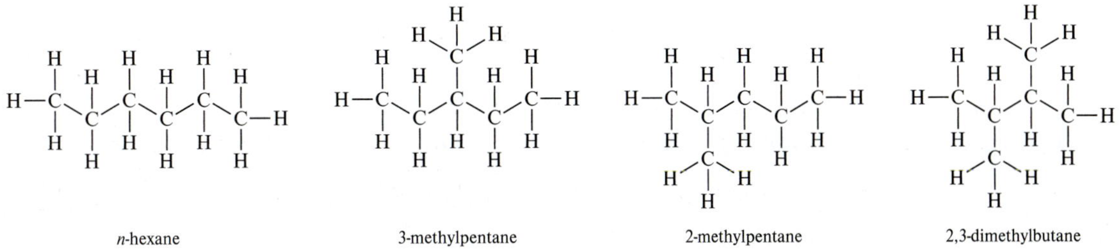

n-hexane   3-methylpentane   2-methylpentane   2,3-dimethylbutane

FIGURE 2–5

Cis- and *trans-*1,2-
dichloroethene.

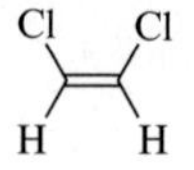

*cis-*1, 2-dichloroethene

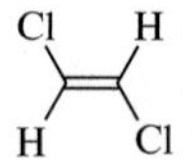

*trans-*1, 2-dichloroethene

$H_3C—C\equiv C—CH_3$ is called dimethylacetylene. The alkynes are highly reactive and dangerously explosive. They are generally of lesser environmental significance because they do not last long in the environment.

Aryl (Aromatic) Compounds

Aromatic compounds are cyclic and have alternating double bonds, in a resonance structure in which all of the carbons in the ring share the π-bond electrons. This sharing may not always be equal, making the aromatic ring slightly polar. Benzene (C_6H_6) is the simplest aromatic compound and is shown in Figure 2–6 with its Kekulé resonance structure. Because of the sharing of the π-bond electrons, creating what can be thought of as a "cloud" of electrons surrounding the carbon atoms, the chemical symbol in Figure 2–7 (known as the Robinson symbol) is often used as a shorthand notation for benzene. The hydrogen atoms are assumed to be bonded to the carbon atoms.

Among the many aromatic compounds of environmental significance, three classes of compounds stand out: the benzenelike hydrocarbons (benzene, toluene, and xylene or BTX), polycyclic aromatic hydrocarbons (PAHs, also known as polynuclear aromatic hydrocarbons—PNAs), and the polychlorinated biphenyls (PCBs). The first class (BTX—shown in Figure 2–8) are found in gasoline and are common in soils contaminated by gasoline products. The next group (PAHs) is a class of fused benzene rings, which are found in many petroleum products and are by-products of the incomplete combustion of other hydrocarbons. The carcinogenic properties of some of these chemicals make them of environmental concern. Several PAHs are shown in Figure 2–9. As shown in Figure 2–10, the biphenyl compounds are composed of two benzenes covalently bonded with a single bond.

Chlorinated biphenyls (PCBs) are biphenyl compounds that have a total of one to five chlorine atoms bonded to each of the aromatic rings (in place of the hydrogen atoms). PCBs were produced in the United States until 1978 and were used extensively in transformer oils, because these compounds are essentially nonflammable. PCBs were also used in such products as carbonless copy paper, electrical insulation, and paints. Although the biphenyl itself does not appear to be toxic, chlorination of the biphenyl increases the toxicity and carcinogenicity of this class of compounds. In general, the greater the degree of chlorination, the more **recalcitrant** (less easily biodegradable) are these compounds. Although their use has been banned for more than 20 years, the presence of PCBs in the environment continues to be a problem.

FIGURE 2–6

Kekulé resonance
structures of benzene.

FIGURE 2–7

Robinson shorthand
notation for aromatic
rings.

Functional Groups and Classes of Compounds

The presence of other elements or groups of elements on the backbone structure of hydrocarbons changes the properties of the chemical considerably. For example, replacing a hydrogen on ethane with a hydroxyl (—OH) functional group changes the chemical from an aqueous-insoluble gas to an aqueous-soluble liquid (ethanol). Functional groups are simply groups of specific bonding arrangements of atoms in organic molecules. Common functional groups are listed in Table 2–7.

FIGURE 2–8

BTX compounds.

benzene   toluene   xylene

FIGURE 2–9

Several examples of polycyclic aromatic hydrocarbons.

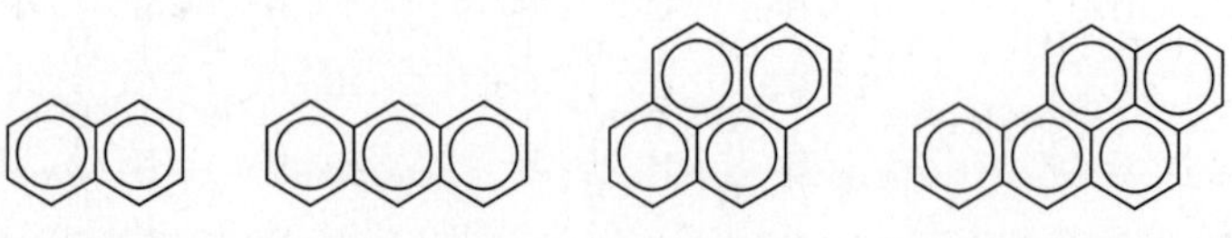

napthalene   anthracene   pyrene   benzo(a)pyrene

FIGURE 2–10

Two covalently bonded
benzene rings. This is the
basic unit of the biphenyl
compounds.

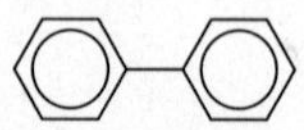

TABLE 2–7	Common Functional Groups and Examples of Compounds			

	Functional Group[a]		Example Compound	
Name	**Chemical Structure**		**Name**	**Structure**
Alcohol	R—OH (on alkyl group)		ethanol	H_3C—CH_2OH
Phenol	R—OH (on aryl group)		phenol	(benzene ring)—OH
Aldehyde	$R-\overset{\overset{\displaystyle O}{\|}}{C}H$		acetaldehyde	$H_3C\overset{\overset{\displaystyle O}{\|}}{C}H$
Ketone	$R-\overset{\overset{\displaystyle O}{\|}}{C}-R'$		methylethylketone	$H_3C-\overset{\overset{\displaystyle O}{\|}}{C}-C_2H_5$
Ester	$R-\overset{\overset{\displaystyle O}{\|}}{C}-O-R'$		methylethylester	$H_3C-\overset{\overset{\displaystyle O}{\|}}{C}-O-C_2H_5$
Ether	R—O—R'		methylethylether	$H_3C-O-C_2H_5$
Amine	$R-C-NH_2$		methylamine	H_3C-NH_2
Amide	$R-\overset{\overset{\displaystyle O}{\|}}{C}-NH_2$		acetamide	$H_3C-\overset{\overset{\displaystyle O}{\|}}{C}-NH_2$
Mercaptan	R—SH		methylmercaptan	CH_3SH
Halides	R—(Cl, Br, I, or F)		chloroform	$CHcl_3$
Sulfonic acid	$R-SO_3H$		benzenesulfonic acid	(benzene ring)—$\overset{\overset{\displaystyle O}{\|}}{\underset{\underset{\displaystyle O}{\|}}{S}}$—OH

[a]*Note:* R represents any functional group or hydrogen.

2–4 WATER CHEMISTRY

Because all environmental systems contain water, the properties of water play a significant role in defining environmental processes.

Physical Properties of Water

The basic physical properties of water relevant to environmental science are density and viscosity. Density is a measure of the concentration of matter and is expressed in three ways:

1. **Mass density, ρ.** Mass density is mass per unit volume and is measured in units of kilograms per cubic meter ($kg \cdot m^{-3}$). Table A–1, in the Appendix, shows the variation of density with temperature for pure water free from air. Dissolved or suspended impurities change the density of water in direct proportion to their concentration and their own density. In most environmental science applications, it is common to ignore the density increase due to impurities in the water. However, environmental scientists and engineers cannot ignore the density of the matter when dealing with high concentrations of solutions and suspensions, such as thickened sludge or stock solutions of reagents such as lime (used in water purification), or in ocean or estuarine waters.

2. **Specific weight, γ.** Specific weight is the weight (force) per unit volume, measured in units of kilonewtons per cubic meter ($kN \cdot m^{-3}$). The specific weight of a fluid is related to its density by the acceleration of gravity, g, which is $9.81 \ m \cdot s^{-2}$.

$$\gamma = \rho g \tag{2-63}$$

3. **Specific gravity, S.** Specific gravity is given by

$$S = \frac{\rho}{\rho_0} = \frac{\gamma}{\gamma_0} \tag{2-64}$$

where the subscript zero denotes the density of water at $3.98°C$, $1000 \ kg \cdot m^{-3}$, and the specific weight of water, $9.81 \ kN \cdot m^{-3}$.

For quick approximations, the density of water at commonly occurring temperatures can be taken as $1000 \ kg \cdot m^{-3}$ (which is conveniently $1 \ kg \cdot L^{-1}$).

4. **Viscosity.** All substances, including liquids, exhibit a resistance to movement, an internal friction. Just think of several liquids such as water, corn syrup, and molasses. The latter two are much more **viscous** than water, that is, they flow much less freely than does water. Alternatively, if you dropped a marble in each of the three fluids, you would find that the marble travels fastest in the water. The higher the viscosity of the fluid, the greater the friction, and the harder it is to pump the liquid. Viscosity is actually a measure of the friction and is presented in one of two ways:

   a. **Dynamic viscosity,** or absolute viscosity, μ, has dimensions of mass per unit length per time, with units of kilograms per meter second ($kg \cdot m^{-1} \cdot s^{-1}$) or pascal per second ($Pa \cdot s^{-1}$).

   b. **Kinematic viscosity, ν,** is defined as the dynamic viscosity divided by the density of the fluid at that temperature.

$$\nu = \frac{\mu}{\rho} \tag{2-65}$$

   It has dimensions of length squared per time with the corresponding units meters squared per second ($m^2 \cdot s^{-1}$).

States of Solution Impurities

Substances can exist in water in one of three classifications: suspended, colloidal, or dissolved. A dissolved substance is one that is truly in solution. The substance is homogeneously dispersed in the liquid. Dissolved substances can be simple atoms or complex molecular compounds. For example, if one adds table salt to water, the dissolved sodium is present as Na^+, its dissociated form of NaCl. On the contrary, if one adds sugar to water, the dissolved form of sugar (glucose) is $C_6H_{12}O_6$. Dissolved substances are present in the liquid, that is, there is only one phase present. The substance cannot be removed from the liquid without accomplishing a phase change such as distillation, precipitation, adsorption, or extraction.

In **distillation,** either the liquid or the substance itself is changed from a liquid phase to a gas phase in order to achieve separation. Distillation can occur in nature when salt water containing sodium chloride salt evaporates, leaving behind the sodium and chloride ions and producing a vapor that is free of sodium chloride.

In **precipitation** the substance in the liquid phase combines with another chemical to form a solid phase, thus achieving separation from the water. This occurs in a water treatment plant when lime (CaO) is added to remove hardness ions (in particular calcium and magnesium).

Adsorption also involves a phase change, wherein the dissolved substance attaches to the surface of solid particles. Attachment can be due to either chemical or physical attractive forces. Adsorption is important in soils, where ions such as nitrate and phosphate can attach, or sorb, to the surfaces of the soil particles.

Liquid extraction can separate a substance from water or a solid by extracting it into another liquid, hence a phase change from water to a different liquid. Liquid extraction is used in some environmental engineering applications, for example, petroleum-based compounds such as polycyclic aromatic hydrocarbons can be removed from soils by extracting them with such organic solvents as hexane.

Under no circumstances can physical methods such as filtration, sedimentation, or centrifugation remove dissolved substances. Activated carbon filters remove dissolved chemicals. They are not true filters, in that they do not "strain out" particles. Activated carbon works by the adsorption and absorption of dissolved substances onto the carbon itself.

Suspended solids are not truly dissolved and are large enough to settle out of solution or be removed by filtration. In this case two phases are present: the liquid water phase and the suspended-particle solid phase. Mixtures of a liquid and suspended particles are referred to as **suspensions**. The lower size range of suspended particles is 0.1–1.0 μm, about the size of bacteria. Suspended particles can range up to about 100 μm in size. Suspended solids are often defined operationally as those solids that can be filtered by a glass fiber filter and are properly called filterable solids. Suspended solids can be removed from water by physical methods such as sedimentation, filtration, and centrifugation.

Colloidal particles are usually defined on the basis of size and are generally from 0.001 to about 1 μm in size. Colloidal particles are kept in suspension by physical and chemical forces of attraction. Milk is an example of a colloidal suspension. The fat molecules are not truly dissolved but are held in suspension by attractive forces to the water. If you add acid (due to natural processes such as fermentation or the addition of vinegar), you will find that a solid precipitate forms. The acid changes the charge on the colloidal particles, allowing the colloidal particles to bind together and come out of suspension. Colloidal particles can be removed from the liquid by physical means such as ultracentrifugation or by filtration through membranes having pore sizes less than 0.45 μm. Colloidal particles cannot be removed by sedimentation unless the particles are first aggregated to form particles large enough to settle.

Concentration Units in Aqueous Solutions or Suspensions

Concentrations of solutions can be given in a variety of units. Chemists tend to use **molarity** or **molality,** but practicing environmental scientists and engineers usually use units of **milligrams per liter (mg · L^{-1}), parts per million (ppm),** or **percent (by weight).** In other cases, it makes more sense to use units of normality.

Weight percent, P, is sometimes employed to express approximate concentrations of commercial chemicals or of solid concentrations of sludges. The term specifies the grams of substance per 100 g of solution or suspension and is mathematically expressed as

$$P = \frac{W}{W + W_0} \times 100\% \qquad \qquad (2\text{–}66)$$

where P = percent of substance by weight
$\quad W$ = mass of substance (grams)
$\quad W_0$ = mass of solute (grams)

Analytical results are often given directly in mass per volume (concentration), and the units are milligrams per liter (mg · L^{-1}). In environmental science and engineering it is often assumed that the substance does not change the density of the water. This is generally true in dilute solutes at constant temperature, but it is not valid for concentrated solutions, in air, or in cases of large temperature fluctuations. However, when this assumption is valid, we can use the density of water (1 g · mL^{-1}) to develop the following conversion:

$$\left(\frac{1 \text{ mg solute}}{\text{L solution}}\right) \left(\frac{1 \text{ L}}{1000 \text{ mL}}\right) \left(\frac{1 \text{ mL}}{\text{g}}\right) \left(\frac{\text{g}}{1000 \text{ mg}}\right) = \frac{1 \text{ mg}}{10^6 \text{ mg}} = 1 \text{ ppm} \qquad (2\text{–}67)$$

or 1 mg $\cdot$ L^{-1} equals 1 part per million (ppm). For very dilute solutions, concentrations of parts per billion (ppb) or parts per trillion (ppt) can be used. Using an analogous conversion to that given in Equation 2–67, we can see that 1 μg $\cdot$ L^{-1} = 1 ppb and 1 ng $\cdot$ L^{-1} = 1 ppt. Using the same assumptions, then a conversion for milligrams per liter to weight percent can be developed.

$$\frac{1 \text{ mg}}{\text{L}} \left(\frac{1 \text{ mL}}{\text{g}}\right) \left(\frac{\text{L}}{1000 \text{ mL}}\right) \left(\frac{\text{g}}{1000 \text{ mg}}\right) = \frac{1 \text{ mg}}{10^6 \text{ mg}} = \frac{1 \text{ mg}}{10^4 (100 \text{ mg})} = 10^{-4} \, P \tag{2-68}$$

or 1 mg $\cdot$ L^{-1} equals $1 \times 10^{-4} \, P$, which can be translated into 1% = 10,000 mg $\cdot$ L^{-1}.

Concentrations may also be reported in units of moles per liter (molarity) or gram-equivalents per liter (normality). When working with chemical reactions, it is necessary to use concentrations of molarity (see pages 41–42) or normality.

Molarity is related to milligrams per liter by

$$\begin{aligned} \text{mg} \cdot \text{L}^{-1} &= \text{molarity} \times \text{molecular weight} \times 10^3 \\ &= (\text{mol} \cdot \text{L}^{-1})(\text{g} \cdot \text{mol}^{-1})(10^3 \text{ mg} \cdot \text{g}^{-1}) \end{aligned} \tag{2-69}$$

A second unit, normality, is frequently used in softening and redox reactions. **Normality** is defined as the number of gram-equivalents of a substance per liter. Gram-equivalents are determined using the equivalent weight of the substance. The **equivalent weight (EW)** is the molecular weight divided by the number (n) of electrons transferred in redox reactions or by the number of protons transferred in acid–base reactions. The value of n depends on how the molecule reacts. In an acid–base reaction, n is the number of hydrogen ions that are transferred. It is easiest to conceptualize this with an example reaction. Let's consider sulfuric acid, which has a chemical formula of H_2SO_4. Sulfuric acid can give up two protons to a base.

$$H_2SO_4 \rightleftharpoons 2H^+ + SO_4^{2-} \tag{2-70}$$

The two moles of protons (H^+), which are released for every mole of sulfuric acid that dissociates, must be accepted by a base. For example,

$$2H^+ + 2NaOH \rightleftharpoons 2H_2O + 2Na^+ \tag{2-71}$$

If we combine Equations 2–70 and 2–71, we can write a third equation.

$$H_2SO_4 + 2NaOH \rightleftharpoons SO_4^{2-} + 2H_2O + 2Na^+ \tag{2-72}$$

in which two moles of protons are transferred to two moles of sodium hydroxide (the base). Thus, for sulfuric acid, the number of equivalents per mole is 2.

When working with acid–base problems, one often uses equivalent weights instead of molecular weights. The equivalent weight is simply the molecular weight divided by the number of protons transferred. Continuing with our use of sulfuric acid as an example, we see that the equivalent weight of sulfuric acid is the molecular weight (98.08 g $\cdot$ mol^{-1}) divided by 2, or 49.04 g $\cdot$ g-equivalent^{-1}.

Determining n for precipitation reactions is simply a special case of that for acid–base reactions. Here the n is equal to the number of hydrogen ions that would be required to replace the cation involved in the precipitation reaction. For example, for $CaCO_3$ it would take two hydrogen ions to replace the calcium, forming H_2CO_3. Therefore, the number of equivalents per mole, n, equals 2.

In oxidation–reduction (redox) reactions, n is equal to the number of electrons transferred in the reaction. For example, in the following reaction:

$$Fe^{2+} + \tfrac{1}{4}O_2 + H^+ \rightleftharpoons Fe^{3+} + \tfrac{1}{2}H_2O \tag{2-73}$$

one electron is transferred and n equals 1. Here the equivalent weight of the ferrous ion is $55.85 \text{g} \cdot \text{g-equivalent}^{-1}$. Obviously, it is impossible to determine the number of equivalence without knowing the reaction.

Normality (N) is the number of gram-equivalents per liter of solution and is related to molarity (M) by

$$N = nM \qquad\qquad\qquad (2\text{–}74)$$

remembering that n is the number of gram-equivalents per mole.

EXAMPLE 2–18   Commercial sulfuric acid (H_2SO_4) is often purchased as a 93 wt% (weight percent) solution. Find the concentration of this solution of H_2SO_4 in units of milligram per liter, molarity, and normality. Sulfuric acid (100%) has a specific gravity of 1.839. Assume that the temperature of the solution is 15°C.

Solution   We can use Equation 2–64 to find the density of a 100% sulfuric acid solution.

$$(1000.0 \text{ g} \cdot \text{L}^{-1})(1.839) = 1839 \text{ g} \cdot \text{L}^{-1}$$

From Table A–1 (Appendix), we find that at 15°C 1.000 L of water weighs 999.103 g. Following this line of reasoning, a 93% solution of H_2SO_4 would have a density of

$$(999.103 \text{ g} \cdot \text{L}^{-1})(0.07) + (1839 \text{ g} \cdot \text{L}^{-1})(0.93) = 1780.2 \text{ g} \cdot \text{L}^{-1} \quad \text{or} \quad 1.8 \times 10^6 \text{ mg} \cdot \text{L}^{-1}$$

The molecular weight of H_2SO_4 is found by looking up the atomic weights in Appendix B.

Number of Atoms (n)	Element	Atomic Weight (AW)	$n \times$ AW
2	hydrogen, H	1.008	2.016
1	sulfur, S	32.06	32.06
4	oxygen, O	15.9994	64.0
		Molecular weight	$98.08 \text{ g} \cdot \text{mol}^{-1}$

The molarity is found by manipulating Equation 2–69 so as to divide the concentration (in grams per liter) by the molecular weight (in grams per mole).

$$\frac{1780.2 \text{ g} \cdot \text{L}^{-1}}{98.08 \text{ g} \cdot \text{mol}^{-1}} = 18.15 \text{ M}$$

The normality is found from Equation 2–74, realizing that H_2SO_4 can donate two hydrogen ions and therefore $n = 2$ g-equivalents $\cdot \text{mol}^{-1}$.

$$N = 18.15 \text{ mol} \cdot \text{L}^{-1} (2 \text{ Eq} \cdot \text{mol}^{-1}) = 36.30 \text{ Eq} \cdot \text{L}^{-1} \qquad \text{or} \qquad 36.3 \text{ N}$$

EXAMPLE 2–19   Find the mass of sodium bicarbonate ($NaHCO_3$) that must be added a 1.00 L volumetric flask containing distilled water to make a 1.0 M solution. Find the normality of the solution.

Solution   The molecular weight of $NaHCO_3$ is 84; therefore, the mass that must be added can be determined by using Equation 2–69.

$$\text{Concentration} = (1.0 \text{ mol} \cdot \text{L}^{-1})(84 \text{ g} \cdot \text{mol}^{-1}) = 84 \text{ g} \cdot \text{L}^{-1}$$

Therefore, 84 g of sodium bicarbonate must be added to 1 L of solution to prepare a 1 M solution. Because HCO_3^- is able to donate or accept only one proton, $n = 1$. Using Equation 2–74, we see that the normality is the same as the molarity.

EXAMPLE 2–20   Determine the equivalent weight of each of the following: Ca^{2+}, CO_3^{2-}, $CaCO_3$.

Solution   Equivalent weight was defined as EW = (Atomic or molecular weight)/n, where n is the oxidation state or the number of electrons or hydrogen ions transferred in the reaction of interest. The units of EW are grams per g-equivalent ($g \cdot g\text{-}Eq^{-1}$) or milligrams per milliequivalent ($mg \cdot mEq^{-1}$).

For calcium, n is equal to its oxidation state in water, so $n = 2$. From the table in the front of the book, the atomic weight of Ca^{2+} is $40.08\ g \cdot mol^{-1}$ The equivalent weight is then

$$EW = \frac{40.08}{2} = 20.04\ g \cdot g\text{-}Eq^{-1} \qquad \text{or} \qquad 20.04\ mg \cdot mEq^{-1}$$

For the carbonate ion (CO_3^{2-}), $n = 2$ because carbonate can accept two protons (H^+). The molecular weight is calculated as shown in the following table.

Number of Atoms (n)	Element	Atomic Weight (AW)	n × AW
1	carbon, C	12.01	12.01
3	oxygen, O	15.9994	48.0
		Molecular weight	60.01 g · mol⁻¹

And the equivalent weight is

$$EW = \frac{60.01\ g \cdot mol^{-1}}{2\ g\text{-}Eq \cdot mol^{-1}} = 30.01\ g \cdot g\text{-}Eq^{-1} \qquad \text{or} \qquad 30.01\ mg \cdot mEq^{-1}$$

With $CaCO_3$, $n = 2$ because two hydrogen ions are required to replace the cation (Ca^{2+}) to form carbonic acid (H_2CO_3). Its molecular weight is the sum of the atomic weight of Ca^{2+} and the molecular weight of CO_3^{2-} and is, therefore, equal to $40.08 + 60.01 = 100.09$. Its equivalent weight is

$$EW = \frac{100.09\ g \cdot mol^{-1}}{2\ g\text{-}Eq \cdot mol^{-1}} = 50.05\ g \cdot g\text{-}Eq^{-1} \qquad \text{or} \qquad 50.05\ mg \cdot mEq^{-1}$$

Buffers

A solution that resists large changes in pH when an acid or base is added or when the solution is diluted is called a **buffer.** A solution containing a weak acid and its salt is an example of a buffer. Atmospheric carbon dioxide (CO_2) produces a natural buffer through the following reactions:

$$CO_2(g) \rightleftharpoons CO_2(aq) + H_2O \rightleftharpoons H_2CO_3^* \rightleftharpoons H^+ + HCO_3^- \rightleftharpoons 2H^+ + CO_3^{2-} \qquad \text{(2–75)}$$

where $H_2CO_3^* =$ "carbonic acid" = true carbonic acid (H_2CO_3) and dissolved carbon dioxide (CO_2 (aq)), which cannot be distinquished analytically

$HCO_3^- =$ bicarbonate ion

$CO_3^{2-} =$ carbonate ion

This is perhaps the most important buffer system in water. We will be referring to it several times in this and subsequent chapters as the carbonate buffer system.

Before we continue looking at the buffering of natural waters, let's review some basic carbonate chemistry. As mentioned earlier, carbonic acid dissociates to form bicarbonate.

$$H_2CO_3^* \rightleftharpoons H^+ + HCO_3^- \qquad K_{a1} = 10^{-6.35} \text{ at } 25°C \qquad \text{(2–76)}$$

Thus, we can write the equilibrium expression

$$\frac{[HCO_3^-][H^+]}{[H_2CO_3^*]} = 10^{-6.35} \qquad \text{(2–77)}$$

Similarly, bicarbonate can act as a Brønsted–Lowry acid and dissociate to form carbonate.

$$HCO_3^- \rightleftharpoons H^+ + CO_3^{2-} \qquad K_{a2} = 10^{-10.33} \text{ at } 25°C \qquad \text{(2–78)}$$

As before we can write an equilibrium expression, this time for bicarbonate.

$$\frac{[CO_3^{2-}][H^+]}{[HCO_3^-]} = 10^{-10.33} \qquad \text{(2–79)}$$

In a closed system, where the concentration of carbonate species is constant.

$$[H_2CO_3^*] + [HCO_3^-] + [CO_3^2] = C_T \qquad \text{(2–80)}$$

Replacing the terms for bicarbonate and carbonate with expressions in terms of carbonic acid, we can write an expression for the concentration of carbonic acid in terms of pH.

$$[H_2CO_3^*] = C_T\left(1 + \frac{K_{a1}}{[H^+]} + \frac{K_{a1}K_{a2}}{[H^+]^2}\right)^{-1} \qquad \text{(2–81)}$$

The following relationships are true:
  (a) pH $<$ p$K_{a1} <$ pK_{a2}; log[$H_2CO_3^*$] $\approx$ log C_T
      d(log[$H_2CO_3^*$])/d(pH) $= 0$ and the line for log[$H_2CO_3^*$] can be represented by a straight line of slope zero.
  (b) p$K_{a1} <$ pH $<$ pK_{a2}; log[$H_2CO_3^*$] $\approx$ pK_{a1} + log C_T − pH
      d(log[$H_2CO_3^*$])/d(pH) $= -1$ and the line for log[$H_2CO_3^*$] can be represented by a straight line of slope −1.
  (c) p$K_{a1} <$ p$K_{a2} <$ pH; log[$H_2CO_3^*$] $\approx$ pK_{a1} + pK_{a2} + log C_T − 2pH
      d(log[$H_2CO_3^*$])/d(pH) $= -2$ and the line for log[$H_2CO_3^*$] can be represented by a straight line of slope −2.

We can develop similar relationships for HCO_3^- and CO_3^{2-} such that the log of the concentrations of each of these species can be plotted. Once we have done this, we can plot these relationships on a log-log graph of log concentration versus pH as shown in Figure 2–11.

FIGURE 2–11

Plot of pC (–log C) versus pH for the carbonate system ($T = 25°C$, $pK_{a1} = 6.35$, $pK_{a2} = 10.33$).

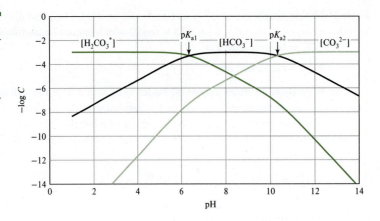

EXAMPLE 2–21 The pH of a water is measured to be 7.5. The concentration of bicarbonate was measured to be 1.3×10^{-3} M. What are the concentrations of carbonate, carbonic acid, and C_T? Assume this system is closed to the atmosphere.

Solution This problem can be solved using the relationships given earlier.

$$pK_{a1} = \frac{[HCO_3^-][H^+]}{[H_2CO_3^*]} = 10^{-6.3}$$

and

$$\frac{[CO_3^{2-}][H^+]}{[HCO_3^-]} = 10^{-10.33}$$

Solving for $[H_2CO_3^*]$, we have

$$[H_2CO_3^*] = \frac{[HCO_3^-][H^+]}{pK_{a1}} = \frac{(1.3 \times 10^{-3})(10^{-7.5})}{10^{-6.3}} = 8.2 \times 10^{-5} \text{ M}$$

Solving for $[CO_3^{2-}]$, we have

$$[CO_3^{2-}] = \frac{pK_{a2}[HCO_3^-]}{[H^+]} = \frac{(10^{-10.33})(1.3 \times 10^{-3})}{10^{-7.5}} = 1.9 \times 10^{-6} \text{ M}$$

$$C_T = [H_2CO_3^*] + [HCO_3^-] + [CO_3^{2-}] = 8.2 \times 10^{-5} + 1.3 \times 10^{-3} + 1.9 \times 10^{-6}$$

$$= 1.384 \times 10^{-3} \text{ M} \approx 1.4 \times 10^{-3} \text{ M}$$

As depicted in Equation 2–75, the CO_2 in solution is in equilibrium with atmospheric CO_2 (g). Any change in the system components to the right of $CO_2(aq)$ causes the $CO_2(g)$ either to be released from solution or to dissolve. If this is true (i.e., we have an open system), the concentration of $H_2CO_3^*$ is constant with pH and C_T changes as pH changes. If this is the case, the plot of log C versus pH looks very different. A typical plot is shown in Figure 2–12.

FIGURE 2–12

pC (–log C)–pH diagram for the carbonate system ($T = 25°C$, $pK_{a1} = 6.35$, $pK_{a2} = 10.33$; $K_H = 10^{-1.5}$ M · atm^{-1}; $P_{CO2} = 10^{-3.5}$ atm).

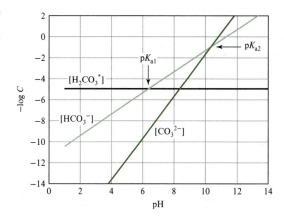

EXAMPLE 2–22   The pH of a water is measured to be 7.5. What are the concentrations of carbonate, bicarbonate, carbonic acid, and C_T? Assume this system is open to the atmosphere. The temperature is 25°C. The Henry's constant for carbon dioxide is $10^{-1.47}$ M · atm^{-1} at this temperature. The partial pressure of carbon dioxide is $10^{-3.53}$ atm.

Solution   The same relationships hold true for this problem as in the previous one. Because we are given the partial pressure of carbon dioxide and the Henry's constant, we can calculate the concentration of carbonic acid using the relationship

$$[H_2CO_3^*] = K_H P_{CO_2} = (10^{-1.47} \text{ M} \cdot \text{atm}^{-1})(10^{-3.53} \text{ atm}) = 10^{-5.0} \text{ M}$$

$$pK_{a1} = \frac{[HCO_3^-][H^+]}{[H_2CO_3^*]} = 10^{-6.3}$$

and

$$\frac{[CO_3^{2-}][H^+]}{[HCO_3^-]} = 10^{-10.33}$$

Solving for $[HCO_3^-]$

$$[HCO_3^-] = \frac{[H_2CO_3^*]pK_{a1}}{[H^+]} = \frac{(1.0 \times 10^{-5})(10^{-6.3})}{10^{-7.5}} = 1.6 \times 10^{-4} \text{ M}$$

Solving for $[CO_3^{2-}]$

$$[CO_3^{2-}] = \frac{pK_{a2}[HCO_3^-]}{[H^+]} = \frac{(10^{-10.33})(1.58 \times 10^{-4})}{10^{-7.5}} = 2.3 \times 10^{-7} \text{ M}$$

$$C_T = [H_2CO_3^*] + [HCO_3^-] + [CO_3^{2-}] = 1.0 \times 10^{-5} + 1.6 \times 10^{-4} + 2.3 \times 10^{-7}$$

$$= 1.70 \times 10^{-4} \text{ M} \approx 1.7 \times 10^{-4} \text{ M}$$

We can examine the character of the buffer system to resist a change in pH by considering the addition of an acid or a base and applying the **law of mass action** (Le Châtelier's principle). For example, if an acid is added to the system, the hydrogen ion concentration will increase, and the system will not be at equilibrium. To achieve equilibrium, the carbonate combines with the

FIGURE 2–13

Behavior of the carbonate buffer system with the addition of acids and bases or the addition and removal of CO_2.

Case I: Open System
Acid is added to carbonate buffer system.[a]

Reaction shifts to the left as $H_2CO_3^*$ is formed when H^+ and HCO_3^- react.[b]
CO_2 is released to the atmosphere.
pH is lowered slightly because the availability of free H^+ (amount depends on buffering capacity).

Case II: Open System
Base is added to carbonate buffer system.

Reaction shifts to the right.
CO_2 from the atmosphere dissolves into solution.
pH is raised slightly because H^+ combines with OH^- (amount depends on buffering capacity).

Case III: Open System
CO_2 is bubbled into carbonate buffer system.

Reaction shifts to the right because $H_2CO_3^*$ is formed when CO_2 and H_2O react.
CO_2 dissolves into solution.
pH is lowered.

Case IV: Open System
Carbonate buffer system is stripped of CO_2.

Reaction shifts to the left to form more $H_2CO_3^*$ to replace that removed by stripping.
CO_2 is removed from solution.
pH is raised.

[a]Refer to Equation 2–75.
[b]The asterisk * in $H_2CO_3^*$ is used to signify the sum of CO_2 and H_2CO_3 in solution.

free protons to form bicarbonate. Bicarbonate reacts to form more carbonic acid, which in turn dissociates to CO_2 and water. The excess CO_2 can be released to the atmosphere in a thermo-dynamically open system. Alternatively, the addition of a base consumes hydrogen ions and the system shifts to the right, with the CO_2 being replenished from the atmosphere. When CO_2 is bubbled into the system or is removed by passing an inert gas such as nitrogen through the liquid (a process called stripping), the pH will change more dramatically because the atmosphere is no longer available as a source or sink for CO_2. Figure 2–13 summarizes the four general responses of the carbonate buffer system. The first two cases (I and II) are common in natural settings when the reactions proceed over a relatively long time. Cases III and IV are not common in natural settings, but may occur in engineered systems. For example in a water treatment plant, we can alter the reactions more quickly than the CO_2 can be replenished from the atmosphere.

Buffering Capacity.   We are often concerned with the ability of a water to resist changes in pH when an acid or base is added. This ability to resist change is referred to as **buffering capacity.** Water chemists use the term **alkalinity** to describe a water's ability to resist changes in pH on the addition of acid, therefore, it is also called **acid-neutralizing capacity. Acidity** describes the ability of a water to resist changes in pH due to the addition of base. Therefore, it is also called **base-neutralizing capacity.**

Alkalinity.   Alkalinity is defined as the sum of all titratable bases to a pH of approximately 4.5. It has units of equivalents per liter or normality (N). It is found by experimentally determining the amount of acid required to reduce the pH of the water sample to 4.5. In most freshwaters the only weak acids or bases that contribute to alkalinity are bicarbonate (HCO_3^-), carbonate (CO_3^{2-}), H^+, and OH^-. In ocean waters, the bromate species also play a significant role in determining alkalinity. The total H^+ that can be neutralized by water containing primarily carbonate species is

$$\text{Alkalinity} = [HCO_3^-] + 2[CO_3^{2-}] + [OH^-] - [H^+] \qquad (2\text{–}82)$$

where [ ] refers to concentrations in moles per liter. In most natural waters (pH 6–8), the OH^- and H^+ are negligible, such that

$$\text{Alkalinity} \approx [HCO_3^-] + 2[CO_3^{2-}] \qquad (2\text{--}83)$$

Note that $[CO_3^{2-}]$ is multiplied by 2 because it can accept two protons. If we were to write Equation 2–83 using *normality* instead of molarity, then Equation 2–82 would become

$$\text{Alkalinity} = (HCO_3^-) + (CO_3^{2-}) + (OH^-) - (H^+) \qquad (2\text{--}84)$$

where ( ) refers to concentrations in units of normality.

The pertinent acid–base reactions are

$$H_2CO_3 \rightleftharpoons H^+ + HCO_3^- \qquad\qquad pK_{a1} = 6.35 \text{ at } 25°C \qquad (2\text{--}76)$$

$$HCO_3^- \rightleftharpoons H^+ + CO_3^{2-} \qquad\qquad pK_{a2} = 10.33 \text{ at } 25°C \qquad (2\text{--}78)$$

From the pK values, some useful relationships can be found. The more important ones are as follows:

1. Below pH of 4.5, the only carbonate species present in appreciable quantity is H_2CO_3 and the concentration of OH^- is negligible. Because carbonic acid does not contribute to alkalinity, in this case the alkalinity is negative (due to the H^+). This water would have no ability to neutralize acids, and any small addition in acid would result in a significant reduction in pH.
2. In the pH range from 7 to 8.3, HCO_3^- predominates over carbonate, and the concentration of H^+ approximately equals the concentration of OH^- (both are also small in comparison with the concentration of HCO_3^-). As such, the alkalinity is approximately equal to the concentration of HCO_3^-.
3. At a pH of greater than 12.3, the predominant carbonate species is CO_3^{2-}, the concentration of H^+ is negligible, and the concentration of OH^- cannot be ignored. Here, the alkalinity equals $2[CO_3^{2-}] + [OH^-]$.

In environmental engineering and science, we need to differentiate between alkaline water and water having high alkalinity. Alkaline water has a pH greater than 7, and a water with high alkalinity has a high buffering capacity. An alkaline water may or may not have a high buffering capacity. Likewise, a water with a high alkalinity may or may not have a high pH.

By convention, alkalinity is not expressed in molarity units as shown in the preceding equations, but rather in **units of milligrams as CaCO$_3$ per liter** or normality $(Eq \cdot L^{-1})$. In order to convert concentrations of the ions to milligrams per liter as $CaCO_3$, multiply milligrams per liter as the species by the ratio of the equivalent weight of $CaCO_3$ to the species equivalent weight (EW).

$$\text{Milligrams per liter as } CaCO_3 = (\text{milligrams per liter as species})\left(\frac{\text{EW of } CaCO_3}{\text{EW of species}}\right) \qquad (2\text{--}85)$$

The alkalinity is then determined as before, except that instead of using normality, the concentrations in units of milligrams per liter as $CaCO_3$ are added or subtracted.

EXAMPLE 2–23   A water contains 100.0 mg $\cdot$ L^{-1} CO_3^{2-} and 75.0 mg $\cdot$ L^{-1} HCO_3^- at a pH of 10 ($T = 25°C$). Calculate the exact alkalinity. Approximate the alkalinity by ignoring the appropriate chemical species.

Solution   First, convert CO_3^{2-}, HCO_3^-, H^+, and OH^- to milligrams per liter as $CaCO_3$. Remember, you are given the concentrations of carbonate and bicarbonate in milligrams per liter as the species and the concentrations of H^+ and OH^- in molar units.

The equivalent weights are

CO_3^{2-}: MW = 60, n = 2, EW = 30

HCO_3^-: MW = 61, n = 1, EW = 61

H^+: MW = 1, n = 1, EW = 1

OH^-: MW = 17, n = 1, EW = 17

and the concentrations of H^+ and OH^- are calculated as follows: pH = 10; therefore $[H^+]$ = 10^{-10} M. Using Equation 2–69,

$$[H^+] = (10^{-10}\ \text{mol} \cdot \text{L}^{-1})(1\ \text{g} \cdot \text{mol}^{-1})(10^3\ \text{mg} \cdot \text{g}^{-1}) = 10^{-7}\ \text{mg} \cdot \text{L}^{-1}$$

Using Equation 2–37, we see that

$$[OH^-] = \frac{K_w}{[H^+]} = \frac{10^{-14}}{10^{-10}} = 10^{-4}\ \text{mol} \cdot \text{L}^{-1}$$

and

$$[OH^-] = (10^{-4}\ \text{mol} \cdot \text{L}^{-1})(17\ \text{g} \cdot \text{mol}^{-1})(10^3\ \text{mg} \cdot \text{g}^{-1}) = 1.7\ \text{mg} \cdot \text{L}^{-1}$$

Now, the concentrations in units of milligrams per liter as $CaCO_3$ are found by using Equation 2–85 and taking the equivalent weight of $CaCO_3$ to be 50.

$$CO_3^{2-} = 100.0 \times \left(\frac{50}{30}\right) = 167\ \text{mg} \cdot \text{L}^{-1}\ \text{as}\ CaCO_3$$

$$HCO_3^- = 75.0 \times \left(\frac{50}{61}\right) = 61\ \text{mg} \cdot \text{L}^{-1}\ \text{as}\ CaCO_3$$

$$H^+ = 10^{-7} \times \left(\frac{50}{1}\right) = 5 \times 10^{-6}\ \text{mg} \cdot \text{L}^{-1}\ \text{as}\ CaCO_3$$

$$OH^- = 1.7 \times \left(\frac{50}{17}\right) = 5.0\ \text{mg} \cdot \text{L}^{-1}\ \text{as}\ CaCO_3$$

The exact alkalinity (in milligrams per liter as $CaCO_3$) is found by

$$\text{Alkalinity} = 61 + 167 + 5.0 - (5 \times 10^{-6}) = 233\ \text{mg} \cdot \text{L}^{-1}\ \text{as}\ CaCO_3$$

2–5   SOIL CHEMISTRY

Although often taken for granted and, seemingly, not as valued as water and air, without soil, life on this planet would not exist. Soil is important for the production of food; the maintenance of carbon, nitrogen, and phosphorus balances; and for the construction of building materials.

Chemically, soil is a mixture of weathered rocks and minerals; decayed plant and animal material (humus and detritus); and small living organisms, including plants, animals, and bacteria. Soil also contains water and air. Typically, soil contains about 95% mineral and 5% organic matter, although the range in composition varies considerably.

The concentrations of chemicals in soil are given in mass units: parts per million, milligrams per kilogram, or micrograms per kilogram. The units vary somewhat based on the magnitude of the mass of chemical present per unit mass (usually kilograms) of soil. For example, when dealing with carbon, the concentration is usually given in percent because carbon generally accounts for about 1 to 25% of soil material. On the contrary, when working with nutrient concentrations (e.g., nitrogen, phosphorus, etc.) units of milligrams per kilogram are used. When working with many hazardous wastes, whose concentrations are usually small, we use units of parts per billion or micrograms per kilogram.

The movement of ionic nutrients such as nitrate, ammonia, and phosphate is governed by ion-exchange reactions. For example, sodium ions may be attached to the soil surface by electrostatic interactions. If water containing calcium is passed through the soil, the calcium will be preferentially exchanged for the sodium according to this reaction:

$$2(Na^+–Soil) + Ca^{2+} \rightleftharpoons Ca^{2+}–(Soil)_2 + 2Na^+ \tag{2–86}$$

By this reaction, two sodium ions are released for every ion of calcium exchanged, thus maintaining the charge balance. Thus, an important characteristic of soil is its exchange capacity. **Exchange capacity** is, essentially, the extent to which a unit mass of soil can exchange a mass of a certain ion of interest. Exchange capacity (reported in units of equivalents of ions per mass of soil) is an important characteristic of soil in terms of its ability to leach ions such as magnesium, calcium, nitrate, and phosphate.

Another important process that occurs in soils is **sorption**. Sorption is essentially the attachment of a chemical to either the mineral or organic portions of soil particles and includes both adsorption and absorption. Van der Waals forces, hydrogen bonding, or electrostatic interactions can result in the attachment of chemicals to the soil surface. In some cases, covalent bonding can actually result, and the chemical is irreversibly bound to the soil.

With low concentrations of pollutants, sorption can be described mathematically by a linear expression.

$$K_d = \frac{C_s \ (\text{mol} \cdot \text{kg}^{-1})}{C_w \ (\text{mol} \cdot \text{L}^{-1})} \tag{2–87}$$

where C_w = the equilibrium concentration of the chemical in the water (mass per volume of water)

K_d = a partition coefficient describing sorption equilibrium of a chemical-distribution ratio = (mass per mass of soil) (mass per volume of water)$^{-1}$

C_s = the equilibrium concentration of the chemical on the soil (mass per mass of soil)

The partition coefficients of various organic pollutants can vary over at least eight orders of magnitude, depending predominately on the chemical characteristics of the pollutant, but also on the nature of the soil itself.

With most neutral organic chemicals, sorption occurs predominately on the organic fraction of the soil itself (as long as the fraction of organic material on the soil is "significant"). In these cases,

$$C_s \approx C_{om} f_{om} \tag{2–88}$$

where C_{om} = concentration of organic chemical in the organic matter of the soil

f_{om} = fraction of organic matter in the soil.

Combining Equations 2–87 and 2–88 yields an equation valid for neutral organic chemicals.

$$K_d = \frac{C_{om} f_{om}}{C_w} \tag{2–89}$$

EXAMPLE 2–24   A soil sample is collected and the soil water is analyzed for the chemical compound 1,2-dichloroethane (DCA). The concentration in the water is found to be 12.5 $\mu g \cdot L^{-1}$. The organic matter content of the soil is 1.0%. Determine the concentration of DCA that would be sorbed to the soil and that associated with the organic matter. DCA has a K_d of 0.724 ($\mu g \cdot kg^{-1}$) ($\mu g \cdot L^{-1}$)$^{-1}$.

Solution   Using Equation 2–87, we see that

$$K_d = \frac{C_s}{C_w}$$

Therefore,

$$C_s = K_d C_w = (0.724\ (\mu g \cdot kg^{-1})(\mu g \cdot L^{-1})^{-1})(12.5\ \mu g \cdot L^{-1}) = 9.05\ \mu g \cdot kg^{-1}$$

Using Equation 2–88,

$$C_{om} = \frac{C_s}{f_{om}} = \frac{(9.05\ \mu g \cdot kg^{-1})}{0.01} = 905\ \mu g \cdot kg^{-1}$$

2–6 ATMOSPHERIC CHEMISTRY

The atmosphere is a thin envelope of gases that surround the Earth's surface, held in place by gravity. As one moves higher in elevation, the Earth's gravitational forces decrease and the density of these gases also decreases. The composition of air varies with location, altitude, anthropogenic sources (e.g., factories and cars), and natural sources (e.g., dust storms, volcanoes, forest fires). The concentrations of some gases vary less than others. The essentially "nonvariable" gases make up approximately 99% (by volume) of the atmosphere. Of the variable gases, water vapor, carbon dioxide, and ozone are the most prevalent. Table 2–8 lists these gases and their volume percents.

The atmosphere is divided into several layers on the basis of temperature. The layer closest to the Earth's surface, the troposphere, extends to approximately 13 km. As shown in Figure 2–14, the temperature in this layer decreases with increasing altitude. It is estimated that 80–85% of the mass of the atmosphere is in the troposphere. The next layer is the stratosphere, which extends to an altitude of approximately 50 km. Within the stratosphere, the temperature increases with increasing altitude until it reaches approximately 0°C at the stratopause (the boundary between the stratosphere and the mesosphere). The temperature increase in the

TABLE 2–8	Composition of the Atmosphere	
Gas	**Percentage by Volume**[a]	
Nonvariable gases		
Nitrogen	78.08	
Oxygen	20.95	
Argon	0.93	
Neon	0.002	
Others	0.001	
Variable gases		
Water vapor	0.1–≈5.0	
Carbon dioxide	0.035	
Ozone	0.000006	
Other gases	Trace	
Particulate matter	Usually trace	

[a]Percentages, except for water vapor, are for dry air.

Data Source: McKinney and Schooch, 1996.

FIGURE 2–14

The Earth's atmosphere.

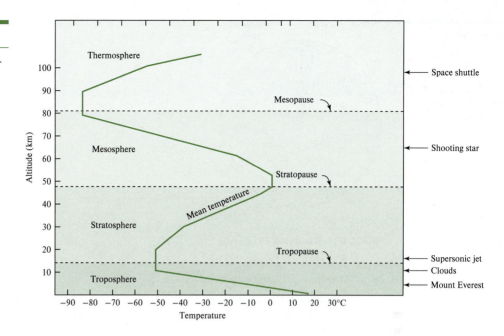

stratosphere is due to the absorption of ultraviolet radiation and the resulting heat given off by the reactions that occur. The troposphere and stratosphere contain approximately 99% of the mass of the atmosphere. In the next layer, the mesosphere, which extends to approximately 80 km, the temperature decreases with increasing altitude until it reaches a temperature of approximately −80°C. The outermost envelope around the Earth is the thermosphere, another region in which the temperature increases with increasing altitude.

One of the major differences between aqueous and atmospheric reactions is the importance of gas-phase and photochemical reactions in the latter. One of the most important gas-phase photochemical reactions that occurs in the troposphere is the formation of ozone from the reaction of ultraviolet radiation, hydrocarbons, and nitrogen oxides (NO_x).

Another critical set of reactions is the absorption of infrared radiation by carbon dioxide (CO_2), methane (CH_4), nitrous oxide (N_2O), and the fluorinated gases (which include hydrofluorocarbons, perfluorocarbons, sulfur hexafluoride, and nitrogen trifluoride). This latter group of chemicals are sometimes used as alternatives for chemicals that deplete stratospheric ozone. Because of the ability of these gases to absorb infrared radiation and therefore warm the troposphere, they are referred to as greenhouse gases. Of the first three chemicals listed above, carbon dioxide accounts for most of the emissions in the U.S. Because of the different extents to which these chemicals absorb infrared radiation, they are often reported in units of CO_2-equivalents (which is defined as the ratio of the accumulated radiative forcing within a specific time horizon caused by emitting 1 kilogram of the gas, relative to that of the reference gas CO_2). Figure 2–15 shows the relative emissions of these gases in the United States. More details about these and other important atmospheric reactions will be presented in Chapter 12 (Air Pollution).

Fundamentals of Gases

Ideal Gas Law.   The behavior of chemicals in air with respect to temperature and pressure can be assumed to be ideal (in the chemical sense) because the concentration of these pollutants are usually sufficiently low. Thus, we can assume that at the same temperature and pressure, different kinds of gases have densities proportional to their molecular masses. This may be written as

$$\rho = \frac{P \times M}{R \times T} \tag{2-90}$$

FIGURE 2–15

Emission of greenhouse gases in the U.S. in 2015 (Total Emissions = 6587 Million Metric Tons of CO_2-equivalents). (*Source:* U.S. Environmental Protection Agency (2017). *Inventory of U.S. Greenhouse Gas Emissions and Sinks: 1990–2015.*)

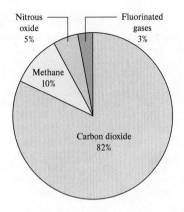

where ρ = density of gas (g · m^{-3})

P = absolute pressure (Pa)

M = molecular mass (g · mol^{-1})

T = absolute temperature (K)

R = ideal gas constant = 8.3143 J · K^{-1} · mol^{-1} (or Pa · m^3 · mol^{-1} · K^{-1})

Because density is defined as mass per unit volume, or the number of moles per unit volume, n/V, the expression may be rewritten in the general form as

$$PV = nRT \tag{2–91}$$

(the ideal gas law) where V is the volume occupied by n moles of gas. At 273.15 K and 101.325 kPa, one mole of an ideal gas occupies 22.414 L.

Dalton's Law of Partial Pressures.   In 1801, the English scientist John Dalton found that the total pressure exerted by a mixture of gases is equal to the sum of the pressures that each type of gas would exert if it alone occupied the container. In mathematical terms,

$$P_t = P_1 + P_2 + P_3 + \cdots + P_n \tag{2–92}$$

where $\qquad P_t$ = total pressure of mixture

P_1, P_2, P_3, P_n = the partial pressure of each gas*

Dalton's law also may be written in terms of the ideal gas law.

$$P_t = n_1 \frac{RT}{V} + n_2 \frac{RT}{V} + n_3 \frac{RT}{V} + \cdots \tag{2–93}$$

$$P_t = (n_1 + n_2 + n_3 + \cdots) \frac{RT}{V} \tag{2–94}$$

Dalton's law is important in air quality assessment because stack and exhaust sampling measurements are made with instruments calibrated with air. Because combustion products have a composition entirely different from air, the readings must be adjusted ("corrected" in sampling parlance) to reflect this difference. Dalton's law forms the basis for the calculation of the correction factor.

Concentrations of Pollutants in Air.   One must be aware that when dealing with concentrations of gases in air, the approximation of 1 ppm = 1 mg · L^{-1} is no longer valid as it is with dilute aqueous solutions. This is because the density of air is not 1 g · mL^{-1} and varies significantly with temperature. With air, concentrations are often reported in units of micrograms per cubic meter or parts per million. With air, the units of parts per million are reported on a

*That is, pressure of each gas if it were in the container alone.

volume–volume basis (unlike that used with aqueous concentrations, which are given on a mass–volume basis). The units of parts per million have the advantage over micrograms per cubic meter in that changes in temperature and pressure do not change the ratio of the volume of pollutant to volume of air. Thus, it is possible to compare concentrations given in parts per million, without considering effects of pressure or temperature. The concentration of particulate matter may be reported only as micrograms per cubic meter. The micrometer unit is used to report particle size.

Converting Micrograms per Cubic Meter to Parts per Million.   The conversion between micrograms per cubic meter and parts per million is based on the fact that at standard conditions (0°C and 101.325 kPa), one mole of an ideal gas occupies 22.414 L. Thus, we may write an equation that converts the mass of the pollutant, M_p, in grams to its equivalent volume, V_p, in liters at standard temperature and pressure (STP).

$$V_p = \frac{M_p}{MW} \times 22.414 \text{ L} \cdot \text{mol}^{-1} \tag{2–95}$$

where MW is the molecular weight of the pollutant in units of grams per mole. For readings made at temperatures and pressures other than standard conditions, the standard volume, 22.414 L · mol^{-1}, must be corrected. We can use the ideal gas law to make the correction.

$$(22.414 \text{ L} \cdot \text{mol}^{-1}) \times \frac{T_2}{273 \text{ K}} \times \frac{101.325 \text{ kPa}}{P_2} \tag{2–96}$$

where T_2 and P_2 are the absolute temperature (in Kelvin) and absolute pressure (in kilopascals) at which the readings were made. Because parts per million is a volume ratio, we may write

$$ppm = \frac{V_p}{V_a + V_p} \tag{2–97}$$

where V_a is the volume of air in cubic meters at the temperature and pressure at which the measurement was taken. We then combine Equations 2–95, 2–96, and 2–97 to yield Equation 2–98.

$$ppm = \frac{(M_p/MW) \times 22.414 \text{ L} \cdot \text{mol}^{-1} \times (T_2/273 \text{ K}) \times (101.325 \text{ kPa}/P_2)}{V_a \times 1000 \text{ L} \cdot \text{m}^{-3}} \tag{2–98}$$

where M_p is the mass of the pollutant of interest in micrograms. The factors converting micrograms to grams and liters to millions of liters cancel one another. Unless otherwise stated, it is assumed that $V_a = 1.00 \text{ m}^3$.

EXAMPLE 2–25   A 1-m^3 sample of air was found to contain 80 μg · m^{-3} of SO_2. The temperature and pressure were 25.0°C and 103.193 kPa when the air sample was taken. What was the SO_2 concentration in parts per million?

Solution   First we must determine the MW of SO_2 from the chart in Appendix B; we find

MW of $SO_2 = 32.06 + 2(15.9994) = 64.06 \text{ g} \cdot \text{mol}^{-1}$

Next we must convert the temperature from Celsius to Kelvin. Thus,

$25°C + 273 \text{ K} = 298 \text{ K}$

Now using Equation 2–98, we find

Concentration

$$= \frac{(80 \text{ μg}/64.06 \text{ g} \cdot \text{mol}^{-1}) \times 22.414 \text{ L} \cdot \text{mol}^{-1} \times (298 \text{ K}/273 \text{ K}) \times (101.325 \text{ kPa}/103.193 \text{ kPa})}{\text{m}^3 \times 10^3 \text{ L} \cdot \text{m}^{-3}}$$

$$= 0.030 \text{ ppm of } SO_2$$

CHAPTER REVIEW

When you have completed studying this chapter, you should be able to do the following without the aid of your textbook or notes:

1. Define the following terms: matter, elements, compounds.
2. State the essence of the law of definite proportions (a.k.a. the law of constant composition).
3. Describe how elements are grouped in the periodic table.
4. Describe the composition of atoms.
5. Define: atomic number, mass number, and atomic weight.
6. Describe isotopes.
7. Describe the types of bonding that hold together atoms and molecules.
8. Define the following terms: molecule, molecular formula, ion, cation, and anion.
9. Describe what a chemical equation represents symbolically.
10. Be able to balance chemical equations.
11. Describe the four principal types of reactions of importance to environmental scientists and engineers.
12. Describe the law of mass action.
13. Describe how ionic strength affects equilibrium constants. Describe how we take into account the effects of ionic strength on equilibrium constants.
14. Describe the common ion effect.
15. Describe what is meant by pH and pOH.
16. Define: weak acid, strong acid, weak base, and strong base.
17. Describe Henry's law.
18. Be able to write the expressions for zero-, first-, and second-order reactions.
19. Define: alkanes, alkenes, alkynes, structural isomers, aryl (aromatic) compounds, alcohols, phenols, aldehydes, ketones, esters, ethers, amines, amides, mercaptans, halides, sulfonic acid compounds.
20. Define: mass density, specific weight, specific gravity, and viscosity.
21. Define: suspended solids and colloidal particles (in terms of the sizes of the particles).
22. Define: buffer, alkalinity, and acidity.
23. List and define three units of measure used to report air pollution data (i.e., ppm, $\mu g \cdot m^{-3}$, and μM).
24. Explain the difference between parts per million in air pollution and parts per million in water pollution.
25. Explain the effect of temperature and pressure on readings made in parts per million.
26. Calculate the concentration of a chemical species in soil in units of parts per million, milligrams per kilogram, and micrograms per kilogram.
27. What is meant by exchange capacity?

With the aid of this text, you should be able do to following:

1. Calculate the molarity or normality of a chemical given the mass added to a particular volume of solution.
2. Calculate the concentration of chemicals that precipitate given the solubility product of the precipitate.
3. Calculate the equilibrium solution concentrations given varying ionic strengths of the background electrolyte.
4. Calculate the solubility of a chemical in the presence of a common ion.
5. Perform dissociation-type calculations.
6. Use Henry's law to calculate the concentration of a gas or the concentration of the gas dissolved in a liquid (given the other).
7. Calculate the mass of chemical remaining after a first-order reaction proceeds for a given time, given the first-order rate constant.

8. Calculate the concentration of a chemical in water in units of percent by weight, molarity, normality, milligrams per liter as the chemical species, milligrams per liter as $CaCO_3$.

9. Calculate the alkalinity of a solution given the concentrations of carbonate, bicarbonate, and the pH.

10. Use the ideal gas law appropriately.

11. Use Dalton's law of partial pressures to calculate the total pressure of a mixture of gases.

12. Calculate the concentration of chemical species in a gas in units of micrograms per cubic meter, parts per million (by volume), and micrograms, as appropriate.

13. Use the distribution coefficient and the equilibrium concentration in water to calculate the equilibrium concentration on the soil.

14. Use the fraction of organic matter to calculate the concentration of the neutral organic chemical in the organic matter.

PROBLEMS

2–1   For each atomic symbol, give the name of the element.
(a) Pb   (b) C   (c) Ca   (d) Zn   (e) O   (f) H   (g) Hg   (h) S   (i) N   (j) Cl
(k) Mg   (l) P

> **Answer:**   (**a**) lead (**b**) carbon (**c**) calcium (**d**) zinc (**e**) oxygen (**f**) hydrogen (**g**) mercury
>
> (**h**) sulfur (**i**) nitrogen (**j**) chlorine (**k**) magnesium (**l**) phosphorus

2–2   The following table gives the number of protons and neutrons in the nucleus of various atoms. (a) Which atom is the isotope of atom A? (b) Which atom has the same mass number as atom A?

	Protons	Neutrons
Atom A	13	15
Atom B	12	15
Atom C	13	16
Atom D	15	17

2–3   Calculate the atomic weight of boron, B, based on the following fractional abundance data.

Isotope	Isotopic Mass (amu)	Fractional Abundance
B-10	10.013	0.1978
B-11	11.009	0.8022

> **Answer:** 10.812

2–4   What element has 17 protons and 18 neutrons in its nucleus?

2–5   A solution of sodium bicarbonate is prepared by adding 45.00 g of sodium bicarbonate to a 1.00-L volumetric flask and adding distilled water until it reaches the 1.00-L mark. What is the concentration of sodium bicarbonate in units of (a) milligrams per liter, (b) molarity, (c) normality, and (d) milligrams per liter as $CaCO_3$?

> **Answer:** (**a**) 4.5×10^4 mg $\cdot$ L^{-1} (**b**) 0.536 M (**c**) 0.536 N (**d**) 2.68×10^4 mg $\cdot$ L^{-1} as $CaCO_3$

2–6   Balance the following chemical equations:
(a) $CaCl_2 + Na_2CO_3 \rightleftharpoons CaCO_3 + NaCl$ (d) $C_4H_{10} + O_2 \rightleftharpoons CO_2 + H_2O$

(b) $C_6H_{12}O_6 + O_2 \rightleftharpoons CO_2 + H_2O$ (e) $Al(OH)_2^- \rightleftharpoons Al^{3+} + OH^-$

(c) $NO_2 + H_2O \rightleftharpoons HNO_3 + NO$

2–7 A magnesium hydroxide solution is prepared by adding 10.00 g of magnesium hydroxide to a volumetric flask and bringing the final volume to 1.00 L by adding water buffered at a pH of 7.0. What is the concentration of magnesium in this solution? (Assume that the temperature is 25°C and the ionic strength is negligible).

> **Answer:** 0.17 M

2–8 For the conditions given in Problem 2–7, what is the concentration of magnesium if the water has an ionic strength of (a) 0.01 M, (b) 0.5 M? (Assume that the temperature of the solution is 25°C.)

2–9 A ferric phosphate solution is prepared by adding 2.4 g of ferric phosphate to a volumetric flask and bringing the final volume to 1.00 L by adding water having a phosphate concentration of 1.0 mg · L^{-1}. What is the concentration of soluble iron in this solution? (Assume that the temperature of the solution is 25°C.)

> **Answer:** 1.20×10^{-17} M

2–10 A solution made up with calcium carbonate is initially supersaturated with Ca^{2+} and CO_3^{2-} ions, such that the concentrations of each are both 1.35×10^{-3} M. When equilibrium is finally reached, what is the final concentration of calcium? (Use the pK_s for aragonite.)

2–11 A solution has an H^+ concentration of 10^{-5} M. (a) What is the pH of this solution? (b) What is the pOH? (Assume that the temperature of the solution is 25°C.)

> **Answer:** **(a)** 5 **(b)** 9

2–12 If 200 mg of HCl is added to water to achieve a final volume of 1.00 L, what is the final pH?

2–13 A solution of acetic acid is prepared in water by adding 11.1 g of sodium acetate to a volumetric flask and bringing the volume to 1.0 L with water. The final pH is measured to be 5.25. What are the concentrations of acetate and acetic acid in solution? (Assume that the temperature of the solution is 25°C.)

> **Answer:** [HA] = 0.033 M   [A$^-$] = 0.102 M

2–14 A 0.21 m^3 drum contains 100 L of a mixture of several degreasing solvents in water. The concentration of trichloroethylene in the headspace (gas phase) above the water was measured to be 0.00301 atm. The Henry's constant for trichloroethylene is 0.00985 atm · m^3 · mol^{-1} at 25°C. What is the concentration of trichloroethylene in the water in units of molarity? (Assume that the temperature of the solution is 25°C.)

2–15 The concentration of a chemical degrades in water according to first-order kinetics. The degradation constant is 0.2 day^{-1}. If the initial concentration is 100.0 mg · L^{-1}, how many days are required for the concentration to reach 0.14 mg · L^{-1}?

> **Answer:** 32.9 days

2–16 The water in a pond is reaerated. Reaeration occurs according to first-order kinetics with a rate of 0.034 day^{-1}. If the temperature of the stream is 15°C and the initial oxygen concentration 2.5 mg · L^{-1} how long (in days) does it take for the oxygen concentration to increase to 6.5 mg · L^{-1}? The oxygen solubility in water at 15°C is 10.15 mg · L^{-1}.

2–17 Hypochlorous acid decays in the presence of ultraviolet radiation. Assume that degradation occurs according to first-order kinetics and the rate of degradation was measured to be 0.12 day^{-1} (at a particular sunlight intensity and temperature). Given this, how long does it take for the concentration of hypochlorous acid to reach nondetectable levels (0.05 mg · L^{-1}) if the initial concentration were 3.65 mg · L^{-1}?

> **Answer:** 35.8 days

2–18 Show that a density of 1 g · mL^{-1} is the same as a density of 1000 kg · m^{-3}. (*Hint:* Some useful conversions are listed inside the back cover of this book.)

2–19 Show that a 4.50% by weight mixture contains 45.0 kg of substance in a cubic meter of water (i.e., 4.50% = 45.0 kg · m^{-3}).

> **Answer:** 45 kg · m^{-3}

2–20 Show that $1 \text{ mg} \cdot \text{L}^{-1} = 1 \text{ g} \cdot \text{m}^{-3}$.

2–21 Calculate the molarity and normality of the following:

(a) $200.0 \text{ mg} \cdot \text{L}^{-1}$ HCl
(b) $150.0 \text{ mg} \cdot \text{L}^{-1}$ H_2SO_4
(c) $100.0 \text{ mg} \cdot \text{L}^{-1}$ $Ca(HCO_3)_2$
(d) $70.0 \text{ mg} \cdot \text{L}^{-1}$ H_3PO_4

Answers:

	Molarity (M)	Normality (N)
(a)	0.005485	0.005485
(b)	0.001529	0.003059
(c)	0.0006168	0.001234
(d)	0.000714	0.00214

2–22 Calculate the molarity and normality of the following:

(a) $80 \text{ μg} \cdot \text{L}^{-1}$ HNO_3
(b) $135 \text{ μg} \cdot \text{L}^{-1}$ $CaCO_3$
(c) $10 \text{ μg} \cdot \text{L}^{-1}$ $Cr(OH)_3$
(d) $1000 \text{ μg} \cdot \text{L}^{-1}$ $Ca(OH)_2$

2–23 Calculate concentration of the following in units of milligrams per liter:

(a) 0.01000 N Ca^{2+}
(b) 1.000 M HCO_3^-
(c) 0.02000 N H_2SO_4
(d) 0.02000 M SO_4^{2-}

Answers: (a) $200.4 \text{ mg} \cdot \text{L}^{-1}$ (b) $61,020 \text{ mg} \cdot \text{L}^{-1}$ (c) $980.6 \text{ mg} \cdot \text{L}^{-1}$ (d) $1921.2 \text{ mg} \cdot \text{L}^{-1}$

2–24 Calculate the concentration of the following in micrograms per liter.

(a) 0.0500 N H_2CO_3
(b) 0.0010 M $CHCl_3$
(c) 0.0300 N $Ca(OH)_2$
(d) 0.0080 M CO_2

2–25 A water initially contains $40 \text{ mg} \cdot \text{L}^{-1}$ of Mg^{2+}. The pH of the water is increased until the concentration of hydroxide ion (OH^-) is 0.001000 M. What is the concentration of magnesium ion in this water at this pH? Give your answer in milligrams per liter. Assume that the temperature of the solution is 25°C.

Answer: $0.4423 \text{ mg} \cdot \text{L}^{-1}$

2–26 Groundwater in Pherric, New Mexico, initially contains $1.800 \text{ mg} \cdot \text{L}^{-1}$ of iron as Fe^{3+}. What must the pH be raised to in order to precipitate all but $0.30 \text{ mg} \cdot \text{L}^{-1}$ of the iron? The temperature of the water is 25°C.

2–27 You made up a saturated solution of calcium sulfate ($CaSO_4$). The temperature is 25°C. You then add $5.00 \times 10^{-3} \text{ M}$ sodium sulfate (Na_2SO_4). What are the concentrations of calcium and sulfate after equilibrium is reached? The pK_s of $CaSO_4$ is 4.58.

Answer: $Ca^{2+} = 0.0032 \text{ M}$, $SO_4^{2-} = 0.0082 \text{ M}$

2–28 The solubility product of calcium fluoride (CaF_2) is 3×10^{-11} at 25°C. Could a fluoride concentration of $1.0 \text{ mg} \cdot \text{L}^{-1}$ be obtained in water that contains $200 \text{ mg} \cdot \text{L}^{-1}$ of calcium? Show your work.

2–29 What amount of NaOH (a strong base), in milligrams, would be required to neutralize the acid in Example 2–11?

Answer: 81.568 or 81.6 mg

2–30 The pH of a finished water from a water treatment process is 10.74. What amount of 0.02000 N sulfuric acid, in milliliters, is required to neutralize 1.000 L of the finished water, assuming that the alkalinity (buffering capacity) of the water is zero?

2–31 How many milliliters of 0.02000 N hydrochloric acid would be required to perform the neutralization in Problem 2–30?

Answer: 27.5 mL

2–32 Calculate the pH of a water at 25°C that contains $0.6580 \text{ mg} \cdot L^{-1}$ of carbonic acid. Assume that $[H^+] = [HCO_3^-]$ at equilibrium and ignore the dissociation of water.

2–33 What is the pH of a water that, at 25°C, contains $0.5000 \text{ mg} \cdot L^{-1}$ of hypochlorous acid? Assume equilibrium has been achieved. Ignore the dissociation of water. Although it may not be justified based on the data available to you, report the answer to two decimal places.

Answer: pH 6.28

2–34 If the pH in Problem 2–33 is adjusted to 7.00, what would the OCl^- concentration in milligrams per liter be at 25°C?

2–35 Convert the following from milligrams per liter as the ion to milligrams per liter as $CaCO_3$.
(a) $83.00 \text{ mg} \cdot L^{-1} \text{ Ca}^{2+}$
(b) $27.00 \text{ mg} \cdot L^{-1} \text{ Mg}^{2+}$
(c) $48.00 \text{ mg} \cdot L^{-1} \text{ CO}_2$ (*Hint:* CO_2 and H_2CO_3 in water are essentially the same: $CO_2 + H_2O \rightleftharpoons H_2CO_3$)
(d) $220.00 \text{ mg} \cdot L^{-1} \text{ HCO}_3^-$
(e) $15.00 \text{ mg} \cdot L^{-1} \text{ HCO}_3^{2-}$

Answers:
(a) $Ca^{2+} = 207.25$ or $207.3 \text{ mg} \cdot L^{-1}$ as $CaCO_3$
(b) $Mg^{2+} = 111.20$ or $111.2 \text{ mg} \cdot L^{-1}$ as $CaCO_3$
(c) $CO_2 = 109.18$ or $109.2 \text{ mg} \cdot L^{-1}$ as $CaCO_3$
(d) $HCO_3^- = 180.41$ or $180.4 \text{ mg} \cdot L^{-1}$ as $CaCO_3$
(e) $CO_3^{2-} = 25.02$ or $25.0 \text{ mg} \cdot L^{-1} \text{ CaCO}_3$

2–36 Convert the following from milligrams per liter as the ion or compound to milligrams per liter as $CaCO_3$.
(a) $200.00 \text{ mg} \cdot L^{-1} \text{ NH}_4^+$
(b) $280.00 \text{ mg} \cdot L^{-1} \text{ K}^+$
(c) $123.45 \text{ mg} \cdot L^{-1} \text{ SO}_4^{2-}$
(d) $85.05 \text{ mg} \cdot L^{-1} \text{ Ca}^{2+}$
(e) $19.90 \text{ mg} \cdot L^{-1} \text{ Na}^+$

2–37 Convert the following from milligrams per liter as $CaCO_3$ to milligrams per liter as the ion or compound.
(a) $100.00 \text{ mg} \cdot L^{-1} \text{ SO}_4^{2-}$
(b) $30.00 \text{ mg} \cdot L^{-1} \text{ HCO}_3^-$
(c) $150.00 \text{ mg} \cdot L^{-1} \text{ Ca}^{2+}$
(d) $10.00 \text{ mg} \cdot L^{-1} \text{ H}_2CO_3$
(e) $150.00 \text{ mg} \cdot L^{-1} \text{ Na}^+$

Answers:
(a) $SO_4^{2-} = 95.98$ or $96.0 \text{ mg} \cdot L^{-1}$
(b) $HCO_3^- = 36.58$ or $36.6 \text{ mg} \cdot L^{-1}$
(c) $Ca^{2+} = 60.07$ or $60.1 \text{ mg} \cdot L^{-1}$
(d) $H_2CO_3 = 6.198$ or $6.20 \text{ mg} \cdot L^{-1}$
(e) $Na^+ = 68.91 \text{ mg} \cdot L^{-1}$

2–38 Convert the following from milligrams per liter as $CaCO_3$ to milligrams per liter as the ion or compound.
(a) $10.00 \text{ mg} \cdot L^{-1} \text{ CO}_2$
(b) $13.50 \text{ mg} \cdot L^{-1} \text{ Ca(OH)}_2$
(c) $481.00 \text{ mg} \cdot L^{-1} \text{ HPO}_4^{2-}$
(d) $81.00 \text{ mg} \cdot L^{-1} \text{ H}_2PO_4^-$
(e) $40.00 \text{ mg} \cdot L^{-1} \text{ Cl}^-$

2–39 What is the "exact" alkalinity of a water that contains $0.6580 \text{ mg} \cdot L^{-1}$ of bicarbonate, as the ion, at a pH of 5.66?

Answer: $0.4302 \text{ mg} \cdot L^{-1}$ as $CaCO_3$

2–40 Calculate the "approximate" alkalinity (in milligrams per liter as $CaCO_3$) of a water containing $120 \text{ mg} \cdot L^{-1}$ of bicarbonate ion and $15.00 \text{ mg} \cdot L^{-1}$ of carbonate ion.

2–41 Calculate the "exact" alkalinity of the water in Problem 2–40 if the pH is 9.43.

Answer: 123.35 mg · L^{-1} as CaCO$_3$

2–42 Calculate the "exact" alkalinity of the water in Problem 2–40 if the pH is 11.03.

2–43 What is the pH of a water that contains 120.00 mg · L^{-1} of bicarbonate ion and 15.00 mg · L^{-1} of carbonate ion?

Answer: 9.43

2–44 What is the density of oxygen at a temperature of 273.0 K and at a pressure of 98.0 kPa?

2–45 Determine the density of nitrogen gas at a pressure of 122.8 kPa and a temperature of 298.0 K.

Answer: 1.39 kg · m^{-3}

2–46 Show that one mole of any ideal gas will occupy 22.414 L at STP. (STP is 273.16 K and 101.325 kPa.)

2–47 What volume would one mole of an ideal gas occupy at 25.0°C and 101.325 kPa?

Answer: 24.46 L

2–48 A sample of air contains 8.583 mol · m^{-3} of oxygen and 15.93 mol · m^{-3} of nitrogen at STP. Determine the partial pressures of oxygen and nitrogen in 1.0 m^3 of the air.

2–49 A 1-m^3 volume tank contains a gas mixture of 18.32 mol of oxygen, 16.40 mol of nitrogen, and 6.15 mol of carbon dioxide. What is the partial pressure of each component in the gas mixture at 25.0°C?

Answer: O$_2$: 45.4 kPa, N$_2$: 40.6 kPa, CO$_2$: 15.2 kPa

2–50 Calculate the volume occupied by 5.2 kg of carbon dioxide at 152.0 kPa and 315.0 K.

2–51 Determine the mass of oxygen contained in a 5.0-m^3 volume under a pressure of 568.0 kPa and at a temperature of 263.0 K.

Answer: 41.56 kg

2–52 One liter of a gas mixture at 0°C and 108.26 kPa contains 250 mg · L^{-1} of H$_2$S gas. What is the partial pressure exerted by this gas?

2–53 A 28-L volume of gas at 300.0 K contains 11 g of methane, 1.5 g of nitrogen, and 16 g of carbon dioxide. Determine the partial pressure exerted by each gas.

Answer: CH$_4$: 61.2 kPa, N$_2$: 4.77 kPa, CO$_2$: 32.4 kPa

2–54 Given the gas mixture of Problem 2–53, how many moles of each gas are present in the 28-L volume?

2–55 The partial pressures of the gases in a 22,414-L volume of air at STP are: oxygen, 21.224 kPa; nitrogen, 79.119 kPa; argon, 0.946 kPa; and carbon dioxide, 0.036 kPa. Determine the gram-molecular weight of air.

Answer: 28.966

2–56 Convert the concentration of SO$_2$ from 80 μg · m^{-3} to units of parts per million, given that the gas has a temperature of 25°C and a pressure of 101.325 kPa.

2–57 Convert the concentration of NO$_2$ from 0.55 ppm to units of micrograms per cubic meter, given that the gas has a temperature of 290 K and pressure of 100.0 kPa.

Answer: 1048.8, or 1050 μg · m^{-3}

2–58 A chemical is placed in a beaker containing 20 g of soil and 500 mL of water. At equilibrium, the chemical is found in the soil at a concentration of 100 mg · kg^{-1} of soil. The equilibrium concentration of this same chemical in the water is 250 μg · L^{-1} What is the partition coefficient for this chemical on this soil?

2–59 A chemical, SpartanGreen, has a partition coefficient of 12,500 $(mg \cdot kg^{-1})(mg \cdot L^{-1})^{-1}$. If the concentration of this chemical in water is found to be 105 $\mu g \cdot L^{-1}$, at equilibrium, what is the concentration on the soil?

Answer: 1312.5 $mg \cdot kg^{-1}$

2–60 Derive the equation given for the half-life of a reactant.

DISCUSSION QUESTIONS

2–1 Would you expect a carbonated beverage to have a pH above, below, or equal to 7.0? Explain why?

2–2 Explain the concentration unit of molarity to a citizen in a community with a hazardous waste site.

2–3 When dealing with acid rain (rainfall having a pH < 4), would a lake having a high or low alkalinity be more or less affected by this form of pollution? Explain your answer.

2–4 When you allow a can of carbonated beverage to sit open so that the carbon dioxide is released to the atmosphere, does the pH of the beverage remain the same, increase, or decrease? Explain your answer.

2–5 Would you expect the mineral kaolinite to dissolve more in a water having a high or a low pH? Explain your answer.

2–6 A gas sample is collected in a special gas-sampling bag that does not react with the pollutants collected but is free to expand and contract. When the sample was collected, the atmospheric pressure was 103.0 kPa. At the time the sample was analyzed the atmospheric pressure was 100.0 kPa. The bag was found to contain 0.020 ppm of SO_2. Would the original concentration of SO_2 be more, less, or the same? Explain your answer. If the concentrations were reported in units of micrograms per cubic meter, would your answer change? How?

2–7 What is the difference between chemical equilibrium and steady-state conditions?

FE EXAM FORMATTED PROBLEMS

2–1 Which statement is correct?
(a) An atom may be separated into elements.
(b) An element can only be a gas or liquid.
(c) An element can be heterogeneous or homogeneous.
(d) A compound can be separated into its elements by physical means.

2–2 Which of the following elements has the largest atomic radius?
(a) Mg
(b) Be
(c) Sr
(d) Ca
(e) Ba

2–3 Which of the following elements has the largest electronegativity?
(a) F
(b) O
(c) C

(d) Li

(e) B

2–4 An ion of an unknown element has an atomic number of 16 and contains 10 electrons. The ion is:

(a) P^{3-}

(b) O^{2-}

(c) Si^{3-}

(d) S^{6+}

(e) F^-

2–5 Which one of the following compounds is classified as an alkene?

(a) Ethanol

(b) Ethylene

(c) Butyne

(d) Butanal

(e) Butanone

2–6 The IUPAC name for the compound $\overset{\text{O}}{\underset{}{\diagup\diagdown\diagup}}\!\!\text{H}$ is:

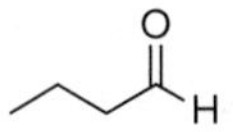

(a) Butanal

(b) Butanol

(c) Butene

(d) Butane

(e) Butyne

2–7 Commercial phosphoric acid (H_3PO_4) is often supplied as a liquid containing 85 wt% (weight percent) solution. The specific gravity of the 85% solution is 1.68 at 20°C.

(a) Determine the concentration of this solution in units of mg/L.

(b) Determine the concentration of this solution in units of molarity (M).

(c) Determine the concentration of this solution in units of normality (N).

2–8 Determine the pH of 10 g/L NaOH solution.

(a) 13.4

(b) 1.6

(c) 10.0

(d) 4.0

2–9 A 20-ounce bottle of a popular sports drink contains 270 mg of sodium. How many moles of sodium does the drink contain?

(a) 0.12 moles

(b) 1.3 moles

(c) 0.012 moles

(d) 10.2 moles

2–10 Assuming that CO_2 behaves as an ideal gas, determine the pressure exerted by the 44 g CO_2 in a 1-L vessel at room temperature (21°C).

(a) 2.41 atm

(b) 24.1 atm

(c) 241. atm

(d) 0.241 atm

REFERENCES

Chang, R. (2002). *Chemistry,* 7th ed. McGraw-Hill, New York.

Great Lakes Binational Toxics Strategy (1999) "Draft Report on Alkyl-lead: Sources, Regulations and Options." October. Accessed May 18, 2017. http://infohouse.p2ric.org/ref/06/05725.pdf.

Henry, J. G., and C. W. Heinke (1989) *Environmental Science and Engineering.* Prentice Hall, Englewood Cliffs, NJ: p. 201.

Kitman, J. L. (2000) "The Secret History of Lead." *The Nation,* March 2.

Kovarik, W. (2005) "How a Classic Occupational Disease Became an International Public Health Disaster." *Int. J. Occup. Environ. Health,* 11: 384–97.

Lead-Franklin. (n.d.) Accessed May 18, 2017. http://corrosion-doctors.org/Elements-Toxic/Lead-Franklin.htm.

Lead Poisoning and Rome (n.d.) Accessed May 18, 2017. http://penelope.uchicago.edu/~grout/encyclopaedia_romana/wine/leadpoisoning.html.

Lewis, W. K., and W. G. Whitman (1924) "Principles of Gas Adsorption," *Ind. Eng. Chem.* 16: 1215.

McKinney, M. L., and R. M. Schooch (1996) *Environmental Science: Systems and Solutions,* Jones and Bartlett Publishers. Sudbury, MA.

Scheer, R., and D. Moss (2012) "Earth Talk: Does the Continued Use of Lead in Aviation Fuel Endanger Public Health and the Environment?" *Scientific American,* September 3, 2012. https://www.scientificamerican.com/article/lead-in-aviation-fuel/

Stumm, W., and J. J. Morgan (1996) *Aquatic Chemistry,* 3rd ed., Wiley and Sons, Inc. New York.

Sunda, W., and R. K. L. Guillard (1976) "The Relationship Between Cupric Ion Activity and the Toxicity of Copper to Phytoplankton." *J. Mar. Res.* 34: 511–29.

The Lead Education and Abatement Design Group (2011) *Chronology-Making_Leaded_Petrol_History.* Dec 23. http://www.lead.org.au/Chronology-Making_Leaded_Petrol_History.pdf.

U.S. EPA (n.d.) Lead Trends. https://www.epa.gov/air-trends/lead-trends.

U.S. Weather Bureau (1976) *U.S. Standard Atmosphere,* U.S. Government Printing Office. Washington, D.C.

White, V. E., and C. J. Knowles (2000) "Effect of Metal Complexation on the Bioavailability of Nitrilotriacetic Acid to *Chelatobacter heintzii* ATCC 29600." *Arch Microbiol.* 173: 373–82.

3

Biology

Case Study

Lake Erie is Dead

In the 1960s, massive algal blooms spread across Lake Erie, the result of industrial pollution, untreated sewage, and agricultural runoff. The phrase "Lake Erie is Dead" was a common refrain then, and Dr. Seuss even mentioned the lake in the 1971 version of the Lorax with the line, "I hear things are just as bad up in *Lake Erie.*" After years of research, scientists discovered that phosphorus was the cause of the algal blooms in Lake Erie and elsewhere. In the 1970s and 1980s, policies and regulations were passed to ban phosphorus from detergents, develop best management practices to reduce phosphorus concentrations in agricultural runoff, and improve sewage treatment to reduce phosphorus levels in the effluent. The quality of Lake Erie water improved, and the problems seemed essentially solved until 2014.

On August 3, 2014, the City of Toledo issued a warning, which effectively cut off the water supply to 400,000 people in Toledo, most of its suburbs, and a few areas in southeastern Michigan. The concentration of the hepatotoxin (liver toxin), microcystin, exceeded drinking water limits, and boiling the water would simply serve to increase the concentration of the toxin. The water was not safe for consumption by humans and pets or for showering and bathing. Governor Kasich of Ohio issued a state of emergency, and the National Guard was brought in to distribute water and deliver water purification systems and ready-to-eat meals to those affected. The community was without water for the better part of three days.

Given all of the efforts of the last four decades, what happened? Part of the problem is geology—the western section of Lake Erie is shallow, with an average depth of approximately 7.3 meters. The water warms quickly and is a perfect breeding ground for the blue-green algae, *Microcystis,* the source of the toxin. The Maumee River, which feeds Lake Erie, flows through the highly productive farms of Indiana and Ohio, carrying with it fertilizers washed from not only the farms, but also golf courses and suburban lawns. Additionally, the type of fertilizers used today contain forms of phosphorus that are more soluble and easily assimilated by plants, including algae. So while we have improved sewage treatment and developed best management practices for agriculture, the sheer volume and form of phosphorus released to Lake Erie exceeds the capacity of the lake to assimilate it without the formation of algae blooms.

Unfortunately, that isn't the entire story—climate change is exacerbating the problem. It isn't the warmer temperatures that are the cause, but extreme precipitation events. As the frequency of these events and the amount of precipitation increases, runoff rates also increase, resulting in sudden pulses of phosphorus being discharged to the lake. And finally, recent research suggests that the combination of nitrogen and phosphorus are a double whammy, further aggravating the problem.

The solution is unclear. Policy change appears to be more difficult than ever. Federal funding to protect the Great Lakes and upgrade wastewater treatment plants is on the chopping block. The Trump Administration threatens to withdraw from the Paris Agreement on Climate Change and severely inhibit or eliminate regulatory agencies. Yet, somehow, solutions must be developed and implemented for millions of people that depend on Lake Erie's water.

3–1   INTRODUCTION

The case study presented above illustrates the importance of biology in public health and environmental engineering. In a 2007 poll conducted by the *British Medical Journal,* sanitation was clearly considered the single greatest medical milestone of the last 150 years. In the first 40 years of the 20th century, mortality rates in the United States decreased by 40%. Life expectancy at birth rose from 47 to 63 over the same period of time. Cutler and Miller (2005) found that treated drinking water and proper sanitation were responsible for nearly half of the total mortality reduction in major cities, three-quarters of the infant mortality reduction, and nearly two-thirds of the child mortality reduction. While our knowledge today is so much greater now than it was in the early 20th century, there is still much to learn and there are many problems to overcome. May you take all the opportunities afforded you to expand your wisdom and knowledge and may you use them for the improvement of the condition of humankind and the world in which we live.

3–2   CHEMICAL COMPOSITION OF LIFE

All living organisms are made of chemical compounds. These molecules all have carbon as their backbone and, as mentioned in Chapter 2, are called organic chemicals. While there are many more organic chemicals found in the cell, we will focus our discussion on the four main classes of large biological molecules, or **macromolecules,** as they are often called. The classes of compounds are carbohydrates, nucleic acids, proteins, and lipids. The first three are **polymers,** chainlike molecules that are composed of many similar or identical smaller molecules that serve as building blocks. These repeated smaller molecules are **monomers.**

Carbohydrates

Organisms use carbohydrates, which include both sugars and their polymers, as sources of energy, as building materials, and as cell markers for identification and communication. Carbohydrates are produced by photosynthetic organisms from carbon dioxide, water, and sunlight. All carbohydrates contain carbon, hydrogen, and oxygen in the ratio 1:2:1 and therefore have an empirical formula CH_2O. There are three groups of carbohydrates: monosaccharides, oligosaccharides, and polysaccharides.

Simple sugars, or **monosaccharides,** are the major nutrients for cells. In cellular respiration, cells extract the energy stored in glucose molecules. Simple sugars also serve as the raw material for the synthesis of amino acids, fatty acids, and other biological compounds. This group of chemicals contains a single chain of carbons, with multiple hydroxyl groups and either a ketone or an aldehyde. Compounds with a ketone are **ketoses,** while those with an aldehyde are **aldoses,** as shown in Figure 3–1a. A sugar with three carbons is a **triose** (Figure 3–1a); those with five carbons are called **pentoses** (Figure 3–1b); and those with six carbons are **hexoses** (Figure 3–1c). While simple sugars have a linear structure in the dry state (as in Figure 3–1), in aqueous solution, chemical equilibrium favors the formation of ring structures as shown in Figure 3–2. When the ring forms, the hydroxyl group at the carbon 1 can end up either within the plane of the ring, resulting in an α-sugar, or above the plane of the ring, forming a β-sugar.

Oligosaccharides contain two or three monosaccharides linked by covalent bonds called glycosidic linkages. The disaccharides maltose and sucrose are shown in Figure 3–3. Sucrose (table sugar) is the most prevalent oligosaccharide. Maltose is the sugar present in milk. Plants use sucrose to transport carbohydrates from leaves to roots.

Polysaccharides can contain polymeric chains of several hundred to several thousand monosaccharides. These chemicals store energy and provide structural support to the cell. Starch contains glucose monomers and stores energy in plants as shown in Figure 3–4a. Humans store energy in muscles and liver cells in the form of glycogen, which can be metabolized into

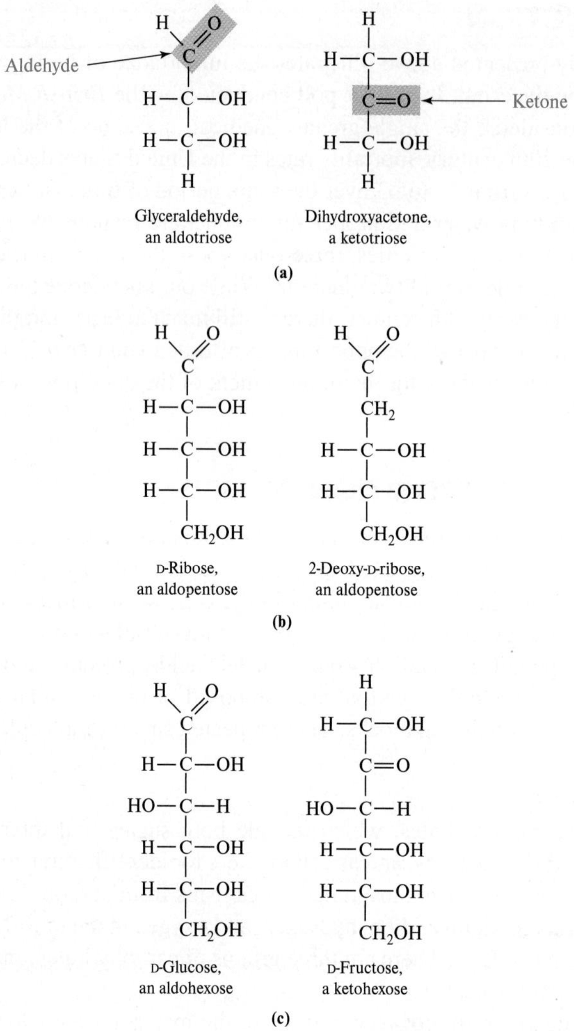

FIGURE 3–1

Simple sugars:
(a) trioses, **(b)** pentoses,
(c) hexoses.

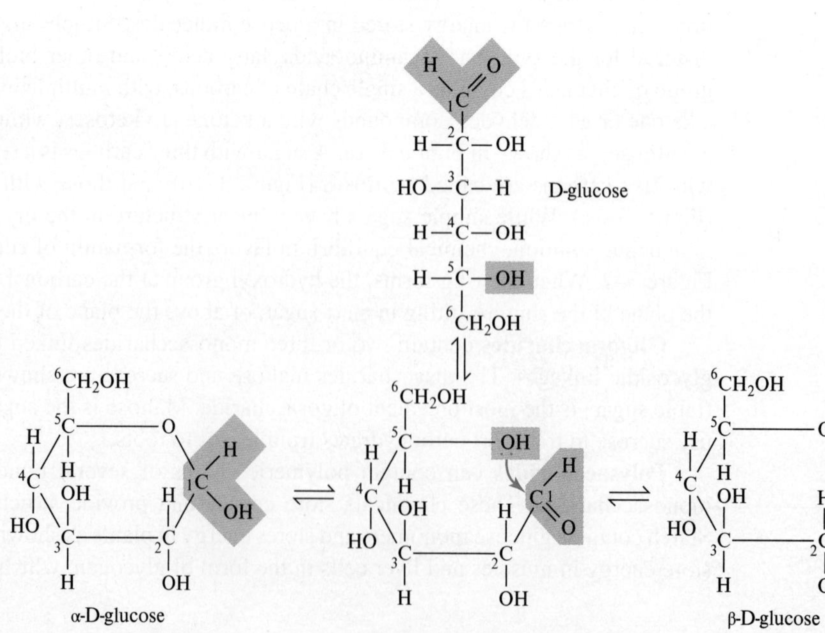

Aldehyde →

H—C=O

H—C—OH

H—C—OH

H

Glyceraldehyde,
an aldotriose

H

H—C—OH

C=O ← Ketone

H—C—OH

H

Dihydroxyacetone,
a ketotriose

(a)

H—C=O

H—C—OH

H—C—OH

H—C—OH

CH_2OH

D-Ribose,
an aldopentose

H—C=O

CH_2

H—C—OH

H—C—OH

CH_2OH

2-Deoxy-D-ribose,
an aldopentose

(b)

H—C=O

H—C—OH

HO—C—H

H—C—OH

H—C—OH

CH_2OH

D-Glucose,
an aldohexose

H

H—C—OH

C=O

HO—C—H

H—C—OH

H—C—OH

CH_2OH

D-Fructose,
a ketohexose

(c)

FIGURE 3–2

When D-glucose dissolves
in water, the hydroxyl
group on carbon 5 reacts
with the aldehyde group
on carbon 1 to form
a closed ring. If the
hydroxyl group on
carbon 1 lies below the
plane of the ring, it is
called α-D-glucose and, if
it lies above the plane, it
is called β-D-glucose.

H—^{1}C=O

H—^{2}C—OH

HO—^{3}C—H   D-glucose

H—^{4}C—OH

H—^{5}C—OH

$^6CH_2OH$

α-D-glucose

β-D-glucose

FIGURE 3–3

Synthesis of the disaccharide, maltose, which is formed from the dehydration of two α-glucose molecules, resulting in an α 1-4 glycosidic linkage. Structure of sucrose, which involves a α 1-2 glycosidic linkage. The decomposition of maltose to glucose occurs by a hydrolysis reaction.

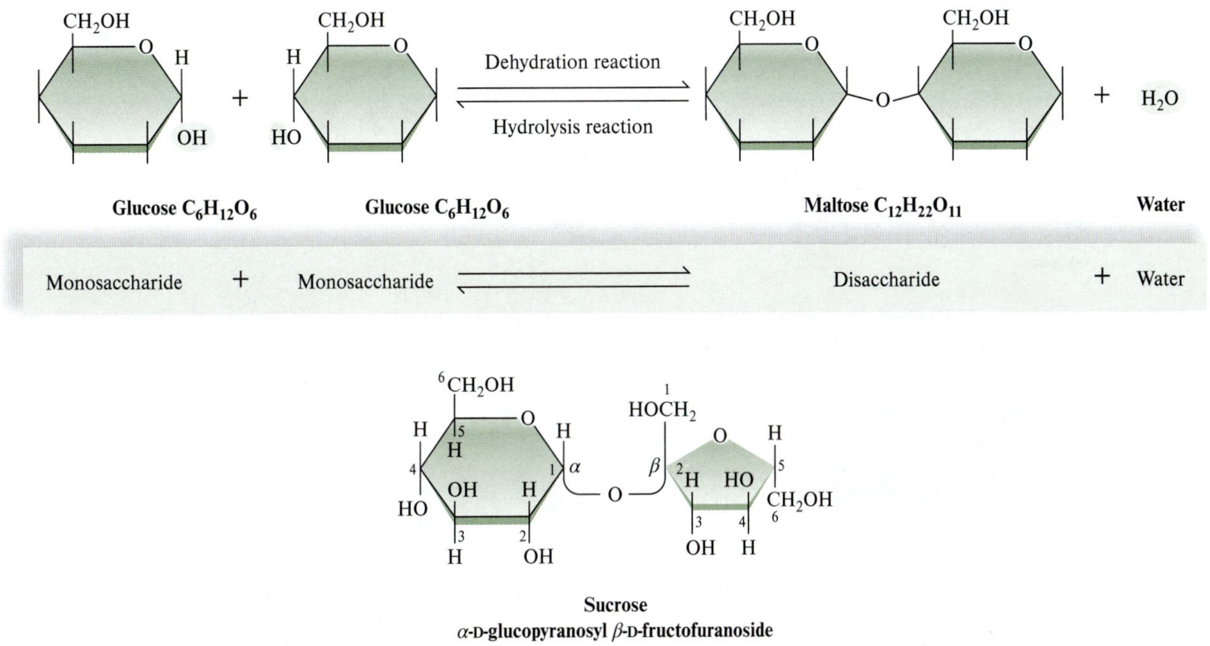

Glucose $C_6H_{12}O_6$ Glucose $C_6H_{12}O_6$ Maltose $C_{12}H_{22}O_{11}$ Water

Monosaccharide + Monosaccharide ⇌ Disaccharide + Water

Sucrose
α-D-glucopyranosyl β-D-fructofuranoside

glucose molecules during physical exercise (Figure 3–4b). Cellulose, shown in Figure 3–4c, is a major component of plant cell walls and provides structural support to the cell. Chitin, another polysaccharide, is used by arthropods to build their exoskeletons and by many fungi in assembling their cell walls. Humans are able to digest starch but not cellulose. Animals, such as rabbits, sheep, and cows, that are able to break down starch actually accomplish this by the presence of symbiotic bacteria and protists that live in their digestive tracts.

Nucleic Acids

Nucleic acids store and transmit hereditary information. They are the only molecules that can produce exact replicas of themselves. In doing so, they allow the organism to reproduce. There are two types of nucleic acids—**deoxyribonucleic acid (DNA)** and **ribonucleic acid (RNA).** DNA provides the directions for its own replication and directs RNA synthesis. RNA controls protein synthesis. RNA and DNA are nucleotide polymers. **Nucleotides** contain a nitrogenous base, a five-carbon (pentose) sugar, and a phosphate group as shown in Figure 3–5a. DNA contains the sugar deoxyribose. RNA contains ribose. The molecular structures are illustrated in Figure 3–5b. As shown in Figure 3–5c, there are five different organic bases that can be found in DNA and RNA: cytosine, thymine, uracil, adenine, and guanine. Cytosine, thymine, and uracil are **pyrimidines,** that is, they contain a six-membered ring of carbon and nitrogen atoms. Adenine and guanine are **purines,** which contain a six-membered ring fused to a five-membered ring of carbon and nitrogen atoms.

DNA contains nucleotides of adenine, guanine, cytosine, and thymine. RNA contains adenine, guanine, cytosine, and uracil, rather than thymine. Nucleotides are attached to one another by phosphodiester linkages between the phosphate group of one nucleotide and the sugar group of the next. RNA is single stranded—that is, it contains a single polymer of nucleotides. As shown in Figure 3–5d, DNA contains two strands, wound around each other in the "double helix"

FIGURE 3–4

Structure and function of **(a)** starch, **(b)** glycogen, and **(c)** cellulose. (a: large photo - ©Steven P. Lynch, insert - ©Freer/Shutterstock; b: large photo - ©Don W. Fawcett/Science Source, insert - ©McGraw-Hlll Education; c: right photo - ©Martin Kreutz/Age Fotostock)

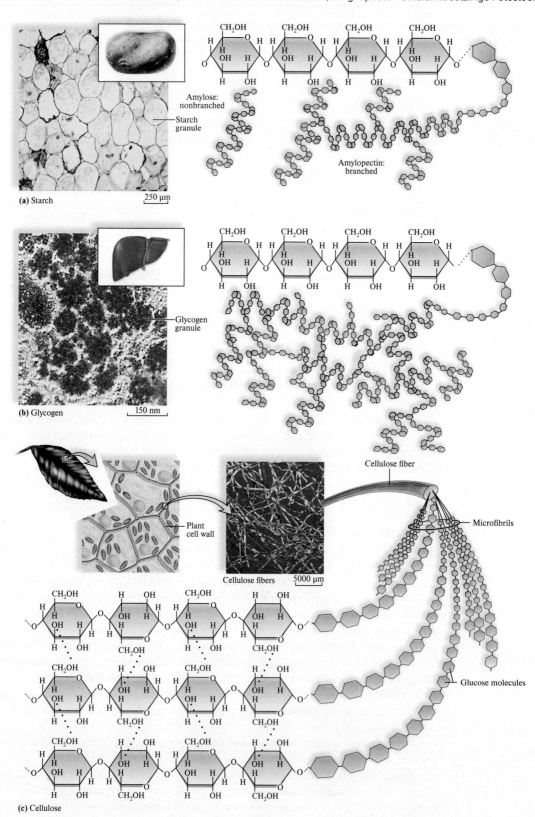

(a) Starch

250 µm

Amylose: nonbranched

Starch granule

Amylopectin: branched

(b) Glycogen

150 nm

Glycogen granule

Cellulose fiber

Plant cell wall

Microfibrils

Cellulose fibers 5000 µm

Glucose molecules

(c) Cellulose

FIGURE 3–5

(a) Structure of a nucleotide. **(b)** In DNA, the sugar is deoxyribose. In RNA, the sugar is ribose. **(c)** There are two classes of nitrogenous compounds, purines and pyrimidines, present in DNA and RNA. The pyrimidine bases are cytosine (C), and thymine (T) and uracil (U). In RNA there is a uracil (U) base instead of thymine which is found in DNA. The purine bases are adenine (A) and guanine (G).
(d) Nucleotides are attached to one another by phosphodiester linkages between the phosphate group of one nucleotide and the sugar of the next. DNA molecules are double stranded. Two strands of DNA are held together by hydrogen bonds.

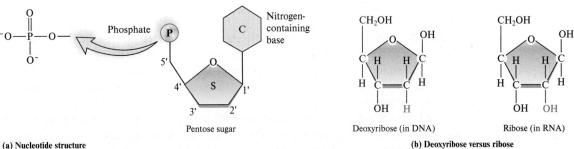

(a) Nucleotide structure

(b) Deoxyribose versus ribose

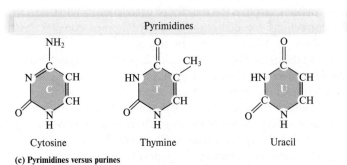

(c) Pyrimidines versus purines

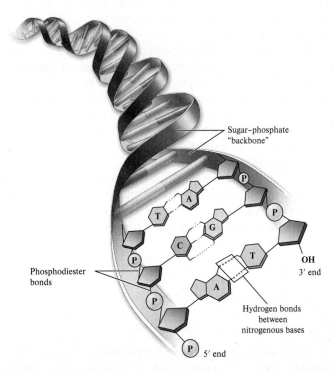

(d)

structure, and held together by hydrogen bonds between the nitrogenous base on one strand and a complementary base on the other strand.

Nucleotides are also important intermediates in the transformation of energy. The nucleotide adenosine triphosphate (ATP) drives virtually all energy transfer reactions in a cell. Nicotinamide adenine dinucleotide (NAD^+) and flavin adenine dinucleotide (FAD^+) are both involved in the production of ATP. Nicotinamide adenine dinucleotide phosphate ($NADP^+$), a molecule similar to NAD^+, is used in photosynthesis.

Proteins

Proteins, a diverse group of compounds, make up more than 50% of the dry weight of cells. They affect almost all cellular functions and are used in structural support, transport of other substances, storage, signaling within cells and from one cell to another, movement, and defense. Enzymes are proteins that act as catalysts. Immunoglobulins are proteins that protect cells. Hemoglobin transports oxygen.

Proteins are polymers of the same set of 20 amino acids (as shown in Figure 3–6) that are folded into specific three-dimensional shapes, determined by the sequence of amino acids in the molecule. Amino acid polymers, called polypeptides, are constructed in the cytoplasm of the cell by protein synthesis. Protein synthesis involves the transfer of genetic information stored in the DNA. RNA carries out the instructions encoded in the DNA. Ribosomes, complex structures formed by the association of ribosomal RNA with specific sets of proteins, move along a messenger RNA molecule to catalyze the assembly of amino acids into protein chains. Proteins can be globular (roughly spherical) or fibrous as shown in Figure 3–7. It is the special conformation (shape) of the protein that determines its function. The structure and function of a protein is highly dependent on the physical and chemical conditions of the environment in which it is found. For example, gastrin, the digestive enzyme found in the stomach, is most effective at pH 2 and "denatures" (changes its shape) at pH values greater than 10. The proteins in a clear egg white are denatured upon exposure to high temperatures, turning the protein an opaque white color.

Enzymes are proteins that catalyze certain reactions and control and direct the reaction pathway. Many reactions are thermodynamically favorable, but do not occur in abiotic systems; however, these same reactions may occur in the cell because of the presence of enzymes. For example, in the absence of microorganisms, a glucose solution is stable. However, as shown in Figure 3–19 in the section on cellular respiration, aerobic organisms take glucose and convert it to carbon dioxide and water, thereby obtaining energy.

The kinetics of enzyme-catalyzed reactions are governed by the same principles as other chemical reactions. The main difference is that the rate enzymatic reactions can be controlled by the concentration of substrate. As shown in Figure 3–8, at very low concentrations of substrate, the rate is directly proportional to the substrate concentration. As a result, the reaction rate can be described by first-order kinetics. As the substrate concentration increases the rate of reaction begins to decrease until it ultimately asymptotes at a maximum rate. At that point, the reaction becomes zero order with respect to the substrate concentration.

The kinetics of enzymatic reactions were first described in 1913 by L. Michaelis and M. L. Menten. In their model the enzyme reacts first with the substrate to form an enzyme-substrate complex:

$$E + S \underset{k_{-1}}{\overset{k_1}{\rightleftharpoons}} ES \tag{3-1}$$

The complex then forms the product and the free enzyme

$$ES \underset{k_{-2}}{\overset{k_2}{\rightleftharpoons}} E + P \tag{3-2}$$

These reactions are assumed to be reversible and the rate constants in the forward direction are denoted by positive subscripts, while those in the reverse direction are given negative subscripts. Using these reaction equations, the Michaelis-Menten equation for a one substrate

FIGURE 3–6

Structural formulas for the 20 amino acids.

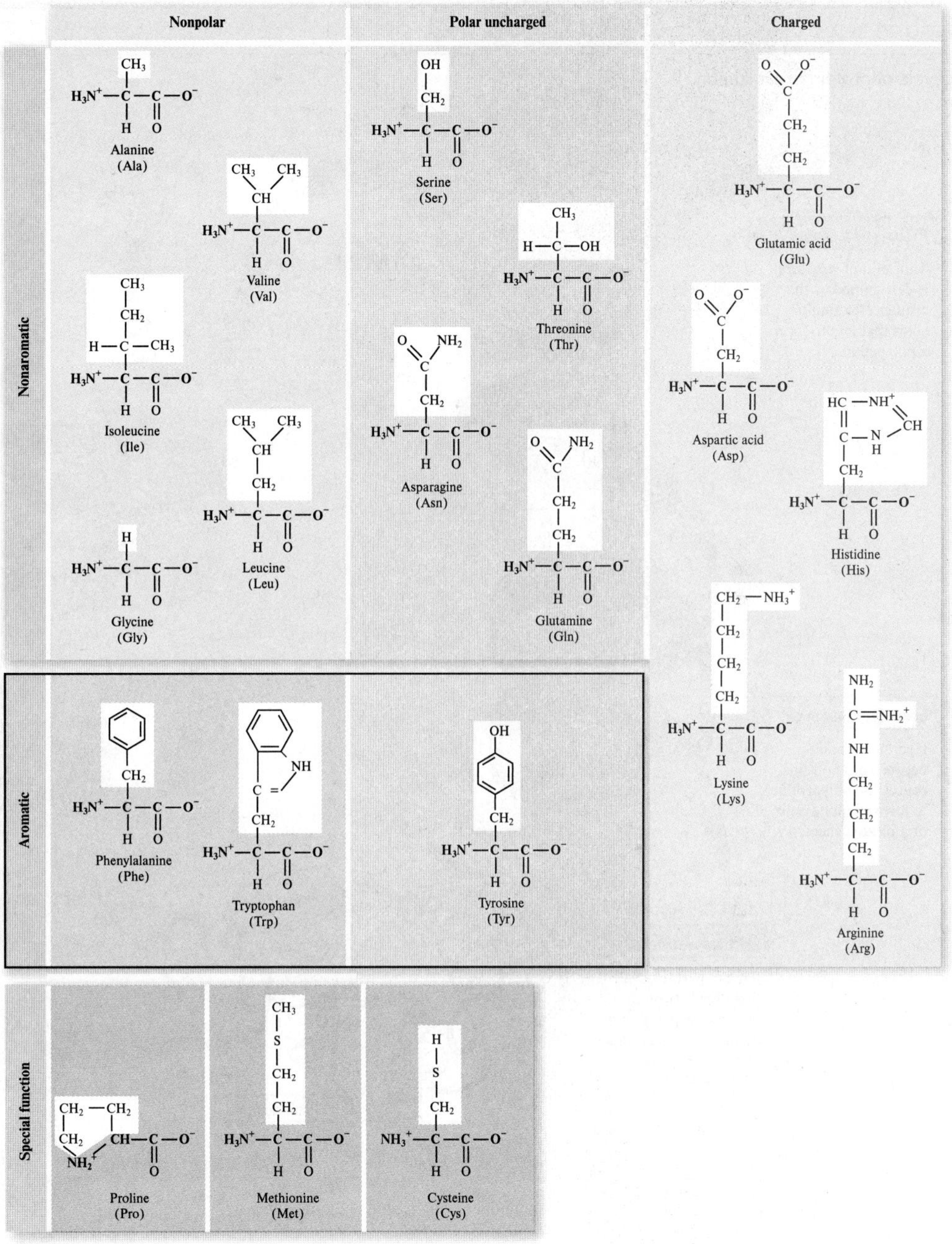

enzyme-catalyzed reaction can be derived. (The derivation is beyond the scope of this text but can be found in most chemistry and biochemistry textbooks.)

$$v_o = \frac{v_{max}[S]}{K_M + [S]} \tag{3-3}$$

where v_o = the initial rate of an enzymatic reaction

v_{max} = the maximum initial velocity

K_M = Michaelis-Menten coefficient

$[S]$ = substrate concentration

FIGURE 3–7

Levels of protein organization.

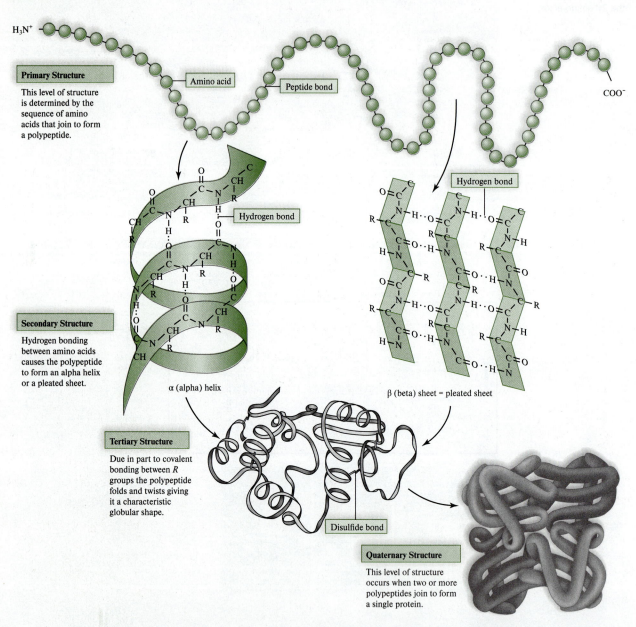

Primary Structure

This level of structure is determined by the sequence of amino acids that join to form a polypeptide.

Amino acid

Peptide bond

Secondary Structure

Hydrogen bonding between amino acids causes the polypeptide to form an alpha helix or a pleated sheet.

Hydrogen bond

α (alpha) helix

Hydrogen bond

β (beta) sheet = pleated sheet

Tertiary Structure

Due in part to covalent bonding between *R* groups the polypeptide folds and twists giving it a characteristic globular shape.

Disulfide bond

Quaternary Structure

This level of structure occurs when two or more polypeptides join to form a single protein.

While it appears that the relationship does not include the enzyme concentration, it is actually incorporated into the term for V_{max}, which is directly proportional to the enzyme concentration. On the contrary, K_M depends on the structure of the enzyme and is independent of the enzyme concentration.

We are interested in the rate of formation of the product, P. The velocity of this reaction, v, is:

$$v = \frac{v_{max}[S]}{K_M + [S]}$$

(3–4)

This equation is known as the Michaelis-Menton equation.

FIGURE 3–8

Effect of substrate concentration on the rate of an enzyme-catalyzed reaction.

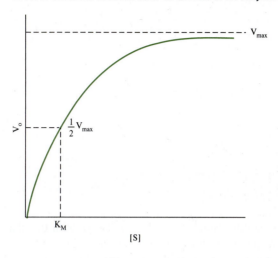

An important relationship exists when the initial velocity is equal to one-half the maximum velocity, that is, when $v_o = \frac{1}{2} v_{max}$.

$$v_o = \frac{v_{max}}{2} = \frac{v_{max}[S]}{K_M + [S]} \qquad\qquad (3\text{–}5)$$

Rearranging this equation yields the equality $K_M = [S]$. So when the initial velocity is one-half the maximum value, the constant K_M is equal to the substrate concentration. This is an important relationship because it allows K_M to be determined by simple experiments using a plot of initial velocity vs. the initial substrate concentration.

The value for K_M can vary greatly. It can be as low as 0.025 mM for glutamate dehydrogenase and the substrate NAD_{ox} to 122 mM for the enzyme chymotrypsin and the substrate glycyltyrosinamide. It also varies with pH and temperature. Although the kinetic behavior of most enzymes are much more complex than is assumed in this one-substrate model, the Michaelis-Menten approach is still useful to quantifying enzymatic activity in tissues and has led to discoveries of new methods for cancer treatment.

EXAMPLE 3–1 The following data were obtained for an enzymatic reaction. Determine V_{max} and K_M.

Initial substrate concentration (mM)	V_o (µg/h)
0.5	40
1.0	75
2.0	139
3.0	179
4.0	213
6.0	255
7.5	280
10.0	313
15.0	350

Solution   To solve this problem, first plot the data presented.

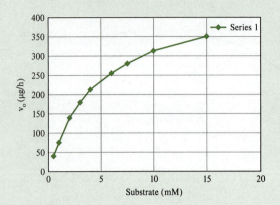

From the plot, one can determine V_{max}. The challenge with this data is that it is difficult to determine where the plot will asymptote. As such, it is difficult to estimate V_{max}. However, the Michaelis-Menten equation can be easily transformed by taking the inverse of both sides of the equation, yielding:

$$\frac{1}{v_o} = \frac{K_M}{V_{max}[S]} + \frac{1}{V_{max}} \qquad (3\text{-}6)$$

This equation is commonly called the Lineweaver–Burk equation. The slope of this plot yields K_M/V_{max}. The y-intercept is $1/V_{max}$. Manipulating the data provided and plotting the reciprocals of v_o and [S] yields the plot:

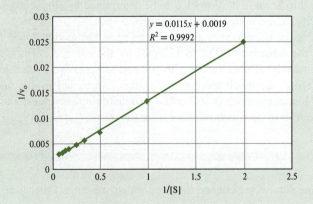

The slope of the plot is 0.0115 and the y-intercept is 0.0019. Therefore, V_{max} is 526 $\mu g \cdot h^{-1}$ and K_M is 6.05 mM.

Lipids

Lipids, which are not polymeric in nature, are hydrophobic molecules made up of carbon, hydrogen, and oxygen. This group is quite varied in nature; the one characteristic they have in common is their repulsion of water molecules. The most important lipids in the cell are fats, phospholipids, and steroids.

Fats are made of glycerol and fatty acids held together by an ester linkage, as shown in Figure 3–9a. Glycerol is an alcohol with three carbons, each bound to a hydroxyl group. A fatty acid usually contains at least 16 carbons and has a carboxyl acid group at one end. Saturated fatty acids contain the maximum number of hydrogen atoms possible; that is, there are no C—C double bonds. Unsaturated fatty acids contain at least one C—C double bond. Most animal fats are saturated. On the contrary, most plant and fish-derived fats are unsaturated.

Fats are the most common energy-storing molecule in the cell. A gram of fat stores about 38 kJ (9 kilocalories) of chemical energy, more than twice that of the same mass of proteins or carbohydrates. Animals are able to convert carbohydrates to fat, which are stored as droplets in the cells of adipose (fat) tissue. This layer of fat serves as insulation against cold temperatures.

Phospholipids are the major component of cell membranes. They are made of a glycerol molecule attached to two fatty acids and a phosphate group, as shown in Figure 3–9b. The "head" of the phospholipid is hydrophilic, while the "tail" is hydrophobic. When in water, phospholipids form spherical micelles, where the hydrophobic tails arrange toward themselves (and the inside of the sphere) and the hydrophilic ends are organized toward the water (and the outside of the sphere). Phospholipids are unique in that this arrangement allows the formation of a membrane (as illustrated in Figure 3–9c), which can selectively pass some molecules but not others.

Steroids are lipids with a carbon skeleton containing four fused hydrocarbon rings. Different steroids have different functional groups attached to the rings, as shown in Figure 3–10. Cholesterol, which has been given a bad name by the popular press, is essential for proper cell function as it is converted into vitamin D and bile salts, and is a precursor of other steroids, including the vertebrate sex hormones.

3–3   THE CELL

Each cell is a fully functional living unit that can create and maintain molecules and structures to sustain life. All living cells obtain food and energy, convert energy, construct and maintain the molecules, carry out chemical reactions, eliminate waste products, reproduce, and maintain **homeostasis** (i.e., balance within itself). All cells are bounded by a plasma membrane, which contains a semifluid substance, cytosol. All cells contain chromosomes, which contain the genetic information in the form of DNA. All cells have ribosomes, the structures that synthesize proteins. The plasma membrane functions as a selective barrier that allows the passage of nutrients and wastes.

Prokaryotes and Eukaryotes

There are two basic types of cells—those of prokaryotes and those of eukaryotes. **Prokaryotic cells** lack a **nucleus,** the region inside a cell containing the DNA as shown in Figure 3–11. Instead, the DNA is encapsulated in a nucleoid, which lacks a membrane to separate it from the rest of the cell. **Eukaryotic cells** contain a nucleus, which is bounded by a membrane. The interior of the eukaryotic cell, between the nucleus and the outer plasma membrane, is the **cytoplasm.** The eukaryotic cell, unlike the prokaryotic cell, contains other specialized membrane-bound structures called **organelles**. The major organelles of an animal cell, along with their function, are shown in Figure 3–12. Figure 3–13 illustrates the major organelles of a plant cell. Eukaryotic cells are usually larger than prokaryotic cells.

Cell Membrane

The structure of the plasma (cell) membrane is very complex. It is only about 8 nm thick, yet it controls all passage of substances into and out of the cell. It exhibits selective permeability, allowing some molecules to pass more easily than others. Membranes are composed predominately of lipids and proteins, although carbohydrates are often also present. Phospholipids, the most prevalent lipids in the membrane, have an important property—they are

FIGURE 3–9

(a) Formation of a fat molecule. **(b)** Structure of a phospholipid and **(c)** schematic showing how phospholipids assemble to form a bilayer.

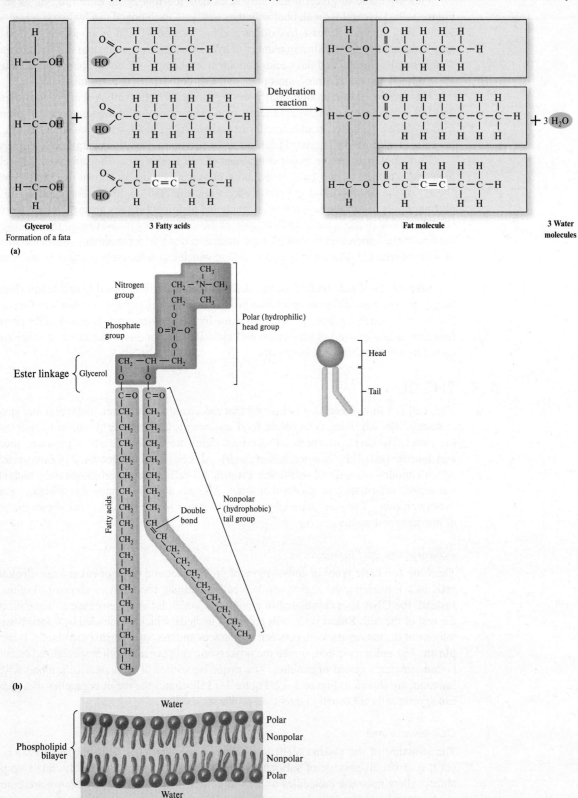

FIGURE 3–10

Cholesterol and testosterone are both members of the sterol family of lipids. Note the structural similarities.

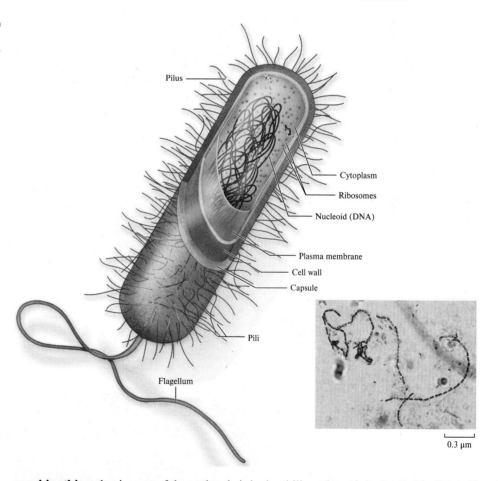

Cholesterol

Testosterone

FIGURE 3–11

Structure of a prokaryotic cell. (Bottom right photo - ©Steven P. Lynch)

Pilus

Cytoplasm

Ribosomes

Nucleoid (DNA)

Plasma membrane

Cell wall

Capsule

Pili

Flagellum

0.3 μm

amphipathic—that is, part of the molecule is hydrophilic and part is hydrophobic. It is believed the membrane has the consistency of a fluid with protein molecules embedded in or attached to a bilayer of phospolipids. This description is referred to as the **fluid mosaic model.**

Animal cell membranes also contain cholesterol molecules, which allow the membrane to function over a wide range of temperatures. The cholesterol molecules reduce the permeability of the membrane to many biological chemicals.

FIGURE 3–12

Structure of an animal cell.

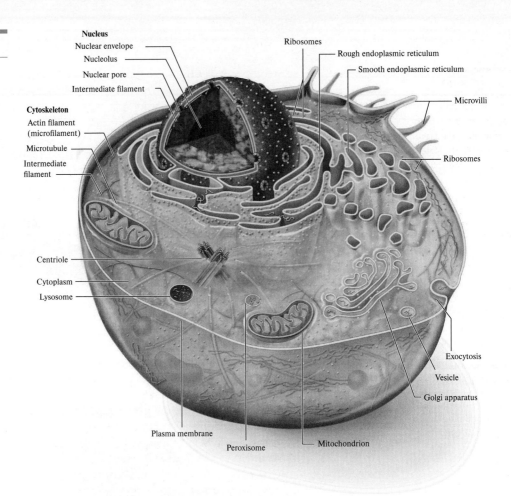

FIGURE 3–13

Structure of a plant cell.

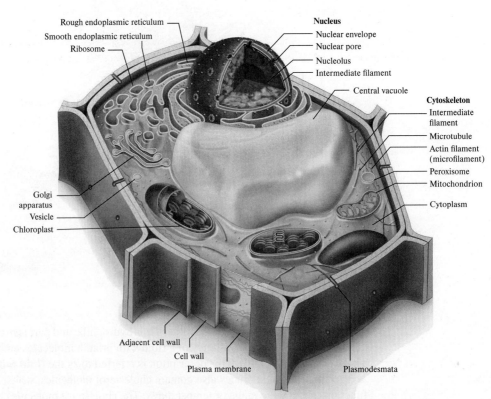

The cell membrane uses several different methods to transport molecules into and out of the cell. Diffusion is the simplest way in which a molecule can move across a membrane. With diffusion, molecules move from a region where they are more concentrated to one where they are less concentrated. The difference in the concentration of the chemical is the driving force. Small, neutral molecules, such as oxygen and carbon dioxide, move through a membrane by diffusion. The diffusion of a molecule across the membrane is called **passive transport** because the cell does not need to expend energy for the process to occur.

Water moves across a biological membrane by a process called osmosis. Osmosis is the diffusion of the *solvent* across a semipermeable membrane that separates the two solutions. If the concentration of water both inside and outside the cell are the same, the net flow of water across the membrane is zero and the conditions are referred to as **isotonic.** If the concentration of water outside the cell is greater than that inside, then the net flow of water will be into the cell and the conditions are **hypotonic.** If, on the other hand, the concentration of water inside the cell is greater than that outside, then the net flow of water will be out of the cell and the conditions are **hypertonic.** Like diffusion, osmosis is a passive process, since the cell does not expend energy for it to occur.

Many polar chemicals and ions cannot pass through a membrane unaided. For example, glucose is too large to pass across a membrane by diffusion. Additionally, it is insoluble in the lipids present in the membrane. To transport such chemicals, the cell has developed specialized transport proteins that reside in the membrane. These transport chemicals are highly selective for the particular solute it is to transport. Many transport proteins have a binding site akin to the active site of an enzyme. This process is called **facilitated diffusion** and biologists are still studying the process to better understand how it works. Like with diffusion and passive transport, facilitated diffusion does not require that the cell expend energy to achieve transport.

As a cell metabolizes nutrients, waste products are produced, which must be removed from the cell. Nutrients from outside of the cell must be transported into the cell for survival. To accomplish this, the cell must expend energy to transport the chemicals from an area of lower concentration to one of higher concentration. The process of moving chemicals against a concentration gradient is called **active transport.** As you are sitting still reading this text, your body is still expending energy. About 40% of the energy used is for active transport. Some cells, such as those in your kidneys, use much more. Renal cells use about 90% of the energy expended on active transport. This is not surprising, since the role of a kidney cell is to pump glucose and amino acids out of the urine and back into the blood and filter toxic waste products from the blood and transport them into the urine.

Active transport occurs with the aid of specific proteins that are embedded in membranes. The energy expended in the process comes from ATP. The transporter protein actively pumps ions across the membrane. In some cases, for example with sodium and potassium ions, the proteins actually contribute to the potential across the membrane, and during the process, energy is stored in the form of voltage. This stored energy can be used at a later time for cellular work.

The energy stored during active transport can be harvested to drive the transport of other solutes across a membrane. For example, in plant cells, this process, called **cotransport,** allows the cell to use the gradient of H^+ generated by its proton pumps to drive the active transport of sugars and amino acids into the cell, against a concentration gradient. This process is analogous to pumping water uphill and then harvesting the energy as it flows downhill to generate electricity.

There are some substances that are too large or too polar (e.g., proteins and polysaccharides) to be transported by the mechanisms discussed previously. For these substances, the cell uses a special mechanism involving a **vesicle,** a membrane-bound sac, to "swallow" or expel the material. The process by which the membrane of the vesicle folds inward, trapping and engulfing the small bit of material from the extracellular fluid, is called **endocytosis.** There are three main types of endocytosis: pinocytosis, phagocytosis, and receptor-mediated endocytosis.

Exocytosis is a similar process but in this case the cell secretes the macromolecule into the extracellular fluid.

With **pinocytosis,** the cell "gulps" small droplets of extracellular fluid, taking with it any dissolved chemicals or particulate matter contained in the fluid. Pinocytosis is nonspecific in nature. This is a very commonly occurring process within a cell.

With **phagocytosis,** or cell "eating," the pseudopodia wrap around a particle, encasing the particle in a membrane-bound sac that is sufficiently large to be classified as a vacuole. The process is quite specific and only occurs in specialized cells such as the macrophages, a type of large blood cell that is involved in immune defense. The single-celled amoeba feeds by phagocytosis.

Unlike pinocytosis and phagocytosis, **receptor-mediated endocytosis** is very specific. Proteins with receptor sites that are specific for a macromolecule are embedded in a membrane. These proteins are exposed to the extracellular fluid and can bind to a macromolecule. Once the protein and receptor molecule bind, the surrounding membrane folds inward and forms a vesicle that contains the two molecules. The vesicle releases its contents inside the cell and then returns to the membrane, turning outward and recycling the receptor molecule and the membrane.

Cell Organelles of Eukaryotes

The nucleus of a cell stores most of the genetic information, which determines the structural characteristics and function of the cell. It is usually the most easily identifiable organelle in a eukaryotic cell and has a diameter of about 5 μm. The nuclear envelope separates the nucleus from the cytoplasm. Within the nucleus is the **chromatin,** a fibrous material that is made up of DNA and proteins. As the cell prepares to divide, the thin chromatin fibers coil up, becoming thick enough to be viewed as separate structures called chromosomes. Each nucleus contains at least one area of chromatin called the **nucleolus.** In the nucleolus, the ribosomal DNA is synthesized and assembled with proteins to form ribosomal subunits, which can be transported through the nuclear pores into the cytoplasm. In the cytoplasm, the subunits assemble to form ribosomes.

Ribosomes are the organelles that synthesize proteins. Each cell contains thousands of these tiny organelles; the number is related to the rates at which the cell synthesizes proteins. Ribosomes can be either "free" (i.e., suspended in the cytosol) or "bound" (i.e., attached to the outside of the endoplasmic reticulum). Bound and free ribosomes are structurally identical and can alternate between the two roles.

The endomembrane system, illustrated in Figure 3–14, contains many of the different membranes of the eukaryotic cell and includes the nuclear envelope, the endoplasmic reticulum, the Golgi apparatus lysosomes, various types of vacuoles, and the plasma membrane. The **endoplasmic reticulum** are made up of numerous folded membranes, which have a very large surface area for chemical reactions to take place. There are two types of endoplasmic reticulum—smooth and rough.

Smooth endoplasmic reticulum (ER) lacks ribosomes on its cytoplasmic surface. This organelle is involved in the synthesis of lipids, including the sex hormones and steroid hormones secreted by the adrenal glands. Enzymes produced by the smooth ER help detoxify drugs and poisons. The smooth ER of muscle cells pump calcium ions from the cytosol, thereby stimulating the muscle cell by a nerve impulse.

Rough ER is involved in the synthesis of secretory proteins, including insulin. It appears rough in micrographs because ribosomes are attached to the cytoplasmic side of the nuclear envelope.

The **Golgi apparatus** (see Figure 3–15) can be thought of as a center of production, storage, sorting, and shipping. Within the Golgi apparatus, macromolecules synthesized in the ER are processed so they become fully functional and are sorted into packages for transport to the appropriate cellular location. The Golgi apparatus also manufactures many polysaccharides that are secreted by the cell. The Golgi apparatus can add molecular identification tags to a macromolecule to aid in identification by other organelles.

Lysosomes are the "composters" of the cell. Within this membrane-bound sac, shown in Figure 3–16, are the hydrolytic enzymes necessary for the digestion of proteins, polysaccharides,

The endomembrane system. (Bottom right photo - ©EM Research Services, Newcastle University)

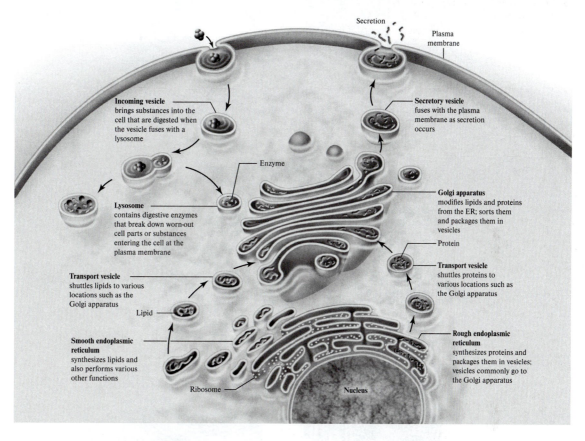

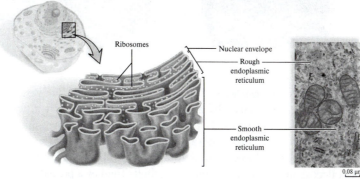

fats, and nucleic acids. The pH within the lysosome is about 5. Cell death can occur if the enzymes from the lysosome are excessively leaked into the cell. **Peroxisomes** are similar in structure to vesicles, but this organelle contains enzymes that generate hydrogen peroxide for various functions, such as breaking down fatty acids into smaller molecules and detoxifying alcohol and other harmful chemicals. The hydrogen peroxide that is produced by the peroxisome is toxic to the cell and must be contained within the membrane to prevent cellular damage. The peroxisome also contains oxidative enzymes that convert hydrogen peroxide to water to protect the cell.

FIGURE 3–15

Structure of the Golgi apparatus. (Bottom right - ©Charles Flickinger, from *Journal of Cell Biology,* 49:221–226, 1971, Fig. 1, p. 224)

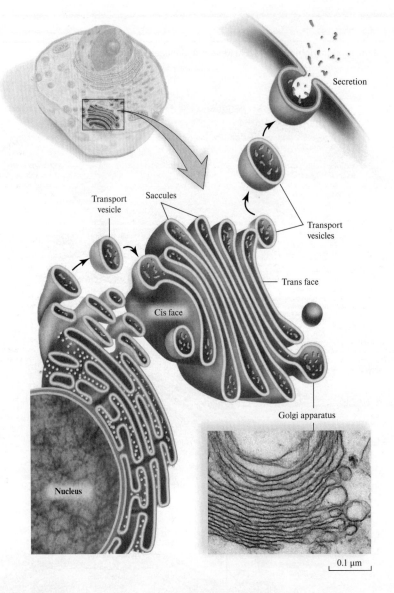

Secretion

Transport vesicle

Saccules

Transport vesicles

Trans face

Cis face

Golgi apparatus

Nucleus

0.1 μm

Vacuoles and **vesicles** are both membrane-bound sacs, but vesicles are smaller than vacuoles. Food vacuoles are formed by a process called **phagocytosis,** in which a cell engulfs a smaller organism or food particles. The food vacuole fuses to a lysosome, and the enzymes digest the organism or food. **Contractile vacuoles** are present in many freshwater protists and pump excess water from the cell. Vesicles form from the Golgi apparatus and travel to the cell membrane where they deposit their content into the extracellular fluid by a process known as exocytosis.

Mitochondria, illustrated in Figure 3–17a, convert energy to a form that the cell can use for work. Within the mitochondria, the energy stored in the form of different macromolecules is transported to a form that can be used by the cell (ATP). Cells that use large amounts of energy (e.g., liver cells) have high concentrations of mitochondria. These organelles contain their own ribosomes and a loop of DNA. Mitochondria self-generate by dividing in the middle to produce two daughter mitochondria, in much the same way that prokaryotic bacteria divide.

The **cytoskeleton** extends from the nucleus to the cell membrane, throughout the cytoplasm, where it organizes the location of the organelles, gives shape to the cell, and allows for movement of parts of the cell. It can be quickly dismantled in one part of the cell and reassembled in another to allow the cell to change shape. The cytoskeleton interacts with proteins called motor

FIGURE 3–16

Structure of the lysosome.
(©Daniel S. Friend)

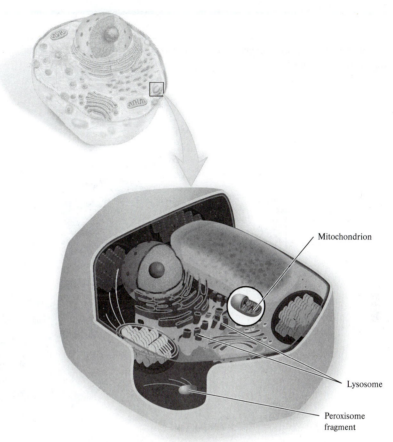

Mitochondrion

Lysosome

Peroxisome
fragment

Mitochondrion and a peroxisome in a lysosome

molecules to bring about the movement of the cilia and flagella, allowing the cell to move. The cytoskeleton also directs the plasma membrane to construct food vacuoles during phagocytosis. The cytoskeleton also appears to help regulate cell function by transmitting mechanical signals to other organelles within the cell. The cytoskeleton is made up of microtubules, microfilaments (actin filaments), and intermediate filaments. Each has a different function, but all are made of proteins and help achieve the shape and function of the cell.

The **centrosome** assembles and directs cell division. The centrosome lacks a membrane. It contains a pair of centrioles, shown in Figure 3–18, which are made of microtubules arranged in a ring. Centrioles also may be involved in the formation of cilia and flagella.

The cilia and flagella are appendages that protrude from the cell and enable locomotion, causing fluid to move over the surface of the cell. Cilia are short cylindrical projections that move in a wavelike motion. They are about 0.25 μm in diameter and about 2–20 μm in length. Flagella, which are about the same diameter as cilia, are about 10–200 μm in length. Flagella move in a rolling, whiplike motion. Both the cilium and flagellum are anchored to the cell by a basal body, which is structurally identical to a centriole.

Cell Organelles of Plant Cells

Plant cells have several unique organelles that are necessary for their survival. These include: the cell wall, the plasmodesmata, central vacuole, tonoplast, and chloroplasts. Other organelles are common to both plant and animal cells, and include the nucleus, mitochondrion, Golgi apparatus, and endoplasmic reticulum. Lysosomes and centrioles are found only in animal cells.

FIGURE 3–17

Structure of the energy-related organelles: **(a)** mitochondrion and **(b)** chloroplast. (a: right - ©EM Research Services, Newcastle University; b: top - ©Dr. Jeremy Burgess/Science Source)

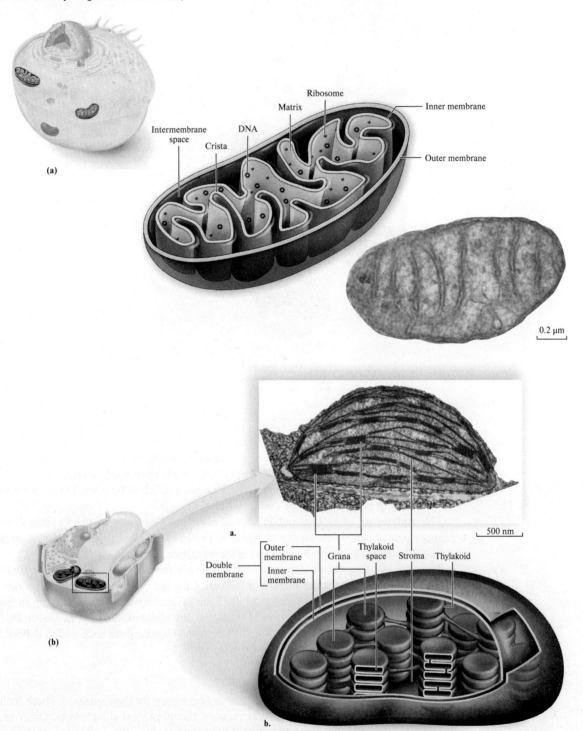

FIGURE 3–18

FIGURE 3–18

Centrioles in the cytoskeleton are composed of microtubules.

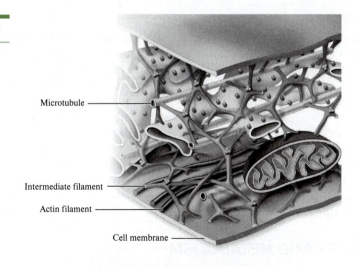

Microtubule

Intermediate filament

Actin filament

Cell membrane

Actin filaments

Microtubules

Intermediate filament

The **cell wall** gives a plant cell its shape, strength, and rigidity, it provides protection, and it prevents the excessive uptake of water by the plant. The cell wall, which is composed mostly of cellulose fibers embedded in a matrix of proteins and polysaccharides, is thicker than the plasma membrane. The **plasmodesmata** are channels through the cell walls that connect the cytoplasm of adjacent cells.

A **central vacuole** is usually present in mature plant cells, while younger cells contain many smaller vacuoles. The organelle provides storage of food, wastes, and various ions. The vacuole is also involved in the breakdown of waste products and in plant growth. The **tonoplast** is the membrane that encloses the central vacuole.

Plastids are found in the cytoplasm of plant cells and photosynthetic protists. All plastids contain stacked internal membrane sacs, which are enclosed in a double membrane. Plastids are able to photosynthesize and store starches, lipids, and proteins. These organelles contain their own DNA and ribosomes. One type of plastid is the chloroplast, illustrated in Figure 3–17b. Chloroplasts give a plant its green color and transfer energy from sunlight into stored energy, in the form of carbohydrates, during photosynthesis. The number of plastids in a cell varies with environmental conditions and plant species.

Cell Organelles of Prokaryotes

As mentioned earlier, prokaryotic cells lack a true nucleus and the other membrane-bound organelles found in eukaryotic cells. The prokaryotic cell is simpler and smaller than eukaryotic cells, often the size of mitochondria, as shown in Figure 3–11.

The **nucleoid** contains a single loop of double-stranded DNA. Some prokaryotes contain **plasmids,** which are small, circular, self-replicating DNA molecules that are separate from the chromosome. A plasmid has only a small number of genes, none of which are required for normal cell survival and reproduction. Plasmids are generally responsible for survival under extreme conditions and for resistance to antibiotics and some toxic chemicals.

All prokaryotes have a cell wall. Some prokaryotic cells have a slime layer called a **capsule,** which surrounds the cell wall. Prokaryotes may also have **flagella,** which rotate like propellers. Others may have **pili,** which are hollow appendages that allow cells to stick to one another or to other surfaces.

3–4 ENERGY AND METABOLISM

All forms of life need energy to function and survive. You needed energy to brush your teeth this morning, to walk to class later in the day, to pick up the fork to eat your dinner, and to get yourself into bed this evening. The energy you expend comes from the food you eat. The same is true for the trees you passed on your way to class, the flowers planted in the gardens on campus, or the algae that float on the surface of a pond. The energy expended in functioning and survival by the plant comes from sunlight.

Cells, Matter, and Energy

Life as we know it would not be possible were it not for the flow of energy into organisms. The sun is the primary source of this energy since all biological life is dependent on the green plants that use sunlight as a source of energy.

Photosynthesis.   The process by which some organisms (namely chlorophyll-containing plants) are able to convert energy from sunlight into chemical energy (in the form of sugars) is called **photosynthesis** and can be represented by the simple equation:

$$6\ CO_2 + 6\ H_2O + 2800\ kJ\ \text{energy from sun} \xrightarrow{\text{chlorophyll}} C_6H_{12}O_6 + 6\ O_2 \qquad (3\text{–}7)$$

Photosynthesis is a very complex process involving over one hundred different chemical reactions. There are two stages, the "photo" stage and the "synthesis" stage. During the photo stage, chemical reactions, driven by light, occur. These chemicals drive the synthesis reactions, in which chemical energy is stored in the form of chemical bonds of the glucose molecule.

All plants contain pigments, which absorb light of specific wavelengths. Most plant leaves contain chlorophyll, which absorbs light in the 450–475 nm and 650–675 nm regions. Chlorophyll also converts the absorbed energy to a form that can drive the synthesis reactions. In order for both light absorption and energy conversion to occur, the chlorophyll must be contained in the chloroplast.

The chloroplast is typically small. In fact, about five thousand chloroplasts could be lined up along a 1-cm row. Yet, a single chloroplast can perform hundreds to thousands of reactions in a single second. Chloroplasts have both an inner and outer membrane. The membranes surround an interior space filled with stroma, a protein-rich semiliquid material. In the stroma are the thylakoids, interconnected membrane-bound sacs. These sacs stack on top of one another to form grana. Photosynthesis occurs in the stroma and thylakoid membrane, which contains the pigments that absorb light and the chemicals involved in the chain of electron transport reactions. The high surface area of the thylakoid membranes greatly increases the efficiency of the photosynthetic reactions. The synthesis reactions occur in the stroma.

Photosynthesis occurs in three stages:

1. Light energy is captured by the pigments.
2. ATP and NADPH are synthesized.
3. Carbon is fixed into carbohydrates by reactions referred to as the Calvin cycle.

In the first stage, the pigments embedded in the thylakoid membranes absorb light and, through the second-stage light reactions, transfer their energy to ADP and $NADP^+$, forming ATP and NADPH. Chlorophyll a is the only pigment that is capable of transferring light energy to drive the carbon fixation reactions. Chlorophyll b and the carotenoids transfer their energy to chlorophyll a to help drive the cycle. In the third stage, in the stroma, sugar is produced from carbon dioxide, using ATP for energy and NADPH for reducing power.

Catabolic Pathways. Energy is stored by organic chemicals in the bonds between atoms. Complex organic chemicals, which are rich in potential energy, are degraded within a cell to smaller compounds that have less energy. These reactions are mediated by enzymes. Some of the energy released by these reactions can be used to do work; the rest is dissipated by heat. The two common catabolic, energy-yielding pathways are: cellular respiration and fermentation.

Cellular respiration is a three-step process involving glycolysis, the Krebs cycle [also known as the citric acid cycle and the tricarboxylic acid (TCA) cycle], and the electron transport chain and oxidative phosphorylation as shown in Figure 3–19. The first two stages are

FIGURE 3–19

Cellular respiration showing the overall yield of energy per molecule of glucose.

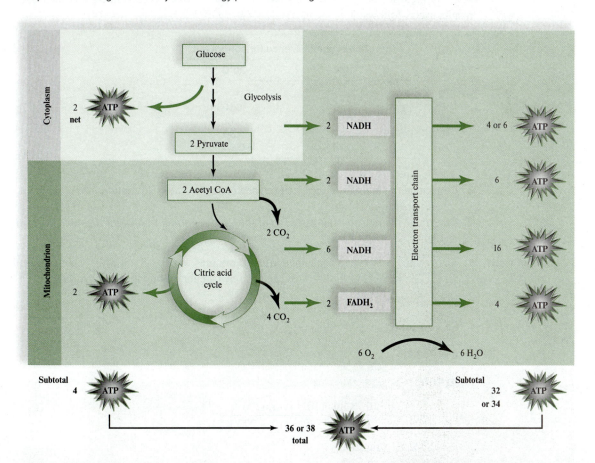

catabolic processes that involve the decomposition of glucose and other organic chemicals. In eukaryotic cells, glycolysis occurs in the cytosol and the Krebs cycle occurs in the mitochondrion. During glycolysis, a molecule of glucose is broken down to two molecules of pyruvate. In addition, glycolysis yields 2 ATP and 2 NADH. During glycolysis and the Krebs cycle, electrons are supplied (via NADH) to the transport chain, thereby driving oxidative phosphorylation. During the Krebs cycle, CO_2 is released, 1 ATP is formed, and electrons are transferred to 3 NAD^+ and 1 FAD. Oxidative phosphorylation, the process by which ATP is synthesized, is driven by the redox reactions that transfer electrons from the substrate (i.e., food) to oxygen. At the end of the chain of reactions, electrons are passed to oxygen, reducing it to H_2O. The process is highly efficient and as many as 38 ATP molecules can be produced from the conversion of a single molecule of glucose to CO_2.

EXAMPLE 3–2 The rate of respiration can be determined by monitoring the rate of oxygen consumption or carbon dioxide production:

$$C_6H_{12}O_6 + 6O_2 \rightleftharpoons 6CO_2 + 6H_2O \qquad (3\text{–}8)$$

In an experiment, the change in gas volume using a respirometer containing 25 germinating pea seeds was measured at two temperatures (10 and 20°C). The experiment was conducted in the dark so that photosynthesis did not occur. The CO_2 produced during cellular respiration was removed by the reaction of CO_2 with potassium hydroxide (KOH) to form solid potassium carbonate (K_2CO_3). In this way, only the oxygen consumed was measured. The results are as follows:

	Corrected difference for control (mL)	
Time (min)	10°C	20°C
0	—	—
5	0.19	0.11
10	0.31	0.19
15	0.42	0.39
20	0.78	0.93

1. Determine the number of moles of oxygen consumed after 20 minutes. Assume the atmospheric pressure is 1 atm.
2. Determine the rate constant for each of the data assuming first-order kinetics. Explain your results.

Solution
1. The volumes of oxygen consumed after 20 minutes were 0.78 mL at 10°C and 0.93 mL at 20°C. The number of moles of oxygen can be calculate using the ideal gas law:

$$PV = nRT$$

Solving for n:

$$n = PV/RT$$

At 10°C:

$$n = \frac{(1 \text{ atm})(0.78 \text{ mL})(10^{-3} \text{ L} \cdot \text{mL}^{-1})}{(0.082 \text{ atm} \cdot \text{L} \cdot \text{mol}^{-1} \cdot \text{K}^{-1})(10 + 273 \text{ K})}$$

$$= 3.4 \times 10^{-5} \text{ moles } O_2$$

At 20 °C:

$$n = \frac{(1 \text{ atm})(0.93 \text{ mL})(10^{-3} \text{ L} \cdot \text{mL}^{-1})}{(0.082 \text{ atm} \cdot \text{L} \cdot \text{mol}^{-1} \cdot \text{K}^{-1})(20 + 273 \text{ K})}$$

$$= 3.9 \times 10^{-5} \text{ moles O}_2$$

2. Using the same approach as employed in (1), the number of moles of oxygen can be calculated for the two data sets:

Time (min)	10°C Vol (mL)	20°C Vol (mL)	10°C n (moles)	20°C n (moles)	10°C ln (n)	20°C ln (n)
0	—	—				
5	0.19	0.11	8.19×10^{-6}	4.58×10^{-6}	−11.7	−12.3
10	0.31	0.19	1.34×10^{-5}	7.91×10^{-5}	−11.2	−11.7
15	0.42	0.3	1.81×10^{-5}	1.62×10^{-5}	−10.9	−11.0
20	0.78	0.93	3.36×10^{-5}	3.87×10^{-5}	−10.3	−10.2

The natural log of the number of moles can then be plotted vs. time to determine the slopes and the rate constant:

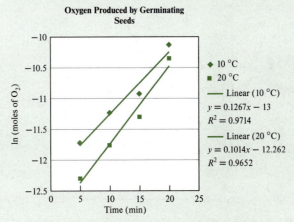

Oxygen Produced by Germinating Seeds

♦ 10 °C
■ 20 °C
— Linear (10 °C)
$y = 0.1267x − 13$
$R^2 = 0.9714$
— Linear (20 °C)
$y = 0.1014x − 12.262$
$R^2 = 0.9652$

Data Source: http://www.scribd.com/doc/7570252/AP-Biology-Lab-Five-Cell-Respiration; http://www.biologyjunction.com/Cell%20Respiration.htm

As shown in the figure the rate constant for the seeds at 10°C was 0.10 min^{-1} and 0.13 min^{-1} at 20°C. The increase is to be expected as the metabolic activity of the seeds would increase with increasing temperature.

Fermentation is the less efficient of the two catabolic pathways. During fermentation, sugars are partially degraded in the absence of oxygen, that is, under anaerobic conditions. Fermentation, an extension of glycolysis (the breakdown of 1 mole of glucose to 2 moles of pyruvate), generates ATP. Under anaerobic conditions, electrons are transferred from NADH to pyruvate or derivatives of pyruvate. The two most common types of fermentation (there are many types) are alcohol fermentation and lactic acid fermentation.

In alcohol fermentation, pyruvate is converted to ethanol in two steps. As shown in Figure 3–20, in the first step, carbon dioxide is eliminated from pyruvate, which is converted

FIGURE 3–20

Yeasts carry out the conversion of pyruvate to ethanol. Muscle cells convert pyruvate into lactate, which is less toxic than ethanol. In each case, the reduction of a molecule of glucose has oxidized NADH back to NAD^+ to allow glycolysis to continue under anaerobic conditions.

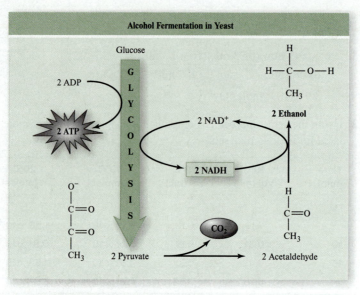

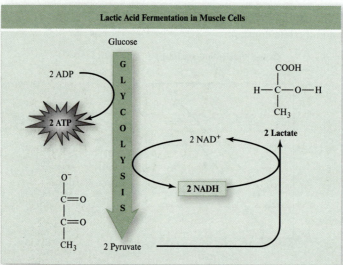

to acetaldehyde ($HCOCH_3$). Acetaldehyde is reduced by NADH to ethanol. Fermentation is important in anaerobic wastewater treatment.

As shown in Figure 3–20, during lactic acid fermentation, pyruvate is reduced to lactate by NADH. Carbon dioxide is not released during this reaction. Lactic acid fermentation, though not widely used in environmental engineering, is important in the production of cheese and yogurt.

EXAMPLE 3–3   Using the two half-reactions given below for the oxidation of acetate to carbon dioxide and the reduction of molecular oxygen to water, determine the amount of energy consumed or released per mole of acetate consumed. Show that the energy production from the oxidation of one mole of glucose is greater than that for acetate.

$$\Delta G°, kJ/e^- \text{ eq}$$

$$1/8 \ CH_3COO^- + 3/8 \ H_2O \rightleftharpoons 1/8 \ CO_2 + 1/8 \ HCO_3^- + H^+ + e^- \qquad -27.40 \qquad (3\text{–}9)$$

$$1/4 \ O_2 + H^+ + e^- \rightleftharpoons 1/2 \ H_2O \qquad -78.72 \qquad (3\text{–}10)$$

Solution When the two half-reactions for the oxidation of acetate are combined, the net reaction is:

$$1/8 \ CH_3COO^- + 1/4 \ O_2 \rightleftharpoons 1/8 \ CO_2 + 1/8 \ HCO_3^- + 1/8 \ H_2O \quad -106.12 \ \Delta G°, kJ/e^- \text{ eq} \qquad (3\text{–}11)$$

As such, the oxidation of acetate to form carbon dioxide and water results in the release of 106.12 kJ of energy per electron transferred. To obtain the energy per mole of acetate oxidized, multiply 106.12 by 8 since the reaction is written for 1/8 mole of acetate. As such, 8(106.12) = 848.96 kJ are released.

The half-reactions for the oxidation of glucose to carbon dioxide and water are:

$$\Delta G°, kJ/e^- \text{ eq}$$

$$1/24 \ C_6H_{12}O_6 + 1/4 \ H_2O \rightleftharpoons 1/4 \ CO_2 + H^+ + e^- \qquad -41.35 \qquad (3\text{–}12)$$

$$1/4 \ O_2 + H^+ + e^- \rightleftharpoons 1/2 \ H_2O \qquad -78.72 \qquad (3\text{–}13)$$

The net reaction is:

$$1/24 \ C_6H_{12}O_6 + 1/4 \ O_2 \rightleftharpoons 1/4 \ CO_2 + 1/4 \ H_2O \qquad -120.07 \qquad (3\text{–}14)$$

Therefore, one mole of glucose yields 2881.68 kJ of energy, significantly more than released from one mole of acetate.

3–5 CELLULAR REPRODUCTION

The ability of an organism to self-replicate is one of the most distinguishing characteristics that separate living things from nonliving matter. Life continues because of this reproduction of cells, or **cell division.**

The Cell Cycle

The cell cycle is made up of two stages: growth and division. In the growth phase, which is called **interphase,** the cell grows and copies its chromosomes in preparation for division. About 90% of the time during the cycle, the cell is in the interphase. During this time, the volume and mass of the cell is increasing as new cellular material is synthesized.

The first part of interphase is referred to as the **G1 phase** or "gap 1." During this time the cell is growing rapidly and metabolic activity is fast. Once the G1 phase is complete, the cell can either enter a rest phase or move to the next stage, called the synthesis or **S phase.** If a cell enters the rest phase, metabolism continues but the cell does not replicate and further progression through the cycle does not occur. During the S phase, DNA is synthesized and chromosomes are replicated. Once the S phase is completed, the cell moves into the second growth phase, called **G2** ("gap 2") **phase.** The cell continues to grow, to produce proteins and cytoplasmic organelles, and to prepare for cell division. During the late interphase, the nucleus is well defined and the nuclear envelope is intact. The chromosomes have duplicated but are still present as loosely packed chromatin fibers. The next stage, called the **mitotic phase,** involves both **mitosis** (division of the nucleus of the cell) and cell division. Cell division involves the division of the cytoplasm of the cell and the formation of two new cells.

FIGURE 3–21

Stages of mitosis in a plant cell.

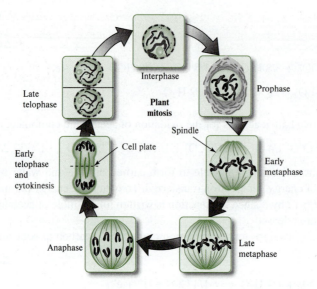

Mitosis is usually divided into five subphases: prophase, prometaphase, metaphase, anaphase, and telophase. As shown in Figure 3–21, during **prophase,** the chromatin fibers become more tightly coiled and discrete chromosomes become visible with a light microscope. Each chromosome is X-shaped, with each half of the X being one copy of the original chromosome. The nucleoli, along with the nuclear membrane, can no longer be seen. Centrioles migrate to opposite ends of the cell. Spindle fibers begin to form between the two centrioles.

The second stage of mitosis is the **prometaphase.** During this phase, the nuclear envelope breaks apart. The microtubules of the spindle move into the nuclear area and intermingle with the chromosomes. Bundles of microtubules extend from each spindle pole toward the center of the cell. A kinetochore forms at the center of each pair of chromatids. Nonkinetochore microtubules network with those from the opposing pole of the cell.

The next phase is the **metaphase,** in which the chromosomes arrange themselves along the metaphase plate, an imaginary plane that divides the cell into two halves. The spindle fibers are attached to the centromere of the replicated chromosomes. A spindle fiber is attached to a chromatid, ensuring that each daughter cell contains the same genetic information.

The fourth phase is called the **anaphase.** During this stage, the centromeres divide and the sister chromatids separate and move to opposite ends of the cell. Each of the sister chromatids is now a chromosome. The poles of the cell move farther apart. By the end of this phase, each half of the cell has a full and identical set of chromosomes.

The last stage is **telophase.** During this stage, the daughter nuclei form at the two poles. Fragments of the nuclear envelope from the parent cell recombine to form a new nuclear envelope. The chromosomes begin to unwind and become less visible. Mitosis is now complete and cytokinesis is well under way.

During cytokinesis, the cytoplasm separates and two new daughter cells form. The spindle fibers disintegrate and disappear. The nucleolus rematerializes. A nuclear membrane forms to encircle each set of chromosomes. Finally, a cell membrane forms. In plant cells, the cell wall develops and two new cells are complete.

Asexual Reproduction

Unlike sexual reproduction, **asexual reproduction** does not involve the formation and fusion of sex cells (gametes). A single individual organism gives rise to offspring, which contain essentially exact copies of the genes from the parent. Each new organism is a clone of the parent. Unicellular organisms, including yeast, reproduce by asexual reproduction.

Prokaryotes, including most bacteria, reproduce by an asexual process called **binary fission.** A single bacterial chromosome carries the genetic material for most bacteria species. The chromosome consists of a circular strand of DNA. Since prokaryotes do not have mitotic spindles, the chromosomes must separate by some other means, although the exact mechanism remains unknown. As the bacterial cell replicates, it continues to grow. By the time replication is complete, the cell is about twice its initial size. Each new cell has a complete genome. The kinetics of binary fission are discussed in Chapter 5.

Some multicellular organisms, including many invertebrates, reproduce asexually. For example the *Hydra,* a relative of the jellyfish, reproduces by **budding.** With budding, the offspring actually grows out of the body of the parent. Other organisms, including sponges, reproduce through the formation and release of a **gemmule** (internal bud), which is a specialized mass of cells that is capable of developing into an offspring. Planarians, a type of carnivorous flatworm, reproduce by fission or **fragmentation,** in which the body of the parent organism can split into distinct pieces, each of which can produce an offspring. Sea stars reproduce by **regeneration,** in which a detached piece of the parent organism is capable of developing into a new individual.

Sexual Reproduction

Sexual reproduction results in the diversity that you observe in your classroom, at the supermarket, and even among your brothers and sisters. With sexual reproduction, the offspring, which originate from two parents, have unique combinations of genes inherited from the parents. Sexual reproduction involves the union of two cells (from different parents) to form a zygote. The zygote contains chromosomes from both parents, but it does not contain twice the number of cells in a somatic (any cell other than a zygote) cell. How can this occur? It is because of a process called **meiosis,** which occurs only in reproductive organs. Meiosis results in the formation of gametes. These reproductive cells (i.e., sperm or eggs) are **haploid** and, as such, contain only one copy of each type of chromosome. In humans, the haploid number is 23, whereas for a dog, n = 39, and for Adder's tongue fern, n = 630! When a haploid sperm cell and a haploid ovum unite during fertilization, the resulting zygote contains the chromosomes from both parents. The zygote contains two sets of chromosomes and is **diploid.** All somatic cells are diploid.

There are three main types of life cycles for organisms that reproduce sexually. Meiosis and fertilization alternate in all of these organisms, although the timing of the two events varies with species.

With animals, including humans, the only haploid cells are the gametes. Meiosis occurs during the formation of gametes, which do not undergo cell division until fertilization. The zygote, which is diploid, divides by mitosis, resulting in the formation of a multicellular organism, with diploid cells.

Most fungi and some algae reproduce by a second type of life cycle. With these organisms, the gametes unite to form a diploid zygote. Before the offspring can develop, meiosis occurs, resulting in a haploid multicellular adult organism. The mature organism produces gametes by mitosis, not by meiosis, as occurs with animals. The zygote is the only diploid cell in these species of organisms.

Plants and some algal species exhibit a complex cycle called **alternation of generations,** which includes both diploid and haploid multicellular stages. The multicellular diploid stage is called the **sporophyte. Spores,** which are haploid, are produced when the sporophyte undergoes meiosis. The spore undergoes mitosis to form a **gametophyte,** which is multicellular and haploid. The gametophyte undergoes mitosis, forming gametes. Two gametes unite during fertilization resulting in a diploid zygote, which develops into a sporophyte generation.

Under conditions unfavorable for asexual reproduction, bacteria can reproduce sexually by a process called **conjugation.** During conjugation, the long tubelike pilus links one bacterium cell to another. One bacterium transfers all or part of its chromosome to the other through this specialized organelle. With the new genetic material, the receiving cell divides by binary fission. This process produces cells with new genetic combinations, and therefore provides additional possibilities that the new cells are better adapted to the changing conditions.

3–6 DIVERSITY OF LIVING THINGS

All plants, animals, and other organisms can be organized into various classifications based on their characteristics. The Greek philosopher Aristotle, who lived in the fourth century BCE, first classified living organisms into two large groups, Plantae and Animalia. He coined the term **kingdom** to describe each group—a term we still use some 2300 years later. Carl von Linné (or Linnaeus) refined the classification in his treatise, *Systema Naturae,* but still he maintained that there were only two kingdoms—the same ones identified by Aristotle. Linnaeus placed great importance on the role of sexual reproduction and classified plants according to the number and arrangement of the reproductive organs. Until the 1960s, biology textbooks referred to only these two kingdoms. The Animalia kingdom included protozoa, and bacteria were placed in the Plantae kingdom. In 1957, Professor Robert J. Whittaker of Cornell University argued that the two-kingdom classification was insufficient. By 1969, the five kingdom system he proposed was widely accepted by the scientific community. The kingdoms are: Animalia, Plantae, Fungi, Protista, and Monera. Eukaryotes are in the first four kingdoms. Prokaryotes were given their own kingdom—Monera.

Whittaker divided the multicellular organisms into kingdoms in part based on their nutritional status. Organisms that are capable of photosynthesizing their own food (autotrophic) were placed in the Plantae kingdom. Those that use organic chemicals as a source of carbon (heterotrophic) and are absorptive were placed in the Fungi kingdom. Most fungi are decomposers that reside on the food they consume; they secrete enzymes capable of digesting organic chemicals and absorb the predigested material. Animals are most commonly heterotrophic and digest their food within specialized organs, and were placed in the Animalia kingdom.

Into the fifth kingdom, Protista, were placed any remaining organisms that did not fit into the other four kingdoms. Most protists are unicellular, although some are multicellular. All protists are aerobic and have mitochondria to perform cellular respiration. Some have chloroplasts and are capable of photosynthesis. Protozoa, slime molds, and algae were placed in this kingdom.

As genetic research and molecular biology advanced, it became increasingly clear that the five-kingdom classification scheme was insufficient. Using molecular techniques, microbiologists learned that there are two groups of bacteria that are genetically and metabolically very different from one another and therefore do not belong in the same kingdom. This and other problems with the five-kingdom classification system gave rise to the three-domain classification system. In the three-domain system, as shown in Figure 3–22, all organisms can be placed in one of three groups or superkingdoms: Bacteria, Archaea, or Eukarya (or Eukaryota). Archaea are said to be "living fossils" from the planet's very early ages, before the Earth's atmosphere contained oxygen. The domain Bacteria includes the true bacteria, including cyanobacteria and enterobacteria. The third domain, Eukarya, includes eukaryotic organisms with a true nucleus, and a wide range of organisms, from protists to humans. Debate continues as to how the domains are related, how many kingdoms or groups are within each of the domains, and how evolutionary history fits into this scheme. Much more research is necessary before there is consensus within the biological community.

3–7 BACTERIA AND ARCHAEA

Environmental engineering and science would not be the fields we know without bacteria and archaea; in fact the world would be very different without these organisms. The collective mass of all bacteria and archaea exceeds that of all eukaryotes by at least an order of magnitude. More bacteria and archaea can be found living in a spadeful of soil than the total number of people who have ever lived. Activated sludge tanks and trickling filters that clean wastewater (see Chapter 11) would not work without these organisms. The carbon, nitrogen, sulfur, and phosphorus cycles would come to a grinding halt without bacteria and archaea. As such, it is imperative that we dedicate a section of this chapter to these important organisms.

FIGURE 3–22

The three domains.

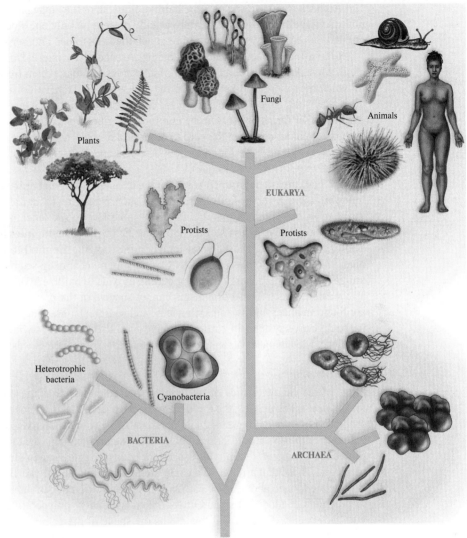

Common ancestor

Archaea

In the traditional five-kingdom classification system, single-celled microorganisms (formally called prokaryotes) are placed in the Monera kingdom. In the three-domain system, these organisms are grouped either as Bacteria or Archaea. As mentioned earlier, Archaea are essentially living fossils, formed within a billion years after the Earth was formed. Archaea are not bacteria (hence the name Archaebacteria is no longer used); their genes differ significantly from those of bacteria. However, both Archaea and Bacteria have one major circular chromosome, although both may have smaller rings of DNA called **plasmids,** which contain only a few genes. The cell membrane of Archaeans are not made of the same lipids as found in other organisms; instead, their membranes are formed from isoprene chains. Archaea are very small, usually less than 1 μm long. Like true bacteria, Archaea come in many shapes. They may be spherical, a form known as **coccus,** and these may be perfectly round or lobed. Others are rod shaped, known as **bacillus,** and may be short bar-shaped rods to long and whiplike. Some unusual Archaea are triangular-shaped or even rectangular like a postage stamp.

Archaea may have one or more hairlike flagella, or they may lack flagella altogether. When multiple flagella are present, they are usually present on only one side of the cell. Archaea may also exude proteins, which are used to attach the cells to one another to form large clusters.

Like all single-celled microorganisms, Archaea lack internal membranes and a true nucleus. Most Archaea have a cell wall, which serves to protect them from the extreme conditions in which they exist. Unlike bacteria, the cell walls of Archaea do not contain peptidoglycan, nor do they contain cellulose as do plant cells. The cell walls of Archaea are chemically distinct from all other forms of life. On the contrary, the ribosomes of Archaea are much more similar to those of eukaryotes than they are to other prokaryotes. Methionine, the amino acid that initiates protein synthesis, is found in both Archaea and Eukarya but is absent in Bacteria (being replaced with formyl-methionine).

Archaea are often grouped based on the extreme environment in which they live. **Extreme halophiles** live in very saline environments, such as the Great Salt Lake and the Dead Sea. They thrive in saline solutions that are 15–20%, some five to six times more saline than seawater. They cannot survive in less-saline conditions. **Extreme thermophiles** live in very hot, and often acidic, environments. Most live at temperatures ranging from 60 to 80°C, although a sulfur-metabolizing species lives in the 105°C waters found near deep-sea hydrothermal vents. **Methanogens** obtain their energy by using carbon dioxide to oxidize hydrogen. Methane is produced as a waste product. These organisms cannot survive in the presence of oxygen. Methanogens are found in anaerobic swamps, marshes, and sediments. Methanogens play an important role in the digestive systems of cattle, termites, and other herbivores whose diet consists mainly of cellulose. Methanogens also play an important role in municipal wastewater treatment.

Bacteria

Some of the characteristics of bacteria are presented above as these organisms are compared to Archaea. Like Archaea, the three main shapes bacteria take are cocci, bacilli, or spirilli (spiral shaped), as shown in Figure 3–23a. Bacteria also grow in distinctive patterns. For those in pairs, the prefix **diplo-** is used. For example, cocci bacteria that aggregate in pairs are referred to as **diplococci.** The prefix **staphylo-** is used for cells that aggregate in clusters resembling bunches of grapes. Those that arrange in a chain are referred to with the prefix **strepto-.**

As mentioned above, most bacteria have a cell wall made of peptidoglycan, which consists of polymers of modified sugars cross-linked with short polypeptides. The polypeptides vary with the species. The cell wall maintains the shape of the cell, protects it from harsh environments, and prevents the cell from bursting in hypotonic solutions.

The Gram stain (see Figure 3–23b) is a very valuable tool for the identification of bacteria and is, therefore, used by environmental engineers and scientists. The stain is used to classify bacteria based on differences in their cell walls. **Gram-positive** bacteria have simpler cell walls that contain relatively high concentrations of peptidoglycan. In contrast, the cell walls of **gram-negative** bacteria are more complex and have less peptidoglycan than gram-positive bacteria. These organisms also contain an outer membrane on the cell wall that is made of lipopolysaccharides, which are often toxic. The outer layer also protects the organism and makes the pathogenic (disease-causing) organisms more resistant to attack by their hosts. Gram-negative bacteria are also more resistant to antibiotics for the same reason.

Many bacteria secrete a thin layer of polysaccharide (or sometimes proteins), which surrounds the bacterial cell to form a capsule. The capsule allows the organism to attach to a host and provides protection against defensive cells, such as white blood cells. Bacteria can also adhere to one another using **pili,** which are hollow, hairlike structures made of protein. Some bacteria form specialized pili, which are used to attach two bacterium cells, during conjugation, at which time DNA is transferred from one cell to another.

As is the case with Archaea, bacteria can also be motile. The most common motile organelle is the **flagella.** Flagella, if present, are thin and hairlike. There may be a single flagella or there many be multiple flagella distributed over the entire surface of the cell, or they may be

FIGURE 3–23

(a) Characteristic shapes of bacteria. **(b)** Gram staining technique. (a: top and middle - ©Janice Haney Carr/CDC, bottom - ©Don Rubbelke/McGraw-HIll Education; b: bottom - ©HBiophoto Associates/Science Source)

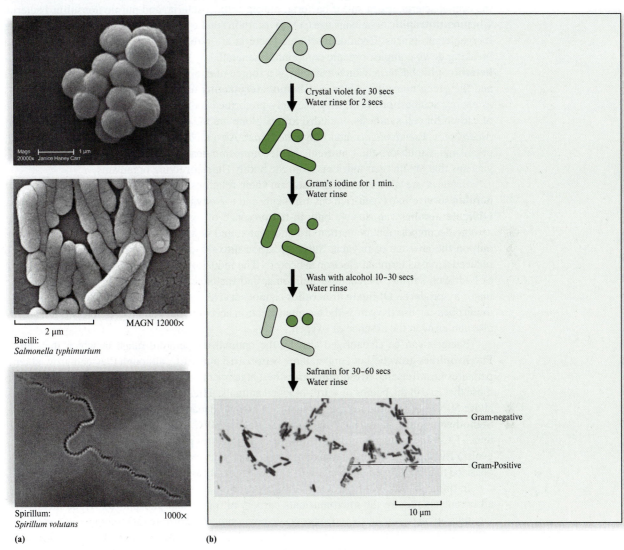

Magn 20000x Janice Haney Carr 1 μm

2 μm   MAGN 12000×

Bacilli:
Salmonella typhimurium

Spirillum: 1000×
Spirillum volutans

Crystal violet for 30 secs
Water rinse for 2 secs

Gram's iodine for 1 min.
Water rinse

Wash with alcohol 10–30 secs
Water rinse

Safranin for 30-60 secs
Water rinse

Gram-negative

Gram-Positive

10 μm

(a)   **(b)**

concentrated at one or both ends of the cell. Some helix-shaped bacteria called spirochetes have two or more helical filaments. Each filament is attached under the outer layer of the cell to a basal apparatus, which acts like a motor. These rotate, making the cell move like a corkscrew. Other bacteria, predominately those that form filamentous chains, have a third type of mechanism for motility. These secrete a slimy thread that anchors the cell to a substratum. The slime allows the organism to glide along other surfaces.

Many bacteria are capable of **taxis,** that is, moving toward or away from a stimulus. Taxis is referred to as positive—toward an object—or negative—away from a substance. Phototaxis is the ability of an organism to respond to light; chemotaxis is the ability to respond to a chemical (e.g., food, oxygen, or a toxin). Barotaxis pertains to pressure and hydrotaxis is related to water.

Bacteria, as is the case with all organisms, can be classified by their nutritional status (i.e., how the organism obtains energy and carbon). The term **trophic** is used to describe the level of nourishment. Trophic levels are discussed in more detail in Chapter 5 (Ecosystems), but its importance warrants some discussion. Organisms that obtain their carbon from inorganic sources such as carbon dioxide (CO_2) or bicarbonate (HCO_3^-) are referred to as **autotrophic.** Those photosynthetic organisms that obtain their carbon from inorganic sources and energy from sunlight are called **photoautotrophic.** Photoautotrophic bacteria are common and include the cyanobacteria, green sulfur bacteria, purple sulfur bacteria, and purple nonsulfur bacteria. **Chemoautotrophic** organisms use CO_2 as a carbon source and obtain their energy by oxidizing inorganic substances. Chemoautotrophic bacteria are rare. **Heterotrophs** derive energy from breaking down complex organic compounds generated by other organisms. Among the **photo-heterotrophic** bacteria, which use light as a source of energy, are the purple nonsulfur bacteria and the green nonsulfur bacteria. **Chemoheterotrophs** use inorganic or organic compounds as energy sources; however, they use only preformed reduced organic chemicals as a source of carbon for cell synthesis. Included in this group are the commonly occurring *Acinetobacter, Alcaligenes, Pseudomonas,* and *Flavobacterium.* Also included in this group are the **saprobes,** organisms that absorb their nutrients by decomposing dead organic matter, and **parasites,** organisms that obtain their nutrients from the bodily fluids of living organisms.

Bacteria are also classified based on their relationship to oxygen. Organisms that are **aerobic** survive in oxygen-rich environments and use oxygen as the terminal electron acceptor. **Obligate aerobes** can survive only in the presence of oxygen. The primary end products of aerobic decomposition are carbon dioxide, water, and new cell tissue. **Anaerobes** can survive only in the absence of oxygen. Sulfate, carbon dioxide, and organic compounds that can be reduced serve as terminal electron acceptors. The reduction of sulfate results in the production of hydrogen sulfide, mercaptans, ammonia, and methane. Carbon dioxide and water are the major by-products. **Obligate anaerobes** cannot survive in the presence of oxygen. **Facultative anaerobes** can use oxygen as the terminal electron acceptor, and, under certain conditions, they also can grow in the absence of oxygen.

Bacteria can be categorized also by the optimal temperature range in which they grow. **Psychrophiles** grow best at temperatures between 15 and 20°C, although they can grow at temperatures near freezing. **Mesophiles** grow best at temperatures between 25 and 40°C and include those that live in the bodies of warm-blooded animals. **Thermophiles** grow best at temperatures above 50°C, at temperatures greater than what would denature key proteins in most organisms. **Stenothermophilies** grow best at temperatures above 50°C and cannot grow at temperatures less than 37°C. **Hyperthermophiles** grow best at temperatures greater than 75°C and include the organism *Pyrodictium,* which is found on geothermally heated areas of the seabed. It must be emphasized that the temperature ranges given above are only approximate and somewhat subjective.

In order for bacteria to grow, a terminal electron acceptor must be available. In addition, all organisms require the micronutrients carbon, nitrogen, and phosphorus. Trace metals and vitamins are necessary as is the appropriate environment (moisture, pH, temperature, etc.).

There are five major groups of bacteria: proteobacteria, chlamydias, spirochetes, gram-positive bacteria, and cyanobacteria. The **proteobacteria** are a large and diverse **clade** (group based on common evolutionary history) of gram-negative bacteria. These bacteria can be photoautotrophic, chemoautotrophic, and heterotrophic. There are both anaerobic and aerobic species. Included in this clade are several species of bacteria that are very important to environmental engineers and scientists. For example, *Nitrosomonas,* a common soil bacteria, oxidizes ammonium to nitrite and therefore plays an important role in nitrogen cycling. *Vibrio cholera* is the organism that causes the waterborne disease cholera. *Escherichia coli,* found in the mammalian intestine, is commonly used as an indicator of fecal pollution. The **chlamydias** are parasites that can survive only in the cells of animals. **Spirochetes** are microscopic, helical heterotrophs that are about 0.25 mm in length. Within this clade is *Borrelia burgdorferi,* the organism that causes Lyme disease. **Gram-positive bacteria** are a large diverse clade that includes *Clostridium botulinum,* the organism that causes the potentially deadly disease botulism. The last clade is the **cyanobacteria,** the only prokaryote that is capable of photosynthesis. Cyanobacteria can be solitary or colonial in nature. They are abundant in both fresh and marine waters and form the basis of many aquatic ecosystems.

3–8 PROTISTS

The third domain is Eukarya, which contains all of the eukaryotes that had been previously classified in the four kingdoms of Protista, Fungi, Animalia, and Plantae. The last three have remained essentially intact in the new domain system. However, the boundaries of the Protista kingdom have crumbled in disarray, with some of those formally classified as Protista being reclassified as Fungi, Animalia, or Plantae, and the rest of the organisms being reclassified into one of about 20 new kingdoms. Nevertheless, the term "protist" is still used by biologists to describe the diverse set of eukaryotic organisms that don't fit neatly into the other three kingdoms.

As mentioned earlier, most protists are unicellular, although some are multicellular and colonial. All are eukaryotes. They may be auto- or heterotrophic. Most protists are aerobic and have mitochondria to perform cellular respiration. Some have chloroplasts and are capable of photosynthesis. Others are **mixotrophs,** which are capable of photosynthesis, but also are heterotrophs. Most are motile and have flagella or cilia at some time during their life cycle. Many protists form resistant cells, called **cysts,** which are capable of surviving harsh environmental conditions. Most do not cause disease, although some are parasitic. They are an important group of organisms as they form the bottom of most aquatic food webs, and as such, will be discussed here. The three major classes of protists are protozoa, algae, and slime and water molds.

Protozoa

Protozoa, meaning "first animals," are single-celled, eukaryotic organisms. Most are aerobic chemoheterotrophs that ingest or absorb their food. The 30,000 or so species of protozoa vary greatly in shape, from the microscopic *Paramecium* to the thumb-sized shell-covered marine forms. Many protozoa, especially those that are parasitic, have complex life cycles. In some cases, the different forms taken during the life cycle have caused biologists to mistakenly believe that the same organism was two or more different species. The protozoa are divided into some 18 phyla, including *Zoomastigina, Dinomastigota, Sarcomastigophora, Labyrinthomorpha, Apicomplexa, Microspora, Ascetospora, Myxozoa,* and *Ciliophora.* These phyla represent four major groups: flagellates, amoebae, ciliates, and sporozoa, illustrated in Figure 3–24. We will focus our discussion herein on these four groups, which are based on modes of locomotion.

Flagellates.   The flagellates are protozoans that have one or more flagella, which move in a whiplike motion. They multiply by binary fission and some species can form cysts. A cytosome may be present. The flagella offer these organisms a competitive advantage as they are able to invade their hosts and adapt to a wide range of environmental conditions. A common freshwater species, *Euglena,* is quite unique in that it is mixotrophic. Among the most common pathogenic flagellates that invade the intestinal tract is *Giardia lamblia.* Giardiasis is a waterborne disease spread by the contamination of water with excreted *G. lamblia* cysts. The cysts have a thick cell wall, allowing them to be resistant to disinfection and able to survive outside the body for several weeks under favorable conditions. The main symptoms are abdominal pain, flatulence, and episodic diarrhea. Parasitic flagellates also cause trypanosomiasis (African sleeping sickness), which is transmitted by the tsetse fly.

Sarcodines (Sarcodina).   Explore a local pond and you're likely to find Sarcodines, more commonly known as amoebas. Amoebas resemble blobs of protoplasm. They are among the 40,000 species of unicellular organisms that are known as Sarcodines. Most amoebas move and engulf their prey by producing limblike extensions of cytoplasm called **pseudopods,** meaning false foot. The internal flow of cytoplasm allows the amoeba to move. Pseudopodia can also be used for feeding by phagocytosis. Under unfavorable environmental conditions, amboebas can form cysts. Amoebic dysentery, a waterborne disease common in developing countries, is caused by the amoeba *Entameba histolytica.*

FIGURE 3–24

Examples of protozoa. **(a)** Flagellates: *Giardia lamblia*. **(b)** Sarcodines, amoeba. **(c)** Ciliates: *Paramecium*. **(d)** Sporozoans: *Cryptosporidium* oocysts purified from murine fecal material. (a: Janice Haney Carr/CDC; b: ©Stephen Durr; c: ©Melba Photo Agency/Alamy Stock Photo; d: ©Michael Abbey/Science Source)

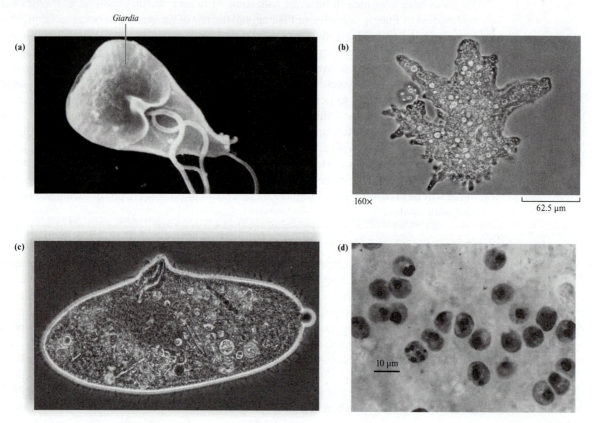

Ciliates. There are more than eight thousand species in the phylum Ciliophora. All move using cilia, which are short, hairlike projections on the cell membrane that beat rhythmically to propel the organism. Many species of ciliates are large and complex and they can grow up to 0.1 mm in length. All reproduce asexually by binary fission and sexually by conjugation. No new cells result from conjugation as only chromosomes are transferred. After conjugation, the cell divides asexually. Ciliates have two kinds of nuclei: a macronucleus and a micronucleus. The macronucleus controls cellular function and asexual reproduction. The micronucleus controls genetic exchange during conjugation. Interestingly, the ciliates are closer evolutionarily to fungi, plants, and animals than to other protozoa. Most live in freshwater—in fact, if you go back to the pond in which you found the amoeba, you are almost certain to find *Paramecium*. *Paramecium* have a protective coating, called a **pellicle,** over their cell membrane. They have hundreds of cilia that beat together to move food particles into their gullet, which leads to a food vacuole. Inside the vacuole, nutrients are extracted. The waste products are excreted through an opening called an **anal pore.**

Sporozoans. Sporozoans, which include those organisms in the phylum Sporozoa, are non-motile. All are parasites and form spores during some point in their complex life cycles. Sporozoans have a number of organelles that help the organism invade their hosts. Among this group is *Cryptosporidium parvum,* a protozoan parasite that infects humans, livestock, pets, and wildlife (including birds, mice, deer, and raccoons). *C. parvum* produces ovoid to spherical oocysts, which contain four sporozoites. Very large numbers (up to 10 million oocysts per gram of feces) of thick-walled oocysts are shed in the feces of an infected animal. When the oocyst is ingested, the sporozoites are released and parasitize the lining of the small intestine. Oocysts are extremely resilient and can survive adverse environmental conditions (near freezing temperatures) for long periods of time and chlorination (at typical doses and retention times). While most individuals can recover from a bout of cryptospiridosis, the disease can kill immunocompromised individuals. It was *C. parvum* that caused more than 100,000 individuals to become ill from an outbreak in Milwaukee, Wisconsin, in 1993, with almost one hundred deaths. Malaria and toxoplamosis are also caused by Sporozoans.

Algae

Algae are photoautotrophic protists that contain **pyrenoids**, organelles that synthesize and store starch. Almost all species are aquatic. Algae may be unicellular, colonial, filamentous, or multicellular. Algae are classified into six phyla based on the type of chloroplasts and pigments they contain, their color (which is related to their pigments), food storage, and the composition of their cell wall. Examples of algae from the four groups are shown in Figure 3–25.

Green Algae.   Green algae contain the same types of chlorophyll and are the same color as most true plants. Like true plants, their cell walls contain cellulose and they store food as starch. Green algae are aquatic and found in damp locations on land. Many unicellular algae, such as *Chlamydomonas,* have flagella. This organism also experiences phototaxis, driven toward light by a red-pigmented eye-spot, a light-sensitive organelle.

Brown Algae.   Brown algae thrive in cool, marine environments. Most are multicellular and are commonly referred to as seaweed. They have cell walls made of cellulose and alginic acid. They have rootlike structures called **holdfasts** that anchor the plants to the rocky seabed and help to prevent them from being washed out to sea. Brown algae have large, flat fronds that are strong enough to bear the constant battering of the waves. On the fronds are air bladders that allow the fronds to float near the surface of the water, where they are able to absorb light and photosynthesize. The species *Macrocystis* (giant kelp) can grow to a length of 100 m.

Red Algae.   Red algae are commonly found in warm seawater. They are smaller and more delicate than brown algae and grow to greater depths than most other algae. Many are branched, with feathery or ribbonlike fronds. Red algae are found on coral reefs.

Diatoms.   Diatoms are the most abundant unicellular algae in the oceans. There are also many freshwater species, and they are a major food supply at the base of the food web. Most diatoms are photosynthetic, although a few are heterotrophic. They have neither cilia nor flagella. Diatoms reproduce both sexually and asexually.

   Diatoms have a rigid cell wall with an outer layer of silica. The remains of these organisms form what is called diatomaceous earth, which is used in the production of detergents, paint removers, fertilizers, and some types of toothpaste. Diatomaceous earth is also used as media in swimming pool filters.

Dinoflagellates.   Most dinoflagellates are unicellular, photosynthetic, and marine. A few are heterotrophic. Like diatoms, they make up a significant portion of the ocean's organic matter. Dinoflagellates have protective cellulose coatings that resemble armor. They are identifiable by their two flagella. One flagellum moves in a spinning motion and propels the organism forward. The second acts as a rudder. Most dinoflagellates are free living or symbiotic, living in the bodies of such invertebrates as sea anemones, mollusks, and coral. In return for protection, they provide their hosts with carbohydrates that they produce through photosynthesis. Some are bioluminescent.

FIGURE 3–25

Examples of algae. (Top middle - ©M.I. Walker/Science Source, top right - ©PhotoLink/Getty Images; middle right - ©Andrew Syred/ Science Source)

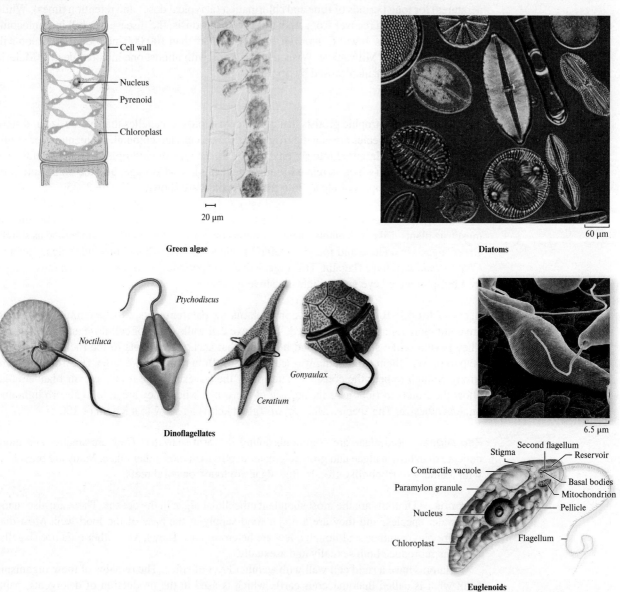

Under some conditions, dinoflagellates reproduce rapidly, causing algae blooms, often referred to as "red tide." The dense populations can produce toxins in sufficient concentrations to kill fish and poison people who consume shellfish that have fed on the algae. Blooms can appear greenish, brown, or reddish orange, depending upon the type of organism, the particular water, and the concentration of the organisms. The term "red tide" is actually a misnomer and "harmful algae blooms" is the preferred term. The extent of the problem is apparent from Figure 3–26, which shows the degree to which the problem has increased in the period of time from 1970 to 2005. Paralytic Shellfish Poisoning (PSP), one of several types of syndromes resulting from the ingestion of algal toxins, is life threatening. The effects are neurological and

FIGURE 3–26

Global distribution of Paralytic Shellfish Poisoning (PSP) toxins in 1970 and 2015. (*Source:* Woods Hole Oceanographic Institute.)

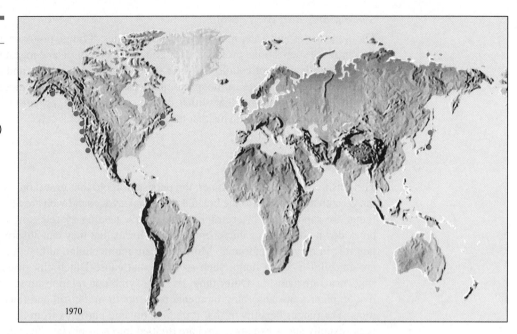

1970

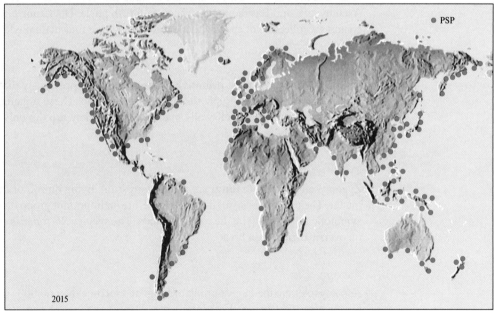

2015

their onset is rapid. With the most severe cases, respiratory arrest can occur. With medical treatment the patient usually fully recovers. Much research needs to be done to better understand which species release toxins and how to prevent such blooms.

Euglenoids.   Euglenoids are small, unicellular, freshwater organisms with two flagella. Usually one flagellum is much shorter than the other. The organisms do not have a rigid cell wall, but, like the *Paramecium,* it is covered with a flexible protein coating called a **pellicle.** Biologists have struggled with classifying this organism since some species have chloroplasts and are photosynthetic, while many are heterotrophs. To complicate things, those with chloroplasts, when raised in the absence of light, will become heterotrophic.

Slime Molds and Water Molds

The slime molds and water molds have characteristics of fungi, protozoa, and plants. Like fungi, they produce spores. Like protozoa, they glide for locomotion and ingest food. Like plants, they have cell walls made of cellulose. Water molds include white rusts and downy mildews. Most live on dead organic matter, although some species are parasitic. Plasmodial slime molds appear as tiny sluglike organisms that slither over moist, decaying, organic material. Cellular slime molds are unicellular organisms that feed on bacteria or yeast cells.

3–9 FUNGI

The mushrooms you ate for dinner, the persistent mold that grows on your shower curtain, and the yeast that was added to the bread used to make your sandwich are all members of this group. Fungi are saprophytic heterotrophic eukaryotes, feeding by releasing digestive enzymes that break down complex organic chemicals into forms that they can absorb. Fungi play an important role in recycling nutrients. Most fungi are multicellular, although a few, including yeasts, are unicellular. Some fungi, such as those that cause Dutch elm disease, athlete's foot, and ringworm, are parasitic. Other fungi live in a symbiotic relationship. For example, some fungi live on plant roots, absorbing inorganic nutrients from the soil and releasing them to the plant roots. The fungus benefits in that it obtains organic nutrients from the plant. Fungi reproduce both sexually and asexually. Fungi are divided into four phyla: Chytridiomycota, Zygomycota, Ascomycota, and Basidiomycota. There is a fifth phyla, Deuteromycota, imperfect fungi, which contains a collection of species that do not fit into the four main phyla.

Chytridiomycota

The **chytrids** are the most primitive of fungi and were previously thought to be Protists. Most are aquatic. Some are saprophytic; others are parasitic. These organisms obtain their nutrients by absorption and have cell walls made of chitin. They are the only fungi with a flagellated stage, the zoospore.

Zygomycota

Zygomycetes or **zygote fungi** are mostly terrestrial, living on soil or decaying plant and animal matter. The mycorrhizae, mentioned above, are an important group of zygote fungi that live in a symbiotic relationship on the roots of plants. The species *Rhizopus* is commonly found growing on overripe fruit or on bread.

Ascomycota

Ascomycotes are the largest group of fungi and include over 60,000 species that inhabit marine, freshwater, and terrestrial environments. They range in size and complexity from unicellular yeasts to intricate cup fungi and morels. Many are saprobes, although some are plant parasites. All species have fingerlike sacs, called **asci,** which contain sexual spores. Ascomycotes reproduce asexually through the production of very large numbers of asexual spores, which can be dispersed by the wind. The spores, called **conidia,** form on the tips of specialized hyphae called **conidiophores**, as shown in Figure 3–27a. Yeasts tend to reproduce asexually by budding, shown in Figure 3–27b.

This group of fungi is important to environmental engineers. Since they require only about half as much nitrogen as do bacteria, they are prevalent in nitrogen-deficient wastewater (McKinney, 1962). They are usually found in environments where the pH is low, including biotowers and trickling filters (see Chapter 11). They can cause "plugging or ponding" in these units. Filamentous fungi can be a cause of bulking of wastewater sludge, resulting in an increase in polymer consumption and making it more difficult to dewater the sludge (see Chapter 11).

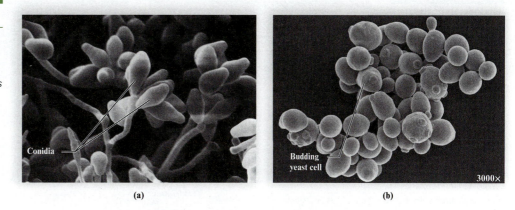

FIGURE 3–27

Asexual reproduction in sac fungi. **(a)** This scanning electron micrograph of *Aspergillus* shows conidia, the spheres at the end of hyphae. **(b)** Scanning electron micrograph showing yeast reproducing by budding. (a: ©Janice Haney Carr/CDC; b: ©Science Photo Library RF/Getty Images)

Recent research has revealed that *Fusarium solani* is capable of both nitrification and denitrification, which makes it an ideal organism for biological wastewater treatment (Guest and Smith, 2002).

Basidiomycota

Included in the phylum Basiodiomycota are the mushrooms that grow on lawns, the bracket fungus that you can find on dead tree branches, and the puffballs found on woodland forest floors. Basidiomycetes are very proficient decomposers of wood and plant material. All of the organisms in this phylum have short-lived, elaborate, sexual reproductive structures (fruiting bodies) called **basidiocarps.** These fruiting bodies produce sexual spores.

The organism *Phanerochaete chyrsosporium* (white rot fungus) has been extensively studied by environmental engineers because of its ability to degrade polycyclic aromatic hydrocarbons (Zhongming and Obbard, 2002), olive oil mill wastewater (Yeşilada, Şik, and Şam, 1999), lignin (Aust, 1995), and other complex organic chemicals.

Deuteromycota

Classification schemes are never perfect, and this one contains the misfits, those species that do not fit into the other four phyla. Deuteromycotes, the imperfect fungi, have no known sexual stages. They reproduce exclusively asexually by producing spores.

One species within this grouping is *Aspergillus fumigatus*. The fungus is a saprophytic aerobe, although it can be parasitic. *Aspergillus fumigatus* is commonly found in compost piles as it is an integral part of the breakdown of compostable materials to a stabilized, finished product. This species is also the most pathogenic of the *Aspergillus* species to humans. It can colonize the air spaces in sinuses, bronchi, or lungs, resulting in acute bronchopulmonary aspergillosis.

3–10 VIRUSES

Viruses are not living organisms; they do not have cytoplasm, organelles, or cell membranes. They do not respire or carry out other life processes. So what are they? Viruses are infectious particles consisting of nucleic acid enclosed in a protein coat, called a **capsid.** The genome of viruses is made up of double-stranded DNA, single-stranded DNA, double-stranded RNA, or single-stranded RNA. Thus, viruses are classified as either DNA viruses or RNA viruses. The smallest virus has only four genes, the largest has several hundred.

Capsids may be helical, polyhedral, or more complex in shape, as illustrated in Figure 3–28. They are made of numerous protein subunits called capsomeres.

FIGURE 3–28

Shapes of viruses.
(a) Adenovirus: DNA virus with a polyhedral capsid and a fiber at each corner. **(b)** T-even bacteriophage: DNA virus with a polyhedral head and a helical tail. **(c)** Tobacco mosaic virus: RNA virus with a helical capsid. **(d)** Influenza virus: RNA virus with a helical capsid surrounded by an envelope with spikes. (Top photos - a: ©Science Photo Library RF/Getty Images; b: ©Eye of Science/ Science Source; c: ©Omikron/Science Source; d: ©Dr. F. A. Murphy/Centers for Disease Control)

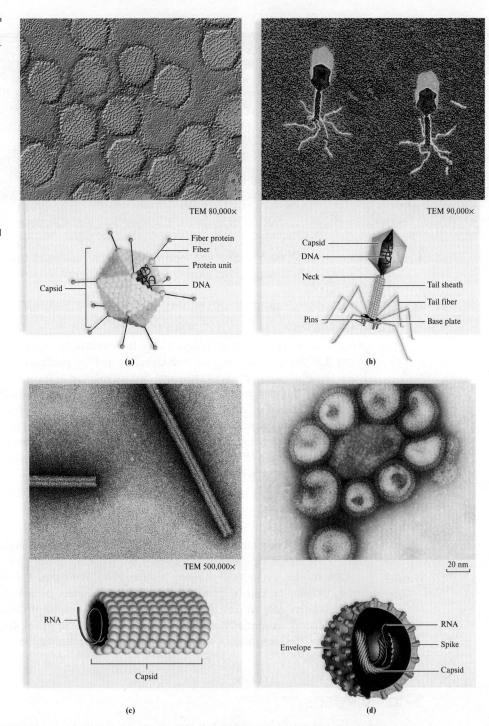

Some viruses have a membranous envelope, called a **viral envelope,** covering the capsid. The membrane contains proteins and phospholipids derived from the host cell, along with similar compounds of viral origin. These envelopes assist the virus to infect its host.

Viruses are obligate intracellular parasites. An isolated virus has no capability to do anything, until it infects a host cell. Each type of virus is limited in the type of host cells it can infect. Some viruses can infect organisms of several species. Others, such as some bacteriophage (a virus that infects bacteria), can infect only the organism *E. coli.*

3–11 MICROBIAL DISEASE

It was not until the late 1800s that western scientists[1] linked pathogenic microorganisms to disease. The first western scientist to make the connection between certain diseases and specific bacteria was Robert Koch, a German physician. He established what are now called **Koch's postulates,** which still guide medical microbiologists. He stated that to establish a specific pathogen as the causal agent of disease, one must (1) identify the same pathogen in each diseased individual or organism tested, (2) isolate the pathogen from the infected individual and grow the organism in a pure culture, (3) induce the disease in an organism using the cultured pathogen, and (4) isolate the same pathogen from the experimental infected organism after the disease had developed. This method works well for most diseases, but does have its limitations. For example, the spirochete *Treponema pallidum* is known to cause syphilis, yet it has never been cultured on artificial media.

Not all microorganisms are pathogenic. In fact, the presence of microorganisms in your large intestine is necessary for the processing of food and the release of vitamins. Microbes that live on your skin provide protection from harmful microorganisms that could otherwise colonize your skin. In order for an organism to be pathogenic, it must be able to invade a host for long enough to cause a detrimental effect. About half of all human diseases is caused by pathogenic prokaryotes.

Some pathogens are **opportunistic,** that is, they do not cause disease unless the defense mechanisms of the host are weakened by such factors as stress, fatigue, poor nutrition, or other diseases. For example, the organism *Streptococcus pneumoniae* is present in the throats of most healthy individuals, but does not cause disease unless the host's defense mechanisms are suppressed. Other examples of opportunistic pathogens are *Legionella pneumophila, Mycobacterium avium,* and *Pseudomonas aeruginosa,* which persist and grow in household plumbing. Individuals with preexisting conditions such as respiratory disease or are immunocompromised are especially susceptible. Exposure is through inhalation, including in showers and from nebulizers (for medication delivery). Although a definitive link has not been found, as shown in Figure 3–29, there was a very significant increase in the number of Legionellosis cases in Flint during the time the city was treating water from the Flint River. (See Case Study in Chapter 10 for more details on the Flint Water Crisis.)

Diseases are referred to as **communicable** or **contagious,** meaning that they are transmitted from one individual to another. Some diseases can cross species, whereas others only affect a single species. Pathogens can also be **virulent,** which means that the microorganisms have a strong ability to cause disease; they are highly pathogenic and effective at overcoming the defense mechanisms of the host. Rabies is caused by a virulent virus; however, the disease is not highly contagious. The virus is considered virulent because exposure to the virus is most certain to cause infection, resulting in the rapid onset of symptoms, including rapid deterioration of the brain and, ultimately, death, if left untreated. However, the disease is not highly contagious because contact with an infected person is not likely to cause disease unless the virus is introduced directly into the blood stream through a bite or open wound. Cholera is an example of a disease that is both highly contagious and virulent. It is responsible for some 200,000 cases of infection worldwide and some five thousand deaths per year.

In some cases, the pathogen causes disease not by invasion of tissue, but by the production and release of toxins. **Exotoxins** are proteins that are emitted by prokaryotes. Disease can result, even in the absence of viable organisms, as long as the toxin is present. For example, botulism is actually caused by an exotoxin produced by the bacterium *Clostridium botulinum,* which is capable of growing on improperly preserved or stored food. The exotoxin is potentially fatal.

---

[1]Medieval medicine was much more advanced in the East as compared to that in the West. In fact, Iranian physician Abu Bakr Muhammad ibn Zakariya' al-Razi (865–925 CE) recognized the connection between parasites and infection. He revolutionized the treatment of smallpox and measles and introduced the use of surgical disinfectants. In fact, his treatises were translated into Latin and became the basis of medical education throughout the West.

FIGURE 3–29

Number of Legionellosis cases in Genesee County, Michigan from January 2010 through December 2015. The City of Flint treated and distributed water from the Flint River from April 2014 through October 2015. Prior to and after that treated water was supplied by the Detroit Water and Sewer Department from Lake Huron. (*Data Source:* https://www.michigan.gov/documents/mdhhs/Updated_5-15_to_11-15_Legionellosis_Analysis_Summary_511707_7.pdf)

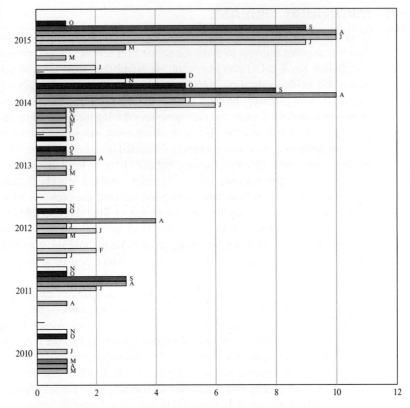

Note: If onset date was not available, referral date was used (i.e., date that the case was referred to public health).
Cases in 2014 include one non-Genesee County resident associated with the 2014 outbreak.

Endotoxins are present in the outer membranes of certain gram-negative bacteria. All members of the genus *Salmonella* produce endotoxins, and *Salmonella*-induced food poisoning is actually not a result of infection but from the release of endotoxins from lysed cells. Cyanobacteria, a group of photosynthetic-planktonic bacteria also referred to as "blue-green algae," produce endotoxins. Microcystin, a hepatotoxin, is one of several toxins produced by this group of organisms. It is thought that low-level exposure to these toxins may cause liver cancer and chronic gastrointestinal disorders. For this reason, the U.S. EPA has placed this compound on the candidate contaminant list, a list of chemicals for which investigations are ongoing to determine if drinking water regulatory limits are necessary.

Infectious diseases can be transmitted by various means. One of the most common is through inhalation of cough or sneeze discharges from another individual. Such infections, including influenza and the common cold, are called **droplet infections.** Other pathogens can enter your body when you ingest contaminated water. These are known as **waterborne infections,** and include typhoid fever and cholera. Some infections, known as **direct contact infections,** are passed from one person to another by direct contact. Many gastrointestinal infections can be spread in this way, as can sexually transmitted diseases. In most cases, the pathogens must enter the body by crossing a mucous membrane. The fourth way infectious diseases can spread is by way of **vectors.** A vector is a disease-carrying animal, often an insect. Bubonic plague (Black Death) spread from China to Europe in 1347 on merchant ships. The disease is caused by a bacterium, *Yersinia pestis,* which was carried on rats and transmitted to humans by flea bites. In this case, the vector is the flea.

The most recently discovered cause of disease is the **prion.** Prions are abnormally folded proteins that can cause normal proteins to take an abnormal shape. Prions enter brain cells and convert the normal cell protein to the prion form. If sufficient numbers of molecules are transformed, the brain of an infected cow does not function properly and the animal succumbs to "mad cow disease" or "transmissible spongiform encephalopathies" (as it is generally known). The animal staggers and appears fearful or crazy. Sheep and goats can be infected with "scrapie," which like mad cow disease affects brain function. The animal becomes uncomfortable and itchy, frantically rubbing against anything it can, and eventually scraping off most of its wool and hair. Scrapie is not a new disease, in fact it was first reported some 250 years ago.

For a long time, scientists did not believe that humans could become infected by prions. However, between 1994 and 1996, 12 people in England succumbed to Creutzfeld-Jakob disease (CJD), a human prion disease with symptoms similar to mad cow disease. All the victims had eaten beef from cattle suspected of having mad cow disease. Autopsies revealed that the brains of 10 of the 12 British patients contained prions similar to those that caused mad cow disease and the individuals had not died of "classical" CJD (Guyer, 1997). Many scientists now believe that the prions originated in sheep suffering from scrapie. Cattle were fed sheep offal, along with their bones and other waste parts of the carcasses. The cattle were exposed to the sheep prions and the prions established themselves in the cattle hosts. People ate flesh from infected cattle, resulting in the infection of humans. Much research is still necessary to determine the exact cause of the disease and ways to prevent transmission.

3–12 MICROBIAL TRANSFORMATIONS

Microorganisms have been used to achieve chemical transformations for thousands of years. Wine, beer, cheese, and yogurt would not be available without microorganisms. In more recent years, microorganisms have been used by sanitary (now called environmental) engineers to produce safe drinking water, treat wastewater, and decontaminate hazardous chemicals.

Water Quality. Engineers and scientists recognized the importance of biological reactions for stream purification as early as 1870, when the Royal Commission on River Pollution in Great Britain concluded that there was no river in England that was long enough to purify its waters of pollutants that entered in the upper reaches. In the winter of 1882–1883, it was noted that when the Schuykill River in Philadelphia, Pennsylvania was covered with ice for long periods of time, anaerobic bacteria took over, resulting in "bad tastes and smells in the river."

The impact of algae on water quality, in terms of malodors, was documented as early as 1917 in a textbook for water supply operators (Folwell, 1917). *Anabaena, Uroglena,* and *Asterionella* were singled out as being among the most common nuisance organisms. Protozoa, spongiana, rotifera, entomostraca (the term formerly used for Crustacea), and mollusks were known to purify reservoir water of organic and mineral impurities.

As mentioned in Chapter 5, microorganisms play significant roles in the cycling of nitrogen, carbon, phosphorus, and sulfur in the environment. The oxidation of organic matter in streams and lakes is described in more detail in Chapter 9, Water Quality Management.

Wastewater Treatment. By 1889, engineers began to use biological wastewater treatment with the development of the trickling filter at the Massachusetts State Board of Health in Lawrence, Massachusetts. By 1908, the first trickling filter was put into operation in Reading, Pennsylvania. In 1917, the first activated sludge plant was put into operation to treat wastewater from the city of Worchester, England (Fuller and McClintock, 1926).

The two major groups of microorganisms involved in biological wastewater treatment are bacteria and eukaryotes (protozoa, crustacean, nematodes, and rotifers). Fungi are rarely present in significant numbers. The primary consumers of the organic material in wastewater are the heterotrophic bacteria, although protozoa can also play an important role. The majority of bacteria present in wastewater are gram-negative and include organisms of the genera *Pseudomonas*, *Arthrobacter, Bacillus, Zoogloea,* and *Nocardia*. The heterotrophic bacteria also make up most of the mass of organisms present. The protozoa, crustaceans, and other organisms are primarily secondary consumers that degrade the byproducts formed by the heterotrophic bacteria along with the dead and lysed bacteria. The role of microorganisms in biological wastewater treatment is discussed in more detail in Chapter 11.

As our understanding of biological processes for wastewater treatment has advanced so has wastewater treatment. Advances in animal waste management have been made and manure digestors are being constructed to use bacterial fermentation of the manure to produce biogas containing methane, carbon dioxide, and small amounts of trace gases, which can be burned to generate electricity. The residual solid material can be used as a soil conditioner and the liquid as a fertilizer.

Drinking Water Treatment.   The role of microorganisms in drinking water treatment is generally thought of more in terms of the protection of human health and the prevention of waterborne diseases. The primary goals of rapid sand filtration and disinfection are the removal and inactivation of pathogens, respectively. By 1917, it had become well accepted that the filtration of drinking water significantly reduced the incidence of typhoid fever. As shown in Figure 10–1, the application of chlorine as a disinfectant resulted in further decreases in the number of deaths due to typhoid fever.

Processes such as slow sand filtration rely on the presence of microorganisms to consume readily biodegradable organics that could otherwise be degraded in the distribution system, resulting in the formation of biofilms on the pipe walls. By the early 1900s, "pipe moss" or "vegetable and animal life" had been observed to grow on the insides of water mains (Folwell, 1917). These filters also effectively remove turbidity and pathogens. By the mid-1800s the practice of using slow sand filtration to treat drinking water had become so well-established in England that in 1852, the Metropolis Water Act was passed, requiring that all water obtained from the River Thames within 5 miles of St. Paul's Cathedral in London had to be filtered before distribution to the public (Huisman and Wood, 1974).

In recent years, there has been a resurgence of interest in the United States in the use of biological treatment to produce potable and palatable drinking water. This has resulted from the observation that the presence of biological instability, that is, electron donors such as biodegradable organic matter, nitrite, ferrous iron, manganese (II), sulfides, and ammonium can foster the growth of bacteria in the distribution system. While these bacteria are seldom pathogenic, they do increase heterotrophic plate counts, result in an increase in turbidity and the formation of taste and odor-causing chemicals, consume dissolved oxygen, and accelerate corrosion. Among the processes that can be used to reduce biological instability are biological activated carbon, ozonation-fluidized bed biological treatment, slow sand filtration, and bank filtration. These processes are discussed in more detail in Chapter 10.

Detoxification of Hazardous Chemicals.   Today, there are more than 70,000 different synthetic organic chemicals on the global market, many of which are now known to be persistent in the environment. However, at the time many of these were first introduced, their fate in the environment was unknown. For example, during World War II there was a lack of animal fats to make soap. As a result, synthetic detergents were developed and manufactured. Their exceptional properties for cleaning made them preferable to soaps. However, their discharge into municipal sewers resulted in excessive foaming in aeration tanks and the receiving bodies of water. Research showed that the cause was the presence of alkyl benzene sulfonate (ABS), which is not readily biodegraded and causes foaming. The replacement of ABS with a linear alkylbenzene sulfonate has solved this problem. Other compounds such as DDT, polychlorinated biphenyls (PCB), halogenated solvents, and the chlorofluorocarbons are also not readily biodegraded and accumulate in the environment. Many of these compounds contain halogens, predominately chlorine, fluorine, and bromine atoms.

Other compounds, such as the aromatic components of gasoline, degrade biologically, albeit slowly and under specific conditions. In 1972, microorganisms were first employed commercially to degrade petroleum compounds from an oil pipeline spill. In the 1980s, groundwater contaminated with toxic chemicals, such as trichloroethylene, tetrachloroethylene, and aviation fuel, were treated using microorganisms. Numerous attempts were made throughout the 1980s and 1990s to genetically engineer microorganisms to obtain enhanced removal of toxic chemicals, however, to date, these efforts have not been as fruitful as originally hoped. Present efforts are focused on giving naturally occurring organisms the competitive advantage so that they effectively degrade the contaminants of interest.

EXAMPLE 3–4

1,1-Dichloroethane (1,1-DCA) is a chlorinated hydrocarbon that is sparingly soluble in water but miscible with most organic solvents. It is used extensively in chemical manufacturing, as a solvent, degreasing agent, as a fumigant in insecticide sprays and fire extinguishers. It is regulated both by the U.S. EPA as a drinking water contaminant and OSHA as an indoor air pollutant.

You are working as an environmental engineer and are designing a groundwater remediation for a site contaminated with 1,1-DCA. You wish to determine if 1,1-DCA has the potential to be degraded microbiologically by sulfate-reducing bacteria. How would you accomplish this?

Solution

To determine if the reaction is thermodynamically feasible we need the half-reactions. The half reaction for the dechlorination reaction of 1,1-DCA to form chloroethane is:

$$1/2 \ C_2H_4Cl_2 + 1/2 \ H^+ + e^- \rightleftharpoons 1/2 \ C_2H_5Cl + 1/2 \ Cl^- \qquad \Delta G° = -49.21 \ kJ/e^- \ eq$$

The half reaction for the oxidation of hydrogen sulfide is:

$$1/8 \ HS^- + 1/2 \ H_2O \rightleftharpoons 1/8 \ SO_4^{2-} + 9/8 \ H^+ + e^- \qquad \Delta G° = -21.23 \ kJ/e^- \ eq$$

Based on these half-reactions, the G° for the dechlorination of 1,1-DCA is -70.44 kJ/e$^-$ eq and the reaction is thermodynamically feasible. Laboratory and field studies have confirmed that bacteria can transform 1,1-DCA to chloroethane.

Clearly, a solid knowledge of biology is absolutely critical to make advances in all of the areas mentioned above, and as such, the environmental engineer and scientist of the future will be expected to have a much greater understanding of biology than those in the past.

CHAPTER REVIEW

When you have completed studying this chapter, you should be able to answer the following questions without your text or notes:

1. List the cell processes necessary for life.
2. Provide examples of a monosaccharide, a disaccharide, and a polysaccharide.
3. What are the two types of nucleic acids? What are the five nucleotides found in DNA and RNA?
4. What are the nucleotides that are important intermediates in the transformation of energy?
5. What are the roles of proteins in cellular function?

6. Define the terms enzymes and immunoglobulins.
7. Describe the importance of enzyme kinetics in environmental engineering.
8. Discuss the basic assumptions of the Michaelis-Menten equation.
   With the aid of this textbook or a list of equations, you should be able to:
   (a) Determine V_{max} and K_M.
   (b) Determine the rate of respiration from experimental data.
   (c) Determine the number of moles of oxygen consumed or the moles of carbon dioxide produced during respiration.
   (d) Using half-reactions, determine if a set of reactions are thermodynamically favorable.
9. Define the following terms: lipids, fats, phospholipids, and steroids. Explain how phospholipids organize themselves into a bilayer in water.
10. Explain the modes by which chemicals can diffuse across a cell membrane.
11. Define the following terms: endocytosis, exocytosis, pinocytosis, phagocytosis, and receptor-mediated endocytosis.
12. State the function of each of the following organelles: chromatin, nucleolus, ribosomes, endoplasmic reticulum, vesicles, Golgi apparatus, lysosomes, vacuoles, mitochondria, cytoskeleton, centrosome. Identify the organelles in an animal cell.
13. State the function of the cell wall, plasmodesmata, central vacuole, tonoplast, and plastid. Identify the organelles in a plant cell.
14. State the function of the nucleoid, flagella, capsule, and pili.
15. Describe how photosynthesis occurs and in which organelles.
16. Describe the three steps in cellular respiration.
17. Explain the cell cycle.
18. Describe mitosis and meiosis.
19. Describe binary fission. Define the following terms: budding, gemmule, fragmentation, regeneration. Give an example of each.
20. What are the differences between a haploid and diploid cell?
21. Describe the different approaches to classifying organisms and how these methods evolved over time.
22. List the basic characteristics of Archaea. List the conditions under which extreme halophiles, extreme thermophiles, and methanogens survive.
23. Identify bacteria that are coccus, bacillus and spirillum. Give examples of each. Define the prefixes: diplo-, staphylo- and strepto-.
24. Explain how the Gram stain is used.
25. What is meant by taxis? Give at least three examples.
26. Define the following terms: autotrophic, chemoautotrophic, photoautotrophic, heterotrophic, chemoheterotrophic, photoheterotrophic, and mixotrophs. Give an example of each.
27. Define the following terms: saprobe, parasite.
28. Define the following terms: anaerobes, obligate aerobes, facultative anaerobes.
29. Define the following terms: psychrophiles, mesophiles, thermophiles, stenothermophiles, hyperhermophiles.
30. What are the five major groups of bacteria?
31. What are the four groups of protozoa? Give an example of each.
32. What are the six phyla of algae? Give an example of each.
33. What are slime molds and water molds?
34. What is the importance of fungi to environmental engineers and scientists?
35. List the five phyla of fungi and give an example of each.
36. What are viruses? Why are they important?
37. What are Koch's postulates?
38. What are opportunistic pathogens? Give an example of each.
39. What is a communicable or contagious disease? What is a virulent pathogen?

40. Define endotoxin and exotoxin.

41. How can infectious diseases be transferred?

42. What is meant by the term "prion"?

PROBLEMS

3–1 The rate of respiration can be determined by monitoring the rate of oxygen consumption or carbon dioxide production:

$$C_6H_{12}O_6 + 6O_2 \rightleftharpoons 6CO_2 + 6H_2O$$

In an experiment, the change in gas volume using a respirometer containing 25 nongerminating pea seeds was measured at two temperatures (10 and 20°C). The experiment was conducted in the dark so that photosynthesis did not occur. The CO_2 produced during cellular respiration was removed by the reaction of CO_2 with potassium hydroxide (KOH) to form solid potassium carbonate (K_2CO_3). In this way, only the oxygen consumed was measured. The results are as follows:

Time (min)	Corrected difference for control (mL)	
	10°C	**20°C**
0	—	—
5	0.005	0.01
10	0.01	0.02
15	0.015	0.035
20	0.027	0.05

(a) Determine the number of moles of oxygen consumed after 20 minutes. Assume the atmospheric pressure is 1 atm.

(b) Determine the rate constant for each of the data assuming first-order kinetics. Compare your results to those obtained for germinating seeds (see Example 3–2).

Answer: (a) 1.2×10^{-6} moles of O_2 at 10°C; 2.1×10^{-6} moles of O_2 at 20°C

(b) 0.108 min^{-1} for 10°C; 0.109 min^{-1} at 20°C

3–2 Calculate the amount of CO_2 produced (in kg) per mole of acetate oxidized during aerobic oxidation. If on the other hand, acetate is oxidized to CO_2 by denitrification, determine the amount of CO_2 produced (in kg) per mole of acetate oxidized. The half reaction for denitrification is:

$$\Delta G°, kJ \cdot (e^- \, eq)$$

$$1/5 \, NO_3^- + 6/5 \, H^+ + e^- \rightleftharpoons 1/10 \, N_2 + 3/5 \, H_2O \qquad -72.2$$

3–3 Hydrogen sulfide (HS$^-$) can be oxidized to SO_4^{2-} by microorganisms that grow inside sewer pipes, resulting in the corrosion of concrete from the acid produced.

(a) If the concentration of HS$^-$ in the sewage is 2.5 mg $\cdot$ L^{-1}, what is the concentration of sulfate that would result, assuming that conversion is 100%?

(b) Assuming that the electron acceptor is acetate, determine the $\Delta G°$ for this reaction and state if it is thermodynamically feasible.

The half-reaction for HS$^-$ to SO_4^{2-};

$$\Delta G°, kJ \cdot (e^- \, eq)^{-1}$$

$$1/8 \, SO_4^{2-} + 9/8 \, H^+ + e^- \rightleftharpoons 1/8 \, HS^- + 1/2 \, H_2O \qquad 21.23$$

Answer: (a) 7.3 mg $\cdot$ L^{-1} SO_4^{2-}

(b) -6.2 kJ(e$^-$ eq)$^{-1}$. The reaction is thermodynamically feasible.

3–4 Calculate the $\Delta G°$ for the oxidation of hydrogen sulfide (HS^-) to SO_4^{2-} if glucose is the electron acceptor. The half-reaction is:

	$\Delta G°$, kJ $\cdot$ (e$^-$ eq)$^{-1}$
$1/24\ C_6H_{12}O_6 + 1/4\ H_2O \rightleftharpoons 1/4\ CO_2 + H^+ + e^-$	-41.35

3–5 You are characterizing a novel enzyme X. You measure the velocity of the reaction with different substrate concentrations and obtain the following data:

Initial substrate concentration (mM)	V_o ($\mu g \cdot h^{-1}$)
3.0	10.4
5.0	14.5
10.0	22.5
30.0	33.8
90.0	40.5
180.0	42.5

(a) Determine V_{max} and K_M.

(b) If the enzyme concentration were decreased to 10% of the amount used in the experiment described above, would the initial reaction velocities change? If so, how?

Answer: (a) $V_{max} = 44.6\ \mu g \cdot h^{-1}$; $K_M = 10.0$ mM

(b) The initial reaction velocities would also decrease to 10% of the original.

3–6 Enzyme X is able to catalyze the breakdown of the toxic red cedar dye to CO_2 and water. In a study of the enzyme you have found that the initial velocities (mM $\cdot h^{-1}$) are dependent on the initial substrate concentration. The data collected are listed below. Plot the data to determine V_{max} and K_M. How would a tenfold decrease in the enzyme concentration affect the observed K_M?

Substrate conc. (mM)	Initial velocity (mM $\cdot h^{-1}$)
2	5
4	9
6	15
8	18
10	23
12	30
16	38
20	44
25	50
30	55
40	60

3–7 Vinyl chloride is a product formed from the anaerobic dehalogenation of tetrachloroethylene to trichloroethylene then to dichloroethylene. Determine if vinyl chloride can be further dechlorinated to produce ethylene either aerobically or anaerobically. Show your work. The pertinent half-reactions are:

	$\Delta G°$, kJ $\cdot$ (e$^-$ eq)$^{-1}$
$1/2\ C_2H_3Cl + 1/2\ H^+ + e^- \rightleftharpoons 1/2\ C_2H_4 + 1/2\ Cl^-$	-35.81
$1/8\ CO_2 + 1/2\ H^+ + e^- \rightleftharpoons 1/8\ CH_4 + 1/4\ H_2O$	23.53
$1/4\ O_2 + H^+ + e^- \rightleftharpoons 1/2\ H_2O$	-78.72

Answer: Vinyl chloride can only be dechlorinated to produce ethylene under anaerobic conditions using CO_2 as an electron acceptor.

3–8 Is the dehalogenation reaction of dichlorophenol to chlorophenol thermodynamically favorable under either aerobic or methanogenic condition? Show your work. The pertinent half-reactions are given below:

	$\Delta G°$, kJ $\cdot$ (e$^-$ eq)$^{-1}$
$1/2\ C_6H_4Cl_2OH + 1/2\ H^+ + e^- \rightleftharpoons 1/2\ C_6H_5ClOH + 1/2\ Cl^-$	-33
$1/8\ CO_2 + 1/2\ H^+ + e^- \rightleftharpoons 1/8\ CH_4 + 1/4\ H_2O$	23.53
$1/4\ O_2 + H^+ + e^- \rightleftharpoons 1/2\ H_2O$	-78.72

DISCUSSION QUESTIONS

3–1 Bacteria and molds typically do not grow in honey or pickle jars, even after their containers have been opened. Explain why.

3–2 Some organisms live in shallow ponds, where much of the water will evaporate during the warm, sunny, summer months. How will the concentrations of solutes change in the water during this time? Will this create problems for the organisms in the pond water? If so, how?

3–3 Cells use matter efficiently. How could we learn from the cell to produce less waste or more efficient landfills?

3–4 Before regulations prohibited it, hazardous chemicals were dumped in water, pumped into the ground, or left in decomposing drums on land. Many old industrial sites have contaminated soil. Explain how the study of bacterial processes provides insight into the development of new technologies to clean up these sites.

3–5 An algae bloom occurs in a lake. What effects might this bloom have on the other aquatic organisms (e.g., protozoans, plants, fish) living in the lake? Explain why.

3–6 How might bacteria mitigate the effects of acid rain?

3–7 What are some precautions that you think should be considered when developing a new biological process that employs bacteria? Explain.

3–8 There are many products on the market that claim antibacterial properties. Do you believe that antibacterial agents should be added to soaps, hand lotions, and other personal care products? Explain your answer in terms of bacterial resistance.

3–9 Design an experiment that could determine the optimal growth conditions for a particular bacteria species.

3–10 Study the problem of harmful algae blooms more thoroughly. Propose several strategies to prevent such blooms and mitigate their effects.

FE EXAM FORMATTED PROBLEMS

3–1 Which of the following is found in eukaryotic plant cells but not eukaryotic animal cells?

(a) Mitochondria

(b) Nucleus

(c) Golgi apparatus

(d) Endoplasmic reticulum

(e) Chloroplasts

3–2 Which of the following is *not* a function of the Golgi apparatus?

    (a) Processing of macromolecules so that they become fully functional

    (b) Sorting of macromolecules for transport to appropriate cellular location

    (c) Synthesis of secretory proteins

    (d) Manufacturing of polysaccharides

    (e) Addition of molecular identification tags to macromolecules

3–3 Methanogenic, halophilic, and thermoacidophilic bacteria are all members of which group?

    (a) Eukaryotes

    (b) Eubacteria

    (c) Archaebacteria

    (d) Plantae

    (e) Protista

3–4 Estimate the rate constant for growth of a population of bacteria during the exponential growth phase. Use the chart given in Figure FEP-3-4.

Use base-10 logarithms.

    (a) $3.75 \ \mathrm{d}^{-1}$

    (b) $0 \ \mathrm{d}^{-1}$

    (c) $2 \ \mathrm{d}^{-1}$

    (d) $10 \ \mathrm{d}^{-1}$

3–5 For Figure FEP-3-5, match the regions of the curve, that is (1) to (5), with the growth phase, that is (a) to (e).

    (a) Exponential growth phase

    (b) Stationary phase

    (c) Lag phase

    (d) Death phase

    (e) Declining growth phase

FIGURE FEP-3-4

Growth of bacteria in a closed system.

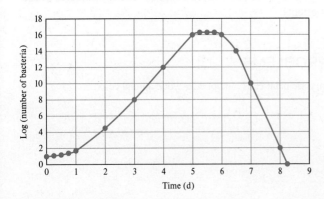

FIGURE FEP-3-5

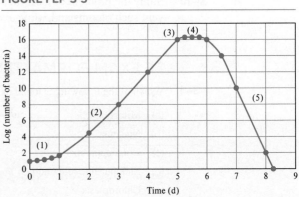

3–6 In the tricarboxylic acid cycle, glucose is oxidized to carbon dioxide. In the process how many molecules of ATP are produced from each molecule of glucose?

    (a) 12

    (b) 20

    (c) 24

    (d) 36

    (e) 42

3–7 Nitrifying bacteria convert ammonia to which of the following:

    (a) Nitrite

    (b) Nitrogen

    (c) Amines

    (d) Nitrate

3–8 Sulfide is converted to sulfate in the presence of oxygen by which organism?

    (a) *Desulfovibrio*

    (b) *Nitrobacter*

    (c) *Nitrosomonas*

    (d) *Thiobacillus*

    (e) *Ferrobacillus*

3–9 During an experiment it was found that *Pisum sativum* (pea plant) has a rate of photosynthesis at 25°C of 0.10 mg glucose/cm^2·hr. Given, that the heat of combustion of glucose is 686 KJ/mole and the molecular weight of glucose is 180 g, calculate the energy, in KJ, that could be theoretically produced over an 18-hr lighted period by a single leaf that has a surface area of 200 cm^2.

    (a) 0.137 kJ

    (b) 1.37 kJ

    (c) 13.7 kJ

    (d) 137 kJ

3–10 A large power plant emits about 40,300 tons of CO_2/d. To counter this, environmentalists plan to plant trees in the area. If on average 500 moles of glucose is synthesized by a tree per day, how many trees need to be planted to reabsorb all the CO_2 released into the atmosphere every day by the factory.

    (a) 3000

    (b) 30,000

    (c) 100,000

    (d) 300,000

    (e) 1,000,000

REFERENCES

Aust, S. D. (1995) "Mechanisms of Degradation by White Rot Fungi," *Environmental Health Perspectives Supplements,* 103(S5).

Cutler, D. M., and G. Miller (2005) "The Role of Public Health Improvements in Health Advances: The 20th Century United States," *Demography,* 42 (1): 1–22.

Folwell, A. P. (1917) *Water Supply Engineering: The Designing and Constructing of Water-Supply Systems,* John Wiley and Sons, New York.

Fuller, G. W., and J. R. McClintock (1926) *Solving Sewage Problems,* McGraw-Hill Book Co., Inc., New York.

Guest, R. K., and D. W. Smith (2002) "A Potential New Role for Fungi in a Wastewater MBR Biological Nitrogen Reduction System," *J. Environ. Eng. Sci.* 1: 433–37. Published on the NRC Research Press website at http://jees.nrc.ca/ on 3 December 2002.

Guyer, R. L. (1997) "Prions: Puzzling Infectious Proteins," National Institutes of Health, Washington, DC. http://science.education.nih.gov/home2.nsf/Educational+Resources/Resource+Formats/Online+Resources/+High+School/D07612181A4E785B85256CCD0064857B

Huisman, L., and W. E. Wood (1974) *Slow Sand Filtration,* World Health Organization, Geneva, Switzerland, pp. 1–89.

Mader, S. S. (2007) *Biology,* 9th ed. McGraw Hill Book Co., New York.

McKinney, R. E. (1962) *Microbiology for Sanitary Engineers,* McGraw-Hill, New York: p. 40.

Raven, P. H., G. B. Johnson, K. A. Mason, J. Losos, and S. Singer (2014) *Biology,* 8th ed., McGraw-Hill, New York.

WHOI (2006) "The Harmful Algae Page," National Office for Harmful Algal Blooms at Woods Hole Oceanographic Institution, Dr. Don Anderson, Director. http://www.whoi.edu/redtide/HABdistribution/PSP_worldmap_1970-2006.jpg

Yeşilada, Ö, S. Şik, and M. Şam (1999) "Treatment of Olive Oil Mill Wastewater with Fungi," *Tr. J. of Biology,* 23: 231–40.

Zhongming, Z., and J. P. Obbard (2002) "Oxidation of Polycyclic Aromatic Hydrocarbons (PAH) by the White Rot Fungus, *Phanerochaete Chrysosporium,*" *Enzyme and Microbial Technology,* 31 (1–2): 3–9.

4

Materials and Energy Balances

4–1 INTRODUCTION

Materials and energy balances are key tools in achieving a quantitative understanding of the behavior of environmental systems. They serve as a method of accounting for the flow of energy and materials into and out of environmental systems. Mass balances provide us with a tool for modeling the production, transport, and fate of pollutants in the environment. Energy balances likewise provide us with a tool for modeling the production, transport, and fate of energy in the environment. Examples of mass balances include prediction of rainwater runoff (Chapter 7), determination of the solid waste production from mining operations (Chapter 8), oxygen balance in streams (Chapter 9), and audits of hazardous waste production (Chapter 14). Energy balances allow us to estimate the efficiency of thermal processes (Chapter 8), predict the temperature rise in a stream from the discharge of cooling water from a power plant (Chapter 9), and study climate change (Chapter 12).

4–2 UNIFYING THEORIES

Conservation of Matter

The **law of conservation of matter** states that (without nuclear reaction) matter can neither be created nor destroyed. This is a powerful theory. It means that if we observe an environmental process carefully, we should be able to account for the "matter" at any point in time. It does not mean that the form of the matter does not change nor, for that matter, the properties of the matter. Thus, if we measure the volume of a fresh glass of water on the counter on Monday and measure it again a week later and find the volume to be less, we do not presume magic has occurred but rather that matter has changed in form. The law of conservation of matter says we ought to be able to account for all the mass of the water that was originally present, that is, the mass of water remaining in the glass plus the mass of water vapor that has evaporated equals the mass of water originally present. The mathematical representation of this accounting system is called a **materials balance** or **mass balance.**

Conservation of Energy

The **law of conservation of energy** states that energy cannot be created or destroyed. Like the law of conservation of matter, this theory means that we should be able to account for the "energy" at any point in time. It also does not mean that the form of the energy does not change. Thus, we should be able to trace the energy of food through a series of organisms from green plants through animals. The mathematical representation of the accounting system we use to trace energy is called an **energy balance.**

Conservation of Matter and Energy

In 1905, Albert Einstein proposed the theory of equivalence between matter and energy, that is $E = (2.2 \times 10^{13})(m)$, where E is in calories, m is in grams, and 2.2×10^{13} is a proportionality constant. This equivalence implies that conversion of 1 g of matter to energy will liberate 2.2×10^{13} calories. The amount of energy does not depend on the nature of the substance that undergoes change. This is a tremendous amount of energy. The conversion of 1 g of matter to energy would raise the temperature of 220 gigagrams (Gg) of water from the freezing point, 0°C, to the boiling point, 100°C. You would have to burn approximately 2.7 megagrams (Mg) of coal to obtain this amount of energy!

The birth of the nuclear age proved Einstein's hypothesis correct, so today we have a combined **law of conservation of matter and energy** that states that the total amount of energy and matter is constant. A nuclear change produces new materials by changing the identity of the atoms themselves. Significant amounts of matter are converted to energy in nuclear explosions. Exchange between mass and energy is not an issue in environmental applications. Thus, there are generally separate balances for mass and energy.

4–3 MATERIALS BALANCES

Fundamentals

In its simplest form a materials balance or mass balance may be viewed as an accounting procedure. You perform a form of material balance each time you balance your checkbook.

$$\text{Balance} = \text{deposit} - \text{withdrawal} \tag{4–1}$$

For an environmental process, the equation would be written

$$\text{Accumulation} = \text{input} - \text{output} \tag{4–2}$$

where accumulation, input, and output refer to the mass quantities accumulating in the system or flowing into or out of the system. The "system" may be, for example, a pond, river, or a pollution control device.

The Control Volume. Using the mass balance approach, we begin solving the problem by drawing a flowchart of the process or a conceptual diagram of the environmental subsystem. All of the known inputs, outputs, and accumulation are converted to the same mass units and placed on the diagram. Unknown inputs, outputs, and accumulation are also marked on the diagram. This helps us define the problem. System boundaries (imaginary blocks around the process or part of the process) are drawn in such a way that calculations are made as simple as possible. The system within the boundaries is called the **control volume.**

We then write a materials balance equation to solve for unknown inputs, outputs, or accumulations or to demonstrate that we have accounted for all of the components by demonstrating that the materials balance "closes," that is, the accounting balances. Alternatively, when we do not have data for all inputs or outputs, we can assume that the mass balance closes and solve for the unknown quantity. The following example illustrates the technique.

EXAMPLE 4–1   Mr. and Mrs. Konzzumer have no children. In an average week they purchase and bring into their house approximately 50 kg of consumer goods (food, magazines, newspapers, appliances, furniture, and associated packaging). Of this amount, 50% is consumed as food. Half of the food is used for biological maintenance and ultimately released as CO_2; the remainder is discharged to the sewer system. The Konzzumers recycle approximately 25% of the solid waste that is generated. Approximately 1 kg accumulates in the house. Estimate the amount of solid waste they place at the curb each week.

Solution   Begin by drawing a mass balance diagram and labeling the known and unknown inputs and outputs. There are, in fact, two diagrams: one for the house and one for the people. However, the mass balance for the people is superfluous for the solution of this problem.

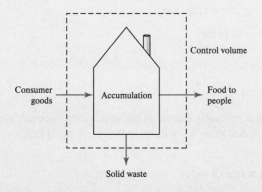

Write the mass balance equation for the house.

Input = accumulation in house + output as food to people + output as solid waste

Now we need to calculate the known inputs and outputs.

One half of input is food = (0.5)(50 kg) = 25 kg

This is output as food to the people. The mass balance equation is then rewritten as

50 kg = 1 kg + 25 kg + output as solid waste

Solving for the mass of solid waste:

Output as solid waste = 50 − 1 − 25 = 24 kg

The mass balance diagram with the appropriate masses may be redrawn as shown below:

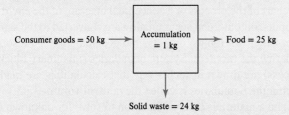

We can estimate the amount of solid waste placed at the curb by performing another mass balance around the solid waste as shown in the following diagram:

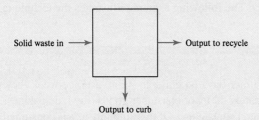

The mass balance equation is

Solid waste in = output to recycle + output to curb

Because the recycled amount is 25% of the solid waste

Output to recycle = (0.25)(24 kg) = 6 kg

Substituting into the mass balance equation for solid waste and solving for output to curb:

24 kg = 6 kg + output to curb

Output to curb = 24 − 6 = 18 kg

Time as a Factor

For many environmental problems time is an important factor in establishing the degree of severity of the problem or in designing a solution. In these instances Equation 4–2 is modified to the following form:

Rate of accumulation = rate of input − rate of output (4–3)

where **rate** is used to mean "per unit of time." In the calculus this may be written as

$$\frac{dM}{dt} = \frac{d(\text{in})}{dt} - \frac{d(\text{out})}{dt} \tag{4–4}$$

where M refers to the mass accumulated and (in) and (out) refer to the mass flowing in or out of the control volume. As part of the description of the problem, a convenient time interval that is meaningful for the system must be chosen.

EXAMPLE 4–2   Truly Clearwater is filling her bathtub, but she forgot to put the plug in. If the volume of water for a bath is 0.350 m^3 and the tap is flowing at 1.32 L · min^{-1} and the drain is running at 0.32 L · min^{-1}, how long will it take to fill the tub to bath level? Assuming Truly shuts off the water when the tub is full and does not flood the house, how much water will be wasted? Assume the density of water remains constant through the control volume.

Solution   The mass balance diagram is shown here.

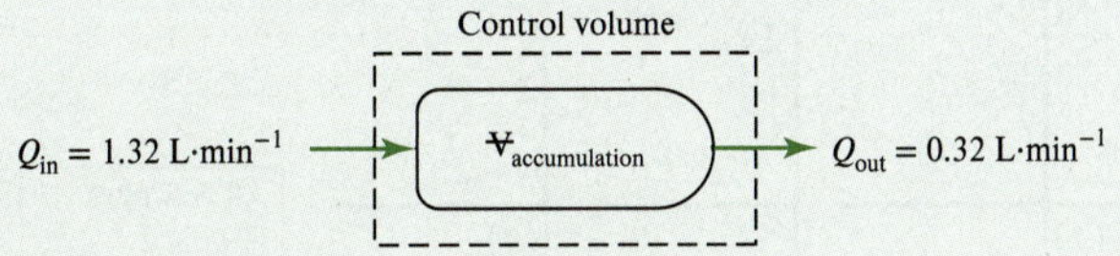

Because we are working in mass units, we must convert the volumes to masses. To do this, we use the density of water.

Mass = (volume)(density) = $(\forall)(\rho)$

where the

Volume = (flow rate)(time) = $(Q)(t)$

So for the mass balance equation, noting that 1.0 m^3 = 1000 L, 0.350 m^3 = 350 L,

Accumulation = mass in − mass out

$$(\forall_{ACC})(\rho) = (Q_{in})(\rho)(t) - (Q_{out})(\rho)(t)$$

Using the assumption of constant density,

$$\forall_{ACC} = (Q_{in})(t) - (Q_{out})(t)$$
$$\forall_{ACC} = 1.32t - 0.32t$$
$$350 \text{ L} = (1.00 \text{ L} \cdot \text{min}^{-1})(t)$$
$$t = 350 \text{ min}$$

The amount of water wasted is

Wasted water = (0.32 L · min^{-1})(350 min) = 112 L

More Complex Systems

A key step in the solution of mass balance problems for systems that are more complex than the previous examples is the selection of an appropriate control volume. In some instances, it may be necessary to select multiple control volumes and then solve the problem sequentially using the solution from one control volume as the input to another control volume. For some complex processes, the appropriate control volume may treat all of the steps in the process as a "black box" in which the internal process steps are not required and therefore are hidden in a black box. The following example illustrates a case of a more complex system and a method of solving the problem.

EXAMPLE 4–3 A storm sewer network in a small residential subdivision is shown in the following sketch. The storm water flows by gravity in the direction shown by the pipes. Storm water only enters the storm sewer on east–west legs of pipe. No storm water enters on the north–south legs. The flow rate for each section of pipe is also shown by the arrows at each section of pipe. The capacity of each pipe is $0.120 \text{ m}^3 \cdot \text{s}^{-1}$. During large rain storms River Street floods below junction number 1 because the flow of water exceeds the capacity of the storm sewer pipe. To alleviate this problem and to provide extra capacity for expansion, it is proposed to build a retention pond to hold the storm water until the storm is over and then gradually release it. Where in the pipe network should the retention pond be built to provide approximately 50% extra capacity ($0.06 \text{ m}^3 \cdot \text{s}^{-1}$) in the remaining system?

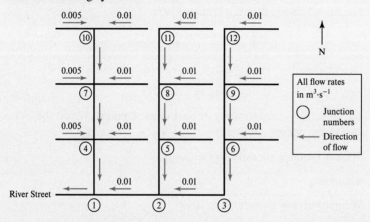

Solution This is an example of a balanced flow problem. That is Q_{out} must equal Q_{in}. Although this problem can almost be solved by observation, we will use a sequential mass balance approach to illustrate the technique. Starting at the upper end of the system at junction number 12, we draw the following mass balance diagram:

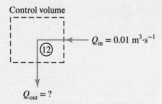

The mass balance equation is

$$\frac{dM}{dt} = \frac{d(\text{in})}{dt} - \frac{d(\text{out})}{dt}$$

Because no water accumulates at the junction

$$\frac{dM}{dt} = 0$$

and

$$\frac{d(\text{in})}{dt} = \frac{d(\text{out})}{dt}$$

$$(\rho)(Q_{in}) = (\rho)(Q_{out})$$

Because the density of water remains constant, we may treat the mass flow rate in and out as directly proportional to the flow rate in and out.

$$Q_{in} = Q_{out}$$

So the flow rate from junction 12 to junction 9 is $0.01 \text{ m}^3 \cdot \text{s}^{-1}$.

At junction 9, we can draw the following mass balance diagram.

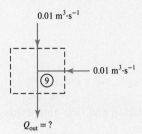

Again using our assumption that no water accumulates at the junction and recognizing that the mass balance equation may again be written in terms of flow rates,

$$Q_{from\ junction\ 9} = Q_{from\ junction\ 12} + Q_{in\ the\ pipe\ connected\ to\ junction\ 9}$$

$$= 0.01 + 0.01 = 0.02 \text{ m}^3 \cdot \text{s}^{-1}$$

Similarly

$$Q_{from\ junction\ 6} = Q_{from\ junction\ 9} + Q_{in\ the\ pipe\ connected\ to\ junction\ 6}$$

$$= 0.02 + 0.01 = 0.03 \text{ m}^3 \cdot \text{s}^{-1}$$

and, noting that storm water enters on east–west legs of pipe

$$Q_{from\ junction\ 3} = Q_{from\ junction\ 6} + Q_{in\ the\ pipe\ connecting\ junction\ 3\ with\ junction\ 2}$$

$$= 0.03 + 0.01 = 0.04 \text{ m}^3 \cdot \text{s}^{-1}$$

By a similar process for all the junctions, we may label the network flows as shown in the following diagram.

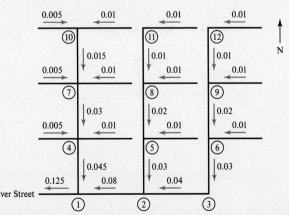

It is obvious that the pipe capacity of $0.12 \text{ m}^3 \cdot \text{s}^{-1}$ is exceeded just below junction 1. By observation we can also see that the total flow into junction 2 is $0.07 \text{ m}^3 \cdot \text{s}^{-1}$ and that a retention pond at this point would require that the pipe below junction 1 carry only $0.055 \text{ m}^3 \cdot \text{s}^{-1}$ This meets the requirement of providing approximately 50% of the capacity for expansion.

Efficiency

The effectiveness of an environmental process in removing a contaminant can be determined using the mass balance technique. Starting with Equation 4–4,

$$\frac{dM}{dt} = \frac{d(\text{in})}{dt} - \frac{d(\text{out})}{dt}$$

The mass of contaminant per unit of time [$d(\text{in})/dt$ and $d(\text{out})/dt$] may be calculated as

$$\frac{\text{Mass}}{\text{Time}} = (\text{concentration})(\text{flow rate})$$

For example,

$$\frac{\text{Mass}}{\text{Time}} = (\text{mg} \cdot \text{m}^{-3})(\text{m}^3 \cdot \text{s}^{-1}) = \text{mg} \cdot \text{s}^{-1}$$

This is called a **mass flow rate.** In concentration and flow rate terms, the mass balance equation is

$$\frac{dM}{dt} = C_{in} Q_{in} - C_{out} Q_{out} \tag{4–5}$$

where dM/dt = rate of accumulation of contaminant in the process

C_{in}, C_{out} = concentrations of contaminant into and out of the process

Q_{in}, Q_{out} = flow rates into and out of the process

The ratio of the mass that is accumulated in the process to the incoming mass is a measure of how effective the process is in removing the contaminant

$$\frac{dM/dt}{C_{in} Q_{in}} = \frac{C_{in} Q_{in} - C_{out} Q_{out}}{C_{in} Q_{in}} \tag{4–6}$$

For convenience, the fraction is multiplied by 100%. The left-hand side of the equation is given the notation η. Efficiency (η) is then defined as

$$\eta = \frac{\text{mass in} - \text{mass out}}{\text{mass in}} (100\%) \tag{4–7}$$

If the flow rate in and the flow rate out are the same, this ratio may be simplified to

$$\eta = \frac{\text{concentration in} - \text{concentration out}}{\text{concentration in}} (100\%) \tag{4–8}$$

The following example illustrates a multistep solution using efficiency as part of the solution technique.

EXAMPLE 4–4   The air pollution control equipment on a municipal waste incinerator includes a fabric filter particle collector (known as a **baghouse**). The baghouse contains 424 cloth bags arranged in parallel, that is 1/424 of the flow goes through each bag. The gas flow rate into and out of the baghouse is 47 m$^3 \cdot$ s^{-1}, and the concentration of particles entering the baghouse is 15 g $\cdot$ m^{-3} In normal operation the baghouse particulate discharge meets the regulatory limit of 24 mg $\cdot$ m^{-3} During preventive maintenance replacement of the bags, one bag is inadvertently not replaced, so only 423 bags are in place.

Calculate the fraction of particulate matter removed and the efficiency of particulate removal when all 424 bags are in place and the emissions comply with the regulatory requirements. Estimate the mass emission rate when one of the bags is missing and recalculate the efficiency of the baghouse. Assume the efficiency for each individual bag is the same as the overall efficiency for the baghouse.

Solution   The mass balance diagram for the baghouse in normal operation is shown here.

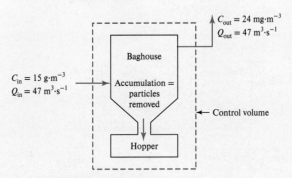

In concentration and flow rate terms, the mass balance equation is

$$\frac{dM}{dt} = C_{in}\, Q_{in} - C_{out}\, Q_{out}$$

The mass rate of accumulation in the baghouse is

$$\frac{dM}{dt} = (15{,}000\ \text{mg} \cdot \text{m}^{-3})(47\ \text{m}^3 \cdot \text{s}^{-1}) - (24\ \text{mg} \cdot \text{m}^{-3})(47\ \text{m}^3 \cdot \text{s}^{-1}) = 703{,}872\ \text{mg} \cdot \text{s}^{-1}$$

The fraction of particulates removed is

$$\frac{703{,}872\ \text{mg} \cdot \text{s}^{-1}}{(15{,}000\ \text{mg} \cdot \text{m}^{-3})(47\ \text{m}^3 \cdot \text{s}^{-1})} = \frac{703{,}872\ \text{mg} \cdot \text{s}^{-1}}{705{,}000\ \text{mg} \cdot \text{s}^{-1}} = 0.9984$$

The efficiency of the baghouse is

$$\eta = \frac{15{,}000\ \text{mg} \cdot \text{m}^{-3} - 24\ \text{mg} \cdot \text{m}^{-3}}{15{,}000\ \text{mg} \cdot \text{m}^{-3}}\ (100\%)$$

$$= 99.84\%$$

Note that the fraction of particulate matter removed is the decimal equivalent of the efficiency.

To determine the mass emission rate with one bag missing, we begin by drawing a mass balance diagram. Because one bag is missing, a portion of the flow (1/424 of Q_{out}) effectively bypasses the baghouse. The bypass line around the baghouse is drawn to show this.

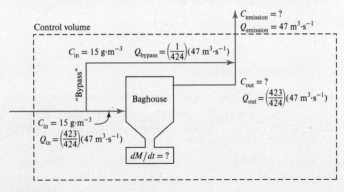

A judicious selection of the control volume aids in the solution of this problem. As shown in the diagram, a control volume around the overall baghouse and bypass flow yields three unknowns: the mass flow rate out of the baghouse, the rate of mass accumulation in the baghouse hopper, and the mass flow rate of the mixture. A control volume around the baghouse alone reduces the number of unknowns to two:

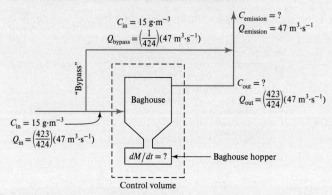

Because we know the efficiency and the influent mass flow rate, we can solve the mass balance equation for the mass flow rate out of the filter.

$$\eta = \frac{C_{in}\, Q_{in} - C_{out}\, Q_{out}}{C_{in}\, Q_{in}}$$

Solving for $C_{out} Q_{out}$

$$C_{out}\, Q_{out} = (1 - \eta)C_{in}\, Q_{in}$$

$$= (1 - 0.9984)(15{,}000\ \text{mg} \cdot \text{m}^{-3})(47\ \text{m}^3 \cdot \text{s}^{-1})(423/424) = 1125\ \text{mg} \cdot \text{s}^{-1}$$

This value can be used as an input for a control volume around the junction of the bypass, the effluent from the baghouse and the final effluent.

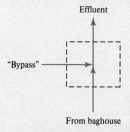

A mass balance for the control volume around the junction may be written as

$$\frac{dM}{dt} = C_{in}\, Q_{in\,from\,bypass} + C_{in}\, Q_{in\,from\,baghouse} - C_{emission}\, Q_{emission}$$

Because there is no accumulation in the junction

$$\frac{dM}{dt} = 0$$

and the mass balance equation is

$$C_{emission}\, Q_{emission} = C_{in}\, Q_{in\,from\,bypass} + C_{in}\, Q_{in\,from\,baghouse}$$

$$= (15{,}000\ \text{mg} \cdot \text{m}^{-3})(47\ \text{m}^3 \cdot \text{s}^{-1})(1/424) + 1125 = 2788\ \text{mg} \cdot \text{s}^{-1}$$

The concentration in the effluent is

$$\frac{C_{emission}\, Q_{emission}}{Q_{out}} = \frac{2788\ \text{mg} \cdot \text{s}^{-1}}{47\ \text{m}^3 \cdot \text{s}^{-1}} = 59\ \text{mg} \cdot \text{m}^{-3}$$

The overall efficiency of the baghouse with the missing bag is

$$\eta = \frac{15{,}000 \text{ mg} \cdot \text{m}^{-3} - 59 \text{ mg} \cdot \text{m}^{-3}}{15{,}000 \text{ mg} \cdot \text{m}^{-3}} (100\%)$$

$$= 99.61\%$$

The efficiency is still very high but the control equipment does not meet the allowable emission rate of 24 mg · m^{-3} It is not likely that a baghouse would ever operate with a missing bag because the unbalanced gas flows would be immediately apparent. However, many small holes in a number of bags could yield an effluent that did not meet the discharge standards but would otherwise appear to be functioning correctly. To prevent this situation, the bags undergo periodic inspection and maintenance and the effluent stream is monitored continuously.

The State of Mixing

The state of mixing in the system is an important consideration in the application of Equation 4–4. Consider a coffee cup containing approximately 200 mL of black coffee (or another beverage of your choice). If we add a dollop (about 20 mL) of cream and immediately take a sample (or a sip), we would not be surprised to find that the cream was not evenly distributed throughout the coffee. If, on the other hand, we mixed the coffee and cream vigorously and then took a sample, it would not matter if we sipped from the left or right of the cup, or, for that matter, put a valve in the bottom and sampled from there, we would expect the cream to be distributed evenly. In terms of a mass balance on the coffee cup system, the cup itself would define the system boundary for the control volume. If the coffee and cream were not mixed well, then the place we take the sample from would strongly affect the value of $d(\text{out})/dt$ in Equation 4–4. On the other hand, if the coffee and cream were instantaneously well mixed, then any place we take the sample from would yield the same result. That is, any output would look exactly like the contents of the cup. This system is called a completely mixed system. A more formal definition is that **completely mixed systems** are those in which every drop of fluid is homogeneous with every other drop, that is, every drop of fluid contains the same concentration of material or physical property (e.g., temperature). If a system is completely mixed, then we may assume that the output from the system (concentration, temperature, etc.) is the same as the contents within the system boundary. Although we frequently make use of this assumption to solve mass balance problems, it is often very difficult to achieve in real systems. This means that solutions to mass balance problems that make this assumption must be taken as approximations to reality.

If completely mixed systems exist, or at least systems that we can approximate as completely mixed, then it stands to reason that some systems are completely unmixed or approximately so. These systems are called **plug-flow systems.** The behavior of a plug-flow system is analogous to that of a train moving along a railroad track (Figure 4–1). Each car in the train must follow the one preceding it. If, as in Figure 4–1b, a tank car is inserted in a train of box cars, it maintains its position in the train until it arrives at its destination. The tank car may be identified at any point in time as the train travels down the track. In terms of fluid flow, each drop of fluid along the direction of flow remains unique and, if no reactions take place, contains the same concentration of material or physical property that it had when it entered the plug-flow system. Mixing may or may not occur in the radial direction. As with the completely mixed systems, ideal plug-flow systems don't happen very often in the real world.

When a system has operated in such a way that the rate of input and the rate of output are constant and equal, then, of course, the mass rate of accumulation is zero (i.e., in Equation 4–4 $dM/dt = 0$). This condition is called **steady state.** In solving mass-balance problems, it is often

FIGURE 4–1

(a) Analogy of a plug-flow system and a train.
(b) Analogy when a pulse change in influent concentration occurs.

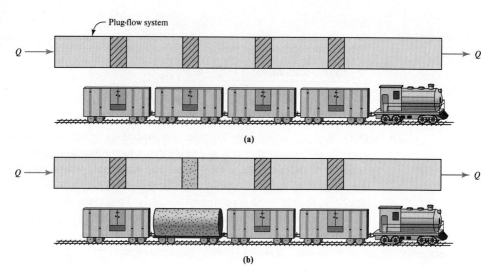

convenient to make an assumption that steady-state conditions have been achieved. We should note that steady state does not imply **equilibrium.** For example, water running into and out of a pond at the same rate is not at equilibrium, otherwise it would not be flowing. However, if there is no accumulation in the pond, then the system is at steady state.

Example 4–5 demonstrates the use of two assumptions: complete mixing and steady state.

EXAMPLE 4–5   A storm sewer is carrying snow melt containing $1.200 \text{ g} \cdot \text{L}^{-1}$ of sodium chloride into a small stream. The stream has a naturally occurring sodium chloride concentration of $20 \text{ mg} \cdot \text{L}^{-1}$. If the storm sewer flow rate is $2000 \text{ L} \cdot \text{min}^{-1}$ and the stream flow rate is $2.0 \text{ m}^3 \cdot \text{s}^{-1}$, what is the concentration of salt in the stream after the discharge point? Assume that the sewer flow and the stream flow are completely mixed, that the salt is a conservative substance (it does not react), and that the system is at steady state.

Solution   The first step is to draw a mass balance diagram as shown here.

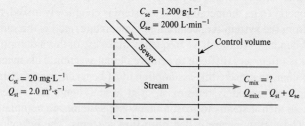

Note that the mass flow of salt may be calculated as

$$\frac{\text{Mass}}{\text{Time}} = (\text{concentration})(\text{flow rate})$$

or

$$\frac{\text{Mass}}{\text{Time}} = (\text{mg} \cdot \text{L}^{-1})(\text{L} \cdot \text{min}^{-1}) = \text{mg} \cdot \text{min}^{-1}$$

Using the notation in the diagram, where the subscript "st" refers to the stream and the subscript "se" refers to the sewer, the mass balance may be written as

Rate of accumulation of salt $= [C_{st}Q_{st} + C_{se}Q_{se}] - C_{mix}Q_{mix}$

where $Q_{mix} = Q_{st} + Q_{se}$

Because we assume steady state, the rate of accumulation equals zero and

$$C_{mix}Q_{mix} = [C_{st}Q_{st} + C_{se}Q_{se}]$$

Solving for C_{mix}

$$C_{mix} = \frac{[C_{st}\,Q_{st} + C_{se}\,Q_{se}]}{Q_{st} + Q_{se}}$$

Before substituting in the values, the units are converted as follows:

$$C_{se} = (1.200\ \text{g} \cdot \text{L}^{-1} \times 1000\ \text{mg} \cdot \text{g}^{-1}) = 1200\ \text{mg} \cdot \text{L}^{-1}$$

$$Q_{st} = (2.0\ \text{m}^3 \cdot \text{s}^{-1})(1000\ \text{L} \cdot \text{m}^{-3})(60\ \text{s} \cdot \text{min}^{-1}) = 120{,}000\ \text{L} \cdot \text{min}^{-1}$$

$$C_{mix} = \frac{[(20\ \text{mg} \cdot \text{L}^{-1})(120{,}000\ \text{L} \cdot \text{min}^{-1})] + [(1200\ \text{mg} \cdot \text{L}^{-1})(2000\ \text{L} \cdot \text{min}^{-1})]}{120{,}000\ \text{L} \cdot \text{min}^{-1} + 2000\ \text{L} \cdot \text{min}^{-1}}$$

$$= 39.34\ \text{or}\ 39\ \text{mg} \cdot \text{L}^{-1}$$

Including Reactions and Loss Processes

Equation 4–4 is applicable when no chemical or biological reaction takes place and no radioactive decay occurs of the substances in the mass balance. In these instances the substance is said to be **conserved.** Examples of conservative substances include salt in water and argon in air. An example of a nonconservative substance (i.e., those that react or settle out) is decomposing organic matter. Particulate matter that is settling from the air is considered a loss process.

In most systems of environmental interest, transformations occur within the system: by-products are formed (e.g., CO_2) or compounds are destroyed (e.g., ozone). Because many environmental reactions do not occur instantaneously, the time dependence of the reaction must be taken into account. Equation 4–3 may be written to account for time-dependent transformation as follows:

Accumulation rate = input rate − output rate ± transformation rate   (4–9)

Time-dependent reactions are called **kinetic reactions.** As discussed in Chapter 2, the rate of transformation, or reaction rate (r), is used to describe the rate of formation or disappearance of a substance or chemical species. With reactions, Equation 4–4 may become

$$\frac{dM}{dt} = \frac{d(\text{in})}{dt} - \frac{d(\text{out})}{dt} + r \qquad (4\text{–}10)$$

The reaction rate is often some complex function of temperature, pressure, the reacting components, and products of reaction.

$$r = -kC^n \qquad (4\text{–}11)$$

where k = reaction rate constant (in s^{-1} or d^{-1})

   C = concentration of substance

   n = exponent or reaction order

The minus sign before the reaction rate constant, k, indicates the disappearance of a substance or chemical species.

In many environmental problems, for example the oxidation of organic compounds by microorganisms (Chapter 9) and radioactive decay (Chapter 16), the reaction rate, r, may be assumed to be directly proportional to the amount of material remaining, that is the value of $n = 1$.

This is known as a **first-order reaction.** In first-order reactions, the rate of loss of the substance is proportional to the amount of substance present at any given time, t.

$$r = -kC = \frac{dC}{dt} \tag{4-12}$$

The differential equation may be integrated to yield either

$$\ln \frac{C}{C_o} = -kt \tag{4-13}$$

or

$$C = C_o e^{-kt} = C_o \exp(-kt) \tag{4-14}$$

where C = concentration at any time t
 C_o = initial concentration
 e = exp = exponential e = 2.7183

For simple completely mixed systems with first-order reactions, the total mass of substance (M) is equal to the product of the concentration and volume ($C\Psi$) and, when Ψ is a constant, the mass rate of decay of the substance is

$$\frac{dM}{dt} = \frac{d(C\Psi)}{dt} = \Psi \frac{d(C)}{dt} \tag{4-15}$$

Because first-order reactions can be described by Equation 4–12, we can rewrite Equation 4–10 as

$$\frac{dM}{dt} = \frac{d(\text{in})}{dt} - \frac{d(\text{out})}{dt} - kC\Psi \tag{4-16}$$

EXAMPLE 4-6 A well-mixed sewage lagoon (a shallow pond) is receiving 430 $\text{m}^3 \cdot \text{d}^{-1}$ of untreated sewage. The lagoon has a surface area of 10 ha (hectares) and a depth of 1.0 m. The pollutant concentration in the raw sewage discharging into the lagoon is 180 mg $\cdot$ L^{-1}. The organic matter in the sewage degrades biologically (decays) in the lagoon according to first-order kinetics. The reaction rate constant (decay coefficient) is 0.70 d^{-1}. Assuming no other water losses or gains (evaporation, seepage, or rainfall) and that the lagoon is completely mixed, find the steady-state concentration of the pollutant in the lagoon effluent.

Solution We begin by drawing the mass-balance diagram.

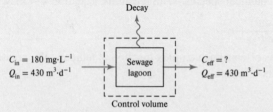

The mass-balance equation may be written as

Accumulation = input rate − output rate − decay rate

Assuming steady-state conditions, that is, accumulation = 0, then

Input rate = output rate + decay rate

This may be written in terms of the notation in the figure as

$$C_{in} Q_{in} = C_{eff} Q_{eff} + kC_{lagoon}\Psi$$

Solving for C_{eff}, we have

$$C_{eff} = \frac{C_{in}\, Q_{in} - kC_{lagoon}\Psi}{Q_{eff}}$$

Now calculate the values for terms in the equation. The input mass rate ($C_{in}Q_{in}$) is

$$(180\ mg \cdot L^{-1})(430\ m^3 \cdot d^{-1})(1000\ L \cdot m^{-3}) = 77{,}400{,}000\ mg \cdot d^{-1}$$

With a lagoon volume of

$$(10\ ha)(10^4\ m^2 \cdot ha^{-1})(1\ m) = 100{,}000\ m^3$$

and the decay coefficient of $0.70\ d^{-1}$, the decay rate is

$$kC\Psi = (0.70\ d^{-1})(100{,}000\ m^3)(1000\ L \cdot m^{-3})(C_{lagoon}) = (70{,}000{,}000\ L \cdot d^{-1})(C_{lagoon})$$

Now using the assumption that the lagoon is completely mixed, we assume that $C_{eff} = C_{lagoon}$. Thus,

$$kC\Psi = (70{,}000{,}000\ L \cdot d^{-1})(C_{eff})$$

Substituting into the mass-balance equation

$$\text{Output rate} = 77{,}400{,}000\ mg \cdot d^{-1} - (70{,}000{,}000\ L \cdot d^{-1} \times C_{eff})$$

or

$$C_{eff}(430\ m^3 \cdot d^{-1})(1000\ L \cdot m^{-3}) = 77{,}400{,}000\ mg \cdot d^{-1} - (70{,}000{,}000\ L \cdot d^{-1} \times C_{eff})$$

Solving for C_{eff}, we have

$$C_{eff} = \frac{77{,}400{,}000\ mg \cdot d^{-1}}{70{,}430{,}000\ L \cdot d^{-1}} = 1.10\ mg \cdot L^{-1}$$

Plug-Flow with Reaction.   As noted in Figure 4–1, in plug-flow systems, the tank car, or "plug" element of fluid, does not mix with the fluid ahead or behind it. However, a reaction can take place in the tank car element. Thus, even at steady-state, the contents within the element can change with time as the plug moves downstream. The control volume for the mass balance is the plug or differential element of fluid. The mass balance for this moving plug may be written as

$$\frac{dM}{dt} = \frac{d(\text{in})}{dt} - \frac{d(\text{out})}{dt} + \Psi\frac{d(C)}{dt} \tag{4–17}$$

Because no mass exchange occurs across the plug boundaries (in our railroad car analogy, there is no mass transfer between the box cars and the tank car), $d(\text{in})$ and $d(\text{out}) = 0$. Equation 4–17 may be rewritten as

$$\frac{dM}{dt} = 0 - 0 + \Psi\frac{d(C)}{dt} \tag{4–18}$$

As noted earlier in Equation 4–12, for a first-order decay reaction, the right-hand term may be expressed as

$$\frac{dC}{dt} = -kC \tag{4–19}$$

The total mass of substance (M) is equal to the product of the concentration and volume ($C\Psi$) and, when Ψ is a constant, the mass rate of decay of the substance in Equation 4–18 may be expressed as

$$\Psi\frac{dC}{dt} = -kC\Psi \tag{4–20}$$

where the left-hand side of the equation $= dM/dt$. The steady-state solution to the mass-balance equation for the plug-flow system with first-order kinetics is

$$\ln \frac{C_{out}}{C_{in}} = -kt_o \qquad (4\text{--}21)$$

or

$$C_{out} = (C_{in})e^{-kt_o} \qquad (4\text{--}22)$$

where k = reaction rate constant (in s^{-1}, min^{-1}, or d^{-1})

$\qquad t_o$ = residence time in plug-flow system (in s, min, or d)

In a plug-flow system of length L, each plug travels for a period $= L/u$, where u = the speed of flow. Alternatively, for a cross-sectional area A, the residence time is

$$t_o = \frac{(L)(A)}{(u)(A)} = \frac{\forall}{Q} \qquad (4\text{--}23)$$

where $\forall$ = volume of the plug-flow system (in m^3)

$\qquad Q$ = flow rate (in $m^3 \cdot s^{-1}$)

Thus, for example, Equation 4–21, may be rewritten as

$$\ln \frac{C_{out}}{C_{in}} = -k\frac{L}{u} = -k\frac{\forall}{Q} \qquad (4\text{--}24)$$

where L = length of the plug-flow segment (in m)

$\qquad u$ = linear velocity (in $m \cdot s^{-1}$)

Although the concentration within a given plug changes over time as the plug moves downstream, the concentration at a fixed point in the plug-flow system remains constant with respect to time. Thus, Equation 4–24 has no time dependence.

Example 4–7 illustrates an application of plug-flow with reaction.

EXAMPLE 4–7   A wastewater treatment plant must disinfect its effluent before discharging the wastewater to a near-by stream. The wastewater contains 4.5×10^5 fecal coliform colony-forming units (CFU) per liter. The maximum permissible fecal coliform concentration that may be discharged is 2000 fecal coliform $CFU \cdot L^{-1}$. It is proposed that a pipe carrying the wastewater be used for disinfection process. Determine the length of pipe required if the linear velocity of the wastewater in the pipe is $0.75 \ m \cdot s^{-1}$. Assume that the pipe behaves as a steady-state plug-flow system and that the reaction rate constant for destruction of the fecal coliforms is $0.23 \ min^{-1}$.

Solution   The mass balance diagram is sketched here. The control volume is the pipe itself.

$C_{in} = 4.5 \times 10^5 \ CFU \cdot L^{-1}$
$u = 0.75 \ m \cdot s^{-1}$

$L = ?$

$C_{out} = 2000 \ CFU \cdot L^{-1}$
$u = 0.75 \ m \cdot s^{-1}$

Using the steady-state solution to the mass-balance equation, we obtain

$$\ln \frac{C_{out}}{C_{in}} = -k\frac{L}{u}$$

$$\ln \frac{2000 \ CFU \cdot L^{-1}}{4.5 \times 10^5 \ CFU \cdot L^{-1}} = -0.23 \ min^{-1} \frac{L}{(0.75 \ m \cdot s^{-1})(60 \ s \cdot min^{-1})}$$

Solving for the length of pipe, we have

$$\ln(4.44 \times 10^{-3}) = -0.23 \text{ min}^{-1} \frac{L}{45 \text{ m} \cdot \text{min}^{-1}}$$

$$-5.42 = -0.23 \text{ min}^{-1} \frac{L}{45 \text{ m} \cdot \text{min}^{-1}}$$

$$L = 1060 \text{ m}$$

A little over 1 km of pipe is needed to meet the discharge standard. For most wastewater treatment systems this would be an exceptionally long discharge and another alternative such as a baffled reactor (discussed in Chapter 10) would be investigated.

Reactors

The tanks in which physical, chemical, and biochemical reactions occur, for example, in water softening (Chapter 10) and wastewater treatment (Chapter 11), are called **reactors.** These reactors are classified based on their flow characteristics and their mixing conditions. With appropriate selection of control volumes, ideal chemical reactor models may be used to model natural systems.

Batch reactors are of the fill-and-draw type: materials are added to the tank (Figure 4–2a), mixed for sufficient time to allow the reaction to occur (Figure 4–2b), and then drained (Figure 4–2c). Although the reactor is well mixed and the contents are uniform at any instant in time, the composition within the tank changes with time as the reaction proceeds. A batch reaction is unsteady. Because there is no flow into or out of a batch reactor

$$\frac{d(\text{in})}{dt} = \frac{d(\text{out})}{dt} = 0$$

For a batch reactor Equation 4–16 reduces to

$$\frac{dM}{dt} = -kC\Psi \qquad (4\text{–}25)$$

As we noted in Equation 4–15

$$\frac{dM}{dt} = \Psi\frac{dC}{dt}$$

So that for a first-order reaction in a batch reactor, Equation 4–25 may be simplified to

$$\frac{dC}{dt} = -kC \qquad (4\text{–}26)$$

FIGURE 4–2

Batch reactor operation. **(a)** Materials added to the reactor. **(b)** Mixing and reaction. **(c)** Reactor is drained. *Note:* There is no influent or effluent during the reaction.

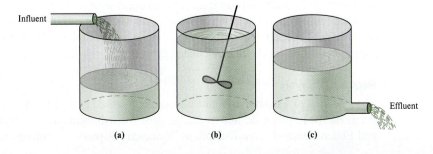

FIGURE 4–3

Schematic diagram of **(a)** completely mixed flow reactor (CMFR) and **(b)** the common diagram. The propeller indicates that the reactor is completely mixed.

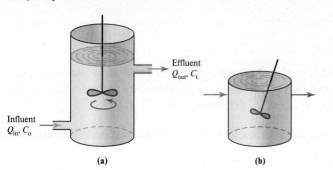

(a)   (b)

FIGURE 4–4

Schematic diagram of a plug-flow reactor (PFR). *Note: $t_3 > t_2 > t_1$.*

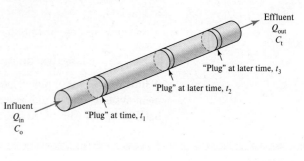

Flow reactors have a continuous type of operation: material flows into, through, and out of the reactor at all times. Flow reactors may be further classified by mixing conditions. The contents of a **completely mixed flow reactor** (CMFR), also called a **continuous-flow stirred tank reactor** (CSTR), ideally are uniform throughout the tank. A schematic diagram of a CMFR and the common flow diagram notation are shown in Figure 4–3. The composition of the effluent is the same as the composition in the tank. If the mass input rate into the tank remains constant, the composition of the effluent remains constant. The mass balance for a CMFR is described by Equation 4–16.

In **plug-flow reactors** (PFR), fluid particles pass through the tank in sequence. Those that enter first leave first. In the ideal case, it is assumed that no mixing occurs in the lateral direction. Although composition varies along the length of the tank, as long as the flow conditions remain steady, the composition of the effluent remains constant. A schematic diagram of a plug-flow reactor is shown in Figure 4–4. The mass balance for a PFR is described by Equation 4–18, where the time element (dt) is the time spent in the PFR as described by Equation 4–23. Real continuous-flow reactors are generally something in between a CMFR and PFR.

For time-dependent reactions, the time that a fluid particle remains in the reactor obviously affects the degree to which the reaction goes to completion. In ideal reactors the average time in the reactor (**detention time** or **retention time** or, for liquid systems, **hydraulic detention time** or **hydraulic retention time**) is defined as

$$t_o = \frac{\forall}{Q} \tag{4–27}$$

where t_o = theoretical detention time (in s)

$\forall$ = volume of fluid in reactor (in m³)

Q = flow rate into reactor (in m³ · s⁻¹)

Real reactors do not behave as ideal reactors because of density differences due to temperature or other causes, short circuiting because of uneven inlet or outlet conditions, and local turbulence or dead spots in the tank corners. The detention time in real tanks is generally less than the theoretical detention time calculated from Equation 4–27.

Reactor Analysis

The selection of a reactor either as a treatment method or as a model for a natural process depends on the behavior desired or recognized. We will examine the behavior of batch, CMFR, and PFR reactors in several situations. Situations of particular interest are the response of the

FIGURE 4–5

Example influent graphs of **(a)** step increase in influent concentration, **(b)** step decrease in influent concentration, and **(c)** a pulse or spike increase in influent concentration. *Note:* The size of the change is for illustration purposes only.

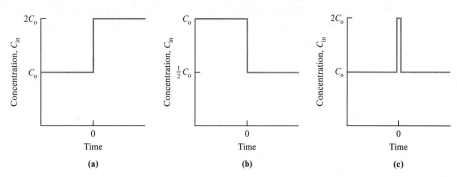

(a)   (b)   (c)

FIGURE 4–6

Batch reactor response to a step or pulse increase in concentration of a conservative substance. C_o = mass of conservative substance/volume of reactor.

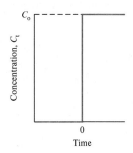

reactor to a sudden increase (Figure 4–5a) or decrease (Figure 4–5b) in the steady-state influent concentration for conservative and nonconservative species (commonly called a step increase or decrease) and the response to pulse or spike change in influent concentration (Figure 4–5c). We will present the plots of the effluent concentration for each of the reactor types for a variety of conditions to show the response to these influent changes.

For nonconservative substances, we will present the analysis for first-order reactions. The behavior of zero-order and second-order reactions will be summarized in comparison at the conclusion of this discussion.

Batch Reactor.   Laboratory experiments are often conducted in batch reactors because they are inexpensive and easy to build. Industries that generate small quantities of wastewater (less than 150 m$^3 \cdot$ d^{-1}) use batch reactors because they are easy to operate and provide an opportunity to check the wastewater for regulatory compliance before discharging it.

Because there is no influent to or effluent from a batch reactor, the introduction of a conservative substance into the reactor either as a step increase or a pulse results in an instantaneous increase in concentration of the conservative substance in the reactor. The concentration plot is shown in Figure 4–6.

Because there is no influent or effluent, for a nonconservative substance that decays as a first-order reaction, the mass balance is described by Equation 4–26. Integration yields

$$\frac{C_t}{C_o} = e^{-kt} \tag{4–28}$$

The final concentration plot is shown in Figure 4–7a. For the formation reaction, where the sign in Equation 4–28 is positive, the concentration plot is shown in Figure 4–7b.

FIGURE 4–7

Batch reactor response for **(a)** decay of a non-conservative substance and **(b)** a formation reaction.

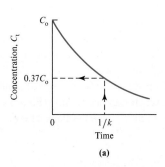

(a)

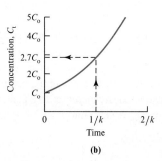

(b)

EXAMPLE 4–8   A contaminated soil is to be excavated and treated in a completely mixed aerated lagoon at a Superfund site. To determine the time it will take to treat the contaminated soil, a laboratory completely mixed batch reactor is used to gather the following data. Assuming a first-order reaction, estimate the rate constant, k, and determine the time to achieve 99% reduction in the original concentration.

Time (d)	Waste Concentration (mg · L^{-1})
1	280
16	132

Solution   The rate constant may be estimated by solving Equation 4–28 for k. Using the 1st and 16th day, the time interval $t = 16 - 1 = 15$ d

$$\frac{132 \text{ mg} \cdot \text{L}^{-1}}{280 \text{ mg} \cdot \text{L}^{-1}} = \exp[-k(15 \text{ d})]$$

$$0.4714 = \exp[-k(15)]$$

Taking the natural logarithm (base e) of both sides of the equation, we obtain

$$-0.7520 = -k(15)$$

Solving for k, we have

$$k = 0.0501 \text{ d}^{-1}$$

To achieve 99% reduction, the concentration at time t must be $1 - 0.99$ of the original concentration

$$\frac{C_t}{C_o} = 0.01$$

The estimated time is then

$$0.01 = \exp[-0.05(t)]$$

Taking the logarithm of both sides and solving for t, we get

$$t = 92 \text{ days}$$

CMFR.   A batch reactor is used for small volumetric flow rates. When water flow rates are greater than 150 m^3 · d^{-1}, a CMFR may be selected for chemical mixing. Examples of this application include equalization reactors to adjust the pH, precipitation reactors to remove metals, and mixing tanks (called **rapid mix** or **flash mix tanks**) for water treatment. Because municipal wastewater flow rates vary over the course of a day, a CMFR (called an **equalization basin**) may be placed at the treatment plant influent point to level out the flow and concentration changes. Some natural systems such as a lake or the mixing of two streams or the air in a room or over a city may be modeled as a CMFR as an approximation of the real mixing that is taking place.

For a step increase in a conservative substance entering a CMFR, the initial level of the conservative substance in the reactor is C_o prior to $t = 0$. At $t = 0$, the influent concentration (C_{in}) instantaneously increases to C_1 and remains at this concentration (Figure 4–8a). With balanced fluid flow ($Q_{in} = Q_{out}$) into the CMFR and no reaction, the mass balance equation for a step increase is

$$\frac{dM}{dt} = C_1 Q_{in} - C_{out} Q_{out} \qquad (4\text{–}29)$$

FIGURE 4–8

Response of a CMFR to **(a)** a step increase in the influent concentration of a conservative substance from concentration C_o to a new concentration C_1. **(b)** Effluent concentration.

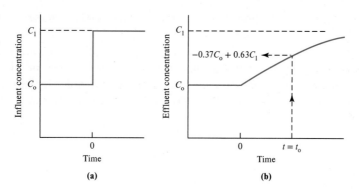

FIGURE 4–9

Flushing of CMFR resulting from **(a)** a step decrease in influent concentration of a conservative substance from C_o to 0. **(b)** Effluent concentration.

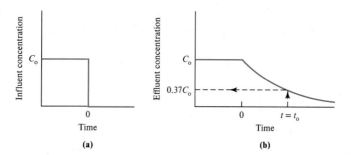

where $M = C\mathrm{V}$. The solution is

$$C_t = C_o \left[\exp\left(-\frac{t}{t_o}\right) \right] + C_1 \left[1 - \exp\left(-\frac{t}{t_o}\right) \right] \tag{4–30}$$

where C_t = concentration at any time t

C_o = concentration in reactor prior to step change

C_1 = concentration in influent after instantaneous increase

t = time after step change

t_o = theoretical detention time = V/Q

Figure 4–8b shows the effluent concentration plot.

Flushing of a nonreactive contaminant from a CMFR by a contaminant-free fluid is an example of a step change in the influent concentration (Figure 4–9a). Because $C_{in} = 0$ and no reaction takes place, the mass balance equation is

$$\frac{dM}{dt} = - C_{out} Q_{out} \tag{4–31}$$

where $M = C\mathrm{V}$. The initial concentration is

$$C_o = \frac{M}{\mathrm{V}} \tag{4–32}$$

Solving Equation 4–31 for any time $t \geq 0$, we obtain

$$C_t = C_o \exp\left(-\frac{t}{t_o}\right) \tag{4–33}$$

where $t_o = \mathrm{V}/Q$ as noted in Equation 4–27. Figure 4–9b shows the effluent concentration plot.

EXAMPLE 4–9   Before entering an underground utility vault to do repairs, a work crew analyzed the gas in the vault and found that it contained 29 mg · m^{-3} of hydrogen sulfide. Because the allowable exposure level is 14 mg · m^{-3}, the work crew began ventilating the vault with a blower. If the volume of the vault is 160 m^3 and the flow rate of contaminant-free air is 10 m^3 · min^{-1}, how long will it take to lower the hydrogen sulfide level to a level that will allow the work crew to enter? Assume the manhole behaves as a CMFR and that hydrogen sulfide is nonreactive in the time period considered.

Solution   This is a case of flushing a nonreactive contaminant from a CMFR. The theoretical detention time is

$$t_o = \frac{\math7V}{Q} = \frac{160 \text{ m}^3}{10 \text{ m}^3 \cdot \text{min}^{-1}} = 16 \text{ min}$$

The required time is found by solving Equation 4–33 for t

$$\frac{14 \text{ mg} \cdot \text{m}^{-3}}{29 \text{ mg} \cdot \text{m}^{-3}} = \exp\left(-\frac{t}{16 \text{ min}}\right)$$

$$0.4828 = \exp\left(-\frac{t}{16 \text{ min}}\right)$$

Taking the logarithm to the base e of both sides

$$-0.7282 = -\frac{t}{16 \text{ min}}$$

$$t = 11.6 \text{ or } 12 \text{ min to lower the concentration to the allowable level}$$

Because the odor threshold for H$_2$S is about 0.18 mg · m^{-3}, the vault will still have quite a strong odor after 12 min.

A precautionary note is in order here. H$_2$S is commonly found in confined spaces such as manholes. It is a very toxic poison and has the unfortunate property of deadening the olfactory senses. Thus, you may not smell it after a few moments even though the concentration has not decreased. Each year a few individuals in the United States die because they have entered a confined space without taking stringent safety precautions.

Because a CMFR is completely mixed, the response of a CMFR to a step change in the influent concentration of a reactive substance results in an immediate corresponding change in the effluent concentration. For this analysis we begin with the mass balance for a balanced flow ($Q_{in} = Q_{out}$) CMFR operating under steady-state conditions with first-order decay of a reactive substance.

$$\frac{dM}{dt} = C_{in}Q_{in} - C_{out}Q_{out} - kC_{out}\math7V \tag{4–34}$$

where $M = C\math7V$ Because the flow rate and volume are constant, we may divide through by $\math7V$ and simplify to obtain

$$\frac{dC}{dt} = \frac{1}{t_o}(C_{in} - C_{out}) - kC_{out} \tag{4–35}$$

where $t_o = \math7V/Q$ as noted in Equation 4–27. Under steady-state conditions, $dC/dt = 0$ and the solution for C_{out} is

$$C_{out} = \frac{C_o}{1 + kt_o} \tag{4–36}$$

FIGURE 4–10

Steady-state response of CMFR to **(a)** a step increase in influent concentration of a reactive substance. **(b)** Effluent concentration. *Note:* Steady-state conditions exist prior to $t = 0$.

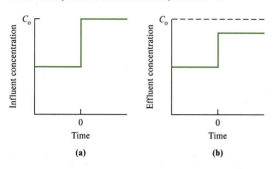

(a)   (b)

FIGURE 4–11

Non-steady-state response of CMFR to **(a)** step decrease from C_o to 0 of influent C_o reactive substance. **(b)** Effluent concentration.

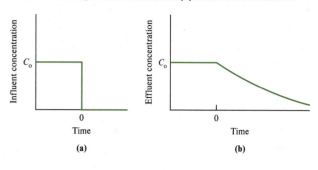

(a)   (b)

where $C_o = C_{in}$ immediately after the step change. Note that C_{in} can be nonzero before the step change. For a first-order reaction where material is produced, the sign of the reaction term is positive and the solution to the mass balance equations is

$$C_{out} = \frac{C_o}{1 - kt_o} \qquad (4\text{–}37)$$

The behavior of the CSTR described by Equation 4–36 is shown diagrammatically in Figure 4–10. The steady-state effluent concentration (C_{out} in Figure 4–10b) is less than the influent concentration because of the decay of a reactive substance. From Equation 4–36, you may note that the effluent concentration is equal to the influent concentration divided by $1 + kt_o$.

A step decrease in the influent concentration to zero ($C_{in} = 0$) in a balanced flow ($Q_{in} = Q_{out}$) CMFR operating under non-steady-state conditions with first-order decay of a reactive substance may be described by rewriting Equation 4–34

$$\frac{dM}{dt} = 0 - C_{out}Q_{out} - kC_{out}\forall \qquad (4\text{–}38)$$

where $M = C\forall$. Because the volume is constant, we may divide through by $\forall$ and simplify to obtain

$$\frac{dC}{dt} = \left(\frac{1}{t_o} + k\right) C_{out} \qquad (4\text{–}39)$$

where $t_o = \forall/Q$ as noted in Equation 4–27. The solution for C_{out} is

$$C_{out} = C_o \exp\left[-\left(\frac{1}{t_o} + k\right)t\right] \qquad (4\text{–}40)$$

where C_o is the effluent concentration at $t = 0$.

The concentration plots are shown in Figure 4–11.

PFR.   Pipes and long narrow rivers approximate the ideal conditions of a PFR. Biological treatment in municipal wastewater treatment plants is often conducted in long narrow tanks that may be modeled as a PFR.

A step change in the influent concentration of a conservative substance in a plug-flow reactor results in an identical step change in the effluent concentration at a time equal to the theoretical detention time in the reactor as shown in Figure 4–12.

Equation 4–21 is a solution to the mass-balance equation for a steady-state first-order reaction in a PFR. The concentration plot for a step change in the influent concentration is shown in Figure 4–13.

FIGURE 4–12

Response of a PFR to **(a)** a step increase in the influent concentration of a conservative substance. **(b)** Effluent concentration.

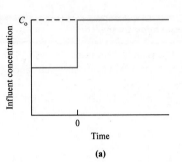

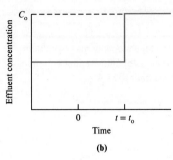

FIGURE 4–13

Response of a PFR to **(a)** a step increase in the influent concentration of a reactive substance. **(b)** Effluent concentration.

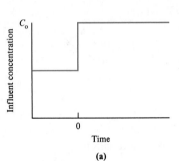

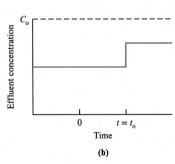

FIGURE 4–14

Passage of a pulse change in influent concentration of a conservative substance through a PFR. u is the linear velocity of the fluid through the PFR.

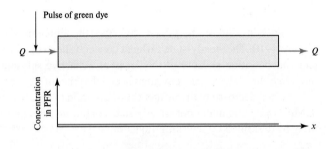

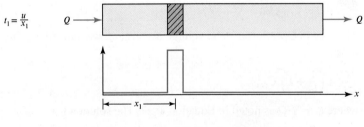

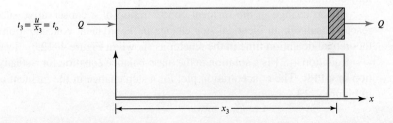

TABLE 4–1 **Comparison of Steady-state Mean Retention Times for Decay Reactions of Different Order[a]**

Reaction Order	r	Ideal Batch	Ideal Plug Flow	Ideal CMFR
Zero[b]	$-k$	$\dfrac{(C_o - C_t)}{k}$	$\dfrac{(C_o - C_t)}{k}$	$\dfrac{(C_o - C_t)}{k}$
First	$-kC$	$\dfrac{\ln(C_o/C_t)}{k}$	$\dfrac{\ln(C_o/C_t)}{k}$	$\dfrac{(C_o/C_t) - 1}{k}$
Second	$-kC^2$	$\dfrac{(C_o/C_t) - 1}{kC_o}$	$\dfrac{(C_o/C_t) - 1}{kC_o}$	$\dfrac{(C_o/C_t) - 1}{kC_t}$

(Equations for Mean Retention Times (t_o))

[a]C_o = initial concentration or influent concentration; C_t = final condition or effluent concentration; units for k: for zero-order reactions – mass · volume^{-1} · time^{-1}; for first-order reactions – time^{-1}; for second-order reactions – volume · mass^{-1} · time^{-1}.

[b]Expressions are valid for $kt_o \le C_o$; otherwise $C_t = 0$.

TABLE 4–2 **Comparison of Steady-state Performance for Decay Reactions of Different Order[a]**

Reaction Order	r	Ideal Batch	Ideal Plug Flow	Ideal CMFR
Zero[b] $t \le C_o/k$	$-k$	$C_o - kt$	$C_o - kt_o$	$C_o - kt_o$
$t > C_o/k$		0		
First	$-kC$	$C_o[\exp(-kt)]$	$C_o[\exp(-kt_o)]$	$\dfrac{C_o}{1 + kt_o}$
Second	$-kC^2$	$\dfrac{C_o}{1 + ktC_o}$	$\dfrac{C_o}{1 + kt_oC_o}$	$\dfrac{(4kt_oC_o + 1)^{1/2} - 1}{2kt_o}$

(Equations for C_t)

[a]C_o = initial concentration or influent concentration; C_t = final condition or effluent concentration.

[b]Time conditions are for ideal batch reactor only.

A pulse entering a PFR travels as a discrete element as illustrated in Figure 4–14 for a pulse of green dye. The passage of the pulse through the PFR and the plots of concentration with distance along the PFR are also shown.

For reaction orders greater than or equal to one, an ideal PFR will always require less volume than a single ideal CMFR to achieve the same percent destruction.

Reactor Comparison. Although first-order reactions are common in environmental systems, other reaction orders may be more appropriate. Tables 4–1 and 4–2 compare the reactor types for zero-, first-, and second-order reactions.

EXAMPLE 4–10 A chemical degrades in a flow-balanced, steady-state CMFR according to first-order reaction kinetics. The upstream concentration of the chemical is 10 mg · L^{-1} and the downstream concentration is 2 mg · L^{-1}. Water is being treated at a rate of 29 m^3 · min^{-1}. The volume of the tank is 580 m^3. What is the rate of decay? What is the rate constant?

Solution From Equation 4–11, we note that for a first-order reaction, the rate of decay $r = -kC$. To find the rate of decay, we must solve Equation 4–34 for kC to determine the reaction rate.

$$\frac{dM}{dt} = C_{in} Q_{in} - C_{out} Q_{out} - kC_{out}\forall$$

For steady state, there is no mass accumulation, so $dM/dt = 0$. Because the reactor is flow-balanced, $Q_{in} = Q_{out} = 29 \text{ m}^3 \cdot \text{min}^{-1}$. The mass balance equation may be rewritten

$$kC_{out}\forall = C_{in} Q_{in} - C_{out} Q_{out}$$

Solving for the reaction rate, we obtain

$$r = kC = \frac{C_{in} Q_{in} - C_{out} Q_{out}}{\forall}$$

$$= kC = \frac{(10 \text{ mg} \cdot \text{L}^{-1})(29 \text{ m}^3 \cdot \text{min}^{-1}) - (2 \text{ mg} \cdot \text{L}^{-1})(29 \text{ m}^3 \cdot \text{min}^{-1})}{580 \text{ m}^3} = 0.4 \text{ mg} \cdot \text{L}^{-1} \cdot \text{min}^{-1}$$

The rate constant, k, can be determined using the equations given in Table 4–1. For a first-order reaction in a CMFR

$$t_o = \frac{(C_o/C_t) - 1}{k}$$

The mean hydraulic detention time (t_o) is

$$t_o = \frac{\forall}{Q} = \frac{580 \text{ m}^3}{29 \text{ m}^3 \cdot \text{min}^{-1}} = 20 \text{ min}$$

Solving the equation from the table for the rate constant, k, we get

$$k = \frac{(C_o/C_t) - 1}{t_o}$$

and

$$k = \frac{\left(\dfrac{10 \text{ mg} \cdot \text{L}^{-1}}{2 \text{ mg} \cdot \text{L}^{-1}} - 1\right)}{20 \text{ min}} = 0.20 \text{ min}^{-1}$$

Reactor Design. Volume is the major design parameter in reactor design. In general, the influent concentration of material, the flow rate into the reactor, and the desired effluent concentration are known. As noted in Equation 4–27, the volume is directly related to the theoretical detention time and the flow rate into the reactor. Thus, the volume can be determined if the theoretical detention time can be determined. The equations in Table 4–1 may be used to determine the theoretical detention time if the decay rate constant, k, is available. The rate constant must be determined from the literature or laboratory experiments.

4–4 ENERGY BALANCES

First Law of Thermodynamics

The **first law of thermodynamics** states that (without nuclear reaction) energy can be neither created nor destroyed. As with the law of conservation of matter, it does not mean that the form of the energy does not change. For example, the chemical energy in coal can be changed to heat and electrical power. **Energy** is defined as the capacity to do useful work. **Work** is done by a force acting on a body through a distance. One **joule** (J) is the work done by a constant force of one newton when the body on which the force is exerted moves a distance of one meter in the direction of the force. **Power** is the rate of doing work or the rate of expanding energy. The first law may be expressed as

$$Q_H = U_2 - U_1 + W \tag{4-41}$$

where Q_H = heat absorbed (in kJ)

U_1, U_2 = internal energy (or thermal energy) of the system in states 1 and 2 (in kJ)

W = work (in kJ)

Fundamentals

Thermal Units of Energy.   Energy has many forms, among which are thermal, mechanical, kinetic, potential, electrical, and chemical. Thermal units were invented when heat was considered a substance (caloric), and the units are consistent with conservation of a quantity of substance. Subsequently, we have learned that energy is not a substance but mechanical energy of a particular form. With this in mind we will still use the common metric thermal unit of energy, the calorie.* One **calorie** (cal) is the amount of energy required to raise the temperature of one gram of water from $14.5°C$ to $15.5°C$. In SI units 4.186 J = 1 cal.

The **specific heat** of a substance is the quantity of heat required to increase a unit mass of the substance one degree. Specific heat is expressed in metric units as $kcal \cdot kg^{-1} \cdot K^{-1}$ and in SI units as $kJ \cdot kg^{-1} \cdot K^{-1}$ where K is kelvins and 1 K = 1°C.

Enthalpy is a thermodynamic property of a material that depends on temperature, pressure, and the composition of the material. It is defined as

$$H = U + P\forall \tag{4-42}$$

where H = enthalpy (in kJ)

U = internal energy (or thermal energy) (in kJ)

P = pressure (in kPa)

$\forall$ = volume (in m^3)

Think of enthalpy as a combination of thermal energy (U) and flow energy ($P\forall$). Flow energy should not be confused with kinetic energy ($\frac{1}{2}Mv^2$). Historically, H has been referred to as a system's "heat content." Because *heat* is correctly defined only in terms of energy transfer across a boundary, this is not considered a precise thermodynamic description and enthalpy is the preferred term.

When a non-phase-change process[†] occurs without a change in volume, a change in internal energy is defined as

$$\Delta U = Mc_v\Delta T \tag{4-43}$$

where ΔU = change in internal energy

M = mass

c_v = specific heat at constant volume

ΔT = change in temperature

When a non-phase-change process occurs without a change in pressure, a change in enthalpy is defined as

$$\Delta H = Mc_p\Delta T \tag{4-44}$$

where ΔH = enthalpy change

c_p = specific heat at constant pressure

---

*In discussing food metabolism, physiologists also use the term *Calorie*. However, the food Calorie is equivalent to a *kilocalorie* in the metric system. We will use the units cal or kcal throughout this text.

[†]Non-phase-change means, for example, water is not converted to steam.

TABLE 4–3   **Specific Heat Capacities for Common Substances**

Substance	c_p (kJ · kg^{-1} · K^{-1})
Air (293.15 K)	1.00
Aluminum	0.95
Beef	3.22
Cement, Portland	1.13
Concrete	0.93
Copper	0.39
Corn	3.35
Dry soil	0.84
Human being	3.47
Ice	2.11
Iron, cast	0.50
Steel	0.50
Poultry	3.35
Steam (373.15 K)	2.01
Water (288.15 K)	4.186
Wood	1.76

Source: Adapted from Guyton (1961), Hudson (1959), Masters (1998), Salvato (1972).

Equations 4–43 and 4–44 assume that the specific heat is constant over the range of temperature (ΔT). Solids and liquids are nearly incompressible and therefore do virtually no work. The change in $P\forall$ is zero, making the changes in H and U identical. Thus, for solids and liquids, we can generally assume $c_v = c_p$ and $\Delta U = \Delta H$, so the change in energy stored in a system is

$$\Delta H = Mc_v \Delta T \tag{4–45}$$

Specific heat capacities for some common substances are listed in Table 4–3.

When a substance changes **phase** (i.e., it is transformed from solid to liquid or liquid to gas), energy is absorbed or released without a change in temperature. The energy required to cause a phase change of a unit mass from a solid to a liquid at constant pressure is called the **latent heat of fusion** or **enthalpy of fusion.** The energy required to cause a phase change of a unit mass from a liquid to a gas at constant pressure is called the **latent heat of vaporization** or **enthalpy of vaporization.** The same amounts of energy are released in condensing the vapor and freezing the liquid. For water the enthalpy of fusion at 0°C is 333 kJ · kg^{-1} and the enthalpy of vaporization is 2257 kJ · kg^{-1} at 100°C. The enthalpy of condensation is 2490 kJ · kg^{-1} at 0°C.

EXAMPLE 4–11   Standard physiology texts (Guyton, 1961) report that a person weighing 70.0 kg requires approximately 2000 kcal for simple existence, such as eating and sitting in a chair. Approximately 61% of all the energy in the foods we eat becomes heat during the process of formation of the energy-carrying molecule adenosine triphosphate (ATP) (Guyton, 1961). Still more energy becomes heat as it is transferred to functional systems of the cells. The functioning of the cells releases still more energy so that ultimately "all the energy released by metabolic processes eventually becomes heat" (Guyton, 1961). Some of this heat is used to maintain the body at a normal temperature of 37°C. What fraction of the 2000 kcal is used to maintain the body temperature at 37°C if the room temperature is 20°C? Assume the specific heat of a human is 3.47 kJ · kg^{-1} · K^{-1}.

Solution   Recognizing that ΔT in °C = ΔT in K, the change in energy stored in the body is

$$\Delta H = (70 \text{ kg})(3.47 \text{ kJ} \cdot \text{kg}^{-1} \cdot \text{K}^{-1})(37°\text{C} - 20°\text{C}) = 4129.30 \text{ kJ}$$

Converting the 2000 kcal to kJ

$$(2000 \text{ kcal})(4.186 \text{ kJ} \cdot \text{kcal}^{-1}) = 8372.0 \text{ kJ}$$

So the fraction of energy used to maintain temperature is approximately

$$\frac{4129.30 \text{ kJ}}{8372.0 \text{ kJ}} = 0.49, \text{ or about } 50\%$$

The remaining energy must be removed if the body temperature is not to rise above normal. The mechanisms of removing energy by heat transfer are discussed in the following sections.

Energy Balances.   If we say that the first law of thermodynamics is analogous to the law of conservation of matter, then energy is analogous to matter because it too can be "balanced." The simplest form of the energy balance equation is

Loss of enthalpy of hot body = gain of enthalpy by cold body   (4–46)

EXAMPLE 4–12   The Rhett Butler Peach, Co. dips the peaches in boiling water (100°C) to remove the skin (a process called blanching) before canning them. The wastewater from this process is high in organic matter and it must be treated before disposal. The treatment process is a biological process that operates at 20°C. Thus, the wastewater must be cooled to 20°C before disposal. Forty cubic meters (40 m³) of wastewater is discharged to a concrete tank at a temperature of 20°C to allow it to cool. Assuming no losses to the surroundings and that the concrete tank has a mass of 42,000 kg and a specific heat capacity of 0.93 kJ · kg⁻¹ · K⁻¹, what is the equilibrium temperature of the concrete tank and the wastewater?

Solution   Assuming the density of the water is 1000 kg · m⁻³, the loss in enthalpy of the boiling water is

$$\Delta H = (1000 \text{ kg} \cdot \text{m}^{-3})(40 \text{ m}^3)(4.186 \text{ kJ} \cdot \text{kg}^{-1} \cdot \text{K}^{-1})(373.15 - T) = 62,480,236 - 167,440T$$

where the absolute temperature is 273.15 + 100 = 373.15 K.

The gain in enthalpy of the concrete tank is

$$\Delta H = (42,000 \text{ kg})(0.93 \text{ kJ} \cdot \text{kg}^{-1} \cdot \text{K}^{-1})(T - 293.15) = 39,060T - 11,450,439$$

The equilibrium temperature is found by setting the two equations equal and solving for the temperature.

$$(\Delta H)_{water} = (\Delta H)_{concrete}$$

$$62,480,236 - 167,440T = 39,060T - 11,450,439$$

$$T = 358 \text{ K or } 85°\text{C}$$

This is not very close to the desired temperature without considering other losses to the surroundings. Convective and radiative heat losses discussed later on also play a role in reducing the temperature, but a cooling tower may be required to achieve 20°C.

For an open system, a more complete energy balance equation is

Net change in energy = energy of mass entering system − energy of mass leaving system

$$\pm \text{ energy flow into or out of system} \qquad (4\text{-}47)$$

For many environmental systems the time dependence of the change in energy (i.e., the rate of energy change) must be taken into account. Equation 4–47 may be written to account for time dependence as follows:

$$\frac{dH}{dt} = \frac{d(H)_{mass\,in}}{dt} + \frac{d(H)_{mass\,out}}{dt} \pm \frac{d(H)_{energy\,flow}}{dt} \qquad (4\text{-}48)$$

If we consider a region of space where a fluid flows in at a rate of dM/dt and also flows out at a rate of dM/dt, then the change in enthalpy due to this flow is

$$\frac{dH}{dt} = c_p M \frac{dT}{dt} + c_p T \frac{dM}{dt} \qquad (4\text{-}49)$$

where dM/dt is the mass flow rate (e.g., in kg · s^{-1}) and ΔT is the difference in temperature of the mass in the system and the mass outside of the system.

Note that Equations 4–47 and 4–48 differ from the mass-balance equation in that there is an additional term: "energy flow." This is an important difference for everything from photosynthesis (in which radiative energy from the sun is converted into plant material) to heat exchangers (in which chemical energy from fuel passes through the walls of the tubes of the heat exchanger to heat a fluid inside). The energy flow into (or out of) the system may be by conduction, convection, or radiation.

Conduction. Conduction is the transfer of heat through a material by molecular diffusion due to a temperature gradient. Fourier's law provides an expression for calculating energy flow by conduction.

$$\frac{dH}{dt} = -h_{tc} A \frac{dT}{dx} \qquad (4\text{-}50)$$

where dH/dt = rate of change of enthalpy (in kJ · s^{-1} or kW)

$\qquad h_{tc}$ = thermal conductivity (in kJ · s^{-1} · m^{-1} · K^{-1} or kW · m^{-1} · K^{-1})

$\qquad A$ = surface area (in m^2)

$\qquad dT/dx$ = change in temperature through a distance (in K · m^{-1})

Note that 1 kJ · s^{-1} = 1 kW. The average values for thermal conductivity for some common materials are given in Table 4–4.

TABLE 4–4	Thermal Conductivity for Some Common Materials[a]
Material	h_{tc} **(W · m^{-1} · K^{-1})**
Air	0.023
Aluminum	221
Brick, fired clay	0.9
Concrete	2
Copper	393
Glass-wool insulation	0.0377
Steel, mild	45.3
Wood	0.126

[a]Note that the units are equivalent to J · s^{-1} · m^{-1} · K^{-1}

Source: Adapted from Kuehn, Ramsey, and Threkeld (1998); Shortley and Williams (1955).

FIGURE 4–15

Sinusoidal Waves

The wavelength (λ) is the distance between two peaks or troughs.

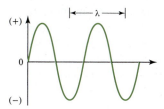

Convection. Forced convective heat transfer is the transfer of thermal energy by means of large-scale fluid motion such as a flowing river or aquifer or the wind blowing. The convective heat transfer between a fluid at a temperature T_f and a solid surface at a temperature T_s can be described by Equation 4–51.

$$\frac{dH}{dt} = h_c A (T_f - T_s) \tag{4–51}$$

where h_c = convective heat transfer coefficient (in $kJ \cdot s^{-1} \cdot m^{-2} \cdot K^{-1}$)
  A = surface area (in m^2)

Radiation. Although both conduction and convection require a medium to transport energy, radiant energy is transported by electromagnetic radiation. The radiative transfer of heat involves two processes: the absorption of radiant energy by an object and the radiation of energy by that object. The change in enthalpy due to the radiative heat transfer is the energy absorbed minus the energy emitted and can be expressed as

$$\frac{dH}{dt} = E_{abs} - E_{emitted} \tag{4–52}$$

Thermal radiation is emitted when an electron moves from a higher energy state to a lower one. Radiant energy is transmitted in the form of waves. Waves are cyclical or **sinusoidal** as shown in Figure 4–15. The waves may be characterized by their wavelength (λ) or their frequency (v). The wavelength is the distance between successive peaks or troughs. Frequency and wavelength are related by the speed of light (c).

$$c = \lambda v \tag{4–53}$$

Planck's law relates the energy emitted to the frequency of the emitted radiation.

$$E = hv \tag{4–54}$$

where h = Planck's constant = $6.63 \times 10^{-34} \, J \cdot s$

The electromagnetic wave emitted when an electron makes a transition between two energy levels is called a **photon.** When the frequency is high (small wavelengths), the energy emitted is high. Planck's law also applies to the absorption of a photon of energy. A molecule can only absorb radiant energy if the wavelength of radiation corresponds to the difference between two of its energy levels.

Every object emits thermal radiation. The amount of energy radiated depends on the wavelength, surface area, and the absolute temperature of the object. The maximum amount of radiation that an object can emit at a given temperature is called **blackbody radiation.** An object that radiates the maximum possible intensity for every wavelength is called a **blackbody.** The term *blackbody* has no reference to the color of the body. A blackbody can also be characterized by the fact that all radiant energy reaching its surface is absorbed.

Actual objects do not emit or absorb as much radiation as a blackbody. The ratio of the amount of radiation an object emits to that a blackbody would emit is called the emissivity (ε). The energy spectrum of the sun resembles that of a blackbody at 6000 K. At normal atmospheric temperatures, the emissivity of dry soil and woodland is approximately 0.90.

Water and snow have emissivities of about 0.95. A human body, regardless of pigmentation, has an emissivity of approximately 0.97 (Guyton, 1961). The ratio of the amount of energy an object absorbs to that which a blackbody would absorb is called absorptivity (α). For most surfaces, the absorptivity is the same value as the emissivity.

Integration of Planck's equation over all wavelengths yields the radiant energy of a blackbody.

$$E_B = \sigma T^4 \tag{4–55}$$

where E_B = blackbody emission rate (in $W \cdot m^{-2}$)
   σ = Stephan–Boltzmann constant = $5.67 \times 10^{-8}\ W \cdot m^{-2} \cdot K^{-4}$
   T = absolute temperature (in K)

For other than blackbodies, the right-hand side of the equation is multiplied by the emissivity.

For a body with an emissivity ε and an absorptivity α at a temperature T_b receiving radiation from its environment that is a blackbody of temperature $T_{environ}$, we can express the change in enthalpy as

$$\frac{dH}{dt} = A(\varepsilon\sigma T_b^4 - \alpha\sigma T_{environ}^4) \tag{4–56}$$

where A = surface area of the body (in m^2)

The solution to thermal radiation problems is highly complex because of the "re-radiation" of surrounding objects. In addition the rate of radiative cooling will change with time as the difference in temperatures changes; initially the change per unit of time will be large because of the large difference in temperature. As the temperatures approach each other, the rate of change will slow. In the following problem we use an arithmetic average temperature as a first approximation to the actual average temperature.

EXAMPLE 4–13   As mentioned in Example 4–12, heat losses due to convection and radiation were not considered. Using the following assumptions, estimate how long it will take for the wastewater and concrete tank to come to the desired temperature (20°C) if radiative cooling and convective cooling are considered. Assume that the average temperature of the water and concrete tank while cooling between 85°C (their combined temperature from Example 4–12) and 20°C is 52.5°C. Also assume that the mean radiant temperature of the surroundings is 20°C, that both the cooling tank and the surrounding environment radiate uniformly in all directions, that their emissivities are the same (0.90), that the surface area of the concrete tank including the open water surface is 56 m^2, and that the convective heat transfer coefficient is 13 $J \cdot s^{-1} \cdot m^{-2} \cdot K^{-1}$.

Solution   The required change in enthalpy for the wastewater is

$$\Delta H = (1000\ kg \cdot m^{-3})(40\ m^3)(4.186\ kJ \cdot kg^{-1} \cdot K^{-1})(325.65 - 293.15) = 5{,}441{,}800\ kJ$$

where the absolute temperature of the wastewater is $273.15 + 52.5 = 325.65$ K.

The required change in enthalpy of the concrete tank is

$$\Delta H = (42{,}000\ kg)(0.93\ kJ \cdot kg^{-1} \cdot K^{-1})(325.65 - 293.15) = 1{,}269{,}450\ kJ$$

For a total of $5{,}441{,}800 + 1{,}269{,}450 = 6{,}711{,}250$ kJ, or 6,711,250,000 J

In estimating the time to cool down by radiation alone, we note that the emissivities are the same for the tank and the environment and that the net radiation is the result of the difference in absolute temperatures.

$$\begin{aligned} E_B &= \varepsilon\sigma\ (T_c^4 - T_{environ}^4) \\ &= \varepsilon\sigma T^4 = (0.90)(5.67 \times 10^{-8}\ W \cdot m^{-2} \cdot K^{-4})[(273.15 + 52.5)^4 - (273.15 + 20)^4] \\ &= 197\ W \cdot m^{-2} \end{aligned}$$

The rate of heat loss is

$(197 \text{ W} \cdot \text{m}^{-2})(56 \text{ m}^2) = 11{,}032 \text{ W, or } 11{,}032 \text{ J} \cdot \text{s}^{-1}$

Using Equation 4–51, the convective cooling rate may be estimated.

$$\frac{dH}{dt} = h_c A (T_f - T_s)$$

$$= (13 \text{ J} \cdot \text{s}^{-1} \cdot \text{m}^{-2} \cdot \text{K}^{-1})(56 \text{ m}^2)[(273.15 + 52.5) - (273.15 + 20)]$$

$$= 23{,}660 \text{ J} \cdot \text{s}^{-1}$$

The time to cool down is then

$$\frac{6{,}711{,}250{,}000 \text{ J}}{11{,}032 \text{ J} \cdot \text{s}^{-1} + 23{,}660 \text{ J} \cdot \text{s}^{-1}} = 193{,}452 \text{ s, or } 2.24 \text{ days}$$

This is quite a long time. If land is not at a premium and several tanks can be built, then the time may not be a relevant consideration. Alternatively, other options must be considered to reduce the time. One alternative is to utilize conductive heat transfer and build a heat exchanger. As an energy-saving measure, the heat exchanger could be used to heat the incoming water needed in the blanching process.

Overall Heat Transfer.   Most practical heat transfer problems involve multiple heat transfer modes. For these cases, it is convenient to use an overall heat transfer coefficient that incorporates multiple modes. The form of the heat transfer equation then becomes

$$\frac{dH}{dt} = h_o A (\Delta T) \tag{4–57}$$

where h_o = overall heat transfer coefficient (in $\text{kJ} \cdot \text{s}^{-1} \cdot \text{m}^{-2} \cdot \text{K}^{-1}$)

ΔT = temperature difference that drives the heat transfer (in K)

Among their many responsibilities, environmental scientists (often with a job title of environmental sanitarian) are responsible for checking food safety in restaurants. This includes ensuring proper refrigeration of perishable foods. The following problem is an example of one of the items, namely the electrical rating of the refrigerator, that might be investigated in a case of food poisoning at a family gathering.

EXAMPLE 4–14   In evaluating a possible food "poisoning," Sam and Janet Evening evaluated the required electrical energy input to cool food purchased for a family reunion. The family purchased 12 kg of hamburger, 6 kg of chicken, 5 kg of corn, and 20 L of soda pop. They have a refrigerator in the garage in which they stored the food until the reunion. The specific heats of the food products (in $\text{kJ} \cdot \text{kg}^{-1} \cdot \text{K}^{-1}$) are hamburger: 3.22; chicken: 3.35; corn: 3.35; beverages: 4.186. The refrigerator dimensions are 0.70 m $\times$ 0.75 m $\times$ 1.00 m. The overall heat transfer coefficient for the refrigerator is 0.43 $\text{J} \cdot \text{s}^{-1} \cdot \text{m}^{-2} \cdot \text{K}^{-1}$. The temperature in the garage is 30°C. The food must be kept at 4°C to prevent spoilage. Assume that it takes 2 h for the food to reach a temperature of 4°C, that the meat has risen to 20°C in the time it takes to get it home from the store, and that the soda pop and corn have risen to 30°C. What electrical energy input (in kilowatts) is required during the first 2 h the food is in the refrigerator? What is the energy input required to maintain the temperature for the second 2 h. Assume the refrigerator interior is at 4°C when the food is placed in it and that the door is not opened during the 4-h period. Ignore the energy required to heat the air in the refrigerator, and assume that all the electrical energy is used to remove heat. If the refrigerator is rated at 875 W, is poor refrigeration a part of the food poisoning problem?

Solution The energy balance equation is of the form

$$\frac{dH}{dt} = \frac{d(H)_{mass\ in}}{dt} + \frac{d(H)_{mass\ out}}{dt} \pm \frac{d(H)_{energy\ flow}}{dt}$$

where dH/dt = the enthalpy change required to balance the input energy

$d(H)_{mass\ in}$ = change in enthalpy due to the food

$d(H)_{energy\ flow}$ = the change in enthalpy to maintain the temperature at 4°C

There is no $d(H)_{mass\ out}$.

Begin by computing the change in enthalpy for the food products.

Hamburger

$$\Delta H = (12\ \text{kg})(3.22\ \text{kJ} \cdot \text{kg}^{-1} \cdot \text{K}^{-1})(20°C - 4°C) = 618.24\ \text{kJ}$$

Chicken

$$\Delta H = (6\ \text{kg})(3.35\ \text{kJ} \cdot \text{kg}^{-1} \cdot \text{K}^{-1})(20°C - 4°C) = 321.6\ \text{kJ}$$

Corn

$$\Delta H = (5\ \text{kg})(3.35\ \text{kJ} \cdot \text{kg}^{-1} \cdot \text{K}^{-1})(30°C - 4°C) = 435.5\ \text{kJ}$$

Beverages
Assuming that 20 L = 20 kg

$$\Delta H = (20\ \text{kg})(4.186\ \text{kJ} \cdot \text{kg}^{-1} \cdot \text{K}^{-1})(30°C - 4°C) = 2176.72\ \text{kJ}$$

The total change in enthalpy = 618.24 kJ + 321.6 kJ + 435.5 kJ + 2176.72 kJ = 3,552.06 kJ

Based on a 2-h period to cool the food, the rate of enthalpy change is

$$\frac{3,552.06\ \text{kJ}}{(2\ \text{h})(3600\ \text{s} \cdot \text{h}^{-1})} = 0.493, \text{ or } 0.50\ \text{kJ} \cdot \text{s}^{-1}$$

The surface area of the refrigerator is

0.70 m × 1.00 m × 2 = 1.40 m^2

0.75 m × 1.00 m × 2 = 1.50 m^2

0.75 m × 0.70 m × 2 = 1.05 m^2

for a total of 3.95 m^2

The heat loss through walls of the refrigerator is then

$$\frac{dH}{dt} = (4.3 \times 10^{-4}\ \text{kJ} \cdot \text{s}^{-1} \cdot \text{m}^{-2} \cdot \text{K}^{-1})(3.95\ \text{m}^2)(30°C - 4°C) = 0.044\ \text{kJ} \cdot \text{s}^{-1}$$

In the first 2 h, the electrical energy required is 0.044 kJ · s^{-1} + 0.50 kJ · s^{-1} = 0.54 kJ · s^{-1}. Because 1 W = 1 J · s^{-1}, the electrical requirement is 0.54 kW, or 540 W. It does not appear that the refrigerator is part of the food poisoning problem.

In the second 2 h, the electrical requirement drops to 0.044 kW, or 44 W.

Note that at the beginning of this example we put "poisoning" in quotation marks because illness from food spoilage may be the result of microbial infection, which is not poisoning in the same sense as, for example, that caused by arsenic.

The result of Example 4–14 is based on an assumption of 100% efficiency in converting electrical energy to refrigeration. This is, of course, not possible and leads us to the second law of thermodynamics.

Second Law of Thermodynamics

The second law of thermodynamics states that energy flows from a region of higher concentration to one of lesser concentration, not the reverse, and that the quality degrades as it is transformed. All natural, spontaneous processes may be studied in the light of the second law, and in all such cases a particular one-sidedness is found. Thus, heat always flows spontaneously from a hotter body to a colder one; gases seep through an opening spontaneously from a region of higher pressure to a region of lower pressure. The second law recognizes that order becomes disorder, that randomness increases, and that structure and concentrations tend to disappear. It foretells elimination of gradients, equalization of electrical and chemical potential, and leveling of contrasts in heat and molecular motion unless work is done to prevent it. Thus, gases and liquids left by themselves tend to mix, not to unmix; rocks weather and crumble; iron rusts.

The degradation of energy as it is transformed means that enthalpy is wasted in the transformation. The fractional part of the heat which is wasted is termed unavailable energy. A mathematical expression called the **change in entropy** is used to express this unavailable energy.

$$\Delta s = Mc_p \ln\frac{T_2}{T_1} \tag{4–58}$$

where Δs = change in entropy
$\quad M$ = mass
$\quad c_p$ = specific heat at constant pressure
$\quad T_1, T_2$ = initial and final absolute temperature
$\quad \ln$ = natural logarithm

By the second law, entropy increases in any transformation of energy from a region of higher concentration to a lesser one. The higher the degree of disorder, the higher the entropy. Degraded energy is entropy, dissipated as waste products and heat.

Efficiency (η) or, perhaps, lack of efficiency is another expression of the second law. Sadi Carnot (1824) was the first to approach the problem of the efficiency of a heat engine (e.g., a steam engine) in a truly fundamental manner. He described a theoretical engine, now called a Carnot engine. Figure 4–16 is a simplified representation of a Carnot engine. In his engine, a material expands against a piston that is periodically brought back to its initial condition so that in any one cycle the change in internal energy of this material is zero, that is $U_2 - U_1 = 0$, and the first law of thermodynamics (Equation 4–41) reduces to

$$W = Q_2 - Q_1 \tag{4–59}$$

where Q_1 = heat rejected or exhaust heat
$\quad Q_2$ = heat input

FIGURE 4–16

Schematic flow diagram of a Carnot heat engine.

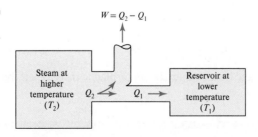

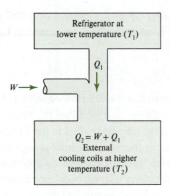

Thermal efficiency is the ratio of work output to heat input. The output is mechanical work. The exhaust heat is not considered part of the output.

$$\eta = \frac{W}{Q_2} \tag{4–60}$$

where W = work output
Q_2 = heat input

or, using Equation 4–59

$$\eta = \frac{Q_2 - Q_1}{Q_2} \tag{4–61}$$

Carnot's analysis revealed that the most efficient engine will have an efficiency of

$$\eta_{max} = 1 - \frac{T_1}{T_2} \tag{4–62}$$

where the temperatures are absolute temperatures (in kelvins). This equation implies that maximum efficiency is achieved when the value of T_2 is as high as possible and the value for T_1 is as low as possible.

A refrigerator may be considered to be a heat engine operated in reverse (Figure 4–17). From an environmental point of view, the best refrigeration cycle is one that removes the greatest amount of heat (Q_1) from the refrigerator for the least expenditure of mechanical work. Thus, we use the **coefficient of performance** (C.O.P.) rather than efficiency.

$$\text{C.O.P.} = \frac{Q}{W} = \frac{Q_1}{Q_2 - Q_1} \tag{4–63}$$

By analogy to the Carnot efficiency,

$$\text{C.O.P.} = \frac{T_1}{T_2 - T_1} \tag{4–64}$$

EXAMPLE 4–15   What is the coefficient of performance of the refrigerator in Example 4–14?

Solution   The C.O.P. is calculated directly from the temperatures.

$$\text{C.O.P.} = \frac{273.15 + 4}{[(273.15 + 30) - (273.15 + 4)]} = 10.7$$

Note that in contrast to the heat engine, the performance increases if the temperatures are close together.

CHAPTER REVIEW

When you have completed studying the chapter, you should be able to do the following without the aid of your textbook or notes:

1. Define the law of conservation of matter (mass).
2. Explain the circumstances under which the law of conservation of matter is violated.
3. Draw a materials-balance diagram given the inputs, outputs, and accumulation or the relationship between the variables.
4. Define the following terms: rate, conservative pollutants, reactive chemicals, steady-state conditions, equilibrium, completely mixed systems, and plug-flow systems.
5. Explain why the effluent from a completely mixed system has the same concentration as the system itself.
6. Define the first law of thermodynamics and provide one example.
7. Define the second law of thermodynamics and provide one example.
8. Define energy, work, power, specific heat, phase change, enthalpy of fusion, enthalpy of evaporation, photon, and blackbody radiation.
9. Explain how the energy-balance equation differs from the materials-balance equation.
10. List the three mechanisms of heat transfer and explain how they differ.
11. Explain the relationship between energy transformation and entropy.

With the aid of this text you should be able to do the following:

1. Write and solve mass-balance equations for systems with and without transformation.
2. Write the mathematical expression for the decay of a substance by first-order kinetics with respect to the substance.
3. Solve first-order reaction problems.
4. Compute the change in enthalpy for a substance.
5. Solve heat transfer equations for conduction, convection, and radiation individually and in combination.
6. Write and solve energy balance equations.
7. Compute the change in entropy.
8. Compute the Carnot efficiency for a heat engine.
9. Compute the coefficient of performance for a refrigerator.

PROBLEMS

4–1 A municipal landfill has available space of 16.2 ha at an average depth of 10 m. Seven hundred sixty-five (765) m^3 of solid waste is dumped at the site 5 days a week. This waste is compacted to twice its delivered density. Draw a mass-balance diagram and estimate the expected life of the landfill in years.

>   **Answer:** 16.25, or 16 years

4–2 Each month the Speedy Dry Cleaning Company buys one barrel (0.160 m^3) of dry cleaning fluid. Ninety percent of the fluid is lost to the atmosphere and 10% remains as residue to be disposed of. The density of the dry cleaning fluid is 1.5940 $g \cdot mL^{-1}$. Draw a mass-balance diagram and estimate the monthly mass emission rate to the atmosphere (in kilograms per month).

4–3 Congress banned the production of the Speedy Dry Cleaning Company's dry cleaning fluid in 2000. They are using a new cleaning fluid. The new dry cleaning fluid has one-sixth the volatility of the former dry cleaning fluid (Problem 4–2). The density of the new fluid is 1.6220 $g \cdot mL^{-1}$. Assume that the amount of residue is the same as that resulting from the use of the old fluid and estimate the mass emission rate to the atmosphere in $kg \cdot mo^{-1}$. Because the new dry cleaning fluid is less volatile, the company will have to purchase less each year. Estimate the annual volume of dry cleaning fluid saved (in $m^3 \cdot y^{-1}$).

4–4 Gasoline vapors are vented to the atmosphere when an underground gasoline storage tank is filled. If the tanker truck discharges into the top of the tank with no vapor control (known as the splash fill method), the emission of gasoline vapors is estimated to be 2.75 kg · m^{-3} of gasoline delivered to the tank. If the tank is equipped with a pressure relief valve and interlocking hose connection and the tanker truck discharges into the bottom of the tank below the surface of the gasoline in the storage tank, the emission of gasoline vapors is estimated to be 0.095 kg · m^{-3} of gasoline delivered (Wark, Warner, and Davis, 1998). Assume that the service station must refill the tank with 4.00 m^3 of gasoline once a week. Draw a mass-balance diagram and estimate the annual loss of gasoline vapor (in kg · y^{-1}) for the splash fill method. Estimate the value of the fuel that is captured if the vapor control system is used. Assume the density of the condensed vapors is 0.800 g · mL^{-1} and the cost of the gasoline is $.80 per liter.

4–5 The Rappahannock River near Warrenton, Virgina, has a flow rate of 3.00 m^3 · s^{-1}. Tin Pot Run (a pristine stream) discharges into the Rappahannock at a flow rate of 0.05 m^3 · s^{-1}. To study mixing of the stream and river, a conservative tracer is to be added to Tin Pot Run. If the instruments that measure the tracer can detect a concentration of 1.0 mg · L^{-1}, what minimum concentration must be achieved in Tin Pot Run so that 1.0 mg · L^{-1} of tracer can be measured after the river and stream mix? Assume that the 1.0 mg · L^{-1} of tracer is to be measured after complete mixing of the stream and Rappahannock has been achieved and that no tracer is in Tin Pot Run or the Rappahannock above the point where the two streams mix. What mass rate (in kilograms per day) of tracer must be added to Tin Pot Run?

Answer: 263.52, or 264 kg · day^{-1}

4–6 The Clearwater water treatment plant uses sodium hypochlorite (NaOCl) to disinfect the treated water before it is pumped to the distribution system. The NaOCl is purchased in a concentrated solution (52,000 mg · L^{-1}) that must be diluted before it is injected into the treated water. The dilution piping scheme is shown in Figure P-4–6. The NaOCl is pumped from the small tank (called a "day tank") into a small pipe carrying a portion of the clean treated water (called a "slip stream") to the main service line. The main service line has a flow rate of 0.50 m^3 · s^{-1}. The slip stream flows at 4.0 L · s^{-1}. At what rate of flow (in L · s^{-1}) must the NaOCl from the day tank be pumped into the slip stream to achieve a concentration of 2.0 mg · L^{-1} of NaOCl in the main service line? Although it is reactive, you may assume that NaOCl is not reactive for this problem.

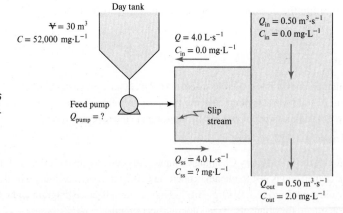

Figure P-4–6
Dilution piping scheme.

4–7 The Clearwater design engineer cannot find a reliable pump to move the NaOCl from the day tank into the slip stream (Problem 4–6). Therefore, she specifies in the operating instructions that the day tank be used to dilute the concentrated NaOCl solution so that a pump rated at 1.0 L · s^{-1} may be used. The tank is to be filled once each shift (8 h per shift). It has a volume of 30 m^3. Determine the concentration of NaOCl that is required in the day tank if the feed rate of NaOCl must be 1000 mg · s^{-1}. Calculate the volume of concentrated solution and the volume of water that is to be added for an 8 hour operating period. Although it is reactive, you may assume that NaOCl is not reactive for this problem.

4–8 In water and wastewater treatment processes a filtration device may be used to remove water from the sludge formed by a precipitation reaction. The initial concentration of sludge from a softening reaction (Chapter 10) is 2% (20,000 mg · L⁻¹) and the volume of the sludge is 100 m³. After filtration the sludge solids concentration is 35%. Assume that the sludge does not change its density during filtration and that liquid removed from the sludge contains no sludge. Using the mass-balance method, determine the volume of sludge after filtration.

4–9 The U.S. EPA requires hazardous waste incinerators to meet a standard of 99.99% destruction and removal of organic hazardous constituents injected into the incinerator. This efficiency is referred to as "four nines DRE." For especially toxic waste, the DRE must be "six nines." The efficiency is to be calculated by measuring the mass flow rate of organic constituent entering the incinerator and the mass flow rate of constituent exiting the incinerator stack. A schematic of the process is shown in Figure P-4–9. One of the difficulties of assuring these levels of destruction is the ability to measure the contaminant in the exhaust gas. Draw a mass balance diagram for the process and determine the allowable quantity of contaminant in the exit stream if the incinerator is burning 1.0000 g · s⁻¹ of hazardous constituent. (Note: the number of significant figures is very important in this calculation.) If the incinerator is 90% efficient in destroying the hazardous constituent, what scrubber efficiency is required to meet the standard?

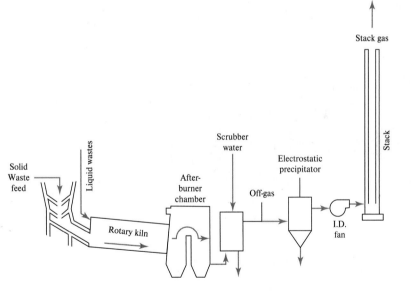

Figure P-4–9
Schematic of hazardous
waste incinerator.

4–10 A new high efficiency air filter has been designed to be used in a secure containment facility to do research on detection and destruction of anthrax. Before the filter is built and installed, it needs to be tested. It is proposed to use ceramic microspheres of the same diameter as the anthrax spores for the test. One obstacle in the test is that the efficiency of the sampling equipment is unknown and cannot be readily tested because the rate of release of microspheres cannot be sufficiently controlled to define the number of microspheres entering the sampling device. The engineers propose the test apparatus shown in Figure P-4-10. The sampling filters capture the microspheres on a membrane filter that allows microscopic counting of the captured particles. At the end of the experiment, the number of particles on the first filter is 1941 and the number on the second filter is 63. Assuming each filter has the same efficiency, estimate the efficiency of the sampling filters. [Note: this problem is easily solved using the particle counts (C_1, C_2, C_3) and efficiency (η).]

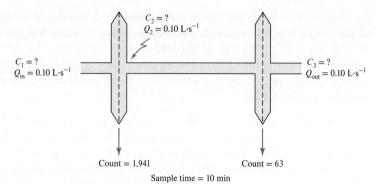

Count = 1,941   Count = 63

Sample time = 10 min

Figure P-4–10
Filtration test apparatus.

4–11   To remove the solution containing metal from a part after metal plating, the part is commonly rinsed with water. This rinse water is contaminated with metal and must be treated before discharge. The Shinny Metal Plating Co. uses the process flow diagram shown in Figure P-4–11. The plating solution contains $85\ g \cdot L^{-1}$ of nickel. The parts drag out $0.05\ L \cdot min^{-1}$ of plating solution into the rinse tank. The flow of rinse water into the rinse tank is $150\ L \cdot min^{-1}$. Write the general mass-balance equation for the rinse tank and estimate the concentration of nickel in the wastewater stream that must be treated. Assume that the rinse tank is completely mixed and that no reactions take place in the rinse tank.

Answer: $C_n = 28.3$, or $28\ mg \cdot L^{-1}$

Figure P-4–11

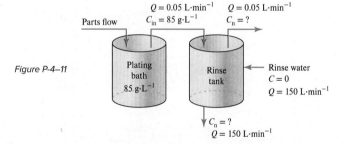

4–12   Because the rinse water flow rate for a nickel plating bath (Problem 4–11) is quite high, it is proposed that the countercurrent rinse system shown in Figure P-4–12 be used to reduce the flow rate. Assuming that the C_n concentration remains the same at $28\ mg \cdot L^{-1}$, estimate the new flow rate. Assume that the rinse tank is completely mixed and that no reactions take place in the rinse tank.

Figure P-4–12

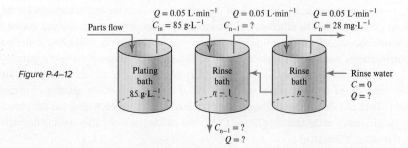

4–13 The Environmental Protection Agency (U.S. EPA, 1982) offers the following equation to estimate the flow rate for countercurrent rinsing (Figure P-4–12):

$$Q = \left[\left(\frac{C_{in}}{C_n} \right)^{1/n} + \frac{1}{n} \right] q$$

where Q = rinse water flow rate, $L \cdot min^{-1}$

$\quad C_{in}$ = concentration of metal in plating bath, $mg \cdot L^{-1}$

$\quad C_n$ = concentration of metal in nth rinse bath, $mg \cdot L^{-1}$

$\quad n$ = number of rinse tanks

$\quad q$ = flow rate of liquid dragged out of a tank by the parts, $L \cdot min^{-1}$

Using the EPA equation and the data from Problem 4–11, calculate the rinse water flow rate for one, two, three, four, and five rinse tanks in series using a computer spreadsheet you have written. Use the spreadsheet graphing function to plot a graph of rinse water flow rate versus number of rinse tanks.

4–14 If biodegradable organic matter, oxygen, and microorganisms are placed in a closed bottle, the microorganisms will use the oxygen in the process of oxidizing the organic matter. The bottle may be treated as a batch reactor, and the decay of oxygen may be treated as a first-order reaction. Write the general mass-balance equation for the bottle. Using a computer spreadsheet program you have written, calculate and then plot the concentration of oxygen each day for a period of 5 days starting with a concentration of $8 \, mg \cdot L^{-1}$. Use a rate constant of $0.35 \, d^{-1}$.

Answer: Day 1 = 5.64, or 5.6 $mg \cdot L^{-1}$; Day 2 = 3.97, or 4.0 $mg \cdot L^{-1}$

4–15 In 1908, H. Chick reported an experiment in which he disinfected anthrax spores with a 5% solution of phenol (Chick, 1908). The results of his experiment are tabulated here. Assuming the experiment was conducted in a completely mixed batch reactor, determine the decay rate constant for the die-off of anthrax.

Concentration of Survivors (number $\cdot mL^{-1}$)	Time (min)
398	0
251	30
158	60

4–16 A water tower containing 4000 m^3 of water has been taken out of service to install a chlorine monitor. The concentration of chlorine in the water tower was 2.0 $mg \cdot L^{-1}$ when the tower was taken out of service. If the chlorine decays by first-order kinetics with a rate constant k = 1.0 d^{-1} (Grayman and Clark, 1993), what is the chlorine concentration when the tank is put back in service 8 hours later? What mass of chlorine (in kg) must be added to the tank to raise the chlorine level back to 2.0 $mg \cdot L^{-1}$? Although it is not completely mixed, you may assume the tank is a completely mixed batch reactor.

4–17 The concept of *half-life* is used extensively in environmental engineering and science. For example, it is used to describe the decay of radioisotopes, elimination of poisons from people, self-cleaning of lakes, and the disappearance of pesticides from soil. Starting with the mass-balance equation, develop an expression that describes the half-life ($t_{1/2}$) of a substance in terms of the reaction rate constant, k, assuming the decay reaction takes place in a batch reactor.

4–18 If the initial concentration of a reactive substance in a batch reactor is 100%, determine the amount of substance remaining after 1, 2, 3, and 4 half-lives if the reaction rate constant is 6 $months^{-1}$.

4–19 Liquid hazardous wastes are blended in a CMFR to maintain a minimum energy content before burning them in a hazardous waste incinerator. The energy content of the waste currently being fed is 8.0 $MJ \cdot kg^{-1}$ (megajoules per kilogram). A new waste is injected in the flow line into the CMFR. It has an energy content of 10.0 MJ · kg^{-1}. If the flow rate into and out of the 0.20 m^3 CMFR is 4.0 L · s^{-1}, how long will it take the effluent from the CMFR to reach an energy content of 9 MJ · kg^{-1}?

Answer: $t = 34.5$, or 35 s

4–20 Repeat Problem 4–19 with a new waste having an energy content of 12 MJ · kg^{-1} instead of 10 MJ · kg^{-1}.

4–21 An instrument is installed along a major water-distribution pipeline to detect potential contamination from terrorist threats. A 2.54 cm diameter pipe connects the instrument to the water-distribution pipe. The connecting pipe is 20.0 m long. Water from the distribution pipe is pumped through the instrument and then discharged to a holding tank for verification analysis and proper disposal. If the flow rate of the water in the sample line is 1.0 L · min^{-1}, how many minutes will it take a sample from the distribution pipe to reach the instruments? Use the following relationship to determine the speed of the water in the sample pipe:

$$u = \frac{Q}{A}$$

where u = speed of water in pipe, m · s^{-1}
$\qquad Q$ = flow rate of water in pipe, m^3 · s^{-1}
$\qquad A$ = area of pipe, m^2

If the instrument uses 10 mL for sample analysis, how many liters of water must pass through the sampler before it detects a contaminant in the pipe?

4–22 A bankrupt chemical firm has been taken over by new management. On the property they found a 20,000-m^3 brine pond containing 25,000 mg · L^{-1} of salt. The new owners propose to flush the pond into their discharge pipe leading to the Atlantic Ocean, which has a salt concentration above 30,000 mg · L^{-1}. What flow rate of fresh water (in m^3 · s^{-1}) must they use to reduce the salt concentration in the pond to 500 mg · L^{-1} within one year?

Answer: $Q = 0.0025$ m^3 · s^{-1}

4–23 A 1900-m^3 water tower has been cleaned with a chlorine solution. The vapors of chlorine in the tower exceed allowable concentrations for the work crew to enter and finish repairs. If the chlorine concentration is 15 mg · m^{-3} and the allowable concentration is 0.0015 mg · L^{-1}, how long must the workers vent the tank with clean air flowing at 2.35 m^3 · s^{-1}?

4–24 A railroad tank car derails and ruptures. It discharges 380 m^3 of pesticide into the Mud Lake drain. As shown in Figure P-4–24, the drain flows into Mud Lake, which has a liquid volume of 40,000 m^3. The water in the creek has a velocity of 0.10 m · s^{-1}, and the distance from the spill site to the pond is 20 km. Assume that the spill is short enough to treat the injection of the pesticide as a pulse, that the pond behaves as a flow-balanced CMFR, and that the pesticide is nonreactive. Estimate the time for the pulse to reach the pond and the time it will take to flush 99% of the pesticide from the pond.

Answer: Time to reach the pond = 2.3 days   Time to flush = 21.3, or 21 days

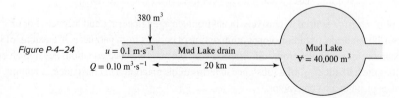

Figure P-4–24

4–25 During a snowstorm the fluoride feeder in North Bend runs out of feed solution. As shown in Figure P-4–25, the rapid mix tank is connected to a 5-km long distribution pipe. The flow rate into the rapid mix tank is 0.44 m^3 · s^{-1}, and the volume of the tank is 2.50 m^3. The velocity in the pipe is 0.17 m · s^{-1}. If the fluoride concentration in the rapid mix tank is 1.0 mg · L^{-1} when the feed stops, how long until the concentration of fluoride is reduced to 0.01 mg · L^{-1} at the end of the distribution pipe? The fluoride may be considered a nonreactive chemical.

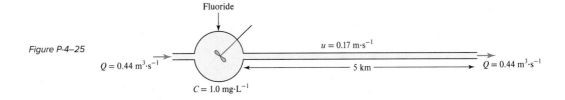

Figure P-4–25

Fluoride

$Q = 0.44$ m^3·s^{-1}

$u = 0.17$ m·s^{-1}

5 km

$Q = 0.44$ m^3·s^{-1}

$C = 1.0$ mg·L^{-1}

4–26 A sewage lagoon that has a surface area of 10 ha and a depth of 1 m is receiving 8640 m^3 · d^{-1} of sewage containing 100 mg · L^{-1} of biodegradable contaminant. At steady state, the effluent from the lagoon must not exceed 20 mg · L^{-1} of biodegradable contaminant. Assuming the lagoon is well mixed and that there are no losses or gains of water in the lagoon other than the sewage input, what biodegradation reaction rate coefficient (d^{-1}) must be achieved for a first-order reaction?

Answer: $k = 0.3478$, or 0.35 day^{-1}

4–27 Repeat Problem 4–26 with two lagoons in series (see Figure P-4–27). Each lagoon has a surface area of 5 ha and a depth of 1 m.

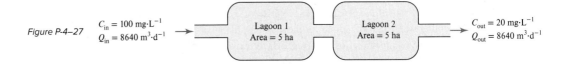

Figure P-4–27

$C_{in} = 100$ mg·L^{-1}
$Q_{in} = 8640$ m^3·d^{-1}

Lagoon 1
Area = 5 ha

Lagoon 2
Area = 5 ha

$C_{out} = 20$ mg·L^{-1}
$Q_{out} = 8640$ m^3·d^{-1}

4–28 Using a spreadsheet program you have written determine the effluent concentration if the process producing sewage in Problem 4–26 shuts down ($C_{in} = 0$). Calculate and plot points at 1-day intervals for 10 days. Use the graphing function of the spreadsheet to construct your plot.

4–29 A 90-m^3 basement in a residence is found to be contaminated with radon coming from the ground through the floor drains. The concentration of radon in the room is 1.5 Bq · L^{-1} (becquerels per liter) under steady-state conditions. The room behaves as a CMFR, and the decay of radon is a first-order reaction with a decay rate constant of 2.09 × 10^{-6} s^{-1}. If the source of radon is closed off and the room is vented with radon-free air at a rate of 0.14 m^3 · s^{-1}, how long will it take to lower the radon concentration to an acceptable level of 0.15 Bq · L^{-1}?

4–30 An ocean outfall diffuser that discharges treated wastewater into the Pacific Ocean is 5000 m from a public beach. The wastewater contains 10^5 coliform bacteria per milliliter. The wastewater discharge flow rate is 0.3 m^3 · s^{-1}. The coliform first-order death rate in seawater is approximately 0.3 h^{-1} (Tchobanoglous and Schroeder, 1985). The current carries the wastewater plume toward the beach at a rate of 0.5 m · s^{-1}. The ocean current may be approximated as a pipe carrying 600 m^3 · s^{-1} of seawater. Determine the coliform concentration at the beach. Assume that the current behaves as a plug flow reactor and that the wastewater is completely mixed in the current at the discharge point.

4–31 For the following conditions determine whether a CMFR or a PFR is more efficient in removing a reactive compound from the waste stream under steady-state conditions with a first-order reaction: reactor volume = 280 m^3, flow rate = 14 m^3 · day^{-1}, and reaction rate coefficient = 0.05 day^{-1}.

Answer: CMFR η = 50%; PFR η = 63%

4–32 Compare the reactor volume required to achieve 95% efficiency for a CMFR and a PFR for the following conditions: steady-state, first-order reaction, flow rate = 14 m^3 · day^{-1}, and reaction rate coefficient = 0.05 day^{-1}.

4–33 The discharge pipe from a sump pump in the dry well of a sewage lift station did not drain properly and the water at the discharge end of the pipe froze. A hole has been drilled into the ice and a 200-W electric heater has been inserted in the hole. If the discharge pipe contains 2 kg of ice, how long will it take to melt the ice? Assume all the heat goes into melting the ice.

Answer: 55.5 or 56 min

4–34 As noted in Examples 4–12 and 4–13, the time to achieve the desired temperature using the cooling tank is quite long. An evaporative cooler is proposed as an alternate means of reducing the temperature. Estimate the amount of water (in m^3) that must evaporate each day to lower the temperature of the 40 m^3 of wastewater from 100°C to 20°C. (Note: While the solution to this problem is straightforward, the design of an evaporative cooling tower is a complex thermodynamic problem made even more complicated in this case by the contents of the wastewater that would potentially foul the cooling system.)

4–35 The water in a biological wastewater treatment system must be heated from 15°C to 40°C for the microorganisms to function. If the flow rate of the wastewater into the process is 30 $m^3 \cdot day^{-1}$, at what rate must heat be added to the wastewater flowing into the treatment system? Assume the treatment system is completely mixed and that there are no heat losses once the wastewater is heated.

Answer: 3.14 GJ $\cdot$ d^{-1}

4–36 The lowest flow in the Menominee River in July is about 40 $m^3 \cdot s^{-1}$. If the river temperature is 18°C and a power plant discharges 2 $m^3 \cdot s^{-1}$ of cooling water at 80°C, what is the final river temperature after the cooling water and the river have mixed? Assume the density of water is constant at 1000 kg $\cdot$ m^{-3}.

4–37 The flow rate of the Seine in France is 28 $m^3 \cdot s^{-1}$ at low flow. A power plant discharges 10 $m^3 \cdot s^{-1}$ of cooling water into the Seine. In the summer, the river temperature upstream of the power plant reaches 20°C. The temperature of the river after the power plant discharge mixes with the river is 27°C (Goubet, 1969). Estimate the temperature of the cooling water before it is mixed with the river water. Ignore radiative and convective losses to the atmosphere as well as conductive losses to the river bottom and banks.

4–38 An aerated lagoon (a sewage treatment pond that is mixed with air) is being proposed for a small lake community in northern Wisconsin. The lagoon must be designed for the summer population but will operate year-round. The winter population is about half of the summer population. Based on these design assumptions, the volume of the proposed lagoon is 3420 m^3. The daily volume of sewage in the winter is estimated to be 300 m^3. In January, the temperature of the lagoon drops to 0°C but it is not yet frozen. If the temperature of the wastewater flowing into the lagoon is 15°C, estimate the temperature of the lagoon at the end of a day. Assume the lagoon is completely mixed and that there are no losses to the atmosphere or the lagoon walls or floor. Also assume that the sewage has a density of 1000 kg $\cdot$ m^{-3} and a specific heat of 4.186 kJ $\cdot$ $kg^{-1} \cdot K^{-1}$.

4–39 Using the data in Problem 4–38 and a spreadsheet program you have written, estimate the temperature of the lagoon at the end of each day for a period of 7 days. Assume that the flow leaving the lagoon equals the flow entering the lagoon and that the lagoon is completely mixed.

4–40 A cooling water pond is to be constructed for a power plant that discharges 17.2 $m^3 \cdot s^{-1}$ of cooling water. Estimate the required surface area of the pond if the water temperature is to be lowered from 45.0°C at its inlet to 35.5°C at its outlet. Assume an overall heat transfer coefficient of 0.0412 kJ $\cdot$ $s^{-1} \cdot m^{-2} \cdot K^{-1}$ (Edinger, Brady, and Graves, 1968). Note: the cooling water will be mixed with river water after it is cooled. The mixture of the 35°C water and the river water will meet thermal discharge standards.

Answer: 174.76 or 175 ha

4–41 Because the sewage in the lagoon in Problem 4–38 is violently mixed, there is a good likelihood that the lagoon will freeze. Estimate how long it will take to freeze the lagoon if the temperature of the wastewater in the lagoon is 15°C and the air temperature is −8°C. The pond is 3 m deep. Although the aeration equipment will probably freeze before all of the wastewater in the lagoon is frozen, assume that the total volume of wastewater freezes. Use an overall heat transfer coefficient of 0.5 kJ $\cdot$ $s^{-1} \cdot m^{-2} \cdot K^{-1}$ (Metcalf & Eddy, Inc., 2003). Ignore the enthalpy of the influent wastewater.

4–42 A small building that shelters a water supply pump measures 2 m × 3 m × 2.4 m high. It is constructed of 1-cm thick wood having a thermal conductivity of 0.126 W · m^{-1} · K^{-1}. The inside walls are to be maintained at 10°C when the outside temperature is −18°C. How much heat must be supplied to maintain the desired temperature? How much heat must be supplied if the walls are lined with 10 cm of glass-wool insulation having a thermal conductivity of 0.0377 W · m^{-1} · K^{-1}? Assume there is no heat loss through the floor. Ignore the wood in the second calculation.

4–43 The radiative heat load on two surfaces of a horizontal leaf is about 1.7 kW · m^{-2} (Gates, 1962). The leaf in turn radiates away a portion of this heat load. If the leaf temperature is near that of the ambient temperature, say 30°C, what fraction of the heat loss is by radiation? Assume the emissivity of the leaf is 0.95.

    Answer: Fraction = 0.2676 or 0.27 or 27%

4–44 Bituminous coal has a heat of combustion* of 31.4 MJ · kg^{-1}. In the United States the average coal-burning utility produces an average of 2.2 kWh of electrical energy per kilogram of bituminous coal burned. What is the average overall efficiency of this production of electricity?

DISCUSSION QUESTIONS

4–1 A piece of limestone rock ($CaCO_3$) at the bottom of Lake Superior is slowly dissolving. For the purpose of calculating a mass balance, you can assume

    (a) The system is in equilibrium.
    (b) The system is at steady state.
    (c) Both of the above.
    (d) Neither of the above.

    Explain your reasoning.

4–2 A can of a volatile chemical (benzene) has spilled into a small pond. List the data you would need to gather to calculate the concentration of benzene in the stream leaving the pond using the mass-balance technique.

4–3 In Table 4–3, specific heat capacities for common substances, the values for c_p for beef, corn, human beings, and poultry are considerably higher than those for aluminum, copper, and iron. Explain why.

4–4 If you hold a beverage glass whose contents are at 4°C, "you can feel the cold coming into your hand." Thermodynamically speaking is this statement true? Explain.

4–5 If you walk barefooted across a brick floor and a wood floor, the brick floor will feel cooler even though the room temperature is the same for both floors. Explain why.

FE EXAM FORMATTED PROBLEMS

4–1 The decay of chlorine in a distribution system follows first-order decay with a rate constant of 0.360 d^{-1}. If the concentration of chlorine in a well mixed water storage tank is 1.00 mg/L at time zero, what will the concentration be one day later. Assume no water flows out of the tank.

    (a) 0.360 mg/L
    (b) 0.368 mg/L
    (c) 0.500 mg/L
    (d) 0.698 mg/L

4–2 A 350 m^3 retention pond that holds rain water from a shopping mall is empty at the beginning of a rain storm. The flow rate out of the retention pond must be restricted to 320 L/min to prevent downstream flooding from a 6-hour storm. What is the maximum flow rate (in L/min) into the pond from a 6-hour storm that will not flood it?

    (a) 5860 L/min
    (b) 321 L/min
    (c) 1320 L/min
    (d) 7750 L/min

---

*Heat of combustion is the amount of energy released per unit mass when the compound reacts completely with oxygen. The mass does not include the mass of oxygen.

4–3   A pipeline carrying 0.50 million gallons per day (MGD) of a 35,000 mg/L brine solution (NaCl) across a creek ruptures. The flow rate of the creek is 2.80 MGD. If the salt concentration in the creek is 175 mg/L, what is the concentration of salt in the creek after the pipeline discharge mixes completely with the creek water?

(a) 1.80×10^4 mg/L

(b) 1.75×10^2 mg/L

(c) 5.45×10^3 mg/L

(d) 6.43×10^3 mg/L

4–4   A wastewater treatment plant has experienced a power outage due to a winter storm. A treatment facility (anaerobic digester) contains 2120 m^3 of wastewater at 37°C. If the wastewater temperature falls below 30°C, the digester will fail. The digester is made of concrete with a thermal conductivity of 2 W/m · K. The surface area of the digester is 989.6 m^2. The concrete walls and ceiling are 30 cm thick. If the outside temperature is 0°C and there is no wind or sunshine, how long will the operator have to get the heating system back into operation before the digester fails? Assume the specific heat capacity of the wastewater is the same as that of water and ignore the lack of mixing.

(a)   16 d

(b)   3 d

(c)   22 min

(d)   2 h

REFERENCES

Chick, H. (1908) "An Investigation of the Laws of Disinfection," *Journal of Hygiene,* p. 698.

Edinger, J. E., D. K. Brady, and W. L. Graves (1968) "The Variation of Water Temperatures Due to Steam-Electric Cooling Operations," *Journal of Water Pollution Control Federation,* 40(9): 1637–39.

Gates, D. M. (1962) *Energy Exchange in the Biosphere,* Harper & Row Publishers, New York, p. 70.

Goubet, A. (1969) "The Cooling of Riverside Thermal-Power Plants," in F. L. Parker and P. A. Krenkel (eds.), *Engineering Aspects of Thermal Pollution,* Vanderbilt University Press, Nashville, TN, p. 119.

Grayman, W. A., and R. M. Clark (1993) "Using Computer Models to Determine the Effect of Storage on Water Quality," *Journal of the American Water Works Association,* 85(7): 67–77.

Guyton, A. C. (1961) *Textbook of Medical Physiology,* 2nd ed., W. B. Saunders Company, Philadelphia, pp. 920–21, 950–53.

Hudson, R. G. (1959) *The Engineers' Manual,* John Wiley & Sons, New York, p. 314.

Kuehn, T. H., J. W. Ramsey, and J. L. Threkeld (1998) *Thermal Environmental Engineering,* Prentice Hall, Upper Saddle River, NJ, pp. 425–27.

Masters, G. M. (1998) *Introduction to Environmental Engineering and Science*, Prentice Hall, Upper Saddle River, NJ, p. 30.

Metcalf & Eddy, Inc. (2003) revised by G. Tchobanoglus, F. L. Burton, and H. D. Stensel, *Wastewater Engineering, Treatment and Reuse,* McGraw-Hill, Boston, p. 844.

Salvato, J. A., Jr. (1972) *Environmental Engineering and Sanitation,* 2nd ed., Wiley-Interscience, New York, pp. 598–99.

Shortley, G., and D. Williams (1955) *Elements of Physics,* Prentice-Hall, Engelwood Cliffs, NJ, p. 290.

Tchobanoglous, G., and E. D. Schroeder (1985) *Water Quality,* Addison-Wesley Publishing Co., Reading, MA, p. 372.

U.S. EPA (1982) *Summary Report: Control and Treatment Technology for the Metal Finishing Industry, In-Plant Changes,* U.S. Environmental Protection Agency, Washington, DC, Report No. EPA 625/8-82-008.

Wark, K., C. F. Warner, and W. T. Davis (1998) *Air Pollution: Its Origin and Control,* 3rd ed., Addison-Wesley, Reading, MA, p. 509.

5

Ecosystems

Case Study

Ecosystems

As discussed in Section 5–5, the population of any species found in the wild depends on factors, such as the availability of food, shelter, concentrations of waste products, disease, and the population density of predators and parasites. Environmental aspects, such as weather, temperature, and flooding also affect population density. The same is true for humans.

From the time the species *Homo sapiens* evolved some 200,000 years ago until approximately 12,000 years ago, the human population on Earth was less than a few million. The human population, as shown in Figure 5–16, increased from 0.6 to 7.3 billion over the period from 1700 to 2015. In this same period of time, we have altered the Earth's atmosphere, with the global carbon dioxide concentration increasing from approximately 280 parts per million (ppm) to slightly greater than 400 ppm. Since 1950, the human population has been increasing at approximately 0.8 billion people every decade. At these rates, the human population and the carbon dioxide level will reach 8 billion and 419 ppm, respectively, by 2025. The trillion dollar questions are, for how long can these rates increase or what is the carrying capacity of the Earth for the human population?

Among the factors that limit growth are the availability of food, which depends on the availability of fertile cropland along with weather conditions. A typical Western diet requires approximately 0.5 hectares per person (Pimentel and Pimentel, 1996). In order to provide such a diet for the human population in 2025, we would need 4 billion hectares of arable land. However, presently the Earth supports only approximately 1.4 billion hectares of cropland and approximately 11% of the world's population is undernourished (United Nations Food and Agriculture Organization, 2013). To make matters worse, the UN Food and Agriculture Organization estimates that the amount of arable land will decrease to about 0.15 ha/person in 2050 (UN United Nations Food and Agriculture Organization, 2010). In addition, climate change is predicted to increase the water stress (withdrawal-to-availability ratio), which will decrease crop productivity.

Another factor mentioned above are wastes. The WHO estimated that in 2012 nearly 25% of total global deaths are the result of living or working in an unhealthy environment (Prüss-Ustün et al., 2016). Environmental risk factors include air, water and soil pollution, chemical exposures, climate change, and ultraviolet radiation. As the global temperature increases as a result of increasing levels of greenhouse gases in the atmosphere, these rates are anticipated to increase due to increasing air pollution, water stress, and endemic diseases.

A third factor is shelter. Climate change and the rising seas will cause mass migration of populations. This, along with increasing water stress, will destabilize countries as populations move as their homelands flood. This is already happening in areas such as Bangladesh; the Pacific Islands, Kiribati, Naura, Tuvalu; and the remote coastal Alaskan village of Shishmaref. A recent study (Hauer, 2017) suggests that migration as a result of rising seas will reshape the U.S. population distribution, potentially stressing land-locked areas unprepared to accommodate this wave of coastal migrants.

So what does an environmental scientist or engineer do? As mentioned in a white paper prepared by Doran (2012), there are numerous ways in which we can respond.

We will need to develop systems that conserve, reuse, treat, and improve the management of our water resources. Environmental engineers and scientists will need to be more effective at communicating to the general public and more involved in the regulatory and legislative process. We will need to find more effective and less energy-intensive ways to treat municipal and industrial wastewater. And finally, we will need to address problems such as antibiotic resistance and climate change. Hopefully, we are up to the challenge.

5–1   INTRODUCTION

Ecosystems

Ecosystems are communities of organisms that interact with one another and with their physical environment, including sunlight, rainfall, and soil nutrients. Organisms within an ecosystem tend to interact with one another to a greater extent than do the organisms between ecosystems. Ecosystems can vary greatly in size. For example, a tidal pool of only about 2 m in diameter could be considered an ecosystem because the plants and animals living in this environment depend on one another and are unique to this type of system. On a larger scale, a tropical rainforest* is also an ecosystem. Even larger is our global biosphere,† which could be considered the "ultimate" Earthbound ecosystem. Within each ecosystem are **habitats,** which are defined as the place where a population‡ of organisms lives.

Ecosystems can be further defined as systems into which matter flows. Matter also leaves ecosystems; however, this flow of matter into and out of the ecosystem is small compared with the amount of matter that cycles within the ecosystem. If we think of a lake as an ecosystem, matter flows into the lake in the form of carbon dioxide that dissolves into the water, nutrients that run off from the land, and chemicals that flow with any streams or rivers feeding the lake. Within the lake, matter flows from one organism to the next in the form of food, excreted material, or respired gases (either oxygen or carbon dioxide). This flow of matter is critical for the existence of an ecosystem.

Another characteristic of an ecosystem is that it can change with time. Later in the chapter we will discuss how lakes change (naturally or anthropogenically) over time, from a system that has very clear water, low levels of nutrients, and low numbers of a large variety of species to one that contains highly turbid water, high levels of nutrients, and large numbers of a few species. Both of these systems (the same lake at different times) are very different ecosystems. Similarly, severe flooding or droughts and extreme temperature changes or other extreme environmental conditions (e.g., volcanic activity or forest fires) can cause significant changes in an ecosystem.

Ecosystems can be natural or artificial. The lake, the tidal pool, or the forest is usually natural (although lakes can be human-made and forests can be cultivated). Constructed wetlands are increasingly being used for the treatment of storm runoff, mining wastes (acid mine drainage), or municipal sewage. Agricultural land is another example of an artificial (or human-made) ecosystem. The criteria explained earlier are met in all ecosystems, whether they are natural or artificial, large or small, long-lasting or temporary.

---

*A tropical rainforest is an example of a biome. **Biomes** are complex communities of plants and animals in a region and a climate. These include deserts, tundra, chaparrals or scrubs, and temperate hardwood forests.

†The sum of all the regions of the earth that support ecosystems is known as the **biosphere.** The biosphere is made up of the atmosphere, the hydrosphere (the water), and the lithosphere (the soil, rocks, and minerals that make up the solid portion of earth).

‡A **population** is defined as a group of organisms of the same species living in the same place at the same time.

5–2 HUMAN INFLUENCES ON ECOSYSTEMS

As environmental engineers and scientists we have a responsibility to protect ecosystems and the life that resides within them. Although ecosystems change naturally, human activity can speed up natural processes by several orders of magnitude (in terms of time).

Seemingly harmless or beneficial activities can wreak havoc on the environment. For example, large-scale agricultural operations, although producing inexpensive food to feed millions, can result in the release of pesticides, fertilizers, and carbon dioxide and other greenhouse gases to the environment. Hydroelectric power is seen as a clean, renewable energy source. However, dam construction can have detrimental effects on river ecosystems, drastically reducing fish populations as well as causing erosion of soil and vegetation during powerful water surges.

Human activity can also change ecosystems through the destruction of species. The loss of habitat can threaten the existence of individual species within an ecosystem. For example, the destruction of the rainforest in Mexico threatens the very existence of the monarch butterfly. If these forests are destroyed to the extent that the monarch loses its winter roosting grounds, **global extinction** of this butterfly could result—the complete and permanent loss of this animal species across the entire planet. However, the localized destruction of the milkweed plant deprives the butterfly of its nesting environment, resulting in **local extinction.**

The destruction of an ecosystem is not the only way humans can affect animal populations. As discussed above, the release of toxic chemicals can also threaten wildlife. The toxic chemicals can be synthetic organic chemicals, such as DDT (**d**ichloro**d**iphenyl**t**richloroethane), petroleum compounds, or heavy metals. Acid rain, resulting from emissions from power plants, automobiles, and industrial operations, also can have a significant effect on ecosystems. This is very apparent in the water bodies in the northeastern United States and northern Europe, where by 1997, hundreds of lakes were devoid of fish (Moyle, 1997). And millions of acres of forest across the world were damaged. However, scientific research led to a solid understanding of the relationship between sulfur dioxide and nitrogen oxide emissions and acid rain, resulting in stringent regulations of power plant emissions. As a result, rain falling in the Northeast today is about half as acidic as it was in the early 1980s and fragile ecosystems are beginning to recover (Willyard, 2010). Agricultural operations in California's San Joaquin Valley have resulted in the mobilization of selenium from soils and the subsequent concentration of the metal in the Kesterson Reservoir. The concentrations are so great that the populations of several species of water birds, including the black-necked stilts, are severely threatened (Moyle, 1997).

A third way species can be threatened is by the introduction of nonnative (**exotic**) species into ecosystems. The introduction of the rabbit onto Norfolk Island, the zebra mussel into the Great Lakes, or the Asian fungus causing Dutch elm disease onto the U.S. east coast has had a significant impact on ecosystems. The introduction of rabbits onto Norfolk Island in 1830 resulted in the loss of 13 species of vascular plants by 1967 (Western and Pearl, 1989). Scientists believe that the zebra mussel (*Dreissena polymorpha*) was introduced into the Great Lakes from the ballast water of a transatlantic freighter that previously visited a port in eastern Europe, where the mussel is common (Glassner-Schwayder, 2000). The zebra mussel can now be found in inland waters in 28 states and the Province of Ontario and Quebec and is thought to be responsible for the reduction of some 80% of the mass of phytoplankton in Lake Erie. Because the zebra mussel is an efficient filterer of water its presence significantly increases water clarity, allowing light to penetrate deeper into the water column, increasing the density of rooted aquatic vegetation, benthic forms of algae, and some forms of insect-like benthic organisms (Glassner-Schwayder, 2000). The mussel has also caused the near extinction of many types of native unionid clams in Lake St. Clair and the western basin of Lake Erie (Glassner-Schwayder, 2000). The zebra mussel attaches itself to the native clams, eventually killing them. One of the more recent invaders of the Great Lakes is the bloody-red shrimp (*Hemimysis anomala*), a half-inch long, bright red shrimp native to the Black and Caspian seas. The shrimp was first spotted by wildlife biologists in Muskegon Lake, which empties into Lake Michigan, in late 2006. Within two years of its discovery in Muskegon Lake, bloody-red shrimp were isolated from all major waterbodies within the Great Lakes-St. Lawrence system, except for Lake Superior (Kestrup and Ricciardi, 2008). The impact of this organism on the Great Lakes is not yet known, but given that it aggressively feeds on the tiny plants and animals, it can be expected that the effects will be significant (U.S. EPA, 2006; Associated Press, 2007). Ricciardi (2012) suggested that predation on *H. anomala* could increase the biomagnification of contaminants in fishes.

The last method by which species can become extinct is through excessive hunting, some legal, others illegal. The manatee whose habitat is the Everglades is threatened by poaching, along with harm due to boat propellers, loss of habitat, and vandalism. The rhinoceros is threatened by poaching, mainly for its horns.

5–3   ENERGY AND MASS FLOW

Ecosystems would not be possible were it not for the flow of energy into them. The sun is the primary source of this energy because all biological life is dependent on the green plants that use sunlight as a source of energy. As such, these sunlight-using organisms are called **primary producers.** Primary producers also obtain their carbon from inorganic sources such as carbon dioxide (CO_2) or bicarbonate (HCO_3^-). As such, they are referred to as **autotrophic.** These photosynthetic organisms that obtain their carbon from inorganic sources are called **photoautotrophic.** *Trophic* is the term used to describe the level of nourishment. Trophic levels are summarized in Table 5–1.

The process by which some organisms (namely chlorophyll-containing plants) are able to convert energy from sunlight into chemical energy (in the form of sugars) is called **photosynthesis** and can be represented by the simple equation

$$6CO_2 + 6H_2O + 2800 \text{ kJ energy from sun} \xrightarrow{\text{chlorophyll}} C_6H_{12}O_6 + 6O_2 \tag{5–1}$$

The chemical compound represented by $C_6H_{12}O_6$ is the simple sugar glucose. The rate of carbon dioxide use and, therefore, glucose production is dependent on sunlight and the number and growth rate of the photoautotrophs, along with other environmental conditions such as temperature and pH. The rate of production of biomass glucose, cells, and other organic chemicals by the primary producers is referred to as **net primary productivity** (NPP). Swamps and tropical forests, for example, have a high NPP, whereas deserts and the arctic tundra do not. The rates

TABLE 5–1	Characteristic Terms for Biological Organisms Based on Energy and Carbon Sources			
	Type	**Energy Source**	**Electron Donor[a]**	**Carbon Source**
	Phototrophs	Light		
	Chemotrophs	Organic or inorganic compounds		
	Lithotrophs (subgroup of chemotrophs)		Reduced inorganic compounds	
	Organotrophs (subgroup of chemotrophs)		Organic compounds	
	Autotrophs			Inorganic compounds (e.g., CO_2)
	Heterotrophs			Organic carbon

[a]Electron donors (reducing agents) are the source of electrons that come from reduced bonds (i.e., C—H bonds). The breaking of these reduced bonds may be coupled directly or indirectly to the production of adenosine triphosphate (ATP) within the cell.

of production can be limited by such factors as sunlight (e.g., the growing season, insolent radiation or sunlight penetration into waters), temperature, water, or the availability of nutrients.

When plants are not photosynthesizing, for example, at night, they are respiring, that is giving off carbon dioxide in a manner similar to the way animals do. **Aerobic respiration** is simply the breakdown of organic chemicals, such as sugars and starches, by molecular oxygen to form gaseous carbon dioxide.

Some organisms are able to obtain energy through photosynthesis but are not capable of reducing carbon dioxide. Thus, they obtain carbon from reduced carbon compounds generated by other organisms. These organisms are known as **photoheterotrophs:** heterotrophic referring to the fact that carbon for cell synthesis is derived from preformed organic compounds usually produced by other organisms. This group of biota include the purple nonsulfur bacteria; the green nonsulfur bacteria; some Chrysophytes including *Chromulina, Chrysochromulina, Dinobryon,* and *Ochromonas;* some Euglenophytes including *Euglena gracilis;* some Cryptophytes; and some Pyrrhophyta (Dinoflagellates) such as *Gymnodinium* and *Gonyaulaux.*

These organisms all obtain their energy from light, and they acquire carbon from either inorganic or organic sources. Similarly, the **chemotrophs** obtain their energy from organic or inorganic carbon rather than from light. Chemotrophs can be either autotrophic, that is, they build cell mass from either inorganic forms of carbon, or heterotrophic, using organic forms of carbon to synthesize new cells and compounds. Chemotrophs can also be either lithotrophs, that is, they obtain energy by breaking inorganic chemical bonds, or organotrophs, which get energy by breaking organic chemical bonds.

Chemotrophs that are autotrophic obtain their energy from organic or inorganic compounds and use inorganic carbon compounds as a carbon source (from reduced carbon bonds). All chemoautotrophs are prokaryotic, archaea, and bacteria. These include the nitrifiers, such as *Nitrosomonas europea.* Other chemoautotrophs are bacteria that live at high temperatures and are found in deep-sea vents. These organisms also use inorganic chemicals, including H_2S, HS^-, S^{2-}, SO_3^{2-}, iron sulfide, Fe^{2+}, NH_3, NO_2^-, hydrogen gas, carbon monoxide, ammonia, or nitrite, as electron acceptors. The redox state of the system considered governs the properties of the system and the predominance of species within the system. Redox chemistry was discussed in Chapter 2.

Chemoheterotrophs use inorganic or organic compounds as energy sources; however, they use only preformed reduced organic chemicals as a source of carbon for cell synthesis.

Examples of chemoheterotrophs include animals, protozoa, fungi, and bacteria. Essentially all cellular pathogens are chemoheterotrophs.

As we move up the trophic levels (Figure 5–1) from the primary producers, we find those organisms that are known as the **primary consumers.** These **chemoheterotrophic** organisms are the herbivores that eat plant material. Although chemoautotrophs may obtain energy from chemicals formed by other organisms, under most circumstances they do not consume the organism to obtain those compounds. Rather the compounds they consume were excreted by the living organism or released during the decay of the dead organisms. These organisms are often referred to as the **decomposers** (as they are a special type of consumers). The **secondary consumers,** also chemoheterotrophic organisms, are carnivores that eat the flesh of animals. In a pond ecosystem, beaver, muskrats, ducks, pond snails, and the lesser water boatman, an aquatic insect, are all primary consumers. Among the secondary consumers living in a pond are the insects (all except the lesser water boatman), leeches, otters, mink, and heron. What we have just described is the **food web,** a description of the complex relationship between organisms in an ecosystem. An example of a food web is provided in Figure 5–2.

FIGURE 5–1

Trophic levels (levels of nourishment).

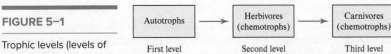

FIGURE 5–2 Simplified Representation of a Food Web, Showing the Main Pathways

Food (energy) moves in the direction of the arrows. The driving force is sunlight. Depictions of the various organisms are not drawn to scale. (*Source:* Fuller, et al., 2006. U.S. EPA Great Lakes National Program Office, Chicago, IL.)

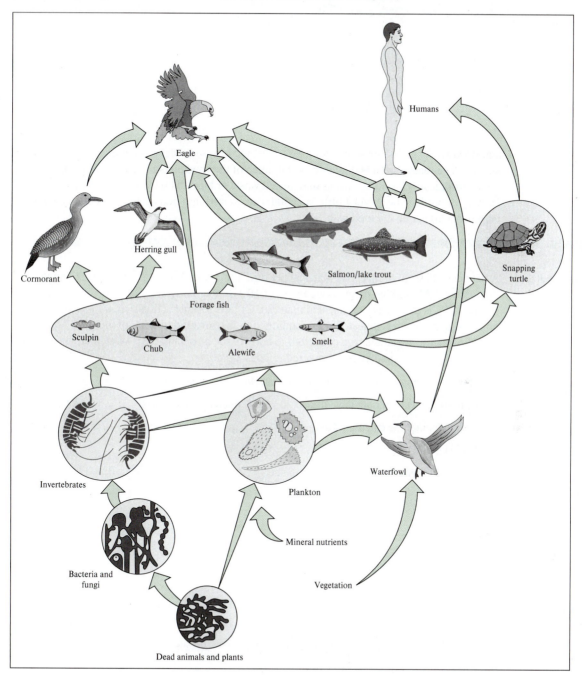

Another term often used is the food, or biomass, pyramid, which attempts to show the quantitative relationships of energy flow by plotting the mass of biomass (all organisms) with trophic level. As we move up the food chain the amount of biomass present decreases (Figure 5–3). Consider, for instance, a meadow as an ecosytem; the vast majority of the biomass would occur as plants. The percentage of primary consumers (as the weight of biomass) would be

FIGURE 5–3

A simplified illustration of the relationship between organisms in an ecosystem is the biomass, or ecological pyramid. This diagram shows both mass and energy flow.

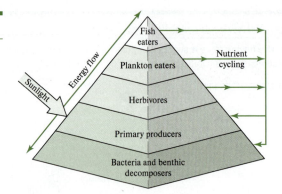

significantly less. The mass of biomass contributed by secondary consumers would be even smaller. The reason for this decrease is that much of the food consumed by an organism higher on the trophic level is either lost as undigested waste or burned up by the organism's metabolic activity to produce heat. Very little is actually converted into body tissue that can be eaten by organisms higher up the food web.

EXAMPLE 5–1 A deer eats 25 kg of herbaceous material per day. The herbaceous matter is approximately 20% dry matter (DM) and has an energy content of $10 \text{ MJ} \cdot (\text{kg DM})^{-1}$. Of the total energy ingested per day, 25% is excreted as undigested material. Of the 75% that is digested, 80% is lost to metabolic waste products and heat. The remaining 20% is converted to body tissue.

How many megajoules are converted to body tissue on a daily basis? Calculate the percentage of energy consumed that is converted to body tissue.

Solution The dry matter content of the herbaceous material is calculated as

$$(25 \text{ kg herbaceous material} \cdot \text{day}^{-1}) \times (0.20 \text{ kg dry matter} \cdot (\text{kg material})^{-1}) = 5.0 \text{ kg DM} \cdot \text{day}^{-1}$$

The energy content is calculated as

$$(10 \text{ MJ} \cdot (\text{kg DM})^{-1})(5.0 \text{ kg DM} \cdot \text{day}^{-1}) = 50 \text{ MJ} \cdot \text{day}^{-1}$$

Now draw a schematic of the energy balance.

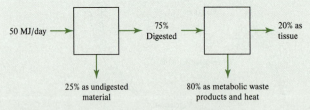

The amount of energy digested is calculated as

$$(0.75) \times 50 \text{ MJ} \cdot \text{day}^{-1} = 37.5 \text{ MJ} \cdot \text{day}^{-1}$$

The amount of this energy that is then turned into tissue is calculated as

$$(0.2) \times 37.5 \text{ MJ} \cdot \text{day}^{-1} = 7.5 \text{ MJ} \cdot \text{day}^{-1}$$

The percentage of "consumed" energy used for body tissue is

$$\left(\frac{7.5 \text{ MJ} \cdot \text{day}^{-1}}{50 \text{ MJ} \cdot \text{day}^{-1}} \right) \times 100 = 15\%$$

This "inefficiency" of conversion of energy is further exemplified in the fact that it is often assumed that only about 10% of the energy consumed by an organism in the form of plant matter is converted into animal tissue (in Example 5–1, it was 15%). Given this, if we consider the mass balance example for lake trout (see Figure 5–2), we can determine the approximate percentage of energy used in building trout tissue.

EXAMPLE 5–2   For every megajoule of energy used by the phytoplankton in Lake Michigan (see Figure 5–2), how many joules of energy are used in building cell tissue in the lake trout? How many in humans? Use the following food web path:

Phytoplankton $\longrightarrow$ zooplankton $\longrightarrow$ alewife $\longrightarrow$ lake trout $\longrightarrow$ humans

Solution   Given the rule of thumb that only 10% of the energy consumed is converted to biomass, then

Phytoplankton $\longrightarrow$ zooplankton $\longrightarrow$ alewife $\longrightarrow$ lake trout $\longrightarrow$ humans
1 MJ   0.1 MJ   0.01 MJ   1000 J   100 J

So far, we have focused our discussion on the trophic levels with no discussion of the type of respiration that occurs in living organisms. Basically, respiration can be aerobic, anaerobic, or anoxic. Organisms that are **aerobic** survive in oxygen-rich environments and use oxygen as the terminal electron acceptor.* Obligate aerobes can survive only in the presence of oxygen. The microorganisms *Bacillus subtilis, Pseudomonas aeruginosa,* and *Thiobacillis ferrooxidans* are examples of obligate aerobes. Humans are also obligate aerobes. The primary end products of aerobic decomposition are carbon dioxide, water and new cell tissue. **Anoxic** environments contain low concentrations (partial pressures) of oxygen. Here nitrate is usually the terminal electron acceptor. The end products of denitrification are nitrogen gas, carbon dioxide, water, and new cells. **Anaerobic respiration** can occur only in the absence of oxygen or nitrate. Obligate anaerobes include *Clostridium* sp. and *Bacteroides* sp. Sulfate, carbon dioxide, and organic compounds that can be reduced serve as terminal electron acceptors. The reduction of sulfate results in the production of hydrogen sulfide, mercaptans, ammonia, and methane. Carbon dioxide and water are the major by-products. *Desulfovibrio desulfuricans* is an example of a sulfur-reducing bacterium.

Facultative anaerobes can use oxygen as the terminal electron acceptor and, under certain conditions, they can also grow in the absence of oxygen. Under anoxic conditions, a group of facultative anaerobes called denitrifiers utilizes nitrites (NO_2^-) and nitrates (NO_3^-) as the terminal electron acceptor. Nitrate nitrogen is converted to nitrogen gas in the absence of oxygen. This process is called **anoxic denitrification.**

Bioaccumulation

So far, in this chapter we have discussed the food web and the ecological pyramid. Bioaccumulation has serious implications for the movement of chemicals in the environment. Chemicals that are hydrophobic (those that don't want to "go into" water) will tend to be lipophilic, that is, "liking lipids." As a result, these chemicals will tend to **partition** (move into) into the fat tissue of animals. This process results in bioaccumulation. **Bioaccumulation** is the total uptake of chemicals by an organism from food items (benthos, fish prey, sediment ingestion, etc.) as well as via mass transport of dissolved chemicals through the gills and epithelium (Schnoor, 1996). When chemicals bioaccumulate, the concentration of a chemical increases over time in

---

*In the oxidation of an electron donor, the electron acceptor (oxidizing agent) is reduced, that is, it accepts electrons.

an organism relative to the chemical's concentration in the environment. For this to occur, these chemicals must be retained in living tissue faster than they are broken down (metabolized) or excreted. For example, if a crustacean or other "lake bottom-dwelling" organism consumes DDT or PCBs from the lake sediments, these chemicals will tend to remain in the fatty tissue of the organism. If we were to put a crustacean from a pristine environment into the contaminated lake, we would notice a corresponding increase in the concentration of the DDT in the crustacean's tissue with residence time in the lake. The problem is compounded when a small fish eats the crustacean because it eats not only the crustacean but the PCB or DDT in the tissue of the crustacean. And the process continues as a larger fish eats the smaller fish. The result of this is that these chemicals tend to biomagnify as we move up the food web.

Biomagnification is the process that results in the accumulation of a chemical in an organism at higher levels than are found in its own food. It occurs when a chemical becomes more and more concentrated as it moves up through a food chain. A typical food chain includes algae eaten by *Daphnia,* a water flea, eaten by an alewife eaten by a lake trout and finally consumed by a kingfisher (or human being). If each step results in bioaccumulation, then an animal at the top of the food chain, through its regular diet, will accumulate a much greater concentration of chemical than was present in organisms lower in the food chain, and biomagnification will occur.

Biomagnification is illustrated by a study of DDT, which showed that where soil levels were 10 ppm, DDT reached a concentration of 141 ppm in earthworms and 444 ppm in the brain tissue of robins (Hunt, 1965). In another study (Figure 5–4), the biomagnification of PCB in the Great Lakes food web can be easily observed. Through biomagnification, the concentration of a chemical in the animal at the top of the food chain may be sufficiently high to cause death or adverse effects on behavior, reproduction, or disease resistance and thus endanger that species, even though contamination levels in the air, water, or soil are low. Fortunately, bioaccumulation does not always result in biomagnification.

Another term often used (and confused) is **bioconcentration**—the uptake of chemicals from the dissolved phase. Through bioconcentration, the concentration of a chemical in an organism becomes greater than its concentration in the air or water in which the organism lives. Although the process is the same for both natural and synthetically made chemicals, the term *bioconcentration* usually refers to chemicals foreign to the organism. For fish and other aquatic animals, bioconcentration after uptake through the gills (or sometimes the skin) is usually the most important bioaccumulation process.

FIGURE 5–4 Persistent Organic Chemicals such as PCBs Biomagnify

This diagram shows the degree of concentration in each level of the Great Lakes aquatic food web for PCBs (in parts per million, ppm). The highest levels are reached in the eggs of fish-eating birds such as herring gulls. (*Source:* Fuller, et al., 2006. U.S. EPA Great Lakes National Program Office, Chicago, IL.)

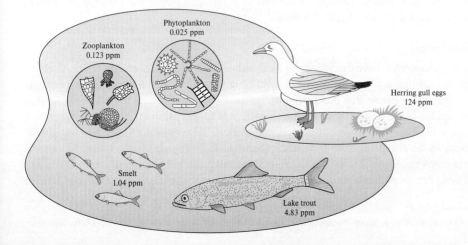

Phytoplankton
0.025 ppm

Zooplankton
0.123 ppm

Herring gull eggs
124 ppm

Smelt
1.04 ppm

Lake trout
4.83 ppm

Bioconcentration factors, the ratio of the concentration of the chemical of interest in the organism to the concentration in water, are used to measure the tendency of a chemical to accumulate in lipid tissue and to relate pollutant concentrations in the water column with that in fish using the following equation:

$$\text{Concentration in fish} = (\text{concentration in water}) \times (\text{bioconcentration factor}) \qquad (5\text{–}2)$$

These factors are important in performing risk assessment calculations for predicting the effect of a chemical on a target species.

EXAMPLE 5-3   The concentration of the pesticide DDT was found to be 5 μg · L^{-1} in the water of a pond. The bioconcentration factor for DDT is 54,000 L · kg^{-1} (U.S. EPA, 1986). What is the expected concentration of DDT in the fish living in the pond?

Solution   Concentration in fish = (5 μg · L^{-1})(54,000 L · kg^{-1}) = 270,000 μg · kg^{-1}, or 270 mg · kg^{-1}

5-4   NUTRIENT CYCLES

Chapter 4 focused on material and energy balances. These same balances, although very complex, hold true globally. The basic elements of which all organisms are composed are carbon, nitrogen, phosphorus, sulfur, oxygen, and hydrogen. The first four of these elements are much more limited in mass and easier to trace than are oxygen and hydrogen. Because these elements are conserved, they can be recycled indefinitely (or cycled through the environment). Because the pathways used to describe the movement of these elements in the environment are cyclic, they are referred to as the carbon, nitrogen, phosphorus, and sulfur cycles.

Carbon Cycle

Although carbon is only the 14th by weight in abundance on earth, it is by far, one of the most important elements on earth as it is the building block of all organic substances and thus, of life itself. Carbon is found in all living organisms, in the atmosphere (predominately as carbon dioxide and bicarbonate), in soil humus, in fossil fuels, and in rock and soils (predominately as carbonate minerals in limestone or dolomite or in shales). Although it was once thought that the largest reservoir of carbon is terrestrial (plants, geological formations, etc.), the ocean actually serves as the greatest reservoir of carbon. As can be seen in Figure 5–5, approximately 85% of world's carbon is found in the oceans.

Photosynthesis is the major driving force for the carbon cycle, shown in Figure 5–6. Plants take up carbon dioxide and convert it to organic matter. Even the organic carbon compounds in fossil fuels had their beginnings in photosynthesis. The "bound," or stored, CO_2 in fossil fuels is released by combustion processes. The cycling of carbon also involves the release of carbon dioxide by animal respiration, fires, diffusion from the oceans, weathering of rocks, and precipitation of carbonate minerals.

The ocean is a major sink of carbon, much of which is found in the form of dissolved carbon dioxide gas, and carbonate and bicarbonate ions. Primary productivity is responsible for the assimilation of inorganic carbon into organic forms. Productivity is limited by the concentrations of nitrogen, phosphorus, silicon, and other essential trace nutrients. The concentrations of CO_2 vary with depth. In shallow waters, photosynthesis is active and there is a net consumption of CO_2. In deeper waters, there is a net production of CO_2 due to respiration and decay processes. Because ocean circulation occurs over such a long time scale, the oceans take up CO_2

FIGURE 5–5

Major reservoirs of carbon (numbers in billions of metric tons). (*Source:* Post, W.M., T.H. Peng, W.R. Emanuel, A.W. King, V.H. Dale, and D.L. DeAngelis (1990). "The Global Carbon Cycle," *American Scientist* vol. 78, p. 310–26.)

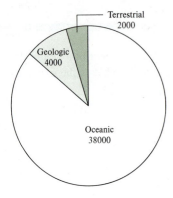

FIGURE 5–6

Cycling of carbon in the environment.

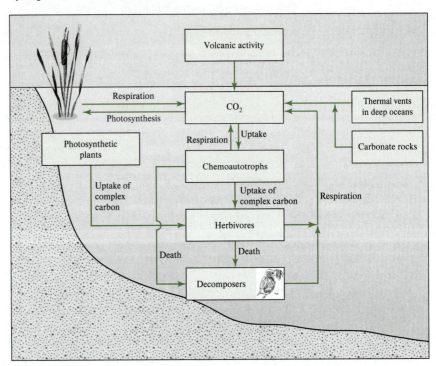

at a slower rate than the rate at which CO_2 from anthropogenic sources is accumulating in the atmosphere. In addition, as the amount of CO_2 dissolved in the ocean increases, the chemical capacity to take up more CO_2 decreases. The rate of uptake of CO_2 is driven by two main cycles: the solubility and biological pumps.

The solubility pump, as it is known, is the net driving force for dissolution of CO_2 into waters. Polar waters are colder at the surface than in deeper regions. As a result of the cold temperatures, CO_2 dissolution is enhanced in colder waters, driving dissolution of CO_2 from the atmosphere into the waters. Because these colder waters are denser than the warmer waters below, the colder waters tend to sink, taking with them CO_2. Because ocean circulation is slow, much of this CO_2 is "lost" to deep waters, keeping surface waters lower in CO_2 and driving dissolution from the atmosphere.

Phytoplankton, zooplankton and their predators, and bacteria make up the biological pump. These organisms take up carbon, resulting in a cycling of much of the carbon and nutrients found in the surface ocean waters. However, as these organisms die, they settle into deeper regions of ocean, taking with them bound CO_2. Additionally, as the dead organisms settle, some of this bound CO_2 finds its way into the ocean depths with the fecal matter of these organisms. Some is carried by currents to deeper regions. Thus, the depths of the ocean become a CO_2 sink, releasing carbon mainly through "upwelling" of water, diffusion across the thermocline,* and seasonal, wind-driven mixing, which brings the deep water to the surface. This mixing of the deeper waters returns nutrients and carbon to the ocean surface, continuing the cycle of photosynthesis and respiration.

Humans have affected significantly the carbon cycle through the combustion of fossil fuel, the large-scale production of livestock, and the burning of forests. Since the Industrial

---

*As discussed in Section 5–6, the thermocline is the layer in a body of water where the temperature decreases sharply with increasing depth. Temperature changes in the thermocline exceed $1°C \cdot m^{-1}$ depth.

Revolution (the 1850s), the concentration of carbon dioxide in the atmosphere has increased from approximately 280 ppm to 383 ppm (by volume) in 2006 (National Oceanic and Atmospheric Administration, 2007). Although there is still some debate, many scientists believe that this increase in carbon dioxide levels has resulted in an increase in global temperatures. These scientists expect that global temperatures will continue to increase, although the extent of this increase is still the subject of some contention. Global warming is discussed in more detail in Chapter 12.

Nitrogen Cycle

Nitrogen in lakes is usually in the form of nitrate (NO_3^-) and comes from external sources by way of inflowing streams or groundwater. When taken up by algae and other phytoplankton, the nitrogen is chemically reduced to amino compounds (NH_2—R) and incorporated into organic compounds. When dead algae undergo decomposition, the organic nitrogen is released to the water as ammonia (NH_3). At normal pH values, this ammonia occurs in the form of ammonium (NH_4^+). The ammonia released from the organic compounds, plus that from other sources such as industrial wastes and agricultural runoff (e.g., fertilizers and manure) is oxidized to nitrate (NO_3^-) by a special group of nitrifying bacteria in a two-step process called nitrification:

$$4NH_4^+ + 6O_2 \rightleftharpoons 4NO_2^- + 8H^+ + 4H_2O \qquad (5\text{–}3)$$

$$4NO_2^- + 2O_2 \rightleftharpoons 4NO_3^- \qquad (5\text{–}4)$$

The first reaction is mediated by the organism *Nitrosomonas* sp., the second by *Nitrobacter* sp.
The overall reaction is

$$NH_4^+ + 2O_2 \rightleftharpoons NO_3^- + 2H^+ + H_2O \qquad (5\text{–}5)$$

As shown in Figure 5–7, nitrogen cycles from nitrate to organic nitrogen, to ammonia, and back to nitrate as long as the water remains aerobic. However, under anoxic conditions, for example, in anaerobic sediments, when algal decomposition has depleted the oxygen supply, nitrate is reduced by bacteria to nitrogen gas (N_2) and lost from the system in a process called **denitrification.** Denitrification reduces the average time nitrogen remains in the lake. Denitrification can also result in the formation of N_2O (nitrous oxide). The denitrification reaction is

$$2NO_3^- + \text{organic carbon} \rightleftharpoons N_2 + CO_2 + H_2O \qquad (5\text{–}6)$$

Some photosynthetic microorganisms can also fix nitrogen gas from the atmosphere by converting it to organic nitrogen and are, therefore, called nitrogen-fixing microorganisms. In lakes the most important nitrogen-fixing microorganisms are photosynthetic bacteria called cyanobacteria, also known as blue-green algae because of their pigments. Because of their nitrogen-fixing ability, cyanobacteria have a competitive advantage over green algae when nitrate and ammonium concentrations are low but other nutrients are sufficiently abundant. Nitrogen fixation also occurs in the soil. The aquatic fern *Azolla* is the only fern that can fix nitrogen. It does so by virtue of a symbiotic association with a cyanobacterium (*Anabaena azollae*). *Azolla* is found worldwide and is sometimes used as a valuable source of nitrogen for agriculture. Similarly, lichens can also contribute to nitrogen fixation. For example, *Lobaria pulmonaria,* a common *nitrogen-fixing* lichen in Pacific northwest forests, fixes nitrogen by a symbiotic relationship with the cyanobacterium *Nostoc.* Lichens such as this are a major source of nitrogen in old growth forests.

Nitrogen fixation is also acomplished in the soils by almost all legumes. Nitrogen fixation occurs in the root nodules that contain bacteria (*Bradyrhizobium* for soybean, *Rhizobium* for most other legumes). The legume family (Leguminosae or Fabaceae) includes many important crop species such as pea, alfalfa, clover, common bean, peanut, and lentil. The reaction describing nitrogen fixation is

$$N_2 + 8e^- + 8H^+ + ATP \longrightarrow 2NH_3 + H_2 + ADP + P_I \qquad (5\text{–}7)$$

where P_I = inorganic phosphate.

FIGURE 5–7

The nitrogen cycle. (*Source:* O'Keefe, et al., 2002.)

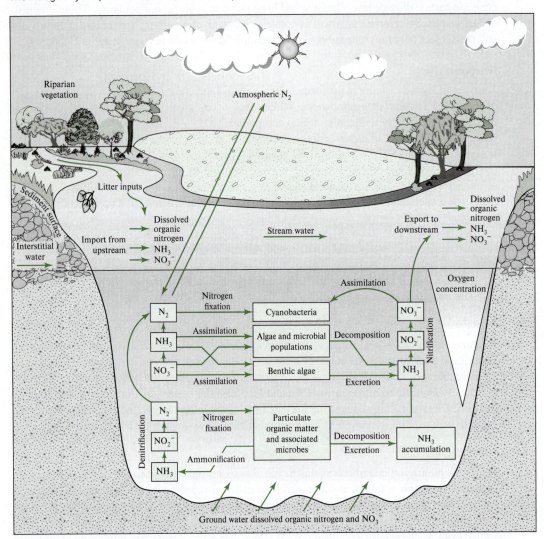

Human influences on the nitrogen cycle have resulted from the manufacture and use of industrial fertilizers, fossil fuel combustion, and large-scale production of nitrogen-fixing crops. Consequently, the release of biologically usable nitrogen from soil and organic matter has increased. Nitrous oxide releases from industrial sources and the combustion of fossil fuels have also increased. The effects of nitrogen releases are significant and range from acid rain and lake acidification to the corrosion of metals and deterioration of building materials. The effects of perturbations on the nitrogen cycle are discussed in more detail later in this book.

Phosphorus Cycle

Phosphorus in unpolluted waters is imported through dust in precipitation or via the weathering of rock. Phosphorus is normally present in watersheds in extremely small amounts, usually existing dissolved as inorganic orthophosphate, suspended as organic colloids, adsorbed onto particulate organic and inorganic sediment, or contained in organic water. In polluted waters, the major source of phosphorus is from human activities. The only significant form of phosphorus available

FIGURE 5–8

The phosphorus cycle. (*Source:* The Michigan Water Research Center, Central Michigan University.)

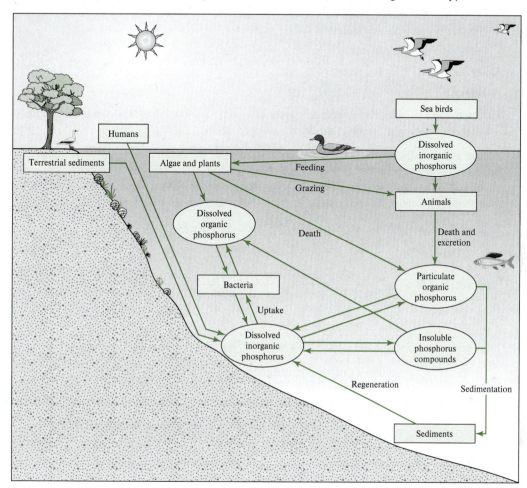

to plants and algae is the soluble reactive inorganic orthophosphate species (HPO_4^{2-}, PO_4^{3-}, etc.) that are incorporated into organic compounds. During algal decomposition, phosphorus is returned to the inorganic form. The release of phosphorus from dead algal cells is so rapid that only a small fraction of it leaves the upper zone of a stratified lake (the epilimnion) with the settling algal cells. However, little by little, phosphorus is transferred to the sediments; some of it in undecomposed organic matter; some of it in precipitates of iron, aluminum, and calcium; and some bound to clay particles. To a large extent, the permanent removal of phosphorus from the overlying waters to the sediments depends on the amount of iron, aluminum, calcium, and clay entering the lake along with phosphorus. The overall cycle can be seen in Figure 5–8.

EXAMPLE 5–4   A farmer has a 7-year rotation of corn, soybeans, and wheat and 4 years of alfalfa. Manure will be applied before corn and wheat and before seeding to alfalfa. The initial soil test results indicate total phosphorus level of 48 kg P · hectare^{-1}. Manure is to be surface-applied in mid-March and disk-incorporated within 2 days of application. One-third of the organic nitrogen and 50% of the nitrogen from NH_4^+ is available to the corn crop. To obtain the desired yield of corn, the local extension agent has told the farmer to apply 100 kg of nitrogen per hectare.

The manure analysis is

- Total N $= 12$ g N $\cdot$ (kg manure)$^{-1}$
- Organic N $= 6$ g N $\cdot$ (kg manure)$^{-1}$
- $(NH_4^+) = 6$ g N $\cdot$ (kg manure)$^{-1}$
- Total $P_2O_5 = 5$ g P $\cdot$ (kg manure)$^{-1}$
- Total $K_2O = 4$ g K $\cdot$ (kg manure)$^{-1}$

Calculate the available nitrogen per kilogram of manure and the kilograms of manure per hectare that must be applied to meet the corn's nitrogen needs.

Solution This is a mass balance problem. Thirty-three percent of the organic nitrogen in the manure is available to the corn, whereas 50% of the inorganic nitrogen (NH_4^+) is available to the corn. There is 6 g organic nitrogen per kilogram of manure and 6 g inorganic nitrogen per kilogram of manure. Therefore, the available N in grams per kilogram can be calculated as

$$\text{Organic N} \times 0.33 + NH_4^+ \times 0.5 = 6 \text{ g N} \cdot kg^{-1} \times 0.33 + 6 \text{ g N} \cdot kg^{-1} \times 0.5$$

$$= 5 \text{ g N} \cdot (\text{kg manure})^{-1}$$

$$\text{Application rate} = \frac{\text{Crop nitrogen needs (in kg N} \cdot \text{hectare}^{-1})}{\text{Available nitrogen (in g N} \cdot (\text{kg manure})^{-1}) \times (10^{-3} \text{ g} \cdot kg^{-1})}$$

$$= \left[\frac{100 \text{ kg N} \cdot \text{hectare}^{-1}}{(5 \times 10^{-3} \text{ kg N} \cdot (\text{kg manure})^{-1})} \right]$$

$$= 20{,}000 \text{ kg manure per hectare}$$

Human activities have led to a release of phosphorus from the disposal of municipal sewage and from concentrated livestock operations. The application of phosphorus fertilizers has also resulted in perturbations in the phosphorus cycle, although these changes are thought to be more localized than the perturbations in the other cycles. Phosphorus releases can have a significant effect on lake and stream ecosystems.

Sulfur Cycle

Until the Industrial Revolution the effect of sulfur on environmental systems was quite small. However, with the Industrial Revolution, our use of sulfur-containing compounds as fertilizers and the release of sulfur dioxide during the combustion of fossil fuels and in metal processing has increased significantly. Mining operations have also resulted in the release of large quantities of sulfur in acid mine drainage. Like the nitrate ion, sulfate is negatively charged and is not adsorbed onto clay particles. Dissolved sulfates thus can be leached from the soil profile by excess rainfall or irrigation. In the environment, sulfur is found predominantly as sulfides (S^{2-}), sulfates (SO_4^{2-}), and in organic forms.

As with the nitrogen cycle, microorganisms play an important role in the cycling of sulfur. Bacteria are involved in the oxidation of pyrite-containing minerals, releasing large quantities of sulfate. In anaerobic environments, sulfate-reducing bacteria (*Desulfovibrio*) reduce sulfate to release hydrogen sulfide. In marine waters, the biological production of dimethylsulfide may occur. The overall cycle is shown in Figure 5–9.

FIGURE 5–9

The sulfur cycle. The lithosphere is the earth's crust and includes rocks and minerals. (*Source:* VanLoan, and W., Duffy, S.J. Environmental Chemistry: A Global Perspective. Oxford University Press, Oxford UK, 2003, p. 345.)

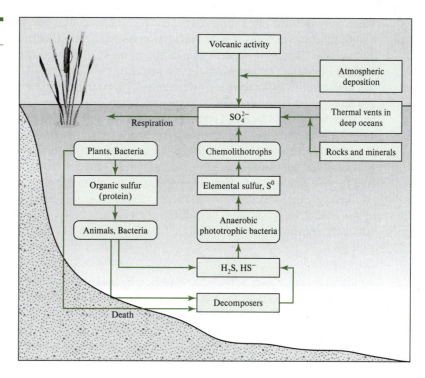

5–5 POPULATION DYNAMICS

Population dynamics is the study of the changes in the numbers and composition of individuals in a population within a study unit and the factors that affect these numbers. The population can be *Escherichia coli,* moose, otters, humans, or any other unit. The study area can be a biological, geographical, political, or an engineered area. For example, if we were studying eastern timber wolves (which travel in packs of five or six), a biological unit we might investigate is the pack. A geographical area could be a mountain range, a valley, or an island. Continuing with our example, we could investigate the numbers of wolves on Isle Royale in Michigan. In this case Isle Royale would be a geographical area. A political unit might be a county, a hunting district, and so on. Lastly, an engineered area might be an aeration tank in a sewage treatment plant in which the population includes rotifers and ciliates.

As environmental scientists and engineers, evaluating population dynamics is critical to (1) understanding how environmental perturbations affect populations, (2) predicting human populations so as to determine water resource needs, (3) predicting bacterial populations in engineered systems, and (4) using populations as indicators of environmental quality. Resource development specialists and wildlife biologists also use population dynamics. They are most concerned with (1) estimating how many animals can be harvested, (2) predicting when a species or population is threatened or endangered with extinction, and (3) understanding how one population might affect another (i.e., competition or predation). Thus, an understanding of population dynamics is necessary for understanding the structure and function of communities and ecosystems.

Factors that cause populations to change may be related to or independent of the number of organisms* in the study area. These factors can be classified as either density-dependent or density-independent. Density refers to the number of organisms per unit area or volume.

---

*We will use the term *organisms* to describe any biological species from bacteria to humans.

We usually measure plants or higher animals in numbers of animals per hectare or per square kilometer. Bacteria, viruses and other aquatic organisms are usually measured in numbers per unit volume. Density-dependent factors are, as implied, a function of density. As the density increases beyond a certain threshold, the population numbers may begin to decline. For example, as the density of bacteria in a reactor increases, they will begin to use up available resources and produce excessive amounts of waste products that may be toxic at high concentrations. The decrease in available food and other resources along with the production of waste will result in a decline in the health of the individual organisms and an increase in the death rate (**mortality**) along with a decrease in the rate of replication or reproduction. For human populations, as the density of the population increases unemployment may increase, causing people to leave the area to seek employment. Thus, the population numbers are observed to decrease as density increases. Density-independent factors are those factors that act on a population independent of the size of the population. Typical density-independent causes of mortality are weather, accidents, and environmental catastrophes (e.g., volcanos, floods, landslides, and fire).

Bacterial Population Growth

As discussed in Chapter 3, bacteria reproduce by binary fission. As such, we can easily model the growth of bacteria in pure cultures using a geometric progression. The models used for growth in mixed cultures are more complicated because of the interactions between the various species. The dynamics of bacterial populations are also relevant to environmental scientists and engineers because of their importance in wastewater treatment and water quality.

Growth in Pure Cultures. As an illustration, let us examine a hypothetical situation in which 1400 bacteria of a single species are introduced into a synthetic liquid medium. Initially nothing appears to happen. The bacteria must adjust to their new environment and begin to synthesize new protoplasm. On a plot of bacterial growth versus time (Figure 5–10), this phase of growth is called the **lag phase.**

At the end of the lag phase the bacteria begin to divide. Because all of the organisms do not divide at the same time, population increase is gradual. This phase is labeled the **accelerated growth phase** on the growth plot.

At the end of the accelerated growth phase, the population of organisms is large enough and the differences in generation time are small enough that the cells appear to divide at a regular rate. Because reproduction is by binary fission (each cell divides to produce two new cells), the increase in population follows in a geometric progression: $1 \rightarrow 2 \rightarrow 4 \rightarrow 8 \rightarrow 16 \rightarrow 32$, and so forth. The population of bacteria (P) after the nth generation is given by the following expression:

$$P = P_o(2)^n \tag{5–8}$$

FIGURE 5–10

Bacterial growth in a pure culture: the log-growth curve.

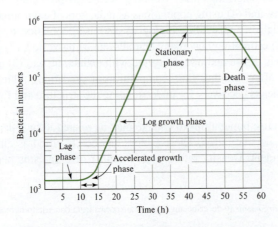

where P_o is the initial population at the end of the accelerated growth phase and n is the number of generations. If we take the log of both sides of Equation 5–8, we obtain the following:

$$\log P = \log P_o + n \log 2 \qquad (5\text{–}9)$$

This means that if we plot bacterial population on a logarithmic scale, this phase of growth would plot as a straight line of slope n. The intercept P_o at t_o is equal to the population at the end of the accelerated growth phase. Thus, this phase of growth is called the **log growth** or **exponential growth phase.** During this phase there are essentially no limitations to cell replication and growth.

In engineered or laboratory settings, the log growth phase tapers off as the substrate becomes exhausted, as toxic by-products build up, as disease takes over, or as space becomes limited. Thus, at some point the population becomes constant either as a result of cessation of fission or a balance in death and reproduction rates. This is depicted by the **stationary phase** on the growth curve. The stationary phase can be rather long as shown in Figure 5–10 or it can be very abrupt.

Following the stationary phase, the bacteria begin to die faster than they reproduce. This **death phase** is due to a variety of causes that are basically an extension of those which lead to the stationary phase. The point at which this decline occurs is referred to as the **carrying capacity.**

EXAMPLE 5–5   If the initial density of bacteria is 10^4 cells per liter at the end of the accelerated growth phase, what is the number of bacteria after 25 generations?

Solution   We are told that $P_o = 10^4$ organisms. Because we are to determine the number of bacteria after 25 generations, $n = 25$. Therefore, we can find the population after 25 generations, P, using Equation 5–8.

$$P = P_o(2)^n$$
$$= 10^4(2)^{25}$$
$$= 3.36 \times 10^{11} \qquad \text{or} \qquad 3.4 \times 10^{11} \text{ cells} \cdot \text{L}^{-1}$$

Growth in Mixed Cultures.   In wastewater treatment, as in nature, **pure cultures*** of microorganisms do not exist. Rather, a mixture of species compete and survive within the limits set by the environment.

The factors governing the dynamics of the various microbial populations are food and shelter limitations, competition and predation by other organisms, and unfavorable physical conditions. The relative success of a pair of species competing for the same substrate is a function of the ability of the species to metabolize the substrate. The more successful species will be the one that metabolizes the substrate more completely. In so doing, it will obtain more energy for synthesis and consequently will achieve a greater mass.

Because of their relatively smaller size and, thus, larger surface area per unit volume, which allows a more rapid uptake of substrate, the density of bacteria will exceed that of fungi. For the same reason, the density of fungi will surpass that of protozoa and the density of filamentous bacteria will exceed that of spherical bacteria (cocci).

*A pure culture is one in which only a single species of microorganism is present.

FIGURE 5–11

Population dynamics in a closed system.

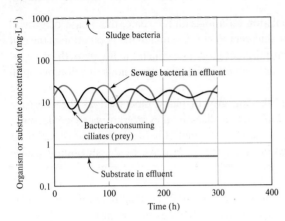

FIGURE 5–12

Population dynamics in an open system.

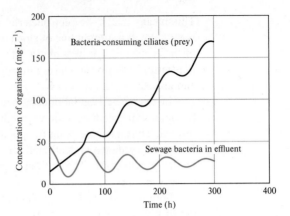

When the supply of soluble organic substrate becomes exhausted, the bacterial population is less successful at replicating and the predator population numbers increase. In a closed system with an initial inoculum of mixed microorganisms and substrate, the populations will cycle as the bacteria give way to higher level organisms, which in turn die for lack of food and are then decomposed by a different set of bacteria (Figure 5–11). In an open system, such as a waste-water treatment plant or a river, with a continuous inflow of new substrate, the predominant populations will change through the length of the plant or river (Figure 5–12*). This condition is known as **dynamic equilibrium.** It is a highly sensitive state, and changes in influent characteristics must be regulated closely to maintain the proper balance of the various populations.

Animal Population Dynamics

The constituents that influence the rate of change in the numbers of a particular species found in the wild include such density-dependent factors as the availability of food, locations to live and build nests for their young, concentrations of toxic waste products, disease, predators, parasites, and so on. Environmental aspects such as weather, temperature, flooding, and snowfall, all of which are density-independent, will also affect population dynamics. As such, population dynamics involve five basic components to which all changes in population can be related: birth, death, gender ratio, age structure, and dispersal.

Clearly population dynamics are affected by the rate at which animals reproduce. A number of components affect a population's birth rate: (1) the amount and quantity of food, (2) age at first reproduction, (3) the birth interval, and (4) the average number of young born per pregnancy. A doubling in the birth (live) rate will more than double the population growth rate.

As mentioned above, several other factors affect population dynamics. **Death,** or **mortality, rate** is defined as the number of animals that die per unit time divided by the number of animals alive at the beginning of that time period. The gender (or sex) ratio will affect mating systems and management. **Gender ratio** is the proportion of males to females within a population. Typically, at birth, this ratio is 50:50. The mating system (monogamous vs. polygamous) will greatly affect population dynamics. In monogamous species, a decline in the population will occur if a

---

*Be careful interpreting this figure. Although at first glance it may appear that it indicates an inverted ecological pyramid (with more biomass in the upper trophic levels than in the lower), this is not the case because the mass of higher organisms measured in the reactor is compared with the mass of bacteria in the effluent. The actual mass of bacteria in the reactor is actually in the thousands of milligrams per liter.

deviation from the 50:50 gender ratio results. In polygamous species, major effects can occur if the population deviates from the 50:50 gender ratio. For example, if all females breed, a population with a ratio of 4 males for every 1 female would result in a 40% lower birth rate than a population with a gender ratio of 50:50. If conversely, the gender ratio were 1 male:4 females and every female bred, then the birth rate would be 160% of that for a population with a gender ratio of 50:50. As expected, age structure will affect population dynamics. This is because of age-specific mortality and pregnancy rates. Dispersal is probably the poorest understood of the factors mentioned. **Dispersal** is defined as the movement of an animal from the location of its birth to a new area where it lives and reproduces. Dispersal usually does not occur until the animal is an adult, and males are usually the gender to disperse.

Population and wildlife biologists use a number of models to describe the rate of change in the numbers of individuals per unit time. The models are similar to those used for bacteria.

The simplest (exponential) model assumes that the resources necessary for population growth are unlimited. Therefore, the population grows at an exponential rate, which is the maximum rate possible for that particular species:

$$\frac{dN}{dt} = rN \tag{5–10}$$

where dN/dt = the change in numbers of animals within a certain population per unit time
r = the specific rate of change
N = number of animals within a certain population

If r is positive, the population size is increasing, if negative, the numbers are decreasing. If $r = 0$, there is no change in the population. If N_o is the number of organisms at time zero, the number of organisms at any particular time, N_t, can be determined by integrating Equation 5–10 over that particular time period:

$$N_t = N_o \exp(rt) \equiv N_o e^{rt} \tag{5–11}$$

The geometric (logistic) model, like the model in Equation 5–11, assumes that resources are unlimited. It would be ridiculous to believe that there is an infinite amount of food and space to support an unlimited number of animals within a particular sized plot of land. As such, we need models that allow us to describe these limits mathematically. This model, however, assumes that growth occurs in discrete intervals described by the term λ.

$$\frac{N(t+1)}{N(t)} = \lambda = e^r \tag{5–12}$$

where $N(t+1)$ = population after $(t+1)$ number in years
$N(t)$ = population after t years
r = the specific growth rate (net new organisms per unit time).

The number of organisms at time, t, can be found from Equation 5–12.

EXAMPLE 5–6

Use the following data along with the exponential model to determine the predicted population of the eastern gray wolf in the state of Wisconsin in the year 2005. Compare that result with that obtained with the geometric model.

Year	1975	1980	1990	1995	1996	1997	1998	1999
Number	8	22	45	83	99	148	180	200

Solution

The value of r in the exponential model can be determined by plotting the natural log of $N(t)/N(0)$ vs. t. In this case, we will use $t = 1975$ as time zero. We can then perform a linear regression to determine the slope. If we do this, we find that the slope is 0.123. Given this, we can then determine the predicted population in 2005, 30 years after the beginning of the data.

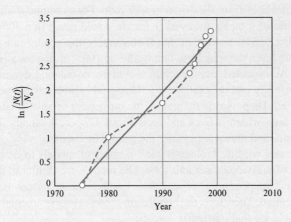

$$N(t) = N(0)e^{0.123(t)} = 8e^{0.123(30)} = 320 \text{ wolves}$$

Now using the geometric model, we can calculate the number in 2005.

Because the geometric model is applied on an interval-by-interval basis, we must develop the general formula to apply it over a time span. This is done for the first interval of 5 years.

For year 0 to year 1 we write: $8 \times \lambda = a$

For year 1 to year 2 we write: $a \times \lambda = b$

For year 2 to year 3 we write: $b \times \lambda = c$

And so on until we get year 4 to year 5: $t \times \lambda = d$.

If we work backward, substituting for the variables a through d, we find the equation to be

$$8 \times \lambda^5 = N(5)$$

where $N(5)$ is the number of individuals 5 years later; and is equal to 22.

So

$$8 \times \lambda^5 = 22$$

Solving for λ yields $\lambda = 1.224$

The general equation applied to get the data in the following table is

$$\lambda = \left[\frac{N(t)}{N(0)}\right]^{1/(t-0)}$$

where t is the year of calculation and 0 is the first year for which data is available.

Year	Number	λ
0	8	
5	22	1.224
15	45	1.122
20	83	1.124
21	99	1.127
22	148	1.142
23	180	1.145
24	200	1.144
	Average	**1.147**

Then, using the geometric growth equation, we rearrange it to solve for $N(30)$

$$N(30) = N(24) \times (1.147^6) = 455 \text{ individuals}$$

Rarely are the resources unlimited. As such, a logistic growth model, which adds a density-dependent term to describe the limitations that exist, is more useful than the simple model described earlier. This model includes a term called the carrying capacity, K, which is simply the numbers of individuals an area can support. As the numbers approach K, the mechanisms (increased mortality, decreased reproduction, increased dispersal) that result in a decrease in the rate of population growth take over. The population change can then be represented by the model:

$$\frac{dN}{dt} = rN \left[\frac{K-N}{K}\right] \qquad (5\text{–}13)$$

The shape of this curve will be S-shaped or sigmoidal.

The number of individuals can be determined by integrating Equation 5–13 to obtain

$$N(t) = \frac{KN_o}{N_o + (K - N_o)e^{-rt}} \qquad (5\text{–}14)$$

EXAMPLE 5–7 Assume that the population of the greater roadrunner in the Guadelope Desert was 200 per hectare at the beginning of 1999. If the carrying capacity, K, is 600 and $r = 0.25 \cdot \text{year}^{-1}$, what is the number of roadrunners one, five and ten years later? What happens when the number of roadrunners equals K?

Solution We substitute the givens into Equation 5–14 for the one year solution:

$$N(1) = \frac{600 \times 200}{200 + (600 - 200)e^{-0.25 \times 1}} = 234 \text{ roadrunners}$$

Doing the same for the five- and ten-year solutions, we get

$N(5) = 381$ roadrunners

$N(10) = 515$ roadrunners

When the number of roadrunners is 600, that is, K, no additional roadrunners can be sustained on this land and the population will remain at 600.

Numerous more complex models exist. These include phenomena known as monotonic damping, damped oscillations, limiting cycles, or chaotic dynamics. A number of these models can also be used to describe plant population dynamics. However, any discussion of these models is beyond the scope of this text. More information on these models can be obtained from *The Economy of Nature* (Ricklef, 2000).

The models discussed thus far are known as single-species models because they describe mathematically a single species of organism. Much more complex models describe the interactions between species by considering predator–prey relationships. These models show how the interactions between two species result in periodic behavior. The models use the following two differential equations to describe the numbers of predators, K, and prey, P:

$$\frac{dP}{dt} = aP - bPK \qquad (5\text{–}15)$$

$$\frac{dK}{dt} = cPK - dK \qquad (5\text{–}16)$$

FIGURE 5–13 Graphical Illustration of the Periodic Nature of Predator–Prey Relationships

The plot with the higher amplitude represents the numbers of prey, in this case, numbers of hares. The lynx is the predator. The model used here is the Lotka–Volterra predation model with initial lynx and hare populations of 1250 and 50,000, respectively. (*Source:* Wilensky, U., & Reisman, K. (2006). Thinking like a wolf, a sheep, or a firefly. Learning biology through constructing and testing computational theories. *Cognition and Instruction,* 24(2), 171–209. (figure 2) and Wilensky, U. (1997). NetLogo Wolf Sheep Predation Model. Evanston, IL: Center for Learning and Computer-Based Modeling, Northwestern University. http://ccl.northwestern.edu/netlogo/models/wolfsheeppredation)

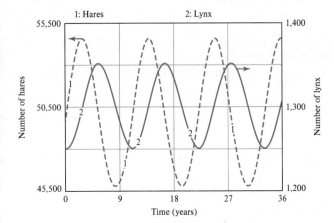

where a = growth rate of the prey
  b = mortality parameter of the prey
  c = growth rate of the predator
  d = mortality parameter of the predator

These equations are often referred to as the Lotka-Volterra model. The cycling nature of these relationships can be seen in Figure 5–13.

Human Population Dynamics

Predicting the dynamics of human populations is important to environmental engineers because it is the basis for the determination of design capacity for municipal water and wastewater treatment systems and for water reservoirs. Population predictions are also important in the development of resource and pollutant management plans. Although several models are used for forecasting human populations, only the exponential model will be discussed here. Human population dynamics also depend on birth, death, gender ratio, age structure, and dispersal. In humans, cultural factors play a significant role. For example, if the average age of first pregnancy in one population is 25, and in another it is 15 and the average age of onset of menopause is 45 in both populations, the birth rates in the two populations will vary significantly. In human populations, dispersal is referred to as immigration and emigration. The effect of cultural differences can be easily illustrated using population pyramids. Population pyramids display the age by gender data of a community at one point in time. Figure 5–14 shows the population pyramids for the United States, Venezuela, Central African Republic, and Spain. Population pyramids are also useful for illustrating changes within a population over time (Figure 5–15).

Assuming an exponential growth rate, the population can be predicted using the equation

$$P(t) = P_o e^{rt} \tag{5–17}$$

where $P(t)$ = the population at time, t
  P_o = population at time, 0
  r = rate of growth
  t = time

Population pyramids for the United States, Venezuela, Central African Republic, and Spain for 2017. (*Source:* U.S. Census Bureau, International Database.)

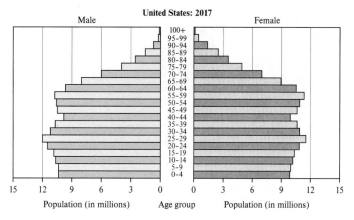

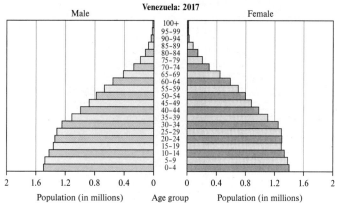

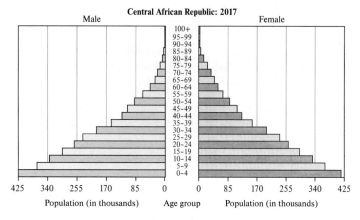

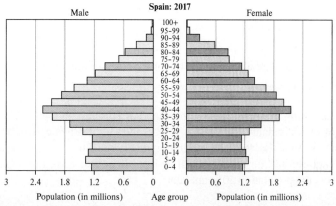

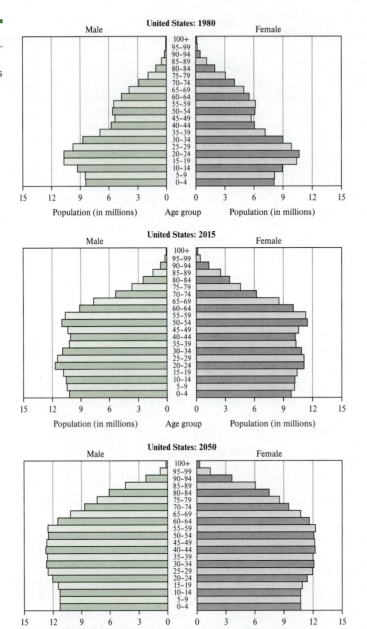

FIGURE 5–15

Population pyramids for the United States in years 1980, 2015, and 2050 illustrating the increase in population and the "graying of America." (*Source:* U.S. Census Bureau, International Database.)

The growth rate can be determined as a function of the birth rate (b), death rate (d), immigration rate (i), and emigration rate (m):

$$r = b - d + i - m \tag{5–18}$$

where the rates are all expressed as some value per unit time.

EXAMPLE 5–8 A population of humanoids on the island of Huronth on the Planet Szacak has a net birth rate (b) of 1.0 individuals/(individual × year) and a net death rate (d) of 0.9 individuals/(individual × year). Assume that the net immigration rate is equal to the net emigration rate. How many years are required for the population to double? If in year zero, the population on the island is 85, what is the population 50 years later?

Solution We must first calculate r. The net growth rate, r, is equal to $b - d$, or $1.0 - 0.9$ or 0.1 individuals/individual $\times$ year. Based on this, we can calculate the projected doubling time.

$$t_{double} = \frac{\ln 2}{r} = \frac{\ln 2}{0.1} = 6.93 \text{ years}$$

The population at year 0, N_o, is 85.

$t = 50$ years

$$r = \frac{0.1 \text{ individuals}}{\text{individual} \times \text{year}}$$

$N_{50} = N_o e^{rt} = 85 \times e^{(0.1)(50)} = 12{,}615$ humanoids. Hopefully the island of Huronth is quite large and able to sustain this population. Note, this model assumes that resources are unlimited. As such, the carrying capacity is not introduced into this equation.

The population of the world can be shown to be exponential (Figure 5–16), although the rate of growth appears to be leveling off (Roser and Ortiz-Ospina, 2018).

One of the most difficult aspects of predicting human populations within a particular region is the accurate prediction of immigration and emigration rates. Another difficult parameter to estimate, and one that is critical for determining world population, is total fertility rate. The **total fertility rate** is the number of children a woman will have over her entire lifetime. The worldwide fertility rate has been decreasing since the 1960s and is now estimated to be about 2.33. Although the fertility rate is decreasing, there are more women in the world to give birth today than there were in the past. This results in a continued increase in population. As shown in Figure 5–17, accurately predicting the future world population depends greatly on correctly predicting the fertility rate. These differences greatly complicate policy decisions.

FIGURE 5–16

Population curve for the world (left) and rate of world population increase (5-year running average) (right). (*Source:* Roser and Ortiz-Ospina (2018).)

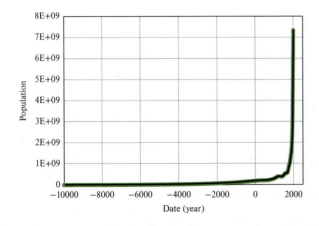
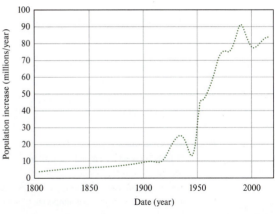

FIGURE 5–17

Prediction of world population using different fertility rates. Total fertility rates of 1.5, 2.1, and 2.6 children are used for low, medium, and high rates, respectively. (*Source:* Population Division of the Department of Economic and Social Affairs of the United Nations Secretariat (2002). *World Population Prospects: The 2000 Revision, vol. III, Analytical Report* (United Nations publication, Sales No. E.01.XIII.20)

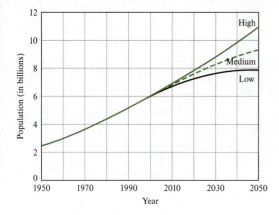

5–6 LAKES: AN EXAMPLE OF MASS AND ENERGY CYCLING IN AN ECOSYSTEM

Lakes provide an excellent example of the cycling of nutrients, mass, and energy within an ecosystem. In this section we will focus on the natural processes that occur in lakes and how human activities have affected these processes. You should note that any of numerous ecosystems could have been chosen as an example of these processes.

Limnology is the study of the ecology of inland waters. The word *limnology* comes from the Greek root *limne* meaning "pool" or "marsh." Limnologists are concerned with the relationships and productivity of freshwater biotic communities with the ever-changing physical, chemical, and biological characteristics of the waters of interest. Although we will focus on lakes, be aware that limnologists study all inland waters, including rivers, creeks, ponds, salt lakes, and even billabongs.*

Stratification and Turnover in Deep Lakes

Nearly all deep lakes in temperate climates† become stratified during the summer and **overturn (turnover)** in the fall due to changes in the water temperature that result from the annual cycle of air temperature changes. In addition, lakes in cold climates undergo winter stratification and spring overturn as well. These physical processes, which are described later on, occur regardless of the water quality in the lake. Many of the lakes in the southern United States are shallow and do not tend to follow the same stratification patterns as those discussed here. Although important, the cycles exhibited by such lakes are beyond the scope of our discussion.

During the summer, the surface water of a lake is heated both indirectly by contact with warm air and directly by sunlight. Warm water, being less dense than cool water, remains near the surface until mixed downward by turbulence from wind, waves, boats, and other forces. Because this turbulence extends only a limited distance below the water's surface, the result is an upper layer of well-mixed, warm water (the **epilimnion**) floating on the lower water (the

---

*Billabong is an aboriginal Australian word for a still body of water that collects first in the lowest areas and grows in size during a rainy season, as the rainfall becomes heavier and more frequent.

†Temperate climates are those without extremes of temperature and rainfall. The British Isles and the northern states of the United States have a temperate climate.

FIGURE 5–18

Temperature and oxygen relationships in temperate eutrophic lakes.

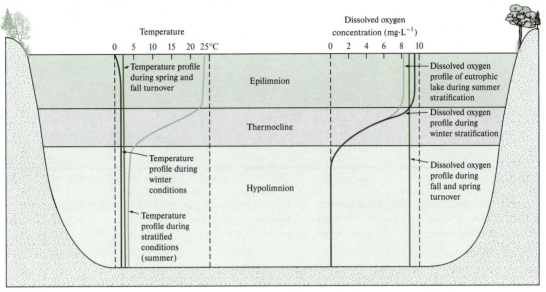

(a) Temperature profile (b) Dissolved oxygen profile

hypolimnion), which is poorly mixed and cool, as shown in Figure 5–18. Because of good mixing, the epilimnion will be aerobic. The hypolimnion will have a lower dissolved oxygen concentration and may become anaerobic or anoxic. The intermediary layer between the epilimnion and hypolimnion is called the **metalimnion.** Within this region the temperature and density change sharply with depth. The **thermocline** may be defined as the region having a change in temperature with depth that is greater than $1°C \cdot m^{-1}$. You may have experienced the thermocline while swimming in a lake. As long as you are swimming horizontally, the water is warm, but as soon as you tread water or dive, the water turns cold. You have penetrated the thermocline. The depth of the epilimnion is related to the size of the lake. It is as shallow as 1 m in small lakes and as deep as 20 m or more in large lakes. The depth of the epilimnion is also related to storm activity in the spring when stratification is developing. A major storm at the right time will mix warmer water to a substantial depth and thus create a deeper than normal epilimnion. Once formed, lake stratification is very stable. It can be broken only by exceedingly violent storms. In fact, as the summer progresses, the stability increases because the epilimnion continues to warm, whereas the hypolimnion remains at a fairly constant temperature.

As shown in Figure 5–19, in the fall, as temperatures drop, the epilimnion cools until it is denser than the hypolimnion. The surface water then sinks, causing overturning. The water of the hypolimnion rises to the surface, where it cools and again sinks. The lake thus becomes completely mixed. If the lake is in a cold climate, this process stops when the temperature reaches 4°C because at this temperature water is the densest. Further cooling or freezing of the surface water results in winter stratification, as shown in Figure 5–19. As the water warms in the spring, it again overturns and becomes completely mixed. Thus, temperate climate lakes have at least one, if not two, cycles of stratification and turnover every year.

Biological Zones

Lakes contain several distinct zones of biological activity, largely determined by the availability of light and oxygen. The most important biological zones, shown in Figure 5–20, are the euphotic, limnetic, littoral, and benthic zones.

FIGURE 5–19

Overturn in stratified lakes. (*Source:* U.S. Environmental Protection Agency, 1995.)

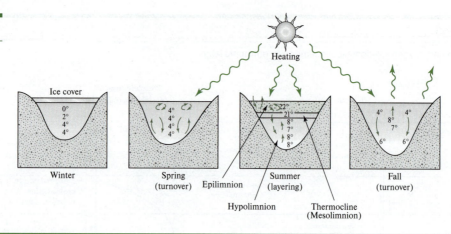

FIGURE 5–20

Biological zones in a temperate lake. (*Source:* WOW. 2004. Water on the Web www.wateronthweb.org—Monitoring Minnesota Lakes on the Internet and Training Water Science Technicians for the Future—A National On-line Curriculum using Advanced Technologies and Real-Time Data. (http://WaterontheWeb.org). University of Minnesota-Duluth, Duluth, MN 55812. Authors: Munson, BH, Axler, RP, Hagley CA, Host GE, Merrick G, Richards C.)

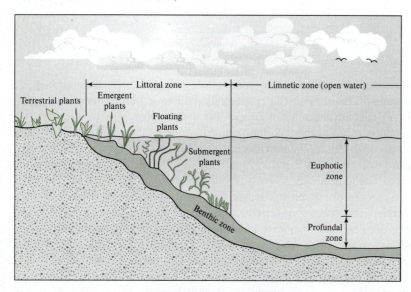

Limnetic Zone. The **limnetic zone** is the layer of open water where photosynthesis can occur. Life in the limnetic zone is dominated by floating organisms (plankton) and actively swimming organisms. The producers in this zone are the planktonic algae. The primary consumers are zooplankton, such as crustaceans and rotifers. The secondary (and higher) consumers are swimming insects and fish.

Euphotic Zone. The upper layer of water through which sunlight can penetrate is called the **euphotic zone.** The depth of the euphotic zone is determined by sunlight penetration and is defined as that region of the lake where light levels are greater than 0.5–1% of that at the surface. At light intensities less than this, algae and macrophytes cannot grow. In most lakes, the euphotic zone exists within the epilimnion. However, in some very clear lakes the euphotic zone can extend well into the hypolimnion. For example, in western Lake Superior, summertime algal growth can occur at depths of up to 25 m, whereas the epilimnion only extends down about 10 m. Similarly, in Lake Tahoe, summertime algal growth has been measured at 100 m depths, whereas, the epilimnion also only extends to about 10 m depth.

TABLE 5–2	**Lake Classification Based on Productivity**			
Lake Classification		Chlorophyll *a* Concentration ($\mu g \cdot L^{-1}$)	Secchi Depth (m)	Total Phosphorus Concentration ($\mu g \cdot L^{-1}$)
Oligotrophic	Average	1.7	9.9	8
	Range	0.3–4.5	5.4–28.3	3.0–17.7
Mesotrophic	Average	4.7	4.2	26.7
	Range	3–11	1.5–8.1	10.9–95.6
Eutrophic	Average	14.3	2.5	84.4
	Range	3–78	0.8–7.0	15–386
Hypereutrophic		>50	<0.5	Often >100

Note: Classification for oligotrophic, mesotrophic, and eutrophic lakes from Wetzel, 1983. Classification for hypereutrophic lakes from Kevern, King, and Ring, 1996.

In deep water, algae are the most important plants, whereas rooted plants grow in shallow water near the shore. In the euphotic zone, plants produce more oxygen by photosynthesis than they remove by respiration. Below the euphotic zone lies the **profundal zone.** The transition between the two zones is called the **light compensation point,** which corresponds roughly to the depth at which the amount of carbon dioxide being converted to sugars by photosynthesis is equal to that being released during respiration.

Littoral Zone.   The shallow water near the shore in which rooted (emergent) water plants (macrophytes) can grow is called the **littoral zone.** The extent of the littoral zone depends on the slope of the lake bottom and the depth of the euphotic zone. The littoral zone cannot extend deeper than the euphotic zone.

Benthic Zone.   The bottom sediments constitute the **benthic zone.** As organisms living in the overlying water die, they settle to the bottom where they are decomposed by organisms living in the benthic zone. Bacteria and fungi are always present. Attached algae may also be present. The presence of higher life forms, such as worms, aquatic insects, molluscs, and crustaceans, depends on the availability of oxygen.

Lake Productivity

The productivity of a lake is a measure of its ability to support aquatic life and is often determined by measuring the amount of algal growth that can be supported by the available nutrients. Although a more productive lake will have a higher biomass concentration than a less productive one, the biomass supported is often undesirable, resulting in taste and odor problems, low dissolved oxygen levels, especially at night, excessive macrophyte growth, loss of water clarity, and proliferation of forage fish and sludge worms. Because of the important role productivity plays in determining water quality, it forms a basis for classifying lakes. Table 5–2 shows how lakes are classified based upon productivity.

Productivity is controlled by the limiting factor, which may be the concentration of nitrogen or phosphorus or the light intensity. This phenomena is known as **Liebig's law of the minimum.*** In many freshwater lakes, phosphorus is often the limiting nutrient. This is because of all the nutrients, only phosphorus is not readily available from the atmosphere or the natural water supply. Therefore, the amount of phosphorus often controls the quantity of algal growth and therefore the productivity of lakes. This can be seen from Figure 5–21 in which the

---

*In 1840, Justin Liebig formulated the idea that the growth of a plant is dependent on the amount of foodstuff that is presented to it in minimum quantity.

FIGURE 5–21

Relationship between summer levels of chlorophyll *a* and measured total phosphorus concentration for 143 lakes. (*Source:* Jones, J.R., Buchmann, R.W., "Prediction of Phosporus and Chlorophyll Levels in Lakes, "Journal of Water Pollution Control Federation, vol. 48, p. 2176, 1976.)

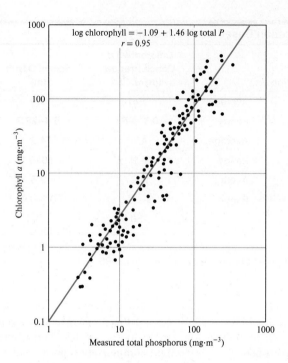

concentration of chlorophyll *a* is plotted against phosphorus concentration. Chlorophyll *a*, one of the green pigments involved in photosynthesis, is found in all algae, so it is used to distinguish the mass of algae in the water from other organic material such as bacteria. It has been estimated that the phosphorus concentration should be below 0.010–0.015 mg $\cdot$ L^{-1} to limit algae blooms (Vollenweider, 1975).

Lakes have a natural life cycle and they change over time, although the times can be many tens of thousands of years. There are young, middle-aged, and old lakes. As lakes age, they slowly shrink as land overtakes the shores of the lake. The trees and other vegetation along the shores of the lake shed organic debris into the lake. This organic matter serves as a carbon source for the organisms in the lake. As the organic matter decays, it can become a new habitat for sedges, grasses, and rushes. This will allow rooted species to flourish where lily pads once floated on the surface. As the lake shrinks and productivity increases, the lake may become anoxic or anaerobic, resulting in a significant change in the lake ecology. This process of succession can continue until the lake becomes a marsh, then a bog, and finally a forest or grassland. This process is illustrated in Figure 5–22.

Oligotrophic Lakes. **Oligotrophic** lakes have a low level of productivity due to a severely limited supply of nutrients to support algal growth. As a result, the water is clear enough that the bottom can be seen at considerable depths. In this case, the euphotic zone often extends into the hypolimnion, which is aerobic. Oligotrophic lakes, therefore, support cold-water game fish. Lake Tahoe on the California–Nevada border, Crater Lake in Oregon, and the blue waters of Lake Superior (Michigan–Wisconsin–Ontario border) are classic examples of oligotrophic lakes. However, the clarity of Lake Tahoe is decreasing due to an increase in the numbers of people residing within the watershed, and a concomitant increase in sewage discharge to the lake.

Eutrophic Lakes. **Eutrophic** lakes have a high productivity because of an abundant supply of nutrients. Because the algae cause the water to be highly turbid, the euphotic zone may extend only partially into the epilimnion. As the algae die, they settle to the lake bottom where they are decomposed by benthic organisms. In a eutrophic lake, this decomposition is sufficient to deplete the hypolimnion of oxygen during summer stratification as shown in Figure 5–23. Because the hypolimnion is anaerobic during the summer, eutrophic lakes support only

FIGURE 5–22

Succession of a lake or
pond.

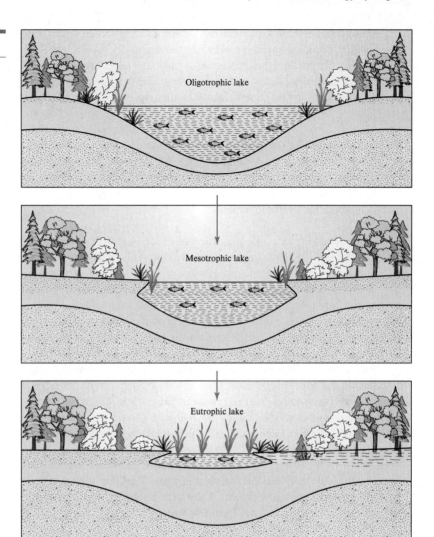

FIGURE 5–23

Seasonal stratification
in lakes will vary with
trophic levels.

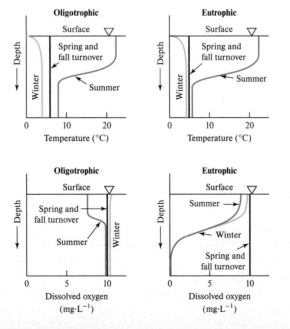

warm-water fish. In fact, most cold-water fish are driven out of the lake long before the hypo-limnion becomes anaerobic because they generally require dissolved oxygen levels of at least 5–6 mg · L^{-1}. Highly eutrophic lakes may also have large mats of floating algae that typically impart unpleasant tastes and odors to the water. Halsted Bay of Lake Minnetonka (Minnesota) and the Neuse River (North Carolina) are examples of eutrophic waters.

Mesotrophic Lakes.   Lakes intermediate between oligotrophic and eutrophic are called **mesotrophic.** Although substantial depletion of oxygen may have occurred in the hypolimnion, it remains aerobic. Lake Ontario, Ice Lake (Minnesota), and Grindstone Lake (Minnesota) are examples of mesotrophic lakes. Lake Michigan and Lake Huron are classified as mesooligo-trophic, meaning the level of productivity is higher than that of an oligotrophic lake but not quite that of a mesotrophic lake.

Dystrophic Lakes.   **Dystrophic** lakes receive a large quantity of organic material from out-side the lake and have low productivity due to low nutrient concentrations. Such lakes are usu-ally surrounded by conifer forests and are small in size. The decomposition of fallen needles results in a significant input of acid to the lake. The lake is typically yellowish brown in color, moderately clear, contains high levels of dissolved organic matter and tannin concentrations, and is acidic. These lakes also have their own characteristic algae, insects, and fish. Sphagnum moss proliferates as a thick mat on the water's surface. Organisms, such as carp, mud min-nows, dragon flies and giant water bugs, which are tolerant of low oxygen levels, replace their predecessors of the eutrophic lake. There are also quantities of emergent vegetation, especially around the edges of the lake. There is zero oxygen on the deep bottom of the lake, thus no fish can survive. Aerobic life exists only in the shallow regions of a dystrophic lake during summer. Examples of dystrophic lakes are Alligator (New) Lake (North Carolina), Swan Creek Lake (North Carolina), Glen Lake, and the bog lakes of northern Michigan.

Hypereutrophic Lakes.   **Hypereutrophic** lakes are extremely eutrophic, with a high algal productivity level and intense algal blooms. They are often relatively shallow lakes with much accumulated organic sediment. They have extensive, dense weed beds and often accumulations of filamentous algae. They have low water clarity and high phosphate and chlorophyll concen-trations. The fish and other aquatic animals in these lakes are subject to extreme shifts in oxygen concentrations; sometimes very high and at other times very low, even depleted. These lakes are often subject to "winter kill" and even "summer kill" during which the depletion of oxygen results in an extensive kill of fish and sometimes other organisms. Recreational use of the wa-ters in hypereutrophic lakes is often impaired. Examples of hypereutrophic lakes are Onondaga Lake (New York) and Upper Klamath Lake (Oregon).

Senescent Lakes.   Very old, shallow lakes in advanced stages of eutrophication are called **senescent** lakes. These lakes have thick organic sediments formed from accumulated aquatic vegetation and dead plant material. Rooted water plants exist in great abundance. These lakes are nearing extinction as a productive lake environment. They will eventually become marshes.

Eutrophication

Eutrophication was once thought to be a natural and inevitable process in which lakes gradu-ally become shallower and more productive through the introduction and cycling of nutrients. However, many oligotrophic lakes have remained such since the last ice age. Ultraoligotrophic lakes such as Lake Tahoe have been unproductive for millions of years. Studies in paleolimnol-ogy also suggest that lakes may undergo natural variations in productivity.

Cultural eutrophication of lakes can occur through the introduction of high levels of nutrients (usually nitrogen and phosphorus, although phosphorus is generally the more limiting of the two). This occurs due to poor management of the watershed and the input of human and

FIGURE 5–24

Simple phosphorus system.

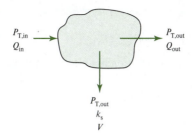

animal wastes. Such changes (from oligotrophic to senescent status) can occur over a period of decades and will cease only if the input of nutrients is halted or significantly reduced. With the control of nitrogen and phosphorus inputs, the process of cultural eutrophication has greatly slowed in both Lake Erie and the Chesapeake Bay. Additional information on the control of cultural eutrophication is presented in Chapter 9.

Because of the importance of phosphorus in enhancing the rates of eutrophication, it is critical that environmental scientists and engineers be able to predict concentrations of phosphorus in lakes. A mass balance approach for a completely mixed lake at steady-state conditions can be used to determine a seasonal/annual estimate. This can be then used to determine the required control measures necessary to halt the process of eutrophication.

Phosphorus enters a lake in surface runoff, from wastewater discharge pipes, and from septic systems. Phosphorus is lost from the system through uptake by organisms, settling of dead biomass (containing phosphorus), and from any outfalls (streams leaving the lake). Figure 5–24 illustrates how we can develop a simple model (Thomann and Mueller, 1987) for the total phosphorus concentration, P_T.

$$\frac{V \, dP_T}{dt} = P_{T,in} Q_{in} - k_s P_{T,out} V - P_{T,out} Q_{out} \qquad (5-19)$$

where V = volume of the hypothetical lake

$P_{T,in}$ = concentration of total phosphorus in the inflow (streams entering the lake, runoff, etc.)

Q_{in} = inflow rate

k_s = phosphorus removal rate (combined settling and biological uptake)

$P_{T,out}$ = concentration of total phosphorus in the lake and in the outflow (under complete mixing conditions)

Q_{out} = outflow rate

At steady state, $dP_T/dt = 0$, therefore

$$0 = P_{T,in} Q_{in} - k_s P_{T,out} V - Q_{out} P_{T,out} \qquad (5-20)$$

$$P_{T,out} = \frac{(P_{T,in} Q_{in})}{(k_s V + Q_{out})} \qquad (5-21)$$

If the removal rate (sometimes called the settling velocity where it does not include biological uptake) is given in units of distance per time (e.g., $m \cdot s^{-1}$), then the velocity should be multiplied by the surface area of the lake, rather than the volume.

EXAMPLE 5–9   Greenlawn Lake has a surface area $2.6 \times 10^6 \, m^2$. The average depth is 12 m. The lake is fed by a stream having a flow rate of $1.2 \, m^3 \cdot s^{-1}$ and a phosphorus concentration of $0.045 \, mg \cdot L^{-1}$. Runoff from the homes along the lake adds phosphorus at an average annual rate of $2.6 \, g \cdot s^{-1}$. The settling rate of the lake is $0.36 \, day^{-1}$. A river flows from the lake at a flowrate of $1.2 \, m^3 \cdot s^{-1}$. What is the steady-state concentration of phosphorus in the lake? What is the trophic state of the lake?

Solution $0 = P_{T,stream} Q_{stream} + P_{T,runoff} - k_s P_{T,out} V - P_{T,out} Q_{out}$

First convert the stream phosphorus concentration to units of grams per cubic meter.

$$P_{stream} = \frac{0.045 \text{ mg} \cdot \text{L}^{-1}(1000 \text{ L} \cdot \text{m}^{-3})}{1000 \text{ mg} \cdot \text{g}^{-1}} = 0.045 \text{ g} \cdot \text{m}^{-3}$$

Therefore,

$$0 = (0.045 \text{ g} \cdot \text{m}^{-3})(1.2 \text{ m}^3 \cdot \text{s}^{-1}) + 2.6 \text{ g} \cdot \text{s}^{-1}$$

$$- (0.36 \text{ day}^{-1}) \left(\frac{1 \text{ day}}{86,400 \text{ s}} \right) (2.6 \times 10^6 \text{ m}^2)(12 \text{ m})(P_{T,out}) - (P_{T,out})(1.2 \text{ m}^3 \cdot \text{s}^{-1})$$

Check each product, the units should be grams per second.

$$0 = 0.054 \text{ g} \cdot \text{s}^{-1} + 2.6 \text{ g} \cdot \text{s}^{-1} - (130 \text{ m}^3 \cdot \text{s}^{-1})(P_{T,out}) - (P_{T,out})(1.2 \text{ m}^3 \cdot \text{s}^{-1})$$

$$P_{T,out} = 0.020 \text{ g} \cdot \text{m}^{-3}, \text{ or } 0.020 \text{ mg} \cdot \text{L}^{-1}$$

$$(0.020 \text{ mg} \cdot \text{L}^{-1})(1000 \text{ µg} \cdot \text{mg}^{-1}) = 20 \text{ µg} \cdot \text{L}^{-1}$$

Therefore, using Table 5–2, you can determine that the lake borders on a eutrophic state.

Eutrophication of marine waters is often controlled by nitrogen concentrations, rather than by phosphorus levels. This is the case in the Massachusetts Bay, where eutrophication was a major problem until quite recently. Combined sewer overflows from the cities of Boston, Cambridge, Chelsea, and Somerville discharged mixed storm water and untreated sewage into Boston Harbor and its tributary, the Charles River. Storm drainage, contaminated with sewage from leaking pipes or illegal sewer connections from buildings, also entered the estuary. Animal waste deposited on the streets contaminated storm water, resulting in an additional input of nutrients into the water. The two wastewater treatment plants serving the city of Boston and surrounding communities, Deer Island and Nut Island Wastewater Treatment Plants (DITP and NITP), were undersized and aging, and provided only minimal (primary) treatment. With the sewage from 48 communities flowing into the Boston Harbor, the estuary was considered one of the most contaminated in the United States.

As shown in Figure 5–25a, eutrophication, as indicated by the high chlorophyll levels, was a problem in much of the inner harbor and estuary (Massachusetts Water Resources Authority, 2002). Bacterial counts were also high in the Inner Harbor, along the shoreline, and in the rivers, resulting in beach closures during much of the warm summer months. Prior to July 1998, ammonia nitrogen levels near the two treatment plant outfalls reached very high levels, up to 100 µM. By late 2000, after the Nut Island Treatment Plant had been shut down, the Deer Island Treatment Plant had been upgraded and the new 15-km long outfall was operational, the ammonia-nitrogen levels decreased to less than the target value of 5 µM. As shown in Figure 5–25b, between 1998 and 2000 (after the treatment plant upgrade and before the outfall was operational), a decrease in phytoplankton was observed. After the outfall was in use, the chlorophyll level in the South Harbor further decreased to even more desirable concentrations. The chlorophyll concentrations in the North Harbor decreased significantly from 2000 to 2005. However, for reasons still unknown, the concentrations increased in Summer 2006 to almost that observed before the outfall was utilized (Massachusetts Water Resources Authority, 2006). Despite the high chlorophyll concentrations, which were mainly due to the proliferation of the diatom *Dactyliosolen fragilissimus,* oxygen levels in the estuary were at desirable levels and the diatom was not harmful or problematic. While the estuary is still not pristine, water quality is much improved compared to that observed before 1998 and the problems related to eutrophication have been abated.

FIGURE 5–25

(a) August 1995–July 8, 1998. While both DITP and NITP were discharging, chlorophyll had a west-to-east decreasing gradient, with the highest levels in Quincy Bay and at the mouth of the Neponset River. **(b)** July 9, 1998–August 2000. With the transfer South System flows to DITP, the shape of chlorophyll contours shifted subtly, with lower chlorophyll levels in the southern harbor and a small, localized increase in chlorophyll at the outer North Harbor. (*Source:* MWRA.)

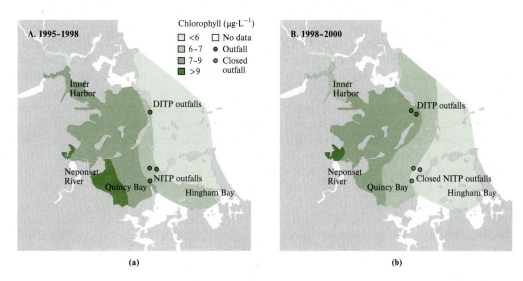

(a)   (b)

5–7   ENVIRONMENTAL LAWS TO PROTECT ECOSYSTEMS

Since the late 1960s and early 1970s, many countries have enacted laws to protect species from extinction. Although numerous laws such as the Lacey Act of 1900 and the Migratory Bird Act of 1929 were promulgated early in the 20th century, the landmark acts protecting species were the Endangered Species Conservation Act of 1969, the Marine Mammal Protection Act of 1972, and the Endangered Species Act of 1973. The Endangered Species Act is important because it declared that endangered species "are of aesthetic, ecological, educational, historical, recreational, and scientific value to the nation and its people." It made the importation, exportation, or selling in interstate or international commerce of any endangered species or any product of an endangered species illegal. It is also illegal to capture, harass, or harm any animal on the endangered species list.

In the United States another law important to the protection of ecosystems and the plants and animals within is the National Environmental Policy Act (NEPA) of 1969. NEPA is significant in that its language justifies the protection of the environment on aesthetic, cultural, and moral grounds.* The U.S. Congress recognized both the profound impact of human activity on natural environment and the critical importance of restoring and maintaining environmental quality to the overall welfare and development of humanity. Through this act, Congress sought to "foster and promote the general welfare to the overall welfare and development of man *(sic),* to create and maintain conditions under which humans and nature can exist in productive harmony, and fulfill the social, economic, and other requirements of present and future generations of Americans." One of the major aspects of the regulations promulgated under this act requires the preparation of environmental impact statements (EISs) on those major actions determined

---

*This act declared a national policy to encourage productive and enjoyable harmony between man (sic) and his environment; to promote efforts which will prevent or eliminate damage to the environment and biosphere and stimulate the health and welfare of man; to enrich the understanding of the ecological systems and natural resources important to the Nation; and to establish a Council on Environmental Quality.

to have a significant impact on the quality of the human environment (40 CFR Part 6; Subpart A). These documents are reviewed by the U.S. EPA and must show the impact of the proposed project on the ecology of the area.

While political boundaries divide people, they do not partition ecosystems that fall at borders. Such is the case at the Canadian–U.S. border, which crosses all of the Great Lakes except Lake Michigan. Under the Boundary Waters Treaty of 1909, the International Joint Commission was set up to prevent and resolve disputes between the United States and Canada and to develop a structure that would allow the two countries to work together to solve common environmental problems. The Great Lakes Water Quality Agreement was first signed in 1972 and renewed in 1978 and commits both the United States and Canada "to restore and maintain the chemical, physical and biological integrity of the Great Lakes Basin Ecosystem" (International Joint Commission, 1978). In accordance with this agreement, the two countries have agreed to eliminate to the greatest extent possible the discharge of any or all persistent toxic substances, provide financial assistance for the construction of publicly owned wastewater treatment plants, and develop and implement best-management practices to protect the Great Lakes. On October 24, 2006, the International Joint Commission recommended that the United States and Canada develop a new agreement that would be more action oriented and would establish achievable goals, timelines, and better provisions for monitoring and reporting of water quality in the Great Lakes.

Clearly, these regulations have had a profound effect on protecting the environment. If we value the environment for its own sake and for the welfare and well-being of humanity, we must work to ensure that these acts and regulations are maintained and all efforts to dilute them are halted.

CHAPTER REVIEW

When you have completed studying this chapter, you should be able to do the following without the aid of your text or notes:

1. Define the terms *ecosystem* and *ecology*. What are the differences in these terms?
2. List the different nutrient levels of organisms. Provide the energy source, electron donor, and carbon source for each classification of organism.
3. Describe primary and secondary productivity. Give examples of producers, consumers, and decomposers.
4. For each type of decomposition (aerobic, anoxic, and anaerobic), list the electron acceptor and important end products.
5. Draw a typical food web for Lake Michigan.
6. Describe the difference between a food web and an ecological pyramid.
7. Describe bioaccumulation, bioconcentration, and biomagnification. How are bioconcentration factors used?
8. Describe a pure culture.
9. Describe how the growth in a mixed culture differs from that in a pure culture.
10. Describe what is meant by dynamic equilibrium.
11. List the models used to describe animal population dynamics. List the attributes and problems associated with each model.
12. Describe what is meant by carrying capacity. Explain how it influences population growth. Do cities have a carrying capacity for humans?
13. Rank the major global pools of carbon (terrestrial vegetation and biomass, atmosphere, surface ocean, and intermediate/deep ocean) by the total mass of carbon in each. What is the predominant form of carbon in each?
14. List the major pathways in the nitrogen cycle, the chemical species that are transformed in each, and the species (give one or two examples) that accomplish each of the transformations.

15. List the major pathways in the phosphorus cycle, the chemical species that are transformed in each, and the species (give one or two examples) that accomplish each of the transformations.

16. List the major pathways in the sulfur cycle, the chemical species that are transformed in each, and the species (give one or two examples) that accomplish each of the transformations.

17. Describe the major difference between the phosphorus cycle and all other nutrient cycles.

18. Sketch and compare the epilimnion and hypolimnion with respect to the following: location in a lake, temperature, and oxygen abundance (i.e., dissolved oxygen).

19. Describe the process of stratification and overturn in lakes.

20. Define the following terms: epilimnion, hypolimnion, thermocline, and mesolimnion.

21. Define the following terms: littoral, limnetic, euphotic, profundal, and benthic zones of lakes.

22. Explain what determines the euphotic zone of a lake and what significance this has for biological growth.

23. Describe the classification of lakes as oligotrophic, mesotrophic, eutrophic, hypereutrophic, or senescent in terms of productivity, clarity, and oxygen levels.

24. Explain the process of eutrophication. Explain the difference between natural and cultural eutrophication.

25. State Leibig's law of the minimum.

26. Name the most common limiting nutrient in lakes and explain why it most commonly limits growth.

27. List three sources of phosphorus that can be controlled to reduce cultural eutrophication.

PROBLEMS

5–1 A population of purple rabbits lives on the island of Zulatop. The rabbits have a net growth rate of 0.09 year^{-1}. At the present time there are 176 rabbits on the island. What is the predicted number of rabbits 5, 10, and 20 years from now? Use the simple exponential growth equation to calculate the number of rabbits.

Answer: $P(5) = 276$ rabbits $P(10) = 433$ $P(20) = 1065$

5–2 Recalculate the number of purple rabbits if the carrying capacity is 386 and you use the logistic equation. Assume the same number of rabbits at the present time and use the same intervals.

5–3 A population of spotted wolves lives on the mountain Hesperides. There were 26 wolves in the year 2054 and 54 wolves counted in 2079. Assuming exponential growth, what is the net growth rate constant?

Answer: 0.029 year^{-1}

5–4 For Problem 5–3, what will the population of wolves be in 2102?

5–5 Using the data presented in Problem 5–3, a net growth rate of 0.04 year^{-1} and a carrying capacity of 159, what will be the predicted population of wolves in the year 2102?

Answer: 91 wolves

5–6 A population of Ladon dragons has a birth rate of 3.3 individuals/(individual × year) and a death rate of 3.2 individuals/(individual × year).

(a) What is r?

(b) Based on your answer from part a, is the population growing, declining, or remaining constant?

(c) Assuming exponential growth, how many years are necessary for the population to double?

(d) On the Wondering Rock Mountain, the population is presently 49. What will the population be in 25 years?

(e) If the carrying capacity were 105 on the Wondering Rock Mountain, what would the population be in 25 years (using the logistic model)?

5–7   The initial density of bacteria is 15,100 cells per liter at the end of the accelerated growth period. What is the density of bacteria (cells per liter) after 28 generations?

Answer: 4.053×10^{12}

5–8   There are 100 bunnies living on an island on November 7, 2006. The net birth rate is 1.2 baby bunnies/bunny-year, 0.85 bunnies/bunny-year go to bunny heaven annually. Numerous bunny tourists decide to stay on the island—so that the net immigration rate is 0.45 bunnies/bunny-year, and 0.12 bunnies/bunny-year leave for better-tasting alfalfa. How many bunnies are there on November 6, 2007? How many bunnies are there on the island on November 6, 2016? Use the Human Population Dynamic Model. State any assumptions inherent in this model.

5–9   Suppose there is a grove of 1334 Asphodel plants on Prometheus Island located in Andarta Lake. The trees are growing with an r of 0.21 individuals/(individual × year). The carrying capacity on the island is 3250. What is the population in 35 years, assuming a logistic growth model applies?

Answer: 3247

5–10   Julana Lake has a surface area of 4.1×10^6 m^2. The average depth of the lake is 15 m. The lake is fed by a stream having a flow rate of 2.02 m$^3 \cdot$ s^{-1} and a phosphorus concentration of 0.023 mg $\cdot$ L^{-1}. A wastewater treatment plant discharges into the lake at a rate of 0.2 m$^3 \cdot$ s^{-1} and a phosphorus concentration of 1.1 mg $\cdot$ L^{-1}. Runoff from the homes along the lake adds phosphorus at an average annual rate of 1.35 g $\cdot$ s^{-1}. The settling rate of phosphorus from the lake averages 0.94 year^{-1}. The river flows from the lake at a flow rate of 2.42 m$^3 \cdot$ s^{-1}. Assume evaporation and precipitation to negate each other. What is the concentration of phosphorus in the river flowing from the lake?

5–11   You have been conducting a water quality study of Lake Arjun, which has a surface area of 8.9×10^5 m^2. The average depth of the lake is 9 m. The lake is fed by a stream having a flow rate of 1.02 m$^3 \cdot$ s^{-1} and a phosphorus concentration of 0.023 mg $\cdot$ L^{-1}. Runoff from the homes along the lake adds phosphorus at an average annual rate of 1.25 g $\cdot$ s^{-1}. The river flows from the lake at a flow rate of 1.02 m$^3 \cdot$ s^{-1}. The average phosphorus concentration in the lake is 13.2 µg $\cdot$ L^{-1}. Assume evaporation and precipitation negate each other. What is the calculated average settling rate of phosphorus?

Answer: 1.19×10^{-5} s^{-1} or   376 year^{-1}

5–12   The concentration of diazinon has been measured to be 23.3 µg $\cdot$ L^{-1} in Lake Pekko. The bioconcentration factor for diazinon is 337 L $\cdot$ kg^{-1}. What is the expected concentration of diazinon in fish living in Lake Pekko?

5–13   The concentration of the pentachlorophenol has been measured to be 42.8 µg $\cdot$ L^{-1} in Adonis Pond. A study of the Matsu fish revealed an average lipid concentration of 30,600 µg $\cdot$ kg^{-1}. What is the bioconcentration factor for this fish?

Answer: 715

5–14   The bioconcentration factor for bis(2-ethylhexyl) phthlate, a commonly used plasticer, in the organism *Daphnia* is 5200 L $\cdot$ kg^{-1}. If the concentration of bis(2-ethylhexyl) phthlate in a lake is 3.6 µg $\cdot$ L^{-1}, estimate the concentration of bis(2-ethylhexyl) phthlate in *Daphnia* in units of µg $\cdot$ kg^{-1}.

5–15   One of the congeners of toxaphene, a persistent pesticide that was used on cotton, is 1,2,3,4,7,7-heptachloro-2-norbornene. The bioconcentration factor for this chemical in fish was determined to be 11,200 L $\cdot$ kg^{-1}. If the concentration is 1.1 ng $\cdot$ L^{-1} in Lake Greenway, determine the estimated concentration in fish in µg $\cdot$ Kg^{-1}.

Answer: 12.3 µg $\cdot$ L

5–16   Farmer Tapio is raising deer. She has 110 female deer age 3–15 months. The deer require 22 MJ of metabolizable energy per day during the spring months. The deer are being fed a mixture of 50% wheat and 50% silage. The wheat contains 85% dry matter (DM) and has 12.5 MJ metabolizable energy per kilogram of DM. The silage has 30% dry matter and 10.5 MJ metabolizable energy per kilogram of DM. How many kilograms of feed are required per day to feed the 110 deer?

5–17 For Problem 5–16 calculate the energy (in megajoules) converted to body tissue on a daily basis. Assume that 19% of the feed consumed is excreted as undigested material. Of the remaining 81% that is digested, 78% is used in generating metabolic waste products and heat. The remaining 22% is incorporated into tissue.

Answer: $3.92 \cdot \text{day}^{-1}$

5–18 Using the data provided in Example 5–4, calculate the mass of phosphorus (in kilograms P_2O_5) and potassium (in kilograms K_2O) per hectare of land.

5–19 In year 2, the farmer mentioned in Example 5–4 plans to plant soybeans. No manure is to be applied during this crop year. The soybean is a legume and can fix adequate atmospheric nitrogen to produce the desired yield of soybeans. Additional nitrogen fertilizer does not produce significant yield increases. Soybeans require $50 \text{ kg N} \cdot (\text{hectare})^{-1}$, $35 \text{ kg P} \cdot (\text{hectare})^{-1}$, and $225 \text{ kg K} \cdot (\text{hectare})^{-1}$ according to the local extension agent.

(a) Determine if the previous year's manure application provides enough phosphorus and potassium to meet the crop nutrient requirements of the soybeans.

(b) Estimate the remaining phosphorus and potassium after the corn is harvested. The corn crop from year 1 would have removed $52 \text{ kg of } P_2O_5 \cdot \text{hectare}^{-1}$ and $38 \text{ kg of } K_2O \cdot \text{hectare}^{-1}$.

Answer: (a) Sufficient nitrogen and phosphorus are provided. Additional potassium would be required.

(b) $77.3 \text{ kg P} \cdot (\text{hectare})^{-1}$

$48.5 \text{ kg K} \cdot (\text{hectare})^{-1}$

5–20 Colette Lake has a surface area of 103 ha and a mean depth of 8 m. The pH of the water is 7.6. The lake receives surface water from a lake upstream, rainfall, and diffuse groundwater input. Atmospheric inorganic N deposition is insignificant. Contaminated groundwater is the major source of inorganic N in the lake. During April, 62 mm of rainfall fell on Colette Lake. During the same month, $4.2 \times 10^6 \text{ m}^3$ of groundwater having an inorganic N concentration of 63 mg/L infiltrated into the lake. Water, having an inorganic N concentration of $8.4 \text{ mg} \cdot \text{L}^{-1}$, flows into the lake at a rate of $21.00 \text{ m}^3 \cdot \text{s–1}$. The sedimentation rate for inorganic N was measured as $49 \text{ mg N} \cdot \text{m}^{-2} \cdot \text{day}^{-1}$. A river drains the lake at a rate of $22.64 \text{ m}^3 \cdot \text{s}^{-1}$. There are no other sources of water into or out of the lake. Calculate the steady state concentration of inorganic N in the lake.

DISCUSSION QUESTIONS

5–1 Rank the major global pools of carbon (terrestrial vegetation and biomass, atmosphere, surface ocean, and intermediate/deep ocean) by the total mass of carbon in each. How is our consumption of fossil fuels and resultant emission of carbon dioxide changing the ratios and predominant forms of carbon in each pool?

5–2 Why is there more standing detritus in tundra soils than in tropical rainforest soils?

5–3 Describe how human activities are altering each of the nutrient cycles.

5–4 Describe how phosphorus availability could control nitrogen uptake.

5–5 How can the university you attend promote biological conservation?

5–6 Choose an endangered species. Provide an argument, both for and against, the protection of this species.

5–7 Is cultural eutrophication ever beneficial? Defend your position.

5–8 Recently, several dams have been removed or breached (removal of the earthen portion while leaving the concrete intact) because they had outlived their intended purpose. Discuss the benefits and negative effects of dam removal.

5–9 What are some environmental impacts of dams? Given the various negative effects dams have on rivers, what is the likelihood that new dams will be built? Discuss your answer.

FE EXAM FORMATTED PROBLEMS

5–1 Organisms that obtain carbon from reduced carbon compounds generated by other organisms and energy from sunlight are known as:

(a) Photoheterotrophs

(b) Photoautotrophs

(c) Chemoheterotrophs

(d) Chemoautotrophs

5–2 For every megajoule (MJ) of energy taken up by grass, how much energy does the snake use to build cell tissue? The food web is:

Grass → Grasshopper → Toad → Snake

(a) 1 MJ

(b) 0.1 MJ

(c) 0.01 MJ

(d) 1000 J

5–3 The concentration of the pesticide malathion was found to be 0.125 mg/L in the water in a pond. The bio-concentration factor for malathion is 1000 L/kg (for brown shrimp). What is the expected concentration of malathion in the brown shrimp living in this pond?

(a) 0.125 µg/kg

(b) 1.25 mg/kg

(c) 12.5 mg/kg

(d) 125 mg/kg

5–4 The process by which ammonia is converted to nitrate by bacteria is called:

(a) Nitrification

(b) Denitrification

(c) Nitrogen fixation

(d) Ammonification

5–5 A community has experienced the following exponential growth in population:

Year	Population
2000	45,000
2010	61,000

If the rate of expected population growth is anticipated to remain equal to the observed growth rate, determine the predicted population in 2030.

(a) 93,000

(b) 112,100

(c) 125,000

(d) 150,000

5–6 The initial density of bacteria is 2×10^3 cells per liter at the end of the accelerated growth phase. What is the number of bacteria after 50 generations?

(a) 2.25×10^9

(b) 2.25×10^{18}

(c) 2.25×10^{50}

(d) 2.25×10^{150}

5–7 The human population on Earth can be described by the equation $P = 2.439 \times 10^{-3} \exp(1.425 \times 10^{-2t})$, where t = year. Determine the predicted population in 2030.

(a) 7.62 billion

(b) 8.99 billion

(c) 11.6 billion

(d) 12.8 billion

5–8 The region in a lake having a change in temperature with depth that is greater than 1°C/m is known as the:

(a) Epilimnion

(b) Hypolimnion

(c) Thermocline

(d) Benthos

5–9 An organism that is autotrophic, photosynthetic, and either unicellular or multicellular is most likely a:

(a) Bacterium

(b) Plant

(c) Algae

(d) Protozoa

5–10 Which of the following is not a likely end product of denitrification?

(a) Nitrogen gas

(b) Carbon dioxide

(c) Water

(d) Ammonia

REFERENCES

Associated Press (2007) *Ravenous, Invasive Shrimp Found in Lake Ontario,* Jan. 18, 2007, http://www.ctv.ca/servlet/ArticleNews/story/CTVNews/20070118/ravenous_shrimp_070118/20070118?hub=SciTech

Doran, M. (2012) "Human Society and Growth."*

Estabrook, V. (2007) *The Phosphorus Cycle in Lakes,* Michigan Water Research Center, http://www.cst.cmich.edu/centers/mwrc/phosphorus%20cycle.htm, accessed April 20, 2007.

Fuller, K., H. Shear, and J. Wittig (2006) *The Great Lakes: An Environmental Atlas and Resource Book,* U.S. EPA Great Lakes National Program Office, Chicago, IL, http://www.epa.gov/glnpo/atlas/images/big05.gif

Glassner-Schwayder, K. (2000) *Zebra Mussels Cause Economic and Ecological Problems in the Great Lakes,* GLSC Fact Sheet 2000-6, August 15, 2000. U.S. Geological Survey, Great Lakes Science Center, Ann Arbor, MI, http://www.glsc.usgs.gov/_files/factsheets/2000-6%20Zebra%20Mussels.pdf

Government of Canada and the U.S. EPA (1995) *The Great Lakes: An Environmental Atlas and Resource Book: Natural Processes in the Great Lakes: Geology,* 3rd ed. Jointly produced by the Government of Canada, Toronto, ON, and the U.S. EPA Great Lakes National Program Office, Chicago, IL, www.epa.gov/grtlakes/atlas/glat-ch2.html#4, accessed April 20, 2007.

Hauer, M. E. (2017) Migration induced by sea-level rise could reshape the US population landscape, *Nature Climate Change,* 7: 321–325. doi:10.1038/nclimate3271

Hunt, L. B. (1965) "Kinetics of Pesticide Poisoning in Dutch Elm Disease," *U.S. Fish, Wildl. Serv. Circ.* 226: 12–13.

*The authors thank Michael Doran, P.E., DEE for sharing his white paper, Human Society and Growth.

International Joint Commission (1978) *What Is the Great Lakes Water Quality Agreement?* Windsor, ON, http://www.ijc.org/rel/agree/quality.html, accessed April 20, 2007.

Jones, J. R., and R. W. Bachmann (1976) "Prediction of Phosphorus and Chlorophyll Levels in Lakes," *Journal of the Water Pollution Control Federation,* 48: 2176.

Kestrup, Å. M., and A. Ricciardi (2008) "Occurrence of the Ponto-Caspian mysid shrimp *Hemimysis anomala* (Crustacea, Mysida) in the St. Lawrence River," *Aquatic Invasions* 3(4): 461–464.

Kevern, N. R., D. L. King, and R. Ring (1996) "Lake Classification System—Part 1," *The Michigan Riparian,* February, 1996, last updated 05/10/2004, http://www.mlswa.org/lkclassif1.htm, accessed April 20, 2007.

Massachusetts Water Resources Authority (2002) *The State of Boston Harbor: Mapping the Harbor's Recovery,* Boston, MA, http://www.mwra.state.ma.us/harbor/html/2002-09.htm

Massachusetts Water Resources Authority (2006) October 25, 2006, letter from Michael J. Hornbrook, Chief Operating Officer to G. Haas, Massachusetts Department of Environmental Protection and Linda Murphy, U.S. EPA, www.mwra.state.ma.us/harbor/pdf/20061025amx.pdf, accessed April 20, 2007.

Moyle, P. B., ed. (1997) *Essays in Wildlife Conservation, a Reader for WFC 10, Wildlife Conservation and Ecology,* Department of Wildlife, Fish, and Conservation Biology, University of California, Davis. Chapter 8, "Wildlife and Pollution," http://www.meer.org/wfc10-a.htm, accessed April 20, 2007.

National Oceanic and Atmospheric Administration (2007) *Trends in Atmospheric Carbon Dioxide,* Earth System Research Laboratory, Global Monitoring Division, http://www.cmdl.noaa.gov/ccgg/trends/index.php#mlo

O'Keefe, T. C., S. R. Elliott, R. J. Naiman, and D. J. Norton (2002) *Introduction to Watershed Ecology,* October 2002, www.epa.gov/watertrain/ecology/rt.html

Pimentel, D., and M. Pimentel (1996) *Food, Energy and Society,* Colorado University Press, Niwot, CO.

Post, W. M., T. H. Peng, W. R. Emanuel, A. W. King, V. H. Dale, and D. L. DeAngelis (1990) "The Global Carbon Cycle," *American Scientist,* 78: 310–26.

Prüss-Ustün, A., Wolf, J., Corvalán, C., Bos, R. and M. Neira (2016) Preventing Disease Through Healthy Environments: A Global Assessment of the Burden of Disease from Environmental Risks. http://apps.who.int/iris/bitstream/handle/10665/204585/9789241565196_eng.pdf;jsessionid=2AC5EFD7A09D0E49EF4653D46E09C4B8?sequence=1

Ricciardi, A., S. Avlijas, and J. Marty (2012) "Forecasting the ecological impacts of the *Hemimysis anomala* invasion in North America: Lessons from other freshwater mysid introductions," *J Great Lakes Res.* 38: 7–13. doi:10.1016/j.jglr.2011.06.007.

Ricklef, R. E. (2000) *The Economy of Nature,* 5th ed., W. H. Freeman and Co., New York.

Roser, M., and E. Ortiz-Ospina (2018) World Population Growth. Published online at OurWorldInData.org. Retrieved from: https://ourworldindata.org/world-population-growth.

Schnoor, J. L. (1996) *Environmental Modeling: Fate and Transport of Pollutants in Water, Air, and Soil,* Wiley and Sons, New York, pp. 342–43.

Sweedler, A. R. (1999) *Energy Issues in the San Diego/Tijuana Region—Briefing Paper,* Center for Energy Studies, San Diego State University, San Diego, CA.

Thomann, R. V., and J. A. Mueller (1987) *Principles of Surface Water Quality Modeling and Control,* Harper and Row Inc., New York.

U.S. Census Bureau (2002) *International Database,* U.S. Census Bureau, International Programs Office, Washington, DC.

U.S. EPA (1986) Superfund Public Health Evaluation Manual, Office of Emergency and Remedial Response, Washington, DC.

U.S. EPA (2006) *Invasive Species in the Great Lakes,* last updated March 9, 2006. http://www.epa.gov/glnpo/monitoring/indicators/exotics/cercopagis.html

United Nations (1998) Population Division, Dept. of Economic and Social Affairs, Geneva, Switzerland.

United Nations (2000) *World Population Prospects, the 2000 Revision,* Vol III, Analytical Report, UN Population Division, Department of Economic and Social Affairs, p. 161, http://www.un.org/esa/population/publications/wpp2000/chapter5.pdf

United Nations Food and Agricultural Organization (2010) Achieving Sustainable Gains in Agriculture. http://www.fao.org/docrep/014/am859e/am859e01.pdf.

United Nations Food and Agricultural Organization (2013) Food Wastage Footprint: Impacts on Natural Resources: Summary Report. http://www.fao.org/docrep/018/i3347e/i3347e.pdf.

VanLoon, G. W., and S. J. Duffy (2000) *Environmental Chemistry: A Global Perspective,* Oxford University Press, New York, p. 345.

Vollenweider, R. A. (1975) "Input-output Models with Special Reference to the Phosphorus Loading Concept in Limnology," *Schweizerische Zeitschrift für Hydrologie,* 37: 53–83.

Western, D., and M. C. Pearl (1989) *Conservation for the Twenty-first Century,* Oxford University Press, New York, p. 55.

Wetzel, R. G. (1983) *Limnology,* W. B. Saunders Co., Philadelphia, PA.

Wilensky, U., and K. Reisman (1998) "Thinking Like a Wolf, a Sheep, or a Firefly: Learning Biology through Constructing and Testing Computational Theories—An Embodied Modeling Approach," Proceedings of the Second International Conference on Complex Systems, Nashua, NH, October 1998.

Willyard, C. (2010) "Smithsonian.com." *Smithsonian Magazine.* 19 Apr. 2010. Accessed March 6, 2012. http://www.smithsonianmag.com/specialsections/ecocenter/air/EcoCenter-Air-Acid-Rain-and-Our -Ecosystem.html?c=y

6

Risk Perception, Assessment, and Management

Case Study

Imposed Risk Versus Assumed Risk

People make an assessment of risk of death for a large number of activities such as crossing a street, riding a bicycle, or flying in an airplane. These are termed *assumed risks*. Drinking contaminated water and breathing contaminated air because there is no alternative are called *imposed risks*. People are more willing to accept risks that are assumed than they are willing to accept risks that are imposed.

The drinking water crisis in Flint, Michigan, as will be discussed in Chapter 10, resulted from an imposed risk. The risk was unknowingly imposed when lead pipe service lines were installed to connect houses to the municipal water treatment system when they were built in the 1930s and 1940s. With adequate water treatment, the risk was minimized. When the water treatment system failed to provide a protective chemical layer on the walls of the lead pipe, lead began leaching into the drinking water. The citizens of Flint were only alerted to the danger when the corrosive water resulted in orange color in the water. Their fear and anger were unmitigated! And rightly so!

6–1   INTRODUCTION

The concepts of risk and hazard are inextricably intertwined. **Hazard** implies a probability of adverse effects in a particular situation. **Risk** is a measure of the probability. In some instances the measure is subjective, or perceived risk. Scientists and engineers use models to calculate an estimated risk. In some instances actual data may be used to estimate the risk. We make estimates of risk for a wide range of environmental phenomena. Examples include the risk of tornadoes, hurricanes, floods, droughts, landslides, and forest fires. We have chosen to limit our discussion to the risk to human health from chemicals released to the environment.

In the last two decades an attempt has been made to bring more rigor to the risk estimation process. Today this process is called **quantitative risk assessment,** or more simply **risk assessment.** The use of the results of a risk assessment to make policy decisions is called **risk management.** Chapters 9–16 discuss alternative measures for managing risk by reducing the amount of contaminants in the environment.

6–2   RISK PERCEPTION

The old political saying, "Perception is reality," is no less true for environmental concerns than it is for politics. People respond to the hazards they perceive. If their perceptions are faulty, risk management efforts to improve environmental protection may be misdirected.

Some risks are well quantified. For example, the frequency and severity of automobile accidents are well documented. In contrast, other hazardous activities such as those resulting from the use of alcohol and tobacco are more difficult to document. Their assessment requires complex epidemiological studies.

When lay people (and some experts for that matter) are asked to evaluate risk, they seldom have ready access to the statistics. In most cases, they rely on inferences based on their experience. People are likely to judge an event as likely or frequent if instances of it are easy to imagine or recall. Also, it is evident that acceptable risk is inversely related to the number of people participating in the activity. In addition, recent events such as a hurricane or earthquake can seriously distort risk judgments.

Figure 6–1 illustrates different perceptions of risk. Four different groups were asked to rate 30 activities and technologies according to the present risk of death from each. Three of the groups, from Eugene, Oregon, included 30 college students, 40 members of the League

FIGURE 6–1

Judgments of perceived risk for experts and lay people plotted against the best technical estimates of annual fatalities for 25 technologies and activities. The lines are the straight lines that best fit the points. The experts' risk judgments are seen to be more closely associated with annual fatality rates than are the lay judgments.

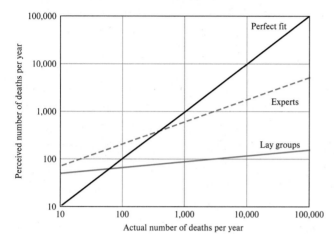

of Women Voters (LOWV), and 25 business and professional members of the "Active Club." The fourth group was composed of 15 people selected from across the United States because of their professional involvement in risk assessment. These groups were asked to estimate the mean fatality rate for a group of activities and technologies given the fact that the annual death toll from motor vehicle accidents in the United States was 50,000. The results are plotted in Figure 6–1. The lines for "experts" and "lay groups" are lines of best fit to the results. The line at 45 degrees depicts an imaginary line of perfect fit. The steeper slope of the line for the experts' risk judgments shows that they are more closely associated with the actual annual fatality rates than those of the lay groups.

Putting risk perception in perspective, we can calculate the risk of death from some familiar causes. To begin, we recognize that we will all die at some time. So, as a trivial example, the lifetime risk of death from all causes is 100%, or 1.0. In the United States in 2014, there were about 2.6 million deaths. Of these, about 591,699 were cancer-related. Without considering age factors, the risk of dying from cancer in a lifetime was about

$$\frac{591,699}{2.6 \times 10^6} = 0.23$$

The annual risk (assuming a 78.8-year life expectancy and again ignoring age factors) is about

$$\frac{0.23}{78.8} = 0.003$$

For comparison, Table 6–1 summarizes the risk of dying from some causes of death.

In developing standards for environmental protection, the EPA often selects a lifetime incremental risk of cancer in the range of 10^{-7} to 10^{-4} as acceptable.

Of course, if the risk of dying in one year is increased, the risk of dying from another cause in a later year is decreased. Because accidents often occur early in life, a typical accident may shorten life by 30 years. In contrast, diseases, such as cancer, cause death later in life, and life is shortened by about 15 years. Therefore, a risk of 10^{-6} shortens life on the average of 30×10^{-6} years, or 15 min for an accident. The same risk for a fatal illness shortens life by about 8 min. It has been noted that smoking a cigarette takes 10 min and shortens life by 5 min (Wilson, 1979).

TABLE 6–1	Annual Risk of Death from Selected Common Human Activities	
Cause of Death	Number of Deaths in Representative Year	Individual Risk per Year
Black lung disease (coal mining)	1,500	1.36×10^{-4} or 1/7,350
Heart attack	614,348	2.97×10^{-3} or 1/337
Cancer	591,699	2.86×10^{-3} or 1/350
Coal mining accident	16	1.14×10^{-4} or 1/9,000
Fire fighting	68	2.57×10^{-6} or 1/390,000
Motorcycle driving	4,668	6.97×10^{-6} or 1/144,000
Motor vehicle	35,398	1.92×10^{-6} or 1/520,000
Truck driving	725	8.68×10^{-7} or 1/1,150,000
Falls	27,000	1.30×10^{-4} or 1/7,700
Football (averaged over participants)		or 1/100,000
Home accidents	136,053	6.57×10^{-4} or 1/1,500
Bicycling	726	2.14×10^{-7} or 1/4,670,000
Air travel (one transcontinental trip/year)		2×10^{-6} or 1/500,000

Sources. CDC, National Center for Health Statistics, 2014; NFPA, June 2016; NHTSA, 2011, 2012.

6–3 RISK ASSESSMENT

In 1989, the EPA adopted a formal process for conducting a baseline risk assessment (U.S. EPA, 1989). This process includes data collection and evaluation, toxicity assessment, exposure assessment, and risk characterization. Risk assessment is considered to be site-specific. Each step is described briefly in the following section.

Data Collection and Evaluation

Data collection and evaluation includes gathering and analyzing site-specific data relevant to human health concerns for the purpose of identifying substances of major interest. This step includes gathering background and site information, the preliminary identification of potential human and ecosystem exposure through sampling, and the development of a sample collection strategy.

When collecting background information, it is important to identify the following:

1. Possible contaminants
2. Concentrations of the contaminants in key sources and media (i.e., air, soil, or water) of interest, characteristics of sources, and information related to the chemical's release potential
3. Characteristics of the environmental setting that could affect the fate, transport, and persistence of the contaminants

The review of the available site information determines basic site characteristics such as air and groundwater movement or soil characteristics. With these data, it is possible to initially identify potential exposure pathways and exposure points important for assessing risk. A conceptual model of pathways and exposure points can be formed from the background data and site information. This conceptual model can then be used to help refine data needs.

Toxicity Assessment

Toxicity assessment is the process of determining the relationship between the exposure to a contaminant and the increased likelihood of the occurrence or severity of adverse effects.

In this discussion we will focus on adverse effects to people, but similar evaluations for plants and animals are also appropriate for evaluation of ecosystem effects. This procedure includes hazard identification and dose-response evaluation. **Hazard identification** determines whether exposure to a contaminant causes increased adverse effects for humans and to what level of severity. Dose-response evaluation uses quantitative information on the dose of the contaminant and relates it to the incidence of adverse reactions in an exposed population. Toxicity values can be determined from this quantitative relationship and used in the risk characterization step to estimate different occurrences of adverse health effects based on various exposure levels.

The single factor that determines the degree of harmfulness of a compound is the dose of that compound (Loomis, 1978). **Dose** is defined as the mass of chemical received by the animal or exposed individual. Dose is usually expressed in units of milligrams per kilogram of body mass ($mg \cdot kg^{-1}$). Some authors use parts per million (ppm) instead of $mg \cdot kg^{-1}$. Where the dose is administered over time, the units may be $mg \cdot kg^{-1} \cdot day^{-1}$. It should be noted that dose differs from the concentration of the compound in the medium (air, water, or soil) to which the animal or individual is exposed.

For toxicologists to establish the "degree of harmfulness" of a compound, they must be able to observe a quantitative effect. The ultimate effect manifested is death of the organism. Much more subtle effects may also be observed. Effects on body weight, blood chemistry, and enzyme inhibition or induction are examples of graded responses. Mortality and tumor formation are examples of **quantal** (all-or-nothing) responses. If a dose is sufficient to alter a biological mechanism, a harmful consequence will result. The experimental determination of the range of changes in a biologic mechanism to a range of doses is the basis of the dose-response relationship.

The statistical relationship of organism response to dose is commonly expressed as a cumulative-frequency distribution known as a **dose-response curve.** Figure 6–2 illustrates the method by which a common toxicological measure, namely the LD_{50}, or lethal dose for 50% of the animals, is obtained. The assumption inherent in the plot of the dose-response curve is that the test population variability follows a Gaussian distribution and, hence, that the dose-response curve has the statistical properties of a Gaussian cumulative-frequency curve.

Toxicity is a relative term. That is, there is no absolute scale for establishing toxicity; one may only specify that one chemical is more or less toxic than another. Comparison of different chemicals is uninformative unless the organism or biologic mechanism is the same and the quantitative effect used for comparison is the same. Figure 6–2 serves to illustrate how a toxicity scale might be developed. Of the two curves in the figure, the LD_{50} for compound B is greater than that for compound A. Thus, for the test animal represented by the graph, compound A is more toxic than compound B as measured by lethality. There are many difficulties in establishing toxicity relationships. Species respond differently to toxicants so that the LD_{50} for a mouse may be very different than that for a human. The shape (slope) of the dose-response curve may differ for different compounds so that a high LD_{50} may be associated with a low "no observed adverse effect level" (NOAEL) and vice versa.

The nature of a statistically obtained value, such as the LD_{50}, tends to obscure a fundamental concept of toxicology: that there is no fixed dose that can be relied on to produce a given biologic effect in every member of a population. In Figure 6–2, the mean value for each test group is plotted. If, in addition, the extremes of the data are plotted as in Figure 6–3, it is apparent that the response of individual members of the population may vary widely from the mean. This implies not only that single-point comparisons, such as the LD_{50}, may be misleading, but that even knowing the slope of the average dose-response curve may not be sufficient if one wishes to protect hypersensitive individuals.

Organ toxicity is frequently classified as an acute or subacute effect. Carcinogenesis, teratogenesis, reproductive toxicity, and mutagenesis have been classified as chronic effects. A glossary of these toxicology terms is given in Table 6–2.

FIGURE 6–2

Hypothetical dose-response curves for two chemical agents (A and B) administered to a uniform population. NOAEL = no observed adverse effect level.

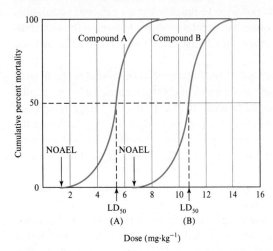

FIGURE 6–3

Hypothetical dose-response relationships for a chemical agent administered to a uniform population.

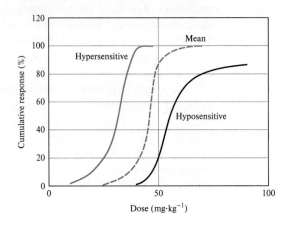

TABLE 6–2		Glossary of Toxicological Terms
	Acute toxicity	An adverse effect that has a rapid onset, short course, and pronounced symptoms.
	Cancer	An abnormal growth process in which cells begin a phase of uncontrolled growth and spread.
	Carcinogen	A cancer-producing substance.
	Carcinomas	Cancers of epithelial tissues. Lung cancer and skin cancer are examples of carcinomas.
	Chronic toxicity	An adverse effect that frequently takes a long time to run its course and initial onset of symptoms may go undetected.
	Genotoxic	Toxic to the genetic material (DNA).
	Initiator	A chemical that starts the change in a cell that irreversibly converts the cell into a cancerous or precancerous state. Needs to have a promoter to develop cancer.
	Leukemias	Cancers of white blood cells and the tissue from which they are derived.
	Lymphomas	Cancers of the lymphatic system. An example is Hodgkin's disease.
	Metastasis	Process of spreading or migration of cancer cells throughout the body.
	Mutagenesis	Mutagens cause changes in the genetic material of cells. The mutations may occur either in somatic (body) cells or germ (reproductive) cells.
	Neoplasm	A new growth. Usually an abnormally fast-growing tissue.
	Oncogenic	Causing cancers to form.
	Promoter	A chemical that increases the incidence to a previous carcinogen exposure.
	Reproductive toxicity	Decreases in fertility, increases in miscarriages, and fetal or embryonic toxicity as manifested in reduced birth weight or size.
	Sarcoma	Cancer of mesodermal tissue such as fat and muscle.
	Subacute toxicity	Subacute toxicity is measured using daily dosing during the first 10% of the organism's normal life expectancy and checking for effects throughout the normal lifetime.
	Teratogenesis	Production of a birth defect in the offspring after maternal or paternal exposure.

It is self-evident that an organ may exhibit acute, subacute, and chronic effects and that this system of classification is not well bounded.

Virtually all of the data used in hazard identification and, in particular, hazard quantification, are derived from animal studies. Aside from the difficulty of extrapolating from one species to another, the testing of animals to estimate low-dose response is difficult. Example 6–1 illustrates the problem.

EXAMPLE 6–1 An experiment was developed to ascertain whether a compound has a 5% probability of causing a tumor. The same dose of the compound was administered to 10 groups of 100 test animals. A control group of 100 animals was, with the exception of the test compound, exposed to the same environmental conditions for the same period of time. The following results were obtained:

Group	Number of Tumors	Group	Number of Tumors
A	6	F	9
B	4	G	5
C	10	H	1
D	1	I	4
E	2	J	7

No tumors were detected in the controls (not likely in reality).

Solution The average number of excess tumors is 4.9%. These results tend to confirm that the probability of causing a tumor is 5%.

If, instead of using 1000 animals (10 groups × 100 animals), only 100 animals were used, it is fairly evident from the data that, statistically speaking, some very anomalous results might be achieved. That is, we might find a risk from 1–10%.

Note that a 5% risk (probability of 0.05) is very high in comparison with the EPA's objective of achieving an environmental contaminant risk of 10^{-7} to 10^{-4}.

Animal studies are only capable of detecting risks on the order of 1%. Toxicologists employ mathematical models to extrapolate data from animals to humans.

One of the most controversial aspects of toxicological assessment is the method chosen to extrapolate the carcinogenic dose-response curve from the high doses administered to test animals to the low doses that humans actually experience in the environment. The conservative worst-case assessment is that one event capable of altering DNA will lead to tumor formation. This is called the one-hit hypothesis. From this hypothesis, it is assumed that there is no threshold dose below which the risk is zero, so that for carcinogens, there is no NOAEL and the dose-response curve passes through the origin.

Many models have been proposed for extrapolation to low doses. The selection of an appropriate model is more a policy decision than a scientific one because no data confirm or refute any model. The one-hit model is frequently used.

$$P(d) = 1 - \exp(-q_o - q_1 d) \tag{6–1}$$

where $P(d)$ = lifetime risk (probability) of cancer
$\quad\quad d$ = dose
q_o and q_1 = parameter to fit data

This model corresponds to the simplest model of carcinogenesis, namely, that a single chemical hit will induce a tumor.

The background rate of cancer incidence, $P(O)$, may be represented by expanding the exponential as

$$\exp(x) = 1 + x + \frac{x^2}{2!} + \cdots + \frac{x^n}{n!} \tag{6–2}$$

For small values of x, this expansion is approximately

$$\exp(x) \approx 1 + x \tag{6–3}$$

Assuming the background rate for cancer is small, then

$$P(O) = 1 - \exp(-q_o) \approx 1 - [1 + (-q_o)] = q_o \tag{6–4}$$

This implies that q_o corresponds to the background cancer incidence. For small dose rates, the one-hit model can then be expressed as

$$P(d) \approx 1 - [1 - (q_o + q_1 d)] = q_o + q_1 d = P(O) + q_1 d \tag{6–5}$$

For low doses, the additional cancer risk above the background level may be estimated as

$$A(d) = P(d) - P(O) = [P(O) + q_1 d] - P(O) \tag{6–6}$$

or

$$A(d) = q_1 d \tag{6–7}$$

This model, therefore, assumes that the excess lifetime probability of cancer is linearly related to dose.

Some authors prefer a model that is based on an assumption that tumors are formed as a result of a sequence of biological events. This model is called the multistage model.

$$P(d) = 1 - \exp[-(q_o + q_1 d + q_2 d^2 + \cdots + q_n d^n)] \tag{6–8}$$

where q_i values are selected to fit the data. The one-hit model is a special case of the multistage model.

EPA has selected a modification of the multistage model for toxicological assessment, called the **linearized multistage model.** This model assumes that we can extrapolate from high doses to low doses with a straight line. At low doses, the slope of the dose-response curve is represented by a **slope factor** (SF) expressed in units of risk per unit dose, or risk (kg $\cdot$ day $\cdot$ mg^{-1}).

The EPA maintains a toxicological data base called IRIS (the **Integrated Risk Information System**) that provides background information on potential carcinogens. IRIS includes suggested values for the slope factor. A list of slope factors for several compounds is shown in Table 6–3.

In contrast to the carcinogens, it is assumed that for noncarcinogens there is a dose below which there is no adverse effect; that is, there is an NOAEL. The EPA has estimated the acceptable daily intake, or reference dose (RfD), that is likely to be without appreciable risk. The RfD is obtained by dividing the NOAEL by safety factors to account for the transfer from animals to humans, sensitivity, and other uncertainties in developing the data. A list of several compounds and their RfD values is given in Table 6–4.

Limitations of Animal Studies.   No species provides an exact duplicate of human response. Certain effects that occur in common lab animals generally occur in people. Many effects produced in people can, in retrospect, be produced in some species. Notable exceptions are toxicities dependent on immunogenic mechanisms. Most sensitization reactions are difficult if not impossible to induce in lab animals. The procedure in transferring animal data to people is then

TABLE 6–3 **Slope Factors and Inhalation Unit Risk for Potential Carcinogens[a]**

Chemical	CPS_o $(kg \cdot day \cdot mg^{-1})$	CPS_i $(kg \cdot day \cdot mg^{-1})$	Inhalation Unit Risk $(per \cdot \mu g \cdot m^{-3})$
Arsenic	1.5	15.1	4.3×10^{-3}
Benzene	0.015	0.029	2.2×10^{-6}
Benzo(a)pyrene	7.3	N/A	N/A
Cadmium	N/A	6.3	1.8×10^{-3}
Carbon tetrachloride	7×10^{-2}	0.0525	6×10^{-6}
Chloroform	0.01	0.08	2.3×10^{-5}
DDT	0.34	0.34	9.7×10^{-5}
Dieldrin	16	16.1	4.6×10^{-3}
Heptachlor	4.5	4.55	1.3×10^{-3}
Hexachloroethane	4×10^{-2}	N/A	
Pentachlorophenol	0.4		
Polychlorinated biphenyls	0.04	N/A	
2,3,7,8-TCDD	1.5×10^{5}	1.16×10^{5}	
Tetrachloroethylene	2.1×10^{-3}	2×10^{-3}	2.6×10^{-7}
Trichloroethylene	4.6×10^{-2}	6×10^{-3}	4.1×10^{-6}
Vinyl chloride	0.72	N/A	4.4×10^{-6}

CPS_o = cancer potency slope, oral; CPS_i = cancer potency slope, inhalation; N/A = not applicable
[a]Values are frequently updated. Refer to IRIS for current data.
Source: U.S. Environmental Protection Agency IRIS data base, January, 2017.

TABLE 6-4 **RfDs for Chronic Noncarcinogenic Effects for Selected Chemicals[a]**

Chemical	Oral RfD $(mg \cdot kg^{-1} \cdot day^{-1})$	Chemical	Oral RfD $(mg \cdot kg^{-1} \cdot day^{-1})$
Acetone	0.9	Pentachlorophenol	5.0×10^{-3}
Barium	0.2	Phenol	0.3
Cadmium	5.0×10^{-4}	PCB	
Chloroform	0.01	Aroclor 1016	7.0×10^{-5}
Chromium VI	3.0×10^{-3}	Aroclor 1254	2.0×10^{-5}
Cyanide	6.3×10^{-4}	Silver	5.0×10^{-3}
Dieldren	5.0×10^{-5}	Tetrachloroethylene	6.0×10^{-3}
1,1-Dichloroethylene	0.05	Toluene	8.0×10^{-2}
Hexachloroethane	7.0×10^{-4}	1,2,4-Trchlorobenzene	0.01
Hydrogen cyanide	6.0×10^{-4}	Xylenes	0.2
Methylene chloride	0.06		

[a]Values are frequently updated. Refer to IRIS for current data.
Source: U.S. Environmental Protection Agency IRIS data base, January, 2017.

TABLE 6–5	Potential Contaminated Media and Corresponding Routes of Exposure	
	Media	**Routes of Potential Exposure**
	Groundwater	Ingestion, dermal contact, inhalation during showering
	Surface water	Ingestion, dermal contact, inhalation during showering
	Sediment	Ingestion, dermal contact
	Air	Inhalation of airborne (vapor phase) chemicals (indoor and outdoor)
		Inhalation of particulates (indoor and outdoor)
	Soil/dust	Incidental ingestion, dermal contact
	Food	Ingestion

to find the "proper" species and study it in context. Observed differences are then often quantitative rather than qualitative.

Carcinogenicity as a result of application or administration to lab animals is often assumed to be transposable to people because of the seriousness of the consequence of ignoring such evidence. However, slowly induced, subtle toxicity—because of the effects of ancillary factors (environment, age, etc.)—is difficult at best to transfer. This becomes even more difficult when the incidence of toxicity is restricted to a small hypersensitive subset of the population.

Limitations of Epidemiological Studies.   Epidemiological studies of toxicity in human populations present four difficulties. The first is that large populations are required to detect a low frequency of occurrence of a toxicological effect. Second, a long or highly variable latency period may be needed between the exposure to the toxicant and a measurable effect. Third, competing causes of the observed toxicological response make it difficult to attribute a direct cause and effect. For example, cigarette smoking; the use of alcohol or drugs; and personal characteristics such as gender, race, age, and prior disease states tend to mask environmental exposures. Fourth, epidemiological studies are often based on data collected in specific political boundaries that do not necessarily coincide with environmental boundaries such as those defined by an aquifer or the prevailing wind patterns.

Exposure Assessment

The objective of this step is to estimate the magnitude of exposure to chemicals of potential concern. The magnitude of exposure is based on chemical intake and pathways of exposure. The most important route (or pathway) of exposure may not always be clearly established. Arbitrarily eliminating one or more routes of exposure is not scientifically sound. The more reasonable approach is to consider an individual's potential contact with all contaminated media through all possible routes of entry. These are summarized in Table 6–5.

The evaluation of all major sources of exposure is known as **total exposure assessment** (Butler et al., 1993). After reviewing the available data, it may be possible to decrease or increase the level of concern for a particular route of entry to the body. Elimination of a pathway of entry can be justified if

1. The exposure from a particular pathway is less than that of exposure through another pathway involving the same media at the same exposure point
2. The magnitude of exposure from the pathway is low, or
3. The probability of exposure is low and incidental risk is not high

There are two methods of quantifying exposure: point estimate methods and probabilistic methods. The EPA uses the point estimate procedure by estimating the **reasonable maximum exposure** (RME). Because this method results in very conservative estimates, some scientists believe probabilistic methods are more realistic (Finley and Paustenbach, 1994). We will limit our consideration to the EPA point estimate technique.

RME is defined as the highest exposure that is reasonably expected to occur and is intended to be a conservative estimate within the range of possible exposures. Two steps are involved in estimating RME: first, exposure concentrations are predicted using a transport model such as the Gaussian plume model for atmospheric dispersion, then pathway-specific intakes are calculated using these exposure concentration estimates. The following equation is a generic intake equation.

$$CDI = C\left[\frac{(CR)(EFD)}{BW}\right]\left(\frac{1}{AT}\right) \tag{6–9}$$

where* CDI = chronic daily intake (in mg $\cdot$ kg body weight^{-1} $\cdot$ day^{-1})

C = chemical concentration, contacted over the exposure period (e.g., in mg $\cdot$ L^{-1} water)

CR = contact rate, the amount of contaminated medium contacted per unit time or event (e.g., in L $\cdot$ day^{-1})

EFD = exposure frequency and duration, describes how long and how often exposure occurs. Often calculated using two terms (EF and ED):

EF = exposure frequency (in days $\cdot$ year^{-1})

ED = exposure duration (in years)

BW = body weight, the average body weight over the exposure period (in kg)

AT = averaging time, period over which exposure is averaged (in days)

For each different medium and corresponding route of exposure, it is important to note that additional variables are used to estimate intake. For example, when calculating intake for the inhalation of airborne chemicals, inhalation rate and exposure time are required. Specific equations for medium and routes of exposure are given in Table 6–6. Standard values for use in the intake equations are shown in Table 6–7.

On April 6, 2016, the U.S. EPA released the 2011 edition of the *Exposure Factors Handbook*. The new handbook makes the standard values shown in Table 6-7 obsolete. The new standard values are reported at a more rigorous level than can be included in a textbook. There are more than 287 values. The 2004 values in Table 6-7 have been retained for the purpose of demonstrating the methodology in estimating the chronic daily intake. For actual risk assessment, the reader should refer to the 2011 edition of the *Exposure Factors Handbook*. It is found online.

The EPA has made some additional assumptions in calculating exposure that are not noted in the tables. In the absence of other data, the exposure frequency (EF) for residents is generally assumed to be 350 days $\cdot$ year^{-1} to account for absences from the residence for vacations. Similarly, for workers, EF is generally assumed to be 250 days $\cdot$ year^{-1} based on a 5 day work-week over 50 weeks per year.

Because the risk assessment process is considered to be an **incremental risk** for cancer, the exposure assessment calculation is based on an assumption that cancer effects are cumulative over a lifetime and that high doses applied over a short time are equivalent to short doses over a long time. Although the validity of this assumption may be debated, the standard risk calculation incorporates this assumption by using a lifetime exposure duration (ED) of 75 years and an averaging time (AT) of 27,375 d (that is 365 days $\cdot$ year^{-1} $\times$ 75 years). For noncarcinogenic effects, the averaging time (AT) is assumed to be the same as the exposure duration (ED) (Nazaroff and Alverez-Cohen, 2001).

---

*The notation in Equation 6–9 and subsequent expressions follows that in EPA guidance documents. The abbreviation "CDI" does **not** imply multiplication of three variables "C," "D," and "I." CDI, CR, EFD, BW, and so on are the notation for the variables. They do not refer to the product of terms.

TABLE 6–6	Residential Exposure Equations for Various Pathways

Ingestion in drinking water

$$CDI = \frac{(CW)(IR)(EF)(ED)}{(BW)(AT)}$$

(6–10)

Ingestion while swimming

$$CDI = \frac{(CW)(CR)(ET)(EF)(ED)}{(BW)(AT)}$$

(6–11)

Dermal contact with water

$$AD = \frac{(CW)(SA)(PC)(ET)(EF)(ED)(CF)}{(BW)(AT)}$$

(6–12)

Ingestion of chemicals in soil

$$CDI = \frac{(CS)(IR)(CF)(FI)(EF)(ED)}{(BW)(AT)}$$

(6–13)

Dermal contact with soil

$$AD = \frac{(CS)(CF)(SA)(AF)(ABS)(EF)(ED)}{(BW)(AT)}$$

(6–14)

Inhalation of airborne (vapor phase) chemicals

$$CDI = \frac{(CA)(IR)(ET)(EF)(ED)}{(BW)(AT)}$$

(6–15)

Ingestion of contaminated fruits, vegetables, fish and shellfish

$$CDI = \frac{(CF)(IR)(FI)(EF)(ED)}{(BW)(AT)}$$

(6–16)

where ABS = absorption factor for soil contaminant (unitless)

AD = absorbed dose (in $mg \cdot kg^{-1} \cdot day^{-1}$)

AF = soil-to-skin adherence factor (in $mg \cdot cm^{-2}$)

AT = averaging time (in days)

BW = body weight (in kg)

CA = contaminant concentration in air (in $mg \cdot m^{-3}$)

CDI = chronic daily intake (in $mg \cdot kg^{-1} \cdot day^{-1}$)

CF = volumetric conversion factor for water = $1 L \cdot 1000 cm^{-3}$

= conversion factor for soil = $10^{-6} kg \cdot mg^{-1}$

CR = contact rate (in $L \cdot h^{-1}$)

CS = chemical concentration in soil (in $mg \cdot kg^{-1}$)

CW = chemical concentration in water (in $mg \cdot L^{-1}$)

ED = exposure duration (in years)

EF = exposure frequency (in $days \cdot year^{-1}$ or $events \cdot year^{-1}$)

ET = exposure time ($h \cdot day^{-1}$ or $h \cdot event^{-1}$)

FI = fraction ingested (unitless)

IR = ingestion rate (in $L \cdot day^{-1}$ or $mg\ soil \cdot day^{-1}$ or $kg \cdot meal^{-1}$)

= inhalation rate (in $m^3 \cdot h^{-1}$)

PC = chemical-specific dermal permeability constant (in $cm \cdot h^{-1}$)

SA = skin surface area available for contact (in cm^2)

Source: U.S. EPA, 1989.

TABLE 6–7	EPA Recommended Values for Estimating Intake	
	Parameter	**Standard Value**
	Average body weight, adult female	65.4 kg
	Average body weight, adult male	78 kg
	Average body weight, child	
	6–11 months	9 kg
	1–5 years	16 kg
	6–12 years	33 kg
	Amount of water ingested daily, adult[a]	2.3 L
	Amount of water ingested daily, child[a]	1.5 L
	Amount of air breathed daily, adult female	11.3 m^3
	Amount of air breathed daily, adult male	15.2 m^3
	Amount of air breathed daily, child (3–5 y)	8.3 m^3
	Amount of fish consumed daily, adult	6 g · day^{-1}
	Water swallowing rate, swimming	50 mL · h^{-1}
	Skin surface available, adult male	1.94 m^2
	Skin surface available, adult female	1.69 m^2
	Skin surface available, child	
	3–6 years (average for male and female)	0.720 m^2
	6–9 years (average for male and female)	0.925 m^2
	9–12 years (average for male and female)	1.16 m^2
	12–15 years (average for male and female)	1.49 m^2
	15–18 years (female)	1.60 m^2
	15–18 years (male)	1.75 m^2
	Soil ingestion rate, children 1–6 years	100 mg · day^{-1}
	Soil ingestion rate, persons > 6 years	50 mg · day^{-1}
	Skin adherence factor, gardeners	0.07 mg · cm^{-2}
	Skin adherence factor, wet soil	0.2 mg · cm^{-2}
	Exposure duration	
	Lifetime	75 years
	At one residence, 90th percentile	30 years
	National median	5 years
	Averaging time	(ED)(365 days · year^{-1})
	Exposure frequency (EF)	
	Swimming	7 days · year^{-1}
	Eating fish and shell fish	48 days · year^{-1}
	Exposure time (ET)	
	Shower, 90th percentile	30 min
	Shower, 50th percentile	15 min

[a]90th percentile.

Source: U.S. EPA, 1989, 1997, 2004.

EXAMPLE 6–2   Estimate the lifetime average chronic daily intake of benzene from exposure to a city water supply that contains a benzene concentration equal to the drinking water standard. The allowable drinking water concentration (maximum contaminant level, MCL) is $0.005 \text{ mg} \cdot \text{L}^{-1}$. Assume the exposed individual is an adult male who consumes water at the adult rate for 63 years*, that he is an avid swimmer and swims in a local pool (supplied with city water) 3 days a week for 30 minutes from the age of 30 until he is 75 years old. As an adult, he takes a long (30 minutes) shower every day. Assume that the average air concentration of benzene during the shower is $5 \text{ µg} \cdot \text{m}^{-3}$ (McKone, 1987). From the literature, it is estimated that the dermal uptake from water is $0.0020 \text{ m}^3 \cdot \text{m}^{-2} \cdot \text{h}^{-1}$. (This is PC in Table 6–6. PC also has units of $\text{m} \cdot \text{h}^{-1}$ or $\text{cm} \cdot \text{h}^{-1}$.) Direct dermal absorption during showering is no more than 1% of the available benzene because most of the water does not stay in contact with skin long enough (Byard, 1989).

Solution   From Table 6–5, we note that five routes of exposure are possible from the drinking water medium: (1) ingestion, dermal contact while (2) showering and (3) swimming, (4) inhalation of vapor while showering, and (5) ingestion while swimming.

We begin by calculating the CDI for ingestion (Equation 6–10):

$$\text{CDI} = \frac{(0.005 \text{ mg} \cdot \text{L}^{-1})(2.3 \text{ L} \cdot \text{day}^{-1})(365 \text{ days} \cdot \text{year}^{-1})(63 \text{ years})}{(78 \text{ kg})(75 \text{ years})(365 \text{ days} \cdot \text{year}^{-1})}$$

$$= 1.24 \times 10^{-4} \text{ mg} \cdot \text{kg}^{-1} \cdot \text{day}^{-1}$$

The chemical concentration (CW) is the MCL for benzene. As noted in the problem statement and footnote, the man ingests water at the adult rate for a duration (ED) equal to his adult years. The ingestion rate (IR) and body weight (BW) were selected from Table 6–7. The exposure averaging time of $365 \text{ days} \cdot \text{year}^{-1}$ for 75 years is the EPA's generally accepted value as discussed on page 243.

Equation 6–12 may be used to estimate absorbed dose while showering:

$$\text{AD} = \frac{(0.005 \text{ mg} \cdot \text{L}^{-1})(1.94 \text{ m}^2)(0.0020 \text{ m} \cdot \text{h}^{-1})(0.50 \text{ h} \cdot \text{event}^{-1})}{(78 \text{ kg})(75 \text{ years})}$$

$$\times \frac{(1 \text{ event} \cdot \text{d}^{-1})(365 \text{ days} \cdot \text{year}^{-1})(63 \text{ years})(103 \text{ L} \cdot \text{m}^{-3})}{(365 \text{ days} \cdot \text{year}^{-1})}$$

$$= 1.04 \times 10^{-4} \text{ mg} \cdot \text{kg}^{-1} \cdot \text{day}^{-1}$$

As in the previous calculation, CW is the MCL for benzene. SA is the adult male surface area. PC is given in the problem statement. A "long shower" is assumed to be the 90th percentile value in Table 6–7 and the 63 years is derived as noted in the previous calculation.

Because only about 1% of this amount is available for adsorption in a shower because of the limited contact time, so the actual adsorbed dose by dermal contact is

$$\text{AD} = (0.01)(1.04 \times 10^{-4} \text{ mg} \cdot \text{kg}^{-1} \cdot \text{day}^{-1}) = 1.04 \times 10^{-6} \text{ mg} \cdot \text{kg}^{-1} \cdot \text{day}^{-1}$$

---

*This is based on Table 6–7 values of a lifetime of 75 years minus a childhood that is assumed to last until age 12.

The adsorbed dose for swimming is calculated in the same fashion:

$$AD = \frac{(0.005 \text{ mg} \cdot \text{L}^{-1})(1.94 \text{ m}^2)(0.0020 \text{ m} \cdot \text{h}^{-1})(0.5 \text{ h} \cdot \text{event}^{-1})}{(78 \text{ kg})(75 \text{ years})}$$

$$\times \frac{(3 \text{ events} \cdot \text{week}^{-1})(52 \text{ weeks} \cdot \text{year}^{-1})(45 \text{ years})(10^3 \text{ L} \cdot \text{m}^{-3})}{(365 \text{ days}) \cdot (\text{year}^{-1})}$$

$$= 3.19 \times 10^{-5} \text{ mg} \cdot \text{kg}^{-1} \cdot \text{day}^{-1}$$

In this case, because there is virtually total body immersion for the entire contact period and because there is virtually an unlimited supply of water for contact, there is no reduction for availability. The value of ET is computed from the swimming time (30 minutes = $0.5 \text{ h} \cdot \text{event}^{-1}$). The exposure frequency is computed from the number of swimming events per week and the number of weeks in a year. The exposure duration (ED) is calculated from the lifetime and beginning time of swimming = 75 years − 30 years = 45 y.

The inhalation rate from showering is estimated from Equation 6–15:

$$CDI = \frac{(5 \text{ μg} \cdot \text{m}^{-3})(10^{-3} \text{ mg} \cdot \text{μg}^{-1})(0.633 \text{ m}^3 \cdot \text{h}^{-1})(0.50 \text{ h} \cdot \text{event}^{-1})}{(78 \text{ kg})(75 \text{ years})}$$

$$\times \frac{(1 \text{ event} \cdot \text{day}^{-1})(365 \text{ days} \cdot \text{year}^{-1})(63 \text{ years})}{(365 \text{ days} \cdot \text{year}^{-1})}$$

$$= 1.70 \times 10^{-5} \text{ mg} \cdot \text{kg}^{-1} \cdot \text{day}^{-1}$$

The inhalation rate (IR) is taken from Table 6–7 and converted to an hourly basis. The values for ET and EF were given in the problem statement. As assumed previously, adulthood occurs from age 12 until age 75.

For ingestion while swimming, we apply Equation 6–11:

$$CDI = \frac{(0.005 \text{ mg} \cdot \text{L}^{-1})(50 \text{ mL} \cdot \text{h}^{-1})(10^{-3} \text{ L} \cdot \text{mL}^{-1})(0.5 \text{ h} \cdot \text{event}^{-1})}{(78 \text{ kg})(75 \text{ years})}$$

$$\times \frac{(3 \text{ events} \cdot \text{week}^{-1})(52 \text{ weeks} \cdot \text{year}^{-1})(45 \text{ years})}{(365 \text{ days} \cdot \text{year}^{-1})}$$

$$= 4.11 \times 10^{-7} \text{ mg} \cdot \text{kg}^{-1} \cdot \text{day}^{-1}$$

The contact rate (CR) is the water swallowing rate. It was determined from Table 6–7. Other values were obtained in the same fashion as those for dermal contact while swimming.

The total exposure would be estimated as:

$$CDI_T = 1.24 \times 10^{-4} + 1.04 \times 10^{-6} + 3.19 \times 10^{-5} + 1.70 \times 10^{-5} + 4.11 \times 10^{-7}$$

$$= 1.74 \times 10^{-4} \text{ mg} \cdot \text{kg}^{-1} \cdot \text{day}^{-1}$$

From these calculations, it becomes readily apparent that, in this case, drinking the water dominates the intake of benzene.

Risk Characterization

In the risk characterization step, all data collected from exposure and toxicity assessments are reviewed to corroborate qualitative and quantitative conclusions about risk. The risk for each media source and route of entry is calculated. This includes the evaluation of compounding effects due to the presence of more than one chemical contaminant and the combination of risk across all routes of entry.

For low-dose cancer risk (risk below 0.01), the quantitative **incremental risk** assessment for a single compound by a single route is calculated as

$$\text{Risk} = (\text{intake})(\text{slope factor}) \tag{6–17}$$

where intake is calculated from one of the equations in Table 6–6 or a similar relationship. The slope factor is obtained from IRIS (see, for example, Table 6–3). For high incremental carcinogenic risk levels (risk above 0.01), the one-hit equation is used.

$$\text{Risk} = 1 - \exp[-(\text{intake})(\text{slope factor})] \tag{6–18}$$

The measure used to describe the potential for noncarcinogenic toxicity to occur in an individual is not expressed as a probability. Instead, EPA uses the noncancer hazard quotient, or hazard index (HI):

$$\text{HI} = \frac{\text{intake}}{\text{RfD}} \tag{6–19}$$

These ratios are not to be interpreted as statistical probabilities. A ratio of 0.001 does not mean that there is a one in one thousand chance of an effect occurring. If the HI exceeds unity, there may be concern for potential noncancer effects. As a rule, the greater the value above unity, the greater the level of concern.

To account for multiple substances in one pathway, EPA sums the risks for each constituent.

$$\text{Risk}_T = \sum \text{risk}_i \tag{6–20}$$

For multiple pathways

$$\text{Total exposure risk} = \sum \text{risk}_{ij} \tag{6–21}$$

where i = the compounds and j = pathways.

In a like manner, the hazard index for multiple substances and pathways is estimated as

$$\text{Hazard index}_T = \sum \text{HI}_{ij} \tag{6–22}$$

In its guidance documents, EPA recommends segregation of the hazard index into chronic, subchronic, and short-term exposure. Although some research indicates that the addition of risks is reasonable (Silva et al., 2002), there is some uncertainty in taking this approach. Namely, should the risk from a carcinogen that causes liver cancer be added to the risk from a compound that causes stomach cancer? The conservative approach is to add the risks.

EXAMPLE 6–3 Using the results from Example 6–2, estimate the risk from exposure to drinking water containing the MCL for benzene.

Solution Equation 6–21 in the form

$$\text{Total Exposure Risk} = \sum \text{Risk}_j$$

may be used to estimate the risk. Because the problem is to consider only one compound, namely benzene, i = 1 and others do not need to be considered. Because the total exposure from

Example 6–2 included both oral and inhalation routes and there are different slope factors for each route in Table 6–3, the risk from each route is computed and summed. Because we do not have a slope factor for contact, we have assumed that it is the same as for oral ingestion. The risk is

$$\text{Risk} = (1.57 \times 10^{-4}\ \text{mg} \cdot \text{kg}^{-1} \cdot \text{day}^{-1})(1.5 \times 10^{-2}\ \text{kg} \cdot \text{d} \cdot \text{mg}^{-1})$$

$$+ (1.70 \times 10^{-5}\ \text{mg} \cdot \text{kg}^{-1} \cdot \text{day}^{-1})(2.9 \times 10^{-2}\ \text{kg} \cdot \text{d} \cdot \text{mg}^{-1})$$

$$= 2.85 \times 10^{-6}\ \text{or}\ 2.9 \times 10^{-6}$$

This is the total lifetime risk (75 years) for benzene in drinking water at the MCL. Another way of viewing this is to estimate the number of people that might develop cancer. For example, in a population of 2 million,

$$(2 \times 10^{6})(2.85 \times 10^{-6}) = 5.7\ \text{or}\ 6\ \text{people might develop cancer}$$

This risk falls within the EPA guidelines of 10^{-4} to 10^{-7} risk. It, of course, does not account for all sources of benzene by all routes. None the less, the risk, compared to some other risks in daily life, appears to be quite small.

6–4 RISK MANAGEMENT

Though some might wish it, it is clear that establishment of zero risk cannot be achieved. All societal decisions, from driving a car to drinking water with benzene at the EPA's regulated concentration, pose a risk. Even banning the production of chemicals, as was done for PCBs, for example, does not remove those that already permeate our environment. Risk management is performed to decide the magnitude of risk that is tolerable in specific circumstances (NRC, 1983). This is a policy decision that weighs the results of the risk assessment against the costs of risk reduction techniques and benefits of risk reduction as well as the public acceptance. The risk manager recognizes that if a very high certainty in avoiding risk (i.e., a very low risk, for example, 10^{-7}) is required, the costs in achieving low concentrations of the contaminant are likely to be high.

To reduce risk, the risk manager's options fall into three categories: change the environment, modify the exposure, or compensate for the effects. Frequently, the final choice is a blend of the options. Equation 6–9 provides a glimpse of the alternatives available. These include modifying the environment by reducing the concentration of the compounds (C) through engineering measures and modifying the exposure by limiting the intake of compounds (CR) by providing warnings and dietary restrictions, or restricting access to contaminated environments and thus reducing the exposure time (EFD). Chapters 9–16 discuss alternative measures for reducing the risk of exposure to toxic compounds. The risk manager uses a benefit–cost analysis to choose the remedy.

Unfortunately, very little guidance can be provided to the risk manager. We know that people are willing to accept a higher risk for things that they expose themselves to voluntarily than for involuntary exposures and, hence, insist on lower levels of risk, regardless of cost, for involuntary exposure. We also know that people are willing to accept risk if it approaches that for disease, that is, a fatality rate of 10^{-6} people per person-hour of exposure (Starr, 1969).

CHAPTER REVIEW

When you have finished studying this chapter, you should be able to do the following without the aid of your textbook or notes:

1. Define and differentiate between risk and hazard.
2. List the four steps in risk assessment and explain what occurs in each step.

3. Define the terms dose, LD_{50}, NOAEL, slope factor, RfD, CDI, IRIS.
4. Explain why it is not possible to establish an absolute scale of toxicity.
5. Explain why the average dose-response curve may not be an appropriate model for developing environmental protection standards.
6. Identify routes of exposure for the release of contaminants in multiple media.
7. Explain how risk management differs from risk assessment and the role of risk perception in risk management.

With the aid of this text, you should be able to do the following:

1. Calculate lifetime risk using the one-hit model.
2. Calculate chronic daily intake or other variables given the medium and values for the remaining variables.
3. Perform a risk characterization calculation for carcinogenic and noncarcinogenic threats by multiple contaminants and multiple pathways.

PROBLEMS

6–1 The recommended time weighted average air concentration for occupational exposure to water soluble hexavalent chromium (Cr VI) is 0.05 mg $\cdot$ m^{-3}. This concentration is based on an assumption that the individual is generally healthy and is exposed for 8 hours per day, 5 days per week, 50 weeks per year, over a working lifetime (that is from age 18 to 65 years). Assuming a body weight of 78 kg and inhalation rate of 15.2 m^3 $\cdot$ d^{-1} over the working life of the individual, what is the lifetime (75 years) CDI?

6–2 A wastewater treatment plant worker spends her day checking wet wells. She is exposed to hydrogen sulfide in this confined space. The average temperature in the wet well is 25°C and the atmospheric pressure is 101.325 kPa. The recommended time weighted average air concentration for occupational exposure to hydrogen sulfide is 10 ppm (v/v). This concentration is based on an assumption that the individual is generally healthy and is exposed for 8 hours per day, 5 days per week, 50 weeks per year, over a working lifetime (that is from age 18 to 65 years). Assuming an average female body weight and inhalation rate over the working life of the individual, what is the lifetime (75 years) CDI?

6–3 The National Ambient Air Quality Standard for sulfur dioxide is 80 μg $\cdot$ m^{-3}. Assuming a lifetime exposure (24 h $\cdot$ day^{-1}, 365 days $\cdot$ year^{-1}) for an adult male of average body weight, what is the estimated lifetime CDI for this concentration? Assume the exposure duration equals the lifetime.

6–4 Children are one of the major concerns of environmental exposure. Compare the CDIs for a 10-month-old child and an adult female drinking water contaminated with 10 mg $\cdot$ L^{-1} of nitrate (as N). Assume a 1-year exposure time, a 1-year averaging time for both the child and adult, and the weight for the child is that of a 6–11 month old.

6–5 Agricultural chemicals such as 2,4-D (2,4-Dichlorophenoxyacetic acid) may be ingested by routes other than food. Compare the CDIs for ingestion of a soil contaminated with 10 mg $\cdot$ kg^{-1} of 2,4-D by a 3-year-old child and an adult male. Assume both the child and adult have an equivalent exposure of 1 day per week for 20 weeks in a year, that the fraction of 2,4-D ingested is 0.10, and that the averaging time equals the exposure time.

6–6 Estimate the chronic daily intake of toluene from exposure to a city water supply that contains a toluene concentration equal to the drinking water standard of 1 mg $\cdot$ L^{-1}. Assume the exposed individual is an adult female who consumes water at the adult rate for 70 years, that she abhors swimming, and that she takes a long (20 minute) bath every day. Assume that the average air concentration of toluene during the bath is 1 μg $\cdot$ m^{-3}. Assume the dermal uptake from water (PC) is 9.0×10^{-6} m $\cdot$ h^{-1} and that direct dermal absorption during bathing is no more than 80% of the available toluene because she is not completely submerged. Use the EPA lifetime exposure of 75 years.

Answer: 3.3×10^{-2} mg $\cdot$ kg^{-1} $\cdot$ day^{-1}

6–7 Estimate the chronic daily intake of 1,1,1-trichloroethane from exposure to a city water supply that contains a 1,1,1-trichloroethane concentration equal to the drinking water standard of $0.2 \text{ mg} \cdot \text{L}^{-1}$. Assume the exposed individual is a child who consumes water at the child rate for 5 years, that she swims once a week for 30 minutes, and that she takes a short (10 minute) bath every day. Assume her average age over the exposure period is 8. Assume that the average air concentration of 1,1,1-trichloroethane during the bath is $2 \text{ μg} \cdot \text{m}^{-3}$. Assume the dermal uptake from water (PC) is $0.0060 \text{ m} \cdot \text{h}^{-1}$ and that direct dermal absorption during bathing is no more than 50 percent of the available l,1,1-trichloroethane because she is not completely submerged.

6–8 Estimate the risk from occupational inhalation exposure to hexavalent chromium. (See Problem 6–1 for assumptions.)

6–9 In its ruling for burning hazardous waste in boilers and industrial furnaces (56 FR 7233, 21 FEB 1991), EPA calculated the doses of various contaminants that would result in a risk of 10^{-5}. Using the standard assumptions in Table 6–7 for an adult male, estimate the dose of hexavalent chromium that results in an inhalation risk of 10^{-5}. Assume the exposure time equals the averaging time.

6–10 Characterize the hazard index for a chronic daily exposure by the water pathway (oral) of $0.03 \text{ mg} \cdot \text{kg}^{-1} \cdot \text{d}^{-1}$ of toluene, $0.06 \text{ mg} \cdot \text{kg}^{-1} \cdot \text{d}^{-1}$ of barium, and $0.3 \text{ mg} \cdot \text{kg}^{-1} \cdot \text{d}^{-1}$ of xylenes.

    Answer: HI $= 1.95$

6–11 Characterize the risk for a chronic daily exposure by the water pathway (oral) of $1.34 \times 10^{-4} \text{ mg} \cdot \text{kg}^{-1} \cdot \text{day}^{-1}$ of tetrachloroethylene, $1.43 \times 10^{-3} \text{ mg} \cdot \text{kg}^{-1} \cdot \text{day}^{-1}$ of arsenic, and $2.34 \times 10^{-4} \text{ mg} \cdot \text{kg}^{-1} \cdot \text{day}^{-1}$ of dichloromethane (methylene chloride).

6–12 In some lakes, the fish are contaminated with methylmercury. As a member of the Department of Environmental Quality, you have been assigned the job of developing an "advisory" for adult males to limit consumption of fish from these lakes. The advisory will be in terms of a recommended limit on the number of fish meals per unit of time. You must determine the unit of time, that is one per day, or one per week, or one per month, etc. In addition to the EPA recommended values for estimating intake, use 1×10^{-6} for the fraction ingested, a 30-year exposure duration, and a 30-year averaging time to estimate the chronic daily intake. Your superiors have set a hazard index of 0.10 as a matter of policy. The RfD for methylmercury is $1 \times 10^{-4} \text{ mg} \cdot \text{kg}^{-1} \cdot \text{day}^{-1}$.

6–13 Your firm has been asked to evaluate the "risk" of removing drums of acid from a waste lagoon that contains cyanide. The engineer assigned to the project has had an emergency appendectomy and is not available, so you have been assigned the task of completing the calculations. The following data have been provided to you:

Volume of air in which occupants live $= 1 \times 10^8 \text{ m}^3$
Wind speed $=$ calm
Number of drums $= 10$
Volume of each drum $= 0.20 \text{ m}^3$
Contents of drums: HCl at a concentration of 10% by mass
Anticipated reaction: $\text{HCl} + \text{NaCN} \rightarrow \text{HCN(g)} + \text{NaCl}$
RfC for HCN $= 3 \times 10^{-3} \text{ mg} \cdot \text{m}^{-3}$

Assuming 100% of the HCl will react to form HCN, estimate the hazard index and indicate whether or not the residents should be evacuated while the drums are being removed.

6–14 A commercial research organization is planning to develop a sensor to detect methyl parathion that may be injected into a water supply by terrorists. The LD_{50} for methyl parathion is reported to be as low as 6 $\text{mg} \cdot \text{kg}^{-1}$ for rats. They are assuming a safety factor of 1×10^5 and a volume of liquid ingested equal to a cup of coffee (about 250 mL) in the design of the instrument. What concentration (in $\text{mg} \cdot \text{L}^{-1}$) must their instrument be able to detect? Using Equation 6–10, the concentration you have calculated, the ingestion rate of 250 mL, an exposure frequency of one day, and exposure duration of one day, an adult male body weight, and an averaging time of 1 day, estimate the hazard index. The RfD for methyl parathion is $2.5 \times 10^{-4}\,\text{mg} \cdot \text{kg}^{-1} \cdot \text{d}^{-1}$.

6–15 Risk is a measure of probability. Probability theory tells us that the probability of a number of independent and mutually exclusive events is the sum of the probabilities of the separate events. Using this concept, estimate the lifetime (75 y) probability of death from the following: driving a motor vehicle or falling or a home accident. Hint: see Table 6–1.

6–16 As noted previously, risk is a measure of probability. EPA has used the 90 percent confidence level to select values for Table 6–7. Probability theory tells us that the probability of two independent events occurring simultaneously or in succession is the product of the individual probabilities. Using this concept, estimate the probability of ingesting 2.3 L of water each day and weighing 75 kg.

DISCUSSION QUESTIONS

6–1 It has been stated that based on LD$_{50}$, 2,3,7,8-TCDD (tetrachlorodibenzo-*p*-dioxin) is the most toxic chemical known. Why might this statement be misleading? How would you rephrase the statement to make it more scientifically correct?

6–2 Which of the following individuals is at greater risk from inhalation of an airborne contaminant: a 1-year-old child; an adult female; an adult male? Explain your reasoning.

6–3 Which of the following individuals is at greater risk from ingestion of a soil contaminant: a 1-year-old child; an adult female; an adult male? Explain your reasoning.

6–4 A hazard index of 0.001 implies

(a) Risk = 10^{-3}

(b) The probability of hazard is 0.001.

(c) The RfD is small compared with the CDI.

(d) There is little concern for potential health effects.

FE EXAM FORMATTED PROBLEMS

6–1 A 4-year-old child has been playing in soil contaminated with 500 mg $\cdot$ kg^{-1} of tetrachloroethylene. It is estimated that he ingested 5 g $\cdot$ d^{-1} of soil over a single 3-month period. The exposure frequency was 7 days per week. What is the estimated chronic daily intake of tetrachloroethylene for the child over the 3-month period?

(a) 0.00016 mg $\cdot$ kg$^{-1} \cdot$ d^{-1}

(b) 0.0756 mg $\cdot$ kg$^{-1} \cdot$ d^{-1}

(c) 0.0312 mg $\cdot$ kg$^{-1} \cdot$ d^{-1}

(d) 0.156 mg $\cdot$ kg$^{-1} \cdot$ d^{-1}

6–2 Characterize the risk of an oral chronic daily intake of 50 parts per billion (ppb) of arsenic.

(a) 0.072

(b) 0.075

(c) 0.033

(d) 0.047

6–3 Characterize the hazard of an oral chronic daily intake of 0.005 mg $\cdot$ L^{-1} of cadmium.

(a) 0.032

(b) 0.031

(c) 10.0

(d) 1.00

6–4 Estimate the concentration of Heptachlor in fish if the concentration in water is 5 ppb. The bioconcentration factor is 15,700.

(a) 78.5 mg · kg^{-1}

(b) 78,500 mg · kg^{-1}

(c) 3140 mg · kg^{-1}

(d) 3.14 mg · kg^{-1}

REFERENCES

Butler, J. P., A. Greenberg, P. J. Lioy, G. B. Post, and J. M. Waldman (1993) "Assessment of Carcinogenic Risk from Personal Exposure to Benzo(a)pyrene in the Total Human Environmental Exposure Study (THEES)," *Journal of the Air & Waste Management Association,* July, 43: 970–77.

Byard, J. L. (1989) "Hazard Assessment of 1,1,1-Trichloroethane in Groundwater," in D. J. Paustenbach (ed.), *The Risk Assessment of Environmental Hazards,* John Wiley & Sons, New York, pp. 331–44.

CDC (2014) National Center for Health Statistics, http://www.cdc.gov/nchs/surveys.htm

Finley, B., and D. Paustenbach (1994) "The Benefits of Probabilistic Exposure Assessment; Three Case Studies Involving Contaminated Air, Water, and Soil," *Risk Analysis,* 14(1): 53–73.

Hutt, P. B. (1978) "Legal Considerations in Risk Assessment," *Food, Drugs, Cosmetic Law J,* 33: 558–59.

Loomis, T. A. (1978) *Essentials of Toxicology,* Lea & Febiger, Philadelphia, p. 2.

McKone, T. E. (1987) "Human Exposure to Volatile Organic Compounds in Household Tap Water: The Indoor Inhalation Pathway," *Environmental Science & Technology,* 21(12): 1194–1201.

National Center for Health Statistics (2004) http://www.cdc.gov/nchs/fastats/deaths

Nazaroff, W. W., and L. Alverez-Cohen (2001) *Environmental Engineering Science,* John Wiley & Sons, Inc., New York, pp. 570–71.

NFPA (2016) *Fire Deaths*, National Fire Protection Association, https//www.nfpa.org/news-and-research

NHTSA (2012) "2011 Annual Assessment," National Highway Traffic Administration, Washington, DC, http://www.nhtsa.gov

NRC (1983) *Risk Assessment in the Federal Government: Managing the Process,* National Research Council, National Academy Press, Washington, DC, pp. 18–19.

Silva, E., N. Rajapakse, and A. Kortenkamp (2002) "Something from "Nothing"—Eight Weak Estrogenic Chemicals Combined at Concentrations Below NOECs Produce Significant Mixture Effects," *Environmental Science & Technology,* 36: 1751–56.

Slovic, P., B. Fischoff, and S. Lichtenstein (1979) "Rating Risk," *Environment,* 21: 1–20, 36–39.

Starr, C. (1969) "Social Benefit Versus Technological Risk," *Science,* 165: 1232–38.

U.S. EPA (1989) *Risk Assessment Guidance for Superfund, Volume I: Human Health Evaluation Manual (Part A),* U.S. Environmental Protection Agency Publication EPA/540/1-89/002, Washington, DC.

U.S. EPA (1994) *Annual Health Effects Assessment Summary Tables* (HEAST), U.S. Environmental Protection Agency Publication No. EPA 510-R-04-001, Washington, DC.

U.S. EPA (1996) National Center for Environmental Assessment—Provisional Value, http//www.epa.gov/ncea

U.S. EPA (1997) *Exposure Factor Handbook,* U.S. Environmental Protection Agency National Center for Environmental Assessment, Washington, DC.

U.S. EPA (2004) *Risk Assessment Guidance Manual for Superfund, Volume I: Human Health Evaluation Manual,* U.S. Environmental Protection Agency Publication EPA/540/R/99/005, Washington, DC.

U.S. EPA (2011) *Exposure Factors Handbook 2011 Edition*, EPA/600/R-09/052F, 2011

U.S. EPA (2017) *IRIS Data Base*, U.S. Environmental Protection Agency, Washington, DC, http://www.epa.gov/iris

Hydrology

Case Study

Potential Failure of the Oroville Dam

In mid-January 2017, after six years of drought, rain began to fall on the Sacramento River watershed, a large basin, approximately 70,000 km² in area, which lies between the Sierra Nevada and Cascade Ranges to the east and Coast Rage and Klamath Mountains to the west. As the largest watershed in California, 31% of the state's total surface water runoff flows in the Sacramento River. The water from the watershed eventually flows into the Pacific Ocean, via the Sacramento-San Joaquin Delta and San Francisco Bay.

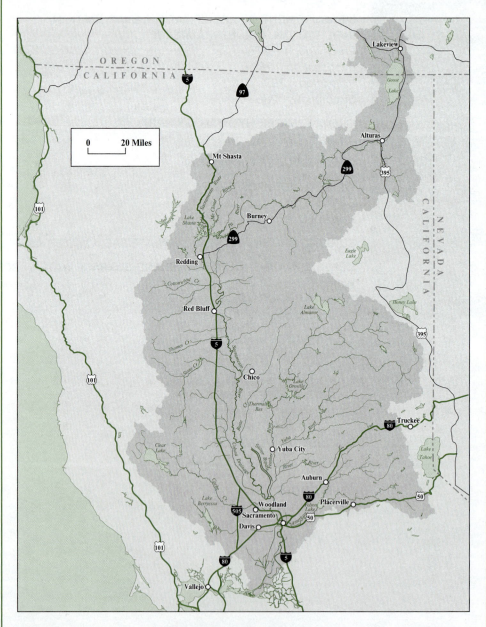

FIGURE 7–1 Sacramento River Watershed Program, Chico, CAReport: Sacramento River Basin.
Source: http://www.sacriver.org/files/documents/roadmap/report/SacRiverBasin.pdf

The watershed is home to numerous aquatic habitats and biota. The river water is used to irrigate approximately 1.5 million hectares of land, most of which supports water-intensive crops such as cotton, grapes, tomatoes, fruit, hay, and rice, which have an annual value of greater than $14 billion. The Sacramento River provides drinking water to Northern and Southern California, along with water for industries, hydroelectric power generation, and recreation and fishing.

By February 12, 2017, the watershed had experienced sequential "atmospheric rivers" and 4.65 inches of rainfall in the previous week. As the level in Lake Oroville peaked at 902.59 ft, 7 inches over the top of the spillway, the 188,000 residents of Butte, Sutter, and Yuba counties awoke to an emergency evacuation order. Authorities feared that a catastrophic failure of the emergency spillway could occur due to the rapid erosion of the "managed" flow that was being allowed to flow over the emergency spillway. Even a partial breach of the spillway would flood the escape route from the village of Oroville within about 35 minutes, leaving its 17,000 to drown.

While the reasons for the failure of the emergency spillway are beyond the scope of this study, the hydrological changes are not. Flow in the Sacramento River watershed is seasonal, with high flows in winter and low flows in late summer. As a result of anthropogenic activities, the winter peak flow now occurs earlier, and there is reduced runoff in the spring. Summer flows are higher than what is naturally occurring due to upstream reservoir releases.

Climate change is anticipated to significantly alter the hydrology of the basin. Decreases in rainfall will likely result in reduced base flow. Increases in precipitation in winter, as was the case with the failure of the Oroville Dam spillway in February 2017, will result in flooding, infrastructure failure, and potential loss of life. Climate change is also expected to reduce water storage in snowpack and snowmelt runoff in the spring. Warmer temperatures will result in oxygen depletion, changes in ecosystems, and potential further reduction in the numbers of Chinook salmon. Much work will need to be done to alleviate these effects that humankind has had on the earth, our only home.

7–1   FUNDAMENTALS OF HYDROLOGY

The availability of water is critical to maintaining ecosystems, as well as for communities, industry, agriculture, and commercial operations. The presence (or lack) of water at sufficient quantity and quality can significantly affect the sustainability of life. It is therefore important for the engineer and environmental scientist to have a solid understanding of our water supply and its distribution in nature.

Hydrology is a multidisciplinary subject that deals with the question of how much water can be expected at any particular time and location. The application of this subject is important to ensuring adequate water for such purposes as drinking, irrigation, and industrial uses, as well as to prevent flooding. Surface water hydrology focuses on the distribution of water on or above the earth's surface. It encompasses all water in lakes, rivers, and streams, on land and in the air. Groundwater hydrology deals with the distribution of water in the earth's subsurface geological materials, such as sand, rock, or gravel.

The Hydrological Cycle

The **hydrological cycle** (Figure 7–2) describes the movement and conservation of water on earth. This cycle includes all of the water present on and in the earth, including salt and fresh water, surface and groundwater, water present in the clouds and that trapped in rocks far below the earth's surface.

FIGURE 7–2

The hydrological cycle. The percentages correspond to the volume in each of the different compartments. (*Source:* Montgomery C., *Environmental Engineering,* 6e, 2003. The McGraw-Hill Companies, Inc.)

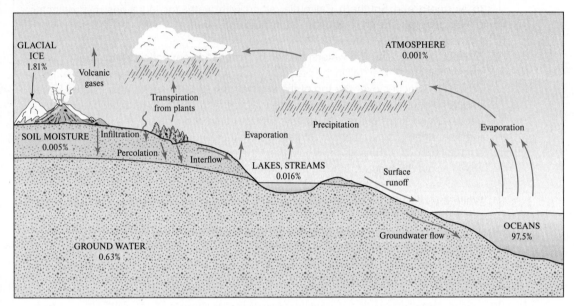

Water is transferred to the earth's atmosphere through two distinct processes: (1) evaporation and (2) transpiration. A third process is derived from the two and is called evapotranspiration. **Evaporation** is the conversion of liquid water from lakes, streams, and other bodies of water to water vapor. **Transpiration** is the process by which water is emitted from plants through the stomata, small openings on the underside of leaves that are connected to the vascular tissue. It occurs predominantly at the leaves while the stomata are open for the passage of carbon dioxide and oxygen during photosynthesis. Because it is often difficult to distinguish between true evaporation and transpiration, hydrologists use the term **evapotranspiration** to describe the combined losses of water due to transpiration and evaporation.

Precipitation is the primary mechanism by which water is released from the atmosphere. Precipitation takes several forms, the most common of which in temperate climates is rain. Additionally, water can fall as hail, snow, sleet, and freezing rain.

As water falls to the earth's surface, the droplets either run over the ground into streams and rivers (referred to as **surface runoff, overland flow,** or **direct runoff**), move laterally just below the ground surface (**interflow**), or move vertically through the soils to form groundwater (**infiltration** or **percolation**). As shown in Figure 7–3, flow in streams (**streamflow** to hydrologists) is generated in a number of ways. Some portion of the flow in a stream (**baseflow**) can originate from groundwater, soil, and springs. This is the portion of the streamflow that would be present even during periods of drought. **Interflow** is that portion of precipitation that infiltrates into the soil and moves horizontally through the shallow soil horizon without ever reaching the water table (**zone of saturation***). **Overland flow** is due to surface runoff and is that portion of precipitation that neither infiltrates into the soil nor evaporates nor evapotranspirates. This water flows down gradient to the nearest channel (river, stream, etc.). The final source of water in a channel is due to **channel precipitation.** This is the rainfall that actually falls into the stream or river.

The movement of water through various phases of the hydrological cycle is extremely complex because it is erratic in both time and space. Here we will take a very simplistic view so

---

*See Aquifers in Section 7–3: Groundwater Hydrology.

FIGURE 7–3

Generation of flow in a channel.

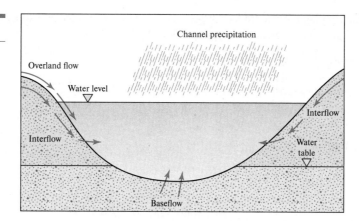

as to develop a water budget. The most important terms for a water budget are evaporation (E), evapotranspiration (E_T), precipitation (P), infiltration (G), interflow (F), and surface runoff (R).

One of the simplest water budgets used by hydrologists is that for a lake. Let's look at how water flows into and out of a lake. Water can flow into the lake by way of any rivers or streams (natural or anthropogenic, including industrial) that flow into the lake, by surface water runoff along the banks of the lake, by precipitation that falls directly into the lake, or by seepage into the bottom sediments of the lake from groundwater. Water can flow out of the lake by way of any streams or rivers that flow from the lake; withdrawal of water for municipal, industrial, or agricultural uses; evaporation; evapotranspiration; or seepage of water through the bottom sediments of the lake. What hydrologists often wish to determine is the net amount (mass) of water that is gained or lost in the lake within a given period. Hydrologists refer to this type of problem as a **storage problem.**

To solve storage problems, we must develop a mass-balance equation for the lake. In this case, the substance is water and the system is the lake. Therefore, the mass-balance equation is simply

Mass rate of accumulation = mass rate in − mass rate out   (7–1)

For the lake, we can write the most general form of this equation as

Mass rate of accumulation = $(Q_{in} + P' + R' + I'_{in} - Q_{out} - E' - E'_T - I'_{out})\rho_{water}$ (7–2)

where Q_{in} = flowrate of stream(s) entering lake (in vol · time^{-1})
  P' = rate of precipitation (in vol · time^{-1})
  R' = rate of runoff (in vol · time^{-1})
  I'_{in} = rate of seepage into lake (in vol · time^{-1})
  Q_{out} = flowrate of streams exiting lake (in vol · time^{-1})
  E' = rate of evaporation from water bodies such as lakes, rivers, and ponds
    (in vol · time^{-1})
  E'_T = rate of evapotranspiration (in vol · time^{-1})
  I'_{out} = rate of seepage out of the lake (in vol · time^{-1})
  ρ_{water} = density of water (in mass · vol^{-1})

Although these problems are not inherently difficult, some confusion arises because Q and R are often given in units of volume per time (e.g., cubic meters per second), whereas rates of precipitation, seepage into and out of the lake, evaporation, and evapotranspiration are often given in units of length per unit time (e.g., centimeters per month or millimeters per hour). One needs to make sure that the units used are consistent (either volume per unit time or length per unit time). Because it is assumed that precipitation, seepage, evaporation, and evapotranspiration all occur over the entire surface area of the lake, we can approximate the volumetric rate by multiplying the rate (in length per unit time) by the surface area of the lake. Therefore, if we define these parameters as is typically done by hydrologists,

P = rate of precipitation (in mm · h^{-1})
I_{in} = rate of seepage into lake (in mm · h^{-1})
E = rate of evaporation (in mm · h^{-1})
E_T = rate of evapotranspiration (in mm · h^{-1})
I_{out} = rate of seepage out of the lake (in mm · h^{-1})

Then Equation 7–2 becomes

$$\text{Mass rate of accumulation} = ((Q_{in} + R^I - Q_{out}) + (P + I_{in} - E - E_T - I_{out}) \times A_s)\rho_{\text{water}}$$

$$\times \left(\frac{1\ \text{m}}{1000\ \text{mm}^{-1}}\right)\left(\frac{1\ \text{h}}{3600\ \text{s}^{-1}}\right) \tag{7–3}$$

where A_s = surface area of the lake (in m^2). For most systems, we can assume that the density of water is constant because the temporal changes in temperature and pressure are small. As such, we can divide both sides of Equations 7–2 or 7–3 by the density of water to yield an equation for the volumetric rate of accumulation.

EXAMPLE 7–1   Sulis Lake has a surface area of 708,000 m^2. Based on collected data, Okemos Brook flows into the lake at an average rate of 1.5 m^3 · s^{-1} and the Tamesis River flows out of Sulis Lake at an average rate of 1.25 m^3 · s^{-1} during the month of June. The evaporation rate was measured as 19.4 cm · month^{-1}. Evapotranspiration can be ignored because there are few water plants on the shore of the lake. A total of 9.1 cm of precipitation fell this month. Seepage is negligible. Due to the dense forest and the gentle slope of the land surrounding the lake, runoff is also negligible. The average depth in the lake on June 1 was 19 m. What was the average depth on June 30th?

Solution   The first step to solving this problem is to determine what we know. We know that the inputs to the lake are

Q_{in} = 1.5 m^3 · s^{-1}
P = 9.1 cm · month^{-1}
I_{in} = 0 (because we were told that seepage is negligible)
R' = 0 (because we were told that runoff is negligible)

We also know that the outputs from the lake are

Q_{out} = 1.25 m^3 · s^{-1}
E = 19.4 cm · month^{-1}
E_T = 0

We also know that surface area of the lake is 708,000 m^2 and the average depth of the lake on June 1 is 19 m.

The following is a picture of the lake as a system

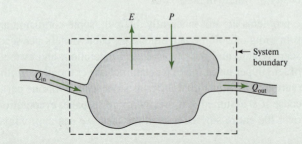

Using the average values given earlier and the most general form of the mass-balance equation (7–2), the mass-balance for this lake can be reduced to

Volumetric rate of accumulation $= Q_{in} - Q_{out} + P - E$

The volumetric rate of accumulation is often referred to as the **change in storage** (ΔS) and

$$\Delta S = Q_{in} - Q_{out} + P - E$$

Because the units for Q and P and E are different, we must ensure that the proper conversions are performed, yielding the same set of units.

Therefore,

$$\Delta S = (1.5 \text{ m}^3 \cdot \text{s}^{-1})(86{,}400 \text{ s} \cdot \text{day}^{-1})(30 \text{ days} \cdot \text{month}^{-1})$$
$$-(1.25 \text{ m}^3 \cdot \text{s}^{-1})(86{,}400 \text{ s} \cdot \text{day}^{-1})(30 \text{ days} \cdot \text{month}^{-1})$$
$$+(9.1 \text{ cm} \cdot \text{month}^{-1})(\text{m} \cdot (100 \text{ cm})^{-1})(708{,}000 \text{ m}^2)$$
$$-(19.4 \text{ cm} \cdot \text{month}^{-1})(\text{m} \cdot (100 \text{ cm})^{-1})(708{,}000 \text{ m}^2)$$

$$= 3{,}888{,}000 \text{ m}^3 \cdot \text{month}^{-1} - 3{,}240{,}000 \text{ m}^3 \cdot \text{month}^{-1}$$
$$+ 64{,}428 \text{ m}^3 \cdot \text{month}^{-1} - 137{,}352 \text{ m}^3 \cdot \text{month}^{-1}$$

Solving the preceding equation, yields

$$\Delta S = 575{,}076 \text{ m}^3 \cdot \text{month}^{-1}$$

Because $\Delta S = 575{,}076 \text{ m}^3 \cdot \text{month}^{-1}$ and the average surface area is $708{,}000 \text{ m}^2$, the change in depth during the month of June is

$$(575{,}076 \text{ m}^3 \cdot \text{month}^{-1})/708{,}000 \text{ m}^2 = 0.81 \text{ m} \cdot \text{month}^{-1}$$

Note that ΔS is positive. As such, the volume in the lake increased during June and, therefore, the depth increases. The new average depth on June 30 would be 19.81 m. Had a negative value for storage been calculated, then the depth of the lake would have decreased.

More Complex Systems. Hydrologists often wish to look at larger systems that include lakes, rivers, the surrounding land, and even the groundwater lying in the geological materials below the land. These systems are called watersheds or basins. The **watershed, or basin,** is defined by the surrounding topography (Figure 7–4). The boundary of the watershed is a **divide** and is the highest elevation surrounding the watershed. All of the water that falls on the inside of the divide has the potential to flow into the streams of the basin contained within the watershed boundary (divide). Water falling outside of the divide is shed to another basin.

Before we begin to develop a water budget for a watershed, let's look at a simpler system: an impervious inclined plane, confined on all sides and having a single outlet. An example of such a system is a small urban parking lot, surrounded by buildings or concrete walls, and sloped in one direction (Figure 7–5). Water can flow only out of the drain.

In this system, water enters the parking lot via rainfall. The water can either remain on the parking lot in the form of puddles (**accumulated storage** or **surface detention** to the hydrologist) or flow from the parking lot (**flow released from storage**). In this case, the hydrological continuity equation (Equation 7–1) becomes

FIGURE 7–4 The Kankakee River Basin above Davis, IN

The arrows indicate that precipitation falling inside the dashed line is in the Davis watershed, whereas that falling outside is in another watershed. The dashed line then "divides" the watersheds.

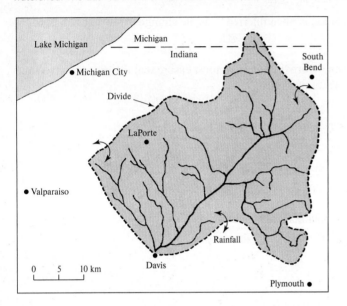

FIGURE 7–5

Schematic of a parking lot.

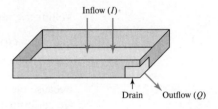

Volumetric accumulation rate = input rate − output rate $\qquad$ (7–4)

or

$$\frac{dS}{dt} = I - Q \qquad (7\text{–}5)$$

Again, we have assumed that the density of water is constant. As such, we can express our mass balance in terms of the volumetric accumulation rate. To a hydrologist, this equation would be written as

Change in storage (in volume · time^{-1})

$\qquad$ = inflow (in volume · time^{-1}) − outflow (in volume · time^{-1}) $\qquad$ (7–6)

At the beginning of the rainstorm, all of the rain accumulates on the surface of the parking lot and none is released from the drain. As rain continues to fall, the amount of water stored on the parking lot surface (surface detention) increases, and eventually water begins to flow from the drain. As the rain stops, water can continue to flow from the drain, resulting in a decrease in the surface detention. Eventually, all of the water can be released from storage as it flows from the drain. In this example, we have ignored the effects of evaporation. Another example of a similar system is a bathtub that is being filled at a rate faster than it is being drained.

The flow and collection of water in the parking lot can be described by a **hydrograph,** a chart in which flowrate is plotted versus time (Figure 7–6). Note that as the water continues to flow into the system, some of that water remains in the system as accumulated storage. Once the flow of water stops (e.g., rainfall ceases), water is then released from storage as outflow. If there is no evaporation or infiltration of water, then the total mass of water flowing into the system must equal that leaving the system by way of the drain.

FIGURE 7–6

Hydrograph showing constant inflow of water (e.g., rainfall) over some duration of time.

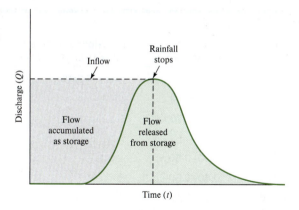

FIGURE 7–7

Effect of the watershed on a hydrograph. $Q_c > Q_b > Q_a$ and $t_c < t_b < t_a$.

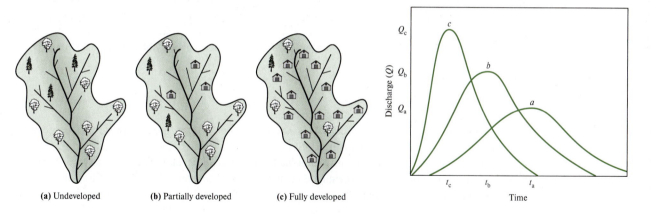

(a) Undeveloped (b) Partially developed (c) Fully developed

The same phenomenon occurs in a watershed, although the system becomes more complicated because water can infiltrate into the ground, flow over land into streams and rivers, and be caught up in puddles and depressions as surface detention. In the watersheds shown in Figure 7–7, we see that all of the water drains to a single point, a stream. The water that flows within the watershed is separated from all other watersheds by the divide. Numerous factors affect the rates at which water flows toward the stream and into the ground. For example, the steeper the slope of the land surrounding the stream, the faster the rate at which the water moves into the stream. Density and type of ground cover will also affect the rate of water transport. The denser the ground cover, the slower the rate of movement. These same factors also affect the volume of water that reaches the stream. This is represented by the runoff coefficient, some of which are given in Table 7–1. The runoff coefficient is defined as R/P or the rate of water that runs off a surface divided by the rate of precipitation.

The effect of some of these factors can be observed by looking at the unit hydrographs shown in Figure 7–7. Shown here are hydrographs for three watersheds, all of which are essentially identical except for the level of development. Watershed (a) is undeveloped, that is, it is covered by dense vegetation. Watershed (b) has been partially developed, so that some

TABLE 7–1		Typical Runoff Coefficients	
Description of Area or Character of Surface	Runoff Coefficient	Description of Area or Character of Surface	Runoff Coefficient
Business		Railroad yard	0.20–0.35
Downtown	0.70–0.95	Natural grassy land	0.10–0.30
Neighborhood	0.50–0.70	Pavement	
Residential		Asphalt, concrete	0.70–0.95
Single-family	0.30–0.50	Brick	0.70–0.85
Multi-units, detached	0.40–0.60	Roofs	0.75–0.95
Multi-units, attached	0.60–0.75	Lawns, sandy soil	
Residential, suburban	0.25–0.40	Flat (<2%)	0.05–0.10
Apartment	0.50–0.70	Average (2–7%)	0.10–0.15
Industrial		Steep (>7%)	0.15–0.20
Light	0.50–0.80	Lawns, heavy soil	
Heavy	0.60–0.90	Flat (<2%)	0.13–0.17
Parks, cemeteries	0.10–0.25	Average (2–7%)	0.18–0.22
Playgrounds	0.20–0.35	Steep (>7%)	0.25–0.35

Source: Joint Committee of the American Society of Civil Engineers and the Water Pollution Control Federation, 1969.

vegetation remains, but numerous roads, driveways, and houses, are also present. The presence of these structures prevents infiltration, resulting in a greater fraction of the rainfall flowing over the land and into the stream. Watershed (c) is heavily developed, similar to an urban environment, where there is little open land through which the water can infiltrate. Most of the water that falls on this watershed flows rapidly into the stream. It should be noted that in watershed (c), the time to reach the maximum stream flow is the shortest. This is due to the fact that there are few trees, grasses and other plants to impede the flow in the developed watershed. As the water falls, it rushes as it falls to the nearest body of water. Also, because there is little open land through which water can infiltrate or on which it can pool, the maximum stream flow is greatest in watershed (c). These high flows can have a significant effect on humans, on stream ecosystems, and on the physical characteristics of the stream because the high flows can result in bank erosion, scouring of stream beds, and even a change in the path of the stream.

The shape of hydrographs also vary on a seasonal and annual basis. For example, the annual cycling of stream flow in Convict Creek draining the 47.2 km^2 watershed near Mammoth Lakes, in Mono County, California, can be seen in Figure 7–8a. Each spring (March to May) snow melt contributes significantly to the streamflow. The dry season (September through April) is indicated on the hydrograph as very low flows. The unseasonably high flow (e.g., in February 1969) was perhaps due to unusually warm weather, resulting in snowmelt or rainfall.

Figure 7–8b provides an example of a much larger watershed (1194 km^2). As noted here, the variations are much less regular, although low flows typically are in winter when precipitation rates are 50–66% of what occurs in April through September. The spike seen in April 2017 was due to the largest rainfall event since 2001 and caused massive flooding in the area.

Having developed the concept of a watershed, let's look at an example of a water budget for a watershed.

FIGURE 7–8a

Ten-year hydrograph for Convict Creek near Mammoth Lakes, California.

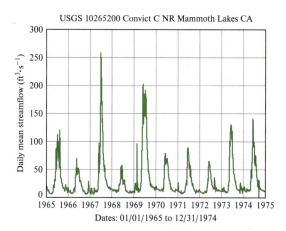

FIGURE 7–8b

Hydrograph for the Red Cedar River at East Lansing, Michigan.

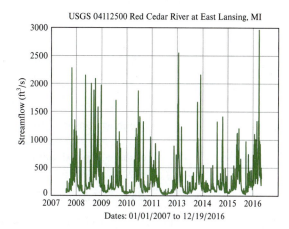

EXAMPLE 7–2 In 1997, the Upper Grand watershed near Lansing, Michigan, with an area of 4530 km^2 received 77.7 cm of precipitation. The average rate of flow measured in the Grand River, which drained the watershed, was 39.6 m$^3 \cdot$ s^{-1}. Infiltration was estimated to occur at an average rate of 9.2×10^{-7} m $\cdot$ s^{-1}. Evapotranspiration was estimated to be 45 cm $\cdot$ year^{-1}. What is the change in storage in the watershed?

Solution To solve this problem, we should draw a picture, list the information we know and that which we are seeking and write the question in symbolic form.

A simple picture of the watershed is shown here.

We know the following information:

$$\text{Area} = 4530 \text{ km}^2$$

$$P = 77.7 \text{ cm} \cdot \text{year}^{-1}$$

$$\text{Infiltration} = G = 9.2 \times 10^{-7} \text{ cm} \cdot \text{s}^{-1}$$

$$E_T = 45 \text{ cm} \cdot \text{year}^{-1}$$

To solve this problem we assume that all of the flow in the river is due to runoff, so that, $R = Q_{out}$.

In words, the mass-balance equation for this system can be written as

Change in storage = rate of precipitation − rate of evapotranspiration − rate of infiltration

− rate of water flowing from the stream.

Symbolically, this can be represented as

$$\Delta S = P - E_T - G - R$$
$$= 77.7 \text{ cm} \cdot \text{year}^{-1} - 45 \text{ cm} \cdot \text{year}^{-1} - (9.2 \times 10^{-7} \text{ cm} \cdot \text{s}^{-1})(60 \text{ s} \cdot \text{min}^{-1})(60 \text{ min} \cdot \text{h}^{-1})$$
$$\times (24 \text{ h} \cdot \text{day}^{-1})(365 \text{ day} \cdot \text{year}^{-1}) - R$$

We must convert R from units of cubic meters per second as given to units of centimeters per year as all other terms are given. To accomplish this, we must divide the flowrate by the area of the watershed it drains and perform all of the necessary unit conversions. Thus, substituting now for R

$$\Delta S = 77.7 \text{ cm} \cdot \text{year}^{-1} - 45 \text{ cm} \cdot \text{year}^{-1} - 29 \text{ cm} \cdot \text{year}^{-1}$$
$$- \frac{(39.6 \text{ m}^3 \cdot \text{s}^{-1})(86,400 \text{ s} \cdot \text{day}^{-1})(365 \text{ day} \cdot \text{year}^{-1})(100 \text{ cm} \cdot \text{m}^{-1})}{(4530 \text{ km}^2)(1000 \text{ m} \cdot \text{km}^{-1})^2}$$

Solving this equation, yields

$$\Delta S = 77.7 - 45 - 29 - 27.6 = -23.9 \text{ cm} \cdot \text{year}^{-1}$$

The negative storage means that there is a net loss of water from the watershed during this period.

We can also calculate the runoff coefficient for this watershed. Remembering that the runoff coefficient equals R/P, then

$$\frac{R}{P} = \frac{27.6 \text{ cm}}{77.7 \text{ cm}} = 0.36$$

This value (from Table 7–1) is typical of what one would observe in a suburban area.

Discharge from relatively small watersheds (less than 13 km^2) is often calculated using the rational method, which simply states that

$$Q = CIA$$

where

Q = peak discharge
C = runoff coefficient
I = rainfall intensity
A = watershed area. The runoff coefficients can be obtained from Table 7-1.

EXAMPLE 7–3 Determine the peak discharge from the grounds of the Spartanite High School grounds during a storm of intensity 2.5 cm/h. The composition of the grounds is

Character of surface	Area (m²)	Runoff Coefficient
Parking lot, asphalt	11,200	0.85
Building	10,800	0.75
Lawns and athletic fields	140,000	0.20

Solution   Since we have three different areas, we can calculate the weighted runoff coefficient as:

$$C' = \frac{\sum_1^n C_n A_n}{\sum_1^n A_n}$$

$$C' = \frac{(11,200 \times 0.85) + (10,800 \times 0.75) + 140,000 \times 0.20}{(11,200 + 10,800 + 140,000)}$$

$$= 0.2816$$

$$Q = CIA$$

$$= (11,200 \text{ m}^2 + 10,800 \text{ m}^2 + 140,000 \text{ m}^2) \times \left(\frac{2.5 \text{ cm}}{\text{h}}\right)\left(\frac{\text{m}}{100 \text{ cm}}\right)\left(\frac{\text{h}}{3600 \text{ s}}\right)$$

$$= 0.32 \text{ m}^3 \cdot \text{s}^{-1}$$

In reality, what this would mean is that a storm sewer draining the site would need to be large enough to handle a flow of $0.32 \text{ m}^3 \cdot \text{s}^{-1}$ to prevent flooding.

7–2   MEASUREMENT OF PRECIPITATION, EVAPORATION, INFILTRATION, AND STREAMFLOW

The development of any hydrologic continuity equation depends on the quality of the data generated. As such, it is important for the environmental scientist and engineer to have a good understanding of these parameters and how each is measured.

Precipitation

Precipitation is the primary input quantity into the hydrologic cycle. Its accurate measurement is essential to the design of successful water resource projects, especially those pertaining to flood control.

Precipitation rates vary greatly on a regional scale. For example, as shown in Figures 7–9 and 7–10, the differences in precipitation can vary significantly, even within a few hundred kilometers.

FIGURE 7–9

Average annual precipitation (in inches) across the state of Washington, for the period 1981–2010. The PRISM model was used to generate the gridded estimates from which the map was generated; the data was obtained from NOAA Cooperative stations and USDA-NRCS SNOTEL stations.
(*Source:* http://www.prism.oregonstate.edu/projects/gallery.php)

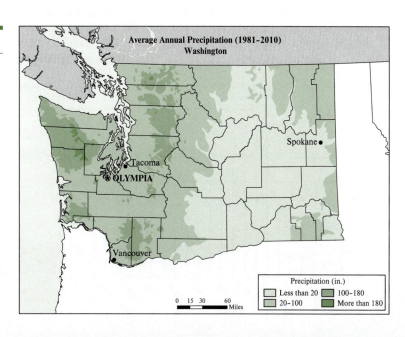

Average Annual Precipitation (1981–2010)
Washington

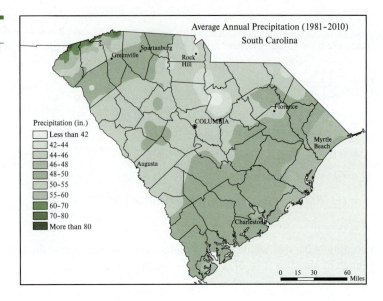

Precipitation tends to decrease with increasing latitude because decreasing temperatures reduce atmospheric moisture. However, there are some exceptions, such as Seattle, with its rainy climate, and San Diego, which is dry. Precipitation also tends to decrease with distance from a body of water, as evidenced by the concentration of precipitation along coastlines (see Figure 7–9) and to some extent to the leeward of the Great Lakes. Mountains are also an important factor in precipitation; heavier precipitation usually occurs along the windward slope of a mountain range, whereas the leeward slope usually lies in the rain shadow. Latitude, annual temperature, and maximum possible water content all contribute to rates of precipitation. However, oceanic currents and global atmospheric patterns are also important factors.

Precipitation not only varies regionally but also temporally. These temporal changes may be more important, from an engineering perspective. Seasonal and annual variations have serious implications for water resource management. As shown in Figure 7–11, monthly precipitation rates for La Crosse, Wisconsin, varied by more than an order of magnitude. This phenomenon is not unusual. Yearly variations can also be significant as shown in Figure 7–12. Here the precipitation rates varied as much as sixteenfold. These yearly variations make it important to design reservoirs that are adequate during years of low rainfall and dams that have the capacity to ensure adequate flood control even in times of high precipitation rates.

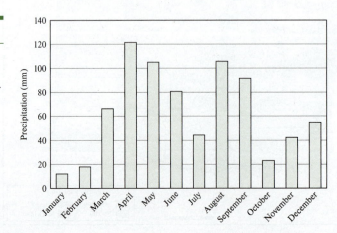

FIGURE 7–12

Annual variations in precipitation in Mojave, CA. (*Source:* Western Regional Climate Center, 2007)

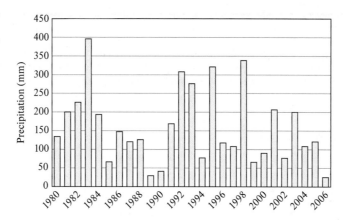

Because floods are among the most frequent and costly natural disasters, the prediction of precipitation rates is very important. Unfortunately, predicting precipitation cycles is extremely difficult, and evidence suggests that some degree of variation may be purely random.

Hydrologists often speak of 50- or 100-year storms. This notation can be confusing because it implies a certain probability. When hydrologists speak of a 100-year storm, they are speaking of a storm with a given intensity and duration that is likely, based on records, to occur once in a 100-year period. Because storms, and therefore, floods, are stochastic* in nature, a 100-year storm can occur 2 years in a row, or even within the same year. However, historical data are becoming less useful because of the effects of climate change. As shown in Figure 7–13a, the total annual precipitation has increased over land areas in the United States. The rate of increase is 0.17 inches per decade. Some parts of the United States have experienced greater increases in precipitation than others, while others are seeing significant decreases in precipitation, as shown in Figure 7–13b.

FIGURE 7–13a

Change in precipitation in the United States, 1901–2015. (*Data source:* NOAA (National Oceanic and Atmospheric Administration), 2016. National Centers for Envronmental Information Accessed February 2016. www.ncei.noaa.gov. For more information, visit U.S. EPA's "Climate Change Indicators in the United States" at www.epa.gov/climate-indicators.)

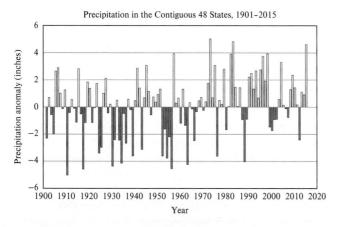

*Stochastic means that the occurrence of floods can be predicted using the probability distribution of an ordered set of flood data that have been collected over time.

FIGURE 7-13b

Precipitation in the contiguous 48 states, 1901–2015. (*Data source:* NOAA (National Oceanic and Atmospheric Administration), 2016. National Centers for Environmental Information Accessed February 2016. www.ncei.noaa.gov. For more information, visit U.S. EPA's "Climate Change Indicators in the United States" at https://www.epa.gov/climate-indicators/climate-change-indicators-us-and-global-precipitation.)

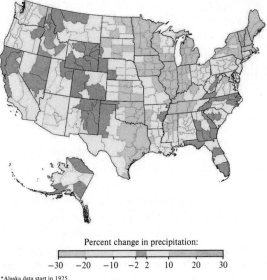

Percent change in precipitation:

−30   −20   −10   −2 2   10   20   30

*Alaska data start in 1925.

Point Precipitation Analysis. Because accurate precipitation data are critical for the determination of flood and drought forecasting, it is important to understand how precipitation rates are determined for a given area. Precipitation can be measured using either gauges, which yield point data (i.e., data for a very limited area, often less than about 20 cm in diameter), or areal data using radar (in which the area over which the rates are averaged is much larger, generally about 2.5 km^2). Each method has its advantages and disadvantages. Although rain gauges can give very accurate data for a very small region, these data must then be extrapolated over much larger regions. Data from a single nearby rain gauge are often sufficiently representative to allow their use in the design of small projects. The analysis of data from a single gauge is called **point precipitation analysis.** Radar may reasonably estimate the precipitation rate if the intensity and duration of the storm are relatively constant over the area in which measurements are taken. However, the location of mountains can interfere with the collection of critical data. Where storms can be very small in area, such as in the southwestern United States, such important data can be completely missed by the radar.

Evaporation

Because evaporation is a significant component of the hydrologic cycle, especially in arid and semiarid climates, the determination or prediction of accurate evaporation rates is important for determining the capacity of humanmade impoundments. Variations in evaporation rates occur both on temporal (Figure 7–14) and spatial scales. Evaporation rates can be estimated using the pan evaporation method, free-water surface evaporation, and lake evaporation. **Pan evaporation** is the rate of evaporation using a standard National Weather Service (NWS) class A pan. Lake evaporation differs from pan evaporation measurements due to the heat storage capacity of the lake and wind currents. A simple method for estimating lake evaporation from class A pan evaporation measurements multiplies the pan value by 0.7 (Farnsworth, Thompson, and Peck, 1982).

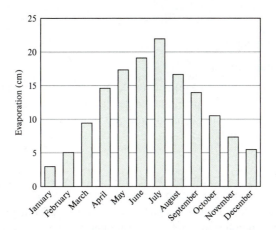

FIGURE 7–14

Average monthly pan evaporation rates for Athens University of Georgia (for 1970–1971). (*Source:* South Carolina State Climatology Office, 2007.)

Another method of estimating evaporation rates is to use Dalton's equation. Dalton showed that the loss of water from the surface of a lake or other body of water is a function of solar radiation, air and water temperature, wind speed, and the difference in vapor pressures at the water surface and in the overlying air (Dalton, 1802). Thus, the relationship

$$E = (e_s - e_a)(a + bu) \qquad (7\text{--}7)$$

where E = evaporation rate (in mm · day^{-1})
   e_s = saturation vapor pressure (in kPa)
   e_a = vapor pressure in overlying air (in kPa)
   a, b = empirical constants
   u = wind speed (in m · s^{-1})

Empirical studies at Lake Hefner, Oklahoma, yielded a similar relationship.

$$E = 1.22(e_s - e_a)u \qquad (7\text{--}8)$$

From these expressions, it is apparent that high wind speeds and low humidities (vapor pressure in the overlying air) result in large evaporation rates. Note that the units for these expressions may not make much sense. This is because these are empirical expressions developed from field data. The constants have implied conversion factors in them. In applying these (and other) empirical expressions, take care to use the same units as those used by the scientists who developed the expression.

EXAMPLE 7–4 Anjuman's Lake has a surface area of 70.8 ha. For April the inflow was 1.5 m^3 · s^{-1}. The dam regulated the outflow (discharge) from Anjuman's Lake to be 1.25 m^3 · s^{-1}. If the precipitation recorded for the month was 7.62 cm and the storage volume increased by an estimated 650,000 m^3, what is the estimated evaporation in cubic meters and centimeters? Assume that no water infiltrates into or out of the bottom of Anjuman's Lake.

Solution Begin by drawing the mass-balance diagram.

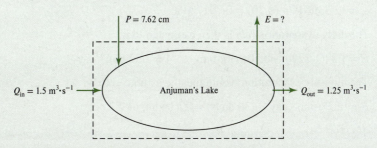

$P = 7.62$ cm $E = ?$

$Q_{in} = 1.5$ m^3·s^{-1} →   Anjuman's Lake   → $Q_{out} = 1.25$ m^3·s^{-1}

The mass-balance equation is

Accumulation = input − output

The accumulation (change in storage) is 650,000 m³. The input consists of the inflow and the precipitation. The product of the precipitation depth and the area on which it fell (70.8 ha) will yield a volume. The output consists of outflow plus evaporation. The change in storage can be represented by the equation

$$\Delta S = Q_{in} + P - E - Q_{out}$$

Make sure that all parameters are in the same units. The flowrates are expressed in cubic meters per second whereas E and P are shown in centimeters. Because we want to calculate the rate of evaporation, we should convert all units to either units of volume per month ($m^3 \cdot month^{-1}$) or units of length per month ($cm \cdot month^{-1}$). Although hydrologists often calculate changes in storage in units of length per unit time, you should recognize that length is not conserved, rather mass is. Since we typically assume that density is constant, one can also assume that volume is also constant. As such, we will solve the problem in units of volume and then calculate the change in depth. Remember also that April has 30 days.

Therefore,

$$\begin{aligned}
650{,}000 \text{ m}^3 = &(1.5 \text{ m}^3 \cdot \text{s}^{-1})(30 \text{ days})(86{,}400 \text{ s} \cdot \text{day}^{-1}) \\
&+ (7.62 \text{ cm})(70.8 \text{ ha})(10^4 \text{ m}^2 \cdot \text{ha}^{-1})(\text{m} \cdot (100 \text{ cm})^{-1}) \\
&- (1.25 \text{ m}^2 \cdot \text{s}^{-1})(30 \text{ days})(86{,}400 \text{ s} \cdot \text{day}^{-1}) - E
\end{aligned}$$

Solving for E, we obtain

$$\begin{aligned}
E &= Q_{in} + P - Q_{out} - \Delta S \\
&= 3.89 \times 10^6 \text{ m}^3 + 5.39 \times 10^4 \text{ m}^3 - 3.24 \times 10^6 \text{ m}^3 - 6.50 \times 10^5 \text{ m}^3 \\
&= 5.39 \times 10^4 \text{ m}^3
\end{aligned}$$

For an area of 70.8 ha, the evaporation rate in units of depth per month is

$$E = \frac{5.39 \times 10^4 \text{ m}^3}{(70.8 \text{ ha})(10^4 \text{ m}^2 \cdot \text{ha}^{-1})} = 0.076 \text{ m} = 7.6 \text{ cm}$$

EXAMPLE 7–5   During April, the wind speed over Anjuman's Lake was estimated to be 4.0 m · s⁻¹. The air temperature averaged 20°C, and the relative humidity was 30%. The water temperature averaged 10°C. Estimate the evaporation rate using Dalton's equation.

Solution   From the water temperature and the values given in Table 7–2, the saturation vapor pressure is estimated as $e_s = 1.2290$ at 10°C. The vapor pressure in the air may be estimated as the product of the relative humidity and the saturation vapor pressure at the air temperature.

$$e_a = (2.3390 \text{ kPa})(0.30) = 0.7017 \text{ kPa}$$

The daily evaporation rate is then estimated to be

$$E = 1.22(1.2290 - 0.7017)(4.0 \text{ m} \cdot \text{s}^{-1}) = 2.57 \text{ mm} \cdot \text{day}^{-1}$$

The monthly evaporation would then be estimated to be

$$E = (2.57 \text{ mm} \cdot \text{day}^{-1})(30 \text{ days}) = 77.1 \text{ mm, or } 7.7 \text{ cm}$$

TABLE 7–2	Water Vapor Pressures at Various Temperatures			
	Temperature (°C)	Vapor Pressure (kPa)	Temperature (°C)	Vapor Pressure (kPa)
	0	0.6104	25	3.1679
	5	0.8728	30	4.2433
	10	1.2290	35	5.6255
	15	1.7065	40	7.3866
	20	2.3390	50	12.4046

Calculated using $e_s \cong 33.8639[(0.00738T + 0.8072)^8 - 0.000019 \mid 1.8T + 48 \mid + 0.001316]$ where T = temperature (°C).

Source: Bosen, J. F. (1960). A Formula for Approximation of the Saturation Vapor Pressure over Water Monthly Weather Review, Aug. 1960: 275–276.

Evapotranspiration.   Evapotranspiration describes the total water removed from an area by transpiration (release of water vapor from plants) and by evaporation of water from soils, snow, and water surfaces. Evapotranspiration is often estimated by subtracting the total outflow for an area from the total input of water. The change in storage must be included in the calculations, unless this change is negligible.

The potential evapotranspiration rate (maximum possible loss) from a well-watered root zone (e.g., that in a golf course) can approximate the rate of evaporation that can occur over a large free-water surface. The available moisture in the root zone will limit the actual evapotranspiration rate, such that, as the root zone dries out the rate of transpiration will decrease. The rate of evapotranspiration is also a function of soil type, plant type, wind speed, and temperature. Plant types may affect evapotranspiration rates dramatically. For example, an oak tree may transpire as much as $160 \, \text{L} \cdot \text{day}^{-1}$, whereas a corn plant may transpire only about $1.9 \, \text{L} \cdot \text{day}^{-1}$. Although empirical models have been developed to attempt to predict rates of evapotranspiration based on the factors mentioned earlier, these models are inherently difficult to calibrate and validate due to the very complex biology and physics that govern evapotranspiration.

Infiltration

Infiltration is the net movement of water into soil. When the rainfall rate exceeds the infiltration rate, water migrates through the surface soil at a rate that generally decreases with time until it reaches a constant value. The rate of infiltration varies with rainfall intensity, soil type, surface condition, and vegetal cover. This temporal decline in the rate is actually due to a filling of the soil pores with water and a reduction in capillary action. The change in infiltration rate with time is shown in Figure 7–15.

Of the numerous equations developed to describe infiltration, Horton's equation (Horton, 1935) is useful for describing the rate. When the rate of rainfall exceeds the rate of infiltration, we can use Horton's equation.

$$f = f_c + (f_o - f_c)e^{-kt} \tag{7–9}$$

where f = infiltration rate (or capacity) (in length $\cdot$ time^{-1})
$\quad f_c$ = equilibrium, critical or final infiltration rate (in length $\cdot$ time^{-1})
$\quad f_o$ = initial infiltration rate (in length $\cdot$ time^{-1})
$\quad k$ = empirical constant (in time^{-1})
$\quad t$ = time

Note that this equation has the same problem as many of the other mass-balance equations used by hydrologists, that is, rates are given in units of length per unit time rather than mass per unit time or volume per unit time. This occurs because obtaining a volumetric rate requires that

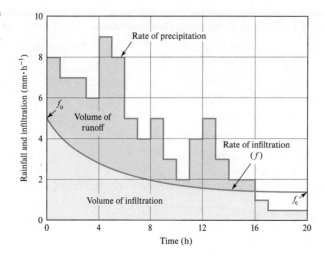

the right-hand side of Equation 7–9 be multiplied by the surface area through which water is infiltrating, thus yielding f in units of volume per unit time. However, because, by convention, hydrologists have used this notation, we have conformed to it.

As mentioned earlier soil type affects the infiltration rate of water. As one might imagine, the sandier the soil, the greater the infiltration rate. The more compacted the soil, or the greater the clay content of the soil, the slower the infiltration rate. Table 7–3 provides data for some common soil types.

Horton's infiltration can be integrated to yield an equation that represents the total volume of water that would infiltrate over a given period. The integrated form of Horton's equation is given in Equation 7–10.

$$\text{Volume} = A_s \int_o^t f\, dt = A_s \int_o^t [f_c + (f_o - f_c)e^{-kt}]\, dt = A_s \left[f_c t + \frac{f_o - f_c}{k}(1 - e^{-kt}) \right] \qquad (7\text{–}10)$$

Although Horton's equation is generally applicable to most soils, several limitations apply. For sandy soils, f_o exceeds most rainfall intensities. As such, all rainfall will infiltrate, and the infiltration rate will equal the precipitation rate. In these cases, Horton's equation will under-predict the infiltration rate. As mentioned previously, the infiltration capacity, f, decreases with cumulative infiltration volume as the pores fill up, not with time. Note that in Horton's equation, infiltration capacity is a function of time, not cumulative infiltration volume.

TABLE 7–3	Parameters for Horton's Equation for Some Typical Soils		
Soil Type	f_c (cm · h^{-1})	f_o (cm · h^{-1})	k (h^{-1})
Alphalpha loamy sand	3.56	48.26	38.29
Carnegie sandy loamy	4.50	35.52	19.64
Dothan loamy sand	6.68	8.81	1.40
Fuquay pebbly loamy sand	6.15	15.85	4.70
Leefield loamy sand	4.39	28.80	7.70
Troop sand	4.57	58.45	32.71

Source: Bedient, Philip; Huber, Wayne C.; Vieux, Baxter E., Hydrology and Floodplain Analysis, 4th Edition, 2008, pg. 67.

EXAMPLE 7–6   A soil has the following characteristics

$$f_o = 3.81 \text{ cm} \cdot \text{h}^{-1} \qquad f_c = 0.51 \text{ cm} \cdot \text{h}^{-1} \qquad k = 0.35 \text{ h}^{-1}$$

What are the values of f at $t = 12$ min, 30 min, 1 h, 2 h, and 6 h? What is the total volume of infiltration over the 6-h period in an area that is 1 m^2?

Solution   Using the stated data, we can calculate the infiltration rates using Horton's equation if $i > f$ (or rate of precipitation exceeds the rate of infiltration). The volume of precipitation that infiltrated can be calculated by integrating Horton's equation over the time interval being considered.

$$\text{Volume} = A_s \int_o^t f \, dt = A_s \int_o^t [f_c + (f_o - f_c)e^{-kt}] \, dt = A_s \left[f_c t + \frac{(f_o - f_c)}{-k} e^{-kt} \Big|_o^t \right]$$

Using Horton's equation, we calculate the infiltration rate for each of the desired time intervals.

Time (h)	Infiltration Rate (cm/h)
0.2	3.58
0.5	3.28
1	2.54
2	2.16
6	0.91

The data calculated can be plotted to show how the infiltration rate decreases with time.

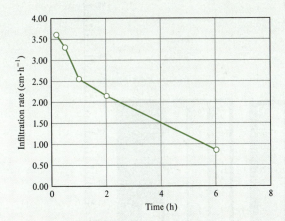

The volume of water that would have infiltrated over the 6 h can be calculated.

$$\text{Volume} = A_s \left[\left\{ (0.51)(6) + \frac{(3.81 - 0.51)}{-0.35} e^{-(0.35)(6)} \right\} \right.$$

$$\left. - \left\{ (0.51)(0) + \frac{(3.81 - 0.51)}{-0.35} e^{-(0.35)(0)} \right\} \right]$$

$$= A_s(11.3 \text{ cm}) = 1 \text{ m}^2 (11.3 \text{ cm})(\text{m} \cdot (100 \text{ cm})^{-1})) = 0.113 \text{ m}^3$$

Streamflow

The actual flowrate in a stream is determined by measuring the velocity and depth (or height of the water above a reference datum) of the water at particular cross sections in the channel. The elevation (stage) readings are calibrated in terms of streamflow. Stage measurements are commonly given in units of feet or meters. **Flow,** or discharge, the total volume of water that flows past a point on the river during some fixed time interval, is also crucial to hydrological measurements. Discharge is a measure of flowrate and can be measured in units of cubic feet per second (cfs), gallons per minute (gpm), or cubic meters per second ($m^3 \cdot s^{-1}$). Flowrate is measured at a location on the channel known as a **stream-gauging station.**

At manual recording stations, readings are made from a marked rod (staff gauge) placed in the stream (Figure 7–16). These systems require that someone visit the station to record the stage elevation.

Using automatic recording stations, the river stage can be continuously monitored and recorded. These data can be transmitted to United States Geological Service (USGS) and National Weather Service (NWS) offices using either telephone lines or satellite radios. This process allows for the remote monitoring of river elevation and prediction of flood conditions.

Automatic recording stations often use a stilling well and shelter for monitoring purposes. A stilling well and a shelter (Figure 7–17) minimize the effects of wave action and protects the piping system and valves from floating logs and other materials. For small streams, a dam with a weir plate (Figure 7–18) may be installed. This system increases the change in elevation for small changes in streamflow and makes readings more precise and accurate.

Rating curves (Figure 7–19) are required to calibrate stage measurements. These curves are constructed by USGS field personnel who periodically visit the gauging station to measure river discharge by monitoring depth and velocity of the river across a cross section of the channel.

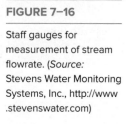

FIGURE 7–16

Staff gauges for measurement of stream flowrate. (*Source:* Stevens Water Monitoring Systems, Inc., http://www .stevenswater.com)

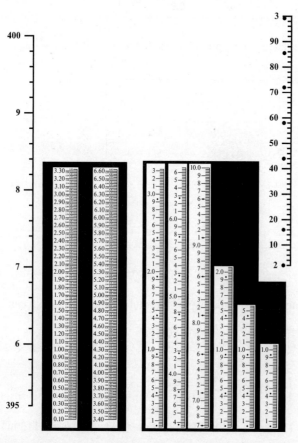

FIGURE 7–17

Schematic of a stilling well and shelter at a stream-gauging station.

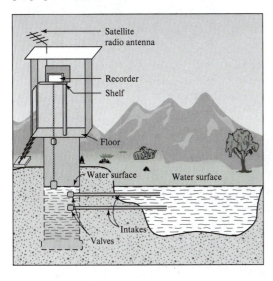

FIGURE 7–18

Weir for stage measurement. (*Source:* Stevens Water Monitoring Systems, Inc., http://www.stevenswater.com.)

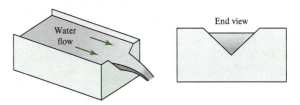

FIGURE 7–19

Typical discharge curve. (*Source:* Wahl, 1995. U.S. Geological Survey Circular 1123, Reston, VA.)

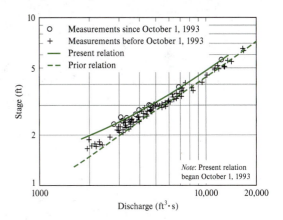

7–3 GROUNDWATER HYDROLOGY

So far, our discussion has focused on surface water. Although surface water is an important natural resource, the importance of groundwater must not be neglected. Groundwater accounts for 25.7% of the total freshwater and 98.4% of the unfrozen freshwater in the hydrosphere (Mather, 1984). Although groundwater is renewable, the rate at which it is replenished is often much slower than the rate at which water is pumped from the water-yielding earthen layer or **aquifer.** Therefore, in locations such as Arizona, eastern Colorado, western Kansas, Texas, Oklahoma, and portions of California, where groundwater is the predominant source of water, the groundwater is being "mined," that is, pumped at a rate that exceeds the rate at which it is replenished. As shown in Figure 7–20, such phenomena as saltwater intrusion and chemical and microbial contamination affect the quality of this important natural resource. Clearly, we must learn to better manage this resource, protecting it for future generations.

Aquifers

As water drains down through the soil, it flows through the root zone and then through a zone referred to as the **unsaturated zone** (also called the **vadose zone** or the **zone of aeration**). As shown in Figure 7–20, the pores of the geologic material in the unsaturated zone are partially filled with water. The remaining portion is filled with air. The water continues to migrate vertically down through the soil until it reaches a level at which all of the openings or voids in the soils are filled with water. This zone is known as the **zone of saturation,** the **saturated zone,** or the **phreatic zone.** The water in the zone of saturation is referred to as **groundwater.** The geologic formation, through which water can flow horizontally and be pumped, is called an **aquifer.**

FIGURE 7–20

Elements of groundwater flow.

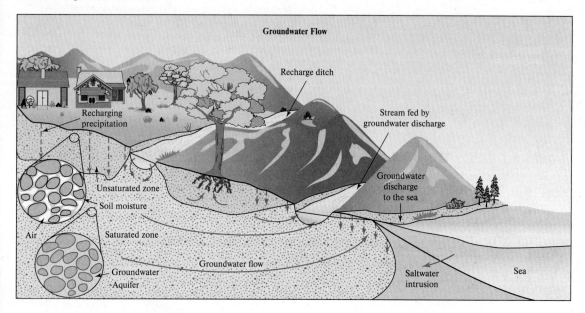

Sand, sandstone, or other sedimentary rocks serve as good aquifers. Aquifers can also be present in other porous geologic materials, such as limestone, fractured basalt, or weathered granite.

Unconfined Aquifers. The upper surface of the zone of saturation in aquifers that are not confined by impermeable geologic material is called the **water table** (Figure 7–21). This type of aquifer is called a **water table aquifer,** a **phreatic aquifer,** or an **unconfined aquifer.**

The smaller void spaces in the geologic material just above the water table may contain water as a result of interactive forces between the water and the soil. The process of soil drawing water above its static level is known as **capillary action.** The zone in which this occurs is referred to as the **capillary fringe.** Although the pores in this region are saturated with water, this water cannot be thought of as a source of supply because the water held in this region will not drain freely by gravity (Figure 7–22).

FIGURE 7–21

Schematic of groundwater aquifers.

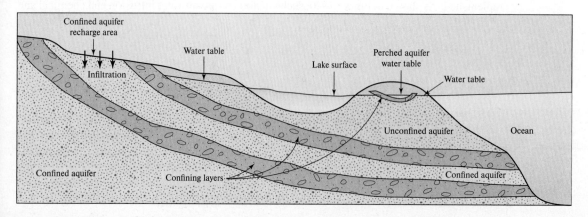

FIGURE 7–22

Schematic drawing showing the zones of aeration and saturation and the water table. Note how the surfaces of the soil particles in the zone of aeration are partially covered by water, whereas in the zone of saturation the pores are completely filled with water. Also note that in the unpumped well, the level of water is the same as the water table. (*Source:* Montgomery C., *Environmental Engineering,* 6e, 2003. The McGraw-Hill Companies, Inc.)

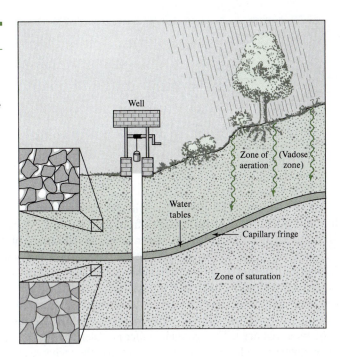

In an unconfined aquifer, the water table can vary significantly with rainfall and seasons. For example, in temperate climates, in the spring when rainfall is usually high and where snow- and icemelt are significant, the water table is nearest to the ground surface. In contrast, when infiltration rates are low, such as during periods of low rainfall or when the ground surface is frozen, the water table is farthest from the surface. This process of infiltration and migration, renewing the supply of groundwater, is referred to as **recharge.**

Perched Aquifers.   A perched aquifer is a lens of water held above the surrounding water table by an impervious geologic layer, such as bedrock or clay. It may cover an area from a few hundred square meters to several square kilometers. Drilling wells into perched aquifers can present problems because the volume of water held in these aquifers is relatively small, resulting in the well "drying out" after a short period of pumping.

Confined Aquifers.   Aquifers bounded both above and below the saturated zone by impermeable layers are referred to as **confined aquifers.** The impermeable layers are called **confining layers.** Confining layers are classified either as **aquicludes** or **aquitards.** Although aquicludes are essentially impermeable to water flow and aquitards are less permeable than the aquifer, but not truly impermeable, the terms are often used interchangeably.

Artesian Aquifer.   The water in a confined aquifer may be under considerable pressure due to the impermeable nature of the confining layers, which restrict flow, or due to elevation differences in the aquifer. The system is analogous to a manometer. When there is no constriction in the manometer, the water level in each leg rises to the same height. This is analogous of the water properties in an unconfined aquifer and is shown in Figure 7–23a. If the water level in the left leg is raised, the increased water pressure in that leg pushes the water up in the right leg until the levels are equal again. As shown in Figure 7–23b, if the right leg is clamped shut, then the water will not rise to the same level. This is analogous to the water properties in a confined aquifer. At the point where the clamp is placed, the water pressure will increase. This pressure is the result of the height of water in the left leg.

FIGURE 7–23

Manometer analogy to water in an aquifer. The manometer in **(a)** is analogous to an unconfined aquifer. The manometer in **(b)** is analogous to a confined aquifer.

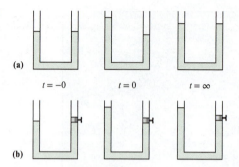

As shown in Figure 7–24, if the water in the aquifer is under pressure, it is called an **artesian aquifer.** The name *artesian* comes from the French province of Artois (*Artesium* in Latin) where, in the days of the Romans, water flowed to the surface of the ground from a well. When the water pressure in the aquifer is sufficiently high to push the water up through the geologic materials of the aquifer and overlying unsaturated zone and out onto the ground surface, the aquifer is known as a **flowing artesian aquifer.**

Water enters an artesian aquifer at some location where the confining layers intersect the ground surface. This is usually in an area of geological uplift. The exposed surface of the aquifer is called the **recharge area.** The artesian aquifer is under pressure for the same reason that the pinched manometer is under pressure, that is, because the recharge area is higher than the bottom of the top confining layer, and, thus, the height of the water above the confining layer causes pressure in the aquifer. The greater the vertical distance between the recharge area and the bottom of the top confining layer, the higher the height of the water, and the higher the pressure.

Springs. Because of the irregularities in underground geologic materials and in surface topography, the water table occasionally intersects the surface of the ground or the bed of a stream, lake, or ocean. At these points of intersection, groundwater flows out of the aquifer, forming lakes, streams, or springs. The location where the water table breaks the ground surface is called a **gravity** or **seepage spring.** Springs can result from either confined or unconfined aquifers.

FIGURE 7–24

Schematic showing piezometric surface of artesian and flowing artesian wells. Note the piezometric surface for the flowing well is at ground surface.

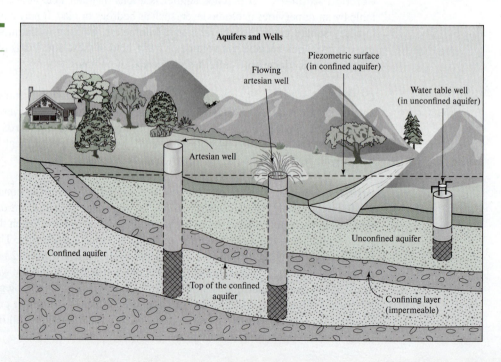

Complexities in Hydrogeology.   The descriptions and schematics presented earlier are great simplifications of what actually occurs in nature. Aquifers are very complex and highly variable geological formations. Variations in groundwater flow occur spatially, both in the vertical and horizontal directions. Lenses of variable geologic material may be present within a formation. For example, it is not uncommon in depositional geological formations to find sand lenses within much more impermeable geologic material, such as silt and clay. Aquifer systems also have divides, analogous to that observed with surface streams. Here the groundwater flow divides into different directions, affecting the extent to which the aquifer can be tapped as a water source.

Piezometric Surfaces and Head.   If we place small tubes (piezometers) vertically into a confined aquifer, the water pressure will cause water to rise in the tubes just as the water in the legs of a manometer rises to a point of equilibrium. The height of the water in the tube, referred to as **piezometric head,** is a measure of the pressure in the aquifer. The piezometric head is measured using the water level in the well. An imaginary plane drawn through the points of equilibrium in several piezometers is called a **piezometric surface.** In an unconfined aquifer, the piezometric surface is the water table.

   If the piezometric surface of a confined aquifer lies above the ground surface, a well penetrating into the aquifer will flow naturally without pumping. In this case the well penetrates into an artesian aquifer. If the piezometric surface is below the ground surface, the well will not flow without pumping.

   The **hydraulic gradient** is the difference in the head at two locations divided by the distance between the locations and can be represented mathematically as

$$\frac{\Delta h}{L} = \frac{h_2 - h_1}{L} \tag{7–11}$$

where $\Delta h/L$ = the hydraulic gradient
   h_2 = the head at location 2
   h_1 = the head at location 1
   L = the linear distance between location 1 and location 2

EXAMPLE 7–7   The head in an unconfined aquifer (Figure 7–25) has been measured at four locations as shown in the following schematic.

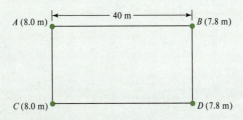

   Using this information, determine the hydraulic gradient.

Solution   The direction of flow is from AC to BD. The hydraulic gradient can be calculated using Equation 7–11

$$\frac{\Delta h}{L} = \frac{h_2 - h_1}{L} = \frac{8.0 - 7.8 \text{ m}}{40 \text{ m}} = 0.005 \text{ m} \cdot \text{m}^{-1}$$

FIGURE 7–25

Schematic showing
piezometric surface and
datum in an unconfined
aquifer.

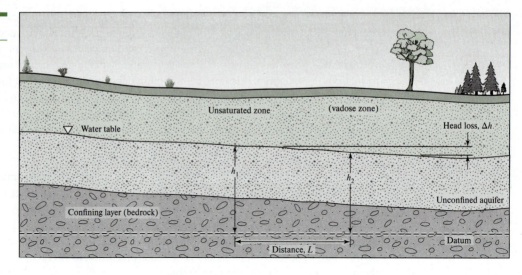

7–4 GROUNDWATER FLOW

Water flows along the piezometric surface from areas of higher head to lower head.

As stated earlier, in unconfined aquifers, the piezometric surface is the water table. The piezometric surface is calculated by subtracting the depth of water below the ground surface from a predefined datum. In many cases the datum is either the height of the top of the confining layer relative to sea level or the depth below ground surface (bgs).

EXAMPLE 7–8   You are working for a construction company and are building a school. In digging the foundation you find water at 7 m bgs. One hundred meters away, you find water at 7.5 m bgs. Choose the datum as the confining layer that is 25 m bgs. What is the piezometric surface at each point, the direction of groundwater flow, and the hydraulic gradient? *Note:* This assumes that the confining layer is parallel to the surface, which may or may not be true; however, assuming this allows us to simplify a complicated problem.

Solution   The first thing we should do is to draw a picture illustrating the problem. Note that at point A, the depth to the water table is 7.0 m, whereas at point B the depth is 7.5 m. Using the datum given (at 25 m bgs), we can calculate the total head of water at each point.

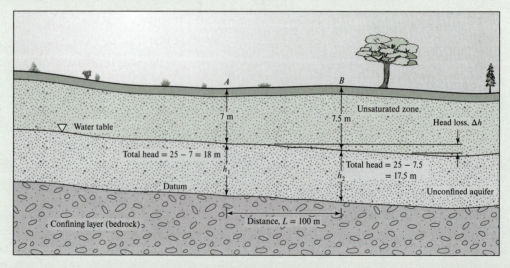

Point A: Total head $= 25 - 7.0$ m $= 18$ m

Point B: Total head $= 25 - 7.5$ m $= 17.5$ m

The groundwater flow is from point A to B, from the higher piezometric surface to the lower. Using these two piezometric surfaces, the hydraulic gradient can be calculated as

$$\frac{\Delta h}{L} = \frac{h_2 - h_1}{L} = \frac{18.0 - 17.5 \text{ m}}{100 \text{ m}} = 0.005 \text{ m} \cdot \text{m}^{-1}$$

So far, we have spoken only of the direction of groundwater flow; however, in many situations, including those where we wish to predict the rate of migration of groundwater contaminants, we need to determine the rate of groundwater flow. The hydrologist Henri Darcy studied the flow of water through columns filled with sand and slanted on their side. He found that the rate of groundwater flow depends on the hydraulic gradient and on a property of the geological material known as the hydraulic conductivity. The **hydraulic conductivity** can be thought of as a measure of how easy it is to obtain flow of water through the porous media (e.g., the sand, gravel, etc.). For example, you might expect that water would flow much more easily through gravel than it would through very fine clay. Because water flows easily through gravel, its hydraulic conductivity would be high; however, because water does not flow readily through clay, its hydraulic conductivity would be low. Hydraulic conductivity depends on the properties of the geological material, including the grain diameter and the porosity. Hydraulic conductivities of typical geological materials are given in Table 7–4.

Hydraulic conductivity is defined as the discharge that occurs through a unit cross section of aquifer (Figure 7–26) under a hydraulic gradient of 1.00. It has units of velocity (meters per second).

TABLE 7–4	Typical Values of Aquifer Parameters		
Aquifer Material		**Porosity (%)**	**Typical Values for Hydraulic Conductivity (m · s⁻¹)**
Clay		55	2.3×10^{-9}
Loam		35	6.0×10^{-6}
Fine sand		45	2.9×10^{-5}
Medium sand		37	1.4×10^{-4}
Coarse sand		30	5.2×10^{-4}
Sand and gravel		20	6.0×10^{-4}
Gravel		25	3.1×10^{-3}
Slate		<5	9.2×10^{-10}
Granite		<1	1.2×10^{-10}
Sandstone		15	5.8×10^{-7}
Limestone		15	1.1×10^{-5}
Fractured rock		5	$1 \times 10^{-8} - 1 \times 10^{-4}$

Source: Davis and Cornwall, 1998; Todd, 1980.

FIGURE 7–26

Illustration of definition of hydraulic conductivity (*K*). (*Source:* Geological Survey water supply paper 1662-D, pp. 74.)

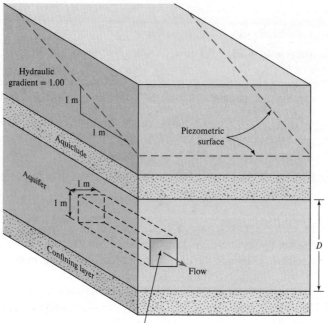

K = discharge that occurs through a unit cross section 1 m square

Darcy found that when groundwater flow is laminar and the aquifer is fully saturated, the velocity of flow is proportional to both the hydraulic gradient and the hydraulic conductivity (Darcy, 1856).

$$v = K\frac{\Delta h}{L} \tag{7–12}$$

where K is the hydraulic conductivity (length per time) and $\Delta h/L$ is the hydraulic gradient (length per length). This equation is not valid for extremely fine materials or for fractured rock. As shown in Figure 7–27, Darcy also found that the flowrate of water, Q, is equal to the specific discharge (also called Darcy velocity), v, times the cross-sectional area through which water flows:

$$Q = vA = \left(K\frac{\Delta h}{L}\right)A \tag{7–13}$$

The actual equation used to accurately model groundwater flow is much more complicated since the hydraulic conductivity varies both in the horizontal and vertical distances and the hydraulic gradient is $\partial h/\partial L$. However, the simplified versions used here provide a basic understanding of groundwater flow.

FIGURE 7–27

This area is filled with solid particles. As shown in Figure 7–28, water can only flow through the pore spaces.

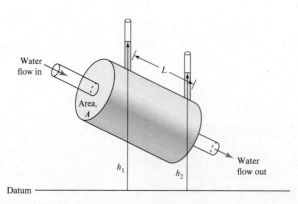

EXAMPLE 7–9 Let's assume that in the previous example the aquifer is coarse sand and that the cross-sectional area of the aquifer through which water flows is 925 m². What is the Darcy velocity of groundwater in this aquifer? What is the specific discharge?

Solution Using Table 7–4, we find that coarse sand has a hydraulic conductivity, K, of 6.9×10^{-4} m · s^{-1}. Because the hydraulic gradient was determined to be 0.005 m · m^{-1}, the Darcy velocity of groundwater flow can be calculated as

$$v = K\left(\frac{\Delta h}{L}\right) = (6.9 \times 10^{-4} \text{ m} \cdot \text{s}^{-1})(0.005 \text{ m} \cdot \text{m}^{-1})(86{,}400 \text{ s} \cdot \text{day}^{-1})$$

$$= 0.298 \text{ m} \cdot \text{day}^{-1}$$

The specific discharge is equal to vA or

$$0.298 \text{ m} \cdot \text{day}^{-1} \times 925 \text{ m}^2 = 275.65 \text{ m}^3 \cdot \text{day}^{-1}$$

In reality, the Darcy velocity is calculated from the flowrate, Q, a measured parameter. As mentioned previously, the specific discharge is a flowrate, and the area used to calculate this value is the cross-sectional area of the flow shown in Figure 7–27. But water does not flow through the entire cross-sectional area since much of this area is filled with solid particles. As shown in Figure 7–28 water can only flow through the pore spaces.

As such, the average linear velocity of the water must be greater than the Darcy velocity. The average linear velocity is calculated as

$$v'_{water} = \frac{v}{\eta} \tag{7–14}$$

where v'_{water} = the average linear velocity of the water (or **seepage velocity**)
v = the Darcy velocity
η = the porosity of the geological material

Porosity is the ratio of the volume of voids (open spaces) in the aquifer material to the total volume. It is a measure of the maximum amount of water that can be stored in the spaces between particles of aquifer material. It does not indicate how much of this water is available to be pumped or drained from this volume of material.

FIGURE 7–28

Cross section of aquifer material showing voids through which water can flow.

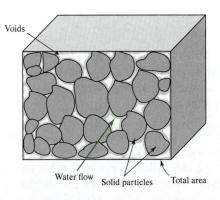

Voids

Water flow   Solid particles   Total area

EXAMPLE 7–10   The geological material in the column shown in Figure 7–27 is coarse sand. The piezometric surfaces $h_1 = 10$ cm and $h_2 = 8.0$ cm. The distance between the two points where h_1 and h_2 were measured is 10.0 cm. The cross-sectional area is 10 cm^2. What is the linear velocity of the water flowing through the column?

Solution   The hydraulic gradient can be calculated as

$$\frac{\Delta h}{L} = \frac{h_2 - h_1}{L} = \frac{10.0 - 8.0 \text{ cm}}{10.0 \text{ cm}} = \frac{2 \text{ cm}}{10.0 \text{ cm}} = 0.2 \text{ cm} \cdot \text{cm}^{-1}$$

From Table 7–4, we see that the hydraulic conductivity, K, of coarse sand is equal to 6.9×10^{-4} m · s^{-1}. The Darcy velocity can be calculated as

$$v = K\frac{\Delta h}{L} = (6.9 \times 10^{-4} \text{ m} \cdot \text{s}^{-1})(0.2 \text{ cm} \cdot \text{cm}^{-1}) = 1.38 \times 10^{-4} \text{ m} \cdot \text{s}^{-1}$$

Assuming that the porosity is 0.3 (as given in Table 7–4), then the linear velocity would be

$$v'_{water} = \frac{v}{\eta} = \frac{1.38 \times 10^{-4} \text{ m} \cdot \text{s}^{-1}}{0.3} = 4.6 \times 10^{-4} \text{ m} \cdot \text{s}^{-1}$$

From this analysis, you can see that the linear velocity of the water in the aquifer is significantly higher than the Darcy velocity.

7–5   WELL HYDRAULICS

Definition of Terms

The aquifer parameters identified and defined in this section are those relevant to determining the available volume of water and the ease of its withdrawal. More general terms have been defined in Sections 7–3 and 7–4.

Specific Yield.   The percentage of water that is free to drain from the aquifer under the influence of gravity is defined as specific yield (Figure 7–29). Specific yield is not equal to porosity because the molecular and surface tension forces in the pore spaces keep some of the water in the voids. Specific yield reflects the amount of water available for development. Values of S for unconfined aquifers range from 0.01 to 0.35 m^3 · m^{-3}. Some average values are shown in Table 7–5.

Storage Coefficient (S).   This parameter is akin to specific yield. The storage coefficient is the volume of available water resulting from a unit decline in the piezometric surface over a unit horizontal cross-sectional area. It has units of m^3 of water · m^{-3} of aquifer. For confined aquifers, the values of S vary from 5×10^{-5} to 5×10^{-3}.

Transmissibility (T).   The coefficient of transmissibility (T) is a measure of the rate at which water will flow through a unit width vertical strip of aquifer extending through its full saturated thickness (Figure 7–30) under a unit hydraulic gradient. It has units of m^2 · s^{-1}. Values of the transmissibility coefficient range from 1.0×10^{-4} to 1.5×10^{-1} m^2 · s^{-1}.

FIGURE 7–29

Specific yield.

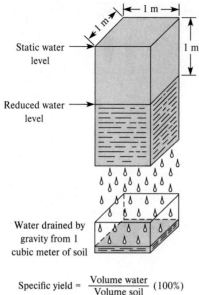

Static water level

Reduced water level

Water drained by gravity from 1 cubic meter of soil

$$\text{Specific yield} = \frac{\text{Volume water}}{\text{Volume soil}} \ (100\%)$$

Cone of Depression

When a well is pumped, the level of the piezometric surface in the vicinity of the well will be lowered (Figure 7–31).

This lowering, or drawdown, causes the piezometric surface to take the shape of an inverted cone called a **cone of depression.** Since the water level in a pumped well is lower than that in the aquifer surrounding it, the water flows from the aquifer into the well. At increasing distances from the well, the drawdown decreases until the slope of the cone merges with the static water table or the original piezometric surface, as shown in Figure 7–32. The distance from the well at which this occurs is called the radius of influence. The radius of influence is not constant but tends to expand with continued pumping.

At a given pumping rate, the shape of the cone of depression depends on the characteristics of the water-bearing formation. Shallow and wide cones will form in aquifers composed

TABLE 7–5   **Typical Values of Specific Yield**

| | Specific Yield (%) | | |
Material	Maximum	Minimum	Average
Coarse gravel	26	12	22
Medium gravel	26	13	23
Fine gravel	35	21	25
Gravelly sand	35	20	25
Coarse sand	35	20	27
Medium sand	32	15	26
Fine sand	28	10	21
Silt	19	3	18
Sandy clay	12	3	7
Clay	5	0	2

Source: Johnson, 1967, as cited by C. W. Fetter, 1994.

FIGURE 7–30

Illustration of definition of
hydraulic conductivity (*K*)
and transmissibility (*T*).

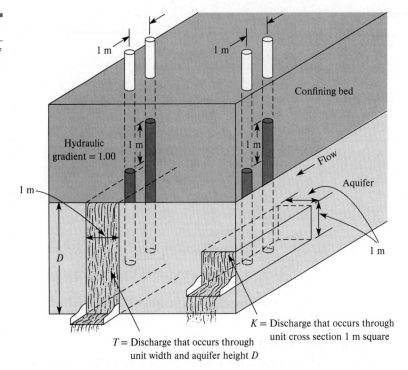

of coarse sands or gravel. Deeper and narrower cones will form in fine sand or sandy clay as shown in Figure 7–33. As the pumping rate increases, the drawdown increases. Consequently, the slope of the cone steepens.

When the cones of depression overlap, the local water table will be lowered (Figure 7–34). This requires additional pumping lifts to obtain water from the interior portion of the group of wells. A wider distribution of the wells over the groundwater basin will reduce the cost of pumping and will allow the development of a larger quantity of water. One rule of thumb is that two wells should be placed no closer together than two times the thickness of the water-bearing strata. For more than two wells, they should be spaced at least 75 meters apart.

The flow of water into the well can be described using Darcy's law. This equation has been solved for steady state and nonsteady or transient flow. Steady state is a condition under which

FIGURE 7–31

Idealized cone of
depression associated
with a homogeneous
aquifer. (*Source:* U.S.
Geological Survey
Circular 1186, Denver, CO,
pp. 86.)

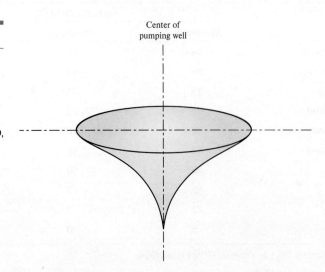

FIGURE 7–32

Schematic showing idealized drawdown in **(a)** an unconfined aquifer and **(b)** a confined aquifer.

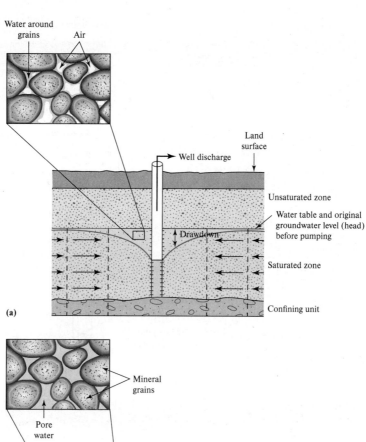

(a)

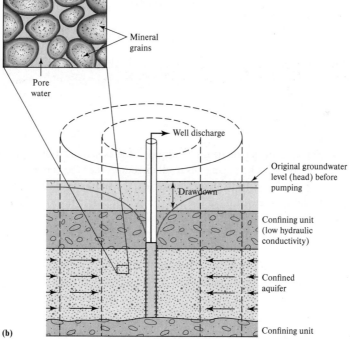

(b)

no changes occur with time. It will seldom, if ever, occur in practice, but may be approached after very long periods of pumping. Transient-flow equations include a factor of time. (Here we will consider only steady flow; the discussion of transient flow is beyond the scope of this text.) The derivation of these equations is based on the following assumptions:

1. The well is pumped at a constant rate.
2. Flow of groundwater toward the well is radial and uniform.
3. Initially, the piezometric surface is horizontal.

FIGURE 7–33

Effect of aquifer material on cone of depression. (*Source:* U.S. Environmental Protection Agency, 1973.)

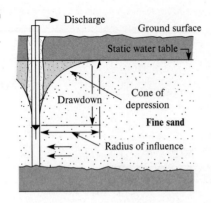 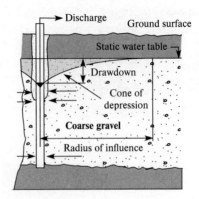

FIGURE 7–34

Effect of overlapping cones of depression. (*Source:* U.S. Environmental Protection Agency, 1973.)

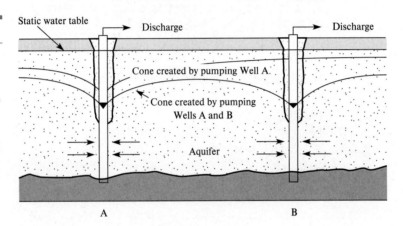

4. The well fully penetrates the aquifer and is screened over the entire height of the aquifer.
5. The aquifer is homogeneous in all directions (i.e., the porosity, conductivity, and other parameters are the same in all directions) and the aquifer is essentially of infinite horizontal extent.
6. Water is released from the aquifer in immediate response to a drop in the piezometric surface.
7. The height of drawdown is small compared to the depth of the aquifer.

Steady Flow in a Confined Aquifer. The equation describing steady, confined aquifer flow was first presented by Dupuit (1863) and subsequently extended by Theim (1906). It may be written as follows:

$$Q = \frac{2\pi T \, (h_2 - h_1)}{\ln(r_2/r_1)}$$

(7–15)

where $T = KD$ = transmissibility, $m^2 \cdot s^{-1}$
$\qquad D$ = thickness of artesian aquifer, m
$\qquad h_1$ = height of the piezometric surface above confining layer at a distance, r_1, from the pumping well, m
$\qquad h_2$ = height of the piezometric surface above confining layer at a distance, r_2, from the pumping well, m
$\qquad r_1, r_2$ = radius from pumping well, m
$\qquad \ln$ = logarithm to base e

Confined aquifers remain completely saturated during pumping by wells.

EXAMPLE 7–11   An artesian aquifer 10.0 m thick with a piezometric surface 40.0 m above the bottom confining layer is being pumped by a fully penetrating well. Steady state drawdowns of 5.00 m and 1.00 m were observed at two nonpumping wells located 20.0 m and 200.0 m, respectively, from the pumped well. The pumped well is being pumped at a rate of 0.016 m³ · s⁻¹. Determine the hydraulic conductivity of the aquifer.

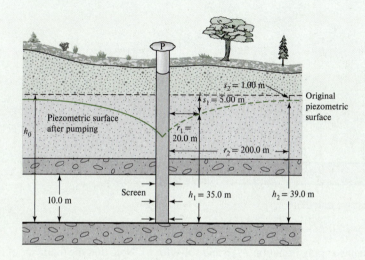

Solution   First we determine h_1 and h_2:

$$h_1 = 40.0 \text{ m} - 5.00 \text{ m} = 35.0 \text{ m}$$

$$h_2 = 40.0 \text{ m} - 1.00 \text{ m} = 39.0 \text{ m}$$

Since,

$$Q = \frac{2\pi KD(h_2 - h_1)}{\ln(r_2/r_1)}$$

we can rearrange this equation to determine K

$$\frac{Q \ln(r_2/r_1)}{2\pi D(h_2 - h_1)} = K$$

Therefore:

$$\frac{0.016 \text{ m}^3 \cdot \text{s}^{-1} \ln(200/20)}{2\pi(10 \text{ m})(39.0 \text{ m} - 35.0 \text{ m})} = 1.50 \times 10^{-4} \text{ m} \cdot \text{s}^{-1}$$

Steady Flow in an Unconfined Aquifer.   For unconfined aquifers, the factor D in Equation 7–15 is replaced by the height of the water table above the lower boundary of the aquifer. The equation then becomes

$$Q = \frac{2\pi K\left(h_2^2 - h_1^2\right)}{\ln(r_2/r_1)} \tag{7–16}$$

EXAMPLE 7–12   A 0.50-m diameter well fully penetrates an unconfined aquifer, which is 30.0 m thick. The drawdown at the pumped well is 10.0 m and the hydraulic conductivity of the gravel aquifer is 6.4×10^{-3} m · s^{-1}. If the flow is steady and the discharge is 0.014 m^3 · s^{-1}, determine the drawdown at a site 100.0 m from the well.

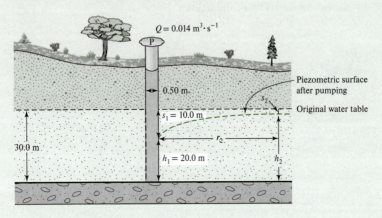

Solution   First we calculate h_1

$$h_1 = 30.0 \text{ m} - 10.0 \text{ m} = 20.0 \text{ m}$$

Then we apply Equation 7–16 and solve for h_2. Note that the diameter of the well was given and r_1 must be calculated as one-half of the diameter or 0.25 m.

$$0.014 \text{ m}^3 \cdot \text{s}^{-1} = \frac{\pi(6.4 \times 10^{-3} \text{ m} \cdot \text{s}^{-1}) \left[h_2^2 - (20.0 \text{ m})^2 \right]}{\ln(100 \text{ m}/0.25 \text{ m})}$$

$$h_2^2 - 400.0 \text{ m}^2 = \frac{(0.014 \text{ m}^3 \cdot \text{s}^{-1})(5.99)}{\pi(6.4 \times 10^{-3} \text{ m} \cdot \text{s}^{-1})}$$

$$h_2 = (4.17 + 400.0)^{1/2}$$

$$h_2 = 20.10 \text{ m}$$

The drawdown is then

$$s_2 = H - h_2 = 30.0 - 20.10 = 9.90 \text{ m}$$

7–6   SURFACE WATER AND GROUNDWATER AS A WATER SUPPLY

Groundwater and surface water provide an important natural resource. Although groundwater is hidden, it is as important as surface water. As shown in Figure 7–35a, approximately 67% of the U.S. population obtains its drinking water from surface water supplies. However, the number of public drinking-water systems that use surface water as a source is only 24% of the total (Figure 7–35b). This is because most small systems use groundwater and, as shown in Figure 7–35c, small systems vastly outnumber medium or large ones. Combined with the fact that large systems supply 58% of the population (Figure 7–35d), we see why groundwater systems account for only 33% of the per capita drinking water even though they account for 76% of the systems.

FIGURE 7–35 Surface Water and Groundwater as a Water Supply (2006 data, U.S. EPA)

(a) Percentage of the population served by drinking-water system source. **(b)** Percentage of drinking-water systems by supply source. **(c)** Percentage of systems by size (~157.4 thousand systems in total in United States). **(d)** Percentage of population served by system size. (*Source:* U.S. Environmental Protection Agency, 2010.)

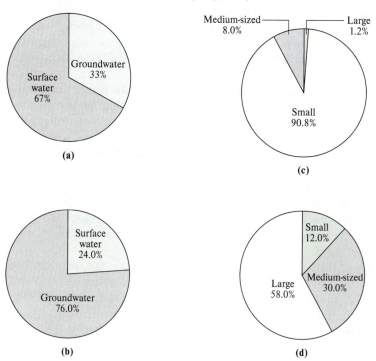

Groundwater has several characteristics that make it desirable as a water supply source over surface water. First, the groundwater system provides natural storage, which eliminates the cost of storage tanks, reservoirs, and other impoundments. Second, because the supply frequently is available at the point of demand, the cost of transmission is reduced significantly. Third, because the natural geological materials filter groundwater, groundwater is clearer and less turbid than surface water. Groundwater is less subject to seasonal fluctuations and has been more protected from pollution; however, unless we provide greater restrictions on such pollutant sources as large-scale agricultural operations, landfills, gasoline stations, and hazardous waste operations, our groundwater supplies will suffer from contamination similar to that affecting our surface water supplies.

7–7   DEPLETION OF GROUNDWATER AND SURFACE WATER

Water Rights

A discussion of the depletion of water must begin with water usage and water rights. In the United States, most water rights are based on two concepts: riparian rights and prior appropriation law. A third concept is a hybrid of riparian rights and prior appropriation law. In the United States, there exists a fourth concept: federal reserved water rights.

The doctrine of riparian rights is based on the "Natural Flow" Doctrine, which states that water rights originate from the ownership of land that abuts a natural water body and that the landowner has the right to water flow, unimpaired in quality and quantity. The doctrine originated in England to protect the interests of the landed gentry. While the landowner has an equal right to the use of water from the source, he/she does not own the water and therefore, water must be used for reasonable, nonconsumptive purposes only. Once the right is established, he/she is entitled to access to unpolluted water and the watercourse and to fish in the water. The landowner does not have the right to unreasonably detain or divert water from the water body, and the water must be returned to the water body from which it was obtained. This set of water rights worked reasonably well in Colonial times in the Eastern part of the United States where water was abundant and the society was agrarian. In more recent times, "reasonable use" has expanded to allow for the consumptive use of water, although what constitutes reasonable use varies widely from state to state and continues to evolve.

Prior appropriation water rights developed in the western part of the United States in the 1800s from laws used to settle mineral rights claims amongst settlers. With appropriation law, the claim must be documented officially and posted, the rights must be used continually or they are lost, and "first in time, first in right" takes precedence to settle disputes over water rights. An appropriation right is independent of land ownership and is based on physical control and beneficial use of a specific amount of water for a specified purpose at a specific location. These rights cannot be sold or transferred. Unlike with riparian rights, water may be diverted and stored, and this was commonly the case. Whether a use is beneficial or not is typically determined by the state. While appropriation law was developed in part to prevent waste, the fact that rights are lost for nonuse for a period of time can and has resulted in "wasteful" use of water simply to prevent the loss of rights.

The third is a hybrid of the riparian rights and prior appropriation law, in which both types of rights are recognized. Usually, the hybrid doctrine is used where riparian rights were recognized first, but the state moved to an appropriation legal system because of limited resources. In such cases, the rights obtained under the riparian system were converted to appropriation rights by allowing riparian land owners to claim a water right and to incorporate that right into the state's prior appropriation system. While the owner did not have to be using the water for a beneficial use at the time of conversion of rights, if the water was not used within a certain number of years or the water claim was not made, the land owner lost his/her riparian rights. California was the first state to move to this approach. Kansas, Nebraska, North and South Dakota, Oklahoma, Oregon, Texas, and Washington also use this hybrid doctrine. The last doctrine is the federal reserved water rights. Federal reserved water rights are those rights guaranteed to the Federal Government or Native American tribes and reservations by tribal treaties with the federal government, executive orders, and statutes. These rights cannot be terminated or forfeited as a result of nonuse. The federal reserved water rights doctrine was established by the U.S. Supreme Court in 1908 in *Winters v. United States* case. With this landmark ruling, the U.S. Supreme Court established that when the federal government created Native American reservations, water rights were reserved in sufficient quantity to meet the purposes for which the reservation was established. The determination of water rights was no longer purely a state matter.

The actual application of these systems varies greatly from state to state. The systems of water rights are complex and diverse. When water bodies cross political and jurisdictional boundaries, conflicts often arise. In the United States, three basic approaches are used to settle such conflicts: (1) litigation before the Supreme Court of the United States; (2) the ratification of legislation by the Congress of the United States; and (3) negotiation and ratification of interstate compacts between states. For example, the Colorado River Compact of 1922 was signed by the states of Colorado, New Mexico, Utah, Wyoming, Nevada, Arizona, and California to settle disputes over the allocation of the waters of the Colorado River and its tributaries. As a result, the Colorado River Basin was divided into two basins, and water rights were allocated to secure the agricultural and industrial development of the Basin, to provide water storage, to promote interstate commerce, and to protect life and property from floods. Over time additional disputes arose, each settled by either ratification of a new law (e.g., The Colorado River Basin Project Act of 1968); Supreme Court decisions (e.g., *Arizona v. California*, 1964); or by the signing of new treaties or compacts (the California Seven Party Agreement of 1931).

Water Use

In 2010, the last date for which comprehensive data is available, approximately 1.34 billion cubic meters of water were withdrawn daily in the United States. Thermoelectric power used the largest percentage of water (45%), followed by irrigation (33%) and public water supplies (almost 12%). The remaining five categories (domestic, livestock, aquaculture, industrial, and mining) account for the remaining (9%) of water withdrawals. Freshwater withdrawals were 86% of the total freshwater and saline-water withdrawals. Withdrawals for thermoelectric-power generation and irrigation have stabilized or decreased since 1980, whereas that for public-supply and domestic uses have increased steadily since then (Maupin et al., 2014).

Water use for irrigation is by far the largest consumptive use of water diverted from streams or withdrawn from aquifers in the United States. Of the total volume of water withdrawn for irrigation, 58% came from surface water sources and 42% from groundwater (Kenny et al., 2009). Approximately 56% of the water withdrawn for irrigation is consumed, whereas only about 16% of the water withdrawn for public water supplies is consumed (Hudson et al., 2004).

As shown in Figure 7-36a, the majority of total U.S. irrigation withdrawals (83%) and irrigated acres (74%) were in 17 Western states. Surface water was the primary source of water in the arid West, except in Kansas, Oklahoma, Nebraska, Texas, and South Dakota, where more groundwater was used, as illustrated in Figures 7-36b and c. In the Eastern states, irrigation

FIGURE 7–36

Irrigation withdrawals by source and state, 2010. (*Source:* U.S. Geological Survey, Georgia Water Science Center http://ga.water.usgs.gov/edu/.)

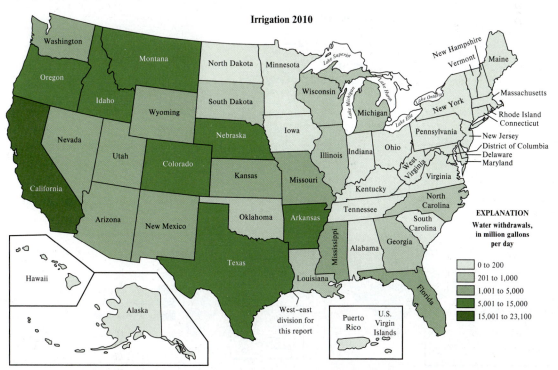

FIGURE 7–36 (*continued*)

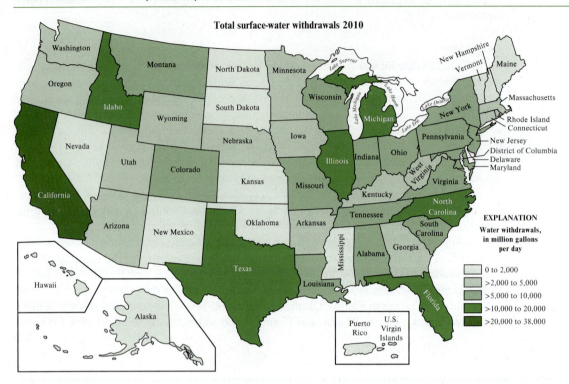

Total surface-water withdrawals 2010

EXPLANATION
Water withdrawals,
in million gallons
per day

0 to 2,000
>2,000 to 5,000
>5,000 to 10,000
>10,000 to 20,000
>20,000 to 38,000

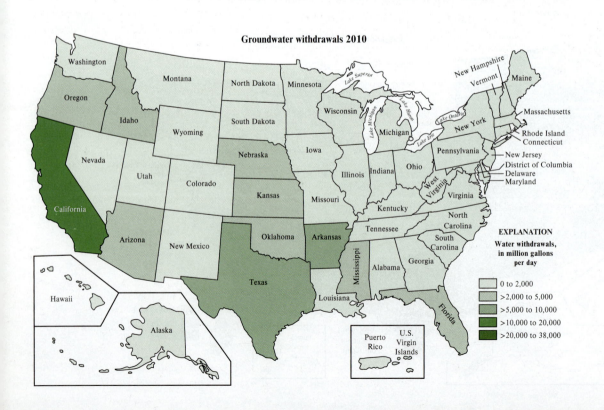

Groundwater withdrawals 2010

EXPLANATION
Water withdrawals,
in million gallons
per day

0 to 2,000
>2,000 to 5,000
>5,000 to 10,000
>10,000 to 20,000
>20,000 to 38,000

TABLE 7-6 Irrigation Withdrawals, Top States, 2010 (percentages calculated from unrounded values)

State	Percentage of Total Withdrawals (%)	Cumulative Percentage of Total Withdrawals (%)
California	20	20
Idaho	12	32
Colorado	8	41
Arkansas	8	48
Montana	6	54

Source: https://water.usgs.gov/watuse/wuir.html.

is used to increase the number of plantings per year and yield per acre, or to supplement any lack of precipitation during periods of drought. Table 7-6 shows the top states for irrigation withdrawals in 2010.

In the West conflicting demands for water supplies are common. California, in particular, has become very concerned about being able to balance the water needs of agriculture, its growing urban population, and the environment. An example of the controversy that can arise over water rights is the recent actions proposed to protect fisheries and wildlife in the Sacramento Delta in California. The watershed of the Delta provides drinking water to two-thirds of the state's population and it supplies some of the state's most productive agricultural areas. The Bay Delta Estuary itself is one of the largest ecosystems for fish and wildlife habitat and production in the United States. Historical and current human activities (e.g., water development, land use, wastewater discharges, introduced species, and harvesting) have degraded the beneficial uses of the Bay Delta Estuary, as evidenced by the declines in the populations of many biological resources of the estuary. In an effort to reverse this decline, the State of California has proposed that some of the water now diverted from the delta be allowed to flow into the delta. This action created considerable controversy because it affects the water rights of other users.

Land Subsidence

Land subsidence, the loss of surface elevation due to removal of subsurface support, occurs in nearly every state. Common causes of land subsidence from human activity are pumping water, oil, and gas from underground reservoirs; the collapse of underground mines; and the drainage of organic soils. Some of the worst problems are seen in the San Joaquin Valley in California where massive groundwater withdrawals have lead to widespread subsidence. As is shown in Figure 7–37 the loss of surface elevation can be dramatic. The sign on the top of the pole shows where the land surface was back in 1925. In this picture, taken in 1977, the land surface has dropped nearly 9 m during that time.

Many problems result from land subsidence, including

- Damage to bridges, roads, drains, well casings, buildings and other structures.
- Flooding and changes in streamflow patterns.
- Loss of groundwater storage. In California more groundwater storage has been lost as a result of subsidence than the volume of all aboveground water storage reservoirs built in the state combined.

The economic damage caused by land subsidence can be significant. For example, the annual cost of subsidence mitigation in the San Joaquin Valley alone is over $180 million.

FIGURE 7–37

Land subsidence in the San Joaquin Valley, 16 km southwest of Mendota, CA. (©U.S. Geological Survey)

7–8   STORMWATER MANAGEMENT

As discussed in Section 7–1, stormwater runoff is generated when precipitation from rain and snowmelt events flows over land or impervious surfaces and does not infiltrate into the ground. As the runoff flows over paved streets, parking lots, building rooftops, fertilized lawns, and other surfaces, it accumulates debris, chemicals, sediment or other pollutants that could adversely affect water quality if the runoff were discharged untreated. Figure 7–7 shows how urbanization, resulting in the rapid flow of storm water into storm drains, sewer systems, and drainage ditches, affects the stream flow hydrograph and causes flooding. This rapid discharge of water can also result in stream bank erosion, increased turbidity in the receiving water, habitat destruction, infrastructure damage, and diminished water quality.

Stormwater discharges, which are considered point sources, were traditionally collected in piped networks and transported offsite as quickly as possible, either directly to a stream or river, to a large stormwater retention or detention basin, or to a combined sewer system flowing to a wastewater treatment plant. Storm water retention or detention basins detain and slow the flow of stormwater, allowing larger and heavier material to settle out and chemicals and smaller particles to be filtered out before the water is discharged into receiving waters. These ponds reduce the likelihood of flooding and reduce the effects of urbanization on water quality and aquatic habitats. However, as these systems do reduce the extent of impervious surfaces, they do not enhance infiltration and, therefore, groundwater recharge. They also consume wildlife habitat and available space for recreation or other needs.

In an attempt to address the problems related to traditional stormwater basins, **low impact development (LID) techniques** and wet weather green infrastructure have been developed and is now mandated by law for federal facilities. The Energy Independence and Security Act of 2007 Sec. 438 mandates that any Federal facility with a footprint that exceeds 46.5 m^3 shall use site planning, design, construction, and maintenance strategies for the property to maintain or restore, to the maximum extent technically feasible, the predevelopment hydrology of the property with regard to the temperature, rate, volume, and duration of flow. LID approaches include strategic site design, measures to control the sources of runoff, and sensible landscape planning. The purpose of LID is to restore natural watershed functions through small-scale treatment at or near the source of runoff. The site should be designed so that it functions hydrologically similar to that before development. **Wet weather green infrastructure** includes approaches and technologies to enhance the infiltration, evapotranspiration, capture, and reuse of stormwater, and thereby maintain or restore the natural hydrology of the site.

Low Impact Development

The goal of LID practices is to reduce runoff volume and enhance the filtration and removal of pollutant from storm water. These practices include bioretention facilities or rain gardens, grass swales and channels, vegetated rooftops, rain barrels, cisterns, vegetated filter strips, and permeable pavements.

Bioretention facilities typically contain these six components: grass buffer strips, sand bed, ponding area, organic layer, planting soil, and vegetation. The grass buffer strips serve to reduce the velocity of the runoff and filter particulate matter from the water. The sand bed helps to aerate the water and drain the planting soil. It also helps to flush pollutants from soil materials. The ponding area provides storage of excess runoff, especially for the initial flush during a rainfall event. It also facilitates the settling of particulate material and the evaporation of water. The organic layer provides a support medium for microbiological growth, which decomposes organic material in the storm water. This layer also serves as a sorbent for heavy metals and other hydrophobic pollutants. The plants, which grow in the planting soil, take up nutrients and aid in the evapotranspiration of water. The soil provides additional water retention and may absorb some pollutants, including hydrocarbons and heavy metals.

Green roofs effectively reduce urban stormwater runoff by reducing the amount of impervious surfaces. They are especially effective in older urban areas, where the percentage of land that is impervious to water flow is high. A green roof is constructed of multiple layers: a vegetative layer, media, a geotextile, and a synthetic drain system. Green roofs also extend the life of the roof, reducing energy costs and conserve valuable land that would otherwise be required for stormwater runoff controls.

Permeable pavements can be used to reduce the percent of impervious surfaces in a watershed. Porous pavements are most appropriate for low traffic areas, such as parking lots and sidewalks. They have been successfully installed in coastal areas with sandy soils and flatter slopes. The infiltration of stormwater into underlying soils promotes pollutant treatment and groundwater recharge.

Other techniques used in LID, including grass swales and channels, rain barrels, cisterns, and vegetated filter strips, serve to redirect runoff from sewers and storm water collection systems. Rain barrels and cisterns allow for the collection of water and its subsequent use for lawn and garden irrigation and nonpotable water uses such as toilet flushing. Such was the case on the campus of Michigan State University, where during the recent renovations of Brody Hall, a cistern was installed. The collected water serves for toilet flushing in the first floor restrooms of this building. Grass swales and channels, along with vegetated filter strips help reduce the amount of impervious surface in development areas and enhance infiltration and groundwater recharge.

Wet Weather Green Infrastructure

As mentioned earlier, wet weather green infrastructure seeks to maintain or restore the natural hydrology of the site. Techniques may include LID approaches discussed above. It may also include **constructed wetlands,** which are designed to mimic natural wetlands so as to retain stormwater and remove pollutants by gravity settling, sorption, biodegradation, and plant uptake. The velocity of water slows down as it flows through a wetland. This allows suspended solids to either become trapped by vegetation and to settle by gravity. Hydrophobic organic pollutants and heavy metals can sorb to plants or soil organic matter. Biodegradable pollutants can be assimilated and transformed by plants or microorganisms. Nutrients are absorbed by wetland soils and taken up by plants and microorganisms. For example, microbes found in wetlands can convert organic nitrogen into useable, inorganic forms (i.e., NO_3^- and/or NH_4^+), which is essential for plant growth. Subsequent reactions (known as denitrification) can convert the nitrate to nitrogen, which can be safely released to the atmosphere. Phosphorus can be assimilated by microorganisms and incorporated into cellular biomass.

Constructed wetlands can also be a cost-effective and technically feasible approach to treating stormwater. Wetlands are often less expensive to build than traditional treatment options, have lower operating and maintenance costs, and can better handle fluctuating water levels. Additionally, they are aesthetically pleasing and can promote water reuse, wildlife habitat, and recreational use. Constructed wetlands should be built on uplands and outside floodplains or floodways in order to prevent damage to natural wetlands. Water control structures need to be installed to ensure the desired hydraulic flow patterns. Where the soils are highly permeable, an impervious, compacted clay liner should be installed to protect underlying groundwater. The original soil can be placed over the liner. Wetland vegetation is then planted or allowed to establish naturally.

There are limitations to constructed wetlands for stormwater management. Highly variable flow conditions may make it difficult to maintain vegetation, although the recycle of water through the wetland can help ameliorate problems associated with low flow conditions. The detained water may act as a heat sink, resulting in the discharge of significantly warmed water to downstream water bodies. In addition, until vegetation is established and during cold seasons the removal of pollutants is likely to be minimal. However, with careful design and proper maintenance constructed wetlands can cost-effectively remove pollutants from stormwater for many years.

CHAPTER REVIEW

When you have completed studying this chapter, you should be able to do the following without the aid of your textbooks or notes:

1. Sketch and explain the hydrological cycle, labeling the parts as in Figure 7–2.
2. List and explain the four factors that reduce the amount of direct runoff.
3. Explain the difference between streamflow that results from direct runoff and that which results from baseflow.
4. Define evaporation, transpiration, runoff, baseflow, watershed, basin, and divide.
5. Write the basic mass-balance equation for the hydrological cycle.
6. Explain why infiltration rates decrease with time.
7. Explain why predicting water levels in streams and lakes is difficult.
8. Explain what is meant by a 100-year flood.
9. Precipitation rates vary with what, and why are these variations important in water resources management?
10. Describe the different types of rainfall.
11. What are the problems associated with extrapolating rainfall gauge data to watersheds?

12. Evaporation of water from water surfaces depends on what?
13. The loss of water due to evapotranspiration depends on what?
14. What factors affect infiltration rates?
15. What method(s) can be used to estimate infiltration rates? What are the limitation(s) of this (these) method(s)?
16. Use Horton's law to estimate infiltration rates and infiltration volumes in a particular time interval.
17. Why is it important to have accurate estimates of infiltration and streamflow?
18. What is meant by streamflow and how is it determined?
19. What factors affect streamflow?
20. Explain what a unit hydrograph is.
21. Explain what factors affect the shape of a hydrograph and show how these factors change its shape.
22. Sketch the groundwater hydrological system, labeling the parts as in Figure 7–21.
23. Define the following terms: aquifer, vadose zone, unsaturated zone, zone of aeration, phreatic zone, aquitard, aquiclude, perched aquifer, confined aquifer, unconfined aquifer, hydraulic conductivity, hydraulic gradient, porosity, piezometric head (surface).
24. Explain why water in an artesian aquifer is under pressure and why water may rise above the surface in some instances and not others.
25. What factors affect the rates of water movement in the subsurface?
26. Use Horton's equation and some form of Dalton's equation to solve complex mass-balance problems.
27. Calculate the hydraulic gradient.
28. Given several hydraulic heads and the locations of the piezometers, sketch a piezometric surface.
29. Compute mass-balances for open and closed hydrological systems.
30. Estimate the volume of water loss through transpiration given the air and water temperature, wind speed, and relative humidity.
31. Determine how long is required for water to travel a certain distance in the subsurface.

With the aid of this text, you should be able to do the following:

1. Sketch a piezometric profile for a single well pumping at a high rate, and sketch a profile for the same well pumping at a low rate.
2. Sketch a piezometric profile for two or more wells located close enough together to interfere with one another.
3. Calculate the drawdown at a pumped well or observation well if you are given the proper input data.

PROBLEMS

7–1 Lake Kickapoo, TX, is approximately 12 km in length by 2.5 km in width. The inflow for the month of April is $3.26 \text{ m}^3 \cdot \text{s}^{-1}$ and the outflow is $2.93 \text{ m}^3 \cdot \text{s}^{-1}$. The total monthly precipitation is 15.2 cm and the evaporation is 10.2 cm. The seepage is estimated to be 2.5 cm. Estimate the change in storage during the month of April.

Answer: $1.61 \times 10^6 \text{ m}^3$

7–2 A 4000-km^2 watershed receives 102 cm of precipitation in a year. The average flow of the river draining the watershed is $34.2 \text{ m}^3 \cdot \text{s}^{-1}$. Infiltration is estimated to be $5.5 \times 10^{-7} \text{ cm} \cdot \text{s}^{-1}$ and evapotranspiration is estimated to be $40 \text{ cm} \cdot \text{year}^{-1}$. Determine the change in storage in the watershed over a year. Compute the runoff coefficient for this watershed.

7–3 Using the values of f_o, f_c, and k for a Dothan loamy sand, find the infiltration rate at times of 12, 30, 60, and 120 min. Compute the total volume of infiltration over 120 min in an area 1 m^2. Assume that the rate of precipitation exceeds the rate of infiltration throughout the storm event.

> **Answer:** Rates are 83, 77, 72, and 68 mm · h^{-1} for times of 12, 30, 60, and 120 min, respectively. The volume = 148 m^3 (for an area of 1 m^2).

7–4 Infiltration data from an experiment yield an initial infiltration rate of 4.70 cm · h^{-1} and a final equilibrium infiltration rate of 0.70 cm · h^{-1} after 60 min of steady precipitation. The value of k was estimated to be 0.1085 h^{-1}. Determine the total volume of infiltration (in an area 1 m^2) for the following storm sequence: 30 mm · h^{-1} for 30 min, 53 mm · h^{-1} for 30 min, 23 mm · h^{-1} for 30 min.

7–5 Using the empirical equation developed for Lake Hefner (Equation 7–8), estimate the evaporation from a lake on a day that the air temperature is 30°C, the water temperature is 15°C, the wind speed is 9 m · s^{-1}, and the relative humidity is 30%.

> **Answer:** 4.7 mm · day^{-1}

7–6 The Dalton-type evaporation equation implies that there is a limiting relative humidity above which evaporation will be negligible regardless of the wind speed. Using the Lake Hefner empirical equation, estimate the relative humidity at which evaporation will be zero if the water temperature is 10°C and the air temperature is 25°C.

7–7 Four monitoring wells have been placed around a leaking underground storage tank. The wells are located at the corners of a 1-ha square. The total piezometric head in each of the wells is as follows: NE corner, 30.0 m; SE corner, 30.0 m; SW corner, 30.6 m: NW corner, 30.6 m. Determine the magnitude and direction of the hydraulic gradient.

> **Answer:** Hydraulic gradient = 6×10^{-3} m · m^{-1}; direction is from west to east.

7–8 After a long wet spell, the water levels in the wells described in Problem 7–7 were measured and found to be the following distances from the ground surface: NE corner, 3.0 m; SE corner, 3.0 m; SW corner, 3.6 m; NW corner, 3.6 m. A confining layer of bedrock exists at a depth of 25 m bgs. Assume that the ground surface is at the same elevation for each of the wells. Determine the magnitude and direction of the hydraulic gradient.

7–9 A gravelly sand has hydraulic conductivity of 6.1×10^{-4} m · s^{-1}, a hydraulic gradient of 0.00141 m · m^{-1}, and a porosity of 20%. Determine the Darcy velocity and the average linear velocity.

> **Answer:** Darcy velocity = 8.6×10^{-7} m · s^{-1}; average linear velocity = 4.3×10^{-6} m · s^{-1}.

7–10 Two piezometers have been placed along the direction of flow in a confined aquifer that is 30.0 m thick. The piezometers are 280 m apart. The difference in piezometric head between the two is 1.4 m. The aquifer hydraulic conductivity is 50 m · day^{-1}, and the porosity is 20%. Estimate the travel time for water to flow between the two piezometers.

7–11 A fully penetrating well in a 28.0-in thick artesian aquifer pumps at a rate of 0.00380 m^3 · s^{-1} for 1941 days (assume to be sufficient to obtain steady state conditions) and causes a drawdown of 64.05 m at an observation well 48.00 m from the pumping well. How much drawdown will occur at an observation well 68.00 m away? The original piezometric surface was 94.05 m above the bottom confining layer. The aquifer material is sandstone. Report your answer to two decimal places.

> **Answer:** $S^2 = 51.08$ m

7–12 An artesian aquifer 5 m thick with a piezometric surface 65 m above the bottom confining layer is being pumped by a fully penetrating well. The aquifer is a mixture of sand and gravel. A steady-state drawdown of 7 m · s^{-1} is observed at a nonpumping well located 10 m away. If the pumping rate is 0.020 m^3 · s^{-1}, how far away is a second nonpumping well that has an observed drawdown of 2 m?

7–13 A contractor is trying to estimate the distance to be expected of a drawdown of 4.81 m from a pumping well under the following conditions:

Pumping rate $= 0.0280 \text{ m}^3 \cdot \text{s}^{-1}$
Pumping time $= 1066 \text{ d}$
Drawdown in observation well $= 9.52 \text{ m}$
Observation well is located 10.00 m from the pumping well
Aquifer material $=$ medium sand
Aquifer thickness $= 14.05 \text{ m}$

Assume that the well is fully penetrating in an unconfined aquifer. Report your answer to two decimal places.

7–14 A well with a 0.25-m diameter fully penetrates an unconfined aquifer that is 20 m thick. The well has a discharge of $0.015 \text{ m} \cdot \text{s}^{-1}$ and a drawdown of 8 m. If the flow is steady and the hydraulic conductivity is $1.5 \times 10^{-4} \text{ m} \cdot \text{s}^{-1}$, what is the height of the piezometric surface above the confining layer at a site 80 inches from the well?

Answer: $S = 1.9 \times 10^{-5}$

DISCUSSION QUESTIONS

7–1 An artesian aquifer is under pressure because of the weight of the overlying geological strata. Is this sentence true or false? If it is false, rewrite the sentence to make it true.

7–2 As a field engineer you have been asked to estimate how long you would have to measure the discharge from a mall parking lot before the maximum discharge would be achieved. What data would you have to gather to make the estimate?

7–3 When a flood has a recurrence interval (return period) of 5 years, it means that the chance of another flood of the same or less severity occurring next year is 5%. Is this sentence true or false? If it is false, rewrite the sentence to make it true.

FE EXAM FORMATTED PROBLEMS

7–1 A given soil has a hydraulic conductivity, K, of $4.8 \times 10^{-6} \text{ m} \cdot \text{s}^{-1}$. The hydraulic gradient is 0.01 m/m. The effective porosity is 0.45. What is the average seepage velocity?

(a) $4.8 \times 10^{-8} \text{ m} \cdot \text{s}^{-1}$
(b) $4.8 \times 10^{-4} \text{ m} \cdot \text{s}^{-1}$
(c) $1.1 \times 10^{-7} \text{ m} \cdot \text{s}^{-1}$
(d) $2.2 \times 10^{-8} \text{ m} \cdot \text{s}^{-1}$

7–2 Rain falls at a maximum rate of $2 \text{ in} \cdot \text{h}^{-1}$ on a watershed of 77 acres. If the runoff coefficient is 0.42, what is the peak discharge?

(a) $5.43 \text{ ft}^3 \cdot \text{s}^{-1}$
(b) $65.2 \text{ ft}^3 \cdot \text{s}^{-1}$
(c) $90.1 \text{ ft}^3 \cdot \text{s}^{-1}$
(d) $782. \text{ ft}^3 \cdot \text{s}^{-1}$

7–3   A property is 500 m long and 250 m wide, with a 3% slope. The runoff coefficient is 0.35. Rainfall occurs at an intensity of 120 mm · h^{-1}. Determine the peak discharge.

   (a)   0.044 $m^3 \cdot s^{-1}$

   (b)   1.46 $m^3 \cdot s^{-1}$

   (c)   2.70 $m^3 \cdot s^{-1}$

   (d)   14.6 $m^3 \cdot s^{-1}$

7–4   Two monitoring wells were constructed in an unconfined aquifer. The wells are separated by a distance of 300 m. The water surface elevations in the up-gradient and down-gradient wells were 101.00 m and 100.85 m, respectively. The aquifer hydraulic conductivity is 1.5 m · day^{-1}. The porosity of the aquifer material is 0.53. The Darcy velocity (m · day^{-1}) in the aquifer is most nearly:

   (a)   0.06

   (b)   0.0014

   (c)   0.00075

   (d)   0.00014

7–5   An artesian well in a fine sand aquifer is 20.0 m thick. The piezometric surface is 40.0 m above the bottom confining layer. The well is fully penetrating and is being pumped at 0.02 $m^3 \cdot s^{-1}$. The inside diameter of the well is 20 cm. The drawdown at the well is 10 m. Determine the drawdown at a distance 100 m from the well. (Note: you will need to look up the hydraulic conductivity in Table 7-4.)

   (a)   2.09 m

   (b)   10.5 m

   (c)   27.3 m

   (d)   37.9 m

7–6   A 10-cm diameter well fully penetrates an unconfined aquifer that is 40.0 m thick. The drawdown at the well is 15.0 m. The hydraulic conductivity of the aquifer is 5.0×10^{-6} m · s^{-1}. Assume steady flow. What is the allowable pumping rate to ensure that the drawdown at a well that is 100.0 m from the pumping well does not exceed 5.0 m?

   (a)   0.0063 $m^3 \cdot s^{-1}$

   (b)   0.063 $m^3 \cdot s^{-1}$

   (c)   0.16 $m^3 \cdot s^{-1}$

   (d)   1.6 $m^3 \cdot s^{-1}$

REFERENCES

Alley, W. M., T. E. Reilly, and O. L. Franke (1999) *Sustainability of Ground-Water Resources,* U.S. Geological Survey Circular 1186, Denver, CO, http://pubs.usgs.gov/circ/circ1186/html/boxa.html

Barber, N. L., J. F. Kenny, K. S. Linsey, D. S. Lumia, and M. A. Maupin (2004) *Estimated Use of Water in the United States in 2000,* U.S. Geological Survey Circular 1268.

Bedient, P. B., and W. C. Huber (1992) *Hydrology and Floodplain Analysis,* 2nd ed., Addison-Wesley, Reading, MA.

Bertoldi, G. L., R. H. Johnston, and K. D. Evenson (1991) *Groundwater in the Central Valley, California. A Summary Report,* U.S. Geological Survey Professional Paper 1401-A.

Dalton, J. (1802) "Experimental Essays on the Constitution of Mixed Gases; on the Force of Steam or Vapor from Waters and Other Liquids, Both in a Torricellian Vacuum and in Air; on Evaporation; and on the Expansion of Gases by Heat," *Mem. Proc. Manchester Lit. Phil. Soc.,* 5: 535–602.

Darcy, H. (1856) *Les Fontaines Publiques de la Ville de Dijon,* Victor Dalmont, Paris, pp. 570, 590–94.

Davis, M. L., and D. A. Cornwell (1998) *Introduction to Environmental Engineering,* 3rd ed., McGraw-Hill, Inc., New York.

Dupuit, J. (1863) *Etudes Theoriques et Pratiques sur le Mouvement des Earax dans les Canaux Deconverts et a Travers les Terrains Permeables,* Dunod, Paris.

Farnsworth, R. K., E. S. Thompson, and E. L. Peck (1982) *Evaporation Atlas for the Contiguous 48 United States,* Washington, DC, U.S. Department of Commerce, National Oceanic and Atmospheric Administration, National Weather Service, NOAA, Technical Report NWS 33, p. 26.

Fetter, C. W. (1994) *Applied Hydrogeology,* 3rd ed. Prentice Hall, Upper Saddle River, NJ.

Glantz, M. H., and I. S. Zonn (2005) *The Aral Sea: Water, Climate and Environmental Change in the Central Asia,* WMO-No. 982, World Meteorological Organization, Geneva, Switzerland.

Horton, R. E. (1935) *Surface Runoff Phenomena: Part I, Analysis of the Hydrograph,* Horton Hydrologic Lab Pub. 101, Edward Bros., Ann Arbor, MI.

Hutson, S. S., N. L. Barber, J. F. Kenny, K. S. Linsey, S. L. Lumina, and M. A. Maupin (2004) *Estimated Use of Water in the United States in 2000,* U.S. Geological Survey Circular 1268, accessed August 11, 2004, at http://water.usgs.gov/pubs/circ/2004/circ1268

Johnson, A. I. (1967) *Specific Yield—Compilation of Specific Yields for Various Materials,* U.S. Geological Survey Water Supply Paper 1662-D.

Johnson Division UOP (1975) *Ground Water and Wells,* Johnson Division UOP, St. Paul, MN.

Joint Committee of the American Society of Civil Engineers and the Water Pollution Control Federation (1969) *Design and Construction of Sanitary and Storm Sewers* (ASCE Manuals and Reports on Engineering Practice No 37), American Society of Civil Engineers, New York, p. 51.

Kenny, J. F., Barber, N. L., Hutson, S. S., Linsey, K. S., Lovelace, J. K., and Maupin, M. A., (2009) *Estimated Use of Water in the United States in 2005,* U.S. Geological Survey Circular 1344, 52 p.

Leake, S. A. (2004) *Land Subsidence from Groundwater Pumping,* U.S. Geological Survey, http://geochange.er.usgs.gov/sw/changes/anthropogenic/subside/

Lide, D. R., editor-in-chief (1995) *CRC Handbook of Chemistry and Physics,* 76th ed. CRC Press, New York, pp. 6–15.

Mather, J. R. (1984) *Water Resources,* V. H. Winston and Sons, Silver Springs, MD.

Maupin, M. A., J. F. Kenny, S. S. Hutson, J. K. Lovelace, N. L. Barber, and K. S. Linsey (2014) *Estimated Use of Water in the United States in 2010,* U.S. Geological Survey Circular 1405, 56 p., https://pubs.usgs.gov/circ/1405/.)

Montgomery, C. (2003) *Environmental Geology,* 6th ed. McGraw-Hill, New York.

National Academies of Science (1991) *Mitigating Losses from Land Subsidence in the United States,* Panel on Land Subsidence, Committee on Ground Failure Hazards Mitigation Research, Division of Natural Hazard Mitigation, National Research Council, National Academy Press, Washington, DC.

NOAA National Weather Service (2007) "2006 Climate for LaCrosse, WI," LaCrosse, WI Weather Forecast Office, LaCrosse, WI 54601, http://www.crh.noaa.gov/arx/?n=lse2006

Spatial Climate Analyis Service (2000) "Average Annual Precipitation: Washington," Oregon State University, http://www.ocs.orst.edu/page_links/comparative_climate/washington/wa.gif

South Carolina State Climatology Office (2007) "Pan Evaporation Records for the South Carolina Area," http://water.dnr.state.sc.us/climate/sco/pan_tables.html

Southeast Regional Climate Center (2000) "South Carolina Normal Annual Precipitation," in *South Carolina Climate Atlas,* http://www.dnr.sc.gov/climate/sco/products/maps/climate_atlas/scatlasannprecip.gif

Theim, G. (1906) *Hydrologische Methoden,* J. M. Gebbart, Leipzig, Germany.

Todd, D. S. (1980) *Groundwater Hydrology,* 2nd ed. John Wiley and Sons, New York.

UNEP/GRID-Arendal (2000) *Water Resources: State of Environment of the Aral Sea Basin,* Sept. 2000, downloaded 31 Jan. 2007, http://enrin.grida.no/aral/aralsea/english/water/water.htm

U.S. Environmental Protection Agency (1973) *Manual of Individual Water Supply Systems,* Publication No. EPA-430-9-73-003, U.S. Government Printing Office, Washington, DC.

U.S. Environmental Protection Agency (2006) *Factoids: Drinking Water and Groundwater Statistics for 2005,* EPA 816-K-03-001, U.S. EPA Office of Water, December 2006, www.epa.gov/safewater/data

United Nations Educational, Scientific and Cultural Organization (2004) *Interdisciplinary Solutions in the Aral Sea,* 3 Sept. 2004, downloaded 31 Jan. 2007, http://portal.unesco.org/shs/en/ev.php-URL_ID=4952&URL_DO=DO_TOPIC&URL_SECTION=201.html

United Nations, Food and Agriculture Organization (1998) "Time to Save the Aral Sea?" *Agriculture 21,* vol. 9, http://www.fao.org/ag/magazine/9809/spot2.htm

USGS (1975) Continuous Record Streamflow Data for Station 10265200 Convict Creek NR; Mammoth Lake, CA.

Wahl, K. L., W. O. Thomas, Jr., and R. M. Hirsch (1995) *Stream-Gaging Program of the U.S. Geological Survey,* U.S. Geological Survey Circular 1123, Reston, Virginia.

Western Regional Climate Center (2007) "Monthly Total Precipitation for Mojave, CA (Data Station 045756)," from National Climatic Data Center Database, www.wrcc.dri.edu/cgi-bin/cliMONtpre .pl?camoja

8

Sustainability

Case Study

A New Precious Metal—Copper!

Present estimates of the natural reserves of copper in North America amount to 113 Tg.* Of this amount, only about 50% can be feasibly extracted. Mass balance modeling studies (Figure 8–1) show that over the last century (1900–1999), 40 Tg of copper were collected and recycled from postconsumer waste, 56 Tg accumulated in landfills or were lost through dissipation, and 29 Tg were deposited in tailings, slag, and waste reservoirs. The residence time model shows a significant rise in the rate at which waste is placed in landfills from postconsumer waste. In 1940, postconsumer copper was placed in landfills at a rate of 270 gigagrams** of copper per year (Gg of Cu · y^{-1}). By 1999, the rate had risen to 2790 Gg of Cu · y^{-1}. In the absence of an efficient collection and processing infrastructure for retired electronics, this rate of landfilling is expected to rise because of the increasing rate of electronic equipment use and corresponding shorter residence times for electronic equipment (Spatari et al., 2005).

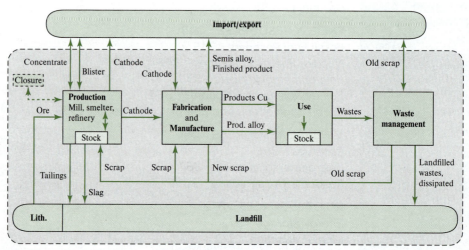

System Boundary "STAF North America"

FIGURE 8–1 Framework for analysis of copper flows within North America.

The United States rate of use of copper is not sustainable based on North American natural reserves. The rise in the price of copper from $2000 per Mg*** to over $8000 per Mg, the highest price since the seventies (that is the 1870s!), signals that the market has already begun to respond to unsustainable use of this natural resource (Freemantle, 2006).

---

*Tg = teragrams, 1×10^{12} g

**Gg = gigagrams, 1×10^9 g

***Mg = megagram, 1×10^6 g

8–1 INTRODUCTION

Sustainability

In the simplest dictionary style definition, sustainability is a method of harvesting or using a resource so that the resource is not depleted or permanently damaged. Beyond this simple definition, there are as many definitions as there are authors that write on the subject. This results from the many perspectives the authors bring to the subject. Here are several perspectives from which one might view sustainability: developing countries, developed countries, ecological, economic, social justice, worldwide, regional, national, local.

The conventional starting point for discussions on sustainability is that published by the World Commission on Environment and Development* (WCED, 1987): *Sustainable development is development that meets the needs of the present without compromising the ability of future generations to meet their own needs.* For our starting point, we prefer to define sustainability in terms of a sustainable economy. A *sustainable economy* is one that produces wealth and provides jobs for many human generations without degrading the environment. There are two fundamental principles of this definition of sustainability:

- Reduction in the use of both renewable and nonrenewable natural resources.
- Provision of solutions that are both long-term and market-based.

In the first instance, emphasis is placed on reduction of natural resources rather than end-of-pipe solutions. In a sustainable economy, development focuses on minimizing resource consumption through increased efficiency, reuse and recycling, and substitution of renewable resources for nonrenewable resources. *Renewable resources* are those that can be replaced within a few human generations. Some examples are timber, surface water, and alternative sources of power such as solar and wind. *Nonrenewable resources* are those that are replaceable only in geologic time scales. Groundwater, fossil fuels (coal, natural gas, and oil) and metal ores are examples of nonrenewable resources.

In the second instance, an effective and cost-efficient approach is for society to provide incentives to use alternatives or reduce the use of water, coal, gasoline, and other substances. Some of these incentives are in the form of technological advances that improve efficiency and some are in the form of sociopolitical changes.

The People Problem

An inherent problem with definitions of sustainability is "The People Problem," that is, the ability of future generations to meet their own needs. To understand the "The People Problem," we need to understand the characteristics of population growth that were discussed in Chapter 5.

A simple way of viewing population growth is by assuming that growth is exponential. Equation 5–17, repeated here for convenience, is basis for discussion:

$$P_t = P_o e^{rt} \tag{8–1}$$

where P_t = population at time t
$\quad P_o$ = population at time $t = 0$
$\quad\quad r$ = rate of growth, individuals/individual · year
$\quad\quad t$ = time

The growth rate is a function of the crude birth rate (b), crude death rate (d), immigration rate (i), and emigration rate (m):

$$r = b - d + i - m \tag{8–2}$$

Example 8–1 provides an illustration of a crude estimate of population growth.

---

*It is also known as the Brundtland Commission Report after its chairman.

EXAMPLE 8–1 Estimate the percent growth of the global population from 2014 to 2050 using the following assumptions: crude birth rate ~23.4 per 1000 people, crude death rate ~7.8 per 1000 people, population ~7,483,000,000 or 7.483×10^9 (WHO, 2014).

Solution Begin by estimating the rate of growth

$$k = \frac{23.4}{1000} - \frac{7.8}{1000} = 0.0234 - 0.0078 = 0.0156$$

The time interval is 2050 to 2014 = 36 years and the population in 2050 is estimated to be

$$P_t = 7.483 \times 10^9 \{\exp(0.0156 \times 36)\}$$
$$= 1.31 \times 10^{10} \text{ people}$$

This is an increase of

$$\left(\frac{1.31 \times 10^{10} - 7.483 \times 10^9}{7.483 \times 10^9}\right)(100\%) = 75.1 \text{ or about } 75\%$$

Comment: Historically, economic and social development has resulted in decreases in the rate of growth. For example, China's rate of growth in 1964 was about 31 per 1000 people per year. In 1990, it was about 12 per 1000 people per year. In 2015, it was about 5.2 per 1000 people per year. As reference points, China's per capita income in 2000 was $950. Current (2016) growth projections estimate per capita income at $8,000 by 2015.

For comparison, the rate of growth in the United States was about 7.5 per 1000 people in 2015; in the United Kingdom and France it was 4.5 per 1000 people in 2015 (United Nations Statistics Division, 2016).

Projections of population using Equation 8–1 are quite crude. More sophisticated analyses take into account the *total fertility rate* (TFR). The TFR is the average number of children that would be born alive to a woman, assuming that current age-specific birth rates remain constant through the women's reproductive years. Three scenarios of global population growth for different TFRs are shown in Figure 5–17 (Haupt and Kane, 1985).

Recent studies (Bremner et al., 2010) reveal that the world's population has reached a transition point. The rapid growth of the second half of the 20th century has slowed. However, factors such as improving mortality and slower than expected declines in birth rate guarantee continued growth. Current projections of world population in 2050 range from 7.5 billion to 10.6 billion (United Nations, 2012). This places the population growth on the "medium fertility rate" curve in Figure 5–17 in Chapter 5. This is a 42% global population increase rather than the 75% increase calculated in Example 8–1.

Given limited resources, both nonrenewable and renewable, there are long-term (>100 years) implications for population growth that cannot be resolved by technological solutions (for example, food for starving people). We make this statement to place the remainder of the discussion in this chapter in context. We will focus on currently available technological improvements.

There Are No Living Dinosaurs

Over the millennia, the climate of the Earth has changed. Natural processes such as variation in solar output, meteorite impacts, and volcanic eruptions cause climate change. These natural phenomena have resulted in ecosystem changes. Animal and plant species have evolved and adapted or have become extinct . . . there are no living dinosaurs. The rate of change has been a major factor in the ability of organisms to adapt successfully.

We inherently refer to the atmosphere when we speak of the climate. But there are also strong interactions with seas and oceans in terms of fluxes of energy, water, and carbon dioxide. While the atmosphere has only a small buffer capacity to resist change, the oceans provide a gigantic buffer capacity for changes in heat, water, and carbon dioxide. Thus, there is a considerable time lag between the causes of climate change and their effects. Because natural changes in climate occur relatively slowly, so also do their effects. Similarly, the effects of human impacts will take some time to be felt on a human time scale.

In this millennium, we are faced with the potential of significant global warming over a geologically short time period for the reasons discussed in Chapter 12. As noted in that chapter, the impacts of global warming on North America are forecast to be a mix of "good news" and "bad news." Exactly how these impacts will be addressed is a formidable challenge.

Vulnerability is a term used to assess the impacts of climate change in general and global warming in particular. To be vulnerable to the effects of global warming, human-environmental systems must not only be exposed to and sensitive to the changes but also unable to cope with the changes (Polsky and Cash, 2005). Conversely, human-environmental systems are **relatively sustainable** if they possess strong adaptive capacity and can employ it effectively. In the United States we have abundant cropland, sufficient natural resources, and a robust economy that gives us a strong adaptive capacity and, hence, a capacity for a relatively sustainable human-environmental system. Whether or not we can employ it effectively is another question. Many other countries do not have these key ingredients and, therefore, do not have a strong adaptive capacity . . . they are vulnerable.

We use the term "relatively sustainable" because there are limits to growth using 20th century economic models based on exploitation of nonrenewable resources. As we will see in later sections of this chapter, the time frame for adaptation on a global basis is on the order of a few generations or less for energy and minerals at current growth rates in consumption. Certain countries, the United States among them, are more richly blessed and have more time. But the time is not infinite.

Go Green

Green engineering is the design, commercialization, and use of processes and products that are feasible and economical while (U.S. EPA, 2010):

- Reducing the generation of pollution at the source.
- Minimizing risk to human health and the environment.

These are not new concepts. In the decades of the 1980s and 1990s numerous publications were devoted to the concept of reducing pollution generation (A&WMA 1988; Freeman, 1990; Freeman, 1995; Higgins, 1989; Nemerow, 1995). What is "new" about green engineering is that it now has a catchy name **and** branches of civil engineering other than environmental engineering as well as other engineering disciplines have taken up the standard.

The concepts of green engineering are, in fact, a demonstration of adaptability to improve sustainability. Many of the changes that we have historically recognized as improvements in efficiency are steps in increasing sustainability. To be sure there are new initiatives to take advantage of the catchword but they all contribute to sustainability . . . and they embody the two principles of sustainability:

- Reducing society's use of natural resources.
- Using market-based solutions.

Of particular note for civil and environmental engineers is the emergence of whole building assessment systems like BRE Environmental Assessment Method (called BREEAM and used in the United Kingdom), Green Globes (used in Canada and the United States), and Leadership in Energy and Environmental Design (called LEED and used in the United States). These programs place considerable emphasis on the selection of green materials or products as an important aspect of sustainability.

In addition to these new assessment systems, an old system called life cycle assessment (LCA) has taken on a new perspective. Traditionally, LCA focused on the costs of building, operating, and closing a facility as a method of comparison of alternatives. The new LCA approach is a methodology for assessing the environmental performance of a product over its full life cycle (Trusty, 2009).

In the following sections of this chapter we will examine the parameters of sustainability for a renewable resource—water, and two nonrenewable resources—energy and minerals. In each of these cases we will give examples of green engineering to demonstrate the contribution of technology to sustainability. Obviously, these are only a sample of the current possibilities.

8-2 WATER RESOURCES

Water, Water, Everywhere

"Water: too much, too little, too dirty" (Loucks et al., 1981). This sentence succinctly summarizes the issues of sustainable water resource management. The discussion of risk and hydrology in Chapters 6 and 7 form a basis for this discussion. A large portion of this text is devoted to environmental engineering measures to prevent water from becoming "too dirty" and cleaning water that is "too dirty." These will be discussed in the context of green engineering examples.

Frequency from Probability Analysis

The relative frequency of an event such as a coin toss is a probability. A rainfall of a given intensity for a given duration is such an **event.** The probability of a single rainfall event, say E_1, is defined as the relative number of occurrences of the event in a long period of record of rainfall events. Thus, $P(E_1)$, the probability of rainfall event E_1, is n_1/N for n_1 occurrences of the same event in a record of N events if N is sufficiently large. The number of occurrences of n_1 is the frequency, and n_1/N is the relative frequency. More formally, we say that the probability of a single event, E_1, is defined as the relative number of occurrences of the event after a long series of trials. Each outcome has a finite probability, and the sum of all the possible outcomes is 1. The outcomes are mutually exclusive. The relative frequency of an event is used as a descriptor of the risk of an activity.

$P(E_1) = 0.10$ implies a 10% chance each year that a rainfall event will "occur." Because the probability of any single, exact value of a continuous variable is zero, "occur" also means the rainfall event will be reached or exceeded. That is, the probability of rainfall event E_1 being exceeded is

$$P(E_1) = \frac{1}{10}$$

In a sufficiently long run of data, the rainfall event would be equaled or exceeded on the average once in 10 years. Hydrologists and engineers often use the reciprocal of annual average probability because it has some temporal significance. The reciprocal is called the average **return period** or average **recurrence interval** (T):

$$T = \frac{1}{\text{Annual average probability}} \tag{8-3}$$

Thus, return period, whether for rainfall events, floods, or droughts, is used as a convenient way to explain the "risk" of a hydrologic event to the public. Unfortunately, it also has led to serious misunderstanding by both the public and, in some cases, design engineers. For emphasis, we remind you of the following:

- A storm (or flood or drought) with a 20-year return period that occurred last year may occur next year (or it may not).
- A storm (or flood or drought) with a 20-year return period that occurred last year may not occur for another 100 years or more.
- A storm (or flood or drought) with a 20-year return period may, on the average, occur 5 times in 100 years.

Using the return period definition for T, the following general probability relations hold where E is the event (storm, flood, drought):

1. The probability that E will be equaled or exceeded in any year is

$$P(E) = \frac{1}{T} \tag{8–4}$$

2. The probability that E will not be exceeded in any year is

$$P(\bar{E}) = 1 - P(E) = 1 - \frac{1}{T} \tag{8–5}$$

3. The probability that E will not be equaled or exceeded in any of n successive years is

$$P(\bar{E})^n = \left(1 - \frac{1}{T}\right)^n \tag{8–6}$$

4. The **risk** (R) that E will be equaled or exceeded at least once in n successive years is

$$R = 1 - \left(1 - \frac{1}{T}\right)^n \tag{8–7}$$

Floods

There are two broad categories of floods: coastal and inland. Inland floods, also called *inundation floods*, as the name implies are the result of a combination of meteorological events that result in inundation of a floodplain. These may be seasonal, such as the historic flooding of the Nile and the rhythmic monsoons that flood great portions of India, or they may be highly irregular with long return periods. Coastal flooding most frequently results from exposure to cyclones or other intense storms. Prime examples are the Bay of Bengal and the Queensland coast of Australia exposed to cyclones, the Gulf and Atlantic coasts of the United States exposed to hurricanes and the coasts of China and Japan affected by typhoons.

A special case of coastal flooding is that due to tsunamis. Earthquakes are a major cause of tsunamis. They most frequently occur along the Pacific Rim because of crustal instability. Wave heights range from barely noticeable to over 10 m with run-ups from less than 5 m to over 500 m.

Floods by themselves are not disasters. Floods occurred long before civilizations arose and will continue long after they have disappeared. Inland floods bring nutrients and contribute to alterations in the river channel that provide nursery habitat for young fish, destroy existing riparian communities, and create new environments for new ecosystems. Both field data and model studies indicate that the most complex and diverse ecosystems are maintained only in riparian environments that fluctuate because of flooding (Power et al., 1995).

Estuaries and natural coastal areas are characterized by many different habitats ranging from sandy beaches to salt marshes, mud flats, and tidal pools that are inhabited by an extraordinary variety of animals and plants. In a similar fashion to inland floods, coastal floods alter the landscape to create new environments for new ecosystems.

Around 3000 B.C.E., the world's first agriculturally based urban civilizations appeared in Egypt and Mesopotamia and maize farming began in Central America. As populations grew people moved to the floodplains to tap the natural resources of water and fish. Over the millennia these urban civilizations matured and grew such that today half of the world's 6.9 billion people live in an urban environment. There are between 18 and 25 major metropolitan cities with a population over 10 million (often cited as *megacities*). Most of these are located near water bodies that, at some point in time, will bring catastrophic floods.

Although only about 7% of the United States' total land area lies in floodplains, more than 20,800 communities are located in flood-prone areas (Hays, 1981). By the mid-1990s 12% of the population occupied more than seven million structures in areas of periodic inundation (Grundfest, 2000). About 50% of the population lives near the coast (Smith and Ward, 1998).

For every city like Chicago or Detroit that are relatively isolated from inland or coastal flooding, there are cities like St. Louis, Los Angeles, and Miami that will probably experience a major flood before the end of this century. The flooding of New Orleans (hurricane Katrina in 2005) and New York (storm surge Sandy in 2012) are harbingers of more to come.

As illustrated in Example 8–2, the probability is not small.

EXAMPLE 8–2 Estimate the risk of a 100-year return period event occurring by the year 2100 if the current year is 2010.

Solution Using Equation 8–6 with $T = 100$ years and $n = 2100 - 2010 = 90$

$$R = 1 - \left(1 - \frac{1}{T}\right)^n$$

$$= 1 - \left(1 - \frac{1}{100}\right)^{90} = 1 - 0.40 = 0.60$$

The estimate is that there is a 60% risk of a 100-year return period event being equaled or exceeded in the next 90 years.

In the United States inland floods tend to be repetitive. From 1972 to 1979, 1900 communities were declared disaster areas by the federal government more than once, 351 were inundated at least three times, 46 at least four times, and four at least five times. Between 1900 and 1980 the coast of Florida experienced 50 major hurricanes, and even as far north as Maryland there is an average of one hurricane per year that has direct or fringe effects on the coast (Smith and Ward, 1998).

Floods and Climate Change. The amount of precipitation falling in the heaviest 1% of the rain events in the United States increased 20% in the past 50 years with eastern events increasing by greater than 60% and western events increasing by 9% (Karl et al., 2009). During this time period, the greatest increases in heavy precipitation have occurred in the Northeast and Midwest. Using the middle 50% of the values from Global Circulation Models (GCMs), annual precipitation forecasts have been made for the time period 2080–2099. They are summarized as follows:

- Western United States: 0 to 9% increase.
- Central United States: precipitation projections cannot be distinguished from natural variability.
- Eastern United States: increase by 5 to 10%.

FIGURE 8–2

Increases in the
amounts of very heavy
precipitation from
1958–2007.
(*Source:* Karl, T. R., J.
M. Melillo, and T. C.
Peterson (2009) Global
Climate Change Impacts
in the United States, U.S.
Global Climate Change
Research Program,
Cambridge University
Press, Cambridge, U.K.)

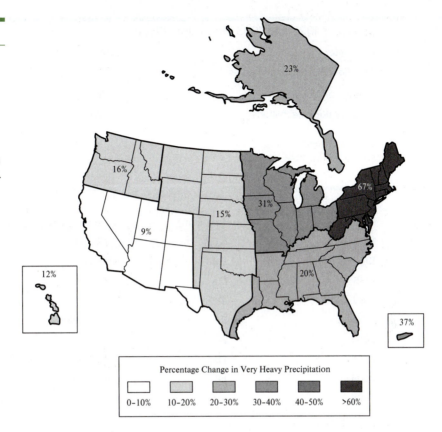

These data (Figure 8–2) and projections imply that higher intensity rainfall is to be expected. With higher intensities comes high river flow rates and the potential for more floods.

As a result of melting of glaciers, global warming models predict an increase in sea level between 0.2 and 0.6 m by 2100 (IPCC, 2007). The sources of sea level rise include thermal expansion, melting of glaciers and ice caps, melting of the Antarctic ice sheet and the Greenland ice sheet. This will result in the loss of about 30% of global coast wetlands and an increase in coastal flooding.

Floods and Sustainability.   Vulnerability, in terms of flood planning, is a measure of the ability to anticipate, cope with, resist, and recover from harm caused by a flood hazard (Blaikie et al., 1994). As we noted in Section 8–1, human-environmental systems are relatively sustainable if they possess strong adaptive capacity and can employ it effectively. The Mississippi River floods regularly (notably so in 1927, 1937, 1947, 1965, 1993, 2008, and 2010). Florida has had hurricane flood surges in the range of 2 to 5 meters on at least five occasions (1935, 1960, 1992, and twice in 2004). The fact that the environs of the Mississippi and Florida remain inhabited and prospering is evidence of sustainability and the adaptability of the people. As shown in Tables 8–1 and 8–2 this has not been without cost.

For a country like Bangladesh, the question of sustainability is moot. Because Bangladesh is a country of rivers, 66% of the country's 144,000 km^2 is either a floodplain or a delta. About 75% of the runoff from the Himalayan mountains drains through the country from July to October. The mean annual rainfall ranges from 1500 mm in the west to over 3000 mm in the southeast. Seventy-five to eighty percent of the rainfall comes from monsoons during the period from June to October. When the Brahmaputra, Ganges, and Meghna rivers peak in August or September approximately one-third of the country floods. Over one-third of the flooded area is under at least one meter of water (Whol, 2000).

TABLE 8-1

Impact of 1927 and 1993 Mississippi Floods

Parameter	1927 Flood	1993 Flood
Area flooded, millions of acres	12.8	20.1
Number of fatalities	246	52
Property damage, billions of dollars[a]	21.6	22.3

[a]In 2012 dollars.
Source: Wright, 1996.

TABLE 8-2

Florida Hurricane Impacts[a]

Year	Surge Height (m)	Economic Loss	Fatalities
1926	4.6	$90 billion	373
1928	2.8	$25 million	1836
1935	n/a	$6 million	408
1960	4	$387 million	50
1992	5	$25 billion	23
2004	2	$16 billion	10
2004	1.8	$8.9 billion	7
2004	1.8	$6.9 billion	3
2005	n/a	$16.8 billion	5

[a]*Source:* NOAA, 2010.
n/a = not available

TABLE 8-3

Bangladesh Cyclone Fatalities

Year	Fatalities
1822	40,000
1876	100,000
1897	175,000
1963	11,468
1965	19,279
1970	300,000
1985	11,000
1991	140,000

Source: Smith and Ward, 1998.

In addition to the inland inundation, Bangladesh suffers recurring lethal coastal storm surges. The surge effects are accentuated by a large astronomical tide, the seabed bathymetry, with shallow water extending to more than 300 km offshore in the northern part of the bay, and the coastal configuration that accentuates the funnel shape of the bay by a right-angle change in the coastline. This produces maximum storm-surge levels that are higher than would be produced on a straight coastline (Smith and Ward, 1998). The human devastation produced by the monsoons is summarized in Table 8-3.

TABLE 8–4	Measures of Adaptive Capacity for Bangladesh and the United States[a]		
	Parameter	Bangladesh	United States
	Population, million	164	310
	Growth rate, %	1.5	0.6
	Net migration, %	−1	+3
	TFR	2.4	2.0
	Infant mortality, deaths/1000	45	1.4
	Population density, people/km^2	1142	32
	GNI,[b] US$/capita	1440	46,970

[a]Source: PRB, 2010.
[b]Gross National Income, 2008.

The economic devastation is equally bad. In 1988–89, nearly half of the nation's development budget was spent to repair flood damage. In 1991, the storm surge caused damage estimated at one-tenth of the nation's gross domestic product.

The comparison of some fundamental parameters in Table 8–4 provides a basis for the very positive outlook for the ability of the United States to adapt in comparison to the daunting prospects for the Bengalis. Although Bangladesh's inundation brings desperately needed nutrients for the soil and replenishes ponds for fish and shrimp spawning, floods bring displacement. Historically, displaced households have moved to new lands. In a country with the world's highest population density, there is precious little "new" land. There is a collective fatalism to this cycle. The Bengalis cope astonishingly well with the floods they've always known. From a long-term perspective it does not appear that Bangladesh has the resources to develop a *sustainable economy* that will produce wealth and provide jobs for many human generations without degrading the environment. They need outside help. Using funds from the National Science Foundation and the Georgia Tech Foundation to supplement sporadic funding from USAID (whose priorities changed from the Clinton to the Bush administration), Webster et al. (2010) were able to develop an exploratory project of 10-day flood forecasts to provide advance warning and allow implementation of emergency response plans. The forecasts and responses were successful for the 2009 flood season. The savings resulting from the forecast was estimated to be US$130–190 for fishery and agriculture income, US$500 per animal, and US$270 per household. Given that the average farmer's income is approximately US$470, the savings in the flooded regions was substantial. To continue this success a better funding base and increased collaboration between federal and international agencies is required.

Floods and Green Engineering.   In the last several decades it has become apparent that intensively engineered, densely populated, biologically impoverished river corridors and coastal areas are not a sustainable response to flood hazards. Unfortunately, most of the world's cultures have a tradition of aggressively interfering and altering natural systems and processes. We have selected two examples of alternatives that have been implemented in the United States. You will note that engineered structures play a minor role in sustainable flood protection. The "green engineering" component is recognizing that traditional structural solutions were bound to fail and that other solutions were required to minimize risk and to provide a *sustainable economy* for many human generations.

Flood-Warning Systems.   Technological advances have made warning systems available to more than 1000 communities in the United States. Coupled with advances in meteorological forecasting, these systems are lifesavers. The reduction in flood and hurricane fatalities can, in part, be directly related to implementation of these systems (see, for example, Tables 8–1 and 8–2). The lack of a warning system, as for example in Bangladesh, inevitably leads to great loss of life (Table 8–3). However, without a plan for disseminating the warning message and implementing an evacuation plan, the warning system is useless. The 1835 fatalities in New Orleans that resulted from Katrina hurricane in 2005 may be attributed to the lack of provision for the poor and handicapped that had no transportation before, during, and after the event as well as remote, safe facilities to accommodate them.

Given the difficulty in evacuating New Orleans, a metropolitan area with a population of 1,235,650 people, the prospects for successful evacuation of even a portion of one of the world's megacities in the event of a forecast 100- or 500-year storm or flood appear slim.

Of course, a warning system is only necessary after poor land use decisions have been made. The fact that the material resources and habitat are destroyed in a flood diminishes the potential for a sustainable economy that produces wealth and provides jobs.

Acquisition and Relocation. One of the most successful, sustainable, long-term solutions to repeated flooding is to move people and structures out of the floodplain. Almost 40 years ago the U.S. federal government authorized a cost sharing program for relocation. One of the most dramatic uses of these funds was for the relocation of the village of Valmeyer, Illinois, from the Mississippi floodplain to a bluff 120 m above the river and two miles away. The village had been flooded in 1910, 1943, 1944, and 1947. Although the U.S. Army Corps of Engineers raised the levee to 14 m after the 1947 flood, it was topped by the 1993 flood that destroyed 90% of the buildings. Relocation and rebuilding began almost immediately. The first business opened in May of 1994 and the first home was occupied in April, 1995. The town's population at the time of the 1993 flood was 900. In 2009, the population was 1168.

Other examples of buyouts and relocation include Hopkinsville, Kentucky, Bismarck, North Dakota, Montgomery, Alabama, and Birmingham, Alabama. Since the 1993 Mississippi flood, nearly 20,000 properties in 36 states have been bought out (Gruntfest, 2000).

From a practical point of view, it is unlikely that more than a few of the 20,800 U.S. communities that are flooded regularly are able or willing to move. However, it is frequently the case that only a portion of the town is flooded. This opens the opportunity for sections of communities to be relocated. The floodway can then become a resource for replenishment of the river ecosystem. In the last few decades, this type of activity has been a means of revitalization of downtown business districts. Predisaster mitigation by acquisition and relocation has resulted in benefit cost ratios between 1.67 and 2.91 (Grimm, 1998). Given the immense cost of floods (Tables 8–1 and 8–2), this seems like an excellent market-based approach.

Droughts

Drought means something different for a climatologist, an agriculturalist, a hydrologist, a public water supply management official, and a wildlife biologist. There are more than 150 published definitions of drought. The literature of the 21st century appears to have adopted the classification by Wilhite and Glantz (1985). They address these different perspectives by classifying drought as a sequence of definitions related to the drought process and disciplinary perspectives: meteorological, agricultural, hydrological, and socioeconomical/political drought. Bruins (2000) adds another useful subdivision: pastoral drought.

In summary the definitions are as follows:

- *Meteorological (climatological) drought:* A deficiency of precipitation for an extended period of time that results from persistent large-scale disruptions in the dynamic processes of the earth's atmosphere.

- *Agricultural drought:* A dry period where *arable* land that can be cultivated does not receive enough precipitation. This may occur even if the total annual precipitation reaches the average amount or above because the precipitation does not occur when the crop needs it. An example is in an early growth stage when seeds are germinating.

- *Pastoral drought:* Land that is *nonarable* but is suitable for grazing may not receive enough moisture to keep natural, indigenous vegetation alive.

- *Hydrological drought:* This type is reflected in reduced stream flow, reservoir drawdown, dry lake beds, and lowering of the piezometric surface in wells.

- *Socioeconomical/political drought:* This type of drought occurs when the government or the private sector creates a demand for more water than is normally available. This may result from attempts to convert nonarable pasture to cultivation or population growth that outstrips the water supply.

The relationship between these various types of droughts, the duration of drought events, and the resultant socioeconomic impacts is shown schematically in Figure 8–3.

The temporal sequence in drought recovery begins when precipitation returns to normal and meteorological drought conditions have abated. Soil water reserves are replenished first, followed by increases in streamflow, reservoir and lake levels, followed by groundwater recovery.

FIGURE 8–3

Relationship between various types of droughts, drought impacts, and drought duration. (*Source:* Davis, M.L. and Cornwell, D.A., *Introduction to Environmental Engineering,* 5e, p.968, 2013. The McGraw-Hill Companies, Inc.)

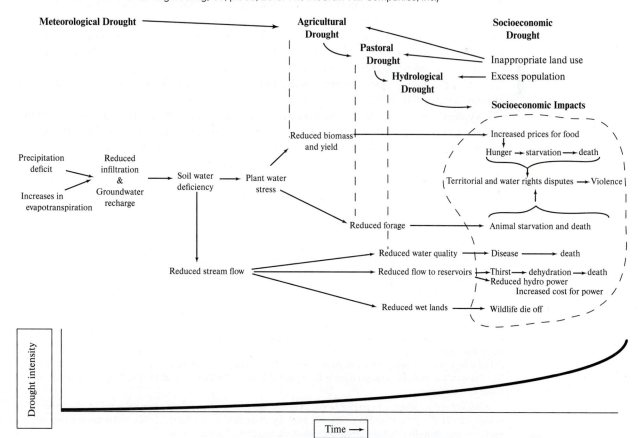

TABLE 8–5

UNESCO Definitions for Bioclimatic Aridity

Climatic Zone	P/ET
Hyperarid	<0.03
Arid	0.03–0.20
Semiarid	0.20–0.50
Subhumid	0.50–0.75

Source: UNESCO, 1979.

Drought impacts may diminish rapidly in the agricultural sector because of its reliance on soil water, but linger for months or years in other sectors dependent on stored surface or subsurface supplies. Groundwater users, often last to be affected by drought onset, may be last to return to normal conditions (Viau and Vogt, 2000).

Drought is a normal part of climate, rather than a departure from normal climate. Drought by itself is not a disaster. Whether it becomes a disaster depends on its impact on local people and the environment. The Nile River which has been the source of life for Egypt from its earliest inhabitation is a source of long-term data that affirms this fact. The annual inundations brought moisture and nutrients to the soil. Low floods meant hardship because there was insufficient soil moisture and nutrients for crops. Hydrological drought in the Nile River has been documented for the period from 3000 B.C.E. to modern times. For the early Dynastic period and the Old Kingdom (circa 3000–2125 B.C.E.) there are 63 annual flood records from eleven different rulers. During this period there was a decrease of the mean discharge of about 30% (Bell, 1970; Butzer, 1976). It has been posited* that between 2250 and perhaps 1950 B.C.E. there was a series of catastrophic low floods (i.e., droughts) that lead to the demise of the Old Kingdom and the start of the First Intermediate period (Bell, 1971; Butzer, 1976). Again in the late Ramessid period (circa 1170–1100 B.C.E.) there is economic evidence[†] of catastrophic failures of the annual flood (Butzer, 1976). Between 622 and 999 B.C.E. there were 102 years of poor floods (Bell, 1975).

Typically, we in the United States often associate drought with arid, semiarid, subhumid regions. In reality, drought occurs in most nations, in both dry and humid regions, and often on a yearly basis. As a basis for planning and management the United Nations Educational, Scientific, and Cultural Organization (UNESCO) has developed a numerical representation of aridity based on the ratio of average precipitation (P) to average annual potential evapotranspiration (ET). This classification system is summarized in Table 8–5.

How Dry Is Dry?   Drought differs from other natural hazards. It is insidious. It takes months or, in some cases, years before it is recognized. This is because of natural aridity and the time it takes for the lack of precipitation to be manifest in agricultural drought, hydrological drought, or socioeconomic drought. Likewise, it is difficult to determine when a drought has ended.

---

*Because there are no meteorological records of precipitation in ancient Egypt (and even if there were records they would invariably show little rain), the droughts were inferred using the definition of socioeconomic/political drought: lack of food implies a hydrological drought which, in turn, implies an agricultural drought. Because the Nile floods are a result of monsoon rains in Ethiopia and Uganda, the meteorological drought had to occur there rather than in Egypt.

[†]The price of emmer wheat with respect to metals rose to between eight and twenty-four times the standard price of earlier times. Prices recovered to predrought levels by about 1070 B.C.E. (Butzer, 1976). Again this is a case of socioeconomic drought reflecting preceding hydrologic and agricultural droughts.

TABLE 8–6	Palmer Drought Severity Index Categories	
	Moisture Category	**PDSI**
	Extreme drought	≤ −4.00
	Severe drought	−3.00 to −3.99
	Moderate drought	−2.00 to −2.99
	Mild drought	−1.00 to −1.99
	Incipient drought	−0.50 to −0.99
	Near normal	+0.49 to −0.49
	Incipient wet	+0.50 to +0.99
	Slightly wet	+1.00 to +1.99
	Moderately wet	+2.00 to +2.99
	Very wet	+3.00 to +3.99
	Extremely wet	≥ +4.00

Source: Based on Palmer, 1965.

Looking back, we have a better understanding of both the beginning and the end. One method of evaluation of a historic drought is the *Palmer Drought Severity Index* (PDSI). The PDSI is a single number representing precipitation, potential evapotranspiration, soil moisture, recharge, and runoff (Palmer, 1965). Implicitly, they are a measure of "agricultural drought."

Numerous authors have criticized the PDSI for a wide variety of reasons. Fundamentally, it is not an operational tool because the drought must end before it can be calculated. Moreover, it requires a substantial amount of sophisticated measurements and calculations to develop the index number. Nonetheless, it has been used extensively to evaluate droughts in the United States. Even here it has limitations because Palmer used data from Iowa and Kansas that are not representative of the western states where snow accumulation and snowmelt are an important component of the water balance. Because of its wide use in the United States and the extensive use of PDSI in comparative data with *dendrochronological* (tree-ring) reconstructions of historical records, we present it in summary form in Table 8–6.

From this table we note that a PDSI less than −0.99 indicates drought.

Historical Reference Droughts.   For a simple reference frame, Fye et al. (2003) classify droughts in the United States into two broad categories: "Dust bowl-like droughts" and "1950s-like droughts." The "Dust bowl-like droughts" were put in this category because of the similarity of their spatial footprint with the 1930s drought. However, the 1930s drought was by far the worst drought over the last 500 years based on regional coverage, intensity, and duration. The 11-year drought from 1946 through 1956 was the second worst drought to impact the United States during the 20th century. Based on PDSI calculations for the 20th century and reconstructions of large-scale decadal tree ring reconstructions of summer PDSI, Fye et al. (2003) developed the maps reproduced in Figures 8–4 and 8–5. Of special note is that these droughts lasted between 6 and 14 years and that shorter drought periods were not considered in the research. Earlier work using dendrochronology identified longer drought periods of severe drought (PDSI < −3.00) circa 900–1100 B.C.E. (200 years) and 1200–1350 B.C.E. (150 years) on the eastern slope of the Sierra Nevada (Stine, 1994). Fye et al. (2003) estimate that the 1950s-like decadal drought has a return period of about 45 years. Shorter drought periods, particularly in the southwest, have much shorter return periods.

FIGURE 8–4

Dust bowl-like droughts. Instrumental PDSI (a) and PDSI analog constructed from tree-ring data (b) for the period 1929–1940. PDSI reconstructed, averaged, and mapped for the period 1527–1865 (c–g). (*Source:* Fye, F.K., D.W. Stahle, and E.R. Cook (2003) "Paleoclimatic Analogs to Twentieth-Century Moisture Regimes Across the United States," *Bulletin of the American Meteorological Society,* vol. 84, no. 7, pp. 901–909.)

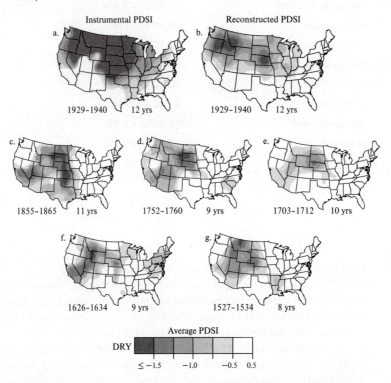

Although the decadal droughts are predominantly located west of the Mississippi, there are instances of incipient and mild drought in areas that we consider to be humid in the dendro-chronological analysis (i.e., the southeastern states and New England).

Regional Water Resource Limitations.   A comparison of the average regional consumptive use and renewable water supply in the United States is shown in Figure 8–6. *Renewable supply* is the sum of precipitation and imports of water, minus the water that is not available for use through natural evaporation and exports. It is a simplified upper limit to the amount of water consumption that could occur in a region on a sustained basis. Because there are requirements to maintain minimum flows in streams and rivers for ecological, navigation, and hydropower, it is not possible to totally develop the renewable supply. However, the ratio of renewable supply to consumptive use is an index of the development of the resource (Metcalf & Eddy, Inc., 2007). Water resource issues for selected regions are summarized in the following paragraphs. These discussions are based on the *National Water Summary—1983* (USGS, 1984) and *Water 2025* (U.S. Department of Interior, 2003).

The West.   Using the map in Figure 8–6, this region consists of the Pacific Northwest, California, and the Great Basin. While the index implies a consumptive use less than 40% of the renewable water supply, rather large local areas in these regions are already stressed to meet demand in regularly occurring meteorological droughts. The issues are (1) the explosive

FIGURE 8–5

1950s-like droughts. Instrumental PDSI (a and c) and PDSI analogs constructed from tree-ring data (b and d). PDSI reconstructed, averaged, and mapped for the period 1542–1883 (e–o). (*Source:* Fye, F.K., D.W. Stahle, and E.R. Cook (2003) "Paleoclimatic Analogs to Twentieth-Century Moisture Regimes Across the United States," *Bulletin of the American Meteorological Society,* vol. 84, no. 7, pp. 901–909.)

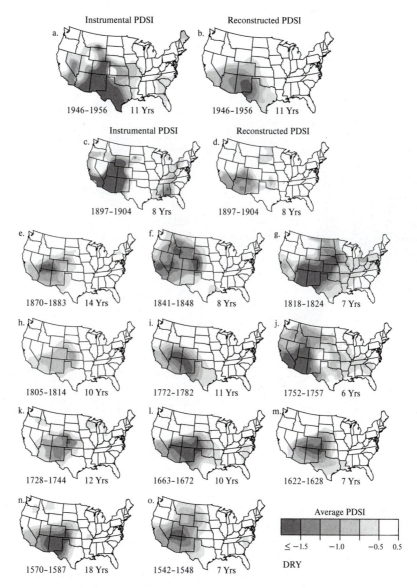

population growth in urban areas, (2) the emerging need for water for environmental and recreational uses, (3) the national importance of the domestic production of food and fiber from western farms and ranches.

In the Pacific Northwest the majority of the precipitation falls on the western slopes of the Rocky Mountains. The Snake and Columbia are the major rivers. Fifteen large dams and more than 100 smaller dams have been built on these rivers and their tributaries. These dams harness 90% of the hydroelectric potential of the region and provide irrigation to over a million

FIGURE 8–6

Comparison of average consumptive use and renewable water supply for the 20 water resources regions of United States (Adapted from USGS, 1984; updated using 1995 estimates of water use). The number in each water resource region is consumptive use/renewable water supply in 10^6 m^3/d, respectively, or consumptive use as a percentage of renewable supply as shown in the legend. (*Source:* Davis, M.L. and Cornwell, D.A., *Introduction to Environmental Engineering,* 5e, p.973, 2013. The McGraw-Hill Companies, Inc.)

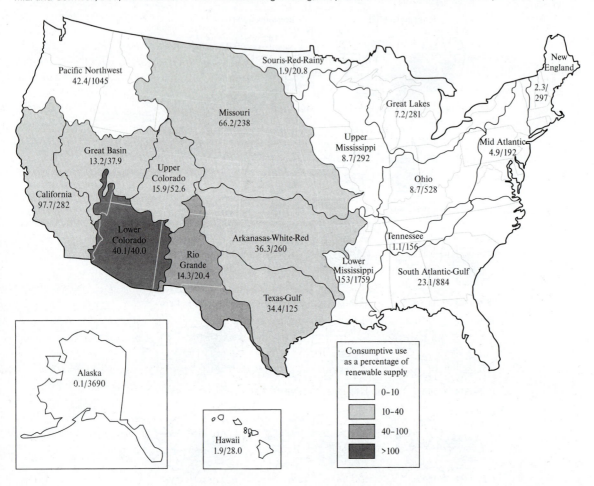

acres. Because of runoff, seepage, evaporation and other losses, almost twice as much water must be delivered as is actually used. As much as 43% of the acreage is in low-value forage and pasture crops that account for 60% of the water consumed. While of great economic benefit to the region, these federally funded projects have decimated the salmon population with a loss of between 5 and 11 million adult salmonids (Feldman, 2007).

California has the most precarious water resource system in the United States. The vast majority of the population is concentrated in the semiarid southern part of the state. Seventy percent of the water supplies are in the north and 80% of the demand is in the midsection and the south. California droughts of two or more years have occurred eight times in the last century. Record breaking droughts occurred in the hydrologic years 1928–1934, 1976–1977, 1987–1992, 2007–2010, and 2012–2017. The most recent drought ended in February 2017 with extraordinary flooding in northern California that caused a major dam failure. Coastal areas in the mid- to southern half of the state experience salt water intrusion due to groundwater pumping.

The Colorado Basins.   The Upper and Lower Colorado basins are the source of water for the Colorado River Compact. The Compact allocates water among seven states: Arizona, California, Colorado, Nevada, New Mexico, Utah, and Wyoming. The Colorado Basin was arbitrarily divided into upper and lower basins at Lee's Ferry for the purpose of water allocation.

Although all of these states have semiarid to arid climates, Arizona is notable for the precariousness of its water balance.* As noted in Figure 8–6, the index shows a consumptive use greater than 100% of the renewable water supply. Surface water is supplied from the following rivers: Colorado, Verde, Gila, and Little Colorado. The vast majority of surface water comes from the Colorado River. The trend in annual flow volume measured at Lee's Ferry from 1895 through 2003 shows a continuous decline of about 6.2×10^8 m^3 per decade. This is due, in part, to upstream water use. Dendrochronological analysis of Utah trees has revealed four droughts lasting more than 20 years, and nine lasting 15 to 20 years (USGS, 2004). Similar data from Mesa Verde, Arizona, reveal a drought lasting 23 years from circa 1276 to 1299 B.C.E. (Haury, 1935). Flow data from Lee's Ferry show three droughts lasting from four to eleven years and an ongoing drought from 2000 through 2003 (USGS, 2004).

Until the late 1970s, Arizona depended almost exclusively on groundwater to supply municipal water needs. The sparse precipitation means that aquifer received virtually no recharge. As a consequence, the aquifer was literally being mined.

The Rio Grand.   Population density in the border lands of the Rio Grand River has quadrupled in Mexico and tripled in the United States since the 1950s (Mumme, 1995). Meanwhile in-stream flows have fallen to 20% of historical levels.

The Central Great Plains.   The Missouri, Arkansas-White-Red river basins and the Texas Gulf region are included in this region. A major transbasin water diversion is from the Colorado River through tunnels drilled through the Rockies to supply Wyoming. As noted in "The West" discussion, the Upper Colorado basin is under severe stress to maintain flow rates to downstream compact members as agreed upon early in the last century.

Although there are major rivers that supply generous amounts of water to multiple communities, irrigated agriculture is a main end use in the region. Farmers in the "High Plains" states of Nebraska, Colorado, Kansas, Oklahoma, New Mexico, and Texas are drawing water from the Ogallala aquifer that underlies the region. In some instances the rate of withdrawal exceeds the average recharge rate (see Figures 8–7 and 8–8). At the current rate of withdrawal it is estimated that the aquifer will be unable to provide fresh water in another 50 to 100 years (KGS/KDA, 2010).

The Eastern Midwest.   This region includes the Souris-Red-Rainy, Upper and Lower Mississippi, Ohio and Tennessee river basins. This region can be said to have plenty of water. Seasonal meteorological droughts occur in some locations with resultant agricultural droughts.

Great Lakes.   The Great Lakes hold 95% of the fresh surface water in the United States. This region includes the 121 watersheds in the states that border the Great Lakes. The surrounding states and Canadian provinces have ratified an agreement to prohibit export of water from this region. Local instances of groundwater depletion in urban areas are of concern.

---

*We have already addressed California's critical situation but would add the note that California has been using half a million acre-feet or 6.2×10^8 m^3 more water than was agreed to in the Compact and affirmed by the U.S. Supreme Court (i.e., 4.4 million acre-feet or 5.43×10^9 m^3). On October 10, 2003, a seven-state agreement called the California 4.4 Plan was signed that gave California until 2017 to reduce its draw on the river to the basic apportionment (Pulwarty et al., 2005).

FIGURE 8–7

Ogallala aquifer. (*Source:* USGS (2007) USGS Fact Sheet 2007–3029, U.S. Geological Survey, Washington, DC.)

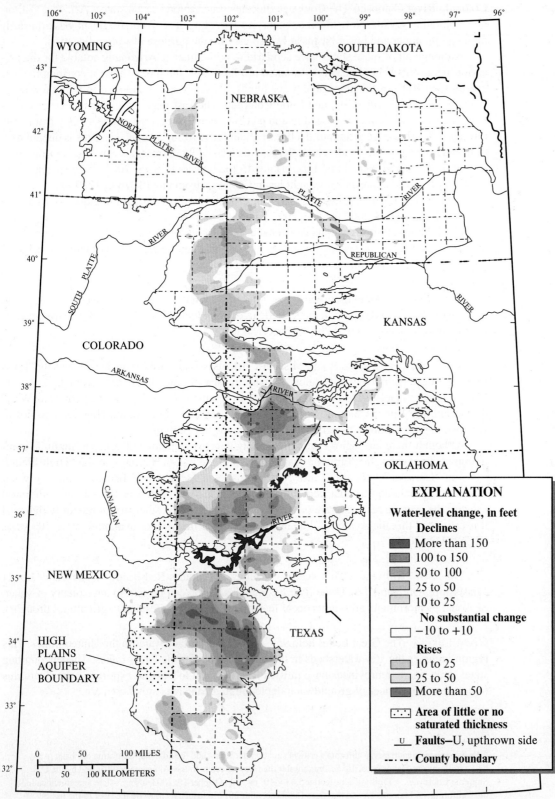

Base from U.S. Geological Survey digital data, 1:2,00,000
Albers Equal-Area projection, Horizontal datum NAD 83,
Standard parallels 29°30′ and 45°30′, central meridian-101°

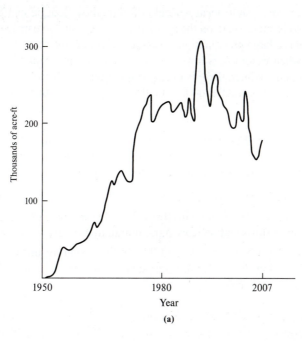

(a)

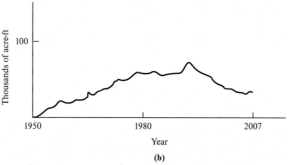

(b)

New England. This region includes the traditional New England states of Maine, New Hampshire, Vermont, Massachusetts, Connecticut, and Rhode Island. Water resources in this region are generally abundant. Population pressure, especially in New Hampshire, Connecticut, and around the Boston metropolitan area, has resulted in dramatic stress on the drinking water supply and virtually irreparable damage to some surface water ecosystems. Even a short-term drought in this region would result in major socioeconomic impact.

Mid-Atlantic. This region includes the states of New York, New Jersey, Pennsylvania, Delaware, and Virginia. This region has experienced severe droughts in the past few decades. The major metropolitan areas rely on water systems that are highly sensitive to climate variation. Like the New England region, population pressure has resulted in dramatic stress on the drinking water supply and virtually irreparable damage to some surface water ecosystems. Even a short-term drought in this region would result in major negative socioeconomic impact.

South Atlantic-Gulf. This region includes North and South Carolina, Georgia, Alabama, and Florida. Development, population growth, demographic preference for the coastal areas, and large seasonal population swings present water management issues that are not easily resolved. A recent (2005–2008) drought in Alabama, Georgia, the Carolinas, and Tennessee has revealed major vulnerability. This was not an unusual event. For example, based on the climatological record, not including short-term agricultural droughts, Georgia can expect a meteorological drought of two or more years on the average about once in 25 years (Stooksbury, 2003).

Florida has a large precipitation excess, but it occurs over a three- to four-month period in the summer when water use is lower because the transient population leaves the state. During high demand in the winter, there is little precipitation.

All along the South Atlantic and Gulf coast, there is a burgeoning problem of saltwater intrusion.

Droughts and Global Warming.   As noted in the discussion of floods and climate change, Global Circulation Model (GCM) projections do not reveal an increase in meteorological drought. However, as noted in Chapter 12, forecast increases in temperature will result in

- Drier crop conditions in the Midwest and Great Plains, requiring more irrigation.
- Warming in the western mountains that will decrease snowpack, cause more winter flooding, reduce summer flows and increase agricultural, hydrological, and socioeconomic drought.
- A rise in sea level between 0.18 and 0.57 m that would result in an increase in severity of saltwater intrusion into drinking water supplies in coastal areas particularly in Florida and much of the Atlantic coast.

Drought and Sustainability.   The determinants of vulnerability to drought impacts include: population growth, settlement patterns, economic development, health infrastructure, mitigation and preparedness, early warning, emergency assistance, and recovery assistance coupled with safe yield of available water resources. As we noted in Section 8–1, human-environmental systems are relatively sustainable if they possess strong adaptive capacity and can employ it effectively.

Safe Yield.   Theoretically, safe yield defines the capability of a water source to sustain a stipulated level of water supply over time as river inflows and groundwater recharge varies both seasonally and annually (Dziegielewski and Crews, 1986). There are two major assumptions in the estimates of safe yield: (1) the data exhibit stationarity and, (2) the required level of water supply remains constant. *Stationarity* means that the data do not exhibit trend or periodicity, or that any trend in the data is explicable and capable of reliable extrapolation into the future (Smith and Ward, 1998). The decline in flow of the Colorado River noted under "Regional Water Resource Limitations" and the forecast of reduction in snowmelt runoff for California are just two examples of the lack of stationarity in the data. Population growth and increased saltwater intrusion because of rising sea levels are examples of the change in the required level of water supply that invalidates the second assumption. Nonetheless, the comparison of safe yield estimates and "current" water consumption shown in Table 8–7 provide an example of a convenient way to benchmark a municipal water supply.

From the perspective of drought mitigation, any water supply system with average demands below the safe yield should not experience water shortages during droughts that are less severe than a "design drought" used to derive the value of the safe yield.

Case Study—California.   Beginning at the end of the 19th century and continuing to this day, numerous audacious water projects have been developed to keep people coming to the state. Early in the 1900s, after shallow aquifers and seasonal rivers could no longer sustain Los Angeles, land was purchased in the Owens Valley, east of the Sierra Nevada. Completion of the Los Angeles Aqueduct sent the entire flow of the Owens River to Los Angeles. Within a decade the lake was a dust bowl and the San Fernado Valley was worth millions (Bourne and Burtynsky, 2010).

In the intervening years, California has built over 3218 km of canals, pipelines and aqueducts to transport water from the north to the south. Because of its dependence on snowmelt for much of its water, over 157 reservoirs have been built to hold the melt water for later

TABLE 8-7	Water Use and Safe Yield for Selected Cities		
City	Current Use, $(10^6 \text{ m}^3 \cdot \text{d}^{-1})$	Safe Yield, $(10^6 \text{ m}^3 \cdot \text{d}^{-1})$	Water Use/ Safe Yield
Binghamton, NY	0.0473	0.174	0.27
Denver, CO	0.749	1.01	0.74
Indianapolis, IN	0.386	0.424	0.91
Merrifield, VA	0.299	0.204	1.47
New York City, NY	5.80	4.88	1.19
Phoenix, AZ	1.03	0.973	1.06
Southern California	12.6	11.6	1.09

Source: Dziegielewski et al., 1991.

distribution. Major pumping stations, located in the Sacramento-San Joaquin Delta, move the water to the south (Bourne and Burtynsky, 2010). The movement of water accounts for nearly 40% of the state's total energy supply (Feldman, 2007).

The Sacramento-San Joaquin Delta, a former 283,300 ha marsh, was drained and diked into islands that became prime farmland and exclusive residential property surrounded by waterways. Today the delta sits 6 m below sea level. Sea level rise combined with more severe storms threaten the dikes.

The delta sits just east of the Hayward Fault in one of the most dangerous earthquake zones in the United States. The forecast is that there is greater than a 60% chance of a major earthquake in the next 30 years. The average island in the delta now has a 90% chance of flooding in the next 50 years (Bourne and Burtynsky, 2010).

Los Angeles County has been pumping "fresh" water into the groundwater aquifer to prevent saltwater intrusion into water supplies since 1947. About 4×10^7 m^3 per year of recycled wastewater are used.

The population of southern California is increasing at a rate of more than 200,000 per year. In 2010, California had a budget deficit of $20 billion and a proposal for $11 billion in water projects.

Notwithstanding the implications of global warming that exacerbate the existing precariousness of the water resource system, from a long-term perspective it does not appear that California has the water resources to develop a *sustainable economy* that will produce wealth and provide jobs for many human generations without degrading the environment. There are too many people for the available water resources and population growth is too high.

Droughts and Green Engineering. As with floods, civil engineering structures play a minor role in sustainable drought protection. The "green engineering" component is recognizing that traditional structural solutions are bound to fail and that other solutions are required to minimize risk and to provide a *sustainable economy* for many human generations. These solutions fall into two broad categories: drought response planning and water conservation planning and implementation.

Drought Response Planning. The basic goal of a response plan is to improve response efforts by enhancing monitoring and early warning, impact assessment, preparedness, and response. Activities in these areas are summarized in the following paragraphs.

- *Early warning:* Parameters that must be monitored to detect the early onset of drought include temperature, precipitation, stream flow, reservoir and groundwater levels, snowpack, and soil moisture.

- *Impact assessment:* Synthesis of the early warning data to describe the magnitude, duration, severity, and spatial extent of the drought requires some type of indicator. The PDSI may be appropriate to evaluate past droughts, but it has been criticized for its inability to make "real time" assessments. Other tools such as the *Standardized Precipitation Index* (SPI), *Surface Water Supply Index* (SWSI), and the *U.S. Drought Monitor* (DM) maps have been proposed. Of these, the DM maps (http://www.drought.unl.edu/dm) are the most recent evolution. They are compiled with input from the National Drought Mitigation Center (NDMC), U.S. Department of Agriculture (USDA), National Oceanic and Atmospheric Administration's (NOAA), Climate Prediction Center (CPC), and the National Climatic Data Center (NCDC).

- *Preparedness:* Plans and facilities to implement them generally require legislative action. Examples include authorization to establish and activate water banks, mandatory conservation, drought surcharges, restrictions on outdoor residential use, and local emergency supply for municipalities and farm animals.

- *Response:* People and animals cannot drink plans. Fish cannot swim in plans. Agencies and individuals must be given **training and facilities** to implement the plans.

In the United States drought response planning is primarily a state responsibility. By 2010, only five states (Alaska, Arkansas, Louisiana, Mississippi, and Vermont) are without a drought response plan.

California has delegated drought response planning to municipalities. San Diego has a vigorous, rigorous drought response plan that includes drought stages and required responses that are enforceable by code. Violators are subject to fines and potential water shutoff. Some examples are shown in Table 8–8. The demand reductions are based on "a reasonable probability that there will be a supply shortage" and that the specified reductions are required to "ensure that sufficient supplies will be available to meet anticipated demands."

The state of Georgia has a comprehensive drought response plan that is triggered by SPI readings for 3, 6, and 12 months. The plan includes several categories of users including municipal-industrial, agricultural, and water quality. Specific restrictions are applied for each of four drought levels. For example, in the municipal category, at the lowest level (Level 1), outdoor water use is limited to scheduled week days. At the highest level (Level 4), outdoor watering is banned (Georgia EPD, 2003). This is a proactive plan that should be able to prevent severe socioeconomic hardship and water rights lawsuits.

In contrast, Tennessee's drought response plan uses qualitative triggers: "Deteriorating water supplies or quality," "conflicts among users," and "very limited resource availability." Local actions are specified qualitatively rather than by prohibitions. At no point are specific actions required. For example, at the "drought alert level" public water suppliers are to "monitor water sources and water use." At the "mandatory restrictions level," restrictions could include "banning of some outdoor water uses, per capita quotas and percent reductions of nonresidential users." At the

TABLE 8–8		San Diego's "Drought Watch Response Levels"
Level	Trigger, % Demand Reduction Required	Typical Restrictions
1	<10	Voluntary restrictions become mandatory Prohibit excessive irrigation Washing of paved surfaces prohibited
2	<20	Landscape irrigation limited to 3 days/week Landscape irrigation is limited to 10 minutes
3	<40	Landscape irrigation limited to two days/week No new potable water services permitted
4	≥40	Landscape irrigation is stopped except for trees and shrubs

"emergency management level" public water suppliers are to "provide bottled water" and "initiate hauling of water." This is a reactive plan that will not be able to prevent severe socioeconomic hardship and water rights lawsuits. The triggers will be too late to initiate preventive strategies. They are based on "hydrologic" drought. In the typical drought sequence, hydrologic drought is in the later stages of drought after meteorological drought and agricultural drought.

Water Conservation Planning and Implementation.   The status of water conservation planning in the United States has been evaluated for each state. The status is reported in four categories (Rashid et al., 2010):

1. Mandate requiring comprehensive program.
2. Recommend conservation planning as part of water supply.
3. Provision of conservation tips.
4. No provisions.

All but ten states fall into one of the first three categories. Those in the first three categories are shown by category number in Figure 8–9. Those in category 4 are Alabama, Alaska, Idaho, Kentucky, Mississippi, North Dakota, Ohio, South Dakota, West Virginia, and Wyoming.

Typical requirements include tiered rate structures, public conservation outreach, voluntary conservation initiatives, and mandatory conservation initiatives.

Case Study—Arizona.   Arizona has, perhaps, the most ambitious and farsighted ongoing conservation plan. After the drought of 1976–1977, the legislature enacted the 1980 Groundwater Management Act (GMA). The act provided for *active management areas* (AMAs) within the state. These are areas where the majority of the population and groundwater overdraft are

---

FIGURE 8–9

Status of state water conservation planning. (1: mandate requiring comprehensive program; 2: recommend conservation planning as part of water supply; 3: provision of conservation tips; 4: no provisions.) (*Source:* Davis, M.L. and Cornwell, D.A., Introduction to Environmental Engineering, 5e, p. 977 and 982 respectively, 2013. The McGraw-Hill Companies, Inc.)

located (Prescott, Phoenix, Pinal, Tucson, and Santa Cruz). The management goal for all of the AMAs is to develop a sustainable water supply. In the case of the major metropolitan areas, this goal is quantified as the safe yield. The AMAs include more than 80% of Arizona's population, 50% of the total water use, and 70% of the state's groundwater overdraft (Pulwarty et al., 2005).

Authorized by an act of Congress in 1969 and built by the United States Bureau of Reclamation, the Central Arizona Project (CAP) is the largest single source of renewable water supply in Arizona. The 540 km system of aqueducts, tunnels, pumping stations, and pipelines was designed to carry 1.4 million acre-feet per year (1.73×10^9 m^3) of Arizona's Colorado River allocation from Lake Havasu to central and southern Arizona. The project cost $4 billion and took 22 years to complete. Under federal guidelines, municipalities have a higher priority for CAP deliveries than agriculture. Deliveries to Phoenix began in 1985. Tucson began receiving water in 1992.

The GMA specifies mandatory reductions in demand for all sectors through conservation and transition to renewable supplies. A key component is the use of Colorado River water through the Assured Water Supply (AWS) program. The AWS requires that all new subdivisions in AMAs demonstrate a 100-year AWS based primarily on renewable water supplies before the subdivision is approved.

Recognizing the precariousness of Arizona's water supplies and the need to utilize its full allocation of the Colorado River flow, in 1996 the state created the Arizona Water Banking Authority (AWBA). The objectives of the AWBA are to:

1. Store water underground that can be recovered for municipal use.
2. Support the management goals of the AMAs.
3. Support Native American water rights.
4. Provide for interstate banking of Colorado River water to assist Nevada and California in meeting water supply while protecting Arizona's entitlement.

The AWS rules can be implemented because developers can access the AWBA through the Central Arizona Groundwater Replenishment District (CAGRD) by committing to replenish groundwater used by its members. With incentive pricing for agriculture and recharge, and the AWS rules in place, Arizona is now utilizing all of its Colorado River allocation. Arizona is banking on artificial recharge as a tool to offset future drought-related shortages.

Green Engineering in Metropolitan Areas. Surveys of water use in urban areas in the southwestern and western United States reveal that a majority of the per capita water use is outdoors (Figure 8–10). Conservation efforts in the metropolitan areas focus first on outdoor use. Some example water conservation measures are summarized below.

- *Drip irrigation:* Using a small pipe with outlets for a plant, tree, or small area of plantings.
- *Smart irrigation:* Using meteorological data and/or soil moisture data to schedule timing and volume of water for irrigation. Real-time meteorological data are used to calculate evapotranspiration (ET is discussed in Chapter 7) and control irrigation. Soil moisture sensor (SMS) controllers measure the soil moisture at the root zone. While ET systems reduce overwatering, they may result in increases for those that have been underwatering (Green, 2010). The SMS systems have an advantage over the ET system in that the SMS system can stop irrigation when enough has been applied by adjusting the run time to maintain the desired moisture (Philpott, 2008). Both systems are designed to avoid watering when it is raining.
- *Rainwater harvesting:* Methods for collecting, holding, and using rainwater (Waterfall, 2004).
- *Water irrigation auditor certification:* A program to train irrigation system installation and maintenance personnel on techniques to conserve water. Homes with inground sprinkler systems use 35% more water than those without inground systems (Feldman, 2007).
- *Water conservation incentives:* A program to credit water bills for a variety of conservation activities. Some examples include: irrigation conservation by replacing worn and leaking equipment ($25 per item), certified irrigation audit ($100), and turf grass removal ($2.70/m^2

FIGURE 8–10

Per capita water use for single families. (*Source:* Western Resource Advocates, 2003.)

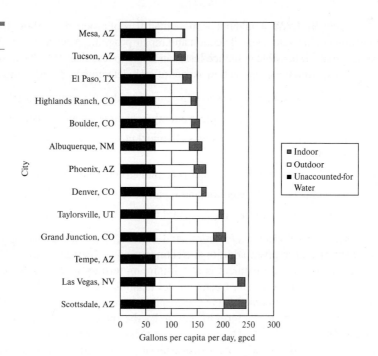

up to $400 for residential). For fiscal years 2006–2010, Prescott, Arizona, estimates a total water savings of over 1.6×10^5 m^3 of water (Prescott, 2010).

- *Low flush toilets:* Water closets that operate on lower volumes of water. Typically a low system will use about 6 L/flush in contrast to a conventional system that uses about 15 L/flush. Prescott, Arizona (2010) saved about 4×10^4 m^3 of water over a four-year period by replacing standard units with low flush units.

- *Tiered rate structures:* Examples include
  - Uniform rate plus seasonal surcharge
  - Inclining block—rate increases with increasing use
  - Water budget rate—an inclining block rate tailored for individual customers
  - See Georgia EPD (2010) for details

- *Water reuse:* Examples include
  - Using *gray water* that comes from bathing and washing facilities that do not contain concentrated human waste for irrigation
  - Using treated municipal wastewater for golf course irrigation

Green Engineering and Water Utilities.   Of the numerous suggestions for water conservation presented in *Water Conservation for Small- and Medium-Sized Utilities* (Green, 2010), two are highlighted here because of the magnitude of their impact on water conservation: leaks and water reuse. *Unaccounted for water* (UAW) is the term used to identify water produced but not otherwise sold. It includes leaks, unauthorized consumption, metering inaccuracies, and data handling errors. It is water that does not produce revenue for the utility.

In a 1997 audit, Lebanon, Tennessee, identified 45% of their water production was UAW (Leauber, 1997). Water audit and leak detection of 47 California water utilities found an average loss of 10% and a range from 5 to 45%. It is estimated that up to 8.6×10^8 m^3 of leakage occurs in California each year (DWR, 2010). Community water systems in the United States lose about 2.3×10^7 m^3 of water per day through leaks (Thornton et al., 2008). It is notable that 10% of all homes account for 58% of household leaks and that households with pools have 55% greater leakage than other households (Feldman, 2007).

Typically, UAW should not be more than 10% of production. Above this level, there is strong incentive to institute a leak detection and repair program. Advances in technology and expertise should make it possible to reduce losses in UAW to less than 10% (Georgia EPD, 2007).

Aside from the conservation aspect of leak detection and repair, there are other benefits:

- Repairing leaks with scheduled maintenance reduces overtime costs of unscheduled repairs.
- Leak detection and repair reduces energy consumption and power costs to deliver water.
- Leak detection and repair reduces chemical costs to treat water that will not be delivered to customers.
- Leak detection and repair reduces potential structural damage to roads.
- UAW audits and leak detection and repair have a very favorable return on investment.
- Little leaks get bigger with age . . . and ultimately result in pipe failure.

Water reuse, *water reclamation*, and *recycled water* are terms used for treated municipal wastewater that is given additional treatment and is distributed for specific, direct beneficial uses. The additional treatment is at a minimum tertiary treatment (Chapter 11) and often includes advanced wastewater treatment (for example, reverse osmosis discussed in Chapter 10). Water reuse has been practiced in California for over a century. In 1910 at least 35 communities were using wastewater for farm irrigation. By the end of 2001, reclaimed water use had reached over 6.48×10^8 m$^3 \cdot$ y^{-1}. Florida has been reclaiming wastewater for over 40 years. Approximately 8.34×10^8 m^3 of reclaimed wastewater was used in 2003 (Metcalf & Eddy, 2007). In 2008, Florida passed a law phasing out ocean outfall disposal of wastewater by 2025. A total of 4.16×10^8 m$^3 \cdot$ y^{-1} will be reclaimed to meet this goal (Greiner et al., 2009).

A list of typical applications is shown in Table 8–9.

TABLE 8–9	Example Applications of Reclaimed Water	
	Category	**Typical Application**
	Agricultural irrigation	Crop irrigation Commercial nurseries
	Groundwater recharge	Groundwater banking Groundwater replenishment Saltwater intrusion control Subsidence control
	Industrial recycling/reuse	Boiler feed water Cooling water Process water
	Landscape irrigation	Cemeteries Golf courses Greenbelts Parks
	Environmental enhancement	Fisheries Lake augmentation Marsh enhancement Stream flow augmentation Snow making
	Nonpotable urban uses	Air conditioning where evaporative cooling is used Fire protection (sprinkler systems) Toilet flushing in recreational areas
	Potable reuse	Blending in water supply reservoirs Blending in groundwater Direct pipe to water supply

Source: Metcalf & Eddy, 2007.

8–3   ENERGY RESOURCES

Until about 300 years ago, the majority of human energy needs were supplied by human and animal labor, water power, wind, wood and other burnable organic matter such as agricultural waste, dung and peat (now lumped together as *biomass*). With the advent of the Industrial Revolution in the late 18th and early 19th centuries, water power remained important but fossil fuels (primarily coal) began to be used on a large scale. Over the course of the 20th century petroleum products assumed a major role in supplying our energy needs.

The United States is both a major producer and major consumer of primary energy in the world. In 2006, the United States produced 21% of the world's primary energy and consumed 26% of the world's primary energy (EIA, 2010a and 2010b). Approximately 85% of our energy consumption is in the form of fossil fuels. Thirty-seven percent of our energy use is for transportation. Although our per capita energy consumption has stabilized in the last three decades, our total energy consumption has grown enormously in the last 40 years (Figures 8–11 and 8–12).

Fossil Fuel Reserves

How do you compare a kilogram of coal with a liter of oil? Or, for that matter, a liter of oil with a liter of natural gas? To make any useful comparison of fossil fuel reserves, we must use an energy basis. For fossil fuels, the net heating value (NHV) serves as a common denominator for comparison. Typical values for NHV of some common fuels are shown in Table 8–10.

Of course, comparison of energy value is not a comparison of the fuel's utility. For example, you cannot replace gasoline with the equivalent NHV of coal or natural gas and put it in the gas tank of your car. The fact that coal cost less than gasoline on a dollar per joule basis is irrelevant in comparing fuels for our current automobile fleet.

The question is not whether or not the world is running out of energy. The question is what are we willing to pay and what environmental impacts will be incurred. Let us first look at the question of "running out." That portion of the identified fossil fuels that can be economically recovered at the time of the calculation is called the *proven commercial energy reserve*. Because of changes in technology and the change in consumer willingness to pay, the reserve may change even though the mass of fuel does not. The estimated U.S. and world reserves are shown in Table 8–11. How long will the reserves last? Several types of estimates are possible. For example, we can estimate the length of time assuming current consumption rates (demand) with no new discoveries or changes in extraction technology. Alternatively, we can assume some

FIGURE 8–11

Per capita energy consumption in the United States, GJ/capita-gigajoules per person.

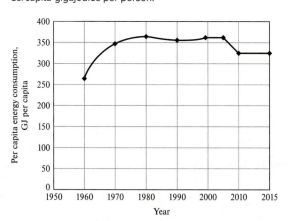

FIGURE 8–12

United States energy consumption, EJ = exajoules.

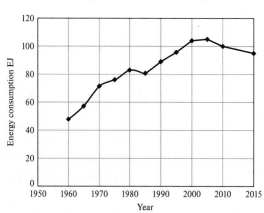

TABLE 8–10	Typical Values of Net Heating Value	
	Material	Net Heating Value, MJ · kg^{-1}
	Charcoal	26.3
	Coal, anthracite	25.8
	Coal, bituminous	28.5
	Fuel oil, no. 2 (home heating)	45.5
	Fuel oil, no. 6 (bunker C)	42.5
	Gasoline (regular, 84 octane)	48.1
	Natural gas[a]	53.0
	Peat	10.4
	Wood, oak	13.3–19.3
	Wood, pine	14.9–22.3

[a]Density take as 0.756 kg · m^{-3}

TABLE 8-11	U.S. and World Proven Commercial Energy Reserves of Fossil Fuels[a]		
	Fuel	United States, EJ[b]	World, EJ
	Coal	11,700	24,000
	Oil	215	12,700
	Natural gas	370	7500
	Total Reserves	**12,285**	**44,200**

[a]Data for oil and gas are for 2015, while data for coal are for 2016. Natural gas data do not include estimates of natural gas that can be recovered from shale formations.
[b]EJ = Exajoules = 10^{18} joules
Source: EIA, 2016 and BP Statistical Review, 2016.

growth in demand with no new discoveries or changes in extraction technology. The length of time that current reserves will last with a constant demand is expressed as

$$T_s = \frac{F}{A} \tag{8–8}$$

where T_s = time until exhaustion (in years)
F = energy reserve (in EJ*)
A = annual demand (in EJ · year^{-1})

The length of time that current reserves will last with a growth in demand is expressed as

$$F = A\left[\frac{(1 + i)^n - 1}{i}\right] \tag{8–9}$$

where i = annual increase in demand as a fraction
n = the number of years to consume the reserve

Example 8–3 shows how to make these two kinds of estimates.

*EJ = exajoule = 1×10^{18} J

EXAMPLE 8-3   In 2015, the international consumption of coal for energy was 160 EJ (British Petroleum, 2016). Assuming the demand remains constant, how long will world reserves last? The estimate of the average world consumption of coal-based energy ranged from a 0.6% increase to a 1.8% decrease in 2015 (EIA, 2015; British Petroleum, 2016). Using the 0.6% increase, estimate how long the world's coal reserves will last.

Solution   If the demand remains constant (Equation 8–8), the world's coal reserves will last:

$$\frac{24,000 \text{ EJ}}{160 \text{ EJ/year}} = 150 \text{ years}$$

If the demand rises at a rate of 0.6% per year, an estimate of the time the world's coal reserves will last may be found using Equation 8–9:

$$24,000 \text{ EJ} = 160 \text{ EJ/y} \left[\frac{(1 + 0.006)^n - 1}{0.006} \right]$$

Solving for n:

$$(150)(0.006) = (1.006)^n - 1$$
$$0.9 + 1 = (1.006)^n$$

Taking the log of both sides

$$\log (1.9) = \log [(1.006)^n] = n \log (1.006)$$
$$2.788 \times 10^{-1} = n (2.598 \times 10^{-3})$$
$$n = 107 \text{ years}$$

Long-range forecasts of the total primary energy supply by resource are illustrated in Figure 8–13. From this analysis it appears that a peak in fossil fuel as a primary energy source will occur shortly. While oil and natural gas will be virtually irrelevant as a fuel source by the end of this century, coal will still have a role to play . . . albeit a significantly shrinking one.

FIGURE 8–13

Peaking of world fossil fuel supply. (*Source:* Davis, M.L. and Cornwell, D.A., Introduction to Environmental Engineering, 5e, p. 987 and 990, 2013. The McGraw-Hill Companies, Inc.)

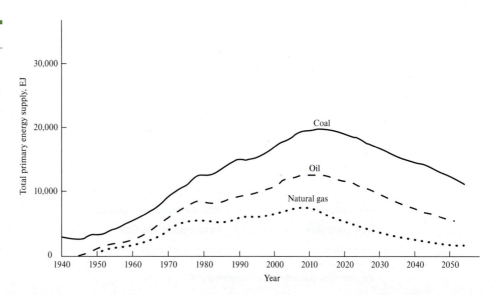

American Electric Power which has 5.4 million customers in 11 states serves as an example of a years-long movement away from coal. The company's coal-generating capacity has fallen from 71% in 2005 to about 47% in 2017. Meanwhile gas capacity has risen from 20% to 27% and renewable energy has jumped from 3% to 13%. The switch is happening even faster across the United States. The U.S. Energy Information Agency estimates that gas on average will provide 34% of the electricity generated in the U.S. compared with 30% for coal (Loveless, 2016).

The data in Table 8–11 and Figure 8–13 do not reflect two more recently developed major sources of oil and natural gas: **"tar sands"** and **"shale gas."**

Tar Sands.   Petroleum may occur in forms that are not fluid. **Tar sands** or **oil sands,** as the name implies, are sands laden with petroleum. They have sufficient petroleum content that, if they occur at shallow depths, they can be mined with surface mining techniques. The Athabasca Field in Alberta, Canada, contains a thick deposit of tar sand that covers an area over 77,000 km^2 (Coates, 1981). About 2.8×10^{10} m^3 of the estimated 3.7×10^{11} m^3 of oil are recoverable with available technology. In Alberta about 20% of the deposits are within 100 m of the surface. These can be excavated by strip mining. The remainder must be recovered from wells. The tar sand is heated and pumped from the wells. About 4 m^3 of water is needed to remove 1 m^3 of oil (Ritter, 2011).

Shale Gas.   In the period 2008 to 2010, the estimates of natural gas energy in shale formations in the United States grew from 400 EJ to 4000 EJ (LaCount and Barcella, 2010). The discovery and exploitation of this new resource began about 2005 with the application of an old technology, hydraulic fracturing (also known as **hydrofracking** or just **fracking**), and the development of horizontal drilling. By 2009, shale gas had grown from 5% of the natural gas supply in the United States to an estimated 20% of the supply.

The fracking process begins with a cased well drilled to about 1800 m depth and horizontal drilling of a bore hole to extend that well to as much as 3000 m in length. Specially formulated fluids such as those shown in Table 8–12 are pumped into the bore hole sufficient rates and pressures to cause the formation to crack. Pumping continues to extend the cracks in the formation. To keep the fracture open, a solid **proppant** consisting of sands or other granular materials are added to the fracture fluids. Once the gas (and/or oil) is released from the formation, it flows back to the surface with the water and additives as "flow-back" where the gas is separated from the flow-back.

TABLE 8–12	Summary of Hydraulic Fracturing Fluid Additives	
Type of Additive	**Function**	**Example Products**
Biocide	Kill microorganisms	Glutaraldehyde carbonate
Breaker	Reduce fluid viscosity	Acid, oxidizer
Buffer	Control pH	Sodium bicarbonate
Clay stabilizer	Prevent clay swelling	Potassium chloride
Fluid loss additive	Improves fluid efficiency	Diesel fuel, fine sand
Friction reducer	Reduce friction	Anionic polymer
Gel stabilizer	Reduce thermal degradation	Sodium thiosulfate
Iron controller	Keep iron in solution	Acetic and citric acid
Surfactant	Lower surface tension	Fluorocarbon

Source: U.S. Department of Energy, 2004.

The well releases flow-back over its lifetime which may be as long as 20 to 40 years. Initially the flow-back was taken to wastewater treatment plants for disposal. Early on it was established that many of the constituents are not amenable to the common biological treatment processes used to treat municipal sewage. By 2012 the consensus method of disposal (for about 90 percent of the flow) was injection of the flow-back into the geologic system below the Bakken or Marcellus formations where most of the shale gas is found (Arnaud, 2015; Ritter, 2014; Vengosh, 2015).

In Texas, Oklahoma, and Ohio, the injection of 10,000 to 20,000 m^3 of flow-back has triggered earthquakes in areas that have never experienced them before. In these local sites, the fracking process has been curtailed until other management techniques can be evaluated. For example, in 2016 Pawnee, Oklahoma, experienced a 5.8 magnitude earthquake. The Oklahoma Geological Survey closed 37 disposal wells as a precautionary response. The U.S. earthquake trend appears to be increasing. In 2005 the U.S. Geologic Survey reported only 3 earthquakes above magnitude 2.5. In 2015, that number skyrocketed to more than 2,700 (Frohlich, 2016).

Nuclear Energy Resources

Fission and fusion are the two potential reactions that may be used to generate nuclear energy. During nuclear fission, a neutron penetrates the nucleus of a fissionable atom (a radioactive isotope of uranium or plutonium) and splits it into daughter products while simultaneously releasing energy. To generate energy by fusion, isotopes of a light element such as hydrogen are fused together to form a heavier element such as helium. In this process energy is released. All commercially operating nuclear reactors are based on fission reactions.

The two main isotopes of uranium are ^{235}U and ^{238}U. Of these, about 99.3% of the naturally occurring uranium is nonfissionable ^{238}U. To induce a sustained fission reaction, the probability that excess neutrons from one fission reaction will cause other fissionable atoms to be split is increased by increasing the concentration of fissionable material or by slowing down the neutrons so they are more likely to be captured by ^{235}U than by ^{238}U.

In the middle of the 20th century, nuclear energy appeared to be a realistic energy alternative to fossil fuel. Many countries began building nuclear reactors. About 440 plants now generate 16% of the world's electric power. Some countries have made nuclear power their primary source of electricity. France, for example, gets 78% of its electricity from fission (Parfit and Leen, 2005).

There are, of course, pros and cons. On the plus side, fission provides abundant power, no global warming CO_2 emissions, and minimal impact on the landscape. On the minus side, in addition to the accidents at Three Mile Island in the United States, the Chernobyl Nuclear Power Plant in the Ukraine, and the Fukushima Daiichi Nuclear Power station in Japan that have made the public skeptical at best, the capital investment in a nuclear power plant is huge compared to a fossil fuel fired power plant and radioactive waste has remained an unresolved problem for over 20 years in spite of the passage of the Nuclear Waste Policy Act of 1987. Although current technology exists to develop renewable nuclear energy, under current U.S. policies, the readily available uranium fuel will last only about 50 years (Parfit and Leen, 2005).

Although ^{238}U is not fissionable, a ^{238}U atom that captures a neutron will be transformed to fissionable plutonium (^{239}Pu). This is the basis of "breeder reactors." The National Academy of Science estimated that the transformation of abundant ^{238}U to ^{239}Pu could satisfy the U.S. electricity requirements for over a hundred thousand years (McKinney and Schoch, 1998). Yet, there are no breeder reactors operating in the United States. Both environmental concerns and the potential for nuclear weapons proliferation have restricted the U.S. development of this energy source. Recently, well-known environmental advocates have come to regard the very real dangers of nuclear accidents and radioactive waste as less of a threat than the dangers of irreversible climate disruption. They now advocate the advancement of solutions to the economic, safety, waste storage and proliferation problems because renewable nuclear energy is sustainable while firing fossil fuels is not (Hileman, 2006).

Environmental Impacts

Waste from Resource Extraction.   Roughly 52% of the coal mined in the United States comes from surface mining. The remainder is from underground mines. Underground coal mining generates relatively small amounts of rock waste. **Strip mining,** in contrast generates large quantities of waste in the form of overburden. Up to 30 m of overburden may be removed to gain access to the coal.

The **spoil banks** from removed overburden or **tailings** from deep mines are sources of dust, excess erosion, and air pollution from fires and water pollution from acid mine drainage. Excess erosion from high-relief terrain can change relatively low sediment yields of 10 Mg $\cdot$ km^{-2} from forested sites to over 11,000 Mg $\cdot$ km^{-2} from excavated sites. The residual, low-grade coal in abandoned deep mines and spoil piles catch fire from spontaneous combustion, lightning, or human activities. These fires are difficult, if not impossible, to control. The poor combustion conditions yield large quantities of particulates and sulfur oxides. Both ground water and surface water pollution, called **acid mine drainage**, result from rainwater leaching through abandoned mines, spoil, and tailings. Although not exclusively associated with coal, sulfide minerals such as iron pyrite (FeS_2) frequently occur in coal beds. Once exposed to the atmosphere and bacterial action, the pyrite is oxidized to form sulfuric acid:

$$4FeS_2 + 15O_2 + 2H_2O \longrightarrow 2Fe_2(SO_4)_3 + 2H_2SO_4$$

This solution leaches into streams and lends them a red or yellow color from the iron precipitates. The acidic water is extremely lethal to aquatic life.

Coal mining, especially in water-poor areas of the west can have adverse effects on the groundwater supply. The water table around Decker, Montana, was lowered 15 m because of strip mining operations (Coates, 1981). Mining operations may also "behead" the recharge areas of aquifers.

Compared with other forms of resource extraction, drilling for oil and natural gas has produced minimal environmental deterioration. Unfortunately, as oil and natural gas exploration intensifies to replace depleting supplies, more environmentally sensitive areas are being explored. The damage from road building and other exploration activities frequently surpasses the actual well-drilling operation.

Oil spills and leaking pipelines are notorious sources of water pollution. Large quantities of gasoline and other petroleum products evaporate into the atmosphere and contribute to the formation of ozone.

As with mineral mining, uranium mining produces rock waste. A major difference, however, is that the uranium tailings are frequently radioactive.

Waste in Energy Production.   Perhaps more insidious than the recovery of fossil and nuclear fuels is the generation of wastes in the production of energy. The burning of fossil fuels releases air pollutants, such as sulfur oxides, nitrogen oxides, and particulates. The sulfur content of the fuel directly affects the mass of sulfur dioxide released. Coal, with the highest concentration of sulfur, contributes the most SO_2. Fuel oil has less sulfur and is frequently chosen as a substitute for coal to reduce SO_2 emissions. Natural gas has virtually no sulfur and, thus, no SO_2 emissions. Because the **ash** content (the unburnable mineral matter) of coal is high, it is also a major source of particulate pollutants. Fuel oil has less ash and less emissions. Natural gas has no ash and no emissions. All three release nitrogen oxides because the reaction chemistry does not depend on the nitrogen content of the fuel. Other pollutants from fossil fuels include toxic metals such as mercury, which occur in trace amounts in coal.

Although nuclear energy production does not release air pollutants, the spent fuel is highly radioactive and poses an as yet unresolved solid waste disposal problem. The sheer volume of coal ash is a major solid waste disposal problem. About 30% of the ash settles out as **bottom ash** in the firing chamber of the power plant. The remaining **fly ash** must be captured by the air pollution control equipment. The following example illustrates how large a volume this ash might occupy for a large coal-fired power plant.

EXAMPLE 8–4   A coal-fired power plant converts about 33% of the coal's energy into electrical energy. For a large 800-MW electrical output, estimate the volume of ash that is produced in a year if the anthracite coal has a NHV of 31.5 MJ $\cdot$ kg^{-1}, an ash content of 6.9%, and the bulk density of the ash is about 700 kg $\cdot$ m^{-3}. Assume that 99.5% of the ash is captured by a combination of the air pollution control equipment and settling in the combustion chamber.

Solution   The solution to this problem begins with an energy balance analysis. The energy-balance diagram is shown here. Because the output is in terms of power (work per unit time), the energy-balance equation is written in the form of power. At steady state, the energy-balance equation is

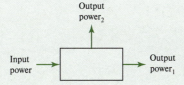

Input power = output power$_1$ + output power$_2$

where output power$_1$ is the power delivered to the electric distribution system and output power$_2$ is waste heat. From the definition of efficiency (see Chapter 4)

$$\eta = \frac{\text{output power}}{\text{input power}}$$

the input energy for the power plant is

$$\text{Input power} = \frac{\text{output power}}{\eta} = \frac{800 \text{ MW}_e}{0.33} = 2424 \text{ MW}_t$$

where MW$_e$ refers to megawatts electrical power and MW$_t$ refers to megawatts thermal power. Converting to units of megajoules per second, we obtain

$$(2424 \text{ MW}_t)\left(\frac{1 \text{ MJ} \cdot \text{s}^{-1}}{\text{MW}}\right) = 2424 \text{ MJ} \cdot \text{s}^{-1}$$

The amount of ash is determined from a mass balance for the coal. The mass-balance diagram is shown here. At steady state, the mass-balance equation is

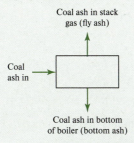

Coal ash in = coal ash in stack gas + coal ash in bottom of boiler

The coal ash entering the boiler is found from the burning rate and the ash content of the coal. Using the NHV of 31.5 MJ $\cdot$ kg^{-1} for anthracite, the amount of coal required is

$$\frac{2424 \text{ MJ} \cdot \text{s}^{-1}}{31.5 \text{ MJ} \cdot \text{kg}^{-1}} = 76.95 \text{ kg} \cdot \text{s}^{-1} \text{ of coal}$$

The ash input is then

$(76.95 \text{ kg} \cdot \text{s}^{-1} \text{ of coal})(0.069) = 5.31 \text{ kg} \cdot \text{s}^{-1}$

With a 99.5% capture efficiency in collecting the ash, the mass of ash to be disposed of is

$(0.995)(5.31 \text{ kg} \cdot \text{s}^{-1}) = 5.28 \text{ kg} \cdot \text{s}^{-1}$

In a year this would amount to

$(5.28 \text{ kg} \cdot \text{s}^{-1})(86{,}400 \text{ s} \cdot \text{day}^{-1})(365 \text{ days} \cdot \text{year}^{-1}) = 1.67 \times 10^8 \text{ kg} \cdot \text{year}^{-1}$

The volume will be

$$\frac{1.67 \times 108 \text{ kg} \cdot \text{year}^{-1}}{700 \text{ kg} \cdot \text{m}^{-3}} = 2.4 \times 10^5 \text{ m}^3 \cdot \text{year}^{-1}$$

This is a very large volume.

The inefficiency of the conversion of fuels into work results in the generation of waste heat. For automobiles and other dispersed energy consumers (light bulbs, refrigerators, and electric motors, for example), the environmental effect is sufficiently dispersed to be of little consequence. However, the generation of power also generates such a large amount of waste heat that the environmental effects may be serious. Thermal pollution of rivers is not limited to fossil fuel power plants but also includes nuclear plants. Example 8–5 illustrates how waste heat affects a stream.

EXAMPLE 8–5   Using the power plant from Example 8–4, and assuming that 15% of the waste heat goes up the stack and that 85% must be removed by cooling water, estimate the flow rate of cooling water required if the change in temperature of the cooling water is limited to 10°C. If the stream has a flow rate of 63 $\text{m}^3 \cdot \text{s}^{-1}$ and a temperature of 18°C above the intake to the power plant, what is the temperature after the cooling water and the stream water have mixed?

Solution   The energy balance diagram for the power plant is shown here.

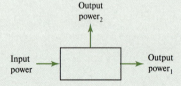

At steady state, the energy balance equation is

Input power = output power$_1$ + output power$_2$

From Example 8–2, input power = 2424 MW$_t$ and output power$_1$ = 800 MW$_e$, so

Output power$_2$ = 2424 MW − 800 MW = 1624 MW$_t$

Of this the stack losses are

$(0.15)(1624 \text{ MW}_t) = 243.6 \text{ MW}_t$

and the river water must store

$$(0.85)(1624 \text{ MW}_t) = 1380.4 \text{ MW}_t$$

Using Equation 4–44, we obtain

$$\frac{dH}{dt} = c_p \Delta T \frac{dM}{dt}$$

where $dH/dt = 1380.4 \text{ MW}_t$, c_p = specific heat of water from Table 4–3 is 4.186, and ΔT = the allowable temperature rise of 10°C. First converting units, we see

$$(1380.4 \text{ MW}_t)\left(\frac{1 \text{ MJ} \cdot \text{s}^{-1}}{\text{MW}}\right) = 1380.4 \text{ MJ} \cdot \text{s}^{-1}$$

then, noting that $\Delta T = 10°C = 10 \text{ K}$, solve Equation 4–44 for dM/dt

$$\frac{dM}{dt} = \frac{1380.4 \text{ MJ} \cdot \text{s}^{-1}}{(4.186 \text{ MJ} \cdot \text{Mg}^{-1} \cdot \text{K}^{-1})(10 \text{ K})} = 32.98 \text{ Mg} \cdot \text{s}^{-1}$$

Using a density of water of $1000 \text{ kg} \cdot \text{m}^{-3}$ (or $1 \text{ Mg} \cdot \text{m}^{-3}$), the required volumetric flow rate is

$$\frac{32.98 \text{ Mg} \cdot \text{s}^{-1}}{1 \text{ Mg} \cdot \text{m}^{-3}} = 32.98, \text{ or } 33 \text{ m}^3 \cdot \text{s}^{-1}$$

To find the increase in stream temperature, we solve Equation 4–44 for ΔT.

$$\Delta T = \frac{1380.4 \text{ MJ} \cdot \text{s}^{-1}}{(4.186 \text{ MJ} \cdot \text{Mg}^{-1} \cdot \text{K}^{-1})(63 \text{ m}^3 \cdot \text{s}^{-1})(1 \text{ Mg} \cdot \text{m}^{-3})} = 5.23 \text{ K}, \text{ or } 5.23°C$$

Because the upstream temperature was 18°C, the downstream temperature will be

18°C + 5.23°C = 23.23, or 23°C.

This will have a dramatically adverse effect on aquatic life in the stream.

Water Use in Energy Production. The production of energy requires large volumes of water. In the United States electricity production from fossil fuels and nuclear energy requires 7.2×10^8 $\text{m}^3 \cdot \text{d}^{-1}$ of water (Torcellini et al., 2003). This is about 39% of all freshwater withdrawals in the United States. Table 8–13 summarizes the freshwater usage for various energy sources.

TABLE 8–13	Freshwater Usage for Various Energy Sources	
	Energy Source	**Estimated Water Consumption, $\text{m}^3 \cdot \text{MW}^{-1} \cdot \text{h}^{-1}$**
	Biomass	1.14–1.51
	Coal	1.14–1.51
	Geothermal	~5.3
	Hydroelectric	5.4
	Natural gas	0.38–0.68
	Nuclear	1.51–2.72
	Solar thermal	2.88–3.48

Source: Torcellini et al., 2003; Desai and Klanecky, 2011; U.S. Department of Energy, 2006.

Terrain Effects.   Coal mining has the most serious effect on the terrain. About 26,000 ha of strip mining is done each year in the United States. In 1998, it was estimated that over 400,000 ha had been strip mined in the United States (McKinney and Schoch, 1998). Underground mines are notorious for land subsidence, including sudden collapses of the shafts that destroy homes, roads, and utilities. More than 800,000 ha of land in the United States has already subsided due to underground mining (Coates, 1981).

Unresolved Environmental Concerns.   Among the many environmental issues in energy production, two stand out as currently unresolved: (1) cleaning the water used in oil extraction from tar sands and reclaiming the land, and (2) potential groundwater contamination from hydrofracking.

The tar sand oil processes recover about 90% of the bitumen from strip-mined sand and about 55% from pumped wells. After recovery of the oil from the extracted tar sand, the slurry of water, bitumen contaminated clay and sand is pumped to tailings ponds where the solids are settled out. The mixture of contaminated clay and water has the consistency of yogurt. If untreated, this mixture would take 40 to 100 years to separate. Addition of proprietary chemicals reduces the time to about 10 years. Of the 650 square kilometers of Alberta that has been mined so far, only about 1 square kilometer has been certified reclaimed with another 73 square kilometers reclaimed but not certified (Ritter, 2011). This issue remains unresolved because of the very tardy response of the oil recovery companies in reclaiming the land.

Hydrofracking has been linked to drinking water contamination in Colorado, Ohio, New York, and Pennsylvania (Easley, 2011). The contaminants include sufficient methane to support combustion from degassing well water. The "flow-back" water that brings methane to the surface of the hydrofracking well is typically disposed of by pumping the wastewater back into a subsurface formation. The wastewater will contain one or more of the hydraulic fracturing fluid additives listed in Table 8–12. The concern is the potential for these contaminants to enter potable groundwater supplies. This issue remains unresolved because there currently is insufficient data to verify the source of the contaminants and the connection to the hydrofracking wells. In addition, there are no applicable appropriate regulations.

Sustainable Energy Sources

Sustainable energy sources, often called **renewables,** include hydropower, biomass or biofuels, geothermal, wave, and solar energy. Although photovoltaics, wind energy, and biofuels now provide only 6% of the world's, they are growing at annual rates of 17 to 29% (Hileman, 2006). Overall, renewable electricity supplied 9% of net generation in the United States in 2005 (computed from EIA, 2004, 2005b). Several of these alternative energy sources along with renewable hydrogen are explored in the following paragraphs. It should be noted that consensus of virtually every energy expert is that there is no "silver bullet," that is, no single alternative is going to solve our long-term demand for energy. Society is going to need all of the energy it can get from all of the available alternatives.

Hydropower.   Hydropower exploits the most fundamental of physics principles by converting the potential energy of stored water to kinetic energy of falling water. The falling water passes through a turbine that drives an electrical generator. Unlike fossil fuels, hydropower is a renewable resource because the water is renewed through the hydrologic cycle. Although naturally occurring water falls have been used to generate hydropower, dams have been the major method of storing the water and raising the elevation to gain potential energy. From the definition of potential energy, the amount of energy available is a function of both the mass of water available and the elevation difference that can be created by damming up the water.

$$E_p = mg(\Delta Z) \tag{8–10}$$

where E_p = potential energy (in J)

m = mass (in kg)

g = acceleration due to gravity

$= 9.81 \text{ m} \cdot \text{s}^{-2}$

ΔZ = head, the difference in elevation between the water surface at the top of the dam and the turbine (in m)

Ignoring the losses in energy from friction, the kinetic energy of the falling water is equal to the potential energy. This provides us with a method for estimating the velocity of the falling water. The definition of kinetic energy (E_k) is

$$E_k = \frac{1}{2}mv^2 \tag{8–11}$$

where v = velocity of water (in m · s^{-1})

If we set the two equations equal to each other, we can solve for the velocity

$$v = [(2g)(\Delta Z)]^{0.5} \tag{8–12}$$

The power available may be estimated from the time rate at which work is done by the falling water. This is the mass flow rate of water traveling through a distance equal to the head.

$$\text{Power} = g(\Delta Z)\frac{dM}{dt} \tag{8–13}$$

where dM/dt = mass flow rate of water (in kg · s^{-1})

$= \rho Q$

ρ = density of water (in kg · m^{-3})

Q = flow rate of water (in m^3 · s^{-1})

Example 8–6 illustrates how the potential energy and power may be estimated.

EXAMPLE 8–6

The Hoover Dam on the Colorado River at the Arizona–Nevada border is the highest dam in the United States. It has a maximum height of 223 m and a storage capacity of about 3.7×10^{10} m^3. What is the potential energy of the Hoover Dam and reservoir? If the maximum discharge is 950 m^3 · s^{-1}, what is the electrical capacity of the generating plant?

Solution

Assuming the density of water is 1000 kg · m^{-3} and applying the potential energy equation:

$$E_p = (3.7 \times 10^{10} \text{ m}^3)(1000 \text{ kg} \cdot \text{m}^{-3})(9.81 \text{ m} \cdot \text{s}^{-2})(111.5 \text{ m})$$

$$= 4.05 \times 10^{16} \text{ J, or } 40.5 \text{ PJ}$$

Note that the final units are

$$\left(\frac{\text{kg} \cdot \text{m}}{\text{s}^2}\right)(\text{m}) = \text{N} \cdot \text{m} = \text{J}$$

and that the average head (223 m/2 = 111.5 m) is used for the calculation.

Of course, all of this energy cannot be recovered. The estimate assumes all of the water is at the average head, that the reservoir is filled to maximum capacity, and that the maximum height is equal to the maximum head when, in fact, the maximum head is the distance between the maximum elevation of the reservoir pool and the turbines. In addition, the efficiency of converting the flowing water to mechanical energy to turn the turbine and generator is less than 100%.

At a flow rate of 950 $m^3 \cdot s^{-1}$, the electrical capacity is

$$Power = (9.81 \text{ m} \cdot s^{-2})(223 \text{ m})(1000 \text{ kg} \cdot m^{-3})(950 \text{ m}^3 \cdot s^{-1}) = 2.08 \times 10^9 \text{ J} \cdot s^{-1}$$
$$= 2.08 \times 10^9 \text{ W} = 2080 \text{ MW}$$

The actual rating of the Hoover Dam is 2000 MW. The average electrical power delivered may be considerably less than this because the average flow rate of water is less.

In 2007, the United States had a hydropower capacity of about 100 GW (gigawatts). This accounted for about 6% of the nation's electrical capacity (EIA, 2006). Most of this is developed at large dams such as Hoover and Glen Canyon in Arizona. It is unlikely, however, that any more large-scale plants will be built in the United States.

The combination of feasible topography for such dams and their negative environmental effects prohibit further expansion of large-scale projects. Dams and the reservoirs they create flood large tracts of land. For example, Lake Mead, the reservoir behind Hoover Dam, covers an area of 640 km^2. Natural habitats of a wide variety of biota are destroyed to accommodate such projects. Villages, homes, farms, and other natural resources are also lost. Frequently, the reservoir traps sediments and nutrients normally supplied to downstream ecosystems while altering the dissolved oxygen, temperature, and mineral composition of the water.

An alternative source of hydropower is the 70,000 small dams already in existence in the United States. These provide an opportunity to develop so called "low-head" hydropower. Because the flow rates are low and the elevation differences are small, these projects are marginally economical.

Biofuels from Biomass. Biomass energy includes wastes, standing forests, and energy crops. In the United States, the "wastes" include wood scraps, pulp and paper scraps, and municipal solid wastes. In developing countries, it may also include animal wastes. In developing countries, up to 90% of the energy may be supplied by biomass. Second only to hydropower in terms of renewable energy utilized in the United States, biomass burning provides about 2.5% of our nation's electrical capacity (computed from EIA, 2010b, 2010c). The environmental impacts are both positive and negative. Certainly, recovering the fuel value from wastes that would otherwise be buried is a positive aspect of burning biomass. The use of standing forests, especially those harvested in an unsustainable manner, results in the creation of a wasteland. Soil erosion and deprivation of the replenishing nutrients have long-term consequences that are not repairable.

In the popular literature, biofuel *is* ethanol. Other biofuels now in commercial use include alkyl esters and 1-butanol. In the United States, the primary feed stock for ethanol production is corn. Fermentation is the major method for production of ethanol for use as a fuel. The starch in the feed stock is hydrolyzed into glucose. The usual method of hydrolysis for fuels is by the use of dilute sulfuric acid and/or fungal amylase enzymes. Certain species of yeast (for example, *Saccaromyces cerevisiae*) anaerobically metabolize the glucose to form ethanol and carbon dioxide:

$$C_6H_{12}O_6 \rightarrow 2CH_3CH_2OH + 2CO_2$$

Ethanol is blended with gasoline. With the recent introduction of **"flex-fuel"** engine design, the allowable ethanol content has been raised from about 15% to 85%—the so called **E85.** At this time (2006), less than 2% of the total U.S. motor vehicle fleet is capable of running on E85, and there are only about 600 E85 service stations nationwide (Hess, 2006a).

Soybean oil serves as the primary feed stock for alkyl esters in the United States. In global use, canola oil provides 84% of the feed stock. Alkyl esters are used as a diesel blend or substitute that is commonly referred to as **biodiesel.** In a modern diesel engine, it can be blended in any percentage from B1 (1% biodiesel and 99% petrodiesel) to B99. Because of incentives from Congress, B20 is the popular blend (Pahl, 2005).

Biobutanol (1-butanol) is being brought to market as a competitor to ethanol. It has several advantages over ethanol. Ethanol attracts water and tends to corrode normal distribution pipelines. Thus, it must be transported by truck, rail, or barge to terminals where it is blended with gasoline. Butanol can be blended at higher concentrations than ethanol without having to retrofit automobile engines. It is also expected to have a better fuel economy than gasoline-ethanol blends (Hess, 2006b). As of 2017, it is still not commercially viable.

A lesser recognized but age-old source of biofuel is the methane generated from anaerobic decomposition of waste material. In Lansing, Michigan, for example, the methane recovered from a municipal solid waste landfill is used to produce power for more than 4500 homes each year.

There are three issues in the use of biofuels as a replacement for petrofuel: the energy balance, the environmental impact, and the availability of land. Critics have questioned the rationale behind the policies that promote ethanol for energy, stating that corn-ethanol has a negative energy value (Pimentel, 1991). That is, according to their estimates, the nonrenewable energy required to grow and convert corn into ethanol is greater than the energy value present in the ethanol fuel. Recent studies have concluded that changes in technology and increases in yields of corn and soybeans have resulted in net energy benefits (Farrell et al., 2006; Hammerschlag, 2006; Hill et al., 2006; Shapouri et al., 2002). These studies revealed that the corn ethanol energy output:input ratios ranged from 1.25 to 1.34. The ratio for soybean biodiesel was even higher, ranging from 1.93 to 3.67.

The use of ethanol is estimated to reduce CO_2 emissions because, for example, of the uptake of CO_2 by corn. E15 could reduce CO_2 emissions from the light duty fleet (all U.S. autos and small trucks) by 39%. Complete conversion to E85 would reduce the emissions on the order of 180% (Morrow, Griffin, and Matthews, 2006). Conversely, the total life cycle emissions of five major air pollutants (carbon monoxide, fine particulate matter, volatile organic compounds, sulfur oxides, and nitrogen oxides) are higher with E85. Low-level biodiesel blends reduce these emissions while reducing green house gas emissions that cause global warming (Hill et al., 2006).

The real issue for biofuels is the limit of agricultural production. "Even dedicating all U.S. corn and soybean production to biofuels would only meet 12% of gasoline demand and 6% of diesel demand," say Hill et al. (2006). The key to biofuels is to identify a nonfood crop that can be produced on marginal land. Switchgrass (*Panicum virgatum*) currently fulfills this requirement. It is a perennial warm-season grass native to the Midwest and Great Plains. It has been grown for decades as a pasture or hay crop on marginal land that is not well suited for conventional row crops. It is tolerant of both wet and dry conditions, requires less fertilizer and pesticides than corn or soybeans, and yields less agricultural waste. The major stumbling block is that the plant material is cellulosic—it cannot be economically converted to ethanol or butanol with our current (2016) technology.

Wind.   Even casual observation of hurricanes and tornadoes reveals that the wind has power. Sails and windmills have been used to harness the energy from lesser winds for centuries.

The force of the wind on a flat plate held normal to it is expressed as

$$F = \frac{1}{2} A\rho v^2 \qquad\qquad\qquad (8\text{–}14)$$

where F = force (in N)
   A = area of plate (in m^2)
   ρ = density of air (in kg $\cdot$ m^{-3})
   v = wind speed (in m $\cdot$ s^{-1})

If the force moves the plate through a distance, then work is done. The product of (distance) $(A)(\rho)$ = mass, which is the definition of kinetic energy (Equation 8–11).

Two characteristics of Equation 8–14 are important in developing wind power. The first, well noted by sailors in "putting on more sail," is that the larger the area, the greater the force captured. The second is less obvious. That is, that the force is proportional to the square of the velocity, and power (kilowatts) is proportional to the cube of velocity (force × velocity). It is a characteristic of the wind that the velocity increases with distance above ground. Thus, a wind-mill will be more effective if it is placed at a higher elevation above the ground than a lower one.

Modern wind turbines for generation of electrical power are placed on towers ranging from 30 to 200 m in height in "wind farms." Individual turbine generating capacity ranges from 750 kW to 8 MW. This is almost a 10-fold increase over the last decade.

Although wind power is environmentally benign, it is not very dependable. Furthermore, the sound of 50 or more whirring propellers is not something most people want in their back-yard. None the less, it has been forecast that up to 12% of the world demand for electricity could be provided by wind in the next two decades. Europe leads the world in wind power with almost 35 GW capacity. Denmark supplies about 20% of its electrical needs from wind power (Parfit and Leen, 2005). Almost 10 GW have been installed in the United States. In the last decade, the cost of wind power has dropped from 18–20 cents per kWh to 4–7 cents per kWh.

Solar.   In 20 days time, the earth receives energy from the sun equal to all of the energy stored as fossil fuels. This resource may be captured directly, as is done with greenhouses; passively, for example with thermal masses to absorb solar radiation; or, actively, through water heaters, photovoltaic cells, or parabolic mirrors.

To be economically realistic solar energy collection is limited to places with a lot of sun-shine; Michigan is not an ideal setting. Furthermore, direct and passive systems are difficult to implement as a retrofit to existing structures. The system needs to be designed together. The active solar collection systems require large areas of land and are very expensive to install.

Photovoltaic (PV) systems for generating electricity have been on the market for about 40 years. Their high cost and the need for backup power or a large array of batteries when the sun doesn't shine have impeded their acceptance. The Japanese have been leaders in im-plementing improvements over the past 20 years. They have reduced the installed cost from $40–50 per W to $5–6 per W. This has resulted in a reduction in electricity cost to 11 to 12 cents per kWh. This is a very favorable rate in comparison to Japanese utility-generated electricity at 21 cents per kWh. In the United States, utility-generated electricity costs about 8.5 cents per kWh, so PV systems are not competitive without a subsidy. California and New Jersey are leaders in providing subsidies. These include not only cash rebates for installation but also regulatory programs that require utilities to purchase power from the PV users when they cannot use all the electricity they generate (Johnson, 2004).

In 2016, worldwide PV system capacity was estimated to have grown to 305,000 MW.

Hydrogen.   A cheap, robust fuel cell is the key to using hydrogen as a fuel. A membrane electrode assembly is the heart of the fuel cell (Figure 8–14). The proton-exchange membrane is a barrier to hydrogen but not protons. The membrane allows a catalyst to strip electrons, which power an electric motor, before the hydrogen reacts with oxygen to form water. The ultimate source of the hydrogen is water. Ideally, the hydrogen is to be split from the water using natural systems such as photovoltaic cells to produce the power.

The automobile industry has done much research on using hydrogen as a fuel replacement for petroleum. Fuel cells and hydrogen fuel are a reality today, but their application to automo-biles has major hurdles to overcome:

- The cost of membranes must be reduced from the current price of $150 per m^2 to less than half that and preferably to about $35 per m^2.
- The life of the membranes must be doubled from 1000 hours to 2000 hours.

FIGURE 8–14

A typical membrane electrode assembly like this one shows two flow field plates that are the heart of the fuel cell.

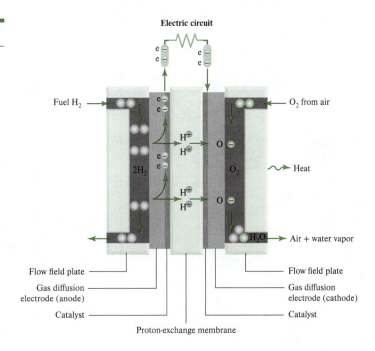

- The operating temperature range must be increased from 80°C to 100°C and the relative humidity range must be increased from 25% to 80%.
- A suitable on-board hydrogen storage system must be devised.
- A hydrogen infrastructure comparable to the petroleum infrastructure must be developed.

These are not insurmountable obstacles, but it will be a while before hydrogen-powered vehicles replace the gasoline-fired engine.

Green Engineering and Energy Conservation

There are numerous examples that may be used to illustrate methods for energy conservation. Not all of these are practical, and some have limited value because of the small segment of the energy demand that they impact. Figure 8–14 reveals the major flows of energy and the areas where conservation effort may provide the most benefit. The several examples selected for discussion here were chosen because of their relevance to civil and environmental engineering practice.

Macroscale Incentives.   Electric motors account for about half of the electricity used in the United States (Masters and Ela, 2008). Oversized motors and constant-speed motors both contribute to inefficient use of electricity. Pumps driven by electric motors are often regulated by adjusting valves instead of the speed of the motor itself. The excess electrical energy is wasted as heat.

Old power plants rarely achieve efficiencies above 35%. In part this occurs because the waste heat is sent to a nearby river or other water body rather than being used elsewhere. **Cogeneration** power plants use the waste heat to heat buildings, thus making much more efficient use of the fuel.

As shown in Figure 8–15, transportation accounts for 27% of the total energy use in the United States. Automobiles and other light-duty vehicles use about 80% of the energy consumed in transportation. Increased efficiency has raised the fuel use of automobiles from 7.7 km · L^{-1} in the mid-1970s to 11.8 km · L^{-1} in 1992. Yet, this increase in efficiency was far outstripped by the growth in the number of vehicle-kilometers traveled. Worldwide, more than 700 million vehicles are on the road, and the growth rate is about 35 million per year.

FIGURE 8–15

Energy Flow, 2005.

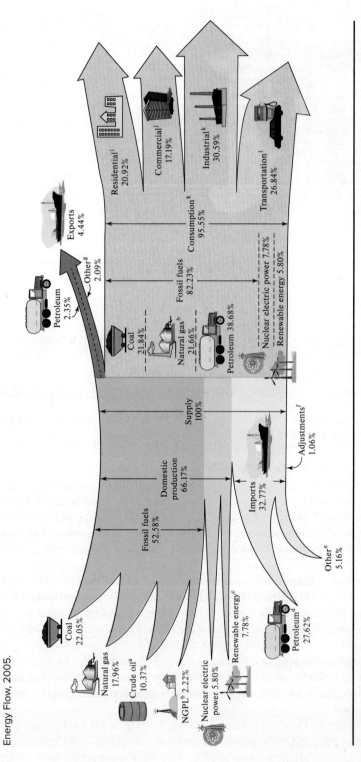

^a Includes lease condensate.

^b Natural gas plant liquids.

^c Conventional hydroelectric power, wood, waste, ethanol blended into motor gasoline, geothermal, solar, and wind.

^d Crude oil and petroleum products. Includes imports into the Strategic Petroleum Reserve.

^e Natural gas, coal, coal coke, and electricity.

^f Stock changes, losses, gains, miscellaneous blending components, and unaccounted-for supply.

^g Coal, natural gas, coal coke, and electricity.

^h Includes supplemental gaseous fuels.

ⁱ Petroleum products, including natural gas plant liquids.

^j Includes 0.04 quadrillion Btu of coal coke net imports.

^k Includes, in quadrillion Btu, 0.34 ethanol blended into motor gasoline, which is accounted for in both fossil fuels and renewable energy but counted only once in total consumption; and 0.08 electricity net imports.

^l Primary consumption, electricity retail sales, and electrical system energy losses, which are allocated to the end-use sectors in proportion to each sector's share of total electricity retail sales.

Notes: Data are preliminary. Values are derived from source data prior to rounding for publication. Total may not equal sum of components due to independent rounding.

Sources: EIA, 2005a, Tables 1.1, 1.2, 1.3, 1.4, and 2.1a.

Some energy conservation techniques are to increase the number of passengers by offering commuter lanes and parking incentives. It has been estimated that if every commuter car carried an average of one more person, 700,000 barrels of oil (about 4.3 PJ of energy) would be saved each day. Mass transit would improve this even more. On a global scale, these techniques are only marginal in reducing the oil depletion. Only a major revolution in transportation modes, for example switching to bicycles, will provide any benefit, and that may come too late.

Green Engineering and Building Construction. In the United States, buildings account for 42% of the energy consumption and 68% of the total electricity consumption (Janes, 2010). About 80% of this energy consumption is used in residential heating, cooling, and lighting systems. Examples of electricity consumption for common household appliances are shown in Table 8–14.

TABLE 8–14	Energy Demand for Common Household Appliances		
Appliance	**Average Demand, W**		**Comment**
Air conditioning			
Central	2000–5000		Function of size of house
Room (window)	750–1200		Function of size of room
Clothes dryer (electric)	4400–5000		
Clothes washer	500–1150		
Computer	200–750		
Dishwasher	1200–3600		Function of water heater and dry cycle (1200 W)
Freezer	335–500		
Furnace fan	350–875		Function of size of house
Lights			
Incandescent equivalent	Compact fluorescent		Note: cfl has a hazardous mercury component
40 W	11		
60 W	16		
75 W	20		
100 W	30		
Microwave oven	1500		
Range			
Oven	2000–3500		Temperature dependent
Small burner element	1200		
Large burner element	2300		
Refrigerator/freezer			
Frost free	400		
Not frost free	300		
Side by side	780		
Television			
CRT 27-inch	170		
CRT 32-inch	200		
LCD 32-inch	125		
Plasma 42-inch	280		
Video game	100		
Water heater (electric)	4000–4500		

Building design teams have become increasingly aware of the need to incorporate green engineering in both new designs and retrofitting old ones. Of particular note for civil and environmental engineers is the emergence of whole building assessment systems like Leadership in Energy and Environmental Design (LEED) and Green Globe. These provide a methodology for assessing the environmental performance of products selected for construction. Environmental performance is measured in terms of a wide range of potential impacts. Examples include human health respiratory effects, fossil fuel depletion, global warming potential, and ozone depletion. Both Green Globes and LEED have adopted a form of Life Cycle Assessment (LCA) that assigns credits for selecting prestudied building assemblies that are ranked in terms of the effects associated with making, transporting, using, and disposing of the products. In the process of selecting material, the LCA of a product also includes the use of other products required for cleaning or maintaining the product. As an interim aid to the design process, a computer program (the Athena EcoCalculator, available at www.athenasmi.org) has been developed to assess alternatives. Ultimately, the EcoCalculator will serve as the basis for inputs to a separate computational system to evaluate whole building LCA (Trusty, 2009).

The first step in using the EcoCalculator is to select an assembly sheet from one of the following categories:

- Columns and beams
- Exterior walls
- Foundations and footings
- Interior walls
- Intermediate floors
- Roofs
- Windows

The number of assemblies in each category varies widely depending on the possible combinations of layers and materials. For example, the exterior wall category includes nine basic types, seven cladding types, three sheathing types, four insulation types, and two interior finish types (Athena Institute, 2010).

Energy conservation is an important element in building design. For example, improved building insulation has a dramatic effect on reduction of energy consumption. From Equation 4–50, we may note that the effectiveness of insulation is a function of both its thermal conductivity and thickness. In the heating and air conditioning market it is more common to refer to the *resistance* (R) of the insulation than its thermal conductivity. The resistance is the reciprocal of the thermal conductivity:

$$R = \frac{1}{h_{tc}} \tag{8–15}$$

where h_{tc} = thermal conductivity

Larger values of R imply better insulating properties. When multiple layers of different materials are used, the combined resistance may be estimated as

$$R_T = R_1 + R_2 + \cdots + R_i \tag{8–16}$$

The resistance form of Equation 4–50 is

$$\frac{dH}{dt} = \frac{1}{R_T}(A)(\Delta T) \tag{8–17}$$

where A = surface area (in m^2)
  ΔT = difference in temperature (in K)
  R_T = resistance (in m$^2 \cdot$ K $\cdot$ W^{-1})

Example 8–7 illustrates the value of additional roofing insulation.

EXAMPLE 8–7   A typical residential construction from the 1950s consisted of the layers shown in the drawing. Estimate the heat loss with the existing insulation scheme and with an additional 20 cm of organic bonded-glass fiber insulation if the indoor temperature is to be maintained at 20°C and the outdoor temperature is 0°C.

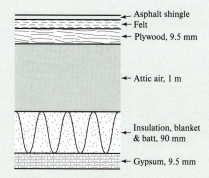

Asphalt shingle
Felt
Plywood, 9.5 mm

Attic air, 1 m

Insulation, blanket & batt, 90 mm

Gypsum, 9.5 mm

Solution   The resistance values are obtained from Table 8–15. Assuming a 1-m^2 surface area, the total resistance for the original construction in units of m$^2 \cdot$ K $\cdot$ W^{-1} is calculated as

R = asphalt + felt + plywood + air + insulation + gypsum

$$= 0.077 + 0.21 + 0.10 + \frac{1000 \text{ mm}}{90 \text{ mm}} (0.4) + 2.29 + 0.056 = 7.18 \text{ m}^2 \cdot \text{K} \cdot \text{W}^{-1}$$

where the ratio 1000/90 is the number of 90 mm air spaces in 1 m of air in the attic.

From Equation 8–17, we get

$$\frac{dH}{dt} = \frac{1}{7.18 \text{ m}^2 \cdot \text{K} \cdot \text{W}^{-1}} (1 \text{ m}^2)(20 - 0)$$

$$= 2.79 \text{ W}$$

The additional insulation will add resistance. To be in consistent units, we must multiply by the thickness of the insulation.

$R = (27.7 \text{ m} \cdot \text{K} \cdot \text{W}^{-1})(0.20 \text{ m}) = 5.54 \text{ m}^2 \cdot \text{K} \cdot \text{W}^{-1}$

The new resistance is then

$R_T = 7.18 + 5.54 = 12.72$

and the heat loss is

$$\frac{dH}{dt} = \frac{1}{12.72 \text{ m}^2 \cdot \text{K} \cdot \text{W}^{-1}} (1 \text{ m}^2)(20 - 0)$$

$$= 1.57 \text{ W}$$

This is about 56% of the original heat loss.

TABLE 8–15	Typical Values of Resistance for Common Building Materials		
Building Material		R (m · K · W^{-1})	R **for Thickness Shown** (m^2 · K · W^{-1})
Building board			
Gypsum, 9.5 mm			0.056
Particle board		7.35	
Building membrane (felt)			
Two layers, 0.73 kg · m^{-2} felt			0.21
Glass			
Single glazing, 3 mm			0.16
Double glazing, 6 mm air space			0.32
Triple glazing, 6 mm air space			0.47
Insulating material			
Blanket and batt			
Mineral fiber from glass,			
≈ 90 mm			2.29
≈ 150 mm			3.32
≈ 230 mm			5.34
≈ 275 mm			6.77
Glass fiber, organic bonded		27.7	
Loose fill milled paper		23	
Spray applied polyurethane foam		40	
Roofing			
Asphalt shingles			0.077
Built-up, 10 mm			0.058
Masonry			
Brick		1.15	
Concrete		0.6	
Siding materials			
Hardboard, 11 mm			0.12
Plywood, 9.5 mm			0.10
Aluminum or steel, over sheathing,			
Hollow-backed			0.11
Insulating board backed, 9.5 mm			0.32
Insulating board backed, 9.5 mm, foil			0.52
Soils		0.44	
Still air, 90 mm			0.4

Source: Data used from ASHRAE, *Handbook of Fundamentals,* American Society of Heating and Air Conditioning Engineers (1993).

Green Engineering and Building Operation.   We tend to think of buildings as inert structures but they are in fact operating. For example, residential buildings have furnaces/air conditioners and refrigerators that operate whether or not anyone is home. Commercial and institutional buildings have heating and ventilating systems (HVAC), area lighting, and computers that operate 24 hours, seven days a week whether or not they are open for business.

Operations in residential buildings can be engineered to reduce energy consumption by the use of energy efficient appliances and regulating the heating/air conditioning systems with programmable thermostats that reduce operational times when the residence is not occupied. Turning off the lights, computers, and televisions when they are not in use also helps.

Commercial/institutional "smart buildings" use computer control systems to regulate area lighting and HVAC systems based on occupancy. Regular preventive maintenance of the HVAC system is essential to conserve energy. Electric motors account for about one-half of the electricity used in the United States (Masters and Ela, 2008). Oversized motors and constant speed motors both contribute to inefficient use of electricity. Out-of-balance fans reduce the efficiency of the HVAC system. Personal computers should be turned off at the end of the work day. Computers draw virtually as much power when they are in the "sleep mode" as they do when they are active.

Installation of "smart" metering systems at residential and commercial buildings provides an opportunity for the owner to obtain real-time data on electricity and natural gas use (McNichol, 2011). These data can be used to develop and implement energy conservation plans on a building by building basis.

Green Engineering and Transportation.   In the United States, transportation accounts for 28% of energy use (EIA, 2010e). The Corporate Average Fuel Economy (CAFE) standards were first enacted in 1975. The objective of the standards is to increase the motor vehicle fuel efficiency. The standards have been adjusted periodically to increase the miles per gallon (mpg). It is estimated that, for cars and light duty trucks, an average fuel economy standard of 15.09 km $\cdot$ L^{-1} was achieved in 2016 models. In addition to energy conservation, the achievement of the CAFE standards will result in the reduction of greenhouse gas carbon dioxide emissions. Yet, this increase in efficiency may be far outstripped by growth in the number of vehicle-kilometers traveled.

Recycling of asphalt and concrete pavement materials not only saves raw material, it also saves energy. When recycled asphalt pavement is incorporated into new pavement, the asphalt cement in the old pavement is reactivated. Recycling in place is another energy-saving step in asphaltic pavement rehabilitation. Other materials including rubber from used tires, glass, and asphaltic roofing can also be recycled into asphaltic pavement (NAPA, 2010).

Concrete pavement is also recyclable. In addition, supplementary cementitious material (SCM) may be used to replace portland cement or as an additive. Common SCMs are fly ash (see Chapter 12), slag cement, ground blast furnace slag, and silica fume (ACPA, 2007).

Green Engineering and Water and Wastewater Supply.   In some regions 30 to 50% of the total operating cost for a drinking water supply is for energy. At wastewater treatment plants, energy accounts for 25 to 40% of operating costs (Feldman, 2007). *Energy Conservation in Water and Wastewater Facilities* (Schroedel and Cavagnaro, 2010) provides numerous techniques to conserve energy. Several of these are highlighted in the following paragraphs.

Much of the cost for energy is for pumping. Pumps driven by electric motors are often regulated by adjusting valves instead of the speed of the motor itself. The excess electrical energy is wasted as heat. Variable frequency drives (vfd) adjust the speed and thus the pumping rate by adjusting the frequency of the electric current. This is a major means to reduce energy use and cost. Replacement of oversized motors and/or pump impellers is another method to reduce energy inefficiency. In wastewater plants, aerators driven by electric motors are another major energy consumer. Changing coarse bubble aerators to fine bubble aerators improves oxygen transfer efficiency and provides the opportunity to reduce motor sizes or implement vfd systems (Herbert, 2010). Using computer control systems to regulate the air flow in proportion to the wastewater flow and strength is another means to reduce energy consumption (Rogers, 2010).

Control of chemical dosing is another method to reduce energy consumption (Truax, 2010). For example, adjusting the dose of lime in softening plants (see Chapter 10) to achieve a final hardness of 130 mg/L as $CaCO_3$ instead of 80 mg/L as $CaCO_3$ will reduce mineral resource consumption, energy in production and transportation, and the energy and costs for sludge disposal.

8–4 MINERAL RESOURCES

Reserves

The estimated United States and world reserves of three metals and phosphorus are shown in Table 8–16. How long will the reserves last? Calculations similar to those in Example 8–3 can be used to make estimates.

EXAMPLE 8–8   In 2002, the international production of iron was 1080 Tg. Assuming the 2002 demand remains constant, how long will world reserves last? World production increased 2.85% from 2001 to 2002. If that rate of increase remains constant, how long will world reserves last?

Solution   If the demand remains constant (Equation 8–1, with F mineral reserve), the world reserve will last

$$\frac{79,000 \text{ Tg}}{1080 \text{ Tg} \cdot \text{year}^{-1}} = 73.15 \text{ or } 73 \text{ years}$$

If the demand rises at a rate of 2.85%, then we can use the following growth expression:

$$79,000 \text{ Tg} = 1080 \text{ Tg} \cdot \text{year}^{-1} \left[\frac{(1 + 0.0285)^n - 1}{0.0285} \right]$$

Solving for n

$$(73.15)(0.0285) = (1.0285)^n - 1$$
$$3.085 = (1.0285)^n$$

Taking the log of both sides

$$\log(3.085) = \log[(1.0285)^n] = n \log[1.0285]$$
$$0.4892 = n(0.0122)$$
$$n = 40.08 \text{ or } 40 \text{ years}$$

TABLE 8-16   **U.S. and World Reserves of Some Common Metals and Phosphorus[a]**

Mineral	United States (Tg[b])	World (Tg[b])
Aluminum	20	28,000
Copper	680	940
Iron	790	82,000
Phosphorus	1100	69,000

[a]Data are for 2016.

[b]Tg = teragrams = 10^{12}

Source: USGS, 2016.

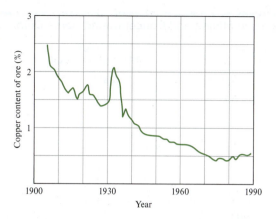

FIGURE 8–16

Since 1905, the quality of copper ore mined in the United States has declined from 2.5% to about 0.5%. (*Source:* U.S. Bureau of Mines 1990.)

This example illustrates the finite nature of our mineral resources. However, both estimates incorporate a number of assumptions that are not realistic. They both assume no changes in technology, no further discoveries of new ores, and, perhaps most importantly, that the grade of the ore must remain the same as it is today to be economically feasible to mine. In general, we have been able to extract lower grade ores with increases in technology as shown in Figure 8–16. It has been estimated that even with marginal grade ores, at current rates of growth, the maximum reserve capacity for aluminum, copper and iron will be exhausted in 160, 33, and 78 years, respectively (Ophuls and Boyan, 1992). In addition, the cost of extracting these ores may be prohibitive.

Phosphorus.   As a component of DNA, phosphorus is an essential element for all living things. With the beginning of the 21st century came a heightened awareness of a potential shortage of this essential element. In the last century the United States produced one-third of the world's production. Seventeen percent of that was exported as phosphate rock. The remainder was converted to fertilizer. Five countries: China, Morocco and Western Sahara, Jordan, South Africa, and the United States control 89% of the world's reserves and are responsible for 70% of its annual production. The United States stopped exporting phosphate rock in 2004. China has begun to restrict its exports by imposing a 130% export tarrif. At current production rates, the reserve life estimate for the United States is approximately 40 years. Like energy and other mineral reserve estimates, the estimates given in Table 8–16 do not include possible new undiscovered deposits and improvements in extraction technology (Vaccari, 2011).

About 17% of the phosphorus in fertilizer enters the human diet (Cordell and White, 2008). Major losses include agricultural erosion and lack of or improper application of animal wastes. Of these, capture of losses from animal waste are the most economically and technically feasible. In the United States about half of the volume of waste from concentrated animal feeding operations is lost either through accumulation in piles or by over application beyond that which the land can assimilate. The loss of phosphorus from fertilizer applications and its environmental impacts are discussed in Chapter 5.

Of the portion of phosphorus that is consumed by humans about 86% is excreted. Of this amount, 40% ends up in landfills and the remainder is discharged to surface waters where it contributes to eutrophication.

Environmental Impacts

Energy.   Although we tend to think of cost in terms of dollars, a more realistic measure of cost for extraction of minerals is the energy required to mine and process them. Table 8–17 gives some examples of the energy expenditures in mining and concentrating some common metallic

TABLE 8–17   **Energy Expenditures in Mining and Concentrating Some Common Metallic Ores**

Ore	Grade (%)	Energy (MJ · kg^{-1} of metal)
Aluminum	25	235
Copper	0.7	74
Iron	30	3.0

Sources: Atkins, Hitter, and Wiloughby, 1991; Hayes, 1976.

ores. These expenditures do not include the energy to fabricate a metallic product such as a beverage can or car bumper.

It is estimated that approximately 1% of the world's energy resource is used for aluminum production each year and over 5% of the world's energy is used in steel production (McKinney and Schoch, 1998).

Waste.   Because the rock from which minerals are mined contains only a fraction of the mineral, a substantial amount of rock waste is generated by mining. In the United States, the annual production of nonfuel mineral mining waste is estimated to be 1.0–1.3 petagrams (Pg) (McKinney and Schoch, 1998). This is about seven times the amount of municipal solid waste generated in U.S. cities. In the following example, a simple mass balance of ore excavation and concentration illustrates why there is so much waste rock.

EXAMPLE 8–9   Estimate the amount of waste rock generated in producing 100 kg of copper from an ore containing 0.5% copper.

Solution   The mass-balance diagram is shown here.

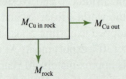

The mass-balance equation is

Accumulation = $M_{Cu\ in\ rock} - M_{Cu\ out}$

At steady state, the accumulation = 0.

At 0.5% concentration, the amount of copper in the ore is

$M_{Cu\ in\ rock} = (0.005)(M_{rock})$

This is also the amount of copper produced ($M_{Cu\ out}$). So

$M_{Cu\ out} = (0.005)(M_{rock})$

Because the $M_{Cu\ out}$ is 100 kg,

$$(M_{rock}) = \frac{100\ \text{kg}}{0.005} = 200{,}000\ \text{kg}$$

This waste rock does not include the soil cover (**overburden**) that is removed to expose the ore.

The waste generated in producing the mineral is not limited to waste rock. Additional waste includes wastewater and air pollutants from the smelting operations. If the ore contained sulfides, then rainwater leaching through the waste rock piles (**tailings**) will become acidified. Erosion of the waste rock piles results in increased loads of suspended solids in nearby streams and rivers. The suspended solids settle and smother hatcheries and clog gills of fish. An estimated 16,000 km of streams have been damaged by leaching from mine tailings.

Smelting operations frequently release sulfur dioxide and volatile heavy metals (e.g., lead and arsenic) to the atmosphere. Smelting operations are estimated to produce 8% of the world-wide sulfur emissions (McKinney and Schoch, 1998).

Terrain Effects.   Although the land directly involved in mining may seem small, it is far from insignificant. In the United States over 1.3 million ha (an area approximately the size of Connecticut) have been disturbed by surface mining (Coates, 1981). In addition to being unsightly, without reclamation, this disturbed land is no longer viable as a support system for renewable resources. In some cases the mined land poses hazards from cave-ins and subsidence of the land surface beneath the roads and cities built above the mines. At some surface mines, landslides are an imminent danger.

Resource Conservation

All methods of resource conservation are not equal. Some yield more or longer term benefits than others. There is a hierarchy of resource conservation. The preferred order is reduced consumption, material substitution, and finally, recycling.

Reduced Consumption.   Since the 1970s, the use of raw materials in the United States and Europe has leveled off or even declined a bit. Part of this trend is explained by the fact that the western industrialized nations have a firmly built infrastructure and major construction of public works is less common than in the past. In addition, the economies of the industrialized nations are reorienting to high-technology goods and consumer services that demand less raw materials. This trend, of course, does not include the developing countries, where the demand for minerals will remain high.

Other, more proactive, means of reducing consumption include product design, process management, and dematerialization. Products that do not wear out as quickly and that are easy to disassemble to recover materials at the end of their usefulness all contribute to reduced consumption. Germany has imposed requirements on their automobile manufacturers that require them to take back and recycle old automobiles. Companies like BMW and Volkswagen have redesigned their cars to be disassembled to recover parts.

Better process management improves efficiency and reduces the amount of waste material generated. This has benefit for production economics as well as reduced consumption of raw materials.

Another strategy, called **dematerialization,** is to reduce the size of the product while performing the intended function. For example, a smaller, lighter car still provides the transportation function but uses a smaller amount of mineral resources. Miniaturization in the electronics industry is a classic example of the savings in materials.

Dematerialization must be considered carefully. In some cases a smaller, lighter product may result in a less durable product that will, over time, use more material than a heavier, more durable product.

Material Substitution.   A simple way to reduce the demand for mineral resources is to produce more durable goods. Frequently, this means substituting a more durable material for a less durable one. A classic example is the substitution of plastic parts for steel on automobiles. The plastic does not corrode, and, thus extends the life of the body.

Alternatively, a material may be substituted because it performs the same function more efficiently. Fiber optic cables in place of copper telephone wire not only reduce the use of a vanishing mineral resource but also provide better communication service.

Recycling. The law of conservation of mass tells us that the minerals we have mined do not disappear. Although they have served their useful life as a product, the mineral content is still there. One of the more obvious solutions to the resource limit problem is recycling. In 1993, 15.6 Tg of metals was disposed of in municipal solid waste systems. Iron accounted for 76% of the mass, and aluminum accounted for 17%. Since the early 1970s the effort to recycle has increased substantially. For example, in 1999/2000 recycled aluminum accounted for 36% of the production of finished metal and recycled copper accounted for 44% of the production (Plunkert, 2001, and Zeltner et al., 1999). The fact that only 35% of the discarded aluminum was recovered implies there is room to grow in this effort.

The advantage of recycling is also reflected in a reduction in energy consumption. For example, the energy to produce aluminum stock from recycled metal is 5.1 MJ · kg^{-1} compared with 235 MJ · kg^{-1} required to produce it from ore.

On the down side, recycling has practical limits. The second law of thermodynamics cannot be violated. Each use results in some deterioration or loss through transformation. Normal wear and tear will result in the loss of some metal. Some loss will be in the form of corrosion, some will be in the form of microscopic abrasive loss, and some will be dissipative loss (e.g., in dyes, paints, inks, cosmetics, and pesticides). In the process of recovery, the crushing, grinding, and remelting will cause some additional loss.

The amount of metal available at the end of each cycle of reclamation may be approximated as a first-order decay.

$$\frac{dM}{dn} = -kM \tag{8-18}$$

where dM/dn = change in mass of metal per cycle of reclamation (in kg · cycle^{-1})
  k = decay constant (cycle^{-1})
  M = mass recovered in previous cycle (in kg)

The solution to this equation yields

$$M = M_o e^{-kn} \tag{8-19}$$

where M = mass recovered (in kg)
  M_o = original mass (in kg)
  k = decay constant (cycle^{-1})
  n = number of cycles

EXAMPLE 8–10 Assume that you could track a single aluminum beverage can (with a mass of 16 g) through several cycles of reclamation and that a 10% "loss" occurred in each recovery cycle. How much new aluminum must be supplied to replace the loss at the end of the third reclamation?

Solution The decay constant is not known but may calculated from the assumptions. For the first cycle, the amount of mass recovered is 90% of the original mass (100 − 10% loss). So

$$\frac{M}{M_o} = 0.90 = e^{-kn}$$

where $n = 1$

Taking the natural logarithm of both sides and solving for k, we obtain

$$\ln(0.90) = \ln[e^{-k(1)}]$$

$$-0.1054 = -k(1)$$

$$k = 0.1054 \text{ cycle}^{-1}$$

The approximate mass remaining at the end of the third cycle is

$$M = M_o e^{-kn} = (16 \text{ g}) \exp[(-0.1054)(3)]$$

$$= (16 \text{ g})(0.7290) = 11.664 \text{ g}$$

The approximate mass of new aluminum that must be supplied to replace the loss at the end of the third reclamation is then

$$16 \text{ g} - 11.664 = 4.336, \text{ or } 4.3 \text{ g}$$

Another way of expressing the limits of recycling is to estimate the equivalent mass of metal if it is recycled an infinite number of times. Or, perhaps, the equivalent mass if it is recycled a given number of times. In the first instance, this may be determined by the sum of a series.

$$\sum_{k=0}^{\infty} M_k = M_o + M_o f + M_o f^2 + \cdots + M_o f^n = \frac{M_o}{1-f} \qquad (8\text{–}20)$$

where M_o = original mass (in kg)
$\qquad$ f = fraction recovered
$\qquad$ n = number of cycles

and $0 < f < 1$ and $n = \infty$

In the second instance, this may be determined by the sum of the series:

$$\sum_{k=0}^{n} M_k = M_o + M_o f + M_o f^2 + \cdots + M_o f^{n-1} = \frac{M_o(f^n - 1)}{f - 1} \qquad (8\text{–}21)$$

where f is not $= 1$ and $n < \infty$.

EXAMPLE 8–11   What is the equivalent mass of aluminum saved if the beverage can in Example 8–10 is recycled an infinite number of times?

Solution   Using the data from Example 8–10 and Equation 8–20, we obtain

$$\sum M_k = \frac{16 \text{ g}}{1 - 0.90} = 160 \text{ g}$$

8–5   SOIL RESOURCES

Energy Storage

The soil is a storage place for energy. The primary source of energy is the sun. Plant photosynthesis captures this energy. It is brought to the soil by litter fall, by plant roots, by animals, and by oxidation of some minerals by microorganisms. The energy input to the soil is used to drive reactions involved in the food chain and in the transformations soil organisms perform on the soil components.

Plant Production

Because of the diversity of soils and the uses we make of them, no single measure comparable to those we used for mineral or energy reserves can be used for soils. However, the intimate relationship between plants and soil as well as our dependence on those plants leads us to use forest land and agricultural production potential as a measure of the soil resource. In doing this, we obviously slight forest potential, ecosystem habitat, and a wide variety of other basis for evaluating the soil resource.

The estimated area of the world land mass covered by forests is about 40×10^6 km^2. Of this area, North America, South America, and Russia each have about 8×10^6 km^2. In 1996, this amounted to about 0.7 ha of forest per person (WCMC, 1998). In the United States, the area of forested land has stabilized at 1920 levels; each year more trees are grown than are harvested (U.S. Forest Service, 1999). In contrast, it is estimated that the world forest resource will be reduced to 0.46 ha per person by 2025 (WCMC, 1998). Unfortunately, the major losses of forest land will be in areas that have the least amount of forest land.

Arable land is land that is or that can be cultivated. Agriculturally productive land consists of arable land and that which is not arable but is suitable for grazing. The total world land area is about 13×10^9 ha (Table 8–18). About half of this land is nonarable. It is desert, swampland, or

---

TABLE 8–18   Population and Cultivated Land on Each Continent

Continent	Population in 2001[a] (millions)	Total[b]	Potentially arable[b]	Cropland[c]	Cropland (ha per capita)
		Area (10⁶ ha)			
Africa	823	2966	733	183	0.22
Asia	3277[d]	2679	627	455	0.14
Australia, New Zealand and Oceania	31	843	154	48	1.55
Europe	729	473	174	140	0.19
North America	486	2139	465	274	0.56
South America	351	1753	680	139	0.40
Russia and Baltic	460	2234	356	232	0.50
Total	6157	13,087	3189	1471	0.24

[a]From Bureau of Census, U.S. Dept. of Commerce, estimated 2002.

[b]From The President's Science Advisory Committee Panel on World Food Supply, 1967.

[c]From World Resources, A report of the International Institute for Environment and Development and World Resources Institute, Washington, D.C., 1987.

[d]Asia does not include Russia and the Baltic states.

mountainous. About 25% of the land supports enough vegetation to provide grazing for animals but otherwise is unsuitable for cultivation (Brady, 1990).

How can we evaluate the reserve capacity of the soil resource? In terms of cultivated land per person (see Table 8–18), the total potentially arable land is more than double that being cultivated. In terms of "consumption," the amount of grain area harvested per person declined from 0.23 ha in 1950 to 0.13 ha in 1998 in spite of a doubling of the world's population. This is a result of the introduction of modern agricultural practices, including mineral fertilization, specially bred varieties, and the application of pesticides and herbicides. In a worst case scenario, if we assume that no further advances are made in improving **yields** (mass of crop $\cdot$ ha^{-1}) and that population growth continues at its current rate, then all of the arable land will have to be put into production by 2050 (Meadows, Meadows, and Randers, 1992).

Compounding the difficulty in assessing the reserve capacity of the soil resource, is the question of sustainability. Can we continue modern agricultural practices without destroying the resource?

8–6   PARAMETERS OF SOIL SUSTAINABILITY

The chief soil factors influencing plant growth are nutrient supply, soil acidity, soil salinity (with special reference to arid and semiarid land), texture and structure, and depth available for rooting. These are discussed in the next sections in the context of sustainability.

Nutrient Cycling

There are 16 elements without which green plants cannot grow: C, H, O, N, P, S, Ca, Mg, K, Cl, Fe, Mn, Zn, Cu, B, and Mo. The first ten are called **macronutrients,** and the last six are called **micronutrients.** Carbon, hydrogen, and oxygen are abundant in the atmosphere and hydrosphere and are not of concern in sustainability. The main limiting factors on plant growth are nitrogen, phosphorous, potassium, and sulfur. Of these, nitrogen is often the most important.

The flow of nutrients is continuous among the soil (inorganic store), the living organisms in the soil (biomass store), and the residues and excreta of living organisms (organic store). Nitrogen is fixed from the atmosphere by soil and root nodule bacteria. It may also be dissolved in rainwater in the form of ammonium. These are the only two external sources of nitrogen. The remainder is supplied in recycled form from the breakdown of dead plant material and animal excreta. This cyclic process is diagrammed in Figure 8–17. If a crop is grown and the plant products such as leaves, stems, roots, and fruit are removed, the nitrogen must be replaced by fertilizer to remain sustainable for plant growth. Crops planted on cleared ground rapidly decrease in productivity after 2–3 years (Courtney and Trudgill, 1984). In the absence of an efficient root network, the rapid leaching of nitrogen makes the addition of nitrogenous fertilizer ineffective as well as being very expensive. The classic example of this problem is that of the tropical rain forests of South America, central Africa, and South East Asia, where subsistence farmers cut and burn the trees to clear land for growing crops. The ash provides some of the nutrients to sustain crops for a few years but the thin layers of soil, the low cation-exchange capacities, and heavy rainfall result in not only the removal of macronutrients but also micronutrients. The subsistence nature of the activity precludes expenditures for mineral fertilizers to replenish the soil.

Cycling processes are also evident for phosphorus (Figure 8–18). Phosphorus is derived from both the decay of organic matter and from the soil minerals. In highly leached soils, cycling may be the only source of phosphorus. The form of phosphorus is as calcium phosphate in alkaline soils and as aluminum or iron phosphates in acid soils. The availability of phosphorus compounds is pH-dependent. They are most available at neutral pH values.

Potassium is also available from mineral weathering. In leached soils, the available potassium is bound up in organic cycles.

FIGURE 8–17

The nitrogen cycle in the soil ecosystem.

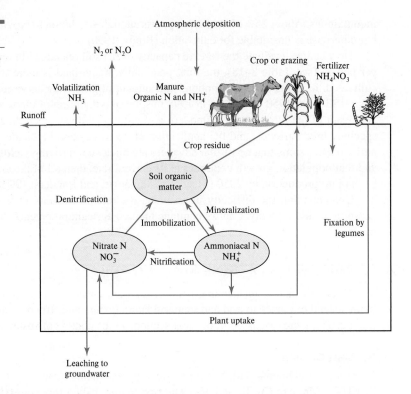

FIGURE 8–18

The phosphorus cycle in the soil ecosystem.

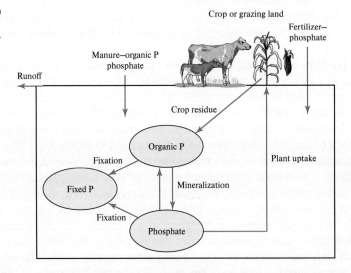

Some similarities between the sulfur and nitrogen cycles is evident (Figure 8–19). Each is largely held in the organic fraction of the soil and each is dependent on microbial action for transformation.

Soil Acidity
Efficient cycling, among other things, depends on a soil pH that favors the soil faunal activity and the breakdown of organic matter by bacteria and fungi. At unfavorable pH values, organic matter will accumulate without releasing its nutrient store. The acidity of leaf litter may have a substantial effect on the soil pH.

FIGURE 8–19

The sulfur cycle.

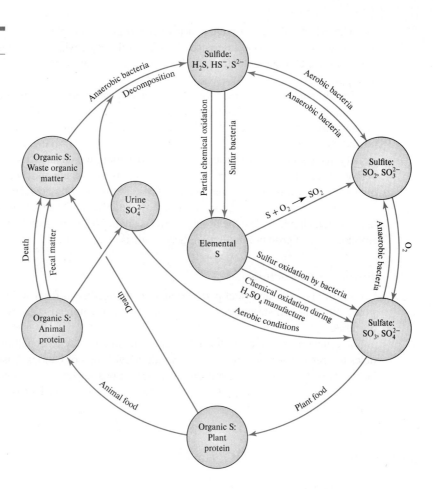

Crops vary in their pH tolerance. For example, alfalfa, sugar beets, some clover, lettuce, peas, and carrots grow best in alkaline soils (pH 7–8); barley, wheat, maize, rye, and oats flourish in circumneutral soils (pH 6.5–7.5); potatoes thrive in acid soil (pH 5) (Courtney and Trudgill, 1984). For many plants, the ability to adapt to varying soil acidity is not well understood. However, tolerance to aluminum toxicity appears to be important in acid soils and the ability to take up iron appears to be important in alkaline soils.

Soil Salinity

Osmosis is defined as the spontaneous transport of a solvent from a dilute solution to a concentrated solution across an ideal semipermeable membrane such as a plant cell wall that impedes passage of the solute but allows the solvent to flow. In most soils this assists the flow of water from the less concentrated soil solution to the more concentrated cell contents.

In saline soils, the soil solution concentration is high and the gradient is reversed; plants wilt because their uptake of water is inhibited. **Saline** soils, sometimes called **white alkali** because of a light-colored surface incrustation that is one of their characteristics, have excess soluble salts that are mostly chlorides and sulfates of sodium, calcium, and magnesium. These can be readily leached out with water that has lower concentrations of these salts. **Saline-sodic** soils contain large quantities of soluble salts and enough adsorbed sodium to seriously affect most plants. Unlike saline soils, leaching will raise the soil pH and the release of sodium will disperse the mineral colloids, which, in turn, results in a tight impervious soil. Leaching of saline-sodic soils leads to sodic soils. **Sodic** soils do not contain large amount of salts. The detrimental effects result from release of sodium, causing both plant toxicity and structural changes to the soil.

The effects of salt on plants varies depending on the plant. Barley, sugar beets, cotton, and sugar cane have good tolerance to saline soils. Rye, wheat, oats, and rice have a moderate tolerance. Orange, grapefruit, beans, and some clovers have poor tolerance (Courtney and Trudgill, 1984).

In arid and semiarid areas, in which evapotranspiration exceeds precipitation, soil moisture supplies are often insufficient to sustain plant growth. Irrigation is an obvious technique to overcome this deficiency. With high evaporation, the salts in the water will tend to be left behind as the water evaporates. Thus, irrigation in areas of high evaporation can lead to high salinity in the soil. Repeated use of water for irrigation, which happens as irrigation from a river percolates through the soil back into the river and is used again and again downstream, leads to increased salt levels in the water and salinization of the soil by downstream users.

Worldwide, crop production is limited by the effects of salinity on about 50% of irrigated lands. In the United States, the problem affects about 30% of irrigated lands (Jacoby, 1999). The sustainability of crop production on these lands is of concern.

Texture and Structure

Groups of clay, silt, and sand particles aggregate together to form soil structures (often called **peds**). Without soil structures, the soil would have few pores for air, water, and plant roots to penetrate.

A crucial process in the formation of structures is clay flocculation. Dispersed clay particles are separated from one another in individual layers of cations—usually sodium. Flocculated clay particles are linked together by other cations, especially calcium. Dispersed clay soils are very dense because the particles are packed closely together. This is why cultivation of soils rich in sodium is difficult.

8–7 SOIL CONSERVATION

Soil Management

The capacity of a soil to consistently produce a crop depends on management of the soil. This goes beyond the mere application of fertilizer. Other soil attributes such as structure, drainage, and organic matter content must also be managed.

Soil Fertility. **Liebig's law of the minimum** states that the growth of a plant depends on the amount of foodstuff presented to it in minimum quantity. For example, a soil with abundant amounts of nitrogen will not be productive if the water available is insufficient. As noted previously, plants must have both macronutrients and micronutrients in sufficient quantity. There is an optimum level for each of these nutrients. Deficiencies will stunt growth. Excesses will be toxic. In natural systems, nutrients extracted from the soil are returned as leaf litter. When the crop is removed, there is no replenishment and the soil's nutrient store will be depleted. To maintain soil fertility nutrients must be replaced by other means.

For centuries, the use of animal manures as fertilizer has been synonymous with successful and stable agriculture. Not only do manures return organic matter and plant nutrients to the soil but they return a high proportion of the solar energy captured by plants (Figure 8–20). Although mineral fertilizers containing nitrogen, phosphorus, and potassium provide the major macronutrients, they alone may not be sufficient. The return of organic matter also aids the maintenance of soil structure.

Crop rotation and laying the land fallow are also recognized techniques for sustained agricultural production. The rotation of legume plants that live symbiotically with the microorganisms that fix nitrogen replenishes the nitrogen supply. Fallow land that produces a grass cover provides a natural replenishment of organic matter that may be plowed back into the soil.

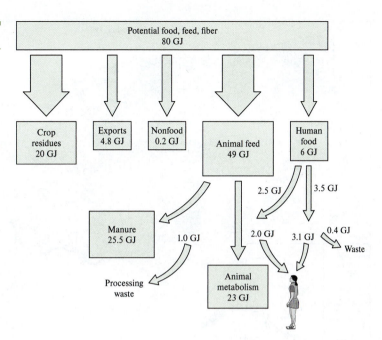

FIGURE 8–20

Diagram of the approximate energy flow for the United States food chain (in billions of joules) showing the high proportion of the energy ultimately found in animal manures.

Liming of the soil may also be required to maintain fertility. Repeated application of nitrate fertilizers results in increased acidity of the soil. Some decay processes also will result in acidification. In addition to pH adjustment, liming reduces the toxicity of acid-soluble trace elements, improves soil structure, and increases the availability of calcium as a nutrient for soil organisms and plants.

Structural Form and Stability. Form determines the packing and pore space in the soil that provides aeration and water retention. Stability determines how the structures will behave when they are plowed.

Organic matter, calcium carbonate, aluminum and iron hydroxides, and silica all act to cement soil grains together and make their structures more durable. If the soil is wet, the "cement" dissolves and the structure becomes unstable. Silt soils are particularly susceptible to structural deterioration because the particles are not cohesive enough nor are they coarse enough to prevent packing.

The addition of manures and lime can help to form more stable structures. In some cases, for example silty clays, drainage may improve their stability. The most significant factor is the timeliness of cultivation—when the surplus water has drained off for the consistency and structure of the soil to withstand the manipulation without breaking down. Limiting tillage is also of value in reducing the compaction by farm equipment. The structure of sandy soils and loam soils with a high organic content are of less concern because they can withstand the manipulation better.

Soil Erosion

Erosion is the transport of soil by water and wind. Although geologic erosion is the mechanism by which sedimentary deposits are formed, short-term erosion due to anthropogenic activities is detrimental to the soil resource. Erosion is damaging to the soil fertility because mainly the nutrient-rich surface soil is removed. "No other soil phenomenon is more destructive worldwide than is soil erosion" (Brady, 1990).

FIGURE 8–21

This dot map shows tons of erosion due to wind and water on cropland and CRP land. Each gray dot represents 181 Gg (200,000 tons) of average annual erosion due to water. Each blue dot represents 181 Gg (200,000 tons) of average annual erosion due to wind.

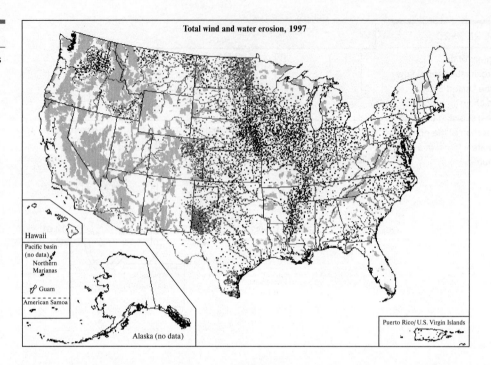

Soil erosion due to overgrazing on pasturelands and rangelands and loss of soil from croplands is still a major problem in the United States. In 1997, an estimated 1.7 Pg of soil were lost from cropland and Conservation Reserve Program (CRP)* land in the United States. This translates into an on-farm economic loss of more than $27 billion each year, of which $20 billion is for replacement of nutrients and $7 billion for lost water and soil.

In addition to the economic effect on farmers, erosion can have serious impacts on the environment, including

- Pollution of lakes and streams by nutrients, particularly phosphorus, and agricultural chemicals that are washed away with the soil.

- Flooding due to silt buildup in drains and waterways.

- Buildup of sediments in wetlands, which can lead to serious environmental problems or even loss of these habitats.

- Siltation in reservoirs, which reduces their capacity.

- Siltation in harbors and waterways. The cost for dredging of harbors and waterways in the United States is estimated to be $1 billion per year.

The average annual soil loss from croplands is $16.4 \text{ Mg} \cdot \text{ha}^{-1}$. The maximum level that can be tolerated if soil productivity is to be maintained is about $11 \text{ Mg} \cdot \text{ha}^{-1}$ (Brady, 1990).

Both water and wind can erode cropland soils. Water erosion results from the removal of soil material by flowing water. When a raindrop strikes the soil surface, it can break up soil aggregates. On a slope, water begins to flow downhill carrying the detached soil grains with it. Wind erosion occurs in regions of low rainfall; it can be widespread, especially during periods of drought. Unlike water erosion, wind erosion is generally not related to slope gradient.

As shown in Figure 8–21, water erosion occurs mainly in areas of the Corn Belt and Southern Plains. Wind erosion takes place mostly in the West, Northern Plains, and Southern

---

*The CRP is a federal program to set aside cropland with the goals of reducing soil erosion, reducing production of surplus commodities, providing income support for farmers, improving environmental quality, and enhancing wildlife habitat.

FIGURE 8–22

Wind and water erosion on crop and CRP land from 1982 to 1997. (*Source:* U.S. Department of Agriculture National Resource Conservation Service.)

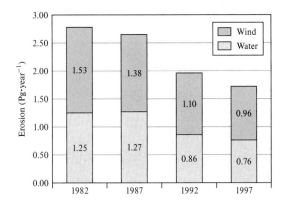

FIGURE 8–23

Kinetic energy of rain in relation to rainfall intensity. (*Source:* U.S. Department of Agriculture National Resource Conservation Service.)

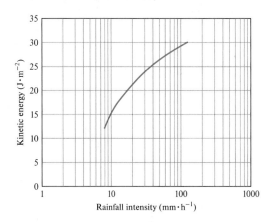

Plains. Both water and wind erosion result in major losses of soil from agricultural land. In 1997, erosion due to water on crop and CRP lands was nearly 1.0 Pg · year^{-1} and about 0.75 Pg · year^{-1} of soil was lost due to wind erosion.

As shown in Figure 8–22, the conservation practices adopted over the past two decades have considerably reduced soil erosion on agricultural lands. The widespread adoption of conservation tillage practices is one of the major contributors to this improvement. Conventional tillage breaks up and buries the crop residue from the previous planting, creating a bare soil surface, which is vulnerable to erosion. Conservation tillage is the practice of leaving some or all of the crop residues on the soil surface to protect it like a mulch. In addition to protecting the soil from erosion, the mulch creates an environment that conserves water in the soil. Conservation tillage has some drawbacks, one of which is an increased dependence on herbicides for weed control. Other practices such as terracing, contour tillage, and crop rotation have helped to reduce water erosion. Diversion of water away from fields or other erosion-sensitive areas can also reduce soil loss.

Measures used to control wind erosion on croplands include conservation tillage, planting windbreaks, and tilling at right angles to the prevailing winds, so that furrows act as small windbreaks to capture blowing soil. Erosion on pastureland and rangeland can be reduced by a number of measures, the single most important of which is control of animal numbers. Animal numbers must be controlled, so that as forage is consumed regrowth has a chance to replace it. For maximum sustained production, plants must be given a period of rest to rebuild these reserves. During droughts the numbers of grazing animals must be tightly controlled, as plants can be injured by a grazing intensity that would not injure them during a less vulnerable season. Proper distribution of animals is also an important grazing management tool. Fences can be used to divide up pastures. On open range land one of the most effective techniques is to provide adequate water facilities that are properly distributed. Ranchers also improve livestock distribution by placing salt and mineral supplies in widespread locations to draw the animals to those areas.

Erosion by Water.   To erode soil, work must be done to disrupt the soil aggregation and to move the soil particles. Energy is supplied by the kinetic energy of the falling raindrops and running water. The energy of falling raindrops is at least 200 times that of running water (White, 1979). Raindrop energy increases with the mass of the drop and the square of its terminal velocity. Both characteristics are a function of rainfall intensity (Figure 8–23).

FIGURE 8–24

Average annual values of the rainfall erosion index in the United States. Note the high values in parts of the east and the generally low values in the west. (*Source:* Wischmeier and Smith, 1978.)

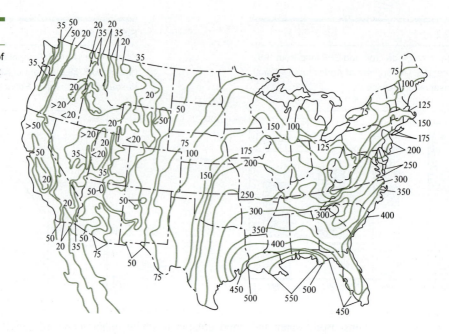

The major factors affecting accelerated erosion are included in the **universal soil loss equation** (USLE):

$$A = (R)(K)(LS)(C)(P) \tag{8–22}$$

where A = predicted soil loss (in $\text{Mg} \cdot \text{ha}^{-1}$)
$\quad R$ = rainfall erosion index
$\quad K$ = soil erodibility factor
$\quad LS$ = topographic factor, a function of
$\qquad L$ = length (in m)
$\qquad S$ = slope (%)
$\quad C$ = crop management factor
$\quad P$ = conservation practice factor

The rainfall erosion index is a function of intensity and total rainfall. Average annual values for the United States are shown in Figure 8–24.

The soil erodibility factor is low for soils in which the water infiltrates readily. Selected values of K are shown in Table 8–19.

TABLE 8–19 Computed K Values for Soils on Erosion Research Stations

Soil	Source of Data	Computed K	Soil	Source of Data	Computed K
Dunkirk silt loam	Geneva, NY	0.69[a]	Mexico silt loam	McCredie, MO	0.28
Keene silt loam	Zanesville, OH	0.48	Cecil sandy loam	Clemson, SC	0.28[a]
Lodi loam	Blacksburg, VA	0.39	Cecil sandy loam	Watkinsville, GA	0.23
Cecil sandy clay loam	Watkinsville, GA	0.36	Tifton loamy sand	Tifton, GA	0.10
Marshall silt loam	Clarinda, IA	0.33	Bath flaggy silt loam		
Hagerstown silty clay loam	State College, PA	0.31[a]	with surface stones		
Austin silt	Temple, TX	0.29	> 5 cm removed	Arnot, NY	0.05[a]

[a]Evaluated from continuous fallow. All others were computed from rowcrop data.
Source: Wischmeier and Smith, 1978.

TABLE 8–20	The Topographic Factor (LS)[a] for Selected Combinations of Slope Length and Steepness

Slope (%)	Slope Length (m)				
	15.35	30.5	45.75	61.0	91.5
2	0.163	0.201	0.227	0.248	0.280
4	0.303	0.400	0.471	0.528	0.621
6	0.476	0.673	0.824	0.952	1.17
8	0.701	0.992	1.21	1.41	1.72
10	0.968	1.37	1.68	1.94	2.37
12	1.280	1.80	2.21	2.55	3.13

[a]Note that the factor increases with both percent slope and length of slope.

Source: Wischmeier and Smith, 1978.

TABLE 8–21	Crop Management or C Values for Different Crop Sequences in Northern Illinois[a]

Crop Sequence[b]	Conventional Tillage[c]		Minimum Tillage		No Tillage	
			Residue Level (kg)		Residue Level (kg)	
	Residue Left	Residue Removed	458–907	908–1816	458–907	908–1816
Continuous soybeans (Sb)	0.49	—	0.33	—	0.29	—
Continuous corn (C)	0.37	0.47	0.31	0.07	0.29	0.06
C–Sb	0.43	0.49	0.32	0.12	0.29	0.06
C–C–Sb	0.40	0.47	0.31	0.12	0.29	0.06
C–C–Sb–G–M	0.20	0.24	0.18	0.09	0.14	0.05
C–Sb–G–M	0.16	0.18	0.15	0.09	0.11	0.05
C–C–G–M	0.12	0.16	0.13	0.08	0.09	0.04

[a]Note the dominating effect of tillage systems and of the maintenance of soil cover. Values would differ slightly in other areas but the principles illustrated would pertain.

[b]Crop abbreviations: C = corn; Sb = soybeans; G = small grain (wheat or oats); M = meadow.

[c]Spring plowed; assumes high crop yields.

Source: Selected data taken from Walker, 1980.

The topographic factor includes the effects of both the slope and the length on the velocity of the water flow. Theoretically, a doubling of the velocity enables the water to move particles 64 times larger, to carry 32 times more mass in suspension, and makes the erosive power four times greater. The topographic factor for selected combinations of length and slope are shown in Table 8–20.

The C values indicate the influence of cropping systems on soil loss, and they depend on the crop being grown, crop stage, tillage, and other management factors. Table 8–21 gives example values for C.

The conservation practice factor reflects the benefits of contouring, strip cropping, and other similar conservation practices (see discussion in the following sections). Example P factors are listed in Table 8–22.

TABLE 8–22   *P* Values for Contour-Farmed Terraced Fields in Relation to Slope Gradient

Slope	Contour Factor	Stripcrop Factor
1–2	0.60	0.30
3–8	0.50	0.25
9–12	0.60	0.30
13–16	0.70	0.35
17–20	0.80	0.40
21–25	0.90	0.45

Source: Wischmeier and Smith, 1978.

EXAMPLE 8–12   Using the USLE estimate, determine the annual soil loss for a farm in central Indiana that has a Marshall silt loam with a slope of 2% and an average slope length of 91.5 m. The land is in continuous corn cultivation, and the farmer uses conventional tillage up and down the slope and leaves the residue.

Solution   From Figure 8–24 select a rainfall erosion index of 175. The value for *K* is 0.33 from Table 8–19. The value of *LS* is 0.280 from Table 8–20, and the value of *C* is 0.37 from Table 8–21. Because the farmer tills up and down hill, the value for *P* is 1.0.

The annual soil loss is then

$$A = (R)(K)(LS)(C)(P)$$

$$= (175)(0.33)(0.280)(0.37)(1.0) = 5.98 \text{ or } 6 \text{ Mg} \cdot \text{ha}^{-1}$$

Erosion by Wind.   Turbulent eddies of the wind lift up on the edges of soil particles. If the effective diameter of the soil particle is less than 0.06 mm in diameter, the particle becomes airborne and remains suspended for a long time. Particles between 0.06 and 0.2 mm in diameter are lifted in a "jump" that is followed by a long flat trajectory back to the ground level, where it either bounces or comes to rest after transmitting its kinetic energy to other particles that may be knocked upward. This process is called **saltation.** Particles between 0.2 and 1 mm frequently creep or roll along. Movement by saltation accounts for more than 50% of wind erosion (White, 1979). Particles larger than 1 mm on the surface protect the smaller particles under them from erosion. Thus, over long periods, as small particles are blown from the surface to expose larger ones below, the amount of erosion by "normal" winds is reduced. This is particularly important in desert environments that lack vegetative cover. It is also a reason why offroad vehicles are particularly harmful to desert soils. They destroy the natural balance by removing the protective layer of large particles.

Conservation Measures.   We can identify several methods for limiting soil erosion by water from the USLE: by reducing the steepness of the slopes cultivated and the length of water runoff by contour plowing (along lines of equal elevation) rather than up and down hill or strip cropping by alternating tilled crops such as corn with untilled crops such as hay and grain; by crop management practices to reduce *C* through minimum tillage or no tillage.

Wind erosion is more difficult to control. Wind breaks help slow the velocity and shorten the fetch* of the wind, but the most effective method is a healthy vegetative cover on a moist soil.

---

*Distance the wind travels without being inhibited by orographic or vegetation effects.

CHAPTER REVIEW

When you have completed studying the chapter, you should be able to do the following without the aid of your textbook or notes:

1. Compare the WCED definition of sustainable development with the definition of sustainable economy.
2. Explain the difference between renewable and nonrenewable resources.
3. Explain "the people problem" in terms that a group of legislators can understand.
4. Define vulnerability in terms of sustainability under conditions of climate change.
5. List three key ingredients to a strong adaptive capacity for sustainability under conditions of climate change.
6. Define green engineering.
7. Explain why floods are a threat to sustainability.
8. Compare the ability of U.S. communities and those in Bangladesh to maintain a sustainable economy in the event of a flood.
9. List two green engineering programs to mitigate loss of life and flood damage.
10. Diagram and explain the relationship between the four definitions of drought.
11. Define PDSI and explain the significance of a PDSI of minus 4.
12. Explain why droughts are a threat to sustainability.
13. Explain the difference between drought response planning and water conservation planning and implementation.
14. Explain why leak detection and repair is an essential component of water conservation.
15. Describe three environmental impacts of coal mining.
16. Define the following terms: acid mine drainage, bottom ash, thermal resistance, cogeneration.
17. Describe the two methods of recovering oil from tar sands.
18. Describe two environmental impacts of recovering oil from tar sands.
19. Describe the process of hydrofracking to recover natural gas.
20. Describe two potential environmental impacts of hydrofracking.
21. Describe four nonfossil fuel alternative energy sources.
22. List three methods of mineral resource conservation and give an example of each.
23. Describe three environmental impacts of mining operations.
24. Define the following terms: weathering, mineral reserve, strip mining, overburden, tailings, spoil banks, dematerialization.
25. Give two examples of soil as a resource.
26. Describe the role of each of the following in determining the sustainability of soil: nutrient cycling, soil acidity, soil salinity, texture, and structure.
27. Explain Liebig's law of the minimum as it pertains to crop production.
28. List three techniques that may be employed to maintain soil fertility.
29. List two techniques to maintain soil structure and form.
30. Describe two measures to reduce soil erosion by water.
31. Define the following terms: arable land, macronutrients, micronutrients, osmosis, saline soils, sodic soils, contour plowing, strip cropping, saltation.

With the aid of this text you should be able to do the following:

32. Estimate a future population or growth rate given appropriate data.
33. Determine one of the following given the required data: the time until exhaustion of an energy reserve with a constant demand, the time until exhaustion of a mineral or energy reserve with a growth in demand, mass of mineral reserve, annual demand.
34. Perform an energy balance on a thermal power plant or other fossil fuel burning facility.
35. Perform an energy balance on a hydropower facility.

36. Perform an energy balance on a heated or cooled structure.

37. Perform mass balance calculations on mining operations.

38. Estimate the mass recovered at the end of each cycle of reclamation of a material.

39. Estimate the equivalent mass for a material recycled an infinite number of times or a given number of times.

40. Estimate the amount of sulfuric acid produced from an ore with a given pyrite concentration.

41. Perform carbon, nitrogen, and phosphorus balances for soil.

42. Estimate the soil loss from water erosion using the Universal Soil Loss Equation (USLE).

PROBLEMS

8–1 It has been estimated that at 2016 consumption rates, the world's petroleum reserve will last 37.5 years. Estimate the world consumption rate in 2016.

      Answer: 227 EJ · year^{-1}

8–2 The conventional coal-fired power plant has an efficiency of about 33%. Assuming that all the coal in Example 8–3 were used to generate electricity (it was not), estimate the time until exhaustion if the efficiency were raised to 40%.

8–3 Using the Internet, determine the pH of the Tiogar River at Mansfield, Pennsylvania. Is this river suffering from acid mine drainage? (HINT: http://waterdata.usgs.gov/nwis. Search under "Real Time" data.)

8–4 It has been proposed that a coal from the Western United States be substituted for the coal in Example 8–4 to reduce the ash disposal volume. The Western coal has an ash content of 4.0% and an NHV of 23.6 MJ · kg^{-1}. Is this a good idea? Estimate the volume of ash produced for the power plant specified in Example 8–4.

      Answer: 1.84×10^5 m^3 · year^{-1}

8–5 It has been proposed that an evaporative cooling tower be used to replace the once-through cooling water in Example 8–5. At what rate must the water be provided from the river to offset the water lost in the cooling tower?

8–6 A house built in the 1950s has 14.86 m^2 of single-glazed windows. Estimate the heat loss with the existing single-glazed window and the loss if the windows are replaced with (a) double-glazed, and (b) triple-glazed windows. Assume the indoor temperature is 20°C and the outdoor temperature is 0°C.

      Answer: Single-glaze 1.86×10^3 W; double-glaze $= 9.29 \times 10^2$ W

8–7 A 25-W compact fluorescent lightbulb (CFL) produces the light equivalent of a 100-W incandescent bulb. The population of North America is estimated to be about 305 million. Estimate the amount of coal needed to light one 100-W incandescent lightbulb for each person for one year and the amount of coal that would be saved if each incandescent bulb were replaced with a CFL. Assume the coal has a NHV of 28.5 MJ · kg^{-1} and the power plants are 33% efficient.

8–8 Repeat Problem 8-7 with oil-fired power plants. Assume the oil is bunker C with a NHV of 42.5 MJ/kg and the power plans are 45% efficient.

8–9 One author has estimated that the time until exhaustion for aluminum is 156 years if the world production remains constant. What is the annual demand based on this estimate?

      Answer: 160 Tg · year^{-1}

8–10 The average increase per year in world production of aluminum increased 3.8% from 1996 to 2002. Use this rate of increase in demand to estimate the annual demand in 2010 if the production in 1996 was 20.8 Tg · year^{-1}.

8–11 In 2004, the United States produced 54.9 Tg of iron from ore with an iron content of 63.0%. Estimate the amount of waste rock generated in mining this ore if the production remains constant until the U.S. reserve is depleted.

      Answer: 3330 Tg or 3.33 Pg

8–12 As the supply of high-grade ores is used up, lower grade ores are used to produce minerals. One author has claimed that the amount of waste rock increases exponentially as the grade of ore decreases.

    (a) Assuming that you are producing 100 kg of metal product, calculate the kg of waste rock per kg of metal for ore containing 50%, 25%, 10%, 5%, and 2.5% metal.

    (b) Write a general expression for the calculation in (a). Is it exponential?

8–13 If the decay constant for recovery of a metal is 0.0202 cycle^{-1}, what is the percent recovery for each cycle?

      Answer: 98%

8–14 In the United States, approximately 1.6 Mg $\cdot$ year^{-1} of aluminum is used in beverage cans. If 63% of this is recovered and recycled each year, what is the equivalent mass if it is recycled an infinite number of times?

8–15 Equation 8–19 may be used to estimate the mass recovered at the end of a given cycle. Equation 8–21 may be used to estimate the equivalent mass after a given number of cycles. Calculate the mass recovered after each of the first three cycles for an aluminum beverage can (Example 8–10) using Equation 8–19 and compare the answer with that from Equation 8–21.

      Answer: For 3rd cycle $M_{TOTAL} = 39.0$ g; $M_K = 43.3$ g

8–16 In 2016, the U.S. Geological Survey estimated that 33% of the demand for copper in the United States was supplied by recycled copper. If the demand for copper remains constant at the 2016 rate of 1.8 Tg per year, how many years will be added to the time until exhaustion of the U.S. reserve?

      Answer: $T_s = 211$ years

8–17 What percent recovery in recycling will be required to double the time to exhaustion for copper if the demand remains constant at the 2016 rate of 1.8 Tg per year?

8–18 What percent recovery in recycling will be required to double the time to exhaustion for copper if the demand increases at a rate 1.0% per year from the 2016 rate of 1.8 Tg per year?

8–19 What is the time until exhaustion for world reserves for copper if the remaining copper is recycled 50% to infinity? Assume the demand remains constant at the 2016 rate of 19.4 Tg per year.

8–20 In Example 8–12, the farmer had the land in continuous corn and used no conservation protection. Estimate the soil loss if a conventional tillage corn-corn-wheat-meadow rotation were employed and the field was contour plowed.

      Answer: 1.16 or 1 Mg $\cdot$ ha^{-1}

DISCUSSION QUESTIONS

8–1 If you are given a choice of bagging your groceries in a paper or a plastic bag, which do you select because it is manufactured from a renewable resource? Explain your choice.

8–2 A bicycle manufacturer is considering changing from steel wheel rims to rims made from either aluminum or titanium alloy. From a resource conservation point of view, which alternative would you recommend? Using the hierarchy of resource conservation, explain your choice.

8–3 A forest area growing on a sandy soil has been harvested to permit agricultural production. Municipal wastewater is available to provide irrigation. Corn is planted but does not yield well despite the abundant application of nitrogen, phosphorus, and potassium. What component of the soil is missing?

8–4 Why is irrigation of farmland in the Southwest potentially damaging to long-term production, but irrigation in the Great Lakes area is not likely to be damaging to the soil?

8–5 As a new home owner, you decide that you would like to have a vegetable garden. The soil is a silty-clay. What can you add to the soil to improve its structure?

FE EXAM FORMATTED PROBLEMS

8–1 Estimate the growth rate of the population of China assuming a population of 1311 million in 2005, a population of 1338 million in 2010, and an exponential growth rate.

    (a) 0.00177 y^{-1}

    (b) $5.4 \times 10^6 \text{ y}^{-1}$

    (c) 0.20 y^{-1}

    (d) 0.00408 y^{-1}

8–2 Estimate the growth rate (individual/individual · y) if the crude birth rate is 14 per 1000, the crude death rate is 8 per 1000, and the net immigration is 3 per 1000.

    (a) 0.0090 y^{-1}

    (b) 0.0060 y^{-1}

    (c) 0.0030 y^{-1}

    (d) 9000 y^{-1}

REFERENCES

ACPA (2007) *Green Highways,* American Concrete Pavement Association, Concrete Pavement Research and Technology Special Report, Skokie, Illinois.

Arnaud, C. H. (2015) "Figuring out Fracking Wastewater," *Chemical & Engineering News,* March 16, pp. 8–9.

ASHRAE (1993) *Handbook of Fundamentals,* American Society of Heating, Refrigerating and Air Conditioning Engineers, Atlanta, Georgia.

Athena Institute (2010) EcoCalculator, www.athenasmi.org/tools/ecoCalculator/index.html.

Atkins, P. R., H. J. Hitter, and D. Wiloughby (1991) "Some Energy and Environmental Impacts of Aluminum Usage," in J. W. Testor, ed., *Energy and Environment in the 21st Century,* MIT Press, Cambridge, MA.

A&WMA (1988) *Waste Minimization,* Air & Waste Management Association, International Specialty Conference Proceedings, October, Baltimore Maryland.

Bell, B. (1970) "The Oldest Records of the Nile Floods," *Geographical Journal,* 136: 569–573.

Bell, B. (1971) "The Dark Ages in Ancient Egypt: I The First Dark Age in Egypt," *American Journal of Archaeology,* 75: 1–26.

Bell, B. (1975) "Climate and History of Egypt," *American Journal of Archaeology,* 79: 223–269.

Blaikie, P., T. Cannon, I. Davis, and B. Wisner (1994) "Bangladesh—A 'Tech-fix' or People's Needs Based Approach to Flooding?" in Blake, P. M. et al. (eds.) *At Risk: Natural Hazards, Peoples Vulnerability, and Disasters,* Routledge, Taylor & Francis Group, New York.

Bourne, J. K., and E. Burtynsky (2010) "California's Pipe Dream," *National Geographic,* 217(4): 132–149.

Brady, N. C. (1990) *The Nature and Properties of Soils,* 9th ed., Macmillan Publishing Company, New York, pp. 264, 431, 432, 549.

Bremner, J., A. Frost, C. Haub, M. Mather, K. Ringheim, and E. Zuehlke (2010) "World Population Highlights: Key Findings from PRB's 2010 World Population Data Sheet," *Population Bulletin,* 65(2).

British Petroleum (2016) Energy Economics–Coal, https://www.bp.com/en/global/corporate/energy-economic/statistical-review-of-world-energy/coal.html

Bruins, H. J. (2000) "Drought Hazards in Israel and Jordan," in Wilhite, D. A. (Ed.) *Drought: A Global Assessment, Volume II,* Routledge, Taylor & Francis Group, New York, pp. 178–193.

Bureau of Census (2002) *Estimated Population,* U.S. Department of Commerce, Washington, DC.

Butzer, K. W. (1976) *Early Hydraulic Civilization in Egypt,* University of Chigaco Press, Chicago, Illinois, pp. 28–29, 55–56.

Coates, D. R. (1981) *Environmental Geology,* John Wiley & Sons, New York, pp. 99, 152, 156, 215.

Cordell, D., and S. White (2008) "The Australian Story of Phosphorus: Sustainability Implications of Global Fertilizer Scarcity for Australia," National workshop on the Future of Phosphorus, Institute for Sustainable Futures, University of Technology, Sydney, Australia, November 14th.

Courtney, F. M., and S. T. Trudgill (1984) *The Soil, An Introduction to Soil Study,* Edward Arnold Publishers, London, pp. 60–61.

Desai, S., and D. A. Klanecky (2011) "Meeting the Needs of the Water-energy Nexus," *Chemical Engineering Progress,* 107(4): 22–27.

Downing, T. E., and K. Bakker (2000) "Drought Risk in a Changing Environment," in Vogt, J. V. and F. Somma (Eds.) *Drought and Drought Mitigation in Europe,* Kluwer Academic Press, Boston, Massachusetts, pp. 79–90.

DWR (2010) California Department of Water Resources, http://www.water.ca.gov/wateruseefficiency/leak

Dziegielewski, B., G. D. Lynne, D. A. Wilhite, and D. P. Sheer (1991) *National Study of Water Management During Drought: A Research Assessment,* Institute of Water Resources Report, IWR Report 91-NDS-3, U.S. Army Corps of Engineers, Fort Belvoir, Virginia.

Dziegielewski, B., and J. E. Crews (1986) "Minimizing the Cost of Coping with Droughts: Springfield, IL," American Society of Civil Engineers, *Journal of Water Resource Management,* 112(4): 419–438.

Easley, P. R. (2011) "Hydrofracking: Is It Worth the Risk," *Opflow,* 37(7): 10–12.

EIA (2004) *Historical Renewable Energy Consumption by Energy Use Sector and Source,* Energy Information Agency, U.S. Department of Energy, www.eia.doe.gov

EIA (2005a) *Table 2.1a Energy Consumption by Sector, Selected Years, 1949–2005,* Energy Information Agency, U.S. Department of Energy, www.eia.doe.gov

EIA (2005b) *Table 1. U.S. Energy Consumption by Energy Source, 2000–2004,* Energy Information Agency, U.S. Department of Energy, www.eia.doe.gov

EIA (2005c) *Table 5b. Historical Renewable Energy Consumption by Energy Use Sector and Energy Source, 2000–2004,* Energy Information Agency, U.S. Department of Energy, www.eia.doe.gov

EIA (2006) *Table 2.2 Existing Capacity by Energy Source,* Energy Information Agency, U.S. Department of Energy, www.eia.doe.gov

EIA (2009) *U.S. Energy Consumption by Energy Source,* Energy Information Agency, U.S. Department of Energy, www.eia.doe.gov

EIA (2010a) *Table E.1 World Primary Energy Consumption by Region,* Energy Information Agency, U.S. Department of Energy, www.eia.doe.gov

EIA (2010b) *Table F.1 World Primary Energy Production by Region,* Energy Information Agency, U.S. Department of Energy, www.eia.doe.gov

EIA (2010c) *Table 8.1. World Crude Oil and Natural Gas Reserves, January 1, 2007,* Energy Information Agency, U.S. Department of Energy, www.eia.doe.gov

EIA (2010d) *Table 8.2. World Estimated Recoverable Coal, December 31, 2005,* Energy Information Agency, U.S. Department of Energy, www.eia.doe.gov

EIA (2010e) *Electric Power Annual,* Energy Information Agency, U.S. Department of Energy, www.eia.doe.gov

EIA (2010f) *Energy in Brief,* Energy Information Agency, U.S. Department of Energy, www.eia.doe.gov

EIA (2015) https://www.eia.gov/todayinenergy/detail.php.id=2930

EIA (2016a) World Primary Energy Consumption by Region, 1995–2016, Table 11.3, Annual Energy Review, Energy Information Administration, U.S. Department of Energy, www.eia.doe.gov

EIA (2016b) Energy Consumption by Sector, Selected Years, 1949–2016, Table 2.1a, Energy Information Administration, U.S. Department of Energy, www.eia.doe.gov

EIA (2016c) U.S. Energy Consumption by Energy Source, 2000–2016, Table 1, Energy Information Administration, U.S. Department of Energy, www.eia.doe.gov

EIA (2016d) Historical Renewable Energy Consumption by Energy Use Sector and Energy Source, 2000–2014, Table 5b, Energy Information Administration, U.S. Department of Energy, www.eia.doe.gov

Farrell, A. E., R. J. Plevin, B. T. Turner, et al. (2006) "Ethanol Can Contribute to Energy and Environmental Goals," *Science,* 311: 506–508.

Feldman, D. L. (2007) *Water Policy for Sustainable Development,* The Johns Hopkins University Press, Baltimore, Maryland, pp. 145–147, 300–301, 304.

Finan, T., C. West, D. Austin, and T. McGuire (2002) "Processes of Adaptation to Climate Variability," *Climate Research,* 21(3): 299–310.

Freeman, H. (1990) *Hazardous Waste Minimization,* McGraw-Hill Publishing Company, New York.

Freeman, H. (1995) *Industrial Pollution Prevention Handbook,* McGraw-Hill, Inc., New York.

Freemantle, M. (2006) "More Than Money's Worth," *Chemical & Engineering News,* 3 Jul, p. 56.

Frohlich, C. (2016) "Oklahoma Curbs Earthquakes by Curbing Wastewater Disposal Wells," *Water Environment and Technology,* 28(11): 24.

Fye, F. K., D. W. Stahle, and E. R. Cook (2003) "Paleoclimatic Analogs to Twentieth-Century Moisture Regimes Across the United States," *Bulletin of the American Meteorological Society,* 84(7): 901–909.

Georgia EPD (2003) *Georgia Drought Management Plan,* Department of Natural Resources, Environmental Protection Division, Atlanta, Georgia.

Georgia EPD (2007) *Water Leak Detection and Repair Program,* Georgia Department of Natural Resources, Environmental Protection Division, Atlanta, Georgia.

Georgia EPD (2010) *Georgia's Water Conservation Implementation Plan,* Georgia Department of Natural Resources, Environmental Protection Division, http://www.ConserveWaterGeorgia.net

Glantz, M. (1994) *Drought Follows the Plow*, Cambridge University Press, Cambridge, United Kingdom.

Glantz, M. (1994) "Drought Follows the Plow," in Wilhite, D. A. (Ed.) *Drought: A Global Assessment, Volume II*, Routledge, Taylor & Francis Group, New York, pp. 285–291.

Green, D. (2010) *Water Conservation for Small- and Medium-Sized Utilities, American Water Works Association,* Denver, Colorado.

Greiner, A. D., J. J. Page, R. H. Cisterna, E. Vadiveloo, and P. A. Davis (2009) "In Search of a Sustainable Future," *Water Environment & Technology,* 21(11): 35–39.

Grimm, M. (1998) "Floodplain Management," *Civil Engineering,* 68(3): 62–64.

Gruntfest, E. (2000) "Nonstructural Mitigation of Flood Hazards," in Wohl, E. E. (Ed.) *Inland Flood Hazards,* Cambridge University Press, Cambridge, UK, pp. 394–395, 398.

Hammerschlag, R. (2006) "Ethanol's Energy Return on Investment: A Survey of the Literature 1990–Present," *Environmental Science & Technology,* 40(6): 1744–1750.

Haupt, A., and T. T. Kane (1985) *Population Reference Handbook,* Population Reference Bureau, Washington, DC.

Hays, W. W. (1981) *Gauging Geological and Hydrological Hazards: Earth-science Considerations,* U.S. Geological Survey Professional Paper 1240-B, Washington, DC.

Hayes, E. T. (1976) "Energy Implications of Materials Recover," in P. H. Abelson and A. L. Hammond, eds., *Materials: Renewable and Nonrenewable Resources,* American Association for the Advancement of Science, Washington, DC.

Haury, E. W. (1935) "Tree Rings—The Archeologist's Time Piece," *American Antiquity,* 1: 98–108.

Hebert, J. (2010) "Process Aeration," Michigan Water Environment Association Process Seminar, East Lansing, MI, December 8.

Hess, G. (2006a) "Push for Biofuels Seen in Farm Bill," *Chemical & Engineering News,* 22 May, pp. 29–31.

Hess, G. (2006b) "BP and DuPont Plan 'Biobutanol,'" *Chemical & Engineering News,* 26 Jun, p. 9.

Higgins, T. (1989) *Hazardous Waste Minimization Handbook*, Lewis Publishers, Chelsea, Michigan.

Hileman, B. (2006) "Heretical Position on Nuclear Power," *Chemical & Engineering News,* 21 Aug, p. 43.

Hill, J., E. Nelson, D. Tilman, S. Polasku, and D. Tiffany (2006) "Environmental, Economic, and Energetic Costs and Benefits of Biodiesel and Ethanol Biofuels," *Proceedings National Academy of Science,* 103(30): 11206–11210.

IIED & WRI (1987) World Resources, International Institute for Environment and Development and World Resources Institute, Washington, DC.

IPCC (2007) *Climate Change 2007: Impacts, Adaptation and Vulnerability—The Physical Science Basis, Summary for Policymakers,* Intergovernmental Panel on Climate Change, Cambridge University Press, Cambridge, United Kingdom, pp. 1–18.

Jacoby, M. (1999) "Botanists Design Plants with a Taste for Salt," *Chemical & Engineering News,* 23 Aug, p. 9.

Janes, D. (2010) "Going for the Green," *EM, Air & Waste Management Association,* January/February, pp. 24–25.

Johnson, J. (2004) "Power from the Sun," *Chemical & Engineering News,* 21 Jun, pp. 25–28.

Karl, T. R., J. M. Melillo, and T. C. Peterson (2009) *Global Climate Change Impacts in the United States, U.S. Global Climate Change Research Program,* Cambridge University Press.

KGS/KDA (2010) "Water Use for Haskell County, Kansas," Kansas Geological Survey and Kansas Department of Agriculture.

LaCount, R., and M. L. Barcella (2010) "Shale Gas, A Game Changer for North American Energy and Environment," *EM,* Air & Waste Management Association, August, pp. 16–19.

Leauber, C. E. (1997) "Leak Detection Cost-effective and Beneficial," *Journal of American Water Works Association,* 89(7): 10.

Loucks, D. P., J. R. Stedinger, and D. A. Haith (1981) *Water Resource Systems Planning and Analysis,* Prentice-Hall, Englewood Cliffs, New Jersey, p. 3.

Loveless, B. (2016) "Coal's Decline May Continue Even Under Trump," USA Today Money, p. 4B.

Masters, G. M., and W. P. Ela (2008) *Introduction to Environmental Engineering and Science, third edition,* Prentice Hall, Upper Saddle River, NJ, p. 612.

McKinney, M. L., and R. M. Schoch (1998) *Environmental Science Systems and Solutions,* Jones and Bartlett Publishers, Sudbury, MA, p. 10.

McNichol, T. (2011) "Rage Against the Machine," *Time*, January 10, p. 62.

Meadows, D. H., D. L. Meadows, and J. Randers (1992) *Beyond the Limits: Confronting Global Collapse, Envisioning a Sustainable Future,* Chelsea Green Publishing Co., Post Mills, VT, p. 51.

Metcalf & Eddy, Inc. (2007) *Water Reuse,* McGraw-Hill, New York, pp. 20–24, 48, 54.

Morrow, W. R., W. M. Griffin, and H. S. Matthews (2006) "Modeling Switchgrass Derived Cellulosic Ethanol Distribution in the United States," *Environmental Science & Technology,* 40(9): 2877–2886.

Mumme, S. P. (1995) "The New Regime for Managing U.S.–Mexican Water Resources," *Environmental Management,* 19(6): 827–835.

NAPA (2010) *Benefits of Asphalt,* www.PaveGreen.com

NCEES (2008) *Fundamentals of Engineering Supplied-Reference Handbook,* National Council of Examiners for Engineering and Surveying, Revised April, 2010, p. 121.

Nemerow, N. L. (1995) *Zero Pollution for Industry*, John Wiley & Sons, Inc., New York.

NOAA (2010) http://www.nch.noaa.gov/HAW2/english/history.shtml

NRCS (1997) National Resource Conservation Service, U.S. Department of Agriculture, http://www.nrcs.usda.gov/technical/land

Ophuls, W., and A. S. Boyan (1992) *Ecology and the Politics of Scarcity Revisited,* W. H. Freeman, New York, p. 76.

Pahl, G. (2005) *Biodiesel, Growing a New Energy Economy,* Chelsea Green Publishing Company, White River Junction, Vermont, p. 47.

Palmer, W. C. (1965) "Meteorological Drought," *Research Paper No. 45,* U.S. Weather Bureau, Washington, DC.

Parfit, M., and S. Leen (2005) "Powering the Future," *National Geographic,* 208(2): 2–31.

Philpott, B. (2008) *Field Guide to Soil Moisture Sensor Use in Florida*, University of Florida, http://www.sjrwmd.com/floridawaterstar/pdfs/SMS_field_guide.pdf

Pimentel, D. (1991) "Ethanol Fuels: Energy Security, Economics and the Environment," *Journal of Agricultural and Environmental Ethics,* 4: 1–13.

Plunkert, P. A. (2001) Aluminum Recycling in the United States in 2000, U.S. Geological Survey Circular 1196-W.

Polsky, C., and D. W. Cash (2005) "Drought, Climate Change, and Vulnerability: The Role of Science and Technology in a Multi-Scale, Multi-Stressor World," in Wilhite, D. A. (Ed.) *Drought and Water Crises: Science, Technology, and Management Issues,* Taylor & Francis, Boca Raton, Florida, p. 218.

Power, M. E., G. Parker, W. E. Dietrich, and A. Sun (1995) "How Does a Floodplain Width Affect Floodplain River Ecology? A Preliminary Exploration Using Simulations," *Geomorphology,* 13: 301–307.

PRB (2010) *2010 World Population Data Sheet,* Population Reference Bureau Washington, DC.

Prescott (2010) *Water Smart,* at http://www.prescott-az.gov

Pulwarty, R. S., K. L. Jacobs, and R. M. Dole (2005) "The Hardest Working River: Drought and Critical Water Problems in the Colorado River Basin," in Wilhite, D. A. (Ed.) *Drought and Water Crises,* Tayor & Francis, New York, pp. 249–285.

Rashid, M. M., W. O. Maddaus, and M. L. Maddus (2010) "Progress in US Water Conservation Planning and Implementation—1990–2009," *Journal of American Water Works Association,* 102(6): 85–99.

Ritter, S. K. (2011) "Water for Oil," *Chemical & Engineering News,* 5 Sep, pp. 56–59.

Ritter, S. K. (2014) "A New Way of Fracking," *Chemical & Engineering News,* May 12, pp. 31–33.

Rogers, R. (2010) "Prime Mover Blower Operations," Michigan Water environment Association Process Seminar, East Lansing, MI, December 8.

Schroedel, R. B., and P. V. Cavagnaro (2010) *Energy Conservation in Water and Wastewater Facilities,* Manual of Practice No. 32, Water Environment Federation, WEF Press, Alexandria, VA.

Shapouri, H., J. A. Duffield, and M. Wang (2002) *The Energy Balance of Corn Ethanol: An Update,* Agricultural Economic Report No. 813, U.S. Department of Agriculture, Washington, DC. 15 pp.

Smith, K., and R. Ward (1998) *Floods: Physical Processes and Human Impacts,* John Wiley & Sons, New York, pp. 142, 153, 157–159, 191.

Spatari, S., M. Betram, R. B. Gordon, K. Henderrson, and T. E. Graedel (2005) "Twentieth Century Copper Stocks and Flows in North America: A Dynamic Analysis," *Ecological Economics,* 54: 37–51.

Stine, S. (1994) "Extreme and Persistent Drought in California and Patagonia During Medieval Time," *Nature,* 369: 546–549.

Stooksbury, D. E. (2003) "Historical Droughts in Georgia and Drought Assessment and Management," *Proceedings of the 2003 Georgia Water Resources Conference,* Apr 23–24, Athens, Georgia.

The President's Science Advisory Committee (1967) Panel on World Food Supply, Washington, DC.

Thornton, J., R. Sturm, and G. Kunkle (2008) *Water Loss Control,* 2nd edition, McGraw-Hill, New York.

Torcellini, P., N. Long, and R. Judkoff (2003) *Consumptive Water Use for U.S. Power Production,* National Renewable Energy Laboratory, U.S. Department of Energy, Golden Colorado, December.

Truax, T. (2010) "Chemical Metering Control Strategy," Michigan Water environment Association Process Seminar, East Lansing, MI, December 8.

Trusty, W. (2009) "Incorporating LCA in Green Building Rating Systems," *EM,* Air & Waste Management Association, December, pp. 19–22.

UNESCO (1979) "Map of the World Distribution of Arid Regions, Explanatory Note," *Man and the Biosphere (MAB) Technical Notes 7,* Paris, UNESCO.

United Nations Statistics Division (2016) Demographic and Social Statistics, https://unstats.un.org/unsd /demographic-social

U.S. Department of Energy (2004) "Hydraulic Fracturing White Paper," Washington, DC, June, Appendix A.

U.S. Department of Energy (2006) *Energy Demands on Water Resources: Report to Congress on the Interdependency of Energy and Water,* Washington, DC.

U.S. Department of Interior (2003) *Water 2025: Preventing Crises and Conflict in the West,* Washington, DC.

U.S. EPA (2010) http:www.epa.gov/oppt/greenengineering

U.S. Forest Service (1999) http://www.fs.fed.us/agenda/

USGS (2016) *Mineral Commodity Summaries, Bauxite and Alumina, Copper, Iron Ore,* U.S. Geological Survey, pp. 33, 57, 87.

USGS (1984) *National Water Summary 1983,* U.S. Geological Survey Water-supply Paper 2250, Washington, DC.

USGS (2004) "Climatic Fluctuations, Drought, and Flow in the Colorado River Basin," fact sheet at http:// pubs.usgs.gov/fs/2004/3062

Vaccari, D. A. (2011) "Sustainability and the Phosphorus Cycle: Inputs, Outputs, Material Flow, and Engineering," *Environmental Engineer,* 17(1): 33–42.

Vengosh, A. (2015) "First Person: Avner Vengosh," *American Scientist,* 103 (5): 312–315.

Viau, A. A., and J. V. Vogt (2000) "Scale Issues in Drought Monitoring," in Vogt, J. V. and F. Somma (Eds.), *Drought and Drought Mitigation in Europe,* Kulwer Academic Publishers, Boston, Massachusetts, pp. 185–193.

Waterfall, P. (2004) *Harvesting Rainwater for Landscape Use,* University of Arizona Cooperative Extension, Pub. No. AZ1344.

Walker, R. D. (1980) "USLE, a Quick Way to Estimate Your Erosion Losses," *Crops and Soils,* 33: 10–13.

WCED (1987) *Our Common Future,* World Commission on Environment and Development, Oxford University Press, Oxford, England.

WCMC (1998) World Conservation Monitoring Centre, Cambridge, UK, http://www.wcmc.org.uk /forest/ data.

Webster, P. J., J. Jian, T. M. Hopson, C. D. Hoyos, P. A. Agudelo, H. Chang, J. A. Curry, et al. (2010) "Extended-Range Probabilistic Forecasts of Ganges and Brahmaputra Floods in Bangladesh," *Bulletin of the American Meteorological Society,* 91(11): 1493–1514.

Western Resources Advocates (2003) "Smart Water: A Comparative Study of Urban Water Use Efficiency Across the Southwest," Water Resource Advocates, Boulder, Colorado.

White, R. E. (1979) *Introduction to the Principles and Practice of Soil Science,* John Wiley & Sons, New York, pp. 24, 34, 152, 156.

Whol., E. E. (Ed.) (2000) *Inland Flood Hazards,* Cambridge University Press, Cambridge, UK, pp. 25–26.

Wilhite, D. A., and W. E. Glantz (1985) "Understanding the Drought Phenomenon: The Role of Definitions," *Water International,* 10: 111–120.

Wilhite, D. A., and M. Buchanan-Smith (2005) "Drought as Hazard: Understanding the Natural and Social Context," in Wilhite, D. A. (Ed.) *Drought and Water Crises,* Tayor & Francis, New York, p. 4.

Wischmeier, W. J., and D. D. Smith (1978) Predicting Rainfall Erosion Loss—A Guide to Conservation Planning, Agriculture Handbook No. 537, Washington, DC, U.S. Department of Agriculture.

Wright, J. M. (1996) "Effects of the Flood on National Policy: Some Achievements, Major Challenges Remain," in Changnon, S. A. (Ed.) *The Great Flood of 1993,* Westview Press, a Division of Harper Collins, Boulder, Colorado, p. 253.

Zeltner, C., H. P. Bader, R. Scheidegger, and P. Baccini (1999) "Sustainable Metal Management Exemplified by Copper in the USA," *Regional Environmental Change,* Nov, pp. 31–46.

Zonn, I., M. H. Glantz, and A. Rubenstein (2000) "The Virgin Lands Scheme in the Former Soviet Union," in Wilhite, D. A. (Ed.) *Drought: A Global Assessment, Volume I,* Routledge, Taylor & Francis Group, New York, pp. 381–388.

9

Water Quality Management

Case Study

Deepwater Horizon: The Largest Oil Spill Disaster in U.S. History

On April 20, 2010, an explosion on the ultradeepwater offshore drilling rig Deepwater Horizon sent crew members flying, propelled shrapnel in all directions, and created a massive fireball that was visible 40 miles away. Eleven crew members died, and this event began what is considered to be one of the worst environmental catastrophes in U.S. history.

Attempts to engage a blowout preventer and a blind shear ram both failed to seal the well. Two days later the rig sank, and for the next 87 days, approximately 4 million barrels (about 170 million gallons) of oil and unknown amounts of methane gushed from the open well causing an oil spill that covered some 600 square miles and impacted approximately 68,000 square miles of ocean (NRDC, 2015) as shown in Figure 9–1. A massive effort began to protect beaches, wetlands, and estuaries from the spreading oil utilizing skimmer ships, floating booms, controlled burns, and dispersant as seen in Figure 9–2. When the oil hit the remote string of barrier islands off the Louisiana coast, it was the peak nesting season for brown pelicans. Adult birds returned from their search for food to find their eggs coated in oil. Diving birds and mangrove roots became coated in oil. First responders became ill as a result of exposure to the oil and the chemicals used to try to contain the oil and prevent further damage.

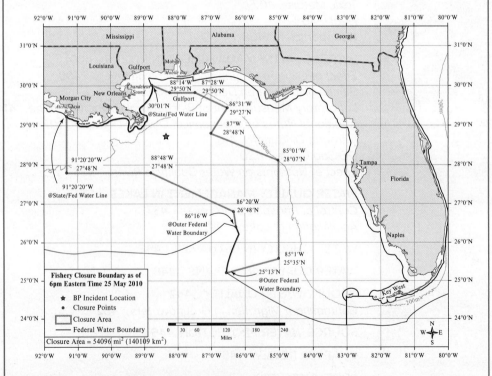

FIGURE 9–1 Map showing the extent to which fisheries were closed in May 2010 as a result of the Deepwater Horizon Oil Spill.

Source: https://upload.wikimedia.org/wikipedia/commons/8/81/Deepwater_Horizon_Oil_Spill_Fishing _Closure_2010-05-25.png.

FIGURE 9–2 Booms surrounding New Harbor Island, Louisiana, during the Deepwater Horizon Oil Spill. (©Eric Gay/AP Photo)

Over the next few years, over 490 miles of coastline along Florida, Alabama, Mississippi, and Louisiana were affected and crews worked to remove thousands of tons of oily material from the Gulf Coast beaches. Nevertheless, in 2015, tar balls could still be found on the beaches of the barrier islands. Seven years later, when the sea is rough, buried oil still resurfaces. The oil that coated the mangrove roots killed the trees. Without the trees to stabilize the beaches, the sand eroded and the islands disappeared (Elliott, 2015).

Numerous studies were conducted and lawsuits convened in the years since the Deepwater Horizon blowout. Among the lessons learned are the importance of strong regulatory oversight and the critical need for communication between the oil companies and their contractors and the U.S. Coast Guard. In a climate that emphasizes short-term profits and deregulation of industry, these lessons have become critically important. We have one Earth, it is ours to protect or destroy.

9–1 INTRODUCTION

How we use the water in lakes, rivers, ponds, and streams is greatly influenced by the quality of the water found in them. Activities such as fishing, swimming, boating, shipping, and waste disposal have very different requirements for water quality. Water of a particularly high quality is needed for drinking water supplies. In many parts of the world, the introduction of pollutants from human activity has seriously degraded water quality, even to the extent of turning pristine trout streams into foul open sewers with few life forms and fewer beneficial uses.

Water quality managers are concerned with controlling pollution from human activity so as to ensure that the water is suitable for its intended uses. Water quality management is also the science of knowing how much waste is too much for a particular water body. To know how much waste can be tolerated (the technical term is **assimilated**) by a water body, water quality managers must know the type of pollutants discharged and the manner in which they affect water quality. They must also know how water quality is affected by natural factors such as the mineral heritage of the watershed, the geometry of the terrain, and the climate of the region. A small, tumbling mountain brook will have a very different assimilative capacity from a sluggish, meandering lowland river, and lakes are different from moving waters.

Originally, the intent of water quality management was to protect the intended uses of a water body while using water as an economic means of waste disposal within the constraints of its assimilative capacity. In 1972, the Congress of the United States enacted the Federal Water Pollution Control Act Amendments of 1972, which established that it was in the national interest to "restore and maintain the chemical, physical, and biological integrity of the nation's waters." In addition to making the water safe to drink, the Congress also established a goal of ensuring "water quality which provides for the protection and propagation of fish, shellfish, and wildlife, and provides for recreation in and on the water." By understanding how pollutants affect water quality, an environmental engineer can properly design the treatment facilities to remove these pollutants to acceptable levels or can help to choose chemicals to be used in a specific process that will not have a deleterious effect on the environment. Traditionally, environmental engineers have focused their efforts on the former, that is, on end-of-pipe treatment, but current thinking demands that the environmental engineer become involved in looking at the entire process to minimize the formation of the pollution in the first place, by choosing chemicals that do not harm the environment, recycling chemicals on site and other means appropriate for the specific process or industry under consideration.

This chapter deals first with the major types of water pollutants and their sources. In the remainder of the chapter, water quality management in rivers and lakes is discussed, placing the emphasis on the categories of pollutants found in domestic wastewaters. For both rivers and lakes, the natural factors affecting water quality will be discussed as the basis for understanding how human activities affect water quality.

9–2 WATER POLLUTANTS AND THEIR SOURCES

The wide range of pollutants discharged to surface waters can be grouped into broad classes.

Point Sources

Domestic sewage and industrial wastes act as **point sources** because they are generally collected by a network of pipes or channels and conveyed to a single point of discharge into the receiving water. Domestic sewage consists of wastes from homes, schools, office buildings, and stores. The term **municipal sewage** includes domestic sewage along with any industrial wastes that are permitted to be discharged into the sanitary sewers. In general, point source pollution can be reduced or eliminated through waste minimization and proper wastewater treatment prior to discharge to a natural water body.

Nonpoint Sources

Urban and agricultural runoff are characterized by multiple discharge points and are called **nonpoint sources.** Often the polluted water flows over the surface of the land or along natural drainage channels to the nearest water body. Even when urban or agricultural runoff waters are collected in pipes or channels, they are generally transported the shortest possible distance for discharge, so that treatment at each outlet is not economically feasible. Much of nonpoint source pollution occurs during rainstorms or spring snowmelt, resulting in large flow rates that make treatment even more difficult. Reduction of agricultural nonpoint source pollution generally requires changes in land use practices and improved education. Urban storm water runoff (including that from streets, parking lots, golf courses, and lawns) can transport pollutants such as nitrogen and phosphorus from fertilizers, herbicides applied to lawns and golf courses, oil, greases, ethylene glycol (used in antifreeze), and cut grass and other organic debris.

Nonpoint pollution from urban storm water and, in particular, storm water collected in **combined sewers** that carry both storm water and municipal sewage may require major engineering work to correct. The original design of combined sewers provided a flow structure that diverted excess storm water mixed with raw sewage (above the design capacity of the wastewater treatment plant) directly to the nearest river or stream. The elimination of **combined sewer overflow** may involve not only the construction of separate storm and sanitary sewers but also the creation of storm water retention basins and expanded treatment facilities to treat the storm water. This is particularly complex and expensive because the combined sewers frequently occur in the oldest, most developed portions of the community, thus disrupting paved streets, utilities, and commercial activities during construction activities. The installation of combined sewers is now prohibited in the United States.

Oxygen-Demanding Material

Anything that can be oxidized in the receiving water resulting in the consumption of dissolved molecular oxygen is termed **oxygen-demanding material.** This material is usually biodegradable organic matter but also includes certain inorganic compounds. The consumption of **dissolved oxygen,** DO (pronounced "dee oh"), poses a threat to fish and other higher forms of aquatic life that must have oxygen to live. The critical level of DO varies greatly among species. For example, brook trout may require about 7.5 mg $\cdot$ L^{-1} of DO, whereas carp can survive at 3 mg $\cdot$ L^{-1}. As a rule, the most desirable commercial and game fish require high levels of dissolved oxygen. Oxygen-demanding materials in domestic sewage come primarily from human waste and food residue. Particularly noteworthy among the many industries that release oxygen-demanding wastes are the food-processing and paper industries. Almost any naturally occurring organic matter, such as animal droppings, crop residues, or leaves, which is released into the water from nonpoint sources, will contribute to the depletion of DO.

Nutrients

Nitrogen and phosphorus, two nutrients of primary concern, are considered pollutants when they become too much of a good thing. All living things require these nutrients for growth. Thus, they must be present in rivers and lakes to support the natural food web.* Problems arise when nutrient levels become excessive and the food web is grossly disturbed, causing some organisms to proliferate at the expense of others. As will be discussed in a later section, excessive nutrients often lead to large growths of algae, which in turn become oxygen-demanding material when they die and settle to the bottom. Some major sources of nutrients are phosphorus-based detergents, fertilizers, food-processing wastes, as well as animal and human excrement. Since agricultural operations are a significant source of nutrient, we will discuss them here.

The release of nutrients from agricultural operations can occur due to the loss of nutrients in runoff or seepage from croplands, rupture of manure storage lagoons, failure of the liner in a manure storage lagoon or the accidental spillage of fertilizers. The loss of nutrients in runoff or seepage from croplands is the major source of nutrient release from agricultural operations.

---

*In simplistic terms, a food web is the collection of interrelated organisms in which the lower levels are the "prey" and the upper levels are the "predators." See Chapter 5 for a more detailed discussion of the food web.

An adequate supply of the nutrients, such as nitrogen, phosphorus, and potassium, and maintenance of proper soil pH are essential to crop growth. Fertilizers have long been used to increase crop yields. The value of silt deposits from flooding rivers in maintaining soil productivity was recognized more than 5000 years ago by the early Egyptians living along the Nile River. Organic manures have been used in Chinese agriculture for over 3000 years. Today farmers employ both chemical fertilizers and manure to increase crop yield. Intensive crop production like that employed in the United States today requires the addition of fertilizers to sustain crop yields. The most widely used fertilizers are lime (to maintain a proper soil pH), nitrogen (N), phosphorus (P), and potassium (K). The use of lime or potassium fertilizers causes few environmental problems; however, widespread problems are associated with the use of nitrogen and phosphorus fertilizers. The primary issues of concern are the eutrophication of surface waters and the contamination of groundwaters with nitrate.

To ensure optimal plant growth nitrogen is usually added to the soil in the form of manure or commercial (chemical) fertilizer. Nitrogen in manure exists primarily in organic and ammoniacal forms. In fresh manure 60–80% of the N exists in the organic form; however, during storage a significant fraction of the organic nitrogen in the manure is converted to ammonia. The most common forms of commercial fertilizer are ammonium nitrate and anhydrous ammonia (NH_3).

The pollution of groundwaters with nitrate is widespread, with 41.4% of domestic wells tested, having a concentration greater than $1 \text{ mg} \cdot L^{-1}$ as N, a level which is considered to result from the effects of human activities. Nitrate concentrations were greater than $10 \text{ mg} \cdot L^{-1}$ as N maximum contaminant level for drinking water in 4.4% of the wells tested. Concentrations were greater than the maximum contaminant level (MCL) in more than 10 percent of wells in several aquifers, including the Basin and Range and Central Valley basin-fill aquifers in the Southwest and in California, the west-central glacial aquifers in the Upper Midwest, the North Atlantic coastal plain aquifers in the central Appalachian region, and the Piedmont crystalline-rock aquifers, also in the central Appalachian region (DeSimone, 2009). A combination of permeable soils, high rates of fertilizer application, and irrigation provides a large source of nitrate and a high potential for nitrate migration to the water table.

The phosphate in fertilizers and manure is initially quite soluble and available to plants. Most phosphate fertilizers have been manufactured by treating rock phosphate (the phosphate-bearing mineral that is mined) with acid to make it more soluble. Manure contains soluble phosphate, organic phosphate, and inorganic phosphate compounds that are quite available. When the fertilizer or manure phosphate comes in contact with the soil, various reactions begin occurring that make the phosphate less soluble and less available. The rates and products of these reactions depend on such soil conditions as pH, moisture content, temperature, and the minerals already present in the soil.

As discussed in Chapter 5, phosphorus is usually the limiting nutrient in freshwater aquatic systems, such as lakes. In contrast to freshwater systems, nitrogen is generally the primary limiting nutrient in the seaward portions of estuarine systems. Estuarine waters may be phosphorus-limited when nitrogen concentrations are high and the N:P ratio > 16:1. In such cases, excess phosphorus may trigger eutrophic conditions.

The effect of agriculture on surface water systems can be illustrated by the example of Chesapeake Bay. The Chesapeake Bay is North America's largest estuary. Its watershed covers 166,000 km^2 and encompasses parts of six states: Delaware, Maryland, New York, Pennsylvania, Virginia, and West Virginia, as well as the District of Columbia. Excess nutrient inputs into the bay have caused eutrophication and periods of hypoxia (dissolved-oxygen concentrations lower than $1.0 \text{ mg} \cdot L^{-1}$), which in turn have killed or stressed living resources in many areas of the bay. The algal blooms from high nutrient inputs and sediment loads also decrease water clarity, which is largely responsible for the decline of submerged aquatic vegetation. Submerged aquatic vegetation, one of the most important components of the ecosystem, provides critical habitat for shellfish and finfish and food for waterfowl. High nutrient levels in the bay's

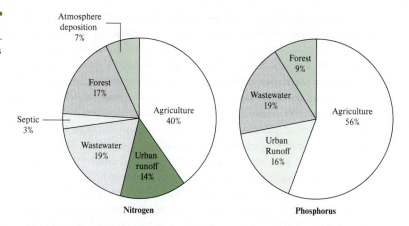

FIGURE 9–3

Nitrogen and phosphorus sources in Chesapeake Bay.

Source: USGS.

waters are thought to be responsible for the proliferation of the single-celled organism, *Pfiesteria piscicida,* which causes fish lesions and fish kills.

The major source of nutrient pollution in the bay's watershed is runoff from agriculture, which contributes about 40% of the nitrogen and about 56% of the phosphorus entering the Chesapeake Bay (Figure 9–3). The major source of nitrogen is the Susquehanna River, which drains some of the most productive agricultural land in the United States. The sources of nutrient pollution from agricultural land are fertilizer and animal waste. Approximately 8.7 Gg of phosphorus entered the bay in 2010. A particular problem within the Chesapeake Bay catchment is that in many areas livestock production is very intensive but crop production is limited. The excessive application of manure to the limited areas of croplands has led to the accumulation of high levels of phosphorus in the fields. For example, a survey conducted in Sussex County, Delaware, which has a high density of livestock operations, found that in 87% of the fields tested the P levels were excessive. By contrast, in New Castle County, Delaware, which has a low density of livestock operations, less than 10% of the field soils tested had excessive P levels. Phosphorus, which is bound to soil particles, can be lost from fields with high P levels in runoff.

Various pollution-reduction strategies were put into place to reduce nutrient pollution in the Chesapeake, including bans on detergents with phosphorus; control of runoff from urban areas, farmland, and pastures; improvements in sewage treatment; and preservation of forest and wetlands, which act as buffers to nutrient-pollution inputs. Controls directed at reducing agricultural and urban erosion also contribute to the lower phosphorus levels. Pollution-reduction strategies are having a positive effect on phosphorus levels in the major tributaries to the bay (the Susquehanna and Potomac Rivers). By contrast, nitrogen levels in the Susquehanna and the Potomac Rivers are increasing. The increase in nitrogen is probably caused by the continued use of nitrogen fertilizer on lawns and cropland, increasing agricultural animal production, and atmospheric deposition of nitrogen from industrial and automotive air pollution.

Pathogenic Organisms

Pathogenic microorganisms found in wastewater include bacteria, viruses, and protozoa excreted by diseased persons or animals. When discharged into surface waters, they make the water unfit for drinking (i.e., nonpotable). If the concentration of pathogens is sufficiently high, the water may also be unsafe for swimming and fishing. Certain shellfish can be toxic because they concentrate pathogenic organisms in their tissues, making the toxicity levels in the shellfish much greater than the levels in the surrounding water. Issues involving the occurrence of the protozoan pathogens *Cryptosporidium* and *Giardia* and resulting outbreaks of gastrointestinal infections will be discussed in more detail in Chapter 10.

Bacteria that have developed immunity to antibiotics are of major concern to environmental professionals. Although the occurrence of these organisms in hospitals is well documented, the presence of antibiotic-resistant bacteria in surface waters was not demonstrated until the late 1990s. Ash and colleagues (1999) sampled 15 U.S. rivers for bacteria resistant to ampicillin, a synthetic penicillin. Antibiotic-resistant bacteria were found in both rural and urban environments with no observable pattern. The Rio Grande was also sampled for antibiotic-resistant bacteria, this time testing for resistance to the antibiotic vancomycin (Sternes, 1999). Antibiotic-resistant bacteria were most commonly found downstream of El Paso; resistance was detected in up to 30% of the bacteria isolated from there. Equally disturbing was the discovery that 30–40% of the bacteria in the streams in rural Dubuque, Iowa, were resistant to tetracycline, and 1% of the bacteria appeared to be immune to this antibiotic (Bennett and Kramer, 1999).

More recent research reveals even more disturbing findings. Pruden et al. (2006) have documented the occurrence of tetracycline and sulfonamide antibiotic resistance genes (ARGs) in irrigation ditches, river sediments, and treated drinking water in northern Colorado. ARGs are portions of DNA that render bacteria resistant to antibiotics. The genes can spread in different ways. Bacteria, for instance, exchange ARGs among themselves. Despite the destruction of cells carrying ARGs, apparently if ARGs are released to the environment the DNA can persist and antibiotic resistance can be transferred to other cells. Clearly the resistance of the bacteria detected reflects our heavy dependence on antibiotics and may significantly alter the future effectiveness of antibiotics (Raloff, 1999). A recent study (LaPara et al., 2011) of the impact of tertiary-treated municipal wastewater indicated that the discharge of highly treated municipal wastewater resulted in a statistically significant increase in the quantities of ARGs in otherwise pristine surface waters, demonstrating that even well-treated municipal wastewater does not effectively remove ARGs.

Suspended Solids

Organic and inorganic particles that are carried by wastewater into a receiving water are termed **suspended solids** (SS). When the speed of the water is reduced by flow into a pool or a lake, many of these particles settle to the bottom as sediment. In common usage, the word *sediment* also includes eroded soil particles that are being carried by water even if they have not yet settled. Colloidal particles, which do not settle readily, cause the turbidity found in many surface waters. Organic suspended solids may also exert an oxygen demand. Inorganic suspended solids are discharged by some industries but result mostly from soil erosion, which is particularly bad in areas of logging, strip mining, and construction. As excessive sediment loads are deposited into lakes and reservoirs, the turbidity increases, light penetration decreases, the bacterial population often increases, and the solids deposit on the bottom of the water body, destroying the habitat for many benthic organisms. Even in rapidly moving mountain streams, sediment from mining and logging operations has destroyed many habitats for aquatic organisms. For example, salmon eggs can only develop and hatch in stream beds of loose gravel. As the pores between the pebbles are filled with sediment, the eggs suffocate and the salmon population is reduced.

Salts

Although most people associate salty water with oceans and salt lakes, all water contains some salt. These salts are often measured by evaporation of a filtered water sample. The salts and other matter that don't evaporate are called **total dissolved solids** (TDS).

Much of the water available in the western United States is saline. In many cases the water is "naturally" saline because salts dissolve in the water as it percolates through the soil or rock formations. However, the widespread use of water for irrigation in arid or semiarid areas has increased salinity problems. Evaporation of water from open reservoirs and canals and during application to plants increases salinity. Also, as plants transpire the salts remain in the soil. Salinity problems can also occur when there is a shallow saline water table. Irrigation raises the water table, which causes upward movement of saline groundwater into the root zone. Salt accumulation in agricultural soils can lead to a reduction in crop yield, particularly in crops that are sensitive to salinity, such as corn, soybeans, rice, lettuce, squash, onion, and bell peppers. Some saline soils can be reclaimed by applying sufficient water to the soil to leach the solutes from the root zone. However, the leaching process may degrade groundwater quality.

Increasing salinity can also cause serious problems in surface waters. As irrigation water percolates through the soil and returns to the river (called return flow or interflow), the salinity increases. The salt concentration increases from the river head waters to the mouth as it is used and reused for irrigation as it flows downstream.

To illustrate this problem we will consider the example of the Salton Sea. The Salton Sea is the largest inland body of water west of the Rockies (surface area of 98,600 ha). It was formed in 1905 when massive flooding caused the Colorado River to break through an irrigation canal headwork and flow freely into the Salton Basin for 18 months. Since then, the sea's existence has been maintained primarily by agricultural return flows from the Imperial, Coachella, and Mexicali valleys. The sea has become a key stop on the Pacific flyway for many species of migratory birds. The Salton Sea National Wildlife Refuge supports over 400 species of birds, among the highest totals of all national refuges. The Salton Sea provides important habitat for several endangered species, including the desert pupfish, Yuma clapper rail, and brown pelican, and it is an important sport fishery.

The Salton Sea is beset by several serious water quality problems. Because it has no outlet, the salt, nutrients, and other pollutants that flow into it are retained within its waters and sediments. Increasing salinity is one of the most serious of these environmental problems. The sea's salinity has been gradually increasing and it has now reached 44 parts per thousand, about 30% higher than ocean water. Many of the native freshwater fish species have disappeared. Beginning in the 1950s, the California Department of Fish and Game introduced marine fish, including orange-mouthed corvina, sargo, gulf croaker, and tilapia, into the Salton Sea. The sea may be becoming too saline for even the marine fish, and the elevated salinity is thought to partially explain the decline in orange-mouthed corvina, the most popular sportfish.

Pesticides

Pesticides are chemicals used by farmers, households, or industry to regulate and control various types of pests or weeds. The major types of pesticides are herbicides, insecticides, and fungicides. Herbicides are used to kill unwanted plants (i.e., weeds). Insecticides are used to kill insects that would otherwise destroy crops, gardens, or structures. Fungicides are employed to control the growth of fungi, many of which cause plant diseases. Table 9–1 shows the most commonly used pesticides along with the mass of chemicals used.

TABLE 9–1	Pesticide Use in the United States in 2012		
	Most Commonly Used	**Total Mass Used in the U.S. (million kg)**	**Percentage of Total Mass of That Type of Pesticide (%)**
Herbicides	Glyphosate, atrazine, metolachlor	308	57
Insecticides	Chloropyrifos, aldicarb, acephate	29.1	5
Fungicides	Chlorothalonil, copper hydroxide, mancozeb	47.7	9
Other*	Metam sodium, dichloropropene, methyl bromide	198	37

*Other includes nematicides, fumigants, and other miscellaneous conventional pesticides, and other chemicals used as pesticides such as sulfur, petroleum oil, and sulfuric acid.

Source: U.S. EPA (2017) *Pesticide Industry Sales and Usage Reports*
https://www.epa.gov/sites/production/files/2017-01/documents/pesticides-industry-sales-usage-2016_0.pdf

The presence of pesticides in surface- and groundwater is ubiquitous in the United States. In a study conducted by the U.S. Geological Survey (USGS) National Water Quality Assessment Program (NAWQA) across the country, it was found that more than 90% of water and fish samples from all streams in developed watersheds contained one or, more often, several pesticides (Figure 9–4). The most commonly occurring pesticides found in water were primarily those currently in use, whereas those found in fish and sediment are the organochlorine insecticides, such as DDT, that were heavily used decades ago. Greater than 50% of the wells sampled contained one or more pesticides, with the highest detection frequencies in shallow groundwater beneath agricultural and urban areas and the lowest frequencies in deeper aquifers (USGS, 2007).

In a study of the western Lake Michigan drainage basins (in Wisconsin and Michigan) by the USGS, 33 pesticides were found in surface waters and 15 in groundwater (Peters et al., 1998). The most commonly detected pesticides were herbicides used on corn, soybeans, small grains, and hay. Atrazine, the herbicide most commonly used on corn in the study area, was detected in all the surface water samples and more than half of the groundwater samples. The herbicides simazine, metolachlor, prometon, and alachlor were found in more than 50% of the stream samples. The good news is that U.S. EPA drinking-water standards were exceeded in only 11% of the surface water samples. However, the EPA has established maximum contaminant levels (MCLs) for only about 20% of the 88 pesticides analyzed in this study and 5% of the 394 pesticides for which the EPA has issued non-enforceable human health benchmarks. The western Lake Michigan basin study found that the concentrations of pesticides were approximately 100 times higher in intensively farmed areas than in nonagricultural or less intensively farmed areas.* Typically, the highest concentrations occurred during the first rainfall events (resulting in runoff) after the application of pesticides. Atmospheric deposition of atrazine was believed to be the cause of the occurrence of this pesticide in two forested streams.

As is illustrated in Figure 9–5, the study also found that the shallow aquifers most vulnerable to contamination were those in intensively cultivated agricultural areas underlain by relatively permeable surficial deposits. Seldom is this shallow groundwater used as a drinking water, lessening the effect of the contamination on human health. Pesticides were also detected in seven drinking-water aquifers. Most of the contaminated water was found in wells drilled in the aquifers not covered by the relatively impermeable confining layer. In conclusion, both USGS studies show the ubiquitous presence of pesticides in the environment. Unfortunately, the presence of pesticides is likely to be underestimated because some of the more difficult to analyze, but widely used, pesticides (e.g., glyphosate, cryolite, and all inorganic pesticides

FIGURE 9–4

The occurrence of pesticides in agricultural, urban and mixed land use areas within NAWQA study units.

Source: USGS.

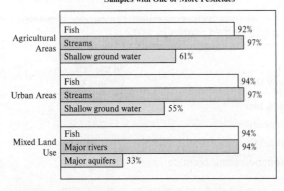

Samples with One or More Pesticides

---

*Intensive farming is defined as the cultivation of row crops such as corn, soy beans, and small grains. Nonintensive farming is grazing, hay production, or the cultivation of less herbicide-intensive crops.

FIGURE 9–5

The occurrence of atrazine in aquifers in eastern Wisconsin. *Source:* Peters et al., 1998.

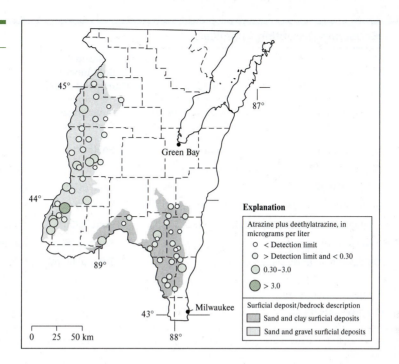

including sulfur and copper) were not included in these studies. Additionally, neither study looked for the presence of the thousands of potential pesticide degradation products that are very rarely included as laboratory analytes.

Pharmaceuticals and Personal Care Products

Pharmaceuticals and personal care products (PPCPs) are a class of compounds that are applied externally or ingested by humans, pets, and other domesticated animals. They are released to the environment through the disposal of expired, unwanted, or excess medications to the sewage system by individuals, pharmacies, or physicians. Another source of PPCPs in the environment is through metabolic excretion—the excretion of the chemically unaltered parent compound and metabolized by-products in urine and feces. PPCPs, such as deodorants and sunscreens, can be washed into our waterways during bathing, washing, and swimming. Some PPCPs are also used as pest-control agents. For example, 4-aminopyridine, an experimental multiple sclerosis drug, is used as an avicide and warfarin is used as both an anticoagulant and a rat poison. PPCPs can also migrate into the environment from poorly engineered landfills and cemeteries (U.S. EPA, 2006; Daughton, 2003).

Most of the PPCPs are polar chemicals, which means that they will tend to migrate to our waterways and not be appreciably sorbed onto soils and sediments. Numerous PPCPs have been detected in muncipal wastewater and surface waters (Blair et al., 2013; Hummel et al., 2006; Buser et al., 1999; Buser, Muller, and Theobald, 1998a; Buser, Poiger, and Muller, 1998b). Conventional muncipal wastewater treatment is not designed to remove PPCPs and, as such, many of these chemicals may pass through the treatment plant unaltered. Some PPCPs, including most painkillers, are readily removed during wastewater treatment (Wang and Wang, 2016; Sedlak, Gray, and Pinkston, 2000). The occurrence of PPCPs in drinking water appears to be less frequent and at lower concentrations (Daughton, 2003). Where necessary, reverse osmosis could be used to remove PPCPs from recycled water (Lin and Lee, 2014; Sedlak et al., 2000).

Little is known about the environmental effects of PPCPs. Fong and Ford (2014) reported that antidepressants affect spawning and larval release in bivalves and disrupt locomotion and

TABLE 9–2	Endocrine-Disrupting Chemicals			
Group 1 Chlorinated Organics	Group 2 Industrial Chemicals	Group 3 Polymers with Molecular Weights ≤ 1000	Group 4 Substances Controlled under the Toxic Substances Control Act	Group 5 Pesticides
2,4,5-Trichlorophenol (S)	Hydrazine (P)	Fluoropolyol	PAHs (K)	Malathion (S)
PCB (K)	Bisphenol A (P)	Methoxypolysiloxane (P)	Triazine	1,2-Dibromo-2-chloropropane
2,3,7,8, TCDD (K)	Transplatin (S)	Polyethylene oxide (s)	Lead (K)	Dicamba (S)
Pentachlorophenol (K)	Benzoflumethiazide	Poly(isobutylene)	Mercury (K)	Aldicarb (P)
Hexachlorobenzene (K)	Propanthelinebromide (S)	Polyurethane (S)	Endosulfans	Aldicarb, nitrofen (S)
	p-Nitrotoluene		Kepone (S)	Aldrin (K)
	Obidoxime chloride (S)		Tributyl tin (K)	DDT and its metabolites DDE and DDD (K)
	Phthalates esters (K)			Carbaryl (K)
	Alkylphenols (K)			

P = probable EDC; S = suspected EDC; K = known EDC; PCB = polychlorinated biphenyl; TCDD = tetrachlordibenzo-*p*-dioxin; PAH = polycyclic aromatic hydrocarbons.
Source: O. A. Sadik and D. M. Witt, 1999; Acerini and Hughes, 2006.

reduce fecundity in snails. In crustaceans, antidepressants affect freshwater amphipod activity patterns, marine amphipod photo- and geotactic behavior, crayfish aggression, and daphnid reproduction and development. Vasquez et al. (2014) provided a review of the literature regarding the effects of pharmaceutical compounds on different biological systems. They discussed the gaps of knowledge that need to be addressed along with future research needs.

Endocrine-Disrupting Chemicals

The class of chemicals known as endocrine disrupters, or EDCs, are a class of chemicals that has received significant interest from the scientific community, regulatory agencies, and the general public. EDCs include the polychlorinated biphenyls, commonly used pesticides such as atrazine and other triazine chemicals, and the phthalates (as shown in Table 9–2). EDCs can mimic estrogens, androgens, or thyroid hormones, or their antagonists, although the structures of many EDCs bear little resemblance to that of natural hormones with which they interfere (see Figure 9–6). They can also interfere with the regulation of reproductive and developmental processes in mammals, birds, reptiles, and fish (Sadik and Witt, 1999; Harries et al., 2000). The chemicals can also alter the normal physiological function of the endocrine system and can affect the synthesis of hormones in the body. EDCs can also target tissues where the hormones exert their effects.

Although numerous studies have indicated the potential effects of these chemicals on wildlife (Acerini and Hughes, 2006), the effects of these chemicals on humans at the low doses usually found in the environment remain under debate. Nevertheless, recent research suggests that there may be reason for concern. For example, in developed countries the incidence of testicular cancer is increasing (Moller, 1998), the quality and quantity of semen is decreasing (Andersen et al., 2000), and the frequency of undescended testes and hypospadias (a birth defect in which the urinary tract opening is not at the tip of the penis) is increasing (Paulozzi, Erickson, and Jackson, 1997). In addition, the rates of male reproductive disorders have increased over the

Example of the chemical structure of suspected endocrine disrupting agents with known estrogenic effects (xenoestrogens) (compared with the natural hormone ligand 17ß-estradiol shown in blue box).

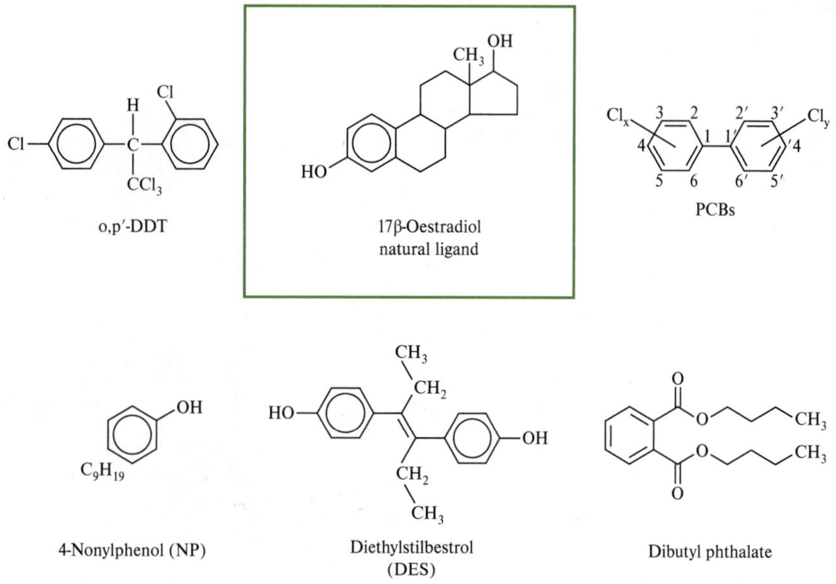

o,p′-DDT

17β-Oestradiol
natural ligand

PCBs

4-Nonylphenol (NP)

Diethylstilbestrol
(DES)

Dibutyl phthalate

last 40 years. There appears to be evidence that exposure to some EDCs has a role in the early onset of puberty in some children (Parent et al., 2003). On the contrary, exposure to PCBs has been implicated in late onset of puberty in girls and boys (Den Hond et al., 2002). Clearly, much research is necessary to better assess and monitor the health risks associated with EDCs.

Other Organic Chemicals

In addition to the groups of organic chemicals discussed previously, there are numerous other organic chemicals whose release into the environment can be problematic. These chemicals include hydrocarbons from combustion processes and oil and gasoline spills, and solvents used in dry cleaning and metal washing. The hydrocarbons formed during combustion include chemicals such as methane, benzene, and a class of compounds called polycyclic aromatic hydrocarbons. The polycyclic aromatic hydrocarbons (PAHs) contain two or more fused benzene rings (as shown in Figure 2–8). Several of the PAHs are known human carcinogens. Since these compounds are lipophilic, they readily bioaccumulate. The PAHs are also released during oil and gasoline spills. Among the solvents used in dry cleaning and metal washing are trichloroethane, tetrachlorethane, trichlorethylene, and tetrachloroethylene (see Figure 2–4). These volatile chemicals are common groundwater contaminants, released from leaking storage tanks or improper disposal. The anaerobic degradation of these compounds can lead to the formation of vinyl chloride, a known human carcinogen. Since many of these chemicals are hydrophobic, when released into lakes and rivers they sorb onto particles and eventually settle and concentrate in the sediments. The mixing of the sediments can result in the resolubilization of these chemicals and release back into the water column.

Arsenic

Arsenic is a naturally occurring element in the environment. Its occurrence in groundwater is largely the result of minerals dissolving naturally from weathered rocks and soils, mainly from iron oxides or sulfide minerals. High arsenic concentrations in groundwater have been documented in parts of Maine, Michigan, Minnesota, South Dakota, Oklahoma, and Wisconsin. Groundwater in many regions of these states has been shown to have arsenic concentrations

FIGURE 9–7

Arsenic concentrations in the groundwater in the U.S. (*Source:* https://water.usgs.gov/nawqa/trace/arsenic/) *Source:* USGS.

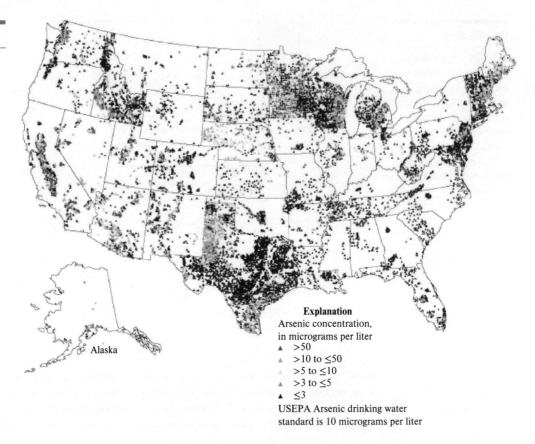

Explanation
Arsenic concentration,
in micrograms per liter
▲ >50
▲ >10 to ≤50
▲ >5 to ≤10
▲ >3 to ≤5
▲ ≤3
USEPA Arsenic drinking water
standard is 10 micrograms per liter

Alaska

exceeding 10 mg · L^{-1} (Welch et al., 1999). Figure 9–7 shows those areas of the United States where arsenic in groundwater is problematic. Arsenic in groundwater is also challenging environmental scientists and engineers in other areas of the world including India, Taiwan, and Bangladesh, where a World Health Organization study has revealed that between 33 and 77 million of Bangladesh's 125 million people are at risk of arsenic poisoning from groundwater.

Arsenic poisoning from groundwater can have numerous health effects, ranging from circulatory disorders, gastrointestinal upsets, diabetes, peripheral neuropathies, and skin lesions (Chen et al., 1992). Epidemiologic studies conducted in several countries (including Taiwan, Japan, England, Hungary, Mexico, Chile, and Argentina) have linked arsenic-contaminated drinking water to skin cancer. Increased mortality from internal cancers of liver, bladder, kidney, and lungs has also been reported (Chen et al., 1992).

On January 22, 2001, the U.S. EPA adopted a new drinking water standard for arsenic. The Final Rule established the MCL at 10 µg · L^{-1}, which became enforceable on January 23, 2006. At the previous MCL of 50 µg · L^{-1}, the lifetime risk for all cancers was calculated to be on the order of 1 in 100. Following the National Research Councils report *Arsenic in Drinking Water* (1999), the EPA proposed an arsenic MCL of 5 µg · L^{-1} A maximum contaminant level goal (MCLG) of zero was proposed. Despite EPA policy, which has been to set the MCL as close as feasible to the MCLG, taking into account technological feasibility and cost-effectiveness, the MCL was not set at 5, but at 10 µg · L^{-1} At a level of 10 µg · L^{-1} the lifetime excess risks for combined lung and bladder cancers were calculated to be 30 per 10,000 people for women and 37 per 10,000 for men (National Research Council, 2001). These risks can be compared with the MCLs set for many other carcinogenic compounds, which are set based on risks of 1:100,000 to 1:1,000,000.

Toxic Metals

Heavy metals enter aquatic environments through the discharge of industrial waste and waste-water treatment plants, storm-water runoff, mining operations, smokestack emissions, and other diffuse sources (such as from vehicles). The most commonly occurring heavy metals are arsenic, cadmium, chromium, copper, nickel, lead, and mercury. As heavy metals persist in the environment, they tend to accumulate in soils, sediments, and biota. Heavy metals can also bioaccumulate and biomagnify.

Mining operations have had a very significant impact on water quality. Acid mine drainage is formed by a series of geochemical and microbial reactions that occur when water comes in contact with the pyrite in coal or the overburden of a mine operation. The resulting water has a low pH and high concentrations of such dissolved metals as copper, lead, and mercury. Acid mine drainage can result in the contamination of drinking water, destruction of ecosystems and biota, and corrosion of infrastructure such as bridges. Contact with water in the most highly polluted streams can result in skin irritations. Contaminated waters are highly colored and can stain river structures red. The riverbed becomes coated with a gelatinous, yellow- or orange-colored, slimy precipitate (ferric hydroxide) called "yellow boy."

One of the earliest instances of heavy metal toxicity that was conclusively determined to result from water pollution and the bioaccumulation of a metal occurred in Minamata, Japan. In May of 1956, it was first reported that people living in the Yatsushiro Sea coastal area, particularly around Minamata Bay, were suffering from some type of neurological disorder. Those affected experienced numbness in their limbs and lips, slurred speech, and vision impairments. Some people fell into unconsciousness or suffered from involuntary movements. People reported that cats seemed to be going insane and jumped off the docks into the bay. Finally, birds were dropping from the sky. By the end of that year, there were 52 human cases of the dreaded disease and 17 people had died. It was later discovered that the disease was caused by exposure to methyl mercury. For years, the Chisso Corporation had been dumping mercury into the bay. In the anoxic sediments, the mercury became methylated. The bottom-feeding organisms, including crabs, shrimp, and shellfish, consumed the methyl mercury, which then moved up the food web to carnivorous fish and birds, then to cats and humans. In 1965, a second outbreak of Minamata disease occurred, this time in the Agano River basin. The number of people actually affected is still unknown, but it is estimated to have been in the thousands.

The California Gold Rush has left its legacy in the Sacramento River watershed: 100 lakes, rivers, and streams are presently on the Clean Water Act Section 303(d) list of impaired water bodies because of mercury contamination. Among the sources of mercury is cinnabar ore (mercury sulfide) that was mined in the Inner Coast Ranges for elemental mercury. The elemental mercury was then transported to the Sierra Nevada and used to extract gold. Several million kilograms of mercury were released during this era. Methyl mercury is especially problematic because in anoxic environments it is converted to methyl mercury, a potent neurotoxin. Methyl mercury is readily absorbed from water and food and is biomagnified in the food chain. As a result, state health managers have issued fish consumption advisories for sensitive populations, such as pregnant women. Unfortunately, little has been done to address the contamination of the watershed.

Another disease caused by metal toxicity is itai-itai. Itai-itai results from chronic exposure to cadmium, resulting in kidney dysfunction followed by osteomalacia. The name means ouch-ouch and comes from the chronic back and bone pain consistent with osteomalacia. In this case, water from the cadmium-polluted Jinzu River was used to irrigate rice. The rice adsorbed the heavy metals, including the cadmium. Exposure to cadmium occurred when people ate the contaminated rice.

Thankfully, we have learned from these incidents and federal and state regulations have been promulgated. This has resulted in a significant reduction in the release of toxic metals from point sources such as industrial operations. More recent regulations have focused on abating the input of metals from nonpoint sources such as urban runoff.

Heat

Although heat is not often recognized as a pollutant, those in the electric power industry are well aware of the problems of disposing of waste heat. Also, waters released by many industrial processes are much warmer than the receiving waters. In some environments an increase in water temperature can be beneficial. For example, production of clams and oysters can be increased in some areas by warming the water. On the other hand, increases in water temperature can have negative effects. Many important commercial and game fish, such as salmon and trout, live only in cool water. In some instances the discharge of heated water from a power plant can completely block salmon migration. Heat is being theorized as one of the potential causes of the feminization of genetically male Chinook salmon in the Columbia River near Hanford, Washington. Apparently, excessively high water temperatures or abnormal temperature fluctuations during egg development can cause a genetically male embryo to develop functioning female sex organs (Nagler et al., 2001). Higher temperatures also increase the rate of oxygen depletion in areas where oxygen-demanding wastes are present. Because temperature increases result in a decrease in oxygen solubility, this increased extent of oxygen depletion along with higher temperatures causes a very pronounced deterioration in the water quality.

Increased temperature results in a decrease in the solubility of carbon dioxide. Since marine waters are a major sink for carbon dioxide, this decrease in solubility could have a significant impact on the concentration of carbon dioxide in the atmosphere. Increases in the temperature of marine waters can also result in coral bleaching, the process whereby the coral polyp expels the algae (zooxanthellae). The exact mechanism is unknown, but it is known that other stresses, such as high light intensity, low salinity, and the presence of pollutants exacerbate the problem. Since the coral and algae have a symbiotic relationship, if the thermal stress lasts for an extended period of time, the corals become bleached and die. Mass coral bleaching began in the 1980s due to steady rise in sea temperatures, pushing reef-building corals closer to their thermal maxima. At the current rate of bleaching, some scientists predict that Australia will lose all or most of its coral reefs by 2050.

Nanoparticles

Nanoparticles are defined as those particles that have a dimension less than 100 nm. Included in this group of chemicals are naturally occurring humic material (derived from plant and animal matter); titania particles used in painkilling creams; fullerene nanotube composites used in the manufacture of tires, tennis rackets, and video screens; fullerene cages used in cosmetics; and protein-based nanomaterials used in the production of soaps, shampoos, and detergents (Wiesner et al., 2006). Fullerenes are a novel material fabricated from carbon, and exist as hollow spheres, ellipsoids, or tubes. Due to their unique properties (including high strength, electrical conductivity, and electron affinity), the commercial applications of these chemicals have grown almost exponentially in a very short time.

Responsible use of these materials necessitates an understanding of how the chemicals behave in the natural environment along with their impact on humans and other organisms. The size, chemical composition, surface structure and chemistry, solubility and shape will affect the interaction of nanoparticles with organisms. Those that are hydrophobic and lipophilic may accumulate in the tissue. As a result of their small size and large specific surface area, nanoparticles can sorb and transport toxic pollutants, which when inhaled can cause a number of pulmonary diseases in mammals, including lung granulomas (Guzman et al., 2006). Metal oxides have been shown to cause pulmonary inflammation in rodents and humans (Wiesner et al., 2006). Inhaled nanoparticles act more like a gas and have the ability to translocate in the body, traveling freely in the blood and reaching organs such as liver or brain. It is clear that much research needs to be accomplished to assess the environmental risks and impacts of these materials so that the field of nanotechnology can grow in a sustainable and responsible way.

9–3 WATER QUALITY MANAGEMENT IN RIVERS

The objective of water quality management is to control the discharge of pollutants so that water quality is not degraded to an unacceptable extent below the natural background level. Controlling waste discharges, however, must be a quantitative endeavor. We must be able to measure the pollutants, predict their effect on water quality, determine the background water quality that would be present without human intervention, and decide the levels acceptable for intended uses of the water.

To most people, the tumbling mountain brook, crystal clear and icy cold, fed by melting snow is safe to drink and the epitome of high water quality.* Certainly a stream in that condition is a treasure, but we cannot expect the Mississippi River to have the same water quality. It never did and never will. Yet both need proper management if the water is to remain usable. The mountain brook may serve as the spawning ground for desirable fish and must be protected from heat and sediment as well as chemical pollution. The Mississippi, however, is already warmed from hundreds of kilometers of exposure to the sun and carries the sediment from thousands of square kilometers of land. But even the Mississippi can be damaged by organic matter and toxic chemicals. Fish do live there, and the river is used as a water supply for millions of people.

The effect of pollution on a river depends both on the nature of the pollutant and the unique characteristics of the individual river.† Some of the most important characteristics include the volume and speed of water flowing in the river, the river's depth, the type of bottom, and the surrounding vegetation. Other factors include the climate of the region, the mineral heritage of the watershed, land use patterns, and the types of aquatic life in the river. Water quality management for a particular river must consider all these factors. Thus, some rivers are highly susceptible to pollutants such as sediment, salt, and heat, whereas others can tolerate large inputs of these pollutants without much damage.

Some pollutants, particularly oxygen-demanding wastes and nutrients, are so common and have such a profound effect on almost all types of rivers that they deserve special emphasis. This is not to say that they are always the most significant pollutants in any one river, but rather that no other pollutant category has as much overall effect on our nation's rivers. For these reasons, the next sections of this chapter will be devoted to a more detailed look at how oxygen-demanding material and nutrients affect water quality in rivers.

Effect of Oxygen-Demanding Wastes on Rivers

The introduction of oxygen-demanding material, either organic or inorganic, into a river depletes the dissolved oxygen in the water. This poses a threat to fish and other higher forms of aquatic life if the concentration of oxygen falls below a critical point. To predict the extent of oxygen depletion, it is necessary to know how much waste is being discharged and how much oxygen will be required to degrade the waste. However, because oxygen is continuously being replenished from the atmosphere and from photosynthesis by algae and aquatic plants, as well as being consumed by organisms, the concentration of oxygen in the river is determined by the relative rates of these competing processes. Organic oxygen-demanding materials are commonly measured by determining the amount of oxygen consumed during degradation in a manner approximating degradation in natural waters. This section begins by considering the factors affecting oxygen consumption during the degradation of organic matter, then moves on to inorganic nitrogen oxidation. Finally, the equations for predicting dissolved oxygen concentrations in rivers from degradation of organic matter are developed and discussed.

---

*Water from this stream may or may not be safe to drink, however appealing it may appear. Runoff containing pathogenic organisms, such as *Cryptosporidum parvum* and *Giardia lambia,* excreted in the feces of wildlife can cause disease in humans. This is a good reason to disinfect all surface waters intended for human consumption.

†Here we will use the word *river* to include streams, brooks, creeks, and any other channel of flowing, freshwater.

Biochemical Oxygen Demand

The amount of oxygen required to oxidize a substance to carbon dioxide and water may be calculated by stoichiometry if the chemical composition of the substance is known. This amount of oxygen is known as the **theoretical oxygen demand** (ThOD).

EXAMPLE 9–1   Compute the ThOD of 108.75 mg $\cdot$ L^{-1} of glucose ($C_6H_{12}O_6$).

Solution   We begin by writing a balanced equation for the reaction.

$$C_6H_{12}O_6 + 6O_2 \longrightarrow 6CO_2 + 6H_2O$$

Next, compute the gram molecular weights of the reactants using the table in Appendix B.

Glucose	Oxygen
6C = 72	(6)(2)O = 192 g $\cdot$ mol^{-1}
12H = 12	
6O = 96	
180 g $\cdot$ mol^{-1}	

Thus, it takes 192 g of oxygen to oxidize 180 g of glucose to CO_2 and H_2O.
The ThOD of 108.75 mg $\cdot$ L^{-1} of glucose is

$$(108.75 \text{ mg} \cdot \text{L}^{-1} \text{ of glucose}) \left(\frac{192 \text{ g} \cdot \text{mol}^{-1} O_2}{180 \text{ g} \cdot \text{mol}^{-1} \text{ glucose}} \right) = 116 \text{ mg} \cdot \text{L}^{-1} O_2$$

In contrast to the ThOD, the **chemical oxygen demand,** COD (pronounced "see oh dee"), is a measured quantity that does not depend on one's knowledge of the chemical composition of the substances in the water. In the COD test, a strong chemical oxidizing agent (chromic acid) is mixed with a water sample and then refluxed. The difference between the amount of oxidizing agent at the beginning of the test and that remaining at the end is used to calculate the COD.

If the oxidation of an organic compound is carried out by microorganisms using the organic matter as a food source, the oxygen consumed is known as **biochemical oxygen demand,** or BOD (pronounced "bee oh dee"). The test is a bioassay that uses microorganisms in conditions similar to those in natural water to measure indirectly the amount of biodegradable organic matter present. **Bioassay** means to measure by biological means. The actual BOD is almost always less than the ThOD due to the incorporation of some of the carbon into new bacterial cells. Thus, a portion of soluble carbon is removed but will not be measured in the BOD test.

To measure the BOD of a water, a water sample is inoculated with bacteria that consume the biodegradable organic matter to obtain energy for their life processes. Because the organisms also use oxygen in the process of consuming the waste, the process is called **aerobic decomposition.** This oxygen consumption is easily measured. The greater the amount of organic matter present, the greater the amount of oxygen used. The BOD test is an indirect measurement of organic matter because we actually measure only the change in dissolved oxygen concentration caused by the microorganisms as they degrade the organic matter. Although not all organic matter is biodegradable and the actual test procedures lack precision, the BOD test is still the most widely used method of measuring organic matter because of the direct conceptual relationship between BOD and oxygen depletion in receiving waters.

FIGURE 9–8

BOD and oxygen-
equivalent relationships.

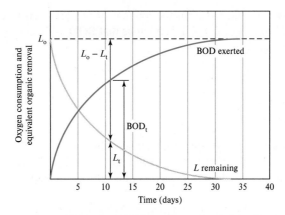

Only under rare circumstances will the ThOD and the COD be equal. If the chemical composition of all of the substances in the water is known and they are capable of being completely oxidized chemically, then the two measures of oxygen demand will be the same. A waste containing only simple sugars would fall into this category. The BOD is never equal to the ThOD or COD because some carbon is always converted to biomass or waste organic compounds and this soluble organic matter, which is removed during the test, is not measured, as discussed earlier.

When a water sample containing degradable organic matter is placed in a closed container and inoculated with bacteria, the oxygen consumption typically follows the pattern shown in Figure 9–8. During the first few days the rate of oxygen depletion is rapid because of the high concentration of organic matter present. As the concentration of organic matter decreases, so does the rate of oxygen consumption. During the last part of the BOD curve, oxygen consumption is mostly associated with the decay of the bacteria that grew during the early part of the test. It is generally assumed that the rate at which oxygen is consumed is directly proportional to the concentration of degradable organic matter remaining at any time. As a result, the BOD curve in Figure 9–8 can be described mathematically as a first-order reaction. Using our definition of reaction rate and reaction order from Chapter 4, this may be expressed as

$$\frac{dL}{dt} = -r_A \qquad (9\text{–}1)$$

where L = oxygen equivalent of the organic chemicals remaining (in mg $\cdot$ L^{-1})
$\quad -r_A = -kL$
$\quad\quad k$ = reaction rate constant (in days^{-1})

Rearranging Equation 9–1 and integrating yields

$$\frac{dL}{L} = -kdt \qquad (9\text{–}2)$$

$$\int_{L_o}^{L_t} \frac{dL}{L} = -k \int_{0}^{t} dt \qquad (9\text{–}3)$$

$$\ln \frac{L_t}{L_o} = -kt \qquad (9\text{–}4)$$

or

$$L_t = L_o e^{-kt} \qquad (9\text{–}5)$$

where L_o = oxygen equivalent of organics at time = 0
$\quad L_t$ = oxygen equivalent of the organic chemicals remaining at time, t (mg $\cdot$ L^{-1})

Rather than L_t our interest is in the amount of oxygen used in the consumption of the organics (BOD$_t$). From Figure 9–8, it is obvious that BOD$_t$ is the difference between the initial value of L_o and L_t, so

$$\text{BOD}_t = L_o - L_t$$
$$= L_o - L_o e^{-kt}$$
$$= L_o(1 - e^{-kt}) \tag{9–6}$$

L_o is often referred to as the **ultimate** BOD, that is, the maximum oxygen consumption possible when the waste has been completely degraded. Equation 9–6 is called the BOD rate equation. Note that lowercase k is used for the reaction rate constant in base e.

EXAMPLE 9–2   If the 3-day BOD (BOD$_3$) of a waste is 75 mg · L^{-1} and the BOD decay constant, k, is 0.345 day^{-1}, what is the ultimate BOD?

Solution   Substitute the given values into Equation 9–6 and solve for the ultimate BOD, L_o.

$$75 = L_o \left(1 - e^{-(0.345\ days^{-1})(3\ days)}\right) = 0.645 L_o$$

or

$$L_o = \frac{75}{0.645} = 116 \text{ mg} \cdot \text{L}^{-1}$$

You should note that the ultimate BOD (L_o) is defined as the maximum BOD exerted by the waste. It is denoted by the horizontal dotted line in Figure 9–8. Because BOD$_t$ approaches L_o asymptotically, it is difficult to assign an exact time to achieve the ultimate BOD. Indeed, based on Equation 9–1, it is achieved only in the limit as t approaches infinity. However, from a practical point of view, we can observe that when the BOD curve is approximately horizontal, the ultimate BOD has been achieved. In Figure 9–8, we would take this to be at about 35 days. In computations, we use a rule of thumb that if BOD$_t$ and L_o agree when rounded to three significant figures, then the time to reach ultimate BOD has been achieved. Given the uncertainties of the BOD test, there are occasions when rounding to two significant figures would be realistic.

Although the ultimate BOD best expresses the concentration of degradable organic matter, it does not, by itself, indicate how rapidly oxygen will be depleted in a receiving water. Oxygen depletion is related to both the ultimate BOD and the BOD rate constant (k). Although the ultimate BOD increases in direct proportion to the concentration of degradable organic matter, the numerical value of the rate constant is dependent on the following:

1. The nature of the waste
2. The ability of the organisms in the system to use the waste
3. The temperature

Nature of the Waste.   There are thousands of naturally occurring organic compounds, and not all of them can be degraded with equal ease. Simple sugars and starches are rapidly degraded and will therefore have a very large BOD rate constant. Cellulose (for example, toilet paper) degrades much more slowly, and compounds such as the higher molecular weight polycyclic aromatic hydrocarbons, highly chlorinated compounds such as DDT, chlorpyriphos, PCBs, caffeine, or many of the estrogenic compounds used in birth control pills are almost

TABLE 9–3	**Typical Values for the BOD Rate Constant**	
Sample		$k\,(20°C)$ (day^{-1})
Raw sewage		0.35–0.70
Well-treated sewage		0.12–0.23
Polluted river water		0.12–0.23

undegradable in the BOD test or in conventional wastewater treatment. In some cases, as with many of the phenolic compounds, the compound is actually toxic to the microorganisms, killing them so that little or no degradation of the waste can occur. Other compounds are intermediate between these extremes. The BOD rate constant for a complex waste depends very much on the relative proportions of the various components. Table 9–3 is a summary of typical BOD rate constants. The lower rate constants for treated sewage compared with raw sewage result from the fact that easily degradable organics are more completely removed than less readily degradable organics during wastewater treatment.

Ability of Organisms to Use Waste.   Any given microorganism is limited in its ability to use organic compounds. As a consequence, many organic compounds can be degraded by only a small group of microorganisms. In a natural environment receiving a continuous discharge of organic waste, that population of organisms that can most efficiently use this waste will predominate. However, the culture used to inoculate the sample used in the BOD test may contain only a very small number of organisms that can degrade the particular organic compounds in the waste. This problem is especially common when analyzing industrial wastes. The result is that the BOD rate constant would be lower in the laboratory test than in the natural water. To avoid this undesirable outcome the BOD test should be conducted with organisms that have been acclimated* to the waste so that the rate constant determined in the laboratory is comparable to that in the river.

Temperature.   Most biological processes speed up as the temperature increases and slow down as the temperature drops. Because oxygen use is caused by the metabolism of microorganisms, the rate of its use is similarly affected by temperature. Ideally, the BOD rate constant should be experimentally determined for the temperature of the receiving water. There are two difficulties with this ideal. Often the temperature of the receiving water changes throughout the year, so a large number of tests would be required to define k. An additional difficulty is the task of comparing data from various locations having different temperatures. Laboratory testing is therefore done at a standard temperature of 20°C, and the BOD rate constant is adjusted to the temperature of the receiving water using the following expression.

$$k_T = k_{20}(\theta)^{T-20} \tag{9–7}$$

where T = temperature of interest (in °C)
  k_T = BOD rate constant at the temperature of interest (in days^{-1})
  k_{20} = BOD rate constant determined at 20°C (in days^{-1})
  θ = temperature coefficient. For typical domestic wastewater, this has a value of 1.135 for temperatures between 4 and 20°C and 1.056 for temperatures between 20 and 30°C. (Schroepfer, Robins, and Susag, 1964)

---

*The word *acclimated* means that the organisms have had time to adapt their metabolisms to the waste or that organisms that can use the waste have been given the chance to predominate in the culture.

EXAMPLE 9–3   A waste is being discharged into a river that has a temperature of 10°C. What fraction of the maximum oxygen consumption has occurred in 4 days if the BOD rate constant, k, determined in the laboratory under standard conditions is 0.115 day^{-1}? (*Note:* All rate constants are base e.)

Solution   Determine the BOD rate constant, k, for the waste at the river temperature using Equation 9–7.

$$k_{10°C} = 0.115(1.135)^{10-20}$$
$$= 0.032 \text{ day}^{-1}$$

Use this value of k in Equation 9–6 to find the fraction of maximum oxygen consumption occurring in 4 days.

$$\frac{\text{BOD}_4}{L_o} = [1 - e^{-(0.032 \text{ days}^{-1})(4 \text{ days})}]$$
$$= 0.12$$

Laboratory Measurement of Biochemical Oxygen Demand

To be as consistent as possible, it is important to standardize testing procedures when measuring BOD. In the paragraphs that follow, the standard BOD test is outlined with emphasis placed on the reason for each step rather than the details. The detailed procedures can be found in *Standard Methods for the Examination of Water and Wastewater* (Eaton et al., 2005), which is the authoritative reference of testing procedures in the water pollution control field.

Step 1. A special 300-mL BOD bottle (Figure 9–9) is completely filled with a sample of water that has been appropriately diluted and inoculated with microorganisms. The bottle is then stoppered to exclude air bubbles. Samples require dilution because the only oxygen available to the

FIGURE 9–9

BOD Bottles

The point on the end of the stopper is to ensure that no air is trapped in the bottle. The bottle in the center is shown with the stopper in place. Water is placed in the small cup formed by the lip. This acts as a seal to further exclude air. The bottle on the right is shown with plastic wrap over the stopper. This is to prevent evaporation of the water seal. (©Harley Seeley of the Instructional Media Center, Michigan State University)

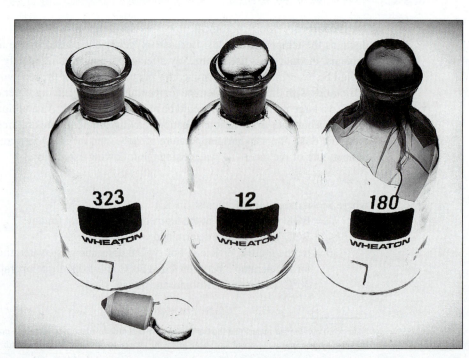

organisms is dissolved in the water. The most oxygen that can dissolve is about 9 mg · L^{-1}, so the BOD of the diluted sample should be between 2 and 6 mg · L^{-1} Samples are diluted with a special dilution water that contains all of the trace elements required for bacterial metabolism so that degradation of the organic matter is not limited by lack of bacterial growth. The dilution water also contains an inoculum of microorganisms so that all samples tested on a given day contain approximately the same type and number of microorganisms.

The ratio of undiluted to diluted sample is called the **sample size,** usually expressed as a percentage, while the inverse relationship is called the **dilution factor.** Mathematically, these are

$$\text{Sample size (\%)} = \frac{\text{volume of undiluted sample}}{\text{volume of diluted sample}} \times 100 \qquad (9\text{–}8)$$

$$\text{Dilution factor} = P = \frac{\text{volume of wastewater sample}}{\text{volume of wastewater plus dilution water}} \qquad (9\text{–}9)$$

The appropriate sample size to use can be determined by dividing 4 mg · L^{-1} (the midpoint of the desired range of diluted BOD) by the estimated BOD concentration in the sample being tested. A convenient volume of undiluted sample is then chosen to approximate this sample size. The dilution factors given earlier are not to be used when the sample size is small compared with the total volume because under such conditions the seed/dilution water will contribute significantly to the BOD. For low dilutions, the more complex equation given in Equation 9–10a should be used.

EXAMPLE 9–4 The BOD of a wastewater sample is estimated to be 180 mg · L^{-1}. What volume of undiluted sample should be added to a 300-mL bottle? Also, what are the sample size and dilution factor using this volume? Assume that 4 mg · L^{-1} BOD can be consumed in the BOD bottle.

Solution 1 Estimate the sample size needed.

$$\text{Sample size} = \frac{4\text{ mg} \cdot \text{L}^{-1}}{180\text{ mg} \cdot \text{L}^{-1}} \times 100 = 2.22\%$$

Estimate the volume of undiluted sample needed because the volume of diluted sample is 300 mL.

Volume of undiluted sample = 0.0222 × 300 mL = 6.66 mL

Therefore a convenient sample volume would be 7.00 mL.

Compute the actual sample size and dilution factor.

$$\text{Sample size} = \frac{7.00\text{ mL}}{300\text{ mL}} \times 100 = 2.33\%$$

$$\text{Dilution factor} = P = \frac{7\text{ mL}}{300\text{ mL}} = 0.0233$$

Solution 2 Although environmental engineers have traditionally used this approach for determining dilution factors for BOD analysis, we could use a more fundamental approach: that of the mass-balance equation (which is inherently what was done above in Solution 1). Let's think about what we are attempting to do. We have a BOD bottle to which we add wastewater and dilution water. The wastewater has an estimated BOD of 180 mg · L^{-1}, and the dilution water has a BOD of zero.

The total volume in the BOD bottle is $V_{ww} + V_{dw}$ or the volume of the wastewater plus the volume of the dilution water. The total mass of BOD in the BOD bottle can be calculated from

$$BOD_{ww} \times V_{ww} + BOD_{dw} \times V_{ww} = (180 \text{ mg} \cdot L^{-1})(V_{ww}) + (0 \text{ mg} \cdot L^{-1})(V_{dw})$$

We also know that the size of a BOD bottle is 300 mL, so that

$$V_{dw} + V_{ww} = 300 \text{ mL} = 0.300 \text{ L}$$

The problem statement indicates that we want a final concentration of BOD of 4 mg · L^{-1} Thus, the final mass of BOD in the bottle is

$$(4 \text{ mg} \cdot L^{-1})(0.300 \text{ L}) = 1.2 \text{ mg}$$

If we equate this to the total BOD we calculated earlier, then

$$BOD_{ww} \times V_{ww} + BOD_{dw} \times V_{ww} = (180 \text{ mg} \cdot L^{-1})(V_{ww}) + (0 \text{ mg} \cdot L^{-1})(V_{dw}) = 1.2 \text{ mg}$$

We can solve for V_{ww} = 1.2 mg/180 mg · L^{-1} = 0.00667 L. Let's multiply this value by 1000 to obtain the answer in milliliters because that is the more appropriate answer: 6.67 mL. This is the same answer we obtained earlier. Again, as stated there, when actually performing a BOD analysis, you would choose a volume that is easier to dispense, such as 7.00 mL.

Step 2. Blank samples containing only the inoculated dilution water are also placed in BOD bottles and stoppered. Blanks are required to estimate the amount of oxygen consumed by the added inoculum in the absence of the sample.

Step 3. The stoppered BOD bottles containing diluted samples and blanks are incubated in the dark at 20°C for the desired number of days. For most purposes, a standard time of 5 days is used. To determine the ultimate BOD and the BOD rate constant, additional times are used. The samples are incubated in the dark to prevent photosynthesis from adding oxygen to the water and invalidating the oxygen consumption results. As mentioned earlier, the BOD test is conducted at a standard temperature of 20°C so that the effect of temperature on the BOD rate constant is eliminated and results from different laboratories can be compared.

Step 4.   After the desired number of days has elapsed, the samples and blanks are removed from the incubator and the dissolved oxygen concentration in each bottle is measured. The BOD of the undiluted sample is then calculated using the following equation:

$$BOD_t = \frac{(DO_{b,t} - DO_{s,t})}{P} \tag{9–10}$$

where $DO_{b,t}$ = dissolved oxygen concentration in blank (blank) after t days of incubation (in mg · L^{-1})

$DO_{s,t}$ = dissolved oxygen concentration in sample after t days of incubation (in mg · L^{-1})

P = dilution factor

The preceding equation is valid only when the BOD of the seed water or the dilution water is negligible. If the BOD of the dilution or seed water is significant, then the following equation must be used.

$$BOD_t = \frac{(DO_{s,i} - DO_{s,t}) - (DO_{b,i} - DO_{b,t})f}{P} \tag{9–10a}$$

where $DO_{b,t}$ and $DO_{s,t}$ are defined earlier and

$DO_{s,i}$ = the initial DO of the sample

$DO_{b,i}$ = the initial DO of the blank (seed) control

f = ratio of seed in diluted sample to seed in seed control

= (% seed in diluted sample)/(% seed in seed control)

= (volume of seed in diluted sample)/(volume of seed in seed control)

EXAMPLE 9–5   What is the BOD$_5$ of the wastewater sample of Example 9–4 if the DO values for the blank and diluted sample after 5 days are 8.7 and 4.2 mg · L^{-1} respectively?

Solution   Substitute the appropriate values into Equation 9–10.

$$BOD_5 = \frac{8.7 - 4.2}{0.0233} = 204.5, \text{ or } 205 \text{ mg} \cdot L^{-1}.$$

Note that because the sample was not seeded, $f = 1$.

Additional Notes on Biochemical Oxygen Demand

Although the 5-day BOD has been chosen as the standard value for most wastewater analysis and for regulatory purposes, ultimate BOD is actually a better indicator of total waste strength. For any one type of waste having a defined BOD rate constant, the ratio between ultimate BOD and BOD$_5$ is constant so that BOD$_5$ indicates relative waste strength. For different types of wastes having the same BOD$_5$, the ultimate BOD is the same only if, by chance, the BOD rate constants are the same. This is illustrated in Figure 9–10 for a municipal wastewater having a $k = 0.345$ day^{-1} and an industrial wastewater having a $k = 0.115$ day^{-1}. Although both wastewaters have a BOD$_5$ of 200 mg · L^{-1}, the industrial wastewater has a much higher ultimate BOD and can be expected to have a greater effect on dissolved oxygen in a river. For the industrial wastewater, a smaller fraction of the BOD was exerted in the first 5 days due to the lower rate constant.

FIGURE 9–10

The effect of k on ultimate BOD for two wastewaters having the same BOD_5.

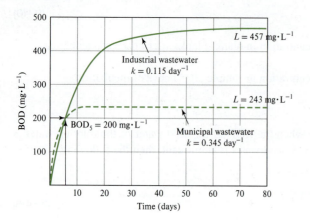

FIGURE 9–11

The effect of k on BOD_5 when the ultimate BOD is constant.

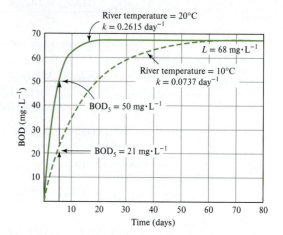

Proper interpretation of BOD_5 values can also be illustrated in another way. Consider a sample of polluted river water for which the following values were determined using standard laboratory techniques: $BOD_5 = 50$ mg $\cdot$ L^{-1}, and $k = 0.2615$ day^{-1}. The ultimate BOD calculated from Equation 9–6 is, therefore, 68 mg $\cdot$ L^{-1}. However, because the river temperature is 10°C, the k value in the river is calculated to be 0.0737 day^{-1} (using Equation 9–7). As shown graphically in Figure 9–11, the laboratory value of BOD_5 seriously overestimates the actual oxygen consumption in the river. Again, a smaller fraction of the BOD is exerted in 5 days when the BOD rate constant is lower.

The 5-day BOD was chosen as the standard value for most purposes because the test was devised by sanitary engineers in England, where the River Thames has a travel time to the sea of less than 5 days, so there was no need to consider oxygen demand at longer times. Because no other time is any more rational than 5 days, this value has become firmly entrenched in the profession.

Nitrogen Oxidation

Until now our unstated assumption has been that only the carbon in organic matter is oxidized. Actually many organic compounds, such as proteins, also contain nitrogen that can be oxidized with the consumption of molecular oxygen. Because the mechanisms and rates of nitrogen oxidation are distinctly different from those of carbon oxidation, the two processes must be considered separately. Logically, oxygen consumption due to oxidation of carbon is called **carbonaceous BOD** (CBOD), and that due to nitrogen oxidation is called **nitrogenous BOD** (NBOD).

The organisms that oxidize the carbon in organic compounds to obtain energy cannot oxidize the nitrogen in these compounds. Instead, the nitrogen is released into the surrounding water as ammonia (NH_3). At normal pH values, this ammonia is actually in the form of the ammonium cation (NH_4^+). The ammonia released from organic compounds, plus that from other sources such as industrial wastes and agricultural runoff (i.e., fertilizers), is oxidized to nitrate (NO_3^-) by a special group of nitrifying bacteria as their source of energy in a process called **nitrification.** The overall reaction for ammonia oxidation (nitrification) is:

$$NH_4^+ + 2O_2 \xrightarrow{\text{microorganisms}} NO_3^- + H_2O + 2H^+ \tag{9–11}$$

From this reaction the theoretical NBOD can be calculated as follows:

$$NBOD = \frac{\text{grams of oxygen used}}{\text{grams of nitrogen oxidized}} = \frac{(2 \text{ moles})(32 \text{ g O}_2 \cdot \text{mol}^{-1})}{(1 \text{ mole})(14 \text{ g N} \cdot \text{mol}^{-1})}$$

$$= 4.57 \text{ g O}_2 \cdot \text{g}^{-1} \text{ N} \qquad\qquad (9\text{–}12)$$

The actual nitrogenous BOD is slightly less than the theoretical value due to the incorporation of some of the nitrogen into new bacterial cells, but the difference is only a few percent.

Because nitrogen can be present in numerous forms (NH_3, NH_4^+, NO_3^-, NO_2^- and various organic compounds), it is often convenient to report concentrations of nitrogenous compounds in units of milligrams per liter as N. For example, it is often convenient for environmental engineers to report ammonia concentrations as "ammonia nitrogen," that is as NH_3–N.

EXAMPLE 9–6   (a) Compute the theoretical NBOD of a wastewater containing 30 mg $\cdot$ L^{-1} of ammonia as nitrogen.
(b) If the wastewater analysis was reported as 30 mg $\cdot$ L^{-1} of ammonia (NH_3), what would the theoretical NBOD be?

Solution   In the first part of the problem, the amount of ammonia was reported as NH_3–N. Therefore, we can use the theoretical relationship developed from Equation 9–11.

Theoretical NBOD = (30 mg NH_3–N $\cdot$ L^{-1})(4.57 mg O$_2$ $\cdot$ mg^{-1} N) = 137 mg O$_2$ $\cdot$ L^{-1}

To answer part (b), we must convert milligrams per liter of ammonia to milligrams per liter as NH_3–N by multiplying by the ratio of gram molecular weights of N to NH_3.

$$(30 \text{ mg NH}_3 \cdot \text{L}^{-1})\left(\frac{14 \text{ g N} \cdot \text{mol}^{-1}}{17 \text{ g NH}_3 \cdot \text{mol}^{-1}}\right) = 24.7 \text{ mg N} \cdot \text{L}^{-1}$$

Now we may use the relationship developed from Equation 9–11.

$$\text{Theoretical NBOD} = (24.7 \text{ mg N} \cdot \text{L}^{-1})\left(\frac{4.57 \text{ mg O}_2}{\text{mg N}}\right) = 113 \text{ mg O}_2 \cdot \text{L}^{-1}$$

The rate at which the NBOD is exerted depends heavily on the number of nitrifying organisms present. Few of these organisms occur in untreated sewage, but the concentration is high in a well-treated effluent. When samples of untreated and treated sewage are subjected to the BOD test, oxygen consumption follows the pattern shown in Figure 9–12. In the case of untreated sewage, the NBOD is exerted after much of the CBOD has been exerted. The lag is due to the time it takes for the nitrifying bacteria to reach a sufficient population for the amount of NBOD exertion to be significant compared with that of the CBOD. In the case of the treated sewage, a higher population of nitrifying organisms in the sample reduces the lag time. Once nitrification begins, however, the NBOD can be described by Equation 9–6 with a BOD rate constant comparable to that for the CBOD of a well-treated effluent ($k = 0.80$ to 0.20 day^{-1}). Because the lag before the nitrogenous BOD is highly variable, BOD$_5$ values are often difficult to interpret. When measurement of only carbonaceous BOD is desired, chemical inhibitors are added to stop the nitrification process. The rate constant for nitrification is also affected by temperature and can be adjusted using Equation 9–7.

FIGURE 9–12

BOD curves showing both carbonaceous (CBOD) and nitrogenous (NBOD) BOD.

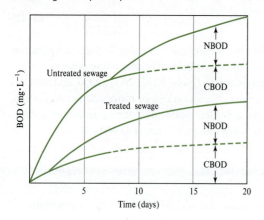

FIGURE 9–13

Typical DO sag curve.

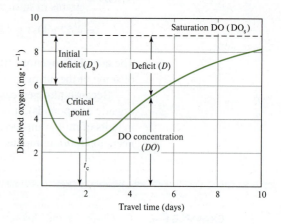

DO Sag Curve

The concentration of dissolved oxygen in a river is an indicator of the general health of the river. All rivers have some capacity for self-purification. As long as the discharge of oxygen-demanding wastes is well within the self-purification capacity, the DO level will remain high, and a diverse population of plants and animals, including game fish, can be found. As the amount of waste increases, the self-purification capacity can be exceeded, causing detrimental changes in plant and animal life. The stream loses its ability to cleanse itself, and the DO level decreases. When the DO drops below about 4 to 5 mg $\cdot$ L^{-1}, most game fish will have been driven out. If the DO is completely removed, fish and other higher animals are killed or driven out, and extremely noxious conditions result. The water becomes blackish and foul-smelling as the sewage and dead animal life decompose under anaerobic conditions (i.e., without oxygen).

One of the major tools of water quality management in rivers is assessing the capability of a stream to absorb a waste load. This is done by determining the profile of DO concentration downstream from a waste discharge. This profile is called the DO sag curve (Figure 9–13) because the DO concentration dips as oxygen-demanding materials are oxidized and then rises again further downstream as the oxygen is replenished from the atmosphere and photosynthesis. As depicted in Figure 9–14, the biota of the stream are often a reflection of the dissolved oxygen conditions in the stream.

To develop a mathematical expression for the DO sag curve, the sources of oxygen and the factors affecting oxygen depletion must be identified and quantified. The only significant sources of oxygen are reaeration from the atmosphere and photosynthesis of aquatic plants. Oxygen depletion is caused by a larger range of factors, the most important being the BOD, both carbonaceous and nitrogenous, of the waste discharge, and the BOD already in the river upstream of the waste discharge. The second most important factor is that the DO in the waste discharge is usually less than that in the river. Thus, the DO at the river is often reduced as soon as the waste is added, even before any BOD is exerted. Other factors affecting dissolved oxygen depletion include nonpoint source pollution, the respiration of organisms living in the sediments (**benthic demand**), and the respiration of aquatic plants. Following the classical approach, the DO sag equation will be developed by considering only initial DO reduction, ultimate BOD, and reaeration from the atmosphere.

FIGURE 9–14

Oxygen sag downstream of an organic source.

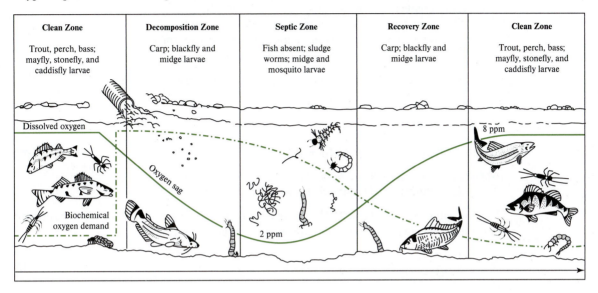

Clean Zone	Decomposition Zone	Septic Zone	Recovery Zone	Clean Zone
Trout, perch, bass; mayfly, stonefly, and caddisfly larvae	Carp; blackfly and midge larvae	Fish absent; sludge worms; midge and mosquito larvae	Carp; blackfly and midge larvae	Trout, perch, bass; mayfly, stonefly, and caddisfly larvae

Mass-Balance Approach.   Simplified mass balances help us understand and solve the DO sag curve problem. Three conservative (those without chemical reaction) mass balances may be used to account for initial mixing of the waste stream and the river. DO, ultimate BOD, and temperature all change as the result of mixing of the waste stream and the river. Once these are accounted for, the DO sag curve may be viewed as a nonconservative mass balance, that is, one with reactions. We can illustrate the mixing of the waste stream and the transport of the waste using the schematic diagram in Figure 9–15.

Through the discharge pipe, the waste stream flows into the river. The rectangle across the river at the location of the discharge pipe becomes the control volume around which we will develop our mass-balance expression. We will assume that the pollutant discharged from the pipe remains in this control volume and the entire volume moves downstream as a single entity. At time zero the volume is located at the pipe. The diagram illustrates the location of the volume at two subsequent times. For this case, we have two inputs and one output. The conservative mass-balance diagram for oxygen (mixing only) is shown in Figure 9–16. The product of the water flow and the DO concentration yields a mass of oxygen per unit of time.

FIGURE 9–15

Schematic diagram of the mixing of the waste stream.

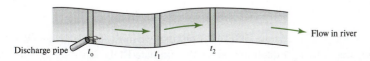

FIGURE 9–16

Mass-balance diagram for BOD and DO mixing.

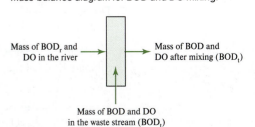

$$\text{Mass of DO in wastewater} = Q_w\text{DO}_w \tag{9-13}$$

$$\text{Mass of DO in river} = Q_r\text{DO}_r \tag{9-14}$$

where Q_w = volumetric flow rate of wastewater (in $m^3 \cdot s^{-1}$)
    Q_r = volumetric flow rate of the river (in $m^3 \cdot s^{-1}$)
    DO_w = dissolved oxygen concentration in the wastewater (in $g \cdot m^{-3}$)
    DO_r = dissolved oxygen concentration in the river (in $g \cdot m^{-3}$)

The mass of DO in the river after mixing equals the sum of the mass fluxes.

$$\text{Mass of DO after mixing} = Q_w\text{DO}_w + Q_r\text{DO}_r \tag{9-15}$$

In a similar fashion for ultimate BOD,

$$\text{Mass of BOD after mixing} = Q_wL_w + Q_rL_r \tag{9-16}$$

where L_w = ultimate BOD of the wastewater (in $mg \cdot L^{-1}$)
    L_r = ultimate BOD of the river (in $mg \cdot L^{-1}$)

The concentrations of DO and BOD after mixing are the respective masses per unit time divided by the total flow rate (i.e., the sum of the wastewater and river flows).

$$\text{DO} = \frac{Q_w\text{DO}_w + Q_r\text{DO}_r}{Q_w + Q_r} \tag{9-17}$$

$$L_a = \frac{Q_wL_w + Q_rL_r}{Q_w + Q_r} \tag{9-18}$$

where L_a = initial ultimate BOD after mixing.

EXAMPLE 9–7   The town of Aveta discharges 17,360 $m^3 \cdot day^{-1}$ of treated wastewater into the Tefnet Creek. The treated wastewater has a BOD_5 of 12 $mg \cdot L^{-1}$ and a BOD decay constant, k, of 0.12 day^{-1} at 20°C. Tefnet Creek has a flow rate of 0.43 $m^3 \cdot s^{-1}$ and an ultimate BOD, L_o, of 5.0 $mg \cdot L^{-1}$. The DO of the river is 6.5 $mg \cdot L^{-1}$ and the DO of the wastewater is 1.0 $mg \cdot L^{-1}$ Compute the DO and initial ultimate BOD, L_o, after mixing.

Solution   The DO after mixing is given by Equation 9–17. To use this equation, we must convert the wastewater flow to compatible units, that is, to cubic meters per second.

$$Q_w = \frac{(17{,}360 \ m^3 \cdot day^{-1})}{(86{,}400 \ s \cdot day^{-1})} = 0.20 \ m^3 \cdot s^{-1}$$

The DO after mixing is then

$$\text{DO} = \frac{(0.20 \ m^3 \cdot s^{-1})(1.0 \ mg \cdot L^{-1}) + (0.43 \ m^3 \cdot s^{-1})(6.5 \ mg \cdot L^{-1})}{0.20 \ m^3 \cdot s^{-1} + 0.43 \ m^3 \cdot s^{-1}} = 4.75 \ mg \cdot L^{-1}$$

Before we can determine the initial ultimate BOD after mixing, we must first determine the ultimate BOD of the wastewater. Solving Equation 9–6 for the ultimate BOD, L_o:

$$L_o = \frac{BOD_5}{(1 - e^{-kt})} = \frac{12 \ mg \cdot L^{-1}}{(1 - e^{-(0.12 \ day^{-1})(5 \ days)})} = \frac{12 \ mg \cdot L^{-1}}{(1 - 0.55)} = 26.6 \ mg \cdot L^{-1}$$

Note that we used the subscript of 5 days in BOD$_5$ to determine the value of t in the equation. Now setting $L_w = L_o$, we can determine the initial ultimate BOD after mixing, L_a, using Equation 9–18.

$$L_a = \frac{(0.20 \text{ m}^3 \cdot \text{s}^{-1})(26.6 \text{ mg} \cdot \text{L}^{-1}) + (0.43 \text{ m}^3 \cdot \text{s}^{-1})(5.0 \text{ mg} \cdot \text{L}^{-1})}{0.20 \text{ m}^3 \cdot \text{s}^{-1} + 0.43 \text{ m}^3 \cdot \text{s}^{-1}} = 11.86, \text{ or } 12 \text{ mg} \cdot \text{L}^{-1}$$

For temperature, we must consider a heat balance rather than a mass balance. This is an application of a fundamental principle of physics:

Loss of heat by hot bodies = gain of heat by cold bodies   (9–19)

The **change in enthalpy** or "heat content" of a mass of a substance may be defined by the following equation:

$$H = mC_p\Delta T \tag{9–20}$$

where H = change in enthalpy (in J)
    m = mass of substance (in g)
    C_p = specific heat at constant pressure (in J $\cdot$ g^{-1} $\cdot$ K^{-1})
    ΔT = change in temperature (in K)

The specific heat of water varies slightly with temperature. A value of 4.19 will be a satisfactory approximation for the range of temperatures usually found in the natural environment. Using our fundamental heat loss = heat gain equation, we may write

$$(m_w)(4.19)\Delta T_w = (m_r)(4.19)\Delta T_r \tag{9–21}$$

The temperature after mixing is found by solving this equation for the final temperature by recognizing that ΔT on each side of the equation is the difference between the final river temperature (T_f) and the starting temperature of the wastewater and the river water, respectively.

$$T_f = \frac{Q_w T_w + Q_r T_r}{Q_w + Q_r} \tag{9–22}$$

Oxygen Deficit.   The DO sag equation has been developed using oxygen deficit rather than dissolved oxygen concentration, to make it easier to solve the integral equation that results from the mathematical description of the mass balance. The oxygen deficit is the difference between the saturation value (at the particular temperature of the water) and the actual dissolved oxygen concentration

$$D = \text{DO}_s - \text{DO} \tag{9–23}$$

where D = oxygen deficit (in mg $\cdot$ L^{-1})
    DO$_s$ = saturation concentration of dissolved oxygen (in mg $\cdot$ L^{-1})
    DO = actual concentration of dissolved oxygen (in mg $\cdot$ L^{-1})

The saturation value of dissolved oxygen is heavily dependent on water temperature; it decreases as the temperature increases. Values of DO$_s$ for freshwater are given in Table A–2 of Appendix A.

Initial Deficit. The beginning of the DO sag curve occurs at the point where a waste discharge mixes with the river. The initial deficit is calculated as the difference between saturated DO and the concentration of the DO after mixing (Equation 9–17).

$$D_a = DO_s - \frac{Q_w DO_w + Q_r DO_r}{Q_{mix}} \tag{9-24}$$

where D_a = initial deficit after river and waste have mixed (in mg · L^{-1})
DO_s = saturation concentration of dissolved oxygen at the temperature of the river after mixing (in mg · L^{-1})

EXAMPLE 9–8   Calculate the initial deficit of the Tefnet Creek after mixing with the wastewater from the town of Aveta (see Example 9–7 for data). The stream temperature is 10°C, and the wastewater temperature is 10°C.

Solution   With the stream temperature, the saturation value of dissolved oxygen (DO_s) can be determined from the table in Appendix A. At 10°C, $DO_s = 11.33$ mg · L^{-1} Because we calculated the concentration of DO after mixing as 4.75 mg · L^{-1} in Example 9–7, the initial deficit after mixing is

$$D_a = DO_s - DO_{mix} = 11.33 \text{ mg} \cdot \text{L}^{-1} - 4.75 \text{ mg} \cdot \text{L}^{-1} = 6.58 \text{ mg} \cdot \text{L}^{-1}$$

Because wastewater commonly has a higher temperature than river water, especially during the winter, the river temperature downstream of the discharge is usually higher than that upstream. Because we are interested in downstream conditions, it is important to use the downstream temperature when determining the saturation concentration of dissolved oxygen.

DO Sag Equation.   Numerous models have been developed to describe the change in BOD in a river or stream with distance (or time) from a wastewater outfall. The level of complexity of these models varies greatly. The simplest model and the basis for all other models is the classical Streeter–Phelps model (Streeter and Phelps, 1925). This model assumes that

1. The river is completely and uniformly mixed in the horizontal direction across the river, and in the vertical direction with depth.
2. There is no dispersion of the pollutant as it moves downstream as shown in Figure 9–17.

This means that in the three-dimensional section shown in Figure 9–17 as box A, the chemical (e.g., DO and BOD) is completely mixed and that the concentration of the chemical is the

FIGURE 9–17

Cross section of stream flow. Each rhomboid represents the location of a control volume as it moves downstream in the river.

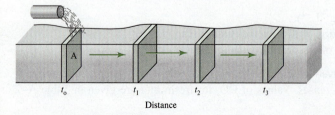

same in all locations in the box. The assumption that no dispersion takes place means that each rectangular cross-section moves down the stream as a packet. The shape of this packet does not change with distance downstream.

The Streeter–Phelps model has essentially two reaction terms: reaeration and deoxygenation. Reaeration describes the rate at which oxygen is replenished and is a function of the stream properties. As you might expect, the greater the turbulence, the more rapids, the greater the rate of reaeration. Reaeration will also depend on the oxygen deficit. The greater the oxygen deficit, the greater the reaeration rate. Thus, the rate at which oxygen is transferred from the air into the stream or river (reaeration) is linearly proportional to the dissolved oxygen deficit (the difference between the concentration of dissolved oxygen in the water and the solubility of oxygen in water at that temperature). The kinetics and modeling of oxygen dissolution are discussed in Chapter 2, Reaction Kinetics. Thus,

$$\text{Rate of reaeration} = k_r(\text{DO}_s - \text{DO}) = k_r D \tag{9–25}$$

where k_r = the reaeration coefficient (in time^{-1})

$\quad$ DO_s = the dissolved oxygen concentration at saturation (in mass · volume^{-1})

$\quad$ D = DO deficit = $(\text{DO}_s - \text{DO})$

The reaeration coefficient is very much dependent on the physical characteristics of the river. The greater the velocity in the stream, the greater the reaeration coefficient will be. O'Connor and Dobbins (1958) found that the reaeration coefficient is related to the stream velocity and depth using the following equation.

$$k_r = \frac{3.9u^{1/2}}{h^{3/2}} \tag{9–26}$$

where k_r = the reaeration coefficient at 20°C (day^{-1})

$\quad$ u = the average stream velocity (in m · s^{-1})

$\quad$ h = the average stream depth (in m)

Note the factor of 3.9 is necessary to make the equation dimensionally correct. The units of velocity must be in meters per second, whereas depth must be in meters. Reaeration coefficients, typically, vary from 0.1 for small ponds to >1.15 for rapids and waterfalls. The reaeration coefficient can be adjusted for temperature changes using the relationship.

$$k_r = k_{r,20}\theta^{(T-20)} \tag{9–27}$$

where $k_{r,20}$ = the reaeration coefficient for a temperature of 20°C

$\quad$ θ = temperature coefficient (1.024)

$\quad$ T = temperature (in °C)

EXAMPLE 9–9 A stream containing no biochemical oxygen demand (this is a hypothetical situation and rarely occurs) has a DO of 5.00 mg · L^{-1} and a flow rate, Q, of 8.70 m^3 · s^{-1}. The temperature of the stream is 18°C. The average velocity in the stream is 0.174 m · s^{-1}. The average depth, H, of the stream is 5 m. Determine the reaeration coefficient, k_r, and the rate of reaeration.

Solution To solve this problem we must first determine the reaeration coefficient at 20°C using Equation 9–26.

$$k_r = \frac{3.9u^{1/2}}{h^{3/2}} = \frac{3.9(0.174 \text{ m} \cdot \text{s}^{-1})^{1/2}}{(5.00 \text{ m})^{1.5}}$$

$$= \frac{3.9(0.417)}{11.18} = 0.146 \text{ day}^{-1}$$

We must also take into account the fact that the stream is not at 20°C but at 18°C. Thus we must use the equation

$$k_r = k_{r,20}\theta^{(T-20)} = (0.146)(1.024)^{(18-20)} = 0.139 \text{ day}^{-1}$$

The oxygen deficit is calculated by taking the difference between the saturation value ($9.54 \text{ mg} \cdot \text{L}^{-1}$—from Appendix A–2) and the dissolved oxygen concentration:

$$D = 9.54 - 5.0 = 4.54 \text{ mg} \cdot \text{L}^{-1}$$

Thus, according to Equation 9–25, the rate of reaeration equals

$$(0.139 \text{ day}^{-1})(4.54 \text{ mg} \cdot \text{L}^{-1}) = 0.632 \text{ mg} \cdot \text{L}^{-1} \cdot \text{day}^{-1}$$

Now, let's consider the rate at which DO disappears from the stream as a result of microbial action (M) or what is commonly referred to as the **rate of deoxygenation.** The rate of deoxygenation at any point in the river is assumed to be proportional to the BOD remaining at that point, so that

$$\text{Rate of deoxygenation} = k_d L_t \tag{9–28}$$

where k_d = the deoxygenation rate (time^{-1})
    L_t = BOD remaining at some time after the wastes enter the stream (mass · time^{-1})

Most models make the assumption that k_d is equal to the constant, k, that is, the rate constant obtained using the BOD test. Thus, using Equation 9–5, we can write Equation 9–28 in terms of the ultimate BOD.

$$\text{Rate of deoxygenation} = k_d L_o e^{-k_d t} \tag{9–29}$$

(Because the rate of deoxygenation is temperature-dependent, k_d must be corrected for temperature by correcting k for temperature.) Although the assumption that $k = k_d$ is reasonable for deep, slow-moving streams, it is a poor assumption for turbulent, shallow, rapidly moving streams. For such streams:

$$k_d = k + \frac{u}{h}\eta \tag{9–30}$$

where k = BOD rate constant (at 20°C)
    u = average velocity of stream flow (in length · time^{-1})
    h = average depth of stream (in length)
    η = bed activity coefficient (ranges from 0.1 for stagnant or deep water to 0.6 for rapidly flowing water)

The deoxygenation coefficient can be corrected for temperature using Equation 9–7 and coefficients used to correct BOD rate constants for temperature.

EXAMPLE 9–10   Determine the deoxygenation rate constant for the reach of Tefnet Creek (Examples 9–7 and 9–8) below the wastewater outfall (discharge pipe). The average speed, u, of the stream flow in the creek is $0.03 \text{ m} \cdot \text{s}^{-1}$. The depth, h, is 5.0 m and the bed-activity coefficient, η, is 0.35. What is the rate of deoxygenation, in units of $\text{mg} \cdot \text{L}^{-1} \cdot \text{day}^{-1}$?

Solution From Example 9–7, the value of the BOD decay constant, k, is 0.12 day^{-1}. Using Equation 9–30, the deoxygenation rate constant, k_d, at 20°C is

$$k_d = 0.12 \text{ day}^{-1} + \frac{0.03 \text{ m} \cdot \text{s}^{-1}}{5.0 \text{ m}} (0.35) = 0.1221, \text{ or } 0.12 \text{ day}^{-1}$$

Note that the units are not consistent. As we have noted before, empirical expressions, such as that in Equations 9–26 and 9–30, may have implicit conversion factors. Thus, you must be careful to use the same units as those used by the author of the equation.

We also note that the deoxygenation rate constant of 0.1221 day^{-1} is at 20°C. In Example 9–8, we noted that the stream temperature was 10°C. Thus, we must correct the estimated k_d value using Equation 9–7.

$$k_d \text{ at } 10°C = (0.1221 \text{ day}^{-1})(1.135)^{10-20} = (0.1221)(0.2819)$$
$$= 0.03442, \text{ or } 0.034 \text{ day}^{-1}$$

From Example 9–7, we know that the ultimate BOD immediately after mixing, L_t, is 12 mg·L^{-1}. Then using Equation 9–28, we obtain

$$\text{Rate of deoxygenation} = k_d L_t$$
$$= 0.034 \text{ day}^{-1} \times 12 \text{ mg} \cdot \text{L}^{-1}$$
$$= 0.408 \text{ mg} \cdot \text{L}^{-1} \cdot \text{day}^{-1}$$

The DO sag curve (Streeter–Phelps model) is simply a mass-balance approach to calculating or predicting the DO level downstream in a river. Using the mass-balance approach introduced in Chapter 4, we can develop the sag curve using a one-dimensional general continuity equation for a constituent in water. Here, the constituent is oxygen and the equation is

$$\frac{\partial C}{\partial t} = \bar{D}_x \frac{\partial^2 C}{\partial x^2} - \bar{v}_x \frac{\partial C}{\partial x} + \sum (\text{reactions}) \tag{9–31}$$

where $\bar{v}_x$ = downstream velocity in the x direction
   $\bar{D}_x$ = dispersion coefficient in the x direction

Here we will ignore the dispersion term and impose steady-state conditions. As such, the first two terms of Equation 9–31 drop out, and the equation is reduced to an ordinary differential equation.

$$\bar{v}_x \frac{dC}{dx} = \sum (\text{reactions}) \tag{9–32}$$

We can obtain the reaction terms by combining Equations 9–25 and 9–28 because the oxygen deficit is a function of the competition between oxygen use and reaeration from the atmosphere. Also if we use the notation used previously, that is, D for deficit, we obtain the equation

$$\bar{v}_x \frac{dD}{dx} = k_d L - k_r D \tag{9–33}$$

where $\dfrac{dD}{dx}$ = the change in oxygen deficit (D) with respect to unit of distance (in mg·L^{-1}·day)
   k_d = deoxygenation rate constant (in day^{-1})
   L = ultimate BOD of river water (in mg·L^{-1})
   k_r = reaeration rate constant (in day^{-1})
   D = oxygen deficit in river water (in mg·L^{-1})

Often, environmental engineers and scientists want to represent the previous equation in terms of travel time downstream. This can be easily accomplished if we recognize that

$$\text{Time} = \frac{x}{\bar{v}_x} \tag{9–34}$$

Therefore, we can write Equation 9–33 as

$$\frac{dD}{dt} = k_d L - k_r D \tag{9–35}$$

By integrating Equation 9–35, and using the boundary conditions (at $t = 0$; $D = D_a$ and $L = L_a$; and at $t = t$, $D = D_t$, and $L = L_t$), the DO sag equation is obtained:

$$D_t = \frac{k_d L_a}{k_r - k_d} (e^{-k_d t} - e^{-k_r t}) + D_a(e^{-k_r t}) \tag{9–36}$$

where D_t = oxygen deficit in river water after exertion of BOD for time, t (in mg · L^{-1})
   L_a = initial ultimate BOD after river and wastewater have mixed (Equation 9–18) (in mg · L^{-1})
   k_d = deoxygenation rate constant (in day^{-1})
   k_r = reaeration rate constant (in day^{-1})
   t = time of travel of wastewater discharge downstream (in days)
   D_a = initial deficit after river and wastewater have mixed (Equation 9–24) (in mg · L^{-1})

When $k_r = k_d$, Equation 9–36 reduces to

$$D_t = (k_d t L_a + D_a)(e^{-k_d t}) \tag{9–37}$$

where the terms are as previously defined.

EXAMPLE 9–11   A city of 200,000 people disposes of 1.05 m^3 · s^{-1} of treated sewage that still has an ultimate BOD (i.e., BOD$_u$) of 28.0 mg · L^{-1} and 1.8 mg · L^{-1} of DO into a river. Upstream from the outfall, the river has a flowrate of 7.08 m^3 · s^{-1} and a velocity of 0.37 m · s^{-1}. At this point, the BOD$_u$ and DO in the river are 3.6 and 7.6 mg · L^{-1} respectively. The saturation value of DO (at the temperature of the river) is 8.5 mg · L^{-1} The deoxygenation coefficient, k_d, is 0.61 day^{-1}, and the reaeration coefficient, k_r, is 0.76 day^{-1}. Assume complete mixing and that the velocity in the river is the same upstream and downstream of the outfall.

1. What is the oxygen deficit and the BOD$_u$ just downstream from the outfall (just after mixing, before any reaction can occur)?
2. What is the DO 16 km downstream?

Solution   1. Using Equation 9–17, we can calculate the concentration of dissolved oxygen in the river after mixing.

$$\text{DO}_{mix} = \frac{(1.8 \text{ mg} \cdot \text{L}^{-1})(1.05 \text{ m}^3 \cdot \text{s}^{-1}) + (7.08 \text{ m}^3 \cdot \text{s}^{-1})(7.6 \text{ mg} \cdot \text{L}^{-1})}{1.05 \text{ m}^3 \cdot \text{s}^{-1} + 7.08 \text{ m}^3 \cdot \text{s}^{-1}} = 6.85 \text{ mg} \cdot \text{L}^{-1}$$

Initial deficit = D_a = 8.5 − 6.85 = 1.6 mg · L^{-1}.

Similarly, using Equation 9–18, we can calculate the concentration of ultimate BOD in the river after mixing.

$$L_{a,mix} = \frac{(28 \text{ mg} \cdot \text{L}^{-1})(1.05 \text{ m}^3 \cdot \text{s}^{-1}) + (3.6 \text{ mg} \cdot \text{L}^{-1})(7.08 \text{ m}^3 \cdot \text{s}^{-1})}{8.13 \text{ m}^3 \cdot \text{s}^{-1}} = 6.75 \text{ mg} \cdot \text{L}^{-1}$$

2. DO 16 km downstream

$$t = \frac{(16 \text{ km})(1000 \text{ m} \cdot \text{km}^{-1})}{(0.37 \text{ m} \cdot \text{s}^{-1})(3600 \text{ s} \cdot \text{h}^{-1})(24 \text{ h} \cdot \text{day}^{-1})} = 0.50 \text{ days}$$

Using Equation 9–36,

$$D_t = \frac{(0.61)(6.75)}{(0.76 - 0.61)} [\exp(-(0.61)(0.50)) - \exp(-(0.76)(0.50))]$$

$$+ 1.6 \exp(-(0.76)(0.50))$$

$$= 2.56 \text{ mg} \cdot \text{L}^{-1}$$

therefore DO $= 8.5 \text{ mg} \cdot \text{L}^{-1} - 2.56 \text{ mg} \cdot \text{L}^{-1} = 5.9 \text{ mg} \cdot \text{L}^{-1}$

To relate travel time to a physical distance downstream, we must also know the average stream velocity. Once D has been found at any point downstream, the DO can be found from Equation 9–36. Note that it is physically impossible for the DO to be less than zero. If the deficit calculated from Equation 9–36 is greater than the saturation DO, then all the oxygen was depleted at some earlier time and the DO is zero. If the result of your calculation yields a negative DO, report it as zero because concentration values cannot be less than zero.

The lowest point on the DO sag curve with respect to dissolved oxygen, which is called the **critical point,** is of major interest because it indicates the worst conditions in the river with respect to dissolved oxygen. The time to the critical point (t_c) can be found by differentiating Equation 9–36, setting it equal to zero, and solving for t using the values for k_r and k_d.

$$t_c = \frac{1}{k_r - k_d} \ln \left[\frac{k_r}{k_d} \left(1 - D_a \frac{k_r - k_d}{k_d L_a} \right) \right] \tag{9–38}$$

or when $k_r = k_d$,

$$t_c = \frac{1}{k_d} \left(1 - \frac{D_a}{L_a} \right) \tag{9–39}$$

and the critical deficit (D_c) is then found by using this critical time in Equation 9–36.

$$D_c = \frac{k_d L_a}{k_r - k_a} (e^{-k_d t_c} - e^{-k_r t_c}) + D_a(e^{-k_r t_c}) \tag{9–40}$$

In some instances there may not be a sag in the DO downstream. The lowest DO may occur in the mixing zone. In these instances Equation 9–38 will not give a useful value.

EXAMPLE 9–12 Using the data presented in Example 9–11,

1. Calculate the critical time and distance.
2. What is the minimum DO?

Solution **1.** Using Equation 9–38, we can calculate the critical time, t_c.

$$t_c = \frac{1}{0.76 - 0.61} \ln \left\{ \frac{0.76}{0.61} \left[1 - \frac{1.6(0.76 - 0.61)}{(0.61)(6.75)} \right] \right\}$$

$$= 1.07 \text{ days}$$

Flow $= 0.37$ m $\cdot$ s^{-1}

Critical distance $= (1.07$ days$)(0.37$ m $\cdot$ s$^{-1})(3600$ s $\cdot$ h$^{-1})(24$ h $\cdot$ day$^{-1})(10^{-3}$ m $\cdot$ km$^{-1})$

$$= 34.2 \text{ km}$$

2. Using the critical time, t_c, for t, we can use Equation 9–36 to calculate the "critical deficit":

$$D = \frac{(0.61)(6.75)}{(0.76 - 0.61)}\{\exp[-(0.61)(1.07)] - \exp[-(0.76)(1.07)]\}$$

$$+ 1.6 \exp[-(0.76)(1.07)]$$

$$= 2.8 \text{ mg} \cdot \text{L}^{-1}$$

therefore, the minimum DO $= 8.5$ mg $\cdot$ L^{-1} $- 2.8$ mg $\cdot$ L^{-1} $= 5.7$ mg $\cdot$ L^{-1}

EXAMPLE 9–13 Determine the DO concentration at a point 5 km downstream from the Aveta discharge into the Tefnet Creek (Examples 9–7, 9–8, 9–10). Also determine the critical DO and the distance downstream at which it occurs.

Solution All of the appropriate data are provided in the three previous examples. With the exceptions of the travel time, t, and the reaeration rate, the values needed for Equations 9–36 and 9–38 have been computed in Examples 9–7, 9–8, and 9–10. The first step then is to calculate k.

$$k_r \text{ at } 20°C = \frac{(3.9)(0.03 \text{ m} \cdot \text{s}^{-1})^{0.5}}{(5.0 \text{ m})^{1.5}} = 0.0604 \text{ day}^{-1}$$

Because k_r is given for 20°C and the stream temperature is at 10°C, Equation 9–7 must be used to correct for the temperature difference.

k_r at $10°C = (0.0604 \text{ day}^{-1})(1.024)^{10-20} = (0.0604)(0.7889) = 0.04766 \text{ day}^{-1}$

Note that the temperature coefficient is the one associated with Equation 9–27.

The travel time t is computed from the distance downstream and the speed of the stream.

$$t = \frac{(5 \text{ km})(1000 \text{ m} \cdot \text{km}^{-1})}{(0.03 \text{ m} \cdot \text{s}^{-1})(86,400 \text{ s} \cdot \text{day}^{-1})} = 1.929 \text{ day}$$

Although it is not warranted by the significant figures in the computation, we have elected to keep four significant figures because of the computational effects of truncating the extra digits. The deficit is estimated using Equation 9–36.

$$D_t = \frac{(0.03442)(11.86)}{0.04766 - 0.03442}\left[e^{-(0.03442)(1.929)} - e^{-(0.04766)(1.929)}\right] + 6.58\left[e^{-(0.04766)(1.929)}\right]$$

$$= (30.83)(0.9358 - 0.9122) + 6.58(0.9122)$$

$$= 6.7299 \text{ or } 6.73 \text{ mg} \cdot \text{L}^{-1}$$

and the dissolved oxygen is

DO $= 11.33 - 6.73 = 4.60$ mg $\cdot$ L^{-1}

The critical time is computed using Equation 9–38.

$$t_c = \frac{1}{0.04766 - 0.03442} \ln \left\{ \left(\frac{0.04766}{0.03442}\right) \left[1 - 6.58 \times \frac{(0.04766 - 0.03442)}{(0.03442)(11.86)}\right] \right\}$$

$$= 6.45 \text{ days}$$

Using t_c for the time in Equation 9–36, we can calculate the critical deficit as

$$D_c = \frac{(0.03442)(11.86)}{0.04766 - 0.03442} \left[e^{-(0.03442)(6.45)} - e^{-(0.04766)(6.45)}\right] + 6.58\left[e^{-(0.04766)(6.45)}\right]$$

$$= 6.85 \text{ mg} \cdot \text{L}^{-1}$$

and the critical DO is

$$DO_c = 11.33 - 6.85 = 4.48 \text{ mg} \cdot \text{L}^{-1}$$

The critical DO occurs downstream at a distance of

$$(6.45 \text{ days})(86{,}400 \text{ s} \cdot \text{day}^{-1})(0.03 \text{ m} \cdot \text{s}^{-1})\left(\frac{1 \text{ km}}{1000 \text{ m}}\right) = 16.7 \text{ km}$$

from the wastewater discharge point. (Remember that $0.03 \text{ m} \cdot \text{s}^{-1}$ is the speed of the stream.)

Management Strategy.   The beginning point for water quality management in rivers using the DO sag curve is to determine the minimum DO concentration that will protect the aquatic life in the stream. This value, called the **DO standard,** is generally set to protect the most sensitive species that exist or could exist in the particular river. For a known waste discharge and a known set of river characteristics, the DO sag equation can be solved to find the DO at the critical point. If this value is greater than the standard, the stream can adequately assimilate the waste. If the DO at the critical point is less than the standard, then additional waste treatment is needed. Usually, the environmental engineer has control over just two parameters: L_a and D_a. By increasing the efficiency of the existing treatment processes or by adding additional treatment steps, the ultimate BOD of the waste discharge can be reduced, thereby reducing L_a. Often a relatively inexpensive method for improving stream quality is to reduce D_a by adding oxygen to the wastewater to bring it close to saturation prior to discharge. To determine whether a proposed improvement will be adequate, the new values for L_a and D_a are used to determine whether the DO standard will be violated at the critical point. As a last resort, mechanical reaeration of rivers can be accomplished to artificially increase k_r and therefore decrease D_a. However, this measure is both costly to install and operate.

When using the DO sag curve to determine the adequacy of wastewater treatment, it is important to use the river conditions that will result in the least DO concentration. Usually these conditions occur in the late summer when river flows are low and temperatures are high. A frequently used criterion is the 10-year, 7-day low flow, which is the recurrence interval of the average low flow for a 7-day period. Low river flows reduce the dilution of the waste entering the river, causing higher values for L_a and D_a. The value of k_r is usually reduced by low river flows because of reduced velocities. In addition, higher temperatures increase k_d more than k_r and also decrease DO saturation, thus making the critical point more severe.

EXAMPLE 9–14 The Flins Company is considering opening one of two possible plants on either the Veles River or on its twin, the Perun River. Among the decisions to be made are what effect the plant discharge will have on each river and which river would be affected less. Effluent data from the Rongo canning plants A and B are considered to be representative of the potential discharge characteristics. In addition, measurements from each river at summer low-flow conditions are available.

Effluent Parameter	Plant A	Plant B
Flow (in $m^3 \cdot s^{-1}$)	0.0500	0.0500
Ultimate BOD at 25°C (in $kg \cdot day^{-1}$)	129.60	129.60
DO (in $mg \cdot L^{-1}$)	0.900	0.900
Temperature (in °C)	25.0	25.0
k at 20°C (in day^{-1})	0.110	0.0693

River Parameter	Veles River	Perun River
Flow (in $m^3 \cdot s^{-1}$)	0.500	0.500
Ultimate BOD at 25°C (in $mg \cdot L^{-1}$)	19.00	19.00
DO (in $mg \cdot L^{-1}$)	5.85	5.85
Temperature (in °C)	25.0	25.0
Speed (in $m \cdot s^{-1}$)	0.100	0.200
Average depth (in m)	4.00	4.00
Bed-activity coefficient	0.200	0.200

Four combinations must be evaluated:

Plant A on the Veles River   Plant B on the Veles River
Plant A on the Perun River   Plant B on the Perun River

Solution Note that for the purpose of explaining the calculations, the number of significant figures given for the data is greater than can probably be measured. The only difference in the combinations is the change in deoxygenation and reaeration coefficients. Thus, we need calculate only one value of L_a and one value of D_a.

We begin by converting the mass flux of ultimate BOD (in $kg \cdot day^{-1}$) to a concentration (in $mg \cdot L^{-1}$). Following our general approach for calculating concentration from mass flow, we divide the mass flux (in $kg \cdot day^{-1}$) by the flow of the water carrying the waste (Q_w, Q_r, or the sum $Q_w + Q_r$):

$$\frac{\text{Mass flux of ultimate BOD discharged (in } kg \cdot day^{-1})}{\text{Flow of water-carrying waste (in } m^3 \cdot s^{-1})}$$

The mass flux units are then converted to milligrams per day and the water flow to liters per day so that the days cancel.

$$\frac{(\text{Mass flux in units of } kg \cdot day^{-1}) \times (1 \times 10^6 \, mg \cdot kg^{-1})}{(\text{Flow rate Q in units of } m^3 \cdot s^{-1}) \times (86,400 \, s \cdot day^{-1})(1 \times 10^3 \, L \cdot m^{-3})}$$

For either plant A or B,

$$L_w = \frac{(129.60 \, kg \cdot day^{-1})(1 \times 10^6 \, mg \cdot kg^{-1})}{(0.0500 \, m^3 \cdot s^{-1})(86,400 \, s \cdot day^{-1})(1 \times 10^3 \, L \cdot m^{-3})}$$

$$= \frac{129.60 \times 10^6 \, mg}{4.320 \times 10^6 \, L}$$

$$= 30.00 \, mg \cdot L^{-1}$$

Now we can compute the mixed BOD using Equation 9–18.

$$L_a = \frac{(0.0500)(30.00) + (0.500)(19.00)}{0.0500 + 0.500}$$

$$= 20.0 \text{ mg} \cdot \text{L}^{-1}$$

From Table A–2 of Appendix A, we find that the DO saturation at 25°C is 8.38 mg · L^{-1} Then using Equation 9–24, we determine the initial deficit:

$$D_a = 8.38 - \frac{(0.0500)(0.900) + (0.500)(5.85)}{0.0500 + 0.500}$$

$$= 8.38 - 5.4$$

$$= 2.98 \text{ mg} \cdot \text{L}^{-1}$$

For the combination of plant A discharging to the Veles River, the reaeration and deoxygenation coefficients are calculated using Equation 9–26 and Equation 9–30, respectively

$$k_d = 0.110 + \frac{0.100 \times 0.200}{4.00}$$

$$= 0.115 \text{ day}^{-1} \text{ at } 20°C$$

and

$$k_r = \frac{3.9(0.100)^{0.5}}{(4.00)^{1.5}}$$

$$= 0.154 \text{ day}^{-1} \text{ at } 20°C$$

Because the temperature of the river is 25°C and the wastewater effluent temperature is also 25°C, we do not have to calculate a temperature after mixing. However, we will have to adjust k_d and k_r to 25°C. For k_d, we use Equation 9–7 with a value of θ of 1.056.

$$k_d = 0.115(1.056)^{25-20}$$

$$= 0.151 \text{ day}^{-1}$$

From the discussion that follows Equation 9–27, we note that θ = 1.024 for reaeration, and thus

$$k_r = 0.154(1.024)^{25-20}$$

$$= 0.173 \text{ day}^{-1}$$

Although perhaps not justified by the coefficients, we round to three significant figures because we will want to calculate travel time to two decimal places.

   Now we have all the information we need to calculate the time to the critical point. Using Equation 9–38, we obtain

$$t_c = \frac{1}{0.173 - 0.151} \ln \left\{ \frac{0.173}{0.151} \left[1 - 2.98 \left(\frac{0.173 - 0.151}{0.151 \times 20.0} \right) \right] \right\}$$

$$= 45.45 \ln\{1.146[1 - 2.98(0.02185)]\}$$

$$= 5.18 \text{ days}$$

Using this value for t in Equation 9–36, we can calculate the deficit at the critical point.

$$D_c = \frac{(0.151)(20.0)}{0.173 - 0.151} \left[e^{-(0.151)(5.18)} - e^{-(0.173)(5.18)} \right] + 2.98 \left[e^{-(0.173)(5.18)} \right]$$

$$= 137.3[(0.0493)] + 2.98[1.224]$$

$$= 6.763 + 1.242$$

$$= 7.99 \text{ mg} \cdot \text{L}^{-1}$$

Using D_c and the appropriate value for the DO saturation that we obtained earlier from Table A–2, we can calculate the DO at the critical point.

$$DO = DO_s - D$$

$$= 8.38 - 7.99 = 0.39 \text{ mg} \cdot \text{L}^{-1}$$

Thus, the lowest DO for the plant A–Veles River combination is 0.39 mg · L^{-1} and it occurs at a travel time of 5.18 days downstream from the plant A outfall. Because the Veles River travels at a speed of 0.100 m · s^{-1}, this would be

$$\frac{(0.100 \text{ m} \cdot \text{s}^{-1})(5.18 \text{ days})(86{,}400 \text{ s} \cdot \text{day}^{-1})}{1000 \text{ m} \cdot \text{km}^{-1}} = 44.8 \text{ km}$$

downstream.

The results of the other combinations are summarized in the following table.

	Plant A		Plant B	
	Veles River	**Perun River**	**Veles River**	**Perun River**
k_d	0.151	0.151	0.104	0.104
k_r	0.173	0.245	0.173	0.245
t_c	5.18	4.11	5.86	4.47
D	7.98	6.62	6.51	5.32
DO	0.40	1.76	1.87	3.06

The best combination is the plant B on the Perun River. This is because of the four options, the deficit is the lowest and the minimum DO the greatest for plant B on the Perun River.

Using a spreadsheet program, we have generated the DO values for a series of times for each of the combinations and plotted the results in Figure 9–18. From this figure we can make the following general observations:

1. Increasing the reaeration rate, while holding everything else as it is, reduces the deficit and decreases the critical time.
2. Decreasing the reaeration rate, while holding everything else as it is, increases the deficit and increases the critical time.
3. Increasing the deoxygenation rate, while holding everything else as it is, increases the deficit and decreases the critical time.
4. Decreasing the deoxygenation rate, while holding everything else as it is, decreases the deficit and increases the critical time.

FIGURE 9–18

Effect of k_d and k_r on DO sag curve. Note that the velocity in the Perun River is twice that in the Veles River.

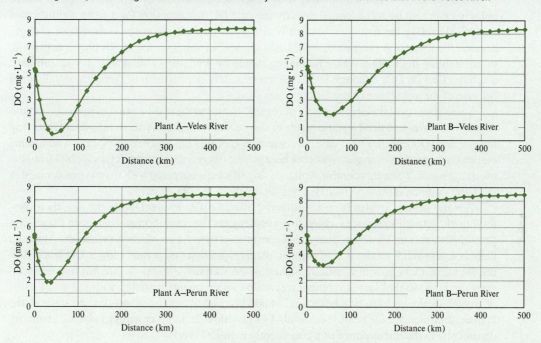

Nitrogenous BOD. Up to this point, only carbonaceous BOD has been considered in the DO sag curve. However, in many cases nitrogenous BOD has at least as much influence on dissolved oxygen levels. Modern wastewater treatment plants can routinely produce effluents with $CBOD_5$ of less than 30 mg · L^{-1}. A typical effluent also contains approximately 30 mg · L^{-1} of nitrogen, which would mean an NBOD of about 137 mg · L^{-1} if it were discharged as ammonia (see Example 9–6). Nitrogenous BOD can be incorporated into the DO sag curve by adding an additional term to Equation 9–36.

$$D = \frac{k_d L_a}{k_r - k_d} (e^{-k_d t} - e^{-k_r t}) + D_a(e^{-k_r t}) + \frac{k_n L_n}{k_r - k_n} (e^{-k_n t} - e^{-k_r t}) \tag{9–41}$$

where k_n = the nitrogenous deoxygenation coefficient (in day^{-1})

L_n = ultimate nitrogenous BOD after waste and river have mixed (in mg · L^{-1})

and the other terms are as previously defined.

It is important to note that with the additional term for NBOD, it is not possible to find the critical time using Equation 9–38. Instead, it must be found by a trial and error solution of Equation 9–41.

Other Factors Affecting DO Levels in Rivers. The classical DO sag curve assumes that there is only one point-source discharge of waste into the river. In reality, this is rarely the case. Multiple point sources can be handled by dividing the river up into reaches with a point source at the head of each reach. A **reach** is a length of river specified by the engineer on the basis of its homogeneity, that is, channel shape, bottom composition, slope, and so on. The oxygen deficit and residual BOD can be calculated at the end of each reach. These values are then used to determine new values of D_a and L_a at the beginning of the following reach. Nonpoint source pollution can also be handled this way if the reaches are made small enough. Nonpoint source pollution can also be incorporated directly into the DO sag equation for a more sophisticated analysis. Dividing the river into reaches is also necessary whenever the flow regime changes because the reaeration coefficient would also change. In small rivers, rapids play a major role in maintaining high DO levels. Eliminating rapids by dredging or damming a river can have a severe effect on DO, although DO levels immediately downstream of dams are usually high because of the turbulence of the falling water.

Some rivers contain large deposits of organic matter in the sediments. These can be natural deposits of leaves and dead aquatic plants or can be sludge deposits from wastewaters receiving little or no treatment. In either case, decomposition of this organic matter places an additional burden on the stream's oxygen resources because the oxygen demand must be supplied from the overlying water. When this benthic demand is significant, compared with the oxygen demand in the water column, it must be included quantitatively in the sag equation.

Aquatic plants can also have a substantial effect on DO levels. During the day, their photosynthetic activities produce oxygen that supplements the reaeration and can even cause oxygen supersaturation. However, plants also consume oxygen for respiration processes. Although there is a net overall production of oxygen, plant respiration can severely lower DO levels during the night. Plant growth is usually highest in the summer when flows are low and temperatures are high, so that large nighttime respiration requirements coincide with the worst cases of oxygen depletion from BOD exertion. In addition, when aquatic plants die and settle to the bottom, they increase the benthic demand. As a general rule, large growths of aquatic plants are detrimental to the maintenance of a consistently high DO level.

Effect of Nutrients on Water Quality in Rivers

Although oxygen-demanding wastes are definitely the most important river pollutants on an overall basis, nutrients can also contribute to deteriorating water quality in rivers by causing excessive plant growth. Nutrients are those elements required by plants for their growth. They include, in order of abundance in plant tissue: carbon, nitrogen, phosphorus, and a variety of trace elements. When sufficient quantities of all nutrients are available, plant growth is possible. By limiting the availability of any one nutrient, further plant growth is prevented.

Some plant growth is desirable because plants form the base of the food chain and thus support the animal community. However, excessive plant growth can create a number of undesirable conditions, such as thick slime layers on rocks and dense growths of aquatic weeds.

The availability of nutrients is not the only requirement for plant growth. In many rivers, the turbidity caused by eroded soil particles, bacteria, and other factors prevents light from penetrating far into the water, thereby limiting total plant growth in deep water. It is for this reason that slime growths on rocks usually occur only in shallow water. Strong water currents also prevent rooted plants from taking hold, and thus limit their growth to quiet backwaters where the currents are weak and the water is shallow enough for light to penetrate.

Effects of Nitrogen.   Nitrogen is detrimental to a receiving body for four reasons:

1. In high concentrations, ammonia in its unionized form is toxic to fish.
2. Ammonia, NH_3, in low concentrations, and nitrate, NO_3^-, serve as nutrients for excessive growth of algae.
3. The conversion of NH_4^+ to NO_3^- consumes large quantities of dissolved oxygen.

4. During the common practice of disinfecting wastewater effluent using chlorination, the chlorine (Cl_2) and hypochlorous acid/hypochlorite ($HOCl/OCl^-$) can react with any ammonia present in the water, forming chloramines. The chloramines, which are not removed during dechlorination prior to discharge, are more toxic than either chlorine and hyporchlorous acid/hypochlorite.

Effects of Phosphorus.   The major harmful effect of phosphorus is that it serves as a vital nutrient for the growth of algae. If the phosphorus availability meets the growth demands of the algae, algae are produced in excess. When the algae die, they become an oxygen-demanding organic material as bacteria seek to degrade them. This oxygen demand frequently overtaxes the DO supply of the water body and, as a consequence, causes fish to die.

Management Strategy.   The strategy for managing water quality problems associated with excessive nutrients is based on the sources for each nutrient. Except under rare circumstances, there is plenty of carbon available for plant growth. Plants use carbon dioxide, which is available from the bicarbonate alkalinity of the water and from the bacterial decomposition of organic matter. As carbon dioxide is removed from the water, it is replenished from the atmosphere. Generally, the major source of trace elements is the natural weathering of rock minerals, a process over which the environmental scientist and engineer has little control. However, because acid rain caused by air pollution accelerates the weathering process, air pollution control can help reduce the supply of trace elements. The removal of trace elements from wastewater is difficult. In addition, such small amounts of trace elements are needed for plant growth that nitrogen or phosphorus is more likely to be the limiting nutrient. Therefore, the practical control of nutrient-caused water quality problems in streams is based on removal of nitrogen or phosphorus from wastewaters before they are discharged.

9–4   WATER QUALITY MANAGEMENT IN LAKES

Control of Phosphorus in Lakes

In Chapter 5 the problem of eutrophication and its effect on ecosystems was discussed. Because phosphorus is usually the limiting nutrient, control of cultural eutrophication must be accomplished by reducing the input of phosphorus to the lake. Once the input is reduced, the phosphorus concentration will gradually fall as phosphorus is buried in the sediment or flushed from the lake. Other strategies for reversing or slowing the eutrophication process, such as precipitating phosphorus with additions of aluminum (alum) or removing phosphorus-rich sediments by dredging, have been proposed. However, if the input of phosphorus is not also curtailed, the eutrophication process will continue. Thus, dredging or precipitation alone can result only in temporary improvement in water quality. In conjunction with reduced phosphorus inputs, these measures can help speed up the removal of phosphorus already in the lake system. Of course, the need to speed the recovery process must be weighed against the potential damage from inundating shoreline areas with sludge and stirring up chemicals buried in the sediment.

To be able to reduce phosphorus inputs, it is necessary to know the sources of phosphorus and the potential for their reduction. The natural source of phosphorus is the weathering of rock. Phosphorus released from the rock can enter the water directly, but more commonly it is taken up by plants and enters the water in the form of dead plant matter. It is exceedingly difficult to reduce the natural inputs of phosphorus. If these sources are large, the lake is generally naturally eutrophic. For many lakes the principal sources of phosphorus are the result of human activity. The most important sources are municipal and industrial wastewaters, seepage from septic tanks, and agricultural runoff that carries phosphorus fertilizers into the water. The relative contributions of various sources of phosphorus are illustrated in the following example.

EXAMPLE 9–15   The ficticious Pinga Lake in Camelot has a surface area of 9.34×10^7 m^2 and depth of 10 m. The lake receives a yearly average of 107 cm of precipitation. The lake receives phosphorus from the following sources:

1. The wastewater treatment plant for the city of Astrid discharges into the lake. The average water usage for the residents of the city is 350 L $\cdot$ capita^{-1} $\cdot$ day^{-1}. There are 54,000 residents of this city. The total phosphate level in the sewage influent is 10 mg $\cdot$ L^{-1} (yearly average). The wastewater treatment plant removes 90% of the phosphorus.

2. The city has recently renovated its sewage collection system, separating the storm water from the sewage, installing a separate sewer system. The storm water is discharged into the lake after treatment through a sand filter that removes 50% of the phosphorus. The storm sewers serve an area that is 9.5 km^2, having a runoff coefficient of 0.40. The phosphorus concentration in the untreated storm water is 0.75 mg $\cdot$ L^{-1}.

3. The lake is fed by a pristine stream, having an average yearly flow rate of 0.65 m^3 $\cdot$ s^{-1}. The stream has an average total phosphorus concentration of 0.05 mg $\cdot$ L^{-1}.

4. Farmland lies to the east of the lake. It has a drainage area of 150 km^2. Manure is applied in early spring before the crops are sown. The phosphorus loading to the land is 0.42 kg $\cdot$ km^{-2} $\cdot$ year^{-1}. The crops remove 60% of the phosphorus applied. The runoff coefficient on this land is 0.30. (This loading to the lake can be averaged throughout the year.)

5. The phosphorus settling rate from the lake is 2.8×10^{-8} s^{-1}.

Calculate the total phosphorus concentration in the lake. What is the trophic state of the lake? You may assume that the rate of evaporation is equal to the rate of precipitation.

Solution   Let's draw a picture of this lake.

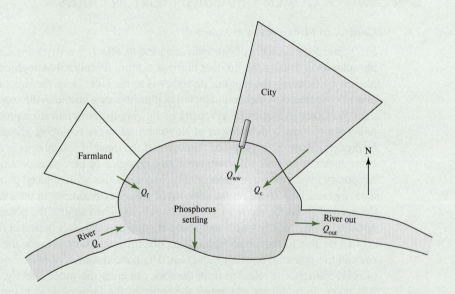

We must first calculate all of our **flow rates:**
The flow rate in the pristine stream, Q_r, is given as 0.65 m^3 $\cdot$ s^{-1}.
The flow rate from the wastewater treatment plant is

$$Q_{ww} = (350 \text{ L} \cdot \text{capita}^{-1} \cdot \text{day}^{-1})(54{,}000 \text{ people})\left(\frac{1 \text{ m}^3}{1000 \text{ L}}\right)\left(\frac{1 \text{ day}}{86{,}400 \text{ s}}\right) = 0.219 \text{ m}^3 \cdot \text{s}^{-1}$$

The runoff for the city must be calculated from the precipitation rate and the drainage area. The precipitation rate is 107 cm · year^{-1}. The drainage area is 9.5 km^2. The runoff rate is 0.40.

$$Q_c = (107 \text{ cm} \cdot \text{year}^{-1})\left(\frac{1 \text{ m}}{100 \text{ cm}}\right)\left(\frac{1 \text{ year}}{(365 \text{ days} \cdot \text{year}^{-1})(86{,}400 \text{ s} \cdot \text{day}^{-1})}\right)$$
$$\times (9.5 \text{ km}^2)\left(\frac{1000 \text{ m}}{1 \text{ km}}\right)(0.40)$$
$$= 0.129 \text{ m}^3 \cdot \text{s}^{-1}$$

The runoff for the farmland is calculated in a similar fashion to that from the city. Here the drainage area is 15 km^2. The runoff rate is 0.30.

$$Q_f = (107 \text{ cm} \cdot \text{year}^{-1})\left(\frac{1 \text{ m}}{100 \text{ cm}}\right)\left(\frac{1 \text{ year}}{(365 \text{ days} \cdot \text{year}^{-1})(86{,}400 \text{ s} \cdot \text{day}^{-1})}\right)$$
$$\times (150 \text{ km}^2)\left(\frac{1000 \text{ m}}{1 \text{ km}}\right)^2(0.3)$$
$$= 1.53 \text{ m}^3 \cdot \text{s}^{-1}$$

Remembering that the rate of evaporation equals the rate of precipitation, the total flow into the lake equals

$$Q_r + Q_{ww} + Q_c + Q_f = (0.65 + 0.219 + 0.129 + 1.53) \text{ m}^3 \cdot \text{s}^{-1} = 2.53 \text{ m}^3 \cdot \text{s}^{-1}.$$

To maintain a mass balance for water, the flow rate out of the lake must equal the flow rate in (this assumes that the volume lost to seepage and evaporation equals the volume in from precipitation). Thus,

$$Q_{out} = 2.53 \text{ m}^3 \cdot \text{s}^{-1}$$

Next we wish to calculate the phosphorus loading.

When the flow rate and concentration are given, then the rate of mass in equals $Q \times C$. This will be the case for all of the flows except that from the farmland.

From the river (inflow)

$$M_r = (0.65 \text{ m}^3 \cdot \text{s}^{-1})(0.05 \text{ mg} \cdot \text{L}^{-1})\left(\frac{1000 \text{ L}}{1 \text{ m}^3}\right)\left(\frac{1 \text{ g}}{1000 \text{ mg}}\right) = 0.033 \text{ g} \cdot \text{s}^{-1}$$

From the wastewater treatment plant, where 90% of the phosphorus is removed

$$M_{ww} = (0.219 \text{ m}^3 \cdot \text{s}^{-1})(10 \text{ mg} \cdot \text{L}^{-1})(1 - 0.90)\left(\frac{1000 \text{ L}}{1 \text{ m}^3}\right)\left(\frac{1 \text{ g}}{1000 \text{ mg}}\right) = 0.219 \text{ g} \cdot \text{s}^{-1}$$

From the city (storm sewer), where 50% is removed by treatment

$$M_c = (0.129 \text{ m}^3 \cdot \text{s}^{-1})(0.75 \text{ mg} \cdot \text{L}^{-1})(1 - 0.50)\left(\frac{1000 \text{ L}}{1 \text{ m}^3}\right)\left(\frac{1 \text{ g}}{1000 \text{ mg}}\right) = 0.048 \text{ g} \cdot \text{s}^{-1}$$

From the farmland, we must use the application rate. The crops use 60% of the phosphorus, so only 40% is available for release.

$$M_f = (0.42 \text{ kg} \cdot \text{km}^{-2} \cdot \text{year}^{-1})(150 \text{ km}^2)(0.40)\left(\frac{1000 \text{ g}}{1 \text{ kg}}\right)\left(\frac{1 \text{ year}}{365 \text{ days}}\right)\left(\frac{1 \text{ day}}{86{,}400 \text{ s}}\right)$$
$$= 7.99 \times 10^{-4} \text{ g} \cdot \text{s}^{-1}$$

Therefore, the total mass of phosphorus into the lake is

$$M_{in} = M_r + M_{ww} + M_c + M_f = (0.033 + 0.219 + 0.048 + 7.99 \times 10^{-4}) \text{ g} \cdot \text{s}^{-1} = 0.301 \text{ g} \cdot \text{s}^{-1}$$

We must now calculate the **mass of phosphorus out.** We will assume steady-state conditions and complete mixing. Phosphorus can leave through settling or from the river leaving the lake.

Remember from Chapter 2 that we can write the degradation terms as $dC/dt = -kC_{lake}\mathbb{V}$, where k is the rate constant, C_{lake} is the concentration in the lake, and $\mathbb{V}$ is the volume. The term for the loss of chemical from the outflow of the lake is $C_{lake}Q_{out}$.

We can now write our **complete mass-balance equation:**

$$\frac{d(\text{Mass})}{dt} = \text{Phosphate in} - \text{phosphate out}$$

or, more specifically,

$$\frac{d(\text{Mass})}{dt} = \text{Phosphate from wastewater} + \text{phosphate from river} + \text{phosphate from farmland}$$
$$+ \text{ phosphate from the city storm sewers} - \text{phosphate lost to settling}$$
$$- \text{ phosphate lost in outflow from lake}$$

We can write this mathematically as we did in Chapter 2.

$$\frac{d(\text{Mass})}{dt} = C_r Q_r + C_{ww} Q_{ww} + C_c Q_c + C_f Q_f - k_s C_{lake}\mathbb{V} - C_{lake}Q_{out}$$

where k_s = settling constant for phosphorus
$\mathbb{V}$ = volume of the lake

We have already calculated the total mass of phosphorus that flows into the river.

$$M_{in} = C_r Q_r + C_{ww} Q_{ww} + C_c Q_c + C_f Q_f = 0.301 \text{ g} \cdot \text{s}^{-1}$$

Therefore,

$$\frac{d(\text{Mass})}{dt} = M_{in} - k_s C_{lake}\mathbb{V} - C_{lake}Q_{out}$$
$$= 0.301 \text{ g} \cdot \text{s}^{-1} - (2.8 \times 10^{-8} \text{ s}^{-1})C_{lake}(9.34 \times 10^7 \text{ m}^2)(10 \text{ m}) - C_{lake}(2.53 \text{ m}^3 \cdot \text{s}^{-1})$$

Setting $d(\text{Mass})/dt = 0$ for steady-state conditions, allows us to solve the above equation for C_{lake}.

$$0.301 \text{ g} \cdot \text{s}^{-1} - (26.15 \text{ m}^3 \cdot \text{s}^{-1})C_{lake} - (2.53 \text{ m}^3 \cdot \text{s}^{-1})C_{lake}$$

$$= \frac{0.301 \text{ g} \cdot \text{s}^{-1}}{28.68 \text{ m}^3 \cdot \text{s}^{-1}} = C_{lake}$$

$$C_{lake} = 0.0105 \text{ g} \cdot \text{m}^{-3}$$

Because we need to compare this to values in units of milligrams per liter, we must convert our answer.

$$C_{lake} = (0.0105 \text{ g} \cdot \text{m}^{-3})\left(\frac{1000 \text{ mg}}{\text{g}}\right)\left(\frac{1000 \text{ µg}}{\text{mg}}\right)\left(\frac{\text{m}^3}{1000 \text{ L}}\right) = 10.5 \text{ µg} \cdot \text{L}^{-1}$$

Therefore, as shown in Table 5–2, the trophic state of the lake is mesotrophic. Even with all of the measures taken, more must be taken to ensure the quality of the lake.

Municipal and Industrial Wastewaters. All municipal sewage contains phosphorus from human excrement. Many industrial wastes are also high in this nutrient. In these cases, the only effective way of reducing phosphorus is through advanced waste treatment processes, which are discussed in Chapter 11. Prior to the late 1980s, municipal wastewater also contained large quantities of phosphorus from detergents containing polyphosphate, which is a chain of phosphate ions (usually three) linked together. The polyphosphate binds with minerals in water to make the detergent a more effective cleaning agent. By the 1970s, phosphorus loading from detergents was approximately twice that from human excrement. Phosphorus from detergents can be removed by advanced waste treatment, but phosphorus can also be removed from detergents so that it never enters the wastewater to begin with. Today's detergents do not contain phosphate because the manufacturers have replaced it with other chemicals. As a direct application of the waste minimization philosophy, by the late 1980s all states that surround the Chesapeake Bay have passed laws banning phosphate detergents as a rapid method of reducing phosphorus inputs to lakes.

Septic Tank Seepage.   The shores of many lakes are dotted with homes and summer cottages, each with its own septic tank and tile field for waste disposal. As treated wastewater moves through the soil toward the lake, phosphorus is adsorbed by soil particles, especially clay. Thus, during the early life of the tile field, very little phosphorus gets to the lake. However, over time, the capacity of the soil to adsorb phosphorus is exceeded and any additional phosphorus will pass on into the lake, contributing to eutrophication. The time it takes for phosphorus to break through to the lake depends on the type of soil, the distance to the lake, the amount of wastewater generated, and the concentration of phosphorus in that wastewater. To prevent phosphorus from reaching the lake, it is necessary to put the tile field far enough from the lake that the adsorption capacity of the soil is not exceeded. If this is not possible, it may be necessary to replace the septic tanks and tile fields with a sewer to collect the wastewater and transport it to a treatment facility.

Agricultural Runoff.   Because phosphorus is a plant nutrient, it is an important ingredient in fertilizers. As rainwater washes off fertilized fields, some of the phosphorus is carried into streams and then into lakes. Most of the phosphorus not taken up by growing plants is bound to soil particles. Bound phosphorus is carried into streams and lakes through soil erosion. Waste minimization can be applied to control phosphorus loading to lakes from agricultural fertilization by encouraging farmers to fertilize more often with smaller amounts and to take effective action to stop soil erosion.

Acidification of Lakes

As shown in Chapter 2, pure rainwater is slightly acidic. As we discussed there, CO_2 dissolves in water to form carbonic acid (H_2CO_3). The equilibrium concentration of H_2CO_3 results in a rainwater pH of approximately 5.6. Thus, acid rain is usually defined to be precipitation with a pH less than 5.6. The phenomena now known as acid rain was first described by Svante Oden (Sweden) in 1968 when he demonstrated that precipitation over the Scandinavian countries was becoming more acidic (Figure 9–19). Thus began a major effort throughout the world to discover exactly what was causing this decrease in pH and how extensive it was. The extent of the problem is illustrated in Figure 9–20a. In 1985, the pH of the precipitation falling over much of the northeastern United States was 4.4, with significant areas receiving rainfall having a pH of <4.1.

FIGURE 9–19

The earliest studies identifying the change in the pH of precipitation were conducted by Svante Oden. The data here show the pH of rainfall in Oslo, Norway, where most of Oden's studies were conducted. *Source:* Svante Oden.

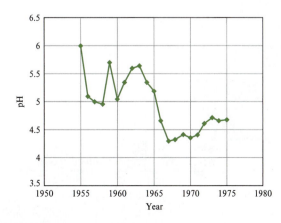

FIGURE 9–20

Hydrogen ion concentrations as pH from measurements of rainwater made at the Central Analytical Laboratory located in Urbana-Champaign, IL. **(a)** 1985 **(b)** 2000 **(c)** 2015.

Source: National Atmospheric Deposition Program/National Trends Network, http://nadp.isws.illinois.edu.

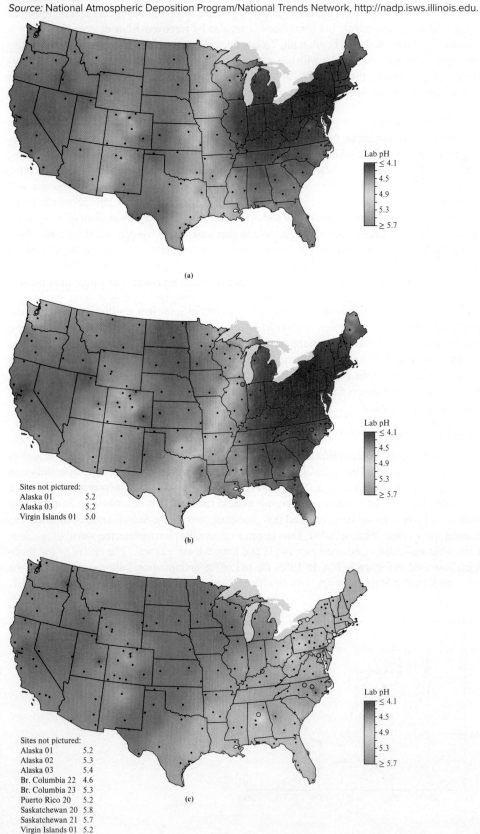

(a)

Sites not pictured:
Alaska 01   5.2
Alaska 03   5.2
Virgin Islands 01 5.0

(b)

Sites not pictured:
Alaska 01   5.2
Alaska 02   5.3
Alaska 03   5.4
Br. Columbia 22   4.6
Br. Columbia 23   5.3
Puerto Rico 20   5.2
Saskatchewan 20   5.8
Saskatchewan 21   5.7
Virgin Islands 01 5.2

(c)

Lab pH
≤ 4.1
4.5
4.9
5.3
≥ 5.7

FIGURE 9–21

Acid rain formation.
Source: U.S. EPA, 2006.

Since its discovery, we have also found that acid rain has caused, among other phenomena

- A complete eradication of some fish species, such as the brook trout, in hundreds of lakes in the Adirondacks
- The acidification of >1350 of mid-Atlantic Highland streams
- The acidification of approximately 580 streams in the mid-Atlantic Coastal Plain
- The acidification of >90% of streams in the New Jersey Pine Barrens

After much research, it was found that these low pH values have been attributed to emissions of sulfur and nitrogen oxides from the combustion of fossil fuels. As shown in Figure 9–21, the release of sulfur and nitrogen oxides into the air results in the formation of nitric and sulfuric acids, which can then deposit as rain, fog droplets, or snow onto plants, structures, lakes, and so on. Because the deposition can occur in both the wet and dry form, acid rain is better termed **acid deposition.**

EXAMPLE 9–16   A chemical analysis has revealed the following composition for the rain falling on the city of Thule.

$$1.38 \text{ mg} \cdot \text{L}^{-1} \text{ HNO}_3$$
$$3.21 \text{ mg} \cdot \text{L}^{-1} \text{ H}_2\text{SO}_4$$
$$0.354 \text{ mg} \cdot \text{L}^{-1} \text{ HCl}$$
$$0.361 \text{ mg} \cdot \text{L}^{-1} \text{ NH}_3$$

What is the pH of the rainwater?

Solution   Nitric, sulfuric, and hydrochloric acids are all strong acids, which would be completely dissociated in water. Ammonia is a weak base.

All waters must be electroneutral, that is, have no excess of either cations or anions. The electroneutrality expression for this water is

$$[\text{H}^+] + [\text{NH}_4^+] = [\text{NO}_3^-] + 2[\text{SO}_4^{2-}] + [\text{Cl}^-] + [\text{OH}^-]$$

where the concentrations are given in moles per liter.

Note: Although hydroxide is not given in the chemical composition, it is present, as it is in *all* water.

We must convert all of the preceding concentrations to units of moles per liter, which can be done by dividing the concentration given by the molecular weight and then converting that number to units of moles per liter.

For example, the molecular weight of nitric acid is

$$(1.008) + (14.01) + 3(16.0) = 63.02 \text{ g} \cdot \text{mol}^{-1}$$

Therefore,

$$\frac{(1.38 \text{ mg HNO}_3 \cdot \text{L}^{-1})\left(\frac{1\,\text{g}}{1000\,\text{mg}}\right)}{63.02 \text{ g} \cdot \text{mol}^{-1}} = 2.19 \times 10^{-5} \text{ M}$$

The remaining calculations are given in the following table:

Chemical Species	Concentration ($mg \cdot L^{-1}$)	Molecular Weight ($g \cdot mol^{-1}$)	Concentration (M)
HNO_3	1.38	63.018	2.1899×10^{-5}
H_2SO_4	3.21	98.076	3.2730×10^{-5}
HCl	0.354	36.458	9.7098×10^{-6}
NH_3	0.361	17.034	2.1193×10^{-5}

$$\sum (\text{anions}) = [NO_3^-] + 2[SO_4^{2-}] + [Cl^-] + [OH^-]$$

We need to make the following approximation to solve this problem.

$$[NO_3^-] + 2[SO_4^{2-}] + [Cl^-] \gg [OH^-]$$

so we can ignore $[OH^-]$. Therefore,

$$\sum (\text{anions}) = 2.19 \times 10^{-5} + (2)(3.28 \times 10^{-5}) + 9.71 \times 10^{-6} \text{ M} = 9.70 \times 10^{-5} \text{ M}$$

Ammonium dissociates to form ammonia. The equilibrium constant is $10^{-9.3}$. Therefore, at acidic pH all of the ammonia would be in the form of the ammonium ion, NH_4^+. Therefore,

$$[H^+] + [NH_4^+] = \sum (\text{anions}) = 9.70 \times 10^{-5}$$
$$[H^+] = \sum (\text{anions}) - [NH_4^+]$$
$$= 9.70 \times 10^{-5} - 2.12 \times 10^{-5} = 7.58 \times 10^{-5} \text{ M}$$
$$pH = -\log(7.58 \times 10^{-5}) = 4.12$$

The assumptions are reasonable because the pH is acidic.

$$[OH^-] = 10^{-(14-4.12)} \text{ M} = 10^{-9.88} \text{ M} = 1.32 \times 10^{-10} \text{ M}$$

This concentration is in fact much less than the concentration of the other anions.

FIGURE 9–22

Age distribution of fish in Patten Lake.

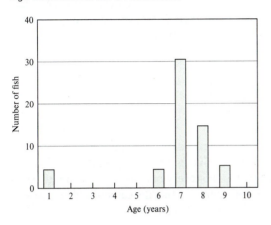

FIGURE 9–23

The relationship between pH and aluminum concentrations in Swedish clearwater lakes.

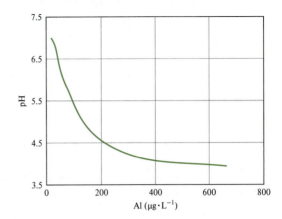

Fish, in particular trout and Atlantic salmon, are very sensitive to low pH levels. Most are severely stressed if the pH drops below 5.5, and few are able to survive if the pH falls below 5.0. If the pH falls below 4.0, cricket frogs and spring peepers experience mortalities in excess of 85%. This is shown in the work of Ryan and Harvey (1980), who found that in Patten Lake, Ontario, Canada, which had become severely acidified, the number of fish ages 2–5 years old was essentially nonexistent (Figure 9–22).

High aluminum concentrations are often the trigger that kills fish. Aluminum is abundant in soil but it is normally bound up in the soil minerals. At conventional pH values aluminum rarely occurs in solution. Acidification of the water releases highly toxic Al^{3+} to the water. This is seen in Figure 9–23, showing the relationship between pH and aluminum concentrations in Swedish clearwater lakes (Dickson, 1980). The lower the pH, the higher the aluminum concentration. Thus, the organisms are stressed not only by the low pH but also by the high aluminum concentrations.

Acid deposition can have numerous other stress effects on aquatic organisms, including

- failure to reproduce
- gill damage resulting in respiratory problems
- failure of eggs to hatch
- interference with Ca uptake (with molluscs)

Most lakes are buffered by the carbonate buffer system (see Chapter 2). To the extent that the buffer capacity of the lake is not exceeded, the pH of the lake will not be appreciably affected by acid rain. If there is a source of bicarbonate to replace that consumed by the acid rain, the buffering capacity can be quite large. Calcareous soils are those containing large quantities of calcium carbonate ($CaCO_3$). The dissolution of calcium carbonate will release carbonate into solution, which can be protonated to form bicarbonate. As such, lakes formed in calcareous soils tend to be resistant to acidification.

EXAMPLE 9–17   Acidic rainfall infiltrates into a geological formation that is predominantly dolomite, a commonly occurring rock-forming mineral. In 1940, the average pH of the groundwater was 6.6. By 1995, the average yearly groundwater pH was 5.6. Assuming that as the groundwater percolates through the dolomite it is at equilibrium, calculate the calcium and magnesium at the two pHs. Assume that the temperature is 25°C and that the concentration of calcium, magnesium, and carbonate in the rainwater is negligible.

Solution   Dolomite has the chemical structure $CaMg(CO_3)_2$ and a pK_s of 17.09. We will assume that the rainfall contains no calcium or magnesium. If the rainfall is at equilibrium with carbon dioxide, then the concentration of carbonic acid is 10^{-5} M (see Chapter 2, we have also assumed that the partial pressure of carbon dioxide in the atmosphere is $10^{-3.5}$ atm).

The equilibrium expression for the dissolution of dolomite is

$$CaMg(CO_3)_2 \rightleftharpoons Ca^{2+} + Mg^{2+} + 2CO_3^{2-}$$

The solubility expression is

$$[Ca^{2+}][Mg^{2+}]\left[CO_3^{2-}\right]^2 = 10^{-17.09}$$

We will assume this because the concentration of calcium and magnesium are zero in the rainfall. Because equimolar amounts of both ions will dissolve into the percolating water, we can assume that the concentration of each cation equals s. For every one mole of calcium that dissolves, two moles of carbonate will also dissolve. Recognizing that the carbonate that dissolves can react to form carbonic acid and bicarbonate, we can write that:

$$C_T = [H_2CO_3^*] + [HCO_3^-] + \left[CO_3^{2-}\right] = 2s$$

We should also remember from Chapter 2 that the equilibrium expression for the dissociation of carbonic acid into carbonate is

$$H_2CO_3^* \rightleftharpoons 2H^+ + CO_3^{2-} \quad K = 10^{-6.3} \times 10^{-10.3} = 10^{-16.6}$$

We should also remember that the equilibrium expression for the dissociation of bicarbonate is

$$HCO_3^- \rightleftharpoons H^+ + CO_3^{2-} \qquad pK = 10.3$$

At a pH of 6.6,

$$\frac{\left[CO_3^{2-}\right][H^+]^2}{[H_2CO_3^*]} = 10^{-16.6} = \frac{\left[CO_3^{2-}\right] 10^{-13.2}}{[H_2CO_3^*]}$$

Thus, $[H_2CO_3^*] = 10^{3.4}\left[CO_3^{2-}\right]$

$$\frac{\left[CO_3^{2-}\right][H^+]}{[HCO_3^-]} = 10^{-10.3}\frac{\left[CO_3^{2-}\right]10^{-6.6}}{[HCO_3^-]}$$

Thus, $[HCO_3^-] = 10^{3.7}\left[CO_3^{2-}\right]$

$$C_T = [H_2CO_3^*] + [HCO_3^-] + \left[CO_3^{2-}\right] = 10^{3.4}\left[CO_3^{2-}\right] + 10^{3.7}\left[CO_3^{2-}\right] + \left[CO_3^{2-}\right]$$

$$= 7.52 \times 10^3 \left[CO_3^{2-}\right] = 2s$$

Therefore,

$$\left[CO_3^{2-}\right] = \frac{2s}{7.52 \times 10^3}$$

From the solubility expression

$$(s)(s)\left(\frac{2s}{7.52 \times 10^3}\right)^2 = 10^{-17.09}$$

Solving for s,

$s = 0.00327$ M $= [Ca^{2+}] = [Mg^{2+}]$ at a pH of 6.6.

At pH 5.6, $s = 0.0207$ M $= [Ca^{2+}] = [Mg^{2+}]$

As you can see, as we decreased the pH from 6.6 to 5.6, the concentration of calcium and magnesium that dissolved in the water increased by approximately sixfold.

FIGURE 9–24

Regions in North America Containing Lakes Sensitive to Acidification by Acid Precipitation

The shaded areas have igneous or metamorphic bedrock geology; the unshaded areas have calcereous or sedimentary bedrock geology. Regions having low alkalinity lakes are concurrent with regions of igneous and metamorphic bedrock geology.

Other factors that affect the susceptibility of a lake to acidification are the permeability and depth of the soil, the make up of the bedrock, the slope and size of the watershed, and the type of vegetation. Thin, impermeable soils provide little time for contact between the soil and the precipitation. This reduces the potential for the soil to buffer the acid precipitation. Likewise, small watersheds with steep slopes reduce the time for buffering to occur. Deciduous foliage tends to decrease acidity. Coniferous foliage tends to yield runoff that is more acidic, than the precipitation itself. Granite bedrock offers little potential to buffer acid rain. Galloway and Cowling (1978) used bedrock geology to predict areas where lakes are potentially most sensitive to acid rain (Figure 9–24). You should note that the predicted areas of sensitivity are also those subjected to very acid precipitation.

As the problems due to acid deposition became increasingly well documented, the U.S. Congress responded by including special provisions in the 1990 Clean Air Act to reduce the emissions of acidic air pollutants. The law set up a market-based system targeting sulfur dioxide emissions, with the goal of reducing sulfur dioxide emissions in 2000 to 40% of their 1980 levels. Phase I of the acid rain reduction program went into effect in 1995, requiring the major sulfur dioxide emitters (10 power plants in 21 Midwest, Appalachian, Southeastern, and Northeastern states) to install pollution control devices to reduce emissions. In 2000, Phase II of the acid rain program went into effect, further reducing the sulfur dioxide releases from both the large coal-burning power plants and other smaller polluters. As can be seen in Figure 9–20b, in 2000, the pH of the rainfall was significantly higher than that observed in 1985. By 2015, the pH of the precipitation falling on much of the northeast is just slightly below the natural pH of rainfall, thereby demonstrating the success of strong environmental regulations.

Further discussion of the control of atmospheric emissions of sulfur and nitrogen oxides occurs in more detail in Chapter 12.

9–5   WATER QUALITY IN ESTUARIES

An **estuary** is formed along the coastline where freshwater from rivers and streams flows into the ocean. It is a place of transition, where the freshwater mixes with the salty seawater. These partially enclosed bodies of water are influenced by tides but are protected from the full force of ocean waves, winds, and storms by the reefs, barrier islands, peninsulas, or the surrounding mainland that define an estuary's seaward boundary (National Estuarine Research Reserve System, 2007; U.S. EPA, 2007).

Estuaries vary greatly in size and shape. They are more commonly known by such names as bays, lagoons, harbors, inlets, or sounds, although not all water bodies by those names are estuaries. The defining feature of an estuary is the mixing of fresh and salt water. San Francisco Bay, Puget Sound, Chesapeake Bay, Boston Harbor, and Tampa Bay are examples of estuaries.

Estuaries are complex and specialized ecosystems formed out of a cycle of wetting and drying, and the submersion and reappearance of aquatic organisms. It is the movement of water that has created the distribution and specialization of organisms in estuaries. The rivers and oceans provide estuaries with an abundance of life-giving nutrients, resulting in an ecosystem that contains more life than a comparably sized plot of agricultural, forested, or grassy land. Within an estuary, many different types of habitats, including shallow open waters, freshwater and salt marshes, sandy beaches, mud and sand flats, rocky shores, oyster reefs, mangrove forests, river deltas, tidal pools, sea grass and kelp beds, and wooded swamps are found. Shore birds, fish, crabs and lobsters, marine mammals, clams and other shellfish, marine worms, sea birds, and reptiles are just some of the animals that make their homes in and around estuaries.

Estuaries could be called the "cradle" of many birds, fish, and other life because the estuary often provides sanctuary to nesting and spawning animals. It is also the ideal spot for tired, resting birds on their long migrations.

In addition to serving as a nursery and sanctuary for much wildlife, the estuary performs many other functions. As the sediment and pollutant-laden water flows from the uplands through fresh and salt marshes, much of the sediments and pollutants are filtered out, creating cleaner and clearer water. Wetland plants and soils buffer the forces of the ocean, absorbing floodwaters and dissipating storm surges. This protects upland organisms as well as valuable real estate from storm and flood damage. Salt marsh grasses and other estuarine plants also help prevent erosion and stabilize the shoreline.

Under siege from excessive nutrients, pathogen contamination, toxic chemicals, alteration of freshwater inflow, loss of habitat, and the introduction of invasive species, our estuaries are in serious danger. In the last 30 or so years, estuaries have experienced declines in fish and wildlife, the introduction of invasive species, deterioration in water quality, and a reduction in overall ecosystem health.

In Chapter 5, we discussed how nutrients such as nitrogen and phosphorus are necessary for the growth of plants and animals and sustaining a healthy aquatic ecosystem. In excess, however, nutrients can contribute to fish disease, "red and brown tides," algae blooms, and low dissolved oxygen. Proliferation of algae can prevent sunlight from penetrating the water, depriving underwater seagrasses of light. Animals that depend on seagrasses for food or shelter leave the area or die. Decreased oxygen levels result in the death of fish and shellfish. In addition, excessive algal growth, resulting in brown and red tides, has been linked to fish kills, manatee deaths, and negative effects on scallops. Increased numbers of algae may also cause foul smells and decreased aesthetic value. Sources of nutrients include point and nonpoint sources, such as sewage treatment plant discharges, stormwater runoff from lawns and agricultural lands, faulty or leaking septic systems, sediment in runoff, animal wastes, atmospheric deposition originating from power plants or vehicles, and groundwater discharges.

Pathogens pose a health threat to swimmers, surfers, divers, and seafood consumers. Fish and filter feeding organisms, such as molluscs and crabs, concentrate pathogens in their tissues and may cause illness to those consuming them. Pathogen contamination can result in the closure of shellfishing areas and bathing beaches. Sources of pathogens include urban and agricultural runoff, boat and marina waste, faulty or leaky septic systems, sewage treatment plant discharges, combined sewer overflows, illegal sewer connections, and waste from pets or wildlife.

Toxic chemicals, such as metals, polycyclic aromatic hydrocarbons (PAHs), polychlorinated biphenyls (PCBs), heavy metals, and pesticides, enter waterways through stormdrains; industrial discharges and runoff from lawns, streets, and farmlands; sewage treatment plants; and from atmospheric deposition. Many toxic contaminants are also found in sediments and are resuspended into the environment by dredging and boating activities. Benthic organisms are exposed to these chemicals and may pose a risk to human health if consumed. As a result fishery and shellfish beds may need to be closed, and consumption advisories may be issued.

One effect of the global nature of our transportation system has been the introduction of nonnative or exotic species. These species are offered free rides in ballast water, on the hulls of ships and can thereby circumvent strict regulations governing the conveyance of nonnative species across international or state boundaries. Some introduced (nonnative) species have contributed to the eradication of some native flora and fauna populations and drastically reduced others, fundamentally altering the food web. As some herbivorous species overpopulate wetlands, the overgrazing of vegetation and the resultant degradation and loss of marsh has resulted. Other effects include (1) modification of nutrient cycles or soil fertility; (2) increased erosion; and (3) interference with navigation, agricultural irrigation, sport and commercial fishing, recreational boating, and beach use.

In an attempt to protect ecosystems, the Chesapeake Bay Agreement and the National Estuary Program (NEP) were created. The Chesapeake Bay, protected under its own federally mandated program, was the first estuary in the United States to be targeted for restoration and protection. In 1983, the governors of Maryland, Virginia, and Pennsylvania, the mayor of the District of Columbia, and the EPA administrator signed the Chesapeake Bay Agreement. This agreement set forth the plans to protect and improve the water quality in the Chesapeake Bay. In 1987, amendments to the Clean Water Act gave rise to the National Estuary Program. The goal of this program is to identify, restore, and protect nationally significant estuaries of the United States, focusing not just on improving water quality but on maintaining the integrity of the whole ecosystem, as well as its economic, recreational, and aesthetic values.

9–6   WATER QUALITY IN OCEANS

Oceans cover about 71% of the Earth's surface. They play an important role in global cycling of carbon dioxide. They also help regulate the cycling of nitrogen and phosphorus. The oceans are also a critical resource, providing food, biochemical compounds, minerals, and recreation.

Coral reefs are fragile, unique, and vibrant ecosystems. Each coral reef supports a different collection of organisms, algae, plants, and animals. These coral reefs support thousands of jobs and billions of dollars in annual revenues from tourism, recreation, and fishing. They are valuable sources of new medicines and biochemical compounds, help prevent shoreline erosion, and provide life-saving protection from storms.

Despite their unique and priceless value, the world's coral reefs are at significant danger from marine pollution, warming temperatures in seas, deposition of sediments, and unsustainable fishing practices. Fertilizers can kill the algae that live within the coral. High nutrient levels from off-shore sewage and agricultural discharges and open ocean dumping can result in algae blooms. The algae consume oxygen, choking the coral reefs and attracting predators, such as sea urchins and crown-of-thorn seastars, which destroy living coral. Warmer oceans are thought to result in coral bleaching in which the entire reef turns white as a result of the death of the algae within the coral that give it its brilliant color. Oil pollution can destroy the coral by coating the surfaces and preventing oxygen uptake. The chemicals in the oil can poison the water in which the coral live.

Another specialized ocean ecosystem is the kelp forest. Kelp forests are found along rocky coastlines at depths of 6–30 m. Within these forests of kelp, a type of brown algae, thrive many

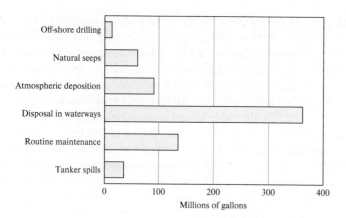

FIGURE 9–25

Sources of oil pollution in marine environments.

Source: Ocean Planet, Smithsonian Institution, 1995.

varieties of fish and other sea creatures, including sea urchins, sea cucumbers, and seastars. Sponges and bryozoans also live along the bottom, as do many varieties of fish and eels. This fragile ecosystem can be disrupted by human activities, such as ocean disposal of municipal sewage and manure from livestock production operations. This can cause an increase in the number of urchins, which will cause widespread destruction of the forests as the urchins dislodge the kelp from its stronghold on the ocean floor.

We have seen how pollution can have devastating effects on ocean ecosystems. Let's step back and look at the sources of this pollution and how as a society we can protect the oceans.

Although large oil spills make headlines, it is apparent from Figure 9–25 that the problem of oil pollution is much larger than just that of tanker spills. Used engine oil spilled on roads and other surfaces can end up in waterways, eventually flowing into the ocean. Bilge cleaning and other ship operations release millions of gallons of oil into navigable waters, in thousands of discharges of just a few gallons each. Hydrocarbons released by automobiles, power plants, and industry are deposited as rainfall or as dry deposition into waterways that flow into the ocean or directly into the ocean. Seepage from the ocean bottom and eroding sedimentary rocks releases oil. Only about 5% of oil pollution in oceans results from major tanker accidents; however, a spill can disrupt sea and shore life for great distances. Finally, spills and operational discharges from offshore oil production can result in ocean pollution.

Until fairly recently, the disposal of chemicals, sewage, and solid waste in the ocean was a common practice. In the United States, the Marine Protection, Research, and Sanctuaries Act Title I (MPRSA) limited ocean dumping. Until 1988, when the Ocean Dumping Ban Act (P.L. 100–688) amended MPRSA, it was permissible to dump sewage sludge and industrial waste in the open ocean. Although ocean dumping is predominantly banned by international law, the United States continues to allow the disposal of sediments and other materials dredged from harbors, marinas, estuaries, and bays. Much of this material contains toxic chemicals. The ocean dumping of mine waste, known as "submarine tailings disposal" is widely practiced in Southeast Asia and the Pacific.

As a result of ocean dumping along with the transport of chemicals from waterways and from the air into the ocean, toxic chemicals can be found in all the earth's oceans. Although more critical in coastal regions, these chemicals can also have a deleterious effect on ocean ecosystems. We have discussed some of the effects of chemicals on coral reefs. Bottom-dwelling fish exposed to toxic compounds released during oil spills are also susceptible. The exposed fish may develop liver disease and reproductive and growth problems. Recent news reports from the FDA warn pregnant women or women desiring to become pregnant to avoid consumption of shark, swordfish, king mackerel, and tilefish. These and other types of fish, such as tuna, can accumulate mercury. When ingested during pregnancy, mercury can damage the central nervous system of the fetus, resulting in slower cognitive development of children (Neegaard, 2001).

TABLE 9–4	An Inventory of the Litter Found on Ducie Atoll		
Item	**Number Found**	**Item**	**Number Found**
Buoys, intact or shards	179	Pieces of plastic pipe	29
Crates (bread, bottle)	14	Pieces of rope	44
Plastic bottles (drinks, toiletries)	71	Fluorescent tubes and light bulbs	12
Glass bottles (from 15 countries)	171	Aerosol cans	7
Jars	18	Food/drink cans	7
Broken plastic pieces	268	Other	34
Bottle tops	74		

Toxins released during algae blooms (resulting from excessive nutrients) have resulted in the death of sea lions and are now thought to harm the already endangered humpback and blue whales. The toxin, demoic acid, has been found in anchovies, sardines, and krill, organisms in the food chain of the whales (Associated Press, 2001).

Improper disposal of solid waste materials can have devastating effects on oceanic wildlife. Animals can drown or be strangled after becoming tangled in discarded or lost fishing gear, or suffer and even die from eating plastics and other garbage. Ducie Atoll, in the Pitcairn Islands, located in the South Pacific almost 300 miles from the nearest inhabited island and over 3000 miles from the nearest continent, is one of the most remote islands in the world. An important breeding area for seabirds, the atoll is, unfortunately, littered with as much trash as some European beaches. Yet when a scientist visited in 1991, he found over 950 pieces of trash in a 1.5-mile (2.4-km) stretch of beach (Benton, 1991). An inventory of the litter found is given in Table 9–4.

The Clean Water Act of 1972 is the principal U.S. federal statute protecting navigable waters and adjoining shorelines from pollution. It forms the foundation for other regulations detailing specific requirements for pollution prevention and response measures. Section 311 of the CWA addresses pollution from oil and hazardous substance releases, providing EPA and the U.S. Coast Guard with the authority to establish a program for preventing, preparing for, and responding to oil spills that occur in navigable waters of the United States. EPA implements provisions of the Clean Water Act through a variety of regulations, including the National Contingency Plan and the Oil Pollution Prevention regulations.

In 1969, a devastating oil spill off the coast of Santa Barbara, California, left beaches covered in an oily coating and killed large numbers of marine animals. In 1972, to protect our oceans and coastlines, Congress passed the Coastal Zone Management Act (CZMA) and the Marine Protection, Research, and Sanctuaries Act (commonly known as the Ocean Dumping Act). Through these acts three programs were created to protect and maintain our nation's marine and coastal environments. These programs are the Coastal Zone Management Program, the National Estuarine Research Reserve System, and the National Marine Sanctuaries Program. The CZMA created a national system to protect and manage coastal and ocean resources, to conserve designated areas, to balance economic, environmental and cultural activities, and to promote science-based resource management decisions. The Marine Protection, Research, and Sanctuaries Act regulates the disposal of all materials into marine waters that are within U.S. jurisdiction.

The National Estuarine Research Reserve System and the National Marine Sanctuaries Program are complementary, one focusing on estuaries and the other on the open ocean. Both programs were developed from the CZMA. The mission of these programs is to study marine ecosystems and develop methods for species and habitat protection, habitat restoration, and pollution control. Through the National Marine Sanctuaries Program, over 46,500 km^2 of ocean and coastal waters from the North Atlantic to the South Pacific are protected. In 2000, the Estuaries and Clean Waters Act was passed by Congress to promote and develop programs for the restoration of estuarine habitats using sound scientific and technical principles. Also in 2000, former President Clinton signed an executive order directing the U.S. EPA to better protect beaches, coasts, and ocean waters from pollution by developing Clean Water Act regulations that strengthen water quality protection for coastal and ocean waters. In developing these regulations, the EPA was to set higher levels of protection in especially valued or vulnerable areas.

On January 14, 2002, the George W. Bush administration overturned stricter environmental standards for the nation's wetlands and streams. The new permitting system is likely to result in an increase in the loss of wetlands and in the destruction of stream quality. On September 19, 2002, the Coastal and Estuarine Land Protection Act was passed by the Senate Committee on Commerce, Science, and Transportation, S.2608. This act amends the CZMA to allow States' National Estuarine Research Reserve units to acquire sensitive or threatened coastal lands through a federal grant program. This act would help protect coastal lands from further development. At present, this act has not been promulgated into law. It is hoped through stricter regulations we can protect our marine environments from the devastation resulting from uncontrolled disposal of our waste materials.

9–7 GROUNDWATER QUALITY

Contaminant Migration in Groundwaters

The contamination of aquifers (see Chapter 7 for a detailed description of groundwater flow and aquifers) with chemicals or microbes can (and has) resulted in the significant deterioration of the quality of groundwater. Contamination can result from a variety of sources, including

- discharge from improperly operated or located septic systems
- leaking underground storage tanks
- improper disposal of hazardous and other chemical wastes
- spills from pipelines or transportation accidents
- recharge of groundwater with contaminated surface water
- leaking dumps and landfills
- leaking retention ponds or lagoons

The types of contaminants that can pollute an aquifer are highly varied. Some of the more common contaminants include

- benzene, toluene, xylene, and ethylbenzene from gasoline spills
- tetrachloroethylene, trichloroethylene, and trichloroethane from dry cleaning and degreasing operations
- radionuclides from weapons manufacturing, research facilities, and hospitals
- polycyclic aromatic hydrocarbons from diesel fuel, crude oil, and other petroleum product spills
- salts from various chemical processes
- fertilizers and pesticides from agricultural operations and the manufacture of these chemicals
- nitrates from agricultural operations
- pathogens from septic systems

The migration of contaminants in soils and aquifers is very much dependent on the properties of the chemical and the geologic material. The more water-soluble a chemical, the more likely it will move vertically down through the soil to the aquifer and migrate with the water,

FIGURE 9–26

Dissolved contamination plume.

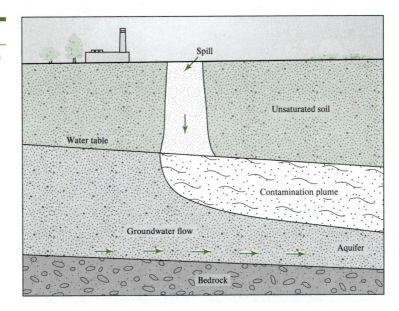

as shown in Figure 9–26. For example, chemicals such as nitrate, methyl tertiary butyl ketone (an additive to gasoline), and methamidophos (a pesticide), which are soluble in water, are less likely to sorb (attach) to the soil or aquifer solids. These chemicals are likely to move through the soil to an underlying aquifer.

Many chemicals are only sparingly soluble in water, and when they migrate in groundwater, they often do so as a separate nonaqueous phase. As such, these chemicals are known as nonaqueous phase liquids (NAPLs). The NAPLs are subdivided into two categories, based on their relative density with water. The light NAPLs (or LNAPLs) are less dense than water and will tend to "float" on the water table as shown in Figure 9–27. Some of the chemical will dissolve into the groundwater, whereas, if the compound is volatile, some volatilize into the pore spaces of the unsaturated zone and move in the gas phase. Examples of LNAPLs include

FIGURE 9–27

Immiscible plume of chemicals that are less dense (LNAPLs) than and insoluble in water.

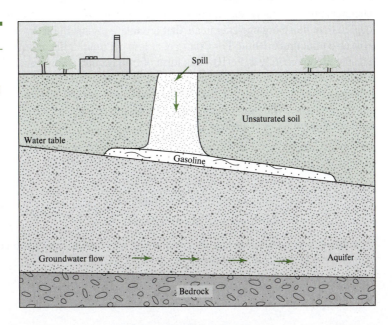

FIGURE 9–28

Immiscible plume of
chemicals that are more
dense (DNAPLs) than and
insoluble in water.

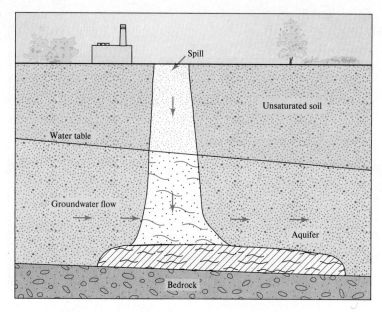

gasoline, and aviation fuel, and some of the chemicals these mixtures contain such as benzene, toluene, and xylene. The dense NAPLs (DNAPLs), on the contrary, have a greater density than that of water and will tend to "sink" in the aquifer until they reach an impervious barrier like bedrock. DNAPLs will tend to "pool" on top of the bedrock, slowly releasing chemicals into the groundwater. Chemicals such as trichloroethylene, tetrachloroethylene, and PCB–laden oils are examples of DNAPLs. The migration of DNAPLs is shown in Figure 9–28.

Chemicals such as chrysene (present in diesel fuel) and PCBs are very insoluble in water and are not likely to move great distances in soils, and therefore, are not likely to contaminate aquifers. To predict contaminant migration, we must revisit Darcy's equation.

Although a few contaminants can move at rates faster than water, most contaminants will move either at the same rate or slower than water. The relative velocities of the water and contaminant are a function of the contaminant and water characteristics. For example, organic chemicals that are more soluble in water will be **retarded** by the geologic material to a lesser degree than those organic chemicals that are relatively insoluble in water. The pH of the water will also affect retardation. For example, at low pH (and in the absence of oxygen) iron will be present predominately as Fe(II). This form of iron is quite soluble and will move with the water. On the contrary, if the pH is high (>6) and oxygen is present, iron will be present as Fe(III), which is much less soluble in water. The Fe(III) will precipitate and therefore, not move with the groundwater. The extent to which chemicals are retarded in water is defined by the **retardation coefficient,** R. R is defined as

$$R = \frac{v'_{water}}{v'_{cont}}$$

(9–42)

where v'_{water} = the linear velocity of the groundwater

v'_{cont} = the linear velocity of the contaminant

The retardation coefficient is a function of the hydrophobicity of the contaminant and on the soil properties. For neutral hydrophobic organic chemicals,

$$R = 1 + \left(\frac{\rho_b}{\eta}\right) K_{oc} f_{oc}$$

(9–43)

TABLE 9–5   Retardation Coefficients of Typical Groundwater Contaminants[a,b]

Compound	Soil Type I $\rho_b = 1.4$ $n = 40\%$ $f_{oc} = 0.002$	Soil Type II $\rho_b = 1.6$ $n = 30\%$ $f_{oc} = 0.005$	Soil Type III $\rho_b = 1.75$ $n = 55\%$ $f_{oc} = 0.05$
Benzene	1.2	1.7	5.3
Toluene	1.7	3.6	17.0
Aniline	1.1	1.2	2.2
Di-*n*-propyl phthalate	5.6	19.0	110.0
Fluorene	23.0	86.0	500.0
n-Pentane	7.0	24.0	140.0

[a]Formulas used for computation of K_{oc} values are from Schwarzenbach, Gschwend, and Imboden, 1993.
[b]Data used in computations are from Schwarzenbach, et al., 1993.
Note: ρ_b = bulk density, $g \cdot cm^{-3}$; n = porosity, %; f_{oc} = fraction of soil that is organic carbon.

where ρ_b = bulk density of the soil

η = the porosity of the soil (as a fraction)

K_{oc} = partition coefficient into the organic carbon fraction of the soil

f_{oc} = fraction of organic carbon in the soil.

A table of retardation coefficients of some typical groundwater contaminants is given in Table 9–5.

EXAMPLE 9–18   Two houses are located adjacent to each other. House A has a septic system located 60 m upgradient from the drinking water well for house B. The owner of house A disposed of a pesticide down the drain, causing the septic field to become contaminated with this pesticide. The linear velocity of the water in the unconfined aquifer used for drinking water is 4.7×10^{-6} m · s^{-1}.

Assuming that the pesticide does not degrade in the soil and that it has a retardation coefficient of 2.4, how many days will it take for the pesticide to reach the well of house B?

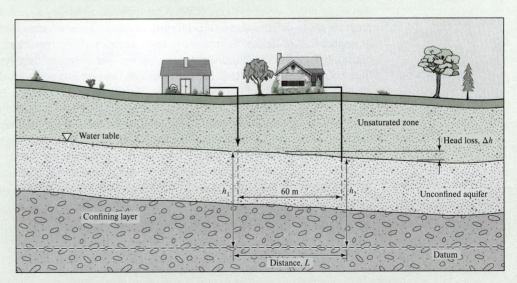

Solution Given that the retardation coefficient is defined as

$$R = \frac{v'_{water}}{v'_{cont}}$$

$$v'_{cont} = \frac{v'_{water}}{R} = \frac{4.7 \times 10^{-6} \text{ m} \cdot \text{s}^{-1}}{2.4} = 1.96 \times 10^{-6} \text{ m} \cdot \text{s}^{-1}$$

Because the distance between the septic field and the well is 60 m, the travel time is

$$\frac{\text{Distance}}{v'} = \left(\frac{60 \text{ m}}{1.96 \times 10^{-6} \text{ m} \cdot \text{s}^{-1}} \right) \left(\frac{\text{days}}{86,400 \text{ s}} \right) = 354 \text{ days}$$

Therefore, it will take 354 days for the pesticide to reach the drinking well. One should note that no mention is made here of the concentration of the chemical. The concentration depends on the mass of contaminant released and on the quantity of water flowing, effectively diluting the chemical.

Note that this is a much simplified solution of a complex problem that involves the use of extensive groundwater models.

As with other aqueous environments, the United States has promulgated numerous laws to protect groundwater resources. For example, the Safe Drinking Act protects sole-source aquifers, that is, those aquifers that are the sole supply of drinking water for a particular region. The Resource Conservation and Recovery Act (RCRA) protects groundwater by regulating the disposal of hazardous chemicals. The Comprehensive Environmental Response, Compensation and Liability Act (CERCLA) helps to protect groundwater resources by mandating and providing funding for the cleanup of abandoned hazardous waste sites. Clearly, groundwater is much too valuable a natural resource to allow it to become damaged by human activities. It is our hope that the financial and technical resources are provided by governments and industries to protect our groundwater for generations to come.

9–8 SOURCE WATER PROTECTION

The Safe Drinking Water Act (SDWA) Reauthorization of 1996 requires states to develop a Source Water Assessment and Protection (SWAP) Program to assess the drinking water sources that serve public water systems (PWSs) for their susceptibility to pollution. The goals of this program are to determine the susceptibility of public water systems to these contaminants and to protect human health. Source water assessments provide important information about the potential risks posed to drinking water supplies. For example, they typically include the delineated protection areas, locations of wells and intakes, inventories and locations of potential contaminant sources, and hydrologic data. The source water protection area is the watershed or groundwater area in which the release of pollutants could contaminate the water supply. For PWSs using surface water, the source water protection area includes the entire watershed upstream of the intake to the State border. If groundwater contributes to the flow of water in the public water supply, then groundwater protection must be included in the SWAP Program.

Where groundwater is the source of drinking water, the Wellhead Protection Program (WHPP) serves as the cornerstone of the state's SWAP Program. The program was developed to help communities:

- Form local teams that will assist with protection of public supply wells in their area.
- Determine the land area that provides water to public supply wells.
- Identify existing and potential sources of contamination.

- Manage potential sources of contamination to minimize their threat to drinking water supplies.
- Develop a contingency plan to prepare for the emergency closing of a well.
- Plan for future water supply needs.

In some cases, it may be determined that the ground water is under the direct influence of surface water (GWUDI). The determination that groundwater is under the direct influence is typically based on site-specific measurements of water quality, information regarding well construction characteristics, and the geology of the area. If a source water is GWUDI, then both the surface and groundwater must be protected.

The first step in a Source Water Assessment (SWA) is the delineation of the boundaries of the area that affects the source water. The second step is to conduct a contaminant source inventory to identify the contaminants of concern and significant potential sources. The contaminants of concern include all contaminants for which there exists an established MCL under the SDWA, those regulated under the Surface Water Treatment Rule, and *Cryptosporidium*. States have the discretion to add additional contaminants that are deemed a threat to human health. The inventory must include descriptions and locations of existing and potential future sources of contamination.

The third part of the SWA is the susceptibility determination, which is an assessment of the susceptibility of the water supply to contamination, based on the source inventory, the hydrology and hydrogeology of the watershed, the characteristics of the contaminants and sources, and other factors such as well intake and integrity. The SWA is then used to develop management plans for the protection of the source water protection area.

The final step in developing a SWAP Program is the dissemination of the plan to the public. The report must be comprehensible, include maps of the delineated source water protection area along with the significant sources of contamination included in the inventory, and be readily available to the public. The EPA has developed databases about local protection efforts and the condition and vulnerability of more than 2000 watersheds that can be found in EPA's "Surf Your Watershed" and the "Watershed Index Online" websites.

Source water protection has numerous benefits. The less polluted a source water is, the less treatment necessary to provide potable drinking water, thereby reducing the cost of treatment. U.S. EPA reports that the cost of addressing contaminated groundwater is on average 30 to 40 times more expensive than preventing contamination. The costs can exceed 200 times, in some cases. In addition, protection can prevent human illness and death, along with eliminating health-related costs due to exposure to contaminated water. The health-related costs of contamination include lost wages and productivity, medical expenses, and, in extreme cases, death. Source water protection has other economic benefits such as maintaining real estate values and safeguarding local tourism and recreation revenues. Finally SWAP Programs help protect our limited water supplies for future generations, build consumer confidence in the water supply, and ensure healthy ecosystems. It is clear that SWAP Programs are well worth the time and cost of development and implementation.

CHAPTER REVIEW

When you have completed studying this chapter, you should be able to do the following without the aid of your text or notes:

1. List the major pollutant categories (there are four) that are produced by each of the four principal sources of wastewater.
2. List the two nutrients of primary concern with respect to a receiving body of water.
3. Define biochemical oxygen demand (BOD).
4. Explain the procedure for determining BOD and specify the nominal values of temperature and time used in the test.

5. List three reasons why the BOD rate constant may vary.
6. Sketch a graph showing the effect of varying rate constant on 5-day BOD if the ultimate BOD is the same, and the effect on ultimate BOD if the 5-day BOD is the same.
7. Using Equation 9–11 in your answer, explain what causes nitrogenous BOD.
8. Sketch a series of curves that show the deoxygenation, reaeration, and DO sag in a river. Show the effect of a change in the deoxygenation or reaeration rate on the location of the critical point and the magnitude of the DO deficit.
9. List three reasons why ammonia nitrogen is detrimental to a receiving body of water and its inhabitants.
10. List three sources of phosphorus that must be controlled to reduce cultural eutrophication of lakes.
11. Define acid rain.
12. Explain why acid rain is of concern.
13. Explain the role of calcareous soils in protecting lakes from acidification.
14. Other than rainwater pH, list six variables that determine the extent of lake acidification and explain how increasing or decreasing the value of each might be expected to change the extent of acidification.

With the aid of this text or a list of equations, you should be able to do the following:

1. Calculate the BOD_5, given the sample size and oxygen consumption, or calculate the sample size, given the allowable oxygen consumption and estimated BOD_5.
2. Calculate the ultimate BOD (L_o), given the BOD exerted (BOD_t) in time, t, and rate constant, or calculate the rate constant, k, given L_o and BOD_t.
3. Calculate a new k for a temperature other than 20°C, given a value at $T°C$.
4. Calculate the BOD rate constant (k) and ultimate BOD (L_o) from experimental data of BOD versus time.
5. Calculate the oxygen deficit, D, in a length of stream (reach), given the required input data.
6. Calculate the critical oxygen deficit, D_c, at the critical point.
7. Calculate the rate of pollutant transport using retardation coefficients.

PROBLEMS

9–1   Glutamic acid ($C_5H_9O_4N$) is used as one of the reagents for a standard to check the BOD test. Determine the theoretical oxygen demand of 150 mg $\cdot$ L^{-1} of glutamic acid. Assume the following reactions apply.

$$C_5H_9O_4N + 4.5O_2 \longrightarrow 5CO_2 + 3H_2O + NH_3$$

$$NH_3 + 2O_2 \longrightarrow NO_3^- + H^+ + H_2O$$

Answer: 212 mg $\cdot$ L^{-1}

9–2   Bacterial cells have been represented by the chemical formula $C_5H_7NO_2$. Compute the theoretical oxygen demand (in grams $O_2 \cdot$ mol^{-1} of cells), assuming the following reactions apply.

$$C_5H_7NO_2 + 5O_2 \longrightarrow 5CO_2 + 2H_2O + NH_3$$

$$NH_3 + 2O_2 \longrightarrow NO_3^- + H^+ + H_2O$$

9–3   The company you work for plans to release a waste stream containing 7.0 mg $\cdot$ L^{-1} of benzaldehyde. Assuming that benzaldehyde degrades according the reaction:

$$C_7H_6O + 8O_2 \longrightarrow 7CO_2 + 3H_2O$$

calculate the theoretical carbonaceous oxygen demand of this waste stream (in mg $\cdot$ L^{-1}).

Answer: 16.906 mg $\cdot$ L^{-1}

9–4 The company you work for plans to release a waste stream containing 10 mg $\cdot$ L^{-1} of phenol. Assuming that phenol degrades according the reaction:

$$C_6H_6O + 7O_2 \longrightarrow 6CO_2 + 3H_2O$$

calculate the theoretical carbonaceous oxygen demand of this waste stream.

9–5 If the BOD$_5$ of a waste is 220.0 mg $\cdot$ L^{-1} and the ultimate BOD is 320.0 mg $\cdot$ L^{-1}, what is the rate constant?

  Answer: $k = 0.233$ day^{-1}

9–6 If the BOD of a municipal wastewater at the end of 7 days is 60.0 mg $\cdot$ L^{-1} and the ultimate BOD is 85.0 mg $\cdot$ L^{-1}, what is the rate constant?

9–7 The BOD rate constant is 0.233 day^{-1} for a municipal wastewater. The BOD$_5$ was measured to be 250 mg $\cdot$ L^{-1}. What is the ultimate BOD?

  Answer: 363 mg $\cdot$ L^{-1}

9–8 The BOD rate constant is 0.433 day^{-1} for a municipal wastewater. The BOD$_5$ was measured to be 272 mg $\cdot$ L^{-1}. What is the ultimate BOD?

9–9 Assuming that the data in Problem 9–5 were taken at 20°C, compute the rate constant at a temperature of 15°C.

  Answer: $k = 0.124$ day^{-1}

9–10 Assuming that the data in Problem 9–6 were taken at 25°C, compute the rate constant at 16°C.

9–11 A sample of municipal sewage is diluted to 1% by volume prior to running a BOD$_5$ analysis. After 5 days the oxygen consumption is determined to be 2.00 mg $\cdot$ L^{-1}. What is the BOD$_5$ of the sewage?

  Answer: BOD$_5$ = 200 mg $\cdot$ L^{-1}

9–12 What sample size (in percent) is required for a BOD$_5$ of 350.0 mg $\cdot$ L^{-1} if the oxygen consumed is to be limited to 6.00 mg $\cdot$ L^{-1}?

9–13 If the ultimate BOD values of two wastes having k values of 0.3800 day^{-1} and 0.220 day^{-1} are 280.0 mg $\cdot$ L^{-1}, what would be the 5-day BOD for each?

  Answer: For $k = 0.38$ day^{-1}, BOD$_5$ = 238 mg $\cdot$ L^{-1}; for $k = 0.22$ day^{-1}, BOD$_5$ = 187 mg $\cdot$ L^{-1}

9–14 If the BOD$_5$ values for two livestock wastes having k values of 0.3800 day^{-1} and 0.240 day^{-1} are 16230.0 mg $\cdot$ L^{-1}, what would be the ultimate BOD for each?

9–15 Using the data from Problem 9–1, calculate the NBOD of glutamic acid. Assume the concentration of glutamic acid is 0.1 mM.

  Answer: CBOD = 14.4 mg O$_2$ $\cdot$ L^{-1} and NBOD = 6.4 g mg O$_2$ $\cdot$ L^{-1}

9–16 Using the data from Problem 9–2, calculate the NBOD of bacterial cells.

9–17 The following data are provided to you:

Seeded wastewater:

   Initial DO = 8.6 mg $\cdot$ L^{-1}
   Final DO (after 5 days) = 2.1 mg $\cdot$ L^{-1}
   Volume of wastewater = 2.5 mL
   Total volume in BOD bottle = 300.0 mL

Seeded dilution water:

   Initial DO = 8.6 mg $\cdot$ L^{-1}
   Final DO (after 5 days) = 7.3 mg $\cdot$ L^{-1}
   Volume of seeded dilution water = 300.0 mL
   T = 20°C

Calculate the BOD of the sample (in units of mg $\cdot$ L^{-1}).

Answer: 625 mg $\cdot$ L^{-1}

9–18 The following data are provided to you:

Seeded wastewater:

> Initial DO = 8.2 mg $\cdot$ L^{-1}
> Final DO (after 5 days) = 2.5 mg $\cdot$ L^{-1}
> Volume of wastewater = 5.0 mL
> Total volume in BOD bottle = 300.0 mL

Seeded dilution water:

> Initial DO = 8.3 mg $\cdot$ L^{-1}
> Final DO (after 5 days) = 7.3 mg $\cdot$ L^{-1}
> Volume of seeded dilution water = 300.0 mL
> T = 20°C

Calculate the BOD of the sample (in units of mg $\cdot$ L^{-1}).

9–19 The Waramurungundi tannery with a wastewater flow of 0.011 m^3 $\cdot$ s^{-1} and a BOD$_5$ of 590 mg $\cdot$ L^{-1} discharges into the Djanggawul Creek. The creek has a 10-year, 7-day low flow of 1.7 m^3 $\cdot$ s^{-1}. Upstream of the Waramurungundi tannery, the BOD$_5$ of the creek is 0.6 mg $\cdot$ L^{-1}. The BOD rate constants (k) are 0.115 day^{-1} for the Waramurungundi tannery and 3.7 day^{-1} for the creek. Calculate the initial ultimate BOD after mixing.

Answer: 9.27 mg $\cdot$ L^{-1}

9–20 A short distance downstream from the Waramurungundi tannery in Problem 9–19, a glue factory and a municipal wastewater treatment plant also discharge into Djanggawul Creek. The wastewater flows and ultimate BOD$_5$ for these discharges are listed in the following table. Determine the initial ultimate BOD after mixing of the creek and the three wastewater discharges.

Source	Flow (m^3 $\cdot$ s^{-1})	Ultimate BOD (mg $\cdot$ L^{-1})
Glue factory	0.13	255
Municipal WWTP	0.02	75

9–21 Using the data given below, calculate the DO concentration just after mixing (in unit of mg $\cdot$ L^{-1}).

	Oz WWTP Effluent	Babbling Run
Flow, m^3 $\cdot$ s^{-1}	0.36	4.8
ult-BOD at 20°C, mg $\cdot$ L^{-1}	27	1.35
DO, mg $\cdot$ L^{-1}	0.9	8.5

(*Note:* WWTP = Wastewater Treatment Plant, Babbling Run: upstream from outfall)

Answer: 7.970 mg $\cdot$ L^{-1}

9–22 Calculate the initial deficit, assuming that the temperature of the Babbling Run is 18°C.

9–23 Compute the deoxygenation rate constant and reaeration rate constant (base e) for the following wastewater and stream conditions.

Source	k (day^{-1})	Temperature (°C)	H (m)	Velocity (m · s^{-1})	η
Wastewater	0.20	20			
Amaunet Stream		20	3.0	0.5	0.4

Answer: $k_d = 0.27$ day^{-1}; $k_r = 0.53$ day^{-1}

9–24 During flood stage, the stream conditions in Problem 9–23 change as shown in the following table. Determine the values of k_d and k_r for flood conditions.

Source	k (day^{-1})	Temperature (°C)	H (m)	Velocity (m · s^{-1})	η
Wastewater	0.20	20			
Amaunet Stream		20	4.0	2.5	0.5

9–25 The initial ultimate BOD after mixing of the Bergelmir River is 12.0 mg · L^{-1}. The DO in the Bergelmir River after the wastewater and river have mixed is at saturation. The river temperature is 10°C. At 10°C, the deoxygenation rate constant (k_d) is 0.30 day^{-1}, and the reaeration rate constant (k_r) is 0.40 day^{-1}. Determine the critical point (t_c) and the critical DO.

Answer: $t_c = 2.88$ days. Critical DO = 7.53 mg · L^{-1}

9–26 Repeat Problem 9–29, assuming the river temperature rises to 15°C so that k_d and k_r change.

9–27 The discharge from the Renenutet sugar beet plant causes the DO at the critical point to fall to 4.0 mg · L^{-1}. The Meskhenet Stream has a negligible BOD, and the initial deficit after the river and wastewater have mixed is zero. What DO will result if the concentration of the waste (L_w) is reduced by 50%? Assume that the flows remain the same and that the saturation value of DO is 10.83 mg · L^{-1} in both cases.

Answer: DO = 7.4 mg · L^{-1}

9–28 The Audhumla town council has asked that you determine whether the discharge of the town's wastewater into the Einherjar River will reduce the DO below the state standard of 5.00 mg · L^{-1} at Gotterdämmerung, 5.79 km downstream, or at any other point downstream. The pertinent data are as follows:

Parameter	Wastewater	Einherjar River
Flow (in m^3 · s^{-1})	0.280	0.877
Ult-BOD at 15°C (mg · L^{-1})	6.44	7.00
DO (in mg · L^{-1})	1.00	6.00
k_d at 28°C (in day^{-1})	N/A	0.199
k_r at 28°C (in day^{-1})	N/A	0.370
Speed (in m · s^{-1})	N/A	0.650
Temperature (in °C)	28	28

Answers: DO at Gotterdämmerung = 4.75 mg · L^{-1}
Critical DO = 4.72 mg · L^{-1} at $t_c = 0.3149$ day

9–29 The town of Edinkira has filed a complaint with the state department of natural resources (DNR) that the city of Quamta is restricting its use of the Umvelinqangi River because of the discharge of raw sewage. The DNR water quality criterion for the Umvelinqangi River is $5.00 \text{ mg} \cdot \text{L}^{-1}$ of DO. Edinkira is 15.55 km downstream from Quamta. What is the DO at Edinkira? What is the critical DO and where (at what distance) downstream does it occur? Is the assimilative capacity of the river restricted? The following data pertain to the 7-year, 10-day low flow at Quamta.

Parameter	Wastewater	Umvelinqangi River
Flow (in $\text{m}^3 \cdot \text{s}^{-1}$)	0.1507	1.08
BOD_5 at 16°C (in $\text{mg} \cdot \text{L}^{-1}$)	128.00	N/A
BOD_u at 16°C (in $\text{mg} \cdot \text{L}^{-1}$)	N/A	11.40
DO (in $\text{mg} \cdot \text{L}^{-1}$)	1.00	7.95
k at 20°C (in day^{-1})	0.4375	N/A
Speed (in $\text{m} \cdot \text{s}^{-1}$)	N/A	0.390
Depth (in m)	N/A	2.80
Temperature (in °C)	28	28
Bed-activity coeff.	N/A	0.200

9–30 Under the provisions of the Clean Water Act, the U.S. Environmental Protection Agency established a requirement that municipalities had to provide secondary treatment of their waste. This was defined to be treatment that resulted in an effluent BOD_5 that did not exceed $30 \text{ mg} \cdot \text{L}^{-1}$. The discharge from Quamta (Problem 9–29) is clearly in violation of this standard. Given the data in Problem 9–29, rework the problem, assuming that Quamta provides treatment to lower the BOD_5 to $30.00 \text{ mg} \cdot \text{L}^{-1}$.

9–31 If the population and water use of Quamta (Problems 9–29 and 9–30) are growing at 5% per year with a corresponding increase in wastewater flow, how many years' growth may be sustained before secondary treatment becomes inadequate? Assume that the treatment plant continues to maintain an effluent BOD_5 of $30.00 \text{ mg} \cdot \text{L}^{-1}$.

> **Answer:** $n = 1.73$ year's growth if we impose the constraint that the critical DO cannot be less than $5.0 \text{ mg} \cdot \text{L}^{-1}$.

9–32 When ice covers a river, it severely limits the reaeration. There is some compensation for the reduced aeration because of the reduced water temperature. The lower temperature reduces the biological activity and, thus, the deaeration rate and, at the same time, the DO saturation level increases. Assuming a winter condition, rework Problem 9–29 with the reaeration reduced to 0 and the river water temperature at 2°C.

9–33 What combination of BOD reduction and wastewater DO increase is required so the Audhumla wastewater in Problem 9–28 does not reduce the DO below $5.00 \text{ mg} \cdot \text{L}^{-1}$ anywhere along the Einherjar River? Assume that the cost of BOD reduction is three to five times that of increasing the effluent DO. Because the cost of adding extra DO is high, limit the excess above the minimum amount such that the critical DO falls between 5.00 and $5.25 \text{ mg} \cdot \text{L}^{-1}$.

> **Answer:** Raising the wastewater DO to $2.7 \text{ mg} \cdot \text{L}^{-1}$ is the most cost-effective remedy.

9–34 What amount of ultimate BOD (in $\text{kg} \cdot \text{day}^{-1}$) may Quamta (Problem 9–29) discharge and still allow Edinkira $1.50 \text{ mg} \cdot \text{L}^{-1}$ of DO above the DNR water quality criteria for assimilation of its waste?

9–35 Assuming that the mixed oxygen deficit (D_a) is zero and that the ultimate BOD (L_r) of the Manco Capac River above the wastewater outfall from Urcaguary is zero, calculate the amount of ultimate BOD (in $\text{kg} \cdot \text{day}^{-1}$) that can be discharged if the DO must be kept at $4.00 \text{ mg} \cdot \text{L}^{-1}$ at a point 8.05 km downstream. The stream deoxygenation rate (k_d) is 1.80 day^{-1} at 12°C, and the reaeration rate (k_r) is 2.20 day^{-1} at 12°C. The river temperature is 12°C. The river flow is $5.95 \text{ m}^3 \cdot \text{s}^{-1}$ with a speed of $0.300 \text{ m} \cdot \text{s}^{-1}$. The Urcaguary wastewater flow is $0.0130 \text{ m}^3 \cdot \text{s}^{-1}$.

> **Answer:** Mass flux $= 1.14 \times 10^4 \text{ kg} \cdot \text{day}^{-1}$ of ultimate BOD

9–36   Calculate the DO at a point 1.609 km downstream from a waste discharge point for the following conditions. Report answers to two decimal places. Rate constants are already temperature adjusted.

Parameters	Stream
k_d	1.911 day^{-1}
k_r	4.49 day^{-1}
Flow	$2.4 \text{ m}^3 \cdot \text{s}^{-1}$
Speed	$0.100 \text{ m} \cdot \text{s}^{-1}$
D_a (after mixing)	0.00
Temperature (in °C)	17.00
ult-BOD (after mixing)	$1100.00 \text{ kg} \cdot \text{day}^{-1}$

9–37   Assume that the Urcaguary wastewater (Problem 9–35) also contains $3.0 \text{ mg} \cdot \text{L}^{-1}$ of ammonia nitrogen with a stream deoxygenation rate of 0.900 day^{-1} at 12°C. What is the amount of ultimate carbonaceous BOD (in $\text{kg} \cdot \text{day}^{-1}$) that Urcaguary can discharge and still meet the DO level of $4.00 \text{ mg} \cdot \text{L}^{-1}$ at a point 8.05 km downstream? Assume also that the theoretical amount of oxygen will ultimately be consumed in the nitrification process.

   Answer: Mass flux $= 6.2 \times 10^3 \text{ kg} \cdot \text{day}^{-1}$

9–38   An aquifer has a hydraulic gradient of 0.00086, a hydraulic conductivity of $200 \text{ m} \cdot \text{day}^{-1}$ and a porosity of 0.23. A chemical with a retardation factor of 2.3 contaminates the aquifer. What is the linear velocity of the contaminant? How long will it take for the contaminant to travel 100 m in the aquifer?

9–39   A chemical contaminates an aquifer. The average linear velocity of water in the aquifer is $2.650 \times 10^{-7} \text{ m} \cdot \text{s}^{-1}$. The aquifer has a porosity of 48.0%. The chemical has a retardation coefficient of 2.65. Calculate the average linear velocity of the pollutant (in $\text{m} \cdot \text{s}^{-1}$).

   Answer: $1.00\text{e-}07 \text{ m} \cdot \text{s}^{-1}$

9–40   A small lake is surrounded by agricultural land. The lake is 150 m long and 120 m wide. The average depth is 30 m. The lake is fed by a small pollution-free stream having an average flow rate of $1.5 \text{ m}^3 \cdot \text{s}^{-1}$. A herbicide, Greatcrop, is detected at a concentration of $45.2 \text{ µg} \cdot \text{L}^{-1}$ in the lake. Greatcrop degrades biologically with a rate constant of 0.22 day^{-1}. Assuming that the lake is at steady-state conditions and completely mixed at the time of sampling, what would have been the mass input of herbicide? Assume that the stream is the only source of water to the lake and that evaporation, precipitation, and seepage can be neglected.

   Answer: $11.2 \text{ kg} \cdot \text{day}^{-1}$

9–41   Blue Lake is 3500 m long and 2800 m wide. The average depth is 20 m. The lake is fed by a stream having an average flow rate of $6.9 \text{ m}^3 \cdot \text{s}^{-1}$. The stream flows through agricultural land, and the average concentration of the herbicide Kill-all in the stream is $33.1 \text{ µg} \cdot \text{L}^{-1}$. Kill-all degrades biologically with a rate constant of 0.18 day^{-1}. Assuming that the lake is at steady-state conditions and completely mixed, what would be the Kill-all concentration in the lake under these conditions?

DISCUSSION QUESTIONS

9–1   Students in a graduate-level environmental engineering laboratory took samples of the influent (raw sewage) and effluent (treated sewage) of a municipal wastewater treatment plant. They used these samples to determine the BOD rate constant (k). Would you expect the rate constants to be the same or different? If different, which would be higher and why?

9–2   If it were your job to set standards for a water body and you had a choice of either BOD_5 or ultimate BOD, which would you choose and why?

9–3 A summer intern has turned in his log book for temperature measurements for a limnology survey. He was told to take the measurements in the air 1 m above the lake, 1 m deep in the lake, and at a depth of 10 m. He turned in the following results but did not record which temperatures were taken where. If the measurements were made at noon in July in Missouri, what is your best guess as to the location of the measurements (i.e., air, 1-m deep, 10-m deep)? The recorded values were: 33°C, 18°C, and 21°C.

9–4 If the critical point in a DO sag curve is found to be 18 km downstream from the discharge point of untreated wastewater, would you expect the critical point to move upstream (toward the discharge point), downstream, or remain in the same place if the wastewater is treated?

9–5 You have been assigned to conduct an environmental study of a remote lake in Canada. Aerial photos and a ground-level survey reveal no anthropogenic waste sources are contributing to the lake. When you investigate the lake, you find a highly turbid lake with abundant mats of floating algae and a hypolimnion DO of 1.0 mg $\cdot$ L^{-1}. What productivity class would you assign to this lake? Explain your reasoning.

9–6 The lakes in Illinois, Indiana, western Kentucky, the lower peninsula of Michigan, and Ohio do not appear to be subject to acidification even though the rainwater pH is 4.4. Based on your knowledge (or what you can discover by research) of the topography, vegetation, and bedrock, explain why the lakes in these areas are not acidic.

9–7 For the community shown below, discuss how you would develop a source water assessment and protection (SWAP) Program. Be sure to

- Identify the boundaries of your source water protection area.

- Identify potential contaminants and sources of concern.

- Determine what information you would collect and analyze to complete the susceptibility determination.

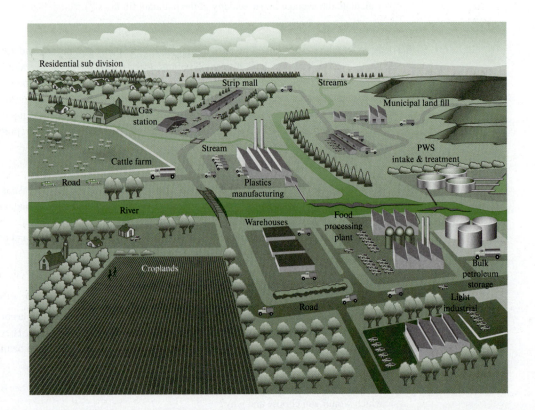

FE Exam Formatted Problems

FE EXAM FORMATTED PROBLEMS

9–1 A sample of water from the River Styx has a BOD rate constant of 0.15 day^{-1} at 20°C. The dissolved oxygen concentration is 8.2 mg/L. The concentration of dissolved oxygen after 5 days is 3.2 mg/L. The ultimate BOD is most nearly:

    (a)   5.0 mg/L

    (b)   9.5 mg/L

    (c)   10.6 mg/L

    (d)   20.2 mg/L

9–2 The BOD$_5$ of a wastewater sample is determined to be 225 mg/L. The BOD rate constant of 0.15 day^{-1} at 20°C. The ultimate BOD is most nearly:

    (a)   225 mg/L

    (b)   426 mg/L

    (c)   476 mg/L

    (d)   525 mg/L

9–3 The nitrification reaction is given by Equation 9–11. An agricultural wastewater has an ammonia concentration of 50 mg/L. The amount of oxygen consumed by nitrification is most nearly:

    (a)   50 mg/L

    (b)   56 mg/L

    (c)   89 mg/L

    (d)   177 mg/L

9–4 A sample of wastewater is diluted by a factor of 25. The diluted wastewater has an initial dissolved oxygen concentration of 8.6 mg/L. After 5 days, the dissolved oxygen is measured at 2.8 mg/L. The BOD$_5$ of the wastewater is most nearly:

    (a)   0.23 mg/L

    (b)   19 mg/L

    (c)   102 mg/L

    (d)   145 mg/L

9–5 A wastewater is discharged to a river that has a summer temperature of 28°C. The BOD rate constant was determined at 20°C in the laboratory was determined to be 0.12 d^{-1}. The percent of the maximum oxygen consumption that would occur in the river after 5 days is most nearly:

    (a)   20%

    (b)   32%

    (c)   60%

    (d)   81%

9–6 The 5-day BOD of a wastewater sample was determined to be 150 mg/L. The 20-day (ultimate) BOD was determined to be 220 mg/L. The BOD rate constant is most nearly:

    (a)   0.06 d^{-1}

    (b)   0.08 d^{-1}

    (c)   0.16 d^{-1}

    (d)   0.23 d^{-1}

9–7 Wastewater with DO concentration of 2 mg/L is discharged to a river. The temperature of the river is 10°C and the river is saturated with oxygen (11.33 mg/L). The flow rate of the river is 3.0 m³/s. The temperature and flow rate of the wastewater are 20°C and 0.3 m³/s, respectively. The temperature of the river after mixing is 11°C. The oxygen saturation level at that temperature is 11.08 mg/L. The oxygen deficit immediately after mixing is most nearly:

(a) 0.20 mg/L

(b) 0.60 mg/L

(c) 0.85 mg/L

(d) 1.23 mg/L

9–8 A waste stream having a flow rate of 0.3 m³/s and a DO concentration of 2 mg/L is discharged to a stream with flow 3.0 m³/s and a DO concentration of 7.5 mg/L. The BOD_5 of the stream is 2 mg/L. The waste stream has BOD_5 of 30 mg/L and a BOD rate constant, k, of 0.2 d^{-1}. The deoxygenation and reaeration rates of the stream are 0.12 d^{-1} and 0.3 d^{-1}, respectively. The DO saturation level of the stream is 8.6 mg/L. The DO after 4 days travel in the stream is most nearly:

(a) 2.4 mg/L

(b) 4.8 mg/L

(c) 6.2 mg/L

(d) 8.1 mg/L

9–9 An industry proposes to discharge a waste stream having a flow rate of 0.3 m³/s and a DO concentration of 1 mg/L to a stream with flow 3.0 m³/s and a DO concentration of 7.5 mg/L. The BOD_5 of the stream is 2 mg/L. The waste stream has BOD_5 of 100 mg/L and a BOD rate constant, k, of 0.2 d^{-1}. The deoxygenation and reaeration rates of the stream are 0.12 d^{-1} and 0.3 d^{-1}, respectively. The DO saturation level of the stream is 8.6 mg/L. The minimum DO downstream from the outfall is most nearly:

(a) 2.0 mg/L

(b) 3.0 mg/L

(c) 4.0 mg/L

(d) 5.0 mg/L

9–10 Excessive phosphorus in a waste stream is most likely to cause:

(a) Acid rain

(b) Cultural eutrophication

(c) Salt water intrusion

(d) Minamata disease

REFERENCES

Acerini, C. L., and I. A. Hughes (2006) "Endocrine Disrupting Chemicals: A New and Emerging Public Health Problem?" *Arch. Dis. Child.*, 91: 633–41.

Andersen, A. G., T. K. Jensen, E. Carlsen, N. Jørgensen, A. M. Andersson, T. Krarup, N. Keiding, and N. E. Skakkebæk (2000) "High Frequency of Sub-Optimal Semen Quality in an Unselected Population of Young Men," *Hum. Reprod.* 15(2): 366–72.

Ash, R. J., B. Mauch, W. Moulder, and M. Morgan (1999) "Antibiotic Resistant Bacteria in U.S. Rivers" (Abstract Q-383), in *Abstracts of the 99th General Meeting of the American Society for Microbiology*, Chicago, May 30–June 3, 1999, p. 607.

Aspelin, A. L. (1994) *Pesticides Industry Sales and Usage: 1992 and 1993 Market Estimates.* Biological and Economic Analysis Division, Office of Pesticide Programs, Office of Prevention, Pesticides, and Toxic Substances, United States Environmental Protection Agency, Washington, DC.

Associated Press (2001) *Toxin Might Hurt Endangered Whales,* January 13, 2001.

Bennett, J., and G. Kramer (1999) "Multidrug Resistant Strains of Bacteria in the Streams of Dubuque County, Iowa" (Abstract Q-86), in *Abstracts of the 99th General Meeting of the American Society for Microbiology,* Chicago, May 30–June 3, 1999, p. 464.

Benton, T. (1991) "Oceans of Garbage," *Nature,* 352: 113.

Blair, B. D., J. P. Crago, C. J. Hedman, and R. D. Klaper (2013) "Pharmaceuticals and Personal Care Products Found in the Great Lakes Above Concentrations of Environmental Concern," *Chemosphere,* 93(9): 2116–2123.

Botkin, D. B., and E. A. Keller (1998) *Environmental Science: Earth as a Living Planet,* 2nd ed., John Wiley and Sons, New York.

Buser, H. R., T. Poiger, and M. D. Muller (1999) "Occurrence and Environmental Behavior of the Chiral Pharmaceutical Drug Ibuprofen in Surface Waters and Waste Water," *Environ. Sci. Technol.,* 33(15): 2529–35.

Buser, H. R., M. D. Muller, and N. Theobald (1998a) "Occurrence of the Pharmaceutical Drug Clofibric Acid and the Herbicide Mecoprop in Various Swiss Lakes and in the North Sea," *Environ. Sci. Technol.,* 32(1): 188–92.

Buser, H. R., T. Poiger, and M. D. Muller (1998b) "Occurrence and Fate of the Pharmaceutical Drug Diclofenac in Surface Waters: Rapid Photodegradation in a Lake," *Environ. Sci. Technol.,* 32(22): 3449–56.

Chen, C. J., C. W. Chen, M. M. Wu, and T. L. Kuo (1992) "Cancer Potential in Liver, Lung, Bladder, and Kidney Due to Inorganic Arsenic in Drinking Water," *British Journal of Cancer,* 66: 888–92.

Cunningham, V. L., M. Buzby, T. Hutchinson, F. Mastrocco, N. Parke, and N. Roden (2006) "Effects of Human Pharmaceuticals on Aquatic Life: Next Steps," *Environ. Sci. Technol. A-Pages,* 40(11): 3456–62.

Daughton, C. G. (2003) "Chemicals from Pharmaceuticals and Personal Care Products," in *Water: Science and Issues,* E. J. Dasch, ed., Macmillan Reference USA, New York: vol. 1, pp. 158–64.

Den Hond, E., H. A. Roels, K. Hoppenbrouwers, T. Nawrot, L. Thijs, C. Vandermeulen, G. Winneke, D. Vanderschueren, and J. A. Staessen (2002) "Sexual Maturation in Relation to Polychlorinated Aromatic Hydrocarbons: Sharpe and Skakkebaek's Hypothesis Revisited," *Environ. Health. Perspect.* 110(8): 771–76.

DeSimone, L. A. (2009) Quality of Water from Domestic Wells in Principal Aquifers of the United States, 1991–2004. U.S. Geological Survey Scientific Investigations Report 2008–5227, 139 p., available online at http://pubs.usgs.gov/sir/2008/5227.

Dickson, W. (1980) "Properties of Acidified Waters," in *Ecological Impact of Acid Precipitation*, D. Drablosand, A. Tollan, eds., SNSF. Oslo, Sandefjord, Norway: pp. 75–83.

Dunphy Guzman, K. A., M. R. Taylor, and J. F. Banfield (2006) "Environmental Risks of Nanotechnology: National Nanotechnology Initiative Funding, 2000–2004," *Environ. Sci. Technol.,* 40(5): 1401–07.

Eaton, A. D., L. S. Clesceri, E. W. Rice, A. E. Greenberg, and M. A. H. Franson, eds. (2005) Standard Methods for the Examination of Water and Wastewater, 21st ed. American Public Health Association (APHA), American Water Works Association (AWWA) & Water Environment Federation (WEF).

Elliott, D. (April 20, 2015) "Five Years After BP Oil Spill, Effects Linger and Recovery Is Slow," NPR National.

Fong, P. P. (1998) "Zebra Mussel Spawning Is Induced in Low Concentrations of Putative Serotonin Reuptake Inhibitors," *Biol. Bull.,* 194(2): 143–49.

Fong, P. P., and A. T. Ford (2014) "The Biological Effects of Antidepressants on the Molluscs and Crustaceans: A Review," *Aquatic Toxicology,* 151: 4–13.

Galloway, J. N., and E. B. Cowling (1978) "The Effects of Precipitation on Aquatic and Terrestrial Ecosystems: A Proposed Precipitation Chemistry Network," *Journal of the Air Pollution Control Association,* 28: 229–35.

Guzman, K. A., M. R. Taylor, and J. F. Banfield (2006) "Environmental Risks of Nanotechnology: National Nanotechnology Initiative Funding, 2000–2004," *Environ. Sci. Technol.,* 40: 1401–07.

Harries, J. E., T. Runnalls, E. Hill, C. A. Harris, S. Maddix, J. P. Sumpter, and C. R. Tyler (2000) "Development of a Reproductive Performance Test for Endocrine Disrupting Chemicals Using Pair-Breeding Fathead Minnows (*Pimephales promelas*)," *Environ. Sci. Technol.,* 34(14): 3003–11.

Hummel, D., D. Löffler, G. Fink, and T. A. Ternes (2006) "Simultaneous Determination of Psychoactive Drugs and Their Metabolites in Aqueous Matrices by Liquid Chromatography Mass Spectrometry," *Environ. Sci. and Technol.,* 40: 7321–28.

LaPara, T. M., T. R. Burch, P. J. McNamara, D. T. Tan, M. Yan, and J. J. Eichmiller (2011) "Tertiary-treated Municipal Wastewater Is a Significant Point Source of Antibiotic Resistance Genes into Duluth-Superior Harbor," *Environ. Sci. Technol.* 45(22): 9543–9549.

Lin, Y. L. and C.-H. Lee (2014) "Elucidating the Rejection Mechanisms of PPCPs by Nanofiltration and Reverse Osmosis Membranes." *Ind. Eng. Chem. Res.,* 53(16): 6798–6806.

Moller, H. (1998) "Trends in Sex-Ratio, Testicular Cancer and Male Reproductive Hazards: Are They Connected?" *APMIS,* 106(1): 232–8, discussion 238–9.

Nagler, J. J., J. Bouma, G. H. Thorgaard, and D. D. Dauble (2001) "High Incidence of a Male-Specific Genetic Marker in Phenotypic Female Chinook Salmon from the Columbia River," *Environ. Health Persp.,* 109(1): 67–69.

National Atmospheric Deposition Program (NRSP/3) National Trends Network (2007) "Isopleth Maps," NADP Program Office, Illinois State Water Survey, Champaign, IL, http://nadp.sws.uiuc.edu/isopleths/.

National Estuarine Research Reserve System (2007) *An Introduction to Estuaries,* http://inlet.geol.sc.edu/nerrsintro.html; http://nerrs.noaa.gov/welcome.html.

National Research Council (1999) *Arsenic in Drinking Water,* National Academy Press, Washington, DC.

National Research Council (2001) "Update: Arsenic in Drinking Water," National Academy Press, Washington, DC.

Neegaard, L. (2001) "FDA: While Pregnant, Limit Fish," *Lansing State Journal,* January 13, 2001.

NRDC (2015) "Summary of Information Concerning the Ecological and Economic Impacts of the BP Deepwater Horizon Oil Spill Disaster," Natural Resources Defense Council, June 2015. *IP,* 15-04-A.

O'Connor, D. J., and W. E. Dobbins (1958) "Mechanism of Reaeration in Natural Streams," *American Society of Civil Engineers Transactions,* 153: 641.

Parent, A. S., G. Teilmann, A. Juul, N. E. Skakkebaek, J. Toppari, and J. P. Bourguignon (2003) "The Timing of Normal Puberty and the Age Limits of Sexual Precocity: Variations Around the World, Secular Trends, and Changes after Migration," *Endocr. Rev.* 24(5): 668–93.

Paulozzi, L. J., J. D. Erickson, and R. J. Jackson (1997) "Hypospadias Trends in Two U.S. Surveillance Systems," *Pediatrics,* 100(5): 831–4.

Peters, C. A., D. M. Robertson, D. A. Saad, D. J. Sullivan, B. C. Scudder, F. A. Fitzpatrick, K. D. Richards, J. S. Stewart, S. A. Fitzgerald, and B. N. Lenz (1998) *Water Quality in the Western Lake Michigan Drainages, Wisconsin, and Michigan, 1992–95,* U.S. Geological Survey Circular 1156, available online at http://water.usgs.gov/pubs/circ1156

Pruden, A., R. Pei, H. Storteboom, and K. H. Carlson (2006) "Antibiotic Resistance Genes as Emerging Contaminants: Studies in Northern Colorado," *Environ. Sci. and Technol.,* 40: 7445–50.

Raloff, J. (1999) "Waterways Carry Antibiotic Resistance," *Science News Online,* June 5, 1999, http://www.sciencenews.org/sn_arc99/6_5_99/fobl.htm.

Ryan, P. M., and H. H. Harvey (1980) "Growth Responses of Yellow Perch, *Perca flavescens* (Mitchill), to Lake Acidification in La Cloche Mountains of Ontario," *Env. Biol. Fish,* 5: 97–108.

Sadik, O. A., and D. M. Witt (1999) "Monitoring Endocrine-Disrupting Chemicals," *Environ. Sci. Technol.,* 33: 368A–74A.

Schroepfer, G. J., M. L. Robins, and R. H. Susag (1964) "Research Program on the Mississippi River in the Vicinity of Minneapolis and St. Paul," *Advances in Water Pollution Research,* 1, pt 1: 145.

Schwarzenbach, R. P., P. M. Gschwend, and D. M. Imboden (1993) *Environmental Organic Chemistry,* John Wiley and Sons, New York.

Sedlak, D. L., J. L. Gray, and K. E. Pinkston (2000) "Understanding Microcontaminants in Recycled Water," *Environ. Sci. Technol.,* 34: 508A–15A.

Simmons, E. (2000) "Making an Impact: A Look at Santa Barbara's Active Role in the Environmental Movement and the Influence of the Environmental Studies Department in Its 30th Anniversary," Daily Nexus, University of California, Santa Barbara, May 19, 2000, http://www.es.ucsb.edu/images/new/ESealcovoredInOil.jpg.

Smithsonian Institution (1995) Ocean Planet: Marine Pollution 1: Oil Pollution, http://seawifs.gsfc.nasa.gov/OCEAN_PLANET/HTML/peril_oil_pollution.html.

Sternes, K. L. (1999) "Presence of High-Level Vancomycin Resistant Enterococci in the Upper Rio Grande" (Abstract Q-63), in *Abstracts of the 99th General Meeting of the American Society for Microbiology,* Chicago, May 30–June 3, 1999, p. 545.

Streeter, H. W., and E. B. Phelps (1925) *A Study of the Pollution and Natural Purification of the Ohio River,* U.S. Public Health Service Bulletin No. 146, Washington, DC.

U.S. EPA (2006) "Origins and Fat of PPCPs in the Environment," U.S. EPA, Office of Research and Development, National Exposure Research Laboratory, Las Vegas, NV, http://www.epa.gov/esd /chemistry/pharma/images/drawing.pdf.

U.S. EPA (2007) *National Estuary Program: About Estuaries,* http://www.epa.gov/owow/estuaries /about1.htm.

U.S. Geological Survey (2007) *The Quality of Our Nation's Waters—Nutrients and Pesticides*, U.S. Geological Survey Circular 1291, Reston, VA.

Vasquez, M. I., A. Lambrianides, M. Schneider, K. Kümmer, and D. Fatta-Kassinos (2014) "Environmental Side Effects of Pharmaceutical Cocktails: What We Know and What We Should Know," *Journal of Hazardous Materials,* 279: 169–189.

Wang, J., and S. Wang (2016) "Removal of Pharmaceuticals and Personal Care Products (PPCPs) from Wastewater: A Review," *Journal of Environmental Management,* 182: 620–640.

Welch, A. H., D. R. Helsel, M. J. Focazio, and S. A. Watkins (1999) "Arsenic in Ground Water Supplies of the United States," *in Arsenic Exposure and Health Effects,* W. R. Chappell, C. O. Abernathy, and R. L. Calderon, eds., Elsevier Science, New York, pp. 9–17, http://webserver.cr.usgs.gov/trace/pubs /segh1998/#AsInGW.

Welsh, A. H., S. A. Watkins, D. R. Helsel, and M. F. Focazio (2000) "Arsenic in Groundwater Resources of the United States," U.S. Geological Survey Fact Sheet 063–00, Reston, VA.

Wiesner, M. R., G. V. Lowry, P. Alvarez, D. Dionysiou, and P. Biswas (2006) "Assessing the Risks of Manufactured Nanomaterials," *Environ. Sci. Technol. A-Pages,* 40(14): 4336–37.

10

Water Treatment

 **Case Study**

The Flint Water Crisis

Flint, Michigan, like many cities in the rustbelt of the United States, has been in decline for quite some time. In its peak in 1960, its population was close to 200,000. It was a company town, with General Motors (GM) employing a large percentage of the population. In the years since, Flint has experienced numerous manufacturing plant closures, depopulation, deindustrialization, and urban decay. Unemployment and poverty skyrocketed, and by 2002, the city of Flint had accrued $30 million in debt. Flint's problems continued, and then in 2011, shortly after Governor Rick Snyder took office, the Republican controlled legislature passed Public Act 4, commonly known as the Emergency Manager Law. As a result of this legislation, in late 2011, the State of Michigan took control of the city's finances and nullified the authority of the mayor and city council (Masten et al., 2016).

The City of Flint began purchasing water from the Detroit Water and Sewerage Department (DWSD) in 1967 when the Flint River could no longer meet the demand of the growing industrial city. The treatment plant, which had been completed in 1952, was shuttered, although it was maintained as an emergency water supply and operated three to four times per year for a few days each time, although complete treatment was not accomplished during these trial runs. However, as per the contract with DWSD, the treated water was discharged back into the Flint River rather than into the distribution system. When the initial 30-year contract expired, the City of Flint continued to purchase water from DWSD on shorter contracts until the last contract was set to expire in 2014.

With the emergency manager in place, the state sought to reduce the city's debt, in part by not renewing the contract with DWSD, and instead provided water from the Flint River via the 1952-era plant. While the plant had undergone some upgrades since being shuttered in 1967, engineering experts estimated that some $50 million of improvements were necessary, and the work would take 52 to 60 months to get the plant ready for use. However, the city did not have these funds, and the emergency manager did not have the time before the contract was set to expire. And so, in April 2014, despite recommendations made and warnings provided by a Michigan Department of Environmental Quality (MDEQ) professional engineer and by the plant operator, the switch was made from purchasing treated water from Lake Huron to treating and distributing treated water from Flint River.

Within three weeks of the switch, Region 5 of the U.S. Environmental Protection Agency (USEPA) received complaints that the treated water was causing rashes. By June of 2014, Flint residents complained about the color (red, blue, yellow) and odor (sewer-like) of the water. On August 14, 2014, the treated water tested positive for the bacteria *Escherichia coli*, and boil-water advisories were issued two days later. Two more boil-water advisories would be issued within the next 21 days. That summer there were 29 cases of Legionnaire's disease, although the community would not learn of this for at least another two years. And then in October of 2014, the GM engine plant announced that it would be switching back to water supplied by DWSD because the treated Flint River water was corroding its engine parts. In December, the MDEQ informed the city that they were in violation of the Disinfection/Disinfection Byproducts (D/DBP) Rule as a result of excessively high total trihalomethane (TTHM) concentrations.

Nevertheless, the city and state insisted that the water was safe to drink, although in January 2015, the state began supplying bottled waters to its employees located in the Flint offices. In February, 2015, LeeAnn Walters, mother of twins and Flint resident, was concerned about the water quality and the health of her children and had the city test her water for lead. The results revealed that the concentration of lead in the two samples taken were 104 and 397 parts per billion (ppb), well above the maximum contaminant level goal (MCLG) of zero and treatment technology–based action level of 15 ppb. As the city dismissed the results as erroneous, she contacted Region 5 USEPA's lead expert Migual del Toral who called the results alarming. By late June, the city received notice of a second violation of the D/DBP Rule and then in late July, the city had the filters rebuilt to install granular activated carbon to control TTHM levels by removing organic matter prior to disinfection. The upgrade cost $1.6M and appeared to have effectively solved the TTHM problem. The residents remained concerned about water quality, and in August 2015, Virginia Tech professor Mark Edwards described the treated Flint River water as "very corrosive" and "causing lead contamination in homes." Of the 120 samples he obtained, he reported that 40.1% contained lead at concentrations in excess of 5 ppb. The next month, Michigan State University professor and pediatrician Mona Hanna-Attisha released a study showing that the number of Flint infants and children with elevated blood levels had increased considerably since the switch to treated Flint River water. The concerns of the public could no longer be dismissed, and on October 16, 2015, nearly 18 months after the initial switch, Flint returned to water from DWSD. As concerns mounted and the story became international news, Flint started adding additional phosphate to bring the concentration from 1 mg/L, as supplied by DWSD, to 2.5 mg/L. It is now more than 20 months since the water was switched back to DWSD, and trust between the residents and governmental officials remains broken. Lead levels in the distribution continue to vary greatly both spatially and temporally, with approximately 5 to 10% of the samples from Flint residences exceeding 5 ppb in any particular week.

What happened in Flint is much more complex than the simple "no corrosion inhibitor" story told in the press. Failures occurred in the treatment and monitoring of regulated parameters. The governmental officials in whom the community placed their trust to ensure that their water was safe ridiculed and mocked the residents and ignored or downplayed analytical results and concerns of the residents. The funds needed for the necessary upgrades were not available until situations became critical. Critical data was withheld from the residents and in some cases other government officials. Government and city officials have been indicted. Engineering firms have been sued. These cases remain in the Courts and may continue to do so for decades.

The situation in Flint is a stark reminder to every one of us: Canon 1 of the ASCE Code of Ethics states that the safety and well-being of the public is paramount. We hope that all reading this will remember this throughout your careers.

10-1   INTRODUCTION

Since ancient times, people have realized the importance of water—for navigation, drinking, recreation, and religious rituals. However, most of the emphasis has been on quantity rather than quality. Where consideration of quality does occur, these discussions have pertained mainly to the aesthetic properties of water. Does it taste good? Is it clear and uncolored? As early as 4000 BCE, the Greeks and southeastern Indians recognized the importance of water treatment for controlling taste and odor. Such methods as filtration of water through charcoal, irradiation by sunlight, and boiling are mentioned in ancient Greek and Sanskrit documents. As early as 1500 BCE, the ancient Egyptians used alum to clarify water (U.S. EPA, 2000). In 1685, the Italian physician Lucas Antonius Portius described a multiple sand filtration method, involving three pairs of sand filters. Water was first strained through a perforated plate. From there, it entered a settling tank. Once large particles were allowed to settle, the water flowed through the filters, which were operated in series. A mixture of sand and pebbles was used to produce good quality water. Around 1703, the French scientist La Hire offered a proposal to the French Academy of Sciences suggesting that every household should have a sand filter to treat rainwater (Jesperson, 2007).

By the early 1800s, Europeans had begun to use slow sand filtration for municipal water treatment (U.S. EPA, 2000). As discussed in Chapter 3, the connection between water quality and disease had not been made until 1854, when John Snow proposed the association between cholera and well water that had been contaminated with sewage. By the later 1800s, western scientists had made the connection between certain diseases and specific bacteria.

With Koch's discovery of disease transmission (see Section 3–11), during the late 19th and early 20th centuries, scientists and engineers focused their concern regarding water quality on turbidity and the presence of microorganisms in public water supplies. As a result, the design of most drinking-water treatment systems used slow sand filtration to reduce turbidity and to remove the microorganisms that caused typhoid, dysentery, and cholera.

Filtration is fairly effective at reducing turbidity and removing bacteria, but the disinfection of public water supplies was the predominant reason for a reduction in the number of waterborne disease outbreaks. In 1893, ozone was first used as a disinfectant in Oudshoom, Holland. In 1897, chlorine was used to disinfect water mains after a typhoid outbreak in England (Cooperative Research Centre for Water Quality and Treatment, 2007). In 1908, chlorine was first used in the United States to disinfect drinking water in Jersey City, New Jersey (U.S. EPA, 2000). As shown in Figure 10–1, the treatment of drinking water in Philadelphia resulted in a significant decrease in the number of deaths due to typhoid. Until 1902, the City of Philadelphia did not filter its water. In 1912, Philadelphia began to chlorinate its water. As a result of disinfection, it is estimated that, in the United States, the rate of deaths due to typhoid and paratyphoid fever decreased from 31.3 (in 1900) to 7.6 (in 1920) per 100,000 people. By 1953, as more public water supplies began to disinfect their water, the rate dropped to less than 0.05 per 100,000 people, and has remained steady since then (Peterson and Calderon, 2003).

FIGURE 10–1

Deaths due to typhoid in Philadelphia, Pennsylvania. (*Source:* City of Philadelphia water dept. May 22, 2007.)

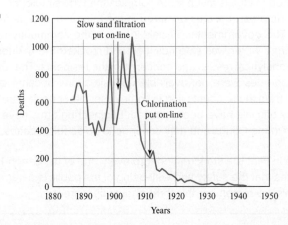

Today, public drinking water systems provide drinking water to about 95% of U.S. residents. In the United States, more than 153,000 water systems are classified as public. Of these, approximately 53,000 community water systems serve 317 million people. In the United States, **public systems** are defined as those systems serving at least 25 persons per day for greater than 60 days out of the year. **Community water systems** supply water to the same population year-round and include, for example, cities, townships, subdivisions, and trailer parks. **Nontransient noncommunity systems** are those that regularly supply to at least 25 of the same people at least 6 months per year but not year-round. These include schools, factories, office buildings, and hospitals that have their own water supply. **Transient noncommunity systems** provide water in locations, such as restaurants, motels, campgrounds, and service stations, where people do not remain for long periods of time. Approximately 2% of the U.S. population is served by nontransient, noncommunity water supplies, while approximately 4.2% of the population is served by transient, noncommunity water systems (U.S. EPA, 2009).

In the United States and in most developed countries, residents can be confident that their health will not be jeopardized because of poor quality water. In the period 1971 through 1996, fewer than 55 waterborne disease outbreaks occurred in the United States each year (Levy et al., 1998). Unfortunately, in many developing nations, clean water is the exception rather than the rule. As former U.S. EPA administrator Russel F. Train remarked,

> Something like 40 percent of the human race does not have adequate access to safe water. Waterborne diseases are estimated to kill more than 25,000 people daily. Schistosomiasis and filariasis, the world's largest causes of blindness, affect—according to one estimate— some 450 million people in more than 70 nations. There are, economist Barbara Ward has said, cities in the developing world where 60 percent of the children born die of infantile gastritis before the age of five. (Train, 1976)

Sadly, the last 30 years has not been kind to the people of many developing countries. While many developed countries and organizations from these countries have committed their technological and financial resources to creating access to safe and adequate drinking water, more than a billion people—almost one-fifth of the world's population—still lack access to safe drinking water. We hope that you and future generations can make a difference and help achieve safe drinking water for all.

Water Quality

As consumers, we expect our water to be both esthetically pleasing and safe to drink. Water that does not impart a taste or odor and is, therefore, pleasant to drink, is called **palatable.** Water that is free of chemicals, microorganisms, and other contaminants, and is, therefore, safe to drink, is called **potable.** We expect our water to be both palatable and potable.

All naturally occurring water supplies are recharged by precipitation. Although generally considered pure, rainwater may, and often does, contain impurities. These impurities, however, are much lower in concentration than that found in water once it reaches the earth's surface. Surface water presents many opportunities for the introduction of mineral and organic substances, microorganisms, and other forms of pollution (contamination).* When water runs over or through the ground surface, it may pick up particles of soil. This is noticeable in the water as cloudiness, or turbidity.† It also picks up particles of organic matter and bacteria. As surface water seeps downward into the soil and through the underlying material to the water

---

*Pollution as used in this text means the presence in water of any foreign substances (organic, inorganic, radiological, or biological) that tend to lower its quality to such a point that it constitutes a health hazard or impairs the usefulness of the water.

†Turbidity, a measure of the decrease in light penetration, is a relative standard. A sample is compared against various standards. The different standards employed are similar, although not exactly the same. For our purposes we will simply refer to the measure of turbidity as a turbidity unit (TU). For a given particle size, the higher the turbidity, the higher the concentration of colloidal particles.

table, most of the suspended particles are filtered out. This natural filtration may be partially effective in removing bacteria and other particulate materials. However, the chemical characteristics of the water may change and vary widely when it comes in contact with mineral deposits. As surface water seeps down to the water table, it dissolves some of the minerals contained in the soil and rocks. Groundwater, therefore, often contains more dissolved minerals than surface water.

The following four categories are used to describe drinking-water quality:

1. Physical: Physical characteristics relate to the quality of water for domestic use and are usually associated with the appearance of water, its color or turbidity, temperature, and, in particular, taste and odor.
2. Chemical: The chemical characterization of drinking water includes the identification of its components and their concentrations.
3. Microbiological: Microbiological agents are important to public health and may also be significant in modifying the physical and chemical characteristics of water.
4. Radiological: Radiological factors must be considered in areas where the water may have come in contact with radioactive substances. The radioactivity of the water is of public health concern.

Consequently, in the development of a water supply system, it is necessary to examine carefully all the factors that might adversely affect the intended use of a water supply source.

Physical Characteristics

As mentioned earlier, water suppliers monitor such physical characteristics as turbidity, color, taste and odor, and temperature. Turbidity is caused by the presence of suspended material such as clay, silt, finely divided organic material, plankton, and other particulate material in water. Although turbidity may not adversely affect health, these particles may harbor microbiological contaminants that are harmful to human health or that decrease the effectiveness of disinfectants.

Dissolved organic material from decaying vegetation and certain inorganic matter may cause color in water. Excessive blooms of algae or the growth of aquatic microorganisms may also impart color. Color can also be caused by inorganic metals such as iron or manganese, or by humic substances, which are formed from the degradation of leaves and other organic debris. Although color itself is not usually objectionable from the standpoint of health, its presence is aesthetically displeasing and suggests that the water needs appropriate treatment.

Foreign matter such as organic compounds, inorganic salts, or dissolved gases can cause taste and odor in water. Certain types of algae, especially the blue-green algae, can also impart foul tastes and odors.

The most desirable drinking waters are consistently cool and do not have temperature fluctuations of more than a few degrees. Groundwater and surface water from mountainous areas generally meet these criteria. As a general rule, water having a temperature between 10–15°C is most palatable.

Chemical Characteristics

Water treatment plants in the United States, Canada, and most of the European Union monitor for a variety of inorganic and organic constituents, including chloride, fluorides, sodium, sulfate, nitrates, and more than 120 organic chemicals. Some of these chemicals are monitored for health reasons; others for more esthetic purposes. For example, iron and manganese rarely pose a health threat, whereas, the presence of nitrate can cause methemoglobinemia (blue baby syndrome) in infants. Nitrates most often originate from the application of agricultural fertilizers, whereas iron and manganese are naturally occurring. However, all three chemical species are viewed as contaminants because of their effects on water quality.

Microbiological Characteristics

Water for drinking and cooking purposes must be made free from disease-producing organisms (**pathogens**), such as viruses, bacteria, protozoa, and helminths (worms). Within the United States, the most common causal agent of waterborne diseases is the protozoal parasite *Giardia lamblia* (Centers for Disease Control, 1990), followed by the bacteria *Legionella*. Some organisms that cause disease in humans originate with the fecal discharges of infected individuals. Others are contained in the fecal discharge of animals.

Unfortunately, the specific disease-producing organisms present in water are not easily identified. The techniques for comprehensive bacteriological examination are complex and time-consuming. As such, it has been necessary to develop tests that indicate the relative degree of contamination in terms of an easily defined quantity.

The most widely used bacteriological assay estimates the number of microorganisms of the coliform group. This grouping includes two genera: *Escherichia coli* and *Aerobacter aerogenes*. The name of the group is derived from the word colon. *E. coli* are gram-negative bacteria that commonly inhabit the intestinal tracts of mammals. *Aerobacter aerogenes* are commonly found in the soil, on leaves, and on grain; on occasion they cause urinary tract infections. The test for these microorganisms, called the **total coliform test,** was selected for the following reasons:

1. As mentioned above, *E. coli* usually inhabits the intestinal tracts of humans and other mammals. Thus, the presence of coliforms is an indication of fecal contamination of the water.
2. Even in acutely ill individuals, the number of *E. coli* excreted in the feces outnumbers the disease-producing organisms by several orders of magnitude. The large numbers of *E. coli* make them easier to culture than pathogenic organisms.
3. The coliform group of organisms survives in natural waters for relatively long periods, but does not reproduce effectively in this environment. Thus, the presence of coliforms in water implies fecal contamination rather than growth of the organism because of favorable environmental conditions. These organisms also survive better in water than most of the bacterial pathogens. This means that the absence of coliforms is a reasonably safe indication that pathogens are not present.
4. The coliform group of organisms is relatively easy to culture. Thus, laboratory technicians can perform the test without expensive equipment.

Current research indicates that testing for *E. coli* specifically may be warranted. Some agencies consider testing for *E. coli* rather than total coliforms as a better indicator of biological contamination. One strain of *E. coli* (0157) produces a potent toxin (vero toxin) that causes cell damage, leading to the development of bloody diarrhea. The potentially fatal toxin can also cause liver and kidney damage. As such, the development of rapid screening techniques specific for this strain of *E. coli* is the focus of much research.

Although molecular tests for pathogens are not commercially available for use in the drinking-water industry at present, much research is being conducted to allow for their use in the future. These techniques use DNA sequencing of genes to identify specific species of organisms. With this method, hundreds of thousands of complementary DNA (cDNA*) strands of known sequence are arrayed on a standard glass microscope slide. RNA is extracted from bacteria that have been isolated from the sample water and then are labeled with a fluorescent dye and bonded to the DNA on the slide. The fluorescent signal of the "spot" on the array is used to help identify the organism.

Radiological Characteristics

Radionuclides may enter water from naturally occurring or anthropogenic sources. They may be found in rainfall, runoff, and water-bearing rocks and soils. Some locations, such as the

---

*Synthetic DNA reverse transcribed from a specific RNA through the action of the enzyme reverse transcriptase.

highlands of western New Jersey, the Reading Prong in Pennsylvania, and the Hudson Highlands in New York, have naturally high concentrations of radionuclides, specifically uranium, in the groundwater. Nuclear testing, the mining and processing of radioactive materials, the operation of power plants, and the use of radionuclides in medicine and scientific studies necessitate the monitoring and analysis of drinking water for radionuclides.

Radioactivity is a health risk because the energy emitted by radioactive materials can damage or kill cells. Exposure to radioactive material can result in acute effects in the form of kidney injury. Chronic effects from long-term low levels of exposure to uranium can produce kidney damage. Uranium interferes with the reabsorption of proteins, although it appears that this effect is reversible. The U.S. EPA has classified all radionuclides as known human carcinogens, and drinking water regulations are based on this premise.

U.S. Water Quality Standards

Prior to 1974, there was no consistent set of drinking-water regulations throughout the United States. In that year Congress enacted the Safe Drinking Water Act (SDWA), requiring the U.S. EPA to set uniform nationwide drinking-water standards. State public health departments are responsible for ensuring that U.S. regulations or more stringent state regulations (if they exist) are met.

With the promulgation of the act, EPA has issued drinking-water standards, or **maximum contaminant levels (MCLs).** These standards limit the amount of each substance that can be present in a treated water. Presently there are more than 100 MCLs.* The limits are determined by assessing the cancer and noncancer risks from exposure to a chemical in drinking water. In setting MCLs for carcinogenic compounds, EPA sets the level so as to limit an individual's added risk of cancer from exposure to that contaminant to between 1:10,000 and 1,000,000 over a lifetime. For microbiological contaminants such as *Giardia lamblia* and chemical contaminants that pose an immediate threat, the standards are set so as to protect human health.

With the act, EPA has promulgated two types of drinking-water standards. The first, described earlier, are the primary standards, which are designed to protect human health and for which MCLs exist. The second type, or secondary drinking-water standards, is based on providing esthetically pleasing water. The secondary MCLs (SMCLs) are recommendations and, as such, are not mandatory.

In 1996, President Clinton signed the Safe Drinking Water Amendments into law. These amendments require that water utilities provide more complete information pertaining to water quality to their customers. Consumer confidence reports documenting the source of each person's drinking water, monitoring results, and providing information concerning health concerns associated with SDWA violations are required. The amendments also encourage more public involvement and emphasize source water protection by focusing more on identifying potential causes of contamination of drinking water source waters before they occur. Federal law also now requires certification of plant operators.

One of the more recently promulgated U.S. regulations is the Ground Water Rule, which was published by the U.S. EPA on November 8, 2006. The purpose of the rule is to protect public water supplies from becoming contaminated with microbial pathogens, especially those due to fecal wastes. The rule also applies to any system that mixes surface and groundwater if the groundwater is added directly to the distribution system and provided to consumers without treatment. A list of the most important U.S. EPA rules for drinking water is provided in Table 10–1.

For some contaminants, EPA has not defined an MCL but instead specifies a specific treatment technique. For those contaminants, the notation TT (treatment technique) is used in place of an MCL. The treatment techniques are specific processes used to treat the water. Some examples include coagulation and filtration, lime softening, and ion exchange. These processes will be discussed in the following sections. EPA now recommends that water be essentially free

---

*A full list of the primary and secondary drinking-water contaminants and their associated MCLs or SMCLs can be found on the U.S. EPA website: http:/www.epa.gov/safewater/contaminants/index.html.

TABLE 10–1   Summary of the Major U.S. EPA Rules for Drinking Water

Rule	Purpose	Target Contaminants
Revised Total Coliform Rule (RTCR) (78 FR 10269, February 13, 2013)	To reduce the number of fecal pathogens to minimal levels by controlling coliform bacteria	Total and fecal coliforms *Escherichia coli (E. coli)*
Surface Water Treatment Rule (SWTR) (40 CFR 141.70-141.75)	To control microbial contaminants	Viruses *Giardia lamblia* Turbidity in filter effluent
Long-term I Enhanced SWTR (67 FR 1812, January 14, 2002, Vol. 67, No. 9)	To control *Cryptosporidium*	*Cryptosporidium sp.*
Long-term 2 Enhanced SWTR (71 FR 654, January 5, 2006, Vol. 71, No. 3)	To protect public health in systems with elevated risks due to high levels of *Cryptosporidium*	*Cryptosporidium sp.* *E. coli* Turbidity in source water
Stage 1 Disinfection and Disinfection By-products Rule (D/DBP1) (63 FR 69390-69476, December 16, 1998, Vol. 63, No. 241)	To reduce exposure to disinfection by-products	Total trihalomethanes Five haloacetic acids Bromate, chlorite, chlorine, chloramines, and chlorine dioxide
Stage 2 Disinfection and Disinfection By-products Rule (D/DBP2) (71 FR388, January 4, 2006, Vol. 71, No. 2)	To further reduce exposure to disinfection by-products	Total trihalomethanes Five haloacetic acids
Lead and Copper Rule (LCR) (56 FR 26460-26564, June 7, 1991, latest revision: Oct. 2007)	To minimize public health risk from lead and copper exposure from drinking water	Water corrosivity Lead Copper
Radionuclides Rule (66 FR 76708, December 7, 2000, Vol. 65, No. 236)	To reduce exposure to radionuclides in drinking water	Combined radium-226 and radium-228 Gross alpha particles Beta particles Photo activity Uranium
Groundwater Rule (GWR) (40 CFR 9, 141, 142, November 8, 2006, Vol. 71, No. 216)	To protect against microbial pathogens in public water systems that use groundwater sources	*E. coli* Viruses
Information Collection Rule (ICR) (40 CFR 141.142 and 141.143, May 14, 1996)	To collect data on the presence of pathogens in drinking water sources (lakes, reservoirs, etc.) the amount of disinfectant and presence of disinfection by-products in treated drinking water, and the effectiveness of certain treatment technologies	Total coliform Fecal coliform *E. coli* Disinfectant concentrations Total trihalomethanes Five haloacetic acids Bromate, chlorite, chlorine, chloramines, and chlorine dioxide
Arsenic and Clarifications to Compliance and New Source Monitoring Rule (66 FR 6976, January 22, 2001)	To protect human health by reducing the allowable concentration of arsenic in drinking water and to clarify how compliance is demonstrated for many inorganic and organic contaminants in drinking water	Arsenic Inorganic contaminants (IOCs) Volatile organic contaminants (VOCs) Synthetic organic contaminants (SOCs)
Unregulated Contaminant Monitoring Rule (UCMR/4) (81 FR92666, Dec. 20, 2016)	To collect data on the presence of 30 contaminants in drinking water sources	Ten cyanotoxins, two metals, eight pesticides, one pesticide-manufacturing product, three brominated haloacetic acids, three alcohols, three other semivolatile chemicals, total organic carbon (TOC), bromide
Aircraft Drinking Water Rule (74 FR53590, Oct . 19, 2009)	To ensure safe and reliable drinking water for crews and passengers.	Coliforms

of some contaminants, such as lead. Other contaminants falling into this category include benzene, carbon tetrachloride, dioxin, trichloroethylene, and vinyl chloride. Standards pertaining to disinfection are discussed in detail in Section 10–6.

As required by the Safe Drinking Water Act, the U.S. EPA must identify and list unregulated contaminants, which may require a national drinking water regulation. The lists of unregulated contaminants (called the Contaminant Candidate List or CCL) are periodically published by the U.S. EPA. The U.S. EPA is then required to decide whether to regulate at least five or more contaminants on the list (called Regulatory Determinations). Regulatory determinations are based on (1) the toxicity of the contaminants and (2) the number of sites at which it is found or is likely to be detected in public waters at a level that is likely to cause public health concern. In order to gather information about unregulated contaminants, the U.S. EPA published the Unregulated Contaminant Monitoring Rule (UCMR). The fourth UCMR (UCMR4) was published in the Federal Register on December 20, 2016. It will require monitoring for 30 contaminants during 2013–2015. This monitoring will provide a basis for future regulatory actions to protect public health.

Other Standards.   Although EPA has set maximum levels of possible contaminants in drinking waters, individual states may require compliance with more stringent regulations. In addition, other agencies such as the American Water Works Association or the World Health Organization have their own set of recommendations or goals (World Health Organization, 2006). In Europe, the European Union sets standards (European Union, 1998).

The design of water treatment systems typically follows criteria set forth in the "Ten State Standards." In 1950, the Great Lakes-Upper Mississippi River Board of State and Provincial Public Health and Environmental Managers created a Water Supply Committee consisting of one associate from each state represented on the Board. The report was first published in 1953, with subsequent revisions, the latest of which was issued in 2012. These reports include policy statements regarding the design and operation of water treatment systems and the protection of water supplies. They also include both interim standards for new treatment processes and recommended standards that are intended to serve as a guide in the design of such facilities.

Water Classification and Treatment Systems

Water Classification by Source.   Although salt water is occasionally used as a drinking-water supply (e.g., in the desert regions along the Persian Gulf and in water-scarce regions, including Southern Florida and California), freshwater is the preferred source. As such, the remaining discussion will focus only on freshwater supplies. Potable water is most conveniently classified as to its source, that is, groundwater or surface water. Groundwater is pumped from wells drilled into aquifers. As mentioned in Chapters 7 and 9, the quantity and quality of water available depends on the type of geological formation forming the aquifer and the properties of the contaminant itself. Drinking-water wells can be shallow (less than 15 m) or deep (greater than 15 m). Although, in general, the deeper the well, the greater the level of protection from contamination, deep wells only provide protection when the wells are properly designed and operated so that surface contamination is prevented. Surface water includes rivers, lakes, and reservoirs. In general, surface and groundwater can be characterized as shown in Table 10–2.

Treatment Systems.   Water treatment facilities in developed countries are typically (1) coagulation plants, (2) softening plants, and (3) limited treatment plants. Coagulation plants (Figure 10–2) are typically used to treat surface water and remove color, turbidity, taste and odors, and bacteria. Most plants use coagulation, flocculation, followed by sedimentation and then filtration and disinfection. However, higher quality (low turbidity and color) surface waters can be treated using direct filtration, in which case, sedimentation is omitted, as shown in Figure 10–3. Softening plants (Figure 10–4) are typically used to treat groundwater that has a high hardness (predominantly calcium and magnesium ions). It often involves chemical addition and reaction, sedimentation, **recarbonation** to reduce the pH, followed by filtration and disinfection. Limited

TABLE 10–2   General Characteristics of Groundwater and Surface Water

Ground	Surface
Constant composition	Varying composition
High mineral content	Low mineral content
Low turbidity	High turbidity
Low or no color	Color
May be bacteriologically safe	Microorganisms present
No dissolved oxygen	Dissolved oxygen
High hardness	Low hardness
H_2S, Fe, Mn	Tastes and odors
Possible chemical toxicity	Possible chemical toxicity

FIGURE 10–2

Flow diagram of a conventional surface-water treatment plant (coagulation plant).

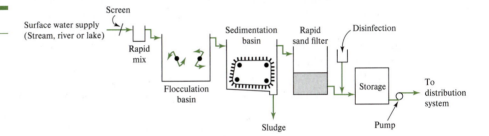

FIGURE 10–3

Flow diagram of a direct filtration plant.

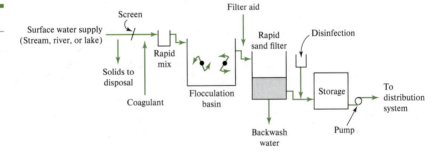

FIGURE 10–4

Flow diagram of a water-softening plant.

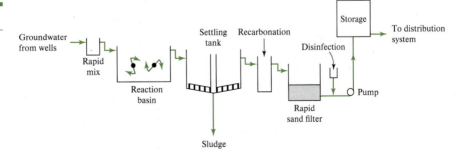

treatment plants generally have a high-quality water source, typically groundwater that is not under the influence of surface water, that is, it is not likely to be "contaminated" with surface water infiltration and, therefore, is little risk of the presence of pathogens. As a result, limited treatment may be all that is required. For example, the Michigan State University treatment plant uses groundwater and treatment consists only of pumping, disinfection, and corrosion control.

FIGURE 10–5

Flow diagram for a membrane filtration plant using microfiltration.

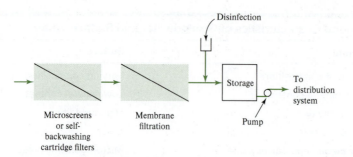

As shown in Figure 10–2, the raw surface water is pumped from a stream, river, lake, or reservoir using low-lift pumps into the coagulation plant. The water is first screened to remove large material including twigs, plant material, and fish. During rapid mixing, chemicals called **coagulants** are added and rapidly dispersed through the water. In the **flocculation basin**, the chemical reacts with the colloidal particles in the water to form larger particles. Once the particles are a sufficient size to precipitate, they are usually removed by gravity in **sedimentation basins**. Following sedimentation, the clarified water is distributed to rapid sand filters to remove residual turbidity. **Disinfection** is then used to reduce the number of pathogenic organisms to levels that will not cause disease. It is usually achieved by the addition of chemicals (chlorine, chloramines, or ozone) or the application of UV radiation. **Fluoridation**, the addition of fluoride to strengthen teeth enamel and reduce the number of cavities, especially in children, can also be a part of the treatment process, as can **corrosion control**, which prevents corrosion of the pipes in the distribution system.

The precipitated chemicals, along with particles, that are removed from the sedimentation basins are present in the sludge that accumulates at the bottom of the tank. This sludge, which contains approximately 94–98% water must be treated further and then disposed of properly. The backwash water from the filters also contains particulate matter and bacteria, along with precipitated chemicals. It, too, must be treated and disposed of.

Some waters may require additional treatment. For example, many groundwaters that contain excessive concentrations of hydrogen sulfide, iron, and manganese may require additional treatment. In such a case, pretreatment can be used to oxidize the reduced iron and manganese and remove the hydrogen sulfide. Tray or packed towers can be used for aeration to achieve hydrogen sulfide removal along with chemical oxidation of iron and manganese. If the waters contain high hardness, then chemical softening would likely follow aeration.

In the last decade, the use of membrane filtration has become more widespread in the United States. Figure 10–5 shows a microfiltration plant used to treat surface water. Screens are used to remove larger material before pumping, followed by microfiltration. Such a treatment plant has the potential to achieve very low turbidity and significant removal of *Giardia, Cryptosporidium,* and viruses. In some cases, coagulation/sedimentation is used prior to membrane filtration.

In the next two sections of this chapter, we discuss the chemistry of coagulation and softening, respectively. The subsequent sections describe the other processes used in water treatment.

10–2 RAPID MIXING, FLOCCULATION, AND COAGULATION

Coagulation is used to remove turbidity, color, and bacteria from drinking waters. The goal of coagulation is to change the surface charge on the particles so they can stick together to form larger particles that will settle by gravity. Although chemists tend to use the term *coagulation* to describe this change in surface charge and subsequent adhesion process, environmental engineers often use the term to describe the entire process by which larger particles and dissolved ions are removed by gravity settling or precipitation. In this chapter we will use the term *coagulation* to describe the removal of particulate matter and the term *softening* for the removal of dissolved hardness ions (predominantly calcium and magnesium).

Colloid Stability and Destabilization

In natural waters, colloids are stable because the surface of these particles are like-charged and they repel one another. In surface water most of the colloidal particles are derived from clays and have a net negative charge. The particles can be thought of as acting like magnets: When the same poles (e.g., north) are pointed at one another, the magnets repel. The goal of coagulation is to ensure that the particles come in contact with one another, "stick together," and form settleable particles called **floc**.

The Physics of Coagulation

There are four mechanisms that cause the destabilization of colloidal particles in natural water suspensions.

- Adsorption and charge neutralization.
- Compression of the electric double layer.
- Adsorption and interparticle bridging.
- Enmeshment in a precipitate.

A charged colloidal particle attracts sufficient counterions (ions of opposite charge) to balance the electrical charge on the particle. As shown in Figure 10–6 a negatively charged particle attracts a layer of cations that are strongly adsorbed to the surface. This layer is called the Stern layer. Beyond the Stern layer there is a layer of loosely bound cations. This layer is termed Gouy-Chapman or diffuse layer. The double layer (Stern and Gouy-Chapman layers) has a net positive charge compared to the bulk solution. There are short-range attractive forces (van der Waals forces) that allow colloidal particles in water to stick together if they can approach close enough to one other. However, as two charged particles approach there will be a repulsive force between them and the interaction of the ions in their double layers may prevent them from sticking together. One way to get the particles to stick together is to reduce the charge on the particle. In natural waters most particles are negatively charged so coagulation can be enhanced by adding cations to the water to induce charge neutralization. Trivalent cations adsorb strongly to negatively charged particles, so they are much more effective coagulants than are divalent or monovalent ions as shown in Figure 10–7a. The greater the concentration of the positive ion added, the greater the extent to which the surface charge is neutralized and the lower the turbidity.

The thickness of the double layer decreases as the ionic strength of the solution increases, so in solutions with a high ionic strength, particles may coagulate as the thickness of the double

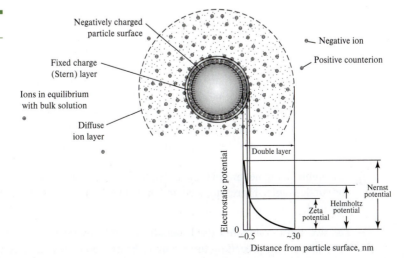

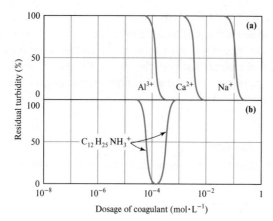

layer may be so small that the particles can approach close enough to each other allowing the van der Waals forces to cause them to stick together.

Hydrolyzed metal salts, prehydrolyzed metal salts, and cationic polymers have a positive charge. They destabilize particles through charge neutralization.

Polymers, such as dodecyl ammonium, shown in Figure 10–7b, absorb on particle surfaces at one or more sites along the polymer chain. The adsorption is the result of coulombic, charge-charge interactions, dipole interactions, hydrogen bonding, and/or van der Waal forces. Other sites on the polymer chain extend into solution and adsorb on the surfaces of other particles, thus creating a "bridge" between particles. These larger particles can be removed by subsequent sedimentation. As shown in Figure 10–7b, the addition of dodecyl ammonium ions, which also have a charge of +1, has a much greater effect on turbidity than does the addition of sodium ions. This occurs because the long chain of the C_{12} group when attached to a particle makes the particle hydrophobic, so it is more likely to interact with another particle that also has a hydrophobic region. If too much dodecyl ammonium is added, however, the turbidity increases. This is because the excess ions adsorbed to the surface in a net positive charge, causing the colloids to again repel. This process is called **restabilization**.

When coagulant doses exceed their solubility, the salts form insoluble precipitates. As the salts precipitate, they "entrap" the colloidal particles within the flocs. These types of precipitates are referred to as sweep floc. Coagulation by the formation of sweep floc is necessary when treating waters having a low turbidity.

Coagulants

As previously mentioned, a coagulant is a chemical that is added to the water to cause the particles to coagulate. A coagulant has three key properties:

1. **Trivalent cation.** As discussed previously, naturally occurring colloids are most commonly negatively charged, hence cations are required to achieve charge neutralization. As shown in Figure 10–7a, trivalent cations are much more effective than monovalent or divalent simple cations, such as sodium and calcium.
2. **Nontoxic.** Obviously, for the production of potable water, the coagulant must be nontoxic.
3. **Insoluble in the neutral pH range.** High concentrations of the coagulant in treated water are undesirable. Therefore, a coagulant is usually relatively insoluble at the pH values desired.

The two most commonly used metallic coagulants are aluminum (Al^{3+}) and ferric iron (Fe^{3+}). Both meet the preceding three requirements, and their reactions are outlined here.

Aluminum Sulfate.   Aluminum sulfate can be purchased as either dry or liquid alum $(Al_2(SO_4)_3 \cdot 14H_2O)$. Commercial alum has an average molecular weight of 594, with approximately 14 waters of hydration.

Alum reacts with alkalinity according to the following reaction:

$$Al_2(SO_4)_3 \cdot 14H_2O + 6HCO_3^- \rightleftharpoons 2Al(OH)_3(s) + 6CO_2 + 14H_2O + 3SO_4^{2-} \tag{10–1}$$

such that each mole of alum added uses six moles of alkalinity and produces six moles of carbon dioxide. The preceding reaction shifts the carbonate equilibrium and decreases the pH. However, as long as sufficient alkalinity is present and $CO_2(g)$ is allowed to evolve, the pH will not be drastically reduced. When sufficient alkalinity is not present to neutralize the sulfuric acid production, the pH may be lowered significantly due to the formation of sulfuric acid.

$$Al_2(SO_4)_3 \cdot 14H_2O \rightleftharpoons 2Al(OH)_3(s) + 3H_2SO_4 + 8H_2O \tag{10–2}$$

If pH control is a problem, lime or sodium carbonate may be added to neutralize the acid, thereby stabilizing the pH.

Two important factors in coagulant addition are pH and dose. The optimum dose and pH must be determined from laboratory jar tests. The optimal pH range for alum is approximately 5.5–6.5, with adequate coagulation possible at pH between 5 and 8 under some conditions.

EXAMPLE 10–1   A water treatment plant with an average flow of $Q = 0.044$ m$^3 \cdot$ s^{-1} treats its water with alum $(Al_2(SO_4)_3 \cdot 14H_2O)$ at a dose of 25 mg $\cdot$ L^{-1}. Alum coagulation is used to remove particulate matter, reduce the concentration of organic matter, and reduce the alkalinity of the water according to Equation 10–1. If the organic matter concentration is reduced from 8 mg $\cdot$ L^{-1} to 3 mg $\cdot$ L^{-1} determine the total mass of alkalinity consumed and the total mass of dry solids removed per day.

Solution   First determine the total amount of alkalinity consumed. We do this by noting that, according to Equation 10–1, for each mole of alum added, six moles of alkalinity (bicarbonate form) are removed. Now convert the alum dose into molar units using its molecular weight of 594.35 g $\cdot$ mol^{-1}

$$\frac{(25 \text{ mg} \cdot \text{L}^{-1})(10^{-3} \text{ g} \cdot \text{mg}^{-1})}{594.35 \text{ g} \cdot \text{mol}^{-1}} = 4.206 \times 10^{-5} \text{ mol} \cdot \text{L}^{-1} \text{ of alum}$$

Thus, with 4.206×10^{-5} mol $\cdot$ L^{-1} of alum added, the amount of HCO_3^- removed is simply six times that, or

$$(6)(4.206 \times 10^{-5}) = 0.000252 \text{ mol} \cdot \text{L}^{-1} \text{ of } HCO_3^- \text{ or } 2.52 \times 10^{-4} \text{ eq} \cdot \text{L}^{-1} \text{ of alkalinity}$$

To find the total alkalinity removed per day, simply multiply the preceding number by the average flow of the plant

$$(0.000252 \text{ eq} \cdot \text{L}^{-1})(0.044 \text{ m}^3 \cdot \text{s}^{-1})(1000 \text{ L} \cdot \text{m}^{-3})(86,400 \text{ s} \cdot \text{day}^{-1}) = 959.4 \text{ eq} \cdot \text{day}^{-1}$$

Converting to a mass basis, using the equivalent weight of bicarbonate to be 61 g $\cdot$ mol^{-1}, yields

$$959.4 \text{ eq} \cdot \text{day}^{-1} \times 61 \text{ g} \cdot \text{eq}^{-1} = 58,526 \text{ g} \cdot \text{day}^{-1}, \text{ or } 58.5 \text{ kg} \cdot \text{day}^{-1}$$

For the second part of the problem we note that for every mole of alum added, two moles of solid precipitate. So, the amount of solid is

$$\left(\frac{2 \text{ mol } Al(OH)_3}{\text{mol alum}}\right)(4.206 \times 10^{-5} \text{ mol} \cdot \text{L}^{-1} \text{ of alum}) = 8.41 \times 10^{-5} \text{ mol} \cdot \text{L}^{-1} \text{ of } Al(OH)_3$$

Again putting this number on a per day basis yields,

$$(8.41 \times 10^{-5} \text{ mol} \cdot \text{L}^{-1})(0.044 \text{ m}^3 \cdot \text{s}^{-1})(1000 \text{ L} \cdot \text{m}^{-3})(86,400 \text{ s} \cdot \text{day}^{-1}) = 319.8 \text{ mol} \cdot \text{day}^{-1}$$

Converting to mass basis, using the molecular weight of aluminum hydroxide ($78 \text{ g} \cdot \text{mol}^{-1}$), we obtain

$$319.8 \text{ mol} \cdot \text{day}^{-1} \times 132 \text{ g} \cdot \text{mol}^{-1} = 42,214 \text{ g} \cdot \text{day}^{-1}, \text{ or } 42.2 \text{ kg} \cdot \text{day}^{-1}$$

The total solids removed also include the settled organic material. We know, based on the influent and effluent levels or $8 \text{ mg} \cdot \text{L}^{-1}$ and $3 \text{ mg} \cdot \text{L}^{-1}$, respectively, that the total organic material removed is simply $5 \text{ mg} \cdot \text{L}^{-1}$. (Note that milligrams per liter of organic material is a very simple approximation and that the normal measurement of such difficult-to-measure substances is normally given as turbidity.) Simply multiply this number by the plant flow to determine the total organic matter settled per day.

$$5 \text{ mg} \cdot \text{L}^{-1} \times 0.044 \text{ m}^3 \cdot \text{s}^{-1} \times 1000 \text{ L} \cdot \text{m}^{-3} \times 86,400 \text{ s} \cdot \text{day}^{-1} = 19,008,000 \text{ mg} \cdot \text{day}^{-1}$$

or

$$19.0 \text{ kg} \cdot \text{day}^{-1}$$

So, by adding the mass fluxes of aluminum hydroxide and settled organic matter sludges, the total amount of dry solids removed per day is calculated.

$$42.2 \text{ kg} \cdot \text{day}^{-1} + 19.0 \text{ kg} \cdot \text{day}^{-1} = 61.2 \text{ kg} \cdot \text{day}^{-1}$$

Iron.   Ferric cations can be supplied by adding either ferric sulfate ($Fe_2(SO_4)_3 \cdot 7H_2O$) or ferric chloride ($FeCl_3 \cdot 7H_2O$). The properties of iron with respect to coagulation efficiency, dose, and pH curves are similar to those of alum. An example of the reaction of $FeCl_3$ in the presence of alkalinity is

$$FeCl_3 \cdot 7H_2O + 3HCO_3^- \rightleftharpoons Fe(OH)_3(s) + 3CO_2 + 3Cl^- + 7H_2O \tag{10-3}$$

and without alkalinity

$$FeCl_3 + 3H_2O \rightleftharpoons Fe(OH)_3(s) + 3H^+ + Cl^- \tag{10-4}$$

If the alkalinity is insufficient, the addition of ferric chloride results in the release of three moles of protons (H^+) for every mole of ferric chloride added. The release of protons lowers the pH. Ferric salts generally have a wider pH range for effective coagulation than aluminum, that is, pH ranges from 4 to 9.

Coagulant Aids.   The three basic types of coagulant aids are activated silica, clay, and polymers.

Activated silica is sodium silicate that has been activated with sulfuric acid, alum, carbon dioxide, or chlorine. When activated silica is added to water, it produces a stable sol (i.e., solid colloidal particles dispersed in a liquid) that has a negative surface charge. The activated silica can react with the positively charged metal hydroxide floc, resulting in a larger, denser floc that settles faster and enhances enmeshment. The addition of activated silica is especially useful for treating highly colored, low-turbidity waters because it increases the density of the floc. However, because activation of silica requires proper equipment and close operational control, many plants are hesitant to use it.

Clays can act much like activated silica in that they have a slight negative charge and can increase the density of the flocs, thus increasing the settling of the floc. Clays are also most useful for treating colored, low-turbidity waters.

The most effective polymers used in water treatment as coagulant aids are the anionic and nonionic polymers. Polymers are long-chain carbon compounds of high molecular weight that have many active sites. Dodecyl ammonium (see Figure 10–7b) is a cationic polymer. The active sites adhere to flocs, joining them together and producing a larger, tougher floc that

FIGURE 10–8

Rapid-mixing devices.

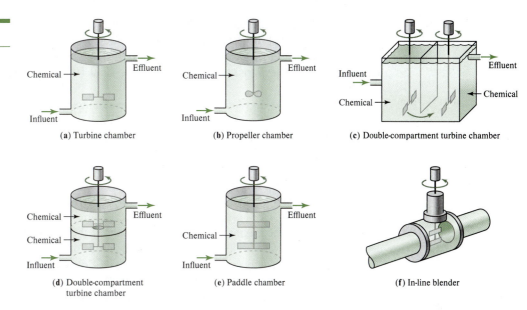

(a) Turbine chamber

(b) Propeller chamber

(c) Double-compartment turbine chamber

(d) Double-compartment turbine chamber

(e) Paddle chamber

(f) In-line blender

settles better. The type of polymer, dose, and point of addition must be determined for each water and requirements may change within a plant on a seasonal, or even daily, basis.

Mixing and Flocculation

Clearly, if the chemical reactions in coagulating and softening a water are going to take place, the chemical must be mixed with the water. In this section we will begin to look at the physical methods needed to produce chemical coagulation and softening.

FIGURE 10–9

Basic impeller styles.

Mixing, or **rapid mixing** as it is called, is the process whereby the chemicals are quickly and uniformly dispersed in the water. Ideally, the chemicals would be instantaneously dispersed throughout the water. During coagulation and softening, the chemical reactions that take place in rapid mixing form precipitates. Either aluminum hydroxide or iron hydroxide forms during coagulation, whereas calcium carbonate and magnesium hydroxide forms during softening. The precipitates created in these processes must be brought into contact with one another so they can form flocs. This contacting process is called **flocculation** and is accomplished by slow, gentle mixing.

Axial flow

Rapid Mix.   Rapid mixing is probably the most important physical operation affecting coagulant dose efficiency. The chemical reaction in coagulation is completed in less than 0.1 s; therefore, it is imperative that mixing be as instantaneous and complete as possible. Rapid mixing can be accomplished within a tank with a vertical shaft mixer or within a pipe using specialized mixing systems. Other methods, such as Parshall flumes, hydraulic jumps, baffled channels, or air mixing, may also be used. Typical tank or inline mixing configurations are shown in Figure 10–8.

Propeller

For dissolution of $CaO/Ca(OH)_2$ mixtures for softening, detention times on the order of 5 to 10 min may be required. Inline mixers are not used to blend softening reagents.

The volume of a rapid mix tank is calculated using the equation

$$\mathcal{V} = Qt_d \tag{10–5}$$

where $\mathcal{V}$ = volume of the reactor
  Q = flow rate
  t_d = detention time

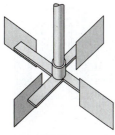

Turbine

The volume seldom exceeds 8 m^3 because of mixing equipment and geometry constraints. The mixing equipment consists of an electric motor, gear-type speed reducer, and either a turbine or axial-flow impeller as shown in Figure 10–9. The turbine impeller provides more turbulence and is preferred for rapid mixing.

Often rules-of-thumb (for example, see Reynolds, 1982) are used to design rapid mix tanks. It has been found that the design liquid depth should be 0.5–1.1 times the basin diameter or width. The impeller diameter should be between 0.30 and 0.50 times the tank diameter or width, and the vertical baffles should extend into the tank about 10% of the tank width or diameter.

Flocculation. Whereas rapid mix achieves successful mixing of the water and the coagulant, flocculation provides the conditions necessary for optimal particle growth so that sedimentation can occur successfully. The objective of flocculation is to allow the particles to collide, adhere to one another, and grow to a size that will settle readily. Mixing must be sufficient so that when collisions occur, they have enough energy for the particles to stick together. On the other hand, too much mixing can shear particles (they break apart), and the resulting particles will be too small for sedimentation to occur effectively. The major design parameter used in flocculation is the velocity gradient, G. This parameter describes the degree of mixing; the higher the G value, the more vigorous the mixing. A detailed discussion of the application of the G value in the design of water and wastewater plants is beyond the scope of this text.

Flocculation is usually accomplished with an axial-flow impeller (see Figure 10–9), a paddle flocculator (Figure 10–10), or a baffled chamber (Figure 10–11). Axial-flow impellers have been recommended over the other types of flocculators (Hudson, 1981) shown in Figure 10–9.

Upflow Solids Contact. Mixing, flocculation, and clarification may be conducted in a single tank such as that shown in Figure 10–12. The influent raw water and chemicals are mixed in the center cone-like structure. The solids flow down under the cone (sometimes called a "skirt").

FIGURE 10–10

Paddle flocculator.

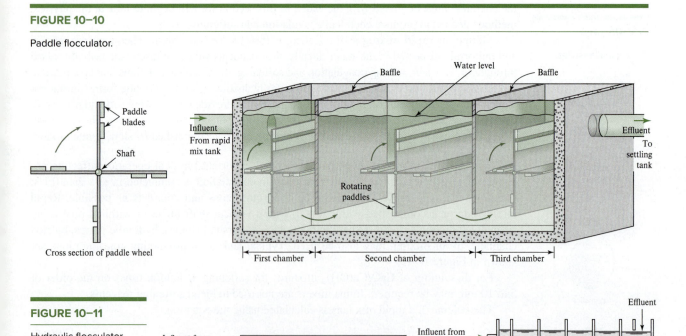

FIGURE 10–11

Hydraulic flocculator.

FIGURE 10–12

Typical upflow solids contact unit. *Source: Water Treatment Plant Design. American Water Works Association, 1969.*

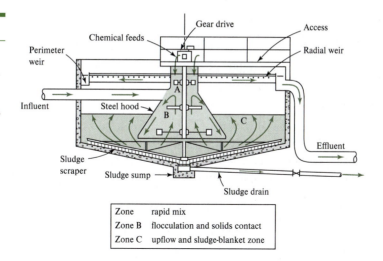

Zone	rapid mix
Zone B	flocculation and solids contact
Zone C	upflow and sludge-blanket zone

As the water flows upward, the solids settle to form a sludge blanket. This design is called an upflow solids-contact basin. The main advantage of this unit is its reduced size. The units are best suited to treating feed water that has a relatively constant quality. It is often favored for softening because the characteristics of groundwater are relatively constant, and the sludge blanket provides further opportunity to drive the precipitation reactions to completion.

EXAMPLE 10–2

The abandoned Catequil Street Water Treatment Plant is to be turned into a research facility treating $0.044 \ \mathrm{m}^3 \cdot \mathrm{s}^{-1}$. A low-turbidity iron coagulation plant has been proposed with the following design parameters:

Rapid mix $t_o = 10 \ \mathrm{s}$ Water temperature $= 18°C$

Size the rapid-mix basin.

Solution

The volume of the rapid-mix tank by Equation 10–5 is

$$\forall = (Q)(t_o) = (0.044 \ \mathrm{m}^3 \cdot \mathrm{s}^{-1})(10 \ \mathrm{s}) = 0.44 \ \mathrm{m}^3$$

Because this volume is less than our guideline of 8 m^3 as the maximum volume for a rapid-mix tank, we can use a tank of 0.44 m^3. To ensure water production even in the event of a major equipment failure, it is necessary to build some redundancy into the water treatment system; therefore, we need two tanks. Because the volume of this tank is quite small, use two 0.44 m^3 tanks and alternate their usage.

Using a depth-to-diameter (or width) ratio of 0.75, we obtain

Depth $= 0.75 \times$ diameter

and remembering that the volume of a cylinder equals

Volume $= \dfrac{\pi d^2}{4} \times$ depth

where $d =$ diameter

we get,

Volume $= \left(\dfrac{\pi d^2}{4} \right)(0.75d) = 0.44 \ \mathrm{m}^3$

Solving for $d = 0.907 \ \mathrm{m}$.

Therefore, the surface area of the tank (to keep the volume near 0.44 m^3) would be

$$\text{Surface area} = \frac{\pi d^2}{4} = \frac{\pi (0.907 \text{ m})^2}{4} = 0.646 \text{ m}^2, \text{ or } 0.65 \text{ m}^2$$

and the depth would be

$$\text{Depth} = \frac{\text{volume}}{\text{surface area}} = \frac{0.44 \text{ m}^3}{0.65 \text{ m}^2} = 0.68 \text{ m}$$

This set of calculations presents a practical problem; we are unlikely to find a tank with these dimensions. Let's pick a tank that is 1 m in diameter and 0.75 m in depth. Now, we must check that this choice does not adversely affect the quality of the design, that is, we must ensure that the tank is large enough to hold the specified volume of 0.44 m^3. The designed tank volume is

$$\text{Volume} = \frac{\pi d^2}{4} \times \text{depth} = \frac{\pi (1)^2}{4} \times 0.75 = 0.59 \text{ m}^3$$

This volume is fine. The detention time is

$$\frac{0.59 \text{ m}^3}{0.044 \text{ m}^3 \cdot \text{s}^{-1}} = 13.4 \text{ min}$$

Thus, the detention time is within suggested guidelines.

10–3 SOFTENING

Hardness

The term *hardness* is used to characterize a water that does not lather well; causes a scum in the bathtub; and leaves hard, white, crusty deposits (scale) on coffee pots, tea kettles, and hot water heaters. The failure to lather well and the formation of scum on bathtubs is the result of the reactions of calcium and magnesium with the soap. For example, the scum on bathtub forms according to the following reaction:

$$Ca^{2+} + 2(Soap)^- \rightleftharpoons Ca(Soap)_2(s) \tag{10–6}$$

As a result of this complexation reaction, soap cannot interact with the dirt on clothing, and the calcium–soap complex itself forms undesirable precipitates. Additionally, hardness minerals result in valves sticking due to the formation of calcium carbonate crystals in the valve and scaling on pipes.

Hardness is defined as the sum of all polyvalent cations (in consistent units). Total hardness (TH) is defined as

$$TH = (Ca^{2+}) + (Mg^{2+}) + (Fe^{3+}) + (Fe^{2+}) + (Ba^{2+}) + (Be^{2+}) + \cdots = \sum_{1}^{i} (X^{n+})_i \tag{10–7a}$$

where $n \geq 2$ and the concentration of each of the ions is expressed in milligrams per liter (mg $\cdot$ L^{-1}) as $CaCO_3$ or milliequivalents per liter (mEq $\cdot$ L^{-1}). Qualitative terms used to describe hardness are listed in Table 10–3. Because many people object to water containing hardness greater than 150 mg $\cdot$ L^{-1} as $CaCO_3$, suppliers of public water have considered it a benefit to soften the water, that is, to remove some of the hardness. A common water treatment goal is to provide water with a hardness in the range of 60 to 120 mg $\cdot$ L^{-1} as $CaCO_3$.

Although all polyvalent cations contribute to hardness, the predominant contributors are calcium and magnesium. Thus, our focus for the remainder of this discussion will be on those two minerals.

TABLE 10-3	Hard Water Classification According to the Water Quality Association	

Term	Concentration Range (mg · L^{-1} as CaCO$_3$)
Soft	<17.1
Slightly hard	17.1–60
Moderately hard	60–120
Hard	120–180
Very hard	>180

The natural process by which water becomes hard is shown schematically in Figure 10–13. As rainwater enters the topsoil, the respiration of microorganisms increases the CO_2 content of the water. As shown in Equation 2–19, CO_2 reacts with water to form H_2CO_3. Limestone, which is made up of solid $CaCO_3$ and $MgCO_3$, reacts with the carbonic acid to form calcium bicarbonate [$Ca(HCO_3)_2$] and magnesium bicarbonate [$Mg(HCO_3)_2$]. Although $CaCO_3$ and $MgCO_3$ are both insoluble in water, the bicarbonates are quite soluble. Gypsum ($CaSO_4$) and $MgSO_4$ may also go into solution to contribute to the hardness.

Because calcium and magnesium usually predominate, it is often convenient in performing softening calculations on these types of waters to *approximate* the total hardness (TH) of a water as the sum of those elements

$$TH \cong Ca^{2+} + Mg^{2+} \tag{10–7b}$$

where the concentrations of each element are in consistent units (mg · L^{-1} as $CaCO_3$ or mEq · L^{-1}). Total hardness is often broken down into two components: (1) that associated with the HCO_3^- anion (called carbonate hardness and abbreviated CH) and (2) that associated with other anions (called noncarbonate hardness and abbreviated NCH).* Total hardness, then, may also be defined as

$$TH = CH + NCH \tag{10–8}$$

Carbonate hardness is defined as the amount of hardness equal to the total hardness or the total alkalinity, whichever is less. Carbonate hardness can be removed by heating the water, because the solubility of calcium and magnesium bicarbonates and carbonates [$CaCO_3$, $MgCO_3$, $Ca(HCO_3)_2$, $Mg(HCO_3)_2$] decreases with increasing temperature. As such, carbonate hardness is often called **temporary hardness** because heating the water removes it.

FIGURE 10-13

Natural process by which water is made hard.

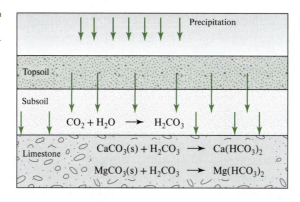

---

*Note that this does not imply that the compounds exist as compounds in solution. They are dissociated.

FIGURE 10–14

Relationships between total hardness, carbonate hardness, and noncarbonate hardness. The pH of the water is between 6.5 and 8.3.

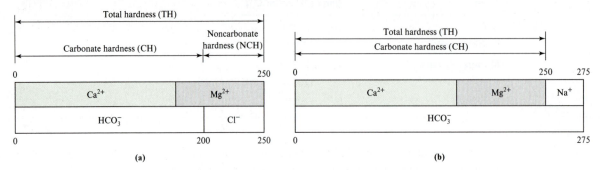

(a)   (b)

Noncarbonate hardness is defined as the total hardness in excess of the alkalinity.* If the alkalinity is equal to or greater than the total hardness, then there is no noncarbonate hardness. Noncarbonate hardness accounts for that portion of the calcium and magnesium (and other polyvalent ions) that is associated with ions other than bicarbonate and carbonate, including sulfate, nitrate, and chloride. Noncarbonate hardness is called **permanent hardness** because it is not removed when water is heated.

The relationships between total hardness, carbonate hardness, and noncarbonate hardness are illustrated in Figure 10–14. In Figure 10–14a, the total hardness is 250 mg · L^{-1} as CaCO$_3$, the carbonate hardness is equal to the alkalinity (HCO$_3^-$ = 200 mg · L^{-1} as CaCO$_3$), and the noncarbonate hardness is equal to the difference between the total hardness and the carbonate hardness (NCH = TH − CH = 250 − 200 = 50 mg · L^{-1} as CaCO$_3$). In Figure 10–14b, the total hardness is again 250 mg · L^{-1} as CaCO$_3$. However, because the alkalinity (HCO$_3^-$) is greater than the total hardness, and because the carbonate hardness cannot be greater than the total hardness (see Equation 10–8), the carbonate hardness is equal to the total hardness, that is, 250 mg · L^{-1} as CaCO$_3$. With the carbonate hardness equal to the total hardness, then all of the hardness is carbonate hardness and there is no noncarbonate hardness. Note that in both cases we have assumed that the pH is between 6.5 and 8.3 because HCO$_3^-$ is the only form of alkalinity present.

---

*When the pH is between 6.5 and 8.3, HCO$_3^-$ is the dominant form of alkalinity, and total alkalinity can be approximated as the concentration of HCO$_3^-$.

EXAMPLE 10–3   A sample of water having a pH of 7.2 has the following concentration of ions.
Calculate the total hardness, the carbonate hardness, the alkalinity and the total dissolved solids for the following ions.

Ion	Concentration (as the ion)	Ion	Concentration (as the ion)
Ca^{2+}	40 mg · L^{-1}	HCO$_3^-$	110 mg · L^{-1}
Mg^{2+}	10 mg · L^{-1}	SO$_4^{2-}$	67.2 mg · L^{-1}
Na$^+$	11.8 mg · L^{-1}	Cl$^-$	11 mg · L^{-1}
K$^+$	7.0 mg · L^{-1}		

Ion	Concentration (mg · L^{-1})	M.W. (mg · mmol^{-1})	$|n|$	Eq. Wt. (mg · mEq^{-1})	Concentration (mEq · L^{-1})	Concentration (mg · L^{-1} as CaCO$_3$)
Ca^{2+}	40.0	40.1				
Mg^{2+}	10.0	24.3				
Na$^+$	11.8	23.0				
K$^+$	7.0	39.1				
HCO$_3^-$	110.0	61.0				
SO$_4^{2-}$	67.2	96.1				
Cl$^-$	11.0	35.5				

Solution   First prepare a table like the one shown.

Next, determine the value of n for each of the ions. In most cases, it will be the oxidation state as given.* Now divide the molecular weight by the number of equivalents per mole, n, to determine the equivalent weight of each of the ions. For example, the molecular weight of calcium can be expressed as 40.1 mg · mmol^{-1} and the number of equivalents is 2 mEq · mmol^{-1}.

Ion	Concentration (mg · L^{-1})	M.W. (mg · mmol^{-1})	$|n|$	Eq. Wt. (mg · mEq^{-1})	Concentration (mEq · L^{-1})	Concentration (mg · L^{-1} as CaCO$_3$)
Ca^{2+}	40.0	40.1	2	20.05		
Mg^{2+}	10.0	24.3	2	12.15		
Na$^+$	11.8	23.0	1	23.0		
K$^+$	7.0	39.1	1	39.1		
HCO$_3^-$	110.0	61.0	1	61.0		
SO$_4^{2-}$	67.2	96.1	2	48.05		
Cl$^-$	11.0	35.5	1	35.5		

The concentration of each of the ions given in milliequivalents per liter can be determined by dividing the concentration given in milligrams per liter by the equivalent weight. For example, for calcium:

$$[Ca^{2+}] = \frac{40.0 \text{ mg} \cdot \text{L}^{-1}}{20.05 \text{ mg} \cdot \text{mEq}^{-1}} = 1.995 \text{ mEq} \cdot \text{L}^{-1}$$

The table now becomes

Ion	Concentration (mg · L^{-1})	M.W. (mg · mmol^{-1})	$|n|$	Eq. Wt. (mg · mEq^{-1})	Concentration (mEq · L^{-1})	Concentration (mg · L^{-1} as CaCO$_3$)
Ca^{2+}	40.0	40.1	2	20.05	1.995	
Mg^{2+}	10.0	24.3	2	12.15	0.823	
Na$^+$	11.8	23.0	1	23.0	0.513	
K$^+$	7.0	39.1	1	39.1	0.179	
HCO$_3^-$	110.0	61.0	1	61.0	1.80	
SO$_4^{2-}$	67.2	96.1	2	48.05	1.40	
Cl$^-$	11.0	35.5	1	35.5	0.31	

*A review of determining the number of equivalents can be found on Section 2–4 in Chapter 2, Chemistry.

Finally, the concentration of each of the ions can be determined in units of milligrams per liter as $CaCO_3$ by multiplying the concentration in milliequivalents per liter by the equivalent weight of $CaCO_3$ (which is 50.0 mg · mEq^{-1}). Therefore, the concentration of calcium in units of milligrams per liter as $CaCO_3$ is

$$(1.995 \text{ mEq} \cdot L^{-1}) \times (50 \text{ mg} \cdot mEq^{-1}) = 99.8 \text{ mg} \cdot L^{-1} \text{ as } CaCO_3$$

Ion	Concentration (mg · L^{-1})	M.W. (mg · $mmol^{-1}$)	$\lvert n \rvert$	Eq. Wt. (mg · mEq^{-1})	Concentration (mEq · L^{-1})	Concentration (mg · L^{-1} as $CaCO_3$)
Ca^{2+}	40.0	40.1	2	20.05	1.995	99.8
Mg^{2+}	10.0	24.3	2	12.15	0.823	41.2
Na^+	11.8	23.0	1	23.0	0.513	25.7
K^+	7.0	39.1	1	39.1	0.179	8.95
HCO_3^-	110.0	61.0	1	61.0	1.80	90.0
SO_4^{2-}	67.2	96.1	2	48.05	1.40	70.0
Cl^-	11.0	35.5	1	35.5	0.31	15.5

The table is thus complete.

In completing this problem, you should check your math and make sure if sampling and analysis is complete and accurate. The sum of the concentrations of cations and ions should be equal to within ±10%. In this case,

$$\sum (\text{cations}) = \sum (\text{anions})$$

$$175.6 \text{ mg} \cdot L^{-1} \text{ as } CaCO_3 = 175.6 \text{ mg} \cdot L^{-1} \text{ as } CaCO_3$$

The exact total hardness is the sum of the multivalent cations

$$[Ca^{2+}] + [Mg^{2+}] = 99.8 + 41.2 = 141 \text{ mg} \cdot L^{-1} \text{ as } CaCO_3$$

The carbonate hardness (CH) is equal to that portion of the hardness associated with carbonate and bicarbonate species, that is, the alkalinity,

$$(HCO_3^-) + (CO_3^{2-}) + (OH^-) - (H^+)$$

Because the pH is between 6.5 and 8.3, the concentration of carbonate is negligible and the concentrations of hydroxide (OH^-) and the hydrogen ions (H^+) essentially cancel each other. Therefore,

$$\text{Alkalinity} = (HCO_3^-) + (\cancel{CO_3^{2-}}) + (\cancel{OH^-}) - (\cancel{H^+}) \approx (HCO_3^-)$$

$$= 90.0 \text{ mg} \cdot L^{-1} \text{ as } CaCO_3$$

Because the total hardness is 141 mg · L^{-1} as $CaCO_3$ and the carbonate hardness is defined as that portion of hardness associated with carbonate and bicarbonate, the concentration of carbonate hardness is equal to the alkalinity, or

$$CH = 90.0 \text{ mg} \cdot L^{-1} \text{ as } CaCO_3$$

The noncarbonate hardness (NCH) is equal to the

$$\text{Total hardness} - CH = 141 - 90.0 = 51.0 \text{ mg} \cdot L^{-1} \text{ as } CaCO_3$$

Finally, the total dissolved solids (TDS) is equal to the sum of all the cations and anions using the concentrations in milligrams per liter as the ion. The reason you can use concentrations as milligrams per liter as the ions is because the TDS is determined by evaporating all of the water and determining the mass of the remaining ions.

$$TDS = 40.0 + 10.0 + 11.8 + 7.0 + 110.0 + 67.2 + 11.0 = 257 \text{ mg} \cdot L^{-1}$$

Bar charts of water composition are often useful in understanding the process of softening. By convention, the bar chart is constructed with cations in the upper bar and anions in the lower bar. In the upper bar, calcium is placed first and magnesium second because calcium concentrations usually exceed that of magnesium. Other cations follow without any specified order. The lower bar is constructed with bicarbonate placed first because in most waters it is the predominant anion. Other anions follow without any specified order. Construction of a bar chart is illustrated in Example 10–4.

EXAMPLE 10–4

Given the following analysis of a groundwater (pH = 7.6), construct a bar chart of the constituents, with concentrations given as milligrams per liter as $CaCO_3$.

Ion	$mg \cdot L^{-1}$ as Ion	Molecular Weight $(g \cdot mol^{-1})$	Number of Equivalents $(Eq \cdot mol^{-1})$	Concentration $(Eq \cdot L^{-1})$	Concentration $(mg \cdot L^{-1}$ as $CaCO_3)$
Ca^{2+}	150.3	40.08	2	7.5	375
Fe^{3+}	1.19	55.85	3	0.064	3.2
Mg^{2+}	14.6	24.31	2	1.20	60
Na^+	100	22.99	1	4.35	217.5
Cl^-	201.4	35.45	1	5.68	284.0
HCO_3^-	450.3	61.02	1	7.38	369.0

Solution

The concentrations of the ions have been converted to $CaCO_3$ equivalents. The results are plotted in Figure 10–15.

The sum of the cation concentrations is 655.7 mg $\cdot L^{-1}$ as $CaCO_3$. The (exact) total hardness is 438.2 mg $\cdot L^{-1}$ as $CaCO_3$. The anions total 653.0 mg $\cdot L^{-1}$ as $CaCO_3$. The carbonate hardness is 369.0 mg $\cdot L^{-1}$ as $CaCO_3$ and the noncarbonate hardness is 69.2 mg $\cdot L^{-1}$ as $CaCO_3$. The discrepancy between the concentrations of cations and anions is the result of other ions that were not measured. If a complete analysis were conducted, and no analytical error occurred, the number of equivalents of cations would equal exactly the number of equivalents of anions. Typically, a complete analysis will vary within ±10% because of analytical errors.

FIGURE 10–15

Bar graph of groundwater constituents.

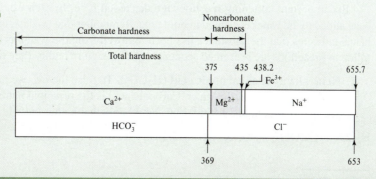

Softening can be accomplished by either the lime–soda process or by ion exchange. Both methods are discussed in the following sections.

Lime–Soda Softening

In lime–soda softening it is possible to calculate the chemical doses necessary to remove hardness. Hardness precipitation is based on the following two solubility reactions:

$$Ca^{2+} + CO_3^{2-} \rightleftharpoons CaCO_3(s) \tag{10–9}$$

and

$$Mg^{2+} + 2OH^- \rightleftharpoons Mg(OH)_2(s) \tag{10–10}$$

The objective is to precipitate the calcium as $CaCO_3$ and the magnesium as $Mg(OH)_2$. To precipitate calcium, the pH of the water must be raised to about 10.3. To precipitate magnesium, the pH must be raised to about 11. If the naturally occurring bicarbonate alkalinity (HCO_3^-) is insufficient for the $CaCO_3(s)$ precipitate to form (i.e., there is noncarbonate hardness), we must add CO_3^{2-} (in the form of Na_2CO_3). Because magnesium is more expensive to remove than is calcium, we leave as much Mg^{2+} in the water as possible. Additionally, it is more expensive to remove noncarbonate hardness than carbonate hardness because we must add another chemical to provide the CO_3^{2-}. Therefore, utilities leave as much noncarbonate hardness in the water as possible.

Softening Chemistry.   The chemical processes used to soften water are a direct application of the law of mass action. We increase the concentration of CO_3^{2-} or OH^- (or both) by the addition of chemicals and drive the reactions given in Equations 10–9 and 10–10 to the right. Insofar as possible, we convert the naturally occurring bicarbonate alkalinity (HCO_3^-) to carbonate (CO_3^{2-}) by the addition of hydroxyl ions (OH^-). Hydroxyl ions cause the carbonate buffer system (Equation 2–75) to shift to the right and, thus, provide the carbonate for the precipitation reaction (Equation 10–9).

The common source of hydroxyl ions is calcium hydroxide ($Ca(OH)_2$). Many water treatment plants find it more economical to buy **quicklime** (CaO), commonly called lime, than hydrated lime ($Ca(OH)_2$). The quicklime is converted to hydrated lime at the water treatment plant by mixing CaO and water to produce a slurry of $Ca(OH)_2$, which is fed to the water for softening. This process is called **slaking.** When carbonate ions must be supplied, the most common chemical chosen is sodium carbonate (Na_2CO_3). Sodium carbonate is commonly referred to as **soda ash** or soda.

Softening Reactions.   The softening reactions are regulated by controlling the pH. First, any free acids are neutralized. Then pH is raised to precipitate the $CaCO_3$; if necessary, the pH is raised further to remove $Mg(OH)_2$. Finally, if necessary, CO_3^{2-} is added to precipitate the noncarbonate hardness.

Six important softening reactions are discussed here. In each case, the chemical that has been added to the water is printed in bold type. Remember that (s) designates the solid form, and hence indicates that the substance has been removed from the water as a solid precipitate. The following reactions are presented sequentially, although in reality they occur simultaneously.

1. **Neutralization of carbonic acid (H_2CO_3).** To raise the pH, we must first neutralize any free acids present in the water. Carbonic acid is the principal naturally occurring acid present in unpolluted water.* You should note that no hardness is removed in this step.

$$H_2CO_3 + \mathbf{Ca(OH)_2} \rightleftharpoons CaCO_3(s) + 2H_2O \tag{10–11}$$

---

*CO_2 and H_2CO_3 in water are essentially the same: $CO_2 + H_2O \rightleftharpoons H_2CO_3$. Thus, the equivalents per mole (n) for CO_2 is 2.

2. **Precipitation of carbonate hardness due to calcium.** As mentioned previously, we must raise the pH to about 10.3 to precipitate calcium carbonate. To achieve this pH, we must convert all of the bicarbonate to carbonate. The carbonate then serves as the common ion for the precipitation reaction.

$$Ca^{2+} + 2HCO_3^- + Ca(OH)_2 \rightleftharpoons 2CaCO_3(s) + 2H_2O \qquad (10\text{–}12)$$

3. **Precipitation of carbonate hardness due to magnesium.** If we need to remove carbonate hardness that results from the presence of magnesium, we must add more lime to achieve a pH of about 11. The reaction may be considered to occur in two stages. The first stage occurs when we convert all of the bicarbonate to carbonate as accomplished in step 2.

$$Mg^{2+} + 2HCO_3^- + Ca(OH)_2 \rightleftharpoons MgCO_3 + CaCO_3(s) + 2H_2O \qquad (10\text{–}13)$$

Note that the hardness of the water did not change because $MgCO_3$ is soluble. The addition of more lime is necessary to remove the hardness due to magnesium.

$$Mg^{2+} + CO_3^{2-} + Ca(OH)_2 \rightleftharpoons Mg(OH)_2(s) + CaCO_3(s) \qquad (10\text{–}14)$$

4. **Removal of noncarbonate hardness due to calcium.** If we need to remove noncarbonate hardness due to calcium, no further increase in pH is required. Instead we must provide additional carbonate in the form of soda ash.

$$Ca^{2+} + Na_2CO_3 \rightleftharpoons CaCO_3(s) + 2Na^+ \qquad (10\text{–}15)$$

5. **Removal of noncarbonate hardness due to magnesium.** If we need to remove noncarbonate hardness due to magnesium, we must add both lime and soda. The lime provides the hydroxyl ion for precipitation of the magnesium.

$$Mg^{2+} + Ca(OH)_2 \rightleftharpoons Mg(OH)_2(s) + Ca^{2+} \qquad (10\text{–}16)$$

Note that although the magnesium is removed, no change in the hardness occurs because the calcium is still in solution. To remove the calcium we must add soda ash.

$$Ca^{2+} + Na_2CO_3 \rightleftharpoons CaCO_3(s) + 2Na^+ \qquad (10\text{–}17)$$

Note that this is the same reaction as the one to remove noncarbonate hardness due to calcium.

EXAMPLE 10–5   The groundwater from the fictitious Apex Aquifer contains 2.3×10^{-5} M CO_2. It is being pumped at a rate of $200 \text{ L} \cdot \text{s}^{-1}$ to serve the residents of the town of Zanidu. The pH of the water is 7.6 and the water analysis is presented in Example 10–4.

(a) Determine the mass (in kilograms) of hydrated lime that must be added each day to neutralize the carbon dioxide present in the water.

(b) Determine the mass (in kilograms) of hydrated lime that must be added each day to precipitate the carbonate hardness due to calcium and magnesium.

(c) Determine the mass (in kilograms) of hydrated lime and of soda ash that must be added each day to precipitate the noncarbonate hardness due to calcium and magnesium.

(d) Determine the total mass of hydrated lime and of soda ash that must be added each day.

Solution   (a) Equation 10–11 shows that we need 1 mole of hydrated lime for every mole of CO_2 present in the water. Therefore, we need 2.3×10^{-5} M of hydrated lime. To calculate the mass needed:

$$(2.3 \times 10^{-5} \text{ M})(200 \text{ L} \cdot \text{sec}^{-1})(86,400 \text{ sec} \cdot \text{day}^{-1})(74.096 \text{ g Ca[OH]}_2 \cdot \text{mol}^{-1}) \times$$
$$(1 \text{ kg} \cdot (1000 \text{ g})^{-1}) = 29.4 \text{ kg} \cdot \text{day}^{-1}$$

(b) Equation 10–12 shows that we need 1 mole of hydrated lime for every mole of calcium re-moved. Using Equations 10–13 and 10–14, it is clear that we need 2 moles of hydrated lime for every 1 mole of magnesium removed.

The ionic analysis given in Example 10–4 states that we have 150.3 mg $\cdot$ L^{-1} of Ca^{2+} and 14.6 mg $\cdot$ L^{-1} of Mg^{2+}. From the calculations presented in this example, we can see that there is more calcium present than alkalinity, so the carbonate hardness is equal to the alkalinity of 7.39 mEq $\cdot$ L^{-1}. From the bar graph, we see that we can assume that the bicarbonate is associated with the calcium. As such, we will assume that there are 7.38 mEq $\cdot$ L^{-1} of noncarbonate hardness, all associated with calcium. The remaining calcium must be noncarbonate hardness and all of the magnesium must be noncarbonate hardness.

The concentration of Ca^{2+} associated with carbonate hardness in molar units is:

$$(7.38 \text{ mEq} \cdot \text{L}^{-1})(\text{mM} \cdot (2 \text{ mEq})^{-1}) = 3.69 \text{ mM}$$

Therefore, we would need 3.69 mM of hydrated lime to remove the carbonate hardness due to calcium. To calculate the mass needed:

$$(3.69 \text{ mM})(200 \text{ L} \cdot \text{sec}^{-1})(86,400 \text{ sec} \cdot \text{day}^{-1})(74.096 \text{ g Ca(OH)}_2 \cdot \text{mol}^{-1})(1 \text{ kg} \cdot (10^6 \text{ mg})^{-1}) = 4725 \text{ kg} \cdot \text{day}^{-1} \text{ to remove the carbonate hardness due to calcium.}$$

(c) The remaining calcium must be associated with noncarbonate hardness. Based on the analysis presented, the concentration of calcium associated with noncarbonate hardness is the total minus that associated with carbonate hardness or:

$$7.5 \text{ mEq} \cdot \text{L}^{-1} - 7.38 \text{ mEq} \cdot \text{L}^{-1} = 0.12 \text{ mEq} \cdot \text{L}^{-1}$$

In molar units, the concentration of Ca^{2+} associated with carbonate hardness is:

$$(0.12 \text{ mEq} \cdot \text{L}^{-1})(\text{mM} \cdot (2 \text{ mEq})^{-1}) = 0.06 \text{ mM}$$

To remove 1 mole of noncarbonate hardness due to calcium, we would need to add 1 mole of soda ash, as shown in Equation 10–15. No additional hydrated lime would need to be added.

Therefore, the mass of soda ash needed is

$$(0.06 \text{ mM})(200 \text{ L} \cdot \text{sec}^{-1})(86,400 \text{ sec} \cdot \text{day}^{-1})(105.99 \text{ g Na}_2\text{CO}_3 \cdot \text{mol}^{-1})(1 \text{ kg} \cdot (10^6 \text{ mg})^{-1}) = 109.9 \text{ kg} \cdot \text{day}^{-1} \text{ to remove the noncarbonate hardness due to calcium.}$$

Since we can assume that all of the magnesium is associated with noncarbonate hardness, we will only need to consider Equations 10–16 and 10–17. For every 1 mole of magnesium, we will need to add 1 mole of hydrated lime and 1 mole of soda ash.

In molar units, the concentration of Mg^{2+} associated with noncarbonate hardness is:

$$(1.20 \text{ mEq} \cdot \text{L}^{-1})(\text{mM} \cdot (2 \text{ mEq})^{-1}) = 0.60 \text{ mM}$$

Therefore, the mass of hydrated lime needed is:

$$(0.60 \text{ mM})(200 \text{ L} \cdot \text{sec}^{-1})(86,400 \text{ sec} \cdot \text{day}^{-1})(74.096 \text{ g Ca(OH)}_2 \cdot \text{mol}^{-1})(1 \text{ kg} \cdot (10^6 \text{ mg})^{-1}) = 76.8 \text{ kg} \cdot \text{day}^{-1}.$$

The mass of soda ash needed is

$$(0.06 \text{ mM})(200 \text{ L} \cdot \text{sec}^{-1})(86,400 \text{ sec} \cdot \text{day}^{-1})(105.99 \text{ g Na}_2\text{CO}_3 \cdot \text{mol}^{-1})(1 \text{ kg} \cdot (10^6 \text{ mg})^{-1}) = 109.9 \text{ kg} \cdot \text{day}^{-1} \text{ to remove the noncarbonate hardness due to calcium.}$$

(d) Therefore, the total mass of hydrated lime that must be added each day is

$$29.4 + 4,725 + 76.8 = 4,831 \text{ kg} \cdot \text{day}^{-1}$$

Therefore, the total mass of soda ash that must be added each day is

$$109.9 + 109.9 = 220 \text{ kg} \cdot \text{day}^{-1}$$

Process Limitations and Empirical Considerations. Lime–soda softening cannot produce a water completely free of hardness because of the solubility of $CaCO_3$ and $Mg(OH)_2$, the physical limitations of mixing and contact, and the lack of sufficient time for the reactions to go to completion. Thus, the minimum calcium hardness that can be achieved is about 30 mg $\cdot$ L^{-1} as $CaCO_3$, and the minimum magnesium hardness is about 10 mg $\cdot$ L^{-1} as $CaCO_3$. Because of the slimy condition that results when soap is used with water that is too soft, we normally strive for a final total hardness on the order of 75–120 mg $\cdot$ L^{-1} as $CaCO_3$.

To achieve reasonable removal of hardness in a practical time period, an excess of $Ca(OH)_2$ greater than the stoichiometric amount is usually provided. Based on our empirical experience, a minimum excess of 20 mg $\cdot$ L^{-1} of $Ca(OH)_2$ expressed as $CaCO_3$ must be provided.

Magnesium in excess of about 40 mg $\cdot$ L^{-1} as $CaCO_3$ forms scales on heat exchange elements in hot water heaters. Because of the expense of removing magnesium, we normally remove only that magnesium that is in excess of 40 mg $\cdot$ L^{-1} as $CaCO_3$. For magnesium removals less than 20 mg $\cdot$ L^{-1} as $CaCO_3$, the basic excess of lime mentioned earlier is sufficient to ensure good results. For magnesium removals between 20 and 40 mg $\cdot$ L^{-1} as $CaCO_3$, we must add an excess of lime equal to the magnesium to be removed. For magnesium removals greater than 40 mg $\cdot$ L^{-1} as $CaCO_3$, the excess lime we need to add is 40 mg $\cdot$ L^{-1} as $CaCO_3$. Addition of excess lime in amounts greater than 40 mg $\cdot$ L^{-1} as $CaCO_3$ does not appreciably improve the reaction kinetics.

EXAMPLE 10–6 The groundwater mentioned in Example 10–5 contains 2.3×10^{-5} M CO_2. It also contains 300 mg $\cdot$ L^{-1} as $CaCO_3$ of carbonate hardness due to calcium and 50 mg $\cdot$ L^{-1} as $CaCO_3$ of carbonate hardness due to magnesium. It is treated at the same rates mentioned in Example 10–5. You are to remove all of the carbonate hardness due to calcium but do not need to remove the magnesium ions. Assume that you remove all but 20 mg $\cdot$ L^{-1} (as $CaCO_3$) of the calcium ions. What mass of calcium carbonate sludge is produced on a daily basis?

Solution Using Equation 10–12 we see that for every mole of calcium removed we need one mole of hydrated lime and we produce two moles of calcium carbonate sludge. Because we remove all but 20 mg $\cdot$ L^{-1} as $CaCO_3$ of the calcium ions, we must remove 280 mg $\cdot$ L^{-1} as $CaCO_3$. We must convert this to a molar basis.

$$\frac{(280 \text{ mg} \cdot \text{L}^{-1} \text{ as } CaCO_3)}{(100.09 \text{ mg} \cdot \text{mmol}^{-1} \text{ } CaCO_3)} = 2.80 \text{ mmol} \cdot \text{L}^{-1} \text{ as } Ca^{2+}, \text{ or } 2.80 \times 10^{-3} \text{ mol} \cdot \text{L}^{-1}$$

Therefore according to the stoichiometry given in Equation 9–12, we produce 5.60×10^{-3} mg $\cdot$ L^{-1} $CaCO_3$ sludge. In units of kilograms per day, the mass of sludge equals

$(5.6 \times 10^{-3} \text{ M})(200 \text{ L} \cdot \text{s}^{-1})(86,400 \text{ s} \cdot \text{day}^{-1})(100.09 \text{ g } CaCO_3 \cdot \text{mol}^{-1})(10^{-3} \text{ kg} \cdot \text{g}^{-1}) =$ 9690 kg $\cdot$ day^{-1} of $CaCO_3$

We also produced 39.8 kg $\cdot$ day^{-1} of $CaCO_3$ sludge in neutralizing the carbon dioxide according to Equation 10–11. Therefore, the total mass of sludge produced each day is 9730 kg.

Ion-Exchange Softening

Ion exchange can be defined as the reversible exchange of an ion on a solid phase with an ion of like charge in an aqueous phase. The most general form of an ion-exchange reaction is

$$nR\text{—}X^+ + Y^{n+} \rightleftharpoons R_n\text{—}Y^{n+} + nX^+ \qquad (10\text{–}18)$$

FIGURE 10–16

Idealized breakthrough curve of a fixed bed adsorber.

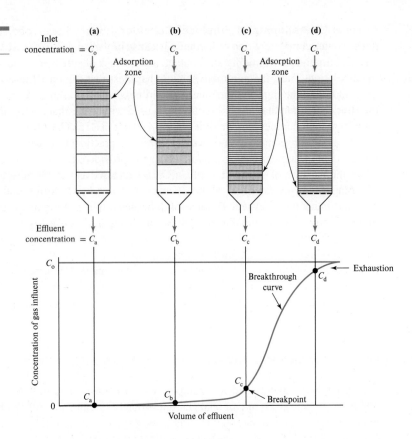

Note that an equivalent number of ions are exchanged in this reaction. In water treatment, ion exchange is most often used for metal removal. Because of its simplicity, ion exchange is most commonly used for in-home water-softening systems. Here the reaction can be represented as

$$Ca^{2+} + 2NaR \rightleftharpoons R_2Ca + 2Na^+ \qquad (10\text{--}19)$$

where R represents the solid ion-exchange material. By this reaction, calcium has been removed from the water and replaced by an equivalent amount of sodium, that is, two sodium ions for each calcium ion. In this reaction, the alkalinity would remain unchanged because there is no reaction or exchange of anions. The exchange results in essentially 100% removal of the hardness from the water until the exchange capacity of the ion-exchange material is reached, as shown in Figure 10–16. When the ion-exchange material becomes saturated, no hardness will be removed. At this point breakthrough is said to have occurred because the hardness ions pass through the ion-exchange adsorber. When the effluent hardness concentration exceeds a predetermined level, breakthrough is said to have occurred. At this point, the ion exchanger is taken out of service, and the ion-exchange material is regenerated. That is, the hardness is removed from the material by passing water containing highly concentrated solution of Na^+ through the ion exchanger. The mass action of having so much Na^+ in the water causes the sodium ions to be preferentially exchanged for the hardness ions.

$$R_2Ca + 2Na^+ + 2Cl^- \rightleftharpoons 2NaR + Ca^{2+} + 2Cl^- \qquad (10\text{--}20)$$

The ion-exchange material can now be used to remove more hardness. The $CaCl_2$ is a waste stream that must be disposed of.

Some large water treatment facilities use ion-exchange softening, but the most common application is for residential water softeners. The ion-exchange material can either be naturally occurring clays, called zeolites, or synthetically made resins. Several manufacturers produce synthetic resins. The resins or zeolites are characterized by the amount of hardness that they remove per volume of resin material and by the amount of salt required to regenerate the resin. The synthetically produced resins have a much higher exchange capacity and require less salt for regeneration.

Hardness Removal with Ion-Exchange Resins. Because the resin removes virtually 100% of the hardness, it is necessary to bypass a portion of the water and then blend to obtain the desired final hardness.

EXAMPLE 10–7 A home water softener has 0.1 m³ of ion-exchange resin with an exchange capacity of $62 \text{ kg} \cdot \text{m}^{-3}$. The four home residents each use water at a rate of $400 \text{ L} \cdot \text{day}^{-1}$. The well water they are using contains $340.0 \text{ mg} \cdot \text{L}^{-1}$ of hardness as $CaCO_3$. It is desirable to soften it to achieve a total hardness of $100 \text{ mg} \cdot \text{L}^{-1}$ as $CaCO_3$. What should the bypass flow rate be?

Solution As always, you first need to draw a picture. The system used is shown in the box with the dotted line.

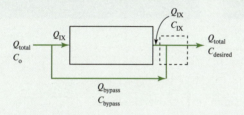

$$Q_{total} = Q_{bypass} + Q_{IX}$$

For the designated system,

$$Q_{IX}C_{IX} + Q_{bypass}C_{bypass} = Q_{total}C_{desired}$$

The hardness concentration after ion exchange treatment can be assumed to be zero. The concentration in the bypass is the same as C_o, or $340 \text{ mg} \cdot \text{L}^{-1}$ as $CaCO_3$. The total flow rate is $(400 \text{ L} \cdot \text{capita}^{-1} \cdot \text{day}^{-1}) \times 4 = 1600 \text{ L} \cdot \text{day}^{-1}$.

$$(Q_{IX})(0) + Q_{bypass}(340 \text{ mg} \cdot \text{L}^{-1}) = (1600 \text{ L} \cdot \text{day}^{-1})(100 \text{ mg} \cdot \text{L}^{-1})$$

$$Q_{bypass} = \frac{(1600 \text{ L} \cdot \text{day}^{-1})(100 \text{ mg} \cdot \text{L}^{-1})}{340 \text{ mg} \cdot \text{L}^{-1}} = 470.6, \text{ or } 471 \text{ L} \cdot \text{day}^{-1}$$

10–4 SEDIMENTATION

Overview

Surface water containing high turbidity may require sedimentation prior to subsequent treatment. Because the detention time of sedimentation basins (also called clarifiers or settling tanks) is usually 2–4 h, only those particles that can settle in this time will be removed. Sedimentation basins (Figure 10–17) are usually rectangular or circular with either a radial or upward water flow pattern. Sludge that is withdrawn from the bottom of sedimentation basins may in some cases be discharged back to the river.

The clarifier is designed so as to remove the settled water from the basin without carrying away any of the floc particles. In ideal reactors, the average velocity of the fluid is equal to its flow rate divided by the area through which it flows, that is,

$$v = \frac{Q}{A_c} \tag{10–21}$$

FIGURE 10–17

A sedimentation basin: **(a)** empty **(b)** full. (©Central Coastal Water Authority)

(a)   (b)

where v = water velocity (in $m \cdot s^{-1}$) ≡ overflow rate

Q = water flow (in $m^3 \cdot s^{-1}$)

A_c = cross-sectional surface area (in m^2)

To reduce the turbulence caused by the rushing water as it leaves the tank, a series of troughs, called weirs, are put in place to provide a large area for the water to flow through and minimize the velocity in the sedimentation tank near the outlet zone. The weirs then feed into a central channel or pipe for transport of the settled water. Figure 10–18 shows various weir arrangements. The length of weir required is a function of the type of solids. The heavier the solids, the harder it is to scour them, and the higher the allowable outlet velocity. Therefore, heavier particles require a shorter length of weir than do lighter particles. The units for weir overflow rates are $m^3 \cdot day^{-1} \cdot m^{-1}$, which is water flow ($m^3 \cdot day^{-1}$) per unit length of weir (m).

EXAMPLE 10–8   A research-scale water treatment plant uses low-turbidity raw water and is designing its overflow weir at a loading rate of 175 $m^3 \cdot day^{-1} \cdot m^{-1}$. If its plant flow rate is 0.044 $m^3 \cdot s^{-1}$, how many linear meters of weir is required?

Solution

$$\frac{(0.044 \; m^3 \cdot s^{-1} \times 86{,}400 \; s \cdot d^{-1})}{175 \; m^3 \cdot day^{-1} \cdot m^{-1}} = 21.7, \text{ or } 22 \text{ m}$$

Determination of Settling Velocity (v_s)

When designing an ideal sedimentation tank, the settling velocity (v_s) of the particle to be removed must first be determined and then the overflow rate (v_o) (determined using Equation 10–21) must be set at some value less than or equal to v_s.

FIGURE 10–18

Weir arrangements: **(a)** rectangular; **(b)** circular. (©Walker Process Equipment)

(a)

(b)

Determination of the particle-settling velocity is different for different types of particles. Settling properties of particles are often categorized into one of three classes:

Type I Sedimentation.   Type I sedimentation is characterized by particles that settle discretely at a constant velocity. They settle as individual particles and do not flocculate or stick to other particles during settling. Examples of these particles are sand and grit material. Generally speaking, the only applications of type I settling are during presedimentation for sand removal, prior to coagulation in a potable water plant, in settling of sand particles during cleaning of rapid sand filters, and in grit chambers (see Section 11–6).

Type II Sedimentation.   Type II sedimentation is characterized by particles that flocculate during sedimentation. Because they flocculate, their size is constantly changing; therefore, the settling velocity is changing. Generally the settling velocity increases with depth and extent of flocculation. These types of particles occur with alum or iron coagulation, in primary sedimentation basins (see Section 11–7), and in settling tanks in trickling filtration (see Section 11–8).

Type III, or Zone, Sedimentation.   In zone sedimentation the particles are at a high concentration (greater than 1000 mg · L^{-1}) such that the particles tend to settle as a mass, and a distinct clear zone and sludge zone are present. Zone settling occurs in lime-softening sedimentation, activated-sludge sedimentation (see Section 11–8), and sludge thickeners (see Section 11–12).

Determination of Overflow Rate (v_o)

Although there are several ways of determining effective particle-settling velocities and consequently of determining overflow rates, the only method we will discuss here is the calculation used for type I sedimentation.

In the case of type I sedimentation, the particle-settling velocity can be calculated and the basin designed to remove a specific size particle. In 1687, Sir Isaac Newton showed that a particle falling in a quiescent fluid accelerates until the frictional resistance, or drag, on the particle is equal to the gravitational force of the particle (Figure 10–19). Some 200 years later, Sir George Gabriel Stokes (1845) found that the terminal settling velocity of a sphere falling under quiescent conditions could be described by the following equation.

$$v_s = \frac{g(\rho_s - \rho)d^2}{18\mu} \tag{10–22}$$

where ρ_s = density of particle (in kg · m^{-3})
  ρ = density of fluid (in kg · m^{-3})
  g = acceleration due to gravity (in m · s^{-2})
  d = diameter of spherical particle (in m)
  μ = dynamic viscosity (in Pa-s)
  18 = a constant

FIGURE 10–19

Forces acting on a free-falling particle in a fluid (F_D = drag force; F_G = gravitational force; F_B = buoyancy force).

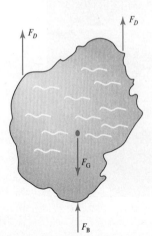

Equation 10–22 is called Stokes' law. It is valid only at laminar flow conditions; that is, where the Reynolds number, R, <1. **Dynamic viscosity** (also called absolute viscosity) is a function of the water temperature. A table of dynamic viscosities (Table A–1) is given in Appendix A.

EXAMPLE 10–9

The design flow and overflow rates for a settling tank are 0.044 m³ · s⁻¹ and 20 m · day⁻¹, respectively. Determine the surface area of this tank. Determine the length of the tank, using conventional wisdom for length-to-width ratios. Determine the tank depth, assuming a detention time of 2 hours.

Solution

1. Find the surface area.
   First change the flow rate to compatible units.

   $$(0.044 \text{ m}^3 \cdot \text{s}^{-1})(86{,}400 \text{ s} \cdot \text{day}^{-1}) = 3801.6 \text{ m}^3 \cdot \text{day}^{-1}$$

   Using Equation 9–21 and an overflow rate of 20 m³ · day⁻¹, the surface area is

   $$A_s = \frac{3801.6 \text{ m}^3 \cdot \text{day}^{-1}}{20 \text{ m} \cdot \text{day}^{-1}} = 190.08 \text{ m}^2, \text{ or } 190 \text{ m}^2$$

   Common length-to-width ratios for settling are between 2:1 and 5:1, and lengths seldom exceed 100 m. A minimum of two tanks is always provided.

   Let's continue with our design, assuming that we can use two tanks, each with a width of 12 m, a total surface area of 190 m² would imply a tank length of

   $$\text{Length} = 190 \text{ m}^2/(2 \text{ tanks})(12 \text{ m wide}) = 7.9 \text{ m, or } 8 \text{ m}$$

   This does not meet our length-to-width ratio of 2:1 through 5:1. The next step would be to choose a smaller width. Let's try 6 m.

   $$\text{Length} = 190 \text{ m}^2/(2 \text{ tanks})(6 \text{ m wide}) = 15.8, \text{ or } 16 \text{ m}$$

   This does meet the length-to-width ratio requirement.

2. Find the tank depth.
   First find the total tank volume from Equation 10–5, using a detention time of 120 min (assumed) based on the rule of thumb that the detention time should be 2–4 h.

   $$\forall = (0.044 \text{ m}^3 \cdot \text{s}^{-1})(120 \text{ min})(60 \text{ s} \cdot \text{min}^{-1}) = 316.8 \text{ m}^3, \text{ or } 320 \text{ m}^3$$

   This would be divided into two tanks as noted earlier. The depth is found as the total tank volume divided by the total surface area.

   $$\text{Depth} = \frac{320 \text{ m}^3}{190 \text{ m}^2} = 1.684 \text{ m, or } 1.7 \text{ m}$$

   This depth would not include the sludge storage zone.
   The final design would then be two tanks, each having the following dimensions: 6 m wide × 16 m long × 1.7 m deep plus sludge storage depth.

10–5 FILTRATION

As the water flows over the weirs and exits the sedimentation tank, it still contains particles that were too small to settle or somehow escaped removal due to fluid patterns. As such, the turbidity of the settling tank effluent ranges from 1 to 10 TU, with a typical value being 3 TU. To reduce this turbidity to 0.3 TU, a filtration process is normally used. In addition, to comply with the Surface

Water Treatment Rule (SWTR) under the Safe Drinking Water Act (SDWA), all surface waters and ground waters that are under the influence of surface water must be treated using filtration.

Filtration is the process by which water flows slowly through a bed of granular media, usually sand, anthracite coal, or garnet. As the water passes through the medium, particles become trapped due to several mechanisms: interception, flocculation, straining, and sedimentation. The densest particles are "strained" like spaghetti is strained through a colander. These particles are simply too large to pass through the pore spaces of the medium and become trapped in the upper depths of the filter. Particles flowing at a sufficiently low velocity are "intercepted" and attach by weak electrostatic forces to the filter medium. If the water is chemically treated prior to filtration, additional flocculation can occur, allowing particle size to grow so that these larger particles can be removed by the other mechanisms. Heavy particles settle out on the filter medium.

As the pore spaces within the filter medium fill up with particles, the velocity of the water increases, causing some of the particles to be sheared off the medium. These particles then flow only to be trapped deeper in the filter. Once the pore spaces are filled to some "capacity," the turbidity level exceeds a critical value and the filter bed needs to be backwashed. Also as the extent to which the pores fill with particles increases, the head loss through the filter continues to increase and it becomes increasingly more difficult to pass water through the filter.

Filters can operate as either slow or rapid sand filters. The first slow sand filter was installed in London, England, in 1829 (Spitzer, 1993). Although the first slow sand filter was installed in the United States in 1872, slow sand filtration never became widely accepted there. Rapid sand or mechanical filtration became the process of choice instead.

Slow sand filters conventionally use only sand (with a gravel support layer), but rapid sand filters use a variety of media. Single-medium beds use only sand, but dual-media filters use sand and anthracite coal. Mixed-media beds use coal, sand, and garnet.

With the promulgation of more stringent regulations for the removal of *Giardia* cysts and *Cryptosporidium* oocysts, filtration is becoming even more widely used in the United States. Slow sand filters are more effective at removing biological particles than are rapid sand filters. This is because the *schmutzdecke* layer, which forms on the surface of sand filters, has smaller pore spaces than do the media used in rapid sand filters, hence achieving more effective filtration. Additionally, the *schmutzdecke* is biologically active, allowing for the biodegradation of a portion of the naturally occurring organic matter that causes problems in distribution systems.

The most important design parameter in sizing filters is the loading rate. The loading rate is the flow rate of water applied per unit area of the filter. It is the velocity of the water approaching the face of the filter.

$$v_a = \frac{Q}{A_s} \tag{10–23}$$

where v_a = face velocity (in m · day^{-1}) = loading rate (in m^3 · day^{-1} · m^{-2})
$\quad Q$ = flow rate onto filter surface (in m^3 · day^{-1})
$\quad A_s$ = surface area of filter (in m^2)

With slow sand filters, water is applied to the sand at a loading rate of 2.9 to 7.6 m^3 · day^{-1} · m^{-2}. As mentioned earlier, filters naturally clog as the pore spaces begin to fill with the particles being removed. Once this happens the filter must be cleaned. With slow sand filters this is accomplished by removing floating material, draining the water until the water level is 4–5 cm below the surface of the sand and then scraping off the top 6–12 mm of sand. Cleaning cycles may be from one to several months. Slow sand filters are well-suited for small communities. Although they require large areas of land, slow sand filters are reliable, require less operator time than rapid sand filters, and have low technical operational demands (Spitzer, 1993).

Rapid sand filters have graded (layered) sand within the bed. The sand grain size distribution is selected to optimize the passage of water while minimizing the passage of particulate matter. Rapid sand filters can also be operated as dual-media filters with sand and anthracite coal. The coal, being lighter and having larger pore spaces than the sand, effectively traps the larger particles in the upper depths of the filter bed.

Rapid sand filters are cleaned in place by forcing water upward through the sand. This operation is called **backwashing.** The washwater flow rate is such that the sand is expanded and the filtered particles are removed from the bed. After backwashing, the sand settles back into place. The largest particles settle first, resulting in a fine sand layer on top and a coarse sand layer on the bottom. The backwash water is either pumped directly to the sanitary sewer for disposal or is treated using sedimentation. The supernatent from the settling ponds is either discharged to the source water or the sanitary sewer and the sludge is landfilled.

Traditionally, rapid sand filters have been designed to operate at a loading rate of $120 \text{ m}^3 \cdot \text{day}^{-1} \cdot \text{m}^{-2}$. Experiments conducted at the Chicago water treatment plant have demonstrated that satisfactory water quality can be obtained with rates as high as $235 \text{ m}^3 \cdot \text{day}^{-1} \cdot \text{m}^{-2}$ (American Water Works Association, 1971). Dual-media filters are operated up to loading rates of $300 \text{ m}^3 \cdot \text{day}^{-1} \cdot \text{m}^{-2}$. Usually, a minimum of two filters are constructed to ensure redundancy. For larger plants ($>0.5 \text{ m}^3 \cdot \text{s}^{-1}$), a minimum of four filters is suggested (Montgomery, 1985). The surface area of the filter tank (often called a filter box) is generally restricted in size to about 100 m^2.

In the mid-1980s, deep-bed monomedia filters came into use. These filters are designed to achieve higher loading rates while producing lower finished water turbidities. The filters typically consist of 1.0–1.5-mm diameter anthracite about 1.5–2.5 m deep. They operate at loading rates up to $800 \text{ m}^3 \cdot \text{day}^{-1} \cdot \text{m}^{-2}$.

EXAMPLE 10–10 As part of a proposed new research treatment plant, Dr. Novella is planning to use rapid sand filtration after sedimentation. Dr. Novella plans to use two banks of sand filters. Each filter bed has a surface area of 3 m × 2 m. The design flow rate to each bank of filters is $0.044 \text{ m}^3 \cdot \text{s}^{-1}$. The design loading rate to each bank of filters is $150 \text{ m}^3 \cdot \text{day}^{-1} \cdot \text{m}^{-2}$.

Determine the number of filter beds in each bank of filters. Determine the loading rate when one filter is out of service.

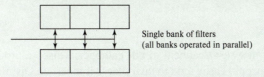

Single bank of filters
(all banks operated in parallel)

Solution The surface area required is the flow rate divided by the loading rate.

$$v_a = \frac{Q}{A_s} = \frac{(0.044 \text{ m}^3 \cdot \text{s}^{-1})(86{,}400 \text{ s} \cdot \text{day}^{-1})}{150 \text{ m}^3 \cdot \text{day}^{-1} \cdot \text{m}^{-2}} = 25.34 \text{ m}^2$$

If the maximum surface area of any one tank is 6 m^2, then the number of filters required is

$$\text{Number} = \frac{25.34}{6} = 4.22$$

Because we cannot build 0.22 filter, we need to round to an integer. Normally, we build an even number of filters to make construction easier and to reduce costs. In this case we would propose to build four filters and check to see that the design loading does not exceed our guideline values. With four filters the loading would be

$$v_a = \frac{Q}{A_s} = \frac{(0.044 \text{ m}^3 \cdot \text{s}^{-1})(86{,}400 \text{ s} \cdot \text{day}^{-1})}{(4 \text{ filters})(6 \text{ m}^2 \cdot \text{filter}^{-1})} = 158.4 \text{ m} \cdot \text{day}^{-1}$$

This is less than the 235 m · day^{-1} recommended maximum loading rate and would be acceptable except that many states require that the filter capacity be sufficient to handle the design flow rate with one filter out of service. Therefore, we must check the loading with three filters in service.

$$v_a = \frac{Q}{A_s} = \frac{(0.044 \text{ m}^3 \cdot \text{s}^{-1})(86{,}400 \text{ s} \cdot \text{day}^{-1})}{(3 \text{ filters})(6 \text{ m}^2 \cdot \text{filter}^{-1})} = 211.4 \text{ m} \cdot \text{day}^{-1}$$

This is less than the 235 m · day^{-1} recommended maximum loading rate and is acceptable. Thus, we would build one bank of four filters.

FIGURE 10–20

Typical cross section of a rapid sand filter. (*Source:* Water Treatment Plant Design, 1969. American Water Works Association.)

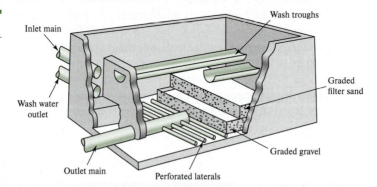

Figure 10–20 shows a cutaway of a rapid sand filter. The bottom of the filter consists of a support medium and collection system. The support medium is designed to keep the sand in the filter and prevent it from leaving with the filtered water. Layers of graded gravel (large on bottom, small on top) traditionally have been used for the support. A perforated pipe is one method of collecting the filtered water. On top of the support medium is a layer of graded sand. The washwater troughs collect the backwash water used to clean the filter. The troughs are placed high enough above the sand layer so that sand will not be carried out with the backwash water.

10–6 DISINFECTION

Disinfection is used in water treatment to kill pathogens (disease-producing microorganisms) present in water that would cause mild to fatal illness if ingested. Disinfection is not the same as sterilization. Sterilization implies the destruction of all living organisms. Drinking water need not be sterile but simply devoid of pathogens at levels that would cause disease.

Four categories of human enteric pathogens are normally of concern: bacteria, viruses, protozoa, and amebic cysts. Effective disinfection must be capable of destroying all four. During the last decade, it has become clear that specific microbial pathogens, such as *Cryptosporidium,* are highly resistant to traditional disinfectants. In 1993, the presence of *Cryptosporidium parvum* oocysts in treated water caused 400,000 people in Milwaukee to experience intestinal illness. More than 4000 were hospitalized, and at least 50 deaths have been attributed to the disease. Cryptosporidiosis outbreaks have also occurred in several other states over the past several years. To protect the general public from waterborne diseases, the U.S. EPA has established maximum contaminant level goals (MCLGs) for viruses, bacteria and protozoa (i.e., *Giardia lamblia* and *Cryptosporidium parvum*). These regulations also include treatment technique requirements for filtered and unfiltered systems that are designed to disinfect against or remove these microbial pathogens.

Under the Long Term 2 Enhanced Surface Water Treatment Rule (LT2 rule), which was promulgated on January 5, 2006, water supplies (surface water and groundwater supplies under the direct influence of surface water) must monitor their source water for *Cryptosporidium,* provide risk-based *Cryptosporidium* treatment, and cover or treat uncovered finished water storage facilities. Based on monitoring results, filtered-water systems will be classified in one of four treatment categories. Most systems are expected to be classified as low risk and no additional treatment will be required. The highest risk waters will require at least 99 or 99.9% (2- or 3-log) inactivation of *Cryptosporidium,* which will likely necessitate additional treatment.

To be of practical service, such water disinfectants must possess the following properties:

1. They must destroy the kinds and numbers of pathogens that may be present in water within a practical period of time over an expected range of water temperatures.
2. They must be effective at the variable compositions, concentrations, and conditions of the waters to be treated.
3. They must be neither toxic to humans and domestic animals nor unpalatable or otherwise objectionable in required concentrations.
4. They must be dispensable at reasonable cost and safe and easy to store, transport, handle, and apply.
5. Their strength or concentration in the treated water must be determined easily, quickly, and (preferably) automatically.
6. They must persist within disinfected water in a sufficient concentration to provide reasonable residual protection against its possible recontamination before use.
7. They must not form toxic by-products due to their reactions with any naturally occurring material in the water.

Disinfection Kinetics

Under ideal conditions, the rate of kill of a type of microorganism can be described by Chick's law (Chick, 1908), which states that the number of organisms destroyed with respect to time is proportional to the number of organisms.

$$\frac{dN}{dt} = -kN \tag{10–24}$$

where N = the number of organisms
k = a first-order rate constant

In its integrated form, the equation is

$$\ln\left(\frac{N}{N_o}\right) = -kt \tag{10–25}$$

where N_o = the number of organisms per unit volume at time zero.

Chick's law does not take into account the concentration of the disinfectant. As such, Watson (1908) proposed a variation of this equation, known as the Chick-Watson law:

$$\ln\left(\frac{N}{N_o}\right) = -k'C^n t \tag{10–26}$$

where $k = k'C^n$
C = disinfectant concentration, mg/L
n = coefficient of dilution

The Chick-Watson law assumes that the disinfectant concentration remains constant.

Under actual conditions, the rate of kill may depart significantly from Chick's law. Increasing rates of kill (with time) may occur because of the time necessary for the disinfectant to damage and inactivate the organism. Decreased rates of kill may occur because of declining concentrations of disinfectant in solution or poor distribution of organisms and disinfectant. To account for these deviations, numerous modifications of Chick's law have been proposed.

In the United States the concept of CT (C = concentration of disinfectant; T = detention time in disinfectant basin) is used to determine the required dose of a disinfectant. The concept is based on Chick's law and data that indicate the required chemical dose necessary to achieve a particular level of treatment for a particular pH and temperature. The time used in this approach is referred to as the t_{10} value, or the time that it takes for 10% of a dye slug to flow from the tank. Tables of CT values have been prepared by the U.S. EPA and are used in designing and operating disinfection systems.

EXAMPLE 10–11   A water treatment plant is to treat 0.1 m^3/s of water. Chlorine is to be used as a disinfectant. At the temperature and pH of the source water, a CT of 200 mg · min · L^{-1} is required. The t_{10} for the contact chamber is 100 min. The ratio of t_{10} to t_o is 0.7. What is the required volume of the reactor? Determine the necessary average chlorine concentration.

1. Calculate t_o, the required hydraulic detention time. The hydraulic detention time and the t_{10} are not equal because the flow in the contact chamber is not ideal. For example, short circuiting occurs at the inlet and exits to the basin, resulting in actual retention times that are less than the theoretical hydraulic detention time.

$$\frac{t_{10}}{t_o} = 0.7$$

$$\frac{100 \text{ min}}{t_o} = 0.7$$

t_o = 142 min or about 145 min

2. From the definition of hydraulic detention time, the volume of the reactor can be calculated.

$$t_o = \frac{V}{Q}$$

$V = (t_o)(Q) = (145 \text{ min})(0.1 \text{ m}^3/\text{s})(60 \text{ s/min})$

$= 870 \text{ m}^3$

3. The average chlorine concentration can be calculated from CT

$$CT = 200 \text{ mg} \cdot \text{min} \cdot L^{-1}$$

$$C = \frac{200 \text{ mg} \cdot \text{min} \cdot L^{-1}}{100 \text{ min}} = 2.0 \text{ mg/L}$$

Disinfectants and Disinfection By-Products

Disinfection By-Products (DBPs) are formed when disinfectants used in water treatment plants react with bromide or naturally occurring organic matter present in the source water. Different disinfectants produce different types or amounts of disinfection by-products. Disinfection by-products for which U.S. EPA drinking-water regulations have been set include trihalomethanes (THMs), haloacetic acids (HAA), bromate, and chlorite.

Trihalomethanes (THMs) are a group of four chemicals that form when chlorine-based disinfectants react with naturally occurring organic matter in water. The THMs are chloroform, bromodichloromethane, dibromochloromethane, and bromoform.

Haloacetic acids (HAAs) are a group of chemicals that form when certain disinfectants react with naturally occurring organic and inorganic matter in water. The regulated haloacetic acids, known as HAA5, are: monochloroacetic acid, dichloroacetic acid, trichloroacetic acid, monobromoacetic acid, and dibromoacetic acid.

Bromate is formed when ozone reacts with naturally occurring bromide found in source water. Chlorite can be formed during the generation of chlorine dioxide.

Bromate (BrO$_3^-$) is a tasteless and colorless inorganic anion, which readily dissolves in water and is stable. It is not found in most natural waters, but is formed when ozone reacts with naturally occurring bromide found in source water. The concentration of bromate formed is dependent on dosage and location of disinfection. Reducing the dosage of ozone or staging ozonation so that ozone is added in multiple but lower doses, reducing the pH, adding ammonia or hydrogen peroxide will reduce bromate formation. While short-term exposure to bromate is not believed to cause any adverse health effects, long-term exposure may increase the risk of cancer. Health Canada has classified bromate as probably carcinogenic to humans (sufficient evidence in animals; no data in humans). The World Health Organization has set a provisional guideline of 10 µg · L^{-1}.

Like bromate, **chlorite** (ClO$_2^-$) is a tasteless and colorless inorganic anion, which readily dissolves in water and is stable—but only in the absence of other reactive chemicals such as free chlorine. It is not found naturally, but is formed during chlorine dioxide (ClO$_2$) generation. Chlorite is also formed during the bleaching of wood pulp by paper mills and the disinfection of municipal wastewater. Chlorite formation can be minimized by careful control of the reactions involved in chloride dioxide generation, reducing oxidant demand, and by minimizing exposure of the water to UV radiation and sunlight. Exposure to chlorite may effect the nervous system in infants and young children, along with fetuses in pregnant women. Chlorite exposure may also result in anemia.

The U.S. EPA first regulated total trihalomethanes (TTHMs) in 1979. The regulations were revised in 1998, when the Stage 1 Disinfectants and Disinfection Byproducts Rule (Stage 1 DBPR) was issued by the U.S. EPA, as part of the 1996 Amendments to the Safe Drinking Water Act. The Stage 1 DBPR set the maximum contaminant level for TTHM and HAA5 at 80 ppb and 60 ppb, respectively. MCLs for bromate and chlorite were also set for systems using ozone and chlorine dioxide, respectively, for treatment. The Stage 2 DBPR, which was proposed in August 2003 and promulgated on December 15, 2005, requires that treatment plants identify locations within the distribution system with high DBP concentrations. These locations will then be used for compliance monitoring. Unlike Stage 1 Rule, this rule requires that the TTHM and HAA5 concentrations be calculated as the locational running annual average (LRAA) for each of these monitoring locations. Under Stage 1 requirements, the compliance was determined by calculating the running annual average of samples from across all monitoring locations in the system. This new method is more protective of human health. A summary of regulations for DBPs under Stage 2 regulations is provided in Table 10–4.

Since the discovery of chlorination by-products in drinking water in 1974, numerous toxicological and epidemiological studies have been conducted. These studies have shown several disinfection by-products to be carcinogenic or to cause reproductive or developmental disorders in laboratory animals. Some considerable uncertainty surrounds the laboratory and epidemiological studies that have been conducted, and extensive debate is ongoing in the scientific community about the significance of these contradictory findings. Taking a conservative approach, EPA believes the weight of evidence presented by the available epidemiological studies on chlorinated drinking water and toxicological studies on individual disinfection by-products supports a potential hazard concern and warrants regulatory action, such as that taken in the Stage 1 D/DBP Rule. Extensive research is currently underway to better understand the potential risks resulting from exposure to disinfection by-products.

Chlorine Reactions in Water

In the United States, chlorine has been the most commonly used disinfectant. When used properly, chlorine is effective, and practical. It has several advantages over other disinfectants. It is long-lasting and, therefore, provides a residual in the distribution system. Chlorine acts to kill

TABLE 10–4	Regulatory Limits and Guidelines for Disinfection By-Products		
Disinfection By-Product	U.S. EPA MCL Set by Under Stage 2 D/DBP Rule	Health Canada	World Health Organization Guidelines
TTHM	80 ppb, annual locational running average	100 ppb	Chloroform: 200 ppb Bromodichloromethane: 60 ppb Dibromochloromethane: 100 ppb Bromoform: 100 ppb
HAA5	60 ppb, annual locational running average	In preparation	Dichloroacetic acid: 50 ppb Trichloroacetic acid: 100 ppb (provisional)
Bromate	10 ppb, annual average	10 ppb	10 ppb (provisional)
Chlorite	1 ppm, monthly average	1 ppm, proposed as of March 2006	0.7 ppm (provisional)

Source: Williams, LeBel, and Benoit, 1995; WHO, 2004.

pathogens by first penetrating through the cell wall, then destroying enzymes within the cytoplasm. It has been used not only as a disinfectant but also to control slime and algae; to destroy unpleasant taste and odor-causing compounds; and to oxidize iron, manganese, and hydrogen sulfide. The major disadvantages of chlorine are the formation of chlorinated by-products due to the reaction of chlorine with naturally occurring organic matter and the lack of effectiveness of chlorine in inactivating *Cryptosporidium* oocysts.

The effectiveness of chlorine depends on several factors, including

- **Dosage** (i.e., concentration). For chlorine to be effective it must be present at a sufficiently high concentration to inactivate the pathogens. Natural waters will contain both inorganic and organic contaminants that are reactive with chlorine and other disinfectants. These include reduced metals and ammonia. Fe(II) and Mn(II) reacts with chlorine to form Fe(III) and Mn(IV), respectively. Ammonia reacts with chlorine to form chloramines. Naturally occurring organic matter, formed from the degradation of plant and animal material, is also reactive with chlorine. For chlorine to be effective, it must be added at a concentration that exceeds the demand exerted by these species.

- **Contact time.** Chlorine must be in physical contact with the pathogens for a sufficient time to achieve inactivation.

- **Turbidity.** The presence of particles (turbidity) essentially hides the pathogen from the disinfectant.

- **Other reactive species.** The presence of other reactive species, such as ammonia, can consume the disinfectant, thereby, reducing the concentration available for inactivation, along with producing toxic chemicals as discussed previously.

- **pH.** Chlorine is most effective at pH values less than 7.5.

- **Water temperature.** As the temperature increases, the rate of disinfection increases; however, chlorine becomes less stable.

Chlorine may be added as the gas (Cl_2), as sodium hypochlorite (NaOCl), or as calcium hypochlorite [$Ca(OCl)_2$]. In water, chlorine gas very rapidly hydrolyzes to form HOCl and hydrochloric acid (HCl).

$$Cl_2(g) + H_2O \rightleftharpoons HOCl + H^+ + Cl^-$$

(10–27)

This reaction is pH-dependent. At the conditions found in most water treatment systems, this reaction is essentially complete, and no dissolved chlorine gas is present. Because hypochlorous acid is a weak acid, it dissociates to form OCl^- according to the reaction

$$HOCl \rightleftharpoons H^+ + OCl^- \tag{10–28}$$

Because the pK_a for this reaction is 7.54 at 25°C, at pH values above this, essentially all of the chlorine is present as OCl^-, whereas below a pH 7.54, hypochlorous acid predominates. This is significant because hypochlorous acid is a much more effective disinfectant than hypochlorite.

Chloramines

Chloramines are formed by the reaction of hypochlorous acid with ammonia as shown below:

$$NH_3 + HOCl \rightleftharpoons NH_2Cl + H_2O \tag{10–29}$$
$$\text{Monochloramine}$$

$$NH_2Cl + HOCl \rightleftharpoons NHCl_2 + H_2O \tag{10–30}$$
$$\text{Dichloramine}$$

$$NH_2Cl_2 + HOCl \rightleftharpoons NCl_3 + H_2O \tag{10–31}$$
$$\text{Trichloramine}$$

The distribution of the reaction products is governed by the rates of formation of monochloramine and dichloramine, which are dependent upon pH, temperature, reaction time, and the initial $HOCl:NH_3$ concentrations. High $HOCl:NH_3$ ratios, low temperatures, and low pH favor the formation of dichloramine. Collectively, chloramines are referred to as **combined chlorine**. The sum of the concentrations of free chlorine and combined chlorine is called **total chlorine**. Chloramines, while effective as a disinfectant, are not as powerful an oxidant as is chlorine or hypochlorous acid. As such, chloramines are usually not used as the primary disinfectant in water treatment systems.

As shown in Figure 10–21, at low chlorine doses, chlorine is consumed by reduced species and natural organic matter (NOM). As the chlorine dose increases, chloramines form along with chlorinated organic compounds. The chlorinated organic compounds form from the reaction of chlorine with organic nitrogenous compounds, such as proteins and amino acids. As a result of these reactions, the chlorine residual increases due to the increased concentration of chloramines (combined chlorine). At higher concentrations, the chlorinated organic compounds and chloramines are oxidized, resulting in a reduction in the chlorine residual. Eventually as those compounds are oxidized, the chlorine residual reaches a minimum value and subsequently

FIGURE 10–21

Diagram showing the effect of chlorine dosage on the chlorine residual in waters containing naturally occurring organic matter and/or ammonia.

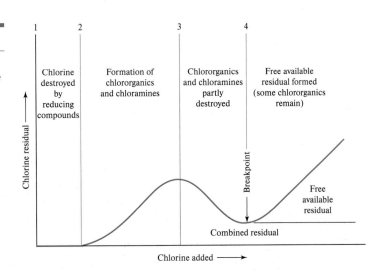

begins to increase. This increase is due to the formation of free chlorine. The point at which the chlorine level reaches a minimum value is known as the breakpoint and the process is called **breakpoint chlorination**.

Chlorine Dioxide

Another very strong oxidant is chlorine dioxide. Chlorine dioxide (ClO_2) must be generated on-site by combining chlorine and sodium chlorite. Chlorine dioxide is often used as a primary disinfectant, inactivating the bacteria and cysts. Chloramine is then used as a secondary disinfectant to provide a residual in the distribution system because chlorine dioxide does not maintain a residual long enough to be useful as a distribution system disinfectant. The advantage of chlorine dioxide over chlorine is that chlorine dioxide does not react with naturally occurring organic matter to form THMs. The major disadvantage of chlorine dioxide is the potential formation of chlorite and chlorate, which are potential human carcinogens.

Ozonation

Ozone is a sweet-smelling, unstable gas. It is a form of oxygen in which three atoms of oxygen are combined to form the molecule O_3. Because of its instability, it is generated at the point of use. Ozone is produced from air or oxygen, and the air exiting the generation equipment can contain up to 13% ozone in oxygen.

Ozone is widely used in drinking-water treatment in Europe and is continuing to gain popularity in the United States. It is a powerful oxidant, more powerful even than hypochlorous acid. Table 10–5 shows the CT values for ozone, chlorine dioxide, and chloramine for 3-log inactivation (99.9% removal) of *Giardia* cysts, indicating the effectiveness of ozone as a disinfectant.

In addition to being a strong oxidant, ozone has the advantage of not forming THMs or any of the chlorinated DBPs. A disadvantage of ozone is that it does form some low-molecular-weight compounds that can react with chlorine-based disinfectants to produce chlorinated aldehydes and ketones. Another disadvantage is that, unless the low-molecular-weight compounds formed from the reaction of ozone with naturally occurring organic matter are removed, the regrowth of bacteria in the distribution system can be problematic. As with chlorine dioxide, ozone does not persist in the water, decaying back to oxygen in minutes. Therefore, a typical flow schematic adds ozone either to the raw water or between the sedimentation basins and filter for primary disinfection, then adds chloramine after the filters as the distribution system disinfectant.

Ultraviolet Radiation

UV radiation can be used as a disinfectant as it has the potential to inactivate pathogens. The inactivation of microorganisms by UV radiation results from a photochemical reaction with nucleic acids and other vital cellular chemicals. This results in injury or death to the exposed organism.

TABLE 10–5	CT Values (mg-min · L^{-1}) for 99.9% *Giardia* Cyst Inactivation at a pH 6.0–9.0					
Temperature	**1°C**	**5°C**	**10°C**	**15°C**	**20°C**	**25°C**
Chlorine dioxide	81	54	40	27	21	14
Chlorine	273	192	144	96	72	48
Ozone	2.9	1.9	1.43	0.95	0.72	0.48
Chloramines	3800	2200	1850	1500	1100	750

Source: U.S. EPA (1999).

Note: Chlorine 1°C actually at 0.5°C; dose for chlorine 1.6 mg · L^{-1}. Other doses unspecified; pH for chlorine 7.5, other oxidants are given as range 6–9.

Ultraviolet radiation has a wavelength in the 200–315 nm range. It is classified as Vacuum UV (100–200 nm), UV-C (200–280 nm), UV-B (280–315 nm), and UV-A (315–400 nm). UV radiation is typically generated by mercury lamps. The optimum UV range for disinfection is between 245 and 285 nm. UV disinfection utilizes either low-pressure lamps that emit maximum energy output at a wavelength of 253.7 nm; medium pressure lamps that emit energy at wavelengths from 180 to 1370 nm; or lamps that emit at other wavelengths in a high intensity "pulsed" manner.

The inactivation of microorganisms by UV radiation is directly related to the UV dose. The UV dose required to inactivate protozoan cysts or oocysts of *Giardia* and *Cryptosporidium* are several times greater than that required for bacteria and viruses (U.S. EPA, 1996). The average UV dose is calculated as:

$$D = IT \tag{10–32}$$

where D = UV dose
$\quad I$ = average intensity, mW $\cdot$ s $\cdot$ cm^{-2}
$\quad T$ = exposure time, s

The UV dose required for effective inactivation is determined by site-specific data relating to the water quality and the required log removal. Since intensity directly affects the dose, germicidal efficacy is dependent on the ability of the UV light to pass through the water to get to the target organism. Thus, the lamps must be free of slime and precipitates and the water should be free of turbidity and color, allowing good transmission of UV light throughout the entire water column.

The rate of inactivation of microorganisms by UV radiation can be described by a modified form of the Chick-Watson law that includes the effects of wavelength of light (Linden and Darby, 1997; MWH, 2005):

$$\left(\frac{dN}{dt}\right)_\lambda = I_\lambda N \tag{10–33}$$

where I_λ = the effective germicidal intensity of UV radiation for wavelength, λ, mW $\cdot$ cm^{-2}.

For multiple wavelengths, the intensity must be integrated over the entire radiation spectrum.

Most conventional UV reactors are either closed vessel or open channel, with closed vessel reactors being more common for drinking water applications. Closed vessel reactors have a small footprint, minimize exposure of UV radiation to plant personnel, avoid the release of airborne contaminants, and have a modular design (U.S. EPA, 1996). The major disadvantage of UV irradiation is that it leaves no residual protection in the distribution system and is generally more expensive than conventional disinfectants.

10–7   OTHER TREATMENT PROCESSES FOR DRINKING WATER

Membrane Processes

Membranes are selective, semipermeable barriers, which exclude some constituents and allow others to pass. The passage of substances across a membrane requires a driving force (i.e., a potential difference across the membrane). With most membrane systems used for drinking-water treatment, pressure serves as the driving force. As shown in Figure 10–22, membrane filtration can be classified on the basis of the size of the particles or ions that are removed by the membrane. Microfiltration and ultrafiltration operate by sieving or size exclusion at low pressures (typically between 20 and 275 kPa). Microfiltration (MF) can be used to remove particles 50 nm and can achieve several log reduction in bacteria, but can only attain a 2–3 log reduction in the number of viruses (U.S. EPA, 2001). Ultrafiltration (UF) removes particles larger than 2 nm and can achieve greater than 4-log removal of viruses (U.S. EPA, 2001).

FIGURE 10–22

Membrane process classification by molecular weight cutoff and membrane pore size. (*Source:* U.S. Environmental Protection Agency, 2001.)

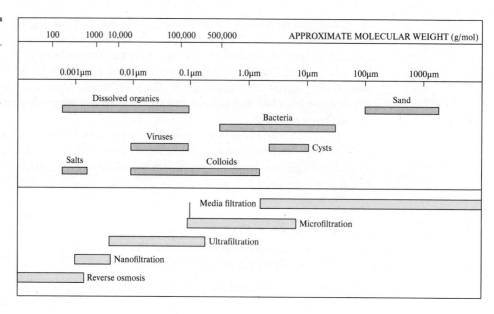

Membrane processes are being used more commonly in water treatment because of their ability to remove small particles including *Cryptosporidium* oocysts and *Giardia* cysts. Compared to conventional treatment methods, membrane processes can provide higher quality water, minimize disinfectant demand, are more compact, provide easier operational control and less maintenance, and generate less sludge.

Ultra- and microfiltration performance is based on both water flux and the rejection of various constituents, typically particulate matter, bacteria, and/or viruses. The pure water flux across a clean membrane is directly proportional to the transmembrane pressure (TMP). It is inversely proportional to the viscosity of water and therefore, directly proportional to the temperature of the water. It is also inversely proportional to the membrane resistance coefficient, which is a measure of the degree to which the flow of water is impeded by the membrane material or the material that deposits on the membrane surface. As such, the volumetric flux, J, is described as:

$$J = \frac{Q_F}{A} = \frac{\Delta P}{\mu R_m}$$

(10–34)

where J = volumetric water flux through the membrane, $\text{m}^3 \cdot \text{m}^{-2} \cdot \text{h}^{-1}$
 Q_F = volumetric flow rate of pure water, $\text{m}^3 \cdot \text{h}^{-1}$
 A = surface area of clean membrane, m^2
 ΔP = transmembrane pressure, kPa
 μ = dynamic viscosity, Pa $\cdot$ s
 R_m = membrane resistance coefficient, m^{-1}

Rejection is a measure of the degree to which constituents in the feed water are prevented from passing through a semipermeable membrane. It is defined by the equation:

$$R = 1 - \left(\frac{C_{permeate}}{C_{feed}} \right)$$

(10–35)

where R = rejection, dimensionless
 $C_{permeate}$ = concentration of the constituent in the permeate, $\text{g} \cdot \text{m}^{-3}$
 C_{feed} = concentration of the constituent in the feed water, $\text{g} \cdot \text{m}^{-3}$

EXAMPLE 10–12   You are to design a membrane system for the community of Red Bull. The design flow rate is 0.100 m$^3 \cdot$ s^{-1}. The temperature of the water ranges from 5°C in winter to 22°C in summer. The membrane you have chosen has maximum TMP of 200 kPa and a resistance coefficient of 4.2×10^{12} m^{-1}. Based on pilot testing, the operating TMP should be no greater than 80% of the maximum TMP. Determine the required membrane area.

Solution   Since the water flux is directly proportional to the temperature of the water, the membrane system should be designed for the winter temperature when the viscosity is the greatest and the flux the lowest.

$$J = \frac{Q_F}{A} = \frac{\Delta P}{\mu R_m}$$

$$A = \frac{(0.100 \text{ m}^3 \cdot \text{s}^{-1})(1.519 \times 10^{-3} \text{ Pa} \cdot \text{s})(4.2 \times 10^{12} \text{ m}^{-1})}{(0.80)(200 \text{ kPa})(1000 \text{ Pa/kPa})} = 3987 \text{ m}^2 \equiv 4000 \text{ m}^2$$

Nanofiltration (NF) and reverse osmosis (RO) can be used for the removal of salts and low-molecular-weight dissolved organic chemicals. With these processes, the operating pressures must exceed the osmotic pressure of the water being treated so as to force the water across a membrane, leaving a concentrated brined solution behind. NF membranes are capable of rejecting high percentages of multivalent ions and divalent cations while allowing monovalent ions to pass. RO is capable of removing ionic substances as small as 1 to 15 Å in diameter.

The fate of salts and particulate matter in membrane systems can be described using a simple mass balance approach. The mass balance for water is given by the equation:

$$Q_F = Q_p + Q_c \tag{10–36}$$

where Q_c = concentrate flow rate in the reject stream, m$^3 \cdot$ h^{-1}
  Q_p = permeate flow rate, m$^3 \cdot$ h^{-1}

Since the salts and/or dissolved material are conservative, a material balance can be written as:

$$Q_F C_F = Q_p C_p + Q_c C_c \tag{10–37}$$

where C_F = concentration of salt or dissolved material in feed water, g $\cdot$ m^{-3}
  C_c = concentration of salt or dissolved material in concentrate, g $\cdot$ m^{-3}
  C_p = concentration of salt or dissolved material in permeate, g $\cdot$ m^{-3}

With reverse osmosis, the water flux is proportional to the difference in the transmembrane pressure, ΔP, and $\Delta \pi$, the difference in the osmotic pressures of the feed and permeate solutions. It is described by the equation:

$$J = k_w (\Delta P - \Delta \pi) \tag{10–38}$$

where k_w = mass transfer coefficient for water flux, m$^3 \cdot$ d$^{-1} \cdot$ m$^{-1} \cdot$ kPa
  $\Delta \pi$ = the difference in the osmotic pressures of the feed and permeate, kPa

The osmotic pressure is determined using the equation:

$$\pi = i\varphi \, CRT$$

where i = the number of ions per mole produced during dissociation of the solute
  φ = osmotic coefficient, unitless
  C = concentration of all solutes, mol $\cdot$ L^{-1}
  R = universal gas constant, 8.314 kPa $\cdot$ m$^3 \cdot$ kg$^{-1} \cdot$ mol$^{-1} \cdot$ K^{-1}
  T = temperature, K

The number of ions per mole, i, would be 2 for NaCl. The osmotic coefficient, φ, depends on the nature of the salt and its concentration. For example, for NaCl it ranges from 0.93 to 1.03 over a concentration range of 10 to 120 g/L of salt. Seawater has an osmotic coefficient that varies from 0.85 to 0.95.

The driving force for the movement of solutes across the membrane is the concentration gradient. The solute flux is

$$J_s = k_s(\Delta C) \tag{10-39}$$

where Js = mass flux of solute, $kg \cdot d^{-1} \cdot m^{-2}$
K_s = mass transfer coefficient for solute flux, $m^3 \cdot d^{-1} \cdot m^{-2}$
ΔC = concentration gradient across the membrane, $kg \cdot m^{-3}$

With reverse osmosis, rejection is defined in the same way as mentioned above with micro- and ultrafiltration. The recovery is the ratio of the permeate flow rate to feed water flux rate:

$$r = \frac{Q_p}{Q_F} \tag{10-40}$$

where Q_p = permeate flow rate, $m^3 \cdot h^{-1}$. For example, if the recovery is 80%, then 80% of the feed flow is produced as permeate and 20% of the feed flow is the reject (concentrate) stream. Using the mass balance approach shown in Equation 10–37, it is apparent that the concentration of salts in the concentrate (reject stream) would be 5 times greater than the concentration in the feed flow if the salt rejection were 100%. It is for this reason that the disposal of the concentrate (brine) stream can be problematic.

EXAMPLE 10–13 A desalination plant treats sea water with a salt concentration of 30,000 mg $\cdot$ L^{-1} at a flow rate of 1.5 m^3 $\cdot$ s^{-1}. The recovery is 70%. The salt rejection is 99.5%. Determine the concentration of salt in the concentrate stream.

Solution Given: $Q_F = 1.5$ m^3 $\cdot$ s^{-1}
With a recovery of 70%,

$$Q_p = (0.70)(1.5 \text{ m}^3 \cdot \text{s}^{-1}) = 1.05 \text{ m}^3 \cdot \text{s}^{-1}$$
$$Q_C = 1.5 \text{ m}^3 \cdot \text{s}^{-1} - 1.05 \text{ m}^3 \cdot \text{s}^{-1} = 0.45 \text{ m}^3 \cdot \text{s}^{-1}$$

Since the salt rejection is 99.5%, the concentration of salt in the permeate stream is

$$C_p = (1 - 0.995)(30,000 \text{ mg} \cdot \text{L}^{-1}) = 150 \text{ mg} \cdot \text{L}^{-1}$$

Using Equation 10–37

$$Q_F C_F = Q_P C_p + Q_c C_c$$

$$(1.5 \text{ m}^3 \cdot \text{s}^{-1})(30,000 \text{ mg} \cdot \text{L}^{-1}) = (1.05 \text{ m}^3 \cdot \text{s}^{-1})(150 \text{ mg} \cdot \text{L}^{-1}) + (0.45 \text{ m}^3 \cdot \text{s}^{-1})(C_c)$$

$$C_c = 99,650 \text{ mg} \cdot \text{L}^{-1} \text{ or about 3.3 times the concentration of salt in the feed water.}$$

Most commercially available membranes used in drinking water applications are poly- meric in nature. Common polymers include polypropylene, polyvinyl difluoride, and cellulose acetate. Each of the membrane materials have different resistance to pH and oxidants and vary- ing surface charge and hydrophobicity. As such, each membrane will have different properties in terms of the sizes and types of particles or ions excluded.

One of the major challenges associated with the operation of membrane filtration is the energy (and therefore, costs) required for its operation as a result of the decrease in the permeate flux due to membrane fouling. Membrane fouling lowers the economic efficiency of membrane systems by reducing the quantity of treated water, shortening membrane life, and increasing the frequency of membrane cleaning. The fouling of membranes in drinking-water systems has been attributed primarily to NOM. Fouling rates are influenced by the concentration and nature of the particulates and solutes present, hydrodynamics, the pore size of the membrane, and the surface characteristics of the membrane. Extensive research is still being conducted in an effort to find a solution to the problem of membrane fouling.

It is important to enhance the performance of membrane systems, so that they can become more affordable. Costs can be decreased by reducing membrane fouling and/or using coarser membranes. While coarser membranes have higher permeabilities or fluxes of water through the membrane, they usually are not as effective at removing microorganisms. They should be used only if other processes are in place to achieve the desired result in terms of microorganism removal or if the membrane is modified to provide more effective treatment.

Advanced Oxidation Processes (AOPs)

AOPs are processes designed to produce hydroxyl radicals (OH). Hydroxyl radicals are highly reactive, nonselective oxidants able to decompose many organic compounds. Two AOPs are the combinations of ozone and hydrogen peroxide and ozone and UV radiation. AOPs are most commonly used for the oxidation of chemicals that cannot be removed by other means.

Carbon Adsorption

Carbon adsorption is essentially the same as that described in Section 5, Chapter 2, with the attachment of chemical compounds onto soil particles. Here, in water treatment, the adsorbent (solid) is activated carbon, either granular (GAC) or powdered (PAC). Activated carbon is most commonly used for the removal of taste- and odor-causing compounds, as well some synthetic organic compounds. The major disadvantage of activated carbon is that the problematic compounds are not destroyed but simply transferred from one matrix, water, to another, carbon surfaces. Additionally, when used for the removal of nonpolar compounds, a GAC filter typically lasts 90–120 days before it loses its adsorptive capacity. Because of its short life, the GAC needs to be regenerated. Regeneration can be accomplished by heating the activated carbon to about 900°C to drive off the adsorbed organic chemicals. Air pollution control devices are necessary to prevent the release of any unburned organic chemicals. This is clearly an expensive proposition. Although a GAC can be used for the removal of THMs, its capacity is very low and the carbon may only last up to 30 days.

Aeration

Aeration is used in drinking-water treatment primarily for the oxidation of iron and for the removal of volatile organic chemicals. Soluble iron can be removed by the following reaction:

$$4Fe(HCO_3)_2 + O_2 + 2H_2O \rightleftharpoons 4Fe(OH)_3 + 8CO_2 \tag{10-41}$$

The rate of oxidation of Fe(II) can be expressed by the rate law (Stumm and Morgan, 1996).

$$\frac{-d[Fe(II)]}{dt} = \frac{k[O_2(aq)]}{[H^+]^2}[Fe(II)] \tag{10-42}$$

where $k = 3 \times 10^{-12}$ $min^{-1} \cdot mol^{-1}$ at 20°C. At pH values greater than 7 the rate of iron oxidation is rapid (>90% removal in less than 10 min).

Manganese oxidation is much slower than that of ferrous iron. At conventional temperatures oxidation of Mn(II) is not extensive at pH values below 9 unless catalytic oxidation is used.

Aeration can also be used for the removal of volatile organic chemicals, such as trichloroethylene and tetrachloroethane. When aeration is used for such means, treatment of the off-gas

FIGURE 10–23

A typical air-stripping unit. (Image ©Raschig-USA. Inc.; *Source:* Remtech Engineers.)

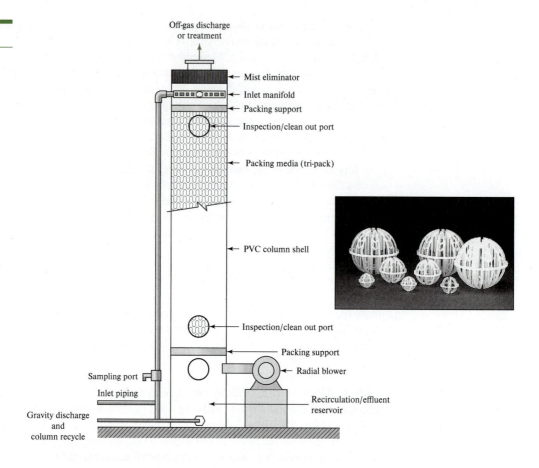

FIGURE 10–23

A typical air-stripping unit. (Image ©Raschig-USA. Inc.; *Source:* Remtech Engineers.)

is necessary to avoid the release of these chemicals into the atmosphere. Countercurrent operation, whereby the stripping gas and the water flow in opposite directions, is often used for maximum removal of the chemicals.

A typical air-stripping unit is shown in Figure 10–23.

10–8 WATER PLANT RESIDUALS MANAGEMENT

Water treatment plants produce a number of waste streams (residuals), which can be divided into five general categories.

1. Settling sludge formed during the removal of turbidity using alum or iron salts
2. Filter backwash water and associated solids
3. Softening sludges
4. Waste streams from ion-exchange regeneration, reverse osmosis reject streams, or spent activated carbon
5. Contaminated air streams generated during aeration processes

The precipitated chemicals and other materials removed from water to make it potable and palatable are termed **sludge.** Satisfactory treatment and disposal of water treatment plant sludge can be the single most complex and costly operation in the plant.

The sludges withdrawn from coagulation and softening plants are composed largely of water (as much as 98%). Thus, for example, 20 kg of solid chemical precipitate is accompanied by 980 kg of water. Assuming equal densities for the precipitate and water (a bad assumption at best), approximately 1 m³ of sludge is produced for each 20 kg of chemicals added to the water.

For even a small plant (say 0.05 m³ · s⁻¹), this might mean up to 800 m³ · year⁻¹ of sludge—a substantial volume to say the least!

The most logical sludge management program attempts to use the following approach in disposing of the sludge:

1. Minimization of sludge generation
2. Chemical recovery of precipitates
3. Sludge treatment to reduce volume
4. Ultimate disposal in an environmentally safe manner

Mass-Balance Analysis

Clarifier sludge production can be estimated by a mass-balance analysis of the sedimentation basin. Because no reaction is taking place, the mass-balance Equation (4–2) reduces to the form

$$\text{Accumulation rate} = \text{input rate} - \text{output rate} \tag{10–43}$$

EXAMPLE 10–14 A coagulation treatment plant with a flow of 0.044 m³ · s⁻¹ is dosing alum at 33.0 mg · L⁻¹. No other chemicals are being added. The raw water/suspended solids concentration is 47.0 mg · L⁻¹. The effluent/suspended solids concentration is measured at 10.0 mg · L⁻¹. The sludge solids content is 1.05%, and the specific gravity of the sludge solids is 2.61. What volume of sludge must be disposed of each day?

Solution The mass-balance diagram for the sedimentation basin is shown here.

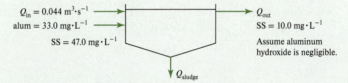

Let's first perform a mass balance on the suspended solids,

$$C_{in} = 47.0 \text{ mg} \cdot \text{L}^{-1}$$

$$C_{out} = 10.0 \text{ mg} \cdot \text{L}^{-1}$$

Therefore, for every liter of water treated (47.0 − 10.0) or 37.0 mg · L⁻¹ suspended solids is removed in the sludge each day. The mass of suspended solids sludge can be calculated as

$$Q_{in}C_{removed} = (0.044 \text{ m}^3 \cdot \text{s}^{-1})(37.0 \text{ mg} \cdot \text{L}^{-1})(1000 \text{ L} \cdot \text{m}^{-3})(86{,}400 \text{ s} \cdot \text{day}^{-1})$$

$$\times (10^{-6} \text{ kg} \cdot \text{mg}^{-1})$$

$$= 140.66 \text{ kg} \cdot \text{day}^{-1}$$

Now let's look at the amount of sludge produced by the addition of alum. Using Equation 10–1, we see that there are two moles of aluminum hydroxide (Al(OH)₃) produced for every mole of alum added. The molecular weight of alum is 594.35 g · mol⁻¹. The molecular weight of aluminum hydroxide is 78.00 g · mol⁻¹. Therefore, using basic stoichiometric relationships, we obtain the alum dose in moles per liter

$$\left(\frac{33.0 \text{ mg} \cdot \text{L}^{-1}}{594.35 \text{ g} \cdot \text{mol}^{-1}} \right) \left(\frac{\text{g}}{1000 \text{ mg}} \right) = 5.55 \times 10^{-5} \text{ mol} \cdot \text{L}^{-1}$$

And the concentration of aluminum hydroxide precipitated

$$= \left(\frac{2 \text{ mol Al(OH)}_3}{\text{mole alum}} \right) (5.55 \times 10^{-5} \text{ mol} \cdot \text{L}^{-1}) = 1.11 \times 10^{-4} \text{ mol} \cdot \text{L}^{-1}$$

Converting the concentration of aluminum hydroxide to units of $mg \cdot L^{-1}$ we obtain

$$(1.11 \times 10^{-4} \text{ mol} \cdot L^{-1})(132 \text{ g} \cdot \text{mol}^{-1})(1000 \text{ mg} \cdot \text{g}^{-1}) = 14.5 \text{ mg} \cdot L^{-1}$$

Thus, the mass of dry aluminum hydroxide sludge produced each day is

$$(0.044 \text{ m}^3 \cdot \text{s}^{-1})(14.5 \text{ mg} \cdot L^{-1})(1000 \text{ L} \cdot \text{m}^{-3})(86,400 \text{ s} \cdot \text{day}^{-1})(10^{-6} \text{ kg} \cdot \text{mg}^{-1})$$
$$= 55.2 \text{ kg} \cdot \text{day}^{-1}$$

Thus the total sludge production is

$$140.66 + 55.2 = 195.9 \text{ kg} \cdot \text{day}^{-1}$$

Because this is a dry mass and the sludge has only 1.05% solids, we must account for the volume of water in estimating the volume to be removed each day. The solids content is defined as the

$$\text{Solids content as percent} = \frac{\text{mass}_{(solids)}}{\text{mass}_{(solids)} + \text{mass}_{(water)}} \times 100$$

$$1.05 = \frac{195.9}{195.9 + \text{mass}_{(water)}} \times 100$$

$$\text{Mass}_{(water)} = 19,380 \text{ kg} \cdot \text{day}^{-1}$$

Now we use the definition of density (mass per unit volume) to find the volume of sludge and water. The specific gravity of the solids is 2.65, and the density of water is $1000 \text{ kg} \cdot \text{m}^{-3}$.

$$\text{Volume} = \frac{\text{mass}}{\text{density}}$$

$$V_T = \text{volume of solids} + \text{volume of water}$$

$$= \frac{195.9 \text{ kg} \cdot \text{day}^{-1}}{2.65 \times 1000 \text{ kg} \cdot \text{m}^{-3}} + \frac{19,380 \text{ kg} \cdot \text{day}^{-1}}{1000 \text{ kg} \cdot \text{m}^{-3}}$$

$$= 0.074 + 19.4$$

$$= 19.5 \text{ m}^3 \cdot \text{day}^{-1}$$

Obviously, the solids account for only a small fraction of the total volume. This is why sludge dewatering is an important part of the water treatment process.

Sludge Treatment

The treatment of solid/liquid wastes produced in water treatment processes involves the separation of the water from the solid constituents to the degree necessary for the selected disposal method. The required degree of treatment is therefore a direct function of the ultimate disposal method.

Several sludge treatment methodologies have been practiced in the water industry. Figure 10–24 shows the most common sludge-handling options available, listed by general categories of thickening, dewatering, and disposal. In choosing a combination of the possible treatment process trains, it is probably best to first identify the available disposal options and their requirements for a final cake solids concentration. Most landfill applications will require a

FIGURE 10–24

Common sludge-handling options.

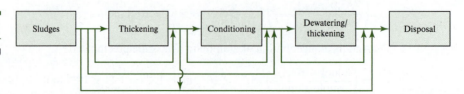

Sludges → Thickening → Conditioning → Dewatering/thickening → Disposal

TABLE 10–6	**Range of Cake Solid Concentrations Obtainable**	
	Lime Sludge (%)	**Coagulation Sludge (%)**
Gravity thickening	15–30	3–4
Basket centrifuge	10–15	
Scroll centrifuge	55–65	10–20
Belt filter press	10–15	
Vacuum filter	45–65	n/a
Pressure filter	55–70	30–45
Sand drying beds	50	20–25
Storage lagoons	50–60	7–15

"handleable" sludge and this may limit the type of acceptable dewatering devices. Methods and costs of transportation may affect the decision of "how dry is dry enough." The criterion should not be to simply reach a given solids concentration but rather to reach a solids concentration of desired properties for the handling, transport, and disposal options available.

Table 10–6 shows a generalized range of results that have been obtained for final solids concentrations from different dewatering devices for coagulant and lime sludges.

To give you an understanding of these solids concentrations, a sludge cake with 35% solids would have the consistency of butter, whereas a 15% sludge would have a consistency much like tar.

After removal of the sludge from the clarifier or sedimentation basin, the first treatment step is usually thickening. Thickening assists the performance of any subsequent treatment, gets rid of much water quickly, and helps to equalize flows to the subsequent treatment device. An approximation for determining sludge volume reduction via thickening is given by

$$\frac{V_2}{V_1} = \frac{P_1}{P_2}$$

(10–44)

where V_1 = volume of sludge before thickening (in m^3)
$\quad\quad V_2$ = volume of sludge after thickening (in m^3)
$\quad\quad P_1$ = percent of solids concentration of sludge before thickening
$\quad\quad P_2$ = percent of solids concentration after thickening

Thickening is usually accomplished by using circular settling basins similar to a clarifier (Figure 10–25). Thickeners can be designed based on pilot evaluations or using data obtained from similar plants. Lime sludges are typically loaded at 100–200 kg · day^{-1} · m^{-2}, and coagulant sludge loading rates are about 15–25 kg · day^{-1} · m^{-2}.

FIGURE 10–25

Continuous-flow gravity thickener.

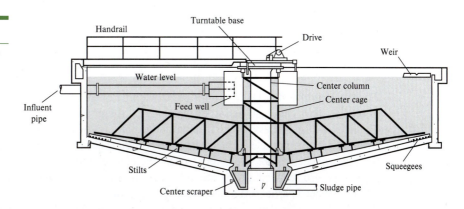

Following thickening of the sludge, dewatering can take place by either mechanical or nonmechanical means. In nonmechanical devices, sludge is spread out, allowing the free water to drain away and the remaining water to evaporate. Sometimes the amount of free water available to drain is enhanced by natural freeze–thaw cycles. In mechanical dewatering, some type of device is used to force the water out of the sludge.

We begin our discussion with the nonmechanical methods and follow with the mechanical methods.

Lagoons.   A **lagoon** is essentially a large pond dug out for containment of the sludge. Lagoons can be constructed as either storage lagoons or dewatering lagoons. **Storage lagoons** are designed to store and collect the solids for some predetermined length of time. They generally have decant capabilities but no underdrain system. Storage lagoons should be equipped with sealed bottoms to protect the groundwater. Once the storage lagoon is full or the decant can no longer meet discharge limitations, it must be abandoned or cleaned. To facilitate drying, the standing water may be removed by pumping, leaving a wet sludge. Coagulant sludges can only be expected to reach a 7–10% solids concentration in storage lagoons. The remaining solids must be either cleaned out wet or allowed to evaporate. Depending on the depth of the wet solids, evaporation can take years. The top layers will often form a crust, preventing evaporation of the bottom layers of sludge.

The primary difference between a dewatering lagoon and a storage lagoon is that a **dewatering lagoon** has a sand and underdrain bottom, similar to a drying bed. Dewatering lagoons can be designed to achieve a dewatered sludge cake. The advantage of a dewatering lagoon over a drying bed is that storage is built into the system to assist in meeting peak solids production or to help in handling sludge during wet weather. Compared with conventional drying beds, bottom sand layers have the disadvantage of plugging up with multiple loadings, thereby increasing the required surface area. Polymer treatment can be useful to prevent clogging.

Sand-Drying Beds.   Sand-drying beds operate on the simple principle of spreading the sludge out and letting it dry. As much water as possible is removed by drainage or decanting, and the rest of the water must evaporate until the desired final solids concentration is reached. Sand-drying beds have been built simply by cleaning an area of land, dumping the sludge, and hoping something happens. At the other end of the spectrum sophisticated automated drying systems have been built.

The dewatering of sludge on sand beds is accomplished by two major methods: drainage and evaporation. The removal of water from sludge by drainage is a two-step process. First, the water is drained from the sludge, into the sand, and out the underdrains. This process may last a few days until the sand is clogged with fine particles or until all the free water has drained away. Further drainage by decanting can occur once a supernatant layer has formed (if beds are provided with a means of removing surface water). Decanting for removal of rain water can also be particularly important with sludges that do not crack.

The water that does not drain or is not decanted must evaporate. Obviously, annual rainfall rates play a role here. Phoenix would be a more efficient area for a sand bed than Seattle!

The filtrate from the sand-drying beds can be recycled, treated, or discharged to a watercourse depending on its quality. Laboratory testing of the filtrate should be performed in conjunction with sand-drying bed pilot testing before a decision is made about what is to be done with it.

Freeze Treatment.   Dewatering sludge by either of the nonmechanical methods may be enhanced by physical conditioning of the sludge through alternate freezing and thawing cycles. The freeze–thaw process dehydrates the sludge particles by freezing the water that is closely associated with the particles. The dewatering process takes place in two stages. In the first stage, the water molecules are selectively frozen, thereby dehydrating the solid particles. The sludge volume is also reduced. In the second stage, the sludge is thawed, and the solid mass forms granular-shaped particles. This coarse material readily settles and retains its new size and shape. This residue sludge dewaters rapidly and makes suitable landfill material.

The supernatant liquid from this process can be decanted, leaving the solids to dewater by natural drainage and evaporation. Pilot-scale systems can be used to evaluate this method's effectiveness and to establish design parameters. Elimination of rain and snow from the dewatering system by the provision of a roof enhances the process considerably.

Centrifuging.   A centrifuge uses centrifugal force to speed up the separation of sludge particles from the liquid. In a typical unit (Figure 10–26), sludge is pumped into a horizontal, cylindrical bowl, rotating at 800–2000 rpm. Polymers used for sludge conditioning are also injected into the centrifuge. The solids are spun to the outside of the bowl, where they are scraped out by a screw conveyor. The liquid, or centrate, is returned to the treatment plant. Centrifuges are very sensitive to changes in the concentration or composition of the sludge, as well as to the amount of polymer applied.

Vacuum Filtration.   A vacuum filter consists of a cylindrical drum, covered with a filtering material or fabric, which rotates partially submerged in a vat of conditioned sludge (Figure 10–27). A vacuum is applied inside the drum to extract water, leaving the solids, or filter cake, on the filter medium. As the drum completes its rotational cycles, a blade scrapes the filter cake from the filter and the cycle begins again.

Continuous Belt Filter Press (CBFP).   The belt filter press operates on the principle that bending a sludge cake contained between two filter belts around a roll introduces shear and compressive forces in the cake, allowing water to work its way to the surface and out of the cake, thereby reducing the cake's moisture content. The device employs double moving belts to continuously dewater sludges through one or more stages of dewatering (Figure 10–28).

FIGURE 10–26

Solid bowl centrifuge.

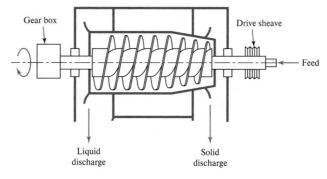

FIGURE 10–27

Vacuum filter.
Source: Komline-Sanderson Engineering Corporation.

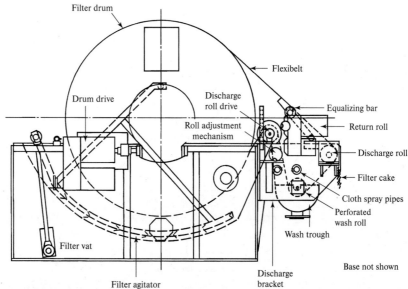

FIGURE 10–28

Continuous belt filter press. (*Source:* U.S. Environmental Protection Agency, 1979.)

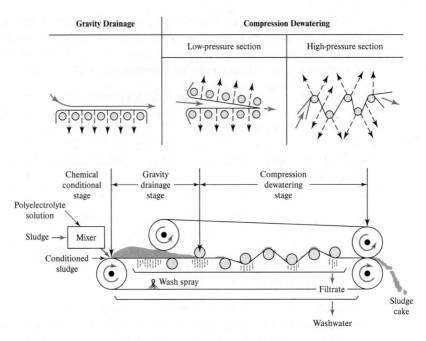

Plate Pressure Filters.   The basic component of a plate filter is a series of recessed vertical plates. Each plate is covered with cloth to support and contain the sludge cake. The plates are mounted in a frame consisting of two bead supports connected by two horizontal parallel bars. A schematic cross section is illustrated in Figure 10–29. Conditioned sludge is pumped into the pressure filter and passes through feed holes in the filter plates along the length of the filter and into the recessed chambers. As the sludge cake forms and builds up in the chamber, the pressure gradually increases to a point where further sludge injection would be counterproductive. At this time the injection ceases.

Ultimate Disposal

After all possible sludge treatment has been accomplished, a residual sludge remains, which must be sent for ultimate disposal. Of the many theoretical alternatives for ultimate disposal, only two are of practical interest:

1. Landfilling
2. Land spreading

FIGURE 10–29

Schematic cross section of a fixed volume recessed plate filter assembly. (*Source:* U.S. Environmental Protection Agency, 1979.)

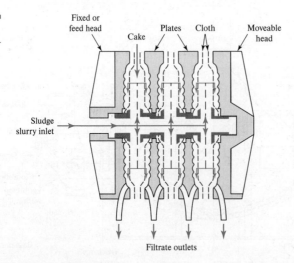

The application of water treatment residuals to land is regulated in the United States under the Resource Conservation and Recovery Act as well as by state and local agencies. Lime softening sludges are among the most commonly land-applied residuals. These sludges are especially beneficial when applied to land treated with nitrogen fertilizers, which typically lower the soil pH. The application of lime softening sludges results in an increase in the pH and greater calcium availability.

Water treatment sludges can be disposed of with biosolids from municipal wastewater treatment plants. This option is beneficial for the facility that produces both potable water and treats wastewater. The two sludges can be permitted and monitored as a single solid waste.

Other beneficial uses include the application on turf grass, in cement and brick manufacturing, as landfill cover, and as road subgrade. Water treatment sludges can also be blended with topsoil and organic matter. The blended product has enhanced nutrient value and water retention.

As discussed in Chapter 13, municipal solid wastes can be disposed of in landfills if there are no other options. Landfills are well-engineered facilities that are located, designed, operated, monitored during operation and after closure, closed, cleaned up when necessary, and financed according to federal regulations. These regulations were promulgated to protect human health and the environment. New landfills are designed to collect and treat potentially harmful landfill gases so that they are not released to the environment. Landfill gases may also be used for energy production. Organic sludges may be incinerated and then applied over land.

Sludges may also be land applied as soil conditioners or amendments. The major environmental concerns regarding such disposal methods are the release of pathogenic organisms and the presence of toxic chemicals. Unlike wastewater sludges, water treatment sludges rarely have sufficiently high organic matter contents to render their putrescible organic compounds problematic.

In the past, ocean disposal of sludges was common; however, concern over the degradation of ocean environments has rendered this method unacceptable. Ocean dumping was banned under the U.S. Ocean Dumping Ban Act of 1988, as described in Chapter 9.

CHAPTER REVIEW

When you have completed studying this chapter, you should be able to do the following without the aid of your textbook or notes:

1. Define pathogen, SDWA, MCL, MCLG, VOC, SOC, DBP, THM, and SWTR.
2. Define potable and palatable and explain why we must provide drinking water that is both potable and palatable.
3. Distinguish among dissolved substances, suspended solids, and colloidal substances based on their size and the mechanism by which they can be removed from water.
4. List the four categories of water quality for drinking water.
5. Identify the microorganism group used as an indicator of fecal contamination of water and explain why it was selected. What are its limitations?
6. Sketch a water-softening plant and a filtration plant, labeling all of the parts and explaining their functions. Show where chemicals are added and residuals produced.
7. Describe the process of coagulation.
8. Differentiate between coagulation and flocculation.
9. Explain the effect of pH on alum and ferric chloride solubility.
10. List the basic types of coagulant aids; explain how each aid works and when it should be employed.
11. Define hardness in terms of the chemical constituents that cause it and the results as seen by the users of hard water.
12. Using diagrams and chemical reactions, explain how water becomes hard.
13. Given the total hardness and alkalinity, calculate the carbonate hardness and noncarbonate hardness.

14. Calculate the theoretical detention time or volume of a tank if you are given the flow rate and either the volume or detention time.
15. Define overflow rate in terms of liquid flow and settling-basin geometry, and state its units.
16. Explain the difference among type I, type II, and type III sedimentation.
17. Compare slow sand filters, rapid sand filters, and dual-media filters with respect to operating procedures and relative loading rates.
18. Explain how a rapid sand filter is cleaned.
19. Explain why a disinfectant that has a residual is preferable to one that does not.
20. Write the equations for the dissolution of chlorine gas in water and the subsequent dissociation of hypochlorous acid.
21. Describe how a membrane filter operates and list the four types of membrane filtration systems.
22. Discuss the challenges associated with membrane filtration.
23. Define the terms thickening, conditioning, and dewatering.
24. List and describe three methods of nonmechanical dewatering of sludge.
25. List and describe four mechanical methods of dewatering of sludge.

With the aid of the text or a list of equations, you should be able to do the following:

1. Calculate the gram equivalent weight of a chemical or compound.
2. Calculate the consumption of alkalinity for a given dose of alum or ferric chloride (given the equations).
3. Calculate the production of CO_2, SO_4^- or acid for a given dose of alum or ferric chloride and the chemical equations.
4. Calculate the amount of lime (as $CaCO_3$) that must be added if insufficient alkalinity is present in order to neutralize the acid produced by the addition of alum or ferric chloride to a given water (given the equations).
5. Calculate the fraction of the "split" for a lime-soda softening system or an ion-exchange softening system.
6. Size rapid-mix and flocculation basins.
7. Size a sedimentation basin and estimate the required weir length.
8. Size a rapid sand filter. Determine how many filters of particular size are needed along with the flow rate through the filters.

PROBLEMS

10–1   The following mineral analysis was reported for a water sample taken from well No. 1 at the Eastwood Manor Subdivision near McHenry Illinois (Woller and Sanderson, 1976a).

Well No. 1, Lab No. 02694, November 9, 1971

(*Note:* All reported as mg · L^{-1} as the ion unless stated otherwise.)

Mineral	Concentration	Mineral	Concentration
Iron (2+)	0.2	Silica (SiO_2)	20.0
Manganese (2+)	0.0	Fluoride	0.35
Ammonium	0.5	Boron	0.1
Sodium	4.7	Nitrate	0.0
Potassium	0.9	Chloride	4.5
Calcium	67.2	Sulfate	29.0
Magnesium	40.0	Alkalinity	284.0 as $CaCO_3$
Barium	0.5	pH (as recorded)	7.6 units

Determine the total, carbonate, and noncarbonate hardness (in mg $\cdot$ L^{-1} as CaCO$_3$) using the predominant polyvalent cation definition in Section 10–3.

Answer:

$$\text{TH} = 332.8 \text{ mg} \cdot \text{L}^{-1} \text{ as CaCO}_3$$

$$\text{CH} = 284.0 \text{ mg} \cdot \text{L}^{-1} \text{ as CaCO}_3$$

$$\text{NCH} = 48.8 \text{ mg} \cdot \text{L}^{-1} \text{ as CaCO}_3$$

10–2 Calculate the total, carbonate, and noncarbonate hardness in Problem 10–1 using all of the polyvalent cations. What is the percent error in using only the predominant cations?

10–3 The following mineral analysis was reported for a water sample taken from well No. 1 at Magnolia, Illinois (Woller and Sanderson, 1976b). Determine the total, carbonate, and noncarbonate hardness (in mg $\cdot$ L^{-1} as CaCO$_3$), using the predominant polyvalent cation definition of hardness.

Well No. 1, Lab No. B109535, April 23, 1973

(*Note:* All reported as mg $\cdot$ L^{-1} as the ion unless stated otherwise.)

Mineral	Concentration	Mineral	Concentration
Iron	0.42	Zinc	0.01
Manganese	0.04	Silica (SiO$_2$)	20.0
Ammonium	11.0	Fluoride	0.3
Sodium	78.0	Boron	0.3
Potassium	2.6	Nitrate	0.0
Calcium	78.0	Chloride	9.0
Magnesium	32.0	Sulfate	0.0
Barium	0.5	Alkalinity	494.035 CaCO$_3$
Copper	0.01	pH (as recorded)	7.7 units

Answer: TH = 327.19 mg $\cdot$ L^{-1} as CaCO$_3$, CH = 327.19 mg $\cdot$ L^{-1} as CaCO$_3$, NCH = 0

10–4 The following mineral analysis was reported for Michigan State University well water (Michigan Department Environmental Quality, 1979). Determine the total, carbonate, and noncarbonate hardness (in mg $\cdot$ L^{-1} as CaCO$_3$).

(*Note:* All units are mg $\cdot$ L^{-1} as the ion unless otherwise stated.)

Mineral	Concentration	Mineral	Concentration
Fluoride	1.1	Silica (SiO$_2$)	13.4
Chloride	4.0	Bicarbonate	318.0
Nitrate	0.0	Sulfate	52.0
Sodium	14.0	Iron	0.5
Potassium	1.6	Manganese	0.07
Calcium	96.8	Zinc	0.27
Magnesium	30.4	Barium	0.2

10–5 Shown below are the results of water quality analyses of the Thames River in London. If the water is treated with 60.00 mg/L of alum to remove turbidity, how much alkalinity will remain? Ignore side reactions with phosphorus and assume all the alkalinity is HCO$_3^-$.

Thames River, London

Constituent	Expressed as	Milligrams per liter
Total hardness	$CaCO_3$	260.0
Calcium hardness	$CaCO_3$	235.0
Magnesium hardness	$CaCO_3$	25.0
Total iron	Fe	1.8
Copper	Cu	0.05
Chromium	Cr	0.01
Total alkalinity	$CaCO_3$	130.0
Chloride	Cl	52.0
Phosphate (total)	PO_4	1.0
Silica	SiO_2	14.0
Suspended solids		43.0
Total solids		495.0
pH[a]		7.4

[a]Not in mg/L.

Answer: Alkalinity remaining = 99.69 or 100 mg/L as $CaCO_3$

10–6   Shown below are the results of water quality analyses of the Mississippi River at Baton Rouge, Louisiana. If the water is treated with 30.00 mg/L of ferric chloride for turbidity coagulation, how much alkalinity will remain? Ignore the side reactions with phosphorus and assume all the alkalinity is HCO_3^-.

Mississippi River, Baton Rouge, Louisiana

Constituent	Expressed as	Milligrams per liter
Total hardness	$CaCO_3$	164.0
Calcium hardness	$CaCO_3$	108.0
Magnesium hardness	$CaCO_3$	56.0
Total iron	Fe	0.9
Copper	Cu	0.01
Chromium	Cr	0.03
Total alkalinity	$CaCO_3$	136.0
Chloride	Cl	32.0
Phosphate (total)	PO_4	3.0
Silica	SiO_2	10.0
Suspended solids		29.9
Turbidity[a]	NTU[b]	12.0
pH[a]		7.6

[a]Not in mg/L.
[b]NTU = nephelometric turbidity units and the pH is dimensionless.

10–7   Shown below are the results of water quality analyses of Crater Lake at Mount Mazama, Oregon. If the water is treated with 40.00 mg/L of alum for turbidity coagulation, how much alkalinity will remain? Assume all the alkalinity is HCO_3^-.

Crater Lake, Mount Mazama, Oregon

Constituent	Expressed as	Milligrams per liter
Total hardness	$CaCO_3$	28.0
Calcium hardness	$CaCO_3$	19.0
Magnesium hardness	$CaCO_3$	9.0
Total iron	Fe	0.02
Sodium	Na	11.0
Total alkalinity	$CaCO_3$	29.5
Chloride	Cl	12.0
Sulfate	SO_4	12.0
Silica	SiO_2	18.0
Total dissolved solids		83.0
pH[a]		7.2

[a]Not in mg/L.

Answer: 9.30 mg/L as $CaCO_3$

10–8 Prepare a bar chart of the water described in Problem 10–1. (*Note:* Valences may be found in Appendix A.) Because all of the constituents were not analyzed, an ion balance is not achieved.

10–9 Prepare a bar chart of the water described in Problem 10–3. Because all of the constituents were not analyzed, an ion balance is not achieved.

10–10 Prepare a bar chart of the water described in Problem 10–4. Because all of the constituents were not analyzed, an ion balance is not achieved.

10–11 Prepare a bar chart of the Lake Michigan water analysis shown below. Because all of the constituents were not analyzed, an ion balance is not achieved. For the estimate of the CO_2 concentration, ignore the carbonate alkalinity.

Lake Michigan at Grand Rapids, MI, Intake

Constituent	Expressed as	Milligrams per liter
Total hardness	$CaCO_3$	143.0
Calcium	Ca	38.4
Magnesium	Mg	11.4
Total iron	Fe	0.10
Sodium	Na	5.8
Total alkalinity	$CaCO_3$	119
Bicarbonate alkalinity	$CaCO_3$	115
Chloride	Cl	14.0
Sulfate	SO_4	26.0
Silica	SiO_2	1.2
Total dissolved solids		180.0
Turbidity[a]	NTU[b]	3.70
pH[a]		8.4

[a]Not in mg/L.
[b]NTU = nephelometric turbidity units and the pH is dimensionless.

10–12 What amount of lime and/or soda ash, in mg/L as $CaCO_3$, is required to soften the Village of Lime Ridge's water to 80.0 mg/L hardness as $CaCO_3$.

Compound	Concentration, mg/L as $CaCO_3$
CO_2	4.6
Ca^{2+}	257.9
Mg^{2+}	22.2
HCO_3^-	248.0
SO_4^{2-}	32.1

10–13 Determine the lime and soda ash dose, in mg/L as $CaCO_3$, to soften the following water to a final hardness of 80.0 mg/L as $CaCO_3$. The ion concentrations reported below are all mg/L as $CaCO_3$.

$$Ca^{2+} = 120.0$$
$$Mg^{2+} = 30.0$$
$$HCO_3^- = 70.0$$
$$CO_2 = 10.0$$

Answers: Total lime addition = 100 mg/L as $CaCO_3$
Total soda ash addition = 40 mg/L as $CaCO_3$

10–14 Two parallel flocculation basins are to be used to treat a water flow of 0.150 $m^3 \cdot s^{-1}$. If the design detention time is 20 min, what is the volume of each tank?

10–15 What is the volume required for a rapid-mix basin that is to be used to treat 0.05 $m^3 \cdot s^{-1}$ of water if the detention time is 60 s?

Answer: 3 m^3

10–16 A water softener is used to treat well water that has a total hardness of 450 $mg \cdot L^{-1}$. The design flow rate is 4.2 $m^3 \cdot s$ with a total hardness of 125 $mg \cdot L^{-1}$. According to the manufacturer, the ion exchange resin used in the water softener has a leakage of 1.5% (which means that the effluent concentration from the softener is 1.5% of the influent concentration). What is the required flow rate through the softener to achieve the desired hardness?

10–17 A water softener is used to treat well water that has a total hardness of 420 $mg \cdot L^{-1}$. The design flow rate is 3.0 $m^3 \cdot s$ with a total hardness of 100 $mg \cdot L^{-1}$. According to the manufacturer, the ion exchange resin used in the water softener has a leakage of 1% (which means that the effluent concentration from the softener is 1% of the influent concentration). What is the required flow rate through the softener to achieve the desired hardness?

Answer: 2.3 $m^3 \cdot s^{-1}$

10–18 An ion exchange softening plant treats water a flow rate of 150 millions gallons per day (MGD). The water has the same characteristics as that listed below. A bypass is used to achieve the desired hardness of 120 $mg \cdot L^{-1}$ as $CaCO_3$. Determine the bypass flow rate assuming that the total hardness is 2.0 $mg \cdot L^{-1}$ as $CaCO_3$ after softening.

Chemical Species or Parameter	Concentration (in mg/L as $CaCO_3$ Unless Otherwise Specified)
Calcium (Ca^{2+})	238.0
Magnesium (Mg^{2+})	90.6
Iron (Fe^{3+})	0.7
Copper (Cu^{2+})	0.02
Sodium (Na^+)	56.2
Bicarbonate (HCO_3^-)	198.9
Sulfate (SO_4^{2-})	86.4
Chloride (Cl^-)	99.7
CO_2	25.0
pH (dimensionless)	7.25

10–19   Determine the terminal settling velocity of particle having a density of 2540 kg $\cdot$ m^{-3} and a diameter of 10 mm in water having a temperature of 10°C.

Answer: 64.2 m $\cdot$ s^{-1}

10–20   Determine the terminal settling velocity of particle having a density of 2650 kg $\cdot$ m^{-3} and a diameter of 100 μm in water having a temperature of 15°C.

10–21   If a 1.0 m^3 $\cdot$ s^{-1} flow water treatment plant uses ten sedimentation basins with an overflow rate of 15 m^3 $\cdot$ day^{-1} $\cdot$ m^{-2}, what should be the surface area (m^2) of each tank?

Answer: 576.0 m^2

10–22   Assuming a conservative value for an overflow rate, determine the surface area (m^2) of each of two sedimentation tanks that together must handle a flow of 0.05162 m^3 $\cdot$ s^{-1} of lime softening floc. Use an overflow rate of 57 m^3 $\cdot$ day^{-1} $\cdot$ m^{-2}.

10–23   Repeat Problem 10–21 for an alum or iron floc with an overflow rate of 20 m^3 $\cdot$ day^{-1} $\cdot$ m^{-2}.

Answer: 111 m^2

10–24   For a flow of 0.8 m^3 $\cdot$ s^{-1}, how many rapid sand filter boxes of dimensions 10 m × 10 m are needed for a loading rate of 110 m^3 $\cdot$ day^{-1} $\cdot$ m^{-2}?

10–25   If a dual-media filter with a loading rate of 300 m^3 $\cdot$ day^{-1} $\cdot$ m^{-2} were built instead of the standard filter in Problem 10–24, how many filter boxes would be required?

Answer: Four filters (rounding to next largest even number)

10–26   The water flow meter at the Westwood water plant is malfunctioning. The plant superintendent tells you the four dual-media filters (each 5.00 m × 10.0 m) are loaded at a velocity of 280 m $\cdot$ day^{-1}. What is the flow rate through the filters (in m^3 $\cdot$ s^{-1})?

10–27   Using Chick's law determine the rate constant for the disinfection of *E. coli*. Initially, the number of organisms is 200 per 100 mL. After 10.0 min, the number is 15 per 100 mL. Assume that the concentration of the disinfectant is constant.

Answer: 0.26 m^{-1}

10–28   A water treatment plant is to treat 0.5 m^3/s of water. Chlorine is to be used as a disinfectant. At the temperature and pH of the source water, a CT of 75 mg $\cdot$ min $\cdot$ L^{-1} is required. The t_{10} for the contact chamber is 30 min. The ratio of t_{10} to t_o is 0.5. What is the required volume of the reactor? Determine the necessary average chlorine concentration.

10–29   A water treatment plant is to treat 0.1 m^3/s of water. Chloramines are to be used as the primary disinfectant. At the temperature and pH of the source water, a CT of 1,250 mg $\cdot$ min $\cdot$ L^{-1} is required to achieve 2.5 log inactivation of *Giardia* cysts. The t_{10} for the contact chamber is 20 min. The ratio of t_{10} to t_o is 0.7. What is the required volume of the reactor? Determine the necessary average chloramine concentration.

Answer: Volume = 180 m^3
Concentration = 62.5 mg $\cdot$ L^{-1}

10–30   Design a longitudinal serpentine chlorine contact chamber for a design flow of 20,000 m^3 $\cdot$ d^{-1}. Chlorine (Cl$_2$) is to be used as the disinfectant. The temperature and pH of the water are 10°C and 8.0, respectively. The chamber is designed to ensure 2 log removal of *Giardia* and have a t_{10}/t_o of 0.5 and a L/W = 60. Assume H = 3W.

(*Note:* The CT value for 2-log inactivation of *Giardia* cysts by free chlorine at 10°C and a pH of 8.0 is 108 (mg $\cdot$ L^{-1})-min at a chlorine concentration of 1 mg $\cdot$ L^{-1}.)

10–31   You are to design a microfiltration membrane system for the community of Lastnight. The design flow rate is 0.960 m^3 $\cdot$ s^{-1}. The temperature of the water is constant throughout the year at 10. The membrane you have chosen has maximum TMP of 230 kPa and a resistance coefficient of 3.5 × 10^{12} m^{-1}. Based on pilot testing, the operating TMP should be no greater than 75% of the maximum TMP. Determine the required membrane area.

> **Answer:** 25,500 m^2

10–32 A reverse osmosis unit is to demineralize Tampa's tertiary treated effluent. Pertinent data are as follows:

> mass transfer coefficient: 0.45 L/(d · m^2 · kPa) at 20°C (winter)
>
> mass transfer coefficient: 0.70 L/(d · m^2 · kPa) at 30°C (summer)
>
> osmotic pressure difference between the feed and product water = 430 kPa,

The unit is designed to treat a flow of 760 m^3/day at a temperature of 20°C (winter conditions, transmembrane pressure = 2800 kPa).

(a) Determine the required membrane area using the winter conditions.

(b) Assuming the same osmotic pressure difference in summer and winter (i.e., identical water quality), determine the transmembrane pressure necessary in summer to maintain the same water flux.

10–33 A water contains 50.40 mg · L^{-1} as CaCO$_3$ of carbon dioxide, 190.00 mg · L^{-1} as CaCO$_3$ of Ca^{2+} and 55.00 mg · L^{-1} as CaCO$_3$ of Mg^{2+}. All of the hardness is carbonate hardness. Using the stoichiometry of the lime soda ash softening equations, what is the daily sludge production (in dry weight, kg · day^{-1}) if the plant treats water at a rate of 2.935 m^3 · s^{-1}? Assume that the effluent water contains no carbon dioxide, 30.0 mg · L^{-1} as CaCO$_3$ of Ca^{2+} and 10.0 mg · L^{-1} as CaCO$_3$ of Mg^{2+}. Be sure to calculate the mass of CaCO$_3$ and Mg(OH)$_2$ sludge produced each day.

> **Answer:** 123409 kg · day^{-1}

10–34 A water contains a suspended solids concentration of 10.30 mg · L^{-1}. Coagulation reduces this to 1.0 mg · L^{-1}. To accomplish this, an alum dose of 13.30 mg · L^{-1} is necessary. What is the dry weight of sludge produced (in kg · day^{-1}) if the plant treats water at a rate of 1.884 m^3 · s^{-1}?

10–35 A water treatment plant is designed at a flow rate of 42.5 L · s^{-1}. The water is to flow into two primary settling tanks operating in parallel. A detention time of 2.5 hours has been determined to be effective. Using a length to width ratio of 2:1 and an effective depth of 3.5 m, calculate the length of the tank in meters.

> **Answer:** 10.5 m

10–36 A water has a pH of 8.00 and a temperature of 10°C. A 3.00 log inactivation of *Giardia* cysts is required. A chlorine concentration of 1.20 mg · L^{-1} is to be used. The water is treated at a rate of 2.409 m^3 · s^{-1}. What is the design volume of the chlorine contacting chamber (in m^3)?

DISCUSSION QUESTIONS

10–1 Explain the word "turbidity" in terms that the mayor of a community could understand.

10–2 Which of the chemicals added to treat a surface water aids in making the water palatable?

10–3 Microorganisms play a role in the formation of hardness in groundwater. True or false? Explain.

10–4 If no bicarbonate is present in a well water that is to be softened to remove magnesium, which chemicals must you add?

10–5 In the United States, chlorine is preferred as a disinfectant over ozone because it has a residual. Why is the presence of a residual important? In recent years ozone has been replacing chlorine as the primary disinfectant in the United States. What are the advantages and disadvantages of this?

10–6 A new water-softening plant is being designed for Lubbock, Texas. The climate is dry and land is readily available at a reasonable cost. What methods of sludge dewatering would be most appropriate? Explain your reasoning.

10–7 Since the attacks on the World Trade Center and the Pentagon on September 11, 2001, it has become apparent that we must provide sufficient security to protect our drinking-water supplies. What measures can be taken to ensure such protection?

10–8 In May 2000, in Walkerton, Ontario, an outbreak of enteric illness occurred due to contamination of the public water supply with *E. coli.* The outbreak killed five people and sickened hundreds. What led to the contamination and what could have prevented it? How could the water treatment plant be redesigned to prevent such an occurrence in the future?

10–9 In Haiti, less than half the population has access to safe water. The poverty level approaches that in sub-Saharan Africa, with an average income in 2002 around $1.40 per day. What measures could be taken to better provide safe drinking water for this population?

10–10 There are plans to build two desalination plants in Mexico that would supply water to the United States. The first, located in Rosarito, would produce 50 million gallons a day. The second in Playas de Rosarito, about 15 miles south of San Diego would produce 100 million gallons a day. Are there ethical issues with building the plants just south of the U.S.-Mexico border to supply water for U.S. residents?

FE EXAM FORMATTED PROBLEMS

10–1 A groundwater contains the following cations:

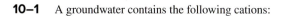

(a) $Na^+ = 35$ mg/L

(b) $Ca^{2+} = 100$ mg/L

(c) $Mg^{2+} = 35$ mg/L

(d) $Fe^{3+} = 1.6$ mg/L

The total hardness (mg/L as $CaCO_3$) is most nearly:

(a) 120

(b) 172

(c) 250

(d) 398

10–2 A rapid mix tank is used to disperse alum into a water stream. Water is to be treated at a rate of 1.5 MGD. Two identical cylindrical tanks are to be used in parallel. The diameter of the tank is 1.8 m and the depth of the water in the tank is 0.8 m. The hydraulic detention time (s) is most nearly:

(a) 50

(b) 100

(c) 150

(d) 200

10–3 A water utility uses rapid sand filtration. The filtration system has four cells, each with dimensions 20 ft × 20 ft and a depth of 6 ft. The plant is designed to treat water at a rate of 8 MGD. If one cell is out of service for backwashing and the plant is treating water at the design flow rate, the hydraulic loading rate (gallons per minute/ft²) is most nearly:

(a) 3.5

(b) 4.6

(c) 11.6

(d) 15.4

10–4   Drinking water leaves the treatment plant with a chlorine concentration of 1.5 mg/L. The chlorine decay rate was found to be 0.27 d^{-1}. Assuming that the distribution system can be modeled as an ideal plug flow reactor, after a hydraulic residence time of 2 d, the chlorine concentration (mg/L) is most nearly:

(a)   0.52

(b)   0.87

(c)   0.97

(d)   1.23

10–5   Which of the following is not a typical characteristic of coagulants used in drinking water treatment?

(a)   Trivalent cation

(b)   Nontoxic

(c)   Monovalent anion

(d)   Insoluble in the neutral pH range

10–6   Noncarbonate hardness accounts for that portion of the hardness that is associated with:

(a)   Carbonate ions

(b)   Bicarbonate ion

(c)   Sulfate ions

(d) HCO_3^-

10–7   A water has the following composition (in mg/L as $CaCO_3$):

(a) $Ca^{2+} = 100$

(b) $Mg^{2+} = 40$

(c) $HCO^{3-} = 90$

If the pH of the water is 8.1, the carbonate hardness is most nearly:

(a)   90

(b)   100

(c)   140

(d)   230

10–8   A water treatment plant is treating water at a flowrate of 0.1 m^3/s. The sedimentation basin has 50 m of weir. The weir loading rate (m^3/d·m) is most nearly:

(a)   2.9

(b)   7.2

(c)   120

(d)   170

10–9   At a pH of 7.5 and a temperature of 5°C, the required CT for disinfection with chlorine is 179 (mg/L)·min. If the t_{10}/t_o ratio is 0.6 and the concentration of chlorine is 1 mg/L, the required detention time (min) in the reactor is most nearly:

(a)   107

(b)   179

(c)   298

(d)   359

10–10 During softening, magnesium is precipitated as $Mg(OH)_2$. The solubility product for $Mg(OH)_2$ is 1.8×10^{-11}. If the pH is 11.5 and equilibrium is attained, the minimum concentration of magnesium (mg/L) is most nearly:

(a) 0.044

(b) 0.44

(c) 4.4

(d) 44.0

REFERENCES

American Water Works Association (1971) *Water Quality and Treatment,* McGraw-Hill, New York, p. 259.

American Water Works Association (1990) *Water Treatment Plant Design,* 2nd ed., McGraw-Hill, New York.

Centers for Disease Control (1990) *Morbidity and Mortality Weekly Reports, March 1, 1990,* vol. 39 (SS-1):1–9, http://www.cdc.gov/MMWR/preview/mmwrhtml/00001596.htm

Chick, H. (1908) "Investigation of the Law of Disinfection," *Journal of Hygiene,* 8:92.

City of Philadelphia Water Department (2007) *Chlorine in Tap Water,* http://www.phillywater.org/WQ/chlorine.htm

Cooperative Research Centre for Water Quality and Treatment (2007) *Keeping Water Safe: Chlorination and Disinfection By-Products,* http://www.waterquality.crc.org.au/dwfacts/dwfact_keep_water_safe.pdf

European Union (1998) *Council Directive on the Quality of Water Intended for Human Consumption,* Issued on 3 November 1998, http://ec.europa.eu/environment/water/waterdrink/index_en.html

Hudson, H. E. (1981) *Water Clarification Processes Practical Design and Evaluation,* Van Nostrand Reinhold, New York, pp. 115–17.

Jaeger Products Inc. (2004) *Plastic Tri-Pak Packing Material,* Houston, TX, http://www.jaeger.com

Jesperson, K. (2007) *Search for Clean Water Continues,* National Environmental Services Center, West Virginia University, Morgantown, WV, http://www.nesc.wvu.edu/ndwc/ndwc_DWH_1.html

Levy, D. A., M. S. Bens, G. F. Craun, R. L. Calderon, and B. L Herwald (1998) *Surveillance for Waterborne- Disease Outbreaks—United States, 1995–1996, Mortality and Morbidity Weekly Reports,* Surveillance Summaries, December 11, 1998, 47(SS-5), 1–34.

Linden, K. G., and J. L. Darby (1997) "Estimating Effective Germicidal Dose from Medium Pressure UV Lamps," *J. Environ. Eng.,* 123(11): 1142–2249.

Masten, S. J., S. H. Davies, and S. P. McElmurry (2016) "Flint Water Crisis: What Happened and Why?" *Journal American Water Works Association,* 108(12): 22-34.

Michigan Department of Environmental Quality (1979) *Annual Data Summary,* Michigan Department of Public Health, Lansing, MI.

Montgomery, J. M. (1985) *Water Treatment Principles and Design,* Wiley Interscience, New York, p. 535.

MWH (2005) *Water Treatment: Principles and Design,* John Wiley & Sons, Hoboken, NJ.

Peterson, C. A., and R. L. Calderon (2003) "Trends in Enteric Disease as a Cause of Death in the United States, 1989–1996," *Am. J. Epidemiol,* 157:58–65.

Remco Engineering (1998) *Air Stripping Towers,* Marietta, GA, http://www.remtech-eng.com/

Reynolds, T. D. (1982) *Unit Operations and Processes in Environmental Engineering,* PWS-Kent Publishing, Boston, MA, pp. 33–54.

Spitzer, E. F. (1993) *Back to Basics Guide to Slow Sand Filtration,* American Water Works Association, Denver, CO.

Stokes, G. (1845) "On the Theories of the Internal Friction of Fluids in Motion, and of the Equilibruim and Motion of Elastic Solids," *Transactions of the Cambridge Philosophical Society,* 8: 287–319.

Stumm, W., and J. J. Morgan (1996) *Aquatic Chemistry: Chemical Equilibria and Rates in Natural Waters,* 3rd. ed. John Wiley and Sons, New York.

Train, R. F. (1976) Excerpted from remarks by former EPA Administrator Russel F. Train, delivered at the Los Angeles World Affairs Forum, December 16, 1976.

U.S. EPA (1979) *Process Design Manual for Sludge Treatment and Disposal,* U.S. EPA Municipal Environmental Research Laboratory, Office of Research and Development, Washington, DC, 625/1–79-011.

U.S. EPA (1996) *Ultraviolet Light Disinfection Technology in Drinking Water Application—An Overview,* U.S. Environmental Protection Agency, Office of Groundwater and Drinking Water, EPA 811-R-96-002, Washington, DC.

U.S. EPA (1999a) *Disinfection Profiling and Benchmarking Guidance Manual,* EPA-815-R-99-013, Washington, DC.

U.S. EPA (1999b) *Alternative Disinfectants and Oxidants Guidance Manual,* U.S. Environmental Protection Agency, Office of Water Publication: EPA 815-R-99-014, Washington, DC. http://www.epa.gov /ogwdw/ mdbp/alternative_disinfectants_guidance.pdf

U.S. EPA (2000) *The History of Drinking Water Treatment,* U.S. EPA Office of Water, 816-F-00-006.

U.S. EPA (2001) *Low Pressure Membrane Filtration for Pathogen Removal: Application, Implementation, and Regulatory Issues,* U.S. EPA Office of Water, 815-C-01-001, Cincinnati, OH.

U.S. EPA (2009) *Factoids: Drinking Water and Groundwater Statistics for 2009,* U.S. EPA Office of Water, EPA 816-K-03-001, December 2006, www.epa.gov/safewater/data

Walker Process Equipment, Inc. (1972) *Bulletin Number 1600-S-107,* Division of Chicago Bridge and Iron Company, Aurora, IL.

Walker Process Equipment, Inc. (1973) *Bulletin Number 9-W-65,* Division of Chicago Bridge and Iron Company, Aurora, IL.

Watson (1908) "A Note on the Variation of the Rate of Disinfection with Change in Concentration of the Disinfecant," *J. Hygiene,* 8: 536.

Williams, T., G. L. LeBel, and F. M. Benoit (1995) *National Survey of Chlorinated Disinfection By-Products in Canadian Drinking Water,* http://www.hc-sc.gc.ca/ewh-semt/pubs/water-eau /byproducts-sousproduits/ index_e.html

Woller, D. M., and E. W. Sanderson (1976a) *Public Water Supplies in McHenry County* (Illinois State Water Survey, publication No. 6019), Urbana, IL.

Woller, D. M., and E. W. Sanderson (1976b) *Public Groundwater Supplies in Putnam County* (Illinois State Water Survey, Publication No. 60-15), Urbana, IL.

World Health Organization (WHO) (2004) *Guidelines for Drinking-water Quality,* 3rd ed., Geneva, Switzerland.

World Health Organization (WHO) (2006) *Guidelines for Drinking-water Quality—First Addendum to Third Edition, Volume 1, Recommendations,* WHO Press, Geneva, Switzerland, http://www.who.int /water_sanitation_health/dwq/gdwq3rev/en/index.html

Case Study

Tenafly Sewage Treatment Plant: A Plant Decades Ahead of Its Time

It was late 1946 and World War II had just ended. With the completion of the George Washington Bridge in 1931, the population of Bergen County, New Jersey, had grown significantly. While most of the towns in the county lacked sewage treatment, Tenafly, a small town of around 7500, was decades ahead of its time.

Built at a time when most large cities treated their wastewater with only sedimentation (primary treatment) and there were no federal wastewater treatment regulations, the 1.75 million gallons per day (MGD) plant employed what would be considered today state-of-the-art treatment. While most plants used primary sedimentation tanks, this plant used fine screens. From the fine screens, the wastewater flowed into activated sludge tanks, with air supplied through diffuser plates. The effluent from the secondary clarifiers was treated through sand filtration and was chlorinated prior to discharge. A portion of the chlorinated plant effluent was recycled into the plant for use in the chlorinators, which reduced water consumption and costs by up to 75%. The remainder of the wastewater effluent was discharged to the Tenakill Brook, which flowed 3.5 miles before discharging into the Oradell Reservoir, the drinking water supply for Tenafly and the surrounding area.

The raw waste activated sludge was conditioned with ferric chloride then dewatered using vacuum filtration. It was then flash-dried, bagged, and marketed as fertilizer. The sludge, "Tenafly Soil Food," had nitrogen and phosphorus contents of 5% and 3%, respectively. It brought a net income of nearly $4000 per year (Adams, 1948).

So two years before the promulgation of the Federal Water Pollution Control Act of 1948, the first major U.S. law to address water pollution, and 26 years before the enactment of the Clean Water Act, the Tenafly plant operated as a model for the nation. It employed innovative technology and equipment, tertiary treatment, and indirect potable water reuse. It also produced sludge (i.e., biosolids) that was applied to soil to supply nutrients and replenish soil organic matter.

11–1   INTRODUCTION

Wastewater Treatment Perspective

Wastewater has historically been considered a nuisance to be discarded in the cheapest, least offensive manner possible. This meant the use of on-site disposal systems such as the pit privy and direct discharge into our lakes and streams. Over the last century it has been recognized that this approach has an undesirable effect on the environment. This led to the variety of treatment techniques that characterize today's municipal treatment systems, which are the focal point of this chapter. As we look forward it becomes obvious that in the interest of sustainability as well as fundamental economic efficiency, we must view wastewater as a raw material to be conserved. Clean water is a scarce commodity and should be treated as such, so it must be conserved and reused. The contents of wastewater are often viewed as pollutants. The abundance of nutrients such as phosphorus and nitrogen are in some treatment schemes (as discussed in Section 11–11) recovered for crop growth. This approach must become more prevalent to achieve a sustainable future. The organic compounds in wastewater are a source of energy. Currently we use the processes described in Section 11–13 to recover some of this energy. Future efforts will focus on improving the efficiency of energy use in the wastewater.

11–2   CHARACTERISTICS OF DOMESTIC WASTEWATER

Physical Characteristics

Fresh, aerobic, domestic wastewater has been said to have the odor of kerosene or freshly turned earth. Aged, septic sewage is considerably more offensive to the olfactory sense. Fresh sewage is typically gray in color. Septic sewage is black. This color results from the precipitation of iron sulfide.

Wastewater temperatures normally range between 10 and 20°C. In general, the temperature of the wastewater will be higher than that of the water supply because of the addition of warm water from households and heating within the plumbing system of the structure.

One cubic meter of wastewater weighs approximately 1,000,000 g and will contain about 500 g of solids. One-half of the solids will be dissolved solids such as calcium, sodium, and soluble organic compounds. The remaining 250 g will be insoluble. The insoluble fraction consists of about 125 g of material that will settle out of the liquid fraction in 30 min under quiescent conditions (these are called **settleable solids**). The remaining 125 g will remain in suspension for a very long time (these are called **suspended solids**). The result is that wastewater is highly turbid.

TABLE 11–1		Typical Composition of Untreated Domestic Wastewater	
Constituent	**Weak**	**Medium** (all mg · L^{-1} except settleable solids)	**Strong**
Alkalinity (as CaCO$_3$)^a	50	100	200
BOD$_5$ (as O$_2$)	100	200	300
Chloride	30	50	100
COD (as O$_2$)	250	500	1000
Suspended solids (SS)	100	200	350
Settleable solids (in mL · L^{-1})	5	10	20
Total dissolved solids (TDS)	200	500	1000
Total Kjeldahl nitrogen (TKN) (as N)	20	40	80
Total organic carbon (TOC) (as C)	75	50	300
Total phosphorus (as P)	5	10	20

^{a}This amount of alkalinity is the contribution from the waste. It is to be added to the naturally occurring alkalinity in the water supply. Chloride is exclusive of contribution from water-softener backwash.

Chemical Characteristics

Because the number of chemical compounds found in wastewater is almost limitless, we normally restrict our consideration to a few general classes. These classes often are better known by the name of the test used to measure them than by what is included in the class. The biochemical oxygen demand (BOD$_5$) test, which we discussed in Chapter 9, is a case in point. Another closely related test is the chemical oxygen demand (COD) test.

The COD test is used to determine the oxygen equivalent of the organic matter that can be oxidized by a strong chemical oxidizing agent (potassium dichromate) in an acid medium. The COD of a waste, in general, will be greater than the BOD$_5$ because more compounds can be oxidized chemically than can be oxidized biologically. BOD$_5$ is typically less than ultimate BOD, which is less than COD, except for totally biodegradable waste.

The COD test can be conducted in about an hour. If it can be correlated with BOD$_5$, it can be used to aid in the operation and control of the wastewater treatment plant (WWTP).

Total Kjeldahl* nitrogen (TKN) is a measure of the total organic and ammonia nitrogen in the wastewater. TKN gives a measure of the availability of nitrogen for building cells, as well as the potential nitrogenous oxygen demand that will have to be satisfied.

Phosphorus may appear in many forms in wastewater, among which are the orthophosphates, polyphosphates, and organic phosphate. For our purpose, we will lump all of these together under the heading "total phosphorus (as P)."

Three typical compositions of untreated domestic wastewater are summarized in Table 11–1. The pH for all of these wastes will be in the range of 6.5 to 8.5, with a majority being slightly on the alkaline side of 7.0.

Characteristics of Industrial Wastewater

Industrial processes generate a wide variety of wastewater pollutants. The characteristics and levels of pollutants vary significantly from industry to industry. The Environmental Protection Agency has grouped the pollutants into three categories: conventional, nonconventional, and priority pollutants. The conventional and nonconventional pollutants are listed in Table 11–2. The priority pollutants are listed in Table 11–3.

*Pronounced "kell dall" after J. Kjeldahl, who developed the test in 1883.

TABLE 11–2	EPA's Conventional and Nonconventional Pollutant Categories	
	Conventional	**Nonconventional**
	Biochemical oxygen demand (BOD$_5$)	Ammonia (as N)
	Total suspended solids (TSS)	Chromium (VI) (hexavalent)
	Oil and grease	Chemical oxygen demand (COD)
	Oil (animal, vegetable)	COD/BOD$_7$
	Oil (mineral)	Fluoride
	pH	Manganese
		Nitrate (as N)
		Organic nitrogen (as N)
		Pesticide active ingredients (PAI)
		Phenols, total
		Phosphorus, total (as P)
		Total organic carbon (TOC)

Sources: Code of Federal Regulations, 2006, 40 CFR parts 413.02, 464.02, 467.02, and 469.12.

TABLE 11–3	EPA's Priority Pollutant List

1. Antimony	27. Dichlorobromomethane	54. Phenol
2. Arsenic	28. 1,1-Dichloroethane	55. 2,4,6-Trichlorophenol
3. Beryllium	29. 1,2-Dichloroethane	56. Acenaphthene
4. Cadmium	30. 1,1-Dichloroethylene	57. Acenaphthylene
5a. Chromium (III)	31. 1,2-Dichloropropane	58. Anthracene
5b. Chromium (VI)	32. 1,1-Dichloropropylene	59. Benzidine
6. Copper	33. Ethylbenzene	60. Benzo(*a*)anthracene
7. Lead	34. Methyl bromide	61. Benzo(*a*)pyrene
8. Mercury	35. Methyl chloride	62. Benzo(*a*)fluoranthene
9. Nickel	36. Methylene chloride	63. Benzo(*ghi*)perylene
10. Selenium	37. 1,2,2,2-Tetrachloroethane	64. Benzo(*k*)fluoranthene
11. Silver	38. Tetrachloroethylene	65. Bis(2-chloroethoxy)methane
12. Thallium	39. Toluene	66. Bis(2-chloroethyl) Ether
13. Zinc	40. 1,2-*trans*-Dichloroethylene	67. Bis(2-chloroisopropyl) Ether
14. Cyanide	41. 1,1,1-Trichloroethane	68. Bis(2-ethylhexyl) Phthalate
15. Asbestos	42. 2,4 Dichlorophenol	69. 4-Bromophenyl phenyl ether
16. 2,3,7,8-TCDD (dioxin)	43. Trichloroethylene	70. Butylbenzyl phthalate
17. Acrolein	44. Vinyl chloride	71. 2-Chloronaphthalene
18. Acrylonitrile	45. 2-Chlorophenol	72. 4-Chlorophenyl phenyl ether
19. Benzene	46. 2,4-Dichlorophenol	73. Chrysene
20. Bromoform	47. 2,4-Dimethylphenol	74. Dibenzo(*a,h*)anthracene
21. Carbon tetrachloride	48. 2-Methyl-4-chlorophenol	75. 1,2-Dichlorobenzene
22. Chlorobenzene	49. 2,4-Dinitrophenol	76. 1,3-Dichlorobenzene
23. Chlorodibromomethane	50. 2-Nitrophenol	77. 1,4-Dichlorobenzene
24. Chloroethane	51. 4-Nitrophenol	78. 3,3-Dichlorobenzidine
25. 2-Chloroethylvinyl ether	52. 3-Methyl-4-chlorophenol	79. Diethyl phthalate
26. Chloroform	53. Pentachlorophenol	80. Dimethyl phthalate

TABLE 11–3	*(continued)*				
81.	Di-*n*-butyl phthalate	97.	*N*-Nitrosodi-*n*-propylamine	113.	β Endosulfan
82.	2,4-Dinitrotoluene	98.	*N*-Nitrosodiphenylamine	114.	Endosulfan sulfate
83.	2,6-Dinitrotoluene	99.	Phenanthrene	115.	Endrin
84.	Di-*n*-octyl phthalate	100.	Pyrene	116.	Endrin aldehyde
85.	1,2-Diphenylhydrazine	101.	1,2,4-Trichlorobenzene	117.	Heptachlor
86.	Fluoranthene	102.	Aldrin	118.	Heptachlor epoxide
87.	Fluorene	103.	α-BHC	119.	PCB-1242
88.	Hexachlorobenzene	104.	β-BHC	120.	PCB-1254
89.	Hexachlorobutadiene	105.	γ-BHC	121.	PCB-1221
90.	Hexachlorocyclopentadiene	106.	δ-BHC	122.	PCB-1232
91.	Hexachloroethane	107.	Chlordane	123.	PCB-1248
92.	Indeno(1,2,3-cd)pyrene	108.	4,4′-DDT	124.	PCB-1260
93.	Isophorone	109.	4,4′-DDE	125.	PCB-1016
94.	Naphthalene	110.	4,4′-DDD	126.	Toxaphene
95.	Nitrobenzene	111.	Dieldrin		
96.	*N*-Nitrosodimethylamine	112.	α Endosulfan		

Source: 40 CFR 131.36, July 1, 1993.

Because of the wide variety of industries and levels of pollutants, we can only present a snapshot view of the characteristics. A sampling of a few industries for two conventional pollutants is shown in Table 11–4.

A similar sampling for nonconventional pollutants is shown in Table 11–5.

TABLE 11–4	Examples of Industrial Wastewater Concentrations for BOD$_5$ and Suspended Solids	
Industry	**BOD$_5$** $(mg \cdot L^{-1})$	**Suspended Solids** $(mg \cdot L^{-1})$
Ammunition	50–300	70–1700
Fermentation	4500	10,000
Slaughterhouse (cattle)	400–2500	400–1000
Pulp and paper (kraft)	100–350	75–300
Tannery	700–7000	4000–20,000

TABLE 11–5	Examples of Industrial Wastewater Concentrations for Nonconventional Pollutants	
Industry	**Pollutant**	**Concentration** $(mg \cdot L^{-1})$
Coke by-product (steel mill)	Ammonia (as N)	200
	Organic nitrogen (as N)	100
	Phenol	2000
Metal plating	Chromium (VI)	3–550
Nylon polymer	COD	23,000
	TOC	8800
Plywood-plant glue waste	COD	2000
	Phenol	200–2000
	Phosphorus (as PO$_4$)	9–15

11–3   WASTEWATER TREATMENT STANDARDS

In Public Law 92-500, the Congress required municipalities and industries to provide **secondary treatment** before discharging wastewater into natural water bodies. The U.S. Environmental Protection Agency (EPA) established a definition of secondary treatment based on three wastewater characteristics: BOD_5, suspended solids, and hydrogen-ion concentration (pH). The definition is summarized in Table 11–6.

PL 92-500 also directed that the EPA establish a permit system called the National Pollutant Discharge Elimination System (NPDES). Under the NPDES program, all facilities that discharge pollutants from any point source into waters of the United States are required to obtain a NPDES permit. Although some states elected to have EPA administer their permit system, most states administer their own program. Before a permit is granted the administering agency will model the response of the receiving body to the proposed discharge to determine if the receiving body is adversely affected (for an example of modeling, see Chapter 9). The permit may require lower concentrations than those specified in Table 11–6 to maintain the quality of the receiving body of water.

In addition, the states may impose additional conditions in the NPDES permit. For example, in Michigan, a limit of 1 $mg \cdot L^{-1}$ of phosphorus is contained in permits for discharges to surface waters that do not have substantial problems with high levels of nutrients. More stringent limits are required for discharges to surface waters that are very sensitive to nutrients.

Carbonaceous biochemical oxygen demand ($CBOD_5$) limits are placed in the NPDES permits for all facilities that have the potential to contribute significant quantities of oxygen-consuming substances. The nitrogenous oxygen demand from ammonia nitrogen is typically the oxygen demand of concern from municipal discharges (see Chapter 9). It is computed separately from the $CBOD_5$ and then combined to establish a discharge limit. Ammonia is also evaluated for its potential toxicity to the stream's biota.

Bacterial effluent limits may also be included in the NPDES permit. For example, municipal wastewater treatment plants in Michigan must comply with limits of 200 fecal coliform bacteria (FC) per 100 mL of water as a monthly average and 400 FC per 100 mL as a 7-day average.

TABLE 11–6	U.S. Environmental Protection Agency Definition of Secondary Treatment		
Characteristic of Discharge	**Units**	**Average Monthly Concentration[a]**	**Average Weekly Concentration[a]**
BOD_5	$mg \cdot L^{-1}$	30[b]	45
Suspended solids	$mg \cdot L^{-1}$	30[b]	45
Hydrogen-ion Concentration	pH units	Within the range 6.0–9.0 at all times[c]	
$CBOD_5$[d]	$mg \cdot L^{-1}$	25	40

Note: Present standards allow stabilization ponds and trickling filters to have higher 30-day average concentrations (45 $mg \cdot L^{-1}$ and 7-day average concentrations (65 $mg \cdot L^{-1}$) of BOD and suspended solids as long as the water quality of the receiving body of water is not adversely affected. Other exceptions are also permitted. The CFR and the *NPDES Permit Writers' Manual* (U.S. EPA, 1996) should be consulted for details on the exceptions.

[a]Not to be exceeded.

[b]Average removal shall not be less than 85%.

[c]Only enforced if caused by industrial wastewater or by in-plant chemical addition.

[d]May be substituted for BOD_5 at the option of the permitting authority.

Source: CFR, 2005.

More stringent requirements are imposed to protect waters that are used for recreation. Total-body-contact recreation waters must meet limits of 130 *Escherichia coli* per 100 mL of water as a 30-day average and 300 *E. coli* per 100 mL at any time. Partial-body-contact recreation is permitted for water with less than 1000 *E. coli* per 100 mL of water.

For thermal discharges such as cooling water, temperature limits may be included in the permit. Michigan rules state that the Great Lakes and connecting waters and inland lakes shall not receive a heat load that increases the temperature of the receiving water more than 1.7°C above the existing natural water temperature after mixing. For rivers, streams, and impoundments the temperature limits are 1°C for cold-water fisheries and 2.8°C for warm-water fisheries. (See Section 4–4 for a discussion of energy balances and Problems 4–37 and 4–40 for typical problems in thermal discharge analysis.)

An example of NPDES limits is shown in Table 11–7. Note that in addition to concentration limits, mass discharge limits are also established.

Pretreatment of Industrial Wastes

Industrial wastewaters can pose serious hazards to municipal systems because the collection and treatment systems have not been designed to carry or treat them. The wastes can damage sewers and interfere with the operation of treatment plants. They may pass through the WWTP untreated or they may concentrate in the sludge, rendering it a hazardous waste.

The Clean Water Act gives the EPA the authority to establish and enforce pretreatment standards for discharge of industrial wastewaters into municipal treatment systems. Specific objectives of the pretreatment program are:

- To prevent the introduction of pollutants into WWTPs that will interfere with their operation, including interference with their use or with disposal of municipal sludge.

- To prevent the introduction of pollutants to WWTPs that will pass through the treatment works or otherwise be incompatible with such works.

- To improve opportunities to recycle and reclaim municipal and industrial wastewaters and sludge.

TABLE 11–7   **NPDES Limits for the City of Hailey, Idaho**

Parameter	Average Monthly Limit	Average Weekly Limit	Instantaneous Maximum Limit
BOD$_5$	30 mg · L^{-1}	45 mg · L^{-1}	N/A[a]
	43 kg · day^{-1}	64 kg · day^{-1}	
Suspended solids	30 mg · L^{-1}	45 mg · L^{-1}	N/A
	43 kg · day^{-1}	64 kg · day^{-1}	
E. coli bacteria	126 per 100 mL	N/A	406 per 100 mL
Fecal coliform bacteria	N/A	200 colonies per 100 mL	N/A
Total ammonia as N	1.9 mg · L^{-1}	2.9 mg · L^{-1}	3.3 mg · L^{-1}
	4.1 kg · day^{-1}	6.4 kg · day^{-1}	7.1 kg · day^{-1}
Total phosphorus	6.8 kg · day^{-1}	10.4 kg · day^{-1}	N/A
Total Kjeldahl nitrogen	25 kg · day^{-1}	35 kg · day^{-1}	N/A

Note: Renewal announcement, 7 February 2001. This table outlines only the quantitative limits. The entire permit is 22 pages long.

[a]N/A = not applicable.

Source: U.S. EPA, 2005.

EPA has established "prohibited discharge standards" (40 CFR 403.5) that apply to all nondomestic discharges to the WWTP and "categorical pretreatment standards" that are applicable to specific industries (40 CFR 405-471). Congress assigned the primary responsibility for enforcing these standards to local WWTPs.

11–4 ON-SITE DISPOSAL SYSTEMS

In less densely populated areas where lot sizes are large and houses are spaced widely apart it is often more economical to treat human waste on-site, rather than use a sewer system to collect the waste and treat it at a centralized location. On-site systems are generally small and may serve individual homes, small housing developments (clusters), or isolated commercial establishments, such as small hotels or restaurants. In the United States about 25% of the population is serviced by on-site wastewater treatment systems. In some states as much as 50% of the population uses on-site systems within rural and suburban communities (U.S. EPA, 1997). As many people chose to move to rural and outer suburban areas the number of decentralized systems is increasing. It is estimated that as much as 40% of new housing construction is taking place in areas that are not connected to municipal sewers.

11–5 MUNICIPAL WASTEWATER TREATMENT SYSTEMS

FIGURE 11–1

Degrees of treatment.

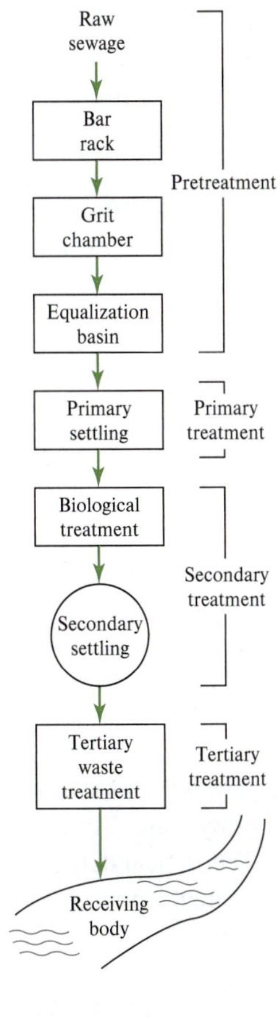

The alternatives for municipal wastewater treatment fall into three major categories (Figure 11–1): (1) primary, (2) secondary, and (3) tertiary treatment. It is commonly assumed that each of the degrees of treatment noted in Figure 11–1 includes the previous steps. For example, primary treatment is assumed to include the pretreatment processes: bar rack, grit chamber, and equalization basin. Likewise, secondary treatment is assumed to include all the processes of primary treatment: bar rack, grit chamber, equalization basin, and primary settling tank.

The purpose of pretreatment is to provide protection to the WWTP equipment that follows. In some municipal plants the equalization step may not be included.

The major goal of primary treatment is to remove from wastewater those pollutants that will either settle or float. Primary treatment will typically remove about 60% of the suspended solids in raw sewage and 35% of the BOD_5. Soluble pollutants are not removed. At one time, this was the only treatment used by many cities. Now federal law requires that municipalities provide secondary treatment. Although primary treatment alone is no longer acceptable, it is still frequently used as the first treatment step in a secondary treatment system. The major goal of secondary treatment is to remove the soluble BOD_5 that escapes the primary process and to provide added removal of suspended solids. Secondary treatment is typically achieved by using biological processes. These provide the same biological reactions that would occur in the receiving water if it had adequate capacity to assimilate the wastewater. The secondary treatment processes are designed to speed up these natural processes so that the breakdown of the degradable organic pollutants can be achieved in a relatively short time. Although secondary treatment may remove more than 85% of the BOD_5 and suspended solids, it does not remove significant amounts of nitrogen, phosphorus, or heavy metals, nor does it completely remove pathogenic bacteria and viruses.

In cases where secondary levels of treatment are inadequate, additional treatment processes are applied to the secondary effluent. These tertiary treatment processes may involve chemical treatment and filtration of the wastewater—much like adding a typical water treatment plant to the tail end of a secondary plant; they may also involve applying the secondary effluent to the land in carefully designed irrigation systems where the pollutants are removed by a soil–crop system. Some of these processes can remove as much as 99% of the BOD_5, phosphorus, suspended solids and bacteria, and 95% of the nitrogen. They can produce a sparkling clean, colorless, odorless effluent. Although these processes and land treatment systems are often applied to secondary effluent for tertiary treatment, they have also been used in place of conventional secondary treatment processes.

Most of the impurities removed from the wastewater do not simply vanish. Some organic compounds are broken down into harmless carbon dioxide and water. Most of the impurities are removed from the wastewater as a solid, that is, sludge. Because most of the impurities removed from the wastewater are present in the sludge, sludge handling and disposal must be carried out carefully to achieve satisfactory pollution control.

11–6 UNIT OPERATIONS OF PRETREATMENT

Several devices and structures are placed upstream of the primary treatment operation to provide protection to the WWTP equipment. These devices and structures are classified as pretreatment because they have little effect in reducing BOD_5. In industrial WWTPs, where only soluble compounds are present, bar racks and grit chambers may be absent. Equalization is frequently required in industrial WWTPs.

Bar Racks

Typically, the first device encountered by the wastewater entering the plant is a **bar rack,** the primary purpose of which is to remove large objects that would damage or foul pumps, valves, and other mechanical equipment. Rags, logs, and other objects that find their way into the sewer are removed from the wastewater on the racks. In modern WWTPs, the racks are cleaned mechanically. The solid material is stored in a hopper and removed to a sanitary landfill at regular intervals.

Bar racks (or bar screens) may be categorized as trash racks, manually cleaned racks, and mechanically cleaned racks. Trash racks have large openings, 40–150 mm, and are designed to prevent very large objects such as logs from entering the plant. These are normally followed by racks with smaller openings. Manually cleaned racks have openings that range from 25 to 50 mm. As mentioned earlier, manually cleaned racks are not frequently employed. They do find application in bypass channels that are infrequently used. Mechanically cleaned racks have openings ranging from 5 to 40 mm. Maximum channel approach velocities range from 0.6 to $1.2 \text{ m} \cdot \text{s}^{-1}$. Regardless of the type of rack, two channels with racks are provided to allow one to be taken out of service for cleaning and repair.

Grit Chambers

Inert dense material, such as sand, broken glass, silt, and pebbles, is called **grit.** If these materials are not removed from the wastewater, they abrade pumps and other mechanical devices, causing undue wear. In addition, they have a tendency to settle in corners and bends, reducing flow capacity and, ultimately, clogging pipes and channels.

Three basic types of grit-removal devices are available: velocity-controlled (also known as a horizontal-flow grit chamber), aerated, and vortex chambers. Only the first can be analyzed by means of the classical laws of sedimentation for discrete, nonflocculating particles (Type I sedimentation). Stokes' law (see Chapter 10) may be used for the analysis and design of horizontal-flow grit chambers if the horizontal liquid velocity is maintained at about $0.3 \text{ m} \cdot \text{s}^{-1}$. Liquid velocity control is achieved by placing a specially designed weir at the end of the channel. A minimum of two channels must be employed so that one can be out of service without shutting down the treatment plant. Cleaning may be either by mechanical devices or by hand. Mechanical cleaning is favored for plants having average flows over $0.04 \text{ m}^3 \cdot \text{s}^{-1}$. Theoretical detention times are set at about 1 min for average flows. Washing facilities are normally provided to remove organic material from the grit.

EXAMPLE 11–1   Will a grit particle with a radius of 0.04 mm and a specific gravity of 2.65 be collected in a horizontal grit chamber that is 13.5 m in length if the average grit-chamber flow is 0.15 m³ · s⁻¹, the width of the chamber is 0.56 m, and the horizontal velocity is 0.25 m · s⁻¹? The wastewater temperature is 22°C.

Solution   Before we can calculate the terminal settling velocity of the particle, we must gather some information from Table A–1 in Appendix A. At a wastewater temperature of 22°C, we find the water density to be 997.774 kg · m⁻³. We will use 1000 kg · m⁻³ as a sufficiently close approximation. Because the particle radius is given to only one significant figure, this approximation is reasonable. From the same table, we find the dynamic viscosity to be 0.955 mPa · s. As noted in the footnote to Table A–1, we must multiply this by 10^{-3} to obtain the viscosity in units of pascal seconds (Pa · s). Using a particle diameter of 0.08×10^{-3} m, we can calculate the terminal settling velocity using Stokes' law (Equation 10–22)

$$v_s = \frac{(9.8 \text{ m} \cdot \text{s}^{-2})(2650 \text{ kg} \cdot \text{m}^{-3} - 1000 \text{ kg} \cdot \text{m}^{-3})(0.08 \times 10^{-3} \text{ m})^2}{18(9.55 \times 10^{-4} \text{ Pa} \cdot \text{s})}$$

$$= 6.02 \times 10^{-3} \text{ m} \cdot \text{s}^{-1}, \text{ or} \approx 6.0 \text{ mm} \cdot \text{s}^{-1}$$

Note that the product of the specific gravity of the particle (2.65) and the density of water is the density of the particle (ρ).

With a flow of 0.15 m³ · s⁻¹ and a horizontal velocity of 0.25 m · s⁻¹, the cross-sectional area of flow may be estimated to be

$$A_c = \frac{0.15 \text{ m}^3 \cdot \text{s}^{-1}}{0.25 \text{ m} \cdot \text{s}^{-1}} = 0.60 \text{ m}^2$$

The depth of flow is then estimated by dividing the cross-sectional area by the width of the channel.

$$h = \frac{0.60 \text{ m}^2}{0.56 \text{ m}} = 1.07 \text{ m}$$

If the grit particle in question enters the grit chamber at the liquid surface, it will take h/v seconds to reach the bottom.

$$t = \frac{1.07 \text{ m}}{6.02 \times 10^{-3} \text{ m} \cdot \text{s}^{-1}} = 178 \text{ s}$$

Because the chamber is 13.5 m in length and the horizontal velocity is 0.25 m · s⁻¹ the liquid remains in the chamber.

$$t = \frac{13.5 \text{ m}}{0.25 \text{ m} \cdot \text{s}^{-1}} = 54 \text{ s}$$

Thus, the particle will not be captured in the grit chamber.

Aerated Grit Chamber.   The spiral roll of the aerated grit chamber liquid "drives" the grit into a hopper that is located under the air diffuser assembly (Figure 11–2). The shearing action of the air bubbles is supposed to strip the inert grit of much of the organic material that adheres to its surface.

Aerated grit chamber performance is a function of the roll velocity and detention time. The roll velocity is controlled by adjusting the air feed rate. Nominal air flow values are in the range

FIGURE 11–2

Aerated grit chamber.

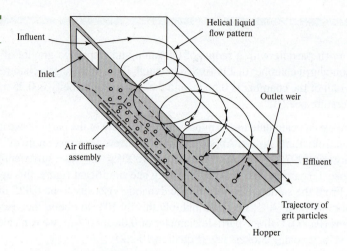

of 0.2 to 0.5 cubic meters per minute of air per meter of tank length ($m^3 \cdot min^{-1} \cdot m^{-1}$). Liquid detention times are usually set to be about three minutes at maximum flow. Length-to-width ratios range from 3:1 to 5:1 with depths on the order of 2 to 5 m.

Grit accumulation in the chamber varies greatly, depending on whether the sewer system is a combined type or a separate type, and on the efficiency of the chamber. For combined systems, 90 m^3 of grit per million cubic meters of sewage ($m^3/10^6$ m^3) is not uncommon. In separate systems you might expect something less than 40 $m^3/10^6$ m^3. Normally the grit is buried in a sanitary landfill.

Vortex Grit Chamber. Wastewater is brought into the chamber tangentially (Figure 11–3). At the center of the chamber, a rotating turbine with adjustable-pitch blades along with the cone-shaped floor produces a spiraling, doughnut-shaped flow pattern. This pattern tends to lift the lighter organic particles and settle the grit into a grit sump. The effluent outlet has twice the width of the influent flume. This results in a lower exit velocity than the influent velocity and thus prevents grit from being drawn into the effluent flow. It should be noted that centrifugal acceleration does not play a significant role in removing the particles. The velocities are too low.

FIGURE 11–3

Vortex grit chamber.
(*Source:* Orenco
Systems®, Inc.)

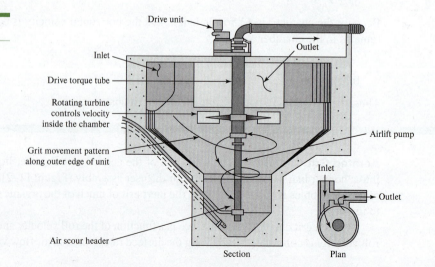

Macerators

Devices that chop up wastewater solids (rags, paper, plastic, and other materials) by revolving cutting bars are called **macerators.** They are used as a replacement for the downstream bar rack but must be installed with a hand-cleaned rack in parallel in case they fail.

Equalization

Flow equalization is not a treatment process per se but a technique that can be used to improve the effectiveness of both secondary and tertiary wastewater treatment processes. Wastewater does not flow into a municipal wastewater treatment plant at a constant rate; the flow rate varies from hour to hour, reflecting the living habits of the area served. In most towns, the pattern of daily activities sets the pattern of sewage flow and strength. Above-average sewage flows and strength occur in midmorning. The constantly changing amount and strength of wastewater to be treated makes efficient process operation difficult. Many treatment units must be designed for the maximum flow conditions encountered, which actually results in their being oversized for average conditions. Other conditions that lead to the inclusion of equalization basins in the process flow scheme include

- Combined sewers that lead to storm flow surges.
- Leaky sewer pipes that allow significant volumes of stormwater to enter the collection system.
- Small plants with very high peaking ratios.
- Industrial wastewater treatment plants that have high cyclic flows or the potential for process malfunctions that discharge large quantities of wastewater or very high strength wastewater.

The purpose of flow equalization is to dampen these variations so that the wastewater can be treated at a nearly constant flow rate. Flow equalization can significantly improve the performance of an existing plant and increase its useful capacity. In new plants, flow equalization can reduce the size and cost of the treatment units (Metcalf & Eddy, 2003).

   Flow equalization is usually achieved by constructing large basins that collect and store the wastewater flow and from which the wastewater is pumped to the treatment plant at a constant rate. These basins are normally located near the head end of the treatment works, preferably downstream of pretreatment facilities such as bar screens, macerators, and grit chambers (Figure 11–4). Adequate aeration and mixing must be provided to prevent odors and deposition of solids. The required volume of an in-line equalization basin is estimated from a mass balance of the flow into the treatment plant with the average flow the plant is designed to treat. From Chapter 4, starting with Equation 4–4

$$\frac{dM}{dt} = \frac{d(\text{in})}{dt} - \frac{d(\text{out})}{dt}$$

and writing in terms of the mass of water

$$\frac{d(\rho \mathbb{V})}{dt} = \frac{d(\rho \mathbb{V})_{\text{in}}}{dt} - \frac{d(\rho \mathbb{V})_{\text{out}}}{dt} \tag{11–1}$$

where ρ = density of water
   Assuming that the density of water is constant and factoring it out, the equation yields

$$\frac{d(\mathbb{V})}{dt} = \frac{d(\mathbb{V})_{\text{in}}}{dt} - \frac{d(\mathbb{V})_{\text{out}}}{dt} \tag{11–2}$$

The term on the left is the change in storage of water in the equalization basin. The term $d(\mathbb{V})_{\text{in}}/dt$ = the flow into the basin, which we normally designate as Q_{in}. Similarly, $d(\mathbb{V})_{\text{out}}/dt$ = the flow out of the basin, which we normally designate as Q_{out}.

FIGURE 11–4

Typical wastewater-treatment plant flow diagram incorporating flow equalization. (a) In-line equalization and (b) off-line equalization. Flow equalization can be applied after grit removal, after primary sedimentation, and after secondary treatment where advanced treatment is used.

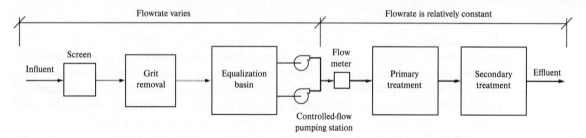

(a)

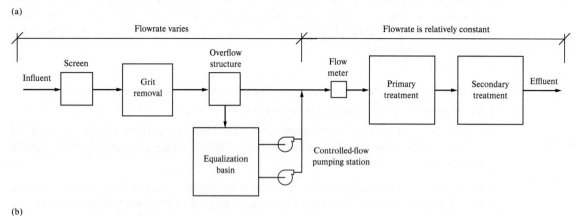

(b)

Thus, in conventional terms we write

$$\frac{d(\text{V})}{dt} = Q_{in} - Q_{out} \tag{11–3}$$

To determine the storage required in a time interval, we rewrite Equation 11–3 as

$$d(\text{V}) = Q_{in}(dt) - Q_{out}(dt) \tag{11–4}$$

For finite time intervals this is

$$\Delta\text{V} = Q_{in}(\Delta t) - Q_{out}(\Delta t) \tag{11–5}$$

This may also be written in terms of volume of wastewater entering and leaving the equalization basin because the product $Q(\Delta t)$ is a volume

$$\Delta\text{V} = \text{V}_{in} - \text{V}_{out} \tag{11–6}$$

To determine the maximum volume required, the sum of ΔV is determined over a cycle that fills and empties the equalization basin. The variable Q_{out} is set to be the average flow over the cycle. The maximum value of ΔV is the required storage.

EXAMPLE 11–2   Design an equalization basin for the following cyclic flow pattern. Provide a 25% excess capacity for equipment, unexpected flow variations, and solids accumulation. Evaluate the effect of equalization on the mass loading of BOD$_5$.

Time (h)	Flow (m$^3 \cdot$ s^{-1})	BOD$_5$ (mg $\cdot$ L^{-1})	Time (h)	Flow (m$^3 \cdot$ s^{-1})	BOD$_5$ (mg $\cdot$ L^{-1})
0000	0.0481	110	1200	0.0718	160
0100	0.0359	81	1300	0.0744	150
0200	0.0226	53	1400	0.0750	140
0300	0.0187	35	1500	0.0781	135
0400	0.0187	32	1600	0.0806	130
0500	0.0198	40	1700	0.0843	120
0600	0.0226	66	1800	0.0854	125
0700	0.0359	92	1900	0.0806	150
0800	0.0509	125	2000	0.0781	200
0900	0.0631	140	2100	0.0670	215
1000	0.0670	150	2200	0.0583	170
1100	0.0682	155	2300	0.0526	130

Solution   Because of the repetitive and tabular nature of the calculations, a computer spreadsheet is ideal for this problem. The spreadsheet solution is easy to verify if the calculations are set up with judicious selection of the initial value. If the initial value is the first flow rate greater than the average after the sequence of nighttime low flows, then the last row of the computation should result in a storage value of zero.

 The first step is to calculate the average flow. In this case it is 0.05657 m$^3 \cdot$ s^{-1}. Next, the flows are arranged in order beginning with the time and flow that first exceeds the average. In this case it is at 0900 h with a flow of 0.0631 m$^3 \cdot$ s^{-1}. The tabular arrangement is shown in the following table. An explanation of the calculations for each column follows.

Time	Flow (m$^3 \cdot$ s^{-1})	$\forall_{in}$ (m^3)	$\forall_{out}$ (m^3)	edS (m^3)	ΣdS (m^3)	BOD$_5$ (mg $\cdot$ L^{-1})	M$_{BOD\text{-in}}$ (kg)	S (mg $\cdot$ L^{-1})	M$_{BOD\text{-out}}$ (kg)
0900	0.0631	227.16	203.65	23.51	23.51	140	31.80	140.00	28.51
1000	0.067	241.2	203.65	37.55	61.06	150	36.18	149.11	30.37
1100	0.0682	245.52	203.65	41.87	102.93	155	38.06	153.83	31.33
1200	0.0718	258.48	203.65	54.83	157.76	160	41.36	158.24	32.23
1300	0.0744	267.84	203.65	64.19	221.95	150	40.18	153.06	31.17
1400	0.075	270	203.65	66.35	288.3	140	37.80	145.89	29.71
1500	0.0781	281.16	203.65	77.51	365.81	135	37.96	140.51	28.62
1600	0.0806	290.16	203.65	86.51	452.32	130	37.72	135.86	27.67
1700	0.0843	303.48	203.65	99.83	552.15	120	36.42	129.49	26.37
1800	0.0854	307.44	203.65	103.79	655.94	125	38.43	127.89	26.04
1900	0.0806	290.16	203.65	86.51	742.45	150	43.52	134.67	27.43
2000	0.0781	281.16	203.65	77.51	819.96	200	56.23	152.61	31.08
2100	0.067	241.2	203.65	37.55	857.51	215	51.86	166.79	33.97
2200	0.0583	209.88	203.65	6.23	863.74	170	35.68	167.42	34.10
2300	0.0526	189.36	203.65	−14.29	849.45	130	24.62	160.69	32.73
0000	0.0481	173.16	203.65	−30.49	818.96	110	19.05	152.11	30.98

(continued)

(continued)

Time	Flow (m³·s⁻¹)	$\forall_{in}$ (m³)	$\forall_{out}$ (m³)	dS (m³)	ΣdS (m³)	BOD_5 (mg·L⁻¹)	$M_{BOD\text{-}in}$ (kg)	S (mg·L⁻¹)	$M_{BOD\text{-}out}$ (kg)
0100	0.0359	129.24	203.65	−74.41	744.55	81	10.47	142.42	29.00
0200	0.0226	81.36	203.65	−122.29	622.26	53	4.31	133.61	27.21
0300	0.0187	67.32	203.65	−136.33	485.93	35	2.36	123.98	25.25
0400	0.0187	67.32	203.65	−136.33	349.6	32	2.15	112.79	22.97
0500	0.0198	71.28	203.65	−132.37	217.23	40	2.85	100.46	20.46
0600	0.0226	81.36	203.65	−122.29	94.94	66	5.37	91.07	18.55
0700	0.0359	129.24	203.65	−74.41	20.53	92	11.89	91.61	18.66
0800	0.0509	183.24	203.65	−20.41	0.12	125	22.91	121.64	24.77

The third column converts the flows to volumes using the time interval (1 h) between flow measurements.

$$\forall = (0.0631 \text{ m}^3 \cdot \text{s}^{-1})(1 \text{ h})(3600 \text{ s} \cdot \text{h}^{-1}) = 227.16 \text{ m}^3$$

The fourth column is the average volume that leaves the equalization basin.

$$\forall = (0.05657 \text{ m}^3 \cdot \text{s}^{-1})(1 \text{ h})(3600 \text{ s} \cdot \text{h}^{-1}) = 203.65 \text{ m}^3$$

The fifth column is the difference between the inflow volume and the outflow volume.

$$\Delta\forall = \forall_{in} - \forall_{out} = 227.16 \text{ m}^3 - 203.65 \text{ m}^3 = 23.51 \text{ m}^3$$

The sixth column is the cumulative sum of the difference between the inflow and outflow. For the second time interval, it is

$$\text{Storage} = dS = 37.55 \text{ m}^3 + 23.51 \text{ m}^3 = 61.06 \text{ m}^3$$

Note that the last value for the cumulative storage is 0.12 m^3. It is not zero because of round-off truncation in the computations. At this point the equalization basin is empty and ready to begin the next day's cycle.

The required volume for the equalization basin is the maximum cumulative storage. With the requirement for 25% excess, the volume would then be

$$\text{Storage volume} = (863.74 \text{ m}^3)(1.25) = 1079.68, \text{ or } 1080 \text{ m}^3$$

The mass of BOD_5 flowing into the equalization basin is the product of the inflow (Q), the concentration of BOD_5 (S_o), and the integration time (Δt).

$$M_{BOD\text{-}in} = (Q)(S_o)(\Delta t)$$

The mass of BOD_5 flowing out of the equalization basin is the product of the average outflow (Q_{avg}), the average concentration (S_{avg}) in the basin, and the integration time (Δt):

$$M_{BOD\text{-}out} = (Q_{avg})(S_{avg})(\Delta t)$$

The average concentration is determined as

$$S_{avg} = \frac{(\forall_i)(S_o) + (\forall_s)(S_{prev})}{\forall_i + \forall_s}$$

where $\forall_i$ = volume of inflow during the time interval Δt (in m³)
   S_o = average BOD_5 concentration during time interval Δt (in g·m⁻³)
   $\forall_s$ = volume of wastewater in the basin at the end of the previous time interval (in m³)
   S_{prev} = concentration of BOD_5 in the basin at the end of the previous time interval
      = previous S_{avg} (in g·m⁻³)

Noting that $1 \text{ mg} \cdot \text{L}^{-1} = 1 \text{ g} \cdot \text{m}^{-3}$, the first row (the 0900 h time) computations are

$$M_{BOD\text{-}in} = (0.0631 \text{ m}^3 \cdot \text{s}^{-1})(140 \text{ g} \cdot \text{m}^{-3})(1 \text{ h})(3600 \text{ s} \cdot \text{h}^{-1})(10^{-3} \text{ kg} \cdot \text{g}^{-1}) = 31.8 \text{ kg}$$

$$S_{avg} = \frac{(227.16 \text{ m}^3)(140 \text{ g} \cdot \text{m}^{-3}) + 0}{227.16 \text{ m}^3 + 0} = 140 \text{ mg} \cdot \text{L}^{-1}$$

$$M_{BOD\text{-}out} = (0.05657 \text{ m}^3 \cdot \text{s}^{-1})(140 \text{ g} \cdot \text{m}^{-3})(1 \text{ h})(3600 \text{ s} \cdot \text{h}^{-1})(10^{-3} \text{ kg} \cdot \text{g}^{-1}) = 28.5 \text{ kg}$$

Note that the zero values in the computation of S_{avg} are valid only at startup of an empty basin. Also note that in this case, $M_{BOD\text{-}in}$ and $M_{BOD\text{-}out}$ differ only because of the difference in flow rates. For the second row (1000 h), the computations are

$$M_{BOD\text{-}in} = (0.0670 \text{ m}^3 \cdot \text{s}^{-1})(150 \text{ g} \cdot \text{m}^{-3})(1 \text{ h})(3600 \text{ s} \cdot \text{h}^{-1})(10^{-3} \text{ kg} \cdot \text{g}^{-1}) = 36.2 \text{ kg}$$

$$S_{avg} = \frac{(241.20 \text{ m}^3)(150 \text{ g} \cdot \text{m}^{-3}) + (23.51 \text{ m}^3)(140 \text{ g} \cdot \text{m}^{-3})}{241.20 \text{ m}^3 + 23.51 \text{ m}^3} = 149.11 \text{ mg} \cdot \text{L}^{-1}$$

$$M_{BOD\text{-}out} = (0.05657 \text{ m}^3 \cdot \text{s}^{-1})(149.11 \text{ g} \cdot \text{m}^{-3})(1 \text{ h})(3600 \text{ s} \cdot \text{h}^{-1})(10^{-3} \text{ kg} \cdot \text{g}^{-1})$$
$$= 30.37 \text{ mg} \cdot \text{L}^{-1}$$

Note that $\mathbf{V}_s$ is the volume of wastewater in the basin at the end of the previous time interval; therefore, it equals the accumulated dS. The concentration of BOD_5 (S_{prev}) is the average concentration at the end of previous interval (S_{avg}) and *not* the influent concentration for the previous interval (S_o).

For the third row (1100 h), the concentration of BOD_5 is

$$S_{avg} = \frac{(245.52 \text{ m}^3)(155 \text{ g} \cdot \text{m}^{-3}) + (61.06 \text{ m}^3)(149.11 \text{ g} \cdot \text{m}^{-3})}{245.52 \text{ m}^3 + 61.06 \text{ m}^3} = 153.83 \text{ mg} \cdot \text{L}^{-1}$$

11–7 PRIMARY TREATMENT

With the screening completed and the grit removed, the wastewater still contains light organic suspended solids, some of which can be removed from the sewage by gravity in a sedimentation tank. These tanks can be round or rectangular. The mass of settled solids is called **raw sludge.** The sludge is removed from the sedimentation tank by mechanical scrapers and pumps (Figure 11–5). Floating materials, such as grease and oil, rise to the surface of the sedimentation tank, where they are collected by a surface-skimming system and removed from the tank for further processing.

Primary sedimentation basins (primary tanks) are characterized by type II flocculant settling. The Stokes' equation cannot be used because the flocculating particles are continually changing in size, shape, and, when water is entrapped in the floc, specific gravity. No mathematical relationship is adequate to describe type II settling. Laboratory tests with settling columns are used to develop design data.

Rectangular tanks with common-wall construction are frequently chosen because they are advantageous for sites with space constraints. Typically, these tanks range from 15 to 100 m in length and 3 to 24 m in width. Circular tanks have diameters from 3 to 90 m.

As in the water treatment clarifier design discussed in Chapter 10, overflow rate is the controlling parameter for the design of primary settling tanks. At average flow, overflow rates typically range from 25 to 60 $\text{m}^3 \cdot \text{m}^{-2} \cdot \text{day}^{-1}$ (or 25–60 $\text{m} \cdot \text{day}^{-1}$). When waste-activated sludge is returned to the primary tank, a lower range of overflow rates is chosen (25–35 $\text{m} \cdot \text{day}^{-1}$). Under peak flow conditions, overflow rates may be in the range of 80 to 120 $\text{m} \cdot \text{day}^{-1}$.

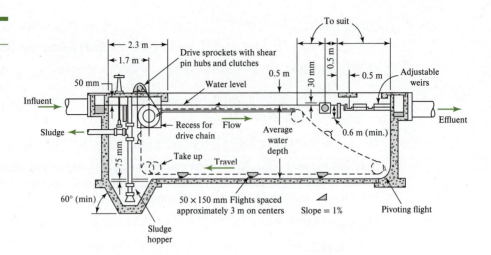

Hydraulic detention time (see Equation 4–27 for definition) in the sedimentation basin ranges from 1.5 to 2.5 h under average flow conditions. A 2.0-h detention time is typical.

The Great Lakes-Upper Mississippi River Board of State Sanitary Engineers (GLUMRB) recommends that peak hourly weir loading (hydraulic flow over the effluent weir) rates not exceed 250 $m^3 \cdot day^{-1}$ of flow per meter of weir length ($m^3 \cdot day^{-1} \cdot m^{-1}$) for plants with average flows less than 0.04 $m^3 \cdot s^{-1}$. For larger flows, the recommended rate is 375 $m^3 \cdot day^{-1} \cdot m^{-1}$ (GLUMRB, 2004). If the side water depths exceed 3.5 m, the weir loading rates have little effect on performance.

As mentioned previously, approximately 50–60% of the raw sewage suspended solids and as much as 30–35% of the raw sewage BOD_5 may be removed in the primary tank.

EXAMPLE 11–3   Evaluate the following primary tank design with respect to detention time, overflow rate, and weir loading.

Design Data

Flow = 0.150 $m^3 \cdot s^{-1}$ Length = 40.0 m (effective)
Influent suspended solids = 280 $mg \cdot L^{-1}$ Width = 10.0 m
Sludge concentration = 6.0%   Liquid depth = 2.0 m
Efficiency = 60%   Weir length = 75.0 m

Solution   The detention time is simply the volume of the tank divided by the flow.

$$t_o = \frac{\Psi}{Q} = \frac{40.0 \text{ m} \times 10.0 \text{ m} \times 2.0 \text{ m}}{0.150 \text{ m}^3 \cdot \text{s}^{-1}} = 5333.33 \text{ s, or } 1.5 \text{ h}$$

This is a reasonable detention time.

The overflow rate is the flow divided by the surface area.

$$v_o = \frac{0.150 \text{ m}^3 \cdot \text{s}^{-1}}{40.0 \text{ m} \times 10.0 \text{ m}} \times 86,400 \text{ s} \cdot \text{day}^{-1} = 32 \text{ m} \cdot \text{day}^{-1}$$

This is an acceptable overflow rate.

The weir loading (WL) is calculated in the same fashion.

$$\text{WL} = \frac{0.150 \text{ m}^3 \cdot \text{s}^{-1}}{75.0 \text{ m}} \times 86,400 \text{ s} \cdot \text{day}^{-1} = 172.8, \text{ or } 173 \text{ m}^3 \cdot \text{day}^{-1} \cdot \text{m}^{-1}$$

This is an acceptable weir loading.

11–8   UNIT PROCESSES OF SECONDARY TREATMENT

Overview

The basic ingredients needed for conventional aerobic secondary biologic treatment are the availability of many microorganisms, good contact between these organisms and the organic material, the availability of oxygen, and the maintenance of other favorable environmental conditions (e.g., favorable temperature and sufficient time for the organisms to work). A variety of approaches have been used in the past to meet these basic needs. The most common approaches for municipalities are the trickling filter process and the activated sludge process. Lagoons are employed when wastewater flows are not large and land space is available.

Role of Microorganisms

The stabilization of organic matter is accomplished biologically using a variety of microorganisms, which convert the colloidal and dissolved carbonaceous organic matter into various gases and into protoplasm. Because protoplasm has a specific gravity slightly greater than that of water, it can be removed from the treated liquid by gravity settling.

It is important to note that unless the protoplasm produced from the organic matter is removed from the solution, complete treatment will not be accomplished because the protoplasm, which itself is organic, will be measured as BOD in the effluent. If the protoplasm is not removed, the only treatment that will be achieved is that associated with the bacterial conversion of a portion of the organic matter originally present to various gaseous end products (Metcalf & Eddy, 2003).

Population Dynamics

The Microbial Ecosystem.   In the discussion of the behavior of bacterial cultures that follows, the inherent assumption is that all the requirements for growth are initially present. The requirements for growth and the dynamic behavior of the microbial ecosystem were discussed extensively in Chapter 5. In wastewater treatment, as in nature, pure cultures of microorganisms do not exist. Rather, a mixture of species compete and survive within the limits set by the environment. **Population dynamics** is the term used to describe the time-varying success of the various species in competition. It is expressed quantitatively in terms of relative mass of microorganisms.

The prime factor governing the dynamics of the various microbial populations is the competition for food. The second most important factor is the predator–prey relationship.

The relative success of a pair of species competing for the same substrate is a function of the ability of the species to metabolize the substrate. The more successful species will be the one that metabolizes the substrate more completely. In so doing, it will obtain more energy for synthesis and consequently will achieve a greater mass.

The Monod Equation.   For the large numbers and mixed cultures of microorganisms found in waste treatment systems, it is convenient to measure biomass rather than numbers of organisms. Frequently, this is done by measuring suspended solids or **volatile suspended solids** (those that burn at $550 \pm 50°C$). When the wastewater contains only soluble organic matter, the volatile suspended solids test is reasonably representative. The presence of organic particles (which is often the case in municipal wastewater) confuses the issue completely. In the log-growth phase (discussed in Chapter 5), the rate of growth of bacterial cells can be defined as

$$r_g = \mu X \tag{11–7}$$

where r_g = growth rate of the biomass (in $mg \cdot L^{-1} \cdot t^{-1}$)
   μ = specific growth rate constant (in t^{-1})
   X = concentration of biomass (in $mg \cdot L^{-1}$)

FIGURE 11–6

Monod growth rate
constant as a function
of limiting food
concentration.

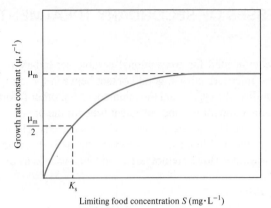

Because $dX/dt = r_g$ for a batch reactor (see Equations 4–12 and 4–26), we may write the following rate expression for biomass increase in a batch reactor

$$\frac{dX}{dt} = \mu X \qquad (11\text{–}8)$$

Because of the difficulty of direct measurement of μ in mixed cultures, Monod developed a model equation that assumes that the rate of food use, and therefore the rate of biomass production, is limited by the rate of enzyme reactions involving the food compound that is in shortest supply relative to its need (Monod, 1949). The Monod equation is

$$\mu = \frac{\mu_m S}{K_s + S} \qquad (11\text{–}9)$$

where μ_m = maximum specific growth rate constant (in t^{-1})

$\quad S$ = concentration of limiting food in solution (in mg $\cdot$ L^{-1})

$\quad K_s$ = half saturation constant (in mg $\cdot$ L^{-1})

$\quad$ = concentration of limiting food when $\mu = 0.5\mu_m$

The relationship between the growth rate constant and limiting food concentration (S) follows a hyperbolic function as shown in Figure 11–6.

Two limiting cases are of interest in the application of Equation 11–9 to wastewater treatment systems. In those cases where there is an excess of the limiting food, then $S \gg K$ and the growth rate constant, μ, is approximately equal to μ_m. Equation 11–8 then becomes zero-order in substrate, S, that is, it is independent of the substrate. At the other extreme, when $S \ll K_s$, the system is food-limited and the growth rate becomes first-order with respect to substrate.

Equation 11–9 assumes only growth of microorganisms and does not take into account natural die-off. It is generally assumed that the death or decay of the microbial mass is a first-order expression in biomass and hence Equations 11–7 and 11–9 are expanded to

$$r_g = \frac{\mu_m S X}{K_s + S} - k_d X \qquad (11\text{–}10)$$

where k_d = endogenous decay rate constant (in t^{-1}).

If all of the food in the system were converted to biomass, the rate of food use, r_{su}, would equal the rate of biomass production. Because of the inefficiency of the conversion process, the rate of food use will be greater than the rate of biomass use, so

$$r_{su} = \frac{1}{Y}(r_g) \qquad (11\text{–}11)$$

where Y = decimal fraction of food mass convened to biomass

$\quad$ = yield coefficient, in milligrams per liter of food used, $\frac{\text{mg} \cdot \text{L}^{-1}\text{ biomass}}{\text{mg} \cdot \text{L}^{-1}\text{ food used}}$

FIGURE 11–7

Conventional activated
sludge plant.

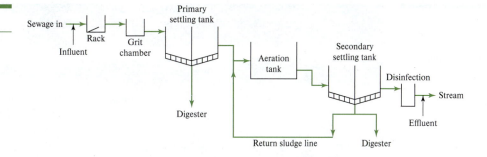

Combining Equations 11–7, 11–9, and 11–11, we obtain

$$r_{su} = \frac{1}{Y} \frac{\mu_m S X}{(K_s + S)} \tag{11–12}$$

Equations 11–10 and 11–12 are a fundamental part of the development of the design equations for wastewater treatment processes.

Activated Sludge

The activated sludge process (Figure 11–7) is a biological wastewater treatment technique in which a mixture of wastewater and biological solids (microorganisms) is agitated and aerated. The biological solids are subsequently separated from the treated wastewater by settling and a fraction is returned to the aeration process as needed. The settling tank that follows the aeration tank is referred to as a **secondary clarifier,** or **final clarifier,** to differentiate it from the sedimentation basin used for primary settling.

The activated sludge process derives its name from the biological mass formed when air is continuously injected into the wastewater. In this process, microorganisms are mixed thoroughly with the organic matter in the wastewater under conditions that stimulate their growth through use of the organic matter as food. As the microorganisms grow and are mixed by the agitation of the air, the individual organisms clump together (flocculate) to form an active mass of microbes (biologic floc) called **activated sludge.**

In practice, wastewater flows continuously into an aeration tank where air is injected to mix the activated sludge with the wastewater and to supply the oxygen needed for the organisms to break down the organic matter. The mixture of activated sludge and wastewater in the aeration tank is called **mixed liquor.** The mixed liquor flows from the aeration tank to a secondary clarifier where the activated sludge is settled out. Most of the settled sludge is returned to the aeration tank (and hence is called **return sludge**) to maintain the high population of microbes that permits rapid breakdown of the organic matter. Because more activated sludge is produced than is desirable in the process, some of the return sludge is diverted to the sludge handling system for treatment and disposal. In conventional activated sludge systems, the wastewater is typically aerated for 6–8 h in long, rectangular aeration basins. About 8 m^3 of air is provided for each cubic meter of wastewater treated. Sufficient air is provided to keep the sludge in suspension (Figure 11–8). The air is injected near the bottom of the aeration tank through a system of perforated pipes or porous stones called **diffusers.** The volume of sludge returned to the aeration basin is typically 20–30% of the wastewater flow.

The activated sludge process is controlled by wasting a portion of the microorganisms each day in order to maintain the proper amount of microorganisms to efficiently degrade the BOD_5. **Wasting** means that a portion of the microorganisms is discarded from the process. The treatment and disposal of the discarded microorganisms are discussed in Section 11–13. The discarded microorganisms are called **waste activated sludge** (WAS). A balance is then achieved between growth of new organisms and their removal by wasting. If too much sludge is wasted, the concentration of microorganisms in the mixed liquor will become too low for

effective treatment. If too little sludge is wasted, a large concentration of microorganisms will accumulate and, ultimately, overflow the secondary tank and flow into the receiving stream.

The **mean cell residence time** θc, also called **solids retention time** (SRT) or **sludge age,** is defined as the average amount of time that microorghanisms are kept in the system. This differs from the **hydraulic detention time** (t_o), which is the average time the wastewater is in the tank.

Many modifications of the conventional activated sludge process have been developed to address specific treatment problems. We have selected the completely mixed flow process for further discussion.

Completely Mixed Activated Sludge Process. The design formulas for the completely mixed activated sludge process are a mass-balance application of the equations used to describe the kinetics of microbial growth. A mass-balance diagram for the completely mixed system (CMFR) is shown in Figure 11–9. The mass-balance equations are written for the system boundary shown by the dashed line. Two mass balances are required to define the design of the reactor: one for biomass and one for food (soluble BOD_5).

Under steady-state conditions, the mass balance for biomass may be written as

Biomass in influent + net biomass growth = biomass in effluent + biomass wasted **(11–13)**

The biomass in the influent is the product of the concentration of microorganisms in the influent (X_o) and the flow rate of wastewater (Q). The concentration of microorganisms in the

FIGURE 11–9

Completely mixed biological reactor with solids recycle.

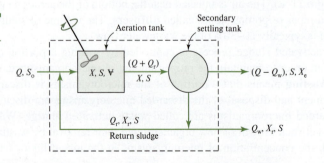

influent is measured as suspended solids (mg · L^{-1}). The biomass that accumulates in the aeration tank is the product of the volume of the tank ($\forall$) and the Monod expression for growth of microbial mass (Equation 11–10)

$$(\forall)\left(\frac{\mu_m S X}{K_s + S} - k_d X\right) \tag{11–14}$$

The biomass in the effluent is the product of flow rate of treated wastewater leaving the plant ($Q - Q_w$) and the concentration of microorganisms that does not settle in the secondary clarifier (X_e). The flow rate of wastewater leaving the plant does not equal the flow rate into the plant because some of the microorganisms must be wasted. The flow rate of wasting (Q_w) is deducted from the flow exiting the plant.

The biomass that is wasted is the product of concentration of microorganisms in the WAS flow (X_r) and the WAS flow rate (Q_r). The mass-balance equation may be rewritten as

$$(Q)(X_o) + (\forall)\left(\frac{\mu_m S X}{K_s + S} - k_d X\right) = (Q - Q_w)(X_e) + (Q_w)(X_r) \tag{11–15}$$

where Q = wastewater flow rate into the aeration tank (in m^3 · day^{-1})

X_o = microorganism concentration (**volatile suspended solids** or VSS*) entering aeration tank (in mg · L^{-1})

$\forall$ = volume of aeration tank (in m^3)

μ_m = maximum growth rate constant (in days^{-1})

S = soluble BOD$_5$ in aeration tank and effluent (in mg · L^{-1})

X = microorganism concentration (**mixed-liquor volatile suspended solids** or MLVSS†) in the aeration tank (in mg · L^{-1})

K_s = half velocity constant

= soluble BOD$_5$ concentration at one-half the maximum growth rate (in mg · L^{-1})

k_d = decay rate of microorganisms (in days^{-1})

Q_w = flow rate of liquid containing microorganisms to be wasted (in m^3 · day^{-1})

X_e = microorganism concentration (VSS) in effluent from secondary settling tank (in mg · L^{-1})

X_r = microorganism concentration (VSS) in sludge being wasted (in mg · L^{-1})

At steady-state, the mass-balance equation for food (soluble BOD$_5$) may be written as

Food in influent + food consumed = food in effluent + food in WAS   **(11–16)**

The food in the influent is the product of the concentration of soluble BOD$_5$ in the influent (S_o) and the flow rate of wastewater (Q). The food that is consumed in the aeration tank is the product of the volume of the tank ($\forall$) and the expression for rate of food use (Equation 11–12)

$$(\forall)\left(\frac{\mu_m S X}{Y(K_s + S)}\right) \tag{11–17}$$

The food in the effluent is the product of flow rate of treated wastewater leaving the plant ($Q - Q_w$) and the concentration of soluble BOD$_5$ in the effluent (S). The concentration of soluble BOD$_5$ in the effluent (S) is the same as that in the aeration tank because we have assumed

---

*Suspended solids means that the material will be retained on a filter, unlike dissolved solids such as NaCl. The amount of the suspended solids that volatilizes at 500 ± 50°C is taken to be a measure of active biomass concentration. The presence of nonliving organic particles in the influent wastewater will cause some error (usually small) in the use of volatile suspended solids as a measure of biomass.

†Mixed-liquor volatile suspended solids is a measure of the active biological mass in the aeration tank. The term *mixed liquor* implies a mixture of activated sludge and wastewater.

that the aeration tank is completely mixed. Because the BOD_5 is soluble, the secondary settling tank will not change the concentration. Thus, the effluent concentration from the secondary settling tank is the same as the influent concentration.

The food in the waste activated sludge flow is the product of the concentration of soluble BOD_5 in the influent (S) and the WAS flow rate (Q_r). The mass-balance equation for steady-state conditions may be rewritten as

$$(Q)(S_o) - (\forall)\left(\frac{\mu_m SX}{Y(K_s + S)}\right) = (Q - Q_w)S + (Q_w)(S) \tag{11-18}$$

where Y = yield coefficient (see Equation 11–11).

To develop working design equations, we make the following assumptions:

1. The influent and effluent biomass concentrations are negligible compared with that in the reactor.
2. The influent food (S_o) is immediately diluted to the reactor concentration in accordance with the definition of a CMFR.
3. All reactions occur in the CMFR.

From the first assumption we may eliminate the following terms from Equation 11–15: QX_o, and $(Q - Q_w)X_e$ because, X_o and X_e are negligible compared with X. Equation 11–15 may be simplified to

$$(\forall)\left(\frac{\mu_m SX}{K_s + S} - k_d X\right) = (Q_w)(X_r) \tag{11-19}$$

For convenience, we may rearrange Equation 11–19 in terms of the Monod equation

$$\left(\frac{\mu_m S}{K_s + S}\right) = \frac{(Q_w)(X_r)}{(\forall)(X)} + k_d \tag{11-20}$$

Equation 11–18 may also be rearranged in terms of the Monod equation

$$\left(\frac{\mu_m S}{K_s + S}\right) = \frac{Q}{\forall}\frac{Y}{X}(S_o - S) \tag{11-21}$$

Noting that the left side of Equations 11–20 and 11–21 are the same, we set the right-hand side of these equations equal and rearrange to give

$$\frac{(Q_w)(X_r)}{(\forall)(X)} = \frac{Q}{\forall}\frac{Y}{X}(S_o - S) - k_d \tag{11-22}$$

Two parts of this equation have physical significance in the design of a completely mixed activated sludge system. The inverse of $Q/\forall$ is the hydraulic detention time (t_o) of the reactor.

$$t_o = \frac{\forall}{Q} \tag{11-23}$$

The inverse of the left side of Equation 11–22 defines the mean cell-residence time (θ_c).

$$\theta_c = \frac{\forall X}{Q_w X_r} \tag{11-24}$$

The mean cell-residence time expressed in Equation 11–24 must be modified if the effluent biomass concentration is not negligible. Equation 11–25 accounts for effluent losses of biomass in calculating θ_c.

$$\theta_c = \frac{(\forall)X}{Q_w X_r + (Q - Q_w)(X_e)} \tag{11-25}$$

From Equation 11–20, it can be seen that once θ_c is selected, the concentration of soluble BOD_5 in the effluent (S) is fixed

$$S = \frac{K_s(1 + k_d \theta_c)}{\theta_c(\mu_m - k_d) - 1} \tag{11-26}$$

TABLE 11–8   **Values of Growth Constants for Domestic Wastewater**

Parameter	Basis	Value[a]	
		Range	Typical
K_s	mg $\cdot$ L^{-1} BOD$_5$	25–100	60
k_d	days^{-1}	0–0.30	0.10
μ_m	days^{-1}	1–8	3
Y	mg VSS $\cdot$ mg^{-1} BOD$_5$	0.4–0.8	0.6

[a]Values at 20°C.

Source: Metcalf and Eddy, Inc., 2003, and Shahriari, Eskicioglu, and Droste, 2006.

Typical values of the microbial growth constants are given in Table 11–8. Note that the concentration of soluble BOD$_5$ leaving the system (S) is affected only by the mean cell-residence time and not by the amount of BOD$_5$ entering the aeration tank or by the hydraulic detention time. It is also important to reemphasize that S is the soluble BOD$_5$ and not the total BOD$_5$. Some fraction of the suspended solids that do not settle in the secondary settling tank also contributes to the BOD$_5$ load to the receiving body. To achieve a desired effluent quality, both the soluble and insoluble fractions of BOD$_5$ must be considered. Thus, using Equation 11–26 to achieve a desired effluent quality (S) by solving for θ_c requires that some estimate of the BOD$_5$ of the suspended solids be made first. This value is then subtracted from the total allowable BOD$_5$ in the effluent to find the allowable S.

$$S = \text{total BOD}_5 \text{ allowed} - \text{BOD}_5 \text{ in suspended solids} \tag{11–27}$$

From Equation 11–22 it is also evident that the concentration of microorganisms in the aeration tank is a function of the mean cell-residence time, hydraulic detention time, and difference between the influent and effluent concentrations.

$$X = \frac{\theta_c(Y)(S_o - S)}{t_o(1 + k_d\theta_c)} \tag{11–28}$$

EXAMPLE 11–4 The town of Gatesville has been directed to upgrade its primary WWTP to a secondary plant that can meet an effluent standard of 30.0 mg $\cdot$ L^{-1} BOD$_5$ and 30.0 mg $\cdot$ L^{-1} suspended solids. They have selected a completely mixed activated sludge system.

Assuming that the BOD$_5$ of the suspended solids may be estimated as equal to 63% of the suspended solids concentration, estimate the required volume of the aeration tank. The following data are available from the existing primary plant.

Existing plant effluent characteristics
Flow = 0.150 m^3 $\cdot$ s^{-1}
BOD$_5$ = 84.0 mg $\cdot$ L^{-1}

Assume the following values for the growth constants: $K_s = 100$ mg $\cdot$ L^{-1} BOD$_5$; $\mu_m = 2.5$ day^{-1}; $k_d = 0.050$ day^{-1}; $Y = 0.50$ mg VSS $\cdot$ mg^{-1} BOD$_5$ removed.

Solution Assuming that the secondary clarifier can produce an effluent with only 30.0 mg $\cdot$ L^{-1} suspended solids, we can estimate the allowable soluble BOD$_5$ in the effluent using the 63% assumption and Equation 11–27.

$$S = \text{BOD}_5 \text{ allowed} - \text{BOD}_5 \text{ in suspended solids}$$
$$= 30.0 - (0.630)(30.0) = 11.1 \text{ mg} \cdot \text{L}^{-1}$$

The mean cell-residence time can be estimated with Equation 11–26 and the assumed values for the growth constants,

$$11.1 = \frac{(100.0 \text{ mg} \cdot \text{L}^{-1}\text{BOD}_5)[1 + (0.050 \text{ day}^{-1})(\theta_c)]}{\theta_c(2.5 \text{ day}^{-1} - 0.050 \text{ day}^{-1}) - 1}$$

Solving for θ_c

$$(11.1)(2.45\,\theta_c - 1) = 100.0 + 5.00\theta_c$$

$$27.20\,\theta_c - 11.1 = 100.0 + 5.00\theta_c$$

$$\theta_c = \frac{111.1}{22.2} = 5.00, \text{ or } 5.0 \text{ days}$$

If we assume a value of 2000 mg $\cdot$ L^{-1} for the MLVSS, we can solve Equation 11–28 for the hydraulic detention time.

$$2000 = \frac{5.00 \text{ days}(0.50 \text{ mg VSS} \cdot \text{mg}^{-1} \text{ BOD}_5)(8.40 \text{ mg} \cdot \text{L}^{-1} - 11.1 \text{ mg} \cdot \text{L}^{-1})}{t_o[1 + (0.050 \text{ day}^{-1})(5.00 \text{ days})]}$$

$$t_o = \frac{2.50(72.9)}{2000(1.25)} = 0.073 \text{ day, or } 1.8 \text{ h}$$

The volume of the aeration tank is then estimated using Equation 11–23.

$$1.8 \text{ h} = \frac{\forall}{(0.150 \text{ m3} \cdot \text{s}^{-1})(3600 \text{ s} \cdot \text{h}^{-1})}$$

$$\forall = 972 \text{ m}^3, \text{ or } 970 \text{ m}^3$$

A commonly used parameter in regulating the performance of the activated sludge process is the food to microorganism ratio (F/M), which is defined as

$$\frac{F}{M} = \frac{QS_o}{VX} \tag{11–29}$$

The units of the F/M ratio are

$$\frac{\text{mg BOD}_5 \cdot \text{day}^{-1}}{\text{mg MLVSS}} = \frac{\text{mg}}{\text{mg} \cdot \text{day}}$$

The F/M ratio is controlled by wasting part of the microbial mass, thereby reducing the MLVSS. A high rate of wasting causes a high F/M ratio. A high F/M yields organisms that are saturated with food. The result is that efficiency of treatment is poor. A low rate of wasting causes a low F/M ratio, which yields organisms that are starved. This results in more complete degradation of the waste.

A long θ_c (low F/M) is not always used, however, because of certain trade-offs that must be considered. A long θ_c means a higher requirement for oxygen and, thus, higher power costs. Problems with poor sludge "settleability" in the final clarifier may be encountered if θ_c is too long. However, because the waste is more completely degraded to final end products and less of the waste is converted into microbial cells when the microorganisms are starved at a low F/M, there is less sludge to handle.

Because both the F/M ratio and the cell detention time are controlled by wasting of organisms, they are interrelated. A high F/M corresponds to a short θ_c, and a low F/M corresponds to a long θ_c. F/M values typically range from 0.1 to 1.0 mg $\cdot$ mg$^{-1} \cdot$ day^{-1} for the various modifications of the activated sludge process.

EXAMPLE 11–5   Two "fill and draw," batch-operated sludge tanks are operated at the "extreme" conditions described below. What is the effect on the operating parameters?

Tank A is settled once each day, and half the liquid is removed with care not to disturb the sludge that settles to the bottom. This liquid is replaced with fresh settled sewage. A plot of MLVSS concentration versus time takes the shape shown in the following figure.

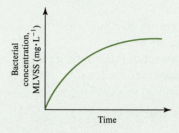

Tank B is not settled; rather, once each day, half the mixed liquor is removed while the tank is being violently agitated. The liquid is replaced with fresh settled sewage. A plot of MLVSS concentration versus time is shown in the following figure.

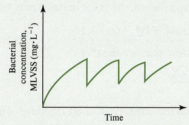

A comparison of the operating characteristics of the two systems is shown in the following table.

Parameter	Tank A	Tank B
F/M	Low	High
θ_c	Long	Short
Sludge produced	None	Much
Oxygen required	High	Low
Power	High	Low

The optimum choice is somewhere between these extremes. A balance must be struck between the cost of sludge disposal and the cost of power to provide oxygen (air).

EXAMPLE 11–6   Compute the F/M ratio for the new activated sludge plant at Gatesville (Example 11–4).

Solution   Using the data from Example 11–4 and Equation 11–29, we obtain

$$\frac{F}{M} = \frac{(0.150 \text{ m}^3 \cdot \text{s}^{-1})(84.0 \text{ mg} \cdot \text{L}^{-1})(86,400 \text{ s} \cdot \text{day}^{-1})}{(970 \text{ m}^3)(2000 \text{ mg} \cdot \text{L}^{-1})}$$

$$= 0.56 \text{ mg} \cdot \text{mg}^{-1} \cdot \text{day}^{-1}$$

This is well within the typical range of F/M ratios.

Sludge Return. The purpose of sludge return is to maintain a sufficient concentration of activated sludge in the reactor basin. The return sludge pumping rate may be determined from a mass balance around the settling tank in Figure 11–10. Assuming that the amount of sludge in the secondary settling tank remains constant (steady-state conditions) and that the effluent suspended solids (X_e) are negligible, the mass balance is

$$\text{Accumulation} = \text{inflow} - \text{outflow} \tag{11–30}$$

$$0 = (Q + Q_r)(X') - (Q_r X_r' + Q_w X_r') \tag{11–31}$$

where Q = wastewater flow rate (in $m^3 \cdot day^{-1}$)
$\quad Q_r$ = return sludge flow rate (in $m^3 \cdot day^{-1}$)
$\quad X'$ = mixed liquor suspended solids (MLSS) (in $g \cdot m^{-3}$)
$\quad X_r'$ = maximum return sludge concentration (in $g \cdot m^{-3}$)
$\quad Q_w$ = sludge wasting flow rate (in $m^3 \cdot day^{-1}$)

Solving for the return sludge flow rate, we get

$$Q_r = \frac{QX' - Q_w X_r'}{X_r' - X'} \tag{11–32}$$

Frequently, the assumption that the effluent suspended solids are negligible is not valid. If the effluent suspended solids are significant, the mass balance may then be expressed as

$$0 = (Q + Q_r)(X') - (Q_r X_r' + Q_w X_r' + (Q - Q_w) X_e) \tag{11–33}$$

Solving for the return sludge flow rate, we see that

$$Q_r = \frac{QX' - Q_w X_r' - (Q - Q_w) X_e}{X_r' - X'} \tag{11–34}$$

Note that X_r' and X' include both the volatile and inert fractions. Thus, they differ from X_r and X by a constant factor. With the volume of the tank and the mean cell-residence time, the sludge wasting flow rate can be determined with Equation 11–24 if the maximum return sludge concentration (X_r') can be determined.

EXAMPLE 11–7   In the continuing saga of the Gatesville plant expansion, we now wish to estimate the return sludge pumping rate. Based on the aeration tank design (Example 11–4) and an informed, reliable source, we have the following data:

Design data
    Flow = $0.150 \, m^3 \cdot s^{-1}$
    MLVSS (X) = $2000 \, mg \cdot L^{-1}$
    MLSS (X') = 1.43 (MLVSS)
    Return sludge concentration (X_r') = $10,000 \, mg \cdot L^{-1}$
    Effluent suspended solids (X_e) may be assumed to be negligible

Solution   The return sludge flow rate may be computed from Equation 11–32. To use this equation we must first estimate the sludge wasting flow rate (Q_w). Solving Equation 11–24 for Q_w, we obtain

$$Q_w = \frac{\forall X}{\theta_c X_r}$$

Note that X_r is the microorganism concentration (MLVSS) in the sludge being wasted. As noted earlier X_r differs from X_r' by a constant factor. In this case the ratio of MLSS/MLVSS is given so we can estimate X_r as

$$X_r = \frac{X_r'}{1.43} = \frac{10,000 \, mg \cdot L^{-1}}{1.43} = 6993 \, mg \cdot L^{-1}$$

Now using the data from Example 11–4, we can compute Q_w.

$$Q_w = \frac{(970 \text{ m}^3)(2000 \text{ mg} \cdot \text{L}^{-1})}{(5 \text{ days})(6993 \text{ mg} \cdot \text{L}^{-1})} = 55.48 \text{ m}^3 \cdot \text{day}^{-1}$$

To compute the return sludge flow rate, we convert the wastewater flow rate to units of cubic meters per day and determine the anticipated concentration of the MLSS using the ratio of MLSS/MLVSS.

$$Q = (0.150 \text{ m}^3 \cdot \text{s}^{-1})(86,400 \text{ s} \cdot \text{day}^{-1}) = 12,960 \text{ m}^3 \cdot \text{day}^{-1}$$

and

$$\text{MLSS} = 1.43(2000 \text{ mg} \cdot \text{L}^{-1}) = 2860 \text{ mg} \cdot \text{L}^{-1}$$

Noting that $1 \text{ mg} \cdot \text{L}^{-1} = 1 \text{ g} \cdot \text{m}^{-3}$, the return sludge concentration is $10,000 \text{ mg} \cdot \text{L}^{-1}$ and ignoring the effluent suspended solids, the return sludge flow rate is then

$$Q_r = \frac{(12,960 \text{ m}^3 \cdot \text{day}^{-1})(2860 \text{ g} \cdot \text{m}^{-3}) - (55.48 \text{ m}^3 \cdot \text{day}^{-1})(10,000 \text{ g} \cdot \text{m}^{-3})}{10,000 \text{ g} \cdot \text{m}^{-3} - 2860 \text{ g} \cdot \text{m}^{-3}}$$

$$= 5113, \text{ or } 5000 \text{ m}^3 \cdot \text{day}^{-1}$$

Sludge Production.   The activated sludge process removes substrate, which exerts an oxygen demand by converting the food into new cell material and degrading this substrate while generating energy. This cell material ultimately becomes sludge, which must be disposed of. Despite the problems in doing so, researchers have attempted to develop enough basic information on sludge production to permit a reliable design basis. Heukelekian and Sawyer both reported that a net yield of 0.5 kg MLVSS $\cdot$ kg^{-1} BOD$_5$ removed could be expected for a completely soluble organic substrate (Heukeleian, Orford, and Maganelli, 1951; Sawyer, 1956). Most researchers agree that, depending on the inert solids in the system and the SRT, 0.40–0.60 kg MLVSS $\cdot$ kg^{-1} BOD$_5$ removed will normally be observed.

The amount of sludge that must be wasted each day is the difference between the amount of increase in sludge mass and the suspended solids (SS) lost in the effluent.

Mass to be wasted = increase in MLSS – SS lost in effluent   **(11–35)**

The net activated sludge produced each day is determined by

$$Y_{obs} = \frac{Y}{1 + k_d\theta_c} \tag{11–36}$$

and

$$P_x = Y_{obs}Q(S_o - S)(10^{-3} \text{ kg} \cdot \text{g}^{-1}) \tag{11–37}$$

where P_x = net waste activated sludge produced each day in terms of VSS (in kg $\cdot$ day^{-1})

Y_{obs} = observed yield (in kg MLVSS $\cdot$ kg^{-1} BOD$_5$ removed)

Other terms are as defined previously.

The increase in MLSS may be estimated by assuming that VSS is some fraction of MLSS. It is generally assumed that VSS is 60–80% of MLVSS. Thus, the increase in MLSS in Equation 11–37 may be estimated by dividing P_x by a factor of 0.6–0.8 (or multiplying by 1.25–1.667). The mass of suspended solids lost in the effluent is the product of the flow rate $(Q - Q_w)$ and the suspended solids concentration (X_e).

EXAMPLE 11–8   Estimate the mass of sludge to be wasted each day from the new activated sludge plant at Gatesville (Examples 11–4 and 11–7).

Solution   Using the data from Example 11–4, calculate Y_{obs}.

$$Y_{obs} = \frac{0.50 \text{ kg VSS} \cdot \text{kg}^{-1} \text{ BOD}_5 \text{ removed}}{1 + [(0.050 \text{ day}^{-1})(5 \text{ days})]} = 0.40 \text{ kg VSS} \cdot \text{kg}^{-1} \text{ BOD}_5 \text{ removed}$$

The net waste activated sludge produced each day is

$$P_x = (0.40)(0.150 \text{ m}^3 \cdot \text{s}^{-1})(84.0 \text{ g} \cdot \text{m}^{-3} - 11.1 \text{ g} \cdot \text{m}^{-3})(86,400 \text{ s} \cdot \text{day}^{-1})(10^{-3} \text{ kg} \cdot \text{g}^{-1})$$

$$= 377.9 \text{ kg} \cdot \text{day}^{-1} \text{ of VSS}$$

The total mass produced includes inert materials. Using the relationship between MLSS and MLVSS in Example 11–7, we obtain

Increase in MLSS $= (1.43)(377.9 \text{ kg} \cdot \text{day}^{-1}) = 540.4 \text{ kg} \cdot \text{day}^{-1}$

The mass of solids (both volatile and inert) lost in the effluent is

$$(Q - Q_w)(X_e) = (0.150 \text{ m}^3 \cdot \text{s}^{-1} - 0.000642 \text{ m}^3 \cdot \text{s}^{-1})(30 \text{ g} \cdot \text{m}^{-3})(86,400 \text{ s} \cdot \text{day}^{-1})$$

$$\times (10^{-3} \text{ kg} \cdot \text{g}^{-1})$$

$$= 387.13, \text{ or } 390 \text{ kg} \cdot \text{day}^{-1}$$

The mass to be wasted is then

Mass to be wasted $= 540.4 - 387.13 = 153.27$, or $150 \text{ kg} \cdot \text{day}^{-1}$

Note that this mass is calculated as dry solids. Because the sludge is mostly water, the actual mass will be considerably larger.

Oxygen Demand.   Oxygen is used in those reactions required to degrade the substrate to produce the energy required for cell synthesis and for respiration. For long SRT systems, the oxygen needed for cell maintenance can be of the same order of magnitude as substrate metabolism. A minimum residual of 0.5 to 2 mg $\cdot$ L^{-1} DO (dissolved oxygen) is usually maintained in the reactor basin to prevent oxygen deficiencies from limiting the rate of substrate removal.

An estimate of the oxygen requirements may be made from the BOD$_5$ of the waste and amount of activated sludge wasted each day. If we assume all of the BOD$_5$ is converted to end products, the total oxygen demand can be computed by converting BOD$_5$ to ultimate BOD (BOD$_L$). Because a portion of waste is converted to new cells that are wasted, the BOD$_L$ of the wasted cells must be subtracted from the total oxygen demand. An approximation of the oxygen demand of the wasted cells may be made by assuming cell oxidation can be described by the following reaction:

$$C_5H_7NO_2 + 5O_2 \rightleftharpoons 5CO_2 + 2H_2O + NH_3 + \text{energy for cells} \tag{11–38}$$

The ratio of gram molecular weights is

$$\frac{\text{Oxygen}}{\text{Cells}} = \frac{5(32)}{113} = 1.42 \tag{11–39}$$

Thus the oxygen demand of the waste activated sludge may be estimated as $1.42(P_x)$

The mass of oxygen required may be estimated as

$$M_{O_2} = \frac{Q(S_o - S)(10^{-3} \text{ kg} \cdot \text{g}^{-1})}{f} - 1.42(P_x) \tag{11–40}$$

where Q = wastewater flow rate into the aeration tank (in $\text{m}^3 \cdot \text{day}^{-1}$)
S_o = influent soluble BOD_5 (in $\text{mg} \cdot \text{L}^{-1}$)
S = effluent soluble BOD_5 (in $\text{mg} \cdot \text{L}^{-1}$)
f = conversion factor for converting BOD_5 to ultimate BOD_L
P_x = waste activated sludge produced (see Equation 11–37)

The volume of air to be supplied must take into account the percent of air that is oxygen and the transfer efficiency of the dissolution of oxygen into the wastewater.

EXAMPLE 11–9 Estimate the volume of air to be supplied ($\text{m}^3 \cdot \text{day}^{-1}$) for the new activated sludge plant at Gatesville (Examples 11–4 and 11–8). Assume that BOD_5 is 68% of the ultimate BOD and that the oxygen transfer efficiency is 8%.

Solution Using the data from Examples 11–4 and 11–8, we see that

$$M_{O_2} = \frac{(0.150 \text{ m}^3 \cdot \text{s}^{-1})(84.0 \text{ g} \cdot \text{m}^{-3} - 11.1 \text{ g} \cdot \text{m}^{-3})(86{,}400 \text{ s} \cdot \text{day}^{-1})(10^{-3} \text{ kg} \cdot \text{g}^{-1})}{0.68}$$

$$- 1.42(377.9 \text{ kg} \cdot \text{day}^{-1} \text{ of VSS})$$

$$= 1389.4 - 536.6 = 852.8 \text{ kg} \cdot \text{day}^{-1} \text{ of oxygen}$$

From Table A–4 in Appendix A, air has a density of $1.185 \text{ kg} \cdot \text{m}^{-3}$ at standard conditions. By mass, air contains about 23.2% oxygen. At 100% transfer efficiency, the volume of air required is

$$\frac{852.8 \text{ kg} \cdot \text{day}^{-1}}{(1.185 \text{ kg} \cdot \text{m}^{-3})(0.232)} = 3101.99, \text{ or } 3100 \text{ m}^3 \cdot \text{day}^{-1}$$

At 8% transfer efficiency, we get

$$\frac{3101.99 \text{ m}^3 \cdot \text{day}^{-1}}{0.08} = 38{,}774.9, \text{ or } 38{,}000 \text{ m}^3 \cdot \text{day}^{-1}$$

Process Design Considerations. The SRT selected for design is a function of the degree of treatment required. A high SRT (or older sludge age) results in a higher quantity of solids being carried in the system and creates a higher degree of treatment. A long SRT also results in the production of less waste sludge.

If industrial wastes are discharged to the municipal system, several additional concerns must be addressed. Municipal wastewater generally contains sufficient nitrogen and phosphorus to support biological growth. The presence of large volumes of industrial wastewater that is deficient in either of these nutrients will result in poor removal efficiencies. Addition of supplemental nitrogen and phosphorus may be required. The ratio of nitrogen to BOD_5 should be 1:32. The ratio of phosphorus to BOD_5 should be 1:150.

Although toxic metals and organic constituents may be at low enough levels that they do not interfere with the operation of the plant, two other untoward effects may result if they are not excluded in a pretreatment program. Volatile organic compounds may be stripped from solution into the atmosphere in the aeration tank. Thus, the WWTP may become a source of air pollution. The toxic metals may precipitate into the waste sludges. Thus, otherwise nonhazardous sludges may be rendered hazardous.

Oil and grease that pass through the primary treatment system will form grease balls on the surface of the aeration tank. The microorganisms cannot degrade this material because it is not in the water where they can physically come in contact with it. Special consideration should be given to the surface-skimming equipment in the secondary clarifier to handle the grease balls.

Trickling Filters

A trickling filter consists of a bed of coarse material (called **media**), such as stones, slats, or plastic materials, over which wastewater is applied. Trickling filters have been a popular biologic treatment process. The most widely used design for many years was simply a bed of stones from 1 to 3 m deep through which the wastewater passed. The wastewater is typically distributed over the surface of the media by a rotating arm (Figure 11–10).

As the wastewater trickles through the bed, a microbial growth establishes itself on the surface of the media in a fixed film. The wastewater passes over the stationary microbial population, providing contact between the microorganisms and the organic compounds in the wastewater.

Trickling filters are not primarily a filtering or straining process as the name implies. The rocks in a rock filter are 25–100 mm in diameter and, hence, have openings too large to strain out solids. They are a means of providing large amounts of surface area where the microorganisms cling and grow in a slime on the rocks as they feed on the organic matter.

Excess growths of microorganisms wash from the rock media and would cause undesirably high levels of suspended solids in the plant effluent if not removed. Thus, the flow from the filter is passed through a sedimentation basin to allow these solids to settle out.

Although rock trickling filters have performed well for years, they have certain limitations. Under high organic loadings, the slime growths can be so prolific that they plug the void spaces between the rocks, causing flooding and failure of the system. Also, the volume of void spaces is limited in a rock filter that restricts the circulation of air and the amount of oxygen available for the microbes. This limitation, in turn, restricts the amount of wastewater that can be processed.

To overcome these limitations, other materials have become popular for filling the trickling filter. These materials include modules of corrugated plastic sheets and plastic rings. These media offer larger surface areas for slime growths (typically 90 m² of surface area per cubic

FIGURE 11–10

Trickling-filter plant with enlargement of trickling filter.

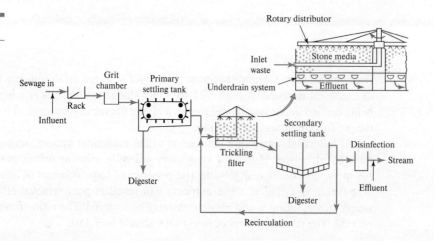

meter of bulk volume, as compared with 40–60 $m^2 \cdot m^{-3}$ for 75-mm rocks) and greatly increase void ratios for increased air flow. The materials are also much lighter than rock, so that the trickling filters can be much taller without facing structural problems. Although rock in filters is usually not more than 3 m deep, synthetic media depths may reach 12 m, thus reducing the overall space requirements for the trickling-filter portion of the treatment plant.

Trickling filters are classified according to the applied hydraulic and organic load. The hydraulic load may be expressed as cubic meters of wastewater applied per day per square meter of bulk filter surface area ($m^3 \cdot day^{-1} \cdot m^{-2}$) or, preferably, as the depth of water applied per unit of time ($mm \cdot s^{-1}$ or $m \cdot day^{-1}$). Organic loading is expressed as kilograms of BOD_5 per day per cubic meter of bulk filter volume ($kg \cdot day^{-1} \cdot m^{-3}$). Common hydraulic loadings range from 1 to 170 $m \cdot day^{-1}$ and organic loadings range from 0.07 to 1.8 kg $BOD_5 \cdot day^{-1} \cdot m^{-2}$ (WEF, 1998).

An important element in trickling filter design is the provision for return of a portion of the effluent to flow through the filter. This practice is called **recirculation.** The ratio of the returned flow to the incoming flow is called the **recirculation ratio.** Recirculation is practiced in stone filters for the following reasons:

1. To increase contact efficiency by bringing the waste into contact more than once with active biological material.
2. To dampen variations in loadings over a 24-h period. The strength of the recirculated flow lags behind that of the incoming wastewater. Thus, recirculation dilutes strong influent and supplements weak influent.
3. To raise the DO of the influent.
4. To improve distribution over the surface, thus reducing the tendency to clog and also reduce filter flies.
5. To prevent the biological slimes from drying out and dying during nighttime periods when flows may be too low to keep the filter wet continuously.

Recirculation may or may not improve treatment efficieny. The more dilute the incoming wastewater, the less likely it is that recirculation will improve efficiency.

Recirculation is practiced for plastic media to provide the desired wetting rate to keep the microorganisms alive. Generally, increasing the hydraulic loading above the minimum wetting rate does not increase BOD_5 removal. The minimum wetting rate normally falls in the range of 25 to 60 $m \cdot day^{-1}$.

Two-stage trickling filters (Figure 11–11) provide a means of improving the performance of filters. The second stage acts as a polishing step for the effluent from the primary stage by

FIGURE 11–11

Two-stage trickling-filter plant.

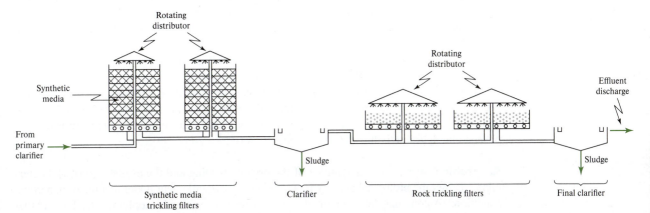

providing additional contact time between the waste and the microorganisms. Both stages may use the same media or each stage may have different media as shown in Figure 11–11. The designer will select the types of media and their arrangement based on the desired treatment efficiencies and an economic analysis of the alternatives.

Secondary Clarifier Design Considerations. Although the secondary settling tank is an integral part of both the trickling filter and the activated sludge process, environmental engineers have focused particular attention on the secondary clarifier used after the activated sludge process. A secondary clarifier is important because of the high solids loading and fluffy nature of the activated sludge biological floc. Also, it is highly desirable that sludge recycle be well thickened. Secondary settling tanks for activated sludge are generally characterized as having type III settling. Some authors would argue that types I and II also occur.

One of the continuing difficulties with the design of secondary settling tanks is accurately predicting effluent suspended solids concentrations as a function of common design and operating parameters. Little theoretical work has been conducted, and empirical correlations have been less than satisfactory. **Computational fluid dynamics,** computer-based computational methods, have shown promise in optimizing the hydraulic performance of sedimentation tanks.

Oxidation Ponds

Treatment ponds have been used to treat wastewater for many years, particularly as wastewater treatment systems for small communities (Benefield and Randall, 1980). Many terms have been used to describe the different types of systems employed in wastewater treatment. For example, in recent years, *oxidation pond* has been widely used as a collective term for all types of ponds. Originally, an **oxidation pond** was a pond that received partially treated wastewater, whereas a pond that received raw wastewater was known as a sewage lagoon. **Waste stabilization pond** has been used as an all-inclusive term that refers to a pond or lagoon used to treat organic waste by biological and physical processes. These processes would commonly be referred to as self-purification if they took place in a stream. To avoid confusion, we will use the following classification (Caldwell, Parker, and Uhte, 1973).

1. **Aerobic ponds:** Shallow ponds, less than 1 m in depth, where DO is maintained throughout the entire depth, mainly by the action of photosynthesis.
2. **Anaerobic ponds:** Deep ponds that receive high organic loadings such that anaerobic conditions prevail throughout the entire pond depth.
3. **Facultative ponds:** Ponds 1–2.5 m deep, which have an anaerobic lower zone, a facultative middle zone, and an aerobic upper zone maintained by photosynthesis and surface reaeration.
4. **Maturation, or tertiary, ponds:** Ponds used for polishing effluents from other biological processes. Dissolved oxygen is furnished through photosynthesis and surface reaeration. This type of pond is also known as a **polishing pond.**
5. **Aerated lagoons:** Ponds oxygenated through the action of surface or diffused air aeration.

Aerobic Ponds. The aerobic pond is a shallow pond in which light penetrates to the bottom, thereby maintaining active algal photosynthesis throughout the entire system. During the daylight hours, large amounts of oxygen are supplied by the photosynthesis process; during the hours of darkness, wind mixing of the shallow water mass generally provides a high degree of surface reaeration. Stabilization of the organic material entering an aerobic pond is accomplished mainly through the action of aerobic bacteria.

Anaerobic Ponds. The magnitude of the organic loading and the availability of DO determine whether the biological activity in a treatment pond will occur under aerobic or anaerobic conditions. A pond may be maintained in an anaerobic condition by applying a BOD_5 load that exceeds oxygen production from photosynthesis. Photosynthesis can be reduced by decreasing the surface area and increasing the depth. Anaerobic ponds become turbid from the presence of reduced metal sulfides. This restricts light penetration to the point that algal growth becomes negligible. Anaerobic treatment of complex wastes involves two distinct stages. In the first stage (known as **acid fermentation**), complex organic materials are broken down mainly to short-chain acids and alcohols. In the second stage (known as **methane fermentation**), these materials are convened to gases, primarily methane and carbon dioxide. The proper design of anaerobic ponds must result in environmental conditions favorable to methane fermentation. Anaerobic ponds are used primarily as a pretreatment process and are particularly suited for the treatment of high-temperature, high-strength wastewaters. However, they have been used successfully to treat municipal wastewaters as well.

Facultative Ponds.   Of the five general classes of lagoons and ponds, facultative ponds are by far the most common type selected as wastewater treatment systems for small communities. Approximately 25% of the municipal wastewater treatment plants in this country are ponds, and about 90% of these ponds are located in communities of 5000 people or fewer. Facultative ponds are popular for such treatment situations because long retention times facilitate the management of large fluctuations in wastewater flow and strength with no significant effect on effluent quality. Also capital, operating, and maintenance costs are less than those of other biological systems that provide equivalent treatment.

A schematic representation of a facultative pond operation is given in Figure 11–12. Raw wastewater enters at the center of the pond. Suspended solids contained in the wastewater settle to the pond bottom, where an anaerobic layer develops. Microorganisms occupying this region do not require molecular oxygen as an electron acceptor in energy metabolism but rather use some other chemical species. Both acid fermentation and methane fermentation occur in the bottom sludge deposits.

The facultative zone exists just above the anaerobic zone. This means that molecular oxygen will not be available in the region at all times. Generally, the zone is aerobic during the daylight hours and anaerobic during the hours of darkness.

Above the facultative zone, an aerobic zone exists that has molecular oxygen present at all times. The oxygen is supplied from two sources. A limited amount is supplied from diffusion across the pond surface; however, the majority is supplied through the action of algal photosynthesis.

FIGURE 11–12

Schematic diagram of facultative pond relationships.

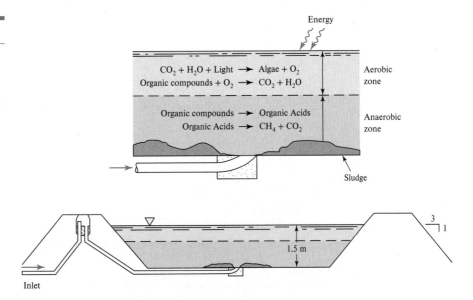

Two rules of thumb commonly used in Michigan in evaluating the design of facultative lagoons are as follows:

1. The BOD_5 loading rate should not exceed 22 kg · ha^{-1} · day^{-1} on the smallest lagoon cell.
2. The detention time in the lagoon (considering the total volume of all cells but excluding the bottom 0.6 m in the volume calculation) should be 6 months.

The first criterion is to prevent the pond from becoming anaerobic. The second criterion is to provide enough storage to hold the wastewater during winter months when the receiving stream may be frozen or during the summer when the flow in the stream might be too low to absorb even a small amount of BOD.

Rotating Biological Contactors

The **rotating biological contactor** (RBC) process consists of a series of closely spaced discs (3–3.5 m in diameter) mounted on a horizontal shaft and rotated while about one-half of their surface area is immersed in wastewater (Figure 11–13). The discs are typically constructed of lightweight plastic. The speed of rotation of the discs is adjustable.

When the process is placed in operation, the microbes in the wastewater begin to adhere to the rotating surfaces and grow there until the entire surface area of the discs is covered with a 1–3-mm layer of biological slime. As the discs rotate, they carry a film of wastewater into the air; this wastewater trickles down the surface of the discs, absorbing oxygen. As the discs complete their rotation, the film of water mixes with the reservoir of wastewater, adding to the

FIGURE 11–13

Rotating biological contactor (RBC) and process arrangement.

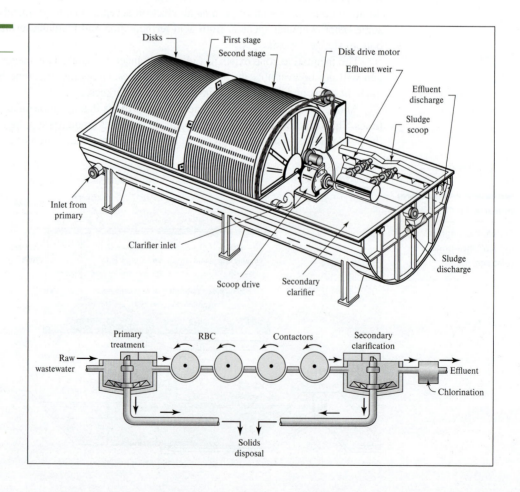

oxygen in the reservoir and mixing the treated and partially treated wastewater. As the attached microbes pass through the reservoir, they absorb other organic compounds for breakdown. The excess growth of microbes is sheared from the discs as they move through the reservoir. These dislodged organisms are kept in suspension by the moving discs. Thus, the discs serve several purposes:

1. They provide media for the buildup of attached microbial growth.
2. They bring the growth into contact with the wastewater.
3. They aerate the wastewater and the suspended microbial growth in the reservoir.

The attached growths are similar in concept to a trickling filter, except the microbes are passed through the wastewater rather than the wastewater passing over the microbes. Some of the advantages of both the trickling filter and activated sludge processes are realized.

As the treated wastewater flows from the reservoir below the discs, it carries the suspended growths out to a downstream settling basin for removal. The process can achieve secondary effluent quality or better. By placing several sets of discs in series, it is possible to achieve even higher degrees of treatment, including biological conversion of ammonia to nitrates.

The history of RBC installations has not been exemplary. Poor mechanical design and lack of understanding of the biological process has resulted in structural failure of shafts, disks, etc. Many of the problems with early installations have been resolved and numerous installations are performing satisfactorily.

The principal advantages of the RBC process are simplicity of operation and relatively low energy costs. They have found application in small communities.

Integrated Fixed-Film Activated Sludge (IFAS)

This category includes any activated sludge system that incorporates a fixed-film media in a suspended growth reactor. The purpose of fixed-film media is to increase the biomass in the reactor. This offers the potential to reduce the basin size or to increase the capacity of an existing basin in a retrofit application. Various types of suspended growth systems have been used. Examples include conventional, modified Ludzack-Ettinger and step denitrification. These processes differ from the moving bed biofilm reactor described below in that they use a return sludge flow.

A number of proprietary media types have been used including rope (no longer in use), sponge, plastic carriers, and a honeycomb polyester fabric called BioWeb™. The media that are fixed in a frame are preferred because they require fewer appurtenances and are less susceptible to hydraulic problems that result from free floating media.

The media frames are placed in conventional aeration tanks above a grid of fine bubble diffusers. The frames vary in size to fit the aeration tank dimensions. For example, an arrangement of two units with dimensions of 3.8 m × 3.8 m × 4 m high is set side by side across the flow path of the reactor.

Moving Bed Biofilm Reactor

This process uses small, plastic elements (on the order of 7 to 22 mm effective diameter) to support the growth of biofilm in the reactor. The suspended growth portion of the hybrid is designed as a complete mix reactor. It is commonly mixed with aeration but may also be mixed with a mechanical mixer. The process does not use a return sludge flow.

The media (typically polyethylene) is formed in a geometry that provides a high surface area (250–515 $m^2 \cdot m^{-3}$). It has a density near that of water (≈ 0.96 g $\cdot$ cm^{-3}). The reactors are normally filled from 1/3 to 2/3 of their empty volume with media. Because of their shape, less than 15% of the water is displaced. A screen across the outlet is used to prevent the media from leaving the aeration tank. Aeration is typically by coarse bubble diffusers.

11–9   DISINFECTION

The last treatment step in a secondary plant is the addition of a disinfectant to the treated wastewater. The addition of chlorine gas or some other form of chlorine is the process most commonly used for wastewater disinfection in the United States. The chemistry of chlorine reactions and by-products discussed in Chapter 10 also apply to wastewater disinfection. Chlorine is injected into the wastewater by automated feeding systems. Wastewater then flows into a basin, where it is held for about 15 min to allow the chlorine to react with the pathogens.

There is concern that wastewater disinfection may do more harm than good. Early U.S. EPA rules calling for disinfection to achieve 200 fecal coliforms per 100 mL of wastewater have been modified to a requirement for disinfection only during the summer season when people may come into contact with contaminated water. There were three reasons for this change. The first was that the use of chlorine and, perhaps, ozone causes the formation of organic compounds that are carcinogenic. The second was the finding that the disinfection process was more effective in killing the predators to cysts and viruses than it was in killing the pathogens themselves. The net result was that the pathogens survived longer in the natural environment because there were fewer predators. The third reason was that chlorine and, in particular, the chloramines formed in reactions with ammonia, are very toxic to fish.

Dechlorination reduces the effects of using chlorine as a disinfectant. Other processes such as ultraviolet light may be used in place of chlorine.

11–10   TERTIARY WASTEWATER TREATMENT

The need for treatment of wastewater beyond that which can normally be accomplished in secondary treatment is based on the recognition of one or more of the following:

1. Increasing population pressures result in increasing loads of organic matter and suspended solids to rivers, streams, and lakes.
2. The need to increase the removal of suspended solids to provide more efficient disinfection.
3. The need to remove nutrients to limit eutrophication of sensitive water bodies.
4. The need to remove constituents that preclude or inhibit water reclamation.

Initially, in the 1970s, these processes were called "advanced wastewater treatment" because they employed techniques that were more advanced than secondary treatment methods. In the last three decades many of these technologies have either been directly incorporated into the secondary processes, for example nutrient removal, or they are so inherent in meeting stringent discharge standards that they have become conventional. These processes include chemical precipitation, granular filtration, membrane filtration, and carbon adsorption. As conventional processes, they are better termed *tertiary treatment* processes rather than an advanced treatment process. In current practice, the employment of air stripping, ion exchange, nanofiltration (NF) or reverse osmosis (RO) treatment, and other similar processes to meet water quality requirements is correctly termed *advanced wastewater treatment*. Advanced wastewater treatment technologies are, fundamentally, those employed to treat water for reuse.

This discussion focuses on the following tertiary treatment processes: granular filtration, membrane filtration, carbon adsorption, chemical phosphorus removal, biological phosphorus removal, and nitrogen control.

Filtration

Secondary treatment processes, such as the activated sludge process, are highly efficient for removal of biodegradable colloidal and soluble organics. However, the typical effluent contains a much higher BOD_5 than would be expected from theory. The typical BOD is approximately 20–50 mg · L^{-1} This is principally because the secondary clarifiers are not perfectly efficient at settling out the microorganisms from the biological treatment processes. These organisms contribute both to the suspended solids and to the BOD_5 because the process of biological decay of dead cells exerts an oxygen demand.

Granular Filtration.   By using a filtration process similar to that used in water treatment plants, it is possible to remove the residual suspended solids, including the unsettled microorganisms. Removing the microorganisms also reduces the residual BOD_5. Conventional sand filters identical to those used in water treatment can be used, but they often clog quickly, thus requiring frequent backwashing. To lengthen filter runs and reduce backwashing, it is desirable to have the larger filter grain sizes at the top of the filter. This arrangement allows some of the larger particles of biological floc to be trapped at the surface without plugging the filter. Multimedia filters accomplish this by using low-density coal for the large grain sizes, medium-density sand for intermediate sizes, and high-density garnet for the smallest size filter grains. Thus, during backwashing, the greater density offsets the smaller diameter so that the coal remains on top, the sand remains in the middle, and the garnet remains on the bottom.

Typically, plain filtration can reduce activated sludge effluent suspended solids from 25 to 10 mg · L^{-1}. Plain filtration is not as effective on trickling filter effluents because these effluents contain more dispersed growth. However, the use of coagulation and sedimentation followed by filtration can yield suspended solids concentrations that are virtually zero. Typically, filtration can achieve 80% suspended solids reduction for activated sludge effluent and 70% reduction for trickling filter effluent.

Membrane Filtration.   The alternative membrane processes have been discussed in Chapter 10. Of the five processes, the one most commonly used in tertiary treatment is microfiltration (MF). It may be used as a replacement for granular filtration or as a subsequent treatment step. MF processes have achieved BOD removals of 75–90% and total suspended solids removals of 95–98%. Performance is highly site-specific. Membrane fouling is of particular concern and on-site pilot testing is highly recommended (Metcalf & Eddy, 2003).

Carbon Adsorption

Even after secondary treatment, coagulation, sedimentation, and filtration, soluble organic materials that are resistant to biological breakdown may persist in the effluent. The persistent materials are often referred to as **refractory organic compounds.** Refractory organic compounds can be detected in the effluent as soluble COD. Secondary effluent COD values are often 30–60 mg · L^{-1}.

The most practical available method for removing refractory organic compounds is by adsorbing them on activated carbon (U.S. EPA, 1979). Adsorption is the accumulation of materials at an interface. The interface, in the case of wastewater and activated carbon, is the liquid–solid boundary layer. Organic materials accumulate at the interface because of physical binding of the molecules to the solid surface. Carbon is activated by heating in the absence of oxygen. The activation process results in the formation of many pores within each carbon particle. Because adsorption is a surface phenomenon, the greater the surface area of the carbon, the greater its capacity to hold organic material. The vast areas of the walls within these pores account for most of the total surface area of the carbon, which makes it so effective in removing organic compounds.

After the adsorption capacity of the carbon has been exhausted, it can be restored by heating it in a furnace at a temperature sufficiently high to drive off the adsorbed organic material. Keeping oxygen at very low levels in the furnace prevents carbon from burning. The organic compounds are passed through an afterburner to prevent air pollution. In small plants where the cost of an on-site regeneration furnace cannot be justified, the spent carbon is shipped to a central regeneration facility for processing.

Chemical Phosphorus Removal

All the polyphosphates (molecularly dehydrated phosphates) gradually hydrolyze in aqueous solution and revert to the ortho form (PO_4^{3-}) from which they were derived. For example, tetrasodium pyrophosphate ($Na_4P_2O_7$) hydrolyzes in aqueous solution to form monohydrogen phosphate (HPO_4^{2-}) typically found in wastewater.

$$Na_4P_2O_7 + H_2O \rightleftharpoons 2Na_2HPO_4 \tag{11-41}$$

The removal of phosphorus to prevent or reduce eutrophication is typically accomplished by chemical precipitation using one of three compounds. The precipitation reactions for each are shown here.

Using ferric chloride:

$$FeCl_3 + HPO_4^{2-} \rightleftharpoons FePO_4 \text{ (s)} + H^+ + 3Cl^- \tag{11-42}$$

Using alum:

$$Al_2(SO_4)_3 + 2HPO_4^{2-} \rightleftharpoons 2AlPO_4 \text{ (s)} + 2H^+ + 3SO_4^{2-} \tag{11-43}$$

Using lime:

$$5Ca(OH)_2 + 3HPO_4^{2-} \rightleftharpoons Ca_5(PO_4)_3OH \text{ (s)} + 3H_2O + 6OH^- \tag{11-44}$$

You should note that ferric chloride and alum reduce the pH, whereas lime increases it. The effective range of pH for alum and ferric chloride is between 5.5 and 7.0. If there is not enough naturally occurring alkalinity to buffer the system to this range, then lime must be added to counteract the formation of H^+.

The precipitation of phosphorus requires a reaction basin and a settling tank to remove the precipitate. When ferric chloride and alum are used, the chemicals may be added directly to the aeration tank in the activated sludge system. Thus, the aeration tank serves as a reaction basin. The precipitate is then removed in the secondary clarifier. This is not possible with lime because the high pH required to form the precipitate is detrimental to the activated sludge organisms. In some wastewater treatment plants, the $FeCl_3$ (or alum) is added before the wastewater enters the primary sedimentation tank. This improves the efficiency of the primary tank but may deprive the biological processes of needed nutrients.

EXAMPLE 11-10   If a wastewater has a soluble orthophosphate concentration of 4.00 mg $\cdot$ L^{-1} as P, what theoretical amount of ferric chloride will be required to remove it completely?

Solution   From Equation 11-42, we see that one mole of ferric chloride is required for each mole of phosphorus to be removed. The pertinent gram molecular weights are as follows:

$FeCl_3 = 162.21$ g
$P = 30.97$ g

With a PO_4-P of 4.00 mg $\cdot$ L^{-1}, the theoretical amount of ferric chloride would be

$$4.00 \times \frac{162.21}{30.97} = 20.95, \text{ or } 21.0 \text{ mg} \cdot L^{-1}$$

Because of side reactions, solubility product limitations, and day-to-day variations, the actual amount of chemical to be added must be determined by jar tests on the wastewater. You can expect that the actual ferric chloride dose will be 1.5–3 times the theoretically calculated amount. Likewise, the actual alum dose will be 1.25–2.5 times the theoretical amount.

Biological Phosphorus Removal

In biological phosphorus removal (BPR or Bio-P) or enhanced biological phosphorous removal (EBPR), as it is sometimes called, the phosphorus in the wastewater is incorporated into cell mass in excess of levels needed for cell synthesis and maintenance. This is accomplished by moving the biomass from an anaerobic to an aerobic environment. The phosphorus contained in the biomass is removed from the process as biological sludge.

Microbiology.   The original work on enhanced Bio-P identified *Acinetobacter* as the responsible genus. Subsequent work has identified Bio-P bacteria in other genera such as *Arthrobacter*, *Aeromonas*, *Nocardia*, and *Pseudomonas*. The Bio-P organisms in these genera are referred to as *phosphorus accumulating organisms* (PAOs).

Based on the work of Comeau et al. (1986), Wentzel et al. (1986) developed a mechanistic model used to explain BPR. This model proposes that (Stephens and Stensel, 1998):

> Chemical oxygen demand (COD) is fermented to acetate by facultative bacteria under anaerobic conditions. The bacteria assimilate acetate in the anaerobic zone and convert it to polyhydroxybutyrate (PHB). Stored polyphosphate is degraded to provide adenosine triphosphate (ATP) necessary for PHB formation, and the polyphosphate degradation is accomplished by the release of orthophosphorus and magnesium, potassium, and calcium. Under aerobic conditions, the PHB is oxidized to synthesize new cells and to produce reducing equivalents needed for ATP formation. Phosphate and inorganic cations are taken up, reforming polyphosphate granules. The amount of phosphate taken up under aerobic conditions exceeds the phosphorus released during anaerobic conditions, resulting in excess phosphorus removal.

Stoichiometry.   Common heterotrophic bacteria in activated sludge have a phosphorus composition of 0.01 to 0.02 g P/g biomass. PAOs are capable of storing phosphorus in the form of phosphates. In the PAOs, the phosphorus content may be as high as 0.2 to 0.3 g P/g biomass.

Acetate (i.e., acetic acid, CH_3COOH) uptake is critical in determining the amount of PAOs and, thus, the amount of phosphorus that can be removed by this pathway. If significant amounts of DO or nitrate enter the anaerobic zone, the acetate will be depleted before it is taken up by the PAOs. Bio-P removal is not used in systems that are designed for nitrification without providing a means of denitrification.

The amount of phosphorus removal can be estimated from the amount of soluble COD in the wastewater influent. The following assumptions are used to evaluate the stoichiometry of biological phosphorus removal: (1) 1.06 g of acetate/g of COD will be produced as the COD is fermented to volatile fatty acids, (2) a cell yield of 0.3 g VSS/g of acetate, and (3) a cell phosphorus content of 0.3 g of P/g VSS. With these assumptions, it is estimated that 10 g of COD is required to remove 1 g of phosphorus (Metcalf & Eddy, 2003). Design principles and practice for Bio-P processes are presented in *Water and Wastewater Engineering* (Davis, 2010).

Nitrogen Control

Nitrogen in any soluble form (NH_3, NH_4^+, NO^- and NO_3^-, but not N_2 gas) is a nutrient and may need to be removed from wastewater to help control algal growth in the receiving body. In addition, nitrogen in the form of ammonia exerts an oxygen demand and can be toxic to fish. Removal of nitrogen can be accomplished either biologically or chemically. The biological process is called **nitrification–denitrification.** The chemical process is called **ammonia stripping.**

Nitrification–Denitrification.   The natural nitrification process can be forced to occur in the activated sludge system by maintaining a cell detention time (θ_c) of 15 days in moderate climates and over 20 days in cold climates. The nitrification step is expressed in chemical terms as follows:

$$NH_4^+ + 2O_2 \xrightleftharpoons{\text{bacteria}} NO_3^- + H_2O + 2H^+ \tag{11-45}$$

Of course, bacteria must be present to cause the reaction to occur. This step satisfies the oxygen demand of the ammonium ion. If the nitrogen level is not of concern for the receiving body, the wastewater can be discharged after settling. If nitrogen is of concern, the nitrification step must be followed by anoxic denitrification by bacteria:

$$2NO_3^- + \text{organic matter} \xrightleftharpoons{\text{bacteria}} N_2 + CO_2 + H_2O \tag{11-46}$$

As indicated by the chemical reaction, organic matter is required for denitrification. Organic matter serves as an energy source for the bacteria. The organic matter may be obtained from within or outside the cell. In multistage nitrogen-removal systems, because the concentration of BOD_5 in the flow to the denitrification process is usually quite low, a supplemental organic carbon source is required for rapid denitrification. (BOD_5 concentration is low because the wastewater previously has undergone carbonaceous BOD removal and the nitrification process.) The organic matter may be either raw, settled sewage or a synthetic material such as methanol (CH_3OH). Raw, settled sewage may adversely affect the effluent quality by increasing the BOD_5 and ammonia content. Design principles and practice for biological nitrogen removal processes are presented in *Water and Wastewater Engineering* (Davis, 2010).

Ammonia Stripping. Nitrogen in the form of ammonia can be removed chemically from water by raising the pH to convert the ammonium ion into ammonia, which can then be stripped from the water by passing large quantities of air through the water. The process has no effect on nitrate, so the activated sludge process must be operated at a short cell detention time to prevent nitrification. The ammonia stripping reaction is

$$NH_4^+ + OH^- \rightleftharpoons NH_3 + H_2O \tag{11-47}$$

The hydroxide is usually supplied by adding lime. The lime also reacts with CO_2 in the air and water to form a calcium carbonate scale, which must be removed periodically. Low temperatures cause problems with icing and reduced stripping ability. The reduced stripping ability is caused by the increased solubility of ammonia in cold water.

11–11 LAND TREATMENT FOR SUSTAINABILITY

An alternative to the previously discussed advanced wastewater treatment processes for producing an extremely high-quality effluent is offered by an approach called land treatment. Land treatment is the application of effluents, usually following secondary treatment, on the land by one of the several available conventional irrigation methods. This approach uses wastewater, and often the nutrients it contains, as a resource rather than considering it as a disposal problem. Treatment is provided by natural processes as the effluent moves through the natural filter provided by soil and plants. Part of the wastewater is lost by evapotranspiration, and the remainder returns to the hydrologic cycle through overland flow or the groundwater system. Most of the groundwater eventually returns, directly or indirectly, to the surface water system.

Land treatment of wastewaters can provide moisture and nutrients necessary for crop growth. In semiarid areas, insufficient moisture for peak crop growth and limited water supplies make this water especially valuable. The primary nutrients (nitrogen, phosphorus, and potassium) are reduced only slightly in conventional secondary treatment processes, so that most of these elements are still present in secondary effluent. Soil nutrients that are consumed each year by crop removal and by losses through soil erosion may be replaced by the application of wastewater. Land application is the oldest method used for treatment and disposal of wastes. Cities have used this method for more than 400 years. Several major cities, including Berlin,

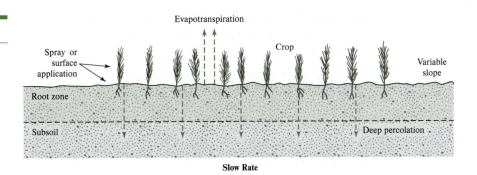

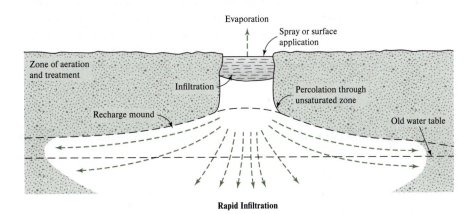

Melbourne, and Paris, have used "sewage farms" for at least 60 years for waste treatment and disposal. Approximately 600 communities in the United States reuse municipal wastewater treatment plant effluent in surface irrigation.

Land treatment systems use one of the three basic approaches (Pound, Crites, and Griffes, 1976):

1. Slow rate
2. Overland flow
3. Rapid infiltration

Each method, shown schematically in Figure 11–14, can produce renovated water of different quality, can be adapted to different site conditions, and can satisfy different overall objectives.

Slow Rate

Irrigation, the predominant land application method in use today, involves the application of effluent to the land for treatment and for meeting the growth needs of plants. The applied effluent is treated by physical, chemical, and biological means as it seeps into the soil. Effluent can be applied to crops or vegetation (including forest land) either by sprinkling or by surface techniques, for purposes such as

1. Avoidance of surface discharge of nutrients
2. Economic return from use of water and nutrients to produce marketable crops
3. Water conservation by exchange when lawns, parks, or golf courses are irrigated
4. Preservation and enlargement of greenbelts and open space

Where water for irrigation is valuable, crops can be irrigated at consumptive use rates (3.5–$10 \text{ mm} \cdot \text{day}^{-1}$, depending on the crop), and the economic return from the sale of the crop can be balanced against the increased cost of the land and distribution system. On the other hand, where water for irrigation is of little value, hydraulic loadings can be maximized (provided that renovated water quality criteria are met), thereby minimizing system costs. Under high-rate irrigation (10–$15 \text{ mm} \cdot \text{day}^{-1}$), water-tolerant grasses with high nutrient uptake become the crop of choice.

Overland Flow

Overland flow is a biological treatment process in which wastewater is applied over the upper reaches of sloped terraces and allowed to flow across the vegetated surface to runoff collection ditches. Renovation is accomplished by physical, chemical, and biological means as the wastewater flows in a thin sheet down the relatively impervious slope.

Overland flow can be used as a secondary treatment process where discharge of a nitrified effluent low in BOD is acceptable or as an advanced wastewater treatment process. The latter objective will allow higher rates of application ($18 \text{ mm} \cdot \text{day}^{-1}$ or more), depending on the degree of advanced wastewater treatment required. Where a surface discharge is prohibited, runoff can be recycled or applied to the land in irrigation or infiltration–percolation systems.

Rapid Infiltration

In infiltration–percolation systems, effluent is applied to the soil at higher rates by spreading in basins or by sprinkling. Treatment occurs as the water passes through the soil matrix. System objectives can include

1. Groundwater recharge
2. Natural treatment followed by pumped withdrawal or the use of underdrains for recovery
3. Natural treatment where renovated water moves vertically and laterally in the soil and recharges a surface watercourse

Where groundwater quality is being degraded by salinity intrusion, groundwater recharge can reverse the hydraulic gradient and protect the existing groundwater. Where existing groundwater quality is not compatible with expected renovated quality, or where existing water rights control the discharge location, a return of renovated water to surface water can be designed, using pumped withdrawal, underdrains, or natural drainage. In Phoenix, Arizona, for example, the native groundwater quality is poor, and the renovated water is withdrawn by pumping, with discharge into an irrigation canal.

Potential Adverse Affects

Land treatment is not without potential adverse environmental affects. Shallow groundwater wells near the land treatment sites may become contaminated with pathogens or high levels of chemicals such as nitrates. The salt concentration in the water may make it unpalatable or, in the worst case, not potable. In addition, return flows to surface waters may raise the salinity to unacceptable levels. For all of these reasons, careful hydrologic analyses must be included in the design and monitoring wells must be provided.

11–12 SLUDGE TREATMENT

In the process of purifying wastewater, another problem is created—sludge. The higher the degree of wastewater treatment, the larger the residue of sludge that must be handled. The exceptions to this rule are where land applications or polishing lagoons are used. Satisfactory treatment and disposal of the sludge can be the single most complex and costly operation in a municipal wastewater treatment system (U.S. EPA, 1979). The sludge is made of materials settled from the raw wastewater and of solids generated in the wastewater treatment processes.

The quantities of sludge involved are significant. For primary treatment, they may be 0.25–0.35% by volume of wastewater treated. When treatment is upgraded to activated sludge, the quantities increase to 1.5 to 2.0% of this volume of water treated. Use of chemicals for phosphorus removal can add another 1.0%. The sludges withdrawn from the treatment processes are still largely water; as much as 97%. Sludge treatment processes, then, are concerned with separating the large amounts of water from the solid residues. The separated water is returned to the wastewater plant for processing.

Sources and Characteristics of Various Sludges

Before we begin the discussion of the various treatment processes, it is worthwhile to recapitulate the sources and nature of the sludges that must be treated.

Grit. The sand, broken glass, nuts, bolts, and other dense material that is collected in the grit chamber is not true sludge in the sense that it is not fluid. However, it still requires disposal. Because grit can be drained of water easily and is relatively stable in terms of biological activity (it is not biodegradable), it is generally trucked directly to a landfill without further treatment.

Primary or Raw Sludge. Sludge from the bottom of the primary clarifiers contains 3–8% solids (1% solids = 1 g solids $\cdot$ 100 mL^{-1} sludge volume), which is approximately 70% organic. This sludge rapidly becomes anaerobic and is highly odiferous.

Secondary Sludge. This sludge consists of microorganisms and inert materials that have been wasted from the secondary treatment processes. Thus, the solids are about 90% organic. When the supply of air is removed, this sludge also becomes anaerobic, creating noxious conditions if not treated before disposal. The solids content depends on the source. Wasted activated sludge is typically 0.5–2% solids, whereas trickling filter sludge contains 2–5% solids. In some cases, secondary sludges contain large quantities of chemical precipitates because the aeration tank is used as the reaction basin for the addition of chemicals to remove phosphorus.

Tertiary Sludges. The characteristics of sludges from the advanced waste treatment processes depend on the nature of the process. For example, phosphorus removal results in a chemical sludge that is difficult to handle and treat. When phosphorus removal occurs in the activated sludge process, the chemical sludge is combined with the biological sludge, making the latter more difficult to treat. Nitrogen removal by denitrification results in a biological sludge with properties very similar to those of waste activated sludge.

Solids Computations

Because most WWTP sludges are primarily water, the volume of the sludge is primarily a function of the water content. Thus, if we know the percent solids and the specific gravity of the solids, we can estimate the volume of the sludge. The solid matter in wastewater sludge is composed of fixed (mineral) solids and volatile (organic) solids. The volume of the total mass of solids may be expressed as

$$\Psi_{solids} = \frac{M_s}{S_s \rho} \qquad \text{(11–48)}$$

where M_s = mass of solids (in kg)

$\quad S_s$ = specific gravity of solids

$\quad \rho$ = density of water = 1000 kg $\cdot$ m^{-3}

Because the total mass is composed of fixed and volatile fractions, Equation 11–48 may be rewritten as

$$\frac{M_s}{S_s\rho} = \frac{M_f}{S_f\rho} + \frac{M_v}{S_v\rho} \tag{11-49}$$

where M_f = mass of fixed solids (in kg)

$\quad M_v$ = mass of volatile solids (in kg)

$\quad S_f$ = specific gravity of fixed solids

$\quad S_v$ = specific gravity of volatile solids

The specific gravity of the solids may be expressed in terms of the specific gravities of the fixed and solid fractions by solving Equation 11–49 for S_s.

$$S_s = M_s\left(\frac{S_f S_v}{M_f S_v + M_v S_f}\right) \tag{11-50}$$

The specific gravity of sludge (S_{sl}) may be estimated by recognizing that, in a similar fashion to the fractions of solids, the sludge is composed of solids and water so that

$$\frac{M_{sl}}{S_{sl}\rho} = \frac{M_s}{S_s\rho} + \frac{M_w}{S_w\rho} \tag{11-51}$$

where M_{sl} = mass of sludge (in kg)

$\quad M_w$ = mass of water (in kg)

$\quad S_{sl}$ = specific gravity of sludge

$\quad S_w$ = specific gravity of water

It is customary to report solids concentrations as percent solids, where the fraction of solids (P_s) is computed as

$$P_s = \frac{M_s}{M_s + M_w} \tag{11-52}$$

and the fraction of water (P_w) is computed as

$$P_w = \frac{M_w}{M_s + M_w} \tag{11-53}$$

Thus, it is more convenient to solve Equation 11–48 in terms of percent solids. If we divide each term in Equation 11–51 by ($M_s + M_w$) and recognize that $M_{sl} = M_s + M_w$, then Equation 11–51 may be expressed as

$$\frac{1}{S_{sl}\rho} = \frac{P_s}{S_s\rho} + \frac{P_w}{S_w\rho} \tag{11-54}$$

If the specific gravity of water is taken as 1.000, as it can be without appreciable error; then solving for S_{sl} yields

$$S_{sl} = \frac{S_s}{P_s + (S_s)(P_w)} \tag{11-55}$$

With these expressions you can calculate the volume of sludge (V_{sl}) with the following equation.

$$V_{sl} = \frac{M_s}{(\rho)(S_{sl})(P_s)} \tag{11-56}$$

EXAMPLE 11–11   Using the following primary settling tank data, determine the daily sludge production.

Operating Data
   Flow = 0.150 m$^3 \cdot$ s^{-1}
   Influent suspended solids = 280.0 mg $\cdot$ L^{-1} = 280.0 g $\cdot$ m^{-3}
   Removal efficiency = 59.0%
   Sludge concentration = 5.00%
   Volatile solids = 60.0%
   Specific gravity of volatile solids = 0.990
   Fixed solids = 40.0%
   Specific gravity of fixed solids = 2.65

Solution   We begin by calculating S_s. We can do this without calculating M_s, M_f, and M_v directly by recognizing that they are proportional to the percent composition.
   With

$$M_s = M_f + M_v$$
$$= 0.400 + 0.600 = 1.00$$

Then Equation 11–50 gives the following:

$$S_s = \frac{(2.65)(0.990)}{[(0.990)(0.400)] + [(2.65)(0.600)]}$$
$$= 1.321, \text{ or } 1.32$$

The specific gravity of the sludge is calculated with Equation 11–55.

$$S_{sl} = \frac{1.321}{0.05 + [(1.321)(0.950)]}$$
$$= 1.012, \text{ or } 1.01$$

   The mass of the sludge is estimated from the incoming suspended solids concentration and the removal efficiency of the primary tank.

$$M_s = 0.59 \times 280.0 \text{ mg} \cdot \text{L}^{-1} \times 0.15 \text{ m}^3 \cdot \text{s}^{-1} \times 86{,}400 \text{ s} \cdot \text{day}^{-1} \times 10^{-3} \text{ kg} \cdot \text{g}^{-1}$$
$$= 2.14 \times 10^3 \text{ kg} \cdot \text{day}^{-1}$$

The sludge volume is then calculated with Equation 11–56.

$$\Psi_{sl} = \frac{2.14 \times 10^3 \text{ kg} \cdot \text{day}^{-1}}{1000 \text{ kg} \cdot \text{m}^{-3} \times 1.012 \times 0.05}$$
$$= 42.29, \text{ or } 42.3 \text{ m}^3 \cdot \text{day}^{-1}$$

Sludge Treatment Processes

The basic processes for sludge treatment are

1. *Thickening:* Separating as much water as possible by gravity or flotation.
2. *Stabilization:* Converting the organic solids to more refractory (inert) forms so that they can be handled or used as soil conditioners without causing a nuisance or health hazard through processes referred to as "digestion." (These are biochemical oxidation processes.)

3. *Conditioning:* Treating the sludge with chemicals or heat so that the water can be readily separated.
4. *Dewatering:* Separating water by subjecting the sludge to vacuum, pressure, or drying.
5. *Reduction:* Convening the solids to a stable form by wet oxidation or incineration. (These are chemical oxidation processes; they decrease the volume of sludge, hence the term reduction.)

Although a large number of alternative combinations of equipment and processes are used for treating sludges, the basic alternatives are fairly limited. The ultimate depository of the materials contained in the sludge must either be land, air, or water. Current policies discourage practices such as ocean dumping of sludge. Air pollution considerations necessitate air pollution control facilities as part of the sludge incineration process.

Thickening.   Thickening is usually accomplished in one of two ways: the solids are floated to the top of the liquid (**flotation**) or are allowed to settle to the bottom (**gravity thickening**). The goal is to remove as much water as possible before final dewatering or digestion of the sludge. The processes involved offer a low-cost means of reducing sludge volumes by a factor of 2 or more. The costs of thickening are usually more than offset by the resulting savings in the size and cost of downstream sludge-processing equipment.

In the flotation thickening process (Figure 11–15) air is injected into the sludge under pressure (275–550 kPa). Under this pressure, a large amount of air can be dissolved in the sludge. The sludge then flows into an open tank where, at atmospheric pressure, much of the air comes out of solution as minute bubbles. The bubbles attach themselves to particles of sludge solids and float them to the surface. The sludge forms a layer at the top of the tank; this layer is removed by a skimming mechanism for further processing. The process typically increases the solids content of activated sludge from 0.5–1% to 3–6%. Flotation is especially effective on activated sludge, which is difficult to thicken by gravity.

Gravity thickening is a simple and inexpensive process that has been used widely on primary sludges for many years. It is a sedimentation process similar to that which occurs in all settling tanks. Sludge flows into a tank that is very similar in appearance to the circular clarifiers used in primary and secondary sedimentation; the solids are allowed to settle to the bottom where a heavy-duty mechanism scrapes them to a hopper from which they are withdrawn for further processing. The type of sludge being thickened has a major effect on performance. The best results are obtained with purely primary sludges. As the proportion of activated sludge increases, the thickness of settled sludge solids decreases. Purely primary sludges can be thickened from 1–3% to 10% solids. The current trend is toward using gravity thickening for primary sludges and flotation thickening for activated sludges, and then blending the thickened sludges for further processing.

FIGURE 11–15

Air flotation thickener.

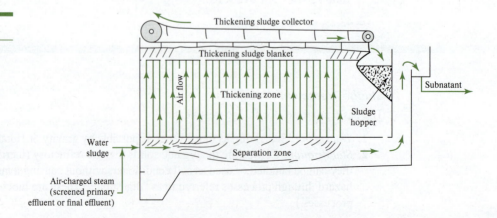

EXAMPLE 11–12   One hundred cubic meters (100.0 m³) of mixed sludge at 4.0% solids is to be thickened to 8.0% solids. Assuming that the specific gravity is not appreciably different from that of water and that it does not change during thickening, what is the approximate volume of the sludge after thickening?

Solution   A 4.0% sludge contains 4.0% by mass of solids and 96.0% water by mass. We can approximate the relationship between volume and percent solids based on a mass balance around the thickener. The mass-balance diagram appears as follows:

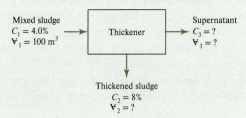

The mass-balance equation is then

$$C_1 \Psi_1 - C_2 \Psi_2 - C_3 \Psi_3 = 0$$

If we assume the supernatant concentration (C_3) is much less than the influent and effluent concentrations (C_1 and C_2), then we can make the assumption that $C_3 = 0$. Then

$$C_1 \Psi_1 = C_2 \Psi_2$$

and

$$\Psi_2 = \frac{C_1 \Psi_1}{C_2}$$

In this case then, the volume of sludge after thickening would be

$$\Psi_2 = \frac{(0.040)(100.0 \text{ m}^3)}{0.080} = 50.0 \text{ m}^3$$

Thus, we can see a substantial reduction in the volume that must be handled by thickening the sludge from 4 to 8% solids.

Stabilization.   The principal purposes of sludge stabilization are to break down the organic solids biochemically so that they are more stable (less odorous and less putrescible) and more dewaterable, and to reduce the mass of sludge (Benefield and Randall, 1980). If the sludge is to be dewatered and burned, stabilization is not used. Two basic stabilization processes are in use. One is carried out in closed tanks devoid of oxygen and is called **anaerobic digestion.** The other approach injects air into the sludge to accomplish **aerobic digestion.**

The aerobic digestion of biological sludges is nothing more than a continuation of the activated sludge process. When a culture of aerobic heterotrophs is placed in an environment containing a source of organic material, the microorganisms remove and use most of this material. A fraction of the organic material removed will be used for the synthesis of new biomass. The remaining material will be channeled into energy metabolism and oxidized to carbon dioxide, water, and soluble inert material to provide energy for both synthesis and maintenance (life-support) functions. Once the external source of organic material is exhausted, however, the microorganisms enter into **endogenous respiration,** where cellular material is oxidized to satisfy the energy of maintenance (i.e., energy for life-support requirements). If this condition is continued over an extended period, the total quantity of biomass will be considerably reduced. Furthermore, that portion remaining will exist at such a low energy state that it can be considered biologically stable and suitable for disposal in the environment. This forms the basic principle of aerobic digestion.

Aerobic digestion is accomplished by aerating the organic sludges in an open tank resembling an activated sludge aeration tank. Like the activated sludge aeration tank, the aerobic digester must be followed by a settling tank unless the sludge is to be disposed of on land in liquid form. Unlike the activated sludge process, the effluent (supernatant) from the clarifier is recycled back to the head end of the plant. This is because the supernatant is high in suspended solids (100–300 mg · L^{-1}), BOD$_5$ (to 500 mg · L^{-1}), TKN (to 200 mg · L^{-1}), and total P (to 100 mg · L^{-1}).

Because the fraction of volatile matter is reduced, the specific gravity of the digested sludge solids will be higher than it was before digestion. Thus, the sludge settles to a more compact mass, and the clarifier underflow concentration can be expected to reach 3%.

The anaerobic treatment of complex wastes is thought to occur in three stages as shown in Figures 11–16 and 11–17 (Speece, 1983; Holland et al., 1987). In the first stage, complex

FIGURE 11–16

Schematic diagram of the patterns of carbon flow in anaerobic digestion.

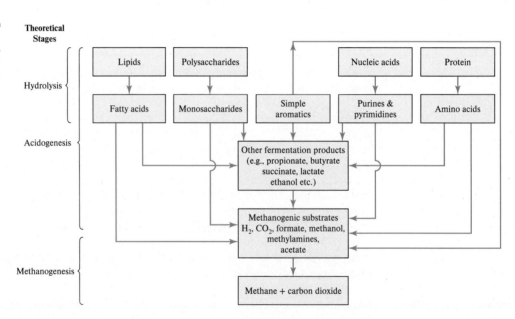

FIGURE 11–17

Steps in anaerobic digestion process with energy flow.

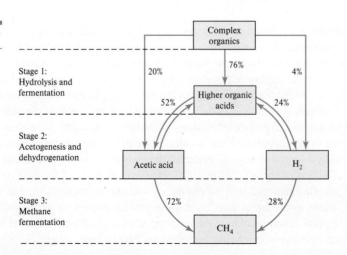

waste components, including fats, proteins, and polysaccharides, are hydrolyzed to their component subunits. This is accomplished by a heterogeneous group of facultative and anaerobic bacteria. These bacteria then subject the products of hydrolysis (triglycerides, fatty acids, amino acids, and sugars) to fermentation and other metabolic processes, leading to the formation of simple organic compounds and hydrogen in a process called **acidogenesis** or **acetogenesis.** The organic compounds are mainly short-chain (volatile) acids and alcohols. In the first two stages little stabilization of BOD or COD is realized. In the third stage, the end products of the second stage are converted to gases (mainly methane and carbon dioxide) by several different species of strictly anaerobic bacteria. Thus, it is here that stabilization of the organic material occurs, a stage generally referred to as **methane fermentation.** Even though the anaerobic process is presented as being sequential in nature, all stages take place simultaneously and synergistically.

The bacteria responsible for acid fermentation are relatively tolerant to changes in pH and temperature and have a much higher rate of growth than the bacteria responsible for methane fermentation. As a result, methane fermentation is generally assumed to be the rate-controlling step in anaerobic waste treatment processes.

Considering 35°C as the optimum temperature for anaerobic waste treatment, Lawrence and Milnes propose that, in the range of 20 to 35°C, the kinetics of methane fermentation of long- and short-chain fatty acids will adequately describe the overall kinetics of anaerobic treatment (Lawrence and Milnes, 1971). Thus, the kinetic equations we presented to describe the completely mixed activated sludge process are equally applicable to the anaerobic process.

The most common design today is a single-stage, high-rate digester (Figure 11–18). It is characterized by heating, auxiliary mixing, uniform feeding, and thickening of the feed stream. Uniform feeding of the sludge is very important to the operation of the digester. For economical anaerobic digestion, a feed concentration of at least 4% total solids is desirable (Shimp et al., 1995). The digestion tanks may have fixed or floating covers. Gas may be stored under the floating cover or in a separate structure. There is no supernatant separation.

The BOD remaining at the end of digestion is still quite high. Likewise, the suspended solids may be as high as 12,000 mg · L^{-1}, whereas the TKN may be on the order of 1000 mg · L^{-1}. Thus, the supernatant from the secondary digester (in the high-rate process) is returned to the head end of the WWTP. The settled sludge is conditioned and dewatered for disposal.

Sludge Conditioning.   Several methods are available for conditioning sludge to facilitate the separation of the liquid and solids. One of the most commonly used is the addition of coagulants such as ferric chloride, lime, or organic polymers. Ash from incinerated sludge has also found

FIGURE 11–18

Schematic of a high-rate anaerobic digester.

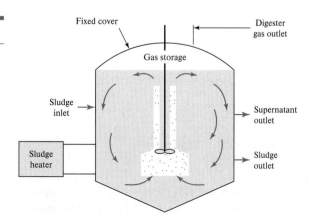

use as a conditioning agent. As happens when coagulants are added to turbid water, chemical coagulants act to clump the solids together so that they are more easily separated from the water. In recent years, organic polymers have become increasingly popular for sludge conditioning. Polymers are easy to handle, require little storage space, and are very effective. The conditioning chemicals are injected into the sludge just before the dewatering process and are mixed with the sludge.

Another conditioning approach is to heat the sludge at high temperatures (175–230°C) and pressures (1000–2000 kPa). Under these conditions, much like those of a pressure cooker, water that is bound up in the solids is released, improving the dewatering characteristics of the sludge. Heat treatment has the advantage of producing a sludge that dewaters better than chemically conditioned sludge. The process has the disadvantages of relatively complex operation and maintenance and the creation of highly polluted cooking liquors that, when recycled to the treatment plant, impose a significant added treatment burden.

Sludge Dewatering.   The most popular method of sludge dewatering in the past has been the use of sludge drying beds. These beds are especially popular in small plants because of their simplicity of operation and maintenance. In 1977, two-thirds of all U.S. wastewater treatment plants used drying beds; one-half of all the municipal sludge produced in the United States was dewatered by this method. Although the use of drying beds might be expected in the warmer, sunny regions, they are also used in several large facilities in northern climates.

Operational procedures common to all types of drying beds involve the following steps:

1. Pump 0.20–0.30 m of stabilized liquid sludge onto the drying bed surface.
2. Add chemical conditioners continuously, if conditioners are used, by injection into the sludge as it is pumped onto the bed.
3. When the bed is filled to the desired level, allow the sludge to dry to the desired final solids concentration. (This concentration can vary from 18 to 60%, depending on several factors, including type of sludge, processing rate needed, and degree of dryness required for lifting. Nominal drying times vary from 10 to 15 days under favorable conditions, to 30–60 days under barely acceptable conditions.)
4. Remove the dewatered sludge either mechanically or manually.
5. Repeat the cycle.

Sand drying beds are the oldest, most commonly used type of drying bed. Many design variations are possible, including the layout of drainage piping, thickness and type of gravel and sand layers, and construction materials. Sand drying beds for wastewater sludge are constructed in the same manner as water treatment plant sludge-drying beds. Sand drying beds can be built with or without provision for mechanical sludge removal and with or without a roof. When the cost of labor is high, newly constructed beds are designed for mechanical sludge removal.

When space is limited or climatic conditions make sand drying beds prohibitively expensive, vacuum filters may be used. A vacuum filter consists of a cylindrical drum covered with a filtering material or fabric, which rotates partially submerged in a vat of conditioned sludge. A vacuum is applied inside the drum to extract water, leaving the solids, or **filter cake,** on the filter medium. As the drum completes its rotational cycle, a blade scrapes the filter cake from the filter and the cycle begins again. In some systems, the filter fabric passes off the drum over small rollers to dislodge the cake. There is a wide variety of filter fabrics, ranging from Dacron to stainless steel coils, each with its own advantages. The vacuum filter can be applied to digested sludge to produce a sludge cake dry enough (15–30% solids) to handle and dispose of by burial in a landfill or by application to the land as a relatively dry fertilizer. If the sludge is to be incinerated, it is not stabilized. In this case, the vacuum filter is applied to the raw sludge to dewater it. The sludge cake is then fed to the furnace to be incinerated.

FIGURE 11–19

Continuous belt filter press. ©Komline-Sanderson Engineering Corporation

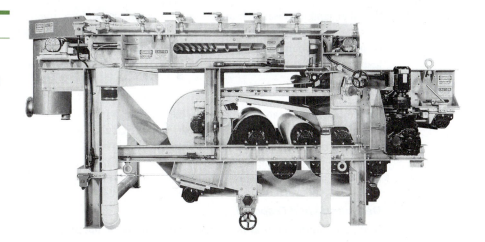

The continuous belt filter presses (CBFP) equipment (Figure 11–19) used in treating waste water sludges is the same as that used for water treatment plant sludges. The CBFP is successful with many normal mixed sludges. Typical dewatering results for digested mixed sludges with initial feed solids of 5% give a dewatered cake of 19% solids at a rate of 32.8 kg · $h^{-1} \cdot m^{-2}$. In general, most of the results with these units closely parallel those achieved with rotary vacuum filters. An advantage of CBFPs is that they do not have the sludge pickup problem that sometimes occurs with rotary vacuum filters. Additionally, they have a lower energy consumption.

Reduction.   If sludge use as a soil conditioner is not practical, or if a site is not available for landfill using dewatered sludge, cities may turn to the alternative of sludge reduction, that is incineration. Incineration completely evaporates the moisture in the sludge and combusts the organic solids to a sterile ash. To minimize the amount of fuel used, the sludge must be dewatered as completely as possible before incineration. The exhaust gas from an incinerator must be treated carefully to avoid air pollution.

11–13   SLUDGE DISPOSAL

Ultimate Disposal

The WWTP process residuals (leftover sludges, either treated or untreated) are the bane of design and operating personnel. Of the five possible disposal sites for residuals, two are feasible and only one is practical. Conceivably, one could ultimately dispose of residues in the following places: in the air, in the ocean, in "outer space," on the land, or in the marketplace. Disposal in the air by burning is in reality not ultimate disposal but only temporary storage until the residue falls to the ground. Although incineration is widely practiced as a sludge treatment device, the ash and residues from air pollution control devices must be disposed of. Disposal of sewage sludge at sea by barging is now prohibited in the United States. "Outer space" is not a suitable disposal site. Thus, we are left with land disposal and use of the sludge to produce a product.

For ease of discussion, we have divided land disposal into three categories: land spreading, landfilling, and dedicated land disposal. We have grouped all of the utilization ideas under one category.

Land Spreading

The practice of applying WWTP residuals for the purposes of recovering nutrients, water, or reclaiming despoiled land such as strip mine spoils is called **land spreading.** In contrast to the other land disposal techniques, land spreading is land-use intensive. Application rates are governed by the character of the soil and the ability of the crops or forests on which the sludge is spread to accommodate it. Land application may be limited by heavy metals (e.g., chromium, lead, and mercury) that poison the soil (See Sludge Disposal Regulations below).

Landfilling

Sludge landfilling can be defined as the planned burial of wastewater solids, including processed sludge, screenings, grit, and ash, at a designated site. The solids are placed into a prepared site or excavated trench and covered with a layer of soil. The soil cover must be deeper than the depth of the plow zone (about 0.20–0.25 m). For the most part, landfilling of screenings, grit, and ash is accomplished with methods similar to those used for sludge landfilling.

Dedicated Land Disposal

Dedicated land disposal is the application of heavy sludge loadings to some finite land area that has limited public access and has been set aside or dedicated for all time to the disposal of wastewater sludge. Dedicated land disposal does not mean in-place use. No crops may be grown. Dedicated sites typically receive liquid sludges. Although application of dewatered sludges is possible, it is uncommon. In addition, disposal of dewatered sludge in landfills is generally more cost-effective.

Utilization

Wastewater solids may sometimes be used beneficially in ways other than as a soil nutrient. Of the several methods worthy of note, composting and cofiring with municipal solid waste are two that have received increasing amounts of interest in the last few years. The recovery of lime and the use of the sludge to form activated carbon have also been in practice to a lesser extent.

Sludge Disposal Regulations

The regulations that govern the use or disposal of sewage sludge are codified as 40 CFR Part 503 and have become known as the "503 regulations." The regulations apply to sewage sludge generated from the treatment of domestic sewage that is land-applied, placed on a surface disposal site, or incinerated in an incinerator that accepts only sewage sludge. The regulations do not apply to sludge generated from treatment of industrial process wastes at an industrial facility, hazardous sewage sludge, sewage sludge with PCB, concentrations of 50 mg $\cdot$ L^{-1} or greater, or drinking water sludge.

11–14 DIRECT POTABLE REUSE

Description

From the time of the earliest habitation of humans, wastewater has been discharged into nearby streams and rivers. From that ancient time until today communities have continued that unsanitary practice. Of course, in more developed countries, the adverse consequences of this practice have been ameliorated by better and better technologies for purifying the wastewater. Thus, today, in the United States and other developed countries where surface water is used to supply drinking water, the water is treated to make it potable. The consequence for users of these surface waters is that they are practicing indirect potable reuse of the water. In many cases, along the major rivers of the United States and other developed countries, indirect potable reuse is practiced multiple times between the river's headwaters and the ultimate discharge into the ocean.

The introduction of wastewater treated to meet the most stringent drinking water quality standards directly into the potable supply distribution system or into the raw water supply immediately upstream of the water treatment plant is called direct potable reuse (DPR). The implementation of DPR is usually the result of extreme circumstances. One example is the case of Chanute, Kansas, in 1956 (Asano et al., 2007). The Neosho River served as the water supply and wastewater disposal site for seven communities upstream of Chanute. In the summer of 1956, the river dried up. To provide water for Chanute, the river was dammed below the wastewater treatment plant. The treated wastewater effluent backed up to the water intake for the drinking water plant. The impoundment pond provided about 17 days of retention. For five months the city used its treated sewage to supply water. It circulated some 8 to 15 times. Because of good treatment, including multiple point chlorination, the treated water met prevailing public health standards. Nonetheless, the treated water gradually developed a pale yellow color and a musty taste and odor. In addition, it had a tendency to foam when agitated and became "rich" in dissolved minerals and organic substances. Bottled water sales flourished. When heavy rains washed out the temporary dam below the sewage outfall, Chanute went back to drinking water that contained treated sewage from the upstream communities. Much has been made of the fact that Chanute reused its municipal wastewater, but little attention has been paid to the fact that Chanute always had some wastewater in its raw water supply.

Public Perception

Over the last decade or two, cities with limited water resources and growing populations have examined a variety of measures to ensure that sufficient potable water is available. Several of these techniques involve the treatment and capture of wastewater. To date, nearly all potable reuse projects have involved indirect potable reuse that depends on aquifer or reservoir storage to further treat, monitor, and store the recycled water before it enters the water distribution system (Pecson et al., 2015). Indirect potable reuse projects in Orange County, California, Los Angeles, California, San Diego, California, and Miami-Dade, Florida, have demonstrated that, with proper education and public outreach, public support for indirect potable reuse can be achieved (Chalmers et al., 2010; Quicho et al., 2012)

Direct potable reuse eliminates the environmental barrier (i.e., aquifer or reservoir storage) by piping the recycled water directly to a drinking water treatment plant. It is the most difficult reuse application for a community to accept. Public health protection is the *sine qua non* of all drinking water supply systems. Public water systems in the United States are considered among the best in the world. Those drinking the water assume it is palatable as well as potable. The experience acquired in implementing indirect potable reuse projects points to the importance of a well thought out plan that includes recognition of the health issues and a technological capability to deal with the threats.

Health Issues

Pathogens and chemicals are the two main health concerns. For contaminants that pose chronic health concerns, short-term fluctuations are less important than average lifetime exposure because brief exposure may be buffered by periods of lower exposure. This holds true for a vast majority but not all of the potential chemical contaminants. Severe health effects can result from a single exposure to pathogens. Thus, pathogens become the primary public health concern in DPR.

The main goal of the design and implementation of a DPR project is to deliver safe water to the customer. C. J. Velz (1970) was one of the early advocates of multiple barrier water treatment systems to provide reliable public health protection. Pecson et al. (2015) summarized a framework for implementing this concept in DPR projects:

1. Reliability = continuous, consistent public health protection
2. Redundancy = measures beyond minimum requirements to ensure treatment goals are reliably met

3. Robustness = employment of a system to address a wide variety of contaminants and that resists catastrophic failure
4. 4. Resilience = use of a treatment train that operators can successfully adapt to recover from failure of a treatment technology component

Technological Capability

Recognizing that long-term growth can be hampered by a lack of adequate potable water, the City of Denver, Colorado, began evaluating alternative water supplies in1968. In 1970 a laboratory scale ($19 \, L \cdot min^{-1}$) advanced wastewater treatment plant process was put into operation. It was operated until 1979. Based on experience with the pilot plant, a $3.8 \times 10^3 \, m^3 \cdot d^{-1}$ DPR demonstration plant was put into operation. The pilot plant was operated over a 13-year period. Whole animal health studies were carried out over a two year period. The conclusion of a rigorous examination of the data was that it is feasible to produce potable water that met all of the current U.S. EPA drinking water standards over a sustained period of time.

CHAPTER REVIEW

When you have completed studying this chapter, you should be able to do the following without the aid of your textbook or notes:

1. List a BOD value for strong, medium, and weak municipal waste.
2. Explain the differences among pretreatment, primary treatment, secondary treatment, and tertiary treatment, and show how they are related.
3. Sketch a graph showing the average variation of daily flow at a municipal WWTP.
4. Define and explain the purpose of equalization.
5. Sketch, label, and explain the function of the parts of an activated sludge plant and a trickling filter plant.
6. Define θ_c, SRT, and sludge age, and explain their use in regulating the activated sludge process.
7. Explain the purpose of the F/M ratio and define F and M in terms of BOD_5 and mixed liquor volatile suspended solids.
8. Explain the relationship between F/M and θ_c.
9. Explain how cell production is regulated using F/M or θ_c.
10. Compare two systems operating at two different F/M ratios.
11. Compare the positive and negative effects of disinfection of wastewater effluents.
12. List four common tertiary processes and the pollutants they remove.
13. Explain why removal of residual suspended solids effectively removes residual BOD_5.
14. Describe refractory organics and the method used to remove them.
15. List three chemicals used to remove phosphorus from wastewaters.
16. Explain biological nitrification and denitrification either in words or with an equation.
17. Explain ammonia stripping either in words or with an equation.
18. Describe the three basic approaches to land treatment of wastewater.
19. State the two major purposes of sludge stabilization.
20. Explain the purpose of each of the sludge treatment steps and describe the major processes used.
21. Describe the locations for ultimate disposal of sludges and the treatment steps needed prior to ultimate disposal.

With the aid of this text, you should be able to do the following:

1. Determine whether a grit particle of given diameter and density will be captured in a given velocity-controlled grit chamber, or determine the minimum diameter that will be captured under a given set of conditions.

2. Determine the required volume of an equalization basin to dampen a given periodic flow.
3. Determine the effect of equalization on mass loading of a pollutant.
4. Evaluate or size primary and secondary sedimentation tanks with respect to detention time, overflow rate, solids loading, and weir loading.
5. Estimate the soluble BOD_5 in the effluent from a completely mixed activated sludge plant, determine the mean cell residence time or the hydraulic detention time to achieve a desired degree of treatment, and determine the wasting flow rate to achieve a desired mean cell residence time or F/M ratio.
6. Calculate the F/M ratio given an influent BOD_5, flow, and detention time, or calculate the volume of the aeration basin given F/M, BOD_5, and flow.
7. Determine return sludge concentration or flow rate.
8. Calculate the required mass of sludge to be wasted from an activated sludge process given the appropriate data.
9. Calculate the theoretical mass of oxygen required and the amount of air required to supply it given the appropriate data.
10. Perform a sludge mass balance, given the separation efficiencies and appropriate mass flow rate.

PROBLEMS

11–1 If a particle having a 0.0170-cm radius and density of $1.95 \text{ g} \cdot \text{cm}^{-3}$ is allowed to fall into quiescent water having a temperature of 4°C, what will be the terminal settling velocity? Assume the density of water = $1000 \text{ kg} \cdot \text{m}^{-3}$. Assume Stokes' law applies.

Answer: $3.82 \times 10^{-1} \text{ m} \cdot \text{s}^{-1}$

11–2 If the terminal settling velocity of a particle falling in quiescent water having a temperature of 15°C is $0.0950 \text{ cm} \cdot \text{s}^{-1}$, what is its diameter? Assume a particle density of $2.05 \text{ g} \cdot \text{cm}^{-3}$ and density of water equal to $1000 \text{ kg} \cdot \text{m}^{-3}$. Assume Stokes' law applies.

11–3 A treatment plant being designed for Cynusoidal City requires an equalization basin to even out flow and BOD variations. The average daily flow is $0.400 \text{ m}^3 \cdot \text{s}^{-1}$. The following flows and BOD_5 have been found to be typical of the average variation over a day. What size equalization basin, in cubic meters, is required to provide for a uniform outflow equal to the average daily flow? Assume the flows are hourly averages.

Time	Flow $(m^3 \cdot s^{-1})$	BOD_5 $(mg \cdot L^{-1})$	Time	Flow $(m^3 \cdot s^{-1})$	BOD_5 $(mg \cdot L^{-1})$
0000	0.340	123	1200	0.508	268
0100	0.254	118	1300	0.526	282
0200	0.160	95	1400	0.530	280
0300	0.132	80	1500	0.552	268
0400	0.132	85	1600	0.570	250
0500	0.140	95	1700	0.596	205
0600	0.160	100	1800	0.604	168
0700	0.254	118	1900	0.570	140
0800	0.360	136	2000	0.552	130
0900	0.446	170	2100	0.474	146
1000	0.474	220	2200	0.412	158
1100	0.482	250	2300	0.372	154

Answer: $\forall = 6110 \text{ m}^3$

11–4 What volume equalization basin is required to even outflow and BOD variations shown in the following data? Assume the flows are hourly averages.

Time	Flow ($m^3 \cdot s^{-1}$)	BOD_5 ($mg \cdot L^{-1}$)	Time	Flow ($m^3 \cdot s^{-1}$)	BOD_5 ($mg \cdot L^{-1}$)
0000	0.0875	110	1200	0.135	160
0100	0.0700	81	1300	0.129	150
0200	0.0525	53	1400	0.123	140
0300	0.0414	35	1500	0.111	135
0400	0.0334	32	1600	0.103	130
0500	0.0318	42	1700	0.104	120
0600	0.0382	66	1800	0.105	125
0700	0.0653	92	1900	0.116	150
0800	0.113	125	2000	0.127	200
0900	0.131	140	2100	0.128	215
1000	0.135	150	2200	0.121	170
1100	0.137	155	2300	0.110	130

11–5 A treatment plant being designed for the village of Excel requires an equalization basin to even outflow and BOD variations. The following flows and BOD_5 have been found to be typical of the average variation over a day. What size equalization basin, in cubic meters, is required to provide for a uniform outflow equal to the average daily flow?

Time	Flow ($m^3 \cdot s^{-1}$)	BOD_5 ($mg \cdot L^{-1}$)	Time	Flow ($m^3 \cdot s^{-1}$)	BOD_5 ($mg \cdot L^{-1}$)
0000	0.0012	50	1200	0.0041	290
0100	0.0011	34	1300	0.0041	290
0200	0.0009	30	1400	0.0042	275
0300	0.0009	30	1500	0.0038	225
0400	0.0009	33	1600	0.0033	170
0500	0.0013	55	1700	0.0039	180
0600	0.0018	73	1800	0.0046	190
0700	0.0026	110	1900	0.0046	190
0800	0.0033	150	2000	0.0044	190
0900	0.0039	195	2100	0.0034	160
1000	0.0047	235	2200	0.0031	125
1100	0.0044	265	2300	0.0020	80

11–6 Compute and plot the unequalized and the equalized hourly BOD mass loadings to the Cynusoidal City WWTP (Problem 11–3). Using the plot and computations, determine the following ratios for BOD mass loading: peak to average (P/A); minimum to average (M/A); and peak to minimum (P/M).

Answers:

	Unequalized	Equalized
P/A	1.97	1.47
M/A	0.14	0.63
P/M	14.05	2.34

11–7 Repeat Problem 11–6 using the data from Problem 11–4.

11–8 Repeat Problem 11–6 using the data from Problem 11–5.

11–9 Using an overflow rate of 26.0 m · day^{-1} and a detention time of 2.0 h, size a primary sedimentation tank for the average flow at Cynusoidal City (Problem 11–3). What would the overflow rate be for the unequalized maximum flow? Assume 15 sedimentation tanks with length to width ratio of 4.7.

> **Answers:**
> Tank dimensions = 15 tanks @ 2.17 m deep × 4.34 m × 20.4 m
> Maximum overflow rate = 39.3 m · day^{-1}

11–10 Determine the surface area of a primary settling tank sized to handle a maximum hourly flow of 0.570 m^3 · s^{-1} at an overflow rate of 60.0 m · day^{-1}. If the effective tank depth is 3.0 m, what is the effective theoretical detention time?

> **Answers:** Surface area = 820.80, or 821 m^2; t_o = 1.2 h

11–11 If an equalization basin is installed ahead of the primary tank in Problem 1–10, the average flow to the tank is reduced to 0.400 m^3 · s^{-1}. What is the new overflow rate and detention time?

11–12 The influent BOD$_5$ to a primary settling tank is 345 mg · L^{-1}. The average flow rate is 0.050 m^3 · s^{-1}. If the BOD$_5$ removal efficiency is 30%, how many kilograms of BOD$_5$ are removed in the primary settling tank each day?

11–13 Using the assumptions given in Example 11–4, the rule of thumb values for growth constants, and the further assumption that the influent BOD$_5$ was reduced by 32.0% in the primary tank, estimate the liquid volume of an aeration tank required to treat the wastewater in Problem 11–3. Assume an MLVSS of 2000 mg · L^{-1}.

> **Answer:** $\mathbf{V}$ = 4032, or 4000 m^3

11–14 Repeat Problem 11–13 using the wastewater in Problem 11–4.

11–15 Repeat Problem 11–13 using the wastewater in Problem 11–5.

11–16 Using a spreadsheet program you have written, rework Example 11–4 using the following MLVSS concentrations instead of the 2000 mg · L^{-1} used in the example: 1000 mg · L^{-1}; and 1500 mg · L^{-1}; 2500 mg · L^{-1}; and 3000 mg · L^{-1}.

11–17 Using a spreadsheet program you have written, determine the effect of MLVSS concentration on the effluent soluble BOD$_5$ (S) using the data in Example 11–4. Assume the volume of the aeration tank remains constant at 970 m^3. Use the same MLVSS values used in Problem 11–16.

11–18 If the *F/M* of a 0.4380 m^3 · s^{-1} activated sludge plant is 0.200 day^{-1}, the influent BOD$_5$ after primary settling is 150 mg · L^{-1}, and the MLVSS is 2200 mg · L^{-1}, what is the volume of the aeration tank?

> **Answer:** Volume = 1.29 × 10^4 m^3

11–19 Two activated sludge aeration tanks at Turkey Run, Indiana, are operated in series. Each tank has the following dimensions: 7.0 m wide × 30.0 m long × 4.3 m effective liquid depth. The plant operating parameters are as follows:

> Flow = 0.0796 m^3 · s^{-1} MLVSS = 1500 mg · L^{-1}
> Soluble BOD$_5$ after primary settling = 130 mg · L^{-1} MLSS = 1.40 (MLVSS)

Determine the following: aeration period, *F/M* ratio.

> **Answers:** Aeration period = 6.3 h; *F/M* = 0.33 mg · mg^{-1} · d^{-1}

11–20 The 500-bed Lotta Hart Hospital has a small activated sludge plant to treat its wastewater. The average daily hospital discharge is 1200 L · day^{-1} per bed, and the average soluble BOD$_5$ after primary settling is 500 mg · L^{-1}. The aeration tank has effective liquid dimensions of 10.0 m wide × 10.0 m long × 4.5 m deep. The plant operating parameters are as follows:

> MLVSS = 2000 mg · L^{-1}
> MLSS = 1.20 (MLVSS)
> Return sludge concentration = 12,000 mg · L^{-1}

Determine the following: Aeration period and *F/M* ratio.

11–21 The Jambalaya shrimp processing plant generates 0.012 $m^3 \cdot s^{-1}$ of wastewater each day. The wastewater is treated in an activated sludge plant. The average BOD_5 of the raw wastewater before primary settling is 1400 $mg \cdot L^{-1}$. The aeration tank has effective liquid dimensions of 8.0 m wide by 8.0 m long by 5.0 m deep. The plant operating parameters are as follows:

> Soluble BOD_5 after primary settling = 966 $mg \cdot L^{-1}$
> MLVSS = 2000 $mg \cdot L^{-1}$
> MLSS = 1.25(MLVSS)
> Settled sludge volume after 30 min = 225.0 $mL \cdot L^{-1}$
> Aeration tank liquid temperature = 15°C

Determine the following: aeration period and *F/M* ratio.

11–22 Using the following assumptions, determine the sludge age, cell wastage flow rate, and return sludge flow rate for the Turkey Run WWTP (Problem 11–19).

Assume:

Suspended solids in the effluent are negligible	Bacterial decay rate = 0.040 day^{-1}
Wastage is from the aeration tank	Effluent BOD_5 = 5.0 $mg \cdot L^{-1}$ (soluble)
Yield coefficient = 0.40	Return sludge concentration = 9130 $mg \cdot L^{-1}$

Answers: θ_c = 11.50 days; Q_w = 0.00182 $m^3 \cdot s^{-1}$; Q_r = 0.0214 $m^3 \cdot s^{-1}$

11–23 Using the following assumptions, determine the solids retention time and the cell wastage flow rate for the Lotta Hart Hospital WWTP (Problem 11–20).

Assume:

Suspended solids in effluent = 30.0 $mg \cdot L^{-1}$	Bacterial decay rate = 0.060 day^{-1}
Wastage is from the return sludge line	Inert fraction of suspended solids = 66.67%
Yield coefficient = 0.60	Allowable BOD in effluent = 30.0 $mg \cdot L^{-1}$

11–24 Using the following assumptions, determine the solids retention time, the cell wastage flow rate, and the return sludge flow rate for the Jambalaya shrimp processing plant WWTP (Problem 11–21).

Assume:

> Allowable BOD_5 in effluent = 25.0 $mg \cdot L^{-1}$
> Suspended solids in effluent = 30.0 $mg \cdot L^{-1}$
> Wastage is from the return sludge line
> Yield coefficient = 0.50 mg VSS/mg BOD_5 removed
> Decay rate of microorganisms = 0.075 d^{-1}
> Inert fraction of suspended solids = 30.0%

11–25 Determine the daily and annual primary sludge production for a WWTP having the following operating characteristics:

Operating data:

Flow = 0.0500 $m^3 \cdot s^{-1}$	Specific gravity of volatile solids = 0.970
Influent suspended solids = 155.0 $mg \cdot L^{-1}$	Fixed solids = 30.0%
Removal efficiency = 53.0%	Specific gravity of fixed solids = 2.50
Volatile solids = 70.0%	Sludge concentration = 4.50%

Answer: V_{sl} = 7.83 $m^3 \cdot day^{-1}$ or 2860 $m^3 \cdot y^{-1}$

11–26 Repeat Problem 11–25 using the following data:

Operating data:

Flow = 2.00 $m^3 \cdot s^{-1}$	Volatile solids = 68.0%
Influent suspended solids = 179.0 $mg \cdot L^{-1}$	Specific gravity of volatile solids = 0.999
Removal efficiency = 47.0%	Fixed solids = 32.0%
Sludge concentration = 5.20%	Specific gravity of fixed solids = 2.50

11–27 Using a computer spreadsheet you have written, and the data in Problem 11–26, determine the daily and annual sludge production at the following removal efficiences: 40%, 45%, 50%, 55%, 60%, and 65%. Plot annual sludge production as a function of efficiency.

11–28 The Pomdeterra wastewater treatment plant produces thickened sludge that has a suspended solids concentration of 3.8%. They are investigating a filter press that will yield a solids concentration of 24%. If they now produce 33 $m^3 \cdot day^{-1}$ of sludge, what annual volume savings will they achieve if they install the press?

> **Answer:** 10,138 or 10,000 $m^3 \cdot year^{-1}$

11–29 Ottawa's anaerobic digester produces 13 $m^3 \cdot day^{-1}$ of sludge with a suspended solids concentration of 7.8%. What volume of sludge must they dispose of each year if their sand drying beds yield a solids concentration of 35%?

DISCUSSION QUESTIONS

11–1 If the state regulatory agency requires tertiary treatment of a municipal wastewater, what, if any, processes would you expect to find preceding the tertiary process?

11–2 What is the purpose of recirculation and how does it differ from return sludge?

11–3 In which of the following cases is the cost of sludge disposal higher?
(a) $\theta_c = 3$ days
(b) $\theta_c = 10$ days

11–4 Would an industrial wastewater containing only NH_4^+ at a pH of 7.00 be denitrified if pure oxygen was bubbled into it? Explain your reasoning.

FE EXAM FORMATTED PROBLEMS

11–1 A 2-liter graduated cylinder was used to determine the sludge volume index (SVI) of an activated sludge sample. The settled volume was 850 mL and the MLSS was 3000 mg/L. What was the SVI?

(a) 283 mL/g
(b) 850 mL/g
(c) 425 mL/g
(d) 142 mL/g

11–2 Estimate the biomass concentration in a CSTR (continuous stirred-tank reactor) aeration tank with the following operating conditions: hydraulic residence time = 3 h; mean cell residence time = 6 d; yield coefficient = 0.6 mg VSS/mg BOD; decay rate = 0.01 d^{-1}; influent soluble BOD = 200 mg/L; effluent BOD = 2.0 mg/L.

(a) 3600 mg/L
(b) 150 mg/L
(c) 210 mg/L
(d) 1200 mg/L

11–3 When the soluble BOD in the influent to the treatment plant rises from 133 mg/L to 222 mg/L, the operator called to ask your advice on a new MLVSS for the plant. It has been operating with an *F/M* of 0.31 mg/mg · d and the operator would like to continue using the same *F/M*. The plant data are as follows: flow rate = 7630 mg/L; aeration tank volume = 1270 m^3; mixed liquor volatile suspended solids = 2600 mg/L. What MLVSS should she use?

(a) 2600 mg/L
(b) 4300 mg/L
(c) 2400 mg/L
(d) 5800 mg/L

11-4 Determine the SRT of an aerobic digester to treat 9500 ft^3 of sludge. The sludge temperature is 10°C and the "503" rules require a 38% reduction in volatile solids. Assume a 25% safety factor in estimating the SRT.

(a) 32 d

(b) 50 d

(c) 40 d

(d) 63 d

REFERENCES

Adams, J. K. (1948) "Operating Experiences at Tenafly, NJ," *Sewage Works Journal,* 20(5): 909–912.

Asano, T., F. L. Burton, H. L. Leverenz, R. Tsuchihashi, and G. Tchobanoglous (2007) *Water Reuse, Issues, Technologies, and Applications,* McGraw-Hill, New York.

Benefield, L. D., and C. W. Randall (1980) *Biological Process Design for Wastewater Treatment,* Prentice Hall, Upper Saddle River, NJ, pp. 322–24, 338–40, 353–54.

Caldwell, D. H., D. S. Parker, and W. R. Uhte (1973) *Upgrading Lagoons,* U.S. Environmental Protection Agency Technology Transfer Publication, Washington, DC.

Chalmers, R. B., M. Patel, and D. Cutler (2010) "Recycled Water Considered a Drought-Proof Water Supply," *Opflow*, 36(4): 14–17.

CFR (1993) *Code of Federal Regulations,* 40 CFR §131.36.

CFR (2005) *Code of Federal Regulations,* 40 CFR §133.102.

CFR (2006) *Code of Federal Regulations,* 40 CFR §413.02, §464.02, §467.02, and §469.12.

Comeau, Y., K. J. Hall, R. E. W. Hancock, and W. K. Oldham (1986) "Biochemical Model for Enhanced Biological Phosphorus Removal," *Water Research*, 20(12): 1511–1522.

Davis, M. L. (2010) *Water and Wastewater Engineering: Design Principles and Practice*, McGraw-Hill, New York.

Ehlers, V. M., and E. W. Steel (1943) *Municipal and Rural Sanitation,* McGraw-Hill, New York, 1943.

Erickson, A. E., B. G. Ellis, J. M. Tiedje, et al. (1974) *Soil Modification for Denitrification and Phosphate Reduction of Feedlot Waste,* U.S. Environmental Protection Agency, EPA Pub. No. 660/2-74-057, Washington, DC.

GLUMRB (2004) *Recommended Standards for Wastewater Facilities,* Great Lakes-Upper Mississippi River Board of State and Provincial Public Health and Environmental Managers, Health Education Service, Inc., Albany, NY, pp. 70–74.

Heukeleian, H., H. Orford, and R. Maganelli (1951) "Factors Affecting the Quantity of Sludge Production in the Activated Sludge Process," *Sewage & Industrial Wastes,* 23: 8.

Holland, K. T., J. S. Knapp, and J. G. Shoesmith (1987) *Anaerobic Bacteria,* Chapman and Hall, New York.

Lawrence, A. W., and T. R. Milnes (1971) "Discussion Paper," *Journal of the Sanitary Engineering Division,* American Society of Civil Engineers, 97: 121.

Metcalf & Eddy, Inc. (2003) *Wastewater Engineering: Treatment and Reuse*, revised by G. Tchobanoglous, F. L. Burton, and H. D. Stensel, McGraw-Hill, New York, 334–335, 1126.

Monod, J. (1949) "The Growth of Bacterial Cultures," *Annual Review of Microbiology,* 3: 371–94.

Pecson, B. M., R. S. Trussell, A. N. Pisarenko, and R. R. Trussell (2015) "Achieving Reliability in Potable Reuse: The Four Rs," *Journal of American Water Works Association*, 107(3): 48–58.

Pound, C. E., R. W. Crites, and D. A. Griffes (1976) *Land Treatment of Municipal Wastewater Effluents, Design Factors I,* U.S. Environmental Protection Agency Technology Transfer Seminar Publication, Washington, DC.

Quicho, J., A. Dorman, M. Steirer, and A. Van (2012) "Sustaining San Diego," *Water Environment and Technology*, 24(5): 36–39.

Sawyer, C. N. (1956) "Bacterial Nutrition and Synthesis," *Biological Treatment of Sewage and Industrial Wastes,* 1: 3.

Shahriari, H., C. Eskicioglu, and R. L. Droste (2006) "Simulating Activated Sludge System by Simple-to-Advanced Models," *Journal of Environmental Engineering,* American Society of Civil Engineers, 132: 42–50.

Shimp, G. F., D. M. Bond, J. Sandino, and D. W. Oerke (1995) "Biosolids Budgets," *Water Environment & Technology*, November, pp. 44–49.

Speece, R. E. (1983) "Anaerobic Biotechnology for Industrial Wastewater," *Environmental Science and Technology,* 17: 416A–27A.

Stephens, H. L., and H. D. Stensel (1998) "Effect of Operating Conditions on Biological Phosphorus Removal," *Water Environment Research*, 70(3): 362–369.

U.S. EPA (1977) *Process Design Manual: Wastewater Treatment Facilities for Sewered Small Communities,* EPA Pub. No. 625/1-77-009, U.S. Environmental Protection Agency, Washington, DC, pp. 8–12.

U.S. EPA (1979) *Environmental Pollution Control Alternatives: Municipal Wastewater,* EPA Pub. No. 625/ 5-79-012, U.S. Environmental Protection Agency, Washington, DC, pp. 33–5, 52–5.

U.S. EPA (1996) *NPDES Permit Writers' Manual,* EPA Pub. No. 833-B-96-003, U.S. Environmental Protection Agency, Washington, DC, pp. 77–8.

U.S. EPA (1997) *Response to Congress on the Use of Decentralized Wastewater Treatment Systems,* EPA Report No. 832-R-97-001b, U.S. Environmental Protection Agency, Washington, DC.

U.S. EPA (2005) http://www.epa.gov Search: Region 10 ⇒ Homepage ⇒ NPDES Permits ⇒ Current NPDES Permits in Pacific Northwest and Alaska ⇒ Current Individual NPDES Permits in Idaho.

Velz, C. J. (1970) *Applied Stream Sanitation*, Wiley-Interscience, New York.

WEF (1998) *Design of Municipal Wastewater Treatment Plants, Vol. 2, Manual of Practice No. 8,* Joint Task Force of the Water Environment Federation and American Society of Civil Engineers, Alexandria, VA, pp. 12–19.

Wentzel, M. C., L. H. Lotter, R. E. Loewenthal, and G. v. R. Marais (1986) "Metabolic Behavior of *Acinetobacter* sp. In Enhanced Biological Phosphorous Removal—A Biochemical Model," *Water,* South African, 12: 209.

12

Air Pollution

Case Study

Killer Smog Blankets Town

It was about 4:30 AM on a November morning in Poza Rica de Hidalgo, in the Veracruz region of Mexico. A dense layer of fog blanketed the valley basin. A light wind blew across the River Cazones. It was 1951, and Poza Rica, which means rich well in Spanish, had been rapidly developing as a city, fueled by oil refineries and natural gas processing plants. It was a small, new city with a population of about 22,000 people.

Sometime between 4:50 and 5:10 AM, a flare on a high sulfur crude oil processing facility went out. There were no alarms and no safety equipment installed on the partially completed, but operative desulfurization unit. With the flare extinguished, the dense gas containing about 16% hydrogen sulfide and 81% carbon dioxide spewed from the stack. About 20 minutes later, refinery workers were able to halt the flow of gas to the stack. The wind carried the gas toward the residential area near the plant.

As residents attempted to flee the area, they were overcome by the fumes. By noon that day, 17 people would be dead. In total, 22 people died, ranging in age from 2 to 50 years. Nine children under the age of 13 died. Three hundred and twenty people were hospitalized. The most frequently reported symptoms were loss of smell, nausea, severe headache, difficulty breathing, and unconsciousness. Approximately 50% of all exposed animals, including pets, birds, and cattle, died. All of the canaries in the area succumbed to the exposure.

While no measurements of hydrogen sulfide were made, it is estimated that its concentration peaked at 1000–2000 parts per million (ppm). Concentrations in excess of 100 ppm are considered immediately dangerous to life or health. This incident demonstrates the need for strict environmental and occupational regulations that protect both public health and welfare.

12–1 INTRODUCTION

Air Pollution Perspective

Air pollution is of public health concern on the micro-, meso-, and macroscales. Indoor air pollution results from products used in construction materials, the inadequacy of general ventilation, and geophysical factors that may result in exposure to naturally occurring radioactive materials. Industrial and mobile sources contribute to mesoscale air pollution that contaminates the ambient air that surrounds us outdoors. Macroscale (or global) effects include transport of ambient air pollutants over large distances. Examples of macroscale effects include acid rain and ozone pollution. Global effects of air pollution result from sources that may potentially change the upper atmosphere, examples of which include depletion of the ozone layer and global warming. Although micro- and macroscale effects are of concern, our focus will predominately be on mesoscale air pollution.

12–2 FUNDAMENTALS

Pressure Relationships and Units of Measure

The fundamental relationships of pressure and the units of measure used in discussing air pollution were presented in Chapter 2: they are **micrograms per cubic meter** ($\mu g \cdot m^{-3}$), **parts per million** (ppm), and the **micrometer** (μm). Micrograms per cubic meter and parts per million are measures of concentration and are both used to indicate the concentration of a gaseous pollutant. Conversion from one of these units to the other was discussed in Chapter 2. The concentration of particulate matter may be reported only as micrograms per cubic meter. The micrometer is used to report particle size.

The advantage of the unit ppm that frequently makes it the unit of choice is that it is a volume-to-volume ratio. (Note that the usage of ppm in air is different from that in water and wastewater, which is a mass-to-mass ratio.) Changes in temperature and pressure do not change the ratio of the volume of pollutant gas to the volume of air that contains it. Thus, it is possible to compare parts per million readings from Denver and Washington, DC, as well as day and night readings in regions with large variations in temperature without further conversion.

Relativity

Before we launch into the esoterics of air pollution, let's take a moment to consider the application of the units ppm and μm in daily life. Four crystals of common table salt in one cup of granulated sugar is approximately equal to 1 ppm on a volume-to-volume basis. Figure 12–1 should help you visualize the size of a micrometer. Note that a hair has an average diameter of approximately 80 μm.

Adiabatic Expansion and Compression

Air pollution meteorology is, in part, a consequence of the thermodynamic processes of the atmosphere. One such process is adiabatic expansion and contraction. An **adiabatic process** is one that takes place with no addition or removal of heat and with sufficient slowness so the gas can be considered to be in equilibrium at all times.

As an example, let us consider the piston and cylinder in Figure 12–2. The cylinder and piston face are assumed to be perfectly insulated. The gas is at pressure, P, a force, F, equal to the product of the pressure and area of the face of the piston, A, must be applied to the piston to maintain equilibrium. If the force is increased and the volume is compressed, the pressure will increase and work will be done on the gas by the piston. Because no heat enters or leaves

FIGURE 12–1

Relative sizes of small particles.

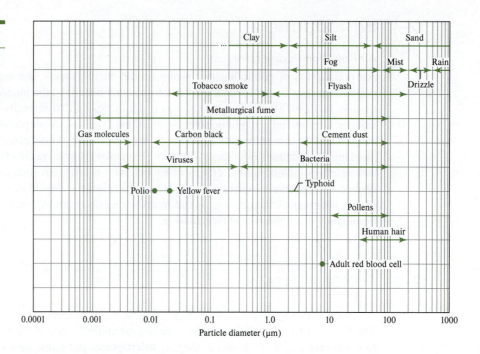

FIGURE 12–2

Work done on gas.

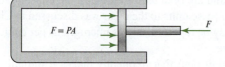

$$F = PA$$

the gas, the work will increase the thermal energy of the gas in accordance with the first law of thermodynamics (Equation 4–41), that is,

(Heat added to gas) = (increase in thermal energy) + (external work done by or on the gas)

Because the left side of the equation is zero (because it is an adiabatic process), the increase in thermal energy is equal to the work done. The increase in thermal energy is reflected by an increase in the temperature of the gas. If the gas is expanded adiabatically, its temperature will decrease.

12–3   AIR POLLUTION STANDARDS

The 1970 Clean Air Act (CAA) required the U.S. Environmental Protection Agency (EPA) to investigate and describe the environmental effects of any air pollutant emitted by stationary or mobile sources that may adversely affect human health or the environment. The EPA used these studies to establish the National Ambient Air Quality Standards (NAAQS). These standards are for the ambient air, that is, the outdoor air that normally surrounds us. EPA calls the pollutants listed in Table 12–1 **criteria pollutants** because they were developed on health-based criteria. The **primary standard** was established to protect human health with an "adequate margin of safety." The **secondary standards** are intended to prevent environmental and property damage. In 1987, the EPA revised the NAAQS. The standard for hydrocarbons was dropped and the standard for total suspended particulates (TSP) was replaced with a particulate standard based on the mass of particulate matter with an aerodynamic diameter less than or equal to 10 μm. This standard is referred to as the PM_{10} standard. In 1997 a new $PM_{2.5}$ standard was added.

TABLE 12–1		National Ambient Air Quality Standards (NAAQS)			
Criteria Pollutant	**Standard Type**	**Concentration**		**Averaging Period or Method**	**Allowable Exceedances**[a]
		($\mu g \cdot m^{-3}$)	**(ppm)**		
CO	Primary	10,000	9	8-hour average	Once per year
	Primary	40,000	35	1-hour average	Once per year
Lead	Primary and secondary	0.15	N/A	Maximum arithmetic mean measured over a calendar quarter	[f]
NO_2	Primary and secondary	100	0.053	Annual arithmetic mean	
NO_2	Primary	189	0.100	1-hour average	[e]
Ozone	Primary and secondary	235	0.12	Maximum hourly average[b]	Once per year
Ozone	Primary and secondary	137	0.070	8-hour average	[c]
Particulate matter (PM_{10})[d]	Primary and secondary	150	N/A	24-hour average	One day per year
Particulate matter ($PM_{2.5}$)	Primary	12.0	N/A	Annual mean	[f]
Particulate matter ($PM_{2.5}$)	Secondary	15.0	N/A	Annual mean	[f]
($PM_{2.5}$)	Primary and secondary	35	N/A	24-hour average	[e]
		15	N/A	Annual arithmetic mean	[f]
SO_2	Primary	80	0.03	Annual arithmetic mean	
	Primary	365	0.14	Maximum 24-hour concentration	Once per year
SO_2	Primary	1,950	0.75	Maximum 1-hour concentration	[g]
SO_2	Secondary	1,300	0.5	Maximum 3-hour concentration	[h]

[a]Allowable exceedances may actually be an average value over a multi-year period.

[b]The 1-hour NAAQS will no longer apply to an area one year after the effective date of the designation of that area for the 8-hour ozone NAAQS. For most areas, the date of designation was June 15, 2004.

[c]Average fourth highest concentration over 3-year period.

[d]Particulate matter standard applies to particles with an aerodynamic diameter ≤ 10 μm.

[e]Three-year average of 98th percentile 24-hour concentration.

[f]Three-year average of weighted annual mean.

[g]Three-year average of 99th percentile of daily maximum 1-hour average.

[h]Not to be exceeded more than once per year.

Source: Code of Federal Regulations 40 CFR 50.4–50.12 and 69 FR 23996 and NAAQS 20 Dec 2016.

States are divided into air quality control regions (AQRs). An AQR that has air quality equal to or better than the primary standard is called an **attainment area.** Those areas that do not meet the primary standard are called **nonattainment areas.**

Under the 1970 CAA, the EPA was directed to establish regulations for **hazardous air pollutants** (HAPs) using a risk-based approach. These were called NESHAPs—national emission standards for hazardous air pollutants. Because EPA had difficulty defining "an ample margin of safety" as required by the law, only seven HAPs were regulated between 1970 and 1990: asbestos, arsenic, benzene, beryllium, mercury, vinyl chloride, and radionuclides. The Clean Air Act Amendments of 1990 directed EPA to establish a HAP emissions control program based on technology for 189 chemicals (Table 12–2). EPA will establish emission

TABLE 12–2	Hazardous Air Pollutants (HAPs)	
Acetaldehyde	Chloroprene	Ethylene dibromide (Dibromoethane)
Acetamide	Cresols/Cresylic acid	Ethylene dichloride (1,2-Dichloroethane)
Acetonitrile	(isomers and mixture)	Ethylene glycol
Acetophenone	o-Cresol	Ethylene imine (Aziridine)
2-Acetylaminofluorene	m-Cresol	Ethylene oxide
Acrolein	p-Cresol	Ethylene thiourea
Acrylamide	Cumene	Ethylidene dichloride
Acrylic acid	2,4-D, salts and esters	(1,1-Dichloroethane)
Acrylonitrile	DDE	Formaldehyde
Allyl chloride	Diazomethane	Heptachlor
4-Aminobiphenyl	Dibenzofurans	Hexachlorobenzene
Aniline	1,2-Dibromo-3-chloropropane	Hexachlorobutadiene
o-Anisidine	Dibutylphthalate	Hexachlorocyclopentadiene
Asbestos	1,4-Dichlorobenzene(p)	Hexachloroethane
Benzene (including benzene	3,3-Dichlorobenzidene	Hexamethylene-1,6-diisocyanate
from gasoline)	Dichloroethyl ether	Hexamethylphosphoramide
Benzidine	[Bis(2-chloroethyl)ether]	Hexane
Benzotrichloride	1,3-Dichloropropene	Hydrazine
Benzyl chloride	Dichlorvos	Hydrochloric acid
Biphenyl	Diethanolamine	Hydrogen fluoride (Hydrofluoric acid)
Bis(2-ethylhexyl)phthalate (DEHP)	N,N-Diethyl aniline (N,N-Dimethylaniline)	Hydroquinone
Bis(chloromethyl)ether	Diethyl sulfate	Isophorone
Bromoform	3,3-Dimethoxybenzidine	Lindane (all isomers)
1,3-Butadiene	Dimethyl aminoazobenzene	Maleic anhydride
Calcium cyanamide	3,3´-Dimethyl benzidine	Methanol
Caprolactam	Dimethyl carbamoyl chloride	Methoxychlor
Captan	Dimethyl formamide	Methyl bromide (Bromomethane)
Carbaryl	1,1-Dimethyl hydrazine	Methyl chloride (Chloromethane)
Carbon disulfide	Dimethyl phthalate	Methyl chloroform
Carbon tetrachloride	Dimethyl sulfate	(1,1,1-Trichloroethane)
Carbonyl sulfide	4,6-Dinitro-o-cresol, and salts	Methyl ethyl ketone (2-Butanone)
Catechol	2,4-Dinitrophenol	Methyl hydrazine
Chloramben	2,4-Dinitrotoluene	Methyl iodide (Iodomethane)
Chlordane	1,4-Dioxane (1,4-Diethyleneoxide)	Methyl isobutyl ketone (Hexone)
Chlorine	1,2-Diphenylhydrazine	Methyl isocyanate
Chloroacetic acid	Epichlorohydrin	Methyl methacrylate
2-Chloroacetophenone	(1-chloro-2,3-epoxypropane)	Methyl tert butyl ether
Chlorobenzene	1,2-Epoxybutane	4,4-Methylene bis(2-chloroaniline)
Chlorobenzilate	Ethyl acrylate	Methylene chloride (Dichloromethane)
Chloroform	Ethyl benzene	Methylene diphenyl diisocyanate (MDI)
Chloromethyl methyl ether	Ethyl carbamate (Urethane)	4,4´-Methylenedianiline
	Ethyl chloride (Chloroethane)	

TABLE 12–2	*(continued)*	
Naphthalene	1,2-Propylenimine (2-Methyl aziridine)	Vinylidene chloride (1,1-Dichloroethylene)
Nitrobenzene	Quinoline	Xylenes (isomers and mixture)
4-Nitrobiphenyl	Quinone	o-Xylenes
4-Nitrophenol	Styrene	m-Xylenes
2-Nitropropane	Styrene oxide	p-Xylenes
N-Nitroso-N-methylurea	2,3,7,8-Tetrachlorodibenzo-p-dioxin	Antimony compounds
N-Nitrosodimethylamine	1,1,2,2-Tetrachloroethane	Arsenic compounds (inorganic, including arsine)
N-Nitrosomorpholine	Tetrachloroethylene (Perchloroethylene)	Beryllium compounds
Parathion	Titanium tetrachloride	Cadmium compounds
Pentachloronitrobenzene (Quintobenzene)	Toluene	Chromium compounds
Pentachlorophenol	2,4-Toluene diamine	Cobalt compounds
Phenol	2,4-Toluene diisocyanate	Coke oven emissions
p-Phenylenediamine	o-Toluidine	Cyanide compounds[a]
Phosgene	Toxaphene (chlorinated camphene)	Glycol ethers[b]
Phosphine	1,2,4-Trichlorobenzene	Lead compounds
Phosphorus	1,1,2-Trichloroethane	Manganese compounds
Phthalic anhydride	Trichloroethylene	Mercury compounds
Polychlorinated biphenyls (Aroclors)	2,4,5-Trichlorophenol	Fine mineral fibers[c]
1,3-Propane sultone	2,4,6-Trichlorophenol	Nickel compounds
beta-Propiolactone	Triethylamine	Polycylic organic matter[d]
Propionaldehyde	Trifluralin	Radionuclides (including radon)[e]
Propoxur (Baygon)	2,2,4-Trimethylpentane	Selenium compounds
Propylene dichloride (1,2-Dichloropropane)	Vinyl acetate	
Propylene oxide	Vinyl bromide	
	Vinyl chloride	

[a]X'CN where X = H' or any other group where a formal dissociation may occur. For example KCN or Ca(CN)$_2$.

[b]Includes mono- and diethers of ethylene glycol, diethylene glycol, and triethylene glycol derivatives R-(OCH$_2$CH$_2$)$_n$-OR' where

n = 1, 2, or 3

R = alkyl or aryl groups

R' = R, H, or groups which, when removed, yield glycol ethers with the structure: R-(OCH$_2$CH)$_n$-OH. Polymers are excluded from the glycol category.

[c]Includes mineral fiber emissions from facilities manufacturing or processing glass, rock, or slag fibers (or other mineral derived fibers) of average diameter 1 micrometer or less.

[d]Includes organic compounds with more than one benzene ring and which have a boiling point greater than or equal to 100°C.

[e]A type of atom which spontaneously undergoes radioactive decay.

Note: For all listings that contain the word *compounds* and for glycol ethers, the following applies: Unless otherwise specified, these listings are defined as including any unique chemical substance that contains the named chemical (i.e., antimony, arsenic, etc.) as part of that chemical's infrastructure.

Source: Public Law 101-549, Nov. 15, 1990.

allowances based on **maximum achievable control technology** (MACT) for 174 categories of industrial sources that potentially emit 9.08 Mg per year of a single HAP or 22.7 Mg per year of a combination of HAPs. A MACT can include process changes, material substitutions, or air pollution control equipment.

Emission standards place a limit on the amount or concentration of one or more specified contaminants that may be emitted from a source. In 1971, the U.S. EPA published final standards for the first of many stationary sources. The initial five industries that were regulated

TABLE 12–3

New Source Performance Standards for Coal-Fired Electric Utility Steam Generating Units of More than 73 MW—Summary*

SO_2 Standard

Emission limit: 90% reduction of potential SO_2 emissions and a limit of SO_2 emissions to 516 g $\cdot$ 10^{-6} kJ (1.2 lb_m/million Btu of heat input)

Particulate Standard

Emission limit: 13 g $\cdot$ 10^{-6} kJ (0.03 lb_m/million Btu heat input)

NO_x Standard

Emission limit for subbituminous coal: 210 g $\cdot$ 10^{-6} kJ (0.50 lb_m/million Btu heat input)

Emission limit for anthracite coal: 260 g $\cdot$ 10^{-6} kJ (0.60 lb_m/million Btu heat input)

Federal Register, 45, February 1980, pp. 8210–8215.

TABLE 12–4

Federal Motor Vehicle Emission Standards in g/mile[a]

Bin	NO_x	NMOG[b]	CO	HCHO[c]	PM[d]
8	0.20	0.125	4.2	0.018	0.02
7	0.15	0.090	4.2	0.018	0.02
6	0.10	0.090	4.2	0.018	0.01
5	0.07	0.090	4.2	0.018	0.01
4	0.04	0.070	2.1	0.011	0.01
3	0.03	0.055	2.1	0.011	0.01
2	0.02	0.010	2.1	0.004	0.01
1	0.00	0.000	0.0	0.000	0.00

[a]For 2004–2009 model years; Full Useful Life
[b]NMOG = nonmethane organic matter
[c]HCHO = formaldehyde
[d]PM = particulate matter
Source: Code of Federal Regulations, 40 CFR 86.1811-04, 2 NOV 2010.

under the New Source Performance Standards (NSPS) included electric steam generating units, Portland cement plants, incinerators, nitric acid plants, and sulfuric acid plants. As an example of the NSPS, those for large electric utility steam generating units are summarized in Table 12–3.

Federal motor vehicle standards are expressed in terms of grams of pollutant per mile of driving. These standards were divided into two tiers. Tier I for 1994–1997 model years and Tier II for the 2004–2009 model years. The emission standards are applicable to light duty vehicles, light duty trucks, and medium duty passenger vehicles. Tier II is subdivided into bins to allow manufacturers to classify their production (called a *fleet*) for the purpose of calculations to meet the standard. By 2008, the original 11 bins had been reduced to the eight shown in Table 12–4. Tier II, Bin 5 roughly defines what the fleet average should be. Because emissions vary as driving conditions change, a standard driving cycle is defined for testing vehicles for compliance. The original driving cycle called the Federal Test Procedure was modified in 1996 by the Supplemental Federal Test Procedure to account for more aggressive driving behavior, the impact of air conditioning, and emissions after the engine is turned off.

Greenhouse gas air pollutant emission standards for passenger automobiles are defined as follows (40 CFR 86.1818-12):

- nitrous oxide $\leq$ 0.010 g/mile
- methane $\leq$ 0.030 g/mile

For carbon dioxide there are target values based on the *footprint* of the vehicle and the model year. For example, a passenger car with a footprint less than or equal to 41 square feet has the following target values by model year:

- 2018 → 202 g/mile
- 2019 → 191 g/mile
- 2020 → 182 g/mile
- 2021 → 172 g/mile
- 2022 → 164 g/mile

where footprint = wheel base × track width.

In March 2017, the Trump administration reversed EPA's January 2017 decision to finalize federal fuel economy standards set for 2022–2025. The proposed standard would have required the automobile industry to meet an estimated combined average of passenger cars and light truck of 54.5 miles per gallon (mpg) of gasoline via fuel economy improvements in model year 2025. This standard is now in abeyance. However, the U.S. auto industry expects to meet the standard for passenger cars and light trucks but, perhaps, not for large pickup trucks.

The industry's reasoning is based on economic considerations. Regulations in the developed countries whose automobile industries are competing for sales require a fuel economy standard of 54.5 mpg in 2025. To compete in the world automotive market, the U.S. passenger vehicles and light trucks must meet the 54.5 mpg standard. The cost and complexity of producing vehicles that differ only in mileage is not warranted.

Outside of the United States, the demand for large pickup trucks is not large enough to force the improvement of mileage. In addition, oil is more abundant and gasoline is relatively cheap. So the U.S auto industry will not spend the money and effort to improve large pickup truck mileage beyond 2022 models.

In addition to the emission standards, Congress enacted the Corporate Average Fuel Economy (CAFE) standard to improve motor vehicle fuel efficiency. These standards were enacted as a national security measure to reduce the United States' dependence on imported oil. While emissions of CO, NMOG, and NO_x are not directly related to fuel efficiency, the emission of carbon dioxide is. As discussed later in this chapter, carbon dioxide contributes significantly to the radiation balance that affects the atmospheric temperature of the earth.

12–4   EFFECTS OF AIR POLLUTANTS

Effects on Materials

Mechanisms of Deterioration.   Five mechanisms of deterioration have been attributed to air pollution: abrasion, deposition and removal, direct chemical attack, indirect chemical attack, and electrochemical corrosion (Yocom and McCaldin, 1968).

Solid particles of large enough size and traveling at high enough speed can cause deterioration by abrasion. With the exception of soil particles in dust storms, most air pollutant particles either are too small or travel at too slow a speed to be abrasive.

Small liquid and solid particles that settle on exposed surfaces do not cause more than aesthetic deterioration. For certain monuments and buildings, such as the White House, this form of deterioration is in itself quite unacceptable. For most surfaces, the cleaning process itself causes the damage. Sandblasting of buildings is an obvious case in point. Frequent washing of clothes weakens their fiber, and frequent washing of painted surfaces dulls their finish.

Solubilization and oxidation–reduction reactions typify direct chemical attack. Frequently, water must be present as a medium for these reactions to take place. Sulfur dioxide and SO_3 in the presence of water react with limestone ($CaCO_3$) to form calcium sulfate ($CaSO_4$) and gypsum ($CaSO_4 \cdot 2H_2O$). Both $CaSO_4$ and $CaSO_4 \cdot 2H_2O$ are more soluble in water than $CaCO_3$, and both are leached away when it rains. The tarnishing of silver by H_2S is a classic example of an oxidation–reduction reaction.

Indirect chemical attack occurs when pollutants are absorbed and then react with some component of the absorbent to form a destructive compound. The compound may be destructive because it forms an oxidant, reductant, or solvent. Furthermore, a compound can be destructive by removing an active bond in some lattice structure. Leather becomes brittle after it absorbs SO_2, which reacts to form sulfuric acid because of the presence of minute quantities of iron in the leather. The iron acts as a catalyst for the formation of the acid. A similar result has been noted for paper.

Oxidation–reduction reactions cause local chemical and physical differences on metal surfaces. These differences, in turn, result in the formation of microscopic anodes and cathodes. Electrochemical corrosion results from the potential that develops in these microscopic batteries.

Factors That Influence Deterioration.   Moisture, temperature, sunlight, and position of the exposed material are among the more important factors that influence the rate of deterioration.

Moisture, in the form of humidity, is essential for most of the mechanisms of deterioration to occur. Metal corrosion does not appear to occur even at relatively high SO_2 pollution levels until the relative humidity exceeds 60%. On the other hand, humidities above 70–90% will promote corrosion without air pollutants. Rain reduces the effects of pollutant-induced corrosion by dilution and washing away of the pollutant.

Higher air temperatures generally result in higher reaction rates. However, when low air temperatures are accompanied by cooling of surfaces to the point where moisture condenses, then the rates may be accelerated.

In addition to the oxidation effect of its ultraviolet wave lengths, sunlight stimulates air pollution damage by providing the energy for pollutant formation and cyclic reformation. The cracking of rubber and the fading of dyes have been attributed to ozone produced by these photochemical reactions.

The position of the exposed surface influences the rate of deterioration in two ways. First, whether the surface is vertical or horizontal or at some angle affects deposition and wash-off rates. Second, whether the surface is an upper or lower one may alter the rate of damage. When the humidity is sufficiently high, the lower side usually deteriorates faster because rain does not remove the pollutants as efficiently.

Effects on Vegetation

Cell and Leaf Anatomy.   Because the leaf is the primary indicator of the effects of air pollution on plants, we shall define some terms and explain how the leaf functions. A typical plant cell has three main components: the cell wall, the protoplast, and the inclusions. Much like human skin, the cell wall is thin in young plants and gradually thickens with age. Protoplast is the term used to describe the protoplasm of one cell. It consists primarily of water, but it also includes protein, fat, and carbohydrates. The nucleus contains the hereditary material (DNA), which controls the operation of the cell. The protoplasm located outside the nucleus is called cytoplasm. Within the cytoplasm are tiny bodies or plastids. Examples include chloroplasts, leucoplasts, chromoplasts, and mitochondria. Chloroplasts contain the chlorophyll that manufactures the plant's food through photosynthesis. Leucoplasts convert starch into starch grains. Chromoplasts are responsible for the red, yellow, and orange colors of the fruit and flowers.

A cross section through a typical mature leaf (Figure 12–3) reveals three primary tissue systems: the epidermis, the mesophyll, and the vascular bundle (veins). Chloroplasts are usually

FIGURE 12–3

Cross section of intact leaf. (*Source:* U.S. Dept. of Health Education and Welfare, Raleigh, N.C., 1970.)

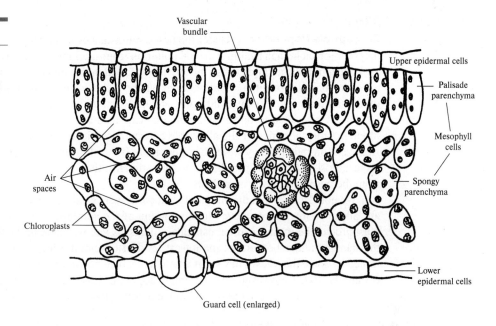

not present in epidermal cells. The opening in the underside of the leaf is called a stoma. (The plural of stoma is stomata.) The mesophyll, which includes both the palisade parenchyma and the spongy parenchyma, contains chloroplasts and is the food production center. The vascular bundles carry water, minerals, and food throughout the leaf and to and from the main stem of the plant.

The guard cells regulate the passage of gases and water vapor in and out of the leaf. When it is hot, sunny, and windy, the processes of photosynthesis and respiration increase. The guard cells open, which allows increased removal of water vapor that otherwise would accumulate because of the increased transport of water and minerals from the roots.

Pollutant Damage.   Ozone injures the palisade cells (Hindawi, 1970). The chloroplasts condense and ultimately the cell walls collapse. This results in the formation of red-brown spots that turn white after a few days. The white spots are called **fleck**. Ozone injury appears to be the greatest during midday on sunny days. The guard cells are more likely to be open under these conditions and thus allow pollutants to enter the leaf.

Plant growth may be inhibited by continuous exposure to 0.5 ppm of NO_2. Levels of NO_2 in excess of 2.5 ppm for periods of 4 h or more are required to produce **necrosis** (surface spotting due to plasmolysis or loss of protoplasm).

Sulfur dioxide injury is also typified by necrosis, but at much lower levels. A concentration of 0.3 ppm for 8 h is sufficient (O'Gara, 1922). Lower levels for longer periods of exposure will produce a diffuse chlorosis (bleaching).

The net result of air pollutant damage goes beyond the apparent superficial damage to the leaves. A reduction in surface area results in less growth and small fruit. For commercial crops this results in a direct reduction in income for the farmer. For other plants the net result is likely to be an early death.

Fluoride deposition on plants not only causes them damage but may result in a second untoward effect. Grazing animals may accumulate an excess of fluoride that mottles their teeth and ultimately causes them to fall out.

Problems of Diagnosis.   Various factors make it difficult to diagnose actual damage from air pollution. Droughts, insects, diseases, herbicide overdoses, and nutrient deficiencies all can cause injury that resembles air pollution damage. Also, combinations of pollutants that alone cause no damage are known to produce acute effects when combined (Hindawi, 1970). This effect is known as **synergism**.

Effects on Health

Susceptible Population. Under the best circumstances it is difficult to assess the effects of air pollution on human health. Personal pollution from smoking results in exposure to air pollutant concentrations far higher than the low levels found in the ambient atmosphere. Occupational exposure may also result in pollution doses far above those found outdoors. Tests on rodents and other mammals are difficult to interpret and apply to human anatomy. Tests on human subjects are usually restricted to those who would be expected to survive. This leads us to a question of environmental ethics. If the allowable concentration levels (standards) are based on results from tests on rodents, they would be rather high. If the allowable concentration levels must also protect those with existing cardiorespiratory ailments, they should be lower than those resulting from the observed effects on rodents.

We noted earlier that the air quality standards were established to protect public health with an "adequate margin of safety." In the opinion of the administrator of the EPA, the standards must protect the most sensitive responders. Thus, as you will note in the following paragraphs, the standards have been set at the lowest level of observed effect. This decision has been attacked by some theorists. They say it would make better economic sense to build more hospitals (Connolly, 1972). However, one also might apply this kind of logic in establishing speed limits for highways, that is, raise the speed limit and build more hospitals, junk yards, and cemeteries!

Anatomy of the Respiratory System. The respiratory system is the primary indicator of air pollution effects in humans. The major organs of the respiratory system are the nose, pharynx, larynx, trachea, bronchi, and lungs (Figure 12–4). The nose, pharynx, larynx, and trachea

---

FIGURE 12–4

The respiratory system.

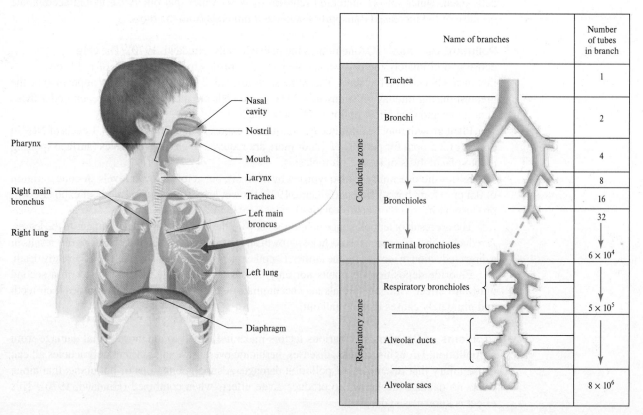

	Name of branches	Number of tubes in branch
Conducting zone	Trachea	1
	Bronchi	2
		4
		8
	Bronchioles	16
		32
	Terminal bronchioles	6×10^4
Respiratory zone	Respiratory bronchioles	5×10^5
	Alveolar ducts	
	Alveolar sacs	8×10^6

together are called the upper respiratory tract (URT). The primary effects of air pollution on the URT are aggravation of the sense of smell and inactivation of the sweeping motion of cilia, which remove mucus and entrapped particles. The lower respiratory tract (LRT) consists of the branching structures known as bronchi and the lung itself, which is composed of grape-like clusters of sacs called alveoli. The alveoli are approximately 300 μm in diameter. The walls of alveoli are lined with capillaries. Carbon dioxide diffuses through the capillary wall into the alveolus, and oxygen diffuses out of the alveolus into the blood cell. The difference in partial pressure of each of the gases causes it to move from the higher to lower partial pressure.

Inhalation and Retention of Particles.   The degree of penetration of particles into the LRT is primarily a function of the size of the particles and the rate of breathing. Particles greater than 5–10 μm are screened out by the hairs in the nose. Sneezing also helps the screening process. Particles in the 1–2-μm size range penetrate to the alveoli. These particles are small enough to bypass screening and deposition in the URT; however, they are big enough that their terminal settling velocity allows them to deposit where they can do the most damage. Particles that are less than 0.5 μm in diameter may diffuse to the alveolus cell wall but do not have a large enough terminal settling velocity to be removed efficiently. Smaller particles diffuse to the alveolar walls. Refer to Figure 12–1 and note the air pollutants that fall in the critical particle size range.

Chronic Respiratory Disease.   Several long-term diseases of the respiratory system are seriously aggravated by and perhaps may be caused by air pollution. Airway resistance is the narrowing of air passages because of the presence of irritating substances. The result is that breathing becomes difficult. **Bronchial asthma** is a form of airway resistance that results from an allergy. An asthma "attack" is the result of the narrowing of the bronchioles because of a swelling of the mucous membrane and a thickening of the secretions. The bronchioles return to normal after the attack. **Chronic bronchitis** is currently defined to be present in a person when excess mucus in the bronchioles results in a cough for 3 months a year for two consecutive years. Lung infections, tumors, and heart disease must be absent. **Pulmonary emphysema** is characterized by a breakdown of the alveoli. The small grape-like clusters become a large non-resilient balloon-like structure. The amount of surface area for gas exchange is reduced drastically. **Cancer of the bronchus** (lung cancer) is characterized by abnormal, disorderly new cell growth originating in the bronchial mucous membrane. The growth closes off the bronchioles. It is usually fatal.

Carbon Monoxide (CO).   This colorless, odorless gas is lethal to humans within a few minutes at concentrations exceeding 5000 ppm. CO reacts with hemoglobin in the blood to form carboxyhemoglobin (COHb). Hemoglobin has a greater affinity for CO than it does for oxygen. Thus, the formation of COHb effectively deprives the body of oxygen. At COHb levels of 5 to 10%, visual perception, manual dexterity, and ability to learn are impaired. A concentration of 50 ppm of CO for 8 h will result in a COHb level of about 7.5%. At COHb levels of 2.5 to 3%, people with heart disease are not able to perform certain exercises as well as they might in the absence of COHb. A concentration of 20 ppm of CO for 8 h will result in a COHb level of about 2.8% (Ferris, 1978). (We should note here that the average concentration of CO inhaled in cigarette smoke is 200–400 ppm!) The sensitive populations are those with heart and circulatory ailments, chronic pulmonary disease, developing fetuses, and those with conditions that cause increased oxygen demand, such as fever from an infectious disease.

Hazardous Air Pollutants.   Most of the information on the direct effects of hazardous air pollutants (also known as air toxics) on human health comes from studies of industrial workers. Exposure to air toxics in the workplace is generally much higher than in the ambient air. We know relatively little about the specific effects of the HAPs at the low levels normally found in ambient air.

The HAPs regulated under the NESHAP program were identified as causal agents for a variety of diseases. For example, asbestos, arsenic, benzene, coke oven emissions, and radionuclides may cause cancer. Beryllium primarily causes lung disease but also affects the liver, spleen, kidneys, and lymph glands. Mercury attacks the brain, kidneys, and bowels. Other potential effects from the HAPs are birth defects and damage to the immune and nervous systems (Kao, 1994).

Mercury has been especially targeted for regulation because it is released during the combustion of coal. Thus, it is one of the few HAPs that is widespread in the environment. Of particular concern are children who are exposed to methylmercury prenatally. They are at increased risk of poor performance on neurobehavioral tasks such as those measuring attention, fine motor function, language skills, visual-spatial abilities, and verbal memory (U.S. EPA, 1997; U.S. EPA, 2004).

Lead (Pb). In contrast to the other major air pollutants, lead is a cumulative poison. A further difference is that it is ingested in food and water, as well as being inhaled. Of that portion taken by ingestion, approximately 5–10% is absorbed in the body. Between 20 and 50% of the inspired portion is absorbed. Those portions that are not absorbed are excreted in the feces and urine. Lead is measured in the urine and blood for diagnostic evidence of lead poisoning.

An early manifestation of acute lead poisoning is a mild anemia (deficiency of red blood cells). Fatigue, irritability, mild headache, and pallor indistinguishable from other causes of anemia occur when the blood level of lead increases to $60–120 \, \mu g \cdot 100 \, g^{-1}$, of whole blood. Blood levels in excess of $80 \, \mu g \cdot 100 \, g^{-1}$ result in constipation and abdominal cramps. When an acute exposure results in blood levels of lead greater than $120 \, \mu g \cdot 100 \, g^{-1}$, acute brain damage (encephalopathy) may result (Goyer and Chilsolm, 1972). Such acute exposure results in convulsions, coma, cardiorespiratory arrest, and death. Acute exposures may occur over a period of one to three weeks.

Canfield et al. (2003) found a decline in intelligence quotient (IQ) of 7.4 points for a lifetime blood lead concentration of up to 10 μg per deciliter. For a lifetime average blood lead concentration ranging from more than 10 μg per deciliter to 30 μg per deciliter, a more gradual decrease of 2.5 IQ points was observed.

Atmospheric lead occurs as a particulate. The particle size range is between 0.16 and 0.43 μm. Before the ban on lead in automobile fuels, nonsmoking residents of suburban Philadelphia exposed to approximately $1 \, \mu g \cdot m^{-3}$ of lead in air had blood levels averaging $1.1 \, \mu g \cdot 100 \, g^{-1}$. At the same time, nonsmoking residents of downtown Philadelphia exposed to approximately $2.5 \, \mu g \cdot m^{-3}$ of lead had blood levels averaging $20 \, \mu g \cdot 100 \, g^{-1}$ (U.S. PHS, 1965). In the early 1990s, 4.4% of U.S. children ages 1 to 5 had elevated lead levels. The percentage dropped to 1.6% by 2002. The U.S. Centers for Disease Control and Prevention attributed this drop to the removal of lead from gasoline as well as other efforts to screen and treat children for lead exposure (U.S. CDC, 2005).

Nitrogen Dioxide (NO_2). The oxides of nitrogen are referred to collectively as NO_x. Exposure to NO_2 concentrations above 5 ppm for 15 min results in cough and irritation of the respiratory tract. Continued exposure may produce an abnormal accumulation of fluid in the lung (pulmonary edema). The gas is reddish brown in concentrated form and gives a brownish yellow tint at lower concentrations. At 5 ppm it has a pungent sweetish odor. The average NO_2 concentration in tobacco smoke is approximately 5 ppm. Slight increases in respiratory illness and decrease in pulmonary function have been associated with concentrations of about 0.10 ppm (Ferris, 1978). You should note that all of these concentrations are very high with respect to the NAAQS value in Table 12–1.

Photochemical Oxidants. Although the photochemical oxidants include peroxyacetyl nitrate, acrolein, peroxybenzoyl nitrates, aldehydes, and nitrogen oxides, the major oxidant is ozone (O_3). Ozone is commonly used as an indicator of the total amount of oxidant present. Oxidant concentrations above 0.1 ppm result in eye irritation. At a concentration of 0.3 ppm, cough and chest discomfort are increased. Those people who suffer from chronic respiratory disease are particularly susceptible.

PM$_{2.5}$.   As noted earlier, large particles are not inhaled deeply into the lungs. This is why EPA switched from an air quality standard based on TSP to one based on particles with an aerodynamic diameter less than 10 µm (PM$_{2.5}$). Studies in the United States, Brazil, and Germany have related higher levels of particulates to increased risk of respiratory, cardiovascular, and cancer-related deaths, as well as pneumonia, lung function loss, hospital admissions, and asthma (Reichhardt, 1995).

Particles 2.5 µm in aerodynamic diameter have been identified as a major contributor to elevated death rates in polluted cities (Pope et al., 1995). One hypothesized biological mechanism is pollution-induced lung damage resulting in declines in lung function, respiratory distress, and cardiovascular disease potentially related to hypoxemia (Pope et al., 1999).

Sulfur Oxides (SO$_x$) and Total Suspended Particulates (TSP).   The sulfur oxides include sulfur dioxide (SO$_2$), sulfur trioxide (SO$_3$), their acids, and the salts of their acids. Rather than try to separate the effects of SO$_2$ and SO$_3$, they are usually treated together. There is speculation that a definite synergism exists whereby fine particulates carry absorbed SO$_2$ to the LRT. The SO$_2$ in the absence of particulates would be absorbed in the mucous membranes of the URT.

Patients suffering from chronic bronchitis have shown an increase in respiratory symptoms when the TSP levels exceeded 350 µg $\cdot$ m^{-3} and the SO$_2$ level was above 0.095 ppm. Studies made in Holland at an interval of 3 years showed that pulmonary function improved as SO$_2$ and TSP levels dropped from 0.10 ppm and 230 µg $\cdot$ m^{-3} to 0.03 ppm and 80 µg $\cdot$ m^{-3} respectively.

Air Pollution Episodes.   The word *episode* is used euphemistically to describe air pollution disasters. Indeed, it was the shock of these disasters that stimulated the first modern legislative action to require control of air pollutants. The characteristics of the three major episodes are summarized in Table 12–5. Careful study of the table will reveal that all of the episodes had

TABLE 12–5   **Three Major Air Pollution Episodes**

	Meuse Valley, Belgium, 1930 (Dec. 1)	Donora, Pennsylvania, 1948 (Oct. 26–31)	London, 1952 (Dec. 5–9)
Population	No data	12,300	8,000,000
Weather	Anticyclone, inversion, and fog	Anticyclone, inversion, and fog	Anticyclone, inversion, and fog
Topography	River valley	River valley	River plain
Most probable source of pollutants	Industry (including steel and zinc plants)	Industry (including steel and zinc plants)	Household coal-burning
Nature of the illnesses	Chemical irritation of exposed membranous surfaces	Chemical irritation of exposed membranous surfaces	Chemical irritation of exposed membranous surfaces
Number of deaths	63	17	4000
Time of deaths	Began after second day of episode	Began after second day of episode	Began on first day of episode
Suspected proximate cause of irritation	Sulfur oxides with particulates	Sulfur oxides with particulates	Sulfur oxides with particulates

Source: World Health Organization, *Air Pollution*, 1961, p. 180.

some things in common. Comparison of these situations and others where no episode occurred (i.e., where the number of dead and ill was considerably less) has revealed that four ingredients are essential for an episode. If one ingredient is omitted, fewer people will get sick and only a few people can be expected to die. The crucial ingredients are: (1) a large number of pollution sources, (2) a restricted air volume, (3) failure of officials to recognize that anything is wrong, and (4) the presence of water droplets of the "right" size (Goldsmith, 1968).

Although a sufficient quantity of any pollutant is lethal by itself, it is generally agreed that some mix is required to achieve the results seen in these episodes. Atmospheric levels of individual pollutants seldom rise to lethal levels without an explosion or transportation accident. However, the proper combination of two or more pollutants will yield untoward symptoms at much lower levels. The sulfur oxides and particulates were the most suspect in the three major episodes.

The meteorology must be such that little air movement occurs, thus preventing the dilution of the pollutants. Although a valley is most conducive to a stagnation effect, the London episode proved that it isn't necessary. The stagnant conditions must persist for several days. Three days appears to be the minimum.

Tragically, each of these hazardous air pollution conditions became lethal because of the failure of city officials to notice anything strange. If they have no measurements of pollution levels or reports from hospitals and morgues, city authorities have no reason to alert the public, shut down factories, or restrict traffic.

The last and, perhaps, most crucial element is fog.* The fog droplets must be of the "right" size, namely, in the 1–2-μm diameter range or, perhaps, in the range below 0.5 μm. As mentioned earlier, these particle sizes are most likely to penetrate into the LRT. Pollutants that dissolve into the fog droplet are thus carried deep into the lung and deposited there.

12–5   ORIGIN AND FATE OF AIR POLLUTANTS

Carbon Monoxide

Incomplete oxidation of carbon results in the production of carbon monoxide. The natural anaerobic decomposition of carbonaceous material by soil microorganisms releases approximately 160 Tg[†] of methane (CH_4) to the atmosphere each year worldwide (IPCC, 1995). The natural formation of CO results from an intermediate step in the oxidation of the methane. The hydroxyl radical (OH·) serves as the initial oxidizing agent. It combines with CH_4 to form an alkyl radical (Wofsy, McConnell, and McElroy, 1972):

$$CH_4 + OH \rightleftharpoons CH_3· + H_2O \tag{12–1}$$

This reaction is followed by a complex series of 39 reactions, which we have oversimplified to the following:

$$CH_3· + O_2 + 2(hv) \rightleftharpoons CO + H_2 + OH· \tag{12–2}$$

This says that $CH_3·$ and O_2 are each zapped by a photon of light energy (hv). The symbol v stands for the frequency of the light. The h is Planck's constant = 6.626×10^{-34} J · Hz^{-1}.

Anthropogenic sources (those associated with the activities of human beings) include motor vehicles, fossil fuel burning for electricity and heat, industrial processes, solid waste disposal, and miscellaneous burning of such things as leaves and brush. Approximately

---

*The word *smog* is a term coined by Londoners before World War I to describe the combination of smoke and fog that accounted for much of their weather. Los Angeles smog is a misnomer because little smoke and no fog is present. In fact, as we shall see later, Los Angeles smog cannot occur without a lot of sunshine. "Photochemical smog" is the correct term to describe the Los Angeles haze.

†One teragram = 1×10^{12} g

600–1250 Tg of CO is released by incomplete combustion of organic carbon in these sources. Motor vehicles account for more than 60% of the emission.

No significant change in the global atmospheric CO level has been observed over the past 20 years. Yet the worldwide anthropogenic contribution of combustion sources has doubled over the same period. Because no apparent change has occurred in the atmospheric concentration, a number of mechanisms (**sinks**) have been proposed to account for the missing CO. The major sink is reaction with hydroxyl radicals to form carbon dioxide. Removal by soil microorganisms and diffusion into the stratosphere are minor routes of removal. It has been estimated that these sinks annually consume an amount of CO that equals or exceeds its production (Seinfeld and Pandis, 1998).

Hazardous Air Pollutants

The EPA has identified 166 categories of major sources and eight categories of area sources for the HAPs listed in Table 12–2 (57 FR 31576, 1992). The source categories represent a wide range of industrial groups: fuel combustion, metal processing, petroleum and natural gas production and refining, surface-coating processes, waste treatment and disposal processes, agricultural chemicals production, and polymers and resins production. There are also a number of miscellaneous source categories, such as dry cleaning and electroplating.

In addition to these direct emissions, air toxics can result from chemical formation reactions in the atmosphere. These reactions involve chemicals emitted to the atmosphere that are not listed HAPs and may not be toxic themselves, but can undergo atmospheric transformations to generate HAPs. For organic compounds present in the gas phase, the most important transformation processes involve photolysis and chemical reactions with ozone, hydroxyl radicals (OH·), and nitrate radicals (Kao, 1994). **Photolysis** is the chemical fragmentation or rearrangement of a chemical upon the adsorption of radiation of the appropriate wavelength. Photolysis is only important during the daytime for those chemicals that absorb strongly within the solar radiation spectrum. Otherwise, reaction with OH· or O_3 is likely to predominate. The HAPs most often formed are formaldehyde and acetaldehyde.

The major removal mechanisms appear to be OH abstraction or addition. The reaction products lead to the formation of CO and CO_2. Eighty-nine of the 189 HAPs have atmospheric lifetimes of less than a day.

Lead

Volcanic activity and airborne soil are the primary natural sources of atmospheric lead. Smelters and refining processes, as well as incineration of lead-containing wastes, are major point sources of lead. Approximately 70–80% of the lead that used to be added to gasoline was discharged to the atmosphere.

Submicrometer-sized lead particles, which are formed by volatilization and subsequent condensation, attach to larger particles or they form nuclei before they are removed from the atmosphere. Once they have attained a size of several microns, they either settle out or are washed out by rain.

Nitrogen Dioxide

Bacterial action in the soil releases nitrous oxide (N_2O) to the atmosphere. In the upper troposphere and stratosphere, atomic oxygen reacts with the nitrous oxide to form nitric oxide (NO).

$$N_2O + O \rightleftharpoons 2NO \tag{12–3}$$

The atomic oxygen results from the dissociation of ozone. The nitric oxide further reacts with ozone to form nitrogen dioxide (NO_2).

$$NO + O_3 \rightleftharpoons NO_2 + O_2 \tag{12–4}$$

The global formation of NO_2 by this process is estimated to be 12 Tg (as N) annually (Seinfeld and Pandis, 1998).

Combustion processes account for 74% of the anthropogenic sources of nitrogen oxides. Although nitrogen and oxygen coexist in our atmosphere without reaction, their relationship is much less indifferent at high temperatures and pressures. At temperatures in excess of 1600 K, they react.

$$N_2 + O_2 \overset{\Delta}{\rightleftharpoons} 2NO \qquad \qquad (12\text{--}5)$$

If the combustion gas is rapidly cooled after the reaction by exhausting it to the atmosphere, the reaction is quenched and NO is the by-product. The NO in turn reacts with ozone or oxygen to form NO_2. The anthropogenic contribution to global emission of NO_x amounted to 32 Tg $\cdot$ y^{-1} (as N) in 1995 (IPCC, 1995). Between 40 and 45% of the NO_x emissions in the United States come from transportation, 30 to 35% from power plants, and 20% from industrial sources (Seinfeld and Pandis, 1998).

The U.S. EPA emission factors provide an example of a method for estimating emissions from coal-fired electric utility boilers. For pulverized coal, dry bottom, wall-fired boilers using bituminous and subbituminous coal:

- Pre-NSPS standards—11 kg $\cdot$ Mg^{-1} (22 lb$_m$ of NO_x/U.S. short ton of bituminous coal)
- Pre-NSPS standards—6 kg $\cdot$ Mg^{-1} (12 lb$_m$ of NO_x/short ton of subbituminous coal)
- After NSPS standards—6 kg $\cdot$ Mg^{-1} (12 lb$_m$ of NO_x/short ton of bituminous coal)
- After NSPS standards—3.7 kg $\cdot$ Mg^{-1} (7.4 lb$_m$ of NO_x/short ton of subbituminous coal)

where a "short ton" is defined as 2000 lb$_m$.

Ultimately, the NO_2 is converted to either NO_2^- or NO^- in particulate form. The particulates are then washed out by precipitation. The dissolution of nitrate in a water droplet allows for the formation of nitric acid (HNO_3). This, in part, accounts for "acid" rain found downwind of industrialized areas.

Photochemical Oxidants

Unlike the other pollutants, the photochemical oxidants result entirely from atmospheric reactions and are not direct emissions from either people or nature; thus, they are called **secondary pollutants.** They are formed through a series of reactions that are initiated by the absorption of a photon by an atom, molecule, free radical, or ion. Ozone is the principal photochemical oxidant. Its formation is usually attributed to the nitrogen dioxide photolytic cycle. Hydrocarbons modify this cycle by reacting with atomic oxygen to form free radicals (highly reactive organic species). The hydrocarbons, nitrogen oxides, and ozone react and interact to produce more nitrogen dioxide and ozone. This cycle is represented in summary form in Figure 12–5. The whole reaction sequence depends on an abundance of sunshine. A result of these reactions is the photochemical "smog" for which Los Angeles is famous.

Sulfur Oxides

Sulfur oxides may be both primary and secondary pollutants. Power plants, industry, volcanoes, and the oceans emit SO_2, SO_3, and SO_4^{2-} directly as primary pollutants. In addition, biological decay processes and some industrial sources emit H_2S, which is oxidized to form the secondary pollutant SO_2. In terms of sulfur, approximately 10 Tg is emitted annually by natural sources. Approximately 75 Tg of sulfur may be attributed to anthropogenic sources each year (Seinfeld and Pandis, 1998).

The most important oxidizing reaction for H_2S appears to be one involving ozone:

$$H_2S + O_3 \rightleftharpoons H_2O + SO_2 \qquad \qquad (12\text{--}6)$$

FIGURE 12–5

Interaction of hydrocarbons with atmospheric nitrogen oxide photolytic cycle. (*Source:* U.S. Dept. of Health, Education and Welfare, Air Quality Criteria for Photochemical Oxidants, National Air Pollution Control Administration Publication No. AP-63; U.S. Government Printing Office, Washington, DC, 1970).

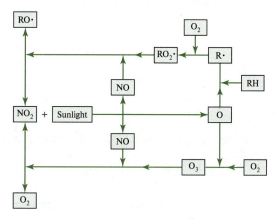

The combustion of fossil fuels containing sulfur yields sulfur dioxide in direct proportion to the sulfur content of the fuel.

$$S + O_2 \rightleftharpoons SO_2 \tag{12–7}$$

This reaction implies that for every gram of sulfur in the fuel, two grams of SO_2 is emitted to the atmosphere. Because the combustion process is not 100% efficient, we generally assume that 5% of the sulfur in the fuel ends up in the ash, that is, 1.90 g SO_2 per gram of sulfur in the fuel is emitted.

EXAMPLE 12–1   An Illinois coal is burned at a rate of $1.00 \text{ kg} \cdot \text{s}^{-1}$. If the analysis of the coal reveals a sulfur content of 3.00%, what is the annual rate of emission of SO_2?

Solution   Using the mass-balance approach, we begin by drawing the following mass-balance diagram.

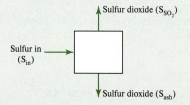

The steady-state mass balance equation may be written as

$$S_{in} = S_{ash} + S_{so_2}$$

From the problem data, the mass of "sulfur in" is

$$S_{in} = 1.00 \text{ kg} \cdot \text{s}^{-1} \times 0.030 = 0.030 \text{ kg} \cdot \text{s}^{-1}$$

In one year,

$$S_{in} = 0.030 \text{ kg} \cdot \text{s}^{-1} \times 86{,}400 \text{ s} \cdot \text{day}^{-1} \times 365 \text{ day} \cdot \text{year}^{-1} = 9.46 \times 10^5 \text{ kg} \cdot \text{year}^{-1}$$

The sulfur in the ash is 5% of the input sulfur.

$S_{ash} = (0.05)(9.46 \times 10^5 \text{ kg} \cdot \text{year}^{-1}) = 4.73 \times 10^4 \text{ kg} \cdot \text{year}^{-1}$

The amount of sulfur available for conversion to SO_2 is

$S_{so_2} = S_{in} - S_{ash} = 9.46 \times 10^5 - 4.73 \times 10^4 = 8.99 \times 10^5 \text{ kg} \cdot \text{year}^{-1}$

The amount of sulfur formed is determined from the proportional weights of the oxidation reaction (Equation 12–7).

$S + O_2 \rightleftharpoons SO_2$

$MW = 32 + 32 = 64$

The amount of sulfur dioxide formed is then 64/32 of the sulfur available for conversion.

$S_{so_2} = \frac{64}{32} \times 8.99 \times 10^5 \text{ kg} \cdot \text{year}^{-1} = 1.80 \times 10^6 \text{ kg} \cdot \text{year}^{-1}$

EPA uses emission factors for estimating emissions from coal-fired electric utility boilers. For pulverized coal, dry bottom, wall-fired boilers using bituminous and subbituminous coal:

- Pre-NSPS standards emission factor for SO_2 from bituminous coal = 38S
- Pre-NSPS standards emission factor for SO_2 from subbituminous coal = 35S
- After NSPS standards emission factor for SO_2 from bituminous coal = 38S
- After NSPS standards emission factor for SO_2 from subbituminous coal = 35S

The quantity "S" is the weight percent sulfur content of the coal. For example, if the fuel is 1.2% sulfur, then S = 1.2 and the emission factor for bituminous coal is $(38)(1.2) = 45.6 \text{ lb}_m/\text{ton}$ where a "ton" is defined as 2000 lb_m.

The ultimate fate of most of the SO_2 in the atmosphere is conversion to sulfate salts, which are removed by sedimentation or by washout with precipitation. The conversion to sulfate is by either of two routes: catalytic oxidation or photochemical oxidation. The first process is most effective if water droplets containing Fe^{3+}, Mn^{2+}, or NH_3 are present.

$$2SO_2 + 2H_2O + O_2 \rightleftharpoons 2H_2SO_4 \qquad \text{(12–8)}$$

At low relative humidities, the primary conversion process is photochemical oxidation. The first step is photoexcitation* of the SO_2.

$$SO_2 + hv \rightleftharpoons \overset{*}{S}O_2 \qquad \text{(12–9)}$$

The excited molecule then readily reacts with O_2 to form SO_3.

$$\overset{*}{S}O_2 + O_2 \rightleftharpoons SO_3 + O \qquad \text{(12–10)}$$

The trioxide is very hygroscopic (readily retains moisture) and consequently is rapidly converted to sulfuric acid:

$$SO_3 + H_2O \rightleftharpoons H_2SO_4 \qquad \text{(12–11)}$$

---

*Photoexcitation is the displacement of an electron from one shell to another, thereby storing energy in the molecule. Photoexcitation is represented in reactions by a star (*).

This reaction in large part accounts for acid rain (i.e., precipitation with a pH value less than 5.6) found in industrialized areas. Normal precipitation has a pH of 5.6 due to the carbonate buffer system.

Particulates

Sea salt, soil dust, volcanic particles, and smoke from forest fires account for about 2.9 petagrams of particulate emissions each year. Anthropogenic emissions from fossil fuel burning and industrial processes account for emissions of 110 Tg per year (Kiehl and Rodhe, 1995). For example the emission factor for a pulverized coal-fired, dry bottom, wall-fired boiler is given as 10A where "A" is the percent ash in the coal. For example, if the fuel is 8% ash, then A = 8 and the emission factor for bituminous or subbituminous coal is (10)(8) = 80 lb_m/ton where a "ton" is defined as 2000 lb_m.

Secondary sources of particulates include the conversion of H_2S, SO_2, NO_x, NH_3, and hydrocarbons. H_2S and SO_2 are converted to sulfates. NO_x, and NH_3 are converted to nitrates. The hydrocarbons react to form products that condense to form particles at atmospheric temperatures. Natural sources of secondary pollutants yield about 240 Tg annually. Anthropogenic sources yield about 340 Tg annually (Kiehl and Rodhe, 1995).

Dust particles that are **entrained** (picked up) by the wind and carried over long distances tend to sort themselves out to the sizes between 0.5 and 50 μm in diameter. Sea salt nuclei have sizes between 0.05 and 0.5 μm. Particles formed as a result of photochemical reactions tend to have very small diameters (< 0.4 μm). Smoke and fly ash particles cover a wide range of sizes from 0.05 to 200 μm or more. Particle mass distributions in urban atmospheres generally exhibit two maxima: one is between 0.1 and 1 μm in diameter; the other is between 1 and 30 μm. The smaller fraction is the result of condensation. The coarse fraction consists of fly ash and dust generated by mechanical abrasion.

Small particles are removed from the atmosphere by accretion to water droplets, which grow in size until they are large enough to precipitate. Larger particles are removed by direct washout by falling raindrops.

EXAMPLE 12–2 Determine whether or not a pulverized coal, dry bottom, wall-fired boilers using bituminous coal at a power plant rate at 61 MW meets the NSPS for SO_2, particulate matter, and NO_x. The power plant burns bituminous coal with a sulfur content of 1.8% and ash content of 6.2%. The coal has a heating value of 14,000 Btu/lb. The boiler efficiency is 35%. Use the EPA emission factors to estimate the emissions. Assume the efficiency of SO_2 control is 85% and the efficiency of particulate control equipment is 99%.

Solution Begin by calculating the coal firing rate for 61 MW at a boiler efficiency of 35%.

$$\frac{61 \text{ MW}}{0.35} = 174.3 \text{ MW or } 174.3 \times 10^6 \text{ W}$$

Using a one-hour time increment, convert W · h to Btu with the conversion factor of 3.4144 Btu/W · h

$$(174.3 \times 10^6 \text{ W})(1 \text{ h})(3.4144 \text{ Btu/W} \cdot \text{h}) = 5.95 \times 10^8 \text{ Btu}$$

The mass of coal burned in an hour is

$$\left(\frac{5.95 \times 10^8 \text{ Btu}}{14,000 \text{ Btu/lb}_m}\right)\left(\frac{1 \text{ ton}}{2000 \text{ lb}_m}\right) = 21.25 \text{ tons}$$

1. Check the SO_2 emission rate.

   Using the EPA emission factor of 38S for bituminous coal:

   Uncontrolled SO_2 emission rate = (38)(1.8) = 68.4 lb_m/ton of coal

The estimated SO_2 emission rate with 85% control is:

(68.4 lb_m/ton of coal)(21.25 tons/h)(1 − 0.85) = 218.03 lb_m

The SO_2 emission rate per million Btu is

$$\frac{218.03 \text{ lb}_m}{5.95 \times 10^8 \text{ Btu}} = 3.66 \times 10^{-7} \text{ lb}_m/\text{Btu}$$

Or on a million Btu basis

$(3.66 \times 10^{-7} \text{ lb}_m/\text{Btu})(10^6) = 0.37 \text{ lb}_m/\text{million Btu}$

This meets the standard of 1.2 lb_m/million Btu but does not meet the 90% reduction requirement.

2. Check the particulate emission rate.

   Using the EPA emission factor of "10A" for a pulverized coal, dry bottom, wall-fired boiler:

   Uncontrolled particulate emission = (10)(6.2) = 62.0 lb_m/ton of coal burned

   The estimated particulate emission with 99% control is

   (62.0 lb_m/ton of coal burned)(21.25 tons)(1 − 0.99) = 13.2 lb_m

   The particulate emission rate per million Btu is

   $$\frac{13.2 \text{ lb}_m}{5.95 \times 10^8 \text{ Btu}} = 2.23 \times 10^{-8} \text{ lb}_m/\text{Btu}$$

   On a million Btu basis

   $(2.23 \times 10^{-8} \text{ lb}_m/\text{Btu})(10^6) = 0.022 \text{ lb}_m/10^6 \text{ Btu}$

   This meets the standard of 0.03 $lb_m/10^6$ Btu.

3. Check the NO_x emission rate

   Using the EPA emission factor of 22 lb_m/ton, the estimated emission is

   (22 lb_m/ton)(21.25 tons) = 467.5 lb_m

   The NO_x emission rate per million Btu is

   $$\frac{467.5 \text{ lb}_m}{5.95 \times 10^8 \text{ Btu}} = 7.86 \times 10^{-7} \text{ lb}_m/\text{Btu}$$

   On a million Btu basis

   $(7.86 \times 10^{-7} \text{ lb}_m/\text{Btu})(10^6) = 0.79 \text{ lb}_m/10^6 \text{ Btu}$

   The standard for bituminous coal is 0.60 $lb_m/10^6$ Btu. The power plant does not meet the NO_x standard.

Comments:

1. The substitution of subbituminous or lignite coal for the bituminous coal is one alternative method to achieve the standard. In general, they have a lower sulfur content and a similar or lower ash content. A coal analysis is required to verify this general assumption.

2. Modification of the burner will be required to meet the NO_x standard.

12–6   MICRO AND MACRO AIR POLLUTION

Air pollution problems may occur on three scales: micro, meso, and macro. Microscale problems range from those covering less than a centimeter to those the size of a house or slightly larger. Mesoscale air pollution problems are those of a few hectares up to the size of a city or county. Macroscale problems extend from counties to states, nations, and in the broadest sense, the globe. Much of the remaining discussion in this chapter is focused on the mesoscale problem. In this section we will address the general microscale and macroscale problems recognized today.

Indoor Air Pollution

People who live in cold climates may spend more than 90% of their time indoors. In the last three decades, researchers have identified sources, concentrations, and effects of air pollutants that arise in conventional domestic residences. The startling results indicate that, in certain instances, indoor air may be substantially more polluted than the outdoor air.

Carbon monoxide from improperly operating furnaces has long been a serious concern. In numerous instances, people have died from furnace malfunction. More recently, chronic low levels of CO pollution have been recognized. Gas ranges, ovens, pilot lights, gas and kerosene space heaters, and cigarette smoke all contribute CO (Table 12–6). Although little or no effort has been exerted to reduce or eliminate the danger from ranges, ovens, and so forth, the public has come to expect that the recreational habits of smokers should not interfere with the quality of the air others breathe. The results of a general ban on cigarette smoking in one office are shown in Table 12–7. Smokers were allowed to smoke only in the designated lounge area. Period 1 was prior to the implementation of the new policy. It is obvious that the new policy

---

TABLE 12–6   **Tested Combustion Sources and Their Emission Rates**

Source	NO	NO₂	NOₓ (as NO₂)	CO	SO₂
			Range of Emission Rates[a] (mg · MJ⁻¹)		

Source	NO	NO$_2$	NO$_x$ (as NO$_2$)	CO	SO$_2$
Range-top burner[b]	15–17	9–12	32–37	40–244	c
Range oven[d]	14–29	7–13	34–53	12–19	c
Pilot light[e]	4–17	8–12	f	40–67	c
Gas space heaters[g]	0–15	1–15	1–37	14–64	c
Gas dryer[h]	8	8	20	69	c
Kerosene space heater[i]	1–13	3–10	5–31	35–64	11–12
Cigarette smoke[j]	2.78	0.73	f	88.43	c

[a]The lowest and highest mean values of emission rates for combustion sources tested in milligrams per megajoule (mg · MJ⁻¹).

[b]Three ranges were evaluated. Reported values are for blue flame conditions.

[c]Combustion source is not emitting pollutant.

[d]Three ranges were evaluated. Ovens were operated for several different settings (bake, broil, self-clean cycle, etc.).

[e]One range was evaluated with all three pilot lights, two top pilot lights and a bottom pilot.

[f]Emission rates not reported.

[g]Three space heaters including one conductive, radiant, and catalytic were tested.

[h]One gas dryer was evaluated.

[i]Two kerosene heaters, including a convective and radiant type, were tested.

[j]One type of cigarette. Reported emission rates are in milligrams per cigarette (800 mg tobacco per cigarette).

Source: Moschandreas, Zabpansky, and Pelta, 1985.

TABLE 12–7	Mean Respirable Particulates (RSP) Carbon Monoxide Levels Measured on the Test Floor	
	RSP ($\mu g \cdot m^{-3}$)	CO ($\mu g \cdot m^{-3}$)
Period 1	26	1908
Period 2	18	1245

Source: Lee et al., 1985.

had a positive effect outside of the lounge. On the other hand, respirable particulate matter was found to increase with one smoker and to rise 56 to 63% with two.

Indoor tobacco smoking is of particular concern because of the carcinogenic properties of the smoke. While *mainstream smoking* (taking a puff) exposes the smoker to large quantities of toxic compounds, the smoldering cigarette in the ashtray (*sidestream smoke*) adds a considerable burden to the room environment. Table 12–8 illustrates the emission rates of mainstream and sidestream smoke.

TABLE 12–8	Emission of Chemicals from Mainstream and Sidestream Smoke	
Chemicals	Mainstream ($\mu g \cdot cigarette^{-1}$)	Sidestream ($\mu g \cdot cigarette^{-1}$)
Particulates		
Aniline	0.36	16.8
Benzo (a) pyrene	20–40	68–136
Methyl naphthalene	2.2	60
Naphthalene	2.8	4.0
Nicotine	100–2500	2700–6750
Nitrosonornicotine	0.1–0.55	0.5–2.5
Pyrene	50–200	180–420
Total phenols	228	603
Total suspended particles	36,200	25,800
Gas and vapor phase		
Acetaldehyde	18–1400	40–3100
Acetone	100–600	250–1500
Acrolein	25–140	55–130
Ammonia	10–150	980–150,000
Carbon dioxide	20,000–60,000	160,000–480,000
Carbon monoxide	1000–20,000	25,000–50,000
Dimethylnitrosamine	10–65	520–3300
Formaldehyde	20–90	1300
Hydrogen cyanide	430	110
Methyl chloride	650	1300
Nitric oxide	10–570	2300
Nitrogen dioxide	0.5–30	625
Nitrosopyrolidine	10–35	270–945
Pyridine	9–93	90–930

Sources: HEW, 1979; Hoegg, 1972; and Wakeham, 1972.

Nitrogen oxide sources are also shown in Table 12–6. NO_2 levels have been found to range from 70 $\mu g \cdot m^{-3}$ in air-conditioned houses with electric ranges to 182 $\mu g \cdot m^{-3}$ in non-air-conditioned houses with gas stoves (Hosein et al., 1985). The latter value is quite high in comparison with the national ambient air quality limits. SO_2 levels were found to be very low in all houses investigated.

Bacteria, viruses, fungi, mites, and pollen are collectively referred to as **bioaerosols.** They require a reservoir (for storage), an amplifier (for reproduction), and a means of dispersal. Most bacteria and viruses in indoor air come from humans and pets. Other microorganisms and pollen are introduced from the ambient air through either natural ventilation or through the intakes of building air-handling systems. Humidifiers, air-conditioning systems, and other places where water accumulates are potential reservoirs for bioaerosols.

Radon is not regulated as an ambient air pollutant but has been found in dwellings at alarmingly high concentrations. We will address the radon issue in depth in Chapter 16. Radon is a radioactive gas that emanates from natural geologic formations and, in some cases, from construction materials. It is not generated from the activities of the householder as is true for the pollutants discussed earlier.

Over 800 volatile organic compounds (VOCs) have been identified in indoor air (Hines et al., 1993). Aldehydes, alkanes, alkenes, ethers, ketones, and polynuclear aromatic hydrocarbons are among them. Although they are not all present all the time, frequently there are several present at the same time. Typical sources of these compounds are listed in Table 12–9.

Between 1979 and 1987, the EPA investigated personal exposures of the general public to VOCs. These studies, titled the Total Exposure Assessment Methodology, revealed that personal exposures exceeded median outdoor air concentrations by a factor of two to five for nearly all

TABLE 12–9	Common Volatile Organic Compounds and Their Sources
Volatile Organic Compounds	**Major Indoor Sources of Exposure**
Acetaldehyde	Paint (water-based), sidestream smoke
Alcohols (ethanol, isopropanol)	Spirits, cleansers
Aromatic hydrocarbons (ethylbenzene toluene, xylenes, trimethylbenzenes)	Paints, adhesives, gasoline, combustion sources
Aliphatic hydrocarbons (octane, decane, undecane)	Paints, adhesives, gasoline, combustion sources
Benzene	Sidestream smoke
Butylated hydroxytoluene (BHT)	Urethane-based carpet cushions
Chloroform	Showering, washing clothes, washing dishes
p-Dichlorobenzene	Room deodorizers, moth cakes
Ethylene glycol	Paints
Formaldehyde	Sidestream smoke, pressed wood products, photocopier
Methylene chloride	Paint stripping, solvent use
Phenol	Vinyl flooring
Styrene	Smoking, photocopier
Terpenes (limonene, -pinene)	Scented deodorizers, polishes, fabric softeners
Tetrachloroethylene	Wearing/storing dry-cleaned clothes
Tetrahydrofuran	Sealer for vinyl flooring
Toluene	Photocopier, sidestream smoke, synthetic carper fiber
1,1,1-Trichloroethane	Aerosol sprays, solvents

Sources: Tucker, 2001, and Wallace, 2001.

of the 19 VOCs investigated. Traditional sources (automobiles, industry, petrochemical plants) contributed only 20 to 25% of the total exposure to most of the target VOCs (Wallace, 2001).

Formaldehyde (CH_2O) has been singled out as one of the more prevalent as well as one of the more toxic, compounds (Hines et al., 1993). Formaldehyde may not be generated directly by the activity of the homeowner. It is emitted by a variety of consumer products and construction materials including pressed wood products, insulation materials (urea-formaldehyde foam insulation in trailers has been particularly suspect), textiles, and combustion sources. In one study, CH_2O concentrations ranged from 0.0455 ppm to 0.19 ppm (Dumont, 1985). Some mobile homes in Wisconsin had concentrations as high as 0.65 ppm. (For comparison, the American Society of Heating, Refrigeration and Air Conditioning Engineers (ASHRAE, 1981) set a guideline concentration of 0.1 ppm.)

Unlike the other air pollution sources that continue to emit as long as there is anthropogenic activity (or in the case of radon, for geologic time), CH_2O is not regenerated unless new materials are brought into the residence. If the house is ventilated over time, the concentration will drop.

The primary source of heavy metals indoors is from infiltration of outdoor air and soil and dust that is tracked into the building. Arsenic, cadmium, chromium, mercury, lead, and nickel have been measured in indoor air. Lead and mercury may be generated from indoor sources such as paint. Old lead paint is a source of particulate lead as it is abraded or during removal. Mercury vapor is emitted from latex-based paints that contain diphenyl mercury dodecenyl succinate to prevent fungus growth.

It is doubtful any regulatory effort will be made to reduce the emissions of these pollutants in the near future. Thus, the house or apartment dweller has little recourse other than to replace gas appliances, remove or cover formaldehyde sources, and put out the smokers.

Acid Rain

Unpolluted rain is naturally acidic because CO_2 from the atmosphere dissolves to a sufficient extent to form carbonic acid (see Chapter 2). The equilibrium pH for pure rainwater is about 5.6. Measurements taken over North America and Europe have revealed lower pH values. In some cases individual readings as low as 3.0 have been recorded. The average pH in rain over the United States in 1994 and 2005 is shown in Figure 9–20.

Chemical reactions in the atmosphere convert SO_2, NO_x, and VOCs to acidic compounds and associated oxidants (Figure 12–6). The primary conversion of SO_2 in the eastern United States is through the aqueous phase reaction with hydrogen peroxide (H_2O_2) in clouds. Nitric acid is formed by the reaction of NO_2 with OH radicals formed photochemically. Ozone is formed and then protected by a series of reactions involving both NO_x and VOCs.

FIGURE 12–6

Acid rain precursors and products.

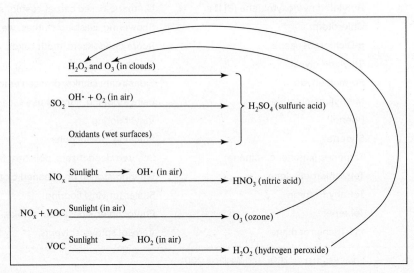

As discussed in Chapter 9, the concern about acid rain relates to potential effects of acidity on aquatic life, damage to crops and forests, and damage to building materials. Lower pH values may affect fish directly by interfering with reproductive cycles or by releasing otherwise insoluble aluminum, which is toxic. Dramatic die back of trees in central Europe has stimulated concern that similar results could occur in North America. It is hypothesized that the acid rain leaches calcium and magnesium from the soil (see Figure 10–13). This lowers the molar ratio of calcium to aluminum, which, in turn, favors the uptake of aluminum by the fine roots, ultimately leading to their deterioration.

In 1980, Congress authorized a 10-year study to assess the causes and effects of acidic deposition. This study was titled the National Acid Precipitation Assessment Program (NAPAP). In September 1987, the NAPAP released an interim report that indicated that acidic precipitation appeared to have no measurable and consistent effect on crops, tree seedlings, or human health, and that a small percentage of lakes across the United States were experiencing pH values lower than 5.0 (Lefohn and Krupa, 1988). On the other hand, oxidant damage was measurable.

Approximately 70% of the SO_2 emissions in the United States are attributable to electric utilities. To decrease the SO_2 emissions, the Congress developed a two-phase control program under Title IV of the 1990 Clean Air Act Amendments. Phase I sets emission allowances for 110 of the largest emitters in the eastern half of the United States. Phase II will include smaller utilities. The utilities may buy or sell allowances. Each allowance is equal to about 1 Mg of SO_2 emissions. If a company does not expend its maximum allowance, it may sell it to another company. As a result of this **market-based system,** utility emissions have decreased by 9 Tg.

In 2003, the EPA reported on the long-term response of surface water chemistry to the 1990 Clean Air Act Amendments (U.S. EPA, 2003a). Eighty-one selected sites in the Northeast and upper Midwest have been monitored for acidity since the early 1980s. The EPA's estimate of changes in the number and proportion of acidic surface waters is summarized in Table 12–10. In two areas, the New England lakes and the Blue Ridge Province streams, there is little evidence of reduction in acidity over the last decade. Sulfate levels have decreased significantly while nitrate levels have not changed appreciably. The widespread decrease in sulfate concentration parallels the general decrease in national emissions of sulfur dioxide since 1980. The EPA concluded from its analysis that surface waters have responded relatively rapidly to the decline in sulfate deposition and that additional reductions in deposition will result in additional declines in sulfate concentration.

In its report, the EPA also noted that, in many cases, sites that are not chronically acidic do undergo short-term episodic acidification during spring snowmelt or during intense rain events. (For updates see: https//fivethirtyeight.com/features/acid-rains-dirty-legacy.)

| TABLE 12–10 | Estimates of Change in Number and Proportion of Acidic Surface Waters in Acid-Sensitive Regions of the North and East | | | |

Region	Population Size	Number Acidic in Past Surveys[a]	Estimated Number Currently Acidic	% Change
New England	6834 lakes	386	374	–3
Adirondacks	1830 lakes	238	149	–38
No. Appalachians	42,426 km	5014	3600	–28
Blue Ridge	32,687 km	1634	1634	0
Upper Midwest	8574 lakes	800	251	–68

[a]Survey dates range from 1984 in upper Midwest to 1993–94 in the northern Appalachians.
Source: Adapted from U.S. EPA, 2003a.

Ozone Depletion

Without ozone, every living thing on the Earth's surface would be incinerated. (On the other hand, as we have already noted, ozone can be lethal.) The presence of ozone in the upper atmosphere (20–40 km and up) provides a barrier to ultraviolet (UV) radiation. The small amounts of UV rays that do seep through provide you with your summer tan. Too much UV, however, will cause skin cancer. Although oxygen also serves as a barrier to UV radiation, it absorbs only over a narrow band centered at a wavelength of 0.2 µm. The photochemistry of these reactions is shown in Figure 12–7. The M refers to any third chemical (usually N_2).

In 1974, Molina and Rowland revealed a potential air pollution threat to this protective ozone shield (Molina and Rowland, 1974). It is noteworthy that they, along with Paul Crutzen, jointly received the Nobel Prize in chemistry for their research. They hypothesized that chlorofluorocarbons (CF_2Cl_2 and $CFCl_3$, often abbreviated as CFC), which are used as aerosol propellants and refrigerants, react with ozone (Figure 12–8). The frightening aspects of this series of reactions are that the chlorine atom removes ozone from the system, and that the chlorine atom is continually recycled to convert more ozone to oxygen. It has been estimated that a 5% reduction in ozone could result in nearly a 10% increase in skin cancer (ICAS, 1975). Thus, CFCs that are rather inert compounds in the lower atmosphere become a serious air pollution problem at higher elevations.

By 1987, the evidence that CFCs destroy ozone in the stratosphere above Antarctica every spring had become irrefutable. In 1987, the ozone hole was larger than ever. More than half of the total ozone column was wiped out and essentially all ozone disappeared from some regions of the stratosphere.

Research confirmed that the ozone layer, on a worldwide basis, shrunk approximately 2.5% in the period from 1976–1986 (Zurer, 1988). Initially, it was believed that this phenomenon was peculiar to the geography and climatology of Antarctica and that the warmer northern hemisphere was strongly protected from the processes that lead to massive ozone losses. Studies of the North Pole stratosphere in the winter of 1989 revealed that this is not the case (Zurer, 1989).

In September 1987, the Montreal Protocol on Substances That Deplete the Ozone Layer was developed. The protocol, which has been ratified by 36 countries and became effective in January 1989, proposed that CFC production first be frozen and then reduced 50% by 1998. Yet, under the terms of the protocol, the chlorine content of the atmosphere would continue to grow because the fully halogenated CFCs have such long lifetimes in the atmosphere. CF_2Cl_2, for example, has a lifetime of 110 years (Reisch and Zurer, 1988). Eighty countries met at Helsinki,

---

FIGURE 12–7

Photoreactions of ozone.

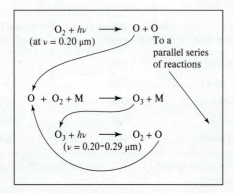

FIGURE 12–8

Ozone destruction by chlorofluoromethane.

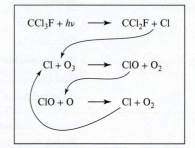

Finland, in the spring of 1989 to assess new information. The delegates gave their unanimous assent to the five-point Helsinki Declaration:

1. All join the 1985 Vienna Convention for the Protection of the Ozone Layer and the follow-up Montreal Protocol
2. Phase out production and consumption of ozone-depleting CFCs no later than 2000
3. Phase out production and consumption as soon as feasible of halons and such chemicals as carbon tetrachloride and methyl chloroform that also contribute to ozone depletion
4. Commit themselves to accelerated development of environmentally acceptable alternative chemicals and technologies
5. Make relevant scientific information, research results, and training available to developing countries (Sullivan, 1989)

The Montreal Protocol was strengthened in 1990, 1992, 1997, and 1999. A ban on halon production took effect in January 1995 (Zurer, 1994). The current terms of the treaty banned production of CFCs, carbon tetrachloride, and methyl chloroform as of January 1996 (Zurer, 1994). As of September 2002, 183 countries had ratified the protocol (UNDP, 2005).

A number of alternatives to the fully chlorinated and, hence, more destructive CFCs have been developed. The two groups of compounds that emerged as significant replacements for the CFCs are hydrofluorocarbons (HFCs) and hydrochlorofluorocarbons (HCFCs). In contrast to the CFCs, HFCs and HCFCs contain one or more C – H bonds. This makes them susceptible to attack by OH radicals in the lower atmosphere. Because HFCs do not contain chlorine, they do not have the ozone-depleting potential associated with the chlorine cycle. Although HCFCs contain chlorine, this chlorine is not transported to the stratosphere because OH scavenging in the troposphere is relatively efficient.

The implementation of the Montreal Protocol appears to be working. The use of CFCs has been reduced to one-tenth of the 1990 levels (UN, 2005). Total tropospheric chlorine from the long- and short-lived chlorocarbons was about 5% lower in 2000 than that observed at its peak in 1992–1994. The rate of change in 2000 was about –22 parts per trillion per year. Total chlorine from CFCs is no longer increasing, in contrast to the slight increase noted in 1998. Total tropospheric bromine from halons continues to increase at about 3% per year, which is about two-thirds of the 1996 rate (UNEP/WHO, 2002).

The issues of ozone depletion and climate change are interconnected. As the atmospheric abundance of CFCs declines, their contribution to global warming will decline. On the other hand, the use of HFCs and HCFCs as substitutes for CFCs will contribute to increases in global warming. Because ozone depletion tends to cool the Earth's climate system, recovery of the ozone layer will tend to warm the climate system (UNEP/WHO, 2002).

Global Warming

Scientific Basis.   The case for global warming has grown very strong over the last three and half decades. As shown in Figure 12–9, the 12-month running mean in 2017 was 1.17°C above the 1880–1920 base period (Hansen, 2018). Mann and Jones (2003) have compiled proxy temperature data from sediments, ice cores, and tree-ring temperature reconstruction over the past two millennia. Their research shows (Figure 12–10) the average global surface temperature has been increasing for the last 100 years and was higher in 2016 than in any time in the past 2000 years.

To understand the physics of global warming, we will use the simplest model of energy balance. It does not take into account location on the planet, time, precipitation, wind, ocean currents, soil moisture, or any of a number of other variables. The model is a simple radiation balance based on the principles described in Chapter 4. It equates the solar energy absorbed by the Earth from the sun with the energy radiated back into space from the Earth.

FIGURE 12–9

Global surface
temperature (1880–1920
base period).

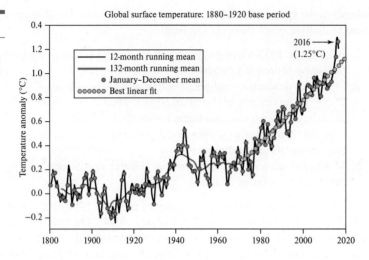

Global surface temperature: 1880–1920 base period

FIGURE 12–10

Global average surface
temperature reconstruc-
tion. Temperature
anomaly is departure
from 1961–1990
instrumental reference
period for which the
average is shown by the
dashed line. (*Source:*
Hansen, J. and M. Sato
Global Surface
Temperatures 1880–1920
Base Period http://
Columbia.edu/2mhs119/
Temperature.

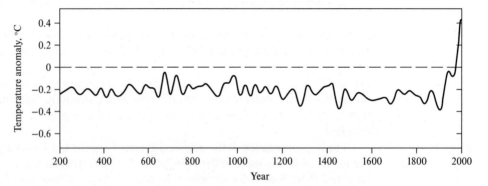

We begin with a definition of the **solar constant.** The solar constant is the average annual intensity of the radiation that is intercepted by the cross section of a sphere equivalent to the Earth's diameter, normal to the incident radiation just outside of the Earth's atmosphere (Figure 12–11). The solar constant has been evaluated over many years, and as a working estimate, it may be taken to be 1400 W · m^{-2} (various sources report values from 1379 to 1396 W · m^{-2}). The rate at which solar energy is radiated on the Earth is the product of the flux of energy (W · m^{-2}) and the area of the intercepting cross section.

$$E = S\pi r^2 \tag{12–12}$$

where E = energy intercepted by Earth (in W)

FIGURE 12–11

Cross-sectional area of
sphere that intercepts
incident radiation from
the Sun outside of the
Earth's atmosphere. The
Earth's radius is *r*.

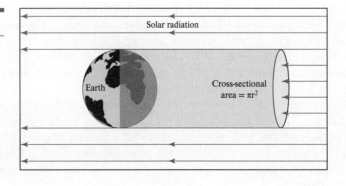

S = solar constant (in W $\cdot$ m^{-2})

r = radius of Earth (in m)

A fraction of the radiation that reaches the Earth is reflected back into space. The ratio of the radiation reflected by an object to that absorbed by it is called the **albedo.** The Earth's albedo is taken to be 0.3. The energy absorbed by Earth is then:

$$E_{abs} = (1 - \alpha)S\pi r^2 \qquad (12\text{–}13)$$

where α = albedo

We assume that the Earth radiates as a blackbody (see Chapter 4) at a temperature T. The energy emitted by a unit area in a unit time is given by Equation 4–55. The total energy emitted by the surface of the Earth is:

$$E_{emit} = \sigma T_e^4 \, 4\pi r^2 \qquad (12\text{–}14)$$

where σ = Stefan-Boltzman constant = 5.67×10^{-8} W $\cdot$ m^{-2} $\cdot$ K^{-4}

$\quad T_e$ = Earth's blackbody temperature (in K)

$\quad 4\pi r^2$ = surface area of sphere

If we assume steady state conditions, that is, that over the millennia the Earth's temperature has not changed appreciably with time, we can assume:

$$E_{abs} = E_{emit} \qquad (12\text{–}15)$$

and solve for the Earth's blackbody temperature:

$$T_e \approx \left[\frac{(1 - \alpha)S}{4\sigma} \right]^{1/4} \qquad (12\text{–}16)$$

$$\approx \left[\frac{(1 - 0.3)(1{,}400 \text{ W} \cdot \text{m}^{-2})}{4(5.67 \times 10^{-8} \text{ W} \cdot \text{m}^{-2} \cdot \text{K}^{-4})} \right]^{1/4}$$

$$\approx 256.3 \text{ or } 256 \text{ K or } -16.6 \text{ or } -17°\text{C}$$

This result is at great variance from the actual value of the Earth's average surface temperature of 288 K (+15°C). The actual temperature differs from the blackbody temperature because of the **greenhouse effect.** To understand the greenhouse effect, we must first review the relationship between the spectrum of wavelengths radiated by an object and its temperature. The wavelength of the energy emitted by a blackbody at a given temperature can be estimated with Wien's displacement equation:

$$\lambda_{max} = \frac{2897.8 \text{ μm K}}{T} \qquad (12\text{–}17)$$

where T = absolute temperature of body (in K)

The Sun is assumed to be a blackbody with a temperature of 6000 K and a peak intensity at about 0.5 μm. The Earth, with a blackbody temperature of 288 K, has its peak at about 10 μm. The blackbody emission spectra for the Sun and Earth are shown in Figure 12–12a. The spectra for the Sun shows incoming "short-wave" radiation. The spectra for the Earth shows outgoing "long-wave" radiation. Note that the abscissa is a logarithmic scale.

As radiant energy enters our atmosphere, it is affected by aerosols and atmospheric gases. Some of the constituents scatter the radiation by reflection, some stop it by adsorption, and some let it pass unchanged. The key phenomenon of interest in causing the greenhouse effect is the ability of gases to absorb radiant energy. As the atoms in gas vibrate and rotate, they absorb and radiate energy in specific wavelengths. If the frequency of the molecular oscillations

FIGURE 12–12

(a) Blackbody radiation curves for the sun (6000 K) and earth (288 K). **(b)** Absorption curves for various gases. The bottom frames show the total atmospheric absorption and the overlay of absorption on the blackbody radiation. The shaded areas depict absorption. The unshaded areas depict transmission. (*Source:* Anthes, R.A., et al. 1981. *The Atmosphere*, Charles E. Merrill Publishing Co., Columbus, OH, p. 89.)

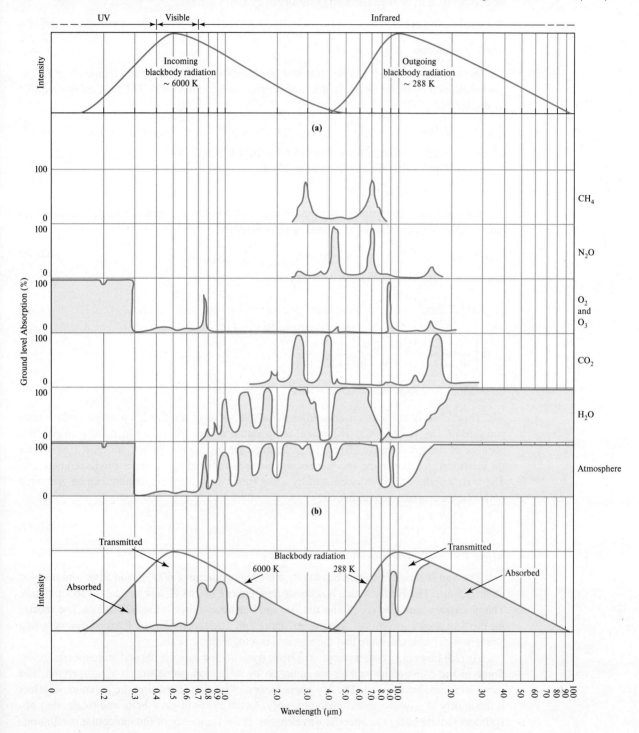

is close to the frequency of the passing radiant energy, the molecule can absorb that energy. The absorption occurs over a limited range of frequencies. It is different for each molecule. A plot of the percent absorption of solar radiation versus wavelength is called the absorption spectrum (Figure 12–12b). The sum of the absorption of the gases at ground level is shown in the bottom frame of Figure 12–12b. The shaded areas show radiation that is absorbed. The unshaded areas show radiation that is transmitted. These areas are often referred to as radiation "windows."

From Figure 12–12b, it is evident that essentially all of the incoming solar radiation at wavelengths in the ultraviolet range (< 0.3 µm) is absorbed by oxygen and ozone. This absorption occurs in the stratosphere. It shields us from harmful ultraviolet radiation.

At the other end of the spectrum, radiatively active gases that absorb at wavelengths greater than 4 µm are called **greenhouse gases** (GHGs). This absorption heats the atmosphere, which radiates energy back to Earth as well as into space. The GHGs act much like the glass on a greenhouse (thus, the name *greenhouse gases*): they let in short-wave (ultraviolet) radiation from the Sun that heats the ground surface, but restrict the loss of heat by radiation from the ground surface. The more GHGs in the atmosphere, the more effective it is in restricting the outflow of long-wave (infrared) radiation. These greenhouse gases act as a blanket that raises the Earth's temperature above the 256 K calculated from the radiation balance.

To elaborate on the radiation balance, we must take into account reflection by clouds and aerosols, evapotranspiration, latent heat release, and convective heat transfer. The simple global mean energy balance shown in Figure 12–13 summarizes the major energy flows.

Increasing levels of GHGs leads to global warming. Unlike ozone, the greenhouse gases are relatively transparent to short-wave ultraviolet light from the Sun. They do, however, absorb and emit long-wave radiation at wavelengths typical of the Earth and atmosphere. CO_2 has been

FIGURE 12–13

The Earth's annual and global mean energy balance. Units are $W \cdot m^{-2}$. Of the incoming solar radiation, 49% (168 $W \cdot m^{-2}$) is absorbed by the surface. That heat is returned to the atmosphere as sensible heat, as evapotranspiration (latent heat), and as thermal infrared radiation. Most of this radiation is absorbed by the atmosphere, which in turn emits radiation both up and down. The radiation lost to space comes from cloud tops and atmospheric regions much colder than the surface. This causes a greenhouse effect.

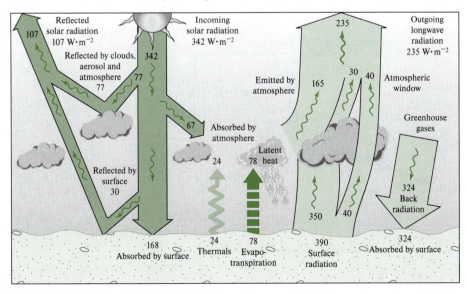

TABLE 12–11   Global Warming Potentials (GWPs) Relative to Carbon Dioxide over 20-Year Time Period

Chemical Species	Lifetime, y	Global Warming Potential, kg of CO_2/kg of gas
Carbon dioxide (CO_2)	30–200	1
Methane (CH_4)	12	62
Nitrous oxide (N_2O)	114	275
CFC-12 (CF_2Cl_2)	100	10,200
HCFC-22 (CHF_3Cl)	12	4800
HFO-1234yf ($CF_3CF=CH_2$)	0.03	<1
Tetrafluoromethane (CF_4)	50,000	3900
Sulfur hexafluoride (SF_6)	3200	15,100

Source: IPCC, 2000; Ritter, S.K., C&EN, 2013.

identified as the major GHG because of its abundance and its strong absorption spectrum in the region where the Earth emits most of its infrared radiation.

Gases other than CO_2 also act as GHGs. In order to allow comparison of the effect of these gases on global warming, we use a weighting factor called the *Global Warming Potential* (GWP). The GWP takes into account three factors:

- Radiative forcing due to the addition a unit mass of each greenhouse gas.

- Estimates of the rate at which the injected unit mass decays over time.

- Estimates of the cumulative radiative forcing that the unit mass addition will have over time.

The GWPs of a few selected chemical species are listed in Table 12–11.

Since the first systematic measurements were made at Mauna Loa in Hawaii in 1958, CO_2 levels have risen from 316 ppm to 406 ppm (Keeling and Whorf, 2005; Pittman, 2011; NOAA, 2017); recent CO_2 levels are shown in Figure 12–14. From analysis of air trapped in ice cores in Greenland and Antarctica, we know that preindustrial levels of CO_2 were about 280 ppm. The ice core records indicate that, over the last 160,000 years, no fluctuations of CO_2 have been larger than 70 ppm (Hileman, 1989) and that the current concentrations are higher than any level attained in the past 650,000 years (Hileman, 2005). It is estimated that the atmospheric CO_2

FIGURE 12–14

High- and low-pressure systems. (*Source:* NOAA March, 2017.)

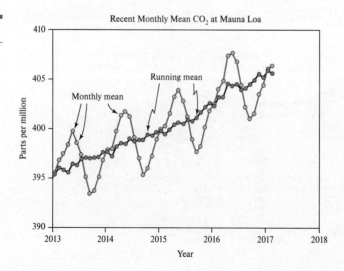

concentration has increased 30% since 1750 and that the present concentration has not been exceeded during the past 420,000 years and likely not during the past 20 million years (IPCC, 2001a). Several gases have been recognized as contributing to the greenhouse effect. Methane (CH_4), nitrous oxide (N_2O), and CFCs are similar to CO_2 in their radiative behavior. Even though their concentrations are much lower than CO_2, these gases are now estimated to trap about 60% as much long-wave radiation as CO_2.

The chemical industry continues to develop replacement refrigerant gases that have lower GWPs. The latest of these, HFO-1234yf, has a GWP of less than 1. Although this is a giant step forward in reducing GWP, the amount of hydrocarbons used as refrigerants is extremely small compared to CO_2 emissions and methane from landfills and fracking activities.

In 2007, the United Nations Intergovernmental Panel on Climate Change* (IPCC) declared (IPCC, 2007a): "Most of the observed increase in globally averaged temperatures since the mid- 20th century is *very likely*[†] due to the observed increase in greenhouse gas concentrations . . . discernable human influences now extend to other aspects of climate, including ocean warming, continental-average temperatures, temperature extremes, and wind patterns."

About three-quarters of the anthropogenic emissions of CO_2 that have been added to the atmosphere over the past 20 years is attributed to the combustion of fossil fuel (IPCC, 2001a). In the 1980s, massive deforestation was identified as a possible contribution. Both the burning of timber and the release of carbon from bacterial degradation contribute. Perhaps more important, deforestation removes a mechanism for removing CO_2 from the atmosphere (commonly referred to as a *sink*). In normal respiration, green plants utilize CO_2 as a carbon source. This CO_2 is fixed in the biomass by photosynthetic processes. A rapidly growing rain forest can fix between 1 and 2 kg per year of carbon per square meter of ground surface. Cultivated fields, in contrast, fix only about 0.2 to 0.4 · m^{-2} —and this amount is recycled by bioconsumption and conversion to CO_2.

Even if atmospheric levels of greenhouse gases are held constant at current levels, Earth will warm about 0.1°C over each of the next few decades because of the long half-life of the greenhouse gases in the atmosphere. Due to the time scale required for removal of CO_2, it will contribute to warming and sea level rise for more than a millennium (IPCC, 2007a).

Impacts.   Attempts to understand the consequences of global warming are based on mathematical models of the global circulation of the atmosphere and oceans. Using best estimates, the IPCC estimates that the globally averaged surface temperature will rise 1.8 to 4.0°C by 2100 (IPCC, 2007a). To date these models have a "good news-bad news" conclusion. Based on the 1.4 to 5.8°C rise in global temperature, the following is predicted for North America (IPCC, 2007b):

1. A decrease in heating costs (partly offset by increased air-conditioning cost).
2. Potential increased food production in areas of Canada and an increase in warm-temperature mixed forest production with modest warming; with severe warming, crop production could possibly become a net loss.
3. Much easier navigation in the Arctic seas.
4. Drier crop conditions in the Midwest and Great Plains, requiring more irrigation.
5. Warming in western mountains is projected to cause decreased snowpack, more winter flooding, and reduced summer flows exacerbating competition for overall water resources.

---

*The IPCC is composed of over 673 scientists and 420 expert reviewers from around the world.

[†]*very likely* is defined by the IPCC as a "9 out of 10 chance of being correct."

TABLE 12–12   **Key Impacts as a Function of Increasing Global Average Temperature Change**

Global Mean Annual Temperature Change Relative to 1980–1999 (°C)

	0	1	2	3	4	5°C

WATER

Increased water availability in moist tropics and high latitudes ▸ (continuing to 5°C)

Decreasing water availability and increasing drought in mid-latitudes and semi-arid low latitudes ▸

Hundreds of millions of people exposed to increased water stress ▸

ECOSYSTEMS

Up to 30% of species at increasing risk of extinction ——— Significant[†] extinctions around the globe ▸

Increased coral bleaching ——— Most corals bleached ——— Widespread coral mortality ▸

Terrestrial biosphere tends toward a net carbon source as: ~15% ——— ~40% of ecosystems affected ▸

Increasing species range shifts and wildfire risk

Ecosystem changes due to weakening of the meridional overturning circulation ▸

FOOD

Complex, localized negative impacts on small holders, subsistence farmers and fishers ▸

Tendencies for cereal productivity to decrease in low latitudes ——— Productivity of all cereals decreases in low latitudes ▸

Tendencies for some cereal productivity to increase at mid- to high latitudes ——— Cereal productivity to decrease in some regions

COASTS

Increased damage from floods and storms ▸

About 30% of global coastal wetlands lost[‡] ▸

Millions more people could experience coastal flooding each year ▸

HEALTH

Increasing burden from malnutrition, diarrheal, cardio-respiratory, and infectious diseases ▸

Increased morbidity and mortality from heat waves, floods, and droughts ▸

Changed distribution of some disease vectors ▸

Substantial burden on health services ▸

[†]Significant is defined here as more than 40%.

[‡]Based on average rate of sea level rise of 4.2 mm · year^{-1} from 2000 to 2080.

Note: The black lines link impacts, dotted arrows indicate impacts continuing with increasing temperature. Entries are placed so that the left hand side of text indicates approximate onset of a given impact. Adaptation to climate change is not included in these estimations. All entries are from published studies recorded in the chapters of the Assessment. Confidence levels for all statements are high.

Source: IPCC, 2007b.

6. Increasing impacts on forests from pests, diseases, and fire.

7. Widespread melting of permanently frozen ground with adverse effects on animal and plant life as well as building technology in Alaska and northern Canada.

8. A rise in sea level between 0.18 and 0.57 m that would result in an increase in the severity of flooding, damage to coastal structures, destruction of wetlands, and saltwater intrusion into drinking water supplies in coastal areas particularly in Florida and much of the Atlantic coast.*

*Contraction of the Greenland ice sheet is projected to contribute to sea level rise after 2100 even if radiative forcing is stabilized at 1.8°C above the current value. The Greenland ice sheet would be virtually eliminated and the resulting contribution to sea level rise would be about 7 m!

As shown in Table 12–12, on a global scale the impacts will range from severe to catastrophic.

Kyoto Protocol.   The framework convention for the protocol was signed in 1992. In 1997, the protocol targets for industrialized countries to reduce their GHG emissions were finalized. To become legally binding two conditions had to be fulfilled:

- Ratification by 55 countries
- Ratification by nations accounting for at least 55% of emissions from 38 industrialized countries plus Belarus, Turkey, and Kazakhstan

The first condition was met in 2002. Following the decision of the United States and Australia not to ratify, Russia's position became crucial to fulfill the second condition. On November 18, 2004, Russia ratified the Kyoto Protocol. It came into force 90 days later on February 16, 2005. At that time, the targets for reducing emissions became binding on the countries that ratified the protocol. The agreement set levels to reduce emissions by 5% from the 1990 base-line level. As of December, 2005, 157 nations had ratified the accord and the United States remained unwilling to make any commitments to reduce greenhouse emissions (AP, 2005a).

Although the United States did not ratify the protocol, 136 U.S. mayors representing more than 30 million people have signed an agreement to meet the goals spelled out in the treaty (AP, 2005b). On December 20, 2005, seven northeastern states (Connecticut, Delaware, Maine, New Hampshire, New Jersey, New York, and Vermont) signed an agreement to establish carbon dioxide emissions caps for electric utilities in their states (Chemical and Engineering News, 2006). In addition, the Community Carbon Exchange (CCX) has signed up more than 50 organizations, including American Electric Power (Columbus, Ohio) and TECO Energy (Tampa, Florida). Several states have also indicated they will work to implement an emission credit program. It is not clear that these credits (either the CCX or the states) will count under any government mandate.

In 2005, Massachusetts petitioned the U.S. EPA to regulate CO_2 emissions from automobiles. EPA declined saying that the Clean Air Act did not authorize it to issue mandatory regulations to address global climate change, that even if it were authorized, it would be unwise because the link between GHG and global warming is not unequivocally established, and that regulation of automobile emissions would be a piecemeal approach and would conflict with the president's comprehensive approach using additional support for technological innovation and nonregulatory programs. On April 2, 2007, the Supreme Court of the United States ruled that the harms associated with climate change are serious and well recognized, and that while reducing automobile emissions may not by itself reverse global warming, it does not follow that the court lacks jurisdiction to decide whether EPA has a duty to take steps to slow or reduce them. The court decided that because greenhouse gases fit well within the act's capacious definition of "air pollutant," EPA has statutory authority to regulate emission of such gases from new motor vehicles. The court further noted that under the act's clear terms, EPA can avoid promulgating regulations only if it determines that greenhouse gases do not contribute to climate change (Supreme Court, 2007).

A Rationale for Action.   While there is still considerable disagreement about the potential for global warming, the consequences of ignoring these trends are sufficiently dramatic that intensive research must continue in the decades to come. Even without the risks of climate change, improvements in energy efficiency to reduce greenhouse gas emissions are amply justified from two points of view: economics and sustainability. Higher energy efficiency will yield economic benefit in reducing the cost of electricity and transportation. Higher efficiency will contribute to sustainability by extending the availability of finite energy resources. The expectation of damages from climate change provides extra incentive for pursuing these programs vigorously.

12–7 AIR POLLUTION METEOROLOGY

The Atmospheric Engine

The atmosphere is somewhat like an engine. It is continually expanding and compressing gases, exchanging heat, and generally creating chaos. The driving energy for this unwieldy machine comes from the Sun. The difference in heat input between the equator and the poles provides the initial overall circulation of the earth's atmosphere. The rotation of the Earth coupled with the different heat conductivities of the oceans and land produce weather.

Highs and Lows.   Because air has mass, it also exerts pressure on things under it. Like water, which we intuitively understand to exert greater pressures at greater depths, the atmosphere exerts more pressure at the surface than it does at higher elevations. The highs and lows depicted on weather maps are simply areas of greater and lesser pressure. The elliptical lines shown on more detailed weather maps are lines of constant pressure, or **isobars.** A two-dimensional plot of pressure and distance through a high- or low-pressure system would appear as shown in Figure 12–15. Figure 12–15a shows a plan view of a high-pressure system in x/y coordinates with the highest pressure (102.8 kPa) in the center. The lower portion of Figure 12–15a shows a plot of pressure versus distance (x) for a cross section through the plan view along the line AA. This cross section has the appearance of a hill with the highest pressure corresponding to the H in the plan view of the high. A similar plot for a low-pressure system is shown in Figure 12–15b. Here a cross section through the line BB shows a valley with the lowest pressure corresponding to the L in the plan view of the low.

The wind flows from the higher pressure areas to the lower pressure areas. On a nonrotating planet, the wind direction would be perpendicular to the isobars (Figure 12–16a). However, because the earth rotates, an angular thrust called the Coriolis effect is added to this motion. The resultant wind direction in the Northern Hemisphere is as shown in Figure 12–16b. The technical names given to these systems are **anticyclones** for highs and **cyclones** for lows. Anticyclones are associated with good weather. Cyclones are associated with foul weather. Tornadoes and hurricanes are the foulest of the cyclones.

Wind speed is in part a function of the steepness of the pressure surface. When the isobars are close together, the pressure **gradient** (slope) is said to be steep and the wind speed relatively high. If the isobars are well spread out, the winds are light or nonexistent.

Turbulence

Mechanical Turbulence.   In its simplest terms, we may consider turbulence to be the addition of random fluctuations of wind velocity (i.e., speed and direction) to the overall average wind velocity. These fluctuations are caused, in part, by the fact that the atmosphere is being sheared. The shearing results from the fact that the wind speed is zero at the ground surface and rises with elevation to near the speed imposed by the pressure gradient. The shearing results in a tumbling, tearing motion as the mass just above the surface falls over the slower moving air at the surface. The swirls thus formed are called **eddies.** These small eddies feed larger ones. As you might expect, the greater the mean wind speed, the greater the mechanical turbulence. The more mechanical turbulence, the easier it is to disperse and spread atmospheric pollutants.

FIGURE 12–15

High- and low-pressure
systems.

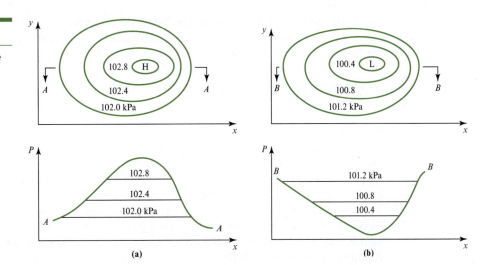

(a)   (b)

Thermal Turbulence.   Like all other things in nature, the rather complex interaction that produces mechanical turbulence is confounded and further complicated by a third party. Heating of
the ground surface causes turbulence in the same fashion that heating the bottom of a beaker full
of water causes turbulence. At some point below boiling, you can see density currents rising off
the bottom. Likewise, if the earth's surface is heated strongly and in turn heats the air above it,
thermal turbulence will be generated. Indeed, the "thermals" sought by glider pilots and hot air
balloonists are these thermal currents rising on what otherwise would be a calm day.

The converse situation can arise during clear nights when the ground radiates its heat away
to the cold night sky. The cold ground, in turn, cools the air above it, causing a sinking density
current.

Stability

The tendency of the atmosphere to resist or enhance vertical motion is termed **stability.** It is
related to both wind speed and the change of air temperature with height (**lapse rate**). For
our purpose, we may use the lapse rate alone as an indicator of the stability condition of the
atmosphere.

FIGURE 12–16

Wind flow due to
pressure gradient.

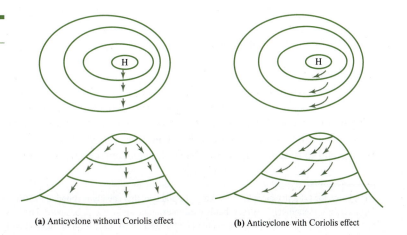

(a) Anticyclone without Coriolis effect   **(b)** Anticyclone with Coriolis effect

There are three stability categories. When the atmosphere is classified as **unstable,** mechanical turbulence is enhanced by the thermal structure. A **neutral** atmosphere is one in which the thermal structure neither enhances nor resists mechanical turbulence. When the thermal structure inhibits mechanical turbulence, the atmosphere is said to be **stable.** Cyclones are associated with unstable air. Anticyclones are associated with stable air.

Neutral Stability.   The lapse rate for a neutral atmosphere is defined by the rate of temperature increase (or decrease) experienced by a parcel of air that expands (or contracts) **adiabatically** (without the addition or loss of heat) as it is raised through the atmosphere. This rate of temperature decrease ($-dT/dz$) is called the **dry adiabatic lapse rate,** designated by the Greek letter gamma (Γ). It has a value of approximately $-1.00°C \cdot 100\ m^{-1}$.* (*Note:* This is not a slope in the normal sense, i.e., it is not dy/dx.) In Figure 12–17a, the dry adiabatic lapse rate of a parcel of air is shown as a dashed line and the temperature of the atmosphere (ambient lapse rate) is shown as a solid line. Because the ambient lapse rate is the same as Γ, the atmosphere is said to have a neutral stability.

Unstable Atmosphere.   If the temperature of the atmosphere falls at a rate greater than Γ the lapse rate is said to be **superadiabatic** and the atmosphere is unstable. Using Figure 12–17b, we can see that this is so. The actual lapse rate is shown by the solid line. If we capture a balloon full of polluted air at elevation A and adiabatically displace it 100 m vertically to elevation B, the temperature of the air inside the balloon will decrease from 21.15 to 20.15°C. At a lapse rate of $-1.25°C \cdot 100\ m^{-1}$, the temperature of the air outside the balloon will decrease from 21.15 to 19.90°C. The air inside the balloon will be warmer than the air outside; this temperature difference gives the balloon buoyancy. It will behave as a hot gas and continue to rise without any further mechanical effort. Thus, mechanical turbulence is enhanced and the atmosphere is unstable. If we adiabatically displace the balloon downward to elevation C, the temperature inside the balloon would rise at the rate of the dry adiabat. Thus, in moving 100 m, the temperature will increase from 21.15 to 22.15°C. The temperature outside the balloon will increase at the superadiabatic lapse rate to 22.40°C. The air in the balloon will be cooler than the ambient air, and the balloon will have a tendency to sink. Again, mechanical turbulence (displacement) is enhanced.

Stable Atmosphere.   If the temperature of the atmosphere falls at a rate less than Γ it is called **subadiabatic,** and the atmosphere is stable. If we again capture a balloon of polluted air at elevation A (Figure 12–17c) and adiabatically displace it vertically to elevation B, the temperature of the polluted air will decrease at a rate equal to the dry adiabatic rate. Thus, in moving 100 m, the temperature will decrease from 21.15 to 20.15°C as before. However, because the ambient lapse rate is $-0.5°C \cdot 100\ m^{-1}$, the temperature of the air outside the balloon will have dropped to only 20.65°C. Because the air inside the balloon is cooler than the air outside the balloon, the balloon will have a tendency to sink. Thus, the mechanical displacement (turbulence) is inhibited.

In contrast, if we displace the balloon adiabatically to elevation C, the temperature inside the balloon would increase to 22.15°C and the ambient temperature would increase to 21.65°C. In this case, the air inside the balloon would be warmer than the ambient air and the balloon would tend to rise. Again, the mechanical displacement would be inhibited.

---

*The value for Γ for dry air is $9.76°C \cdot km^{-1}$. In practice, this usually is rounded to $10°C \cdot km^{-1}$. The dry adiabatic lapse rate is a theoretical construct that is used to examine atmospheric behavior. The U.S. Standard Atmosphere describes a more realistic average mean temperature profile. A discussion of the Standard Atmosphere may be found in Chapter 2.

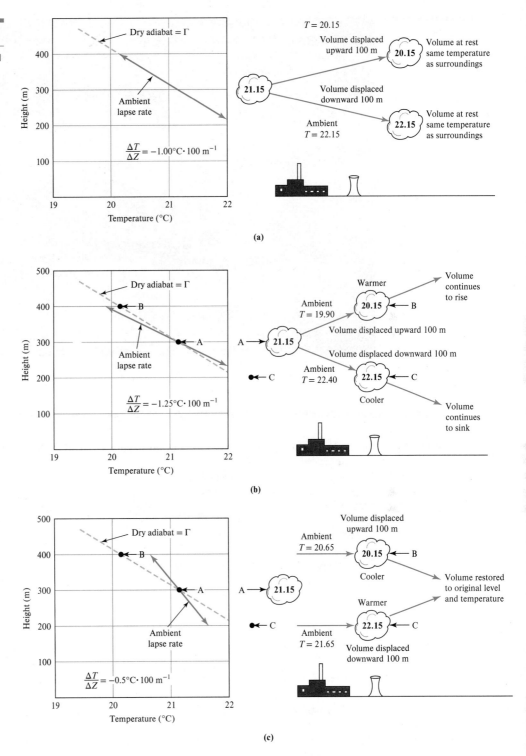

FIGURE 12–17

Lapse rate and displaced air volume. (*Source:* U.S. Atomic Energy Commission, 1955.)

There are two special cases of subadiabatic lapse rate. When there is no change of temperature with elevation, the lapse rate is called **isothermal.** When the temperature increases with elevation, the lapse rate is called an **inversion.** The inversion is the most severe form of a stable temperature profile. It is often associated with restricted air volumes that cause air pollution episodes. Figure 12–18 is a dramatic illustration of an inversion.

FIGURE 12–18

Inversion at Four Corners power plant. The photo was taken during shakedown at the start up of the plant before the air pollution control equipment was operational. (*Source:* U. S. National Park Service)

EXAMPLE 12–3   Given the following temperature and elevation data, determine the stability of the atmosphere.

Elevation (m)	Temperature (°C)
2.00	14.35
324.00	11.13

Solution   Begin by determining the existing lapse rate.

$$\frac{\Delta T}{\Delta Z} = \frac{T_2 - T_1}{Z_2 - Z_1}$$

$$= \frac{11.13 - 14.35}{324.00 - 2.00} = \frac{-3.22}{322.00} = -0.0100°\text{C} \cdot \text{m}^{-1} = -1.00°\text{C} \cdot 100 \text{ m}^{-1}$$

Now we compare this with Γ and find that they are equal. Thus, the atmospheric stability is neutral.

Comment: Temperature measurements are typically not measured to two decimal places.

Terrain Effects

Heat Islands.   A heat island results from a mass of material, either natural or anthropogenic, that absorbs and reradiates heat at a greater rate than the surrounding area. This causes moderate to strong vertical convection currents above the heat island. The effect is superimposed on the prevailing meteorological conditions and is nullified by strong winds. Large industrial complexes and small to large cities are examples of places that would have a heat island.

Because of the heat island effect, atmospheric stability will be less over a city than it is over the surrounding countryside. Depending upon the location of the pollutant sources, this can be either good news or bad news. First, the good news: For ground level sources such as automobiles, the bowl of unstable air that forms will allow a greater air volume for dilution of the pollutants. Now the bad news: Under stable conditions, plumes from tall stacks would be carried out over the countryside without increasing ground level pollutant concentrations. Unfortunately, the instability caused by the heat island mixes these plumes to the ground level.

Land–Sea Breezes.   Under a stagnating anticyclone, a strong local circulation pattern may develop across the shoreline of large water bodies. During the night, the land cools more rapidly than the water. The relatively cooler air over the land flows toward the water (a land breeze, Figure 12–19). During the morning the land heats faster than water. The air over the land becomes relatively warm and begins to rise. The rising air is replaced by air from over the water body (a sea or lake breeze, Figure 12–20).

The effect of the lake breeze on stability is to impose a surface-based inversion on the temperature profile. As the air moves from the water over the warm ground, it is heated from below. Thus, for stack plumes originating near the shoreline, the stable lapse rate causes a fanning plume that does not disperse close to the stack (Figure 12–21). The lapse condition grows to the height of the stack as the air moves inland. At some point inland, unstable conditions develop and the plume rapidly descends to the ground causing a "fumigation" at the ground surface.

FIGURE 12–19

Land breeze during the night.

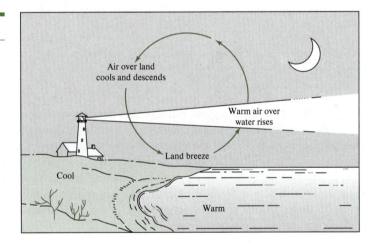

FIGURE 12–20

Lake breeze during the day.

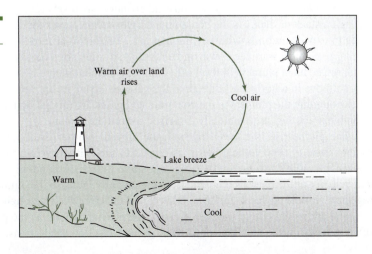

FIGURE 12–21

Effect of lake breeze on plume dispersion.

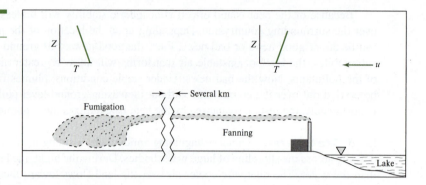

Valleys. When the general circulation imposes moderate to strong winds, valleys that are oriented at an acute angle to the wind direction channel the wind. The valley effectively peels off part of the wind and forces it to follow the direction of the valley floor.

Under a stagnating anticyclone, the valley will set up its own circulation. Warming of the valley walls will cause the valley air to be warmed. It will become more buoyant and flow up the valley. At night the cooling process will cause the wind to flow down the valley.

Valleys oriented in the north–south direction are more susceptible to inversions than level terrain. The valley walls protect the floor from radiative heating by the sun. Yet the walls and floor are free to radiate heat away to the cold night sky. Thus, under weak winds, the ground cannot heat the air rapidly enough during the day to dissipate the inversion that formed during the night.

12–8 ATMOSPHERIC DISPERSION

Factors Affecting Dispersion of Air Pollutants

The factors that affect the transport, dilution, and dispersion of air pollutants can generally be categorized in terms of the emission point characteristics, the nature of the pollutant material, meteorological conditions, and the effects of terrain and anthropogenic structures. We have discussed all of these except the source conditions. Now we wish to integrate the first and third factors to describe the qualitative aspects of calculating pollutant concentrations. We shall follow this with a simple quantitative model for a point source. More complex models for point sources (in rough terrain, in industrial settings, or for long time periods), area sources, and mobile sources are left for more advanced texts.

Source Characteristics. Most industrial effluents are discharged vertically into the open air through a stack or duct. As the contaminated gas stream leaves the discharge point, the plume tends to expand and mix with the ambient air. Horizontal air movement will tend to bend the discharge plume toward the downwind direction. At some point between 300 and 3000 m downwind, the effluent plume will level off. While the effluent plume is rising, bending, and beginning to move in a horizontal direction, the gaseous effluents are being diluted by the ambient air surrounding the plume. As the contaminated gases are diluted by larger and larger volumes of ambient air, they are eventually dispersed toward the ground.

The plume rise is affected by both the upward inertia of the discharge gas stream and by its buoyancy. The vertical inertia is related to the exit gas velocity and mass. The plume's buoyancy is related to the exit gas mass relative to the surrounding air mass. Increasing the exit velocity or the exit gas temperature will generally increase the plume rise. The plume rise, together with the physical stack height, is called the **effective stack height.**

The additional rise of the plume above the discharge point as the plume bends and levels off is a factor in the resultant downwind ground level concentrations. The higher the plume rises initially, the greater distance there is for diluting the contaminated gases as they expand and mix downward.

For a specific discharge height and set of plume dilution conditions, the ground level concentration is proportional to the amount of contaminant materials discharged from the stack outlet for a specific period. Thus, when all other conditions are constant, an increase in the pollutant discharge rate will cause a proportional increase in the downwind ground level concentrations.

Downwind Distance.   The greater the distance between the point of discharge and a ground level receptor downwind, the greater will be the volume of air available for diluting the contaminant discharge before it reaches the receptor.

Wind Speed and Direction.   The wind direction determines the direction in which the contaminated gas stream will move across local terrain. Wind speed affects the plume rise and the rate of mixing or dilution of the contaminated gases as they leave the discharge point. An increase in wind speed will decrease the plume rise by bending the plume over more rapidly. The decrease in plume rise tends to increase the pollutant's ground level concentration. On the other hand, an increase in wind speed will increase the rate of dilution of the effluent plume, tending to lower the downwind concentrations. Under different conditions, one or the other of the two wind speed effects becomes the predominant effect. These effects, in turn, affect the distance downwind of the source at which the maximum ground level concentration will occur.

Stability.   The turbulence of the atmosphere follows no other factor in power of dilution. The more unstable the atmosphere, the greater the diluting power. Inversions that are not ground-based, but begin at some height above the stack exit, act as a lid to restrict vertical dilution.

Dispersion Modeling

General Considerations and Use of Models.   A dispersion model is a mathematical description of the meteorological transport and dispersion process that is quantified in terms of source and meteorologic parameters during a particular time. The resultant numerical calculations yield estimates of concentrations of the particular pollutant for specific locations and times.

To verify the numerical results of such a model, actual measured concentrations of the particular atmospheric pollutant must be obtained and compared with the calculated values by means of statistical techniques. The meteorological parameters required for use of the models include wind direction, wind speed, and atmospheric stability. In some models, provisions may be made for including lapse rate and vertical mixing height. Most models will require data about the physical stack height, the diameter of the stack at the emission discharge point, the exit gas temperature and velocity, and the mass rate of emission of pollutants.

Models are usually classified as either short-term or climatological models. Short-term models are generally used under the following circumstances: (1) to estimate ambient concentrations where it is impractical to sample, such as over rivers or lakes, or at great distances above the ground; (2) to estimate the required emergency source reductions associated with periods of air stagnation under air pollution episode alert conditions; and (3) to estimate the most probable locations of high, short-term, ground-level concentrations as part of a site selection evaluation for the location of air monitoring equipment.

Climatological models are used to estimate mean concentrations over a long period or to estimate mean concentrations that exist at particular times of the day for each season over a long period. Long-term models are used as an aid in understanding atmospheric transport over long distances. We will be concerned only with short-term models in their most simple application.

Basic Point Source Gaussian Dispersion Model.   The basic Gaussian diffusion equation assumes that atmospheric stability is uniform throughout the layer into which the contaminated gas stream is discharged. The model assumes that turbulent diffusion is a random activity and

FIGURE 12–22

Plume dispersion
coordinate system.
(*Source:* Turner, D.B.
(1967) *Workbook of
Atmospheric Dispersion
Estimates*, U.S.
Department of Health,
Education, and Welfare,
U.S. Public Health Service
Publication No. 999-AP-28.)

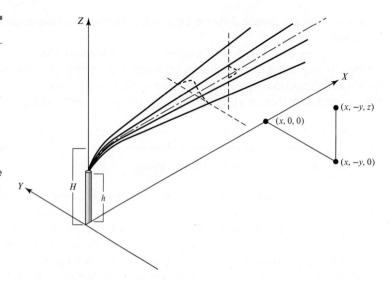

hence the dilution of the contaminated gas stream in both the horizontal and vertical direction can be described by the Gaussian or normal equation. The model further assumes that the contaminated gas stream is released into the atmosphere at a distance above ground level that is equal to the physical stack height plus the plume rise (ΔH). The model assumes that the degree of dilution of the effluent plume is inversely proportional to the wind speed (u). The model also assumes that pollutant material that reaches ground level is totally reflected back into the atmosphere like a beam of light striking a mirror at an angle. Mathematically, this ground reflection is accounted for by assuming a virtual or imaginary source located at a distance of $-H$ with respect to ground level, and emitting an imaginary plume with the same source strength as the real source being modeled. The same general idea can be used to establish other boundary layer conditions for the equations, such as limiting horizontal or vertical mixing.

The Model. We have selected the model equation in the form presented by D. B. Turner (1967). It gives the ground level concentration (χ) of pollutant at a point (coordinates x and y) downwind from a stack with an effective height (H) (Figure 12–22). The standard deviation of the plume in the horizontal and vertical directions is designated by s_y and s_z, respectively. The standard deviations are functions of the downward distance from the source and the stability of the atmosphere. The equation is as follows:

$$\chi\,(x, y, 0, H) = \left[\frac{E}{\pi s_y s_z u}\right]\left[\exp\left[-\frac{1}{2}\left(\frac{y}{s_y}\right)^2\right]\right]\left[\exp\left[-\frac{1}{2}\left(\frac{H}{s_z}\right)^2\right]\right] \tag{12–18}$$

where $\chi\,(x, y, 0, H)$ = downwind concentration at ground level (in $g \cdot m^{-3}$)

$\qquad E$ = emission rate of pollutant (in $g \cdot s^{-1}$)

$\qquad s_y, s_z$ = plume standard deviations (in m)

$\qquad u$ = wind speed (in $m \cdot s^{-1}$)

$\qquad x, y, z,$ and H = distances (in m)

$\qquad$ exp = exponential e such that terms in brackets immediately following are powers of e, that is, $e^{[\ ]}$, where $e = 2.7182$

The value for the effective stack height is the sum of the physical stack height (h) and the plume rise ΔH.

$$H = h + \Delta H \tag{12–19}$$

Note that the model assumes flat terrain, that is, $z = 0$.

ΔH may be computed from Holland's formula as follows (Holland, 1953):

$$\Delta H = \frac{v_s d}{u}\left[1.5 + \left(2.68 \times 10^{-2}(P)\left(\frac{T_s - T_a}{T_s}\right)d\right)\right] \tag{12–20}$$

where v_s = stack velocity (in m · s^{-1})

$\quad d$ = stack diameter (in m · s^{-1})

$\quad u$ = wind speed (in m

$\quad P$ = pressure (in kPa)

$\quad T_s$ = stack temperature (K)

$\quad T_a$ = air temperature (K)

The values of s_y and s_z depend on the turbulent structure or stability of the atmosphere. Figures 12–23 and 12–24 provide graphical relationships between the downwind distance x in kilometers and values of s_y and s_z in meters. The curves on the two figures are labeled A through F. The label A refers to very unstable atmospheric conditions, B to unstable atmospheric conditions, C to slightly unstable conditions, D to neutral conditions, E to stable conditions, and F to very stable atmospheric conditions. Each of these stability parameters represents an averaging time of approximately 3–15 min.

Other averaging times may be approximated by multiplying χ by empirical constants, for example, 0.36 for 24 h. Turner presented a table (Table 12–13) and discussion that allows an estimate of stability based on wind speed and the conditions of solar radiation.

FIGURE 12–23

Horizontal dispersion coefficient. (*Source:* Turner, 1967.)

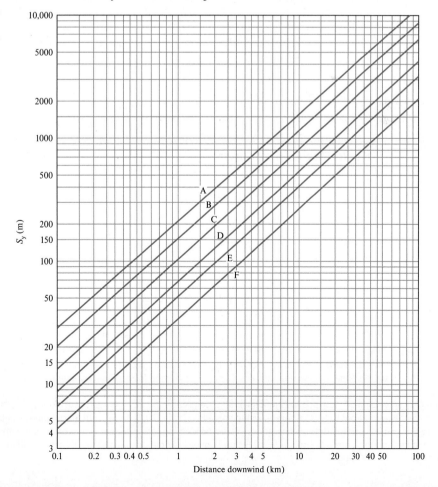

FIGURE 12–24

Vertical dispersion coefficient. (*Source:* Turner, 1967.)

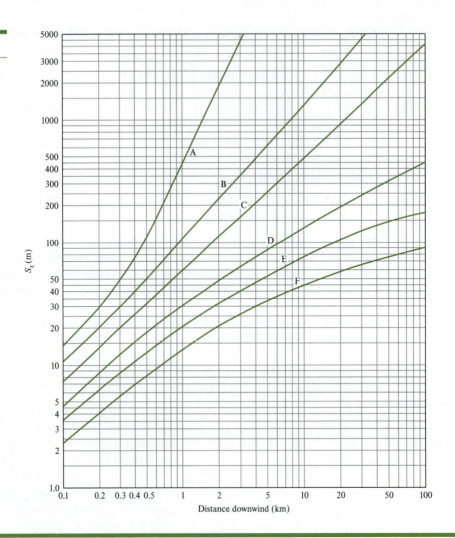

TABLE 12–13   **Key to Stability Categories**

Surface Wind Speed (at 10 m) (m · s⁻¹)	Day[a] Incoming Solar Radiation			Night[a]	
	Strong	**Moderate**	**Slight**	**Thinly Overcast or $\geq \frac{4}{8}$ Low Cloud**	**$\leq \frac{3}{8}$ Cloud**
<2	A	A–B	B		
2–3	A–B	B	C	E	F
3–5	B	B–C	C	D	E
5–6	C	C–D	D	D	D
>6	C	D	D	D	D

[a]The neutral class, D, should be assumed for overcast conditions during day or night. Note that "thinly overcast" is not equivalent to "overcast."

Notes: Class A is the most unstable and class F is the most stable class considered here. Night refers to the period from one hour before sunset to one hour after sunrise. Note that the neutral class, D, can be assumed for overcast conditions during day or night, regardless of wind speed.

   "*Strong*" incoming solar radiation corresponds to a solar altitude greater than 60 degrees with clear skies; "slight" insolation corresponds to a solar altitude from 15 to 35 degrees with clear skies. Table 170, Solar Altitude and Azimuth, in the Smithsonian Meteorological Tables, can be used in determining solar radiation. Incoming radiation that would be strong with clear skies can be expected to be reduced to moderate with broken $\left(\frac{5}{8} \text{ to } \frac{7}{8} \text{ cloud cover}\right)$ middle clouds and to slight with broken low clouds.

Source: Turner, 1967.

TABLE 12–14 Values of *a, c, d,* and *f* for Calculating s_y and s_z

Stability Class	x < 1 km				x ≥ 1 km		
	a	c	d	f	c	d	f
A	213	440.8	1.941	9.27	459.7	2.094	−9.6
B	156	100.6	1.149	3.3	108.2	1.098	2
C	104	61	0.911	0	61	0.911	0
D	68	33.2	0.725	−1.7	44.5	0.516	−13
E	50.5	22.8	0.678	1.3	55.4	0.305	−34
F	34	14.35	0.74	−0.35	62.6	0.18	−48.6

Source: Martin, 1976.

TABLE 12–15 Exponent *p* Values for Rural and Urban Regimes

Stability Class	Rural	Urban	Stability Class	Rural	Urban
A	0.07	0.15	D	0.15	0.25
B	0.07	0.15	E	0.35	0.30
C	0.10	0.20	F	0.55	0.30

Source: User's Guide for ISC3 Dispersion Models, Vol. II, EPA-454/B-95-003b, U.S. Environmental Protection Agency, September, 1995.

For computer solutions of the dispersion model, it is convenient to have an algorithm to express the stability class lines in Figures 12–23 and 12–24, D. O. Martin has developed the following equations that provide an approximate fit (Martin, 1976).

$$s_y = ax^{0.894} \tag{12–21}$$

$$s_z = cx^d + f \tag{12–22}$$

where the constants *a, c, d,* and *f* are defined in Table 12–14. These equations were developed to yield s_y and s_z in meters for downwind distance *x* in kilometers.

As noted earlier, the wind speed varies with height. Unless the wind speed at the effective height of the plume (*H*) is known, the wind speed must be corrected to account for the change in speed with elevation. For elevations up to a few hundred meters a power law expression of the following form may be used to estimate the wind speed at heights other than that of the measurement.

$$u_2 = u_1 \left(\frac{z_2}{z_1}\right)^p \tag{12–23}$$

where u_2 = windspeed at elevation z_2

u_1 = windspeed at elevation z_1

p = function of the terrain roughness and the stability

EPA's recommended values for *p* are shown in Table 12–15.

EXAMPLE 12–4 It has been estimated that the emission of SO_2 from a coal-fired power plant is 1656.2 g · s^{-1}. At 3 km downwind on an overcast summer afternoon, what is the centerline concentration of SO_2 if the wind speed at the top of the stack (120.0 m) is 4.50 m · s^{-1}? (*Note:* "centerline" implies $y = 0$.)

Stack parameters:
Height = 120.0 m
Diameter = 1.20 m
Exit velocity = 10.0 m · s⁻¹
Temperature = 315°C

Atmospheric conditions:
Pressure = 95.0 kPa
Temperature = 25.0°C

Solution We begin by determining the effective stack height (H).

$$\Delta H = \frac{(10.0 \text{ m} \cdot \text{s}^{-1})(1.20 \text{ m})}{4.50 \text{ m} \cdot \text{s}^{-1}} \left\{ 1.5 + \left[2.68 \times 10^{-2}(95.0) \left(\frac{588 \text{ K} - 298 \text{ K}}{588 \text{ K}} \right) (1.20 \text{ m}) \right] \right\}$$

$$= 8.0 \text{ m}$$

$$H = 120.0 + 8.0 = 128.0 \text{ m}$$

Next, we must determine the atmospheric stability class. The footnote to Table 12–13 indicates that the D class should be used for overcast conditions.

Because the wind speed measurement was taken at the top of the stack and the effective stack height (H) is not significantly different from the stack height, we will ignore the power law correction for wind speed.

From Equations (12–21) and (12–22) we can determine that at 3 km downwind with a D stability, the plume standard deviations are

$$s_y = 68(3)^{0.894} = 181.6 \text{ m}$$

$$s_z = 44.5(3)^{0.516} - 13 = 65.4 \text{ m}$$

Thus,

$$\chi = \left[\frac{1656.2}{\pi(181.6)(65.4)(4.50)} \right] \left\{ \exp \left[-\frac{1}{2} \left(\frac{0}{181.6} \right)^2 \right] \right\} \left\{ \exp \left[-\frac{1}{2} \left(\frac{128.0}{65.4} \right)^2 \right] \right\}$$

$$= 1.45 \times 10^{-3} \text{ g} \cdot \text{m}^{-3}, \text{ or } 1.5 \times 10^{-3} \text{ g} \cdot \text{m}^{-3} \text{ of } SO_2$$

Comments:

1. The plume standard deviations shown in Figures 12–22 and 12–23 are based on field measurements using a 10-minute sampling time. The field measurements were conducted in relatively open country. Application of the equations to other settings is not recommended.

2. This concentration is about 0.56 ppm. Several authors have proposed rules of thumb to correct for the averaging time difference between the standards and the estimates provided by Equation 12–18 and the standard deviation figures. These corrections range from 0.33 to 0.63 for an hourly average. Using these rules, the maximum one hour concentration would be estimated to be between 0.19 ppm and 0.35 ppm. These estimates would imply that the NAAQS one-hour standard would not be exceeded. Extrapolation of the rules of thumb to longer time periods is hard to justify because the wind seldom maintains "a direction" for more than a few minutes.

3. The stack temperature used for this example is a bit high for coal-fired power plants. For brick and lined stacks the temperature drop along the height of the stack is about 0.9°C · m⁻¹ (0.5°F/ft). The temperature of the gas entering the stack is on the order of 320°C (Fryling, 1967). Thus, a more realistic temperature would have been about 210°C.

Inversion Aloft. When an inversion is present, the basic diffusion equation must be modified to take into account the fact that the plume cannot disperse vertically once it reaches the inversion layer. The plume will begin to mix downward when it reaches the base of the inversion

FIGURE 12–25

Effect of elevated inversion on dispersion.

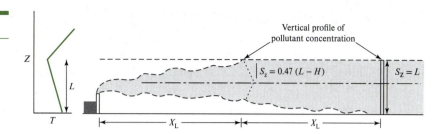

layer (Figure 12–25). The downward mixing will begin at a distance X_L downwind from the stack. The X_L distance is a function of the stability in the layer below the inversion. It has been determined empirically that the vertical standard deviation of the plume can be calculated with the following formula at the distance X_L.

$$s_z = 0.47(L - H) \qquad (12\text{–}24)$$

where L = height to bottom of inversion layer (in m)

H = effective stack height (in m)

When the plume reaches twice the distance to initial contact with the inversion base, the plume is said to be completely mixed throughout the layer below the inversion. Beyond a distance equal to $2X_L$ the centerline concentration of pollutants may be estimated by using the following equation.

$$\chi = \frac{E}{(2\pi)^{1/2}s_y u L} \qquad (12\text{–}25)$$

Note that s_y is determined by the stability of the layer below the inversion and the distance to the receptor using either Figure 12–24 or Equation 12–21. We call this the "inversion" or "short form" of the dispersion equation.

EXAMPLE 12–5 Determine the distance downwind from a stack at which we must switch to the inversion form of the dispersion model given the following meteorologic situation:

Effective stack height: 50 m	Cloud cover: none
Inversion base: 350 m	Time: 1130 h
Wind speed: 7.3 m · s⁻¹ at the effective stack height	Season: summer

Wind speed: 7.3 m · s^{-1} at the effective stack height

Solution Determine the stability class using Table 12–13. At > 6 m · s^{-1} with strong radiation, the stability class is C.

Calculate the value of s_z.

$$s_z = 0.47(350 - 50 \text{ m}) = 141 \text{ m}$$

Using Figure 12–25, find X_L. With $s_z = 141$, draw a horizontal line to stability class C. Drop a vertical line to the "distance downwind." Find $X_L = 2.5$ km.

Therefore, at any distance equal to or greater than 5 km downwind ($2X_L$), use the inversion form of the equation (Equation 12–25).

For distances less than 5 km, we use Equation 12–18 with s_z determined from the distance to the point of interest and the stability. Thus, in no case do we use s_z computed from Equation (12–24) to calculate χ.

12-9 INDOOR AIR QUALITY MODEL

If we envision a house or room in a house or other enclosed space as a simple box (Figure 12–26), then we can construct a simple mass-balance model to explore the behavior of the indoor air quality as a function of infiltration of outdoor air, indoor sources and sinks, and leakage to the outdoor air. If we assume the contents of the box are well mixed, then

Rate of pollutant increase in box = rate of pollutant entering box from outdoors

+ rate of pollutant entering box from indoor emissions

− rate of pollutant leaving box by leakage to outdoors

− rate of pollutant leaving box by decay   **(12–26)**

or

$$\mathrm{V}\frac{dC}{dt} = QC_a + E - QC - kC\mathrm{V} \tag{12-27}$$

where V = volume of box (in m^3)

C = concentration of pollutant ($g \cdot m^{-3}$)

Q = rate of infiltration of air into and out of box (in $m^3 \cdot s^{-1}$)

C_a = concentration of pollutant in outdoor air (in $g \cdot m^{-3}$)

E = emission rate of pollutant into box from indoor source (in $g \cdot s^{-1}$)

k = pollutant reaction rate coefficient (in s^{-1})

Emission factors for selected indoor air pollutants are listed in Table 12–16.

Reaction rate coefficients for a selected list of pollutants are given in Table 12–17. The general solution for Equation 12–27 is

$$C_t = \frac{(E/\mathrm{V}) + C_a(Q/\mathrm{V})}{(Q/\mathrm{V}) + k}\left\{1 - \exp\left[-\left(\frac{Q}{\mathrm{V}} + k\right)t\right]\right\} + C_0\exp\left[-\left(\frac{Q}{\mathrm{V}} + k\right)t\right] \tag{12-28}$$

The steady-state solution for Equation 12–28 may be found by setting $dC/dt = 0$ and solving for C.

$$C_\infty = \frac{QC_a + E}{Q + k\mathrm{V}} \tag{12-29}$$

When the pollutant is conservative and does not decay with time or have a significant reactivity, $k = 0$. In the special case when the pollutant is conservative and the ambient concentration is negligible and the initial indoor concentration is zero, Equation 12–27 reduces to

$$C_t = \frac{E}{Q}\left\{1 - \exp\left[-\left(\frac{Q}{\mathrm{V}}\right)t\right]\right\} \tag{12-30}$$

FIGURE 12–26

Mass-balance model for indoor air pollution.

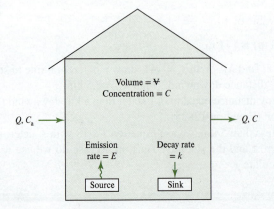

Volume = V
Concentration = C

$Q, C_a \longrightarrow$ $\longrightarrow Q, C$

Emission rate = E Decay rate = k

Source   Sink

TABLE 12–16   Emission Factors for Selected Indoor Air Pollution Sources and Pollutants

Pollutant	Emission Factor $\mu g \cdot h^{-1} \cdot m^{-2}$, at Various Times After Being Put into Use				
	1 h	1 Day	1 Week	1 Month	1 Year
Floor Materials: Carpet, Synthetic Fiber					
Formaldehyde	15	10	5	2	1
Styrene	50	20	6	3	2
Toluene	300	40	20	10	1
TVOC[a]	600	80	20	10	5
Paints and Coatings: Solvent-Based Paint					
Decane	200,000	2000	0	0	0
Nonane	100,000	100	0	0	0
Pentylcyclohexane	10,000	3000	0	0	0
Undecane	100,000	10,000	0	0	0
m,p Xylenes	50,000	5	0	0	0
TVOC	3×10^6	200,000	0	0	0
Paints and Coatings: Water-Based Paints					
Acetaldehyde	100	10	2	1	0
Ethylene glycol	20,000	20,000	15,000	4,000	0
Formaldehyde	40	100	2	1	0
TVOC	50,000	40,000	20,000	200	20

Photocopiers: Dry-Process, $\mu g \cdot h^{-1}$ per Machine

	Machine in Standby Mode	Machine Making Copies
Ethylebenzene	10	30,000
Styrene	500	7,000
m,p Xylenes	200	20,000

Formaldehyde Emission Factors, $\mu g \cdot d^{-1} \cdot m^{-2}$

Medium density fiberboard	17,600–55,000
Particleboard	2000–25,000
Paper products	260–280
Fiberglass products	400–470
Clothing	35–570

[a]TVOC = Total volatile organic compounds.
Sources: Godish, 2001, and Tucker, 2001.

TABLE 12–17   Reaction Rate Coefficients for Selected Pollutants

Pollutant	k (s^{-1})	Pollutant	k (s^{-1})
CO	0.0	Particulates ($<$0.5 μm)	1.33×10^{-4}
HCHO	1.11×10^{-4}	Radon	2.11×10^{-6}
NO	0.0	SO$_2$	6.39×10^{-5}
NO$_x$ (as N)	4.17×10^{-5}		

Source: Traynor, Allen, and Apte, 1982.

EXAMPLE 12–6 An unvented kerosene heater is operated for 1 h in an apartment having a volume of 200 m³. The heater emits SO_2 at a rate of 50 μg · s⁻¹. The ambient air concentration (C_a) and the initial indoor air concentration (C_o) of SO_2 are 100 μg · m⁻³. If the rate of ventilation is 50 L · s⁻¹ and the apartment is assumed to be well mixed, what is the indoor air concentration of SO_2 at the end of 1 h?

Solution The concentration may be determined using the general solution form of the indoor air quality model (Equation 12–28). The decay rate for SO_2 from Table 12–14 is 6.39×10^{-5} s⁻¹ and 50 L · s⁻¹ is equivalent to 0.050 m³ · s⁻¹.

$$C_t = \frac{(50 \text{ μg} \cdot \text{s}^{-1}/200 \text{ m}^3) + 100 \text{ μg} \cdot \text{m}^{-3}(0.050 \text{ m}^3 \cdot \text{s}^{-1}/200 \text{ m}^3)}{(0.050 \text{ m}^3 \cdot \text{s}^{-1}/200 \text{ m}^3) + 6.39 \times 10^{-5} \text{ s}^{-1}}$$

$$\times \left\{ 1 - \exp\left[-\left(\frac{0.050 \text{ m}^3 \cdot \text{s}^{-1}}{200 \text{ m}^3} + 6.39 \times 10^{-5} \text{ s}^{-1} \right)(3600 \text{ s}) \right] \right\}$$

$$+ (100 \text{ μg} \cdot \text{m}^{-3}) \exp\left[-\left(\frac{0.050 \text{ m}^3 \cdot \text{s}^{-1}}{200 \text{ m}^3} + 6.39 \times 10^{-5} \text{ s}^{-1} \right)(3600 \text{ s}^{-1}) \right]$$

$$= 876.08[1 - \exp(-1.13)] + 100 \exp(-1.13)$$

$$= 876.08(1 - 0.323) + 100(0.323)$$

$$= 593.09 + 32.3 = 625.39, \text{ or } 630 \text{ μg} \cdot \text{m}^{-3}$$

12–10 AIR POLLUTION CONTROL OF STATIONARY SOURCES

Gaseous Pollutants

Absorption. Control devices based on the principle of absorption attempt to transfer the pollutant from a gas to a liquid phase. This is a mass-transfer process in which the gas dissolves in the liquid. The dissolution may or may not be accompanied by a reaction with an ingredient of the liquid. **Mass transfer** is a diffusion process wherein the pollutant gas moves from points of higher concentration to points of lower concentration. The removal of the pollutant gas takes place in three steps:

1. Diffusion of the pollutant gas to the surface of the liquid
2. Transfer across the gas–liquid interface (dissolution)
3. Diffusion of the dissolved gas away from the interface into the liquid

Structures such as spray chambers (Figure 12–27) and towers or columns (Figure 12–28) are two classes of devices employed to absorb pollutant gases. In scrubbers, which are a type of spray chamber, liquid droplets are used to absorb the gas. In towers, a thin film of liquid is used as the absorption medium. Regardless of the type of device, the solubility of the pollutant in the liquid must be relatively high. If water is the solute, this generally limits the application to a few inorganic gases such as NH_3, Cl_2, and SO_2. Scrubbers are relatively inefficient absorbers but have the advantage of being able to simultaneously remove particulates. Towers are much more efficient absorbers but they become plugged by particulate matter.

FIGURE 12–27

Spray chamber.

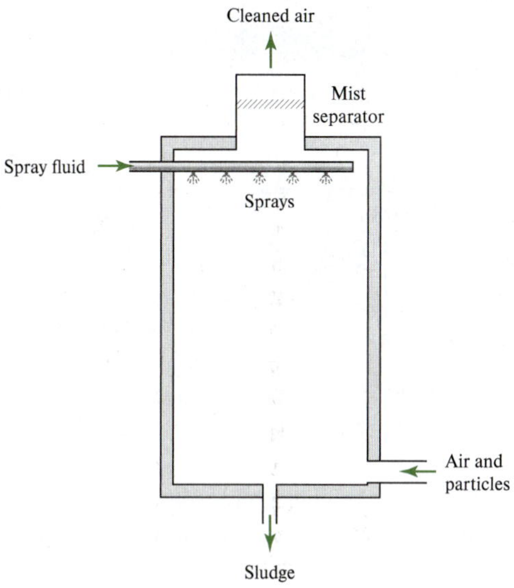

FIGURE 12–28

Absorption systems.

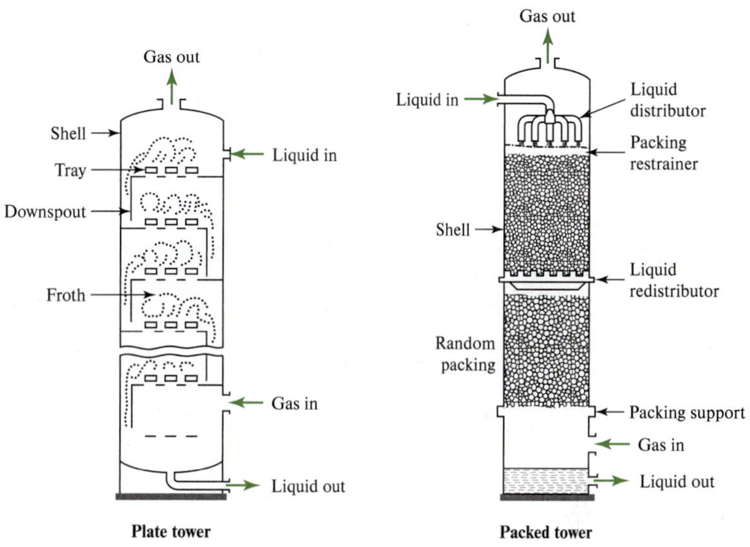

Plate tower

Packed tower

The amount of absorption that can take place for a nonreactive solution is governed by the partial pressure of the pollutant. For dilute solutions, as we have in pollution control systems, the relationship between partial pressure and the concentration of the gas in solution is given by Henry's law.

$$P_g = K_H C_{equil} \qquad (12\text{–}31)$$

where P_g = partial pressure of gas in equilibrium with liquid (in kPa)

K_H = Henry's law constant (in kPa · m^3 · g^{-1})

C_{equil} = concentration of pollutant gas in the liquid phase (in g · m^{-3})

Equation 12–31 implies that the partial pressure of the gas must increase as the liquid accumulates more pollutant or else it will come out of solution. Because the liquid is removing pollutant from the gas phase, the partial pressure is decreasing as the gas is cleaned. This is just the reverse of what we want to happen. The easiest way to get around this problem is to run the gas and liquid in opposite directions, called **countercurrent flow.** In this manner, the high concentration gas is absorbed into a liquid with a high pollutant concentration. The lower concentration gas is absorbed by liquid with no pollutants in it.

Adsorption.   Adsorption is a mass-transfer process in which the gas is bonded to a solid. It is a surface phenomenon. The gas (the adsorbate) penetrates into the pores of the solid (the adsorbent) but not into the lattice itself. The bond may be physical or chemical. Electrostatic forces hold the pollutant gas when physical bonding is significant. Chemical bonding is by reaction with the surface. Pressure vessels having a fixed bed are used to hold the adsorbent (Figure 12–29). Active carbon (activated charcoal), molecular sieves, silica gel, and activated alumina are the most common adsorbents. Active carbon is manufactured from nut shells (coconuts are great) or coal subjected to heat treatment in a reducing atmosphere. Molecular sieves are dehydrated zeolites (alkali-metal silicates). Sodium silicate is reacted with sulfuric acid to make silica gel. Activated alumina is a porous hydrated aluminum oxide. The common property of these adsorbents is a large "active" surface

FIGURE 12–29

Adsorption system.

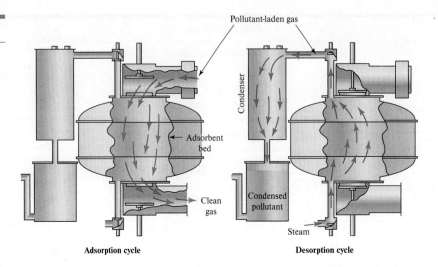

Pollutant-laden gas

Condenser

Adsorbent bed

Clean gas

Condensed pollutant

Steam

Adsorption cycle

Desorption cycle

area per unit volume after treatment. They are very effective for treating hydrocarbon pollutants. In addition, they can capture H_2S and SO_2. One special form of molecular sieve can also capture NO_2. With the exception of the active carbons, adsorbents have the drawback that they preferentially select water before any of the pollutants. Thus, water must be removed from the gas before it is treated. All of the adsorbents are subject to destruction at moderately high temperatures (150°C for active carbon, 600°C for molecular sieves, 400°C for silica gel, and 500°C for activated alumina). They are very inefficient at these high temperatures. In fact, their activity is regenerated at these temperatures!

The relation between the amount of pollutant adsorbed and the equilibrium pressure at constant temperature is called an **adsorption isotherm.** The equation that best describes this relation for gases is the one derived by Langmuir (1918).

$$W = \frac{aC_g^*}{1 + bC_g^*} \tag{12–32}$$

where W = amount of gas per unit mass of adsorbent (in kg · kg^{-1})

a, b = constants determined by experiment

C_g^* = equilibrium concentration of gaseous pollutant (in g · m^{-3})

In the analysis of experimental data, Equation 12–32 is rewritten as follows:

$$\frac{C_g^*}{W} = \frac{1}{a} + \frac{b}{a} C_g^* \tag{12–33}$$

In this arrangement, a plot of (C_g^*/W) versus C_g^* should yield a straight line with a slope of (b/a) and an intercept equal to ($1/a$).

In contrast to absorption towers where the collected pollutant is continuously removed by flowing liquid, the collected pollutant remains in the adsorption bed. Thus, while the bed has sufficient capacity, no pollutants are emitted. At some point in time, the bed will become saturated with pollutant. As saturation is approached, pollutant will begin to leak out of the bed, a process called **breakthrough.** When the bed capacity is exhausted, the influent and effluent concentration will be equal. A typical breakthrough curve is shown in Figure 12–30. To allow for continuous operation, two beds are provided; while one is collecting pollutant, the other is being regenerated. The concentrated gas released during regeneration is usually returned to the process as recovered product.

Combustion. When the contaminant in the gas stream is oxidizable to an inert gas, combustion is a possible alternative method of control. Typically, CO and hydrocarbons fall into this

FIGURE 12–30

Adsorption wave and breakthrough curve.

FIGURE 12–31

Direct flame incineration.

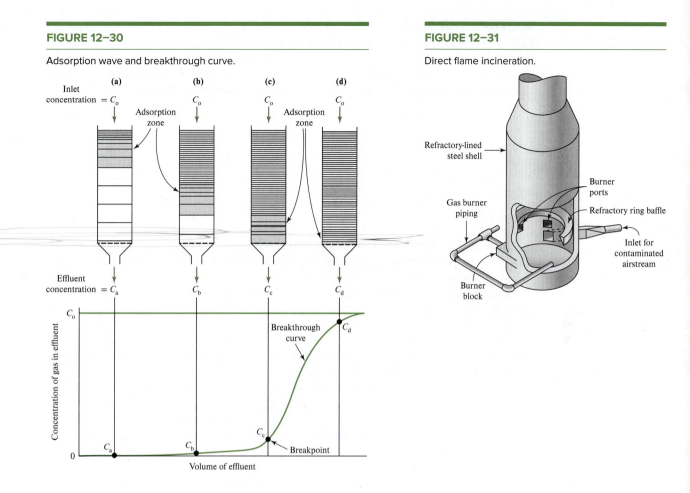

category. Both direct flame combustion by afterburners (Figure 12–31) and catalytic combustion have been used in commercial applications.

Direct flame incineration is the method of choice if two criteria are satisfied. First, the gas stream must have an energy concentration greater than $3.7 \text{ MJ} \cdot \text{m}^{-3}$. At this energy concentration, the gas flame will be self-supporting after ignition. Below this point, supplementary fuel is required. The second requirement is that none of the by-products of combustion be toxic. In some cases the combustion by-product may be more toxic than the original pollutant gas. Direct flame incineration has been successfully applied to varnish-cooking, meat-smokehouse, and paint bake-oven emissions.

Some catalytic materials enable oxidation to be carried out in gases that have an energy content of less than $3.7 \text{ MJ} \cdot \text{m}^{-3}$. Conventionally, the catalyst is placed in beds similar to adsorption beds. Frequently, the active catalyst is a platinum or palladium compound. The supporting lattice is usually a ceramic. Aside from expense, a major drawback of catalysts use is their susceptibility to poisoning by sulfur and lead compounds in trace amounts. Catalytic combustion has successfully been applied to printing-press, varnish-cooking, and asphalt-oxidation emissions.

Flue Gas Desulfurization

Flue gas desulfurization (FGD) systems fall into two broad categories: nonregenerative and regenerative. In **nonregenerative** systems the reagent used to remove the sulfur oxides from the gas stream is used and discarded. In **regenerative** systems the reagent is recovered and reused. In terms of the number and size of systems installed, nonregenerative systems dominate.

Nonregenerative Systems.   There are nine commercial nonregenerative systems (Hance and Kelly, 1991). All have reaction chemistries based on lime (CaO), caustic soda (NaOH), soda ash (Na_2CO_3), or ammonia (NH_3).

The SO_2 removed in a lime/limestone-based FGD system is converted to sulfite. The overall reactions are generally represented by (Karlsson and Rosenberg, 1980).

$$SO_2 + CaCO_3 \rightleftharpoons CaSO_3 + CO_2 \tag{12-34}$$

$$SO_2 + Ca(OH)_2 \rightleftharpoons CaSO_3 + H_2O \tag{12-35}$$

when using limestone and lime, respectively. Part of the sulfite is oxidized with the oxygen content in the flue gas to form sulfate.

$$CaSO_3 + \tfrac{1}{2}O_2 \rightleftharpoons CaSO_4 \tag{12-36}$$

Although the overall reactions are simple, the chemistry is quite complex and not well defined. The choice between lime and limestone, the type of limestone, and method of calcining and slaking can influence the gas–liquid–solid reactions taking place in the absorber.

The principal types of absorbers used in the wet scrubbing systems include Venturi scrubber–absorbers, static packed scrubbers, moving-bed absorbers, tray towers, and spray towers (Black and Veatch, 1983).

Spray dryer-based FGD systems consist of one or more spray dryers and a particulate collector.* The reagent material is typically a slaked lime slurry or a slurry of lime and recycled material. Although lime is the most common reagent, soda ash has also been used. The reagent is injected in droplet form into the flue gas in the spray dryer. The reagent droplets absorb SO_2 while being dried. Ideally, the slurry or solution droplets are completely dried before they hit the wall of the dryer vessel. The flue gas stream becomes more humidified in the process of evaporation of the reagent droplets, but it does not become saturated with water vapor. This is the single most significant difference between spray dryer FGD and wet scrubber FGD. The humidified gas stream and a significant portion of the particulate matter (fly ash, FGD reaction products, and unreacted reagent) are carried by the flue gas to the particulate collector located downstream of the spray dryer vessel (Cannell and Meadows, 1985). Generally, larger units firing high-sulfur coals use wet FGD. Smaller units use spray dryers.

Control Technologies for Nitrogen Oxides

Almost all nitrogen oxide (NO_x) air pollution results from combustion processes. They are produced from the oxidation of nitrogen bound in the fuel, from the reaction of molecular oxygen and nitrogen in the combustion air at temperatures above 1600 K (see Equation 12–5), and from the reaction of nitrogen in the combustion air with hydrocarbon radicals. Control technologies for NO_x are grouped into two categories: those that prevent the formation of NO_x during the combustion process and those that convert the NO, formed during combustion into nitrogen and oxygen (Prasad, 1995).

Prevention.   The processes in this category employ the fact that reduction of the peak flame temperature in the combustion zone reduces NO_x formation. Nine alternatives have been developed to reduce flame temperature.

1. Minimizing operating temperatures
2. Fuel switching
3. Low excess air
4. Flue gas recirculation
5. Lean combustion

---

*Historically, from a mass-transfer point of view, spray drying refers to the evaporation of a solvent from an atomized spray. Simultaneous diffusion of a gaseous species into the evaporating droplet is not true spray drying. Nonetheless, many authors have adopted the term *spray drying* as synonymous with dry scrubbing.

6. Staged combustion
7. Low NO_x burners
8. Secondary combustion
9. Water–steam injection

Routine burner tune-ups and operation with combustion zone temperatures at minimum values reduce the fuel consumption and NO_x formation. Converting to a fuel with a lower nitrogen content or one that burns at a lower temperature will reduce NO_x formation. For example, petroleum coke has a lower nitrogen content and burns with a lower flame temperature than coal. On the other hand, natural gas has no nitrogen content but burns at a relatively high flame temperature and, thus, produces more NO_x than coal.

Low excess air and flue gas recirculation work on the principle that reduced oxygen concentrations lower the peak flame temperatures. In contrast, in lean combustion, additional air is introduced to cool the flame.

In staged combustion and low NO_x burners, initial combustion takes place in a fuel-rich zone that is followed by the injection of air downstream of the primary combustion zone. The downstream combustion is completed under fuel-lean conditions at a lower temperature.

Staged combustion consists of injecting part of the fuel and all of the combustion air into the primary combustion zone. Thermal NO_x production is limited by the low flame temperatures that result from high excess air levels.

Water–steam injection reduces thermal NO_x emissions by lowering the flame temperature.

Postcombustion.   Three processes may be used to convert NO_x to nitrogen gas: selective catalytic reduction (SCR), selective noncatalytic reduction (SNCR), and nonselective catalytic reduction (NSCR).

The SCR process uses a catalyst bed (usually vanadium-titanium, or platinum-based and zeolite) and anhydrous ammonia (NH_3). After the combustion process, ammonia is injected upstream of the catalyst bed. The NO_x reacts with the ammonia in the catalyst bed to form N_2 and water.

In the SNCR process ammonia or urea is injected into the flue gas at an appropriate temperature (870–1090°C). The urea is converted to ammonia, which reacts to reduce the NO_x to N_2 and water.

NSCR uses a three-way catalyst similar to that used in automotive applications. In addition to NO_x control, hydrocarbons and carbon monoxide are convened to CO_2 and water. These systems require a reducing agent similar to CO and hydrocarbons upstream of the catalyst. Larger boilers that have postcombustion NO_x controls are generally equipped with SCR.

Typical reduction capabilities of the NO_x techniques range from 30 to 60% for the "prevention" methods, 30 to 50% for SNCR, and 70 to 90% for the SCR systems (Srivastava, Staudt, and Josewicz, 2005).

Particulate Pollutants

Cyclones.   For particle sizes greater than about 10 μm in diameter, the collector of choice is the cyclone (Figure 12–32). This is an inertial collector with no moving parts. The particulate-laden gas is accelerated through a spiral motion, which imparts a centrifugal force to the particles. The particles are hurled out of the spinning gas and impact on the cylinder wall of the cyclone. They then slide to the bottom of the cone. Here they are removed through an airtight valving system.

As the diameter of the cyclone is reduced, the efficiency of collection is increased; however, the pressure drop also increases. This increases the power requirements for moving the gas through the collector. Because an efficiency increase will result, even if the tangential velocity remains constant, the efficiency may be increased without increasing the power consumption by using multiple cyclones in parallel (**multiclones**).

Cyclones are quite efficient for particles larger than 10 μm. Conversely, they are not very efficient for particles 1 μm or less in diameter. Thus, they are employed only for coarse dusts. Some applications include controlling emissions of wood dust, paper fibers, and buffing fibers.

FIGURE 12–32

Reverse flow cyclone.

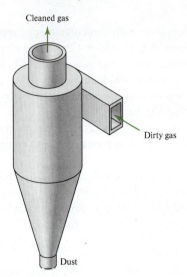

FIGURE 12–33

Baghouse.

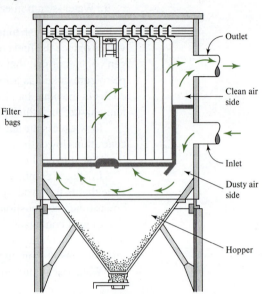

Filters. When high efficiency control of particles smaller than 5 μm is desired, a filter may be selected as the control method. Two types are in use: (1) the deep bed filter and (2) the baghouse (Figure 12–33). The deep bed filter resembles a furnace filter. A packing of fibers is used to intercept particles in the gas stream. For relatively clean gases and low volumes, such as air conditioning systems, these are quite effective. For dirty industrial gas with high volumes, the baghouse is preferable.

The fundamental mechanisms of collection by filters include screening or sieving because the particles are larger than the openings between the fibers, interception by the fibers themselves, and electrostatic attraction because of the difference in static charge on the particle and fiber. Once a dust cake begins to form on the fabric, sieving is probably the dominant mechanism.

The bags are made of either natural or synthetic fibers. Synthetic fibers are widely used as filtration fabrics because of their low cost, better temperature- and chemical-resistance characteristics, and small fiber diameter. Bag life varies between 1 and 5 years. Two years is considered normal. Bag diameters range from 0.1 to 0.35 m. Their lengths vary between 2 and 10 m. The bags are suspended from the toe and fastened by a collar at the open end. They are arranged in groups in separate compartments.

Reverse-air baghouses operate by directing the dirty gas into the inside of the bag. The particulate matter is collected on the inside of the bag in much the same manner as a vacuum cleaner bag. The bags are cleaned by isolating a compartment and reversing the gas flow. The reverse flow combined with the inward collapse of the bag causes the collected dust cake to fall into the hopper below.

Pulse-jet baghouses are designed with frame structures, called cages, that support the bags. The particulate matter is collected on the outside of the bag. The dust cake is removed by directing a pulsed jet of compressed air into the bag. This causes a sudden expansion of the bag. Dust is removed primarily by inertial forces as the bag reaches maximum expansion.

Baghouses have found a wide variety of applications. Examples include the carbon black industry, cement crushing, feed and grain handling, gypsum, limestone crushing, and sanding machines. Baghouse application to boiler flue gas is finding wide acceptance currently, in contrast to past practice. This is because of the better thermal properties of the bag material. Cotton and wool fiber bags, for example, cannot be used for sustained temperatures above 90–100°C.

Glass fiber bags, however, can be used at temperatures up to 260°C. Of all of the particulate control devices, only the filtration method has potential to include the addition of adsorption media to facilitate concurrent removal of gas phase contaminants. Baghouse sizing is based on the ratio of gas flow rate to filter area ($m^3 \cdot s^{-1} \cdot m^{-2}$ of cloth area). Note that this air-to-cloth ratio is a velocity ($m^3 \cdot s^{-1}$). An average value would be about 0.01 $m^3 \cdot s^{-1} \cdot m^{-2}$ for a woven fabric in a conventional baghouse.

Liquid Scrubbing.   When the particulate matter to be collected is wet, corrosive, or very hot, the fabric filter may not work. Liquid scrubbing might. Typical scrubbing applications include control of emission of talc dust, phosphoric acid mist, foundry cupola dust, and open hearth steel furnace fumes.

Liquid scrubbers vary in complexity. Simple spray chambers are used for relatively coarse particle sizes. For high-efficiency removal of fine particles, the combination of a Venturi scrubber followed by a cyclone would be selected (Figure 12–34). The underlying principle of operation of the liquid scrubbers is that a differential velocity between the droplets of collecting liquid and the particulate pollutant allows the particle to impinge onto the droplet. Because the droplet–particle combination is still suspended in the gas stream, an inertial collection device is placed downstream to remove it. Because the droplet enhances the size of the particle, the collection efficiency of the inertial device is higher than it would be for the original particle without the liquid drop.

Electrostatic Precipitation (ESP).   High-efficiency, dry collection of particles from hot gas streams can be obtained by electrostatic precipitation of the particles. The ESP is usually constructed of alternating plates and wires (Figure 12–35). A large direct current potential (30–75 kV) is established between the plates and wires, which results in the creation of an ion field between the wire and plate (Figure 12–36a). As the particle-laden gas stream passes between the wire and the plate, ions attach to the particles, giving them a net negative charge (Figure 12–36b). The particles then migrate toward the positively charged plate, where they stick (Figure 12–36c). The plates are rapped at frequent intervals and the agglomerated sheet of particles falls to a hopper.

Unlike the baghouse, the gas flow between the plates is not stopped during cleaning. The gas velocity through the ESP is kept low (less than 1.5 $m \cdot s^{-1}$) to allow particle migration. Thus, the terminal settling velocity of the sheet is sufficient to carry it to the hopper before it exits the precipitator.

One operational problem of ESPs is of particular note. **Fly ash** is a generic term used to describe the particulate matter carried in the effluent gases from furnaces burning fossil fuels. ESPs often are used to collect fly ash. The strongest force holding fly ash to the collection plate is electrostatic and is caused by the flow of current through the fly ash. The fly ash acts like a resistor and, hence, resists the flow of current. This resistance to current flow is called the **resistivity of the fly ash.** It is measured in units of ohms per centimeter ($\Omega \cdot cm^{-1}$). If the resistivity is too low (less than $10^4 \, \Omega \cdot cm^{-1}$), not enough charge will be retained to produce a strong

FIGURE 12–34

Venturi scrubber.

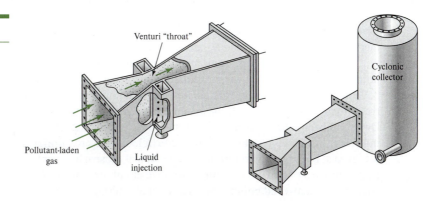

FIGURE 12–35

Electrostatic precipitator with **(a)** wire in tube, **(b)** wire and plate. (*Source:* NCAPC Cincinnati.)

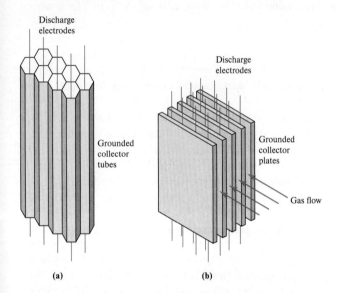

(a)   (b)

FIGURE 12–36

Particle charging and collection in ESP. (*Source:* NCAPC Cincinnati.)

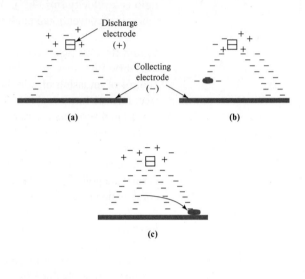

force and the particles will not "stick" to the plate. Conversely, and often more importantly, if the resistivity is too high (greater than $10^{10}\ \Omega \cdot cm^{-1}$), there is an insulating effect. The layer of fly ash breaks down locally, and a local discharge of current (**back corona**) from the normally passive collection electrode occurs. This discharge lowers the sparkover voltage and produces positive ions that decrease particle charging and, hence, collection efficiency.

The presence of SO_2 in the gas stream reduces the resistivity of the fly ash. This makes particle collection relatively easy. However, the mandate to eliminate SO_2 emissions has frequently been satisfied by switching to low-sulfur coal. The result has been increased particulate emissions. This problem can be resolved by adding conditioners such as SO_3 or NH_3 to reduce the resistivity or by building larger precipitators.

Electrostatic precipitators have been used to control air pollution from electric power plants, Portland cement kilns, blast furnace gas, kilns and roasters for metallurgical processes, and mist from acid production facilities.

Control Technologies for Mercury

During combustion, the mercury in coal is volatilized and converted to Hg° vapor. As the flue gas cools, a series of complex reactions converts Hg° to Hg^{2+} and particulate Hg compounds (Hg_p). The presence of chlorine favors the formation of mercuric chloride. In general, the majority of gaseous mercury in bituminous coal-fired boilers is Hg^{2+}. The majority of gaseous mercury in subbituminous and lignite-fired boilers is Hg°.

Existing boiler control equipment achieves some ancillary removal of mercury compounds. Hg_p is collected in particulate control equipment. Soluble Hg^{2+} compounds are collected in FGD systems. Particulate control equipment has achieved a range of emission reductions from 0 to 90%. Of these units, fabric filters obtained the highest levels of control. Dry scrubbers achieve average total mercury (particulate plus compounds) removal ranging from 0 to 98%. Wet FGD scrubber efficiencies were similar. Higher efficiencies were achieved at boilers using bituminous coal than at those using subbituminous and lignite coal. EPA estimates that existing controls remove about 36% of the 75 Mg of mercury input with coal in U.S. coal-fired boilers (Srivastava et al., 2005).

There are two broad approaches being developed to control mercury emissions: powdered activated carbon (PAC) injection and enhancement of existing control devices. The leading candidates for top efficiency (>90%) are PAC with pulse-jet fabric filters and FGD (wet or dry) with fabric filters (U.S. EPA, 2003b). In 2013, bromine emerged as a major component of mercury control strategies for coal-burning power plants. Treating PAC with bromine enhances the capture of mercury to achieve 90% or more removal with fabric filters (Reisch, 2015).

12–11   AIR POLLUTION CONTROL OF MOBILE SOURCES

Engine Fundamentals

Before we examine some cures for the pollution from the common gasoline automobile engine, it may be useful to compare the three familiar types of engines: gasoline, diesel, and jet.

The Gasoline Engine.   In the typical automobile engine with no air pollution controls, a mixture of fuel and air is fed into a cylinder and is compressed and ignited by a spark from the spark plug. The explosive energy of the burning mixture moves the pistons. The pistons' motion is transmitted to the crankshaft that drives the car. The burnt, spent mixture passes out of the engine and out through the tail pipe.

The ratio of air-to-fuel is the single most important factor in determining emissions from a four-stroke internal combustion engine. An estimate of the theoretical mass of air required to burn the fuel may be made using C_7H_{13} to represent the blend of hydrocarbons that we call gasoline. The complete combustion of C_7H_{13} in pure oxygen may be expressed by the following stoichiometric equation:

$$C_7H_{13} + 10.25O_2 \rightarrow 7CO_2 + 6.5H_2O \tag{12–37}$$

Because air rather than pure oxygen is used, we use the molar ratio of nitrogen to air, i.e., 3.76 moles of N_2 for each mole of O_2. Thus, Equation 9–71 is rewritten as

$$C_7H_{13} + 10.25O_2 + 38.54N_2 \rightarrow 7CO_2 + 6.5H_2O + 38.54N_2 \tag{12–38}$$

The oxidation of nitrogen to nitrogen oxides has been ignored in this reaction. The calculation of the stoichiometric air-to-fuel ratio for this reaction is illustrated in Example 12–7.

EXAMPLE 12–7   Determine the stoichiometric air-to-fuel ratio for C_7H_{13}. Ignore constituents other than oxygen and nitrogen in air and ignore the oxidation of nitrogen to nitrogen oxides.

Solution   From Equation 12–38, 10.25 moles of oxygen and 38.54 moles of nitrogen react with each mole of C_7H_{13}. Calculate the molar masses of each constituent as follows:

$$1 \text{ mole of } C_7H_{13} = (7 \text{ moles} \times 12 \text{ g/mole}) + (13 \text{ moles} \times 1 \text{ g/mole}) = 97 \text{ g}$$

$$10.25 \text{ moles of } O_2 = 10.25 \times 2 \text{ moles} \times 16 \text{ g/mole} = 328 \text{ g}$$

$$38.54 \text{ moles of } N_2 = 38.54 \times 2 \text{ moles} \times 14 \text{ g/mole} = 1079 \text{ g}$$

The stoichiometric air-to-fuel ratio is

$$\frac{\text{Air}}{\text{Fuel}} = \frac{(328 \text{ g} + 1079 \text{ g})}{97 \text{ g}} = 14.5$$

FIGURE 12–37

Diurnal variation of NO, NO_2, and O_3 concentrations in Los Angeles on July 19, 1965. (*Source:* NAPCA, 1970.)

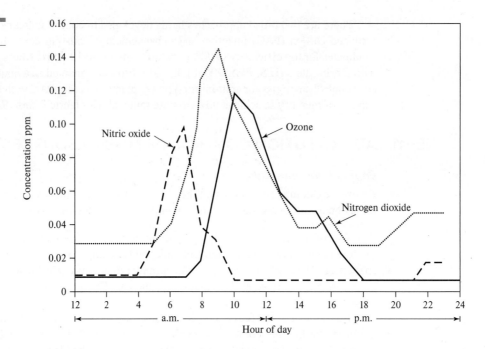

FIGURE 12–37

Diurnal variation of NO, NO_2, and O_3 concentrations in Los Angeles on July 19, 1965. (*Source:* NAPCA, 1970.)

For maximum power, however, the proportion of air to fuel must be less. Most driving takes place at less than the 15-to-1 air-to-fuel ratio. Combustion is incomplete, and substantial amounts of material other than carbon dioxide and water are discharged through the tail pipe. One result of having an inadequate supply of air is the emission of carbon monoxide instead of carbon dioxide. Other by-products are unburned gasoline and hydrocarbons.

Because of the high temperatures and pressures that exist in the cylinder, copious amounts of NO_x are formed (see Equation 12–5).

The role of automobile emissions in the formation of ozone is illustrated in Figure 12–37. As the work day begins and automobiles take to the road, NO concentrations increase. NO is oxidized to NO_2 that increases in concentration while the NO concentration declines. Photolysis by sunlight decomposes NO_2 to NO plus O. The atomic oxygen combines with diatomic oxygen in the atmosphere to form ozone. Ozone can then convert NO back to NO_2. These reactions are summarized in the following equations:

During combustion: $N_2 + O_2 \rightarrow 2NO$ (12–39)

Oxidation of NO in atmosphere: $2NO + O_2 \rightarrow 2NO_2$ (12–40)

Photolysis: $NO_2 + h\nu \rightarrow NO + O$ (12–41)

Formation of ozone: $O + O_2 + M \rightarrow O_3 + M$ (12–42)

Conversion of NO back to NO_2: $O_3 + NO \rightarrow NO_2 + O_2$ (12–43)

where $h\nu$ represents a photon and M represents another molecule such as nitrogen. These reaction products in combination with reaction products from hydrocarbons and ozone are the constituents of photochemical smog.

The Diesel Engine.   The diesel engine differs from the four-stroke gasoline engine in two respects. First, the air supply is unthrottled; that is, its flow into the engine is unrestricted. Thus, a diesel normally operates at a higher air-to-fuel ratio than does a gasoline engine. Second, there is no spark ignition system. The air is heated by compression. That is, the air in the engine cylinder is squeezed until it exerts a pressure high enough to raise the air temperature to about 540°C, which is enough to ignite the fuel oil as it is injected into the cylinder.

FIGURE 12–38

Effect of air-to-fuel ratio **(a)** on emissions and **(b)** on power and economy. (*Source:* John Wiley & Sons, Inc.)

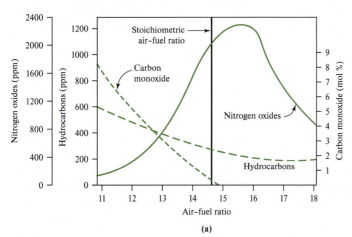

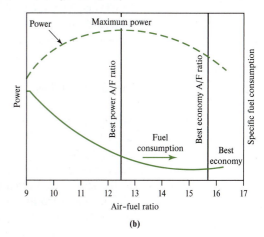

A well-designed, well-maintained, and properly adjusted diesel engine will emit less CO and hydrocarbons than the four-stroke engine because of the diesel's high air-to-fuel ratio. The higher operating temperatures, however, lead to substantially higher NO_x emissions.

The Jet Engine.   Large commercial aircraft that use the thrust of compressed gases for propulsion may contribute significant amounts of particulates and NO_x to urban atmospheres.

Propulsion is obtained when air drawn into the front of the engine is compressed and then heated by burning fuel. The expanding gas passes through turbine blades, which drive the compressor. The gas then exits the engine through an exhaust nozzle.

Effect of Design and Operating Variables on Emissions.   The list of variables that affect internal combustion (automobile) emissions includes the following (Patterson and Henein, 1972):

1. Air-to-fuel ratio
2. Load or power level
3. Speed
4. Spark timing
5. Exhaust back pressure
6. Valve overlap
7. Intake manifold pressure
8. Combustion chamber deposit buildup
9. Surface temperature
10. Surface-to-volume ratio
11. Combustion chamber design
12. Stroke-to-bore ratio
13. Displacement per cylinder
14. Compression ratio

A discussion of all of these items is beyond the scope of an introductory text such as this. Therefore, we shall restrict ourselves to a few of the variables that serve to illustrate the kinds of problems encountered in trying to design pollution out of an internal combustion engine.

The air-to-fuel ratio (A/F) is fairly easy to regulate. As we noted previously, it has a direct effect on all three emissions. As shown in Figure 12–38a, the A/F of 14.6 is the **stoichiometric** mixture for complete combustion.* At lower ratios, both CO and HC emissions increase. At higher ratios, to about 15.5, NO_x emissions increase. At very lean mixtures (high A/F ratios), the NO_x emission begins to decrease.

---

*Note that *stoichiometric* (stoi-chio-met-ric) means "combined in exactly the proper proportions according to their molecular weight."

Combustion chamber design has changed dramatically over the last three decades. For example, the *extra-lean-burn engine* cylinder has an A/F ratio that is richer at the spark plug and leaner elsewhere. In this engine, the A/F ratio is as high as 25. The result is a very low emission of CO and VOCs. Coupled with a higher drive ratio, this engine also gives improved gasoline efficiency. As with all lean-burn engine modifications, this one has the drawback of higher NO_x formation (Cooper and Alley, 2002).

Retarding the timing of the spark relative to the stroke of the piston decreases the hydro-carbon emissions by reducing the amount of unburned fuel. NO_x emissions also decrease with increased retarding. Little or no change occurs in CO emissions.

Hybrid automobiles.   The hybrid utilizes both gasoline and electricity for energy to drive the vehicle. Fuel efficiency is increased by harnessing small amounts electricity generated during braking and coasting to recharge batteries that power the vehicle at lower speeds. The current hybrids on the market can achieve 25 kilometers per liter of fuel ($km \cdot L^{-1}$) compared to the standard internal combustion engine rate of 6–10 $km \cdot L^{-1}$. Research vehicles and "home" modified hybrids that use more powerful batteries or just more batteries have been demonstrated to achieve up to 100 $km \cdot L^{-1}$. These "plug-in" versions must be recharged from a standard electric source. Although the plug-ins are not yet cost efficient, the increased fuel efficiency of the standard hybrid, which is cost efficient, effectively reduces the automobile's emissions by 60%.

Control of Automobile Emissions

Blowby.   The flow of air past the moving vehicle is directed through the crankcase to rid it of any gas–air mixture that has blown past the pistons, any evaporated lubricating oil, and any escaped exhaust products. The air is drawn in through a vent and emitted through a tube extending from the crankcase at a rate that depends on the speed of the car. About 20–40% of the car's total hydrocarbon emissions are sent into the atmosphere from the crankcase. These emissions are called crankcase blowby. All vehicles manufactured after 1963 are required to have a positive crankcase ventilation valve to eliminate blowby emissions.

Fuel Tank Evaporation Losses.   Evaporation of VOCs from the fuel tank is controlled by one of two systems. The simplest system is to place an activated charcoal adsorber in the tank vent line. Thus, as the gasoline expands during warm weather and forces vapor out of the vent, the VOCs are trapped on the activated carbon.

An alternative method is to vent the tank to the crankcase, but with this method, it is more difficult to achieve 100% control than with the activated charcoal system.

Carburetor Evaporation Losses.   During engine operation, the hydrocarbon vapors generated in the carburetor are vented internally to the engine intake system. After the engine is shut off, the gasoline in the float bowl continues to evaporate because of the high temperature in the engine compartment. This phenomenon is called **hot soak.** These losses may be controlled by the use of an activated carbon adsorption system (called a **canister**) or by venting the vapors to the crankcase. The canister system is preferred. Modern fuel injection systems do not have carburetors and thus avoid evaporation losses.

Engine Exhaust.   The number of techniques for reducing engine exhaust emissions far exceeds the list of engine variables that contribute to the production of emissions. In general, the control strategies can be grouped into three categories: engine modifications, fuel system modifications, and exhaust treatment devices. To some extent all three techniques have been employed. The stringent requirements of the Clean Air Act have led to the general adoption of catalytic converter exhaust treatment systems.

A three-way catalyst is currently employed as the most effective means of reducing emissions. It provides for the simultaneous oxidation of hydrocarbons and CO and the reduction of NO_x. The catalyst materials are either noble metals or base metals deposited on an inert support material.

The major problems with the catalysts are their susceptibility to "poisoning" by lead, phosphorus, and sulfur, and their poor wear characteristics under thermal cycling. The poisoning problem is solved by removing the lead, phosphorus, and sulfur from the fuel.

Another approach being implemented is fuel modification. The use of lead in fuels was completely phased out by January 1996. In addition, diesel fuel refining is being changed so that it will contain less sulfur and emit 20% less VOCs. Lowering the gasoline vapor pressure (called the **Reid vapor pressure**) reduces hydrocarbon emissions. **Oxyfuel** is yet another alternative. Oxyfuel is one with more oxygen to allow the fuel to burn more efficiently. Other alternatives include alcohols, liquified petroleum gas, and natural gas.

Inspection–Maintenance (I/M) Programs.   The devices installed by automobile manufacturers are extremely successful in minimizing the pollution from the exhaust and from evaporating fuel. However, as with other aspects of running an automobile, these devices wear out and fail. Because their failure does not inhibit the operation of the automobile, they are not likely to be repaired by the owner. In those areas that have exceeded the NAAQS (nonattainment areas), inspection–maintenance programs have been implemented to ensure that the control devices are in good working order. These programs require periodic checks of the exhaust and, in some instances, the evaporative controls. If the vehicle fails the inspection, the owner is required to provide the required maintenance and have the vehicle reinspected. Failure to pass the inspection may be cause to deny the issuance of license plates or tags.

12–12 WASTE MINIMIZATION FOR SUSTAINABILITY

The best and first step in any air pollution control strategy should be to minimize the production of pollutants in the first place. Because a large proportion of air pollutants results from the combustion of fossil fuels, an obvious approach to waste minimization is to conserve energy. Modern technology has yielded more efficient furnaces that improve fuel use, but simple measures such as turning off the lights in unoccupied rooms, turning down the heat at night and, in factories, during weekends and holidays, can have a dramatic effect. Because of the interrelationship between energy consumption and water supply, water conservation also reduces air pollution. In a similar manner, building smaller, lighter automobiles reduces air pollution because less fuel is burned to propel them, but alternatives such as mass transit, walking, and bicycles can contribute significantly to reduced fuel consumption. The introduction of the hybrid automobile is a major step in improving fuel economy. Alternative sources of energy such as solar, wind, and nuclear also reduce air pollution emissions. (Nuclear power, of course, has a series of pollution problems that may outweigh the benefits of reduced air pollution.)

The chlorofluorocarbon destruction of the ozone layer can only be resolved by waste minimization. Preventing the escape of CFCs from refrigeration systems, the use of alternative propellants for spray cans, and similar measures are the only ones that will be successful because control devices make no sense. Waste minimization is, in fact, the method of control specified by the Montreal Protocol. In a similar fashion, the production of ozone in the lower atmosphere can only be reduced by minimizing the release of precursor hydrocarbons and the production of NO_x. Reduced use of solvents and the substitution of water-based paints for solvent-based paints are examples of methods to reduce hydrocarbon release.

The fundamental method for minimizing anthropogenic inputs to global warming is to minimize the anthropogenic emissions of GHGs. As noted earlier, more efficient use of energy is a major method of reducing GHGs. In and of itself, efficient energy use makes economic sense. The fact that efficient energy use will contribute immensely to the reduction of air pollution and GHGs while enhancing sustainability makes it an ideal candidate for waste minimization.

FIGURE 12–39

(*Source:* U.S. Department of Energy; Worland, 2017.)

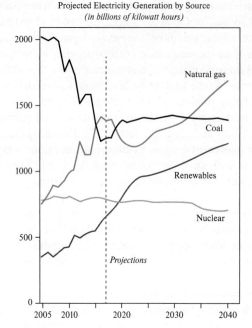

Projected Electricity Generation by Source
(in billions of kilowatt hours)

In 2015 the U.S. EPA entered a consent agreement that set the timeline for the agency to designate certain areas of the United States as "nonattainment" for the SO_2 NAAQS. Twenty-nine SO_2 nonattainment areas were established in 2013. Most of these areas included coal-fired power plants.* EPA identified 68 power plants in the 2015 areas included in the consent agreement.

On August 3, 2015 President Obama unveiled the Clean Power Plan. The Clean Power Plan was designed to limit global warming from CO_2 by switching to natural gas, solar, and wind for energy. Natural gas produces half of the CO_2 emissions that coal-firing produces and none of the SO_2. Solar and wind energy produce virtually no global warming gases beyond those resulting from manufacturing the equipment and building the structures.

On March 28, 2017 President Trump signed an Executive Order that is expected to effectively scrap the Clean Power Plan and roll back regulations on coal-fired power plant emissions. The intent was to protect the coal-burning power plants from closure forced by the requirement for expensive modifications and, thus, bring back jobs in the coal mining industry.

The reality is that coal has lost its grip on the power industry. Coal is no longer the fuel of choice for the electric power generation in the United States. In 2006 burning coal provided 49% of the country's electricity. In 2016 that figure had declined to 30%. The U.S. Energy Information Agency predicts that natural gas and renewable power (solar and wind) will account for 58% of the nation's energy by 2040 (Figure 12–39). Looking at a 50-year timeline, the electric power industry's planning is to grow renewable power and, in the meantime, to rely on natural gas because it is both cleaner and cheaper than coal (Davenport, 2017; Loveless, 2016, Worland and U.S. Department of Energy, 2017).

President Trump's Executive Order will not provide sustainable jobs in the coal mining industry. The economic reality is that solar and wind energy sources are sustainable and will be the future energy source for the United States.

*Arkansas, Colorado, Georgia, Hawaii, Iowa, Illinois, Indiana, Kansas, Kentucky, Louisiana, Maryland, Michigan, Missouri, Mississippi, North Carolina, North Dakota, Nebraska, New York, Ohio, Oklahoma, South Dakota, Tennessee, Texas, Wisconsin, and Wyoming.

CHAPTER REVIEW

When you have completed studying this chapter, you should be able to do the following without the aid of your textbook or notes:

1. List the six criteria air pollutants for which the U.S. EPA has designated NAAQS.
2. List and define three units of measure used to report air pollution data (i.e., ppm, $\mu g \cdot m^{-3}$, and μm).
3. Explain the difference between ppm in air pollution and ppm in water pollution.
4. Explain the effect of temperature and pressure on readings made in ppm.
5. Explain the influence of moisture, temperature, and sunlight on the severity of air pollution effects on materials.
6. Differentiate between acute and chronic health effects from air pollution.
7. State which particle sizes are more important with respect to alveolar deposition and explain why.
8. Explain why it is difficult to define a causal relationship between air pollution and health effects.
9. List three potential chronic health effects of air pollution.
10. List four common features of air pollution episodes and identify the locations of three "killer" episodes.
11. Discuss the natural and anthropogenic origin of the six criteria air pollutants and identify the likely mechanisms for their removal from the atmosphere.
12. Identify one indoor air pollution source for each of the following pollutants: CH_2O, CO, NO_x, Rn, respirable particulates, and SO_x.
13. Define the term *acid rain* and explain how it occurs.
14. Discuss the photochemistry of ozone in the upper atmosphere using the pertinent chemical reactions. Discuss the hypothesized effect of chlorofluorocarbons on these reactions.
15. Explain the term *greenhouse effect,* its hypothesized cause, and why it is being debated, pro and con.
16. Determine the stability (ability to dissipate pollutants) of the atmosphere from vertical temperature readings.
17. Explain why valleys are more susceptible to inversions than is flat terrain.
18. Explain why lake breezes and land breezes occur.
19. Explain how a lake breeze adversely affects the dispersion of pollutants.
20. State the theoretical principle on which each of the following air pollution control devices operates: (a) absorption column (either a packed tower or plate tower), (b) adsorption column, (c) either afterburner or catalytic combustor, (d) cyclone, (e) baghouse, (f) Venturi scrubber, and (g) electrostatic precipitator.
21. Select the correct air pollution control device for a given pollutant and source.
22. Discuss the pros and cons of FGD and the problem of fly ash resistivity.
23. Explain the difference between prevention and postcombustion techniques for reduction of nitrogen oxide emissions and give one example of each.
24. Sketch the relationship between air-to-fuel ratio and emission of CO, HC, and NO_x from automobiles.

With the aid of this text, you should be able to do the following:

1. Solve gas law problems.
2. Convert parts per million (ppm) to micrograms per cubic meter ($\mu g \cdot m^{-3}$) and vice versa.
3. Calculate the amount of SO_2 that will be released from burning coal or fuel oil with a given sulfur content in percent.
4. Calculate the ground level concentration of air pollutants released from a stationary elevated source or the emission rate (E) for a given ground level concentration.
5. Calculate the stoichiometric air-to-fuel ratio for a given fuel chemical composition.

PROBLEMS

12–1 Show by calculation that, in Figure 12–15, the incoming energy (in $W \cdot m^2$) and outgoing energy (in $W \cdot m^2$) are in balance (1) at the top of the atmosphere and (2) at the Earth's surface.

12–2 In simplified form, radiative forcing of temperature change may be expressed as

$$\Delta T_e = \gamma \Delta F$$

Where ΔT_e = surface temperature change
  γ = climate sensitivity factor
  $\Delta \gamma \, F$ = radiative forcing
    $= \Delta E_{emit} - E_{abs}$

The climate sensitivity factor is between 0.34 and 1.03°C $\cdot$ m^2 $\cdot$ W^{-1} (IPCC, 1996).

Estimate the range in change in temperature of the Earth's surface (in °C) if the albedo of the Earth decreases by 0.01 because of the melting of glaciers. Assume that energy emitted remains the same.

12–3 Given the following temperature profiles, determine whether the atmosphere is unstable, neutral, or stable. Show all work and explain choices.

(a) Z (m)	T (°C)	(b) Z (m)	T (°C)	(c) Z (m)	T (°C)
2	–3.05	10	6.00	18	14.03
318	–6.21	202	3.09	286	16.71

Answers: (a) Neutral; **(b)** Unstable; **(c)** Stable (inversion)

12–4 Determine the atmospheric stability for each of the following temperature profiles. Show all work and explain choices.

(a) Z (m)	T (°C)	(b) Z (m)	T (°C)	(c) Z (m)	T (°C)
1.5	–4.49	12	28.05	8	18.55
349	0.10	279	19.67	339	17.93

12–5 Determine the atmospheric stability for each of the following temperature profiles. Show all work and explain choices.

(a) Z (m)	T (°C)	(b) Z (m)	T (°C)	(c) Z (m)	T (°C)
2.00	5.00	2.00	5.00	2.00	–21.01
50.00	4.52	50.00	5.00	50.00	–25.17

12–6 Given the following observations, use the Key to Stability Categories (Table 12–13) to determine the stability.

(a) Clear winter morning at 9:00 A.M.; wind speed of 5.5 m $\cdot$ s^{-1}

(b) Overcast summer afternoon at 1:30 P.M.; wind speed of 2.8 m $\cdot$ s^{-1}

(c) Clear winter night at 2:00 A.M.; wind speed of 2.8 m $\cdot$ s^{-1}

(d) Summer morning at 11:30 A.M.; wind speed of 4.1 m $\cdot$ s^{-1}

Answer: (a) D; **(b)** D; **(c)** F; **(d)** B

12–7 Determine the atmospheric stability category of the following observations.

(a) Clear summer afternoon at 1:00 P.M.; wind speed of 5.6 m $\cdot$ s^{-1}

(b) Clear summer night at 1:30 A.M.; wind speed of 2.1 m $\cdot$ s^{-1}

(c) Overcast winter afternoon at 2:30 P.M.; wind speed of 6.6 m $\cdot$ s^{-1}

(d) Summer afternoon at 1:00 P.M. with broken low clouds; wind speed of 5.2 m $\cdot$ s^{-1}

12–8 Determine the atmospheric stability category of the following observations.

(a) Clear summer afternoon at 1:00 P.M.; wind speed of 1.6 m · s^{-1}

(b) Overcast summer night at 1:30 A.M.; wind speed of 2.1 m · s^{-1}

(c) Overcast winter morning at 9:30 A.M.; wind speed of 6.6 m · s^{-1}

(d) Thinly overcast winter night at 8:00 P.M.; wind speed of 2.4 m · s^{-1}

12–9 A power plant in a college town is burning coal on a cold, clear winter morning at 8:00 A.M. with a wind speed of 2.6 m · s^{-1} measured at 30 m elevation and an inversion layer with its base at a height of 697 m. The effective stack height is 30 m. Calculate the distance downwind X_L at which the plume released will reach the inversion layer and begin to mix downward.

Answer: 5.8 km

12–10 A factory releases a plume into the atmosphere on an overcast summer afternoon. At what distance downwind will the plume begin mixing downward if an inversion layer exists at a base height of 414 m and the windspeed is 1.8 m · s^{-1} measured at the effective stack height is 45 m?

12–11 Given the same power plant and conditions as were found in Example 12–4, determine the concentration of SO_2 at a point 4 km downwind and 0.2 km perpendicular to the plume centerline ($y = 0.2$ km) if there is an inversion with a base height of 328 m.

Answer: 1.16×10^{-3} g · m^{-3}

12–12 Calculate the downwind concentration at 30 km ($y = 0$) in g · m^{-3} resulting from an emission of 1976 g · s^{-1} of SO_2 into a 2.5 m · s^{-1} wind at 1:00 A.M. on a clear winter night. Assume an effective stack height of 85 m and an inversion layer at 185 m. Identify the stability class and show all work.

12–13 On a clear summer afternoon with a wind speed of 3.20 m · s^{-1} measured at 100 m elevation, the TSP concentration was found to be 1520 μg · m^{-3} at a point 2 km downwind and 0.5 km perpendicular to the plume centerline from a coal-fired power plant. Given the following parameters and conditions, determine the TSP emission rate of the power plant:

Stack parameters:

Height = 75.0 m
Diameter = 1.50 m
Exit velocity = 12.0 m · s^{-1}
Temperature = 322°C

Atmospheric conditions:

Pressure = 100.0 kPa
Temperature = 28.0°C

12–14 Determine the required ventilation rate for a particle board manufacturing facility operating 24 hours per day. The facility must meet a formaldehyde (HCHO) concentration of 0.10 ppm. It generates 25,000 μg · d^{-1} · m^{-2} of HCHO. The facility work space is 30 m × 10 m × 5 m. The exhaust from the building is scrubbed and vented through a 10-tall stack. The ambient air intake contains no HCHO.

12–15 Potential buyers of a house ask for an inspection prior to the purchase. The inspector notes that the geology underlying the house is known to have "hot spots" of radon emissions. Upon investigation the house is found to be on a hot spot that emits 2.0 pCi · m^{-2} · s^{-1} of radon gas. The ambient concentration outside of the house is negligible. The house basement and first floor each have a floor space of 56 m^2. The ceiling height in both the basement and the first floor is 2.3 m. Does the concentration exceed the action level of 4 pCi · L^{-1} of air?

12–16 The concentration of a pollutant downwind of a line source of air pollution such as an interstate highway or a wild fire may be estimated with the following equation if one approximates the formula as a straight infinite line with the wind blowing perpendicular to the line. The modeling equation is

$$X(x) = \frac{2E}{(2\pi)^{0.5} s_z u}$$

The federal emission standard for NO_x for automobiles is 0.4 g/mi or 0.25 g – km^{-1}. Using the line source equation, estimate the ground level concentration of NO_x 200 m from a highway with cars passing the point of interest at a rate of 20 vehicles per second. The wind speed is 2 m · s⁻¹ and the stability class is D. Does the estimated concentration exceed the NAAQS 1-hour primary standard for NO_x?

12–17 The federal emission standard for NO_x for heavy trucks is 2.5 g/mi or 1.6 g – km⁻¹. Using the line source equation from Problem 12–16, estimate the ground level concentration of NO_x 200 m from a highway with heavy trucks passing the point of interest at a rate of 20 vehicles per second. The wind speed is 2 m · s⁻¹ and the stability class is D. Does the estimated concentration exceed the NAAQS 1-hour primary standard for NO_x?

12–18 Rework Example 12–4 with a stack temperature of 210°C.

12–19 Rework Example 12–4 with an inversion base at 310 m.

DISCUSSION QUESTIONS

12–1 A gas sample is collected in a special gas sampling bag that does not react with the pollutants collected but is free to expand and contract. When the sample was collected, the atmospheric pressure was 103.0 kPa. At the time the sample was analyzed, the atmospheric pressure was 100.0 kPa. The bag was found to contain 0.020 ppm of SO_2. Would the original concentration of SO_2 be more, less, or the same? Explain.

12–2 Under which of the following conditions would you expect the strongest inversion (largest positive lapse rate) to form?

(a) Foggy day in the fall after the leaves have fallen   (b) Clear winter night with fresh snow on the ground
(c) Clear summer morning just before sunrise

Explain why.

12–3 Cement dust is characterized by very fine particulates. The exhaust gas temperatures from a cement kiln are very hot. Which of the following air pollution control devices would appear to be appropriate? Explain the reasoning for your selection.

(a) Venturi scrubber   (b) Baghouse   (c) Electrostatic precipitation

12–4 Photochemical oxidants are not directly attributable to either people or natural sources. Why, then, are automobiles singled out as the major cause of the formation of ozone?

12–5 Explain why the $PM_{2.5}$ standard is more appropriate than a TSP for protection of human health.

FE EXAM FORMATTED PROBLEMS

12–1 A power plant emits 600 g of SO_2 · s⁻¹ from an effective stack height of 300 m. The wind speed at this height is 5.0 m/s. The stability class is C. What is the maximum downwind concentration?

(a) 2.4×10^{-4} g/m³
(b) 1.5×10^{-6} g/m³
(c) 1.8×10^{-4} g/m³
(d) 1.9×10^{-5} g/m³

12–2 Select an appropriate air to cloth ratio for fly ash and estimate the number of bags required for a gas flow rate of 15 m³/s. Assume each bag is 15 cm in diameter and 5 m in length and that bag cleaning is by pulse jet.

(a)   255 bags
(b)   800 bags
(c)   1700 bags
(d)   50 bags

12–3   Estimate the stoichiometric air to fuel ratio for a 100% ethanol (CH_3CH_2OH) fuel. One mole of CH_3CH_2OH = 46 g; one mole of O_2 = 32 g; one mole of N_2 = 28 g. Assume the following reaction:

$$CH_3CH_2OH + 3O_2 + 11.1N_2 \rightarrow 2CO_2 + 3H_2O + 11.1N_2$$

(a)   25:1

(b)   11.1:1

(c)   15:1

(d)   18:1

12–4   Calculate the rate of emission of SO_2 in g/s that results in a centerline ($y = 0$) concentration at ground level of 1.412×10^{-3} g/m^3 one kilometer from the stack. The time of the measurement was 1 P.M. on clear summer afternoon. The wind speed was 1.8 m/s measured at a height of 10 m. The effective stack height is 94 m. No inversion is present.

(a)   790 g/s

(b)   860 g/s

(c)   280 g/s

(d)   440 g/s

REFERENCES

AEC (1968) *Meteorology and Atomic Energy, 1968,* U.S. Atomic Energy Commission, USAEC Division of Technical Information Extension, Oak Ridge, TN.

Anthes, R. A., J. J. Cahir, A. B. Fraser, H. A. Panofsky (1981) *The Atmosphere,* Charles E. Merrill Publishing Co., Columbus, OH, p. 89.

AP (2005a) Associated Press, *Lansing State Journal,* December 18, p. 14A.

AP (2005b) Associated Press, http://www.myrtlebeachonline.com

ASHRAE (1981) *Ventilation for Acceptable Air Quality,* American Society of Heating, Refrigerating, and Air Conditioning Engineers, Standard 62-1981, Atlanta, GA.

Black & Veatch Consulting Engineers (1983) *Lime FGD Systems Data Book*—Second Edition, EPRI Publication No. CS-2781.

Canfield, R. L., C. R. Henderson, D. A. Cory-Slechta, et al. (2003) "Intellectual Impairment in Children with Blood Lead Concentrations Below 10 μg per Deciliter," *The New England Journal of Medicine,* 348: 1517–26.

Cannell, A. L., and M. L. Meadows (1985) "Effects of Recent Operating Experience on the Design of Spray Dryer FGD Systems," *Journal of the Air Pollution Control Association,* 35(7): 782–89.

CFR (2007) *Code of Federal Regulations,* 40 CFR §50.4-50.12.

Chemical and Engineering News (2006) "Government Concentrates: Seven States Agree to Cut CO_2 Emissions," January 2, p. 16.

Connolly, C. H. (1972) *Air Pollution and Public Health,* Holt, Rinehart & Winston, New York, p. 7.

Cooper, C. D., and F. C. Alley (2002) *Air Pollution Control: A Design Approach,* Waveland Press, Long Grove, IL, p. 547.

Davenport, C. (2017) "Coal Has Lost Its Grip on Power," *New York Times,* April 6, 2017.

Davis, M. L., and D. A. Cornwell (2008) *Introduction to Environmental Engineering,* 4th ed., McGraw-Hill, New York.

Dumont, R. S. (1985) "The Effect of Mechanical Ventilation on Rn, NO_2, and CH_2O Concentrations in Low Leakage Houses and a Simple Remedial Measure for Reducing Rn Concentration," *Transactions, Indoor Air Quality in Cold Climates, Hazards and Abatement Measures,* Air Pollution Control Association, Pittsburgh, PA, pp. 90–104.

FR (1992) *Federal Register,* 57 FR 31576.

Ferris, B. G. (1978) "Health Effects of Exposure to Low Levels of Regulated Air Pollutants," *Journal of the Air Pollution Control Association,* 28: 482–97.

Fryling, G. R. (1967) *Combustion Engineering,* Combustion Engineering, Inc., New York, pp. 21–28.

Godish, T. (2001) "Aldehydes," in J. D. Spengler, J. M. Samet, and J. F. McCarthy (eds.) *Indoor Air Quality Handbook,* McGraw-Hill, New York, pp. 32.1–32.22.

Goldsmith, J. R. (1968) "Effects of Air Pollution on Human Health," in A. C. Stern, ed., *Air Pollution,* Academic Press, New York, pp. 554–57.

Goody, R. M., and Y. L. Yung (1989) *Atmospheric Radiation: Theoretical Basis,* Oxford University Press, New York, p. 4.

Goyer, R. A., and J. J. Chilsolm (1972) "Lead," in D. K. K. Lee, ed., *Metallic Contaminants and Human Health,* Academic Press, New York, pp. 57–95.

Hance, S. B., and J. L. Kelly (1991) "Status of Flue Gas Desulfurization Systems," Paper No. 91-157.3, 84th Annual Meeting of the Air and Waste Management Association, San Francisco, CA.

Hansen, J. (2018) "Global Temperature in 2017," https://www.climatescienceawarenesssolutions.org/blog-1/2018/1/18/global-temperature-in-2017.

Hansen, J., and M. Sato (2004) "Temperature Trends: 2004 Summation," http://www.giss.nasa.gov/data/update/gistemp/2004/

HEW (1979) *Smoking and Health: A Report of the Surgeon General of the Public Health Service,* U.S. Department of Health, Education, and Welfare, Pub. No. 79-50066, Washington, DC.

Hileman, B. (1989) "Global Warming," *Chemical and Engineering News,* March 13, pp. 25–44.

Hileman, B. (2005) "Ice Core Record Extended," *Chemical and Engineering News,* November 28, p. 7.

Hindawi, I. (1970) *Air Pollution Injury to Plants,* U.S. Department of Health, Education, and Welfare, National Air Pollution Control Administration Publication No. AP-71, Washington, DC, p. 13.

Hines, A. L., T. K. Ghosh, S. K. Loyalka, and R. C. Warder (1993) *Indoor Air Quality & Control,* PTR Prentice Hall, Englewood Cliffs, NJ, pp. 21, 22, 34.

Hoegg, V. R. (1972) "Cigarette Smoke in Closed Spaces," *Environmental Perspectives,* 2: 117.

Holland, J. Z. (1953) *A Meteorological Survey of the Oak Ridge Area,* U.S. Atomic Energy Commission Report No. ORO-99, Washington, DC, p. 540.

Hosein, R., F. Silverman, P. Coreg, et al. (1985) "The Relationship Between Pollutant Levels in Homes and Potential Sources," *Transactions, Indoor Air Quality in Cold Climates, Hazards and Abatement Measures,* Air Pollution Control Association, Pittsburgh, PA, pp. 250–60.

ICAS (1975) *The Possible Impact of Fluorocarbons and Hydrocarbons on Ozone,* Interdepartmental Committee for Atmospheric Sciences, Federal Council for Science and Technology, National Science Foundation Publication No. ICAS 18a-FY 75, Washington, DC, p. 3.

IPCC (1995) *Climate Change 1994: Radiative Forcing of Climate Change and an Evaluation of the IPCC 1992 Emission Scenarios,* Intergovernmental Panel on Climate Change, Cambridge University Press, Cambridge, UK.

IPCC (1996) *Climate Change 1995: Climate Change 1995, The Science of Climate Change,* Intergovernmental Panel on Climate Change, Cambridge University Press, Cambridge, UK.

IPCC (2000) *IPCC Special Report: Emissions Scenarios,* Intergovernmental Panel on Climate Change, Cambridge University Press, Cambridge, UK.

IPCC (2001a) *Climate Change 2001: The Scientific Basis, Summary for Policymakers,* Intergovernmental Panel on Climate Change, Cambridge University Press, Cambridge, UK, pp. 1–18.

IPCC (2007a) *Climate Change 2007: The Physical Science Basis, Summary for Policymakers,* Intergovernmental Panel on Climate Change, Cambridge University Press, Cambridge, UK, pp. 1–18.

IPCC (2007b) *Climate Change 2007: Impacts, Adaptation and Vulnerability—The Physical Science Basis, Summary for Policymakers,* Intergovernmental Panel on Climate Change, Cambridge University Press, Cambridge, UK, pp. 1–18.

Kao, A. S. (1994) "Formation and Removal Reactions of Hazardous Air Pollutants," *Journal of the Air Pollution Control Association,* 44: pp. 683–96.

Karlsson, H. T., and H. S. Rosenberg (1980) "Technical Aspects of Lime/Limestone Scrubbers for Coal-fired Power Plants, Part It Process Chemistry and Scrubber Systems," *Journal of the Air Pollution Control Association,* 30(6): 710–14.

Keeling, C. M., and T. P. Whorf (2005) "Atmospheric Carbon Dioxide Record from Mauna Loa," http://www.mlo.noaa.gov

Kiehl, J. T., and H. Rodhe (1995) "Modeling Geographic and Seasonal Forcing Due to Aerosols," in R. J. Charlson and J. Heintzenberg, eds., *Aerosol Forcing of Climate,* John Wiley & Sons, New York, pp. 281–96.

Kiehl, J. T., and K. E. Trenberth (1997) "Earth's Annual Global Mean Energy Budget," *Bulletin of the American Meteorological Society,* 78: 197–208.

Langmuir, I. (1918) "The Adsorption of Gaseson Plane Surfaces of Glass, Mica and Platinum," *Journal of the American Chemical Society,* 40: 1361.

Lee, H. K., T. K. McKenna, L. N. Renton, et al. (1985) "Impact of a New Smoking Policy on Office Air Quality," *Indoor Air Quality in Cold Climates, Transactions of the Air Pollution Control Association,* Pittsburgh, PA, pp. 307–22.

Lefohn, A. S., and S. V. Krupa (1988) *Acidic Precipitation, a Technical Amplification of NAPAP's Findings,* Proceedings of an APCA International Conference, Pittsburgh, PA, p. 1.

Loveless, B. (2016) "Coal's Decline May Continue Even Under Trump," *USA Today,* November 21, 2016, p. B4.

Mann, M. E., and P. D. Jones (2003) "Global Surface Temperatures over the Past Two Millennia," *Geophysical Research Letters,* 30(15): CLM 5-1–CLM 5-4.

Martin, D. O. (1976) "Comment on the Change of Concentration Standard Deviations with Distance," *Journal of the Air Pollution Control Association,* 26: 145–46.

McCabe, L. C. and G. D. Clayton (1952) Air Pollution by Hydrogen Sulfide in Poza Rica, Mexico; an Evaluation of the Incident of Nov. 24, 1950. *A.M.A. Archives of Industrial Hygiene and Occupational Medicine* 6(3):199–213.

Molina, M. J., and F. S. Rowland (1974) "Stratospheric Sink for Chlorofloromethanes; Chlorine Atom Catalysed Destruction of Ozone," *Nature,* 248: 810–12.

Moschandreas, D. J., J. D. Zabpansky, and S. D. Pelta (1985) *Characteristics of Emissions from Indoor Combustion Sources,* Gas Research Institute Report No. 85/0075, Chicago, IL.

NAPCA (1970) *Air Quality Criteria for Photochemical Oxidants,* National Air Pollution Control Administration Publication AP-63, Department of Health, Education, and Welfare, Washington, DC.

NCAPC (1967) *Training Course Manual in Air Pollution, Control of Particulate Emissions,* National Center for Air Pollution Control, Department of Health, Education and Welfare, Cincinnati, OH.

Noticias (2004) Noticias.info, agencia International de noticias, http://www.noticias.info/Archivo 08/11/ 2004 notas_de_prensa_archivo

NOAA (2017) "Recent Monthly Mean CO_2 at Mauna Loa," http://www.esrl.noaa.gov/gmd/ccgg/trends.

O'Gara, P. J. (1922) "Sulfur Dioxide and Fume Problems and Their Solutions," *Industrial Engineering Chemistry,* 14: 744.

Patterson, D. J., and N. A. Henein (1972) *Emissions from Combustion Engines and Their Control,* Ann Arbor Science, Ann Arbor, MI, p. 143.

Peixoto, J. P., and A. H. Oort (1992) *Physics of Climate,* American Institute of Physics, New York.

Pittman, D. (2011) "Monitoring a Troubling Trend," *Chemical and Engineering News,* January 10, p. 26.

Pope, C. A., D. W. Dockery, R. E. Kanner, G. M. Villegas, and J. Schwartz (1999) "Oxygen Saturation, Pulse Rate, and Particulate Air Pollution: A Daily Time-series Panel Study," *American Journal of Respiratory and Critical Care Medicine,* 159: 365–72.

Pope, C. A., M. J. Thun, M. M. Namboodri et al. (1995) "Particulate Air Pollution as a Predictor of Mortality in a Prospective Study of U.S. Adults," *American Journal of Respiratory and Critical Care Medicine,* 151: 669–74.

Prasad, A. (1995) "Air Pollution Control Technologies for Nitrogen Oxides," *The National Environmental Journal,* May/June, pp. 46–50.

Public Law 101-549 (1990).

Reichhardt, T. (1995) "Weighing the Health Risks of Airborne Particulates," *Environmental Science and Technology,* 29: 360A–364A.

Reisch, M., and P. S. Zurer (1988) "CFC production: DuPont Seeks Total Phaseout," *Chemical and Engineering News,* April 4, p. 4.

Reisch, M. S. (2015) "Bromine Bails out Big Power Plants," *Chemical and Engineering News,* March 16, pp. 17, 21.

Ritter, S. (2013) "Halocarbons Reassessed," *Chemical and Engineering News,* p. 27.

Seinfeld, J. H. (1975) *Air Pollution, Physical and Chemical Fundamentals,* McGraw-Hill, *New York,* p. 71.

Seinfeld, J. H., and S. N. Pandis (1998) *Atmospheric Chemistry and Physics,* John Wiley & Sons, Inc., New York, pp. 59 & 71.

Srivastava, R. K., J. E. Staudt, and W. Josewicz (2005) "Preliminary Estimates of Performance and Cost of Mercury Emission Control Technology Applications on Electric Utility Boilers: An Update," *Environmental Progress,* 24(2): 198–213.

Sullivan, D. A. (1989) "International Gathering Plans Ways to Safeguard Atmospherric Ozone," *Chemical and Engineering News,* June 26, 33–36.

Supreme Court (2007) *Massachusetts et al. v. Environmental Protection Agency et al.,* Supreme Court of the United States, syllabus No. 05-1120, argued November 29, 2006—Decided April 2, 2007.

Traynor, G. W., J. R. Allen, and M. G. Apte (1982) *Indoor Air Pollution from Portable Kerosene-fired Space Heaters, Woodburning Stoves and Woodburning Furnaces,* Lawrence Berkeley Laboratory Report No. LBL-14027.

Treybal, R. E. (1988) *Mass Transfer Operations,* McGraw-Hill, New York, pp. 253 & 535.

Tucker, W. G. (2001) "Volatile Organic Compounds," in J. D. Spengler, J. M. Samet, and J. F. McCarthy (eds.) *Indoor Air Quality Handbook,* McGraw-Hill, New York, pp. 31.1–31.20.

Turner, D. B. (1967) *Workbook of Atmospheric Dispersion Estimates,* U.S. Department of Health, Education, and Welfare, U.S. Public Health Service Publication No. 999-AP-28, Washington, DC.

UN (2005) *The Millennium Development Goals Report: 2005,* United Nations, New York, p. 32.

UNDP (2005) *The Montreal Protocol,* http://www.undp.org/seed/eap/montreal/montreal.htm

UNEP/WHO (2002) "Executive Summary," *Scientific Assessment of Ozone Depletion: 2002,* United Nations Environmental Programme/World Health Organization, New York.

U.S. CDC (2005) *Third National Report on Human Exposure to Environmental Chemicals,* U.S. Centers for Disease Control and Prevention, National Center for Environmental Health, NCEH Pub. No. 05-0570, pp. 41, 74–75.

U.S. DHEW (1970) *Air Quality Criteria for Photochemical Oxidants,* National Air Pollution Control Administration Publication No. AP-63, U.S. Government Printing Office, Washington, DC.

U.S. EPA (1995) *User's Guide for ISC3 Dispersion Models, Vol. II,* U.S. Environmental Protection Agency, Report No. EPA-454/B-95-003b.

U.S. EPA (1997) *1997 Mercury Study Report to Congress,* U.S. Environmental Protection Agency, http://www.epa.gov

U.S. EPA (2003a) *Response of Surface Water Chemistry to the Clean Air Act Amendments of 1990,* U.S. Environmental Protection Agency, Report No. 620/R-03/001, Research Triangle Park, NC, pp. 59–62.

U.S. EPA (2003b) *Performance and Cost of Mercury Emission Control Technology Applications on Electric Utility Boilers,* Report No. 600/R-03/1100, Research Triangle Park, NC.

U.S. EPA (2004) *EPA Fact Sheet,* U.S. Environmental Protection Agency, http://www.epa.gov

U.S. PHS (1965) *Survey of Lead in the Atmosphere of Three Urban Communities,* U.S. Department of Health, Education, and Welfare, U.S. Public Health Service Publication No. 999-AP-12, Washington, DC.

Wakeham, H. (1972) "Recent Trends in Tobacco Smoke Research," in I. Schmelz (ed.) *The Chemistry of Tobacco Smoke Research,* Plenum Press, New York.

Wallace, L. A. (2001) "Assessing Human Exposure to Volatile Organic Compounds," in J. D. Spengler, J. M. Samet, and J. F. McCarthy (eds.) *Indoor Air Quality Handbook,* McGraw-Hill, New York, pp. 33.1–33.35.

WHO (1961) *Air Pollution,* World Health Organization Regional Office for Europe, European Series 23, Copenhagen, pp. 180, 210–20.

Wofsy, S. C., J. C. McConnnell, and M. B. McElroy (1972) "Atmospheric CH_4, CO, and CO_2," *Journal of Geophysical Research,* 67: 4477–93.

Worland, J. and U.S. Department of Energy (2017) "Coal's Last Kick," *Time,* April 17, p. 43.

Yocom, J. E., and R. O. McCaldin (1968) "Effects on Materials and the Economy," in A. C. Stern (ed.), *Air Pollution,* vol. I, 2nd ed., Academic Press, New York, pp. 617–54.

Zurer, P. S. (1988) "Studies on Ozone Destruction Expand Beyond Antarctic," *Chemical and Engineering News,* May 30, pp. 18–25.

Zurer, P. S. (1989) "Scientists Find Arctic May Face Ozone Hole," *Chemical and Engineering News,* February 27, p. 5.

Zurer, P. S. (1994) "Scientists Expect Ozone Loss to Peak About 1998," *Chemical and Engineering News,* September 12, p. 5.

13

Solid Waste Engineering

Case Study

Too Much Waste, Too Little Space

The disposal of waste products is an age-old problem for society. Prehistoric people disposed of their wastes, including bones, tools, and clothes, in waste heaps called middens. Ancient civilizations often recycled materials from conquered cities and pillaged buildings. As cities grew and the human population increased in number, problems related to solid waste disposal have increased. Landfill space is limited and an increase in affluence has resulted in an increase in the amount of waste produced on a per capita basis. Clearly, something must be done.

Communities have taken different approaches to dealing with the problems related to solid waste. In Denmark, in the 1980s, per capita solid waste generation exceeded that of most of the rest of Europe and landfill space was rapidly being depleted. Incineration of waste was not considered an option as the Danes were concerned with the air pollution that would result from incineration. Denmark responded by adopting a strict waste management program, which included a substantial waste tax, which more than doubled the cost of disposal. Tight restrictions on waste management were promulgated, making recycling mandatory and more convenient. The money generated from taxes has been used to develop programs to reduce waste generation and to increase recycling and the reuse of household, industrial, and construction materials. In the period of time from 1987 to 1997, the amount of waste collected decreased by almost 26% and the amount of material recycled and reused increased significantly. The recycling and reuse of construction materials rose by more than 100% in the period from 1991 to 1995. The composting of organic household waste increased some 580% in the period from 1990 to 1994. Between 1986 and 1995, the recycling of paper and cardboard increased by greater than 77%. Clearly, the use of national taxes can effectively reduce the amount of waste discarded and increase the reuse and recycling of waste material (Andersen, 1998).

Germany has dealt with its problems related to solid waste management by placing regulations on packaging, putting much of the responsibility for minimizing solid waste on the manufacturers. Manufacturers are required to accept back the packaging of their goods, which must be reused or recycled. Since companies found that they needed help to meet the regulated quotas, the Duales System Deutschland (DSD) was created. Companies pay to become a member and DSD puts its green dot trademark on the company's packaging. This guarantees that the packaging material will be recycled upon collection. Drop off and curbside collection for "green dot" packaging is available to make recycling easy and convenient. DSD has regularly exceeded the government-set recycling targets of 75% for glass packaging, 70% for paper and steel packaging, and 60% of aluminum, plastics, and composite (cartons) packaging (Cartledge, 2004; U.S. EPA, 1998).

In contrast, Hamilton, Ontario, responded to its solid waste problem with voluntary programs. The City of Hamilton set goals of 42% and 65% waste reduction by 2005 and 2008, respectively. Recycling of solid waste is encouraged through various educational programs in the schools and through the local media. Weekly curbside pickup of plastics, paper, cardboard, and glass makes recycling easy and convenient. Residents need only sort recyclable items into two blue bins: one for dry paper products and another for plastic, glass, metal, aluminum containers, and other recyclable items. Leaf and yard waste is also collected several times in the fall and spring. In 2006, the City of Hamilton began the city-wide source separated organics program, whereby organic material is placed curbside in green carts. This organic waste is collected in special "split body" trucks that have a cart tipper to lift and empty the green carts (see Figure 13–1). These trucks collect green carts and garbage at the same time. A wall

FIGURE 13–1 Dual compartment truck used in the City of Hamilton, Ontario. Conventional garbage is collected in compartment 1, while organic wastes are collected in compartment 2. (©City of Hamilton, 2007)

inside the truck keeps organic waste separate from garbage and each side of the truck can be emptied at different places in order to keep the materials separate. The compost from the leaf and yard waste collection program ends up in city parks with some being given back to Hamilton residents. Compost from the green cart program is marketed by the facility operator. Hamilton's diversion rate has been steadily increasing, with a total of 40% at the end of 2006. Prior to the launch of the green cart program, Hamilton had a waste diversion rate of 30% at the end of 2005.

These three case studies provide three very different approaches to dealing with the problem of solid waste management. There is clearly no right way, and any approach taken must take into consideration the social and cultural norms and expectations of the particular society. However, it is clear that we must develop successful strategies to minimize the amount of waste material disposed of in the limited landfill space we have available to us.

13–1 INTRODUCTION

Solid waste is a generic term used to describe the things we throw away. It includes objects the lay audience commonly calls garbage, refuse, and trash. The U.S. EPA regulatory definition is broader in scope. It includes any discarded item: things destined for reuse, recycle, or reclamation; sludges; and hazardous wastes. The regulatory definition specifically excludes radioactive wastes and in situ mining wastes.

We have limited the discussion in this chapter to solid wastes generated from residential and commercial sources. Sludges were discussed in Chapters 10 and 11. Hazardous waste will be discussed in Chapter 14 and radioactive waste in Chapter 16.

Magnitude of the Problem

Solid waste disposal creates a problem primarily in highly populated areas. In general, the more concentrated the population, the greater the problem becomes, although some very populated areas have developed creative solutions to minimize the problems. Various estimates have been made of the quantity of solid waste generated and collected per person per day. Since 1960, when the U.S. EPA began reporting data, the total mass of municipal solid waste (MSW) generated per year has increased from $79.9 \text{ Tg} \cdot \text{yr}^{-1}$ to $225.9 \text{ Tg} \cdot \text{yr}^{-1}$ in 2014. The generation rate peaked in 2005 at $229.2 \text{ Tg} \cdot \text{yr}^{-1}$. The per capita MSW generation rate peaked, in 2000, at a rate of $2.14 \text{ kg} \cdot \text{person}^{-1} \cdot \text{day}^{-1}$. For 2014, the last year for which data is available, the rate was $2.03 \text{ kg} \cdot \text{person}^{-1} \cdot \text{yr}^{-1}$. EPA's goals for the nation were to recycle 35% of the MSW by 2005. The U.S. continues to fall short of that goal and in 2014, 34.1% of the MSW was recycled or composted (U.S. EPA 2011).

Social customs result in significant variations in the mass of waste generated. For example, more frequent collections tend to result in an increase in the total amount of material collected. Increasing use of in-home garbage disposal units tends to decrease the mass of food waste disposed as MSW. As the extent of usage of packaged and prepared food increases, so does the mass of packaging waste, whereas the amount of raw food waste disposed decreases. The total mass of waste is lower in lower income areas, whereas the percentage of food waste increases.

Residential locations (including multifamily dwellings) generate approximately 55–65% of all MSW. Averages are subject to adjustment depending on many local factors. Studies show wide differences in amounts collected by municipalities because of differences in climate, living standards, time of year, education, location, and collection and disposal practices.

As observed in Figure 13–2, on both a total mass and a mass per capita basis, the United States generates more solid waste per year than any other industrialized nation. This difference would be significantly greater if the United States were not also a leader in the recycling of municipal waste, as shown in Figure 13–3. This data can be used to begin to size MSW landfills and determine numbers of trucks necessary to collect MSW as is accomplished in Example 13–1.

FIGURE 13–2

Variability of the masses of MSW generated by major country.

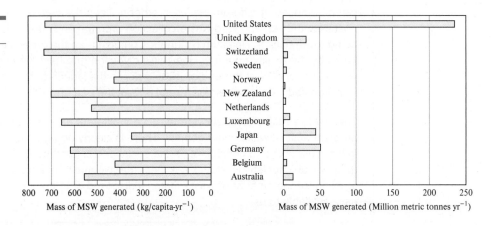

FIGURE 13–3

Variability of recycling patterns of MSW by major countries.

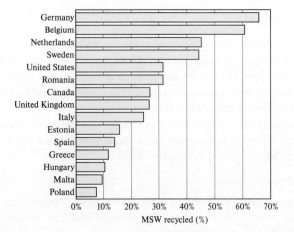

13–2   CHARACTERISTICS OF SOLID WASTE

The terms *refuse* and *solid waste* are used more or less synonymously, although the latter term is preferred. Solid waste can be classified in several different ways. The point of origin is important in some cases, so classification, as shown in Table 13–1, as domestic, institutional, commercial, industrial, street, demolition, or construction may be useful. The nature of the material may be important, so classification can be made on the basis of organic, inorganic, combustible, noncombustible, putrescible, and nonputrescible fractions. The classification of solid waste is usually used for choosing the treatment, collection, recycling, and disposal options.

Putrescible waste is the animal and vegetable waste resulting from the handling, preparation, cooking, and serving of food. It is composed largely of degradable organic matter and moisture. It also includes small amounts of free liquids. Putrescible wastes originate primarily in home kitchens, stores, markets, restaurants, and other places where food is stored, prepared, or served. This type of waste decomposes rapidly, particularly in warm weather, and may quickly produce undesirable odors. Putrescible wastes have some commercial value as animal food and as a base for commercial feeds; however, this use may be precluded by health considerations.

Municipal solid waste, commonly known to the lay audience as garbage, is a subset of solid waste and is defined as durable goods (e.g., appliances, tires, batteries), nondurable goods,

TABLE 13–1	Sources of Solid Wastes Within a Community	
Source	**Typical Facilities, Activities, or Locations Where Wastes Are Generated**	**Types of Solid Wastes**
Residential	Single family and multifamily detached dwellings, low-, medium-, and high-rise apartments, etc.	Food wastes, paper, cardboard, plastics, textiles, leather, yard wastes, wood, glass, tin cans, aluminum, other metals, ashes, street leaves, special wastes (including bulky items, consumer electronics, white goods, yard wastes collected separately, batteries, oil, and tires), household hazardous wastes
Commercial	Stores, restaurants, markets, office buildings, hotels, motels, print shops, service stations, auto repair shops, etc.	Paper, cardboard, plastics, wood, food waste, glass, metals, special wastes (see above), hazardous wastes, etc.
Institutional	Schools, hospitals, prisons, governmental centers	As above in commercial
Construction and demolition	New construction sites, road repair/renovation sites, razing of buildings, broken pavement	Wood, steel, concrete, dirt, etc.
Municipal services (excluding treatment facilities)	Street cleaning, landscaping, catch basin cleaning, parks and beaches, other recreational areas	Special wastes, rubbish, street sweepings, landscape and tree trimmings, catch basin debris, general wastes from parks, beaches, and recreational areas
Treatment plant sites; municipal incinerators	Water, wastewater, and industrial treatment processes, etc.	Treatment plant wastes, principally composed of residual sludges
Municipal solid waste[a]	All of the above	All of the above
Industrial	Construction, fabrication, light and heavy manufacturing, refineries, chemical plants, power plants, demolition, etc.	Industrial process wastes, scrap materials, etc. Nonindustrial wastes including food wastes, rubbish, ashes, demolition and construction wastes, special wastes, hazardous wastes
Agricultural	Field and row crops, orchards, vineyards, dairies, feedlots, farms, etc.	Spoiled food wastes, agricultural wastes, rubbish, hazardous wastes

[a]The term *municipal solid waste* (MSW) normally is assumed to include all of the wastes generated in a community with the exception of industrial process wastes and agricultural solid wastes.

Source: Tchobanoglous, Theisen, and Vigil, 1993.

FIGURE 13–4

Typical composition of MSW. In 2014, U.S. residents, businesses, and institutions produced more than 234 million metric tonnes, which amounts to approximately 2.03 kg of waste per person per day. Other includes miscellaneous inorganic wastes and electrolytes from urine and feces in disposable diapers.

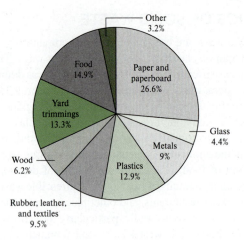

containers and packaging, food wastes, yard trimmings and miscellaneous organic wastes from residential, commercial, and industrial nonprocess sources.

The average municipal solid waste composition in the United States in 2014 is shown in Figure 13–4.

As shown in Figure 13–5, the composition of MSW varies across the globe. Composition of MSW depends on numerous factors, including climate, frequency of collection, usage patterns of in-home garbage disposal units, social customs, per capita income, use and density of food and other material packaging, recycling patterns (e.g., implementation of curbside recycling).

FIGURE 13–5

Variable composition of MSW across the globe. (*Source:* OECD Environmental Data Compendium: 2002 http://www.oecd.org/env /indicators-modelling-outlooks/oecdenvironmentaldatacompendium.htm.)

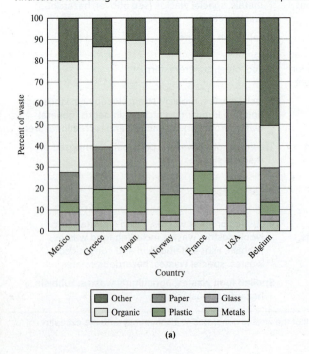

(a)

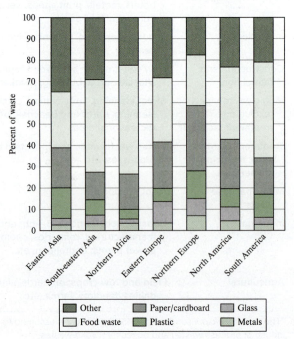

(b)

The density of solid waste is the mass of solid waste per unit volume. The density of loose combustible refuse is approximately $115 \text{ kg} \cdot \text{m}^{-3}$, and the density of collected solid waste is 180 to $450 \text{ kg} \cdot \text{m}^{-3}$. In the landfill, compacted solid waste can have a density ranging from 350 to $500 \text{ kg} \cdot \text{m}^{-3}$, and the density of well-compacted solid waste can range from 600 to $750 \text{ kg} \cdot \text{m}^{-3}$. Within a landfill, the density will depend on compaction and depth of the waste, the age of the waste, the composition of the waste and its moisture content. Unfortunately, where density is reported, rarely do these reports provide information to properly interpret the meaning of the density value given.

EXAMPLE 13–1 A town of 20,000 in Germany generates $0.95 \text{ kg} \cdot (\text{capita})^{-1} \cdot \text{day}^{-1}$ of MSW. A town of the same size in the United States generates $1.9 \text{ kg} \cdot (\text{capita})^{-1} \cdot \text{day}^{-1}$ of MSW.

1. How much MSW is generated in each town?
2. How many trucks would be needed to collect the waste twice weekly? The trucks each have a capacity of 4.5 t (metric tonnes) and operate 5 days per week. Assume that the trucks average two loads per day at 75% capacity.
3. If each of the towns recycles waste in percentages given in Figure 13–4, determine the mass of MSW that enters the landfill. If the density of the waste is $280 \text{ kg} \cdot \text{m}^{-3}$, what is the volume of MSW?

Solution

1. The amount of MSW generated is simply the number of people multiplied by the per capita MSW generation rate.

   For the German town:

   $$20{,}000 \text{ people} \times 0.95 \text{ kg} \cdot (\text{person})^{-1} \cdot \text{day}^{-1} = 19{,}000 \text{ kg} \cdot \text{day}^{-1}$$

   For the U.S. town:

   $$20{,}000 \text{ people} \times 1.9 \text{ kg} \cdot (\text{person})^{-1} \cdot \text{day}^{-1} = 38{,}000 \text{ kg} \cdot \text{day}^{-1}$$

2. To determine the number of collection trucks needed by each town, first calculate the total amount generated by each town on a weekly basis:

   For the German town:

   $$19{,}000 \text{ kg} \cdot \text{day}^{-1} \times 7 \text{ days} = 133{,}000 \text{ kg} \cdot \text{week}^{-1}$$

   For the U.S. town:

   $$38{,}000 \text{ kg} \cdot \text{day}^{-1} \times 7 \text{ days} = 266{,}000 \text{ kg} \cdot \text{week}^{-1}$$

   Next, determine the capacity of each truck.

   $$2 \text{ loads} \cdot \text{day}^{-1} \, (0.75 \times 4.5 \text{ t per truck load})(1000 \text{ kg} \cdot \text{t}^{-1}) = 6750 \text{ kg} \cdot \text{day}^{-1} \cdot \text{truck}^{-1}$$

   Next, determine the total collection demand that must be met by an unknown number of trucks. Two collections occur (effectively splitting the demand in half) for each residence during the 5-day workweek.

   For the German town:

   Using the MSW generation rates calculated earlier, each week the German town generates 133,000 kg of MSW. Because the waste is collected twice a week, each truck collects

   66,500 kg every 2.5 days (during a 5-day work week)

   Similarly, for the U.S. town, each truck collects,

   133,000 kg every 2.5 days (during a 5-day work week)

   Finally, calculate how many trucks are needed to fill this demand by dividing the total collection demand by the capacity for a single truck.

For the German town:

$$\frac{(66,500 \text{ kg})}{(2.5 \text{ days})(6750 \text{ kg} \cdot \text{day}^{-1} \cdot \text{truck}^{-1})} = 3.94, \text{ or 4 trucks}$$

For the U.S. town:

$$\frac{(133,000 \text{ kg})}{(2.5 \text{ days})(6750 \text{ kg} \cdot \text{day}^{-1} \cdot \text{truck}^{-1})} = 7.88, \text{ or 8 trucks}$$

3. The mass of waste that enters the landfill is simply the total amount of MSW produced minus the amount recycled. The volume is then calculated by dividing the mass by the density.

For the German town:

$$19,000 \text{ kg} \cdot \text{day}^{-1} - (0.16 \times 19,000 \text{ kg} \cdot \text{day}^{-1}) = 15,960 \text{ kg} \cdot \text{day}^{-1}$$

$$\frac{(15,960 \text{ kg} \cdot \text{day}^{-1})}{(280 \text{ kg} \cdot \text{m}^{-3})} = 57 \text{ m}^3 \cdot \text{day}^{-1}$$

For the U.S. town:

$$38,000 \text{ kg} \cdot \text{day}^{-1} - (0.23 \times 38,000 \text{ kg} \cdot \text{day}^{-1}) = 29,260 \text{ kg} \cdot \text{day}^{-1}$$

$$\frac{(29,260 \text{ kg} \cdot \text{day}^{-1})}{(280 \text{ kg} \cdot \text{m}^{-3})} = 104.5 \text{ m}^3 \cdot \text{day}^{-1}$$

13–3   SOLID WASTE MANAGEMENT

In discussing solid waste management, we must consider the waste from the point of generation to the point of final disposal. As shown in Figure 13–6, solid waste management is a complex process, involving multiple steps.

As indicated in Figure 13–6, the first step in solid waste management is the generation of the waste. Once a material no longer has value to its owner, it is considered waste. As discussed earlier in this chapter, the generation of waste varies by country, socioeconomic status, and as a result of many other practices.

Once the waste is generated on site, it must be processed in some way. This processing may include washing, separation, and storage so as to recycle some portion of the waste. Public law and education significantly affect this step. For example, in some communities it is illegal to discard lawn clippings and other similar biomass in the regular trash collection. This material must be disposed of during separate trash collections. Educating the public as to the importance of recycling will affect this step also.

Waste collection is the next step in the management process. Collection includes picking up solid wastes and emptying containers into suitable vehicles for transport. This step also includes the collection of recyclable material. As will be discussed in the next section, collection and transport of waste represents a significant fraction of the total cost of waste management.

The collected waste can be transferred to a central storage facility or to a processing facility. If processing occurs, it usually includes mass and volume reduction, along with separation into the various components that can be reused. The separated waste may at this point become a valuable commodity. In effect, it is no longer a waste. The organic portion of the waste can be transformed into heat by chemical means (usually incineration) or into fuel gas or compost (by biologically mediated reactions).

FIGURE 13–6

Elements of a solid waste management system.

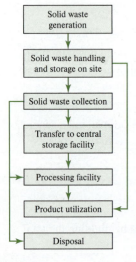

The final steps include transport and disposal. The most common means of final disposal is landfilling.

13–4 SOLID WASTE COLLECTION

The solid waste collection policies of a city begin with decisions made by elected representatives about whether collection is to be made by (1) city employees (municipal collection), (2) private firms that contract with city government (contract collection), or (3) private firms that contract with private residents (private collection).

Elected officials may also determine what type of solid wastes are to be collected and from whom. In some municipalities broad classes of solid wastes (e.g., yard waste) are not accepted for general collection. In others, certain materials (e.g., tires, furniture, or dead animals) may be excluded. Hazardous wastes are excluded from regular collections. The nature of the service may be governed by limitations of disposal facilities or by the opinion of the legislative body as to what service should be performed. Almost all municipal systems collect residential waste, but only about one-third collect industrial waste.

The final decision concerning collection, which is made by the elected officials, is the frequency of collection. The proper frequency for the most satisfactory and economical service is governed by the amount of solid waste that must be collected and by climate, cost, and public requests. For the collection of solid waste that contains putrescible waste, the maximum period should not be greater than

1. The normal time for the accumulation of the amount that can be placed in containers of reasonable size
2. The time it takes for fresh putrescible waste to decay and emit foul odors under average storage conditions
3. The length of the fly-breeding cycle, which, during the hot summer months, is less than 7 days

In the last three decades the prevailing frequency of collection has changed from twice a week pickup to once a week. The increased use of once per week service is due to two factors. First, unit costs are reduced when frequency is cut from twice to once per week. Second, the increased percentage of paper and decreased volume of putrescible wastes permit longer periods of acceptable storage.

Once policy has been set, the actual method of collection is determined by engineers or managers. Major considerations include how the solid waste will be collected, how the crews will be managed, and how the trucks will be routed. A more complete discussion of collection methods, estimation, and routing is beyond the scope of this text.

13–5 WASTE AS RESOURCE

Background and Perspective

The Earth's prime mineral deposits are limited. As high-quality ores are depleted, lower grade ores must be used. Lower grade ores require proportionately greater amounts of energy and capital investment to extract. In a broad economic context, the long-term reasonableness of a market-accounting system that applies only current development costs to our use of depletable, nonrenewable natural resources such as aluminum, copper, iron, and petroleum should be viewed with concern. High rates of virgin raw material extraction result in high rates of solid waste production. In the United States, blatant mispricing—including the "depletion allowance" on minerals and unreasonably low rail rate fares on ores in contrast to scrap—is in no small way responsible for the exploitation of mineral deposits and the large amounts of solid waste generated. Furthermore, our high-waste, low-recycle lifestyle is inherently wasteful of a bountiful endowment of natural resources.

Our renewable resources, primarily timber, are also under siege. Our preference for over-packaged items, in combination with a wanton lack of care in our forests, has strained nature's capacity for growth and replenishment. Europe, India, and Japan have long been faced with a want of timber. We in the United States should learn from their predicaments.

The prevention of waste generation (resource conservation) and the productive use of waste material (resource recovery) represent means of alleviating some of the problems of solid waste management. At one time in our history, resource recovery played an important role in our industrial production. Until the mid-20th century, salvage (reprocessing of wastes to recover an original raw material, now called *recycling*) from household wastes was an increasingly important source of materials. In the 5 years preceding 1939, recycled copper, lead, aluminum, and paper supplied 44, 39, 28, and 30%, respectively, of the total raw materials shipments to fabricators in the United States (National Center for Resource Recovery, 1974). Ultimately, it became more economical to process virgin materials than to use recovered ones.

In principle, processable MSW could provide 95% and 73% of our nation's needs in glass and paper, respectively. However, in the categories of composites, plastics, and wood, only a fraction of the annual needs of the country is recoverable from MSW. While recycling involves returning the material to its raw form, re-use is sometimes more energy efficient. The percentage of metals potentially recycled varies greatly and is dependent on the cost of recycling and raw materials, the ease by which the material is recycled, and the market for both the purchase of recycled material and virgin product. The costs for recycling copper, brass, and aluminum are approximately the same, however, approximately 85% of copper and aluminum can be potentially recycled whereas the potential for brass, an alloy, is only about 45%. Despite high costs, composites are difficult to recycle as most are based on thermosets, polymers that do not melt when heated. Ceramics are almost never recycled. In terms of energy production, if all the MSW in the United States were converted to energy, it would supply 3–5% of the nation's need. This would be enough to supply all of our residential and commercial lighting requirements (Vence and Powers, 1980). Although some communities might be able to generate 7–10% or more of their need, nationally the actual potential is probably on the order of 0.3%.

In light of these facts, it is not difficult to see why "recovery for recovery's sake" or "energy recovery for energy conservation" receive scant attention except in times of national emergency or failure of the local sanitary landfill to accommodate the waste being generated. The fact that, under our current market-accounting system, resource conservation and resource recovery, except in locally favorable situations, costs more than landfilling is a further disincentive to the conservation of our natural resources. Why, then, should any governing body even consider resource conservation and resource recovery? The answer, simply stated, is "to protect our environment." In the same context that we have attempted to control air and water pollution, we have an obligation to leave our descendants something better than the accumulated litter of our squandering habits. Although there will be some expected cost, that cost should be minimized. EPA has set as a national goal that 35% of all solid waste should be recycled by 2005. The greatest rates of recycling are in the west coast states (EPA Regions 9 and 10), whereas the lowest rates are in the western states (EPA Region 8) and the southwest (EPA Region 6).

Individual states have enacted over 500 laws on recycling, ranging from purchasing preferences to comprehensive recycling goals. More than 6000 reuse centers around the country recycle materials ranging from specialized programs for building materials or unneeded materials in schools to local programs such as Goodwill and the Salvation Army, according to the Reuse Development Organization. As of 2016, 10 states have bottle deposit regulations. Twenty-four states have yard waste bans (Miller, 2006). In the 50 states and the District of Columbia, there are over 3780 yard waste compositing programs and over 9500 curbside recycling programs. The EPA has estimated the increases in recovery for recyclables as shown in Table 13–2. It estimates that overall recovery increased from 6.4% in 1960 to 16.2% in 1990 to 28% in 1997 to 34.1% in 2011 (U.S. EPA, 2011). The 7.7 Tg of MSW recycled and composted provides an annual benefit of greater than 189 Tg of CO_2-equivalent emission reductions, equivalent to more than 36 million passenger vehicles.

TABLE 13–2	Recycling Rates for Some MSW Components					
Material	1990 Recycling Rate (%)[a]	1995 Recycling Rate (%)[b]	2000 Recycling Rate (%)[b]	2005 Recycling Rate (%)[c]	2010 Recycling Rate (%)[d]	2014 Recycling Rate (%)[e]
Paper and paperboard	27.8	40.0	45.4	50	62.5	64.7
Glass	20.0	24.5	23.0	21.6	27.1	26.0
Total Metals	26.1	38.9	35.4	36.8	35.1	34.0
Aluminum	35.9	34.6	27.4	21.5	19.9	19.8
Plastics	2.2	5.3	5.4	5.7	8.2	9.5
Yard trimmings	12.0	30.3	56.9	61.9	57.5	61.1
Rubber and leather	6.4	8.8	12.2	14.3	15.0	17.5
Wood	3.3	9.6	3.8	9.4	14.5	15.9
Clothing and other textiles	11.5	12.2	13.5	15.3	15.0	16.2

[a]U.S. EPA, 1997.
[b]U.S. EPA, 2002.
[c]U.S. EPA, 2006.
[d]U.S. EPA, 2011.
[e]U.S. EPA, 2014.

Green Chemistry and Green Engineering

"Green chemistry" and "green engineering" are changing the way engineers design and fabricate. As we begin to look at the big picture of how engineering decisions affect the environment and our planet's life, we are finally beginning to try to solve environmental problems before they happen. Rather than using "end-of-the-pipe" treatment techniques, engineers are attempting to select materials, design manufacturing processes, and enhance energy conservation to minimize their effects on the environment. Where appropriate, the effect of these decisions on MSW management are considered in the following sections.

Recycling

Recycling can also be described as closed-loop or open-loop. **Closed-loop, or primary, recycling** is the use of recycled products to make the same or similar products. Examples of primary recycling include the use of glass bottles to make other types of glass bottles or aluminum cans to make new aluminum cans. **Secondary recycling** is the use of recycled materials to make new products with different characteristics than the originals. For example, milk jugs made of high-density polyethylene (HDPE) can be recycled into toys or drainage pipes. **Tertiary recycling** is the recovery of chemicals or energy from postconsumer waste materials. Many electronic companies, for instance, recover the solvents used in their manufacturing processes and distill them for reuse.

At the lowest and most appropriate technological level, recycled materials are separated at the source by the consumer (**source separation**). This is the most appropriate level because it requires the minimum expenditure of energy. With stringent goals for recycling, municipalities are looking at detailed recycling options.

Methods of Collecting Recyclable Materials.   Generally, the collection options available to a municipality for residential use include curbside collection and drop-off centers. Recycled materials can be collected by a community and then taken to a material-processing facility or to material transfer stations, where the material is sorted and separated via mechanical means.

The primary method of collecting dry recyclables in the United States today is curbside collection. This has the advantage of being easier on the resident than having to drive to a recycling center. There are two basic types of curbside collection for recycling. In the first, the homeowner is given a number of bins or bags. The homeowner separates the refuse as it is used, placing it in the appropriate bin. On collection day the container is placed on the curb. The primary disadvantage of supplying home storage containers is the cost, which can represent a significant investment. A second method of curbside recycling is to provide the homeowner with only one bin, into which is placed all the recyclable materials. Curbside personnel then separate material as it is being picked up, placing each type of material into a separate compartment in the vehicle.

A second collection alternative is a drop-off center. Because recycling is a community-specific operation, a drop-off system must be designed around and in consideration of conditions particular to the area of involvement. To evaluate and select the most appropriate drop-off system, we must consider critical factors such as location, materials handled, population, number of centers, operation, and public information. When drop-offs are used to supplement curbside programs, fewer and smaller drop-off sites may be required than necessary when curbside programs are not implemented. When drop-off sites are the only, or primary, recycling system in a community, the system must provide for increased capacity. Careful planning to accommodate traffic flow, as well as storage and collection of materials, must be part of the siting activity.

The convenience of a drop-off center will directly affect the amount of citizen participation. Strategically locating a drop-off center in an area of high traffic flow, where the center is highly visible, will encourage a greater level of participation. Even rural areas with widely scattered populations can provide good locations for drop-offs. Rural homeowners have certain common travel patterns that bring them to a few locations at regular intervals—to a grocery store, church, or post office, to name a few.

Recycling of Materials.   Several types of materials are commonly recycled, including glass, metals (Al, Fe, Cu, steel), paper, and plastics. Paper (especially newspaper and cardboard) is easily recycled. Recycling paper involves the removal of inks, glue, and coatings. The paper is then reconverted to pulp that can be pressed into new paper. Because this process results in the breakdown of some of the paper fibers and a reduction in strength of the paper, new pulp material must be added to the recycled paper. By 2014, 64.7% of the paper and paperboard produced in the United States was being recycled. Old corrugated cardboard boxes account for more than 61% of the paper that is recycled (U.S. EPA, 2016).

In 1990, some 25 billion kg of plastic materials were produced in the United States. Of the various forms of plastics that are produced and can be recycled, high-density polyethylene is the most common. Known by the Society of Plastics Industry code, #2-HDPE, this type is used to produce milk bottles, detergent bottles, and film products such as produce bags. Recycled HDPE is also used for the production of protective wrap, grocery bags, and molded products such as toys and pails. Polyethylene terephthlate (#1-PETE) is most commonly used originally in the production of carbonated soft drink bottles. Some states have ensured the recycling of a large percentage of this material through "bottle bills," requiring deposits on soft drink bottles. Although this has been successful in the recycling of soft drink containers, as the consumption of bottled water has increased, so has the disposal of water bottles as MSW because they are not always covered by bottle bill regulations.

Glass is one of the easiest materials to recycle. Almost all of the glass that is recycled is made into new glass containers and bottles. A small amount is used to make other products, such as glass wool, fiberglass insulation, paving material, and building products. Nevertheless, in the United States, in 2014, only 26.0% of the glass produced was recycled. From Figure 13–7, it is clear that while significant strides have been made in the recycling of bottles containing liquids such as beer, soft drinks, and wine, much work still needs to be done to ensure that food and other jars and bottles are recycled. Bottle laws can have a significant impact on glass recycling. For example, in Michigan, where all beer and soft drinks have a 10-cent deposit, the redemption rate is around 95%.

One area where research is ongoing is the use of waste glass as a partial replacement for cement in the production of cement. Not only is this beneficial in terms of glass recycling, but

Glass products in MSW.

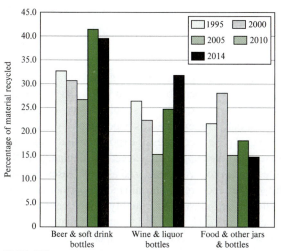

U.S. EPA, 1997.
U.S. EPA, 2002.
U.S. EPA, 2006.
U.S. EPA, 2011.
U.S. EPA, 2014

Percentage of aluminum beverage cans recycled.

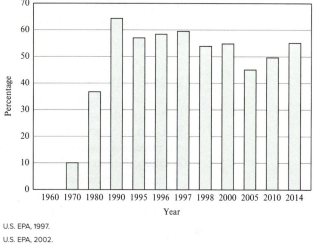

U.S. EPA, 1997.
U.S. EPA, 2002.
U.S. EPA, 2006.
U.S. EPA, 2011.
U.S. EPA, 2014

it has the potential to reduce CO_2 emissions as each tonne of cement emits one tonne of carbon dioxide (CO_2) to the atmosphere. Additionally, the production of cement is very energy intensive; the production of a tonne of cement consumes almost 5.5 million kJ of energy. Researchers at Michigan State University have discovered that the chemical composition and the reactivity of waste glass, milled to the particle size of cement, make it a suitable supplement to cement in the production of concrete. As a result, Michigan State University is pioneering large-scale introduction of recycled glass concrete into construction projects on its campus.

The most commonly recycled metals are iron and steel. MSW contains about 7% metal waste (before recycling). "Tin" cans are made almost entirely of steel. In fact, one tonne of "tin" cans contains only 1.9 kg of tin. Approximately 97% of all metal food containers are made of steel. Because of its relative ease of collection and processing, 70.7% of all steel cans produced in 2014 were recycled (U.S. EPA, 2014). The major impediment to recycling of cans is the high cost of transportation. For recycling of this material to be economical, the cost of collection, transportation, and processing must be less than the cost of producing new steel.

Other metals that can be recycled include aluminum, copper, lead, nickel, tin, and zinc. Virtually all nonferrous metals can be recycled if they are sorted and free of foreign products. Although this is not a problem with aluminum cans, this restriction discourages the recycling of many nonferrous metals. Today, 100% of all soda and beer cans are made of aluminum and the average aluminum beverage cans contain about 51% postconsumer recycled aluminum. As can be seen from Figure 13–8, the percentage of aluminum cans recycled each year rose steadily until 1994. Unfortunately, the percentage of cans recycled reached a maximum of about 64% and has decreased in recent years. More needs to be done to increase the percentage of this important resource that is recycled.

Improving the Market for Recyclables.   Over the last decade the market and prices for recycled materials have fluctuated widely. Among the factors that affect the market and prices are

- Economic conditions
- Overall demand
- Demand for products specifically made partially or entirely from recovered materials
- Quality of recovered materials (contamination problems)

- Capacity to use recovered material in a process
- Excess capacity to produce virgin material
- Transportation costs
- Export markets
- Differences between supply and demand
- Legislation

The market for recycled paper products varies greatly and is highly sensitive to economic conditions. During a recession, the demand for newspapers, books, and other paper products decreases significantly, reducing the demand for all paper; thus the supply of recovered paper will exceed demand. Export markets also vary greatly; for example, as paper recovery in Europe and Asia increases, their demand for U.S. recycled paper decreases. If export demand decreases, the price of recovered material also decreases, and the supply will exceed the demand. On the contrary, if the export demand increases, the price of recovered material also increases, and the demand will exceed the supply. The quality of the recovered material is an important factor in determining desirability of the material to the paper industry. The quality can be enhanced by better separation of the particular types of paper from each other and from other materials such as plastics or glass. The cost of virgin paper will also directly affect paper recovery. Strong legislature governing recycling will enhance the recovery of paper; weakening regulations will have the opposite effect. Thus, the market for recovered paper is enhanced by

- A strong economy
- Export demand
- Separation of paper from other materials
- High cost for virgin material
- Strong regulations

Most recovered glass is used by glass container manufacturers. Some recovered glass is used by fiberglass manufacturers and in making glass aggregates and abrasives. As with paper, the recovery of glass can be enhanced by the separation of glass from other materials to decrease the level of contamination. The market for recovered glass depends on the tolerance for colored glass contamination in the production process. For example, green glass fabrication can accept up to 50% different colors, whereas contamination must be less than 5% in flint (colorless) glass production. The distance to market, and therefore, transportation costs, will play a significant role in recycling. These costs can be decreased by attempting to increase the density of material when transported to market.

Aluminum containers have been the most sought-after recovered material because the conversion of scrap aluminum into new material offers considerable cost savings over production from newly mined and processed bauxite. Because the aluminum market can absorb more recovered material, an increase in the extent of aluminum recovery is feasible. Recovery could be enhanced by

- The reduction in contamination by separating aluminum from other metals
- Strong "bottle laws"
- The availability of a high street price for used beverage containers

The recovery of steel from residential use in MSW depends greatly on the willingness of the steel industry to use recycled material. In 2012, about 66% of the steel processed in the United States was recovered material. As the steel industry has been actively promoting recovery and recycling steel, the market appears positive.

HDPE and PETE materials account for about 80% (by mass) of the recovered plastic in MSW. The market for recovered plastic varies greatly with plastic type. As long as the supply of virgin HDPE remains low, the price for recovered HDPE will be favorable. An oversupply of virgin PETE material has kept the price of and demand for recovered PETE low. The regulations governing the recycling of plastic materials have also been weakened. Consequently, the market for recovered plastics can be enhanced by

- Strengthening recycling regulations
- Reducing production of virgin PETE

Compost, described in detail in the next subsection, can be produced from components of MSW such as yard trimmings, food wastes, and soiled paper. These products can be mixed with non-MSW feed stocks, including livestock manure, industrial food waste, and wastewater sludges. The market for compost material can be enhanced through

- The production of high-quality and consistent material that has general customer acceptability
- The production of products with specific nutrient content depending on individual customer needs and demands
- The education of consumers as to the benefits of using compost
- Strong regulations governing the disposal of yard trimmings

Composting

Composting is the controlled decomposition of organic materials, such as leaves, grass, and food scraps, by microorganisms. The result of this decomposition process is compost, a crumbly, earthy-smelling, soil-like material. Because MSW in the United States contains up to about 25% yard trimmings and food scraps, composting of this material can greatly reduce the amount of waste that ends up in landfills or incinerators.

The material to be composted contains both carbon and nitrogen sources. Leaves, straw, and woody materials serve as a major carbon source. Grass and food scraps provide significant nitrogen. A ratio of 30 parts carbon to 1 part nitrogen is necessary to achieve optimal microorganism growth and energy production. Odor and other problems can occur if the carbon to nitrogen ratio is not 30:1.

The decomposition of organic materials in composting involves both physical and chemical processes. During decomposition, organic materials are broken down through the metabolic activities of various invertebrates present in composting material, including mites, millipedes, beetles, sowbugs, earwigs, earthworms, slugs, and snails. These organisms need adequate moisture and oxygen to degrade the organic materials in the most efficient manner.

Microbes in the pile create considerable heat and essentially "cook" the compost. Temperatures between 25 and 55°C are common in properly maintained large-scale compost piles. Backyard compost piles may not reach these levels. These high temperatures are necessary for rapid composting as well as for destroying weed seeds, insect larvae, and potentially harmful bacteria. When the compost is finished, it has a crumbly texture throughout the pile.

More than 68% of the MSW generated in the United States can be composted. When MSW is to be composted, recyclables, such as glass and aluminum, and noncompostable material must be removed prior to composting. The remaining material, which is predominately organic in nature, can then be composted. Aerated windrows are the most common method. In-vessel composting, in which the material is left to decompose while enclosed in a temperature- and moisture-controlled chamber, can also be used. Composting of MSW using aerobic digestors is used in some European countries. Prior to digestion, the MSW can be combined with livestock manure or food-processing wastes. The finished compost can be sold, given away, or used by the company or municipality in local landscaping or agricultural projects.

Source Reduction

If our standard of living is to be sustainable, we must find ways to better reuse the materials we create. We simply cannot continue to dispose of 2 kg of refuse per day per person. Thus, it is necessary that we focus our efforts on "source reduction" or preventing the formation of waste in the first place. We need to design, manufacture, purchase, and use materials in ways that minimize the amount and toxicity of the waste we generate. By practicing source reduction, we also reduce the costs of and problems associated with recycling, composting, landfilling, or incinerating waste. We also conserve precious natural resources and reduce pollution. Source reduction can be accomplished in numerous ways.

Waste Reduction and Process Modification.   Source reduction can be accomplished by reducing the quantity of materials generated before they are ever ready for recycling or reuse. This includes the redesign of products or packages to reduce the quantity of material used. For example, in 1977 the mass of a 2-L soft drink bottle was 68 g. Today, the same size bottle has a mass of only 51 g. Although the amount of plastic in each bottle may seem inconsequential, this reduction has kept 115 million kg of plastic per year out of the waste stream. Companies can also use packaging that reduces the amount of damage or spoilage, thereby reducing the mass of material that must be discarded.

Nontoxic materials can be substituted for toxic ones. For example, soy-based or water-based inks can be used. Lead foil on wine bottles can be replaced with aluminum, which can be easily recycled. Freon can be replaced with non-ozone-consuming chemicals. High-temperature hydrocarbons and silicone oil have replaced polychlorinated biphenyls (PCBs) in transformers and other electrical equipment. Transformers have also been redesigned and retrofitted to allow for the use of lubricants and insulators other than PCBs.

The mass of waste material can be reduced if manufacturers design products that have longer lives and are easily repaired. Assuming that more durable goods will cost more, consumers must also be willing to have damaged equipment repaired.

Businesses and individuals can modify their practices to reduce waste. For example, e-mail can be used to distribute messages instead of letters. Reports can be copied on both sides of a page. Additionally, mailing lists can be routinely updated to ensure that mail is not being sent astray. Individuals can request to be removed from mailing lists to reduce the amount of unwanted and waste mail. Products can be purchased in bulk quantities to reduce waste.

Reuse of Products.   Other ways of source reduction are the reuse of durable goods (e.g., appliances, furniture, computers) by repairing them when they are broken or refurbishing them when they are out-dated, donating them to charity and community groups, or selling them to others that have a direct use for them. Ideally, products should be used more than once, either for the same purpose or for a different one. Reusing, when possible, is preferable to recycling because the item does not need to be reprocessed before it can be used again. Some ways that items can be reused are

- Using durable coffee mugs rather than disposable cups
- Using cloth napkins or towels rather than paper napkins or towels
- Refilling water bottles from the tap rather than buying disposable water bottles
- Donating old magazines or surplus equipment to charity
- Reusing boxes for other purposes
- Using plastic containers for storage rather than using disposable plastic bags
- Turning empty jars into containers for leftover food or other items
- Purchasing refillable pens and pencils
- Participating in a paint collection and reuse program

Education and Legislation.   Source reduction may be implemented by educating people about the benefits of such actions or by enforcement, for example through the use of fees or deposits. Legislation requiring mandatory refunds or deposits on both returnable and nonreturnable beverage containers has been and will continue to be hotly contested by the beverage and beverage container industries. In those states that have enacted mandatory refund or deposit legislation (e.g., California, Connecticut, Hawaii, Iowa, Maine, Massachusetts, Michigan, New York, Oregon, and Vermont), there is strong evidence to show that the claims of lost jobs and lost business are either unfounded or are offset by increases in jobs and business in other sectors.

These programs have been a great success. Typically, between 70 and 85% of the beverage containers are returned. In Oregon, a reduction in total roadside litter of 39% by item count and 47% by volume was reported after the second year of implementation of its law. Furthermore, for glass containers there is a significant energy savings in that a glass bottle reused ten times consumes less than a third of the energy of a single-use container. Average reuse cycles vary from 10 to 20 times per container.

Other materials such as residual ash and solids from incineration can be used as source material. For example, industrial waste consisting of fly ash from power plants has been used as a raw material for manufacturing building materials. Concrete-like blocks have been produced from oil shale and coal fly ash. Fly ash has also been used in the construction of roads. Chicken manure has been incinerated to produce a source material for phosphorus production, as an additive/basic material for fertilizer production, and as a basic road-building material.

13–6 SOLID WASTE REDUCTION

Combustion Processes

Fundamentals of Combustion.   Combustion is a chemical reaction in which the elements in the fuel are oxidized in the presence of excess oxygen. The major oxidizable elements in the fuel, in this case, the solid waste, are carbon and hydrogen. To a lesser extent sulfur and nitrogen are also present. With complete oxidation, carbon is oxidized to carbon dioxide, hydrogen to water, and sulfur to sulfur dioxide. Some fraction of the nitrogen may be oxidized to nitrogen oxides.

The combustion reactions are a function of oxygen, time, temperature, and turbulence. Sufficient excess of oxygen must be available to drive the reaction to completion in a short time. As shown in Figure 13–9, the oxygen is most frequently supplied by forcing air into the combustion chamber. Over 100% excess air may be provided to ensure sufficient excess. Sufficient time must be provided for the combustion reactions to proceed to completion. The amount of time is a function of the combustion temperature and the turbulence in the combustion chamber. Some minimum temperature must be exceeded to initiate the combustion reaction (i.e., to ignite the waste). Higher temperatures also yield higher quantities of nitrogen oxide emissions, so there is a tradeoff in destroying the solid waste and forming air pollutants. Mixing of the combustion air and the combustion gases is essential for the completion of the reaction.

As the solid waste enters the combustion chamber and its temperature increases, volatile materials are driven off as gases. Rising temperatures cause the organic components to

FIGURE 13–9

Combustion chambers in a mass-fired incinerator. **(a)** Martin grate (Courtesy of Ogden Martin Systems, Inc.) and **(b)** Dusseldorf grate. (*Source:* **(a)** Ogden Martin Systems, Inc.; **(b)** American Ref-Fuel, Inc.)

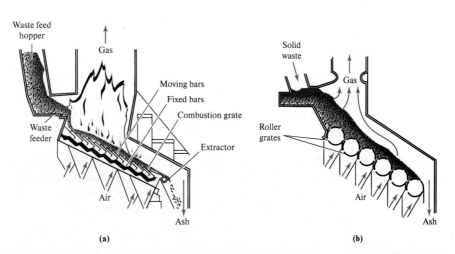

thermally "crack" and form gases. When the volatile compounds are driven off, fixed carbon remains. When the temperature reaches 700°C, the carbon is ignited. To achieve destruction of all the combustible material (**burn out**), it is necessary to achieve 700°C throughout the bed of waste and ash (Pfeffer, 1992).

In the **flame zone** the hot volatilized gases mix with oxygen. This reaction is very rapid, going to completion within 1–2 s if the air and turbulence are sufficient.

The evolution of solid waste combustion has led to higher temperatures both for destroying toxic compounds and increasing the opportunity to use the waste as an energy source by producing steam.

Heating Value of Waste. The heating value of waste is measured in kilojoules per kilogram (kJ · kg^{-1}) and is determined experimentally using a bomb calorimeter. A dry sample is placed in a chamber and burned. The heat released at a constant temperature of 25°C is calculated from a heat balance. Because the combustion chamber is maintained at 25°C, combustion water produced in the oxidation reaction remains in the liquid state. This condition produces the maximum heat release and is defined as the **higher heating value** (HHV).

In actual combustion processes, the temperature of the combustion gas remains above 100°C until the gas is discharged into the atmosphere. Consequently, the water from combustion processes is always in the vapor state. The heating value for combustion is termed the **lower heating value** (LHV). The following equation gives the relationship between HHV and LHV:

$$LHV = HHV - [(\Delta H_v)(9H)] \tag{13–1}$$

where ΔH_v = heat of vaporization of water = 2420 kJ · kg^{-1}
   H = hydrogen content of combusted material

The factor of 9 results because 1 gram mole of hydrogen will produce 9 gram moles of water (i.e., 18/2). Note that this water is only that resulting from the combustion reaction. If the waste is wet, the free water must also be evaporated. The energy required to evaporate this water may be substantial, resulting in a very inefficient combustion process from the point of view of energy recovery. The ash content also reduces the energy yield because it reduces the proportion of dry organic matter per kilogram of fuel and because it retains some heat when it is removed from the furnace.

Types of Incinerators

Conventional (Mass-Fired) Incineration. Mass-fired incinerators are the most common form of MSW incineration. The basic arrangement of the conventional incinerator is shown in Figure 13–10. These systems can accept refuse that has received little pretreatment other than the removal of oversized items, such as kitchen stoves and mattresses. Local programs for the removal of potentially hazardous chemicals such as pesticides and other hazardous household chemicals are necessary to prevent environmental damage. Because the refuse is generally not pretreated, although it may have some heat value, it is normally quite wet and is not **autogenous** (self-sustaining in combustion) until it is dried. Conventionally, auxiliary fuel is provided for the initial drying stages. Because of the large amount of particulate matter generated in the combustion process, some form of air pollution control device is required. Normally, electrostatic precipitators (discussed in Chapter 12) are chosen. Bulk volume reduction in incinerators is about 90%. Thus, about 10% of the material still must be carried to a landfill.

Refuse-Derived Fuel Facilities. **Refuse-derived fuel** (RDF) is the combustible portion of solid waste that has been separated from the noncombustible portion through processes such as shredding, screening, and air classifying (Vence and Powers, 1980). By processing municipal solid waste (MSW), refuse-derived fuel containing 12–16 MJ · kg^{-1} can be produced from 55–85% of the refuse received. This system is also called a **supplemental fuel system** because

Cross-sectional drawing of a conventional mass-fired incinerator. (©Covanta Energy Corp., Inc.)

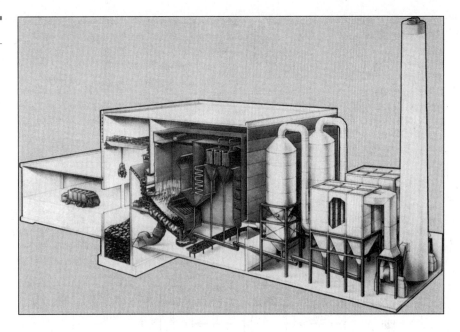

the combustible fraction is typically marketed as a fuel to outside users (utilities or industries) as a supplement to coal or other solid fuels in their existing boilers.

In a typical system, MSW is fed into a trommel, or rotating screen, to remove glass and dirt, and the remaining fraction is conveyed to a shredder for size reduction. Shredded wastes may then pass through an air classifier to separate the light fraction (plastics, paper, wood, textiles, food wastes, and smaller amounts of light metals) from the heavy fraction (metals, aluminum, and small amounts of glass and ceramics).

The light fraction, after being routed through a magnetic system to remove ferrous metals, is ready for fuel use. The heavy fraction is conveyed to another magnetic removal system for recovery of ferrous metals. Aluminum may also be recovered. The remaining glass, ceramics, and other nonmagnetic materials from the heavy fraction are then sent to the landfill.

The first full-scale plant to prepare RDF has been in operation in Ames, Iowa, since 1975. Subsequently, other plants using similar technology have been designed and constructed. Figure 13–11 shows the process flow diagram for the Southeastern Virginia Public Service Authority's RDF plant.

Modular Incinerators.   **Modular incinerators** are usually prefabricated units with capacities of 4.5 to 107 tonnes of solid waste per day. Most produce steam as the sole energy product. Most modular incinerators use a system involving two combustion chambers. The gases produced in the first chamber flow to the second chamber where they are burned more completely. The second chamber often serves as the only air pollution control device. Other modular units employ additional air pollution control equipment and are, thus, able to control emissions as effectively as mass-burn facilities. Although many modular units plan to retrofit existing controls to meet more stringent standards, others will close due to the expense associated with upgrading units to meet these stringent air emission standards. As a result, interest in modular incinerators is decreasing.

Fluidized-Bed Incinerators.   Fluidized bed incinerators have been used extensively in Japan and are becoming more common elsewhere in the world. In a **fluidized bed incinerator,** sand is heated to about 800°C by oil or gas. The sand is blown around, or "fluidized," in the incinerator by a blower that sends air from the bottom upward. As the sludge enters the bottom of the incinerator, the heated fluidized sand hits the sludge, both breaking it apart and burning it.

FIGURE 13–11

Southeastern Virginia Public Service Authority's RDF plant.

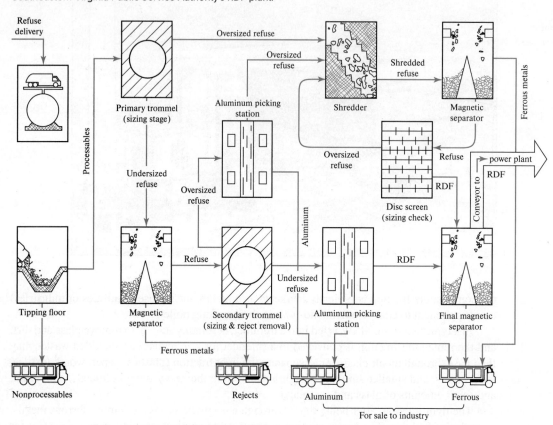

This type of incinerator requires front-end preprocessing of waste. Glass and metals must be removed from the refuse to be incinerated. Fluidized bed incinerators can, however, successfully burn wastes of widely variable moisture and heat content, such as paper and wood. Combustion and heat recovery in fluidized-bed incinerators are very efficient, and the level of pollutant emissions is low.

Fluidized bed systems appear to be more consistent in their operation than mass-burn incinerators and may be more effectively operated to control emissions, reduce residual ash production, and achieve high energy conversion efficiencies. Fluidized bed incinerators are able to co-combust fuels, meaning that, for example, they can burn municipal waste with coal or propane. These type of incinerators appear to be more effective than small sized mass-burn incinerators.

Public Health and Environmental Issues

The combustion of MSW can result in the emission of particulate matter, acid gases (SO_x, HCl, HF), NO_x (primarily NO and NO_2), carbon monoxide, organics, and heavy metals. Carbon dioxide emissions are not significant; if all MSW were incinerated, then CO_2 emissions would contribute <2% of that produced during each year in the United States.

The heavy metals are distributed in the bottom ash, in fly ash, and in particulate matter released in gases. Lead and cadmium will be distributed between the bottom* and fly ashes. Many

---

*Bottom ash** is the solid residue that falls through the grate of the incinerator. It is considerably larger in size than fly ash.

of the other metals will be found in the fly ash.* Until 1998, this ash could be disposed of in a Subtitle C landfill because the EPA deemed the ash exempt from hazardous waste regulations. However, in 1998, the U.S. Supreme Court ruled that if the ash generated by MSW incinerators fails federal toxicity tests, it must be treated as a hazardous waste.

NO_x, which is produced by the reaction of air during combustion at high temperatures, is often present in the off-gases from MSW incinerators. NO_x is known to have adverse effects on respiratory function and contribute to acid rain. HCl and SO_2 will also be produced in the off-gas. Particulate matter is known to have adverse public health and environmental effects as described in Chapter 12 (Air Pollution). Any mercury in the waste feed will be volatilized into the off-gases. MSW incinerators are the largest single source of dioxins, furans, and dioxin-like PCBs[†] (U.S. EPA, 2000). The EPA has estimated that 1100 g TEq $\cdot$ year^{-1} of these chemicals were released in 1995. Emissions standards and guidelines have been promulgated for new and existing large (>220 tonnes $\cdot$ day^{-1}) municipal waste combustion (MWC) facilities and proposed for small MWC facilities (31–220 tonnes $\cdot$ day^{-1}). EPA estimates that when full compliance with the MWC rules is attained, the annual emissions resulting from MSW incinerators will decline significantly to about 24 g TEq $\cdot$ year^{-1}. The 2000 National Research Council report (2000) suggests that maximal achievable control technology (MACT) standards will still not adequately address the release of dioxins along with mercury and lead. The dioxin and dioxin-like compounds are persistent in the environment. When these chemicals enter the aquatic environment, they pass through the food web and accumulate in fish, wildlife, and, ultimately, in people.

Other organic chemicals emitted during the incineration of plastics include the polycyclic aromatic hydrocarbons (PAHs). Polystyrenes appear to yield the highest concentrations of PAHs. PAHs are of concern because many are known or potential carcinogens. Because the PAHs are hydrophobic, they are known to bioaccumulate and move up the food web.

Human Health Effects. Few epidemiological studies of the effects of MSW incinerators on human health have been conducted. A French study (Zmirou, Parent, and Potelon, 1984) found that the purchase of respiratory medications decreased with distance away from a refuse incinerator; however, no causal relationships were discovered. A 3-year study of acute respiratory effects on humans as a result of MSW landfills found no difference in the prevalence of respiratory symptoms among residents in communities near a MSW incinerator and those in a comparison community (Shy et al., 1995). Several studies have indicated that residents living in the vicinity of incinerators have higher incidences of respiratory disease than those living in areas at a distance from the incinerator. A comprehensive study of the incidence of cancer in 14 million people living in Great Britain near MSW incinerators found no causal relationship (Elliott et al., 1996). Although the incidence rates increased with proximity to MSW incinerators, the excess cancers could be attributed to higher unemployment, overcrowding, and lower socioeconomic class.

Studies of MSW incinerator workers show a greater risk of adverse health effects due to greater exposure to pollutants than is true for residents living near the MSW incinerator.

---

*Fly ash** is the fine particulate in the smoke stream. It is collected in precipitators, filters, and scrubbers.

[†]Dioxin is a collective name for 75 different polychlorinated dibenzo-*p*-dioxins. There are 135 different polychlorinated-dibenzo furans. These compounds can be extremely toxic substances to some animal species. The most toxic is 2,3,7,8-tetrachlorodibenzo-*p*-dioxin (2,3,7,8-TCDD), in which chlorine atoms are present at positions 2, 3, 7, and 8 in the dioxin molecule. This dioxin is the most commonly studied because it's extremely toxic for some laboratory animals, but its human health effects at very low concentrations and exposures have been debated considerably. 2,3,7,8-TCDD was the dioxin present in Agent Orange used as a defoliant in Vietnam and was also the major toxic chemical in the Seveso (Italy) disaster in 1976. On July 10, 1976, a plant producing a disinfectant (2,4,5-trichlorophenol) released an estimated 2 kg of TCDD-equivalent dioxins over 20 min on approximately 20 km^2 of land. A few days later wild and domestic animals (chickens, birds, rabbits, and some dogs) died. Children who were playing outdoors at the time suffered skin burns known as chloracne. Over 700 people were evacuated from the most highly contaminated areas. A 1993 study published in the *Journal of Epidemiology* suggests that residents of Seveso were beginning to exhibit excess cancers, at the time of the study (Bertazzi et al., 1993). More recent studies, however, suggest that dioxin is not a human carcinogen (Cole et al., 2003).

Lead, cadmium, and mercury appear to pose the greatest risk to workers. The NRC report, *Waste Incineration and Public Health,* suggests that proposed and recently promulgated regulations governing the use of MACTs will have little effect on the health of incinerator workers (National Research Council, 2000).

Other Thermal Treatment Processes

In addition to incineration, MSW can be processed by pyrolysis or gasification. **Pyrolysis** is the thermal processing of a material in the absence of oxygen. **Gasification** is the partial combustion in which a fuel is burned with less than a stoichiometric amount of oxygen. It is an energy-efficient technique for the reduction in waste volume and for recovering energy. Both processes convert the solid waste into gaseous, liquid, and solid fuels. The main difference in the two systems is that pyrolysis uses an external heat source to drive the endothermic reactions, whereby gasification reactions are self-sustaining.

13–7 DISPOSAL BY SANITARY LANDFILL

Although source reduction, reuse, recycling, and composting can divert large portions of MSW from landfill disposal, some waste still must be placed on land. Before 1979, MSW could be disposed of in what was known as a "dump" or, in many cases, an open and large pile. Dumping garbage in large open piles created several problems. As the garbage decomposed, it was unsightly and foul-smelling and attracted insects, gulls, rats, and other rodents. These animal "vectors" were problematic to human health because of the diseases they can carry. Uncontrolled fires, either set or spontaneously combusting, beleaguered open dumps. As rain seeped through the garbage, it carried with it harmful bacteria and hazardous chemicals into the groundwater and nearby lakes or streams. As a result of these problems, open dumps were banned by the EPA in 1979 and have been replaced by sanitary landfills.

Today, about 55% of the MSW generated is disposed of at sanitary landfills. A **sanitary landfill** is a land disposal site employing an engineered method of disposing of solid wastes on land in a manner that minimizes environmental hazards by spreading the solid wastes to the smallest practical volume, and applying and compacting cover material at the end of each day. According to the EPA, modern landfills are well-engineered facilities that are located, designed, operated, monitored, closed, cared for after closure, cleaned up when necessary, and financed to ensure compliance with federal regulations. Others, however, believe that these views provide a false sense of security. Some believe that groundwater contamination by landfills is inevitable, that the landfills will require periodic maintenance in perpetuity, and that we are simply transferring the economical and public health burdens associated with disposal of MSW to future generations (Lee and Sheehan, 1996).

Federal regulations governing MSW disposal, namely the Subtitle D of the Resource Conservation and Recovery Act of 1991 (Federal Register, 1991), were established to protect human health and the environment. In addition, Subtitle D landfills,* built under these regulations, can collect potentially harmful landfill gas emissions and convert the gas into energy.

Since 1988, the number of landfills has decreased from about 8000 to 2314 in 1998. However, the average size of the landfill has increased. Although, some years ago, the major concern was that we would have already exceeded the capacity of existing landfills, this does not appear to be the case today. The number of years remaining for disposal of MSW varies greatly by region. As shown in Figure 13–12 most of the states are estimated to have more than 10 years of landfill capacity remaining. This is in contradiction to what has been estimated elsewhere, that is, the landfills in the Northeast have essentially no time remaining, whereas those in the intermountain West have some 50 years. The landfills in the rest of the country have somewhere between 7 and 11 years (Scarlett, 1995).

---

*Because municipal landfills are regulated under Subtitle D of RCRA, they are often referred to as "Subtitle D landfills."

FIGURE 13–12

Years of remaining U.S. landfill capacity, by state. (*Source:* Glenn, J. 1999 "State of Garbage in the U.S., 1998" *Biocycle* 40(4): 60–71.)

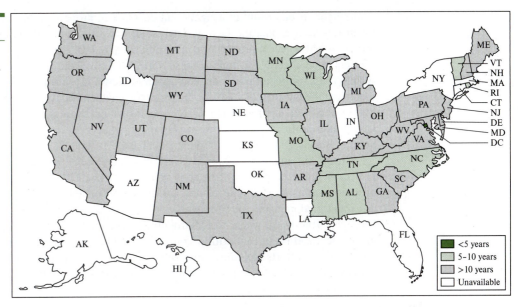

Site Selection

Site location is perhaps the most difficult obstacle to overcome in the development of a sanitary landfill. Opposition by local citizens eliminates many potential sites. In choosing a location for a landfill, consideration should be given to the following variables:

- Restricted locations such as wetlands, flood plains, and seismic impact areas
- Public opposition
- Proximity to major roadways
- Load limits on roadways and bridges
- Underpass limitations
- Traffic patterns and congestion
- Location of the groundwater table and sole-source aquifers
- Soil conditions and topography
- Availability of cover material
- Climate (e.g., floods, mud slides, snow)
- Zoning requirements
- Buffer areas around the site (e.g., high trees on the site perimeter)
- Location of historic buildings, endangered species, and similar environmental factors

In October of 1991, under Subtitle D of RCRA, the EPA promulgated new federal regulations for landfills. These included siting criteria that include restrictions on distances from airports, flood plains, and fault areas, as well as limitations on construction in wetlands, seismic impact areas, and other areas of unstable geology such as landslide areas and those susceptible to sink holes. Other restrictions may apply. For example, a landfill that is less than a mile from a drinking-water intake must comply with the specific groundwater-monitoring regulations under 40 CFR 258.51–258.55 (October 9, 1994). Landfills should also be greater than 10,000 ft from airport runways used by turbojet equipment.

In siting landfills the concerns of nearby neighbors must be addressed. Due to the nature of MSW landfills and the concomitant truck traffic, MSW landfills are often considered to be poor neighbors and result in a significant decrease in property values for distances of several kilometers surrounding the facility. The use of adequate land buffers between the landfilling operation and adjacent properties can significantly reduce the landfill's effect on the value of adjacent properties. Nevertheless, these buffers do not avoid other effects of MSW landfills, such as truck traffic, air pollution, and impaired views. In many cases it is necessary to provide more than the fair market value for the property. Payment can be made by those generating the waste (who often don't want the waste in their own backyards) along with those operating the landfill and profiting from the waste. On the positive site, landfills do bring some economic benefits, mainly in the way of jobs and a tax base.

One of the major sociological issues associated with landfills is that of environmental justice. A growing number of persons believe that landfill and other solid waste-processing operations, such as recycling plants, are disproportionately located in lower socioeconomic areas and predominately in communities populated by people of color. Consider the case of Chicago's 10th ward, where the Metropolitan Sanitary District of Greater Chicago proposed to enlarge a 260-hectare landfill by converting 120 hectares of wetlands into landfill. Through organizational efforts including large public meetings, demonstrations, and political meetings with the then candidate for Mayor Harold Washington, a local action group succeeded in blocking the landfill (Buntin, 1995).

Operation

The most commonly used method of operation at sanitary landfills is called the **area method** (Figure 13–13). The area method uses a three-step process: spreading the waste, compacting it, and covering it with soil.

In the area method, the solid waste is deposited on the surface, compacted, then covered with a layer of compacted soil, called the **daily cover,** at the end of the working day. Use of the area method is seldom restricted by topography; flat or rolling terrain, canyons, and other types of depressions are all acceptable.

A profile view of a typical landfill is shown in Figure 13–14. The waste and daily cover placed in a landfill during one operational period form a **cell.** The operational period is usually 1 day. The waste is dumped by the collection and transfer vehicles onto the working **face.** It is spread in 0.4–0.6-m layers and compacted. At the end of each day **cover** material is placed over the cell. The cover material may be native soil or other approved materials. Recommended depths of cover for various exposure periods are given in Table 13–3. The dimensions of a cell are determined by the amount of waste and the operational period. A **lift** may refer to the placement of a layer of waste or the completion of the horizontal active area of the landfill. In Figure 13–14 a lift is shown as the completion of the active area of the landfill. The first lift is

FIGURE 13–13

The area method.

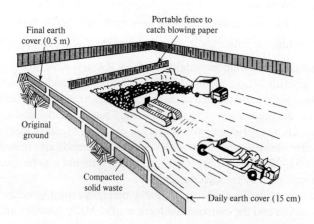

FIGURE 13–14

Sectional view through a sanitary landfill.

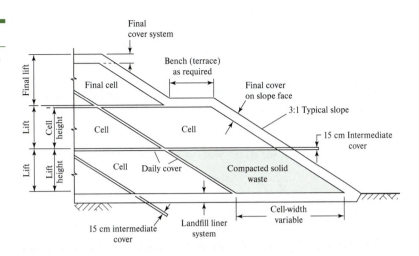

TABLE 13–3 Recommended Depths of Cover

Type of Cover	Minimum Depth (m)	Exposure Time (days)
Daily	0.15	<7
Intermediate	0.30	7–365
Final	0.60	>365

called a **fluffy lift** because the waste is not compacted until 2 m of waste is deposited. This is done to prevent damage to the liner as a result of the heavy compaction equipment. An extra layer of intermediate cover may be provided if the lift is exposed for long periods. The active area may be up to 300 m in length and width. Trenches vary in length from 30 to 300 m with widths of 5 to 15 m. The trench depth may be 3 to 9 m (Tchobanoglous et al., 1993).

Benches are used where the height of the landfill exceeds 15–20 m. They are used to maintain the slope stability of the landfill, for the placement of surface water drainage channels, and for the location of landfill gas collection piping.

Final cover is applied to the entire landfill site after all landfilling operations are complete. A modern final cover will contain several different layers of material to perform very specific functions.

Additional considerations in the operation of the landfill are those required by the 1991 Subtitle D regulations promulgated by EPA. These require exclusion of hazardous waste, use of cover materials, disease vector control, explosive gas control, air quality measurements, access control, runoff and run-on controls, surface water and liquids restrictions, and groundwater monitoring, as well as record keeping (Federal Register, 1991).

EXAMPLE 13–2 Determine the area required for a new landfill site with a projected life of 30 years for a population of 250,000 generating $2.02 \text{ kg} \cdot (\text{capita})^{-1} \cdot \text{day}^{-1}$. The density of the compacted waste is $470 \text{ kg} \cdot \text{m}^{-3}$. The height of the landfill cannot exceed 15 m.

Solution This is really just a mass-balance problem. Assuming that what goes into the landfill compacted remains in the landfill for perpetuity (which is not too bad an assumption, given that food has been found virtually intact some 25–30 years after it was placed in landfills), the volume of landfill necessary can be calculated as

$$\frac{(250{,}000 \text{ people})[2.02 \text{ kg} \cdot (\text{capita}^{-1}) \cdot \text{day}^{-1}]}{(470 \text{ kg} \cdot \text{m}^{-3})} = 1074 \text{ m}^3 \cdot \text{day}^{-1}$$

Therefore, over 30 years, the volume required is

$$(1074 \text{ m}^3 \cdot \text{day}^{-1})(365 \text{ days} \cdot \text{year}^{-1})(30 \text{ years}) = 11{,}760{,}300 \text{ m}^3$$

If we have a 15-m height constraint, then we need an area of

$$\frac{11{,}760{,}300 \text{ m}^3}{15 \text{ m}} = 784{,}020 \text{ m}^2, \text{ or } 78.4 \text{ ha}$$

Note that for simplicity, the solution did not include the volume of the daily cover that would have been used. The solution also does not take into consideration settlement of the waste due to decomposition and consolidation. Dealing with these issues is a task for subsequent upper-division courses in solid waste management.

Environmental Considerations

Disease vectors (carriers of disease) and water and air pollution should not be a problem in a properly operated and maintained landfill. Good compaction of the waste, daily covering with good compaction of the cover, and good housekeeping are musts to control flies, rodents, and fires.

Burning, which may cause air pollution, is never permitted at a sanitary landfill. If accidental fires should occur, they must be extinguished immediately using soil, water, or chemicals.

Odors, dust, and wind-blown litter can be controlled by covering the wastes quickly and carefully, and by sealing any cracks that may develop in the cover. Dust can also be controlled by using typical dust mitigation techniques such as the application of water or biodegradable oils to the soils being excavated.

Leachate

Leachate, liquid that passes through the landfill, extracts dissolved and suspended matter from the waste material. The liquid enters the landfill from external sources such as rainfall, surface drainage, groundwater, and the liquid in and produced from the decomposition of the waste.

Solid wastes placed in a sanitary landfill may undergo a number of biological, chemical, and physical changes. Aerobic and anaerobic decomposition of the organic matter results in both gaseous and liquid end products. Some materials are chemically oxidized. Some solids are dissolved in water percolating through the fill. A range of leachate compositions is listed in Table 13–4. The volatile organic chemicals produced by this decomposition in the landfill gas often contribute to contamination of groundwater because they dissolve in the leachate as it passes through the landfill.

Quantity. The amount of leachate generated from a landfill site may be estimated using a hydrologic mass balance for the landfill. Those portions of the global hydrologic cycle (see Chapter 7) that typically apply to a landfill site include precipitation, surface runoff, evaporation, transpiration (when the landfill cover is completed), infiltration, and storage. Similar to the mass balance equations developed in Chapter 7, we can write that

$$\Delta S = P - E - R + G' - F \tag{13–2}$$

TABLE 13–4 — Typical Data of the Composition of Leachate from New and Mature Landfills

Constituent	New Landfill (< 2 years) Range	New Landfill (< 2 years) Typical	Mature Landfill (>10 years)
BOD$_5$ (5-day biochemical oxygen demand)	2000–30,000	10,000	100–200
TOC (total organic carbon)	1500–20,000	6000	80–160
COD (chemical oxygen demand)	3000–60,000	18,000	100–500
Total suspended solids	200–2000	500	100–400
Organic nitrogen	10–800	200	80–120
Ammonia nitrogen	10–800	200	20–40
Nitrate	5–40	25	5–10
Total phosphorus	5–100	30	5–10
Ortho phosphorus	4–80	20	4–8
Alkalinity as CaCO$_3$	1000–10,000	3000	200–1000
pH (no units)	4.5–7.5	6	6.6–7.5
Total hardness as CaCO$_3$	300–10,000	3500	200–500
Calcium	200–3000	1000	100–400
Magnesium	50–1500	250	50–200
Potassium	200–1000	300	50–400
Sodium	200–2500	500	100–200
Chloride	200–3000	500	100–400
Sulfate	50–1000	300	20–50
Total iron	50–1200	60	20–200

Source: Tchobanoglous et al., 1993.

where ΔS = change in storage
P = precipitation
E = evaporation
R = surface water runoff
G' = infiltration from groundwater
F = percolation from landfill as leachate

Precipitation may be estimated in the conventional fashion from climatological records. Surface runoff or run-on may be estimated using the various engineering techniques that are beyond the scope of this book. Evaporation and transpiration are often lumped together as *evapotranspiration* (E). Evaporation rates may be estimated from regional data such as that provided by the U.S. Geological Survey (2000). Transpiration data may be obtained from plant science textbooks and other sources (Robbins, Weir, and Stocking, 1959; Kozlowski, 1943), from direct measurements (Nobel, 1983; Salisbury and Ross, 1992) and from modeling efforts (Kaufmann, 1985). Infiltration (and exfiltration) may be estimated using Darcy's law (Equation 7–12). Until the landfill becomes saturated, some of the water that infiltrates will be stored in both the cover material and the waste. The quantity of water that can be held against the pull of gravity is referred to as **field capacity.** Theoretically, when the landfill reaches its field capacity, leachate will begin to flow. Then, the potential quantity of leachate is the amount of moisture within the landfill in excess of the field capacity. In reality, the field capacity does not have to be exceeded, and leachate will begin to flow almost immediately because of channeling in the waste through the waste material.

The time that it takes for the leachate to penetrate through a liner can be estimated the following:

$$t = \frac{d^2\eta}{K(d + h)}$$

(13–3)

where t = breakthrough time
$\quad\quad d$ = thickness (m)
$\quad\quad K$ = hydraulic conductivity (m/y)
$\quad\quad \eta$ = porosity
$\quad\quad h$ = hydraulic head (m)

Although field capacity is often used, it does have limitations. For example, the criteria for deciding when leachate just begins to flow is difficult and depends on the frequency of sampling and the accuracy with which moisture content can be measured.

As mentioned previously, the amount of leachate to leave (exfiltrate from) the landfill can be determined using Darcy's law. Leachate systems are designed to have hydraulic conductivities on the order of 10^{-9} m $\cdot$ s^{-1}. When using compacted clay liners, the thickness should be approximately 1 m.

EXAMPLE 13–3   Calculate the volumetric flow rate of leachate through a compacted clay liner if the area of the landfill is 15 ha and the liner thickness is 1 m. The hydraulic conductivity is 7.5×10^{-10} m $\cdot$ s^{-1}. Assume that the head of water is 0.6 m.

Solution   First determine the Darcy velocity for the leachate through the clay layer using Equation 7–12.

$$v = K\left(\frac{\Delta h}{L}\right) = (7.5 \times 10^{-10} \text{ m} \cdot \text{s}^{-1})(0.6 \text{ m/1 m}) = 4.5 \times 10^{-10} \text{ m} \cdot \text{s}^{-1}$$

To determine the flow, simply apply the continuity equation ($Q = vA$)

$$Q = (4.5 \times 10^{-10} \text{ m} \cdot \text{s}^{-1})(15 \text{ ha})(10^4 \text{ m}^2 \cdot \text{ha}^{-1}) = 6.75 \times 10^{-5} \text{ m}^3 \cdot \text{s}^{-1}$$

This solution assumes that there is no collection system. However, because we add a collection system in landfills, thus decreasing the hydraulic head, the infiltration of leachate out of the liner and to groundwater is minimized.

The EPA and the Army Engineer Research and Development Center (formerly Waterways Experiment Station) developed a computer model of the hydrologic balance called the Hydrologic Evaluation of Landfill Performance (HELP) (Schroeder et al., 1984). The program contains extensive data on the characteristics of various soil types, precipitation patterns, and evapotranspiration–temperature relationships as well as the algorithms to perform a routing of the moisture flow through the landfill. Although simple to use, the HELP model overestimates leachate production, often by a factor of two to three. This can be a costly mistake because it could result in a significant overdesign of the leachate collection and treatment system.

Control of Leachate.   To prevent the seepage of leachate from the landfill and thus, to avoid groundwater contamination, strict leachate control measures are required. Under the 1991 Subtitle D rules (40 CFR 258) promulgated by EPA, new landfills must be lined in a specific manner or meet maximum contaminant levels for the groundwater at the landfill boundary. The specified liner system includes a synthetic membrane (**geomembrane**) at least 30 mils (0.76 mm) thick supported by a compacted soil liner at least 0.6 m thick. The soil liner must have a hydraulic conductivity of no more than 1×10^{-7} cm $\cdot$ s^{-1}. Flexible membrane liners consisting of HDPE must be at least 60 mils thick (Federal Register, 1991). A schematic of the EPA specified liner system is shown in Figure 13–15.

FIGURE 13–15

A composite liner and leachate collection system.

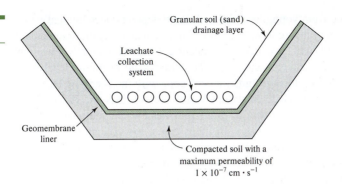

Granular soil (sand) drainage layer

Leachate collection system

Geomembrane liner

Compacted soil with a maximum permeability of 1×10^{-7} cm · s^{-1}

Several geomembrane materials are available, examples of which include PVC, HDPE, chlorinated polyethylene (CPE), and ethylene propylene diene monomer (EPDM). Designers show a strong preference for PVC and especially for HDPE. Recently, geosynthetic clay liners have been used as bottom liners and final covers. These are not as thick as liners that use compacted clay yet are fast and easily installed and have low hydraulic conductivities. The use of bentonite clays allows the liner to be self-sealing in the event of damage.

Although the geomembranes are highly impermeable (hydraulic conductivities are often less than 1×10^{-12} cm · s^{-1}), they can be easily damaged or improperly installed. Damage may occur during construction by construction equipment, by failure due to tensile stress generated by the overburden, tearing as a result of differential settling of the supporting soil, puncture from sharp objects in the overburden, puncture from coarse aggregate in the supporting soil, and tearing by landfill equipment during operation. Installation errors primarily occur during seaming when two pieces of geomembrane must be attached or when piping must pass through the liner. A liner placed with adequate quality control should have fewer than three to five defects per hectare.

The soil layer under the geomembrane acts as a foundation for the geomembrane and as a backup for control of leachate flow to the groundwayer. Compacted clay generally meets the requirement for a hydraulic conductivity of less than 1×10^{-7} cm · s^{-1}. In addition to having a low permeability, it should be free of sharp objects greater than 1 cm in diameter, graded evenly without pockets or hillocks, compacted to prevent differential settlement, and free of cracks.

Leachate Collection. Under the 1991 Subtitle D rules (Federal Register, 1991), the leachate collection system must be designed so that the depth of leachate above the liner does not exceed 0.3 m. The leachate collection system is designed by sloping the floor of the landfill to a grid of underdrain pipes* that are placed above the geomembrane. A 0.3-m-deep layer of granular material (e.g., sand) with a high hydraulic conductivity (EPA recommends greater than 1×10^{-2} cm · s^{-1}) is placed over the geomembrane to conduct the leachate to the underdrains. In addition to carrying the leachate, this layer also protects the geomembrane from mechanical damage from equipment and solid waste. In some instances a **geonet** (a synthetic matrix that resembles a miniature chain link fence), with a **geofabric** (an open-weave cloth) protective layer to keep out the sand, is placed under the sand and above the geomembrane to increase the flow of leachate to the pipe system.

Leachate Treatment. The composition of landfill leachate varies greatly but is generally highly contaminated with ammonia, organic contaminants measured as chemical oxygen demand (COD) and biological oxygen demand (BOD), halogenated hydrocarbons, and heavy metals. In addition, leachate usually contains high concentrations of inorganic salts—mainly sodium chloride, carbonate, and sulfate. Thus, it must be collected and treated before discharge.

*__Underdrain pipes__ are perforated pipes designed to collect the leachate.

Leachate from young landfills with large deposits of municipal waste, for example, will often contain very high concentrations of ammonia and organic chemicals. In contrast, leachate from old landfills will have high salt concentrations with distinctly lower ammonia, heavy metals, and organic chemical concentrations. Therefore, state-of-the-art leachate treatment plants should be designed as multistage units that can cope with the changing leachate characteristics over time.

Landfill leachate treatment varies greatly. In some cases, the collected leachate is sent without any pretreatment to municipal wastewater treatment plants, where it is treated along with municipal wastewater. In other cases, the collected leachate is pretreated by chemical flocculation followed by sedimentation and filtration before it is sent to the municipal wastewater system. In other cases, it is necessary to treat the leachate separately on site before discharge. One such possibility of a treatment system is to use a two-stage activated sludge process—nitrification–denitrification—for ammonia, COD, and BOD removal. This may not yield sufficient treatment, and further processing may be required to remove residual COD. Several options include activated carbon and ozone. However, none of these treatment processes will remove the high salt content. In cases where salts must be removed, reverse osmosis (RO) can be employed for desalination. A common treatment process combination is biological pretreatment of the leachate followed by an RO stage. On-site treatment of the leachate can be achieved by recycling the flow through the landfill, allowing the microorganisms additional opportunity to degrade the biodegradable material in the leachate.

Methane and Other Gas Production

The principal gaseous products emitted from a landfill (methane and carbon dioxide) are the result of microbial decomposition. Typical concentrations of landfill gases and their characteristics are summarized in Table 13–5. During the early life of the landfill, the predominant gas is carbon dioxide, which is produced from aerobic decomposition of the waste. During this time the nitrogen (N_2) gas concentration is also high. As the landfill matures, the gas is composed almost equally of carbon dioxide and methane. In the methanogenic phase, the concentration of nitrogen will decrease. Once methanogenesis is complete, methane generation will cease.

Methane is explosive in the presence of oxygen when it is present at concentrations greater than 40%. Methane release is also a concern because it is a greenhouse gas and can damage crops planted in the vicinity of the landfill. As such, its movement must be controlled. The heat content of this landfill gas mixture ranges from 16,000–20,000 kJ · m^{-3}. Although not as

TABLE 13–5		Typical Constituents Found in MSW Landfill Gas	
Component	**Percent (dry volume basis)**	**Characteristic**	**Value**
Methane	45–60	Temperature (°C)	35–50
Carbon dioxide	40–60	Specific gravity	1.02–1.05
Nitrogen	2–5	Moisture content	Saturated
Oxygen	0.1–1.0	High heating value (in kJ · m^{-3})	16,000–20,000
Sulfides, disulfides, mercaptans, etc.	0–1.0		
Ammonia	0.1–1.0		
Hydrogen	0–0.2		
Carbon monoxide	0–0.2		
Trace constituents	0.01–0.06		

Source: Tchobanoglous et al., 1993.

substantial as methane alone (37,750 kJ · m^{-3}), landfill gas has sufficient economic value that many landfills have been tapped with wells to collect and use methane for energy production.

Typically, landfill gas contains numerous other compounds besides carbon dioxide, methane, and ammonia. Because of the toxicity of the trace gases emitted from landfills, these gases are of concern. More than 150 compounds have been measured at various landfills. Many of these nonmethane organic chemicals (NMOC) may be classified as hazardous air pollutants (HAPs) or volatile organic chemicals (VOCs). The occurrence of significant VOC concentrations is often associated with older landfills that previously accepted industrial and commercial wastes containing these compounds. The concentrations of 13 compounds measured in landfill gases from several California sites are shown in Table 13–6. Other emissions may include nitrogen oxides (NO$_x$), sulfur dioxide, hydrogen chloride (HCl), and particulate matter.

The rate of emissions is governed by the rates of gas production and transport. Gas production is controlled by biological and chemical decomposition along with volatilization. The transport of the gases is controlled by the rates of gas diffusion, advection, and convection.

Landfill gas (LFG) collection systems can be active or passive. **Active systems** provide a pressure gradient to force the gas from the cells. Mechanical blowers or compressors are used. **Passive systems** allow a natural pressure gradient to build up, which causes the gas to move from the cells into the collection devices.

LFG can be purified to produce a gas that can be recovered for energy generation. Alternatively, open combustion using flares can be employed to burn off LFG. Thermal incinerators operate at a sufficiently high temperature to oxidize the VOCs in the LFG to carbon dioxide and water. Purification techniques can use adsorption, absorption, or membranes. Because gas collection systems are not 100% efficient, emissions of methane and other LFGs will nevertheless occur.

TABLE 13–6   Concentrations of Specified Air Contaminants Measured in Landfill Gasses (in ppb)

	Landfill Site						
Compound	Yolo Co.	City of Sacramento	Yuba Co.	El Dorado Co.	L.A.-Pacific (Ukiah)	City of Clovis	City of Willits
Vinyl chloride	6900	1850	4690	2200	<2	66,000	75
Benzene	1860	289	963	328	<2	895	<18
Ethylene dibromide	1270	<10	<50	<1	<1	<1	<0.5
Ethylene dichloride	nr	nr	nr	<20	0.2	<20	4
Methylene chloride	1400	54	4500	12,900	<1	41,000	<1
Perchloroethylene	5150	92	140	233	<0.2	2850	8.1
Carbon tetrachloride	13	<5	<7	<5	<0.2	<5	<0.2
1,1,1-TCA[a]	1180	6.8	<60	3270	0.52	113	0.8
TCE[b]	1200	470	65	900	<0.6	895	8
Chloroform	350	<10	<5	120	<0.8	1200	<0.8
Methane	nr	nr	nr	nr	0.11%	17%	0.14%
Carbon dioxide	nr	nr	nr	nr	0.12%	24%	<0.1%
Oxygen	nr	nr	nr	nr	nr	10%	21%

nr: Not reported by operator.

[a]1,1,1-TCA: 1,1,1-trichloroethane, methyl chloroform.

[b]TCE: Trichloroethane, trichloroethylene.

Source: California Air Resources Board, 1988.

EXAMPLE 13–4 A landfill with a gas collection system is in operation and serves a population of 200,000. MSW is generated at a rate of 1.95 kg · (capita)$^{-1}$ · day^{-1}. Gas is produced at an annual rate of 6.2 L · kg^{-1} of MSW delivered to the landfill. The gas contains 55% methane. Gas recovery is 15% of that generated. The heat content of the landfill gas is approximately 17,000 kJ · m^{-3} (a value lower than the theoretical value because of dilution of the methane with air during recovery). The landfill company and a developer have proposed to build a subdivision in the vicinity of the landfill and pipe the methane generated to the homes to be used for heating. The homes are estimated to use an average of 110×10^6 kJ of heat energy each year. Peak usage during winter is 1.5 times the average usage. How many homes can be built in the subdivision?

Solution This is essentially a mass and energy-balance problem, in which solid waste is being disposed of in a landfill. Methane is generated during the decomposition of the waste. The methane is used by the homes for heating. The energy generated in the landfill must equal the energy consumed by the homeowners.

The amount of methane generated is equal to
(Amount of waste disposed in landfill per person on yearly basis)
× (Number of people served by landfill)
× (Rate of gas production per mass of solid waste)
× (Fraction of methane in gas)

$= [1.95 \text{ kg} \cdot (\text{capita})^{-1} \cdot \text{day}^{-1}](365 \text{ days} \cdot \text{year}^{-1})(200{,}000 \text{ persons})(6.2 \text{ L gas} \cdot \text{kg}^{-1})(0.55)$
$= 4.85 \times 10^8 \text{ L methane} \cdot \text{year}^{-1}$

But only 15% is recovered.

$(4.85 \times 10^8 \text{ L methane} \cdot \text{year}^{-1})(0.15) = 7.28 \times 10^7 \text{ L methane} \cdot \text{year}^{-1}$

The heat content of the methane recovered is

$(17{,}000 \text{ kJ} \cdot \text{m}^{-3})(7.28 \times 10^7 \text{ L} \cdot \text{year}^{-1})(10^{-3} \text{ m}^3 \cdot \text{L}^{-1}) = 1.24 \times 10^9 \text{ kJ} \cdot \text{year}^{-1}$

Energy to heat the homes during times of peak demand (i.e., winter) must be supplied:

Peak demand = Averaged usage × 1.5
$= (110 \times 10^6 \text{ kJ} \cdot \text{year}^1)(1.5) = 1.65 \times 10^8 \text{ kJ} \cdot \text{home}^{-1} \cdot \text{year}^{-1}$

$$\text{Number of homes that can be heated} = \frac{\text{energy generated}}{\text{energy required per home}}$$

$$= \frac{1.24 \times 10^9 \text{ kJ} \cdot \text{year}^{-1}}{1.65 \times 10^8 \text{ kJ} \cdot \text{home}^{-1} \cdot \text{year}^{-1}} = 7.5 \text{ homes}$$

Because we can't heat a fraction of a home, then we can build seven homes.

Leachate Monitoring. Under EPA regulations, all landfill operators must monitor for both gaseous and liquid contaminants that have the potential to be released from the landfill. The goal of any groundwater monitoring plan is to have reliable and representative information on aquifer characteristics, groundwater flow directions, and chemical and physical characteristics of groundwater being monitored. The goals of surface water monitoring are to assess the characteristics of surface waters that have the potential to be affected by leachate releases, contaminated groundwater seepage, or contaminated surface water run-off. The monitoring of surface and groundwater allows the landfill operators to demonstrate that the landfill is performing as designed, to provide data that the leachate is not infiltrating into the groundwater, to identify risks associated with improved operation of the landfill, and to identify when the landfill no longer poses a risk to human or environmental health.

Landfill Design

The design of the landfill has many components, including site preparation, buildings, monitoring wells, size, liners, leachate collection system, final cover, and gas collection system. In the following discussion we will limit ourselves to introductory consideration of the design of the size of the landfill, the selection of a liner system, the design of a leachate collection system, and a discussion of the final cover system.

To estimate the volume required for a landfill, it is necessary to know the amount of refuse being produced and the density of the in-place, compacted refuse. The volume of refuse differs markedly from one city to another because of local conditions.

Salvato (1972) recommends a formula of the following form for estimating the annual volume required.

$$\Psi_{LF} = \frac{PEC}{D_c} \tag{13–4}$$

where Ψ_{LF} = volume of landfill (in m^3)
P = population
E = ratio of cover (soil) to compacted fill = $(\Psi_{sw} + \Psi_c)/\Psi_{sw}$
Ψ_{sw} = volume of solid waste (in m^3)
Ψ_c = volume of cover (in m^3)
C = average mass of solid waste collected per capita per year (in kg · person^{-1})
D_c = density of compacted fill (in kg · m^{-3})

The density of the compacted fill depends somewhat on the equipment used at the landfill site and the moisture content of the waste. Compacted solid waste densities vary from 300 to 700 kg · m^{-3}. Nominal values are generally in the range of 475 to 600 kg · m^{-3}.

EXAMPLE 13–5   How much landfill space does Apocatequil County require for 20 years of operation? You may assume that 142,000 persons are being served by this landfill and that each person generates 2.0 kg · day^{-1} of waste. The density of the uncompacted waste is 106 kg · m^{-3} and a compaction ratio of 4.2 can be used. The ratio of cover to compacted fill is 1.9.

Solution   Apply Equation 13–4.

$$\Psi_{LF} = \frac{PEC}{D_c} = \frac{(142,000 \text{ persons})(2.0 \text{ kg} \cdot \text{day}^{-1})(365 \text{ days} \cdot \text{year}^{-1})(1.9)}{(106 \text{ kg} \cdot \text{m}^{-3})(4.2)} = 442,400 \text{ m}^3$$

Multiply by 20 years.

$$442,400 \times 20 = 8,848,000 = 8,900,000 \text{ m}^3$$

Had we not considered the volume of cover material, we would have determined the volume to be $4.66 \times 10^6 \text{ m}^3$.

Landfill Closure

The major function of the final cover is to prevent moisture from entering the finished landfill. If no moisture enters, then at some time the leachate production will reach minimal proportions and the chance of groundwater contamination will be minimized.

Modern final cover design consists of a surface layer, biotic barrier, drainage layer, hydraulic barrier, foundation layer, and gas control. The surface layer is to provide suitable soil for plants to grow. This minimizes erosion. A soil depth of about 0.3 m is appropriate for grass. The biotic barrier is to prevent the roots of the plants from penetrating the hydraulic barrier. At this time, there does not seem to be an optimal material for this barrier. The drainage layer serves the same function here as in the leachate collection system—that is, it provides an easy flow path to a grid of perforated pipes. This collection piping system is subject to differential settling and may fail because of it. Some designers do not recommend installing it because they prefer to use the funds to develop a thicker hydraulic barrier. The hydraulic barrier serves the same function as the liner by preventing movement of water into the landfill. The EPA recommends a composite liner consisting of a geomembrane and a low hydraulic conductivity soil that also serves as the foundation for the geomembrane. This soil also protects the geomembrane from the rough aggregate in the gas control layer. The gas control layer is constructed of coarse gravel that acts as a vent to carry the gases to the surface. If the gas is to be collected for its energy value, a series of gas recovery wells is installed. A negative pressure is placed on these wells to draw the gas into the system.

Completed landfills generally require maintenance because of uneven settling. Maintenance consists primarily of regrading the surface to maintain good drainage and filling in small depressions to prevent ponding and possible subsequent groundwater pollution. The final soil cover should be about 0.6 m deep.

Completed landfills have been used for recreational purposes, such as parks, playgrounds, or golf courses. Parking lots and storage areas are other final uses. Because of the characteristic uneven settling and gas evolution from landfills, construction of buildings on completed landfills should be avoided.

On occasion, one-story buildings and runways for light aircraft might be constructed. In such cases, it is important to avoid concentrated foundation loading, which can result in uneven settling and cracking of the structure. The designer must provide the means for the gas to dissipate into the atmosphere and not into the structure.

CHAPTER REVIEW

When you have completed studying this chapter, you should be able to do the following without the aid of your textbook or notes:

1. State the average mass of solid waste produced per capita per day in the United States in 1998.
2. Define the following terms: refuse, solid waste, putrescible waste.
3. List the elements of a solid waste management system.
4. Explain how policy makers decide how and how often solid waste is to be collected.
5. Define the following terms: resource conservation, resource recovery, closed-loop recycling, open-loop recycling, composting.
6. List the methods of collecting recyclable materials.
7. List the common materials that can be recycled. Be able to discuss the specific issues pertaining to each material.
8. Discuss how the market for recyclable materials can be improved.
9. Define what is meant by reuse of durable goods. Describe several methods of reuse.
10. Discuss how education and legislation affect source reduction.
11. Define the following terms: combustion, pyrolysis, gasification. What is the difference among these processes?
12. List the different types of incinerators. Explain how an incinerator operates.
13. List the major public health and environmental issues dealing with incinerators.

14. List and discuss the factors pertinent to the selection of a landfill site.
15. Describe the methods of constructing a sanitary landfill.
16. Explain the purpose of daily cover in a sanitary landfill and state the minimum desirable depth of daily cover.
17. Define leachate and explain why it occurs.
18. Sketch a sanitary landfill that includes proper cover and a leachate collection system.
19. Define or explain the following terms: HHV, LHV, RDF, source separation.
20. Explain the effect of source separation on the heating value of solid waste and on the potential for hazardous air pollution emissions.

With the aid of this text or a list of equations, you should be able to do the following:

1. Determine the volume and mass of solid waste from various establishments.
2. Estimate the volume and area requirements for a landfill.
3. Compute the LHV given the HHV and the chemical formula for a compound to be burned.
4. Estimate the volumetric flow rate of leachate.
5. Estimate methane generation and energy production rates.

PROBLEMS

13–1 The Bailey Stone Works employs six people. Assuming that the density of uncompacted waste is $480 \text{ kg} \cdot \text{m}^{-3}$, determine the annual volume of solid waste produced by the stone works assuming a waste generation rate of $1 \text{ kg} \cdot \text{capita}^{-1} \cdot \text{day}$.

Answer: $4.6 \text{ m}^3 \cdot \text{year}^{-1}$

13–2 The student population of a high school is 881. The school has 30 standard classrooms. Assuming a 5-day school week with solid waste pickups on Wednesday and Friday before school starts in the morning, determine the size of storage container required. Assume waste is generated at a rate of $0.11 \text{ kg} \cdot (\text{capita})^{-1} \cdot \text{day}$ plus 3.6 kg per room and that the density of uncompacted solid waste is $120.0 \text{ kg} \cdot \text{m}^{-3}$. Standard container sizes are as follows (all in m^3): 1.5, 2.3, 3.0, and 4.6.

13–3 The Nairyosangha landfill located in Peshdadians serves a population of 562,400 people generating MSW at a rate of $1.89 \text{ kg} \cdot \text{capita}^{-1} \cdot \text{day}$. The volume of the landfill is 11,240,000 m^3. At the present time, 63% of the landfill is used. The ratio of cover to compacted fill is 1.9. Determine projected life remaining for the landfill. Assume the density of the compacted waste to be $490 \text{ kg} \cdot \text{m}^{-3}$.

Answer: 2.76 years

13–4 The Sibzianna landfill serves a population of 253,000. Each week 325 trucks bring a total of 2180 t of MSW to the landfill. Each truck averages two loads per day. The trucks each have a capacity of 4.5 t and operate 5 days per week.

(a) How much MSW is generated per person? Give your answer in $\text{kg} \cdot \text{capita}^{-1} \cdot \text{day}$.

(b) What is the average load (in metric tonnes) in each truck when they arrive at the landfill? At what capacity are the trucks operating?

(c) If, the density of the waste is $265 \text{ kg} \cdot \text{m}^{-3}$, what is the volume of MSW disposed of at the landfill each week?

13–5 Professor Dexter has made measurements of her household solid waste, shown in the following table. If the container volume is 0.0757 m^3, what is the average density of the solid waste produced in her household during the period of time from March 18th through April 8th? Assume that the mass of each empty container is 3.63 kg.

Date	Can Number	Gross Mass[a] (kg)	Date	Can Number	Gross Mass[a] (kg)
March 18	1	7.26	April 8	1	6.35
	2	7.72		2	8.17
March 25	1	10.89		3	8.62
	2	8.17			
	3	7.26			

[a]Container plus solid waste.

Answer: Average density = 58.4 kg · m^{-3}

13–6 A sanitary landfill is being designed to handle solid waste generated by Binford, Vermont, at a rate of 50 Mg · day^{-1}. It is expected that the waste will be delivered by compactor truck on a 5 days · week^{-1} basis. The density as spread is 122 kg · m^{-3}. It will be spread in 0.50-m layers and compacted to 0.25 m. Assuming three such lifts per day and a daily cover of 0.15 m, determine the following: (a) annual volume of landfill consumed in cubic meters, and (b) daily horizontal area covered by the solid waste. Ignore the soil volume between stacks.

13–7 Calculate the volumetric flow rate of leachate through a compacted clay liner if the area of the landfill is 21 ha and the liner thickness is 1.3 m. The hydraulic conductivity is 2.5 × 10^{-10} m · s^{-1}. Assume that the head of water is 0.8 m.

Answer: 2.8 m^3 · day^{-1}

13–8 A developer plans to build a completely self-contained city on the Planet Zaqar. The city will have a population of 555,000. The homes in the city are to be heated with methane gas generated at the landfill. The landfill will have a gas collection system. Through education and conservation, MSW generation will be reduced to a rate of 0.45 kg · capita^{-1} · day. It is anticipated that gas can be produced at an annual rate of 25 L gas · kg^{-1} of MSW delivered to the landfill and that the gas will contain 58% methane. Gas recovery is 25% of that which is generated. The heat content of the methane gas is approximately 18,000 kJ · m^{-3} (a value lower than the theoretical value because of dilution of the methane with air during recovery). The homes are estimated to use an average 40 × 10^6 kJ of heat energy · year^{-1}. Peak usage during winter is 1.5 times the average usage. Is this proposal feasible from the standpoint of generating sufficient heat for domestic use? What are alternative sources of "waste" materials that could be used as a resource to generate heat?

13–9 A landfill that is 12 ha in area has a liner of thickness 0.9 m. Each year 1700 m^3 of leachate is collected. The hydraulic conductivity of the liner is 3.9 × 10^{-10} m · s^{-1}. What is the head of water above the liner?

Answer: 1.04 m

13–10 The city of Lidköping, Sweden, incinerates its biowaste, along with that of seven surrounding municipalities, for energy production. The city operates two 17 MW biowaste-fueled boilers, which burn 70,000 tonnes of waste per year (Åström, 1999). The overall efficiency of the system is 88%. Calculate the mass of coal (in tonnes) that would need to be burned each day, if coal were used in place of the biowaste. The coal has a heat content of 33,500 kJ · kg^{-1}.

13–11 The concentrations of vinyl chloride, benzene, and methylene chloride in the landfill gas for the City of Sacramento are given in Table 13–6. Determine the concentration of these gases in μg · L^{-1}.

13–12 You are to design a leachate treatment system for the landfill in Saint-Etienne, France (Baig et al., 1997). The leachate has the following properties:

BOD_5 = 170 mg · L^{-1}
COD = 1400 mg · L^{-1}

Ammonia nitrogen $= 690$ mg $\cdot$ L^{-1} as N

Suspended solids $= 270$ mg $\cdot$ L^{-1}

Flow rate $= 20$ m$^3 \cdot$ h^{-1}

The leachate is pretreated with chemical coagulation using ferric chloride, flocculation with an anionic polymer, and subsequent flotation. The pretreatment results in the removal of 48% COD, 58% BOD$_5$, and 10% nitrogen. The suspended solids concentration after pretreatment is 35 mg $\cdot$ L^{-1}. You plan to use extended diffused aeration to remove the BOD and nitrogen. If additional COD removal is required, you plan to use either activated carbon or ozonation. Based on the stoichiometry of the reaction of ammonia with oxygen (Equation 9–11), determine the mass of oxygen that must be supplied daily to oxidize the ammonia. Determine the mass of oxygen (in kg $\cdot$ day^{-1}) necessary to oxidize the BOD$_5$.

DISCUSSION QUESTIONS

13–1 Which of the following soil types would be suitable for (a) liner, (b) drainage layer, (c) gas venting?

(1) Gravel (>2.5 cm diameter)

(2) Glacial till ($K = 10^{-7}$ cm $\cdot$ s^{-1})

(3) Clay ($K = 1 \times 10^{-9}$ cm $\cdot$ s^{-1})

(4) Silt ($K = 1 \times 10^{-6}$ cm $\cdot$ s^{-1})

(5) Coarse sand ($K = 0.1$ cm $\cdot$ s^{-1})

(6) Fine sand ($K = 0.001$ cm $\cdot$ s^{-1})

13–2 Although the market value of compost is negligible, many communities have implemented yard waste composting systems. Explain why.

13–3 Discuss the characteristics of waste that are favorable for biological treatment, thermal processing, recycling, and landfilling.

13–4 Discuss at least two aspects of concern of a municipal waste landfill.

13–5 Discuss why biological treatment of solid waste is used extensively in Western Europe and Palestine but not in the United States.

13–6 What is the effect of biowaste composting on gas production from a landfill?

13–7 The biowaste to energy plant for the city of Lidköping, Sweden, is discussed in Problem 13–10. This plant uses fluidized bed technology with cross flow air injection. The exhaust gases contain large particles, heavy metals, SO$_2$, and HCl. Discuss the unit operations (see Chapter 12) you would use to treat the gases before release to the atmosphere.

FE EXAM FORMATTED PROBLEMS

13–1 A clay liner for a waste pond is 0.9 m thick. The effective porosity of the liner is 0.15. The hydraulic conductivity is 2×10^{-9} m/s. The height of the liquid waste above the linear is 3 m. The breakthrough time (y) for the liquid to penetrate the liner is most nearly:

(a) 0.25

(b) 0.50

(c) 1.0

(d) 1.5

13–2 A town of 50,000 generates 1.0 kg/capita-day of MSW. MSW is collected weekly. The trucks each have a capacity of 5 metric tonnes and operate 5 days per week. Each truck can haul 3 loads per day at 80% capacity. How many trucks are needed?

(a) 3

(b) 6

(c) 12

(d) 18

13–3 A new landfill with a design life of 35 years is to be constructed. The landfill is to serve a population of 200,000 generating 2.0 kg/capita-day. The density of the waste is 240 kg/m³. At the landfill the waste is compacted so that its volume is decreased in half. The landfill cannot exceed 15 m. Determine the volume required.

(a) 5.0×10^6 m³

(b) 11×10^6 m³

(c) 15×10^6 m³

(d) 20×10^6 m³

REFERENCES

Andersen, M. S. (1998) "Assessing the Effectiveness of Denmark's Waste Tax," *Environment,* 40(4): 10–15, 38–41.

Åström, J. (1999) Waste to Energy Plant, Sycon Energikonsult AB, IEB Bioenergy Task 23, December.

Baig S., E. Thiéblin, F. Zuliani, R. Jenny, and C. Coste (1997) "Landfill Leachate Treatment: Case Studies," *International Conference on Ozonation and Related Oxidation Processes in Water and Liquid Waste Treatment* (Berlin, April 21–23, 1997), pp. V.4.1–V.4.16.

Bertazzi, A., A. C. Pesatori, D. Consonni, A. Tironi, M. T. Land, and C. Zocchetti (1993) "Cancer Incidence in a Population Accidentally Exposed to 2,3,7,8-Tetrachlorodibenzo-para-dioxin," *Epidemiology,* 4(5): 398–406.

Buntin, S. (1995) *Environmental Liberty and Social Justice for All: How Advocacy Planning Can Help Combat Environmental Racism,* November 1995.

California Air Resources Board (1988) *The Landfill Gas Testing Program: A Report to the Legislature,* Sacramento, CA, June 1988.

Cartledge, J. (2004) *Profits Warning: Why Germany's Green Dot Is Selling Up,* (25.11.04) http://www.letsrecycle.com/features/dsd.jsp

City of Hamilton (2005) *2005 Annual Report Waste Management Division Public Works Department,* City of Hamilton, Ontario, http://www.myhamilton.ca/NR/rdonlyres/54F35506-DB93-4EBB-8682-F030C3E0CC6F/0/2005WasteManagementAnnualReportFinal.pdf

City of Hamilton Waste Management Division (2007) May 26, 2007, http://www.myhamilton.ca/myhamilton/ CityandGovernment/CityDepartments/PublicWorks/WasteManagement

Cole, P., D. Trichopoulos, H. Pastides, T. Starr, and J. S. Mandel (2003) "Dioxin and Cancer: A Critical Review," *Regul. Toxicol, Pharmacol.,* 38(3): 378–88.

Elliott, P. G., G. Shaddick, I. Kleinschmidt, D. Jolley, P. Walls, J. Beresford, and C. Grundy (1996) "Cancer Incidence Near Municipal Solid Waste Incinerators in Great Britain," *Br. J. Cancer,* 72(5): 702–10.

Federal Register (1991) *Code of Federal Regulations,* 40 CFR 257 and 258, and "Solid Waste Disposal Facility Criteria: Final Rule," 9 October 1991.

Glenn, J. (1999) "State of Garbage in the U.S., 1998," *Biocycle,* 40(4): 60–71.

Kaufmann, M. R. (1985) *Annual Transpiration in Subalpine Forests: Large Differences Among Four Tree Species,* USDA Forest Service, Rocky Mountain Forest and Range Experiment Station, Fort Collins, CO, 80526-2098.

Kozlowski, T. T. (1943) "Transpiration Rates of Some Forest Tree Species During the Dormant Species," *Plant Physiol.,* 18(2): 252–60.

Lee, G. F., and W. Sheehan (1996) "Landfills Offer False Sense of Security," *Biocycle,* 37(9): 8.

Miller, C. (2006) "Yard Waste," *Waste Age,* May 1, 2006, http://wasteage.com/mag/waste_yard_waste_4/

National Center for Resource Recovery (1974) *Resource Recovery from Municipal Solid Waste,* Lexington Books, Lexington, MA.

National Research Council (2000) *Waste Incineration and Public Health,* Committee on Health Effects of Waste Incineration, Board of Environmental Studies and Technology, Commission of Life Sciences, National Research Council, National Academy Press, Washington, DC.

Nobel, P. S. (1983) *Biophysical Plant Physiology and Ecology,* W. H. Freeman and Co., San Francisco.

Pfeffer, J. T. (1992) *Solid Waste Management Engineering,* Prentice Hall, Upper Saddle River, NJ, p. 172.

Robbins, W., T. Weir, and C. Stocking (1959) *Botany, an Introduction to Plant Science,* John Wiley and Sons, New York.

Salisbury, F. S., and C. W. Ross (1992) *Plant Physiology,* 4th ed., Belmont, CA: Thomson Brooks/Cole.

Salvato, J. A. (1972) *Environmental Engineering and Sanitation,* Wiley and Sons, New York, p. 427.

Scarlett, L. (1995) *Solid Waste Recycling Costs: Issues and Answers,* Reason Public Policy Institute Policy Study No. 193 (Los Angeles: Reason Public Policy Institute, August 1995), p. 2.

Schroeder, P. R., J. M. Morgan, T. M. Walski, and A. C. Gibbon (1984) *The Hydrologic Evaluation of Landfill Performance (HELP) Model Documentation, User's Guide,* U.S. Environmental Protection Agency, Washington, DC, EPA/530/SW-84-009.

Shy, C. M., D. Degnan, D. L. Fox, S. Mukerjee, M. J. Hazucha, B. A. Boehlecke, D. Rothenbacher, P. M. Briggs, R. B. Devlin, D. D. Wallace, R. K. Stevens, and P. A. Bromberg (1995) "Do Waste Incinerators Induce Adverse Respiratory Effects? An Air Quality and Epidemiological Study of Six Communities," *Environ. Health Perspectives,* 103(7-8): 714–24.

Tchobanoglous, G., H. Theisen, and S. A. Vigil (1993) *Integrated Solid Waste Management: Engineering Principles and Management Issues,* McGraw Hill Inc., New York.

U.S. EPA (1997) *Characterization of Municipal Solid Waste in the United States: 1996 Update,* United States Environmental Protection Agency, Washington, DC, May 1997, EPA530-R-97-015.

U.S. EPA (1998) *Environmental Labeling Issues, Policies and Practices Worldwide,* Office of Prevention, Pesticides, and Toxic Substances, Washington, DC, EPA 742-R-98-009 Dec. 1998.

U.S. EPA (1999) *Characterization of Municipal Solid Waste in the United States: 1998 Update,* United States Environmental Protection Agency, Washington, DC, 1998.

U.S. EPA (2000) *Municipal Solid Waste Generation, Recycling, and Disposal in the United States: Facts and Figures for 1998,* Washington, DC, EPA 530-F-00-024, April 2000.

U.S. EPA (2002) *Municipal Solid Waste in the United States: 2000 Facts and Figures,* United States Environmental Protection Agency, Washington, DC, June 2002, EPA530-R-02-001.

U.S. EPA (2006) *Municipal Solid Waste in the United States: 2005 Facts and Figures,* United States Environmental Protection Agency, Washington, DC, October 2006, EPA530-R-06-011.

U.S. EPA (2011) *Municipal Solid Waste in the United States: 2010 Facts and Figures,* United States Environmental Protection Agency, Washington, DC, December 2011, EPA530-F-11-005.

U.S. EPA (2016) *Advancing Sustainable Materials Management: 2014 Fact Sheet,* United States Environmental Protection Agency, November 2016.

U.S. EPA (2016) *Advancing Sustainable Materials Management: 2014 Fact Sheet,* https://www.epa.gov /sites/production/files/2016-11/documents/2014_smmfactsheet_508.pdf

U.S. Geological Survey (2000) *Ground Water Atlas,* Water Resources Information Service, U.S. Geological Survey, Washington, DC.

Vence, T. D., and D. L. Powers (1980) "Resource Recovery Systems, Part l: Technological Comparison," *Solid Wastes Management/Resource Recovery Journal,* 5, pp. 26–93.

Zmirou, D., B. Parent, and J. L. Potelon (1984) "Etude Épidémiologique Des Effets Sur La Santé Des Rejets Atmosphériques D'une Usine D'incinération De Déchets Industriels Et Ménagers," *Rev. Epidém. Santé Publ.,* 32: 391–97.

14

Hazardous Waste Management

Case Study

Don't Cry over Spilled Milk but Do Cry over Contaminated Milk

It was autumn of 1973 in rural Michigan when dairy farmers began noticing that milk production was decreasing rapidly. There was a significant increase in the incidence of miscarriages and stillborn calves. Calves and chicks were born with significant deformities, including disfigured hooves and talons. The problems were reported to the Department of Agriculture and were dismissed as "bad husbandry." Results of feed samples that were sent to State of Michigan laboratories came back "normal."

In April 1974, the results of cattle samples that were sent by veterinarian Alpha Clark to laboratories outside of Michigan came back with high amounts of polybrominated biphenyl (PBB), a chemical that was commonly used as a flame retardant in household products. The State responded by suing Clark for trafficking cattle out of the state. While the lawsuit was eventually dropped, it took several more months before the State determined the cause. In the meantime, farmers continued to watch their herds dwindle and die. Once PBB was identified as the cause, greater than 500 farms were quarantined and more than 30,000 cattle were killed and buried in pits. In addition, 1.5 million chickens and thousands of pigs, sheep, and rabbits were killed. In the meantime, dairy, chicken, and pork products that were tainted with PBB had been sold to and consumed by consumers across the state.

What resulted in an environmental and economic disaster began with a packaging mistake at the Velsicol Chemical plant, in St. Louis, Michigan. The company had been adding magnesium oxide to cattle feed in an attempt to enhance milk production. The product was called NutriMaster. The same company also produced PBB, which was sold under the trade name FireMaster. It is still unclear what happened next, but bags containing PBB were sent to feed mills across Michigan where the product was mixed with cattle feed. The same machinery was also used to process feed for other animals, resulting in cross-contamination of that feed with PBB.

PBB is no longer manufactured in the U.S. but the results remain. According to a study conducted by the Emory Rollins School of Public Health (Jacobson et al., 2017), some 60% of Michigan residents have elevated blood PBB levels. Women who were exposed to PBB were more likely to give birth to infants who scored low on Apgar tests and have a higher incidence of breast cancer. Daughters of women with high exposure to PBB were more likely to miscarry. The water supply for the city of St. Louis, Michigan, is contaminated with chemicals from the Velsicol Chemical plant, which include PBB. Despite decades of remediation efforts costing millions of dollars, the legacy of a careless mistake remains.

14–1   INTRODUCTION

A **hazardous waste** is any waste or combination of wastes that poses a substantial danger, now or in the future, to human, plant, or animal life and that therefore must be handled or disposed of with special precautions.

Dioxins and PCBs

We would like to elaborate on two particular hazardous wastes that have achieved national prominence: dioxins and PCBs. Because of their newsworthiness, we provide you with a brief summary of the chemistry of these compounds, where they come from, and their environmental effects.

Dioxins are found as over 20 different isomers of a basic chlorodioxin structure (some examples are shown in Figure 14–1). The most common form, 2,3,7,8-tetrachlorodibenzo-*p*-dioxin (TCDD), has become recognized as probably the most poisonous of all synthetic chemicals. Dioxins are a contaminant by-product that may be thermally generated during the manufacture or burning of chlorophenols; pesticides such as 2,4,5-T; Agent Orange, a defoliant made of a 50/50 mix of 2,4-D and 2,4,5-T; algae-controlling herbicides; insecticides; and preservatives. Dioxins are not manufactured for any commercial purpose. They occur only as a contaminant by-product. To date, no dioxin has been found to be formed naturally in the environment. Widespread TCDD contamination has been reported in particulate matter from commercial and domestic combustion processes. Additional background dioxin contamination (0.1–10 ppm) may persist and bioaccumulate following the field application of herbicides.

TCDD is a crystalline solid at room temperature. It is only slightly soluble in water (0.2–0.6 ppb). TCDD is considered to be a highly stable compound. It is thermally degraded at temperatures over 700°C. It is photochemically degraded under ultraviolet light in the presence of a hydrogen-donating solvent such as a solution of olive oil in cyclohexanone.

TCDD contamination was found at levels of parts per million in 2,4,5-T and 2,4-D used for weed control in the United States and as a defoliant in Vietnam; in wastes at the Love Canal disposal sites; in orthochlorophenol crude oil spill residues in the Sturgeon, Missouri, train derailment; and in fallout from an explosion at a chlorophenol manufacturing plant spill in Seveso, Italy. At this last site engineers and scientists were challenged to develop environmentally safe control strategies.

The environmental health effects of dioxin in people are not well documented. However, alleged birth defects in newborns in South Vietnam caused researchers to begin animal toxicological investigations. TCDD is known to cause severe skin disorders, such as chloracne. In test animals it is a carcinogen, teratogen, mutagen, and embryo toxin, and is known to affect immune responses in mammals. It is considered persistent, and it bioaccumulates in aquatic organisms. At this date (2017) no deaths have been directly correlated with low-level TCDD

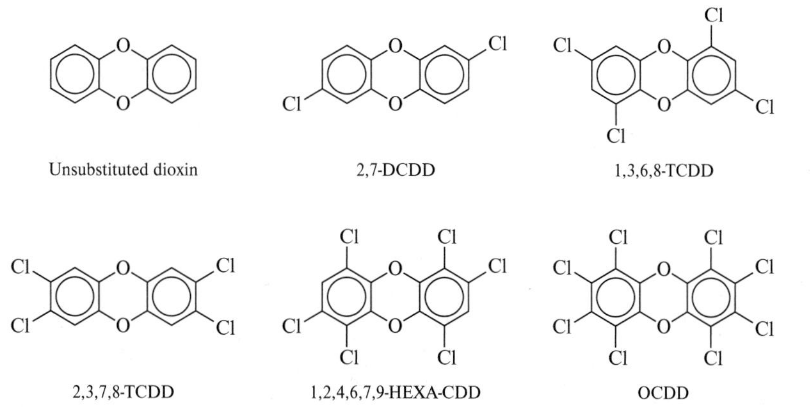

FIGURE 14–1

Some examples of dioxins.

Unsubstituted dioxin   2,7-DCDD   1,3,6,8-TCDD

2,3,7,8-TCDD   1,2,4,6,7,9-HEXA-CDD   OCDD

exposure. Nor have epidemiological findings shown any increased incidence of carcinogenesis, teratogenesis, mutagenesis or newborn defects, miscarriages, or similar adverse health effects in people. In May, 2000, the U.S. EPA released a draft report that presented evidence that dioxin, even in trace amounts, may cause adverse human health effects (Hileman, 2000). The draft report states that TCDD is a known human carcinogen and that other dioxins are likely human carcinogens. EPA believes that dioxins may cause a wide range of other effects, including disruption of regulatory hormones, reproductive and immune system disorders, and abnormal fetal development. Levels of dioxins in the environment were negligible until about 1930, peaked about 1970, and have been declining since then. Concentrations of dioxins in human lipid tissue have declined since 1980.

The term **PCB (polychlorinated biphenyl)** refers to a class of organic chemicals produced by the chlorination of a biphenyl molecule. It is composed of ten possible forms and, theoretically, more than 200 isomers. These forms arise from a specified number of chlorine substitutions on the biphenyl molecule and correspond to the chemical nomenclatures monochlorobiphenyl, dichlorobiphenyl, trichlorobiphenyl, and so on. Several isomers for each PCB molecule are possible, the number depending on available substitution sites on each biphenyl portion (2-6, 2'-6') of the molecule. However, not all possible isomers are likely to be formed during the manufacturing processes. In general, the most common ones are those that have either an equal number of chlorine atoms on both rings or a difference of only one chlorine atom between rings. Some examples are shown in Figure 14–2.

Commercial PCB mixtures were manufactured under a variety of trade names. The chlorine content of any product varied from 18 to 79%, depending on the extent of chlorination during the manufacturing process or on the amount of isomeric mixing engaged in by individual producers. Each company had a specific system for identifying the chlorine content of its product. For example, Aroclor 1248, 1254, and 1260 indicate 48, 54, and 60% chlorine, respectively; Clophen A60, Phenochlor DP6, and Kaneclor 600 designate that these products contain mixtures of hexachlorobiphenyls.

The only important U.S. producer of PCBs was Monsanto Industrial Chemicals Co., which had plants at Anniston, Alabama, where production of PCBs ended in 1970, and Sauget, Illinois, where production ceased in 1977. Sold under Monsanto's registered trademark of Aroclors, mixtures of PCBs had been used originally as a coolant–dielectric for transformers and capacitors, as heat transfer fluids, and as protective coatings for woods when low flammability was essential or desirable. Producers and users alike, apparently unaware of any potential hazards from exposure to PCBs, initially operated in accordance with earlier results of toxicity tests that indicated no effects (Penning, 1930). The expansion of open-ended applications between 1930 and 1960, incorporating PCBs into such commodities as paints, inks, dedusting agents, and pesticides led to the widespread dissemination of which we are now aware. By 1937, toxic effects were noted in occupationally exposed workers, and threshold limit values were imposed at manufacturing sites.

The general pattern of release of PCBs to the environment changed significantly during the early 1970s. Until then, essentially no restrictions were imposed either on the use or disposal of PCBs. After evidence became available in 1969 and 1970 that chronic exposure could result in hazards to human health and the environment, Monsanto voluntarily banned sales of PCBs, and the release rate from industrial use was reduced through stringent control measures.

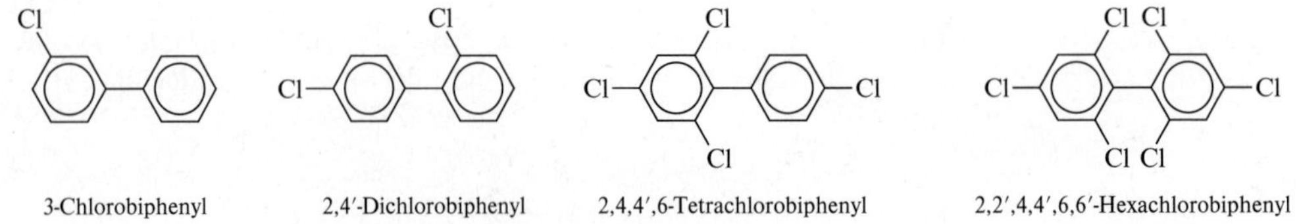

FIGURE 14–2

Molecular structure and names of a few selected polychlorinated biphenyls.

3-Chlorobiphenyl   2,4'-Dichlorobiphenyl   2,4,4',6-Tetrachlorobiphenyl   2,2',4,4',6,6'-Hexachlorobiphenyl

However, significant reservoirs of mobile PCBs (those available for transport among environmental media and biota) still exist along with even larger, currently immobile reservoirs. The latter include those materials containing PCBs that are still in service and those deposited in landfills and dumps. The major factor affecting future release of PCBs from these sources will be government regulations controlling storage and disposal of the chemical.

14–2 EPA'S HAZARDOUS WASTE DESIGNATION SYSTEM

EPA designates a waste material to be hazardous in two ways (40 CFR 260):* (1) by its presence on EPA-developed lists or (2) by evidence that the waste exhibits ignitable, corrosive, reactive, or toxic characteristics.

The list of hazardous wastes includes spent halogenated and nonhalogenated solvents; electroplating baths; wastewater treatment sludges from many individual production processes; and heavy ends, light ends, bottom tars, and side-cuts from various distillation processes.

Some commercial chemical products are also listed as being hazardous wastes when discarded. These include "acutely hazardous" wastes such as arsenic acid, cyanides, and many pesticides as well as "toxic" wastes such as benzene, toluene, and phenols.

EPA has designated five hazardous waste categories. Each hazardous waste is given an EPA hazardous waste number, often referred to as the **hazardous waste code.** Each of the five categories may be identified by the prefix letter assigned by EPA. The five categories may be described as follows:

1. Specific types of wastes from nonspecific sources; examples include halogenated solvents, nonhalogenated solvents, electroplating sludges, and cyanide solutions from plating batches. There are 28 listings in this category. These wastes have a waste code prefix letter F.
2. Specific types of wastes from specific sources; examples include oven residue from the production of chrome oxide green pigments and brine purification muds from the mercury cell process in chlorine production where separated, prepurified brine is not used. There are 111 listings in this category. These wastes have a waste code prefix letter K.
3. Any commercial chemical product or intermediate, off-specification product, or residue that has been identified as an acute hazardous waste. Examples include potassium silver cyanide, toxaphene, and arsenic oxide. There are approximately 203 listings in this category. These wastes have a waste code prefix letter P.
4. Any commercial chemical product or intermediate, off-specification product, or residue that has been identified as hazardous waste. Examples include xylene, DDT, and carbon tetrachloride. There are approximately 450 listings in this category. These wastes have a waste code prefix letter U.
5. Characteristic wastes, which are wastes not specifically identified elsewhere, that exhibit properties of ignitability, corrosivity, reactivity, or toxicity. These wastes have a waste code prefix letter D.

The wastes that appear on one of the lists specified in items one through four are called **listed wastes.** Those wastes that are declared hazardous because of their general properties are called **characteristic wastes.** The characteristics of ignitability, corrosivity, and reactivity may be referred to as ICR. The toxicity characteristic may be referred to as TC.

Figure 14–3 shows a generalized flow scheme for determining if a waste is hazardous according to EPA definitions. Of particular importance in the scheme are those things that are not included in the RCRA regulations. For example, domestic sewage; certain nuclear materials, household wastes, including toxic and hazardous materials; and small quantities (less than $100 \text{ kg} \cdot \text{month}^{-1}$)

---

*Note: CFR means Code of Federal Regulations. The prefix number refers to the volume. The suffix number is the paragraph number that spells out the regulation.

FIGURE 14–3

Flow scheme for determining if the waste is hazardous.

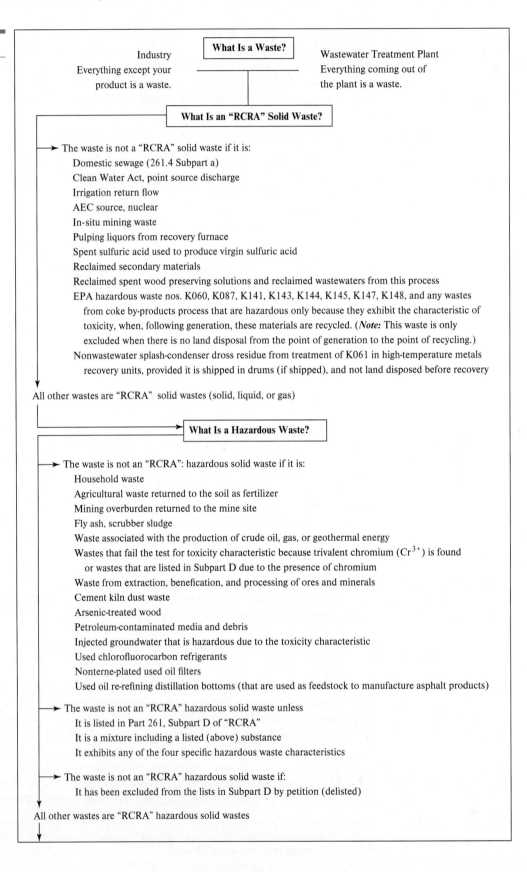

What Is a Waste?

Industry
Everything except your product is a waste.

Wastewater Treatment Plant
Everything coming out of the plant is a waste.

What Is an "RCRA" Solid Waste?

The waste is not a "RCRA" solid waste if it is:
Domestic sewage (261.4 Subpart a)
Clean Water Act, point source discharge
Irrigation return flow
AEC source, nuclear
In-situ mining waste
Pulping liquors from recovery furnace
Spent sulfuric acid used to produce virgin sulfuric acid
Reclaimed secondary materials
Reclaimed spent wood preserving solutions and reclaimed wastewaters from this process
EPA hazardous waste nos. K060, K087, K141, K143, K144, K145, K147, K148, and any wastes from coke by-products process that are hazardous only because they exhibit the characteristic of toxicity, when, following generation, these materials are recycled. (*Note:* This waste is only excluded when there is no land disposal from the point of generation to the point of recycling.)
Nonwastewater splash-condenser dross residue from treatment of K061 in high-temperature metals recovery units, provided it is shipped in drums (if shipped), and not land disposed before recovery

All other wastes are "RCRA" solid wastes (solid, liquid, or gas)

What Is a Hazardous Waste?

The waste is not an "RCRA": hazardous solid waste if it is:
Household waste
Agricultural waste returned to the soil as fertilizer
Mining overburden returned to the mine site
Fly ash, scrubber sludge
Waste associated with the production of crude oil, gas, or geothermal energy
Wastes that fail the test for toxicity characteristic because trivalent chromium (Cr^{3+}) is found or wastes that are listed in Subpart D due to the presence of chromium
Waste from extraction, benefication, and processing of ores and minerals
Cement kiln dust waste
Arsenic-treated wood
Petroleum-contaminated media and debris
Injected groundwater that is hazardous due to the toxicity characteristic
Used chlorofluorocarbon refrigerants
Nonterne-plated used oil filters
Used oil re-refining distillation bottoms (that are used as feedstock to manufacture asphalt products)

The waste is not an "RCRA" hazardous solid waste unless
It is listed in Part 261, Subpart D of "RCRA"
It is a mixture including a listed (above) substance
It exhibits any of the four specific hazardous waste characteristics

The waste is not an "RCRA" hazardous solid waste if:
It has been excluded from the lists in Subpart D by petition (delisted)

All other wastes are "RCRA" hazardous solid wastes

FIGURE 14–3

(continued)

What Wastes Are Subject to Regulation?

The "RCRA" hazardous solid waste is currently not subject to Subtitle C regulations if:
  The total combined "RCRA" hazardous waste generated at the site is less than 100 kg·month^{-1}.
  It is intended to be legitimately reclaimed or reused, (261.6). However it is subject to RCRA reporting
    requirements regarding storage and transportion if it is a sludge or contains a Part 261 listed substance.
      The "RCRA" hazardous solid waste is temporarily exempt from certain regulations if:
  It is a hazardous waste that is generated in a product or raw material storage tank, a product or raw
    material transport vehicle or vessel, a product or raw material pipeline, or in a manufacturing process
      unit or an associated nonwaste-treatment-manufacturing unit.

All other "RCRA" hazardous solid wastes are subject to Subtitle C of RCRA regulation with respect to
disposal, transport, and storage.

Requirements for Recyclable Materials

Hazardous wastes that are not subject to requirements for generator, transporters, and storage facilities:
Regulated under Subparts C through H (261.6):
  Recyclable materials used in a manner constituting disposal
  Hazardous wastes burned for energy recovery in boilers and industrial furnaces
  Recyclable materials from which precious metals are reclaimed
  Spent lead-acid batteries that are being reclaimed
Not subject to regulation or to the notification requirements of RCRA:
  Industrial ethyl alcohol that is reclaimed
  Used batteries returned to a battery manufacturer for regeneration
  Scrap metal
  Fuels produced from the refining of oil bearing hazardous wastes
  Oil reclaimed from hazardous waste resulting from normal petroleum refining, production, and
    transportation practices
  Hazardous waste fuel produced from oil-bearing hazardous wastes
  Petroleum coke produced from petroleum refinery hazardous wastes
  Used oil that is recycled and is also a hazardous waste solely because it exhibits a hazardous characteristic
    is not subject to the requirements of Parts 260–268 of this chapter, but is regulated under Part 279 of
    this chapter.

---

are excluded from the RCRA regulations. This does not mean that these wastes are not regulated at all. In fact they are regulated under other statutes and, thus, do not need to be regulated under RCRA.

Some waste streams do not come under the purview of RCRA but are, nonetheless, considered hazardous. These special wastes include, for example, PCBs and asbestos. PCBs and asbestos are regulated under the Toxic Substances Control Act (abbreviated TSCA and pronounced "tas-kah").

14–3   RCRA AND HSWA

Congressional Actions on Hazardous Waste

In 1976 Congress passed the Resource Conservation and Recovery Act (abbreviated RCRA and pronounced "rick-rah") directing the EPA to establish hazardous waste regulations. RCRA was amended in 1984 by the Hazardous and Solid Waste Amendments (abbreviated HSWA and pronounced "hiss-wah"). RCRA and HSWA were enacted to regulate the generation and disposal of hazardous wastes. These acts did not address abandoned or closed waste disposal sites or spills. The Comprehensive Environmental Response, Compensation, and Liability Act (abbreviated by CERCLA and pronounced "sir-klah"), commonly referred to as "Superfund," was enacted in 1980 to address these problems. SARA, the Superfund Amendments and Reauthorization Act of 1986, extended the provisions of CERCLA. In the following sections, we shall attempt to tell you about the who, what, where, and how of RCRA, HSWA, CERCLA, and SARA.

Cradle-to-Grave Concept

The EPA's cradle-to-grave hazardous waste management system is an attempt to track hazardous waste from its generation point (the "cradle") to its ultimate disposal point (the "grave"). The system requires generators to attach a **manifest** (itemized list describing the contents) form to their hazardous waste shipments. This procedure is designed to ensure that wastes are directed to, and actually reach, a permitted disposal site.

Generator Requirements

Generators of hazardous waste are the first link in the cradle-to-grave chain of hazardous waste management established under RCRA. Generators of more than 100 kg of hazardous waste or 1 kg of acutely hazardous waste per month must (with a few exceptions) comply with all of the generator regulations.

The regulatory requirements for hazardous waste generators include (U.S. EPA, 1986):

1. Obtaining an EPA identification (ID) number
2. Handling of hazardous waste before transport
3. Manifesting of hazardous waste
4. Record keeping and reporting

EPA assigns each generator a unique identification number. Without this number the generator is barred from treating, storing, disposing or transporting, or offering for transportation any hazardous waste. Furthermore, the generator is forbidden from offering the hazardous waste to any transporter, or treatment, storage, or disposal (TSD) facility that does not also have an EPA ID number.

Pretransport regulations are designed to ensure safe transportation of a hazardous waste from origin to ultimate disposal. In developing these regulations, EPA adopted those used by the Department of Transportation (DOT) for transporting hazardous wastes off-site (49 CFR Parts 172, 173, 178, and 179).

In addition to adopting these DOT regulations, EPA also developed pretransport regulations that cover the accumulation of waste prior to transport. A generator may accumulate hazardous waste on site for 90 days or less as long as they are properly stored, there is an emergency plan, and personnel are trained in the proper handling of hazardous waste.

The 90-day period allows a generator to collect enough waste to make transportation more cost-effective, that is, instead of paying to haul several small shipments of waste, the generator can accumulate waste until there is enough for one big shipment. If the generator accumulates hazardous waste on site for more than 90 days, it is considered an operator of a storage facility and must comply with requirements for such facilities.

The uniform hazardous waste manifest (the **manifest**) is the key to cradle-to-grave waste management (Figure 14–4). Through the use of a manifest, generators can track the movement of hazardous waste from the point of generation to the point of ultimate treatment, storage, or disposal.

HSWA requires that each manifest certify that the generator has in place a program to reduce the volume and toxicity of the waste to the degree that is economically practicable, as determined by the generator, and that the TSD method chosen by the generator is the best practicable method currently available that minimizes the risk to human health and the environment.

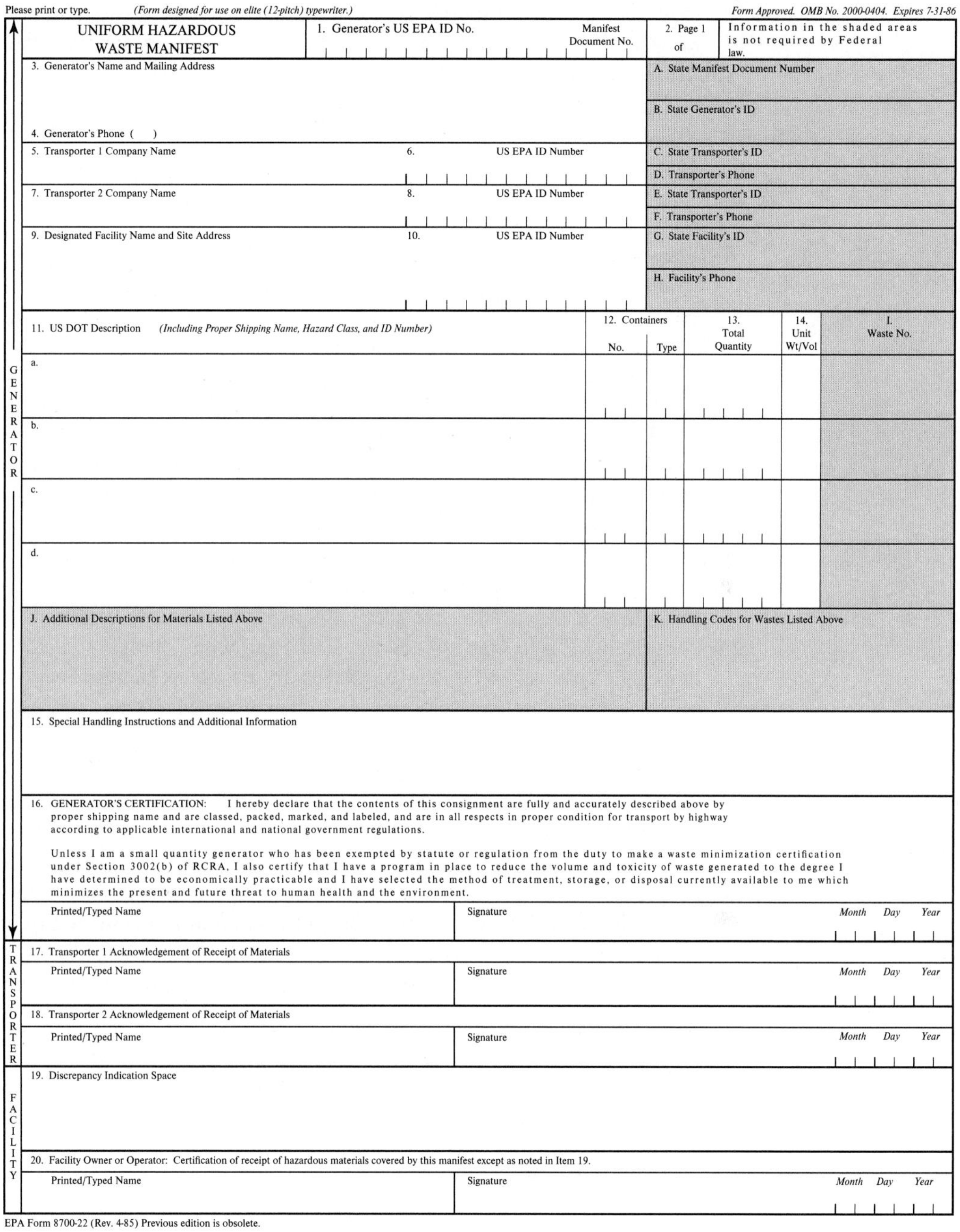

FIGURE 14–4

Uniform hazardous waste manifest. (*Source:* EPA.)

Please print or type.   *(Form designed for use on elite (12-pitch) typewriter.)*   Form Approved. OMB No. 2000-0404. Expires 7-31-86

UNIFORM HAZARDOUS WASTE MANIFEST	1. Generator's US EPA ID No. Manifest Document No.	2. Page 1 of	Information in the shaded areas is not required by Federal law.

3. Generator's Name and Mailing Address

A. State Manifest Document Number

B. State Generator's ID

4. Generator's Phone (   )

5. Transporter 1 Company Name   6.   US EPA ID Number

C. State Transporter's ID

D. Transporter's Phone

7. Transporter 2 Company Name   8.   US EPA ID Number

E. State Transporter's ID

F. Transporter's Phone

9. Designated Facility Name and Site Address   10.   US EPA ID Number

G. State Facility's ID

H. Facility's Phone

11. US DOT Description *(Including Proper Shipping Name, Hazard Class, and ID Number)*	12. Containers No. Type	13. Total Quantity	14. Unit Wt/Vol	I. Waste No.
a.				
b.				
c.				
d.				

J. Additional Descriptions for Materials Listed Above

K. Handling Codes for Wastes Listed Above

15. Special Handling Instructions and Additional Information

16. GENERATOR'S CERTIFICATION: I hereby declare that the contents of this consignment are fully and accurately described above by proper shipping name and are classed, packed, marked, and labeled, and are in all respects in proper condition for transport by highway according to applicable international and national government regulations.

Unless I am a small quantity generator who has been exempted by statute or regulation from the duty to make a waste minimization certification under Section 3002(b) of RCRA, I also certify that I have a program in place to reduce the volume and toxicity of waste generated to the degree I have determined to be economically practicable and I have selected the method of treatment, storage, or disposal currently available to me which minimizes the present and future threat to human health and the environment.

Printed/Typed Name   Signature   Month Day Year

17. Transporter 1 Acknowledgement of Receipt of Materials

Printed/Typed Name   Signature   Month Day Year

18. Transporter 2 Acknowledgement of Receipt of Materials

Printed/Typed Name   Signature   Month Day Year

19. Discrepancy Indication Space

20. Facility Owner or Operator: Certification of receipt of hazardous materials covered by this manifest except as noted in Item 19.

Printed/Typed Name   Signature   Month Day Year

(Left margin vertical labels: GENERATOR / TRANSPORTER / FACILITY)

EPA Form 8700-22 (Rev. 4-85) Previous edition is obsolete.

The manifest is part of a controlled tracking system. Each time the waste is transferred, that is, from a transporter to the designated facility or from a transporter to another transporter, the manifest must be signed to acknowledge receipt of the waste. A copy of the manifest is retained by each link in the transportation chain. Once the waste is delivered to the designated facility, the owner or operator of the facility must send a copy of the manifest back to the generator. This system ensures that the generator has documentation that the hazardous waste has reached its ultimate destination.

If 35 days pass from the date on which the waste was accepted by the initial transporter and the generator has not received a copy of the manifest from the designated facility, the generator must contact the transporter or the designated facility to determine the whereabouts of the waste. If 45 days pass and the manifest still has not been received, the generator must submit an exception report.

The record keeping and reporting requirements for generators provide EPA and states with a method for tracking the quantities of waste generated and the movement of hazardous wastes.

Transporter Regulations

Transporters of hazardous waste are the critical link between the generator and the ultimate off-site treatment, storage, or disposal of hazardous waste. The transporter regulations were developed jointly by EPA and the DOT to avoid contradictory requirements coming from the two agencies (U.S. EPA, 1986). Although the regulations are integrated, they are not contained under the same act. A transporter must comply with the regulations under 49 CFR 171–179 (The Hazardous Materials Transportation Act) as well as those under 40 CFR Part 263 (Subtitle C of RCRA).

Even if generators and transporters of hazardous waste comply with all appropriate regulations, transporting hazardous waste can still be dangerous. There is always the possibility that an accident will occur. To deal with this possibility, the regulations require transporters to take immediate action to protect health and the environment if a release occurs by notifying local authorities or diking off the discharge area.

The regulations also give officials special authority to deal with transportation accidents. Specifically, if a federal, state, or local official, with appropriate authority, determines that the immediate removal of the waste is necessary to protect human health or the environment, the official can authorize waste removal by a transporter who lacks an EPA ID and without the use of a manifest.

Treatment, Storage, and Disposal Requirements

Treatment, storage, and disposal (TSD) facilities are the last link in the cradle-to-grave hazardous waste management system. All TSDs handling hazardous waste must obtain an operating permit and abide by the TSD regulations. The TSD regulations establish performance standards that owners and operators must apply to minimize the release of hazardous waste into the environment.

A TSD facility may perform one or more of the following functions (U.S. EPA, 1986):

1. *Treatment:* Any method, technique, or process, including neutralization, designed to change the physical, chemical, or biological character or composition of any hazardous waste so as to neutralize it or render it nonhazardous or less hazardous; to recover it; make it safer to transport, store, or dispose of; or make it amenable for recovery, storage, or volume reduction.
2. *Storage:* The holding of hazardous waste for a temporary period, at the end of which the hazardous waste is treated, disposed, or stored elsewhere.
3. *Disposal:* The discharge, deposit, injection, dumping, spilling, leaking, or placing of any solid waste or hazardous waste into or on any land or water so that any constituent thereof may enter the environment or be emitted into the air or discharged into any waters, including groundwaters.

The act establishes standards that consist of administrative–nontechnical requirements and technical requirements.

The purpose of the administrative–nontechnical requirements is to ensure that owners and operators of TSDs establish the necessary procedures and plans to operate a facility properly and to handle any emergencies or accidents. They cover the subject areas shown in the following table:

Subpart	Subject
A	Who is subject to the regulations
B	General facility standards
	Waste analysis, security, inspections, training
	Ignitable, reactive, or incompatible wastes
	Location standards (permitted facilities)
C	Preparedness and prevention
D	Contingency plans and emergency procedures
E	Manifest system, record keeping, and reporting

The objective of the technical requirements is to minimize the potential for threats resulting from hazardous waste treatment, storage, and disposal at existing facilities waiting to receive an operating permit. There are two groups of requirements: general standards that apply to several types of facilities and specific standards that apply to a waste management method.

The general standards cover three areas:

1. Groundwater monitoring requirements
2. Closure, postclosure requirements
3. Financial requirements

Groundwater monitoring is only required of owners or operators of a surface impoundment, landfill, land treatment facility, or some waste piles used to manage hazardous waste. The purpose of these requirements is to assess the effect of a facility on the groundwater beneath it. Monitoring must be conducted for the life of the facility except at land disposal facilities, which must continue monitoring for up to 30 years after the facility has closed.

The groundwater monitoring program outlined in the regulations requires a monitoring system of four wells to be installed: one upgradient from the waste management unit and three downgradient. The downgradient wells must be placed so as to intercept any waste migrating from the unit, should such a release occur. The upgradient wells must provide data on groundwater that is not influenced by waste coming from the waste management unit (called background data). If the wells are properly located, comparison of data from upgradient and downgradient wells should indicate if contamination is occurring.

Once the wells have been installed, the owner or operator monitors them for 1 year to establish background concentrations for selected chemicals. These data form the basis for all future data comparisons. There are three sets of parameters for which background concentrations are established: drinking water, groundwater quality, and groundwater contamination.

Closure is the period when wastes are no longer accepted, during which owners or operators of TSD facilities complete TSD operations: apply final covers to or cap landfills and dispose of or decontaminate equipment, structures, and soil. Postclosure, which applies only to disposal facilities, is the 30-year period after closure during which owners or operators of disposal facilities conduct monitoring and maintenance activities to preserve and look after the integrity of the disposal system.

Financial requirements were established to ensure that funds are available to pay for closing a facility, for rendering postclosure care at disposal facilities, and to compensate third parties for bodily injury and property damage caused by sudden and nonsudden accidents related to the facility's operation (states and federal governments are exempted from abiding by these requirements). There are two kinds of financial requirements: financial assurance for closure–postclosure and liability coverage for injury and property damage.

Land Ban. HSWA significantly expanded the scope of RCRA. HSWA was created, in large part, in response to strongly voiced citizen concerns that existing methods of hazardous waste disposal, particularly land disposal, were not safe. Section 3004 of HSWA sets restrictions on land disposal of specific wastes, commonly called the "land ban," or **land disposal restrictions** (LDR). As specifically required by section 3004(m), the agency established levels or methods of treatment, if any, which substantially reduce the likelihood of migration of hazardous constituents from waste so that short-term and long-term threats to human health and the environment are minimized. On September 18, 1994 the universal treatment standards (UTS) were published by EPA to streamline the process (59 FR 47980, 1994).

Underground Storage Tanks

An **underground storage tank** (UST) system includes an underground storage tank, connected piping, underground ancillary equipment, and containment system, if any. On September 23, 1988, the EPA promulgated the final rules for underground storage tanks (Bair, 1989).

The new regulations include a number of exclusions:

- Hazardous waste UST systems
- Regulated wastewater treatment facilities
- Any equipment or machinery that contains regulated substances for operational purposes such as hydraulic lift tanks and electrical equipment tanks
- Any UST system of less than 41.5 L
- Any UST system containing a **de minimis** (negligible) concentration of regulated substances
- Any emergency spill or overflow containment system that is expeditiously emptied after use

All UST systems must have corrosion protection, which can be accomplished in one of three ways: (1) construction of fiberglass-reinforced plastic, (2) steel- and fiberglass-reinforced plastic composite, or (3) a coated steel tank with cathodic protection. Cathodic protection systems must be regularly tested and inspected. All owners and operators must also provide spill and overfill prevention equipment and a certificate of installation to ensure that the methods of installation were in compliance with the regulations.

Release (leak) detection must be instituted for all USTs. Several different methods are allowed for petroleum UST systems. However, some systems have specific requirements, for instance, a pressurized delivery system must be equipped with an automatic line leak detector and have an annual line tightness test. All new or upgraded UST systems storing hazardous substances must have secondary containment with interstitial monitoring.

When release is confirmed, owners and operators must begin corrective action. Immediate corrective action measures include mitigation of safety and fire hazards, removal of saturated soils and floating free product, and an assessment of further corrective action needed. As with any remediation situation, a corrective action plan may be required for long-term cleanups of contaminated soil and groundwater.

14–4 CERCLA AND SARA

The Superfund Law

The Comprehensive Environmental Response, Compensation, and Liability Act (CERCLA) of 1980, better known as "Superfund," became law "to provide for liability, compensation, cleanup and emergency response for hazardous substances released into the environment and the cleanup of inactive hazardous waste disposal sites." CERCLA was generally intended to give EPA authority and funds to clean up abandoned waste sites and to respond to emergencies related to hazardous waste. The law provides for both response and enforcement mechanisms. The four major provisions of the law establish

1. A fund (the "Superfund") to pay for investigations and remediation at sites where the responsible people cannot be found or will not voluntarily pay;

2. A priority list of abandoned or inactive hazardous waste sites for cleanup (the national priority list);
3. The mechanism for action at abandoned or inactive sites (the national contingency plan);
4. Liability for those responsible for cleaning up.

Initially the trust fund was supported by taxes on producers and importers of petroleum and 42 basic chemicals. In its first 5-year period, Superfund collected about $1.6 billion, with 86% of that money coming from industry and the remainder from federal government appropriations. In 1986 the Superfund Amendments and Reauthorization Act (SARA) greatly expanded the money available to remediate Superfund sites. The fund was raised to $8.6 billion for a 5-year period by taxing petroleum products ($2.75 billion), business income ($2.5 billion), and chemical feedstocks ($1.4 billion). The remainder was from general revenues.

The National Priority List

The **national priority list** (NPL) serves as a tool for the EPA to use in identifying sites that appear to present a significant risk to public health or the environment and that may merit use of Superfund money. First published in 1982, it is updated three times a year. In January 2012, the list contained 1298 sites (U.S. EPA, 2012).

The first NPL was formulated from notification procedures and existing information sources. Subsequently, a numeric ranking system known as the hazard ranking system was developed. Sites with high HRS scores may be added to the list. Sites on the NPL are eligible for Superfund money. Those with lower scores are not likely to be eligible.

The Hazard Ranking System

The **hazard ranking system** (HRS) is a procedure for ranking uncontrolled hazardous waste sites in terms of the potential threat based upon containment of the hazardous substances, route of release, characteristics and amount of the substances, and likely targets. The methodology of the HRS provides a quantitative estimate that represents the relative hazards posed by a site and takes into account the potential for human and environmental exposure to hazardous substances. The HRS score is based on the probability of contamination from four pathways—groundwater, surface water, soil, and air—on the site in question. The groundwater and air migration pathways are evaluated for ingestion and inhalation, respectively. The surface water migration and soil exposure pathways are evaluated for multiple intake routes. Surface water is evaluated for exposure of (1) drinking water, (2) human food chain, and (3) the environment (contact) to hazardous wastes. These exposures are evaluated for two separate migration components—overland/flood migration and groundwater to surface water migration. Soil is evaluated for potential hazardous waste exposure to the (1) resident population and (2) nearby population (40 CFR 300).

Use of the HRS requires considerable information about the site and its surroundings, the hazardous substances present, and the geology of the aquifers and the intervening strata. The factors that most affect an HRS site score are the proximity to a densely populated area or source of drinking water, the quantity of hazardous substances present, and the toxicity of those hazardous substances. The HRS methodology has been criticized for the following reasons:

1. There is a strong bias toward human health effects, with only a slight chance of a site in question receiving a high score if it represents a threat or hazard only to the environment.
2. Because of the human health bias, there is an even stronger bias in favor of highly populated affected areas.
3. The air emission migration route must be documented by an actual release, while groundwater and surface water routes have no such documentation requirement.
4. The scoring for toxicity and persistence of chemicals may be based on site containment, which is not necessarily related to a known or potential release of the toxic chemicals.

5. A high score for one migration route can be more than offset by low scores for the other migration routes.

6. Averaging of the route scores creates a bias against a site that has only one hazard, even though that one hazard may pose extreme threat to human health and the environment.

The HRS scores range from 0 to 100, with a score of 100 representing the most hazardous site. Occasional exceptions have been made in the HRS priority ranking to meet the CERCLA requirement that a site designated by a state as its top priority be included on the NPL.

The National Contingency Plan

The **national contingency plan** (NCP) provides detailed direction on the action to be taken at a hazardous waste site, including initial assessment to determine if an emergency or imminent threat exists, creation of emergency response actions, and a method to rank sites (the HRS) and establish priority for future action. When sufficient indication points to a site posing a potential risk to the environment, a detailed study is required (O'Brien and Gere Engineers Inc., 1988; 40 CFR 400.300).

The NCP describes the steps to be taken for the detailed evaluation of the risks associated with a site. Such an evaluation is termed a **remedial investigation** (RI). The process of selecting an appropriate remedy is termed the **feasibility study** (FS). The remedial investigation and the feasibility study are often combined into a single measure, known popularly as a remedial investigation/feasibility study (**RI/FS**). The requirements of the RI/FS are usually outlined in a written work plan, which must be approved by the relevant federal and state agencies before it may be implemented.

A remedial investigation includes the development of detailed plans that address the following items:

1. *Site characterization:* A description of the hydrogeological and geophysical sampling and analytical procedures to be applied to discover the nature and extent of the waste materials, the physical characteristics of the site, and any receptors that could be affected by the wastes at the site.

2. *Quality control:* The guidelines to be enforced to ensure that all the data collected from the characterization program are valid and satisfactorily accurate.

3. *Health and safety:* The procedures to be employed to protect the safety of the individuals who will work at the site and perform the site characterization.

The RI activities and subsequent evaluation of the data gathered are termed a **risk assessment** or an endangerment assessment. The remedial investigation report documents the evaluation.

The remedial investigation report serves as a basis for the feasibility study, which evaluates various remedial alternatives. The review criteria include overall protection of human health and the environment; compliance with applicable or relevant and appropriate regulations; long-term effectiveness; reduction in toxicity, mobility, or volume; short-term effectiveness; technical and administrative implementability; cost; state acceptance; and community acceptance. All the remedies selected must be capable of reducing the risk at the hazardous waste site to an acceptable level. And, in general, the lowest cost alternative that achieves this objective is chosen as the course of action. The results of the feasibility study are presented in a written report, called the **record of decision** (ROD). This document serves as a preliminary basis for the design of the selected alternative.

One of the keys to the NCP is that it specifies that the degree of cleanup be selected in accordance with several criteria, including the degree of hazard to the "public health, welfare and the environment." Therefore, no predetermined level of remediation can be required or must be achieved at any site. Rather, the degree of correction is established on a site-by-site basis. What is acceptable in one location may not necessarily be acceptable in another.

On completion and approval of the RI/FS, the next step is the preparation of plans and specifications for the selected remedy—the **remedial design** (RD). To complete the process, the construction and other activities are undertaken in accordance with the plans and specifications.

Liability

Perhaps the most far-reaching provision of CERCLA that has stood the test of the courts was the establishment of *strict joint, and several liability* for cleanup of an NPL site. Those identified by EPA as **potentially responsible parties** (PRPs) may include generators, present owners, or former owners of facilities or real property where hazardous wastes have been stored, treated, or disposed of, as well as those who accepted hazardous waste for transport and selected the facility. PRPs have strict liability; that is, liability without fault. Neither care nor negligence, neither good nor bad faith, neither knowledge nor ignorance, can be claimed as a defense. Congress correctly predicted that there would be instances where the PRPs would contest their contribution to the problem and would, then, be unwilling to share the costs or the responsibility. The strict liability provision orders that the PRP is liable even if the method of disposal was in accordance with prevailing standards, laws, and practice at the time of disposal. In other words, CERCLA is a "pay now, argue later" statute (O'Brien and Gere Engineers Inc., 1988).

Although the language specific to "joint and several" liability was removed from CERCLA, the courts have interpreted the law as though the language were included. This means that if a PRP contributed any wastes to a site, that PRP can be held accountable for all costs associated with the cleanup. This concept was strongly reaffirmed in SARA. If the PRP refuses to pay, the federal government can sue to recover costs. These actions have been successful. In certain instances where those liable fail, without sufficient cause, to properly provide for cleanup, they may be liable for triple damages!

Superfund Amendments and Reauthorization Act

SARA reaffirmed and strengthened many of the provisions and concepts of the CERCLA program. In SARA, Congress clearly expressed a preference but not a requirement, for remedies such as incineration or chemical treatment that render a waste nonhazardous rather than transport to another disposal site or simple containment on site.

Another aspect of SARA is that the level of cleanup should achieve compliance with **applicable or relevant and appropriate requirements** (ARARs). ARARs are environmental standards from programs other than CERCLA and SARA. For example, if a state has a regulation regarding atmospheric emissions from incinerators, then a Superfund cleanup using incineration must meet those applicable standards. Furthermore, if a similar standard appears to be relevant and appropriate, then EPA may elect to apply it. For example, if drums of waste found on an uncontrolled hazardous waste site have contents that appear to have the same constituents as F001-F005 spent solvent, then the UTS standards for RCRA waste may be considered relevant and appropriate even though there is no specific evidence to identify the origin of the waste.

SARA significantly strengthens the requirement to consider damages to natural resources, especially those off-site. CERCLA also required such observance, but few sites included this factor in practice. SARA provides a mechanism to include the issue in future investigations and remediation activities.

Title III. SARA includes a major addition to the provisions of CERCLA, namely title III—emergency planning and community right to know. Under the emergency planning provisions, facilities must notify the state emergency response commission if they have quantities of extremely hazardous substances that exceed EPA specified **threshold planning quantities**. In addition, communities must establish local emergency planning committees (LEPCs) to develop a chemical emergency response plan. This plan must include identification of regulated facilities, emergency response and notification procedures, training programs, and evacuation plans in case of a chemical release.

If a facility accidentally releases chemicals that are on one of two lists (i.e., EPA's extremely hazardous substance list or the CERCLA section 103(a) list), in regulated quantities (RQ), and the release has the potential for exposure off-site, they must notify the LEPC immediately. The law also requires a report on response actions taken, known or anticipated health risks, and advice on medical attention for exposed individuals.

Perhaps the most revolutionary provision of title III is the establishment of the **community's right to know** amounts of chemicals and their location in facilities in their community. Thus, information about potential hazards from chemicals is available to the public. In addition, each year those facilities that release chemicals above specified threshold amounts must submit a **toxic release inventory** on an EPA-specified form (**form R**). This inventory includes both accidental and routine releases as well as off-site shipments of waste. The publication of these data has resulted in strenuous efforts by industry to control their previously unregulated and, hence, uncontrolled emissions because of the public outcry at the large quantities of materials being dumped into the environment.

14–5 HAZARDOUS WASTE MANAGEMENT

A logical priority in managing hazardous waste would be to

1. Reduce the amount of hazardous wastes generated in the first place.
2. Stimulate "waste exchange." (One factory's hazardous wastes can become another's feedstock; for instance, acid and solvent wastes from some industries can be used by others without processing.)
3. Recycle metals, the energy content, and other useful resources contained in hazardous wastes.
4. Detoxify and neutralize liquid hazardous waste streams by chemical and biological treatment.
5. Reduce the volume of waste sludges generated in item four by dewatering.
6. Destroy combustible hazardous wastes in special high-temperature incinerators equipped with proper pollution control and monitoring systems.
7. Stabilize or solidify sludges and ash from items five and six to reduce leachability of metals.
8. Dispose of remaining treated residues in specially designed landfills.

Waste Minimization

The key elements necessary for the success of a waste minimization program include (Fromm, Bachrach, and Callahan, 1986)

Top-level organizational commitment
Financial resources
Technical resources
Appropriate organization, goals, and strategy

The commitment of senior management is the first element that must be in place. Efforts to establish the other elements can follow. The organizational structure adopted should promote communication and feedback from participants. Often, the best ideas come from line operators who work with the processes every day.

Some firms set quantitative waste minimization goals. Other firms are more qualitative in their goal setting.

Waste Audit. An important first step in establishing a strategy for waste minimization is to conduct a waste audit. The audit should proceed stepwise:

1. Identify waste streams
2. Identify sources

3. Establish priority of waste streams for waste minimization activity
4. Screen alternatives
5. Implement
6. Track
7. Evaluate progress

The key question at the outset of a waste audit is "Why is this waste being generated?" You must first establish the primary cause(s) of waste generation before attempting to find solutions. The audit should be waste stream-oriented to produce a list of specific minimization options for additional evaluation or implementation. Once the causes are understood, solution options can be formulated. An efficient materials and waste-tracking system that allows computation of mass balances is useful in establishing priorities. Knowing how much material is going in and how much of it is ending up as waste allows you to decide which process and which waste to address first.

EXAMPLE 14–1 A manufacturing company has, as part of its first audit, gathered the following data. Estimate the potential annual air emissions in kilograms of volatile organic compounds (VOCs) from the company. Note that 1 U.S. barrel of liquid = 0.12 m^3

Purchasing Department Records

Material	Purchase Quantity (barrels)
Methylene chloride (CH_2Cl_2)	228
Trichloroethylene (C_2HCl_3)	505

Wastewater (ww) Treatment Plant Influent

Material	Average Concentration (mg · L^{-1})
CH_2Cl_2	4.04
C_2HCl_3	3.25

Average flow into treatment plant is 0.076 m^3 · s^{-1}.

Hazardous Waste (hw) Manifests

Material	Barrels	Concentration (%)
CH_2Cl_2	228	25
C_2HCl_3	505	80

Unused Barrels at End of Year

CH_2Cl_2	8
C_2HCl_3	13

Solution The materials balance diagram will be the same for each waste.

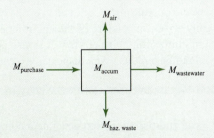

The mass balance equation is

$$M_{purchase} = M_{air} + M_{ww} + M_{hw} + M_{accum}$$

Solving this equation for M_{air} gives us the estimated VOC emission.

First, we calculate the mass purchased. The density of each compound is found in Appendix A.

Mass Purchased

$$M_{(CH_2Cl_2)} = (228 \text{ barrels} \cdot \text{year}^{-1})(0.12 \text{ m}^3 \cdot \text{barrel}^{-1})(1327 \text{ kg} \cdot \text{m}^{-3})$$

$$= 36,306.72 \text{ kg} \cdot \text{year}^{-1}$$

$$M_{(C_2HCl_3)} = (505 \text{ barrels} \cdot \text{year}^{-1})(0.12 \text{ m}^3 \cdot \text{barrel}^{-1})(1476 \text{ kg} \cdot \text{m}^{-3})$$

$$= 89,445.60 \text{ kg} \cdot \text{year}^{-1}$$

Now we calculate the mass received at the wastewater treatment plant. (Note that $1.0 \text{ mg} \cdot \text{L}^{-1} = 1.0 \text{ g} \cdot \text{m}^{-3}$.)

$$M_{(CH_2Cl_2)} = (4.04 \text{ g} \cdot \text{m}^{-3})(0.076 \text{ m}^3 \cdot \text{s}^{-1})(86,400 \text{ s} \cdot \text{day}^{-1})(365 \text{ days} \cdot \text{year}^{-1})(10^{-3} \text{ kg} \cdot \text{g}^{-1})$$

$$= 9682.81 \text{ kg} \cdot \text{year}^{-1}$$

$$M_{(C_2HCl_3)} = (3.25 \text{ g} \cdot \text{m}^{-3})(0.076 \text{ m}^3 \cdot \text{s}^{-1})(86,400 \text{ s} \cdot \text{day}^{-1})(365 \text{ days} \cdot \text{year}^{-1})(10^{-3} \text{ kg} \cdot \text{g}^{-1})$$

$$= 7789.39 \text{ kg} \cdot \text{year}^{-1}$$

The mass shipped to the hazardous waste disposal facility is calculated next.

$$M_{(CH_2Cl_2)} = (228)(0.12)(1327)(0.25) = 9076.68 \text{ kg} \cdot \text{year}^{-1}$$

$$M_{(C_2HCl_3)} = (505)(0.12)(1476)(0.80) = 71,556.48 \text{ kg} \cdot \text{year}^{-1}$$

Accumulated is then

$$M_{(CH_2Cl_2)} = (8)(0.12)(1327) = 1273.92 \text{ kg} \cdot \text{year}^{-1}$$

$$M_{(C_2HCl_3)} = (13)(0.12)(1476) = 2302.56 \text{ kg} \cdot \text{year}^{-1}$$

The estimated air emission for each compound is then

$$M_{(CH_2Cl_2)} = 36,306.72 - 9682.81 - 9076.68 - 1273.92$$

$$= 16,273.31, \text{ or } 16,000 \text{ kg} \cdot \text{year}^{-1}$$

$$M_{(C_2HCl_2)} = 89,445.60 - 7789.39 - 71,556.48 - 2302.56$$

$$= 7797.17, \text{ or } 7800 \text{ kg} \cdot \text{year}^{-1}$$

Note that we round to two significant figures because the volume of the barrels is known to only two significant figures. From this analysis, to reduce the mass of air pollutants emitted, the company should attack the methylene chloride source first. We should also point out that simply counting "barrels in" from the purchasing record and "barrels out" from the hazardous waste manifest would give a highly erroneous picture of the environmental effect of this company's emissions. From a waste minimization point of view, it is also apparent that C_2HCl_3, at 80% concentration in barrels going to hazardous waste disposal, is a candidate for recycling.

The first four steps of the waste audit allow you to generate a comprehensive set of waste management options following the hierarchy of source reduction first, waste exchange second, recycling third, and treatment last.

The screening of options begins with source control. The source control investigation should focus on changes in (1) input materials, (2) process technology, and (3) the human aspect of production. Input material changes can be classified into three separate elements: purification, substitution, and dilution.

Purification of input materials is performed to avoid the introduction of inert materials or impurities into the production process. Such an introduction results in waste because the process inventory must be purged to prevent the undesirable accumulation of impurities. Examples of purification of feed materials to lower waste generation include the use of deionized rinse water in electroplating or the use of oxygen instead of air in oxychlorination reactors for production of ethylene dichloride.

Substitution is the replacement of a toxic material with one characterized by lower toxicity or higher environmental desirability. Examples include using phosphates in place of dichromates as cooling water corrosion inhibitors or the use of alkaline cleaners in place of chlorinated solvents for degreasing. These, and other changes, are examples of an approach called "green chemistry" that industries are beginning to use in their processes (U.S. EPA, 2017).

Dilution is a minor component at input material changes and is exemplified by use of more dilute plating solutions to minimize **dragout** (material carried out of one tank and into another).

Technology changes are those made to the physical plant. Examples include process changes; equipment, piping, or layout changes; changes to process operational settings; additional automation; energy conservation; and water conservation.

Procedural or institutional changes consist of improvements in the ways people affect the production process. Also referred to as "good operating practices" or "good housekeeping," these include operating procedures, loss prevention, waste segregation, and material handling improvements.

Waste Exchange

Waste minimization by consignment of excess unused materials to an independent party for resale to a third party saves both in waste production and in the cost (environmental and financial) of production from new raw materials. In essence "one person's trash becomes another person's treasure." The difference between a manufacturing by-product, which is costly to treat or dispose, and a usable or salable by-product involves opportunity, knowledge of processes outside the generator's immediate production line, and comparative pricing of virgin material. Waste exchanges serve as information clearinghouses through which the availability and need for various types of materials can be established.

Recycling

Under RCRA and HSWA, EPA has carefully defined recycling to prohibit bogus recyclers that are really TSDs from taking advantage of more lenient rules for recycling. The definition says that a material is recycled if it is used, reused, or reclaimed (40 CFR 261.1 (c)(7)). A material is "used or reused" if it is either (1) employed as an ingredient (including its use as an intermediate) to make a product (however, a material will not satisfy this condition if distinct components of the material are recovered as separate end products, as when metals are recovered from metal-containing secondary materials) or (2) employed in a particular function as an effective substitute for a commercial product (40 CFR 261.1 (c)(5)). A material is reclaimed if it is processed to recover a useful product or if it is regenerated. Examples include the recovery of lead from spent batteries and the regeneration of spent solvents (40 CFR 261.1 (c)(4)) (U.S. EPA, 1988a).

Distillation processes can be used to recover spent solvent. The principal characteristics that determine the potential for recovery are the boiling points of the various useful constituents and the water content. The more dilute the waste solvent, the less economical it is to recover. Recovered solvents can be reused by the generator or sold for at least a substantial fraction of the cost of virgin material, and the credit for recovered solvent can more than offset the cost of recovery.

There are several technologies for recovery of metals from metal-plating rinse water. Most are applicable only to waste streams containing a single metal constituent. Examples include ion exchange, electrodialysis, evaporation, and reverse osmosis.

In October 1988 a federal appeals court struck down an EPA policy not to list used oil collected for recycling as a hazardous waste. Prior to that ruling, PCB-contaminated oils, petroleum industry sludges, and leaded tank bottoms were the only oils regulated. The majority of oil and oily wastes generated were not classified as hazardous under EPA regulations. These wastes are amenable either to recovery for use as fuel or to refinement for use as lubricants. Although all waste oil is now deemed hazardous, those oils that were formerly recovered may still be recovered, but the requirements for tracking them are more stringent.

14–6 TREATMENT TECHNOLOGIES

The wastes that remain after the implementation of waste minimization must be detoxified and neutralized. A large number of treatment technologies are available to accomplish this, many of which are applications of processes we have discussed in earlier chapters. Examples include biological oxidation (Chapter 11), chemical precipitation, ion exchange, and oxidation–reduction (Chapter 2), and carbon adsorption (Chapter 12). Here we will discuss these as they apply to hazardous waste treatment, and we will introduce some new technologies.

Biological Treatment

In contrast to naturally occurring compounds, **anthropogenic compounds** (those created by human beings) are relatively resistant to biodegradation (Kobayashi and Rittman, 1982). One reason is that the organisms that are naturally present often cannot produce the enzymes necessary to bring about transformation of the original compound to a point at which the resultant intermediates can enter into common metabolic pathways and be completely mineralized.

Many environmentally important anthropogenic compounds are halogenated, and halogenation is often implicated as a reason for their persistence. The list of halogenated organic compounds includes pesticides, plasticizers, plastics, solvents, and trihalomethanes. Chlorinated compounds are the best known and most studied because of the highly publicized problems associated with DDT and other pesticides and numerous industrial solvents. Hence, chlorinated compounds serve as the basis for most of the information available on halogenated compounds.

Some of the characteristics that appear to confer persistence to halogenated compounds are the location of the halogen atom, the halide involved, and the extent of halogenation. The first step in biodegradation, then, is sometimes dehalogenation, for which there are several biological mechanisms.

Simple generalizations do not appear to be applicable. For example, until recently, oxidative pathways were mostly believed to be the typical means by which halogenated compounds were dehalogenated. Anaerobic, reductive dehalogenation, either biological or nonbiological, is now recognized as the critical factor in the transformation or biodegradation of certain classes of compounds. Compounds that require reductive dechlorination are common among the pesticides, as well as halogenated one- and two-carbon aliphatic compounds.

Reductive dehalogenation involves the removal of a halogen atom by oxidation–reduction. In essence, the mechanism involves the transfer of electrons from reduced organic substances via microorganisms or a nonliving (**abiotic**) mediator, such as inorganic ions (e.g., Fe^{3+}) and biological products [e.g., NAD(P), flavin, flavoproteins, hemoproteins, porphyrins, chlorophyll, cytochromes, and glutathione]. The mediators are responsible for accepting electrons from reduced organic substances and transferring them to the halogenated compounds. The major requirements for the process are believed to be available free electrons and direct contact between the donor, mediator, and acceptor of electrons. Significant reductive dechlorination usually occurs only when the oxidation–reduction potential of the environment is 0.35 V and lower; the exact requirements appear to depend on the compound involved.

Although simple studies using pure cultures of microorganisms and single substrates are valuable, if not essential, for determining biochemical pathways, they cannot always be used to predict biodegradability or transformation in more natural situations. The interactions among environmental factors, such as dissolved oxygen, oxidation–reduction potential, temperature, pH, availability of other compounds, salinity, particulate matter, competing organisms, and concentrations of compounds and organisms, often control the feasibility of biodegradation. The compound's physical or chemical characteristics, such as solubility, volatility, hydrophobicity, and octanol–water partition coefficient contribute to the compound's availability in solution. Often compounds not soluble in the water are not readily available to organisms for biodegradation. There are some exceptions. For example, DDT, which is only slightly soluble in water, may be degraded by the white rot fungus found on decaying trees. This is because the enzymes involved in the white rot reaction are secreted from the cell.

Simple culture studies are similarly inadequate for predicting the fate of substances in the environment if many interactions occur between different organisms. First, substances that cannot be changed significantly in pure culture studies often will be degraded or transformed under mixed culture conditions. A good example of this type of interaction is **cometabolism,** in which a compound, the nongrowth substrate, is not metabolized as a source of carbon or energy, but is incidentally transformed by organisms using other compounds as growth substrates. The growth substrates provide the energy needed to cometabolize the nongrowth substrates. Second, products of the initial transformation by one organism may subsequently be broken down by a series of different organisms until compounds that can be metabolized by normal metabolic pathways are formed. An example is the degradation of DDT, which is reportedly mineralized directly by only one organism, a fungus; other organisms studied appear to degrade DDT only through cometabolism, resulting in numerous transformation products that subsequently can be used by other organisms. For example, *Hydrogenomonas* can metabolize DDT only as far as *p*-chlorophenylacetic acid (PCPA), whereas *Arthrobacter* species can then remove the PCPA.

Table 14–1 demonstrates that members of almost every class of anthropogenic compound can be degraded by some microorganism. The table also illustrates the wide variety of microorganisms that participate in environmentally significant biodegradation.

TABLE 14–1 Examples of Anthropogenic Compounds and Microorganisms That Can Degrade Them

Compound	Organism	Compound	Organism
Aliphatic (nonhalogenated)		Polycyclic aromatics (nonhalogenated)	
Acrylonitrile	Mixed culture of yeast, mold, protozoan bacteria	Benzo(*a*)pyrene, naphthalene, Benzo(*a*)anthracene	*Cunninghamella elegans* *Pseudomonas*
Aliphatic (halogenated)		Polycyclic aromatics (halogenated)	
Trichloroethane, trichloroethylene, methyl chloride, methylene chloride	Marine bacteria, soil bacteria, sewage sludge	PCBs 4-Chlorobiphenyl	*Pseudomonas, Flavobacterium* Fungi
Aromatic compounds (nonhalogenated)		Pesticides	
Benzene, 2,6-dinitrotoluene, creosol, phenol	*Pseudomonas* sp., sewage sludge	Toxaphene	*Corynebacterium pyrogenes*
		Dieldrin	*Anacystic nidulans*
		DDT	Sewage sludge, soil bacteria
Aromatic compounds (halogenated)		Kepone	Treatment lagoon sludge
1,2-; 2,3-; 1,4-dichlorobenzene, hexachlorobenzene, trichlorobenzene, pentachlorophenol	Sewage sludge Soil microbes	Nitrosamines Dimethylnitrosamines	*Rhodopseudomonas*
		Phthalate esters	*Micrococcus* 12B

Source: Extracted from Table I of Kobayashi and Rittman, 1982.

After almost 20 years of research, the use of novel microorganisms for biological treatment of anthropogenic compounds is still a developing concept. A number of advancements are required before large-scale application is possible, among the most important of which is an improved knowledge of metabolic pathways for the biodegradation of specific compounds by different organisms. The metabolic capabilities of many microorganisms, in particular algae and oligotrophic bacteria, are poorly understood. Such knowledge is necessary if limiting reactions are to be determined and the proper types of organisms are to be selected for specific applications. More information about appropriate types of microorganisms to be selected and maintained in real-world treatment systems is needed, especially for the more novel microbial cultures. To develop special-purpose organisms by genetic manipulation, major advances in the understanding of the genetic structure of the many different types of organisms in nature are needed.

Conventional biological treatment processes such as activated sludge and trickling filters have been used to treat hazardous wastes. The major modification to the activated sludge processes has been to extend the mean cell residence time from the conventional values of 4 to 15 days to much longer periods of 3 to 6 months. In a similar fashion, trickling filter loading rates are much lower than those employed in municipal treatment systems. One innovation that has been adopted by TSD facilities is the sequencing batch reactor (SBR). The SBR is a periodically operated, fill-and-draw reactor (Herzbnin, Irvine, and Malinowski, 1985). Each reactor in an SBR system has five discrete periods in each cycle: fill, react, settle, draw, and idle. Biological reactions are initiated as the raw wastewater fills the tank. During the fill and react phase, the waste is aerated in the same fashion as an activated sludge unit. After the react phase, the mixed liquor suspended solids are allowed to settle. The treated supernatant is discharged during the draw phase. The idle stage, the time between the draw and fill, may be zero or may be a few days depending on wastewater flow demand. The SBR has a major advantage in that wastes may be tested for completeness of treatment before discharge.

Chemical Treatment

Chemical detoxification is a treatment technology, either employed as the sole treatment procedure or used to reduce the hazard of a particular waste prior to transport, incineration, and burial.

It is important to remember that a chemical procedure cannot magically make a toxic chemical disappear from the **matrix** (wastewater, sludge, etc.) in which it is found, but can only convert it to another form. Thus, it is vital to ensure that the products of a chemical detoxification step are less of a problem than the starting material. It is equally important to remember that the reagents for such a reaction can be hazardous.

The spectrum of chemical methods includes complexation, neutralization, oxidation, precipitation, and reduction. An optimum method would be fast, quantitative, inexpensive, and leave no residual reagent, which itself would be a pollution problem. The following paragraphs describe a few of these techniques.

Neutralization. Solutions are neutralized by a simple application of the law of mass balance to bring about an acceptable pH. Sulfuric or hydrochloric acid is added to basic solutions, and caustic (NaOH) or slaked lime [$Ca(OH)_2$] is added to acidic solutions. Though a waste is hazardous at pH values less than 2 or greater than 12.5, and it would seem that simply bringing the pH into the range 2–12.5 would be adequate, good treatment practice requires that final pH values be in the range 6–8 to protect natural biota.

Oxidation. The cyanide molecule is destroyed by oxidation. Chlorine is the oxidizing agent most frequently used. Oxidation must be conducted under alkaline conditions to avoid the generation of hydrogen cyanide gas. Hence, this process is often referred to as **alkaline chlorination.** In chlorine oxidation, the reaction is carried out in two steps:

$$NaCN + 2NaOH + Cl_2 \rightleftharpoons NaCNO + 2NaCl + H_2O \qquad (14\text{--}1)$$

$$2NaCNO + 5NaOH + 3Cl_2 \rightleftharpoons 6NaCl + CO_2 + N_2 + NaHCO_3 + 2H_2O \qquad (14\text{--}2)$$

In the first step, the pH is maintained above 10 and the reaction proceeds in a matter of minutes. In this step, great care must be taken to maintain relatively high pH values because at lower pHs there is a potential for the evolution of highly toxic hydrogen cyanide gas. The second reaction step proceeds most rapidly around a pH of 8, but it is not as rapid as the first step. Higher pH values may be selected for the second step to reduce chemical consumption in the following precipitation steps. This increases the reaction time. Often the second reaction is not carried out because the CNO^- (i.e., cyanate anion) is considered nontoxic by current regulations.

Ozone also may be used as the oxidizing agent. Ozone has a higher redox potential than chlorine, thus there is a higher driving force toward the oxidized state. When ozone is used, the pH considerations are similar to those discussed for chlorine. Ozone cannot be purchased. It must be made on site as part of the process.

This technology can be applied to a wide range of cyanide wastes: copper, zinc, and brass plating solutions; cyanide from cyanide salt heating baths; and passivating solutions. The process has been practiced on an industrial scale since the early 1940s. For extremely high cyanide concentrations (>1%), oxidation may not be desirable. Cyanide complexes of metals, particularly iron and, to some extent, nickel, cannot be decomposed easily by cyanide oxidation techniques.

Electrolytic oxidation of cyanide is carried out by anodic electrolysis at high temperatures. The theoretical basis of the process is that cyanide reacts with oxygen in solution in the presence of an electric potential to produce carbon dioxide and nitrogen gas. Normally, the destruction is carried out in a closed cell. Two electrodes are suspended in the solution and a DC current is applied to drive the reaction. The bath temperature must be maintained in the range of 50 to 95°C.

This technology is used for the destruction of cyanide in concentrated spent stripping solutions; in plating solutions for copper, zinc, and brass; in alkaline descalers; and in passivating solutions. It has been more successful for wastes containing high concentrations of cyanide ($50,000–100,000 \text{ mg} \cdot \text{L}^{-1}$), but it has also been successfully used for concentrations as low as $500 \text{ mg} \cdot \text{L}^{-1}$.

Chemical oxidation methods for treating organic compounds in wastewater have received extensive study. In general, they apply only to dilute solutions and often are considered expensive compared with biological methods. Some examples include wet air oxidation, hydrogen peroxide, permanganate, chlorine dioxide, chlorine, and ozone oxidation. Of these, wet air oxidation and ozonation have shown promise as a pretreatment step for biological processes.

Wet air oxidation, also known as the Zimmerman process, operates on the principle that most organic compounds can be oxidized by oxygen given sufficient temperature and pressure. Wet air oxidation may be described as the aqueous phase oxidation of dissolved or suspended organic particles at temperatures of 175 to 325°C and sufficiently high pressure to prevent excessive evaporation. Air is bubbled through the liquid. The process is fuel-efficient; once the oxidation reaction has started, it is usually self-sustaining. Because this method is not limited by reagent cost, it is potentially the most widely applicable of all chemical oxidation methods. The method has been shown to be of use in destroying a wide range of organic compounds, including some pesticides. Although wet oxidation can provide acceptable levels of destruction for many hazardous compounds, it is generally not as complete as incineration. In many instances, the addition of metal salt catalysts can increase the destruction efficiency or allow the process to be run at lower temperature or pressure.

Precipitation.   Metals are often removed from plating rinse waters by precipitation. This is a direct application of the solubility product principle (see Chapter 2). By raising the pH with lime or caustic, the solubility of the metal is reduced (Figure 14–5) and the metal hydroxide precipitates. Optimum removal is achieved by selecting the optimum pH as shown in Figure 14–5. Though there is an optimum for each metal, in many cases, the metals are mixed and the lowest value for an individual metal may not be achievable for the mixture.

FIGURE 14–5

Solubilities of metal
hydroxides as a function
of pH. (*Source:* U.S.
Environmental Protection
Agency, 1981.)

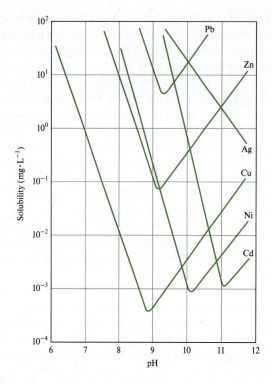

EXAMPLE 14–2   A metal plating firm is installing a precipitation system to remove zinc. They plan to use a pH meter to control the feed of hydroxide solution to the mixing tank. What pH should the controller be set at to achieve a zinc effluent concentration of $0.80 \text{ mg} \cdot \text{L}^{-1}$? The K_{sp} of $Zn(OH)_2$ is 7.68×10^{-17}.

Solution   From Table A–8 in Appendix A we find that the zinc hydroxide reaction is

$$Zn^{2+} + 2OH^- \rightleftharpoons Zn(OH)_2$$

As shown in Chapter 2, we can write the solubility product equation as

$$K_{sp} = [Zn^{2+}][OH^-]^2$$

Because we want the zinc concentration to be no greater than $0.80 \text{ mg} \cdot \text{L}^{-1}$, we calculate the moles per liter of zinc.

$$[Zn^{2+}] = \frac{0.80 \text{ mg} \cdot \text{L}^{-1}}{(65.38 \text{ g} \cdot \text{mol}^{-1})(1000 \text{ mg} \cdot \text{g}^{-1})} = 1.224 \times 10^{-5} \text{ mol} \cdot \text{L}^{-1}$$

Now we solve for the hydroxide concentration.

$$[OH^-]^2 = \frac{7.68 \times 10^{-17}}{1.224 \times 10^{-5}} = 6.275 \times 10^{-12}$$

Therefore,

$$[OH^-] = (6.275 \times 10^{-12})^{0.5} = 2.505 \times 10^{-6}$$

The pOH is

$$pOH = -\log(2.505 \times 10^{-6}) = 5.601$$

And the pH set point for the controller is

$$pH = 14 - pOH = 14 - 5.601 = 8.399, \text{ or } 8.4$$

Reduction.   Although most heavy metals readily precipitate as hydroxides, hexavalent chromium used in plating solutions must be reduced to trivalent chromium before it will precipitate. Reduction is usually done with sulfur dioxide (SO_2) or sodium sulfite ($NaHSO_3$). With SO_2 the reaction is

$$3SO_2 + 2H_2CrO_4 + 3H_2O \rightleftharpoons Cr_2(SO_4)_3 + 5H_2O \qquad (14\text{–}3)$$

Because the reaction proceeds rapidly at low pH, an acid is added to control the pH between 2 and 3.

Physical/Chemical Treatment

Several treatment processes are used to separate hazardous waste from aqueous solution. The waste is not detoxified but only concentrated for further treatment or recovery.

Carbon Adsorption.   Adsorption is a mass-transfer process in which gas vapors or chemicals in solution are held to a solid by intermolecular forces (e.g., hydrogen bonding and van der Waals' interactions). It is a surface phenomenon. Pressure vessels having a fixed bed are used to hold the adsorbent (see Chapter 12). Activated carbon, molecular sieves, silica gel, and activated alumina are the most common adsorbents. The active sites become saturated at some point in time. When the organic material has commercial value, the bed is then regenerated by passing steam through it. The vapor-laden steam is condensed and the organic fraction is separated from the water. If the organic compounds have no commercial value, the carbon may be either incinerated or shipped to the manufacturer for regeneration. Carbon systems for recovery of vapor from degreasers and for polishing wastewater effluents have been in commercial application for more than 40 years.

Distillation.   The separation of more volatile materials from less volatile ones by a process of vaporization and condensation is called distillation. When a liquid mixture of two or more components is brought to the boiling point of the mixture, a vapor phase is created above the liquid phase. If the vapor pressures of the pure components are different (which is usually the case), then the constituent(s) having the higher vapor pressure will be more concentrated in the vapor phase than the constituent(s) having the lower vapor pressure. If the vapor phase is cooled to yield a liquid, a partial separation of the constituents will result. The degree of separation depends on the relative differences in the vapor pressures. The larger the differences, the more efficient the separation. If the difference is large enough, a single separation cycle of vaporization and condensation is sufficient to separate the components. If the difference is not large enough, multiple cycles (stages) are required. Four types of distillation may be used: batch distillation, fractionation, steam stripping, and thin film evaporation.

Both batch distillation and fractionation are well proven technologies for recovery of solvents. Batch distillation is particularly applicable for wastes with high solids concentrations. Fractionation is applicable when multiple constituents must be separated and when the waste contains minimal suspended solids.

Air Stripping.   When the volatility of the organic compound is relatively high and the concentration relatively low, some form of stripping may be appropriate. Air stripping has been used to purge large quantities of contaminated groundwater of small concentrations of volatile organic matter. The behavior of the process is the inverse of absorption discussed in Chapter 12. Air and contaminated liquid are passed countercurrently through a packed tower. The volatile compounds evaporate into the air, leaving a clean liquid stream. The contaminated air stream must then be treated to avoid an air pollution problem. Frequently this is accomplished by passing the air through an activated carbon column. The carbon is then incinerated. Air stripping has been used to remove tetrachloroethylene, trichloroethylene, and toluene from water (Gross and TerMaath, 1987).

The air stripper design equation may be written as follows:

$$Z_T = \frac{L}{A} \times \frac{\ln\{(C_1/C_2) - (L\,RT_g/G\,H_c)[(C_1/C_2) - 1]\}}{K_L a[1 - (L\,RT_g/G\,H_c)]} \tag{14-4}$$

where Z_T = packed tower depth (in m)

$\quad\quad L$ = water flow (in $m^3 \cdot min^{-1}$)

$\quad\quad A$ = cross-sectional area of tower (in m^2)

$\quad\quad G$ = air flow (in $m^3 \cdot min^{-1}$)

$\quad\quad H_c$ = Henry's constant (in $atm \cdot m^3 \cdot mol^{-1}$)

$\quad\quad R$ = universal gas constant = 8.206×10^{-5} $atm \cdot m^3 \cdot mol^{-1} \cdot K^{-1}$

$\quad\quad T_g$ = temperature of air (in K)

$\quad C_1, C_2$ = influent and effluent organic concentration in the water (in $mol \cdot m^{-3}$)

$\quad\quad K_L$ = liquid mass transfer coefficient (in $mol \cdot min^{-1} \cdot m^{-2} \cdot mol^{-1} \cdot m^3$)

$\quad\quad a$ = effective interfacial area of packing per unit volume for mass transfer (in $m^2 \cdot m^{-3}$)

EXAMPLE 14-3   Well 12A at the City of Tacoma is contaminated with 350 $\mu g \cdot L^{-1}$ of 1,1,2,2-tetrachloroethane. The water must be cleaned to the detection limit of 1.0 $\mu g \cdot L^{-1}$. Design a packed tower stripping column to meet this requirement using the following design parameters.

Henry's law constant = 5.0×10^{-4} $atm \cdot m^3 \cdot mol^{-1}$ Temperature = 25°C

$K_L a = 10 \times 10^{-3}$ s^{-1} Column diameter may not

Air flow rate = 13.7 $m^3 \cdot s^{-1}$ exceed 4.0 m

Liquid flow rate = 0.044 $m^3 \cdot s^{-1}$ Column height may not exceed 6.0 m

Solution   The Henry's law constants given in Appendix A are in $kPa \cdot m^3 \cdot mol^{-1}$. To convert these to $atm \cdot m^3 \cdot mol^{-1}$, divide by the atmospheric pressure at standard conditions, that is, 101.325 $kPa \cdot atm^{-1}$.

The stripper equation is then solved for $Z_T A$, the column volume.

$$Z_T A = (0.044) \frac{\ln\{(350/1) - [(0.044)(8.206 \times 10^{-5})(298)/(13.7)(5.0 \times 10^{-4})][(350/1) - 1]\}}{10 \times 10^{-3}\{1 - [(0.044)(8.206 \times 10^{-5})(298)/(13.7)(5.0 \times 10^{-4})]\}}$$

$$= (0.044)(674.74) = 29.69 \ m^3$$

Any number of solutions are now possible within the boundary conditions of 4-m diameter and 6-m height. For example:

Diameter (m)	Tower Height (m)
4.00	2.36
3.34	3.39
2.58	5.68

Steam Stripping.   For gases of lower volatility or higher concentration (>100 ppm) **steam stripping** may be employed. The physical arrangement of the process is much like that of an air stripper except that steam is introduced instead of air. The addition of steam enhances the stripping process by decreasing the solubility of the organic compound in the aqueous phase and by increasing the vapor pressure. Steam stripping has been used to treat aqueous waste contaminated with chlorinated hydrocarbons, xylenes, acetone, methyl ethyl ketone, methanol, and pentachlorophenol. Concentrations treated range from 100 ppm to 10% organic chemical (U.S. EPA, 1987).

Evaporation. Recovery of plating metals by evaporation is accomplished by boiling off sufficient water from the collected rinse stream to allow the concentrate to be returned to the plating bath. The condensed steam is recycled for use as rinse water. The boil-off rate, or evaporator duty, is set to maintain the water balance of the plating bath. Evaporation is usually performed under vacuum to prevent thermal degradation of additives and to reduce the amount of energy required for evaporation of the water.

There are four types of evaporators: rising film, flash evaporators using waste heat, submerged tube, and atmospheric pressure. Rising film evaporators are built so that the evaporative heating surface is covered by a waste water film and does not lie in a pool of boiling wastewater. Flash evaporators are of similar configuration, but the plating solution is continuously recirculated through the evaporator along with the wastewater. This allows the use of waste heat in the plating bath to augment the evaporation process. In the submerged tube design, the heating coils are submerged in the wastewater. Atmospheric evaporators do not recover the distillate for reuse, and they do not operate under vacuum.

Ion Exchange. Metals and ionized organic chemicals can be recovered by ion exchange. Ion-exchange chemistry was discussed in Chapter 10. In ion exchange, the waste stream containing the ion to be removed is passed through a bed of resin. The resin is selected to remove either cations or anions. In the exchange process, ions of like charge are removed from the resin surface in exchange for ions in solution. Typically, either hydrogen or sodium is exchanged for cations (metal) in solution. When the bed becomes saturated with the exchanged ion, it is shut down and the resin is regenerated by passing a concentrated solution containing the original ion (hydrogen or sodium) back through the bed. The exchanged pollutant is forced off the bed in a concentrated form that may be recycled. A typical ion-exchange column is shown in Figure 14–6. A prefilter is required to remove suspended material that would hydraulically foul the column. It also removes organic compounds and oils that would foul the resin.

FIGURE 14–6

Typical ion-exchange resin column.

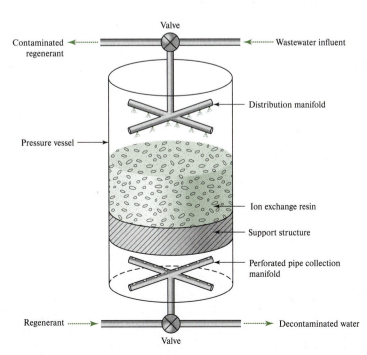

As a rule, ion-exchange systems are suitable for chemical recovery applications in which the rinse water feed has a relatively dilute concentration (<1000 mg $\cdot$ L^{-1}) and a relatively low concentration is required for recycle. Ion exchange has been demonstrated commercially for recovery of plating chemicals from acid–copper, acid–zinc, nickel, tin, cobalt, and chromium plating baths.

In full-scale operation, the resin bed is not allowed to reach saturation because the concentration of the solute will exceed most discharge standards before this occurs. Normal operation then requires either an operating cycle that will allow regeneration of the spent resin during nonworking hours or, in the case of 24-h, 7-day per week schedules, multiple beds so that one may be taken off-line.

During ion exchange, the normal flow pattern is downward through the bed. The hydraulic loading may range from 25 to 600 m^3 $\cdot$ day^{-1} $\cdot$ m^{-2}. Lower hydraulic loadings result in longer contact periods and better exchange efficiency. Because the surface of the bed acts like a filter, regeneration is often done using countercurrent, that is, the regenerating solution is pumped into the bottom of the column. This results in a cleansing of the column much like the backwashing of a rapid sand filter. Regeneration hydraulic loadings range from 60 to 120 m^3 $\cdot$ day^{-1} $\cdot$ m^{-2}.

Electrodialysis. The electrodialysis unit uses a membrane to selectively retain or transmit specific molecules. The membranes are thin sheets of ion-exchange resin reinforced by a synthetic fiber backing. The construction of the unit is such that anion membranes are alternated with cation membranes in stacks of cells in series (Figure 14–7). An electric potential is applied across the membrane to provide the motive force for ion migration. Cation membranes permit passage of only positively charged ions, and anion membranes permit passage of only negatively charged ions. The flow is directed through the membrane in two hydraulic circuits (Figure 14–8). One circuit is ion-depleted and the other is ion-concentrated. The degree of

---

FIGURE 14–7

Electrodialysis. (Cations in the feed water show the same behavior as sodium (Na$^+$) and anions show the same behavior as chloride (Cl$^-$). Under the action of an electric field, cation-exchange membranes permit passage only of positive ions, and anion-exchange membranes permit passage only of negatively charged ions.)

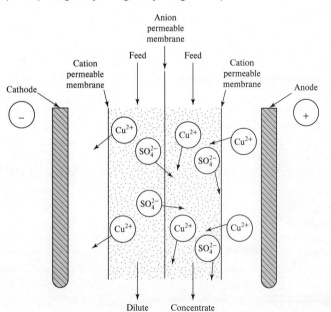

FIGURE 14–8

Electrodialysis unit flow schematic.

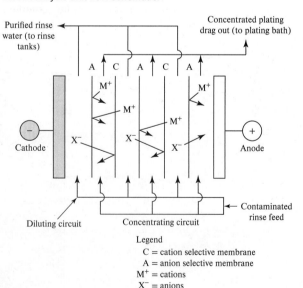

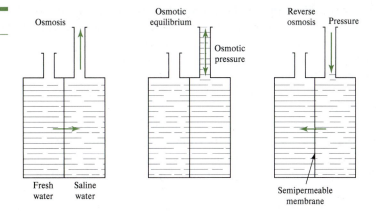

FIGURE 14–9

Direct and reverse osmosis.

purification achieved in the dilute circuit is set by the electric potential. The ability to pass the charge is proportional to the concentration of the ionic species in the dilute stream. Because ion migration is proportional to electric potential, the optimum system is a trade-off between energy requirements and degree of contaminant removal.

Electrodialysis has been in commercial operation for more than four decades in the production of potable water from brackish water. It has also been used in deashing of sugars, desalting of food products such as whey, and to recover waste developer in the photo-processing industry and nickel from a metal plating rinse water. Typically, electrodialysis can separate a waste stream containing 1000–5000 mg · L^{-1} inorganic salts into a dilute stream that contains 100–500 mg · L^{-1} salt and a concentrated stream that contains up to 10,000 mg · L^{-1} salt.

Reverse Osmosis.   **Osmosis** is the spontaneous transport of a solvent from a dilute solution to a concentrated solution across an ideal semipermeable membrane that impedes passage of the solute but allows the solvent to flow. Solvent flow can be reduced by exerting pressure on the solution side of the membrane, as shown in Figure 14–9. If the pressure is increased above the osmotic pressure on the solution side, the flow reverses. Pure solvent will then pass from the solution into the solvent. As applied to metal finishing wastewater, the solute is the metal and the solvent is pure water.

Many configurations of the membrane are possible. The driving pressure is on the order of 2800 to 5500 kPa. No commercially available membrane polymer has demonstrated tolerance to all extreme chemical factors such as pH, strong oxidizing agents, and aromatic hydrocarbons. However, selected membranes have been demonstrated on nickel, copper, zinc, and chrome baths.

Solvent Extraction.   Solvent extraction is also called **liquid extraction** and **liquid–liquid extraction.** Contaminants can be removed from a waste stream using liquid–liquid extraction if the wastewater is contacted with a solvent having a greater solubility for the target contaminants than the wastewater. The contaminants will tend to migrate from the wastewater into the solvent. Although predominately a method for separating organic materials, it may also be applied to remove metals if the solvent contains a material that will react with the metal. **Liquid ion exchange** is one kind of these reactions.

In the solvent extraction process, the solvent and the waste stream are mixed to allow mass transfer of the constituent(s) from the waste to the solvent. The solvent, immiscible in water, is then allowed to separate from the water by gravity. The solvent solution containing the extracted contaminant is called the **extract.** The extracted waste stream with the contaminants removed is called the **raffinate.** As in distillation, the separation may need to be done in one or more stages. In general, more stages result in a cleaner raffinate. The degree of complexity of the apparatus varies from simple mixer–settlers to more exotic contacting devices. If the extract is sufficiently enriched, it may be possible to recover useful material. Distillation is often employed to recover the solvent and reusable organic chemicals. For metal recovery, the ion-exchange material is regenerated by the addition of an acid or alkali. The process has found wide application in the ore processing industry, in food processing, in pharmaceuticals, and in the petroleum industry.

Incineration

In an incinerator, chemicals are decomposed by oxidation at high temperatures (800°C or greater). The waste, or at least its hazardous components, must be combustible to be destroyed. The primary products from combustion of organic wastes are carbon dioxide, water vapor, and inert ash; however, a multitude of other products can be formed.

Products of Combustion.   The percentages of carbon, hydrogen, oxygen, nitrogen, sulfur, halogens, and phosphorus in the waste, as well as the moisture content, need to be known to determine stoichiometric combustion air requirements and to predict combustion gas flow and composition. Actual incineration conditions generally require excess oxygen to maximize the formation of **products of complete combustion** (POCs) and minimize the formation of **products of incomplete combustion** (PICs).

The incineration of halogenated organic compounds results in the formation of halogenated acids, which require further treatment to ensure environmentally acceptable air emissions from the incineration process. Chlorinated organic compounds are the most common halogenated hydrocarbons found in hazardous waste. The incineration of chlorinated hydrocarbons with excess air results in the formation of carbon dioxide, water, and hydrogen chloride. An example is the following reaction for the incineration of dichloroethylene:

$$2C_2H_2Cl_2 + 5O_2 \rightleftharpoons 4CO_2 + 2H_2O + 4HCl \tag{14-5}$$

The hydrogen chloride must be removed before the carbon dioxide and steam can be safely exhausted into the atmosphere.

Hazardous waste may contain either organic or inorganic sulfur compounds. When these wastes are incinerated, sulfur dioxide is produced. For example, the destruction of ethyl mercaptan results in the following reaction:

$$2C_2H_5SH + 9O_2 \rightleftharpoons 4CO_2 + 6H_2O + 2SO_2 \tag{14-6}$$

The sulfur dioxide produced by the incineration of sulfur-containing wastes must not exceed air quality standards.

Excess air must be provided to ensure complete combustion. However, the amount of the excess can only be determined emperically. For example, a highly volatile, clean, hydrocarbon waste would probably require much less excess air than would a heavy hydrocarbon sludge within a high solids content. Incineration of sludges and solids may require as much as two to three times excess air above stoichiometric equivalents. Too much excess air should be avoided because it increases the fuel required to heat the waste to destruction temperatures, reduces residence time for the hazardous wastes to be oxidized, and increases the volume of air emissions to be handled by the air pollution control equipment.

By-products from the incineration of hazardous wastes may also result from incomplete combustion as well as from the products of combustion. PICs include carbon monoxide, hydrocarbons, aldehydes, ketones, amines, organic acids, and polycyclic aromatic hydrocarbons. In a well-designed incinerator, these products are insignificant in amount. However, in poorly designed or overloaded incinerators, PICs may pose environmental concerns. PCBs, for instance, decompose under such conditions into highly toxic chlorinated dibenzo furans. The hazardous material, hexachlorocyclopentadiene, found in many hazardous wastes, is known to decompose into the even more hazardous compound hexachlorobenzene (HCB) (Oppelt, 1981).

Suspended particulate emissions are also produced during incineration. These include particles of mineral oxides and salts from the mineral constituents in the waste material, as well as fragments of incompletely burned combustibles.

Last, but not least, ash is a product of combustion. The ash is considered a hazardous waste. Metals not volatilized end up in the ash. Unburned organic compounds may also be found in the ash. When organic compounds remain, the ash may simply be incinerated again. The metals must be treated prior to land disposal.

Design Considerations. The most important factors for proper incinerator design and operation are combustion temperature, combustion gas residence time, and the efficiency of mixing the waste with combustion air and auxiliary fuel (Oppelt, 1981).

Chemical and thermal dynamic properties of the waste that are important in determining its time and temperature requirements for destruction are its elemental composition, net heating value, and any special properties (e.g., explosive properties) that may interfere with incineration or require special design considerations.

In general, higher heating values are required for solids versus liquids or gases, for higher operating temperatures, and for higher excess air rates if combustion is to be sustained without auxiliary fuel consumption. Although sustained combustion (**autogenous combustion**) is possible with heating values as low as 9.3 MJ · kg^{-1}, in the hazardous waste incineration industry it is common practice to blend wastes (and fuel oil, if necessary) to obtain an overall heating value of 18.6 MJ · kg^{-1} or greater.

Blending is also used to limit the net chlorine content of chlorinated hazardous waste to a maximum of roughly 30% by weight to reduce chlorine concentrations in the combustion gas. The chlorine and, especially, hydrogen chloride that forms from the chlorine, are very corrosive. They oxidize the fire brick in the incinerator, which causes it to fail.

Hazardous waste incinerators must be designed to achieve a 99.99% destruction and removal efficiency (**DRE**) of the principal organic hazardous components (**POHCs**) in the waste. This is commonly referred to as "four 9s DRE"; higher DREs may be referred to as five 9s, six 9s, that is, 99.999 and 99.9999% DRE, respectively. Because of the complexity of the wastes being burned, little success has been achieved in predicting the time and temperature requirements for achieving the 99.99% DRE. Empirical tests (**trial burns**) are required to demonstrate compliance. Experience has demonstrated that highly halogenated materials are more difficult to destroy than those with low halogen content.

Incinerator Types. Two technologies dominate the incineration field: liquid injection and rotary kiln incinerators (Oppelt, 1981). Over 90% of all incineration facilities use one of these technologies. Of these, more than 90% are liquid injection units. Less commonly used incinerators include fluidized beds and starved air/pyrolysis systems.

Horizontal, vertical, and tangential liquid injection units are used. The majority of the incinerators for hazardous wastes inject liquid hazardous waste at 350–700 kPa through an atomizing nozzle into the combustion chamber. These liquid incinerators vary in size from 300,000 to 9 MJ of heat released per second. An auxiliary fuel such as natural gas or fuel oil is often used when the waste is not autogenous. The liquid wastes are atomized into fine droplets as they are injected. A droplet size in the range 40–100 μm is obtained with atomizers or nozzles. The droplet volatilizes in the hot gas stream and the gas is oxidized. Efficient destruction of liquid hazardous wastes requires minimizing unevaporated droplets and unreacted vapors.

Residence time, temperature, and turbulence (often referred to as the "three Ts") are optimized to increase destruction efficiencies. Typical residence times are 0.5–2 s. Incinerator temperatures usually range between 800 and 1600°C. A high degree of turbulence is desirable for achieving effective destruction of the organic chemicals in the waste. Depending on whether the liquid incinerator flow is axial, radial, or tangential, additional fuel burners and separate waste injection nozzles can be arranged to achieve the desired temperature, turbulence, and residence

FIGURE 14–10

Rotary kiln incinerator.

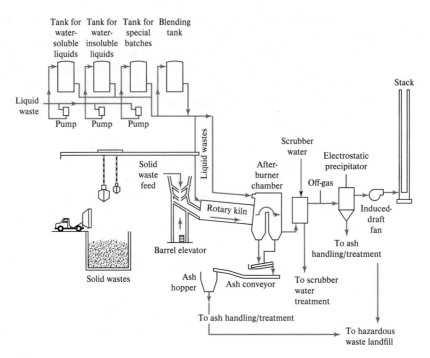

time. Vertical units are less likely to experience ash buildup. Tangential units have a much higher heat release and generally superior mixing.

The rotary kiln is often used in hazardous waste disposal systems because of its versatility in processing solid, liquid, and containerized wastes. Waste is incinerated in a refractory-lined rotary kiln, as shown in Figure 14–10. The shell is mounted at a slight incline from the horizontal plane to facilitate mixing the waste materials with circulating air. Solid wastes and drummed wastes are usually fed by a conveyor system or a ram. Liquids and pumpable sludges are injected through a nozzle. Noncombustible metal and other residues are discharged as ash at the end of the kiln.

Rotary kilns are typically 1.5–4 m in diameter and range in length from 3 to 10 m. Rotary kiln incinerators usually have a length-to-diameter ratio (L/D) of between 2 and 8. Rotational speeds range from 0.5 to 2.5 cm · s⁻¹, depending on kiln periphery. High L/D ratios, along with slower rotational speeds, are used for wastes requiring longer residence times. The feed end of the kiln has airtight seals to adequately control the initial incineration reactions.

Residence times for solid wastes are based on the rotational speed of the kiln and its angle. The residence time to volatilize waste is controlled by the gas velocity. The retention time of solids in the incinerator can be estimated from the following, where the coefficient 0.19 is based on limited experimental data:

$$t_o = \frac{0.19L}{NDS} \tag{14–7}$$

where t_o = retention time (in min)

L = kiln length (in m)

N = kiln rotational speed (in rev · min⁻¹)

D = kiln diameter (in m)

S = kiln slope (in m · m⁻¹)

Rotary kiln systems typically include secondary combustion chambers or afterburners to ensure complete destruction of the hazardous waste. Kiln operating temperatures range from 800 to 1600°C. Afterburner temperatures range from 1000 to 1600°C. Liquid wastes are often injected into the secondary combustion chamber. The volatilized and combusted wastes leave the kiln and enter the secondary chamber, where additional oxygen is provided. High heating value liquid waste or fuel may be introduced into the secondary chamber. Both the secondary combustion chamber and the kiln are usually equipped with an auxiliary fuel firing system for startup.

Cement kilns are very efficient at destroying hazardous waste. Their long residence times and high operating temperatures exceed the requirements for destruction of most wastes. Hydrochloric acid generated from chlorinated hydrocarbon wastes is neutralized by the lime in the kiln while slightly lowering the alkalinity of the cement products. Although cement plants can save energy by incinerating liquid wastes, the expense of obtaining permits and public resistance have inhibited use of this process.

Air Pollution Control (APC). Typical APC equipment on an incinerator will include an afterburner, liquid scrubber, demister, and fine particulate control device. Afterburners are used to control emission of unburned organic by-products by providing additional combustion volume at an elevated temperature. Scrubbers are used to physically remove particulate matter, acid gases, and residual organic compounds from the combustion gas stream. Metals, of course, are not destroyed in the incineration process. Some are volatilized and then collected in the air pollution control device. The large liquid droplets that escape from the scrubber are captured in a mist collector. The final stage in gas cleaning is to remove the fine particles that remain. Electrostatic precipitators have been used for this step. Scrubber water and residues from other APC devices are still considered hazardous and must be treated before ultimate land disposal.

Permitting for Hazardous Waste Incinerators. The permitting for hazardous waste incinerators is a complex, multifaceted program conducted simultaneously on federal, state, and local levels. Because of the variety of state and local regulations for the handling, transportation, treatment, and disposal of hazardous wastes, as well as those concerning the operation of incinerators, each startup facility has a unique set of permit requirements.

Generally speaking, hazardous waste incinerators require permits under the following laws: federal RCRA, state RCRA, for PCBs the Toxic Substances and Control Act (TSCA), state and federal wastewater discharge, and state and federal air pollution control. A variety of local permits may also be necessary. Each of these require data substantiating an incinerator's operation at or above performance levels determined by environmental legislation. Each requires a public hearing and discussion of environmental effects as well.

Hazardous waste incinerators must meet three performance standards (Theodore and Reynolds, 1987):

1. *Principal organic hazardous constituents* (POHC). The DRE for a given POHC is defined as the mass percentage of the POHC removed from the waste. The POHC performance standard requires that the DRE for each POHC *designated* in the permit be 99.99% or higher. The DRE performance standard implicitly requires sampling and analysis to measure the amounts of the designated POHC(s) in both the waste stream and the stack effluent gas during a trial burn. (The term *designated POHC* is described in more detail later in this section.) The DRE is determined for each designated POHC from a mass balance of the waste introduced into the incinerator and in the stack gas*:

$$\text{DRE} = \frac{(W_{in} - W_{out})}{W_{in}} \times 100\% \tag{14–8}$$

---

*Note that this is not a mass balance around the incinerator. Hazardous waste that ends up in the scrubber water, APC residue, and ash are not counted. Hence, the oxidation can be very poor and the incinerator can still meet the 99.99% rule if the scrubber is efficient or the waste ends up in the ash. This is one reason that residues are considered hazardous and must be treated before land disposal.

where W_{in} = mass feed rate of one POHC in the waste stream

W_{out} = mass emission rate of the same POHC present in exhaust emissions prior to release to the atmosphere

2. *Hydrochloric acid.* An incinerator burning hazardous waste and producing stack emissions of more than 1.8 kg · h^{-1} of hydrogen chloride (HCl) must control HCl emissions such that the rate of emission is no greater than the larger of either 1.8 kg · h^{-1} or 1% of the HCl in the stack gas prior to entering any pollution control equipment.

3. *Particulates.* Stack emissions of particulate matter are limited to 180 mg · dscm^{-1} (milligrams per dry standard cubic meter) for the stack gas corrected to 7% oxygen.

This adjustment is made by calculating a corrected concentration:

$$P_c = P_m \frac{14}{(21 - Y)} \tag{14–9}$$

where P_c = corrected concentration of particulate (in mg · dscm^{-1})

P_m = measured concentration of particulate (in mg · dscm^{-1})

Y = percent oxygen in the dry flue gas

In this way, a decrease in the particulate concentration due solely to increasing air flow in the stack is not rewarded, and an increase in the particulate concentration due solely to reduction in the air flow in the stack is not penalized.

Compliance with these performance standards is documented by a **trial burn** of the facility's waste streams. As part of the RCRA permit application, a trial burn plan detailing waste analysis, an engineering description of the incinerator, sampling and monitoring procedures, test schedule and protocol, as well as control information, must be developed. If EPA determines that the design is adequate, a temporary or draft permit is issued, allowing the owner or operator to build the incinerator and initiate the trial burn procedure.

The temporary permit covers four phases of operation. During the first phase, immediately following construction, the unit is operated for **shake-down** purposes to identify possible mechanical deficiencies and to ensure its readiness for the trial burn procedures. This phase of the permit is limited to 720 h of operation using hazardous waste feed. The trial burn is conducted during the second phase.

To verify compliance with the POHC performance standard during the trial burn, it is not required that the incinerator DRE be measured for every POHC identified in the waste. The POHCs with the greatest potential for a low DRE, based on the expected difficulty of thermal degradation (incinerability) and the concentration of the POHC in the waste, become the *designated POHCs* for the trial burn. The EPA permit review personnel work with the owners or operators of the incinerator facility in determining which POHCs in a given waste should be designated for sampling and analysis during the trial burn.

If a wide variety of wastes are to be treated, a difficult-to-incinerate POHC at high concentration may be proposed for the trial burn. The substitute POHC is referred to as a **surrogate POHC.** The surrogate POHC does not have to be actually present in the normal waste. It does, however, have to be considered more difficult to incinerate than any POHC found in the waste.

The third phase consists of completing the trial burn and submitting the results. This phase can last several weeks to several months, during which the incinerator is allowed to operate under specified conditions.

Provided that performance standards are met in the trial burn, the facility can begin its fourth and final phase, which continues through the duration of the permit. In the event that the trial burn results do not demonstrate compliance with standards, the temporary permit must be modified to allow for a second trial burn.

EXAMPLE 14–4   A test burn waste mixture consisting of three designated POHCs (chlorobenzene, toluene, and xylene) is incinerated at 1000°C. The waste feed rate and the stack discharge are shown in the following table. The stack gas flow rate is 375.24 dscm · min^{-1} (dry standard cubic meters per minute). Is the unit in compliance?

Compound	Inlet (kg · h^{-1})	Outlet (kg · h^{-1})	Compound	Inlet (kg · h^{-1})	Outlet (kg · h^{-1})
Chlorobenzene (C_6H_5Cl)	153	0.010	HCl		1.2
Toluene (C_7H_8)	432	0.037	Particulates at 7% O_2		3.615
Xylene (C_8H_{10})	435	0.070			

Outlet concentrations were measured in the stack after APC equipment.

Solution   We begin by calculating the DRE for each of the POHCs.

$$DRE = \frac{W_{in} - W_{out}}{W_{in}} \times 100$$

$$DRE_{chlorobenzene} = \frac{153 - 0.010}{153} \times 100 = 99.993\%$$

$$DRE_{toluene} = \frac{432 - 0.037}{432} \times 100 = 99.991\%$$

$$DRE_{xylene} = \frac{435 - 0.070}{435} \times 100 = 99.984\%$$

The DRE for each designated POHC must be at least 99.99%. In this case, the designated POHC xylene fails to meet the standard. The other POHCs exhibit a DRE of greater than 99.99%.

Now we check compliance for the HCl emission. The HCl emission may not exceed 1.8 kg · h^{-1} or 1% of the HCl prior to the control equipment, whichever is greater. It is obvious that the 1.2 kg · h^{-1} emission meets the 1.8 kg · h^{-1} limit. This would be sufficient to demonstrate compliance, but we will calculate the mass emission rate prior to control for the purpose of comparison. To do this, we assume all the chlorine in the feed is converted to HCl. The molar feed rate of chlorobenzene (M_{CB}) is

$$M_{CB} = \frac{W_{CB}}{(MW)_{CB}} = \frac{(153 \text{ kg} \cdot \text{h}^{-1})(1000 \text{ g} \cdot \text{kg}^{-1})}{112.5 \text{ g} \cdot \text{mol}^{-1}} = 1360 \text{ mol} \cdot \text{h}^{-1}$$

where M_{CB} = molar flow rate of chlorobenzene
$(MW)_{CB}$ = molecular weight of chlorobenzene

Each molecule of chlorobenzene contains one atom of chlorine. Therefore,

$$M_{HCl} = M_{CB}$$

$$= 1360 \text{ mol} \cdot \text{h}^{-1}$$

$$W_{HCl} = (\text{GMW of HCl})(\text{in mol} \cdot \text{h}^{-1})$$

$$= (36.5 \text{ g} \cdot \text{mol}^{-1})(1360 \text{ mol} \cdot \text{h}^{-1})$$

$$= 49,640 \text{ g} \cdot \text{h}^{-1}, \text{ or } 49.64 \text{ kg} \cdot \text{h}^{-1}$$

This is the HCl emission prior to control. The emission of $1.2 \text{ kg} \cdot \text{h}^{-1}$ is greater than 1% of the uncontrolled emission, that is,

$$1\% \text{ of uncontrolled} = (0.01)(49.64)$$

$$= 0.4964 \text{ kg} \cdot \text{h}^{-1}$$

However, the incinerator passes the HCl limits because the HCl emission is less than $1.8 \text{ kg} \cdot \text{h}^{-1}$.

The particulate concentration was measured at 7% O_2 and, therefore, does not need to be corrected. The outlet loading (W_{out}) of the particulates is

$$W_{out} = \frac{(3.615 \text{ kg} \cdot \text{h}^{-1})(10^6 \text{ mg} \cdot \text{kg}^{-1})}{(375.24 \text{ dscm} \cdot \text{min}^{-1})(60 \text{ min} \cdot \text{h}^{-1})} = 160 \text{ mg} \cdot \text{dscm}^{-1}$$

This is less than the standard of $180 \text{ mg} \cdot \text{dscm}^{-1}$ and is, therefore, in compliance with regard to particulates. However, because the incinerator fails the DRE for xylene, the unit is out of compliance.

Regulations for PCBs. Incineration of PCBs is regulated under TSCA rather than RCRA. Thus, some of the permit conditions for incineration of PCBs are different from those for other RCRA hazardous wastes.

The conditions for incineration of liquid PCBs may be summarized as follows:

1. *Time and temperature.* Either of two conditions must be met. The residence time of the PCBs in the furnace must be 2 s at 1200°C ± 100°C with 3% excess oxygen in the stack gas or, alternatively, the furnace residence time must be 1.5 s at 1600°C ± 100°C with 2% excess oxygen in the stack gas.

   The EPA has interpreted these conditions to require a liquid PCB DRE ≥ 99.9999%.

2. *Combustion efficiency.* The combustion efficiency shall be at least 99.99%, computed as follows:

$$\text{Combustion efficiency} = \frac{C_{CO_2}}{C_{CO_2} + C_{CO}} \times 100\% \tag{14–10}$$

   where C_{CO_2} = concentration of carbon dioxide in stack gas
   C_{CO} = concentration of carbon monoxide in stack gas

3. *Monitoring and controls.* In addition to these permitted limits, owners or operators of incinerators are required to monitor and control the variables that affect performance. The rate and quantity of PCBs fed to the combustion system must be measured and recorded at regular intervals of no longer than 15 min. The temperatures of the incineration process must be continuously measured and recorded. The flow of PCBs to the incinerator must stop automatically whenever one of the following occurs: the combustion temperature drops below the temperatures specified, that is, 1200 or 1600°C; when there is a failure of monitoring operations; when the PCB rate and quantity measuring and recording equipment fails; or when excess oxygen falls below the percentage specified. Scrubbers must be used for HCl removal during PCB incineration.

   In addition, a trial burn must be conducted and the following exhaust emissions must be monitored:

Oxygen (O_2)	Total chlorinated organic content
Carbon monoxide (CO)	PCBs
Oxides of nitrogen (NO_x)	Total particulate matter
Hydrogen chloride (HCl)	

An incinerator used for incinerating nonliquid PCBs, PCB articles, PCB equipment, or PCB containers must comply with the same rules as those for liquid PCBs, and the mass air emissions from the incinerator must be no greater than 0.001 g PCB per kilogram of the PCB introduced into the incinerator, that is, a DRE of 99.9999%.

Stabilization–Solidification

Because of their elemental composition, some wastes, such as nickel, cannot be destroyed or detoxified by physical or chemical means. Thus, once they have been separated from aqueous solution and concentrated in ash or sludge, the hazardous constituents must be bound up in stable compounds that meet the LDR restrictions for leachability.

The terminology for this treatment technology has evolved in the last decade. In the early to mid 1980s "chemical fixation, encapsulation," and "binding" were often used interchangeably with solidification and stabilization. With the promulgation of the LDR restrictions, the EPA established a more precise definition for solidification–stabilization and discouraged the use of the other terms to describe the technology (U.S. EPA, 1988b). EPA linked solidification and stabilization because the resultant material from the treatment must be both stable and solid. **Stability** is determined by the degree of resistance of the mixture of the hazardous waste and additive chemical to leaching in the **Toxicity Characteristic Leaching Procedure** (TCLP) (55 FR 22530, 1990). In the EPA definition, then, solidification–stabilization refers to chemical treatment processes that chemically reduce the mobility of the hazardous constituent.

Reduced leachability is accomplished by the formation of a lattice structure or chemical bonds that bind the hazardous constituent and thereby limit the amount of constituent that can be leached when water or a mild acid solution comes into contact with the waste matrix. There are two principal solidification–stabilization processes: cement-based and lime-based. The cement or lime additive is mixed with the ash or sludge and water. It is then allowed to cure to form a solid. The correct mix proportions are determined by trial-and-error experiments on waste samples. In both techniques the stabilizing agent may be modified by other additives such as silicates. In general, this technology is applicable to wastes containing metals with little or no organic contamination, oil, or grease.

14–7 LAND DISPOSAL

Deep Well Injection

Deep well injection consists of pumping wastes into geologically secure formations. The general technical requirements of a suitable hazardous waste injection well include (Warner, 1998)

1. A saline-water-bearing formation that is large enough and permeable enough to accept the waste.
2. Overlying and underlying strata (confining layers) that are sufficiently impermeable to confine the waste to the injection strata.
3. The absence of solution-collapse features, faults, joints, and abandoned wells that might permit the escape of the waste.

Obviously, these requirements imply a rigorous understanding of the geologic and hydrologic characteristics of the subsurface. In addition, it is often necessary to pretreat the waste before injection. The purpose of the pretreatment is to modify the waste characteristics so that it is compatible with the injection equipment and the geologic strata. Some waste characteristics of importance are the presence of microorganisms, suspended solids, oils, and entrained or dissolved gases.

Pumping of wastes into these formations has been practiced primarily in Louisiana and Texas. In promulgating the final third of the LDR restrictions, the EPA allowed disposal of waste in class I injection wells for wastes disposed under Clean Water Act regulations (55 FR 22530, 1990).

Land Treatment

Land treatment is sometimes called **land farming** of the waste. In this practice, waste was incorporated with soil material in the manner that fertilizer or manure might be. Microorganisms in the soil degraded the organic fraction of the waste. Under the LDR restrictions, this practice is prohibited.

The Secure Landfill

Although far from ideal, the use of land for the disposal of hazardous wastes is a major option for the foreseeable future. Furthermore, we recognize that incinerator ash, scrubber bottoms, and the results of biological, chemical, and physical treatment leave residues of up to 20% of the original mass. These residues must be secured in an economical fashion. At this juncture, the secure landfill is the only option.

The basic physical problem with land disposal of hazardous waste stems from the movement of water, originating as precipitation into and through the disposal site. The dissolution of waste material results in contaminants being transported from the waste site to larger regions of the soil zone and, too often, to an underlying aquifer. Problems of groundwater pollution frequently lead to the condemnation of wells and to the contamination of surface water bodies fed by the associated aquifer. In many instances, well contamination is not detected until years after land disposal of waste has begun, owing primarily to the slow movement of the conveying groundwater. For some chemical species, adsorption to soil particles may also act to retard the movement of the contaminant plume.

Water pollution, caused by a hazardous waste facility, may evolve in a variety of ways. Leachate from landfills may drain out of the side of the landfill and appear as surface runoff. Alternatively, it may seep down slowly through the unsaturated zone, if one exists, and enter an underlying aquifer. Breaks in liners of holding ponds or cracks in the bottom of storage tanks also lead to a downward migration of contaminants toward the water table. Sometimes this migration is impeded by a relatively impermeable geologic barrier, such as a clay layer.

Without the institution of remedial measures, buried waste usually acts as a continuing source of pollution. The waste constituents continue to be transported in the subsurface by infiltrating precipitation. Thus, it is generally recommended that sites that handle hazardous wastes be located above a natural barrier, as well as an applied liner. Moreover, the site should be instrumented to continuously monitor the condition of any associated aquifers. If leachate generation is anticipated, there should be a system for the collection and treatment of the leachate.

The technology of the secure landfill may be divided into two phases: siting and construction. The following discussion is drawn primarily from E. F. Wood and colleagues (Wood et al., 1984).

Landfill Siting.   In siting a hazardous waste landfill, the four main considerations are air quality, groundwater quality, surface water quality, and subsurface migration of gases and leachates. Aside from the sociopolitical aspects, the last three components are the major factors to be considered in siting the landfill.

Air quality must be considered to prevent adverse effects to the air caused by volatilization, gas generation, gas migration, and wind dispersal of landfilled hazardous wastes. Generally, these can be controlled by proper construction techniques and do not inhibit the siting.

The hydrogeologic siting problem can be divided into four main areas: hydrology, climate, geology, and soil.

Hydrologic considerations in locating a hazardous waste landfill include distance to the groundwater table, the hydraulic gradient, the proximity of wells, and the proximity to surface waters. The proximity of a site to water supplies and the type of natural materials that occur between a site and a water supply influence contaminant migrations. When the distance from the surface to the groundwater table is short, contaminant travel time is also short, allowing for little attenuation before pollutants disperse laterally in the saturated zone. It is desirable to have the average distance to the groundwater table large enough so that contaminants may be significantly attenuated. This also facilitates monitoring of the saturated zone and will permit detection of leaks and implementation remedial action to be undertaken in a timely fashion.

A hydraulic gradient that slopes away from local groundwater supplies is desired. The steeper the hydraulic gradient, the lower the attenuation time and the faster the water movement. Therefore, a moderate hydraulic gradient may be most acceptable.

The distance from the disposal site to water supply wells and surface waters must be as large as possible to protect them from potential contamination in case the landfill leaks. Furthermore, the proximity to surface waters must take into account the potential for flooding. Site flooding will weaken the structure of a land emplacement facility, causing it to fail and leak wastes. Therefore, it is essential that the facility not be built on a floodplain or area subject to local flooding.

Climate is considered a driving force in contaminant migration. It may be excluded when considering potential sites within the same region, where climate is unlikely to vary significantly.

Structural integrity of host rock is important in terms of seismic risk zones, dipping, and cleavage. Seismic risk zones are areas where earthquakes have occurred or have a high probability of occurring. They indicate the presence of geologic faults and fractures. The presence of faults and fractures is extremely important because they provide a natural pathway for the flow of contaminants, even in low hydraulic conductivity and low-porosity rock. Future seismic activity could damage the landfill cells and storage tanks of a land disposal facility during or after the construction, filling, and closing of the site, unless it is structurally designed to withstand ground motion.

Transport capacity refers to a soil's ability to allow migration of contaminants. Thus, the greater the soil transport capacity, the greater the migration of contaminants, which is undesirable. A soil with a low hydraulic conductivity can lengthen the flow period and act as a natural defense by retarding the movement of contaminants. Glacial outwash plains and deltaic sands are both well-sorted sand and gravel beds with high hydraulic conductivity. Thus, they allow wastes to move faster and further. Clays and silts have lower hydraulic conductivities and, thus, inhibit the movement of wastes.

Sorption capacity depends on the organic content, predominant minerals, pH, and soil. Sorption includes both absorption and adsorption of contaminants. Sorption is important in limiting the movement of metals, phosphorus, and organic chemicals. **Cation-exchange capacity** (CEC) is a measure of the ability of the soil to trade cations in the soil for those in waste. The higher the CEC, the more cations will be retained. The capacity of soil to retard contaminant migration also depends on the presence of numerous hydrous oxides, particularly iron oxides, and other compounds such as phosphates and carbonates. These compounds precipitate heavy metals from solution.

The hydrogen ion concentration (pH) of soil influences the dominant removal mechanism for metal cations. The dominant removal mechanism for metal cations when pH <5 is exchange or adsorption; when pH >6, it is precipitation.

Landfill Construction. A secure landfill means, in essence, that no leachate or other contaminant can escape from the fill and adversely affect the surface water or groundwater (Josephson, 1981). Leakage from the site is not acceptable during or after operations. Neither is any external or internal displacement, which could be brought about by slumping, sliding, and flooding. Wastes must not be allowed to migrate from the site.

It is nearly impossible to create an impervious burial vault for hazardous wastes and guarantee its integrity forever. Landfill design and operation is regulated to minimize migration of wastes from the site. The current EPA rules (40 CFR 264.300) for hazardous waste landfills require a minimum of (1) two or more liners, (2) a leachate collection system above and between

FIGURE 14–11

Minimum technology landfill liner design and recommended final cover design. (*Source:* U.S. Environmental Protection Agency, 1989.)

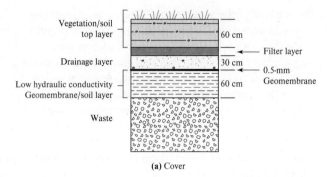

(a) Cover

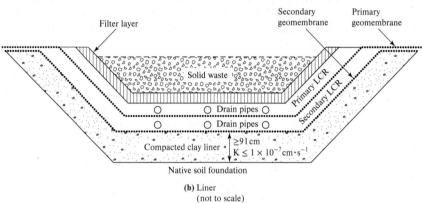

(b) Liner
(not to scale)

the liners, (3) surface run-on and run-off control to collect and control at least the water volume resulting from a 24-h, 25-year storm, (4) monitoring wells, and (5) a "cap" (Figure 14–11).

The liner system must include (57 FR 3462, 1992)

1. A top liner designed and constructed of materials (e.g., a geomembrane) to prevent migration of hazardous constituents into the liner during the active life and postclosure care period.
2. A composite bottom liner consisting of at least two components. The upper component must be designed and constructed of materials (e.g., a geomembrane) to prevent migration of hazardous constituents into the liner during the active life and postclosure care period. The lower component must be designed and constructed of materials to minimize migration of hazardous constituents if a breach in the upper component were to occur. The lower component must be constructed of at least 91 cm of compacted soil material with a hydraulic conductivity of no more than 1×10^{-7} cm · s^{-1}.

The leachate collection and removal system (LCR) immediately above the top liner must be designed, constructed, operated, and maintained to collect and remove leachate so that the leachate depth over the liner does not exceed 30 cm. The LCR between the liners and immediately above the bottom liner is also a leak detection system. The LCR must, at a minimum, be

1. Constructed with a bottom slope of 1% or more.
2. Constructed of a granular drainage material with a hydraulic conductivity of 1×10^{-2} cm · s^{-1} or more and a thickness of 30 cm or more; or be constructed of synthetic or geonet drainage materials with a transmissivity of 3×10^{-5} m^2 · s^{-1} or more.
3. Constructed of sufficient strength to prevent collapse and be designed to prevent clogging.

FIGURE 14–12

Definition of hydraulic
gradient for landfill liner.

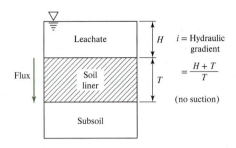

The leachate collection system must include pumps of sufficient size to remove the liquids to prevent leachate from backing up into the drainage layer. The leachate must be treated to meet discharge limits. The treated leachate may be discharged into the municipal wastewater treatment system or into a waterway.

The amount of leachate may be estimated using Darcy's law (see Chapter 7). The hydraulic gradient for a liner is defined as shown in Figure 14–12. The flow rate cannot exceed the amount of water available, that is, the product of the precipitation rate and the area of the landfill. The travel time of a contaminant through a soil layer may be estimated as the linear length of the flow path (T) divided by the seepage velocity (Equation 7–14).

EXAMPLE 14–5   How long will it take for leachate to migrate through a 0.9-m clay liner with a hydraulic conductivity of 1×10^{-7} cm · s^{-1} if the depth of leachate above the clay layer is 30 cm and the porosity of the clay is 55%?

Solution   The Darcy velocity is found using Equation 7–12.

$$v = K\left(\frac{dh}{dr}\right)$$

where the hydraulic gradient (dh/dr) is defined as in Figure 14–12.

$$\frac{dh}{dr} = \frac{0.30 \text{ m} + 0.9 \text{ m}}{0.9 \text{ m}} = 1.33$$

The Darcy velocity is then

$$v = (1 \times 10^{-7} \text{ cm} \cdot \text{s}^{-1})(1.33) = 1.33 \times 10^{-7} \text{ cm} \cdot \text{s}^{-1}$$

From Equation 7–14, the seepage velocity is

$$v' = \frac{K\,(dh/dr)}{\eta} = \frac{1.33 \times 10^{-7} \text{ cm} \cdot \text{s}^{-1}}{0.55} = 2.42 \times 10^{-7} \text{ cm} \cdot \text{s}^{-1}$$

The travel time is then

$$t = \frac{T}{v'} = \frac{(0.9 \text{ m})(100 \text{ cm} \cdot \text{m}^{-1})}{2.42 \times 10^{-7} \text{ cm} \cdot \text{s}^{-1}} = 3.71 \times 10^{8} \text{ s, or about 12 years}$$

The site operator must keep careful records of the location and dimensions of each cell and must depict each cell on a map keyed to permanently surveyed vertical and horizontal markers. Records must show the contents of each cell and the approximate location of each hazardous waste type within the cell.

The purpose of groundwater monitoring is to ensure that programs for managing runon, runoff, and leachates are functioning properly so that groundwater remains uncontaminated. If contamination is occurring, early warning can be given and countermeasures taken. The site owner or operator has to place a sufficient number of monitoring wells around the limits of the facility to be able to describe the background (upgradient) and downgradient water quality. The regulations set forth, in detail, how the monitoring wells must be sunk, screened, sealed, sampled, and located, with special emphasis on location of the downgradient wells.

General groundwater quality, especially the suitability of the uppermost aquifer for use as a drinking-water source, must meet EPA's primary drinking-water standards. The flow rate for each sump must be calculated weekly during the active life and closure period, and monthly during the postclosure care period. If the landfill is leaking to the groundwater, the site operator must file an assessment plan with the EPA that shows how the problem is to be remedied.

14–8   GROUNDWATER CONTAMINATION AND REMEDIATION

The Process of Contamination

Hazardous waste landfills are, of course, not the only source of groundwater contamination. Other sources include municipal landfills, septic tanks, mining and agricultural activities, "midnight dumping," and leaking underground storage tanks. It has been estimated that as many as 5000 gasoline storage tanks at local service stations are leaking.

The threat of contamination to groundwater depends on the specific geologic and hydrologic conditions of the site. The process of contamination and the process of transport of contaminants was discussed in Chapter 9.

EPA's Groundwater Remediation Procedure

The federal program for cleanup of contaminated sites follows a procedural sequence as shown in Figure 14–13. Each of these steps is discussed in the following paragraphs.

Preliminary Assessment.   EPA involvement usually begins with the identification of a potential hazardous waste site. The initial information can come from a variety of sources, including

FIGURE 14–13

Steps involved in the Superfund cleanup process. (*Source:* U.S. Environmental Protection Agency, 2005.)

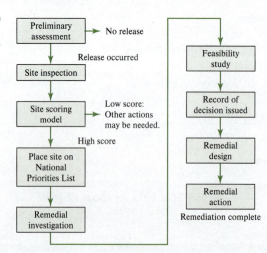

local citizens and officials, state environmental agencies, the site owners themselves, or simply from awareness of potential problems associated with particular industries.

EPA has developed an inventory system called the comprehensive environmental response, compensation, and liability information system (CERCLIS) to document all of the sites in the United States that may be candidates for remedial action. This is a continuing program that identifies sites as information about them becomes available. The growth in the number of CERCLIS sites has been dramatic and is expected to continue for the foreseeable future as additional abandoned and contaminated sites are discovered. As of September, 2005, 1239 sites were on the NPL and 12,031 sites were in the CERCLIS inventory.

A **preliminary assessment** (PA) is the first step in identifying the potential for contamination from a particular site. The primary objectives of the PA are to determine if a contaminant has been released to the environment, if there is immediate danger to persons living or working near the site, and whether a site inspection is necessary. Samples for environmental analysis are generally not taken during the PA. Following the preliminary assessment, EPA or the designated state agency might determine that an immediate threat to residents or employees at the site requires an immediate removal action. Otherwise, on the basis of the preliminary assessment, the site is classified by EPA into one of the three following categories:

1. No further action is needed, because there is no threat to human health or the environment.
2. Additional information is required to complete the preliminary assessment.
3. Inspection of the site is necessary.

Site Inspection.   Site inspection requires sampling to determine the types of hazardous substances present and to identify the extent of contamination and its migration. The actual site inspection includes preparation of a work plan and an on-site safety plan. The site assessment has three objectives:

1. To determine which releases pose no threat to public health and the environment.
2. To determine if there is any immediate threat to persons living or working near the release.
3. To collect data to determine whether or not a site should be included on the NPL.

HRS, NPL, RI/FS, and ROD.   The next series of steps in the EPA's procedure include calculations to complete the HRS; inclusion on the NPL if the score is sufficiently high; conduct of a RI/FS; and issuance of an ROD. These steps were discussed in detail in Section 14–4.

Remedial Design and Remedial Action.   EPA-funded remedial actions may be taken only at those sites that are on the NPL. This ranking helps ensure that the Superfund dollars are used in the most cost-effective manner and where they will yield the greatest benefit.

Before a remedial action can be taken at a site, a number of questions must be answered. These can be classified as problem definition, design alternatives, and policy.

1. *Problem definition questions.* What are the contaminants and how much contamination is present? How large is the surface area of the contaminated site? What is the size of the contaminated groundwater plume? Where is the exact location of the plume and in what direction is it moving?
2. *Design questions.* Based on the alternatives available, what is the best way to clean up the site? How should these alternatives be implemented? What products will be produced during treatment? How long will it take to complete the remediation and what will it cost?
3. *Policy questions.* What level of protection is adequate? In other words, how clean is clean?

The answers to the first two sets of questions require scientific and engineering background that is supported by extensive sampling of the contaminated site area. The last question cannot be answered objectively; rather it is a subjective and often political question.

The NCP defines three types of responses for incidents involving hazardous substances. In these responses removal is differentiated from remediation. **Removal** is, as its name suggests, the physical relocation of the waste—usually to a secure hazardous waste landfill. **Remediation** means that the waste is to be treated to make it less toxic or less mobile or the site is to be contained to minimize further release. Remediation can take place on site or at a TSD facility. The three types of responses are

1. *Immediate removal* is a prompt response to prevent immediate and significant harm to human health or the environment. Immediate removals must be completed within 6 months.
2. *Planned removal* is an expedited removal when some response, not necessarily an emergency response, is required. The same 6-month limitation also applies to planned removal.
3. *Remedial response* is intended to achieve a site solution that is a permanent remedy for the particular problem involved.

Immediate removals are done to prevent an emergency involving hazardous substances. These emergencies might include fires; explosions; direct human contact with a hazardous substance; human, animal, or food-chain exposure; or contamination of drinking-water sources. An immediate removal involves cleaning up the hazardous site to protect human health and life, containing the hazardous release, and minimizing the potential for damage to the environment. For example, a truck, train, or barge spill could involve an immediate removal determination by EPA to get the spill cleaned up.

Immediate removal responses may include activities such as sample collection and analysis, containment or control of the release, removal of the hazardous substances from the site, provision of alternative water supplies, installation of security fences, evacuation of threatened citizens, or general deterrent of the spread of the hazardous contaminants.

A planned removal involves a hazardous site that does not present an immediate emergency. Under Superfund, EPA may initiate a planned removal if the action will minimize the damage or risk and is consistent with a more effective long-term solution to the problem. Planned removals are carried out by EPA if the responsible party is either unknown or cannot or will not take timely and appropriate action. The state in which the cleanup is located must be willing to match at least 10% of the costs of the removal action as well as agree to nominate the site in question for the NPL.

Mitigation and Treatment

Because the spread of contaminants is usually confined to a plume, only localized areas of an aquifer need to be reclaimed and restored. Cleanup of a contaminated aquifer, however, is often troublesome, time-consuming, and costly. The original source of contamination can be eliminated, but the complete restoration of the groundwater is fraught with additional problems, such as defining the site's subsurface environment, locating potential contamination sources, defining potential contaminant transport pathways, determining contaminant extent and concentration, and choosing and implementing an effective remedial process (Griffin, 1988).

Although not a simple task, cleanup is possible. Various methods are being used and have proven successful in certain cases. These efforts have ranged from containment to destruction of the contaminants, either in their original position in the aquifer or by withdrawing the groundwater. Examples of these remedial methods include installing pumping wells to remove contaminated water, building trenches to arrest only the contaminated flow, and stimulating biodegradation of groundwater contaminants.

In many circumstances, the most reasonable and economic remedial approach is to treat the water to attain the necessary quality for a specific use. This treated water may then be used or returned to the aquifer. Certainly, combinations of barriers and treatment methods should be considered. Source control (removal or remediation of the source), physical control, and treatment methods all will have their part in mitigating groundwater contamination problems. Legal implications may also dictate possible strategies.

FIGURE 14–14

Effect of a recharge well on groundwater flow patterns.

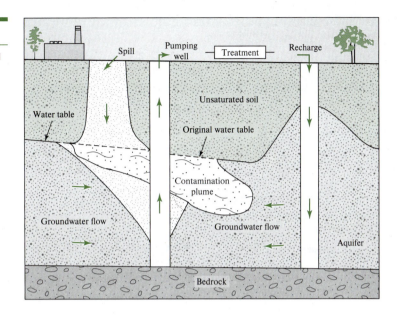

Well Systems.   A well system serves as an example of a common remedial process for treating contaminated groundwater. Well systems manipulate the subsurface hydraulic gradient through injection or withdrawal of water. They are designed to control the movement of the groundwater directly and of the subsurface pollutants indirectly. All require installation of wells at selected sites. It is necessary to first conduct a hydrogeologic study to determine the characteristics of the contaminant plume (width, length, depth, and general shape), the hydraulic gradient across the plume, and the hydrogeologic characteristics of the aquifer. There are three major types of well systems: well point, deep well, and pressure ridge systems.

Both well point and deep well systems withdraw water. The former uses closely spaced, shallow wells, each connected to a main pipe (header) that is connected to a centrally located pump. The well point system is used only for shallow water table aquifers. Deep well systems are used for greater depths and are most often pumped individually. Both well point and deep well systems should be designed so that the radius of influence of the system completely intercepts the contaminant plume.

The principle of pressure ridge systems is the inverse of that of a well system. Clean, uncontaminated water is injected to form an upconing of the water table that acts as a barrier to groundwater flow. The injection well is also called a recharge well. The upconing effect of a recharge well is shown in Figure 14–14.

Nonaqueous phase liquids (NAPLs) such as gasoline are referred to as "product" because their recovery may have some commercial value. When the NAPL floats on the groundwater table, special recovery techniques may be employed to recover it. Product recovery systems to recover NAPL use wells that terminate in the NAPL plume rather than in the aquifer. Because all hydrocarbons are slightly soluble in water, the product recovery system is usually accompanied by a groundwater pumping system to remove and treat the contaminated groundwater. A typical system is shown in Figure 14–15.

Treatment Technologies.   The treatment methods used to clean the contaminated groundwater are fundamentally the same as those discussed in Section 14–6. Each must be tailored to the local situation.

Pump-and-Treat.   One technology that is substantially different from those discussed in Section 14–6 is pump-and-treat. The objectives of a pump-and-treat system include hydraulic

FIGURE 14–15

Lockheed Aeronautical Systems Company's Aqua-Detox groundwater treatment system. (*Source:* Hazmat World, November 1989, EHS Publishing.)

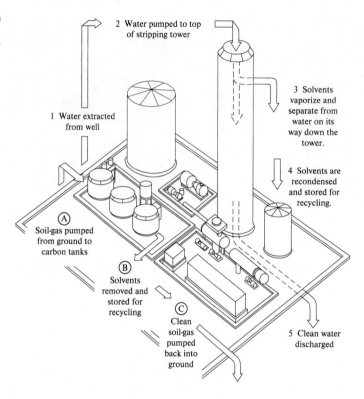

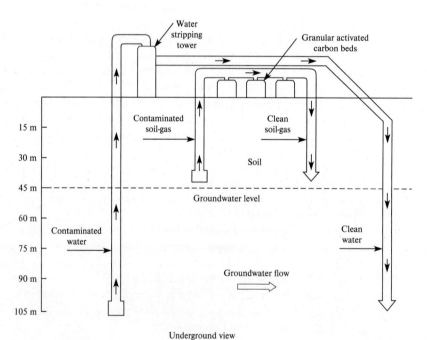

containment of the contaminated plume and removal of the contaminant from the groundwater. The design of the well system for a pump-and-treat remediation is an application of well hydraulics described in Chapter 7.

The **capture zone** of an extraction well is that portion of the groundwater that will discharge into the well. The capture zone is not necessarily coincident with the cone of depression

FIGURE 14–16

Groundwater flow lines influenced by pumping well.

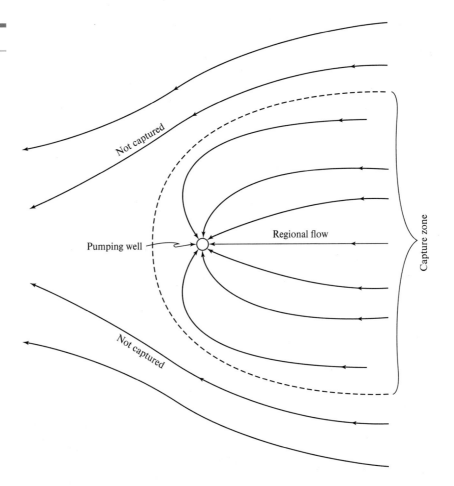

(Section 7–5) because the groundwater flow lines can be diverted by the influence of the pumping well without being captured by the well (Figure 14–16). Under steady-state conditions, the extent of the cone of depression largely depends on the transmissivity and pumping rate. The extent of the capture zone depends on the regional hydraulic gradient as well as the transmissivity and pumping rate.

Three parameters are used to delineate the capture zone: (1) the width of the capture zone at an infinite distance up-gradient from the pumping well, (2) the width of the capture zone at the location of the pumping well, and (3) the location of the down-gradient distance of the capture zone from the pumping well (called the **stagnation point**). These parameters are shown in Figure 14–17.

Javandel and Tsang (1986) developed a highly idealized model of the capture zone that can be used to examine the relationship between some of the important variables. The model assumes a homogeneous, isotropic aquifer uniform in cross section and infinite in width. The aquifer may be either confined or unconfined. However, in the case of the unconfined aquifer, drawdown must be insignificant with respect to the total thickness of the aquifer. The extraction wells are assumed to be fully penetrating.

With a single well located at the origin of the coordinate system shown in Figure 14–17, Javandel and Tsang (1986) developed the following equation to describe the y coordinate of the capture zone envelope:

$$y = \pm \frac{Q}{2\,Dv} - \frac{Q}{2\pi Dv}\tan^{-1}\frac{y}{x} \qquad \text{(14–11)}$$

FIGURE 14–17

Type curve for analytical solution to capture zone analysis for a single extraction well.

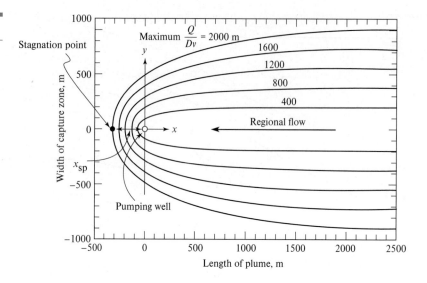

where x, y = distances from the origin (in m)

Q = well pumping rate (in $m^3 \cdot s^{-1}$)

D = aquifer thickness (in m)

v = Darcy velocity (in $m \cdot s^{-1}$)

Note that the "$\pm$" allows computation of the y coordinate above and below the x-axis. Masters (1998) has shown that this equation may be rewritten in terms of the angle ϕ (in radians) drawn from the origin to the x, y coordinate of interest on the line describing the capture zone curve.

That is

$$\tan \phi = \frac{y}{x} \tag{14–12}$$

so that, for $0 \le \phi \ge 2\pi$, Equation 14–12 may be rewritten as

$$y = \pm \frac{Q}{2\,Dv} - \left(1 - \frac{\phi}{\pi}\right) \tag{14–13}$$

This equation allows us to examine some important fundamental relationships:

- The width of the capture zone is directly proportional to the pumping rate.
- The width of the capture zone is inversely proportional to the Darcy velocity.
- As x approaches infinity, $\phi = 0$ and $y = Q/(2\,Dv)$. This sets the maximum total width of the capture zone at $2[Q/(2\,Dv)] = Q/(Dv)$ as shown in Figure 14–17.
- For $\phi = \pi/2$, $x = 0$ and y is equal to $Q/(4\,Dv)$. Thus, the width of the capture zone at $x = 0$ is $2[Q/(4\,Dv)] = Q/(2\,Dv)$.

The distance to the stagnation point down-gradient of the extraction well (x_{sp}) may be estimated from the following equation (LeGrega, Buckingham, and Evans, 2001):

$$X_{sp} = \frac{Q}{2\pi\,Dv} \tag{14–14}$$

Javandel and Tsang prepared a series of "type" curves for various well configurations (one, two, three, or four wells) and several widths of the capture zone at $x = \infty$. The suggested approach to using the capture zone technique is summarized as follows:

1. Prepare a site map with the plume shape at the same scale as the type curves.
2. Superimpose the site map on the one-well type curve with the direction of regional flow parallel to the x-axis. Place the leading edge of the plume just beyond the location of the extraction well. Select the capture zone curve that completely captures the plume. This defines the required value of Q/Dv at $x = \infty$.
3. Determine the required pumping rate by multiplying Q/Dv by Dv. If the required pumping rate can be achieved by the use of one well, then the problem is solved. If one well does not produce the required pumping rate, then go to step 4.
4. Repeat step 2 using the two-, three-, or four-well type curves as required to achieve an acceptable pumping rate. Each well in the multiple-well scenarios is assumed to pump at the same rate.

The capture zone of multiple extraction wells must overlap to prevent the groundwater flow from passing between them. If the distance between the extraction wells is less than or equal to $Q/\pi Dv$, the capture zones will overlap. Assuming that the wells are located symmetrically around the x-axis, the optimum spacing may be calculated using the following:

- For two wells space at $Q/\pi Dv$
- For three wells space at $1.26\ Q/\pi Dv$
- For four wells space at $1.2\ Q/\pi Dv$

The question of whether or not the required pumping rate can be achieved is determined, in a confined aquifer, by the available drawdown that will not lower the piezometric surface into the aquifer. This can be calculated using the methods discussed in Section 7–5. For an unconfined aquifer, the restriction noted above, that the drawdown must be insignificant with respect to the total thickness of the aquifer, requires a judgment decision.

EXAMPLE 14–6   The drinking water well at village of Oh Six is threatened by a contaminant plume in the aquifer. The confined aquifer is 28.7 m thick. It has a hydraulic conductivity of 1.5×10^{-4} m $\cdot$ s^{-1}, a storage coefficient of 3.7×10^{-5}, and a regional hydraulic gradient of 0.003. The contaminant plume is 300 m wide at its widest point. The maximum allowable pumping rate based on the allowable drawdown is 0.006 m^3 $\cdot$ s^{-1}. Locate a single extraction well so that the stagnation point is 100 m from the drinking-water well and so that the capture zone encompasses the plume. A sketch-map of the drinking-water well and the plume relationship is shown below.

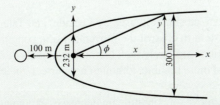

Solution   Determine the Darcy velocity using Equation 7–12:

$$v = K\frac{dh}{dr} = (1.5 \times 10^{-4}\ \text{m} \cdot \text{s}^{-1})(0.003) = 4.5 \times 10^{-7}\text{m} \cdot \text{s}^{-1}$$

The width of the capture zone at an infinite distance up-gradient is

$$\frac{Q}{Dv} = \frac{0.006 \ \text{m}^3 \cdot \text{s}^{-1}}{(28.7 \ \text{m})(4.5 \times 10^{-7} \ \text{m} \cdot \text{s}^{-1})} = 464.58 \ \text{or} \ 465 \ \text{m}$$

The width of the capture zone at the extraction well will be

$$\frac{Q}{2 \ Dv} = \frac{0.006 \ \text{m}^3 \cdot \text{s}^{-1}}{2(28.7 \ \text{m})(4.5 \times 10^{-7} \ \text{m} \cdot \text{s}^{-1})} = 232.29 \ \text{or} \ 232 \ \text{m}$$

The stagnation point will be

$$x_{sp} = \frac{0.006 \ \text{m}^3 \cdot \text{s}^{-1}}{2\pi(28.7 \ \text{m})(4.5 \times 10^{-7} \ \text{m} \cdot \text{s}^{-1})} = 73.94 \ \text{or} \ 74 \ \text{m}$$

down-gradient from the extraction well.

The distance down-gradient from the lead edge of the plume that the extraction well must be placed is determined using Equation 14–13. With $y = 150 \ \text{m}$

$$150 \ \text{m} = (232.29 \ \text{m}) \left(1 - \frac{\phi}{\pi}\right)$$

Solving for the angle (in radians) from the extraction well to the point where the plume just touches the capture zone is

$$\phi = 0.35 \ \pi \ \text{rad}$$

Using the geometry shown in the sketch-map, the distance is therefore

$$x = \frac{y}{\tan \phi} = \frac{150 \ \text{m}}{\tan (0.35 \ \pi)} = \frac{150 \ \text{m}}{1.96} = 76.4 \ \text{or} \ 76 \ \text{m}$$

This solution is highly idealized. The lead edge of the plume is conveniently of a geometry that allows us to locate it using Equation 14–13. A more ellipsoid plume geometry would project the lead edge in advance of the tangent point. This solution technique would then lead to a very erroneous positioning of the extraction well.

While the highly idealized situation used by Javandel and Tsang is useful in understanding the behavior of a well system to control the movement of a contaminant plume, in actual field sites the boundary conditions are rarely met. Computer models that have been calibrated to the local conditions will yield more reliable, although not perfect, understanding of the behavior of a proposed pump-and-treat system.

Because the rate of removal of contaminant decreases exponentially over time, and because the concentrations in the water rebound over time due to diffusion and desorption from the soil, pump-and-treat systems are limited in their value as a mass removal technology.

CHAPTER REVIEW

When you have finished studying this chapter, you should be able to do the following without the aid of your textbook or notes:

1. Sketch the chemical structure of 2,3,7,8-TCDD.
2. Explain how 2,3,7,8-TCDD occurs or when it is found in nature.
3. Sketch the chemical structure of the PCB 2,4′-dichlorobiphenyl.
4. Explain the origin of PCBs.
5. Define hazardous waste.
6. List the five ways a waste can be found to be hazardous and briefly explain each.
7. Explain why dioxin and PCB are hazardous wastes.
8. State how long generators may store their waste.
9. Define the abbreviations CFR, FR, RCRA, HSWA, CERCLA, and SARA.
10. Explain the major difference (objective) between RCRA/HSWA and CERCLA/SARA.
11. Define the terms *cradle to grave* and *manifest system*.
12. Explain what *land ban,* or LDR, means.
13. Define the abbreviations TSD and UST.
14. List the four major provisions of CERCLA.
15. Define the following abbreviations: NPL, NCP, HRS, RI/FS, ROD, and PRP.
16. Explain why it is important for a site to be placed on the NPL.
17. Explain the concept of "joint and several liability" and the implications to those with wastes found in an abandoned hazardous waste site.
18. List and explain four hazardous waste management techniques.
19. List the objectives of a waste audit.
20. Differentiate among waste minimization, waste exchange, and recycling.
21. List six disposal technologies for hazardous wastes.
22. Explain why seismic risk is important in landfill siting.
23. Explain how hydraulic conductivity and sorption capacity of soil limit the migration of hazardous wastes.
24. Explain what hydrologic features are important in siting a landfill.
25. List the minimum EPA requirements for a hazardous waste landfill and sketch a landfill that meets these.
26. Explain the difference between deep well injection and land treatment.
27. Define the following acronyms: PIC, POHC, and DRE as they apply to incineration.
28. List the most important factors for proper incinerator design and operation.
29. List the two types of incinerators most commonly used for destroying hazardous waste.
30. Explain the terms *designated POHC* and *surrogate* as they apply to a trial burn.
31. Outline the steps in EPA's remediation procedures.
32. Differentiate between "remediation" and "removal" as they pertain to a CERCLA/SARA cleanup.

With the aid of this text, you should be able to do the following:

1. Determine whether or not a waste is an EPA hazardous waste based on its composition, source, or characteristics.
2. Perform a mass balance to identify waste sources or waste minimization opportunities.
3. Write the reactions for oxidation or reduction of chemical contaminants to mineralized form.
4. Perform solubility product calculations to estimate treatment doses for precipitation or the concentration of contaminants that remain in solution.
5. Determine the dimensions of an air stripping column and air or liquid flow rate, given the values for remaining variables.

6. Evaluate a chemical feed to an incinerator to determine whether or not the chlorine content is acceptable and design a mix of waste feeds to achieve a desired chlorine feed rate.

7. Evaluate the operating variables for an incinerator to determine regulatory compliance for DRE, HCl emissions, and particulate emissions.

8. Estimate the hydraulic conductivity of a liner material based on laboratory measurements.

9. Estimate the quantity of leachate given the precipitation rate, area, hydraulic gradient, and hydraulic conductivity.

10. Estimate the seepage velocity and travel time of a contaminant through a soil given the hydraulic gradient, hydraulic conductivity, porosity, and length of the flow path.

11. Estimate the required pumping rate for a single extraction well for a specified aquifer thickness, hydraulic conductivity, hydraulic gradient, and capture zone width.

PROBLEMS

14–1   Determine whether the following is a RCRA hazardous waste: Municipal wastewater containing 2.0 mg $\cdot$ L^{-1} of selenium.

   Answer: Not a hazardous waste

14–2   Determine whether the following is a RCRA hazardous waste: An empty pesticide container that a home-owner wishes to discard.

14–3   A dry cleaner accumulates 10 kg $\cdot$ month^{-1} of a hazardous waste solvent. To save shipping cost he would like to accumulate 6 months' worth before he ships it to a TSD facility. Can he do this? Explain. (*Note:* This is a "trick" question that requires searching the applicable regulations in the CFR.)

   Answer: Yes

14–4   The town of What Cheer has set up a recycling center to collect old fluorescent lightbulbs. They anticipate collecting about 250 kg $\cdot$ mo^{-1} of fluorescent bulbs. What is the maximum time the fluorescent bulbs can be stored before they must be disposed? (*Hint:* Use the Internet to access the appropriate CFR.)

14–5   A vapor degreaser uses 590 kg $\cdot$ week^{-1} of trichloroethylene (TCE). It is never dumped. The incoming parts have no TCE on them and the exiting parts drag out 3.8 L $\cdot$ h^{-1} of TCE. The sludge removed from the bottom of the degreaser each week has 1.0% of the incoming TCE in it. The plant operates 8 h $\cdot$ d^{-1} for 5 d $\cdot$ week^{-1}. Draw the mass balance diagram for the degreaser and estimate the loss due to evaporation (in kg $\cdot$ week^{-1}). The density of TCE is 1.460 kg $\cdot$ L^{-1}.

   Answer: $M_{evap.}$ = 362.18 or 360 kg $\cdot$ week^{1}

14–6   Using the data in the following table and Figure P-14–6, use the mass-balance technique to determine the mass flow rate (in kilograms per day) of organic compounds to the condensate collection tank (sample location 4 in Figure P-14–6).

Sample Location	Flow Rate (L $\cdot$ min^{-1})	Total Volatile Organic	Temperature (°C)
1	40.5	5858. mg $\cdot$ L^{-1}	25
2	44.8	0.037 mg $\cdot$ L^{-1}	80
3	57. (vapor)	44.13%	20

Notes:

(a) % is volume percent

(b) Vapor flow rate is corrected to 1 atm and 20°C.

(c) Liquid organic density may be assumed to be 0.95 kg $\cdot$ L^{-1}.

(d) Assume the molecular weight of the organic vapor is equal to that of methylene chloride.

(e) Steam mass flow rate is 252 kg $\cdot$ h^{-1} at 106°C.

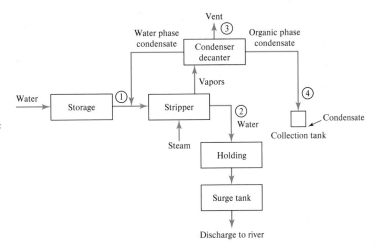

Figure P-14–6

14–7 What is the efficiency of the condenser-decanter in Problem 14–6?

Answer: $\eta = 62.5\%$

14–8 Given the waste constituent and concentration shown in the following table, determine the quantity (in kilograms per day) of hydrated lime ($Ca(OH)_2$) required to neutralize the waste. Estimate the total dissolved solids (TDS) after neutralization. Report your answer in milligrams per liter.

Constituent	Concentration $(mg \cdot L^{-1})$	Flow $(L \cdot min^{-1})$
HCl	100	5

Answers: Lime $= 0.730 \text{ kg} \cdot d^{-1}$; TDS $= 152 \text{ mg} \cdot L^{-1}$

14–9 Given the constituents and concentrations shown in the following table, determine the quantity (in kilograms per day) of sulfuric acid required to neutralize the waste. Estimate the total dissolved solids (TDS) after neutralization. Report your answer in milligrams per liter.

Constituent	Concentration $(mg \cdot L^{-1})$	Flow $(L \cdot min^{-1})$
NaOH	15	200

14–10 It has been proposed to mix a 1500 L bath containing 5.00% by volume of H_2SO_4 with a 1500 L bath containing 5.00% by weight of NaOH. The specific gravity of the acid added is 1.841 and its purity is 96%. The base added is 100% pure. Estimate the final pH (to two decimal places) and the final TDS (in $mg \cdot L^{-1}$) of the mixture of the two baths. (*Note:* The pH is very low.)

14–11 Write the reaction equation to oxidize sodium cyanide using sodium hypochlorite (NaOCl).

14–12 Write the reaction equation to oxidize sodium cyanide using ozone (O_3).

14–13 Write the reaction equation to reduce hexavalent chromium in chromic acid ($H_2Cr_2O_7$) to trivalent chromium using $NaHSO_3$.

14–14 A metal plating solution contains $50.00 \text{ mg} \cdot L^{-1}$ of copper. Determine the concentration, in moles per liter, to which the hydroxide concentration must be raised to precipitate all but $1.3 \text{ mg} \cdot L^{-1}$ of the copper using lime. The K_{sp} of copper hydroxide is 2.00×10^{-19}. Estimate the final pH.

14–15 A plating rinse water flowing at $100 \text{ L} \cdot \text{min}^{-1}$ contains $50.0 \text{ mg} \cdot \text{L}^{-1}$ of Zn. Calculate the theoretical pH required to achieve the EPA's pretreatment standard for existing dischargers of $2.6 \text{ mg} \cdot \text{L}^{-1}$ and estimate the theoretical dose of hydrated lime to remove only the required amount of Zn to achieve the standard (i.e., $50 \text{ mg} \cdot \text{L}^{-1}$ minus the standard). Assume the lime is 100% pure.

14–16 A metal plating sludge as removed from a clarifier has a solids concentration of 4%. If the volume of sludge is $1.0 \text{ m}^3 \cdot \text{day}^{-1}$, what volume will result if the sludge is processed in a filter press to a solids concentration of 30%? If the pressed sludge is dried to 80% solids, what volume will result?

Answer: Press $V_2 = 0.133 \text{ m}^3 \cdot \text{day}^{-1}$; dryer $V_2 = 0.05 \text{ m}^3 \cdot \text{day}^{-1}$

14–17 In Problem 14–16, ferrocyanide is found in the clarifier sludge at a concentration of $400 \text{ mg} \cdot \text{kg}^{-1}$ (4% solids). Assuming that the ferrocyanide is part of the precipitate and that none escapes from the filter press, what concentration would be expected in filter cake?

14–18 A drinking-water supply at Oscoda, Michigan, has been contaminated by trichloroethylene. The average concentration in the water is estimated to be $6000 \text{ μg} \cdot \text{L}^{-1}$. Using the following design parameters, design a packed-tower stripping column to reduce the water concentration to the State of Michigan discharge limit of $1.5 \text{ μg} \cdot \text{L}^{-1}$. Note that more than one column in series may be required for reasonable tower heights.

Henry's law constant $= 6.74 \times 10^{-3} \text{ m}^3 \cdot \text{atm} \cdot \text{mol}^{-1}$ Temperature $= 25°C$
$K_L a = 0.720 \text{ min}^{-1}$ Column diameter may not exceed 4.0 m
Air flow rate $= 60 \text{ m}^3 \cdot \text{min}^{-1}$ Column height may not exceed 6.0 m
$G/L = 18$

Answer: $Z_T = 6 \text{ m}$; $d = 3.15 \text{ m}$

14–19 Well 13 at Watapitae is contaminated with $440 \text{ μg} \cdot \text{L}^{-1}$ of tetrachloroethylene (perchloroethylene). The water must be remediated to achieve a concentration of $0.2 \text{ μg} \cdot \text{L}^{-1}$ (the detection limit). Design a packed-tower stripping column to meet this requirement using the following design parameters. Note that more than one column in series may be required for reasonable tower heights.

Henry's law constant $= 100 \times 10^{-4} \text{ m}^3 \cdot \text{atm} \cdot \text{mol}^{-1}$ Temperature $= 20°C$
$K_L a = 13.5 \times 10^{-3} \text{ s}^{-1}$ Column diameter may not exceed 4.0 m
Air flow rate $= 15 \text{ m}^3 \cdot \text{s}^{-1}$ Column height may not exceed 6.0 m
Liquid flow rate $= 0.22 \text{ m}^3 \cdot \text{s}^{-1}$

14–20 An incinerator operator receives the following shipments of waste for incineration. Can the operator mix these wastes to achieve 30% by mass of chlorine in the feed?

Trichloroethylene $= 18.9 \text{ m}^3$ Toluene $= 21.3 \text{ m}^3$
1,1,1-Trichloroethane $= 5.3 \text{ m}^3$ o-Xylene $= 4.8 \text{ m}^3$

Answer: No

14–21 An incinerator operator receives the following shipments of waste for incineration. What volume of methanol (CH_3OH) must the operator mix to achieve 30% by mass of chlorine in the feed? Assume the density of methanol is $0.7913 \text{ g} \cdot \text{mL}^{-1}$.

Carbon tetrachloride $= 12.2 \text{ m}^3$ Hexachlorobenzene $= 15.3 \text{ m}^3$ Pentachlorophenol $= 2.5 \text{ m}^3$

14–22 A hazardous waste incinerator is being fed methylene chloride at a concentration of $5858 \text{ mg} \cdot \text{L}^{-1}$ in a aqueous stream at a rate of $40.5 \text{ L} \cdot \text{min}^{-1}$. Calculate the mass flow rate of the feed in units of grams per minute.

Answer: $237.25 \text{ g} \cdot \text{min}^{-1}$

14–23 Methylene chloride was measured in the flue gas of a hazardous waste incinerator at a concentration of $211.86 \text{ μg} \cdot \text{m}^{-3}$. If the flow rate of gas from the incinerator was $597.55 \text{ m}^3 \cdot \text{min}^{-1}$, what was the mass flow rate of methylene chloride in grams per minute?

14–24 Assuming that the same incinerator is being evaluated in Problems 14–22 and 14–23, what is the DRE for the incinerator?

> **Answer:** DRE = 99.947%

14–25 Xylene is fed into an incinerator at a rate of 481 kg · h⁻¹. If the mass flow rate at the stack is 72.2 g · h⁻¹, is the unit in compliance with the EPA rules?

14–26 1,2-Dichlorobenzene is being burned in an incinerator under the following conditions:

Operating temperature = 1150°C   Stack gas flow rate = 6.70 m³ · s⁻¹ at standard conditions
Feed flow rate = 173.0 L · min⁻¹   Stack gas concentrations after APC equipment
Feed concentration = 13.0 g · L⁻¹   Dichlorobenzene = 338.8 μg · dscm⁻¹
Residence time = 2.4 s   HCl = 77.2 mg · dscm⁻¹
Oxygen in stack gas = 7.0%   Particulates = 181.6 mg · dscm⁻¹

Assume all of the chlorine in the feed is converted to HCl. Does the incinerator comply with the EPA rules?

> **Answer:** Incinerator fails to comply with HCl limits and with particulate limits.

14–27 The POHCs from a trial burn are shown in the table below. The incinerator was operated at a temperature of 1100°C. The stack gas flow rate was 5.90 dscm · s¹ with 10.0% oxygen. Assuming that all the chlorine in the feed is converted to HCl, is the unit in compliance if the emissions are measured downstream of the APC equipment?

Compound	Inlet (kg · h⁻¹)	Outlet (kg · h⁻¹)
Benzene	913.98	0.2436
Chlorobenzene	521.63	0.0494
Xylenes	1378.91	0.5670
HCl	n/a	4.85
Particulates	n/a	10.61

14–28 During a trial burn, an incinerator was fed a mixed feed containing trichloroethylene, 1,1,1-trichloroethane, and toluene in an aqueous solution. Each component accounted for 5.0% of the feed solution on a volume basis. The feed rate was 40 L · min⁻¹. The incinerator was operated at a temperature of 1200°C. The stack gas flow rate was 9.0 dscm · s⁻¹ with 7% oxygen. Assuming that all the chlorine in the feed is converted to HCl, is the unit in compliance with the following emissions measured after the APC equipment?

Trichloroethylene = 170 μg · dscm⁻¹   HCl = 83.2 mg · dscm⁻¹
1,1,1-Trichloroethane = 353 μg · dscm⁻¹   Particulates = 123.4 mg · dscm⁻¹
Toluene = 28 μg · dscm⁻¹

14–29 During a trial burn, an incinerator was fed a mixed feed containing hexachlorobenzene (HCB), pentachlorophenol (PCP), and acetone (ACET) in an aqueous solution. Each component accounted for 9.3% of the feed solution on a volume basis, that is, HCB = 9.3%, PCP = 9.3%, and ACET = 9.3%. The feed rate was 140 L · min⁻¹. The incinerator was operated at a temperature of 1200°C. The stack gas flow rate was 28.32 dscm · s⁻¹ with 14% oxygen. Assuming that all the chlorine in the feed is converted to HCl, is the unit in compliance if the following emissions are measured downstream of the APC equipment?

Hexachlorobenzene = 170 μg · dscm⁻¹   HCl = 83.2 μg · dscm⁻¹
Pentachlorophenol = 353 μg · dscm⁻¹   Particulates = 123.4 mg · dscm⁻¹
Acetone = 28 μg · dscm⁻¹

> **Answer:** Incinerator complies with all emission limits except particulates.

14–30   A standard permeameter is being considered for testing a clay for a hazardous waste landfill base. If the clay must have a hydraulic conductivity of 10^{-7} cm $\cdot$ s^{-1} and the dimensions of the permeameter are as shown here, how long will the test take if a minimum of 100.0 mL of liquid must be collected for an accurate measurement? See Figure P-14–30 for notation and permeameter equation.

$L = 10$ cm

$h = 1$ m

Diameter of sample $= 5.0$ cm

Figure P-14–30

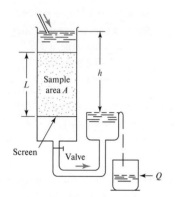

Standard constant head permeameter equation:

$$K = \frac{QL}{hAt}$$

where $K =$ hydraulic conductivity
$\quad Q =$ quantity of discharge
$\quad L =$ length of sample
$\quad h =$ hydraulic head
$\quad A =$ cross-sectioned area of sample
$\quad t =$ time

14–31   A soil sample has been tested to determine hydraulic conductivity using a falling-head permeameter (Figure P-14–31). The following data were recorded:

Diameter of a $= 1$ mm   Initial head $= 1.0$ m
Diameter of $A = 10$ cm   Final head $= 25$ cm
Length, $L = 25$ cm   Duration of test $= 14$ days

Figure P-14–31

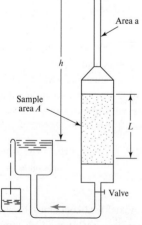

Falling head permeameter equation:

$$K = 2.3 \frac{aL}{At} \log\left(\frac{h_o}{h_1}\right)$$

where $K =$ hydraulic conductivity
$\quad a =$ cross-sectional area of stand pipe
$\quad A =$ cross-sectional area of sample
$\quad L =$ length of sample
$\quad t =$ time
$\quad h_o, h_1 =$ head at beginning of test
$\qquad\quad$ and at time, t, respectively

From these data, calculate the hydraulic conductivity of the sample. Assuming the sample is representative of the landfill site, is this a good soil for a hazardous waste landfill base?

Answer: 2.86×10^{-9} cm $\cdot$ s^{-1}

14–32   An old 100 ha hazardous waste landfill was built on a 10-m deep clay liner. An aquifer lies immediately below the clay layer. The clay layer through which the leachate must pass has a hydraulic conductivity of 1×10^{-7} cm $\cdot$ s^{-1}. If the liquid level (leachate) is 1.0 m deep above the clay layer, how much leachate (in cubic meters per day) will reach the aquifer when the clay layer becomes saturated? Assume Darcy's law applies.

14–33 The three soil layers described here lie between the bottom of a hazardous waste landfill and the underlying aquifer. The depth of leachate above the top soil layer is 0.3 m. How long will it take (in years) for the leachate to migrate to an aquifer located at the bottom of soil C?

Soil A
    Depth = 3.0 m
    Hydraulic conductivity = 1.8×10^{-7} cm · s^{-1}
    Porosity = 55%

Soil B
    Depth = 10 m
    Hydraulic conductivity = 2.2×10^{-5} m · s^{-1}
    Porosity = 25%

Soil C
    Depth = 12.0 m
    Hydraulic conductivity = 5.3×10^{-5} mm · s^{-1}
    Porosity = 35%

Answer: 28 years

14–34 The practical quantitation limit (PQL) for the solvent trichloroethylene (TCE) is 5 µg · L^{-1}. If a barrel (approximately 0.12 m^3) of spent solvent leaked into an aquifer, approximately how many cubic meters of water would be contaminated at the PQL? Assume the barrel container 100% TCE.

14–35 A drum (0.12 m^3) of carbon tetrachloride has leaked into a sandy soil. The soil has a hydraulic conductivity of 7×10^{-4} m · s^{-1} and a porosity of 0.38. The groundwater table is 3 m below grade and has a hydraulic gradient of 0.002. The aquifer is 28 m thick. A single well intercept system is proposed using a well pumping at 0.014 m^3 s^{-1}. Estimate the width of the capture zone at the well.

DISCUSSION QUESTIONS

14–1 What was the outcome of a hazardous waste episode in 1971 at Times Beach, Missouri? (*Hint:* You will need to do an Internet search.)

14–2 Does the "land ban" actually ban the disposal of hazardous waste on the land? Explain.

14–3 A multimillion dollar company has just learned that one drum out of several hundred found at an abandoned waste disposal site has been identified as theirs. Their attorney explains that they may be potentially responsible for cleanup of all the drums at the site if no other former owners of the drums can be identified. Is this correct? Why or why not?

14–4 Your boss has proposed that your company institute a recycling program to minimize the generation of waste. Is recycling the best first step to investigate in a waste minimization program? If not, what others would you suggest and in what order?

14–5 A metal plater is proposing to treat his waste sludge to recover the nickel from it. Would this be

(a) Recycling?   (b) Reusing?   (c) Reclaiming?

    State the correct answer(s) and explain why you made your choice(s).

14–6 It is not necessary to measure every POHC in an incinerator trial burn. True or false? Explain your answer.

FE EXAM FORMATTED PROBLEMS

14–1 Calculate the incinerator destruction and removal efficiency for methylene chloride using the following measurements:

Aqueous influent
Flow rate = 53.0 L · m^{-1}
Concentration = 7120 mg · L^{-1}

Stack gas effluent
Flow rate = 606.5 m^3/min
Concentration = 188.3 µg/m^3

(a) 99.97%
(b) 69.74%

    (c) 97.36%

    (d) 99.77%

14–2 What is the combustion efficiency of a PCB incinerator if the stack effluent concentration of CO = 100 ppm and the concentration of CO_2 = 100,000 ppm?

    (a) 0.999%

    (b) 99.0%

    (c) 10.0%

    (d) 99.9%

14–3 If the pH controller for feeding hydroxide to a zinc precipitation tank is set at pH = 8.4, what will be the effluent concentration of zinc after settling of the precipitate? The K_{sp} of $Zn(OH)_2$ is 7.68×10^{-17}.

    (a) $1.99 \text{ mg} \cdot \text{L}^{-1}$

    (b) $0.80 \text{ mg} \cdot \text{L}^{-1}$

    (c) $0.0013 \text{ mg} \cdot \text{L}^{-1}$

    (d) $0.082 \text{ mg} \cdot \text{L}^{-1}$

14–4 Estimate the soil partition coefficient of benzene for a soil having the following properties: bulk density = 1.4 g/cm^3, porosity = 0.40, organic carbon fraction = 0.002, and retardation coefficient = 1.2.

    (a) 143

    (b) 171

    (c) 314

    (d) 14.2

REFERENCES

Bair, J. B. (1989) "Underground Storage Tanks (UST)," *Michigan Environmental Law Journal,* 8(2): 2.

CFR (2007) *Code of Federal Regulations,* 40 CFR §260.

CFR (2007) *Code of Federal Regulations,* 40 CFR §300.

FR (1990) *Federal Register,* 55 FR 22530.

FR (1992) *Federal Register,* 57 FR 3462.

FR (1992) *Federal Register,* 59 FR 47980.

Fromm, C. H., A. Bachrach, and M. S. Callahan (1986) "Overview of Waste Minimization Issues, Approaches and Techniques," in E. T. Oppelt, B. L. Blaney, and W. F. Kewner, eds., *Transactions of an APCA International Specialty Conference on Peformance and Costs of Alternatives to Land Disposal of Hazardous Waste,* Air Pollution Control Association, Pittsburgh, PA, pp. 6–20.

Griffin, R. D. (1988) *Principles of Hazardous Materials Management,* Lewis Publishers, Ann Arbor, MI.

Gross, R. L., and S. G. TerMaath (1987) "Packed Tower Aeration Strips Trichloroethylene from Groundwater," *Environmental Progress,* 4: 119–24.

Hazmat World (1989) "Lockheed Aeronautical Systems Aqua-Detox Groundwater Treatment System."

Herzbnin, P. A., R. L. Irvine, and K. C. Malinowski (1985) "Biological Treatment of Hazardous Waste in Sequencing Batch Reactors," *Journal of the Water Pollution Control Federation,* 57: 1163–67.

Hileman, B. (2000) "Leaked Dioxin Report Stokes Controversy," *C&E News,* May 29, p. 13.

Jacobson, M. H. et al. (2017) "Serum Polybrominated Biphenyls (PBBs) and Polychlorinated Biphenyls (PCBs) and Thyroid Function among Michigan Adults Several Decades after the 1973–1974 PBB Contamination of Livestock Feed Environ," *Health Perspect.,* 125(9): 097020.

Javandel, I., and C. Tsang (1986) "Capture Zone Type Curves: A Tool for Cleanup," *Ground Water,* 24(5): 616–25.

Josephson, J. (1981) "Hazardous Waste Landfill," *Environmental Science and Technology,* 15(3): 250.

Kobayashi, H., and B. P. Rittman (1982) "Microbial Removal of Hazardous Organic Compounds," *Environmental Science and Technology,* 16: 170A–172A.

LaGrega, M. D., P. L. Buckingham, and J. C. Evans (2001) *Hazardous Waste Management,* McGraw-Hill, Boston, 1014–1016.

Masters, G. M. (1998) *Introduction to Environmental Engineering and Science,* Prentice Hall, Upper Saddle River, NJ, p. 240.

O'Brien and Gere Engineers Inc. (1988) *Hazardous Waste Site Remediation,* Van Nostrand Reinhold, New York, pp. 11–13.

Oppelt, E. T. (1981) "Thermal Destruction Options for Controlling Hazardous Wastes," *Civil Engineering ASCE,* September, pp. 72–75.

Penning, C. H. (1930) "Physical Characteristics and Commercial Possibility of Chlorinated Diphenyl," *Industrial & Engineering Chemistry,* 22: 1180–83.

Theodore, L., and J. Reynolds (1987) *Introduction to Hazardous Waste Incineration,* John Wiley and Sons, New York, pp. 76–85.

U.S. EPA (1981) *Development Document for Effluent Limitations: Guideline and Standards for the Metal Finishing Point Source Category,* U.S Environmental Protection Agency Publication No. EPA/440/1-83-091, Washington, DC.

U.S. EPA (1986) *RCRA Orientation Manual,* U.S. Environmental Protection Agency Office of Solid Waste, Publication No. EPA/530-SW-86-001, Washington, DC.

U.S. EPA (1987) *A Compendium of Technologies Used in the Treatment of Hazardous Waste,* U.S. Environmental Protection Agency Publication No. EPA/625/8-87/014, Cincinnati, OH.

U.S. EPA (1988a) *Waste Minimization Opportunity Assessment Manual,* U.S. Environmental Protection Agency Publication No. EPA/625/7-88/003), Cincinnati, OH.

U.S. EPA (1988b) *Best Demonstrated Available Technology (BDAT) Background Document for F006,* U.S. Environmental Protection Agency Publication No. EPA/530-SW-88-0009-I, Washington, DC.

U.S. EPA (1989) "Requirements for Hazardous Waste Landfill Design, Construction, and Closure," *Seminar Publication,* U.S. Environmental Protection Agency Publication EPA 625/4-89/022, Washington, DC.

U.S. EPA (1991) *Design and Construction of RCRA/CERCLA Final Covers,* U.S. Environmental Protection Agency Publication EPA 625/4-91/025, Washington, DC.

U.S. EPA (2012) CERCLIS Database, http://cfpub.epa.gov/superrcpad/cursites

U.S. EPA (2017) Green Chemistry, https://www.epa.gov/greenchemistry

Warner, D. L. (1998) "Subsurface Injection of Liquid Hazardous Wastes," in H. M. Freeman, ed., *Standard Handbook of Hazardous Waste Treatment and Disposal,* 2nd ed., McGraw-Hill, New York, pp. 10.46–10.58.

Wood, E. F., R. A. Ferrara, W. G. Gray, and G. F. Pinder (1984) *Groundwater Contamination from Hazardous Waste,* Prentice Hall, Englewood Cliffs, NJ, pp. 2–4, 145–58.

Case Study

The Noise, The Noise, The Noise!

Most of us have been frustrated by noise of some kind, but can you imagine life in one of the noisiest cities in the world? Among the loudest cities is New Delhi, a city of 25 million inhabitants. Within the city, the noise level can exceed 100 decibels (dB), well above the recommended levels of 50–55 dB for residential areas. The average noise level was reported to be 80 dB, which is about as loud as a freight train (at 15 meters). The single largest contributor to noise in Delhi are vehicles, followed by neighborhoods and generators. Religious and political events and household industries were also reported to contribute to noise levels (Firdaus and Ahmed, 2010).

The predominant effects of the noise levels were annoyance and interference with communication. Also reported were headache, nausea, giddiness, and fatigue; increased blood pressure; increased heartbeat, breathing rates, and sweating; and depression, mood swings, indigestion, and hypertension. Residents of Delhi are reported to have significant hearing loss.

Efforts are underway to control noise levels in the city of Delhi. These include attempts to limit the use of vehicular horns, revised noise ordinances, improved sound insulation measures, changes in zoning to separate residential areas from commercial and industrial zones, and the addition of greenbelt areas to absorb the noise (Garg and Maji, 2016).

15–1   INTRODUCTION

Noise, commonly defined as unwanted sound, is an environmental phenomenon to which we are exposed before birth and throughout life. Noise can also be considered an environmental pollutant, a waste product generated in conjunction with various anthropogenic activities. Under the latter definition, noise is any sound—independent of loudness—that can produce an undesired physiological or psychological effect in an individual and that may interfere with the social ends of an individual or group. These social ends include all of our activities—communication, work, rest, recreation, and sleep.

As waste products of our way of life, we produce two general types of pollutants. The general public has become well aware of the first type—the mass residuals associated with air and water pollution—that remain in the environment for extended periods. However, only recently has attention been focused on the second general type of pollution, the energy residuals such as the waste heat from manufacturing processes that creates thermal pollution of our streams. Energy in the form of sound waves constitutes yet another kind of energy residual, but, fortunately, one that does not remain in the environment for a long time. The total amount of energy dissipated as sound throughout the earth is not large compared with other forms of energy; it is only the extraordinary sensitivity of the ear that permits such a relatively small amount of energy to adversely affect us.

It has long been known that noise of sufficient intensity and duration can induce temporary or permanent hearing loss, ranging from slight impairment to nearly total deafness. In general, a pattern of exposure to any source of sound that produces high enough levels can result in temporary hearing loss. Exposure persisting over time can lead to permanent hearing impairment. It has been estimated that 1.7 million workers in the United States between 50 and 59 years of age have enough hearing loss to be awarded compensation. In 2012 dollars, the potential cost to industry could be in excess of $1 billion (Olishifski and Harford, 1975). Short-term, but frequently serious, effects include interference with speech communication and the perception of other auditory signals, disturbance of sleep and relaxation, annoyance, interference with an individual's ability to perform complicated tasks, and general diminution of the quality of life.

Beginning with the technological expansion of the Industrial Revolution and continuing through a post-World War II acceleration, environmental noise in the United States and other industrialized nations has been gradually and steadily increasing, with more geographic areas becoming exposed to significant levels of noise. Where once noise levels sufficient to induce some degree of hearing loss were confined to factories and occupational situations, noise levels approaching such intensity and duration are today being recorded on city streets and, in some cases, in and around the home.

There are valid reasons why widespread recognition of noise as a significant environmental pollutant and potential hazard or, as a minimum, a detractor from the quality of life, has been slow in coming. In the first place, noise, if defined as unwanted sound, is a subjective experience. What is considered noise by one listener may be considered desirable by another.

Secondly, noise has a short decay time and thus does not remain in the environment for long, as do air and water pollution. By the time the average individual is spurred to action to abate, control, or, at least, complain about sporadic environmental noise, the noise may no longer exist.

Thirdly, the physiological and psychological effects of noise on us are often subtle and insidious, appearing so gradually that it becomes difficult to associate cause with effect. Indeed, to those persons whose hearing may already have been affected by noise, it may not be considered a problem at all.

Furthermore, the typical citizen is proud of this nation's technological progress and is generally happy with the things that technology delivers, such as rapid transportation, labor-saving appliances, and new recreational devices. Unfortunately, many technological advances have been associated with increased environmental noise, and large segments of the population have tended to accept the additional noise as part of the price of progress.

In the last three decades, the public has begun to demand that the price of progress not fall to them. They have demanded that the environmental impact of noise be mitigated. The cost of mitigation is not trivial. The average cost of soundproofing each of six hundred suburban houses around the Chicago O'Hare airport was about $27,500 in 1997 (Sylvan, 2000). Through 2001, the Boston Logan airport had spent about $99 million and the Los Angeles International airport had allocated about $119 million for soundproofing and land acquisition. At the end of 2001, the total amount spent in the United States for noise mitigation exceeded $5.2 billion (De Neufville and Odoni, 2003). The cost to retrofit and replace airplanes to reduce noise probably exceeds $3.6 billion (in 2005 dollars) (Achitoff, 1973). Traffic noise reduction programs have been in place since the first noise barrier was built in 1963. As of 2001, 44 state departments of transportation and the Commonwealth of Puerto Rico had constructed more than 2900 linear kilometers of noise barriers at a cost of more than $2.8 billion (in 2005 dollars) (FHWA, 2005).

The engineering and scientific community has already accumulated considerable knowledge concerning noise, its effects, and its abatement and control. In that regard, noise differs from most other environmental pollutants. Generally, the technology exists to control most indoor and outdoor noise. As a matter of fact, this is one instance in which knowledge of control techniques exceeds the knowledge of biological and physical effects of the pollutant.

Properties of Sound Waves

Sound waves result from the vibration of solid objects or the separation of fluids as they pass over, around, or through holes in solid objects. The vibration or separation causes the surrounding air to undergo alternating compression and rarefaction, much in the same manner as a piston vibrating in a tube (Figure 15–1). The compression of the air molecules causes a local increase in air density and pressure. Conversely, the rarefaction causes a local decrease in density and pressure. These alternating pressure changes are the sound detected by the human ear.

Let us assume that you could stand at point A in Figure 15–1. Also let us assume that you have an instrument that will measure the air pressure every 0.000010 s and plot the value on a graph. If the piston vibrates at a constant rate, the condensations and rarefactions will move down the tube at a constant speed. That speed is the speed of sound (c). The rise and fall of pressure at point A will follow a cyclic or wave pattern over a period (Figure 15–2). The wave pattern is called sinusoidal. The time between successive peaks or between successive troughs of the oscillation is called the **period** (P). The inverse of this, that is, the number of times a peak arrives in one second of oscillations, is called the **frequency** (f). Period and frequency are then related as follows:

$$P = \frac{1}{f} \tag{15–1}$$

Because the pressure wave moves down the tube at a constant speed, you would find that the distance between equal pressure readings would remain constant. The distance between adjacent crests or troughs of pressure is called the **wavelength** (λ). Wavelength and frequency are then related as follows:

$$\lambda = \frac{c}{f} \tag{15–2}$$

The **amplitude** (A) of the wave is the height of the peak or depth of the trough measured from the zero pressure line (see Figure 15–2). From Figure 15–2 we can also note that the

---

FIGURE 15–1

Alternating compression and rarefaction of air molecules resulting from a vibrating piston.

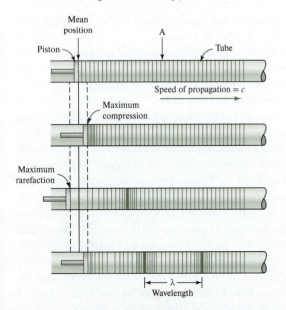

FIGURE 15–2

Sinusoidal wave that results from alternating compression and rarefaction of air molecules. The amplitude is shown as A and the period is P.

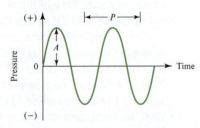

average pressure could be zero if an averaging time was selected that corresponded to the period of the wave. This would result regardless of the amplitude! This, of course, is not an acceptable state of affairs. The **root mean square** (rms) **sound pressure** (p_{rms}) is used to overcome this difficulty.* The rms sound pressure is obtained by squaring the value of the amplitude at each instant in time, summing the squared values, dividing the total by the averaging time, and taking the square root of the total. The equation for rms is

$$\bar{p}_{rms} = (p^2)\frac{1}{2}\left[\frac{1}{t_m} \int p^2(t) dt\right]^{1/2} \tag{15–3}$$

where the overbar refers to the time-weighted average and t_m is the time period of the measurement.

Sound Power and Intensity

Work is defined as the product of the magnitude of the displacement of a body and the component of force in the direction of the displacement. Thus, traveling waves of sound pressure transmit energy in the direction of propagation of the wave. The rate at which this work is done is defined as the **sound power** (W).

Sound **intensity** (I) is the time-weighted average sound power per unit area normal to the direction of propagation of the sound wave. Intensity and power are related as follows:

$$I = \frac{W}{A} \tag{15–4}$$

where A is a unit area perpendicular to the direction of wave motion. Intensity, and hence, sound power, is related to sound pressure in the following manner:

$$I = \frac{(p_{rms})^2}{\rho c} \tag{15–5}$$

where I = intensity (in $W \cdot m^{-2}$)
  p_{rms} = root mean square sound pressure (in Pa)
  ρ = density of medium (in $kg \cdot m^{-3}$)
  c = speed of sound in medium (in $m \cdot s^{-1}$)

Both the density of air and speed of sound are a function of temperature. Given the temperature and pressure, the density of air may be determined from Table A–4 in Appendix A. The speed of sound in air at 101.325 kPa may be determined from the following equation:

$$c = 20.05\sqrt{T} \tag{15–6}$$

where T is the absolute temperature in kelvins (K) and c is in meters per second.

Levels and the Decibel

The sound pressure of the faintest sound that a normal healthy individual can hear is about 0.00002 Pa. The sound pressure produced by a Saturn rocket at liftoff is greater than 200 Pa. Even in scientific notation this is an "astronomical" range of numbers.

To cope with this problem, a scale based on the logarithm of the ratios of the measured quantities is used. Measurements on this scale are called levels. The unit for these types of measurement scales is the **bel,** which was named after Alexander Graham Bell:

$$L' = \log\frac{Q}{Q_o} \tag{15–7}$$

---

*Sound pressure = (total atmospheric pressure) − (barometric pressure).

where L' = level, bels

Q = measured quantity

Q_o = reference quantity

log = logarithm in base 10

A bel turns out to be a rather large unit, so for convenience it is divided into 10 subunits called **decibels** (dB). Levels in decibels are computed as follows:

$$L = 10 \log \frac{Q}{Q_o} \tag{15–8}$$

The decibel does not represent any physical unit. It merely indicates that a logarithmic transformation has been performed.

Sound Power Level. If the reference quantity (Q_o) is specified, then the decibel takes on physical significance. For noise measurements, the reference power level has been established as 10^{-12} W. Thus, sound power level may be expressed as

$$L_w = 10 \log \frac{W}{10^{-12}} \tag{15–9}$$

Sound power levels computed with Equation 15–9 are reported as decibels re: 10^{-12} W.

Sound Intensity Level. For noise measurements, the reference sound intensity (Equation 15–4) is 10^{-12} W · m^{-2}. Thus the sound intensity level is given as

$$L_I = 10 \log \frac{I}{10^{-12}} \tag{15–10}$$

Sound Pressure Level. Because sound-measuring instruments measure the p_{rms}, the sound pressure level (SPL) is computed as follows:

$$L_p = 10 \log \frac{(p_{rms})^2}{(p_{rms})_o^2} \tag{15–11}$$

which, after extracting the squaring term, is given as

$$L_p = 20 \log \frac{(p_{rms})}{(p_{rms})_o} \tag{15–12}$$

The reference pressure has been established as 20 μPa (micropascals). A scale showing some common SPLs is shown in Figure 15–3.

Combining Sound Pressure Levels. Because of their logarithmic heritage, decibels don't add and subtract the way apples and oranges do. Remember: adding the logarithms of numbers is the same as multiplying them. If you take a 60-dB noise (re: 20 μPa) and add another 60-dB noise (re: 20 μPa) to it, you get a 63-dB noise (re: 20 μPa). If you're strictly an apple-and-orange mathematician, you may take this on faith. For skeptics, this can be demonstrated by converting the decibels to sound power level, adding them, and converting back to decibels. Figure 15–4 provides a graphical solution for this type of problem. For noise pollution work, results should be reported to the nearest whole number. When several levels are to be combined, combine them two at a time, starting with lower valued levels and continuing two at a time with each successive pair until one number remains. Henceforth, in this chapter we will assume levels are all "re: 20 μPa" unless stated otherwise.

FIGURE 15–3

Relative scale of sound pressure levels.

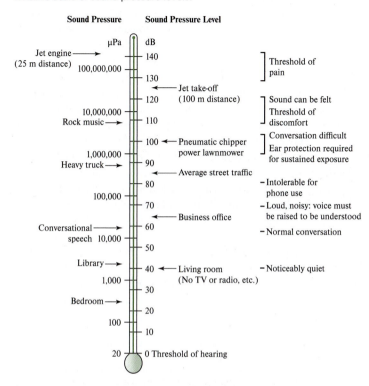

FIGURE 15–4

Graph for solving decibel addition problems.

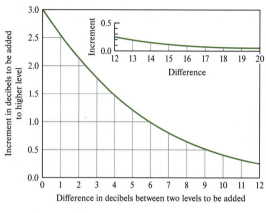

EXAMPLE 15–1

What sound power level results from combining the following three levels: 68 dB, 79 dB, and 75 dB?

Solution

This problem can be worked by converting the readings to sound power level, adding them, and converting back to decibels.

$$L_w = 10 \log \sum 10^{(68/10)} + 10^{(75/10)} + 10^{(79/10)}$$

$$= 10 \log(117{,}365{,}173)$$

$$= 80.7 \text{ dB}$$

Rounding off to the nearest whole number yields an answer of 81 dB re: 20 μPa.

An alternative solution technique using Figure 15–4 begins by selecting the two lowest levels: 68 dB and 75 dB. The difference between the values is $75 - 68 = 7.00$. Using Figure 15–4, draw a vertical line from 7.00 on the abscissa to intersect the curve. A horizontal line from the intersection to the ordinate yields about 0.8 dB. Thus, the combination of 68 dB and 75 dB results in a level of 75.8 dB. This, and the remainder of the computation, is shown in the following diagram.

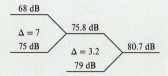

Characterization of Noise

Weighting Networks. Because our reasons for measuring noise usually involve people, we are ultimately more interested in the human reaction to sound than in sound as a physical phenomenon. Sound pressure level, for instance, can't be taken at face value as an indication of loudness because the frequency (or pitch) of a sound has quite a bit to do with how loud it sounds. For this and other reasons, it often helps to know something about the frequency of the noise you're measuring. This is where weighting networks come in (GRC, 1972). **Weighting networks** are electronic filtering circuits built into the meter to attenuate certain frequencies. They permit the sound level meter (Figure 15–5) to respond more to some frequencies than to others with a prejudice something like that of the human ear. Writers of the acoustical standards have established three weighting characteristics: A, B, and C. The chief difference among them is that very low frequencies are filtered quite severely by the A network, moderately by the B network, and hardly at all by the C network. Therefore, if the measured sound level of a noise is much higher on C weighting than on A weighting, much of the noise is probably of low frequency. If you really want to know the frequency distribution of a noise (and most serious noise measurers do), it is necessary to use a sound analyzer. But if you are unable to justify the expense of an analyzer, you can still find out something about the frequency of a noise by the shrewd use of the weighting networks of a sound level meter.

Figure 15–6 shows the response characteristics of the three basic networks as prescribed by the American National Standards Institute (ANSI) specification number Sl.4-1971. When a weighting network is used, the sound level meter electronically subtracts or adds the number of decibels shown at each frequency shown in Table 15–1 from or to the actual SPL at

FIGURE 15–5

Precision sound level meter.
(©Larson Davis, Inc.)

FIGURE 15–6

Response characteristics of the three basic weighting networks.

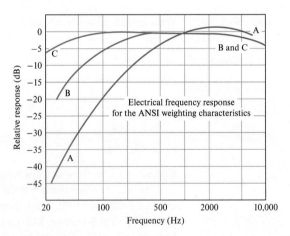

TABLE 15–1   Sound Level Meter Network Weighting Values

Frequency (Hz)	Curve A (dB)	Curve B (dB)	Curve C (dB)	Frequency (Hz)	Curve A (dB)	Curve B (dB)	Curve C (dB)
10	−70.4	−38.2	−14.3	500	−3.2	−0.3	0
12.5	−63.4	−33.2	−11.2	630	−1.9	−0.1	0
16	−56.7	−28.5	−8.5	800	−0.8	0	0
20	−50.5	−24.2	−6.2	1000	0	0	0
25	−44.7	−20.4	−4.4	1250	0.6	0	0
31.5	−39.4	−17.1	−3.0	1600	1.0	0	−0.1
40	−34.6	−14.2	−2.0	2000	1.2	−0.1	−0.2
50	−30.2	−11.6	−1.3	2500	1.3	−0.2	−0.3
63	−26.2	−9.3	−0.8	3150	1.2	−0.4	−0.5
80	−22.5	−7.4	−0.5	4000	1.0	−0.7	−0.8
100	−19.1	−5.6	−0.3	5000	0.5	−1.2	−1.3
125	−16.1	−4.2	−0.2	6300	−0.1	−1.9	−2.0
160	−13.4	−3.0	−0.1	8000	−1.1	−2.9	−3.0
200	−10.9	−2.0	0	10,000	−2.5	−4.3	−4.4
250	−8.6	−1.3	0	12,500	−4.3	−6.1	−6.2
315	−6.6	−0.8	0	16,000	−6.6	−8.4	−8.5
400	−4.8	−0.5	0	20,000	−9.3	−11.1	−11.2

that frequency. It then sums all the resultant numbers by logarithmic addition to give a single reading. Readings taken when a network is in use are said to be "sound levels" rather than "sound pressure levels." The readings taken are designated in decibels in one of the following forms: dB(A); dBa; dBA; dB(B); dBb; dBB; and so on. Tabular notations may refer to L_A, L_B, L_C.

EXAMPLE 15–2   A new type 2 sound level meter is to be tested with two pure tone sources that emit 90 dB. The pure tones are at 1000 Hz and 100 Hz. Estimate the expected readings on the A, B, and C weighting networks.

Solution   From Table 15–1 at 1000 Hz, we note that the relative response (correction factor) for each of the weighting networks is zero. Thus for the pure tone at 1000 Hz we would expect the readings on the A, B, and C networks to be 90 dB.

From Table 15–1 at 100 Hz, the relative response for each weighting network differs. For the A network, the meter will subtract 19.1 dB from the actual reading, for the B network, the meter will subtract 5.6 dB from the actual reading, and for the C network, the meter will subtract 0.3 dB. Thus, the anticipated readings would be

A network: 90 − 19.1 = 70.9, or 71 dB(A)
B network: 90 − 5.6 = 84.4, or 84 dB(B)
C network: 90 − 0.3 = 89.7, or 90 dB(C)

EXAMPLE 15–3   The following sound levels were measured on the A, B, and C weighting networks:

Source 1: 94 dB(A), 95 dB(B), and 96 dB(C)
Source 2: 74 dB(A), 83 dB(B), and 90 dB(C)

Characterize the sources as "low frequency or mid/high frequency."

Solution   From Figure 15–6, we can see that readings on the A, B, and C networks will be close together if the source emits noise in the frequency range above about 500 Hz. This range may be classified mid/high frequency because we cannot distinguish between "mid" and "high" frequency using a type 2 sound level meter. Likewise, we can see that below 200 Hz (low frequency), readings on the A, B, and C scale will be substantially different. The readings from the A network will be lower than the readings from the B network, and readings from both the A and B networks will be lower than those from the C network.

Source 1: Note that the sound levels on each of the weighting networks differ by 1 dB. From Figure 15–6, it appears that the sound level will be in the mid/high-frequency range.
Source 2: Note that the sound levels on each of the weighting networks differ by several decibels and that the reading from the A network is lower than that from the B network and both are below that from the C network. From Figure 15–6, it appears that the sound level will be in the low-frequency range.

Octave Bands.   To completely characterize a noise, it is necessary to break it down into its frequency components, or spectra. Normal practice is to consider 8 to 11 octave bands.* The standard octave bands and their geometric mean frequencies (center band frequencies) are given in Table 15–2. Octave analysis is performed with a combination precision sound level meter and an octave filter set.

Although octave band analysis is frequently sufficient for community noise control (i.e., identifying violators), more refined analysis is required for corrective action and design. One-third octave band analysis provides a slightly more refined picture of the noise source than the full octave band analysis. This improved resolution is usually sufficient for determining corrective action for community noise problems. Narrow band analysis is highly refined and may imply band widths down to 2 Hz. This degree of refinement is only justified in product design and testing or in troubleshooting industrial machine noise and vibration.

Averaging Sound Pressure Levels.   Because of the logarithmic nature of the decibel, the average value of a collection of SPL measurements cannot be computed in the normal fashion. Instead, the following equation must be used:

$$\bar{L}_p = 20 \log \frac{1}{N} \sum_{j=1}^{N} 10^{(L_j/20)} \qquad (15\text{–}13)$$

where $\bar{L}_p$ = average SPL (in dB re: 20 μPa)
N = number of measurements
L_j = the jth SPLs (in dB re: 20 μPa)
$j = 1, 2, 3, \ldots N$

This equation is equally applicable to sound levels in decibels(A). It may also be used to compute average sound power levels if the factors of 20 are replaced with 10s.

---

*An octave is the frequency interval between a given frequency and twice that frequency. For example, given the frequency 22 Hz, the octave band is from 22 to 44 Hz. A second octave band would then be from 44 to 88 Hz.

TABLE 15–2   Octave Bands

Octave Frequency Range (Hz)	Geometric Mean Frequency (Hz)	Octave Frequency Range (Hz)	Geometric Mean Frequency (Hz)
22–44	31.5	1400–2800	2000
44–88	63	2800–5600	4000
88–175	125	5600–11,200	8000
175–350	250	11,200–22,400	16,000
350–700	500	22,400–44,800	31,500
700–1400	1000		

EXAMPLE 15–4   Compute the mean sound level from the following four readings (all in decibels): 38, 51, 68, and 78.

Solution   First we compute the sum.

$$\sum_{j=1}^{4} = 10^{(38/20)} + 10^{(51/20)} + 10^{(68/20)} + 10^{(78/20)}$$

$$= 1.09 \times 10^4$$

Now we complete the computation.

$$\bar{L}_p = 20 \log \frac{1.09 \times 10^4}{4}$$

$$= 68.7, \text{ or } 69 \text{ dBA}$$

Straight arithmetic averaging would yield 58.7, or 59 dB.

Types of Sounds. Patterns of noise may be qualitatively described by one of the following terms: **steady-state, or continuous; intermittent;** and **impulse, or impact. Continuous noise** is an uninterrupted sound level that varies less than 5 dB during the period of observation. An example is the noise from a household fan. **Intermittent noise** is a continuous noise that persists for more than 1 s that is interrupted for more than 1 s. A dentist's drilling would be an example of an intermittent noise. **Impulse noise** is characterized by a change of sound pressure of 40 dB or more within 0.5 s with a duration of less than 1 s.* The noise from firing a weapon would be an example of an impulsive noise.

    Two types of impulse noise generally are recognized. The type A impulse is characterized by a rapid rise to a peak SPL followed by a small negative pressure wave or by decay to the background level (Figure 15–7). The type B impulse is characterized by a damped (oscillatory) decay (Figure 15–8). When the duration of the type A impulse is simply the duration of the initial peak, the duration of the type B impulse is the time required for the envelope to decay to 20 dB below the peak. Because of the short duration of the impulse, a special sound-level meter must be employed to measure impulse noise. You should note that the peak SPL is different from the impulse sound level because of the time-averaging used in the latter.

---

*The Occupational Safety and Health Administration (OSHA) classifies repetitive events, including impulses, as steady noise if the interval between events is less than 0.5 s.

FIGURE 15–7

Type A impulse noise.

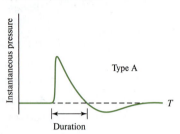

FIGURE 15–8

Type B impulse noise.

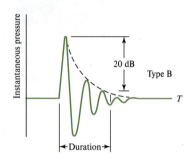

15–2 EFFECTS OF NOISE ON PEOPLE

For the purpose of our discussion, we have classified the effects of noise on people into the following two categories: auditory effects and psychological–sociological effects. **Auditory effects** include both hearing loss and speech interference. **Psychological–sociological effects** include annoyance, sleep interference, effects on performance, and acoustical privacy.

The Hearing Mechanism

Before we can discuss hearing loss, it is important to outline the general structure of the ear and how it works.

Anatomically, the ear is separated into three sections: the outer ear, the middle ear, and the inner ear (Figure 15–9). The outer and middle ear serve to convert sound pressure to

FIGURE 15–9

Anatomical divisions of the ear.

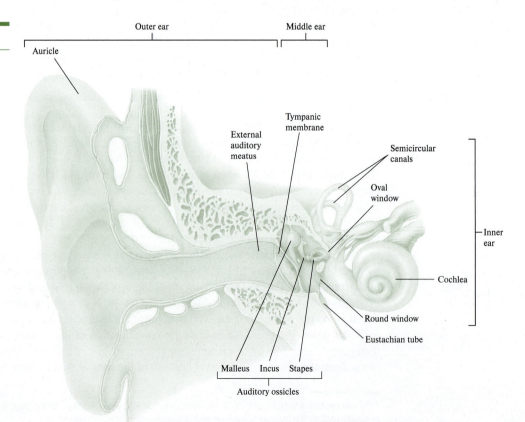

FIGURE 15–10

The sound transducer mechanism housed in the middle ear.

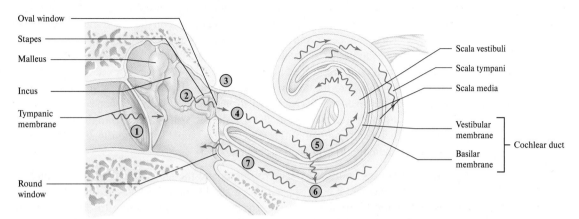

1. Sound waves strike the tympanic membrane and cause it to vibrate.

2. Vibration of the tympanic membrane causes the three bones of the middle ear to vibrate.

3. The foot plate of the stapes vibrates in the oval window.

4. Vibration of the foot plate causes the perilymph in the scala vestibuli to vibrate.

5. Vibration of the perilymph causes displacement of the basilar membrane. Short waves (high pitch) cause displacement of the basilar membrane near the oval window, and longer waves (low pitch) cause displacement of the basilar membrane some distance from the oval window. Movement of the basilar membrane is detected in the hair cells of the spiral organ, which are attached to the basilar membrane.

6. Vibrations of the perilymph in the scala vestibuli and of the endolymph in the cochlear duct are transferred to the perilymph of the scala tympani.

7. Vibrations in the perilymph of the scala tympani are transferred to the round window, where they are dampened.

vibrations. In addition, they perform the protective role of keeping debris and objects from reaching the inner ear. The Eustachian tube extends from the middle ear space to the upper part of the throat behind the soft palate. The tube is normally closed. Contraction of the palate muscles during yawning, chewing, or swallowing opens the tubes. This allows the middle ear to ventilate and equalize pressure. If external air pressure changes rapidly, for example, by a sudden change in elevation, the tube is opened by involuntary swallowing or yawning to equalize the pressure.

The sound transducer mechanism is housed in the middle ear.* It consists of the **tympanic membrane** (eardrum) and three **ossicles** (bones) (Figure 15–10). The ossicles are supported by ligaments and may be moved by two muscles or by deflection of the tympanic membrane. The muscle movement is involuntary. Loud sounds cause these muscles to contract, which stiffens and diminishes the movement of the ossicular chain (Borg and Counter, 1989). This offers some protection for the delicate inner ear structure from physical injury. The discussion on the middle ear that follows is excerpted from Clemis (1975).

The primary function of the middle ear in the hearing process is to transfer sound energy from the outer to the inner ear. As the eardrum vibrates, it transfers its motion to the malleus. Since the bones of the ossicular chain are connected to one another, the movements of the malleus are passed on to the incus, and finally to the stapes, which is imbedded in the oval window.

As the stapes moves back and forth in a rocking motion, it passes the vibrations into the inner ear through the oval window. Thus, the mechanical motion of the eardrum is effectively transmitted through the middle ear and into the fluid of the inner ear.

*A **transducer** is a device that transmits power from one system to another. In this case, sound power is converted to mechanical displacement, which is later measured and interpreted by the brain.

The sound-conducting transducer amplifies sound by two main mechanisms. First, the large surface area of the drum as compared to the small surface area of the base of the stapes (footplate) results in a hydraulic effect. The eardrum has about 25 times as much surface area as the oval window. All of the sound pressure collected on the eardrum is transmitted through the ossicular chain and is concentrated on the much smaller area of the oval window. This produces a significant increase in pressure.

The bones of the ossicular chain are arranged in such a way that they act as a series of levers. The long arms are nearest the eardrum, and the shorter arms are toward the oval window. The fulcrums are located where the individual bones meet. A small pressure on the long arm of the lever produces a much stronger pressure on the shorter arm. Since the longer arm is attached to the eardrum and the shorter arm is attached to the oval window, the ossicular chain acts as an amplifier of sound pressure. The magnification effect of the entire sound-conducting mechanism is about 22-to-1.

The inner ear houses both the balance receptors and the auditory receptors. The auditory receptors are in the **cochlea,** a bone shaped like a snail coiled two and one-half times around its own axis (see Figure 15–9). A cross section through the cochlea (Figure 15–11) reveals three compartments: the **scala vestibuli,** the **scala media,** and the **scala tympani.** The scala vestibuli and the scala tympani are connected at the apex of the cochlea. They are filled with a fluid called **perilymph** in which the scala media floats. The hearing organ, the **organ of Corti,** is housed in the scala media. The scala media contains a different fluid, endolymph, which bathes the organ of Corti.

The scala media is triangular in shape and is about 34 mm in length. As shown in Figure 15–11, there are cells growing up from the **basilar membrane.** They have a tuft of hair at one end and are attached to the hearing nerve at the other. A gelatinous membrane (**tectorial membrane**) extends over the hair cells and is attached to the **limbus spiralis.** The hair cells are embedded in the tectorial membrane.

Vibration of the oval window by the stapes causes the fluids of the three scalae to develop a wave-like motion. The movement of the basilar membrane and the tectorial membrane in opposite directions causes a shearing motion on the hair cells. The dragging of the hair cells sets up electrical impulses in the auditory nerves, which are transmitted to the brain.

The nerve endings near the oval and round windows are sensitive to high frequencies. Those near the apex of the cochlea are sensitive to low frequencies.

FIGURE 15–11

Cross section through the cochlea. (*Source:* U.S. Environmental Protection Agency, 1974.)

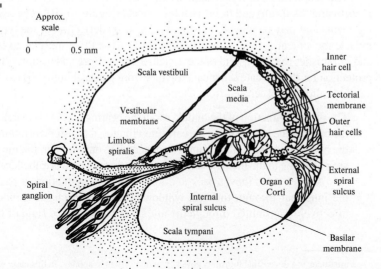

Normal Hearing

Frequency Range and Sensitivity.   The ear of the young, audiometrically healthy, adult male responds to sound waves in the frequency range of 20 to 16,000 Hz. Young children and women often have the capacity to respond to frequencies up to 20,000 Hz. The speech zone lies in the frequency range of 500 to 2000 Hz. The ear is most sensitive in the frequency range from 2000 to 5000 Hz. The smallest perceptible sound pressure in this frequency range is 20 μPa.

A sound pressure of 20 μPa at 1.000 Hz in air corresponds to a 1.0-nm displacement of the air molecules. The thermal motion of the air molecules corresponds to a sound pressure of about 1 μPa. If the ear were much more sensitive, you would hear the air molecules crashing against your ear like waves on the beach!

Loudness.   In general, two pure tones having different frequencies but the same SPL will be heard as different loudness levels. Loudness level is a psychoacoustic quantity.

In 1935, Fletcher and Munson conducted a series of experiments to determine the relationship between frequency and loudness (Fletcher and Munson, 1935). A reference tone and a test tone were presented alternately to the test subjects, who were asked to adjust the sound level of the test tone until it sounded as loud as the reference. The results were plotted as SPL in decibels versus the test tone frequency. The curves are called the **Fletcher–Munson, or equal loudness, contours.** The reference frequency is 1000 Hz. The curves are labeled in **phons,** which are the SPLs of the 1000-Hz pure tone in decibels. The lowest contour (dashed line) represents the "threshold of hearing." The actual threshold may vary by as much as ±10 dB between individuals with normal hearing.

Audiometry.   Hearing tests are conducted with a device known as an **audiometer.** Basically, it consists of a source of pure tones with variable SPL output into a pair of earphones. If the instrument also automatically prepares a graph of the test results (an **audiogram**), then it will include a weighting network called the **hearing threshold level** (HTL) scale.

The HTL scale is one in which the loudness of each pure tone is adjusted by frequency such that "0" dB is the level just audible for the average normal young ear. Two reference standards are in use: ASA-1951 and ANSI-1969. The ANSI reference values are shown in Figure 15–12. Note the similarity to the Fletcher–Munson contours. The initial audiogram prepared for an individual may be referred to as the **baseline HTL** or simply as the HTL.

The audiogram shown in Figure 15–13 reflects excellent hearing response. The average normal response may vary ±10 dB from the "0" dB value. As noted on the audiogram, this test was conducted with the ANSI-1969 weighting network.

You may have noted that we keep stressing young in our references to normal hearing. This is because hearing loss occurs due to the aging process, a type of loss called **presbycusis.**

FIGURE 15–12

The ANSI reference values for hearing threshold level.

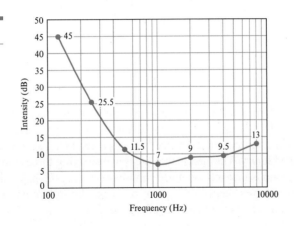

FIGURE 15–13

An audiogram illustrating excellent hearing response.

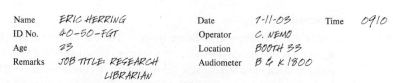

Name ERIC HERRING   Date 7-11-03   Time 0910
ID No. 40-50-FGT   Operator C. NEMO
Age 23   Location BOOTH 33
Remarks JOB TITLE: RESEARCH LIBRARIAN   Audiometer B & K 1800

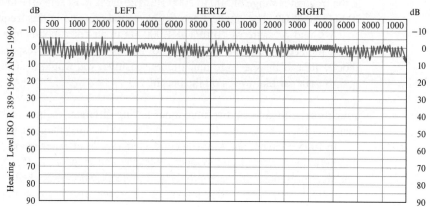

FIGURE 15–14

Various degrees of injury to the hair cells. (*Source:* U.S. Environmental Protection Agency, 1974.)

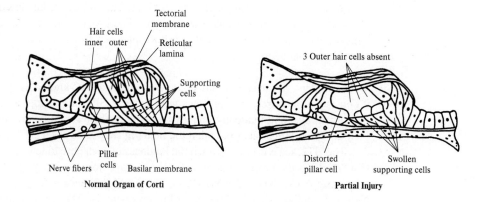

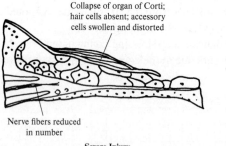

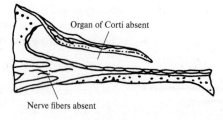

Hearing Impairment

Mechanism. With the exception of eardrum rupture from intense explosive noise, the outer and middle ear rarely are damaged by noise. More commonly, hearing loss is a result of neural damage involving injury to the hair cells (Figure 15–14). Two theories are offered to explain noise-induced injury. The first is that excessive shearing forces mechanically damage the hair cells. The second is that intense noise stimulation forces the hair cells into high metabolic activity, which overdrives them to the point of metabolic failure and consequent cell death. Once destroyed, hair cells are not capable of regeneration.

Measurement.   Because direct observation of the organ of Corti in persons having potential hearing loss is impossible, injury is inferred from losses in their HTL. The increased SPL required to achieve a new HTL is called **threshold shift.** Obviously, any measurement of threshold shift depends on having a baseline audiogram taken before the noise exposure.

Hearing losses may be either temporary or permanent. Noise-induced losses must be separated from other causes of hearing loss such as age (presbycusis), drugs, disease, and blows on the head. **Temporary threshold shift** (TTS) is distinguished from **permanent threshold shift** (PTS) by the fact that in TTS removal of the noise over stimulation will result in a gradual return to baseline hearing thresholds.

Factors Affecting Threshold Shift.   Important variables in the development of temporary and permanent hearing threshold changes include the following (NIOSH, 1972).

1. *Sound level.* Sound levels must exceed 60–80 dBA before the typical person will experience TTS.
2. *Frequency distribution of sound.* Sounds having most of their energy in the speech frequencies are more potent in causing a threshold shift than sounds having most of their energy below the speech frequencies.
3. *Duration of sound.* The longer the sound lasts, the greater the amount of threshold shift.
4. *Temporal distribution of sound exposure.* The number and length of quiet periods between periods of sound influences the potentiality of threshold shift.
5. *Individual differences in tolerance.* These differences may vary greatly among individuals.
6. *Type of sound* (*steady-state, intermittent, impulse, or impact*). The tolerance to peak sound pressure is greatly reduced by increasing the duration of the sound.

Temporary Threshold Shift (TTS).   TTS is often accompanied by a ringing in the ear, muffling of sound, or discomfort of the ears. Most of the TTS occurs during the first 2 h of exposure. Recovery to the baseline HTL after TTS begins within the first hour or two after exposure. Most of the recovery that is going to be attained occurs within 16 to 24 h after exposure.

Permanent Threshold Shift (PTS).   There appears to be a direct relationship between TTS and PTS. Noise levels that do not produce TTS after 2–8 h of exposure will not produce PTS if continued beyond this time. The shape of the TTS audiogram will resemble the shape of the PTS audiogram.

Noise-induced hearing loss generally is first characterized by a sharply localized dip in the HTL curve at the frequencies between 3000 and 6000 Hz. This dip commonly occurs at 4000 Hz (Figure 15–15). This is the **high-frequency notch.** The progress from TTS to PTS with continued noise exposure follows a fairly regular pattern. First, the high-frequency notch broadens and spreads in both directions. Although substantial losses may occur above 3000 Hz, the individual will not notice any change in hearing. In fact, the individual will not notice any hearing loss until the speech frequencies between 500 and 2000 Hz average more than a 25-dB increase in HTL on the ANSI-1969 scale. The onset and progress of noise-induced permanent hearing loss is slow and insidious. The exposed individual is unlikely to notice it. Total hearing loss from noise exposure has not been observed.

Acoustic Trauma.   The outer and middle ear rarely are damaged by intense noise. However, explosive sounds can rupture the tympanic membrane or dislocate the ossicular chain. The permanent hearing loss that results from very brief exposure to a very loud noise is termed **acoustic trauma** (Davis, 1958). Damage to the outer and middle ear may or may not accompany acoustic trauma.

Protective Mechanisms.   Although the extent and mechanisms are unclear, it appears that the structures of the middle ear offer some protection to the delicate sensory organs of the

FIGURE 15–15

An audiogram illustrating hearing loss at the high-frequency notch.

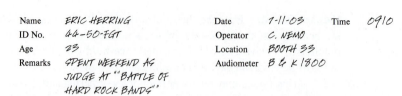

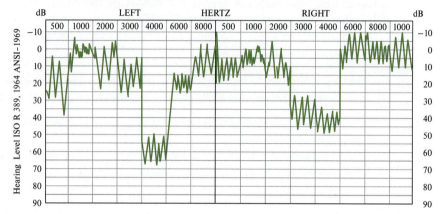

inner ear (Davis, 1957). One mechanism of protection is a change in the mode of vibration of the stapes. As noted earlier, there is some evidence that the muscles of the middle ear contract reflexively in response to loud noise. This contraction results in a reduction in the amplification that this series of levers normally produces. Changes in transmission may be on the order of 5 to 10 dB. However, the reaction time of the muscle–bone structure is on the order of 10 ms. Thus, this protection is not effective against steep acoustic wave fronts that are characteristic of impact or impulsive noise.

Damage-Risk Criteria

A **damage-risk criterion** specifies the maximum allowable exposure to which a person may be exposed if risk of hearing impairment is to be avoided. The American Academy of Ophthalmology and Otolaryngology has defined **hearing impairment** as an average HTL in excess of 25 dB (ANSI-1969) at 500, 1000, and 2000 Hz. This is called the **low fence. Total impairment** is said to occur when the average HTL exceeds 92 dB. Presbycusis is included in setting the 25-dB ANSI low fence. Two criteria have been set to provide conditions under which nearly all workers may be repeatedly exposed without adversely affecting their ability to hear and understand normal speech.

Continuous or Intermittent Exposure.   The National Institute for Occupational Safety and Health (NIOSH) has recommended that occupational noise exposure be controlled so that no worker is exposed in excess of the limits defined by line B in Figure 15–16. In addition, NIOSH recommends that new installations be designed to hold noise exposure below the limits defined by line A in Figure 15–16. The Walsh–Healey Act, which was enacted by Congress in 1969 to protect workers, used a damage-risk criterion equivalent to the line A criterion.

Speech Interference

As we all know, noise can interfere with our ability to communicate. Many noises that are not intense enough to cause hearing impairment can interfere with speech communication. The interference, or **masking,** effect is a complicated function of the distance between the speaker and listener and the frequency components of the spoken words. The speech interference level

FIGURE 15–16

NIOSH occupational noise exposure limits for continuous or intermittent noise exposure.

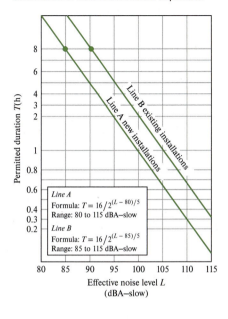

FIGURE 15–17

Quality of speech communication as a function of sound level and distance. (*Source:* U.S. Environmental Protection Agency, 1971.)

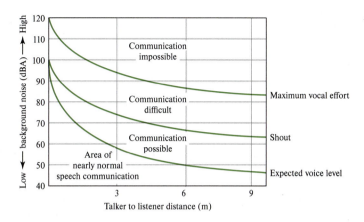

(SIL) was developed as a measure of the difficulty in communication that could be expected with different background noise levels. It is now more convenient to talk in terms of A-weighted background noise levels and the quality of speech communication (Figure 15–17).

EXAMPLE 15–5   Consider the problem of a speaker in a quiet zone who wishes to speak to a listener operating a 4.5-Mg truck 6.0 m away. The sound level in the truck cab is about 73 dBA.

Solution   Using Figure 15–17, we can see that she is going to have to shout very loudly to be heard. However, if she moved to within about 1.0 m, she would be able to use her "expected" voice level, that is, the unconscious slight rise in voice level that one would normally use in a noisy situation.

It can be seen that at distances not uncommon in living rooms or classrooms (4.5–6.0 m), the A-weighted background level must be below about 50 dB for normal conversation.

Annoyance

Annoyance by noise is a response to auditory experience. Annoyance has its base in the unpleasant nature of some sounds, in the activities that are disturbed or disrupted by noise, in the physiological reactions to noise, and in the responses to the meaning of "messages" carried by the noise (Miller, 1971). For example, a sound heard at night may be more annoying than one heard by day, just as one that fluctuates may be more annoying than one that does not. A sound that resembles another sound that we already dislike and that perhaps threatens us may be especially annoying. A sound that we know is mindlessly inflicted and will not be removed soon may be more annoying than one that is temporarily and regretfully inflicted. A sound, the source of which is visible, may be more annoying than one with an invisible source. A sound that is new may be less annoying. A sound that is locally a political issue may have a particularly high or low annoyance (May, 1978).

The degree of annoyance and whether that annoyance leads to complaints, product rejection, or action against an existing or anticipated noise source depend on many factors. Some of these factors have been identified, and their relative importance has been assessed. Responses to aircraft noise have received the greatest attention. There is less information available concerning responses to other noises, however, such as those of surface transportation and industry, and those from recreational activities (Miller, 1971). Many of the noise-rating or forecasting systems now in existence were developed in an effort to predict annoyance reactions.

Sonic Booms.   One noise of special interest with respect to annoyance is called **sonic boom** or, more correctly as we shall see, sonic booms.

The flow of air around an aircraft or other object whose speed exceeds the speed of sound (supersonic) is characterized by the existence of discontinuities in the air known as **shock wave.** These discontinuities result from the sudden encounter of an impenetrable body with air. At subsonic speeds, the air seems to be forewarned; thus, it begins its outward flow before the arrival of the leading edge. At supersonic speeds, however, the air in front of the aircraft is undisturbed, and the sudden impulse at the leading edge creates a region of overpressure (Figure 15–18) where the pressure is higher than atmospheric pressure. This overpressure region travels outward with the speed of sound, creating a conically shaped shock wave called the **bow wave** that changes the direction of airflow. A second shock wave, the **tail wave,** is produced by the tail of the aircraft and is associated with a region where the pressure is lower than normal. This underpressure discontinuity causes the air behind the aircraft to move sideways.

Major pressure changes are experienced at the ear as the bow and tail shock waves reach an observer. Each of these pressure deviations produces the sensation of an explosive sound (Minnix, 1978). Note that the pressure wave and, hence, the sonic boom exist whenever the aircraft is at supersonic speed and not "just when it breaks the sound barrier."

Both the loudness of the noise and the startling effect of the impulse (it makes us "jump") are found to be very annoying. Apparently we can never get used to this kind of noise. Supersonic flight by commercial aircraft is forbidden in the airspace above the United States. Supersonic flight by military aircraft is restricted to sparsely inhabited areas.

FIGURE 15–18

Sonic booms resulting from bow wave and tail wave set in motion by supersonic flight.

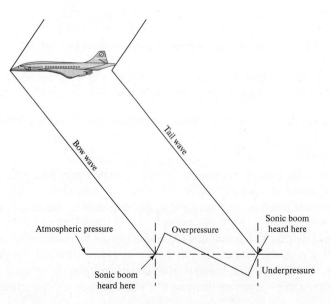

Sleep Interference

Sleep interference is a special category of annoyance that has received a great deal of attention and study (Miller, 1971). Almost all of us have been wakened or kept from falling asleep by loud, strange, frightening, or annoying sounds. It is commonplace to be wakened by an alarm clock or clock radio. But it also appears that one can get used to sounds and sleep through them. Possibly, environmental sounds only disturb sleep when they are unfamiliar. If so, disturbance of sleep would depend only on the frequency of unusual or novel sounds. Everyday experience also suggests that sound can help to induce sleep and, perhaps, to maintain it. The soothing lullaby, the steady hum of a fan, or the rhythmic sound of the surf, can serve to induce relaxation. Certain steady sounds can serve as an acoustical shade and mask disturbing transient sounds.

Common anecdotes about sleep disturbance suggest an even greater complexity. A rural person may have difficulty sleeping in a noisy urban area. An urban person may be disturbed by the quiet when sleeping in a rural area. And how is it that a parent may wake to a slight stirring of his or her child, yet sleep through a thunderstorm? These observations all suggest that the relations between exposure to sound and the quality of a night's sleep are complicated.

The effects of relatively brief noises (about 3 min or less) on a person sleeping in a quiet environment have been studied the most thoroughly. Typically, presentations of the sounds are widely spaced throughout a sleep period of 5 to 7 h. A summary of some of these observations is presented in Figure 15–19. The dashed lines are hypothetical curves that represent the percent of awakenings under conditions in which the subject is a normally rested young adult male who has been adapted for several nights to the procedures of a quiet sleep laboratory. He has been instructed to press an easily reached button to indicate that he has awakened, and had been moderately motivated to awake and respond to the noise.

While in light sleep, subjects can awake to sounds that are about 30–40 dB above the level at which they can be detected when subjects are conscious, alert, and attentive. While in deep sleep, the stimulus may have to be 50–80 dB above the level at which they can be detected by conscious, alert, attentive subjects before they will awaken the sleeping subject.

The solid lines in Figure 15–19 are data from questionnaire studies of persons who live near airports. The percentage of respondents who claim that flyovers wake them or keep them from falling asleep is plotted against the A-weighted sound level of a single flyover. These curves are for the case of approximately 30 flyovers spaced over the normal sleep period of 6 to

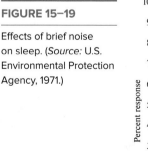

FIGURE 15–19

Effects of brief noise on sleep. (*Source:* U.S. Environmental Protection Agency, 1971.)

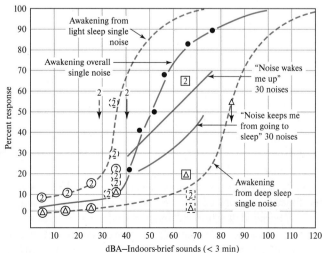

8 h. The filled circles represent the percentage of sleepers that awake to a 3-min sound at each A-weighted sound level (dBA) or lower. This curve is based on data from 350 persons, each tested in his or her own bedroom. These measures were made between 2:00 and 7:00 A.M. It is reasonable to assume that most of the subjects were roused from a light sleep.

Effects on Performance

When a task requires the use of auditory signals, speech or nonspeech, then noise at any intensity level sufficient to mask or interfere with the perception of these signals will interfere with the performance of the task.

Where mental or motor tasks do not involve auditory signals, the effects of noise on their performance have been difficult to assess. Human behavior is complicated, and it has been difficult to discover exactly how different kinds of noises might influence different kinds of people doing different kinds of tasks. Nonetheless, the following general conclusions have emerged. Steady noises without special meaning do not seem to interfere with human performance unless the A-weighted noise level exceeds about 90 dB. Irregular bursts of noise (intrusive noise) are more disruptive than steady noises. Even when the A-weighted sound levels of irregular bursts are below 90 dB, they may sometimes interfere with performance of a task. High-frequency components of noise, above about 1000–2000 Hz, may produce more interference with performance than low-frequency components of noise. Noise does not seem to influence the overall rate of work, but high levels of noise may increase the variability of the rate of work. "Noise pauses" may occur followed by compensating increases in work rate. Noise is more likely to reduce the accuracy of work than to reduce the total quantity of work. Complex tasks are more likely to be adversely influenced by noise than are simple tasks.

Acoustic Privacy

Without opportunity for privacy, either everyone must conform strictly to an elaborate social code or everyone must adopt highly permissive attitudes. Opportunity for privacy avoids the necessity for either extreme. In particular, without opportunity for acoustical privacy, one may experience all of the effects of noise previously described, and, in addition, one is constrained because one's own activities may disturb others. Without acoustical privacy, sound, like a faulty telephone exchange, reaches the "wrong number." The result disturbs both the sender and the receiver.

15–3 RATING SYSTEMS

Goals of a Noise-Rating System

An ideal noise-rating system is one that allows measurements by sound level meters or analyzers to be summarized succinctly and yet represent noise exposure in a meaningful way. In our previous discussions on loudness and annoyance, we noted that our response to sound is strongly dependent on the frequency of the sound. Furthermore, we noted that the type of noise (continuous, intermittent, or impulsive) and the time of day that it occurred (night being worse than day) were significant factors in annoyance.

Thus, the ideal system must take frequency into account. It should differentiate between daytime and nighttime noise. And, finally, it must be capable of describing the cumulative noise exposure. A statistical system can satisfy these requirements.

The practical difficulty with a statistical rating system is that it would yield a large set of parameters for each measuring location. A much larger array of numbers would be required to characterize a neighborhood. It is literally impossible for such an array of numbers to be used effectively in enforcement. Thus, considerable effort has been made to define a single number measure of noise exposure. The following paragraphs describe three of the systems now being used.

FIGURE 15–20

Cumulative distribution curve.

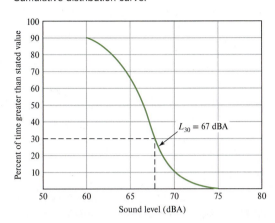

FIGURE 15–21

Probability distribution plot.

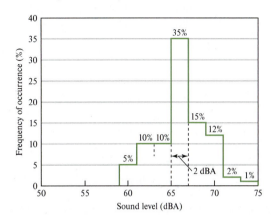

The L_N Concept

The parameter L_N is a statistical measure that indicates how frequently a particular sound level is exceeded. If, for example, we write $L_{30} = 67$ dBA, then we know that 67 dB(A) was exceeded for 30% of the measuring time. A plot of L_N against N where $N = 1\%, 2\%, 3\%$, and so forth, would look like the cumulative distribution curve shown in Figure 15–20.

Allied to the cumulative distribution curve is the probability distribution curve. A plot of this will show how often the noise levels fall into certain class intervals. In Figure 15–21 we can see that 35% of the time the measured noise levels ranged between 65 and 67 dBA; for 15% of the time they ranged between 67 and 69 dBA; and so on. The relationship between this picture and the one for L_N is really quite simple. By adding the percentages given in successive class intervals from right to left, we can arrive at a corresponding L_N, where N is the sum of the percentages and L is the lower limit of the left-most class interval added, thus, L_{30}

$$L(1 + 2 + 12 + 15) = 67 \text{ dBA}$$

The L_{eq} Concept

The equivalent continuous equal energy level (L_{eq}) can be applied to any fluctuating noise level. It is that constant noise level that, over a given time, expends the same amount of energy as the fluctuating level over the same period. It is expressed as follows:

$$L_{eq} = 10 \log \frac{1}{t} \int_{o}^{t} 10^{L(t)/10} \, dt \tag{15–14}$$

where t = the time over which L_{eq} is determined
$\quad L(t)$ = the time varying noise level in dBA

Generally speaking, there is no well-defined relationship between $L(t)$ and time, so a series of discrete samples of $L(t)$ have to be taken. This modifies the expression to

$$L_{eq} = 10 \log \sum_{i=1}^{i=n} 10^{L_i/10} t_i \tag{15–15}$$

where n = the total number of samples taken
$\quad L_i$ = the noise level in dBA of the ith sample
$\quad t_i$ = fraction of total sample time

EXAMPLE 15–6 Consider the case where a noise level of 90 dBA exists for 10 min and is followed by a reduced noise level of 70 dBA for 30 min. What is the equivalent continuous equal energy level for the 40-min period? Assume a 5-min sampling interval.

Solution If the sampling interval is 5 min, then the total number of samples (n) is 8, and the fraction of total sample time (t_i) for each sample is $1/8 = 0.125$. With these preliminary calculations, we may now compute the sum.

$$\sum_{t=1}^{2} = (10^{90/10})(0.250) + (10^{70/10})(0.750)$$

$$= (2.50 \times 10^8) + (7.50 \times 10^6) = 2.58 \times 10^8$$

And finally, we take the log to find

$$L_{eq} = 10 \log(2.58 \times 10^8) = 84.11, \text{ or } 84 \text{ dBA}$$

The example calculation is depicted graphically in Figure 15–22. From this you may note that great emphasis is put on occasional high noise levels.

The equivalent noise level was introduced in 1965 in Germany as a rating specifically to evaluate the effect of aircraft noise on the neighbors of airports (Burck et al., 1965). It was almost immediately recognized in Austria as appropriate for evaluating the effect of street traffic noise in dwellings and schoolrooms. It has been embodied in the national test standards of Germany for rating the subjective effects of fluctuating noises of all kinds, such as from street and road traffic, rail traffic, canal and river ship traffic, aircraft, industrial operations (including the noise from individual machines), sports stadiums, playgrounds, and the like.

The L_{dn} Concept

The L_{dn} is the L_{eq} computed over a 24-hour period with a "penalty" of 10 dBA for a designated nighttime period. Thus, it is a day-night average and the subscript "dn" is assigned instead of "eq." In applications to airport noise, the L_{dn} may be referred to as DNL. The nighttime period is from 10 P.M. to 7 A.M. The L_{dn} equation is derived from the L_{eq} equation with the time increment specified as 1 second. Because the time over which the L_{dn} is computed is a day, the total time period is 86,400 seconds. Equation 15–15 is then written as

$$L_{dn} = 10 \log\left[\frac{1}{86,400} \sum 10^{L_i/10} t_i + \sum 10^{(L_j+10)/10} t_i\right] \tag{15–16}$$

FIGURE 15–22

Graphical illustration of L_{eq} computation given in Example 15–6.

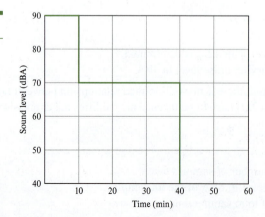

Because 10(log 86,400) ≈ 49.4, the day-night average sound level may be written as

$$L_{dn} = 10 \log \left[\sum 10^{L_i/10} t_i + \sum 10^{(L_j+10)/10} t_i \right] - 49.4 \qquad (15\text{–}17)$$

15–4   COMMUNITY NOISE SOURCES AND CRITERIA

It is not our intent to provide a detailed discussion of the noise characteristics of all community noise sources. Likewise, we have not attempted to provide a comprehensive list of noise criteria. Rather, we have selected a few examples to provide you with a feeling for the magnitude and range of the numbers.

Transportation Noise

Aircraft Noise.   The noise spectra of a wide body fan jet (e.g., the Boeing 747) reveal that SPLs are higher on takeoff than during the approach to land. This is typical of all aircraft. With the notable exception of the turbo jets, smaller aircraft have lower SPLs.

The annoyance criteria for aircraft operations are based on extensive field measurements and opinion surveys. The results of annoyance surveys at nine airports in the United States and Great Britain are summarized in Figure 15–23. L_{dn} is the L_{eq} for a 24-h period with a 10-dB penalty added to the sound levels that occur during the night which is defined as 10 P.M. to 7 A.M.

Highway Vehicle Noise.   For most automobiles, exhaust noise constitutes the predominant source for normal operation below about 55 km · h⁻¹ (Figure 15–24). Although tire noise is much less of a problem in automobiles than in trucks, it is the dominant noise source at speeds above 80 km · h⁻¹. Although not as noisy as trucks, the total contribution of automobiles to the noise environment is significant because of the very large number in operation.

Diesel trucks are 8–10 dB noisier than gasoline-powered ones. At speeds above 80 km · h⁻¹, tire noise often becomes the dominant noise source on the truck. The "crossbar" tread is the noisiest.

Motorcycle noise is highly dependent on the speed of the vehicle. The primary source of noise is the exhaust. The noise spectra of two-cycle and four-cycle engines are of somewhat different character. The two-cycle engines exhibit more high-frequency spectra energy content.

The U.S. Federal Highway Administration (FHA) has developed the traffic noise standards shown in Table 15–3. The levels are above those that would be expected to yield no problems but are below those of many existing highways.

FIGURE 15–23

Relationship between exposure to aircraft noise and annoyance. (*Source:* U.S. Environmental Protection Agency, 1971.)

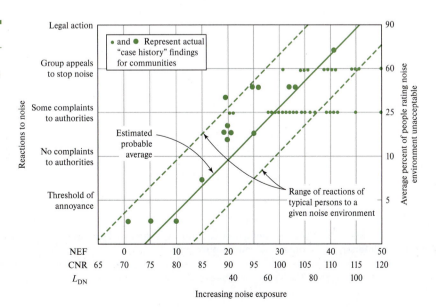

FIGURE 15–24

Typical noise spectra of automobiles. (*Source:* U.S. Environmental Protection Agency, 1971.)

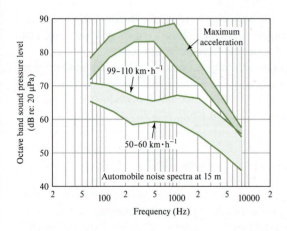

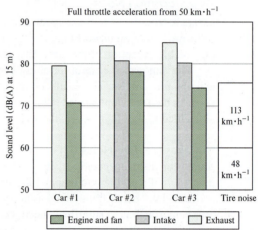

TABLE 15–3 FHA Noise Standards for New Construction

Land Use Category	Exterior Design Noise Level[a] (dBA)		Description of Land Use Category
	L_{eq}	L_{10}	
A	57	60	Tracts of lands in which serenity and quiet are of extraordinary significance and serve an important public need, and where the preservation of those qualities is essential if the area is to continue to serve its intended purpose. For example, such areas could include amphitheaters, particular parks or portions of parks, or open spaces, which are dedicated or recognized by appropriate local officials for activities requiring special qualities of serenity and quiet.
B	67	70	Residences, motels, hotels, public meeting rooms, schools, churches, libraries, hospitals, picnic areas, recreation areas, playgrounds active sports areas, and parks.
C	72	75	Developed lands, properties or activities not included in categories A and B.
D	Unlimited	Unlimited	Undeveloped lands.
E	52 (interior)	55 (interior)	Public meeting rooms, schools, churches, libraries, hospitals, and other such public buildings.

[a]Either L_{eq} or L_{10} may be used, but not both. The levels are to be based on a 1-h sample.

Source: FHWA, 1973.

TABLE 15–4	**Summary of Noise Characteristics of Internal Combustion Engines**				
Source	A-Weighted Noise Energy[a] (kW · h · day^{-1})	Typical A-Weighted Noise Level at 15.2 m (dBA)	8-h Exposure Level[b] (dBA)		Typical Exposure Time (h)
			Average	Maximum	
Lawn mowers	63	74	74	82	1.5
Garden tractors	63	78	N/A	N/A	N/A
Chain saws	40	82	85	95	1
Snow blowers	40	84	61	75	1
Lawn edgers	16	78	67	75	0.5
Model aircraft	12	78	70[c]	79[c]	0.25
Leaf blowers	3.2	76	67	75	0.25
Generators	0.8	71	—	—	—
Tillers	0.4	70	72	80	1

[a]Based on estimates of the total number of units in operation per day.

[b]Equivalent level for evaluation of relative hearing damage risk.

[c]During engine trimming operation.

Source: U.S. EPA, 1971.

Other Internal Combustion Engines

Because of their ubiquitous nature and the general interest they stimulate, the combustion engines listed in Table 15–4 are included at this point. "In general, these devices are not significant contributors to average residential noise levels in urban areas. However, the relative annoyance of most of the equipment tends to be high (U.S. EPA, 1971a)." The 8-h exposure level is in reference to the equipment operator.

Construction Noise

The range of sound levels found for 19 common types of construction equipment is shown in Figure 15–25. Although the sample was limited, the data appear to be reasonably accurate. The noise produced by the interaction of the machine and the material on which it acts often contributes greatly to the sound level.

It is difficult, at best, to quantify the annoyance that results from construction noise. The following generalizations appear to hold.

1. Single-house construction in suburban communities will generate sporadic complaints if the boundary line 8-h L_{eq} exceeds 70 dBA.
2. Major excavation and construction in a normal suburban community will generate threats of legal action if the boundary line 8-h L_{eq} exceeds 85 dBA.

Zoning and Siting Considerations

The U.S. Department of Housing and Urban Development (HUD) set out guideline criteria for noise exposure at residential sites for new construction (Table 15–5). The Federal Aviation Administration (FAA) has also specified noise levels for land use compatibility. These guidelines, and those given earlier for traffic noise (see Table 15–3), if followed in zoning and siting, will minimize annoyance and complaints.

Levels to Protect Health and Welfare

In accordance with the directive from Congress, the EPA published noise criteria levels that it deemed necessary to protect the health and welfare of U.S. citizens (Table 15–6) (U.S. EPA,

FIGURE 15–25

Range of sound
levels from various
types of construction
equipment (based on
limited available data
samples). (*Source:* U.S.
Environmental Protection
Agency, 1972.)

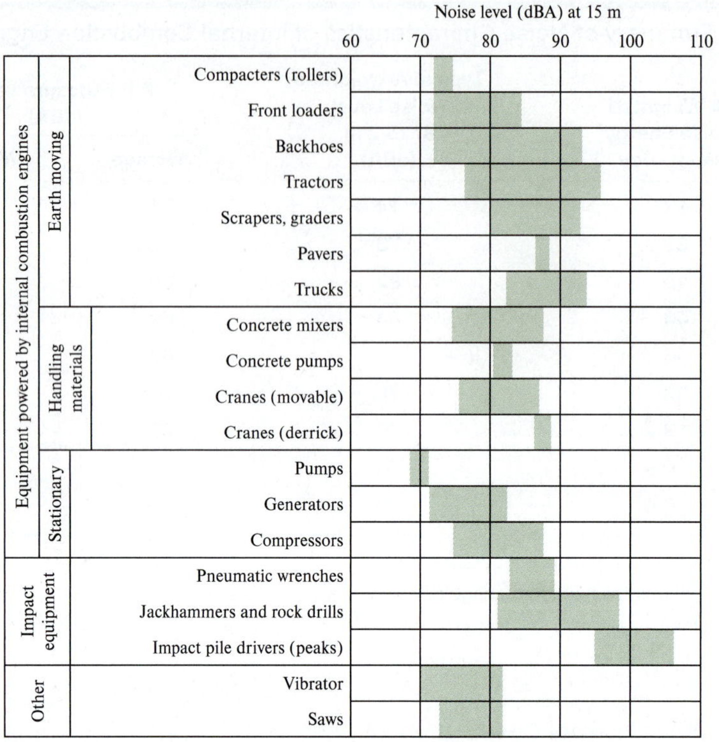

Noise level (dBA) at 15 m

TABLE 15–5   **HUD Noise Assessment Criteria for New Residential Construction**

General External Exposures	Assessment
Exceeds 89 dBA 60 min per 24 hours	Unacceptable
Exceeds 75 dBA 8 hours per 24 hours	
Exceeds 65 dBA 8 hours per 24 hours	Discretionary: normally unacceptable
Loud repetitive sounds on site	
Does not exceed 65 dBA more than 8 hours per 24 hours	Discretionary: normally acceptable
Does not exceed 45 dBA more than 30 min per 24 hours	Acceptable

1974). The EPA maintained that a quiet residential environment is necessary in both urban and rural areas to prevent activity interference and annoyance and to permit the hearing mechanism an opportunity to recuperate if it is exposed to high levels during the day. The L_{dn} of 45 provides a fair margin of safety.

15–5   TRANSMISSION OF SOUND OUTDOORS

Inverse Square Law

If a sphere of radius δ vibrates with a uniform radial expansion and contraction, sound waves radiate uniformly from its surface. If the sphere is placed such that no sound waves are reflected back in the direction of the source, and if the product $\kappa\delta$, where κ is the wave number,* is much

---

* $\kappa = 2\pi/\lambda$ where λ, wavelength, κ has units of reciprocal length, m^{-1}.

TABLE 15–6   Yearly Energy Average L_{eq} Identified as Requisite to Protect the Public Health and Welfare with an Adequate Margin of Safety

	Measure	Indoor			Outdoor		
		Activity Interference	Hearing Loss Consideration	To Protect Against Both Effects[b]	Activity Interference	Hearing Loss Consideration	To Protect Against Both Effects[b]
Residential with outside space and farm residences	L_{dn}	45		45	55		55
	$L_{eq(24)}$		70			70	
Residential with no outside space	L_{dn}	45		45			
	$L_{eq(24)}$		70				
Commercial	$L_{eq(24)}$	a	70	70[c]	a	70	70[c]
Inside transportation	$L_{eq(24)}$	a	70	a			
Industrial	$L_{eq(24)}$[d]	a	70	70[c]	a	70	70[c]
Hospitals	L_{dn}	45		45	55		55
	$L_{eq(24)}$		70			70	
Educational	$L_{eq(24)}$	45		45	55		55
	$L_{eq(24)}$[d]		70			70	
Recreational areas	$L_{eq(24)}$	a	70	70[c]	a	70	70[c]
Farm land and general unpopulated land	$L_{eq(24)}$				a	70	70[c]

[a]Because different types of activities appear to be associated with different levels, identification of a maximum level for activity interference may be difficult except when speech communication is a critical activity.

[b]Based on lowest level.

[c]Based only on hearing loss.

[d]An $L_{eq(8)}$ of 75 dB may be identified in these situations so long as the exposure over the remaining 16 h per day is low enough to result in negligible contribution to the 24-h average, that is, no greater than an L_{eq} of 60 dB.

Note: Explanation of identified level for hearing loss: The exposure period that results in hearing loss at the identified level is a period of 40 years.

Source: U.S. EPA, 1974.

less than 1, then the sound intensity at any radial distance r from the sphere is inversely proportional to the square of distance, that is,

$$I = \frac{W}{4\pi r^2} \tag{15–18}$$

where I = sound intensity (in $W \cdot m^{-2}$)
W = sound power of source (in W)

This is the **inverse square law.** It explains that portion of the reduction of sound intensity with distance that is due to wave divergence (Figure 15–26). For a line source such as a roadway or a railroad, the reduction of sound intensity is inversely proportional to r rather than r^2. From a practical standpoint it is difficult if not impossible to measure the sound power of the source. In such instances we measure the SPL at some known distance from the source and then use the inverse square law or radial dependence relationships to estimate the SPL at some other distance. For example, using the inverse square law, the sound pressure level, L_{p2}, at a distance r_2 from the source may be determined if the sound pressure level, L_{p1}, at some closer point, r_1, is known:

$$L_{p2} = L_{p1} - 10 \log\left(\frac{r_2}{r_1}\right)^2 \tag{15–19}$$

For a line source, the sound pressure level, L_{p2}, at a distance r_2 from the source, may be determined by a similar equation.

$$L_{p2} = L_{p1} - 10 \log\left(\frac{r_2}{r_1}\right) \tag{15–20}$$

Radiation Fields of a Sound Source

The character of the wave radiation from a noise source will vary with distance from the source (Figure 15–27). At locations close to the source, the **near field,** the particle velocity is not in phase with the sound pressure. In this area, L_p fluctuates with distance and does not follow the inverse square law. When the particle velocity and sound pressure are in phase, the location of the sound measurement is said to be in the **far field.** If the sound source is in free space, that is, there are no reflecting surfaces, then measurements in the far field are also **free field measurements.** If the sound source is in a highly reflective space, for example, a room with steel walls, ceiling, and floor, then measurements in the far field are also **reverberant field**

FIGURE 15–26

Illustration of inverse square law.

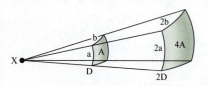

FIGURE 15–27

Variation of sound-pressure level in an enclosure along radius r from a noise source.

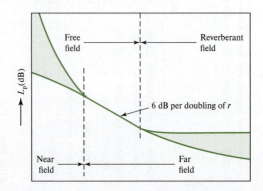

measurements. The shaded area in the far field of Figure 15–27 shows that L_p does not follow the inverse square law in the reverberant field.

Directivity

Most real sources do not radiate sound uniformly in all directions. If you were to measure the SPL in a given frequency band at a fixed distance from a real source, you would find different levels for different directions. If you plotted these data in polar coordinates, you would obtain the directivity pattern of the source.

The **directivity factor** is the numerical measure of the directivity of a sound source. In logarithmic form the directivity factor is called the **directivity index.** For a spherical source it is defined as follows:

$$DI_\theta = L_{p\theta} - L_{ps} \tag{15–21}$$

where $L_{p\theta}$ = sound pressure level measured at distance r' and angle θ from a directive source radiating power W into an echo-free (**anechoic**) space (in dB)

   L_{ps} = sound pressure level measured at distance r' from a nondirective source radiating power W into anechoic space (in dB). This is the same source as the directive source, but acting in the ideal fashion that we assumed in developing the inverse square law.

For a source located on or near a hard, flat surface, the directivity index takes the following form:

$$DI_\theta = L_{p\theta} - L_{ps} + 3 \tag{15–22}$$

The 3-dB addition is made because the measurement is made over a hemisphere instead of a sphere. That is, the intensity at a radius, r, is twice as large if a source radiates into a hemisphere rather than the ideal sphere we have used up to this point. Each directivity index is applicable only to the angle at which L_p was measured and only for the frequency at which it was measured.

We assume that the directivity pattern does not change its shape regardless of the distance from the source. This allows us to apply the inverse square law to directive sources simply by measuring the SPL at a distance r_1 along the directive angle of interest.

Airborne Transmission

It has become standard practice to restrict prediction of airborne transmission of noise to conditions favorable for propagation. These are specified as (ISO, 1989, 1990):

1. The wind direction is within an angle of 45° of the direction connecting the center of the sound source and the center of the specified area, with the wind blowing from the source to the receiver.
2. The wind speed is between approximately 1 and 5 m · s^{-1} measured at a height of 3–11 m.
3. Propagation in any near-horizontal direction under a well-developed ground-based inversion.

Effects of Atmospheric Conditions.   Sound energy is absorbed in quiet isotropic air by molecular excitation and relaxation of oxygen molecules and, at very low temperatures, by heat conduction and viscosity in the air. Molecular excitation is a complex function of the frequency of noise, humidity, and temperature. In general, we may say that as the humidity decreases, sound absorption increases. As the temperature increases to about 10–20°C (depending on the noise frequency), absorption increases. Above 25°C, absorption decreases. Sound absorption is higher at higher frequencies.

The vertical temperature profile greatly alters the propagation paths of sound. If a super-adiabatic lapse rate exists, sound rays bend upward and noise shadow zones are formed. If an inversion exists, sound rays are bent back toward the ground. This results in an increase in the sound level. These effects are negligible for short distances but may exceed 10 dB at distances over 800 m.

In a similar fashion, wind speed gradients alter the way noise propagates. Sound traveling with the wind is bent down, while sound traveling against the wind is bent upward. When sound waves are bent down, there is little or no increase in sound levels. But when sound waves are bent upward, sound levels can be noticeably reduced.

Basic Point Source Model.   A point source is one for which $\kappa\delta \ll 1$ and for which Equation 15–18 holds. According to Magrab,

> In practice most noise sources cannot be classified as simple point sources. However, the sound field of a complicated sound source will look as if it were a point source if the following two conditions are met: (1) $r/\delta \gg 1$, that is, the distance from the source is large compared to its characteristic dimension, and (2) $\delta/\lambda \ll r/\delta$, that is, the ratio of the size of the source to the wavelength of sound in the medium is small compared to the ratio of the distance from the source to its characteristic dimension. Recall that $r/\delta \gg 1$ from the first condition. A value of $r/\delta > 3$ is a sufficient approximation; therefore, $\delta\lambda \ll 3$ (Magrab, 1975).

The basic point source equation is

$$L_p \cong L_w - 20 \log r - 11 - A_e \qquad\qquad\text{(15–23)}$$

where L_p = the desired SPL (re: 20 μPa) at angle θ and distance r from source dB

$\quad\ L_w$ = the measured sound power level (re: 10^{-12} W) at angle θ, dB

$\quad\ A_e$ = attenuation for the distance r dB

With the exception of the last term (A_e), it is the inverse square law. The A_e term is the excess attenuation beyond wave divergence. It is caused by environmental conditions and has units of dB.

The A_e term may be further divided into five terms as follows:

A_{e1} = attenuation by absorption in the air, dB

A_{e2} = attenuation by the ground, dB

A_{e3} = attenuation by barriers, dB

A_{e4} = attenuation by foliage, dB

A_{e5} = attenuation by houses, dB

Because of the introductory nature of this text, we have chosen to limit the following discussion to consideration of the basic point source equation. For detailed examination of the other cases, we recommend that you consult Piercy and Daigle (1991).

EXAMPLE 15–7   The sound power level (re: 10^{-12} W) of a compressor is 124.5 dB at 1000 Hz. Determine the SPL 200 m downwind on a clear summer afternoon if the wind speed is 5 m · s^{-1}, the temperature is 20°C, the relative humidity is 50%, and the barometric pressure is 101.325 kPa. At these conditions, the attenuation by air absorption (A_{e1}) is 0.94 dB. The total ground attenuation (A_{e2}) is −2.21 dB, that is reflection increases the noise level.

Solution   Using the basic point source model (Equation 15–23), the SPL at the receiver is

$$L_p = 124.5 - 20 \log (200) - 11 - 0.94 - (-2.21)$$

$$= 124.5 - 46 - 11 - 0.94 + 2.21 = 68.77 \text{ or } 69 \text{ dB at } 1000 \text{ Hz}$$

15–6 TRAFFIC NOISE PREDICTION

L_{eq} Prediction

The Ontario Ministry of Transportation and Communications developed a traffic noise prediction equation based on the L_{eq} concept (Hajek, 1977). The empirical equation they developed is as follows:

$$L_{eq} = 42.3 + 10.2 \log(V_c + 6V_t) - 13.9 \log D + 0.13S \tag{15-24}$$

where L_{eq} = energy equivalent sound level during 1 h (in dBA)
   V_c = volume of automobiles (four tires only) (in vehicles $\cdot$ h^{-1})
   V_t = volume of trucks (six or more tires) (in vehicles $\cdot$ h^{-1})
   D = distance from edge of pavement to receiver (in m)
   S = average speed of traffic flow during 1 h (in km $\cdot$ h^{-1})

This equation does not account for barriers. A nomograph is available that takes barriers into account.

L_{dn} Prediction

The Ontario method is a direct extension of the L_{eq} methodology that enables the calculation of L_{dn}. The modified model has the following form:

$$L_{dn} = 31.0 + 10.2 \log[ADDT + \%T\ ADDT/20] - 13.9 \log D + 0.13S \tag{15-25}$$

where L_{dn} = equivalent A-weighted sound level during 24-h period with 10-dBA weighting applied to 2200 − 0700 h (in dBA)
 ADDT = annual average daily traffic (in vehicles $\cdot$ day^{-1})
  $\%T$ = average percentage of trucks during a typical day (%)

This equation, like Equation 15–24, does not take into account barriers.

15–7 NOISE CONTROL

Source-Path-Receiver Concept

If you have a noise problem and want to solve it, you have to find out something about what the noise is doing, where it comes from, how it travels, and what can be done about it. A straightforward approach is to examine the problem in terms of its three basic elements: that is, sound arises from a source, travels over a path, and affects a receiver or listener (Berendt, Corlis, and Ojalvo, 1976).

The source may be one or any number of mechanical devices that radiate noise or vibratory energy. Such a situation occurs when several appliances or machines are in operation at a given time in a home or office.

The most obvious transmission path by which noise travels is simply a direct line-of-sight air path between the source and the listener. For example, aircraft flyover noise reaches an observer on the ground by the direct line-of-sight air path. Noise also travels along structural paths. Noise can travel from one point to another via any one path or a combination of several paths. Noise from a washing machine operating in one apartment may be transmitted to another apartment along air passages such as open windows, doorways, corridors, or duct work. Direct physical contact of the washing machine with the floor or walls sets these building components into vibration. This vibration is transmitted structurally throughout the building, causing walls in other areas to vibrate and to radiate noise.

The receiver may be, for example, a single person, a classroom of students, or a suburban community.

Solution of a given noise problem might require alteration or modification of any or all of these three basic elements:

1. Modifying the source to reduce its noise output
2. Altering or controlling the transmission path and the environment to reduce the noise level reaching the listener
3. Providing the receiver with personal protective equipment

Control of Noise Source by Design

Reduce Impact Forces.   Many machines and items of equipment are designed with parts that strike forcefully against other parts, producing noise. Often, this striking action or impact is essential to the machine's function. Several steps can be taken to reduce noise from impact forces. The particular remedy to be applied will be determined by the nature of the machine in question.

A few of the more obvious design modifications are

1. Reduce the weight, size, or height of fall of the impacting mass.
2. Cushion the impact by inserting a layer of shock-absorbing material between the impacting surfaces. In some situations, you could insert a layer of shock-absorbing material behind each of the impacting heads or objects to reduce the transmission of impact energy to other parts of the machine.
3. Whenever practical, one of the impact heads or surfaces should be made of nonmetallic material to reduce resonance (ringing) of the heads.
4. Substitute the application of a small impact force over a long period for a large force over a short period to achieve the same result.

Reduce Speeds and Pressures.   Reducing the speed of rotating and moving parts in machines and mechanical systems results in smoother operation and lower noise output. Likewise, reducing pressure and flow velocities in air, gas, and liquid circulation systems lessens turbulence, resulting in decreased noise radiation. Some specific suggestions that may be incorporated in design are discussed the following sections.

Reduce Frictional Resistance.   Reducing friction between rotating, sliding, or moving parts in mechanical systems frequently results in smoother operation and lower noise output. Similarly, reducing flow resistance in fluid distribution systems results in less noise radiation. Four of the more important factors that should be checked to reduce frictional resistance in moving parts are alignment, polish, balance, and eccentricity (out-of-roundness).

Reduce Radiating Area.   Generally speaking, the larger the vibrating part or surface, the greater the noise output. The rule of thumb for quiet machine design is to minimize the effective radiating surface areas of the parts without impairing their operation or structural strength. This can be done by making parts smaller, removing excess material, or by cutting openings, slots, or perforations in the parts. For example, replacing a large, vibrating sheet-metal safety guard on a machine with a guard made of wire mesh or metal webbing might result in a substantial reduction in noise because of the drastic reduction in surface area of the part.

Reduce Noise Leakage.   In many cases, machine cabinets can be made into rather effective soundproof enclosures through simple design changes and the application of some sound-absorbing treatment. Substantial reductions in noise output may be achieved by adopting some of the following recommendations:

1. All unnecessary holes or cracks, particularly at joints, should be caulked.
2. All electrical or plumbing penetrations of the housing or cabinet should be sealed with rubber gaskets or a suitable nonsetting caulk.

3. If practical, all other functional or required openings or ports that radiate noise should be covered with lids or shields edged with soft rubber gaskets to achieve an airtight seal.
4. Other openings required for exhaust, cooling, or ventilation purposes should be equipped with mufflers or acoustically lined ducts.
5. Openings should be directed away from the operator and other people.

Isolate and Dampen Vibrating Elements.

In all but the simplest machines, the vibrational energy from a specific moving part is transmitted through the machine structure, forcing other component parts and surfaces to vibrate and radiate sound—often with greater intensity than that generated by the originating source itself.

Generally, vibration problems can be considered in two parts. First, we must prevent energy transmission between the source and surfaces that radiate the energy. Second, we must dissipate or attenuate the energy somewhere in the structure. The first part of the problem is solved by isolation. The second part is solved by damping.

The most effective method of vibration isolation involves the resilient mounting of the vibrating component on the most massive and structurally rigid part of the machine. Damping material or structures are those that have some viscous properties. They tend to bend or distort slightly, thus consuming part of the noise energy in molecular motion. The use of spring mounts on motors and laminated galvanized steel and plastic in air-conditioning ducts are two examples.

Provide Mufflers or Other Silencers.

There is no real distinction between mufflers and silencers. They are often used interchangeably. They are, in effect, acoustical filters and are used when fluid flow noise is to be reduced. The devices can be classified into two fundamental groups: absorptive and reactive mufflers. An **absorptive muffler** is one whose noise reduction is determined mainly by the presence of fibrous or porous materials, which absorb the sound. A **reactive muffler** is one whose noise reduction is determined mainly by geometry. It is shaped to reflect or expand the sound waves with resultant self-destruction.

Although several terms are used to describe the performance of mufflers, the most frequently used appears to be insertion loss (IL). **Insertion loss** is the difference between two SPLs that are measured at the same point in space before and after a muffler has been inserted. Because each muffler IL is highly dependent on the manufacturer's selection of materials and configuration, we will not present general IL prediction equations.

Noise Control in the Transmission Path

After you have tried all possible ways of controlling the noise at the source, your next line of defense is to set up devices in the transmission path to block or reduce the flow of sound energy before it reaches your ears. This can be done in several ways: (1) absorb the sound along the path, (2) deflect the sound in some other direction by placing a reflecting barrier in its path, or (3) contain the sound by placing the source inside a sound-insulating box or enclosure.

Selection of the most effective technique will depend on various factors, such as the size and type of source, intensity and frequency range of the noise, and the nature and type of environment.

Separation.

We can make use of the absorptive capacity of the atmosphere, as well as divergence, as a simple, economical method of reducing the noise level. Air absorbs high-frequency sounds more effectively than low-frequency sounds. However, if enough distance is available, even low-frequency sounds will be absorbed appreciably.

If you can double your distance from a point source, you will have succeeded in lowering the SPL by 6 dB. It takes about a 10-dB drop to halve the loudness. If you have to contend with a line source such as a railroad train, the noise level drops by only 3 dB for each doubling of distance from the source. The main reason for this lower rate of attenuation is that line sources radiate sound waves that are cylindrical in shape. The surface area of such waves only increases twofold for each doubling of distance from the source. However, when the distance from the train becomes comparable to its length, the noise level will begin to drop at a rate of 6 dB for each subsequent doubling of distance.

Indoors, the noise level generally drops only from 3 to 5 dB for each doubling of distance in the near vicinity of the source. However, farther from the source, reductions of only 1 or 2 dB occur for each doubling of distance due to the reflections of sound off hard walls and ceiling surfaces.

Absorbing Materials.   Noise, like light, will bounce from one hard surface to another. In noise control work, this is called **reverberation.** If a soft, spongy material is placed on the walls, floors, and ceiling, the reflected sound will be diffused and soaked up (absorbed).

Sound-absorbing materials such as acoustical tile, carpets, and drapes placed on ceiling, floor, or wall surfaces can reduce the noise level in most rooms by about 5–10 dB for high-frequency sounds, but only 2–3 dB for low-frequency sounds. Unfortunately, such treatment provides no protection to an operator of a noisy machine who is in the midst of the direct noise field. For greatest effectiveness, sound-absorbing materials should be installed as close to the noise source as possible.

Because of their light weight and porous nature, acoustical materials are ineffectual in preventing the transmission of either airborne or structure-borne sound from one room to another. In other words, if you can hear people walking or talking in the room or apartment above, installing acoustical tile on your ceiling will not reduce the noise transmission.

Acoustical Lining.   Noise transmitted through ducts, pipe chases, or electrical channels can be reduced effectively by lining the inside surfaces of such passageways with sound-absorbing materials. In typical duct installations, noise reductions on the order of 10 dB $\cdot$ m^{-1} for an acoustical lining 2.5 cm thick are well within reason for high-frequency noise. A comparable degree of noise reduction for the lower frequency sounds is considerably more difficult to achieve because it usually requires at least a doubling of the thickness or the length of acoustical treatment.

Barriers and Panels.   Placing barriers, screens, or deflectors in the noise path can be an effective way of reducing noise transmission, provided that the barriers are large enough in size, and depending on whether the noise is high or low frequency. High-frequency noise is reduced more effectively than low frequency.

The effectiveness of a barrier depends on its location, its height, and its length. Referring to Figure 15–28, we can see that the noise can follow four different paths.

First, the noise follows a direct path to receivers who can see the source well over the top of the barrier. The barrier does not block their line of sight and therefore provides no attenuation. No matter how absorptive the barrier is, it cannot pull the sound downward and absorb it.

FIGURE 15–28

Noise paths from a source to a receiver. (*Source:* National Cooperative Highway Research Program Report 174, Washington, DC.)

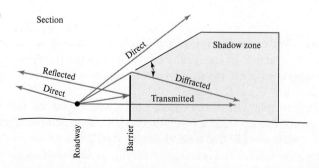

Second, the noise follows a diffracted path to receivers in the shadow zone of the barrier. The noise that passes just over the top edge of the barrier is diffracted (bent) down into the apparent shadow zone in the figure. The larger the angle of diffraction, the more the barrier attenuates the noise in this shadow zone. In other words, less energy is diffracted through large angles than through smaller ones.

Third, in the shadow zone, the noise transmitted directly through the barrier may be significant in some cases. For example, with extremely large angles of diffraction, the diffracted noise may be less than the transmitted noise. In this case, the transmitted noise compromises the performance of the barrier. It can be reduced by constructing a heavier barrier. The allowable amount of transmitted noise depends on the total barrier attenuation desired.

The fourth path shown in Figure 15–28 is the reflected path. After reflection, the noise is of concern only to a receiver on the opposite side of the source. For this reason, acoustical absorption on the face of the barrier may sometimes be considered to reduce this reflected noise; however, this treatment will not benefit any receivers in the shadow zone. It should be noted that in most practical cases the reflected noise does not play an important role in barrier design. If the source of noise is represented by a line of noise, another short-circuit path is possible. Part of the source may be unshielded by the barrier. For example, the receiver might see the source beyond the ends of the barrier if the barrier is not long enough. This noise from around the ends may compromise, or short-circuit, barrier attenuation. The required barrier length depends on the total net attenuation desired. When 10–15-dB attenuation is desired, barriers must, in general, be very long. Therefore, to be effective, barriers must not only break the line of sight to the nearest section of the source but also to the source far up and down the line.

Of these four paths, the noise diffracted over the barrier into the shadow zone represents the most important parameter from the barrier design point of view. Generally, the determination of barrier attenuation or barrier noise reduction involves only calculation of the amount of energy diffracted into the shadow zone.

Enclosures.   Sometimes it is much more practical and economical to enclose a noisy machine in a separate room or box than to quiet it by altering its design, operation, or component parts. The walls of the enclosure should be massive and airtight to contain the sound. Absorbent lining on the interior surfaces of the enclosure will reduce the reverberant buildup of noise within it. Structural contact between the noise source and the enclosure must be avoided, so that the source vibration is not transmitted to the enclosure walls, thus short-circuiting the isolation.

Control of Noise Source by Redress

The best way to solve noise problems is to design them out of the source. However, we are frequently faced with an existing source that, either because of age, abuse, or poor design, is a noise problem. The result is that we must redress, or correct, the problem as it currently exists. The following identify some measures that might apply if you are allowed to tinker with the source: balance rotating parts; reduce frictional resistance; apply damping materials; seal noise leaks; and perform routine maintenance to repair mufflers, rough road surfaces, and so forth.

Protect the Receiver

When All Else Fails.   When exposure to intense noise fields is required and none of the measures discussed so far is practical, as, for example, for the operator of a chain saw or pavement breaker, then measures must be taken to protect the receiver. The following two techniques are commonly employed.

Alter Work Schedule.   Limit the amount of continuous exposure to high noise levels. In terms of hearing protection, it is preferable to schedule an intensely noisy operation for a short interval of time each day over a period of several days rather than a continuous 8-h run for a day or two.

FIGURE 15–29

Attenuation of ear protectors at various frequencies. (*Source:* National Bureau of Standards Handbook, Berendt et al., p. 119, 1976.)

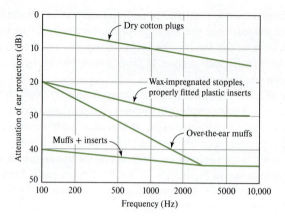

In industrial or construction operations, an intermittent work schedule would benefit not only the operator of the noisy equipment, but also other workers in the vicinity. If an intermittent schedule is not possible, then workers should be given relief time during the day. They should take their relief time at a low-noise-level location and should be discouraged from trading relief time for dollars, paid vacation, or an "early out" at the end of the day!

Inherently noisy operations, such as street repair, municipal trash collection, factory operation, and aircraft traffic, should be curtailed at night and early morning to avoid disturbing the sleep of the community. Remember: operations between 10 P.M. and 7 A.M. are effectively 10 dBA higher than the measured value.

Ear Protection.   Molded and pliable earplugs, cup-type protectors, and helmets are commercially available as hearing protectors. Such devices may provide noise reductions ranging from 15 to 35 dB (Figure 15–29). Earplugs are effective only if they are properly fitted by medical personnel. As shown in Figure 15–29, maximum protection can be obtained when both plugs and muffs are employed. Only muffs that have a certification stipulating the attenuation should be used.

These devices should be used only as a last resort, after all other methods have failed to lower the noise level to acceptable limits. Ear protection devices should be used while operating lawn mowers, mulchers, and chippers, and while firing weapons at target ranges. It should be noted that protective ear devices do interfere with speech communication and can be a hazard in some situations where warning calls may be a routine part of the operation (e.g., TIMBERRRR!). A modern ear-destructive device is a portable digital music player that uses earphones. In this "reverse" muff, high noise levels are directed at the ear without attenuation. If you can hear someone else's radio–recorder, that person is subjecting himself to noise levels in excess of 90–95 dBA!

CHAPTER REVIEW

When you have completed studying this chapter, you should be able to do the following without the aid of your textbook or notes:

1. Define frequency, based on a sketch of a harmonic wave you have drawn, and state its units of measure (namely, hertz, Hz).
2. State the basic unit of measure used in measuring sound energy (namely, the decibel, dB) and explain why it is used.
3. Define sound pressure level in mathematical terms, that is,

$$SPL = 20 \log \frac{P_{rms}}{(P_{rms})_o}$$

4. Explain why a weighting network is used in a sound level meter.

5. List the three common weighting networks and sketch their relative frequency response curves. (Label frequencies, that is, 20, 1000, and 10,000 Hz; and relative response, that is, 0, −5, −20, and −45 dB, as in Figure 15–6.)

6. Differentiate between a mid/high-frequency noise source and a low-frequency source on the basis of A, B, and C scale readings.

7. Explain the purpose of octave band analysis.

8. Differentiate among continuous, intermittent, and impulsive noise.

9. Sketch the curves and label the axes of the two typical types of impulsive noise.

10. Sketch a Fletcher–Munson curve, label the axes, and explain what the curve depicts.

11. Define *phon*.

12. Explain the mechanism by which hearing damage occurs.

13. Explain what hearing threshold level (HTL) is.

14. Define *presbycusis* and explain why it occurs.

15. Distinguish between temporary threshold shift (TTS), permanent threshold shift (PTS), and acoustic trauma with respect to cause of hearing loss, duration of exposure, and potential for recovery.

16. Explain why impulsive noise is more dangerous than steady-state noise.

17. Explain the relationship between the allowable duration of noise exposure and the allowable level for hearing protection, that is, damage-risk criteria.

18. List five effects of noise other than hearing damage.

19. List the three basic elements that might require alteration or modification to solve a noise problem.

With the aid of this text, you should be able to do the following:

1. Calculate the resultant sound pressure level from a combination of two or more sound pressure levels.

2. Determine the A-, B-, and C-weighted sound levels from octave band readings.

3. Compute the mean sound level from a series of sound level readings.

4. Compute the following noise statistics if you are provided the appropriate data: L_N, L_{eq}, L_{dn}.

5. Determine whether or not a noise level will be acceptable given a series of measurements and the criteria listed in Tables 15–3, 15–5, and 15–6.

6. Calculate the sound level at a receptor site after transmission through the atmosphere.

7. Estimate the noise level (L_{eq} or L_{dn}) that might be expected for a given roadway configuration and traffic pattern.

PROBLEMS

15–1 A building located near a road is 6.92 m high. How high is the building in terms of wavelengths of a 50.0-Hz sound? Assume that the speed of sound is 346.12 m · s^{-1}.

   Answer: One wavelength

15–2 Repeat Problem 15–1 for a 500-Hz sound if the temperature is 25.0°C.

15–3 Determine the sum of the following sound levels (all in dB): 68, 82, 76, 68, 74, and 81.

   Answer: 85.5 or 86 dB

15–4 A motorcyclist is warming up his racing cycle at a racetrack approximately 200 m from a sound level meter. The meter reading is 56 dBA. What meter reading would you expect if 15 of the motorcyclist's friends join him with motorcycles having exactly the same sound emission characteristics? You may assume that the sources may be treated as ideal point sources located at the same point.

15–5 A sound power level reading of 127 dB was taken near a construction site where chippers were being used. When all but one of the chippers stopped working, the sound power level reading was 120 dB. Estimate the number of chippers in operation when the reading of 127 was obtained. You may assume that the sources may be treated as ideal point sources located at the same point.

15–6 A law enforcement officer has taken the following readings with her sound level meter. Is the noise source a predominantly low- or middle-frequency emitter? Readings: 80 dBA, 84 dBB, and 90 dBC.

    Answer: Predominantly low frequency

15–7 The following readings have been made outside the open stage door of the opera house: 109 dBA, 110 dBB, and 111 dBC. Is the singer a bass or a soprano? Explain how you arrived at your answer.

15–8 Convert the following octave band measurements to an equivalent A-weighted sound level.

Band Center Frequency (Hz)	Band Level (dB)	Band Center Frequency (Hz)	Band Level (dB)
31.5	78	1000	80
63	76	2000	80
125	78	4000	73
250	82	8000	65
500	81		

    Answers: 85.5 or 86 dBA

15–9 The following noise spectrum was obtained from a jet aircraft flight overhead at an altitude of 250 m. Compute the equivalent A-weighted sound level using sound power level addition in a spreadsheet program you have written.

Band Center Frequency (Hz)	Band Level (dB)	Band Center Frequency (Hz)	Band Level (dB)
125	85	1000	100
250	88	2000	104
500	96	4000	101

15–10 Using the typical noise spectrum for automobiles traveling at 50 to 60 km · h^{-1}, determine the equivalent A-weighted level using sound power level addition in a spreadsheet program you have written. The following band levels were estimated from Figure 15–24.

Band Center Frequency (Hz)	Band Level (dB)	Band Center Frequency (Hz)	Band Level (dB)
63	67	1000	59
125	64	2000	55
250	58	4000	51
500	59	8000	45

15–11 You have been asked to evaluate the A-weighted sound level of a new-model lawn mower and make a recommendation on an acceptable noise spectrum to achieve 74 dBA. Three approaches are being considered by the manufacturer: (1) an improved muffler that will reduce the sound level 3 dB in each frequency band; (2) a reduction in the speed of the mower, which will reduce the sound level 5 dB in each frequency band; and (3) an engine redesign that will reduce the sound level 15 dB in the five highest frequency bands. Using a computer spreadsheet program you have written, compute the A-weighted sound level for the sound spectrum shown below and develop a recommended noise spectrum based on the manufacturer's alternatives that results in a sound level of less than 74 dBA. Assume that each of the alternative reductions may be added together (by decibel addition) in each frequency band in which it is applicable.

Band Center Frequency (Hz)	Band Level (dB)	Band Center Frequency (Hz)	Band Level (dB)
63	78	1000	79
125	76	2000	80
250	76	4000	78
500	77	8000	70

15–12 Compute the average SPL of the following readings by simple arithmetic averaging and by logarithmic averaging (Equation 15–13) (all readings in dB): 42, 50, 65, 71, and 47. Does arithmetic averaging underestimate or overestimate the SPL?

Answers: $x = 55.00$ or 55 dB; $L_p = 61.57$ or 62 dB

15–13 Repeat Problem 15–12 for the following data (all in dB): 76, 59, 35, 69, and 72.

15–14 The following noise record was obtained in the front yard of a home. Is this a relatively quiet or a relatively noisy neighborhood? Determine the equivalent continuous equal energy level.

Time (h)	Sound Level (dBA)	Time (h)	Sound Level (dBA)
0000–0600	42	1500–1700	50
0600–0800	45	1700–1800	47
0800–0900	50	1800–0000	45
0900–1500	47		

Answers: It is a quiet neighborhood. $L_{eq} = 46.2$ or 46 dBA

15–15 A developer has proposed putting a small shopping mall next to a very quiet residential area in Nontroppo, Michigan. Based on the measurements given below, which were taken at a similar size mall in a similar setting, should the developer expect complaints or legal action? Calculate L_{eq}.

Time (h)	Sound Level (dBA)	Time (h)	Sound Level (dBA)
0000–0600	42	2000–2200	68
0600–0800	55	2200–0000	57
0800–1000	65	1800–0000	45
1000–2000	70		

15–16 The U.S. EPA (1974) estimated that the following was a typical noise exposure pattern for a factory worker living in an urban area. Estimate the L_{dn} for the exposure shown.

Time (h)	Sound Level (dBA)	Time (h)	Sound Level (dBA)
0000–0500	52	1200–1530	90
0500–0700	78	1530–1800	52
0700–1130	90	1800–2200	60
1130–1200	70	2200–0000	52

15–17 The U.S. EPA (1974) estimated that the following was a typical noise exposure pattern for a middle school student living in an urban area. Estimate the L_{dn} for the exposure shown.

Time (h)	Sound Level (dBA)	Time (h)	Sound Level (dBA)
0000–0700	52	1500–1700	75
0700–0900	82	1700–1800	90
0900–1200	60	1800–2100	60
1200–1300	65	2100–0000	52
1300–1500	60		

15–18 Two oil-fired boilers for a 600 megawatt (MW) power plant produce a sound power level of 139 dB (re: 10^{-12} W) at 4000 Hz, from the induced draft fans. Determine the SPL 408.0 m downwind on a clear winter night when the wind speed is 4.50 m · s^{-1}, the temperature is 0.0°C, the relative humidity is 30.0%, and the barometric pressure is 101.3 kPa. The air attenuation at these conditions is 28.15 dB. The height of the boiler is 12 m. The ground attenuation is −2.86 dB.

Answer: SPL at 408.0 m = 50.50 or 50 dB at 4000 Hz

15–19 The 125-Hz sound power level (re: 10^{-12} W) from a jet engine test cell is 149 dB. What is the 125-Hz SPL 1200 m downwind on a clear summer morning during an inversion when the wind speed is 1.50 m · s^{-1}? The temperature is 25.0°C, the relative humidity is 70.0%, and the barometric pressure is 101.3. The air attenuation under these conditions is 0.36 dB. The ground attenuations is 1.79 dB.

15–20 Consider an ideal single lane of road that carries 1200 vehicles per hour uniformly spaced along the road and determine the following:

(a) The average center-to-center spacing of the vehicles for an average traffic speed of 40.0 km · h^{-1}

(b) The number of vehicles in a 1-kilometer length of the lane when the average speed is 40.0 km · h^{-1}

(c) The sound level (dBA) 60.0 m from a 1-kilometer length of this roadway with automobiles emitting 71 dBA at the edge of an 8.0-m-wide roadway. Assume that the autos travel at a speed of 40.0 km · h^{-1}, that the sound radiates ideally from a hemisphere, and that contributions of less than 0.3 dBA may be ignored.

Answers: (a) Average center-to-center spacing = 33.3 m.
(b) Number of vehicles in a kilometer length = 30 veh · km^{-1}.
(c) $L_p = 47.47$ or 48 dBA.

15–21   Repeat Problem 15–20 if the vehicle speed is increased to 80.0 km · h^{-1} and the spacing is decreased or increased appropriately to maintain 1200 vehicles per hour.

15–22   In preparation for a public hearing on a proposed interstate bypass at Nontroppo, Michigan, the county road commission has requested that you prepare an estimate of the potential for violation of FHA noise standards 75 meters from the interstate.

Data for I-481 at Pianissimo Avenue estimated traffic:
Automobiles: 7800 per hour at 88.5 km · h^{-1}
Medium trucks: 520 per hour at 80.5 km · h^{-1}
Heavy trucks: 650 per hour at 80.5 km · h^{-1}

Compute the unattenuated L_{eq}, at the receiver for autos only.

Answer: $L_{eq} = 70$ dBA

15–23   Determine the potential for violation of FHA noise standards at the north side of Fermata School 123.17 m from the edge of the interstate highway.

Data for I-481 at Fermata School estimated traffic:
Same as at Pianissimo Avenue (Problem 15–22)

Compute the unattenuated L_{eq}, at the school for the combined traffic flow.

DISCUSSION QUESTIONS

15–1   Classify each of the following noise sources by type, that is, continuous, intermittent, or impulse. (Not all sources fit these three classifications.)

(a) Electric saw   (b) Air conditioner   (c) Alarm clock (bell type)   (d) Punch press

15–2   Is the following statement true or false? If it is false, correct it in a nontrivial manner.

"A sonic boom occurs when an aircraft breaks the sound barrier."

15–3   Is the following statement true or false? If it is false, correct it in a nontrivial manner.

"Excessive continuous noise causes hearing damage by breaking the stapes."

15–4   As the safety officer of your company, you have been asked to determine the feasibility of reducing exposure time as a method of reducing hearing damage for the following situation:

The worker is operating a high-speed grinder on steel girders for a high-rise building. The effective noise level at the operator's ear is 100 dBA. She cannot wear protective ear devices because she must communicate with others.

What amount of exposure time would you set as the limit?

FE EXAM FORMATTED PROBLEMS

15–1   Assuming the basic point source model applies (i.e., ignore attenuation factors), at what distance is a 100 dBA sound pressure level measured 2 m from a source attenuated to 70 dBA?

(a)   15 m

(b)   200 m

(c)   80 m

(d)   5000 m

15–2   What sound power level results from combining the following three levels: 72 dB, 88 dB, and 90 dB?

(a)   96 dB

(b)   250 dB

(c)   110 dB

(d)   92 dB

15–3   Compute the mean sound pressure level of the following three readings: 36 dBA, 76 dBA, and 83 dBA.

(a)   77 dBA

(b)   65 dBA

(c)   86 dBA

(d)   80 dBA

15–4   The noise level outside of a school should not exceed 70 dBA. If the sound pressure level 5 m from the edge of a roadway is 80 dBA, what is the sound pressure level outside of a school that is 30 m from the edge of the roadway? Assume that the roadway is a continuous line source.

(a)   64 dBA

(b)   72 dBA

(c)   88 dBA

(d)   62 dBA

REFERENCES

Achitoff, L. (1973) "Aircraft Noise—A Threat to Aviation," *Journal of Water, Air and Soil Pollution,* 2(3): 357–63.

Berendt, R. D., E. L. R. Corlis, and M. S. Ojalvo (1976) *Quieting: A Practical Guide to Noise Control,* National Bureau of Standards Handbook 119, U.S. Department of Commerce, Washington, DC, pp. 16–41.

Borg, E., and S. A. Counter (1989) "The Middle-Ear Muscles," *Scientific American,* August, pp. 74–79.

Burck, W. et al. (1965) "Gutachten erstatet im Auftrag des Bundesministers für Gesundheitswesen," *Fluglärm,* Göttingen.

Clemis, J. D. (1975) "Anatomy, Physiology, and Pathology of the Ear," in J. B. Olishifski and E. R. Harford, eds., *Industrial Noise and Hearing Conservation,* National Safety Council, Chicago, IL, p. 213.

Davis, H. (1957) "The Hearing Mechanism," in C. M. Harris, ed., *Handbook of Noise Control,* McGraw-Hill, New York, pp. 4–6.

Davis, H. (1958) "Effects of High Intensity Noise on Navy Personnel," *U.S. Armed Forces Medical Journal,* 9: 1027–47.

De Neufville, R., and A. R. Odoni (2003) *Airport Systems: Planning, Design, and Management,* McGraw-Hill, New York, p. 198.

Firdaus, G., and A. Ahmed (2010) "Noise Pollution and Human Health: A Case Study of Municipal Corporation of Delhi," *Indoor Built Environ.,* 19:648–656.

FHWA (1973) *Policy and Procedure Memorandum 90-2, Noise Standards and Procedures,* U.S. Department of Transportation, Washington, DC, http://www.fhwa.dot.gov/environment

FHWA (2005) "Priority, Market-Ready Technologies and Innovations, FHWA Traffic Noise Model®, Version 2.1," Federal Highway Administration, U.S. Department of Transportation, Washington, DC.

Fletcher, H., and W. A. Munson (1935, Oct.) "Loudness, Its Definition, Measurement and Calculation," *Journal of Acoustic Society of America,* 5: 82–105.

Garg, N., and S. Maji (2016) "A Retrospective View of Noise Pollution Control Policy in India: Status, Proposed Revisions and Control Measures," *Current Science,* 111:29–38.

GRC (1972) *A Primer of Noise Measurement,* General Radio Company, Concord, MA.

Hajek, J. J. (1977) "L_{eq} Traffic Noise Prediction Method," *Environmental and Conservation in Transportation: Energy, Noise and Air Quality,* Transportation Research Board, National Academy of Science, Transportation Research Record No. 648, Washington, DC, pp. 48–73.

ISO (1989) *Acoustics—Attenuation of Sound During Propagation Outdoors, Part 2, A General Method of Calculation,* International Organization for Standardization, ISO/DIS 9613-2, Geneva, Switzerland.

ISO (1990) *Acoustics—Attenuation of Sound During Propagation Outdoors, Part 1, Calculation of Absorption of Sound by the Atmosphere,* International Organization for Standardization, ISO/DIS 9613-1, Geneva, Switzerland.

Kryter, K. D., et al. (1971) *Nonauditory Effects of Noise,* Report of WG-63, National Academy of Science, Washington, DC.

Kugler, B. A., D. E. Commins, and W. J. Galloway (1976) *Highway Noise: A Design Guide for Protection and Control,* National Cooperative Highway Research Program Report 174, Washington, DC.

Magrab, E. B. (1975) *Environmental Noise Control,* John Wiley & Sons, New York, pp. 4 & 6.

May, D. (1978) *Handbook of Noise Assessment,* Van Nostrand Reinhold, New York, p. 5.

Miller, J. D. (1971) *Effects of Noise on People,* U.S. Environmental Protection Agency Publication No. NTID 300.7, Washington, DC, pp. 59, 67–69, 93.

Minnix, R. B. (1978) "The Nature of Sound," D. M. Lipscomb and A. C. Taylor, eds., *Noise Control Handbook of Principles and Practices,* pp. 29–30, Van Nostrand Reinhold Company, New York.

NIOSH (1972) *Criteria for a Recommended Standard: Occupational Exposure to Noise,* National Institute for Occupational Safety and Health, U.S. Department of Health, Education, and Welfare, Washington, DC.

Olishifski, J. B., and E. R. Harford, eds. (1975) *Industrial Noise and Hearing Conservation,* National Safety Council, Chicago, IL.

Piercy, J. E., and G. A. Daigle (1991) "Sound Propagation in the Open Air," in C. M. Harris (ed.) *Handbook of Acoustical Measurement and Noise Control,* McGraw-Hill, New York, pp. 3.1–3.26.

Sylvan, S. (2000) *Best Environmental Practices in Europe and North America,* County Administration of Vastra Gotaland, Sweden.

U.S. EPA (1971a) *Transportation Noise and Noise from Equipment Powered by Internal Combustion Engines,* U.S. Environmental Protection Agency Publication No. NTID 300.13, Washington, DC, p. 230.

U.S. EPA (1971b) *Report to the President and Congress on Noise,* U.S. Environmental Protection Agency, NRC 500.1, 1971.

U.S. EPA (1974) *Information on Levels of Environmental Noise Requisite to Protect Public Health and Welfare with an Adequate Margin of Safety,* U.S. Environmental Protection Agency, Publication No. 550/9-74-004, Washington, DC, pp. 29, B9, B10.

16

Ionizing Radiation

Kristin Erickson, Radiation Safety Officer, Office of Radiation, Chemical, and Biological Safety, Michigan State University, contributed to this chapter.

Case Study

Chernobyl: A Land Lost Forever?

April 26, 1986 began like just about any other day in the old city of Chernobyl, which lies about 130 km north of Kiev, Ukraine, and about 20 km south of the border with Belarus. Unbeknownst to the people of Chernobyl and the nearby city of Pripyat, operators at the nuclear power plant had been testing the facility to determine how long the turbines would operate following a loss of main electrical power supply. However, by the morning of the 26th, the reactor was in an extremely unstable condition that resulted in a massive power surge followed by several explosions that would immediately kill two operators.

The accident resulted in the largest uncontrolled radioactive nonmilitary release into the environment ever recorded. Large quantities of radioactive substances, particularly iodine-131 and cesium-137, were released into the air for about 10 days. Within the next few weeks, another 28 individuals died as a result of exposure to radiation.

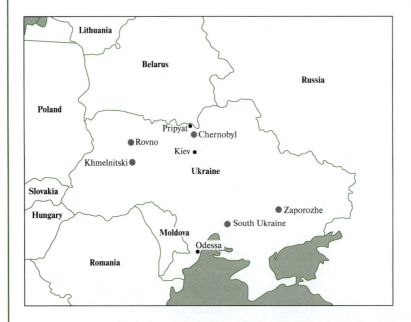

The accident resulted in the relocation of hundreds of thousands of people along with significant economic losses. People were worried and confused as a result of the evacuations and warnings. Shortly after the accident, the Soviet Union dissolved, resulting in political and economic instability. Together, this resulted in high levels of stress and anxiety, along with a lack of health care. Rates of alcohol and tobacco abuse remain high in the affected areas.

In the decades since the Chernobyl accident, the most common health impact has been a large increase in the incidence of thyroid cancer among people who were young children and adolescents at the time of the accident and lived in the most contaminated areas of Belarus, the Russian Federation, and Ukraine. In the days immediately after the accident, radioactive iodine contaminated pastures grazed by cows. The radioactive iodine, which became concentrated in the cow's milk, accumulated in the thyroids of the children who drank the contaminated milk. However, by monitoring the affected population for thyroid disease, the mortality rate due to thyroid cancer has been minimized (World Health Organization, 2006).

The Chernobyl accident has resulted in significant improvements in nuclear reactor safety in the former Soviet states. Increased collaboration between these states and the West has resulted in substantial improvements of reactors, along with a culture of safety. The World Association of Nuclear Operators was formed in 1989, and this group has helped foster cooperation between operators in more than 30 countries. In 1994, the Convention on Nuclear Safety was adopted. More than 100,000 people have been relocated to affected areas in Belarus, where low-level radioactive contamination occurred. With this move, the infrastructure for gas, water, and power had to be rebuilt. The use of local wood was banned and areas with high level contamination have been reforested. The actual Chernobyl site is now a tourist attraction, although it remains with a 30 km exclusion zone (World Nuclear Association, 2018).

16–1 FUNDAMENTALS*

Atomic Structure

We assume that you are familiar with the Bohr model of atomic structure. In this model the atom is described as consisting of a central nucleus surrounded by a number of electrons in closed orbits about the nucleus. The orbital electrons are grouped in shells.

The nucleus itself can be considered as composed of two distinct kinds of particles: protons, which carry a positive unit charge, e^+, and neutrons, which are uncharged. In a particular atom there are Z electrons, each carrying a charge e^-, orbiting around the nucleus, and a nucleus composed of N neutrons and P protons. The condition of electrical neutrality for the atom as a whole yields $Pe - Ze = 0$, that is, the number of protons in the nucleus is equal to the number of orbital electrons.

The number Z is the atomic charge or atomic number of the atom, and $Z + N$ is the atomic mass number, usually denoted by A. The parameters A and Z completely define a particular atomic species, this being known as a **nuclide.**

The masses of nuclides are measured in terms of the **unified atomic mass unit,** with the symbol u. This is defined as the unit of mass equal to one-twelfth the mass of an atom of carbon of atomic mass number 12. This gives 1 u as 1.6606×10^{-27} kg. On this scale, the mass of the neutron is 1.0088665 u, the mass of the proton 1.0088925 u, and the mass of the electron 0.0005486 u.

From the definition of the mass scale, giving proton and neutron masses of the order unity, it is clear that the **atomic mass number** will be a whole number approximation to the nuclidic mass in u. For example, a nuclide of magnesium which contains 12 protons and 12 neutrons has $A = 24$ and a nuclidic mass of 23.985045 u. The difference between the nuclidic mass and the atomic mass number is called the **mass excess.**

The chemical properties of the atom, and hence its designation as a particular element, depend on the number of orbital electrons, that is, on the atomic number Z. Given Z, the element is uniquely defined. As an example, if a given atom has two orbital electrons, it must be helium (assuming that the atom is not ionized or in some similar nonequilibrium state). Similarly an atom with eight electrons must be oxygen.

A particular nuclide is denoted by ^A_ZX, where X takes the place of the element symbol. But as Z determines the element, Z and X denote the same thing. Thus, the shorthand can be amended to ^{A}X. For example, carbon has six neutrons and six protons. Therefore, this nuclide can be written ^{12}C, or carbon-12.

---

*This discussion follows Coombe (1968).

FIGURE 16-1

Three isotopes of
hydrogen. (*Source:*
Coombe, R.A., *An
Introduction to
Radioactivity for
Engineers*, pp. 1–37, 1968,
MacMillan.)

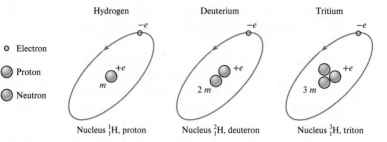

For each element (determined only by Z) several nuclides (determined by Z and A) have the same Z value but different values of A. These different nuclides of the same element are called **isotopes.** Hydrogen with $Z = 1$ has three isotopes with atomic mass numbers of 1, 2, and 3. As Z must remain constant at 1, this means that they have zero, one, and two neutrons, respectively. This is illustrated in Figure 16–1. These isotopes all act chemically as hydrogen, but their nuclidic masses are different. The nuclidic mass of ^{1}H is 1.007825 u, that of ^{2}H (known as deuterium) is 2.014102 u, and that of ^{3}H (known as tritium) is 3.016049 u.

The atomic weight of an element is defined as the combined nuclidic masses of all the isotopes, weighted according to their natural relative abundances. It is denoted by A. In the case of hydrogen it follows that the atomic weight is

$$1.007825(0.9844) + 2.014102(0.0156) + 3.016049(0) = 1.00797$$

The masses of the hydrogen isotopes are not obtained by simple addition of neutron masses. For example, the nuclidic mass of ^{1}H plus a neutron is 2.016490 u, whereas the mass of deuterium is 2.014102 u. This difference of 0.002388 u is called the **mass defect.** This is because when a proton and a neutron are brought together to form a deuteron (the nucleus of deuterium), energy is released to bind them together. Conversely, energy must be supplied to split them apart. This required energy, the **binding energy,** is obtained from Einstein's equation for the conversion of mass into energy,

$$E = \Delta mc^2 \tag{16–1}$$

where Δm is the mass defect and c is the speed of light.

All energies of emitted radiation and particles, as well as the various atomic and nuclear energy levels, are quoted in terms of the **electron volt,** eV. This is the energy that would be acquired by an electron in falling through a potential difference of one volt. From this definition the following equivalent units of energy can be established:

$$1\ \text{eV} = 1.602 \times 10^{-12}\ \text{erg} = 1.602 \times 10^{-19}\ \text{J}$$

For nuclear energy levels and radiation energies, the electron volt is usually an inconveniently small unit. The units MeV and KeV are then used for 10^6 eV and 10^3 eV, respectively. Using Equation 16–1, with the information that $c = 2.99793 \times 10^8$ m · s^{-1}, and 1 u = 1.6606×10^{-27} kg, then the energy equivalent of 1 u is 931.634 MeV. In other words, this means that if an electron of mass 0.0005486 u were completely annihilated, the energy released would be approximately 0.511 MeV.

Radioactivity and Radiation

By definition, isotopes have different ratios of neutrons to protons in the nucleus. Some ratios give rise to unstable conditions. This is usually because the neutron-to-proton ratio is too large. Because of this instability, the nucleus changes its state to attain equilibrium, and in so doing emits either a particle or electromagnetic radiation to carry off the excess energy. This phenomenon of nuclear disintegration is known as **radioactivity,** and an isotope that displays such activity is known as a **radioisotope.**

There are three types of isotopes. Some are stable, others are naturally radioactive, and the third group can be artificially produced and are also radioactive. These artificially produced radioisotopes are the isotopes most used in industrial application.

Three major types of decay product carry off the surplus energy when a radioisotope decays: alpha particles, beta particles, and gamma radiation.

Alpha-Particle Emission.   Conceptually, the source of the instability of the heavy elements is their size; their nuclei are too large. How can they become smaller? One method would be to eject protons or neutrons. Rather than doing this singly, the heavy-element atoms expel them in "packages" containing two protons and two neutrons. This "package" is called an **alpha particle** (α). An alpha particle is equivalent to the nucleus of the helium-4 atom, that is, it is simply a body consisting of two protons and two neutrons bound together. Consequently, if an alpha particle is emitted, the nucleus must change to one that has a charge $2e$ less and a mass approximately 4 u less. The general expression is

$$_{Z}^{A}X \rightleftharpoons _{Z-2}^{A-4}X + _{2}^{4}He \tag{16–2}$$

Atoms that eject the helium "package" are said to decay through emission of alpha radiation. Alpha-particle emission occurs mainly in radioisotopes whose atomic number is greater than 82. With increasing atomic number, the occurrence of alpha-particle decay increases rapidly, and it is a characteristic of the very heavy elements. It is particularly in evidence in the main decay chains of the natural radioactive isotopes.

You should note that an atom undergoing alpha-particle decay changes into a new element. It is a new element because the product nucleus (often called a **daughter**) contains two fewer protons than the parent atom. Through emission of an alpha particle, uranium becomes thorium. Similarly, radium becomes radon.

Beta-Particle Emission.   The instability that is the cause of beta-particle emission arises from the fact that the neutron-to-proton ratio in the nucleus is too high (there are too many neutrons in the nucleus). To achieve stability, a neutron in the nucleus can decay into a proton and an electron. The proton remains in the nucleus so that the neutron-to-proton ratio is decreased, and the electron is ejected. This ejected electron is known as a **beta particle** (β). The general expression for the decay is

$$_{Z}^{A}X \rightleftharpoons _{Z+1}^{A}X + \beta^{-} \tag{16–3}$$

Note that we use the β^{-} to represent an electron of nuclear origin to differentiate it from electrons from other sources. The negative sign is used with the β if there is any chance of ambiguity because a similar particle, called a positron, also exists that carries a positive charge.

Again, as in alpha radiation, emission of a beta particle changes the parent atom into a new element because the number of protons in the nucleus increases by one. If the daughter product also is radioactive, it will, in turn, emit a beta particle, becoming another new element, and so on, until finally a stable neutron-to-proton ratio is reached. Through such a series of changes, for example, the fission-product element krypton becomes rubidium that, in turn, becomes strontium, which finally converts to stable yttrium.

EXAMPLE 16–1   Identify the particles that are emitted in each step of the decay chain represented by

$$_{36}^{89}Kr \rightarrow _{37}^{89}Rb \rightarrow _{38}^{89}Sr \rightarrow _{39}^{89}Y$$

Solution   Because Z increases by 1 in each case, the particle emitted in each step is a beta particle.

Gamma-Ray Emission. Either alpha or beta particles may be accompanied by gamma radiation. Whereas alpha or beta radiation brings about a change in the size of the nucleus or the number of a particular type of particle therein, the emission of gamma radiation represents only a release of energy. This is the energy that remains in the newly formed nucleus after emission of the alpha or beta particle. Electromagnetic radiation in the form of gamma rays is emitted when a nucleus in an excited state transfers to a more stable state. The nucleus thus retains its original composition, the excess energy being radiated away. If the frequency of the radiation is ν, and the nucleus changes from a state of energy E_1 to a state of energy E_2, then the two energies are related by the equation

$$E_1 - E_2 = h\nu \tag{16-4}$$

where h is Planck's constant, having a value of 6.624×10^{-27} ergs. The energy of the emitted gamma ray is thus $h\nu$. In equations, the gamma ray is represented by the Greek letter gamma (γ).

X-Rays. Gamma rays are similar to x-rays. Their difference lies only in their source. Gamma rays originate from a nucleus transferring from one nuclear excited state to another, whereas x-rays originate from electrons transferring from a higher to a lower atomic energy state. As atomic energy levels are in general spaced much closer in terms of energy than nuclear levels, it follows from Equation 16–4 that the frequencies of x-rays are much less than those of gamma rays. As far as industrial applications are concerned, the only difference between them is the penetrating power. Because penetrating power increases with frequency, gamma rays have more penetrating power than x-rays.

Multiple Emissions. In the preceding discussions, only single emission has been considered. In practice, two or more different types of emission are possible, and in a great many cases several particles of the same type but of different energies are emitted. This latter effect is due to the multiplicity of nuclear energy levels both in the original isotope nucleus and in the nucleus formed by particle emission.

Radioactive Decay

Each unstable (radioactive) atom will eventually achieve stability by ejecting an alpha or beta particle. This shift to a more stable form is called **decay.** Each radioactive decay process is characterized by the fact that only a fraction of the unstable nuclei in a given sample will decay in a given time. The probability that a particular nucleus in the sample will decay during a time interval dt is $-\lambda dt$ where λ is the radioactive decay constant. It is defined as the probability that any particular nucleus will disintegrate in unit time.

For a large number of like nuclei together, we make the assumption that λ is independent of the age of a nucleus and is the same for all nuclei. This means that λ is a constant. If N is the number of nuclei present at a time t, then the number of decays occurring in a time dt can be written $\lambda N dt$. As the number of nuclei decreases by dN in this time, we can write

$$dN = -\lambda N dt \tag{16-5}$$

The negative sign denotes that N is decreasing with time. Equation 16–5 shows that the rate of decay is proportional to the number of nuclei present, that is, it is a first-order reaction.

Equation 16–5 can be rearranged and integrated.

$$\int_{N_o}^{N} \frac{dN}{N} = \int_{0}^{t} \lambda \, dt$$

$$\ln \frac{N}{N_o} = -\lambda t$$

or

$$N = N_o \exp(-\lambda t) \tag{16-6}$$

TABLE 16-1   **Some Radioisotope Half-Lives**

Radioisotope	Half-Life	Radioisotope	Half-Life
Polonium-212	3.04×10^{-7} s	Calcium-45	165 days
Carbon-10	19.3 s	Cobalt-60	5.27 years
Oxygen-15	2.05 min	Tritium	12.5 years
Carbon-11	20.4 min	Strontium-90	28 years
Radon-222	3.825 days	Cesium-137	30 years
Iodine-131	8.06 days	Radium-226	1622 years
Phosphorus-32	14.3 days	Carbon-14	5570 years
Polonium-210	138.4 days	Potassium-40	1.4×10^{9} years

where N_o is the number of radioactive nuclei present at time $t = 0$. Equation 16–6 shows that radioactive decay follows an exponential form. In particular, the time taken for a given number of nuclei to decay to half that number, $T_{1/2}$, is obtained from Equation 16–6 as

$$\ln \frac{N_o/2}{N_o} = -\lambda T_{1/2}$$

Solving for $T_{1/2}$ yields

$$T_{1/2} = \frac{\ln 2}{\lambda} = \frac{0.693}{\lambda} \tag{16-7}$$

This equation relates two important parameters of a radioactive species: λ and the half-life, $T_{1/2}$. These quantities are characteristic properties of a particular species. Half-lives of radioisotopes cover an enormous range of values, from microseconds to millions of years. To illustrate this, some values are given in Table 16–1.

EXAMPLE 16–2   Kal Karbonate must dispose of a vial containing 2.0 μCi $\cdot$ L^{-1} of ^{45}Ca. How long must the radioisotope be held to meet an allowable sewer discharge standard of 2.0×10^{-4} μCi $\cdot$ mL^{-1}?

Solution   In Table 16–1, see that the half-life of ^{45}Ca is 165 d.

Calculate the value of λ using Equation 16–7

$$\lambda = \frac{0.693}{165 \text{ d}} = 4.20 \times 10^{-3} \text{ d}^{-1}$$

Calculate the holding time using Equation 16–6.

$$2.0 \times 10^{-4} \text{ } \mu\text{Ci} \cdot \text{mL}^{-1} = (2.0 \text{ } \mu\text{Ci} \cdot \text{L}^{-1})(10^{-3} \text{ L} \cdot \text{mL}^{-1}) \exp\left[(-4.20 \times 10^{-3})(t)\right]$$

$$0.10 = \exp\left[(-4.20 \times 10^{-3})(t)\right]$$

Taking the logarithm of both sides of the equation:

$$\ln(0.10) = \ln\{\exp\left[(-4.20 \times 10^{-3})(t)\right]\}$$

$$-2.30 = (-4.20 \times 10^{-3})(t)$$

$$t = 548.23 \text{ or } 550 \text{ days.}$$

Specific Activity and the Becquerel.   The quantity N is called the **activity** of a sample. In SI units the becquerel (Bq) is the unit used for activity. One **becquerel** of radioactive material is that quantity of unstable atoms whose frequency of decay is one disintegration per second. This definition covers all modes of disintegration for both single isotopes and mixtures.

For many years the unit used for activity was the curie. One **curie** of radioactive material is that quantity of unstable atoms whose frequency of decay is 3.700×10^{10} disintegrations per second. One becquerel is equal to 2.7×10^{-11} Ci. The curie is quite a large unit for a lot of purposes. Millicuries ($1 \text{ mCi} = 10^{-3}$ Ci) or microcuries ($1 \text{ } \mu\text{Ci} = 10^{-6}$ Ci) and even picocuries ($1 \text{ pCi} = 10^{-12}$ Ci) were chosen as more manageable units to work with.

The specific activity of a radioisotope is the activity per gram of the pure radioisotope. The number of atoms of a pure radioisotope in one gram is given by

$$N = \frac{N_A}{A} \tag{16-8}$$

where N_A is Avogadro's number (6.0248×10^{23}) and A is the nuclidic mass. The specific activity S of a particular radioisotope is an intrinsic property of that radioisotope.

$$S = \frac{\lambda N_A}{A} \text{ disintegrations} \cdot \text{s}^{-1} \tag{16-9}$$

Growth of Subsidiary Products.   In the process of decay, a new nuclide is formed, the daughter product. If the daughter product is stable, its concentration will gradually increase as the parent decays. On the other hand, if the daughter product is itself radioactive, the variation in concentrations of parent, daughter, and granddaughter products will very much depend on the relative rates of decay.

In several cases a radioactive isotope decays into another nuclide that is itself radioactive. This can continue for a large number of nuclides, resulting in a decay chain. The characteristics of any particular chain depend largely on the relative decay constants of its various members.

The simplest case is the growth of a radioactive daughter product from the parent atoms. Let us assume we begin with N_1 parent atoms of decay constant λ_1 and N_2 daughter atoms of decay constant λ_2. The rate at which the daughter product is increasing is then the difference between the rate at which it is produced by its parent and the rate at which it decays. This can be written as

$$\frac{dN_2}{dt} = \lambda_1 N_1 - \lambda_2 N_2 \tag{16-10}$$

The rate of production of the daughter is simply the decay rate of the parent.

Using Equation 16–6 with the notation that N_1 is the number of nuclei of the parent and N_{10} is the initial number

$$N_1 = N_{10} \exp(-\lambda_1 t)$$

Substituting in Equation 16–10, we obtain

$$\frac{dN_2}{dt} = \lambda_1 N_{10} \exp(-\lambda_1 t) - \lambda_2 N_2 \tag{16-11}$$

Rearranging, we get

$$\frac{dN_2}{dt} + \lambda_2 N_2 = \lambda_1 N_{10} \exp(-\lambda_1 t) \tag{16-12}$$

This equation can readily be solved by multiplying throughout by the factor $e^{\lambda_2 t}$. Thus,

$$\exp(\lambda_2 t)\frac{dN_2}{dt} + \exp(\lambda_2 t)\lambda_2(N_2) = \lambda_1 N_{10} \exp(-\lambda_1 t) \exp(\lambda_2 t) \tag{16-13}$$

and

$$\frac{dN_2 e^{\lambda_2 t}}{dt} = \lambda_1 N_{10} \exp[(\lambda_2 - \lambda_1)t] \tag{16–14}$$

On integration this yields

$$N_2 e^{\lambda_2 t} = \frac{\lambda_1 N_{10}}{\lambda_2 - \lambda_1} \exp[(\lambda_2 - \lambda_1)t] + C \tag{16–15}$$

The integration constant C is determined from the boundary conditions. For this case, at $t = 0$, there was no daughter product present, that is, $N_2 = 0$ at $t = 0$. Using these boundary conditions, Equation 16–15 reduces to

$$N_2 = \frac{\lambda_1 N_{10}}{\lambda_2 - \lambda_1} (e^{-\lambda_1 t} - e^{-\lambda_2 t}) \tag{16–16}$$

Characteristics of Daughter Products. In the derivation of Equation 16–16, it was assumed that N_2 was zero at zero time. Because the daughter nuclide itself decays, then at an infinite time, N_2 will again be zero. Between these two times when $N_2 = 0$, there will be a time, say t', when N_2 will reach a maximum. At this time, the rate of increase will be passing through a turning point, that is, $dN_2/dt = 0$. Using this fact, together with Equation 16–16, it can be shown that

$$t' = \frac{\ln \lambda_2 - \ln \lambda_1}{\lambda_2 - \lambda_1} \tag{16–17}$$

Secular Equilibrium. A limiting case of radioactive equilibrium in which $\lambda_1 \ll \lambda_2$ and in which the parent activity does not decrease measurably during many daughter half-lives is known as **secular equilibrium.** An example of this is ^{238}U decaying to ^{234}Th. In this case, a useful approximation of the value of N_2 after a large number of half-lives is

$$N_2 = N_{10} \frac{\lambda_1}{\lambda_2} \tag{16–18}$$

Continuous Production of Parent. The previous calculations assumed that at zero time a certain number of parent atoms were present and then decayed. In many cases of interest the parent is continuously replenished. Such cases occur for instance in nuclear reactors, where the parent nuclides are continuously being created by neutron bombardment. Another case is the continuous production of carbon-14 by cosmic rays incident on the nuclei present in the upper atmosphere.

End Products. Any radioactive decay chain must finally arrive at a nuclide that is stable. The relevant equations can readily be obtained, for any stable nuclide has $\lambda = 0$. For example, consider the case of a radioisotope whose daughter is stable. For this, Equation 16–16 can be used with $\lambda_2 = 0$. Thus,

$$N_2 = N_{10} (1 - e^{-\lambda_1 t}) \tag{16–19}$$

Similar modifications can be made to other equations concerned with longer decay chains.

Radioisotopes

Naturally Occurring Radioisotopes. Most of the 50 naturally occurring radioisotopes are associated with three distinct series: the thorium series, the uranium series, and the actinium series. Each one of these series starts with an element of high atomic mass (uranium-238, thorium-232, and uranium-235, respectively) and then decays by a long series of alpha- and beta-particle emissions to reach a stable nuclide (lead-206, lead-208, and lead-207, respectively).

The three chains are associated with the heavy elements, and very few naturally occurring radioisotopes are found with atomic masses less than 82.

The half-lives of the naturally occurring radioisotopes are very long. Presumably they were constituents of the earth at its formation and their activity has not yet died away beyond detection.

Two other important isotopes that occur in the natural environment but that are not strictly naturally occurring are hydrogen-3 (tritium) and carbon-14. These radioisotopes are artificially produced by cosmic rays bombarding the upper atmosphere of the earth. At present the quantities of these isotopes are in equilibrium, their production rate by cosmic radiation being balanced by their natural decay rate. Because of this phenomenon, these isotopes are of particular use in archaeological dating.

Artificially Produced Radioisotopes.   The artificial production of radioisotopes is mainly carried out either by nuclear reactors or by particle accelerators. The cyclotron is the accelerator in most general use because the required bombarding particle energies are easily obtained and the output is reasonably high. The transmutation of a stable isotope to a radioactive one is effected by bombarding a target nucleus with a suitable projectile, either electromagnetic or a particle, to produce the required isotope from the resultant nuclear reaction.

When an accelerator is used, the bombarding particles are usually protons, deuterons, or alpha particles. In the nuclear reaction brought about by the bombardment of zinc-64 with energetic deuterons from a cyclotron, the deuteron and zinc-64 nucleus combine to form a new element. The new element has a charge of $30e + e$ and an atomic mass number of $64 + 2$. This compound nucleus is then ^{66}Ga, gallium-66. This intermediate nucleus disintegrates almost immediately by one of several possible modes of decay. If a proton is emitted, for example, the final nucleus must be left with a nuclear charge of $32e$ and an atomic mass number of 65, so it is ^{65}Zn. This isotope does not occur in nature.

For the production of radioisotopes for industrial application, the most common nuclear reactions used are those from thermal neutrons. A target sample, in a suitable container, is inserted into the core of a reactor and left there for varying amounts of time. In the core of a reactor there is a copious supply of thermal neutrons. These interact with the target nucleus to produce the required radioisotope, a process known as **neutron activation.**

Fission

A **nuclear reactor** is an assembly of fissionable material (such as uranium-235, plutonium-239, or uranium-233) arranged in such a way that a self-sustaining **chain reaction** is maintained. When these nuclei are bombarded with neutrons of the appropriate energy, they split up, or **fission,** into fission fragments and neutrons. For the nuclear reaction to continue, at least one of the neutrons produced must be available to produce another fission instead of escaping from the assembly or being used up in some other nuclear reaction. Thus, there is a minimum **(critical)** mass below which the reaction cannot be self-sustaining. Actual reactors are built with an excess mass to make large amounts of neutrons available. The excess neutron production is controlled by the use of **moderators.** The moderators are made of materials with large neutron-capture cross sections, such as boron, cadmium, or hafnium. These are formed into **control rods** that are moved in and out of the reactor to moderate the excess neutrons.

The fission chain reaction is characterized by an enormous release of heat. This heat must be carried away by an efficient cooling system to prevent mechanical failure of the reactor assembly **(meltdown)** and, ultimately, an uncontrolled fission. The ultimate uncontrolled reaction is, of course, an atomic explosion.

The fission fragments are simply lower mass elements. There are, most commonly, two fission fragments from each nucleus with an energy of the order of 200 MeV shared between them. The uranium nucleus does not split into the same two fragments each time. The breakup is far from symmetrical and can occur in more than 30 different ways. The most commonly produced isotopes are grouped around the mass numbers 95 and 139.

The fragments produced from the fission process have very large neutron-to-proton ratios so that they are highly unstable. Many transitions have to occur before a stable nucleus is finally achieved. These successive decays give rise to a decay chain.

Fission fragments, because of their high mass and very high initial charge, have extremely short ranges in matter. Hence, they are contained within the fuel element when a uranium nucleus fissions. The spent nuclear reactor fuel elements thus provide a very intense radioactive source that presents many problems in the subsequent handling and processing. Fission fragments themselves can sometimes be used as a radioactive source for industrial application.

The Production of X-Rays*

X-rays were discovered in 1895 by Wilhelm Conrad Roentgen. During the course of some studies, he covered a cathode ray tube with a black cardboard box and observed fluorescence on a screen coated with barium platinocyanide near the tube. After further investigation of this phenomenon, he concluded that the effect was caused by the generation of new invisible rays capable of penetrating opaque materials and producing visible fluorescence in certain chemicals. He called these new invisible rays **x-rays.** Because of their discoverer, x-rays are also sometimes referred to as Roentgen rays.

As pointed out previously, x-rays are electromagnetic waves and occupy the same portion of the electromagnetic spectrum as gamma rays. Like gamma rays, x-rays can pass through solid material. The mode of interaction of x-rays with matter is the same, as are the biological and photographic effects.

Whereas gamma rays come from within the nucleus of the atom, x-rays are generated outside the nucleus by the interaction of high-speed electrons with the atom. For this reason, there is a difference in the energy distribution of x- and gamma rays. Gamma rays from any single radionuclide consist only of rays of one or several discrete energies. X-rays consist of a broad, continuous spectrum of energies. The continuous spectrum will be discussed in detail later.

The X-Ray Tube.   X-rays are produced whenever a stream of high-speed electrons strikes a substance. This is caused by their sudden stoppage or deflection by atoms within the target material. The x-ray tube (Figure 16–2) is designed to provide the high-speed electrons and the interacting material. Essential components of an x-ray tube are (1) a highly evacuated glass envelope containing the cathode and anode; (2) a source of electrons proceeding from a cathode; and (3) a target (or anode) placed in the path of the electron stream.

FIGURE 16–2

Typical x-ray tube in self-rectified circuit.

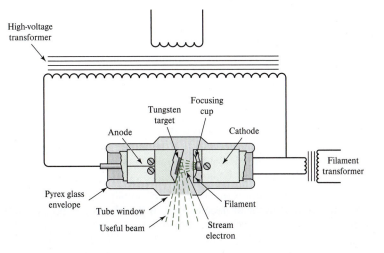

*This discussion follows U.S. PHS (1968).

The development of the hot filament tube by William D. Coolidge in 1913 was a major advance. Most x-ray tubes in use today are of this type. Here, the free electrons are "boiled out" of an incandescent filament within an evacuated tube and given their velocity by accelerating them through an electric field. In the hot filament tube, the quality and intensity of radiation can be controlled independently by simple electrical means. The intensity of radiation is directly proportional to the current and is proportional to the square of the voltage. This allows a much wider range of wavelengths and intensities, while the characteristics of the tube remain reasonably constant throughout its useful life.

The high voltages required for x-ray tube operation are best obtained by step-up transformers, whose output is always AC. Because the electrons must flow only from cathode to anode within the tube some means of rectification is necessary. A self-rectified tube acts as its own rectifier. When an alternating voltage is applied to such a tube, electrons flow only from the cathode to anode as long as the anode remains cool. If the anode becomes hot, the flow of electrons reverses during the second half-cycle and the cathode is damaged. Thus the self-rectified tube is limited to low currents and short periods of operation. The use of "valves" (rectifiers) in the power supply circuit eliminates the inverse voltage on the x-ray tube. Thus, more power can be handled by the x-ray tube, the radiation output is increased, and the time of exposure is shortened.

X-Ray Production Efficiency. On average, the fraction of the electron energy emitted as electromagnetic radiation increases with the atomic number of the atoms of the target and the velocity of the electrons. This fraction is very small and can be represented by the following empirical equation:

$$F = 1.1 \times 10^{-9} \, ZV \tag{16-20}$$

where F = fraction of the energy of the electrons converted into x-rays

$\quad\quad Z$ = atomic number of the target

$\quad\quad V$ = energy of the electrons (in volts)*

Typically, less than 1% of the electrical power supplied is converted into x-ray energy. The remaining energy (over 99%) appears as heat produced at the target (largely through ionization and excitation). As a result, electron bombardment of the target raises it to a high temperature and, if the heat produced is not dissipated fast enough, the target will melt. This heat production is a serious factor in limiting the capacity of an x-ray tube.

A suitable target must have the following characteristics:

1. A high atomic number because efficiency is directly proportional to Z
2. A high melting point because of the high temperatures involved
3. A high thermal conductivity to dissipate the heat
4. Low vapor pressure at high temperatures to prevent target evaporation

The Continuous Spectrum. When high-speed electrons are stopped by a target, the radiation produced has a continuous distribution of energies (wavelengths). As the fast-moving electrons enter the surface layers of the target, they are abruptly slowed down by collision with the strong Coulomb field of the nucleus and are diverted from their original direction of motion. Each time the electron suffers an abrupt change of speed, a change in direction, or both, energy in the form of x-rays is radiated. The energy of the x-ray photon emitted depends on the degree of deceleration. If the electron is brought to rest in a single collision,

---

*The electron energy is generally expressed in terms of the voltage applied across the tube.

the energy of the resulting photon corresponds to the kinetic energy of the electron stopped and will be a maximum. If the electron suffers a less drastic collision, a lower energy photon is produced. Because a variety of types of collisions will be occurring, photons of all energies up to the maximum will be produced. This accounts for the continuous distribution of an x-ray spectrum. The maximum intensity (peak of the curve) occurs at a wavelength about 1.5 times the minimum wavelength. The total intensity of radiation from a given x-ray tube is represented by the area under the spectral curve. The intensity has been found, as might be expected, to be directly proportional to the electron current (number of electrons striking the target).

Radiation Dose*

Fundamentally, the harmful consequences of ionizing radiations to a living organism are due to the energy absorbed by the cells and tissues of the organism. This absorbed energy (or dose) produces chemical decomposition of the molecules present in the living cells. The mechanism of the decomposition appears to be related to ionization and excitation interactions between the radiation and atoms within the tissue. The amount of ionization or number of ion pairs produced by ionizing radiations in the cells or tissues provides some measure of the amount of decomposition or physiological damage that might be expected from a given quantity or dose. The ideal basis for radiation dose measurement could be, therefore, the number of ion pairs (or ionizations) taking place within the medium of interest. For certain practical reasons, the medium chosen for defining exposure dose is air.

Exposure Dose—the Roentgen. The exposure dose of x- or gamma radiation within a specific volume is a measure of the radiation based on its ability to produce ionization in air. The unit used for expressing the exposure to x- or gamma radiation is the roentgen (R). Its merit lies in the fact that the magnitude of the exposure dose in roentgens can usually be related to the absorbed dose, which is of importance in predicting or quantifying the expected biological effect (or injury) resulting from the radiation.

The **roentgen** is an exposure dose of x- or gamma radiation such that the associated corpuscular emission per 0.001293 g of air[†] produces, in air, ions carrying one electrostatic unit of quantity of electricity of either sign. Because the ionizing property of radiation provides the basis for several types of detection instruments and methods, such devices may be used to quantify the exposure dose. Note that this is a unit of exposure dose based on ionization of air; it is not a unit of ionization, nor is it an absorbed dose in air.

Absorbed Dose—the Gray. The absorbed dose of any ionizing radiation is the energy imparted to matter by ionizing radiations per unit mass of irradiated material at the place of interest. The SI unit of absorbed dose is the gray (Gy). One **gray** is equivalent to the absorption of $1 \text{ J} \cdot \text{kg}^{-1}$ (Joule per kilogram). The former unit of absorbed dose was the rad. One **rad** is equivalent to the absorption of $100 \text{ ergs} \cdot \text{g}^{-1}$. One Gy = 100 rads. It should be emphasized that although the roentgen unit is strictly applicable only to x- or gamma radiation, the gray unit may be used regardless of the type of ionizing radiation or the type of absorbing medium.

To make a conversion from roentgens to grays two things must be known: the energy of the incident radiation and the mass absorption coefficient of the absorbing material. This is demonstrated in Example 16–3.

---

*This discussion follows U.S. PHS (1968).

[†]One cubic centimeter of air at standard temperature and pressure (STP) has a mass of 0.001293 g.

EXAMPLE 16–3   A dose of 1.0 R of gamma radiation was measured in air. From empirical studies, it is known that, on the average, 34 eV of energy is transferred (or absorbed) in the process of forming each ion pair in air. What is the equivalent absorbed dose in 1.0 cubic centimeter of air?

Solution   To form 1 esu per 0.001293 g of air (mass of 1 cubic centimeter at STP), the radiation must produce 1.61×10^{12} ion pairs when absorbed in air. Thus, using the empirical estimate, the total energy absorbed is

$$[34 \text{ eV} \cdot (\text{ion pair})^{-1}](1.61 \times 10^{12} \text{ ion pairs} \cdot \text{g}^{-1}) = 5.48 \times 10^{13} \text{ eV} \cdot \text{g}^{-1}$$

In ergs rather than electron volts,

$$(5.48 \times 10^{13} \text{ eV} \cdot \text{g}^{-1})(1.602 \times 10^{-12} \text{ erg} \cdot \text{eV}^{-1}) = 87 \text{ ergs} \cdot \text{g}^{-1}$$

Because $1 \text{ erg} = 1 \times 10^{-7} \text{ J}$, 1 R of exposure dose to 1.0 cm^3 of air at standard conditions results in the absorbed dose of

$$(87 \text{ ergs} \cdot \text{g}^{-1})(10^{-7} \text{ J} \cdot \text{erg}^{-1})(10^3 \text{ g} \cdot \text{kg}^{-1}) = 8.7 \times 10^{-3} \text{ J} \cdot \text{kg}^{-1} = 8.7 \times 10^{-3} \text{ Gy}.$$

Relative Biological Effectiveness (Quality Factor).   Although all ionizing radiations are capable of producing similar biological effects, the absorbed dose, measured in grays, that will produce a certain effect may vary appreciably from one type of radiation to another. The difference in behavior, in this connection, is expressed by means of a quantity called the **relative biological effectiveness** (RBE) of the particular radiation. The RBE of a given radiation may be defined as the ratio of the absorbed dose (grays) of gamma radiation (of a specified energy) to the absorbed dose of the given radiation required to produce the same biological effect. Thus, if an absorbed dose of 0.2 Gy of slow neutron radiation produces the same biological effect as 1 Gy of gamma radiation, the RBE for slow neutrons would be

$$\text{RBE} = \frac{1 \text{ Gy}}{0.2 \text{ Gy}} = 5$$

The value of the RBE for a particular type of nuclear radiation depends on several factors, such as the energy of the radiation, the kind and degree of the biological damage, and the nature of the organisms or tissue under consideration.

Tissue Weighting Factor (W_T).   The **tissue weighting factor** (W_T) is a modifying factor used in dose calculations to correct for the fact that different tissues and organs have varying degrees of radiosensitivity depending on the radioisotope and the chemical form of the radioisotope. Some tissues and organs are very sensitive; others are not radiosensitive at all. For example, because iodine is easily incorporated in thyroid tissue, the thyroid gland is very sensitive to the radioiodines. The W_T is, therefore, high for the radioiodines. When the tissue or organ is not radiosensitive, the value of W_T may be very small or zero for that tissue.

The Sievert.   With the concept of the RBE in mind, it is now useful to introduce another SI unit, known as the sievert (Sv). One **sievert** equals the radiation dose having the same biological effect as a gray of gamma radiation. This was formerly know as the **rem,** an abbreviation of "roentgen equivalent man" (1 Sv = 100 rem). The gray is a convenient unit for expressing energy absorption, but it does not take into account the biological effect of the particular nuclear radiation absorbed. The sievert, however, which is defined by

Dose in Sv = RBE × dose in grays × W_T

provides an indication of the extent of biological injury (of a given type) that would result from the absorption of nuclear radiation. Thus, the sievert is a unit of biological dose.

16–2   BIOLOGICAL EFFECTS OF IONIZING RADIATION*

The fact that ionizing radiation produces biological damage has been known for many years. The first case of human injury was reported in the literature just a few months following Roentgen's original paper in 1895 announcing the discovery of x-rays. As early as 1902, the first case of x-ray induced cancer was reported in the literature.

Early human evidence for harmful effects as a result of exposure to radiation in large amounts existed in the 1920s and 1930s based on the experience of early radiologists, persons working in the radium industry, and other special occupational groups. The long-term biological significance of smaller, chronic doses of radiation, however, was not widely appreciated until the 1950s, and most of our current knowledge of the biological effects of radiation has been accumulated since World War II.

Sequential Pattern of Biological Effects

The sequence of events following radiation exposure may be classified into three periods: a latent period, a period of demonstrable effect, and a recovery period.

Latent Period.   Following the initial radiation event, and often before the first detectable effect occurs, there is a time lag referred to as the **latent period.** There is a vast time range possible in the latent period. In fact, the biological effects of radiation are arbitrarily divided into short-term, or acute, and long-term, or delayed, effects on this basis. Those effects that appear within a matter of minutes, days, or weeks are called **acute effects** and those which appear years, decades, and sometimes generations later are called **delayed effects.**

Demonstrable Effects Period.   During or immediately following the latent period, certain discrete effects can be observed. One of the phenomena seen most frequently in growing tissues exposed to radiation is the cessation of mitosis or cell division. This may be temporary or permanent, depending on the radiation dosage. Other effects observed are chromosome breaks, clumping of chromatin, formation of giant cells or other abnormal mitoses, increased granularity of cytoplasm, changes in staining characteristics, changes in motility or ciliary activity, cytolysis, vacuolization, altered viscosity of protoplasm, and altered permeability of the cell wall. Many of these effects can be duplicated individually with other types of stimuli. The entire gamut of effects however, cannot be reproduced by any single chemical agent.

Recovery Period.   Following exposure to radiation, recovery can and does take place to a certain extent. This is particularly manifest in the case of the acute effects, that is, those appearing within a matter of days or weeks after exposure. There is, however, a residual damage from which no recovery occurs, and it is this irreparable injury that can give rise to later delayed effects.

Determinants of Biological Effects

The Dose-Response Curve.   For any biologically harmful agent, it is useful to correlate the dosage administered with the response or damage produced. "Amount of damage" in the case of radiation might be the frequency of a given abnormality in the cells of an irradiated animal, or the incidence of some chronic disease in an irradiated human population. In plotting these two variables, a dose-response curve is produced. With radiation, an important question has been the nature and shape of this curve. Two possibilities are illustrated in Figures 16–3 and 16–4.

---

*This discussion follows U.S. PHS (1968).

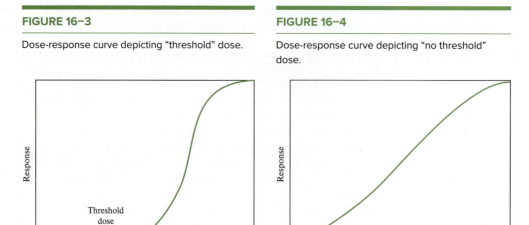

FIGURE 16–3

Dose-response curve depicting "threshold" dose.

FIGURE 16–4

Dose-response curve depicting "no threshold" dose.

Figure 16–3 is a typical "threshold" curve. The point at which the curve intersects the abscissa is the threshold dose, that is, the dose below which there is no response. If an acute and easily observable radiation effect, such as reddening of the skin, is taken as "response," then this type of curve is applicable. The first evidence of the effect does not occur until a certain minimum dose is reached.

Figure 16–4 represents a linear, or nonthreshold, relationship, in which the curve intersects the abscissa at the origin. Here any dose, no matter how small, involves some degree of response. There is some evidence that the genetic effects of radiation constitute a nonthreshold phenomenon, and one of the underlying (and prudent) assumptions in the establishment of radiation protection guidelines and in radiation control activities in public health programs has been the assumption of a nonthreshold effect. Thus, some degree of risk is assumed when large populations of people are exposed to even very small amounts of radiation. This assumption often makes the establishment of guidelines for acceptable radiation exposure an enormously complex task because the concept of "acceptable risk" comes into play in which the benefit to be accrued from a given radiation exposure must be weighed against its hazard.

Rate of Absorption.   The rate at which the radiation is administered or absorbed is most important in the determination of what effects will occur. Because a considerable degree of recovery occurs from the radiation damage, a given dose will produce less of an effect if divided (thus allowing time for recovery between dose increments) than if it were given in a single exposure.

Area Exposed.   Generally when an external radiation exposure is referred to without qualification as to the area of the body involved, whole-body irradiation is assumed. The portion of the body irradiated is an important exposure parameter because the larger the area exposed, other factors being equal, the greater the overall damage to the organism. Even partial shielding of the highly radiosensitive blood-forming organs such as the spleen and bone marrow can mitigate the total effect considerably. An example of this phenomenon is in radiation therapy, in which doses that would be lethal if delivered to the whole body are commonly delivered to very limited areas, such as to tumor sites.

Variation in Species and Individual Sensitivity.   There is a wide variation in the radiosensitivity of various species. Lethal doses for plants and microorganisms, for example, are usually hundreds of times larger than those for mammals. Even among different species of rodents, it is not unusual for one to demonstrate three or four times the sensitivity of another.

Within the same species, biological variability accounts for a difference in sensitivity among individuals. For this reason the lethal dose for each species is expressed in statistical terms. The LD_{50} for that species, or the dose required to kill 50% of the individuals in a large population, is the standard statistical measure. For people, the LD_{50} is estimated to be approximately 450 R.

Variation in Cell Sensitivity.   Within the same individual, a wide variation in susceptibility to radiation damage exists among different types of cells and tissues. In general, cells that are rapidly dividing or have a potential for rapid division are more sensitive than those that do not divide. Furthermore, nondifferentiated cells (i.e., primitive, or nonspecialized) are more sensitive than highly specialized cells. Within the same cell families then, the immature forms, which are generally primitive and rapidly dividing, are more radiosensitive than the older, mature cells, which have specialized in structure and function and have ceased to divide.

Acute Effects

An acute dose of radiation is one delivered to a large portion of the body during a very short time. If the amount of radiation involved is large enough, acute doses may result in effects that can manifest themselves within a period of hours or days. Here the latent period, or time elapsed between the radiation insult and the onset of effects, is relatively short and grows progressively shorter as the dose level is raised. These short-term radiation effects are composed of signs and symptoms collectively known as acute **radiation syndrome.**

The stages in acute radiation syndrome may be described as follows:

1. *Prodrome.* This is the initial phase of the syndrome and is usually characterized by nausea, vomiting, and malaise. It may be considered analogous to the prodrome state in acute viral infections in which the individual is subject to nonspecific systemic reactions.
2. *Latent stage.* During this phase, which may be likened to the incubation period of a viral infection, the subjective symptoms of illness may subside, and the individual may feel well. Changes, however, may be taking place within the blood-forming organs and elsewhere that will subsequently give rise to the next aspect of the syndrome.
3. *Manifest illness stage.* This phase reflects the clinical picture specifically associated with the radiation injury. Among the possible signs and symptoms are fever, infection, hemorrhage, severe diarrhea, prostration, disorientation, and cardiovascular collapse. Which, if any, of the foregoing phenomena are observed in a given individual largely depend on the radiation dose received.
4. *Recovery or death.*

Relation of Dose to Type of Acute Radiation Syndrome

As mentioned earlier, each kind of cell has a different sensitivity to radiation. At relatively low doses, for example, the most likely cells to be injured are those with greatest sensitivity, such as the immature white blood cells of lymph nodes and bone marrow. At low doses the observable effects during the manifest illness stage would be in these cells. Thus, you would expect to observe fever, infection, and hemorrhage. This is known as the **hematopoietic form** of the acute radiation syndrome.

At higher doses, usually over 6 Gy, cells of somewhat lower sensitivity will be injured. Of particular importance are the epithelial cells lining the gastrointestinal tract, for when these are destroyed a vital biological barrier is broken down. As a result, fluid loss may occur as well as overwhelming infection and severe diarrhea in the **gastrointestinal form** of the acute radiation syndrome.

In the **cerebral form,** which may result from doses of 100 Gy or more, the relatively resistant cells of the central nervous system are damaged, and the affected individual undergoes a rapid illness, characterized by disorientation and shock.

Considering the large degree of individual variation that exists in the manifestation of radiation injury, it is difficult to assign a precise dose range to each of these forms of the syndrome. The following generalizations, however, may serve to provide a rough indication of the kinds of doses involved. At 0.5 Gy or less, ordinary laboratory or clinical methods will show no indications of injury. At 1 Gy, most individuals show no symptoms, although a small percentage may show mild blood changes. At 2 Gy, most persons show definite signs of injury; this dose level may prove fatal to those individuals most sensitive to the effects of radiation. At 4.5 Gy, the mean lethal dose has been reached, and 50% of exposed individuals will succumb. Approximately 6 Gy usually marks the threshold of the gastrointestinal form of the acute radiation syndrome, with a very poor prognosis for all individuals involved. A fatal outcome may well be certain at 8–10 Gy.

Delayed Effects

Long-term effects of radiation are those that may manifest themselves years after the original exposure. The latent period, then, is much longer than that associated with acute radiation syndrome. Delayed radiation effects may result from previous acute, high-dose exposures or from chronic, low-level exposures over a period of years.

No unique disease is associated with the long-term effects of radiation. These effects manifest themselves in human populations simply as a statistical increase in the incidence of certain already existing conditions. Because of the low incidence of these conditions, it is usually necessary to observe large populations of irradiated persons to measure these effects. Biostatistical and epidemiological methods are then used to indicate relationships between exposure and effect. In addition to the large numbers of people needed for human studies of delayed radiation effects, the situation is further complicated by the latent period. In some cases, a radiation-induced increase in a disease may go unrecorded unless the study is continued for many years.

Also note that although it is possible to perform true experiments with animal populations, in which all factors with the exception of radiation exposure are kept identical in study populations, human data are limited to "secondhand" information accrued from populations irradiated for reasons other than radiobiological information. Often a special characteristic of irradiated human populations is the presence of some preexisting disease that makes it extremely difficult to draw meaningful conclusions when these groups are compared with nonirradiated ones.

Despite these difficulties, many epidemiologic investigations of irradiated human beings have provided convincing evidence that ionizing radiation may indeed result in an increased risk of certain diseases long after the initial exposure. This information supplements and corroborates that gained from animal experimentation that demonstrates these same effects.

Among the delayed effects thus far observed have been somatic damage, which may result in an increased incidence of cancer, embryological defects, cataracts, lifespan shortening, and genetic mutations. With proper selection of animal species and strains, and of dose, ionizing radiation may be shown to exert an almost universal carcinogenic action, resulting in tumors in a great variety of organs and tissues. There is human evidence as well that radiation may contribute to the induction of various kinds of neoplastic diseases (cancers).

Human Evidence.   Both empirical observations and epidemiologic studies of irradiated individuals have more or less consistently demonstrated the carcinogenic properties of radiation. Some of these findings are summarized here.

Early in this century, when delayed radiation effects were little recognized, luminous numerals on watches and clocks were painted by hand with fine sable brushes, dipped first in radium-containing paint and then often "tipped" on the lips or tongue. Young women commonly were employed in this occupation. Years later, studies of these individuals who had ingested radium paint have disclosed an increased incidence of bone sarcomas and other malignancies resulting from the burdens of radium that had accumulated in their bones.

Some early medical and dental users of x-rays, largely unaware of the hazards involved, accumulated considerable doses of radiation. As early as the year 1910, there were reports of cancer deaths among physicians, presumably attributable to x-ray exposure. Skin cancer was a notable finding among these early practitioners. Dentists, for example, developed lesions on the fingers with which they repeatedly held dental films in their patients' mouths.

Early in this century, certain large mines in Europe were worked for pitchblende, a uranium ore. Lung cancer was highly prevalent among the miners as a result of the inhalation of large quantities of airborne radioactive materials. It was estimated that the risk of lung cancer in the pitchblende miners was at least 50% higher than that of the general population.

One of the strongest supports for the concept that radiation is a leukemogenic agent in people comes from the epidemiologic studies of the survivors of the atomic bombing in Hiroshima, Japan. Survivors exposed to radiation above an estimated dose of approximately 1 Sv showed a significant increase in the incidence of leukemia. In addition, leukemia incidence correlated well with the estimated dose (expressed as distance from the detonation point), thus strengthening the hypothesis that the excess leukemia cases were indeed attributable to the radiation exposure. There is also some indication of an increase in thyroid cancer among the heavily irradiated survivors.

A pioneering study of children of mothers irradiated during pregnancy purported to show an increased risk of leukemia among young children if they had been irradiated in utero as a result of pelvic x-ray examination of the mother. Mothers of leukemic children were questioned as to their radiation histories during pregnancy with the child in question, and these responses were compared with those of a control group, consisting of mothers of healthy playmates of the leukemic children. Originally this work received much criticism, based partly on the questionnaire technique used to elicit the information concerning radiation history. It was believed that differences in recall between the two groups of mothers might have biased the results. A larger subsequent study designed to correct for the objections to the first one corroborated its essential findings and established the leukemogenic effect on the fetus of prenatal x-rays.

Considering the fact that immature, undifferentiated, and rapidly dividing cells are highly sensitive to radiation, it is not surprising that embryonic and fetal tissues are readily damaged by relatively low doses of radiation. It has been shown in animal experiments that deleterious effects may be produced with doses of only 0.10 Gy delivered to the embryo. There is no reason to doubt that the human embryo is equally susceptible.

The majority of the anomalies produced by prenatal irradiation involve the central nervous system, although the specific type of damage is related to the dose and to the stage of pregnancy during which irradiation takes place. In terms of embryonic death, the very earliest stages of pregnancy, perhaps the first few weeks in human beings, are most radiosensitive. From the standpoint of practical radiation protection, this very early sensitivity is of great significance, because it involves a stage in human embryonic development in which pregnancy may well be unsuspected. For this reason, the International Committee on Radiological Protection has recommended that routine nonemergency diagnostic irradiation involving the pelvic area of women in the childbearing years be limited to the 10-day interval following the onset of menstruation. Such precautions would virtually eliminate the possibility of inadvertently exposing a fertilized egg.

The period from approximately the second through the sixth week of human gestation, when pregnancy could still be unsuspected, is the most sensitive for the production of congenital anomalies in the newborn. During this period, embryonic death is less likely than in the extremely early stage, but the production of morphological defects in the newborn is a major consideration.

During later stages of pregnancy, embryonic tissue is more resistant to gross and easily observable damage. However, functional changes, particularly those involving the central nervous system, may result from such late exposures. These would be difficult to measure or evaluate at birth. They usually involve subtle alterations in such phenomena as learning patterns and development and may have a considerable latent period before they manifest themselves. There is some evidence that the decreasing sensitivity of the fetus to gross radiation damage as pregnancy progresses may not apply for the leukemogenic effects of prenatal irradiation. Another important factor to be considered in evaluating the radiation hazard during late pregnancy is that irradiation may produce true genetic mutations in the immature germ cells of the fetus for which no threshold dose has been established.

Lifespan Shortening.   In a number of animal experiments, radiation has been demonstrated to have a lifespan-shortening effect. The aging process is complex and largely obscure, and the exact mechanisms involved in aging are as yet uncertain. Irradiated animals in these investigations appear to die of the same diseases as the nonirradiated controls, but they do so at an earlier age. How much of the total effect is due to premature aging and how much to an increased incidence of radiation-induced diseases is still unresolved.

Genetic Effects

Background.   The fertilized egg is a single cell resulting from the union of sperm and egg; millions of cell divisions develop it into a complete new organism. The information that produces the characteristics of the new individual is carried in the nucleus of the fertilized egg on rod-shaped structures called chromosomes, arranged in 23 pairs. In each pair, one member is contributed by the mother and the other by the father. With each cell division that the rapidly developing embryonic tissue undergoes, all of this information is faithfully duplicated, so that the nucleus in each new cell of the developing organism contains essentially all of the information. This, of course, includes the germ cells in the new organism, which are destined to become sperm or eggs, and thus the information is transmitted from one generation to the next. This hereditary information is often likened to a template or to a code, which is reproduced millions of times over with remarkable accuracy. It is possible to damage the hereditary material in the cell nucleus by means of external influences, and when this is done the garbled or distorted genetic information will be reproduced just as faithfully when the cell divides as was the original message. When this kind of alteration occurs in those cells of the testes or ovaries that will become mature sperm or eggs, it is referred to as **genetic mutation;** if the damaged sperm or egg cell is then used in conception, the defect is reproduced in all of the cells of the new organism that results from this conception, including those that will become sperm or eggs, and thus whatever defect resulted from the original mutation can be passed on for many generations.

Most geneticists agree that the great preponderance of genetic mutations are harmful. By virtue of their damaging effects, they can be gradually eliminated from a population by natural means because individuals afflicted with this damage are themselves less likely to reproduce successfully than are normal individuals. The more severe the defect produced by a given mutation, the more rapidly it will be eliminated and vice versa; mildly damaging mutations may require a great many generations before they gradually disappear.

As a balance to this natural elimination of harmful mutations, fresh ones are constantly occurring. A large number of agents have mutagenic properties, and it is probable that our current knowledge includes just a fraction of these. In addition, mutations can arise within the germ cells of an organism without external insult. Among the various external influences found to be mutagenic are a wide variety of chemicals, certain drugs, and physical factors such as elevated temperatures and ionizing radiation. Natural background radiation probably accounts for a small proportion of naturally occurring mutations. For people, it has been estimated that background radiation probably produces less than 10% of these. Anthropogenic radiation, of course, if delivered to the gonads, can also produce mutations over and above those that occur spontaneously. Radiation, it should be noted, is not unique in this respect and is probably one of a number of environmental influences capable of increasing the mutation rate.

Animal Evidence.   The mutagenic properties of ionizing radiation were first discovered in 1927, using the fruit fly as the experimental animal. Since that time, experiments have been extended to include other species, and many investigations have been carried out on mice. Animal experimentation remains our chief source of information concerning the genetic effects of radiation, and as a result of the intensive experimentation, certain generalizations may be made. Among those of health significance are (1) there is no indication of a threshold dose for the genetic effects of radiation, that is, a dose below which genetic damage does not occur, and (2) the degree of mutational damage that results from radiation exposure seems to be dose-rate dependent, so that a given dose is less effective in producing damage if it is protracted or fractionated over a long period.

Human Evidence.   A major human study on genetic effects was made with the Japanese who survived the atomic bomb in 1945. As the index of a possible increase of the mutation rate, the sex ratio in the offspring of certain irradiated groups (families, for example, in which the mother had been irradiated but the father had not) was observed. Assuming that some of the mutational damage in the mothers would be recessive, lethal, and sex-linked, a shift in the sex ratio among these families might be expected in the direction of fewer male births than in completely nonirradiated groups, and this seemed to be the case in early reports. Later evaluation of more complete data, however, did not bear out the original suggestion of an effect on the sex ratio.

The preconception radiation histories of the parents of leukemic children compared with those of normal children was a part of the subject of another investigation. From the results, it would appear that there is a statistically significant increase in leukemia risk among children whose mothers had received diagnostic x-rays during this period. The effect here is apparently a genetic rather than an embryonic one because the irradiation occurred prior to the conception of the child.

A somewhat similar study ascertained the radiation exposure histories of the parents of children with Down syndrome. Most of this exposure was prior to the conception of the child. A significantly greater number of the mothers of children with Down syndrome reported receiving fluoroscopy and x-ray therapy than did mothers of the normal children in the control group.

The findings of these two studies may provide evidence that ionizing radiation is a mutational agent in people. However, the findings should be viewed with some reservations because there could be significant differences to begin with between populations of people requiring x-rays and those who do not. These differences alone might account for a slightly higher incidence of leukemia or Down syndrome in the offspring of the former group, irrespective of the radiation received. To date, there has been no incontrovertible evidence found of genetic effects in humans from radiation exposure.

16–3   RADIATION STANDARDS

Two population groups are given distinctly different treatment in the establishment of exposure–dose guidelines and rules. Standards are set for those occupationally engaged in work requiring ionizing radiation and for the general public. Although there are many standard-setting bodies, in general, the limits are consistent between groups. The Nuclear Regulatory Commission (NRC) has published guidelines in the *Code of Federal Regulations* (10 CFR 20) that serve as the standard in the United States. The dose guidelines are in addition to the natural background dose.

The allowable dose for occupational exposure is predicated on the following assumptions: the exposure group is under surveillance and control; it is adult; it is knowledgeable of its work and the associated risks; its exposure is at work, that is, $40 \text{ h} \cdot \text{week}^{-1}$; and it is in good health. On this basis, no individual is to receive more than 0.05 Sv per year of radiation exposure.

For the population at large, the allowable whole body dose in one calendar year is 0.001 Sv. This dose does not include medical and dental doses that, for diagnostic and therapeutic reasons, may far exceed this amount.

In addition to these dose rules, the NRC has set standards for the discharge of radionuclides into the environment. Table 16–2 is an extract from that list. These concentrations are measured above the existing background concentration and are annual averages. Discharges must be limited such that the amounts shown are not exceeded in ambient air or natural waters. If a mixture of isotopes is released into an unrestricted area, the concentrations shall be limited so that the following relationship exists:

$$\frac{C_A}{\text{MPC}_A} + \frac{C_B}{\text{MPC}_B} + \frac{C_C}{\text{MPC}_C} \leq 1 \tag{16–21}$$

TABLE 16-2 Selected Maximum Permissible Concentrations of Radionuclides in Air and Water Above Background

Radionuclide	Class	Occupational Values			Effluent Concentrations		Releases to Sewers
		Oral Ingestion ALI (μCi)[b]	Inhalation ALI (μCi)	Inhalation DAC (μCi)[c]	Air (μCi · mL⁻¹)	Water (μCi · mL⁻¹)	Monthly Average Conc. (μCi · mL⁻¹)
Barium-131	D[a], all compounds	3×10^{3}	8×10^{3}	3×10^{-6}	1×10^{-8}	4×10^{-5}	4×10^{-4}
Beryllium-7	W, all compounds except those given for Y	4×10^{4}	2×10^{4}	9×10^{-6}	3×10^{-8}	6×10^{-4}	6×10^{-3}
	Y, oxides, halides and nitrates	—	2×10^{4}	8×10^{-6}	3×10^{-8}	—	—
Calcium-45	W, all compounds	2×10^{3}	8×10^{2}	4×10^{-7}	1×10^{-9}	2×10^{-5}	2×10^{-4}
Carbon-14	Monoxide	—	2×10^{6}	7×10^{-4}	2×10^{6}	—	—
	Dioxide	—	2×10^{5}	9×10^{-5}	3×10^{-7}	—	—
	Compounds	2×10^{3}	2×10^{3}	1×10^{-6}	3×10^{-9}	3×10^{-5}	3×10^{-4}
Cesium-137	D, all compounds	1×10^{2}	2×10^{2}	6×10^{-8}	2×10^{-10}	1×10^{-6}	1×10^{-5}
Iodine-131	D, all compounds	3×10^{1} Thyroid (9×10^{1})	5×10^{1} Thyroid (2×10^{3})	2×10^{-8}	—	—	—
Iron-55	D, all compounds except those given for W	9×10^{3}	2×10^{3}	8×10^{-7}	3×10^{-9}	1×10^{-6}	1×10^{-5}
	W, oxides, hydroxides and halides	—	4×10^{3}	2×10^{-6}	6×10^{-9}	1×10^{-4}	1×10^{-3}
Phosphorus-32	D, all compounds except those given for W	6×10^{2}	9×10^{2}	4×10^{-7}	1×10^{-9}	9×10^{-6}	9×10^{-5}
	W, phosphates of Zn²⁺, S³⁺, Mg²⁺, Fe³⁺, Bi³⁺ and lanthanides	—	4×10^{2}	2×10^{-7}	5×10^{-10}	—	—
Radon-222	With daughters removed	—	1×10^{4}	4×10^{-6}	1×10^{-8}	—	—
	With daughters present	—	1×10^{2} (or 4 working level months)	3×10^{-8} (or 0.33 working level)	1×10^{-10}	—	—
Strontium-90	D, all soluble compounds except SrTiO₃	3×10^{1} Bone surface 4×10^{1}	2×10^{1} Bone surface 2×10^{1}	8×10^{-9}	3×10^{-11}	5×10^{-7}	5×10^{-6}
	Y, all insoluble compounds and SrTiO₃	—	4	2×10^{-9}	6×10^{-12}	—	—
Zinc-65	Y, all compounds	4×10^{2}	3×10^{2}	1×10^{-7}	4×10^{-10}	5×10^{-6}	5×10^{-5}

[a] D, W, and Y are classes denoting the time of retention in the body, days, weeks and years, respectively.
[b] ALI is the Annual Limit of Intake.
[c] DAC is the derived air concentration.

Source: Excerpted from title 10, *CFR*, part 20, Appendix B.

where C_A, C_B, C_C = concentrations of radionuclides A, B, and C, respectively
(in $\mu Ci \cdot mL^{-1}$)

MPC_A, MPC_B, MPC_C = maximum permissible concentrations of radionuclides A, B, and C
from Table II of Appendix B, Part 20 of the *CFR* (10 CFR 20)

Radon.   Unlike the standards for exposure and releases to the environment, those for radon in indoor air are established by the EPA. This is because radon is not the result of anthropogenic activity but rather occurs naturally. The EPA guidelines suggest that the annual average radon exposure be limited to 4 $pCi \cdot L^{-1}$ of air.

16–4   RADIATION EXPOSURE

External and Internal Radiation Hazards

External radiation hazards result from exposure to sources of ionizing radiation of sufficient energy to penetrate the body and cause harm. Generally speaking, it requires an alpha particle of at least 7.5 MeV to penetrate the 0.07 mm protective layer of the skin. A beta particle requires 70 keV to penetrate the same layer (U.S. PHS, 1970). Unless the sources of alpha or beta radiation are quite close to the skin, they pose only a small external radiation hazard. X-rays and gamma rays constitute the most common type of external hazard. When of sufficient energy, both are capable of deep penetration into the body. As a result no radiosensitive organ is beyond the range of their damaging power.

Radioactive materials may gain access to the body by ingestion, by inhalation of air containing radioactive materials, by absorption of a solution of radioactive materials through the skin, and by absorption of radioactive material into the tissue through a cut or break in the skin. The danger of ingesting radioactive materials is not necessarily from swallowing a large amount at one time, but rather from the accumulation of small amounts on the hands, on cigarettes, on foodstuffs, and other objects that bring the material into the mouth.

Any radioactive material that gains entry into the body is an internal hazard. The extent of the hazard depends on the type of radiation emitted, its energy, the physical and biological half-life of the material, and the radiosensitivity of the organ where the isotope localizes. Alpha and beta emitters are the most dangerous radionuclides from the standpoint of internal hazard because their specific ionization is very high. Radionuclides with half-lives of intermediate length are the most dangerous because they combine fairly high activity with a half-life sufficiently long to cause considerable damage. Polonium is an example of a potentially very serious internal hazard. It emits a highly ionizing alpha particle of energy 5.3 MeV and has a half-life of 138 days.

Natural Background

People are exposed to natural radiation from cosmic, terrestrial, and internal sources. Typical gonadal exposures from natural background are summarized in Figure 16–5. Cosmic radiation is that originating outside of our atmosphere. This radiation consists predominately, if not entirely, of protons whose energy spectrum peaks in the range of 1 to 2 GeV. Heavy nuclei are also present. The impact of primary and very high energy secondary cosmic rays produces violent nuclear reactions in which many neutrons, protons, alpha particles, and other fragments are emitted. Most of the neutrons produced by cosmic rays are slowed to thermal energies and, by n, p (neutron–proton) reaction with ^{14}N, produce ^{14}C. The lifetime of carbon-14 is long enough that it becomes thoroughly mixed with the exchangeable carbon at the Earth's surface (carbon dioxide, dissolved bicarbonate in the oceans, living organisms, etc.). Some of the cosmic radiation penetrates to the Earth's surface and contributes directly to our whole body dose. Terrestrial radiation exposure comes from the 50 naturally occurring radionuclides found in the Earth's crust. Of these, radon has come to have the most significance as a common environmental hazard to the general public.

FIGURE 16–5

Average dose per year to person living in the United States. (*Source:* U.S. Department of Energy.)

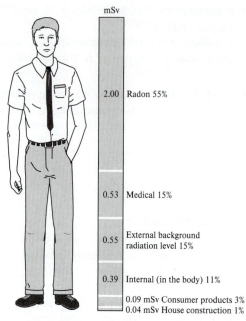

mSv

2.00 | Radon 55%

0.53 | Medical 15%

0.55 | External background radiation level 15%

0.39 | Internal (in the body) 11%

0.09 mSv Consumer products 3%
0.04 mSv House construction 1%

Total = 3.60 mSv · year^{-1}

Radon is the product of the radioactive decay of its parent, radium. Radium is produced from each of the three major series: ^{235}U, ^{238}U, and ^{232}U. The radon isotopes produced are ^{222}Rn, ^{220}Rn, and ^{219}Rn. These have half-lives of 3.8 days, 55.6 s, and 3.92 s, respectively. ^{222}Rn, because of its longer half-life and the abundance of its parent uranium in geologic materials, is generally more abundant and, hence, is considered the greater environmental hazard. Because the half-life of radium and its parents is so long, the source is essentially undiminished over human time scales.

The hazard of radon does not come from radon itself but from its radioactive decay products (^{218}Po, ^{214}Po, ^{214}Bi). The decay products are charged atoms of heavy metals that readily attach themselves to airborne particulates. The main health problem stems from the inhalation of unattached decay products and these particulates. The decay products and particulates become lodged in the lung. As they continue to decay, they release small bursts of energy in the form of alpha, beta, and gamma radiation that damage the lung tissue and could ultimately lead to lung cancer (Kuennen and Roth, 1989).

Radon is a gas. It is colorless, odorless, and generally chemically inert like other noble gases such as helium, neon, krypton, and argon. It does not sorb, hydrolyze, oxidize, or precipitate. Thus, its movement through the ground is not inhibited by chemical interaction with the soil.

Migration of radon occurs by two mechanisms: diffusion as a gas through the pore spaces in the soil and by dissolution and transport in the groundwater. The rate of diffusion or transport is a function of the emanation rate, porosity, structural channels, moisture content, and hydrologic conditions. These migration routes lead to two mechanisms of effect on people. Buildings constructed in areas of high radon emanation may have radon gas penetrate the structure through natural construction openings such as floor drains or joints (Table 16–3) or through structural failures such as cracks that develop from foundation settlement. In areas where the public water supply is drawn from an aquifer that has radon emanation, shower water may release radon. One rule of thumb is that a radon concentration of 10,000 pCi · L^{-1} of water, when heated and agitated, will produce about l pCi · L^{-1} of air (Murane and Spears, 1987).

X-Rays

X-ray machine use is widespread in industry, medicine, and research. All such uses are potential sources of exposure.

TABLE 16–3   Radon Gas Measurements in the Floor Drains and in the Basement Air of Seven Houses

House No.	Radon Concentration in Floor Drain (pCi · L^{-1})	Radon Concentration in Basement Air (pCi · L^{-1})	Ratio Drain/Basement
1	169.3	2.51	67.5
2	98.4	2.24	43.9
3	91.4	1.43	63.9
4	413.3	1.87	221
5	255.4	3.95	64.7
6	173.4	3.02	57.4
7	52.1	9.63	5.4
Average	179.0	3.52	

Medical and Dental Use.   In addition to the 300,000–400,000 medical-technical personnel that are occupationally exposed to radiation in the use of these machines, a considerable portion of the general population is also exposed. A large portion of the 2,500,000 persons seen daily by physicians have some x-ray diagnostic procedure performed on them.

Industrial Uses.   Industrial x-ray devices include radiographic and fluoroscopic units used for the determination of defects in castings, fabricated structures, and welds, and fluoroscopic units used for the detection of foreign material in, for example, airline luggage. Use of these units may result in whole body exposure to the operators and people who are nearby.

Research Use.   High-voltage x-ray machines are becoming familiar features of research laboratories in universities and similar institutions. Other x-ray equipment used in research includes x-ray diffraction units used for crystal analysis, electron microscopes, and particle accelerators.

Radionuclides

Naturally Occurring.   Thousands of becquerels of radium is in use in the medical field. In this use, many individuals besides the patient, including other patients, nurses, technicians, radiologists, and physicians, are potentially exposed to radiation.

Static eliminators, employing polonium or radium as the radioactive source, have been widely used in industry. Typical industries where they may be found are the textile and paper trades, printing, photographic processing, and telephone and telegraph companies.

Artificially Produced.   Over 6000 universities, hospitals, and research laboratories in the United States are using radionuclides for medical, biological, industrial, agricultural, and scientific research and for medical diagnosis and therapy. Over a million people in the United States receive radiotherapy treatment each year. Possible exposure from such radionuclides is involved with their preparation, handling, application, and transportation. Exposures, internal and external, might also arise through contamination of the environment by wastes originating from the use of these materials.

Nuclear Reactor Operations

Sources of radiation exposure associated with nuclear reactor operations include the reactor itself; its ventilation and cooling wastes; procedures associated with the removal and reprocessing of its "spent" fuel and the resulting fission product wastes; and procedures associated with the mining, milling, and fabrication of new fuels.

Radioactive Wastes

There are three principal sources of radioactive wastes: reactors and chemical processing plants, research facilities, and medical facilities. Regulations for the handling and disposal of radioactive wastes are designed to minimize exposure to the general public, but the regulations obviously provide less protection to those handling the waste.

16–5   RADIATION PROTECTION*

The principles discussed here are generally applicable to all types or energies of radiation. Their application will vary however, depending on the type, intensity, and energy of the source. For example, beta particles from radioactive materials require different shielding from that for high-speed electrons from an accelerator. Ideally, we would like to provide protection that results in a radiation exposure of zero. In actuality, technical and economic limitations force us to compromise so that the risks are small compared with the benefits obtained. The radiation standards set the limit above which the risk is deemed to be too great.

Reduction of External Radiation Hazards

Three fundamental methods are employed to reduce external radiation hazards: distance, shielding, and reduction of exposure time.

Distance.   Distance is not only very effective, but also in many instances the most easily applied principle of radiation protection. Beta particles of a single energy have a finite range in air. Sometimes the distance afforded by the use of remote control handling devices will supply complete protection.

The inverse square law for reduction of radiation intensity applies for point sources of x-, gamma, and neutron radiation. The inverse square law states that radiation intensity from a point varies inversely as the square of the distance from the source.

$$\frac{I_1}{I_2} = \frac{(R_2)^2}{(R_1)^2} \qquad (16\text{–}22)$$

where I_1 is the radiation intensity at distance R_1 from the source, and I_2 is the radiation intensity at distance R_2 from the source. Inspection of this formula will show that increasing the distance by a factor of 3, for example, reduces the radiation intensity to one-ninth of its value. The inverse square law does not apply to extended sources or to radiation fields from multiple sources.

X-ray tubes act sufficiently like point sources so that reduction calculations by this law are valid. Gamma ray sources whose dimensions are small in comparison with the distances involved may also be considered point sources, as can capsule neutron sources.

Shielding.   Shielding is one of the most important methods for radiation protection. It is accomplished by placing some absorbing material between the source and the person to be protected. Radiation is attenuated in the absorbing medium. When so used, "absorption" does not imply an occurrence such as a sponge soaking up water, but rather absorption here refers to the process of transferring the energy of the radiation to the atoms of the material through which the radiation passes. X- and gamma radiation energy is lost by three methods: photoelectric effect, Compton effect, and pair production.

The **photoelectric effect** is an all-or-none energy loss. The x-ray, or photon, imparts all of its energy to an orbital electron of some atom. This photon, because it consisted only of energy in the first place, simply vanishes. The energy is imparted to the orbital electron in the form of kinetic energy of motion, and this greatly increased energy overcomes the attractive force of the

---

*This discussion follows U.S. PHS, 1968.

nucleus for the electron and causes the electron to fly from its orbit with considerable velocity. Thus, an ion pair results. The high-velocity electron (which is called a **photoelectron**) has sufficient energy to knock other electrons from the orbits of other atoms, and it goes on its way producing secondary ion pairs until all of its energy is expended.

The **Compton effect** provides a means of partial energy loss for the incoming x- or gamma ray. Again the ray appears to interact with an orbital electron of some atom, but in the case of Compton interactions, only a part of the energy is transferred to the electron, and the x- or gamma ray "staggers on" in a weakened condition. The high-velocity electron, now referred to as a Compton electron, produces secondary ionization in the same manner as does the photoelectron, and the weakened x-ray continues on until it loses more energy in another Compton interaction or disappears completely via the photoelectric effect. The unfortunate aspect of Compton interaction is that the direction of flight of the weakened x- or gamma ray is different from that of the original. In fact, the weakened x- or gamma ray is frequently referred to as a "scattered" photon, and the entire process is known as Compton scattering. By this mechanism of interaction, the direction of photons in a beam may be randomized, so that scattered radiation may appear around corners and behind shields although at a lesser intensity.

Pair production, the third type of interaction, is much rarer than either the photoelectric or Compton effect. In fact, pair production is impossible unless the x- or gamma ray possesses at least 1 MeV of energy. (Practically speaking, it does not become important until it possesses 2 MeV of energy.) **Pair production** may be thought of as the lifting of an electron from a negative to a positive energy state. The pair is a positron–electron pair that results from the photon ejecting an electron and leaving a "hole" the positron. If there is any excess energy in the photon above the 1 MeV required to create two electron masses, it is simply shared between the two electrons as kinetic energy of motion, and they fly out of the atom with great velocity. The negative electron behaves in exactly the ordinary way, producing secondary ion pairs until it loses all of its energy of motion. The positron also produces secondary ionization so long as it is in motion, but when it has lost its energy and slowed almost to a stop, it encounters a free negative electron somewhere in the material. The two are attracted by their opposite charges, and, upon contact, annihilate each other, converting both their masses into pure energy. Thus, two gamma rays of 0.51 MeV arise at the site of the annihilation. The ultimate fate of the annihilation gammas is either photoelectric absorption or Compton scattering followed by photoelectric absorption.

Because the energy of the photon must be greater than 1 MeV for pair production to occur, this process is not a factor in the absorption of x-rays used in dental and medical radiography. The energies of x-rays used in this type of radiography are rarely more than 0.1 MeV.

The predominating mechanism of interaction with the shielding material depends on the energy of the radiation and the absorbing material. The photoelectric effect is most important at low energies, the Compton effect at intermediate energies, and pair production at high energies. As x- and gamma ray photons travel through an absorber, their decrease in number caused by the above-mentioned absorption processes is governed by the energy of radiation, the specific absorber medium, and the thickness of the absorber traversed. The general attenuation may be expressed as follows:

$$\frac{dI}{dx} = -uI_o \tag{16–23}$$

where dI = reduction of radiation

$\quad\quad I_o$ = incident radiation

$\quad\quad u$ = proportionality constant

$\quad\quad dx$ = thickness of absorber traversed

Integrating yields

$$I = I_o \exp(-ux) \tag{16–24}$$

FIGURE 16–6

Transmission through lead of gamma rays from radium; cobalt-60; cesium-137; gold-198; iridium-192; tantalum-182; and sodium-24.

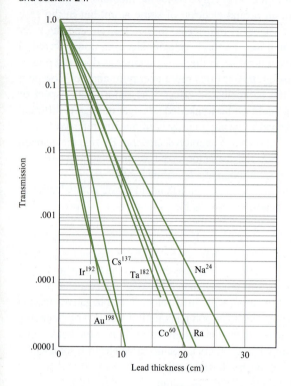

FIGURE 16–7

Transmission through concrete (density 2.35 Mg · m^{-3}) of gamma rays from radium; cobalt-60, cesium-137, gold-198; iridium-192.

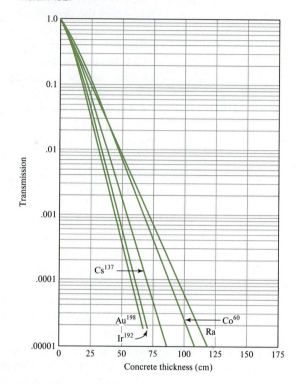

Using this formula, it is easy to calculate the radiation intensity behind a shield of thickness x, or to calculate the thickness of absorber necessary to reduce radiation intensity to a desired level, if the factor u is known. This factor is called the **linear absorption coefficient** when x is a linear dimension. The value of u depends on the energy of the radiation and the absorbing medium. The ratio I/I_o is sometimes called the **transmission.** Tables and graphs are available that give values of u determined experimentally or that give transmission values for varying thickness or different shielding materials (Figures 16–6 through 16–9).

If the radiation being attenuated does not meet narrow-beam conditions, or thick absorbers are involved, the absorption equation becomes

$$I = BI_o \exp(-ux) \tag{16–25}$$

where B is the buildup factor that takes into account an increasing radiation intensity due to scattered radiation within the absorber.

For alpha and beta emissions from radionuclides (not accelerators), substantial attenuation can be achieved with modest shielding. The amount of shielding required is, of course, a function of the particle energy. For example, a 10-MeV alpha particle has a range of 1.14 m in air, whereas a 1-MeV particle has a range of 2.28 cm. Virtually any solid material of any substance can be used to shield alpha particles. Beta particles can also be shielded relatively easily. For example, a ^{32}P beta at 1.71 MeV can be attenuated 99.8% by 0.25 cm of aluminum. However, materials with high atomic numbers, such as metals, should not be used for high-energy beta shielding due to the production of **Bremsstrahlung radiation** (radiation produced by stopping another kind of radiation). In materials with high atomic numbers, the beta particle is absorbed, but the excess "trapped" energy is released in the form of an x-ray. For this reason, Plexiglas or Lucite, typically 6–12 mm thick, is often used.

FIGURE 16–8

Transmission through iron of gamma rays from radium; cobalt-60, cesium-137; iridium-192.

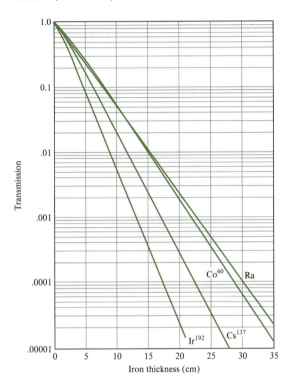

FIGURE 16–9

Transmission through lead of x-rays.

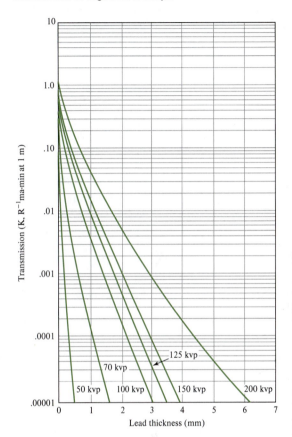

Fast neutrons are poorly absorbed by most materials. Therefore, it is necessary to slow them down for efficient absorption. Because the greatest transfer of energy takes place in collisions between particles of equal mass, hydrogenous materials are most effective for slowing down fast neutrons. Water, paraffin, and concrete are all rich in hydrogen, and thus, important in neutron shielding. Once the neutrons have been reduced in energy, they may be absorbed by either boron or cadmium. When a boron atom captures a neutron, it emits an alpha particle, but because of the extremely short range of alpha particles, no additional hazard results. Neutron capture by cadmium results in the emission of gamma radiation. Lead or a similar gamma absorber must be used as a shield against these gammas. A complete shield for a capsule-type neutron source may consist of, first, a thick layer of paraffin to slow down the neutrons, then a surrounding layer of cadmium to absorb the slow neutrons, and finally, an outer layer of lead to absorb both the gammas produced in the cadmium and those emanating from the capsule.

Some care must be exercised in using shielding to reduce exposure. People outside the "shadow" cast by the shield are not necessarily protected. A wall or partition is not necessarily a safe shield for persons on the other side. Their allowable dose may be less than conceived in the design of the barrier. Radiation can "bounce around corners" because it can be scattered.

Scattered radiation is present to some extent whenever an absorbing medium is in the path of radiation. The absorber then acts as a new source of radiation. Frequently, room walls, the floor, and other solid objects are near enough to a source of radiation to make scatter appreciable. When a point source is used under these conditions, the inverse square law is no longer completely valid for computing radiation intensity at a distance. Measurement of the radiation is then necessary to determine the potential exposure at any point.

Reduction of Exposure Time.   By limiting the duration of exposure to all radiation sources and by providing ample recuperative time between exposures, the untoward effects of radiation can be minimized. Recognition of the zero threshold theory of damage warrants that exposures, no matter how small, be minimized. The standards established by the NRC are upper bounds to be avoided and not goals to be achieved.

In emergency situations it may occasionally be necessary to work in areas of very high dose rates. This can be done with safety by limiting the total exposure time so that the average permissible value for a day based on the radiation protection guide dose of 1×10^{-3} Gy (0.1 rad) per week is not exceeded. This does not imply that a worker should be allowed to extend this practice beyond receiving 1×10^{-3} Gy in a short period of time, that is, a dose of 1×10^{-3} Gy one day and no dose for 6 days would comply with the rule but would be considered excessive. Repetitions of this cycle would be unacceptable. Emergency situations may require that work be done in relays of several people in the same job so that the value of the radiation protection guide is not exceeded by any one person.

Reduction of Internal Radiation Hazards

Occupational.   The prevention and control of contamination is the most effective way to reduce internal hazards in the workplace. The use of protective devices and good handling techniques affords a large measure of protection. Dust should be kept to a minimum by elimination of dry sweeping. Laboratory operations should be carried out in a hood. The exhaust air from the hood must be filtered with a high-efficiency filter. The filter must be replaced regularly in an approved manner. Protective clothing should be worn so that normal street clothes do not become contaminated. Respirators should be worn during emergency operations or when dust is generated. Eating must not be permitted in areas where radioactive materials are handled. Proper training in the care and handling of radioactive materials is, perhaps, the most important method for reducing the potential for internal radiation exposure in the workplace.

Radon.   Aside from cigarette smoke, the most likely nonoccupational internal radiation hazard is from radon in private dwellings. Because the radon primarily originates in the soil beneath the house, control efforts are aimed at the basement or crawl space.

The EPA suggests two major approaches for new construction: reduction of the pathways for radon entry and reduction of the draft of the house on surrounding and underlying soil. The methods to reduce the pathways for entry are summarized in Figure 16–10. Of particular concern are penetrations into the foundation such as floor drains (see Table 16–3) and cracks in the floor. The use of a polyethylene sheet below the slab is particularly effective for controlling leaks that result from slab cracks that develop as the house settles. Because heat in the upper floors tends to rise, creating a draft much like a chimney, the house has a tendency to create a negative pressure on the basement and, hence, "suck in" radon from the soil pore spaces. Figure 16–11 shows some techniques to minimize the draft effect (Murane and Spears, 1987).

FIGURE 16–10

Methods to reduce pathways for radon entry.

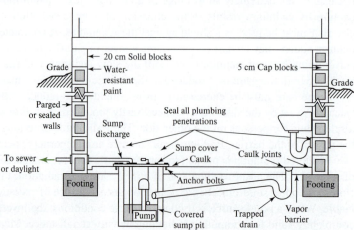

FIGURE 16–11

Methods to reduce the vacuum.

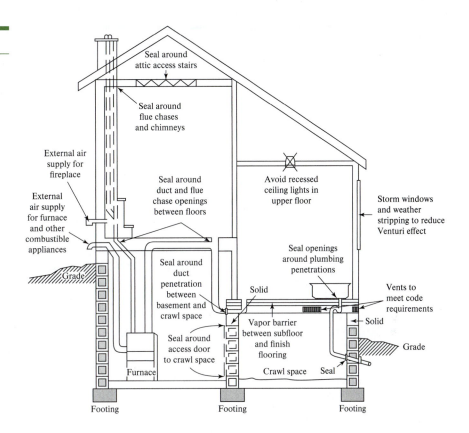

For existing structures, the remedies are more difficult to install, will be expensive, and may not yield satisfactory results. If drain tiles are present around the outside or inside of the perimeter footings, these are ideally located to permit vacuum to be drawn near some of the major soil gas entry routes (the joint between the slab and the foundation wall and the footing region where the radon can enter the voids in the block walls). Other efforts have included drilling holes in the slab itself and creating a vacuum system beneath the whole slab. Several suction points (three to seven) are required for this technique to work (Henschel and Scott, 1987). One demonstration project showed that jacking the house off the foundation and sealing the block walls was effective. In addition, a proprietary epoxy coating was applied to the floor and walls (Figure 16–12) (Ibach and Gallagher, 1987).

16–6   RADIOACTIVE WASTE

Types of Waste

No single scheme is satisfactory for classifying radioactive waste in a quantitative way. Usage has led us to categorize wastes into "levels." **High-level wastes** are those with activities measured in curies per liter; **intermediate-level wastes** have activities measured in millicuries per liter; **low-level wastes** have activities measured in microcuries per liter. Other classifications skip the intermediate-level wastes and use the terms high-level, **transuranic,** and low-level. The high-level wastes (HLW) are those resulting from reprocessing of spent fuel or the spent fuel itself from nuclear reactors. Transuranic wastes are those containing isotopes above uranium in the periodic table. They are the by-products of fuel assembly, weapons fabrication, and reprocessing. In general their radioactivity is low but they contain long-lived isotopes (those with half-lives greater than 20 years). The bulk of low-level wastes (LLW) has relatively little radioactivity. Most require little or no shielding and may be handled by direct contact.

FIGURE 16–12

Interior membrane linings and sealants to prevent radon gas infiltration. (*Source:* Ibach and Gallagher (1987). Retrofit and Preoccupancy Radon Mitigation Program for Homes, "Indoor Radon II Proceeding of the Second APCA International Specialty Conference," Cherry Hill, NJ, Air Pollution Control Association, pp. 172–182.)

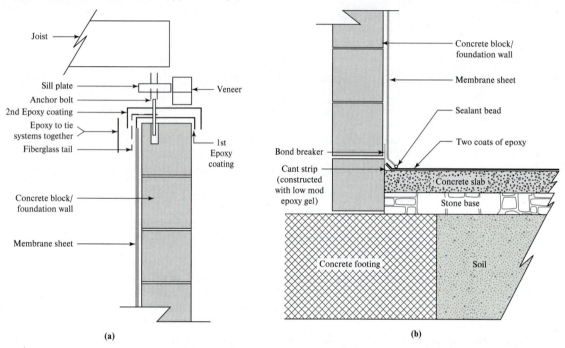

Management of High-Level Radioactive Waste

In 2005, there were about 104 operating reactors in the United States (EIA, 2005). Roughly 10 m³ of spent fuel is generated annually from each of these reactors. The construction of the fuel assembly results in considerably less fission product waste. Approximately 0.1 m³ of the 10 m³ is fission product waste. Of course, it is evenly distributed throughout the assembly and cannot be easily separated. The management choices are (1) store it indefinitely in the form in which it was removed from the reactor, (2) reprocess it to extract the fission products and recycle the other materials, or (3) dispose of it by burial or other isolation technique.

Under the Nuclear Waste Policy Act of 1987, Congress has prescribed that a storage facility be constructed that will not become permanent. The NRC has detailed the rules for the site in the *Code of Federal Regulations* (10 CFR 60). Some of the important provisions are summarized here (Murray, 1989):

1. The design and operation of the facility should not pose an unreasonable risk to the health and safety of the public. The radiation dose limit is a small fraction of that due to natural background.
2. A multiple barrier is to be used.
3. A thorough site study must be made. Geologic and hydrologic characteristics of the site must be favorable.
4. The repository must be located where there are no attractive resources, be far from population centers, and be under federal control.
5. High-level wastes are to be retrievable for up to 50 years from the start of operations.
6. The waste package must be designed to take into account all of the possible effects from earthquakes to accidental mishandling.

7. The package is to have a design life of 300 years.
8. Groundwater travel time from the repository to the source of public water is to be at least 1000 years.
9. The annual release of radionuclides must be less than 1/1000 of a percent of the amount of the radioactivity that is present 1000 years after the repository is closed.

Waste Isolation Pilot Plant

The waste isolation pilot plant (WIPP) project was authorized by Congress in 1979. After much political negotiation, the WIPP was authorized as a military transuranic waste facility exempt from licensing by the NRC. Under construction since 1983, the facility consists of 16 km of shafts and tunnels 650 m below ground in southeast New Mexico. The geologic material is a Permian salt basin. It was supposed to begin accepting waste in 1989 but has not done so. Certain risk analyses and compliance with RCRA requirements have delayed its use. The facility was conceived to demonstrate design concepts and to conduct some experiments with military high-level wastes that will later be retrieved. Many design questions about gas generation in the decaying waste and potential contamination scenarios are still being worked out.

Management of Low-Level Radioactive Waste

Historical Perspective.   Between 1962 and 1971, six commercial waste disposal sites were licensed. Three were subsequently closed because they failed. The three sites (Maxey Flats, Kentucky; Sheffield, Illinois; and West Valley, New York) all experienced similar problems. They used shallow land burial to dispose of the waste. This was accomplished by excavating a trench about 3–6 m deep and placing the drums and other containers (often cardboard boxes) of radionuclides in the trench and covering them with excavated soil. The completed trench was covered with a mound of earth and seeded.

Water seeped through the cover material and animals burrowed through it. The heavy clay sites chosen precisely to limit passage to the groundwater system served as holding ponds for the rainwater and ultimately accelerated the corrosion of the drums. At West Valley, when increased radioactivity called attention to this phenomenon, the trenches were opened and pumped to the nearest stream! Concurrently, it was discovered that the drums were often 30–50% empty. This, combined with the fact that the backfill material was heavy clay that did not completely fill the void spaces between the drums, allowed significant settlement of the cover material. This enhanced the collection of precipitation that contributed to the corrosion and failure of the drums.

These episodes led to a major rethinking at how we should manage our radioactive wastes. One result was that in 1980, Congress enacted the Low-Level Waste Policy Act. It says that each state is responsible for providing for the availability of capacity either within or outside of the state for disposal of low-level radioactive waste (LLRW) generated within its borders. The law provided for the formation of **compacts** between states to allow a regional approach to management. As of March 2004, the compact organization was as shown in Figure 16–13. The compacts decide what facilities are required and which states will serve as hosts. Although the compacts were supposed to begin accepting waste in 1986, the negotiation process has taken longer than expected and the deadline has been extended to beyond the year 2010. Many compacts have yet to select sites, let alone begin construction. The three currently available sites will soon run out of capacity, so there is some urgency to solve the problem.

Waste Minimization.*   As with all waste problems we have dealt with in this text, the first step in managing low-level radioactive waste is to minimize its production. Since 1980, considerable strides have been made in reducing the volume of LLRW (Figure 16–14). A number of procedures can be effectively employed.

---

*This discussion follows NRC (1976).

FIGURE 16–13

Low-level radioactive waste compacts (data as of March 2004). Lines between states (for example New England and Texas) join members of the same compact.

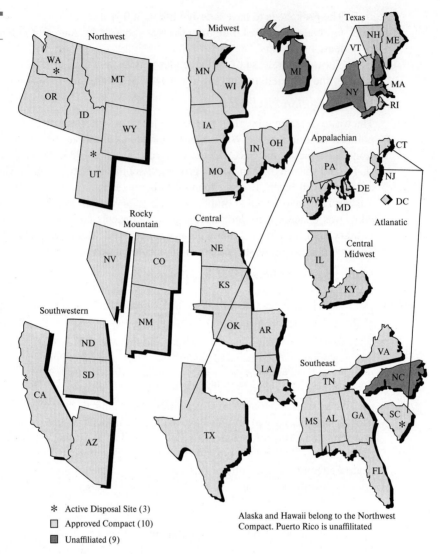

FIGURE 16–14

Low-level radioactive waste disposal.

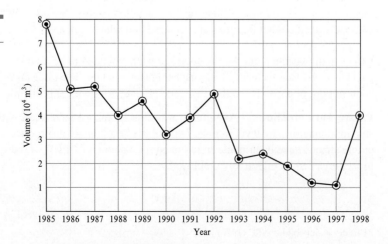

Immediate sorting of solid radioactive waste from nonradioactive waste is an essential initial step in any scheme for the reduction of the volume of that waste and for the recovery of radionuclides from uranium and transuranium waste. It is optimistic to expect much reduction of that volume of waste by sorting out uncontaminated waste unless it is done at the point of origin. Training plant personnel to do this work at the point of origin has been reasonably successful. To ask radioactive waste management personnel to do the sorting of an unknown mixture of wastes at a subsequent time and place creates an unacceptable hazard of exposure to radiation by inhalation, injury, or ambient external exposure.

Often material only suspected of being radioactively contaminated is labeled and disposed of as such without necessarily actually being radioactive. Much of the so-called "radioactive" waste fits into such a category merely because of the place where it was generated. The cost of assaying such suspected low-level solid wastes to determine their true radioactive content is such that it is often cheaper to combine suspected waste with known radioactive waste than to separate it. This suspicious but not always radioactive waste takes up burial space unnecessarily. Time, effort, and money are needlessly expended in putting these nonradioactive wastes in the special radioactive waste landfills.

It has been a general practice to assume that all waste is radioactive if it has been generated in a laboratory using radioactive materials or by a radiochemical or similar processing activity. It is termed "radiation zone" or "contaminated area" wastes. Thus, waste that is suitable for disposal in a municipal landfill is mixed with contaminated waste. The burden of proof that the waste is not radioactive is on the person certifying or releasing the waste. Testing of the waste is time-consuming and is often omitted.

Probably the method most likely to succeed in reducing the amount of nonradioactive waste is a careful delineation and reduction of the so-called radiation zone and contamination areas. It is now common to define such areas rather broadly and to include certain zones and areas from which it should be obvious that the waste would not be radioactive. An example would be the office and administrative areas within a radiation zone. Such areas produce much nonradioactive waste that is often included for convenience in the low-level solid radioactive waste from the technical areas. In laboratory situations where nonradioactive wastes are generated alongside the radioactive waste, point source segregation can result in minimal radioactive waste generation.

Separation of combustible or compactible waste at the point of origin both improves waste handling and reduces volume. By sorting, wastes that are not compatible for incineration do not have to be handled at the incinerator. Because the volume reduction in an incinerator is greater than that in a compactor, the more wastes that are capable of incineration that reach the incinerator rather than the compactor, the greater the volume reduction.

Volume Reduction by Compression.

Compression of solid low-level radioactive waste is suitable for about half the waste generated. There are three kinds of compression devices: compactors, balers, and baggers.

Compactors force material into the final storage, shipping, or disposal container. A favorite container is the 0.210-m^3 drum. Some space saving is possible. A variant of the compactor is called the **packer.** In this device, the material is compressed into a reusable container. At the burial grounds, the compacted material is dumped directly without any effort to retain its compacted form. Space saving is minimal with packer systems.

Balers compress the waste into bales that are wrapped, tied, or banded and then stored, shipped, or disposed of in burial grounds. Considerable space saving is possible with balers.

Baggers compress waste into a predetermined shape that is injected into round or rectangular bags, boxes, or drums before storage, shipment, or disposal. Some space saving is possible with this method of compaction.

These three techniques may be suited to general and sometimes even to unique situations. Unfortunately, such treatment does not reduce the possibility of burning while in storage, and only certain materials are suitable for compaction. These include paper, cloth, rubber, plastics, wood, glass, and small light metal objects. Large, rigid metal objects must be excluded because they are usually relatively incompressible and can damage the container and compressing machinery. Moisture (free or absorbed in large quantities by blotting paper or rags) has to be avoided because of its potential forcible release under high pressure, creating a great hazard to operators. Obviously, corrosive, pyrophoric, and explosive waste must be excluded from such processing, whether it is organic or inorganic.

The compression machinery must be economical, reliable, and easy to operate. Many commercial devices are available, but all must be modified by providing air containment, off-gas ventilation, often filtration, and, if necessary, shielding.

Volume Reduction by Incineration.   Reduction of volumes of solid radioactive waste by incineration has interested managers of low-level radioactive waste, particularly in those parts of the world where land area is at a premium and costs are high. Under these conditions, the advantages of volume reduction are so great that the drawbacks seem only obstacles to be surmounted. In Europe, where land is scarce and more revered, the incineration of solid combustible radioactive waste is a common and apparently satisfactory method of pretreatment before final disposal.

There are certain advantages, such as volume reductions of 80–90%, reported for selected burnable waste. This may be a high estimate if such factors as residues from off-gas treatment and refractory changes are considered. This would represent a considerable saving in land used for burial, in transportation, and in long-term monitoring. In addition, it would free us from the nagging worry about the possible problem of long-burning subterranean fires. Special attention should be given to the problems of burning organic matter (solvents, ion-exchange resins, etc.) and putrescible biological material (animal cadavers, excreta, etc.). Incineration of radioactive waste must be carried out under controlled conditions to prevent the formation of radioactive aerosols and must comply with both RCRA and NRC rules if the wastes are RCRA wastes as well as being radioactive.

Long-Term Management and Containment

Site Selection.   One concern in the burial of radioactive waste is that groundwater or infiltrating surface water will leach the waste and mobilize the radioactive materials. The radionuclides would be carried by this water back to the surface as a part of natural groundwater discharge or through a water well. Because of this concern, hydrogeologic and hydrochemical considerations in site selection become paramount.

The types of hydrogeologic and hydrochemical data that may be needed to determine whether or not a site is adequate include (Papadopulos and Winograd, 1974)

1. Depth to water table, including perched water tables, if present
2. Distance to nearest points of groundwater, spring water, or surface water usage (including well and spring inventory, and, particularly, wells available to the public)
3. Ratio of pan evaporation to precipitation minus runoff (by month for a period of at least 2 years)
4. Water table contour map
5. Magnitude of annual water table fluctuation
6. Stratigraphy and structure to base of shallowest confined aquifer
7. Baseflow data on perennial streams traversing or adjacent to storage site
8. Chemistry of water in aquifers and aquitards and of leachate from the waste trenches
9. Laboratory measurements of hydraulic conductivity, effective porosity, and mineralogy of core and grab samples (from trenches) of each lithology in unsaturated and saturated (to base of shallowest confined aquifer) zone-hydraulic conductivity to be measured at different water contents and tensions
10. Neutron moisture meter measurements of moisture content of unsaturated zone measurements to be made in specially constructed holes (at least 2 years' record needed)

11. In situ measurements of soil moisture tension in upper 4.5–9 m of unsaturated zone (at least 2 years' record needed)
12. Three-dimensional distribution of head in all saturated hydrostratigraphic units to base of shallowest confined aquifer
13. Pumping, bailing, or slug tests to determine transmissivity and storage coefficients
14. Definition of recharge and discharge areas for unconfined and shallowest confined aquifers
15. Field measurements of dispersivity coefficients
16. Laboratory and field determination of the distribution coefficient for movement of critical nuclides through all hydrostratigraphic units
17. Rates of denudation or slope retreat

These data are necessary for a complete definition of flow and nuclide transport through both the unsaturated and saturated zones.

It is not possible to immobilize a radioactive contaminant in a burial site for long periods of geologic time (i.e., for millions of years) with complete certainty. However, there appear to be hydrogeologic environments in which these contaminants can be kept below the surface and away from people until they have decayed to acceptable levels.

The problem is not merely a matter of ensuring optimum confinement, but also one of ensuring confinement for a minimum but specified time or describing and predicting the performance of these radioactive contaminants in the subsurface until this period has elapsed. For this reason, burial sites having complex hydrogeology in which such predictions are difficult or impossible are probably not suitable for storing radioactive waste.

From a geological standpoint, there appear to be two basic approaches to the long-term control of buried radioactive waste. The simplest approach is to prevent water from reaching the waste and thereby to eliminate the possibility of contaminants in the waste being mobilized. In arid climates, where there is little or no infiltration, this appears to be feasible.

In humid climates, where there is infiltration, some sort of engineered container or facilities that would isolate the waste from the water for hundreds of years is necessary. Whether or not such a facility can be designed, constructed, and demonstrated remains to be seen.

The second approach to long-term control involves burying the waste in a hydrogeologic environment that can be demonstrated to be safe despite the fact that radioactive contaminants can and will be mobilized. Demonstrating that such sites are, in fact, safe requires a quantitative evaluation of the factors influencing contaminant movement. Such an evaluation may be quite difficult but appears to be our only option if we wish to bury radioactive waste in humid climates or in climates where infiltration is capable of mobilizing or leaching the buried waste.

It is also important to give attention to the possible biological and microbiological environment of a burial site. Soil microorganisms, earthworms, larger burrowing animals, and the deep taproots of plants seeking water and nourishment (particularly in desert areas) can all be factors in moving components of waste out of a burial place into the biosphere. Some organisms can release organic compounds into the soil that can serve as complexing agents to mobilize otherwise insoluble contaminants. Some organisms can concentrate radionuclides by surprisingly high factors from their environment and so can change both the biochemical availability and the distribution of a radionuclide.

Site Selection Criteria.   Michigan's site selection criteria serve to illustrate the factors that need to be considered in selecting disposal sites.

The first objective is to avoid population centers and conflicts with human activities. Michigan established an isolation distance of 1 km and required that projected population growth must not infringe to the extent that it would interfere with health and safety performance objectives of environmental monitoring.

Areas within 1.6 km of a fault where tectonic movement has occurred within the last 10,000 years are excluded as candidate sites. Likewise excluded are areas where significant earthquake intensity has been measured and flood plains exist. Mass wasting, erosion, and similar geologic processes are to be evaluated for possible damage to the facility.

Areas where groundwater flows from sites more than 30 m in 100 years or where groundwater could reach an aquifer in less than 500 years are excluded. The criteria also exclude areas over sole source aquifers and areas where groundwater discharges to the surface within 1 km. The facility may not be built within 16 km of the Great Lakes.

The criteria specify that the safest transportation net will be used. Highways with low accident rates located away from population centers are favored.

The site must have no complex meteorological characteristics and must avoid resource development conflicts. Likewise, environmentally sensitive areas such as wetlands and shorelands must be avoided. Areas that have formally proposed or approved development plans as of January 1, 1988, are excluded.

These criteria are extremely rigorous. Because of these constraints and the more serious problem of public opposition, no new sites have been finalized in the United States. Some compacts have had severe problems and conflicts that have resulted in the expulsion of one of the states. For example, after being selected as the host state, Michigan failed to identify an acceptable site and was expelled from the Midwest Compact.

A few compacts are proceeding quite well. These compacts are involving the public, community officials, regulators, and generators in joint efforts to identify sites, complete licensing applications, secure contractors, and construct the site. The most successful approach appears to be one of identification of actual candidate sites followed by a volunteer applicant.

The two currently operating sites in the United States are at Hanford, Washington, and Barnwell, South Carolina. These sites are accepting low-level radioactive waste from across the United States. There is a tremendous financial advantage to them in doing so. In November 1995, the total cost of disposing of a 0.210-m^3 drum was about $3000. Many generators have been storing this waste for several years (e.g., in Michigan, 55 generators have been storing waste for 5 years), and they are willing to pay these prices because of the lack of space.

CHAPTER REVIEW

When you have completed studying this chapter, you should be able to do the following without the aid of your textbook or notes:

1. Explain what an isotope is.
2. Explain why some isotopes are radioactive and others are not.
3. Explain how alpha, beta, x-ray, and gamma ray emissions occur and how they differ.
4. Define the unit becquerel.
5. Explain the process of fission in a nuclear reactor.
6. Explain how x-rays are produced in an x-ray machine.
7. Define the concept of radiation dose and the units of roentgen, rad, Gy, Sv, and rem.
8. Explain the concepts of RBE and W_T.
9. List the pattern of biological effects of radiation.
10. Discuss the determinants of biological effects.
11. Discuss the difference between acute and delayed biological effects of radiation.
12. List three possible delayed effects of radiation exposure.
13. State the acceptable occupational and nonoccupational dose of radiation as established by the NRC.
14. Explain the difference between internal and external radiation hazard.
15. Select a material and its thickness to protect against alpha or beta radiation.
16. Describe the sources of background radiation.
17. Explain why radon is a hazard and the mechanism by which the hazard is realized.
18. List three fundamental methods of reducing external radiation hazard.

19. Explain how to reduce occupational exposure to internal radiation hazards.

20. Describe how radon enters a house and give some techniques that may be used to inhibit radon entry.

21. List and describe the three types of radioactive waste (HLW, transuranic, and LLW).

22. Describe how each type of radioactive waste is to be disposed of.

23. Discuss waste minimization practice in reducing the volume of LLW.

With the aid of this text, you should be able to do the following:

1. Determine what particles are emitted in a given decay chain.

2. Determine the activity of a radioisotope given the original activity and the time interval.

3. Determine the activity resulting from the growth of a daughter product from a parent radionuclide.

4. Determine the time to achieve maximum activity of a daughter product.

5. Apply the inverse square law to determine radiation intensity.

6. Determine whether a combination of radionuclides exceeds the permissible concentrations.

7. Calculate the radiation intensity behind a shielding material or the desired thickness of a shielding material to achieve a reduction of radiation intensity.

PROBLEMS

16–1 What are the elements $^{40}_{18}X$ and $^{14}_{7}X$?

    Answer: argon and nitrogen

16–2 What are the elements $^{8}_{4}X$ and $^{238}_{92}X$?

16–3 What particle is emitted in the decay chain represented by

$$^{14}_{6}C \longrightarrow {}^{14}_{7}N?$$

    Answer: beta

16–4 What particle is emitted in the decay chain represented by

$$^{32}_{15}P \longrightarrow {}^{32}_{16}S?$$

16–5 What particles are emitted in each step in the decay chain represented by

$$^{226}_{88}Ra \longrightarrow {}^{222}_{86}Rn \longrightarrow {}^{218}_{84}Po \longrightarrow {}^{214}_{82}Pb?$$

16–6 What particles are emitted in each step in the decay chain represented by

$$^{214}_{82}Pb \longrightarrow {}^{214}_{83}Bi \longrightarrow {}^{214}_{84}Po \longrightarrow {}^{210}_{82}Pb \longrightarrow {}^{210}_{83}Bi \longrightarrow {}^{210}_{84}Po \longrightarrow {}^{206}_{82}Pb?$$

16–7 What particles are emitted in each step in the decay chain represented by

$$^{238}_{92}U \longrightarrow {}^{234}_{90}Th \longrightarrow {}^{234}_{91}Pa \longrightarrow {}^{234}_{92}U?$$

16–8 Show that if a positron and electron are annihilated, then an energy of 1.02 MeV is released.

16–9 A laboratory solution containing 0.5 $\mu Ci \cdot L^{-1}$ of ^{32}P is to be disposed of. How long must the radioisotope be held to meet the allowable discharge activity?

16–10 An accident has contaminated a laboratory with ^{45}Ca. The radiation level is ten times the tolerance level. How long must the room be isolated before the tolerance level is reached?

16–11 A hospital waste containing 100 $\mu Ci \cdot L^{-1}$ of ^{131}I is to be disposed of. How long must the radioisotope be held to meet the allowable discharge activity?

16–12 If in August 1911 Mme. Curie prepared an international standard containing 20.00 mg of $RaCl_2$, what will the radium content of this standard be in August 2010?

16–13 What is the mass of a 50 μCi sample of pure ^{131}I?

> **Answer:** 4.04×10^{-10} g

16–14 By emitting an alpha particle, ^{210}Po decays to ^{206}Pb. If the half-life of ^{210}Po is 138.4 d, what volume of ^{4}He will be produced in 1 year from 50 Ci of ^{210}Po? Assume the gas is at standard temperature and pressure.

16–15 Using a spreadsheet program you have written, calculate and plot the growth curve of ^{222}Rn from an initially pure sample of ^{226}Ra. Assume no ^{222}Rn is present initially.

16–16 When an x-ray unit is operated at 70 kV and 5 mA, it produces an intensity of D R · min^{-1} at 1.0 m from the source. What intensity will it produce 2.0 m from the source?

16–17 If the source of x-rays in Problem 16–16 is operated at 15 mA, what intensity will be produced 2.0 m from the source?

> **Answer:** 0.75 D

16–18 What thickness (in cm) of lead is required to shield a ^{60}Co source so that the transmission is reduced 99.6%?

16–19 What is the equivalent thickness (in cm) of concrete to accomplish the same attenuation as the lead in Problem 16–18?

> **Answer:** ~55 cm

16–20 An existing concrete wall that is 25 cm thick is to be used to shield a ^{60}Co source so that the transmission is reduced 99.6%. What additional thickness (in cm) of lead is required to achieve this transmission reduction?

16–21 Determine the proportionality constant (u) for lead when it is used to shield ^{137}Cs.

16–22 Determine the proportionality constant (u) for iron when it is used to shield ^{137}Cs.

> **Answer:** $u = 0.391$

DISCUSSION QUESTIONS

16–1 Explain why an archaeological artifact such as wood or bone may be dated by measuring its concentration of carbon-14.

16–2 Would you expect the tissue weighting factor (W_T) for x-rays to the big toe to be greater than, less than, or the same as that for radioiodine to the thyroid? Explain your choice.

16–3 What kind of radionuclide emitter (alpha, beta, gamma, or x-ray) is most dangerous from an internal hazard point of view? Explain why.

16–4 A laboratory worker has requested your advice on a shield for work she is doing with high-energy beta particles. What would you recommend?

16–5 You have an opportunity to purchase an older home with a basement that is serviced by a floor drain. What measures might you request to limit the migration of radon into the basement?

16–6 What is the status of the proposed Yucca mountain disposal site? How much money has been spent to determine if this site is acceptable?

REFERENCES

CFR (1989) Code of Federal Regulations, 40 CFR 60.

CFR (2006) Code of Federal Regulations, 40 CFR, part 20.

Coombe, R. A. (1968) *An Introduction to Radioactivity for Engineers,* Macmillan/St. Martin's Press, New York, pp. 1–37.

EIA (2005) Energy Information Administration, http://www.eia.doe.gov/cneaf/nuclear/page/nuc_ reactors/reactsum.html

Henschel, D. B., and A. G. Scott (1987) "Testing of Indoor Radon Reduction Techniques in Eastern Pennsylvania: An Update," *Indoor Radon II, Proceedings of the Second APCA International Specialty Conference, Cherry Hill, NJ,* Air Pollution Control Association, Pittsburgh, PA, pp. 146–59.

Ibach, M. T., and J. H. Gallagher (1987) "Retrofit and Preoccupancy Radon Mitigation Program for Homes," *Indoor Radon II Proceedings of the Second APCA International Specialty Conference, Cherry Hill, NJ,* Air Pollution Control Association, Pittsburgh, PA, pp. 172–82.

Kuennen, W., and R. C. Roth (1989) "Reduction of Radon Working Level by a Room Air Cleaner," presented at the 82nd Annual Meeting of the Air & Waste Management Association, Anaheim, CA, June 1989.

Murane, D. M., and J. Spears (1987) "Radon Reduction in New Construction," *Indoor Radon II, Proceedings of the Second APCA International Specialty Conference, Cherry Hill, NJ,* Air Pollution Control Association, Pittsburgh, PA, pp. 183–94.

Murray, R. L. (1989) *Understanding Radioactive Waste,* Battelle Press, Richland, WA, pp. 137–38.

NRC (1976) *The Shallow Land Burial of Low-Level Radioactively Contaminated Solid Waste,* National Academy of Science, Washington, DC.

Papadopulos, S. S., and I. J. Winograd (1974) *Storage of Low-Level Radioactive Wastes in the Ground: Hydrogeological and Hydrochemical Factors,* U.S. Environmental Protection Agency Report No. 520/3-74-009, Washington, DC.

U.S. PHS (1968) *Introduction to Medical X-Ray Protection, Training and Manpower Development Program,* U.S. Public Health Service, Rockville, MD.

U.S. PHS (1970) *Radiological Health Handbook,* PHS Publication No. 2016, U.S. Public Health Service, Rockville, MD, p. 204.

U.S. NRC (2007) U.S. Nuclear Regulatory Commission, http://www.nrc.gov/reading-nm/basic–ref/glossary/ exposure.html

World Health Organization (2006) *Health Effects of the Chernobyl Accident: An Overview.* Last edited April 2006. http://www.who.int/ionizing_radiation/chernobyl/backgrounder/en/

World Nuclear Association (2018) *Chernobyl Accident 1986.* Last edited April 2018. http://www.world-nuclear.org/information-library/safety-and-security/safety-of-plants/chernobyl-accident.aspx

Properties of Air, Water, and Selected Chemicals

Appendix A

| TABLE A–1 | Physical Properties of Water at 1 atm | | | |

Temperature (°C)	Density, ρ (kg $\cdot$ m^{-3})	Specific Weight, γ (kN $\cdot$ m^{-3})	Dynamic Viscosity, μ (mPa $\cdot$ s)	Kinematic Viscosity, ν (μm^2 $\cdot$ s^{-1})
0	999.842	9.805	1.787	1.787
3.98	1000.000	9.807	1.567	1.567
5	999.967	9.807	1.519	1.519
10	999.703	9.804	1.307	1.307
12	999.500	9.802	1.235	1.236
15	999.103	9.798	1.139	1.140
17	998.778	9.795	1.081	1.082
18	998.599	9.793	1.053	1.054
19	998.408	9.791	1.027	1.029
20	998.207	9.789	1.002	1.004
21	997.996	9.787	0.998	1.000
22	997.774	9.785	0.955	0.957
23	997.542	9.783	0.932	0.934
24	997.300	9.781	0.911	0.913
25	997.048	9.778	0.890	0.893
26	996.787	9.775	0.870	0.873
27	996.516	9.773	0.851	0.854
28	996.236	9.770	0.833	0.836
29	995.948	9.767	0.815	0.818
30	995.650	9.764	0.798	0.801
35	994.035	9.749	0.719	0.723
40	992.219	9.731	0.653	0.658
45	990.216	9.711	0.596	0.602
50	988.039	9.690	0.547	0.554
60	983.202	9.642	0.466	0.474
70	977.773	9.589	0.404	0.413
80	971.801	9.530	0.355	0.365
90	965.323	9.467	0.315	0.326
100	958.366	9.399	0.282	0.294

Pa $\cdot$ s = (mPa $\cdot$ s) $\times$ 10^{-3}

m^2 $\cdot$ s^{-1} = (μm^2 $\cdot$ s^{-1}) $\times$ 10^{-6}

TABLE A–2   **Saturation Values of Dissolved Oxygen in Fresh Water Exposed to a Saturated Atmosphere Containing 20.9% Oxygen Under a Pressure of 101.325 kPa[a]**

Temperature (°C)	Dissolved Oxygen (mg · L⁻¹)	Saturated Vapor Pressure (kPa)
0	14.62	0.6108
1	14.23	0.6566
2	13.84	0.7055
3	13.48	0.7575
4	13.13	0.8129
5	12.80	0.8719
6	12.48	0.9347
7	12.17	1.0013
8	11.87	1.0722
9	11.59	1.1474
10	11.33	1.2272
11	11.08	1.3119
12	10.83	1.4017
13	10.60	1.4969
14	10.37	1.5977
15	10.15	1.7044
16	9.95	1.8173
17	9.74	1.9367
18	9.54	2.0630
19	9.35	2.1964
20	9.17	2.3373
21	8.99	2.4861
22	8.83	2.6430
23	8.68	2.8086
24	8.53	2.9831
25	8.38	3.1671
26	8.22	3.3608
27	8.07	3.5649
28	7.92	3.7796
29	7.77	4.0055
30	7.63	4.2430
31	7.51	4.4927
32	7.42	4.7551
33	7.28	5.0307
34	7.17	5.3200
35	7.07	5.6236
36	6.96	5.9422
37	6.86	6.2762
38	6.75	6.6264

[a]For other barometric pressures, the solubilities vary approximately in proportion to the ratios of these pressures to the standard pressures.

Source: Calculated by G. C. Whipple and M. C. Whipple from measurements of C. J. J. Fox, *Journal of the American Chemical Society,* vol. 33, p. 362, 1911.

TABLE A–3	Viscosity of Dry Air at Approximately 100 kPa[a]			
	Temperature (°C)	Dynamic Viscosity (μPa · s)	Temperature (°C)	Dynamic Viscosity (μPa · s)
	0	17.1	55	20.1
	5	17.4	60	20.3
	10	17.7	65	20.6
	15	17.9	70	20.9
	20	18.2	75	21.1
	25	18.5	80	21.4
	30	18.7	85	21.7
	35	19.0	90	21.9
	40	19.3	95	22.2
	45	19.5	100	22.5
	50	19.8	150	25.2

[a]$\mu = 17.11 + 0.0536\,T + (P/8280)$ where T is in degrees Celsius and P is in kilopascals.

TABLE A–4	Properties of Air at Standard Conditions[a]		
	Molecular weight	M	28.97
	Gas constant	R	$287\ \mathrm{J \cdot kg^{-1} \cdot K^{-1}}$
	Specific heat at constant pressure	c_p	$1005\ \mathrm{J \cdot kg^{-1} \cdot K^{-1}}$
	Specific heat at constant volume	c_v	$718\ \mathrm{J \cdot kg^{-1} \cdot K^{-1}}$
	Density	ρ	$1.185\ \mathrm{kg \cdot m^{-3}}$
	Dynamic viscosity	μ	$1.8515 \times 10^{-5}\ \mathrm{Pa \cdot s}$
	Kinematic viscosity	ν	$1.5624 \times 10^{-5}\ \mathrm{m^2 \cdot s^{-1}}$
	Thermal conductivity	k	$0.0257\ \mathrm{W \cdot m^{-1} \cdot K^{\cdot 1}}$
	Ratio of specific heats, c_p/c_v	k	1.3997
	Prandtl number	Pr	0.720

[a]Measured at 101.325 kPa pressure and 298 K temperature.

TABLE A–5	Properties of Saturated Water at 298 K		
	Molecular weight	M	18.02
	Gas constant	R	$461.4\ \mathrm{J \cdot kg^{-1} \cdot K^{-1}}$
	Specific heat	c	$4181\ \mathrm{J \cdot kg^{-1} \cdot K^{-1}}$
	Prandtl number	Pr	6.395
	Thermal conductivity	k	$0.604\ \mathrm{W \cdot m^{-1} \cdot K^{-1}}$

TABLE A–6	Frequently Used Constants		
	Standard atmospheric pressure	P_{atm}	101.325 kPa
	Standard gravitational acceleration	g	$9.8067\ \mathrm{m \cdot s^{-2}}$
	Universal gas constant	R_u	$8314.3\ \mathrm{J \cdot kg^{-1} \cdot mol^{-1} \cdot K^{-1}}$
	Electrical permittivity constant	ϵ_0	$8.85 \times 10^{-12}\ \mathrm{C \cdot V^{-1} \cdot m^{-1}}$
	Electron charge	q_e	$1.60 \times 10^{-19}\ \mathrm{C}$
	Boltzmann's constant	k	$1.38 \times 10^{-23}\ \mathrm{J \cdot K^{-1}}$

TABLE A–7   Properties of Selected Organic Compounds

Name	Formula	M.W.	Density $(g \cdot mL^{-1})$	Vapor Pressure (mm Hg)	Henry's Law Constant $(kPa \cdot m^3 \cdot mol^{-1})$
Acetone	CH_3COCH_3	58.08	0.79	184	0.01
Benzene	C_6H_6	78.11	0.879	95	0.6
Bromodichloromethane	$CHBrCl_2$	163.8	1.971		0.2
Bromoform	$CHBr_3$	252.75	2.8899	5	0.06
Bromomethane	CH_3Br	94.94	1.6755	1300	0.5
Carbon tetrachloride	CCl_4	153.82	1.594	90	3
Chlorobenzene	C_6H_5Cl	112.56	1.107	12	0.4
Chlorodibromomethane	$CHBr_2Cl$	208.29	2.451	50	0.09
Chloroethane	C_2H_5Cl	64.52	0.8978	700	0.2
Chloroethylene	C_2H_3Cl	62.5	0.912	2550	4
Chloroform	$CHCl_3$	119.39	1.4892	190	0.4
Chloromethane	CH_3Cl	50.49	0.9159	3750	1.0
1,2-Dibromoethane	$C_2H_2Br_2$	187.87	2.18	10	0.06
1,2-Dichlorobenzene	$1,2\text{-}Cl_2\text{--}C_6H_4$	147.01	1.3048	1.5	0.2
1,3-Dichlorobenzene	$1,3\text{-}Cl_2\text{--}C_6H_4$	147.01	1.2884	2	0.4
1,4-Dichlorobenzene	$1,4\text{-}Cl_2\text{--}C_6H_4$	147.01	1.2475	0.7	0.2
1,1-Dichloroethylene	$CH_2{=}CCl_2$	96.94	1.218	500	15
1,2-Dichloroethane	$ClCH_2CH_2Cl$	98.96	1.2351	700	0.1
1,1-Dichloroethane	CH_3CHCl_2	98.96	1.1757	200	0.6
Trans-1,2-Dichloroethylene	$CHCl{=}CHCl$	96.94	1.2565	300	0.6
Dichloromethane	CH_2Cl_2	84.93	1.327	350	0.3
1,2-Dichloropropane	$CH_3CHClCH_2Cl$	112.99	1.1560	50	0.4
Cis-1,3-Dichloropropylene	$ClCH_2CH{=}CHCl$	110.97	1.217	40	0.2
Ethyl benzene	$C_6H_5CH_2CH_3$	106.17	0.8670	9	0.8
Formaldehyde	$HCHO$	30.05	0.815		
Hexachlorobenzene	C_6Cl_6	284.79	1.5691		
Pentachlorophenol	Cl_5C_6OH	266.34	1.978		
Phenol	C_6H_5OH	94.11	1.0576		
1,1,2,2-Tetrachloroethane	$CHCl_2CHCl_2$	167.85	1.5953	5	0.05
Tetrachloroethylene	$Cl_2C{=}CCl_2$	165.83	1.6227	15	3
Toluene	$C_6H_5CH_3$	92.14	0.8669	28	0.7
1,1,1-Trichloroethane	CH_3CCl_3	133.41	1.3390	100	3.0
1,1,2-Trichloroethane	$CH_2ClCHCl_2$	133.41	1.4397	25	0.1
Trichloroethylene	$ClHC{=}CCl_2$	131.29	1.476	50	0.9
Vinyl chloride	$H_2C{=}CHCl$	62.50	0.9106	2200	50
o-Xylene	$1,2\text{-}(CH_3)_2C_6H_4$	106.17	0.8802	6	0.5
m-X...	$1,3\text{-}(CH_3)_2C_6H_4$	106.17	0.8642	8	0.7
	$1,4\text{-}(CH_3)_2C_6H_4$	106.17	0.8611	8	0.7

...lene; ethyl chloride = chloroethane; ethylene chloride = 1,2-dichloroethane; ethylidene chloride = 1,1-dichloroethane; ...luene; methyl chloride = chloromethane; methyl chloroform = 1,1,1-trichloroethane; methylene chloride = ...ichloromethane = carbon tetrachloride; tribromomethane = bromoform.

TABLE A–8 Typical Solubility Product Constants

Equilibrium Equation	K_{sp} at 25°C	Equilibrium Equation	K_{sp} at 25°C
$AgCl \rightleftharpoons Ag^+ + Cl^-$	1.76×10^{-10}	$Fe(OH)_2 \rightleftharpoons Fe^{2+} + 2OH^-$	4.79×10^{-17}
$Al(OH)_3 \rightleftharpoons Al^{3+} + 3OH^-$	1.26×10^{-33}	$FeS \rightleftharpoons Fe^{2+} + S^{2-}$	1.57×10^{-19}
$BaSO_4 \rightleftharpoons Ba^{2+} + SO_4^{2-}$	1.05×10^{-10}	$PbCO_3 \rightleftharpoons Pb^{2+} + CO_3^{2-}$	1.48×10^{-13}
$Cd(OH)_2 \rightleftharpoons Cd^{2+} + 2OH^-$	5.33×10^{-15}	$Pb(OH)_2 \rightleftharpoons Pb^{2+} + 2OH^-$	1.40×10^{-20}
$CdS \rightleftharpoons Cd^{2+} + S^{2-}$	1.40×10^{-29}	$PbS \rightleftharpoons Pb^{2+} + S^{2-}$	8.81×10^{-29}
$CdCO_3 \rightleftharpoons Cd^{2+} + CO_3^{2-}$	6.20×10^{-12}	$Mg(OH)_2 \rightleftharpoons Mg^{2+} + 2OH^-$	1.82×10^{-11}
$CaCO_3 \rightleftharpoons Ca^{2+} + CO_3^{2-}$	4.95×10^{-9}	$MgCO_3 \rightleftharpoons Mg^{2+} + CO_3^{2-}$	1.15×10^{-5}
$CaF_2 \rightleftharpoons Ca^{2+} + 2F^-$	1.61×10^{-10}	$MnCO_3 \rightleftharpoons Mn^{2+} + CO_3^{2-}$	2.23×10^{-11}
$Ca(OH)_2 \rightleftharpoons Ca^{2+} + 2OH^-$	7.88×10^{-6}	$Mn(OH)_2 \rightleftharpoons Mn^{2+} + 2OH^-$	2.04×10^{-13}
$Ca_3(PO_4)_2 \rightleftharpoons 3Ca^{2+} + 2PO_4^{3-}$	2.02×10^{-33}	$NiCO_3 \rightleftharpoons Ni^{2+} + CO_3^{2-}$	1.45×10^{-7}
$CaSO_4 \rightleftharpoons Ca^{2+} + SO_4^{2-}$	3.73×10^{-5}	$Ni(OH)_2 \rightleftharpoons Ni^{2+} + 2OH^-$	5.54×10^{-16}
$Cr(OH)_3 \rightleftharpoons Cr^{3+} + 3OH^-$	6.0×10^{-31}	$NiS \rightleftharpoons Ni^{2+} + S^{2-}$	1.08×10^{-21}
$Cu(OH)_2 \rightleftharpoons Cu^{2+} + 2OH^-$	2.0×10^{-19}	$SrCO_3 \rightleftharpoons Sr^{2+} + CO_3^{2-}$	5.60×10^{-10}
$CuS \rightleftharpoons Cu^{2+} + S^{2-}$	1.28×10^{-36}	$Zn(OH)_2 \rightleftharpoons Zn^{2+} + 2OH^-$	7.68×10^{-17}
$Fe(OH)_3 \rightleftharpoons Fe^{3+} + 3OH^-$	2.67×10^{-39}	$ZnS \rightleftharpoons Zn^{2+} + S^{2-}$	2.91×10^{-25}
$FeCO_3 \rightleftharpoons Fe^{2+} + CO_3^{2-}$	3.13×10^{-11}		

TABLE A–9 Typical Valences of Elements and Compounds in Water

Element or Compound	Valence	Element or Compound	Valence
Aluminum	3^+	Manganese	2^+
Ammonium (NH_4)	1^+	Nickel	2^+
Barium	2^+	Oxygen	2^-
Boron	3^+	Nitrogen	$3^+, 5^+, 3^-$
Cadmium	2^+	Nitrate (NO_3)	1^-
Calcium	2^+	Nitrite (NO_2)	1^-
Carbonate (CO_3)	2^-	Phosphorus	$5^+, 3^-$
Carbon dioxide (CO_2)	a	Phosphate (PO_4)	3^-
Chloride (*not* chlorine)	1^-	Potassium	1^+
Chromium	$3^+, 6^+$	Silver	1^+
Copper	2^+	Silica	b
Fluoride (*not* fluorine)	1^-	Silicate (SiO_4)	4^-
Hydrogen	1^+	Sodium	1^+
Hydroxide (OH)	1^-	Sulfate (SO_4)	2^-
Iron	$2^+, 3^+$	Sulfide (S)	2^-
Lead	2^+	Zinc	2^+
Magnesium	2^+		

[a]Carbon dioxide in water is essentially carbonic acid:
$CO_2 + H_2O \rightleftharpoons H_2CO_3$
As such, the equivalent weight = GMW/2.
[b]Silica in water is reported as SiO_2. The equivalent weight is equal to the gram molecular weigh

TABLE A–10   Values of K_{ow}, Water Solubilities, and Henry's Law Constants for Selected Organic Compounds

Compound	log K_{ow}	Water Solubility (mg · L^{-1})	K_H (atm · M^{-1})	Compound	log K_{ow}	Water Solubility (mg · L^{-1})	K_H (atm · M^{-1})
Data from Yaws for 25°C				**Aromatic compounds, continued**			
				Napthalene	3.3	32	0.46 (20°C)
Halogenated aliphatic compounds				Phenanthrene	4.46	1.18	
Methanes				Anthracene	4.45	0.053	
Chloromethane	0.91	5900	8.2	Fluorene	4.18	1.89	
Dichloromethane	1.25	19,400	2.5				
Chloroform	1.97	7500	4.1	**Other aromatic compounds**			
Bromoform	2.4	3100	0.59	Chlorobenzene	2.84	390	4.5
Carbon tetrachloride	2.83	790	29	1,2-Dichlorobenzene	3.43	92	2.8
Dichlorodifluoromethane	2.16	18,800	390	1,3-Dichlorobenzene	3.53	123	3.4
				1,4-Dichlorobenzene	3.44	80	
Ethanes				1,2,4-Trichlorobenzene			3.0
Chloroethane	1.43	9000	6.9	Hexachlorobenzene	5.73	0.0047	
1,1-Dichloroethane	1.79	5000	5.8	Nitrobenzene	1.85	1940	0.021
1,2-Dichloroethane	1.48	8700	1.18	3-Nitrotoluene	2.45	500	0.075
1,1,1-Trichloroethane	2.49	1000	22	Phenol	1.46	80,000	0.00076
1,1,2-Trichloroethane	1.89	4400	0.92	Diethyl phthalate	2.47	1000	0.00014
Hexachloroethane	3.91	8	25	2-Chlorophenol	2.15	25,000	0.037
				3-Chlorophenol	2.5	25,000	0.00204
Ethenes				Dibenzofuran	4.12		
Vinyl chloride	1.62	2700	22				
1,1-Dichloroethene	2.13	3400	23	**Other aliphatic compounds**			
1,2-cis-Dichloroethene	1.86	3500	7.4	Methyl t-butyl ether	0.94	51,000	0.54
1,2-trans-Dichloroethene	2.09	6300	6.7	Methyl ethyl ketone	0.29	250,000	0.030
Trichloroethene	2.42	1100	11.6				
Tetrachloroethene	3.4	150	26.9	**Data from Schnoor et al. for 20°C**			
Aromatic compounds							
Hydrocarbons				2-Nitrophenol	1.75	2100	
Benzene	2.13	1760	5.6	Benzo(a)pyrene	6.06	0.0038	0.00049
Toluene	2.73	540	6.4	Acrolein	0.01	210,000	0.0038
Ethylbenzene	3.15	165	8.1	Alachlor	2.92	240	
Styrene	2.95	322	2.6	Atrazine	2.69	33	
o-Xylene	3.12	221	4.2	Pentachlorophenol	5.04	14	
m-Xylene	3.2	174	6.8	DDT	6.91	0.0055	0.038
p-Xylene	3.15	200	6.2	Lindane	3.72	7.52	0.0048
1,2,3-Trimethylbenzene	3.66	36	7.4	Dieldrin	3.54	0.2	0.0002
1,2,4-Trimethylbenzene	4.02	35		2,4-D	1.78	900	0.00000172

Sources: C. L. Yaws, "Chemical Properties Handbook," McGraw-Hill, New York, 1999. Schnoor et al., "Processes, Coefficients, and Models for Simulating Toxic Organics and Heavy Metals in Surface Waters," U.S. Environmental Protection Agency, EPA/600/3-87/015, June 1987.

TABLE A–11   Henry's Law Constants for Common Gases Soluble in H_2O

Temperature (°C)	$K_H \times 10^{-4}$ (atm)							
	Air	N_2	O_2	CO_2	CO	H_2	H_2S	CH_4
0	4.32	5.29	2.55	0.073	3.52	5.79	0.027	2.24
10	5.49	6.68	3.27	0.104	4.42	6.36	0.037	2.97
20	6.64	8.04	4.01	0.142	5.36	6.83	0.048	3.76
30	7.71	9.24	4.75	0.186	6.20	7.29	0.061	4.49
40	8.70	10.4	5.35	0.233	6.96	7.51	0.075	5.20

Source: G. Kiely, *Environmental Engineering.* 1996 The McGraw-Hill Education Companies, New York.

List of Elements with Their Symbols and Atomic Masses

List of the Elements with Their Symbols and Atomic Masses*

Element	Symbol	Atomic Number	Atomic Mass†	Element	Symbol	Atomic Number	Atomic Mass†
Actinium	Ac	89	(227)	Lanthanum	La	57	138.9
Aluminum	Al	13	26.98	Lawrencium	Lr	103	(257)
Americium	Am	95	(243)	Lead	Pb	82	207.2
Antimony	Sb	51	121.8	Lithium	Li	3	6.941
Argon	Ar	18	39.95	Lutetium	Lu	71	175.0
Arsenic	As	33	74.92	Magnesium	Mg	12	24.31
Astatine	At	85	(210)	Manganese	Mn	25	54.94
Barium	Ba	56	137.3	Meitnerium	Mt	109	(266)
Berkelium	Bk	97	(247)	Mendelevium	Md	101	(256)
Beryllium	Be	4	9.012	Mercury	Hg	80	200.6
Bismuth	Bi	83	209.0	Molybdenum	Mo	42	95.94
Bohrium	Bh	107	(262)	Neodymium	Nd	60	144.2
Boron	B	5	10.81	Neon	Ne	10	20.18
Bromine	Br	35	79.90	Neptunium	Np	93	(237)
Cadmium	Cd	48	112.4	Nickel	Ni	28	58.69
Calcium	Ca	20	40.08	Niobium	Nb	41	92.91
Californium	Cf	98	(249)	Nitrogen	N	7	14.01
Carbon	C	6	12.01	Nobelium	No	102	(253)
Cerium	Ce	58	140.1	Osmium	Os	76	190.2
Cesium	Cs	55	132.9	Oxygen	O	8	16.00
Chlorine	Cl	17	35.45	Palladium	Pd	46	106.4
Chromium	Cr	24	52.00	Phosphorus	P	15	30.97
Cobalt	Co	27	58.93	Platinum	Pt	78	195.1
Copper	Cu	29	63.55	Plutonium	Pu	94	(242)
Curium	Cm	96	(247)	Polonium	Po	84	(210)
Darmstadtium	Ds	110	(269)	Potassium	K	19	39.10
Dubnium	Db	105	(260)	Praseodymium	Pr	59	140.9
Dysprosium	Dy	66	162.5	Promethium	Pm	61	(147)
Einsteinium	Es	99	(254)	Protactinium	Pa	91	(231)
Erbium	Er	68	167.3	Radium	Ra	88	(226)
Europium	Eu	63	152.0	Radon	Rn	86	(222)
Fermium	Fm	100	(253)	Rhenium	Re	75	186.2
Fluorine	F	9	19.00	Rhodium	Rh	45	102.9
Francium	Fr	87	(223)	Roentgenium	Rg	111	(272)
Gadolinium	Gd	64	157.3	Rubidium	Rb	37	85.47
Gallium	Ga	31	69.72	Ruthenium	Ru	44	101.1
Germanium	Ge	32	72.59	Rutherfordium	Rf	104	(257)
Gold	Au	79	197.0	Samarium	Sm	62	150.4
Hafnium	Hf	72	178.5	Scandium	Sc	21	44.96
Hassium	Hs	108	(265)	Seaborgium	Sg	106	(263)
Helium	He	2	4.003	Selenium	Se	34	78.96
Holmium	Ho	67	164.9	Silicon	Si	14	28.09
Hydrogen	H	1	1.008	Silver	Ag	47	107.9
Indium	In	49	114.8	Sodium	Na	11	22.99
Iodine	I	53	126.9	Strontium	Sr	38	87.62
Iridium	Ir	77	192.2	Sulfur	S	16	32.07
Iron	Fe	26	55.85	Tantalum	Ta	73	180.9
Krypton	Kr	36	83.80	Technetium	Tc	43	(99)

(continued)

Element	Symbol	Atomic Number	Atomic Mass[†]	Element	Symbol	Atomic Number	Atomic Mass[†]
Tellurium	Te	52	127.6	Uranium	U	92	238.0
Terbium	Tb	65	158.9	Vanadium	V	23	50.94
Thallium	Tl	81	204.4	Xenon	Xe	54	131.3
Thorium	Th	90	232.0	Ytterbium	Yb	70	173.0
Thulium	Tm	69	168.9	Yttrium	Y	39	88.91
Tin	Sn	50	118.7	Zinc	Zn	30	65.39
Titanium	Ti	22	47.88	Zirconium	Zr	40	91.22
Tungsten	W	74	183.9				

[*]All atomic masses have four significant figures. These values are recommended by the Committee on Teaching of Chemistry, International Union of Pure and Applied Chemistry.

[†]Approximate values of atomic masses for radioactive elements are given in parentheses.

Periodic Table of Chemical Elements

Legend (sample cell):

9	(Atomic number)
F	
Fluorine	
19.00	(Atomic mass)

Main table (group numbers shown as IUPAC 1–18 / U.S. A–B notation):

1 / 1A	2 / 2A	3 / 3B	4 / 4B	5 / 5B	6 / 6B	7 / 7B	8 / 8B	9 / 8B	10 / 8B	11 / 1B	12 / 2B	13 / 3A	14 / 4A	15 / 5A	16 / 6A	17 / 7A	18 / 8A
1 **H** Hydrogen 1.008																	2 **He** Helium 4.003
3 **Li** Lithium 6.941	4 **Be** Beryllium 9.012											5 **B** Boron 10.81	6 **C** Carbon 12.01	7 **N** Nitrogen 14.01	8 **O** Oxygen 16.00	9 **F** Fluorine 19.00	10 **Ne** Neon 20.18
11 **Na** Sodium 22.99	12 **Mg** Magnesium 24.31											13 **Al** Aluminum 26.98	14 **Si** Silicon 28.09	15 **P** Phosphorus 30.97	16 **S** Sulfur 32.07	17 **Cl** Chlorine 35.45	18 **Ar** Argon 39.95
19 **K** Potassium 39.10	20 **Ca** Calcium 40.08	21 **Sc** Scandium 44.96	22 **Ti** Titanium 47.88	23 **V** Vanadium 50.94	24 **Cr** Chromium 52.00	25 **Mn** Manganese 54.94	26 **Fe** Iron 55.85	27 **Co** Cobalt 58.93	28 **Ni** Nickel 58.69	29 **Cu** Copper 63.55	30 **Zn** Zinc 65.39	31 **Ga** Gallium 69.72	32 **Ge** Germanium 72.59	33 **As** Arsenic 74.92	34 **Se** Selenium 78.96	35 **Br** Bromine 79.90	36 **Kr** Krypton 83.80
37 **Rb** Rubidium 85.47	38 **Sr** Strontium 87.62	39 **Y** Yttrium 88.91	40 **Zr** Zirconium 91.22	41 **Nb** Niobium 92.91	42 **Mo** Molybdenum 95.94	43 **Tc** Technetium (98)	44 **Ru** Ruthenium 101.1	45 **Rh** Rhodium 102.9	46 **Pd** Palladium 106.4	47 **Ag** Silver 107.9	48 **Cd** Cadmium 112.4	49 **In** Indium 114.8	50 **Sn** Tin 118.7	51 **Sb** Antimony 121.8	52 **Te** Tellurium 127.6	53 **I** Iodine 126.9	54 **Xe** Xenon 131.3
55 **Cs** Cesium 132.9	56 **Ba** Barium 137.3	57 **La** Lanthanum 138.9	72 **Hf** Hafnium 178.5	73 **Ta** Tantalum 180.9	74 **W** Tungsten 183.9	75 **Re** Rhenium 186.2	76 **Os** Osmium 190.2	77 **Ir** Iridium 192.2	78 **Pt** Platinum 195.1	79 **Au** Gold 197.0	80 **Hg** Mercury 200.6	81 **Tl** Thallium 204.4	82 **Pb** Lead 207.2	83 **Bi** Bismuth 209.0	84 **Po** Polonium (210)	85 **At** Astatine (210)	86 **Rn** Radon (222)
87 **Fr** Francium (223)	88 **Ra** Radium (226)	89 **Ac** Actinium (227)	104 **Rf** Rutherfordium (257)	105 **Db** Dubnium (260)	106 **Sg** Seaborgium (263)	107 **Bh** Bohrium (262)	108 **Hs** Hassium (265)	109 **Mt** Meitnerium (266)	110 **Ds** Darmstadtium (269)	111 **Rg** Roentgenium (272)	112	(113)	114	(115)	116	(117)	(118)

Lanthanides:

58 **Ce** Cerium 140.1	59 **Pr** Praseodymium 140.9	60 **Nd** Neodymium 144.2	61 **Pm** Promethium (147)	62 **Sm** Samarium 150.4	63 **Eu** Europium 152.0	64 **Gd** Gadolinium 157.3	65 **Tb** Terbium 158.9	66 **Dy** Dysprosium 162.5	67 **Ho** Holmium 164.9	68 **Er** Erbium 167.3	69 **Tm** Thulium 168.9	70 **Yb** Ytterbium 173.0	71 **Lu** Lutetium 175.0

Actinides:

90 **Th** Thorium 232.0	91 **Pa** Protactinium (231)	92 **U** Uranium 238.0	93 **Np** Neptunium (237)	94 **Pu** Plutonium (242)	95 **Am** Americium (243)	96 **Cm** Curium (247)	97 **Bk** Berkelium (247)	98 **Cf** Californium (249)	99 **Es** Einsteinium (254)	100 **Fm** Fermium (253)	101 **Md** Mendelevium (256)	102 **No** Nobelium (254)	103 **Lr** Lawrencium (257)

Key:
- Metals
- Metalloids
- Nonmetals

The 1–18 group designation has been recommended by the International Union of Pure and Applied Chemistry (IUPAC) but is not yet in wide use. In this text we use the standard U.S. notation for group numbers (1A–8A and 1B–8B). No names have been assigned for elements 112, 114, and 116. Elements 113, 115, 117, and 118 have not yet been synthesized.

Appendix D

Useful Unit Conversion and Prefixes

Useful Conversion Factors

Multiply	By	To Obtain
atmosphere (atm)	101.325	kilopascal (kPa)
Calorie (international)	4.1868	Joules (J)
centipoise	10^{-3}	$Pa \cdot s$
centistoke	10^{-6}	m^2/s
cubic meter (m^3)	35.31	cubic feet (ft^3)
cubic meter	1.308	cubic yard (yd^3)
cubic meter	1,000.00	liter (L)
cubic meter/s	15,850.0	gallons/min (gpm)
cubic meter/s	22.8245	million gal/d (MGD)
cubic meter/m^2	24.545	gallons/sq ft (gal/ft^2)
cubic meter/d $\cdot$ m	80.52	gal/d $\cdot$ ft (gpd/ft)
cubic meter/d $\cdot$ m^2	24.545	gal/d $\cdot$ ft^2 (gpd/ft^2)
cubic meter/d $\cdot$ m^2	1.0	meters/d (m/d)
days (d)	24.00	hours (h)
days (d)	1,440.00	minutes (min)
days (d)	86,400.00	seconds (s)
dyne	10^{-5}	Newtons (N)
erg	10^{-7}	Joules (J)
grains (gr)	6.480×10^{-2}	grams (g)
grains/U.S. gallon	17.118	mg/L
grams (g)	2.205×10^{-3}	pounds mass (lb_m)
hectare (ha)	10^4	m^2
hectare (ha)	2.471	acres
Hertz (Hz)	1	cycle/s
Joule (J)	1	N $\cdot$ m
J/m^3	2.684×10^{-5}	Btu/ft^3
kilogram/m^3 (kg/m^3)	8.346×10^{-3}	lb_m/gal
kilogram/m^3	1.6855	lb_m/yd^3
kilogram/ha (kg/ha)	8.922×10^{-1}	lb_m/acre
kilogram/m^2 (kg/m^2)	2.0482×10^{-1}	lb_m/ft^2
kilometers (km)	6.2150×10^{-1}	miles (mi)
kilowatt (kW)	1.3410	horsepower (hp)
kilowatt-hour	3.600	megajoules (MJ)
liters (L)	10^{-3}	cubic meters (m^3)
liters	1,000.00	milliliters (mL)
liters	2.642×10^{-1}	U.S. gallons
megagrams (Mg)	1.1023	U.S. short tons
meters (m)	3.281	feet (ft)
meters/d (m/d)	2.2785×10^{-3}	ft/min
meters/d	3.7975×10^{-5}	meters/s (m/s)
meters/s (m/s)	196.85	ft/min
meters/s	3.600	km/h
meters/s	2.237	miles/h (mph)
micron (μ)	10^{-6}	meters
milligrams (mg)	10^{-3}	grams (g)
milligrams/L	1	g/m^3
milligrams/L	10^{-3}	kg/m^3
Newton (N)	1	kg $\cdot$ m/s^2

(*continued*)

Multiply	By	To Obtain
Pascal (Pa)	1	N/m^2
Poise (P)	10^{-1}	$Pa \cdot s$
square meter (m^2)	2.471×10^{-4}	acres
square meter (m^2)	10.7639	sq ft (ft^2)
square meter/s	6.9589×10^6	gpd/ft
Stoke (St)	10^{-4}	m^2/s
Watt (W)	1	J/s
Watt/cu meter (W/m^3)	3.7978×10^{-2}	hp/1,000 ft^3
Watt/sq meter $\cdot$ °C ($W/m^2 \cdot$ °C)	1.761×10^{-1}	Btu/h $\cdot$ $ft^2 \cdot$ °F

SI Unit Prefixes

Amount	Multiples and Submultiples	Prefixes	Symbols
1,000,000,000,000,000,000	10^{18}	exa	E
1,000,000,000,000,000	10^{15}	peta	P
1,000,000,000,000	10^{12}	tera	T
1,000,000,000	10^9	giga	G
1,000,000	10^6	mega	M[a]
1,000	10^3	kilo	k[a]
100	10^2	hecto	h
10	10	deka	da
0.1	10^{-1}	deci	d
0.01	10^{-2}	centi	c[a]
0.001	10^{-3}	milli	m[a]
0.000,001	10^{-6}	micro	μ[a]
0.000,000,001	10^{-9}	nano	n
0.000,000,000,001	10^{-12}	pico	p
0.000,000,000,000,001	10^{-15}	femto	f
0.000,000,000,000,000,001	10^{-18}	atto	a

[a]Most commonly used.

Greek Alphabet

Greek Alphabet

A	α	Alpha	N	ν	Nu
B	β	Beta	Ξ	ξ	Xi
Γ	γ	Gamma	O	o	Omicron
Δ	δ	Delta	Π	π	Pi
E	ε	Epsilon	P	ρ	Rho
Z	ζ	Zeta	Σ	σ	Sigma
H	η	Eta	T	τ	Tau
Θ	θ	Theta	Υ	υ	Upsilon
I	ι	Iota	Φ	ϕ	Phi
K	κ	Kappa	X	χ	Chi
Λ	λ	Lambda	Ψ	ψ	Psi
M	μ	Mu	Ω	ω	Omega

Index

879